International Conference of Computational Methods
in Sciences and Engineering 2004 (ICCMSE 2004)

LECTURE SERIES ON
COMPUTER AND COMPUTATIONAL SCIENCES 1

INTERNATIONAL CONFERENCE OF COMPUTATIONAL METHODS IN SCIENCES AND ENGINEERING 2004 (ICCMSE 2004)

EDITED BY

THEODORE SIMOS[*] AND GEORGE MAROULIS[**]

[*] DEPARTMENT OF COMPUTER SCIENCE AND TECHNOLOGY, UNIVERSITY OF THE PELOPONNESE, TRIPOLIS, GREECE
[**] DEPARTMENT OF CHEMISTRY, UNIVERSITY OF PATRAS, PATRAS, GREECE

Utrecht • Boston, 2004

VSP   Tel: +31 30 692 5790
an imprint of Brill Academic Publishers   Fax: +31 30 693 2081
P.O. Box 346   vsppub@brill.nl
3700 AH Zeist   www.brill.nl
The Netherlands   www.vsppub.com

First published in 2004

ISBN 90-6764-418-8
Series ISSN 1573-4196

A C.I.P. record for this book is available from the Library of Congress

Printed in The Netherlands by Ridderprint bv, Ridderkerk.

VSP International
Science Publishers
P.O. Box 346, 3700 AH Zeist
The Netherlands

*Lecture Series on Computer
and Computational Sciences*
Volume 1, 2004, pp. i-iii

Preface for the Proceedings of the

International Conference of Computational Methods in Sciences and Engineering 2004 (ICCMSE 2004)

Recognised Conference by the European Society of Computational Methods in Sciences and Engineering (ESCMSE)

The *International Conference of Computational Methods in Sciences and Engineering 2004 (ICCMSE 2004)* is taken place at the Hotel Armonia between 19th and 23rd November 2004.

The aim of the conference is to bring together computational scientists from several disciplines in order to share methods and ideas.

Topics of general interest are: Computational Mathematics, Theoretical Physics and Theoretical Chemistry. Computational Engineering and Mechanics, Computational Biology and Medicine, Computational Geosciences and Meteorology, Computational Economics and Finance, Scientific Computation. High Performance Computing, Parallel and Distributed Computing, Visualization, Problem Solving Environments, Numerical Algorithms, Modelling and Simulation of Complex System, Web-based Simulation and Computing, Grid-based Simulation and Computing, Fuzzy Logic, Hybrid Computational Methods, Data Mining, Information Retrieval and Virtual Reality, Reliable Computing, Image Processing, Computational Science and Education etc.

The International Conference of Computational Methods in Sciences and Engineering (ICCMSE) is unique in its kind. It regroups original contributions from all fields of the traditional Sciences, Mathematics, Physics, Chemistry, Biology, Medicine and all branches of Engineering. It would be perhaps more appropriate to define the ICCMSE as a Conference on Computational Science and its applications to Science and Engineering. Based on the universality of mathematical reasoning the ICCMSE favours the interaction of various fields of Knowledge to the benefit of all. Emphasis on the multidisciplinary character of the Conference and the opening to new forms of interaction was central to the ICCSME 2003 held at Kastoria, in the north of Greece.

It would suffice to give here a powerful example of the new spirit championed by the ICCSME. In Quantum Chemistry one applies Quantum Physics to the study of the electronic structure, properties and the interactions of atoms, molecules and more complex systems. In order to obtain explicit results one needs efficient Mathematical methods and algorithms. The application of these methods and algorithms depends almost exclusively on the use of high-performance Computers. What is more, with the aid of modern technology it is possible to create particularly performant visualization tools. Thus we have been able to study closely and to visualize fundamental chemical phenomena. Our ability to simulate the behaviour of molecules in condensed phases has opened new horizons in Medicine and Pharmacology. One sees perceives easily in interaction Chemistry, Physics, Mathematics, Computer Science and Technology, Medicine and Pharmacology. Computational Science occupies a privileged position in this continuum. The principal ambition of the ICCSME is to promote the exchange of novel ideas through the close interaction of research groups from all Sciences and Engineering.

In addition to the general programme the Conference offers an impressive number of Symposia. The purpose of this move is to define more sharply new directions of expansion and progress for Computational Science.

We note that for ICCMSE there is a co-sponsorship by American Chemical Society

More than 490 extended abstracts have been submitted for consideration for presentation in ICCMSE 2004. From these extended abstracts we have selected 289 extended abstracts after international peer review by at least two independent reviewers. These accepted papers will be presented at ICCMSE 2004.

After ICCMSE 2004 the participants can send their full papers for consideration for publication in one of the nine journals that have accepted to publish selected Proceedings of ICCMSE 2004. We would like to thank the Editors-in-Chief and the Publishers of these journals. The full papers will be considered for publication based on the international peer review by at least two independent reviewers.

We would like also to thank:

- The Scientific Committee of ICCMSE 2004 (see in page iv for the Conference Details) for their help and their important support. We must note here that it is a great honour for us that leaders on Computational Sciences and Engineering have accepted to participate the Scientific Committee of ICCMSE 2004.

- The Symposiums' Organisers for their excellent editorial work and their efforts for the success of ICCMSE 2004.

- The invited speakers for their acceptance to give keynote lectures on Computational Sciences and Engineering.

- The Organising Committee for their help and activities for the success of ICCMSE 2004.

- Special thanks for the Secretary of ICCMSE 2004, Mrs Eleni Ralli-Simou (which is also the Administrative Secretary of the European Society of Computational Methods in Sciences and Engineering (ESCMSE)) for her excellent job.

Prof. Theodore Simos
President of ESCMSE
Chairman ICCMSE 2004
Editor of the Proceedings
Department of Computer Science
and Technology
University of the Peloponnese
Tripolis
Greece

Prof. George Maroulis
Co-Editor of the Proceedings
Department of Chemistry
University of Patras
Patras
Greece

September 2004

VSP International
Science Publishers
P.O. Box 346, 3700 AH Zeist
The Netherlands

*Lecture Series on Computer
and Computational Sciences*
Volume 1, 2004, pp. iv-v

Conference Details

International Conference of Computational Methods in Sciences and Engineering 2004 (ICCMSE 2004), Hotel Armonia, 19-23 November, 2004.

Recognised Conference by the European Society of Computational Methods in Sciences and Engineering (ESCMSE)

Chairman and Organiser

Professor T.E. Simos, President of the European Society of Computational. Methods in Sciences and Engineering (ESCMSE). Active Member of the European Academy of Sciences and Arts and Corresponding Member of the European Academy of Sciences, Department of Computer Science and Technology, Faculty of Sciences and Technology, University of Peloponnese, Greece.

Scientific Committee

Prof. H. Ågren, Sweden
Prof. H. Arabnia, USA
Prof. J. Vigo-Aguiar, Spain
Prof. Dž. Belkić, Sweden
Prof. K. Belkić, USA
Prof. E. Brändas, Sweden
Prof. G. Maroulis, Greece
Prof. R. Mickens, USA
Dr. Psihoyios, UK
Prof. B. Wade, USA
Prof. J. Xu, USA
Prof. Risto M. Nieminen, Finland

Invited Speakers

Prof. J. Vigo-Aguiar, Spain
Prof. Dž. Belkić, Sweden
Prof. K. Belkić, USA
Prof. Paul G. Mezey, Canada
Dr B. Champagne, Belgium
Prof. V. Barone, Italy
Prof. S.Farantos, Greece
Prof. R. Fournier, Canada

Prof. M. Kosmas, Greece
Prof. P. Kusalik, Canada
Dr S. Pal, India
Dr M.G. Papadopoulos, Greece
Prof. C. Pouchan, France
Prof. B.M. Rode, Austria
Prof. K. Szalewicz, USA
 Prof A.J. Thakkar, Canada

VSP International
Science Publishers
P.O. Box 346, 3700 AH Zeist
The Netherlands

Lecture Series on Computer
and Computational Sciences
Volume 1, 2004, pp. vi-vii

European Society of Computational Methods in Sciences and Engineering (ESCMSE)

Aims and Scope

The *European Society of Computational Methods in Sciences and Engineering (ESCMSE)* is a non-profit organization. The URL address is: http://www.uop.gr/escmse/

The aims and scopes of *ESCMSE* is the construction, development and analysis of computational, numerical and mathematical methods and their application in the sciences and engineering.

In order to achieve this, the *ESCMSE* pursues the following activities:

• Research cooperation between scientists in the above subject.
• Foundation, development and organization of national and international conferences, workshops, seminars, schools, symposiums.
• Special issues of scientific journals.
• Dissemination of the research results.
• Participation and possible representation of Greece and the European Union at the events and activities of international scientific organizations on the same or similar subject.
• Collection of reference material relative to the aims and scope of *ESCMSE.*

Based on the above activities, *ESCMSE* has already developed an international scientific journal called **Applied Numerical Analysis and Computational Mathematics (ANACM)**. This is in cooperation with the international leading publisher, **Wiley-VCH**.

ANACM is the official journal of *ESCMSE*. As such, each member of *ESCMSE* will receive the volumes of **ANACM** free of charge.

Categories of Membership

European Society of Computational Methods in Sciences and Engineering (ESCMSE)

Initially the categories of membership will be:

• **Full Member (MESCMSE):** PhD graduates (or equivalent) in computational or numerical or mathematical methods with applications in sciences and engineering, or others who have contributed to the advancement of computational or numerical or mathematical methods with applications in sciences and engineering through research or education. Full Members may use the title MESCMSE.

• **Associate Member (AMESCMSE):** Educators, or others, such as distinguished amateur scientists, who have demonstrated dedication to the advancement of computational or numerical or mathematical methods with applications in sciences and engineering may be elected as Associate Members. Associate Members may use the title AMESCMSE.

• **Student Member (SMESCMSE):** Undergraduate or graduate students working towards a degree in computational or numerical or mathematical methods with applications in sciences and engineering or a related subject may be elected as Student Members as long as they remain students. The Student Members may use the title SMESCMSE

• **Corporate Member:** Any registered company, institution, association or other organization may apply to become a Corporate Member of the Society.

Remarks:

1. After three years of full membership of the European Society of Computational Methods in Sciences and Engineering, members can request promotion to Fellow of the European Society of Computational Methods in Sciences and Engineering. The election is based on international peer-review. After the election of the initial Fellows of the European Society of Computational Methods in Sciences and Engineering, another requirement for the election to the Category of Fellow will be the nomination of the applicant by at least two (2) Fellows of the European Society of Computational Methods in Sciences and Engineering.

2. All grades of members other than Students are entitled to vote in Society ballots.

3. All grades of membership other than Student Members receive the official journal of the ESCMSE Applied Numerical Analysis and Computational Mathematics (ANACM) as part of their membership. Student Members may purchase a subscription to ANACM at a reduced rate.

We invite you to become part of this exciting new international project and participate in the promotion and exchange of ideas in your field.

VSP International
Science Publishers
P.O. Box 346, 3700 AH Zeist
The Netherlands

Lecture Series on Computer
and Computational Sciences
Volume 1, 2004, pp. viii-xxi

Table of Contents

VSP International
Science Publishers
P.O. Box 346, 3700 AH Zeist
The Netherlands

*Lecture Series on Computer
and Computational Sciences*
Volume 1, 2004, pp. 1-5

Bending Analysis of Thick Rectangular Plates with Various Boundary Conditions Using Extended Kantorovich Method

M. M. Aghdam[1], J.P. Vafa

Department of Mechanical Engineering,
Amirkabir University of Technology,
Hafez Ave., Tehran, Iran

Received 7 August, 2004; accepted in revised form 30 August, 2004

Abstract: Bending of thick rectangular plates with various boundary conditions is studied using Extended Kantorovich Method (EKM). The governing equations based on the Reissner shear deformation theory include a system of eight first order partial differential equations (PDEs) with eight unknowns. Application of EKM to the governing equations yields to a double set of eight ordinary differential equations (ODEs) in terms of x and y. These equations were then solved iteratively until a level of prescribed convergence is achieved. It is shown that the convergence of the method is very fast. Results demonstrate good agreement with those of finite element analysis.

Keywords: Bending analysis; Reissner plate theory; Extended Kantorovich Method; Iterative procedure

1. Introduction and theory background

One of the approximate methods to solve the governing equations of plates is the EKM [1-7]. It was mainly used for various analyses of thin isotropic and orthotropic plates where the governing equation is a single forth order PDE [2-3] as well as eigenvalue problems [4]. However, there are also studies [5-7] dedicated to thick plates where a system of three second order PDE was solved. In most of these studies, clamped boundary conditions were considered due to the nature of the governing equations and ease of applying clamped boundary conditions in the EKM.

In this study, the first order shear deformation theory of Reissner for thick plates [8] is employed. In order to apply various combinations of BC's, the governing equations in their complete form which consist of eight first orders PDE with eight unknowns are considered. This offers a simple procedure to apply any combination of boundary conditions for the plate. For a rectangular thick plate ($b/h<10$) with length, a, width, b and thickness h, the governing equations in terms of unknown deflection and rotations (w, φ_x, φ_y), bending and twisting moments (M_{xx}, M_{yy}, M_{xy}) and transverse shear forces (Q_x, Q_y) include a system of eight first orders PDE as:

$$\varphi_x + w_{,x} = BQ_x; \qquad \varphi_y + w_{,y} = BQ_y; \qquad M_{xx} = D(\varphi_{x,x} + v\varphi_{y,y} + Cq);$$

$$M_{yy} = D(\varphi_{y,y} + v\varphi_{x,x} + Cq); \qquad M_{xy} = (D/2)(1-v)(\varphi_{x,y} + \varphi_{y,x});$$

$$M_{xx,x} + M_{xy,y} = Q_x; \qquad M_{xy,x} + M_{yy,y} = Q_y; \qquad Q_{x,x} + Q_{y,y} = -q(x,y); \qquad (1)$$

where comma denotes differentiation with respect to x or y, $q(x, y)$ is applied load, E, v are Young's modulus and Poisson's ratio of the plate and plate constants, $D = Eh^3/12(1-v^2)$, $B = 12(1+v)/5Eh$ and $C = 6v(1+v)/5Eh$. The boundary conditions of the plate for one edge, say $x=0$ or $x=a$, are:

---

[1] Corresponding author. E-mail: aghdam@aut.ac.ir, Tel: +98 21 6454 3429, Fax: +98 21 641 9736.

For Clamped edge :
$$w = \varphi_x = \varphi_y = 0$$
For Simply Supported edge :
$$w = M_{xx} = \varphi_y = 0 \qquad (2)$$
For free edge :
$$M_{xx} = M_{xy} = Q_x = 0$$

Boundary conditions along $y=0$ or $y=b$ can be obtained by interchanging subscripts x and y in equations (2).

2. Application of the EKM

Based on the EKM, all unknowns should be considered as products of separable dimensionless functions of x and y. as:

$$
\begin{aligned}
&(M_{xx})_{ij} = (D/b)\delta_{1i}(x)\psi_{1j}(y); && (M_{yy})_{ij} = (D/b)\delta_{2i}(x)\psi_{2j}(y); && (M_{xy})_{ij} = (D/b)\delta_{3i}(x)\psi_{3j}(y) \\
&(Q_x)_{ij} = (D/b^2)\delta_{4i}(x)\psi_{4j}(y); && (Q_y)_{ij} = (D/b^2)\delta_{5i}(x)\psi_{5j}(y); && \\
&(\varphi_x)_{ij} = \delta_{6i}(x)\psi_{6j}(y); && (\varphi_y)_{ij} = \delta_{7i}(x)\psi_{7j}(y); && (w)_{ij} = b\delta_{8i}(x)\psi_{8j}(y)
\end{aligned}
\qquad (3)
$$

in which, subscripts i and j denote number of iterations in x and y, respectively, $\delta_{ki}(x)$ and $\psi_{kj}(y)$ are unknown dimensionless functions of x and y to be determined. Introducing (3) into (1) and (2) yields to the new forms of the governing equations and boundary conditions (along $x=0$ or $x=a$) in terms of δ_k, ψ_k and their first derivatives, δ'_k, ψ'_k as:

$$
\begin{aligned}
&\delta_1\psi_1 - b\delta'_6\psi_6 - \upsilon b\delta_7\psi'_7 = bCq \ ; \ \ \delta_2\psi_2 - \upsilon b\delta'_6\psi_6 - b\delta_7\psi'_7 = bCq \ ; \ \ \delta_3\psi_3 - g\delta_6\psi'_6 - g\delta'_7\psi_7 = 0 \ ; \\
&f\delta_4\psi_4 - \delta_6\psi_6 - b\delta'_8\psi_8 = 0 \qquad ; \qquad f\delta_5\psi_5 - \delta_7\psi_7 - b\delta_8\psi'_8 = 0 \qquad ; \qquad b\delta'_1\psi_1 + b\delta_3\psi'_3 - \delta_4\psi_4 = 0 \\
&b\delta'_3\psi_3 + b\delta_2\psi'_2 - \delta_5\psi_5 = 0 \ ; \qquad \qquad \delta'_4\psi_4 + \delta_5\psi'_5 = -b^2 q/D
\end{aligned}
\qquad (4)
$$

For Clamped edge :
$$\delta_8(0) = \delta_7(0) = \delta_6(0) = 0$$
For Simply Supported edge :
$$\delta_8(0) = \delta_1(0) = \delta_7(0) = 0 \qquad (5)$$
For free edge :
$$\delta_1(0) = \delta_3(0) = \delta_4(0) = 0$$

where in (4), $f = h^2/5(1-v)$ and $g = (1-v)/2$.

Based on the general procedure of the weighted residual method, it is necessary to multiply each of equations (4) by an appropriate weighting function. These functions would be $\psi_1(y), \psi_2(y) \cdots \psi_8(y)$ for the first, second... eighth equation, respectively. According to the EKM, assuming prescribed functions for $\psi_i(y)$, as initial guesses, and integrating over the entire length of the plate in y direction, b one can obtain a system of ODE as:

$$
\begin{bmatrix}
F_{11} & 0 & 0 & 0 & 0 & -bF_{16}d & -\upsilon bF'_{17} & 0 \\
0 & F_{22} & 0 & 0 & 0 & -\upsilon bF_{26}d & -bF'_{27} & 0 \\
0 & 0 & F_{33} & 0 & 0 & -gF'_{36} & -gF_{37}d & 0 \\
0 & 0 & 0 & fF_{44} & 0 & -F_{46} & 0 & -bF_{48}d \\
0 & 0 & 0 & 0 & fF_{55} & 0 & -F_{57} & -bF'_{58} \\
bF_{61}d & 0 & bF'_{63} & -F_{64} & 0 & 0 & 0 & 0 \\
0 & bF'_{72} & bF_{73}d & 0 & -F_{75} & 0 & 0 & 0 \\
0 & 0 & 0 & F_{84}d & F_{85} & 0 & 0 & 0
\end{bmatrix}
\begin{bmatrix}
\delta_1 \\ \delta_2 \\ \delta_3 \\ \delta_4 \\ \delta_5 \\ \delta_6 \\ \delta_7 \\ \delta_8
\end{bmatrix}
=
\begin{bmatrix}
bCM_1 \\ bCM_2 \\ 0 \\ 0 \\ 0 \\ 0 \\ 0 \\ -\dfrac{b^2}{D}M_8
\end{bmatrix}
\qquad (6)
$$

where $d \equiv d/dx$, $F_{ij} = \int_0^b \psi_i\psi_j \, dy$, $F'_{ij} = \int_0^b \psi_i\psi'_j dy$ and $M_i = \int_0^b q\psi_i dy$.

Solving the system of ODE (6) with appropriate boundary conditions (5) at both edges, $x=0$ and $x=a$, it is possible to determine the first estimate for the dimensionless functions $\delta_i(x)$. Now the procedure can be continued by multiplying the obtained $\delta_i(x)$ to the equations (4) with the sequence just described for

$\psi_i(y)$ and integrating over the length of the plate in the x direction, a. This yields to another system of ODE in $\psi_i(y)$ as:

$$
\begin{bmatrix}
H_{11} & 0 & 0 & 0 & 0 & -bH'_{16} & -vbH_{17}d & 0 \\
0 & H_{22} & 0 & 0 & 0 & -vbH'_{26} & -bH_{27}d & 0 \\
0 & 0 & H_{33} & 0 & 0 & -gH_{36}d & -gH'_{37} & 0 \\
0 & 0 & 0 & fH_{44} & 0 & -H_{46} & 0 & -bH'_{48} \\
0 & 0 & 0 & 0 & fH_{55} & 0 & -H_{57} & -bH_{58}d \\
bH'_{61} & 0 & bH_{63}d & -H_{64} & 0 & 0 & 0 & 0 \\
0 & bH_{72}d & bH'_{73} & 0 & -H_{75} & 0 & 0 & 0 \\
0 & 0 & 0 & H'_{84} & H_{85}d & 0 & 0 & 0
\end{bmatrix}
\begin{bmatrix}
\psi_1 \\ \psi_2 \\ \psi_3 \\ \psi_4 \\ \psi_5 \\ \psi_6 \\ \psi_7 \\ \psi_8
\end{bmatrix}
=
\begin{bmatrix}
bCN_1 \\ bCN_2 \\ 0 \\ 0 \\ 0 \\ 0 \\ 0 \\ -\dfrac{b^2}{D}N_8
\end{bmatrix}
\tag{7}
$$

where $d \equiv d/dy$, $H_{ij} = \int_0^a \delta_i \delta_j dx$, $H'_{ij} = \int_0^a \delta_i \delta'_j dx$ and $N_i = \int_0^a q \delta_i dx$.

By solving system of equations (7) and obtaining $\psi_i(y)$, the first iteration will be completed. This yields to the first approximation for the eight unknowns. The procedure should be continued by solving equations (6) using the new obtained functions, $\psi_i(y)$. Equations (6) and (7) should be solved in every iteration until a level of prescribed convergence is achieved. It should be noted that the solution for systems of ODEs (6) and (7) can be obtained using either analytical or numerical methods. It is also worth nothing that the convergence of the method is very fast and normally three to four iterations are sufficient to get the final results.

3. Results and discussions

The procedure described in previous section is applied to an isotropic square plate (40×40) subjected to a uniformly distributed load with various boundary conditions and different aspect ratios, b/h. One of the advantages of the EKM, unlike other weighted residual methods, is that arbitrary dimensionless functions can be used as the initial guess for $\psi_i(y)$. It is not even necessary for these functions to satisfy the boundary conditions [5-7]. This is due to the fact that the boundary conditions will be satisfied in the subsequent iterations as equations (6) and (7) are solved iteratively. Therefore, a simple and similar function is used as initial guess for all eight functions, $\psi_i(y)$ as:

$$
\psi_i(y) = Sin(\pi y / b) \qquad i = 1 \text{ to } 8 \tag{8}
$$

Using initial guess (8), one can determine all constants (F_{ij}, F'_{ij} and M_i) in equations (6). Then by solving equations (6) the first approximation for $\delta_i(x)$ can be obtained. The new functions, $\delta_i(x)$ are employed to calculate all constants (H_{ij}, H'_{ij} and N_i) in equations (7). Finally, solutions to equations (7) yield to new and updated $\psi_i(y)$. This completes the first iteration. Normally three to four iterations are enough to get final convergence.

Maximum deflections of the square plate (40*40) subjected to uniformly distributed load with different boundary conditions are tabulated in Table 1. Three iterations were carried out. In Table one, C, S, and F are used for Clamped, Simply supported and Free edges, respectively. Included in the table are also results obtained from commercial finite element code ABAQUS [9]. Dimensionless deflections of the square plate along centerline (a/2, y/b) for CFFF, CSCS and SFSF boundary conditions are depicted in Figures 1-3, respectively. As can be seen in the figures, the convergence of the method is fast and various BC's, particularly free BC do not influence the convergence rate of the method.

4. Conclusions

Bending of thick rectangular plates is studied using EKM. The governing equations in their complete form which consist of eight first orders PDE with eight unknowns are considered. Hence, it is shown that applying various combinations of boundary conditions, rather than all clamped edges is straightforward. Results also reveal that the method is a simple iterative process with rapid convergence. Finally, it should be noted that the method leads to approximate closed form solution for the problem as long as one can determine particular solutions for the system of ODE (6) and (7).

#	B.C	b/h	w(max)	
			ABAQUS	EKM
1	CCCC	8	1.218E-04	1.264E-04
2	CCCC	4	2.477E-05	2.596E-05
3	SSSS	8	3.590E-04	3.283E-04
4	CSCS	5	5.645E-05	5.573E-05
5	SFSF	8	1.192E-03	1.178E-03
6	SFSF	5	3.109E-04	3.051E-04
7	CFFF	8	9.867E-03	9.705E-03
8	CFFF	5	2.484E-03	2.449E-03

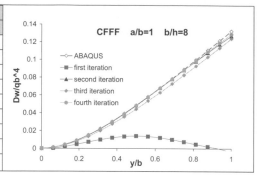

Table 1: Maximum deflection of thick square plates with various boundary conditions

Figure 1: Convergence of dimensionless deflection of the CFFF plate along centerline ($a/2$, y/b)

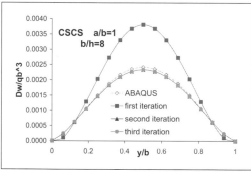

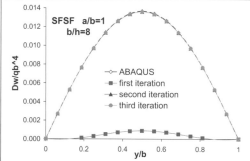

Figure 2: Convergence of dimensionless deflection of the CSCS plate along centerline ($a/2$, y/b)

Figure 3: Convergence of dimensionless deflection of the SFSF plate along centerline ($a/2$, y/b)

References

[1] A.D.Kerr, An extension of the Kantorovich method, *Quart Appl Math* **26** 219-229(1968).

[2] A.D.Kerr and H. Alexander, An application of the extended Kantorovich method to the stress analysis of a clamped rectangular plate, *Acta Mech* **6** 180–196(1968).

[3] M. Dalaei and A.D.Kerr, Analysis of clamped rectangular orthotropic plates subjected to a uniform lateral load, *Int J Mech Sci* **37** 527–535(1995).

[4] A.D. Kerr, An extended Kantorovich method for the solution of eigenvalue problems, *Int J Solids Struct* **5** 559–572(1969).

[5] M.M. Aghdam, M. Shakeri and S.J. Fariborz, Solution to the Reissner plate with clamped edges, *ASCE J Eng Mech* **122** 679–682(1996).

[6] S. Yuan, Y. Jin and F.W. Williams, Bending analysis of Mindlin plates by extended Kantorovich method, *ASCE J Eng Mech* **124** 1339–1345(1998).

[7] M.M. Aghdam and S.R. Falahatgar, Bending analysis of thick laminated plates using extended Kantorovich method, *Composite Structures* **62** 279–283 (2003).

[8] Reissner E, The effect of transverse shear deformation of the bending of elastic plates, *J. Appl. Mech.*, **12** 69-77(1945).

[9] Hibbit, Karlsson and Sorensen, *ABAQUS User's manual* Version 5.7,1997.

VSP International
Science Publishers
P.O. Box 346, 3700 AH Zeist
The Netherlands

*Lecture Series on Computer
and Computational Sciences*
Volume 1, 2004, pp. 6-9

Application of the Extended Kantorovich Method to Bending Analysis of Sector Plates

M. M. Aghdam[1], V. Erfanian

Department of Mechanical Engineering,
Amirkabir University of Technology,
Hafez Ave., Tehran, Iran

Received 7 August, 2004; accepted in revised form 30 August, 2004

Abstract: The Extended Kantorovich Method (EKM) is used to study bending of sector plates with clamped edges subjected to non-uniform loading. The governing equation includes a forth order partial differential equation (PDE) in cylindrical coordinates. The solution procedure based on the EKM is to convert the governing equation to two separate ordinary differential equations in terms of r and θ. The obtained ODE systems are then solved in an iterative manner until convergence was achieved. It is shown that similar to other application of EKM in Cartesian coordinate system the convergence of the method is vary fast. Comparison of the deflection of the plate at various points of the plate shows very good agreement with results of finite element code ANSYS.

Keywords: Extended Kantorovich Method; Bending analysis; sector plates; Non-uniform loading

1. Introduction and theory

Since its first introduction [1], EKM has been extensively used to obtain approximate solution for various two dimensional elasticity problems in Cartesian coordinate system, see for example [2-7]. These studies includes torsion of rectangular bars [1], eigenvalue, buckling and vibration problems [2-4], bending of thick isotropic [5] and orthotropic [6] plates and free-edge analysis of laminates [7]. Application of this method in other coordinate systems is not found in the open literature.

In this study, the EKM is successfully employed to the bending of sector plates with clamped edges subjected to any type of non-linear loading. Approximate closed form solutions are obtained for the governing equations of the plate. The governing equation for a thin sector plate, shown in Fig. 1, is a forth order PDE as [8]:

$$\nabla^4 w = \left(\frac{\partial^2}{\partial r^2} + \frac{1}{r}\frac{\partial}{\partial r} + \frac{1}{r^2}\frac{\partial^2}{\partial \theta^2} \right)\left(\frac{\partial^2}{\partial r^2} + \frac{1}{r}\frac{\partial}{\partial r} + \frac{1}{r^2}\frac{\partial^2}{\partial \theta^2} \right)w = q(r,\theta)\Big/D \tag{1}$$

where w is deflection, $q(r,\theta)$ is applied load, $D=Et^3/12(1-v^2)$ is flexural rigidity and E, v, t are Young's modulus, Poisson's ratio and thickness of the plate. For clamped boundary conditions, deflection and its first derivative with respect to the normal direction of the boundary should be vanished, i.e.:

$$w = dw/dr = 0 \qquad \text{for } r=r_1 \text{ and } r=r_2 \qquad -\tau \le \theta \le \tau$$
$$w = dw/d\theta = 0 \qquad \text{for } \theta = \pm\tau \qquad r_1 \le r \le r_2 \tag{2}$$

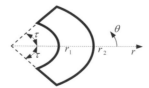

Figure 1: Geometry of sector plate and coordinate system

---

[1] Corresponding author. E-mail: aghdam@aut.ac.ir, Tel: +98 21 6454 3429, Fax: +98 21 641 9736.

2. Iterative Solution by the EKM

The general procedure of the weighted residual methods suggests assuming a multi-term expression for the deflection of the plate. However, it was shown that even a single-term assumption for deflection can provide very good results in the EKM procedure [4-6]. Therefore, a single-term expression is used for the deflection of the plate as:

$$w(r,\theta) = w_{ij}(r,\theta) = f_i(r)g_j(\theta) \tag{3}$$

where f_i and g_j are functions to be determined. According to the Galerkin weighted residual method, for a function, $w(r,\theta)$ that satisfies boundary conditions (2) the Galerkin equation is:

$$\int_{-\tau}^{+\tau}\int_{r_1}^{r_2}\left(D\nabla^4 w - q\right)\delta w\, r\, dr\, d\theta = 0 \tag{4}$$

Substitution of (3) into (4) and assuming a prescribed function for $g_j(\theta)$ leads to the following expression for equation (4):

$$\int_{r_1}^{r_2}\left[\int_{-\tau}^{+\tau}\left(D\nabla^4 w_{ij} - q\right)g_j\, d\theta\right]\delta f_i\, r\, dr = 0 \tag{5}$$

According to the variational method rules, equation (5) is satisfied if the expression in the bracket is vanished, hence:

$$\int_{-\tau}^{+\tau}\left(D\nabla^4 f_i g_j - q\right)g_j\, d\theta = 0 \tag{6}$$

Using a prescribed function for $g_j(\theta)$, as initial guess, and integrating with respect to θ, equation (6) turns into a forth order ODE as:

$$A_1 d^4 f_i + 2A_1 \frac{1}{r}d^3 f_i + A_2 \frac{1}{r^2}d^2 f_i - A_2 \frac{1}{r^3}df_i + A_3 \frac{1}{r^4}f_i = E \tag{7}$$

in which $d \equiv d/dr$, A_i and E are:

$$A_1 = \int_{-\tau}^{+\tau}g_j^2 d\theta; \qquad\qquad A_2 = \int_{-\tau}^{+\tau}\left(2\frac{d^2 g_j}{d\theta^2} - g_j\right)g_j\, d\theta$$

$$A_3 = \int_{-\tau}^{+\tau}\left(\frac{d^4 g_j}{d\theta^4} + 4\frac{d^2 g_j}{d\theta^2}\right)g_j\, d\theta; \qquad E = \int_{-\tau}^{+\tau}q(r,\theta)g_j d\theta \tag{8}$$

It should be noted that by using (3) to define deflection of the plate, the boundary conditions (2) are also changed to:

$$f_i = df_i/dr = 0 \qquad \text{for } r=r_1 \text{ and } r=r_2 \qquad -\tau \le \theta \le \tau$$
$$g_j = dg_j/d\theta = 0 \qquad \text{for } \theta = \pm\tau \qquad r_1 \le r \le r_2 \tag{9}$$

Solving equation (7) with boundary data (9) leads to the first estimate of the function $f_i(r), i=1$. Now it is possible to continue the procedure by introducing the obtained function to equation (4). This yields to the new form of the Galerkin equation as:

$$\int_{-\tau}^{+\tau}\left[\int_{r_1}^{r_2}\left(D\nabla^4 f_i g_j - q\right)f_i\, dr\right]\delta g_j\, d\theta = 0 \tag{10}$$

Again, putting (10) equal to zero and integrating with respect to r leads to the second iterative ODE with respect to θ as:

$$B_1 d^4 g_j + B_2 d^2 g_j + B_3 g_j = F \tag{11}$$

Where in (11), $d \equiv d/d\theta$ and other constants are:

$$B_1 = \int_{r_1}^{r_2}\frac{f_i^2}{r^3}dr; \qquad B_2 = \int_{r_1}^{r_2}\left(\frac{4f_i}{r^3} - \frac{2}{r^2}\frac{df_i}{dr} + \frac{2}{r}\frac{d^2 f_i}{dr^2}\right)f_i\, dr$$

$$B_3 = \int_{r_1}^{r_2}\left(r\frac{d^4 f_i}{dr^4} + 2\frac{d^3 f_i}{dr^3} - \frac{1}{r}\frac{d^2 f_i}{dr^2} + \frac{1}{r^2}\frac{df_i}{dr}\right)f_i\, dr; \qquad F = \int_{r_1}^{r_2}q(r,\theta)f_i dr \tag{12}$$

The process is continued by solving ODE (11) together with boundary conditions (9) and obtaining the new prediction for $g_j(\theta)$. This finishes the first iteration. Equations (7) and (11) should be solved iteratively and new updated estimates for functions $f_i(r)$ and $g_j(\theta)$ are determined. Iterations are continued until a level of convergence is achieved.

3. Results and discussions

The procedure described in previous section is applied to a sector plate with clamped edges subjected to non-uniform distributed load. Plate constants and applied load are :

$$r_1 = 0.3;\ r_2 = 0.4;\ h = 0.002\,\text{m};\qquad E = 2.7 \times 10^{11};\quad v = 0.3;$$

$$q(r,\theta) = -5 \times 10^3 \left(\frac{(r - r_1)(r - r_2)}{(r_2 - r_1)^2} \right) \cos\!\left(\frac{\pi\theta}{2\tau} \right) \tag{13}$$

To start iterations, the initial guess for $g_j(\theta), j = 0$ is assumed as:

$$g_0 = (1 - \theta^2/\tau^2) \tag{14}$$

Using this function as initial guess in conjunction with (8), one can determine constants A_i in (7). This equation is then solved to obtain $f_1(r)$ which is used in (12) to determine constants B_i of equation (11). New and updated function $g_1(\theta)$ can be calculated by solving equation (11). This finishes the first iteration. This procedure is employed to update functions until convergence is reached. It is worth mentioning that closed form solutions exist for both ODEs (7) and (11) although they are not appeared in this manuscript. Therefore, all mechanical terms, i.e. defection, rotations, forces and moments may be obtained using well known expressions which can be found elsewhere [8].

Figures 2 and 3 represent convergence of the functions $f_i(r)$ and $g_j(\theta)$, respectively. As can be seen in the figures, the convergence for both functions is very fast even in comparison with rectangular plates where EKM was extensively employed [5-6]. According to (3), multiplication of these two functions provides expression for deflection of the plate. In order to examine the validity of the results, another analysis was also carried out to determine deflection of the plate using the commercial finite element code ANSYS. Relative difference between results of this study and those of the ANASYS for the deflection of the plate is determined. The relative difference along centerlines of the sector plate, i.e. lines $(r,0)$ and $\left(\frac{(r_1 + r_2)}{2}, \theta \right)$ are depicted in figures 4 and 5, respectively. As figures indicate, maximum relative difference between results of this study and those of the ANASYS is about 1.5%. It is also worth noting that this difference is less than 0.1% for the maximum deflection of the plate which in this case occurs at the center of the plate.

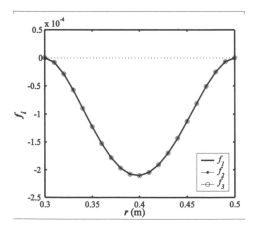

Figure 2: Convergence of $f_i(r)$ for sector plate

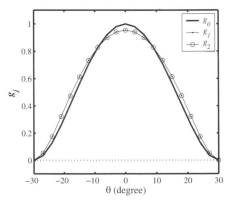

Figure 3: Convergence of $g_i(\theta)$ for sector plate

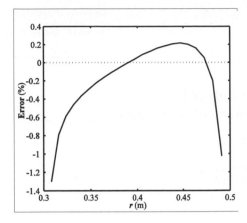
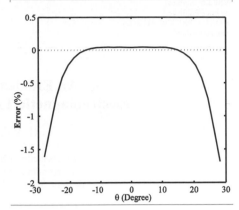

Figure 4: Relative difference between ANSYS and EKM predictions for deflection of the plate along centerline $(r, 0)$

Figure 5: Relative difference between ANSYS and EKM predictions for deflection of the plate along centerline $(0.4, \theta)$

4. Concluding remarks

The applicability of the extended Kantorovich method in cylindrical coordinates is examined by solving a simple elasticity problem, i.e. bending of sector plate subjected to non-uniform distributed load. The governing equation of the problem is a forth order partial differential equation. Application of the EKM converts the governing equation to a double set of ordinary differential equations in r and θ. By assuming an initial guess for one of the functions, these equations are then solved iteratively until convergence is achieved. Although not appeared in the text, closed form solutions can be found for both ODEs. It is revealed that the method provides vary fast convergence to the final results. The method also offers reasonably accurate results in comparison with those of finite element code ANSYS.

References

[1] A.D.Kerr, An extension of the Kantorovich method, *Quart Appl Math* **26** 219-229(1968).

[2] A.D. Kerr, An extended Kantorovich method for the solution of eigenvalue problems, *Int J Solids Struct* **5** 559–572(1969).

[3] S. Yuan and Y. Jin, Computation of elastic buckling loads of rectangular thin plates using the extended Kantorovich method, *Computers & Structures* **66** 861–867(1998).

[4] M. Dalaei and A.D.Kerr, Natural vibration analysis of clamped rectangular orthotropic plates, *J. Sounds & Vibration* **189** 399-406(1996).

[5] M.M. Aghdam, M. Shakeri and S.J. Fariborz, Solution to the Reissner plate with clamped edges, *ASCE J Eng Mech* **122** 679–682(1996).

[6] M.M. Aghdam and S.R. Falahatgar, Bending analysis of thick laminated plates using extended Kantorovich method, *Composite Structures* **62** 279–283 (2003).

[7] H.S. Kim M. Cho and G.I. Kim, Free-edge strength analysis in composite laminates by the extended Kantorovich method, *Composite Structures* **49** 229–235 (2000).

[8] A.P. Boresi and O.M. Sidebottom: *Advanced Mechanics of Materials.* John Wiley & Sons, New York, 1985.

VSP International
Science Publishers
P.O. Box 346, 3700 AH Zeist
The Netherlands

*Lecture Series on Computer
and Computational Sciences*
Volume 1, 2004, pp. 10-13

An Efficient Solver for
Electromagnetic Transient Simulation

G. Ala[1], F. Viola

Dipartimento di Ingegneria Elettrica
Università degli Studi di Palermo
Viale delle Scienze, I-90128 Palermo, Italia

E. Francomano[2], A. Tortorici, E. Toscano

Dipartimento di Ingegneria Informatica
Università degli Studi di Palermo
Viale delle Scienze, I-90128 Palermo, Italia

Received 2 August, 2004; accepted in revised form, 16 August 2004

Abstract: In this paper a mesh-free method for electromagnetic transient simulation is presented. This method enables to simulate electromagnetic problems avoiding the use of a regular grid in the discretized model. The proposed method is a particular mesh-free method based on *particles*, that is no initial regular grid but only an arbitrary particles distribution is required. Thus the particles can be thickened only in little spots, where the electromagnetic field varies rapidly or in those regions in which particular objects have to be simulated. Maxwell's equations with the assigned boundary and initial conditions in time domain have been numerically solved by means of the proposed method. The first application results referred to a simple canonical case has shown that the method has a great potentiality with a reduced computational complexity .

Keywords: mesh-free method, smoothed particle hydrodynamics method, Maxwell's equation.
PACS: 02.60,41.20,02.70.B

1.   Introduction

The grid based methods are of great interest in solving electromagnetic (EM) problems. As well known [1], [2], [3] the accuracy of numerical approximation strictly depends on the shape and dimension of elementary cell of the mesh and on the connectivity laws on the grid nodes. Therefore, by working with complicated and irregular geometry, the pre-processing construction of the grid could be cumbersome. Advantages over the traditional numerical methods can be obtained by using the so called mesh-free methods. In particular, the mesh-free particle methods (MPMs) seem to be very promising in EM simulation. Among the MPMs, the smoothed particle hydrodynamics (SPH) [4]-[8], has some special advantages such as its intrinsic adaptive nature. This adaptability is achieved at the earlier stage of the field variable approximation that is performed at each time step based on a current local set of arbitrarily distributed particles. SPH method has been firstly developed in astrophysics and then extended to mechanical dynamics, fluid dynamics, molecular biology, etc. [8]. In the authors opinion SPH can be applied in an efficient and flexible way also in computational electromagnetics [9]. To this aim, the particles are fixed in the Eulerian space as interpolation points for the field functions. An explicit finite difference method (FDM) is used for time integration. The proposed method has been named smoothed particle electromagnetics (SPEM).

The problem domain is discretized by particles without use of pre-defined mesh and a connectivity law among the nodes; field functions and their derivatives are approximated by using the information belonging to the particles placed in the influence domain of the considered one. Thus, the proposed method enables to model irregular problems geometry with generalized non-homogeneous media only

---

[1] Corresponding author E-mail: ala@diepa.unipa.it

[2] Corresponding author E-mail: elisa@cere.pa.cnr.it

with an initial discretization; moreover the particle discretization can be improved during the process time stepping. These features lead to a great reduction of the computational complexity.

2. The smoothed particle interpolation method

In this section a brief overview of the smoothed particle interpolation method [10] is provided. By considering the governing problem equations with proper boundary and/or initial conditions, the problem domain Ω is discretized by a set of arbitrarly distributed particles.

The field functions $f(r)$ are approximated by means of the integral representation named as kernel approximation. Namely, by introducing a smoothing kernel $W(r - r', h)$, the kernel approximation operator $<>$ is generated:

$$< f(r) >= \int_{\Omega} f(r')W(r - r', h)dr' \tag{1}$$

where r, r' are the position vectors, h is the smoothing length defining the influence area of the smoothing kernel W. A second order of accuracy can be obtained if the smoothing kernel verify the following properties:

$$\int_{\Omega} W(r - r', h)dr' = 1 \tag{2}$$

$$\lim_{h \to 0} W(r - r', h) = \delta(r - r') \tag{3}$$

$$W(r - r', h) = 0 \quad for \quad |r - r'| > \alpha h \tag{4}$$

where α is a constant related to the smoothing kernel for the point at r. In order to solve Maxwell differential time domain equations, the curl operator has to be approximated. Thus, following [9] the approximation of the spatial derivative $\nabla \times f(r)$ of the field function is obtained by (1). Namely:

$$< \nabla \times f(r) >= \int_{\Omega} \nabla \times [f(r')W(r - r', h)]dr' - \int_{\Omega} \nabla'W(r - r', h) \times f(r')dr' \tag{5}$$

where the apex in ∇' indicates that the derivative is referred to r'.

By using the curl theorem, equation (5) can be expressed as follows:

$$< \nabla \times f(r) >= \int_{S} n \times [f(r')W(r - r', h)]dS - \int_{\Omega} \nabla'W(r - r', h) \times f(r')dr' \tag{6}$$

where S is the surface boundary of the domain Ω, and n is the external unity vector perpendicular to S. Since the smoothing function W is defined to have compact support, when the support domain is located inside the problem domain, the surface integral in (6) is zero. On the other hand, if the support domain overlaps with the geometry boundary the smoothing kernel W is truncated and the surface integral in non-zero. If it is the case, some modifications have to be introduced on the problem boundary if the surface integral is considered to be zero anyway. Therefore, for the points whose support domain is within the problem boundary, (6) is rewritten as follows:

$$< \nabla \times f(r) >= -\int_{\Omega} \nabla'W(r - r', h) \times f(r')dr' \tag{7}$$

It is clear that, under the previous made hypothesis the differential operator on a function is transmitted on the smoothing kernel.

By introducing a number of particles covering the problem domain, the integral representation can be converted to a numerical formulation by summing the contribution over all the nearest neighboring particles (NNP) [8] by performing the so called particle approximation.

The particle approximation for a field function set over particle i can be written using the average of those values of the function at all NNP of particle i weighted by the smoothing function:

$$< f(r_i) >= \sum_{j=1}^{N_i} f(r_j)W(r_i - r_j, h)\omega_j \tag{8}$$

where r_i and r_j are position vectors of particles i and j, N_i is the number of NNP of particle i and ω_j are coefficients of quadrature formulae.

For the spatial derivative of the field function, the following particle approximation is obtained:

$$< \nabla \times f(r_i) >= \sum_{j=1}^{N_i} \nabla_i W(r_i - r_j, h) \times f(r_j) \omega_j \tag{9}$$

The support domain for a field point *P(r)* is the domain where the known values of the neighboring points inside this domain are used to obtain the unknown information at *P(r)*. A local support domain is considered. Only the particles that are within this local region of a point are used in order to approximate the field variables and its derivatives at that point, i.e. NNP. The support domain for a particle is closely related to the smoothing length *h*. In fact, the extension of this domain is determined by the product of *h* by the factor α. This factor is selected in relation with the behavior of the smoothing kernel and it determines the number of NNP [8]. Even if the smoothing length can vary both temporarily or spatially, in the present paper a unique smoothing length is considered, so the support domain has the same extension for each particle.

In order to perform the function approximation based on a set of nodes placed in an arbitrary way without using a predefined mesh that provides the connectivity of the nodes, the smoothing kernel is of primary importance. In fact the $W(r - r', h)$ function determines the pattern for the kernel approximation, defines the extension of the support domain, and determines the consistency and the accuracy of both the kernel and particle approximation. One of the most used smoothing kernel is the Gaussian one [6], [8]. This kernel is smooth enough also for high orders derivatives and is very stable and accurate especially for disordered particles. However, it is not really compact from a theoretical point of view even if it approaches to zero very fast and so it is with compact support, practically.

3.   SPEM and Maxwell' equations

Let us consider the time-dependent Maxwell's curl equations in free space:

$$\frac{\partial E}{\partial t} = \frac{1}{\varepsilon_0} \nabla \times H, \quad \frac{\partial H}{\partial t} = \frac{-1}{\mu_0} \nabla \times E \tag{10}$$

E, H are space vectors in three dimensions. In order to exploit the features of the SPEM method, firstly the one dimensional case is considered. The electric field is supposed to be oriented in the *x* direction, the magnetic field in the *y* direction, and the space variation is accounted for the *z* direction. Equation (10) become:

$$\frac{\partial E_x}{\partial t} = -\frac{1}{\varepsilon_0} \frac{\partial H_y}{\partial z}, \quad \frac{\partial H_y}{\partial t} = -\frac{1}{\mu_0} \frac{\partial E_x}{\partial z}. \tag{11}$$

By using the previous developed concepts for spatial derivatives, the following discrete equations hold:

$$\frac{\partial E_x(z_i)}{\partial t} = \frac{-1}{\varepsilon_0} \sum_j^{N_i} H_y(z_j) \frac{\partial W(z_i - z_j, h)}{\partial z_i} \omega_j \tag{12}$$

$$\frac{\partial H_y(z_i)}{\partial t} = \frac{-1}{\mu_0} \sum_j^{N_i} E_x(z_j) \frac{\partial W(z_i - z_j, h)}{\partial z_i} \omega_j \tag{13}$$

where the summations are extended to all the NNP of particle z_i.

By taking the central difference approximation for the time derivative (considering n as $n \Delta t$) and after few manipulations the following formulations hold:

$$E_x^{n+1/2}(z_i) = E_x^{n-1/2}(z_i) - \frac{\Delta t}{\varepsilon_0} \sum_j^{N_i} H_y^n(z_j) \frac{\partial W(z_i - z_j, h)}{\partial z_i} \omega_j \tag{14}$$

$$H_y^{n+1}(z_i) = H_y^n(z_i) - \frac{\Delta t}{\mu_0} \sum_j^{N_i} E_x^{n+1/2}(z_j) \frac{\partial W(z_i - z_j, h)}{\partial z_i} \omega_j \tag{15}$$

The explicit time integration scheme is subjected to the Courant-Friedrichs-Levy (CFL) condition [2] for the stability. This CFL condition requires the time step to be proportional to the smallest spatial particle resolution, which in SPEM formulation is represented by the smallest smoothing length. In the previous equations Δt respects the CFL condition, by choosing the spatial step equals to the smoothing length.

In order to validate the proposed numerical scheme, a comparison with a standard finite difference time domain (FDTD) method has been carried out. The problem domain is one meter length, and the pulse is generated in its central point. In fig. 1 the space profile of the propagating pulse is shown, by comparing

SPEM computations with the FDTD results at the same time step. A regular particles distribution is not needed, the particles are randomly placed near the position of the nodes of the FDTD grid.

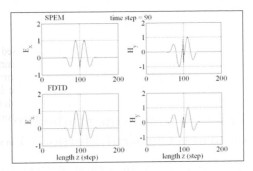

Fig. 1 – Propagation of the pulse generated in the center of the spatial domain after 90 time steps.

4.   Conclusions

In the paper a new meshless method for transient electromagnetic analysis is proposed. The reported one dimensional example can be illustrative of a more complex case in which, in order to better reproduce the geometry of an object, irregular particles distribution is necessary. It is important to underline that this problem in a FDTD scheme constitute a hard duty, which needs a generalized refinement of the computational lattice or the implementation of complex sub-grid processes with overlapping mesh; these tasks require additional computational resources which can become very cumbersome for large scale simulation.

Acknowledgments

The author wishes to thank the anonymous referees for their careful reading of the manuscript and their fruitful comments and suggestions.

REFERENCES

[1]   D. M. Sullivan, Electromagnetic simulation using the FDTD method, *IEEE press, 2000.*

[2]   A. Taflove and S. Hagness, Computational Electrodynamics: The Finite-Difference Time-Domain Method, *Artech House, Boston, MA, 2000.*

[3]   M. N. O. Sadiku, Numerical techniques in electromagnetics, *CRC Press, 1992.*

[4]   L. B. Lucy, A numerical approach to the testing of the fission hypothesis, *Astronomy Journal* **82** 1013-1024 (1977).

[5]   R. A. Gingold and J. J. Monaghan, Smoothed particle hydrodynamics: theory and application to nonspherical stars, *Mon. Not. R. Astronomical Society* **181** 375-389 (1977).

[6]   J. J. Monaghan, An introduction to SPH, *Computer Physics Communications* **48**, 89-96 (1988).

[7]   J. J. Monaghan, Smoothed particle hydrodynamics, *Annual Rev. Astron. Astrophys.,* **30**, 543-574 (1992).

[8]   G. R. Liu and M. B. Liu, Smoothed particle hydrodynamics - a meshfree particle method, *World Scientific Publishing, 2003.*

[9]   G. Ala, A. Spagnuolo, F. Viola: *An advanced gridless method for electromagnetic transient simulation.* Proceedings of EMC Europe 2004, September 6-10, 2004, Eindhoven, The Netherlands.

[10]   P. Laguna, Smoothed particle interpolation, *The Astrophysical Journal* **439** 814-821 (1994).

VSP International
Science Publishers
P.O. Box 346, 3700 AH Zeist
The Netherlands

*Lecture Series on Computer
and Computational Sciences*
Volume 1, 2004, pp. 14-15

Brittle Fracture in Silicon Studied by an Hybrid Quantum/Classical Method (LOTF)

T. Albaret(1) [1], A. De Vita(2), P. Gumbsch(3)

(1)Laboratoire de Physique de la Matière Condensée et Nanostructures,
Bâtiment Léon Brillouin, Campus de La Doua
Université Claude Bernard - Lyon 1
69622 Villeurbanne , France
(2) Physics Department, Kings College London,
Strand, London WC2R 2LS, United Kingdom
(3) IZBS, University of Karlsruhe, Kaiserstraße 12 76131 Karlsruhe, Germany

Received 10 September, 2004; accepted in revised form 20 September, 2004

Brittle fracture is a nice example of intrinsically multi-level problems in material science[1, 2]. On the large scale one has to correctly take into account the driving force of the crack that depends on the elastic properties of the whole system. But simultaneously it is also necessary to accurately describe the bond disruption processes that take place at the atomic scale [4, 5].

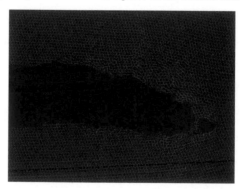

Figure 1: Snapshot of a (111)[1-10] crack tip obtained with a Stillinger-Weber model after a $\simeq$ 7 ps dynamics

In strongly covalent materials the quantum precision in the vicinity of the crack tip is particularly crucial. In the case of diamond silicon the use of purely classical potentials leads to qualitatively wrong descriptions of the crack dynamics showing for example plastic deformations, high fracture energy release rates, unphysical crack tip structures, large surface roughness and

---

[1]Corresponding author. E-mail: tristan.albaret@lpmcn.univ-lyon1.fr

Figure 2: Snapshot of a (111)[1-10] crack tip obtained with the LOTF method after a $\simeq$ 7 ps dynamics

even crack blunting [3] and wrong crack propagation direction. On the other hand full quantum calculations remain practically untractable on the several thousands of atoms needed to integrate the slowly decaying crack elastic field. To face this problem we used the hybrid method LOTF in which the parameters of an adaptable potential vary with the time to capture the quantum forces evaluated in some selected regions of the system [6, 8, 7]. The method was applied to (110)[1-10] and (111)[1-10] cracks upon mode I opening, each simulation box containing around 180 000 atoms. The results evidence that the local accurate treatment in the crack tip region largely influences the structure and roughness of the opened surfaces (see figure 1 and 2) as well as the fracture propagation direction.

References

[1] F.F. Abraham, N. Bernstein, J.Q. Broughton, D. Hess, MRS bull. **25**, 27 (2000)

[2] N. Bernstein and D.W. Hess Phys. Rev. Lett. **91**, 025501 (2003)

[3] D. Holland and M. Marder Adv. Mater. **11**, 793, (1999)

[4] R. Pérez and P. Gumbsch Phys. Rev. Lett. **84**, 5347 (2000)

[5] R. Pérez and P. Gumbsch Acta. Mater. **48**, 4517 (2000)

[6] G. Csanyi, T. Albaret, M. Payne and A. De Vita , to be published in Phys. Rev. Lett.

[7] A. De Vita and R. Car, Mat. Res. Soc. Symp. Proc. **491**, 473 (1998).

[8] G. Trimarchi, T. Albaret, P. Gumbsch, S. Meriani, J. Barth and A. De Vita CIMTEC proceedings, **1 G** 35 (2002)

VSP International
Science Publishers
P.O. Box 346, 3700 AH Zeist
The Netherlands

*Lecture Series on Computer
and Computational Sciences*
Volume 1, 2004, pp. 16-19

Preliminary Application of CFX as Tool in the Aerodynamic Study of Combustion Chamber for Gas Micro Turbine

Harley Souza Alencar[†]
Marco Antonio do Nascimento[‡]
Helcio Villanova[§]

Federal University of Engineering of Itajubá - UNIFEI
Depto. of Mechanical Engineering.
ZIP CODE 37500-000, Itajubá, Minas Gerais, Brazil

Received July 31, 2004; accepted in revised form December, 2004

Abstract: The use of the gas microturbines has been a promising alternative energy to assist the crescent demand of regional energy in Brazil, where more than 10 % of electrical energy is from thermoelectrical system and, besides, there are some programs of environmental regulation in energy, that detach the control of the emission of the pollutant agents of the types NOx, SOx and CO. In spite of these has been operated with fuels of low calorific power, there are limitations as for the instability of the fire; to the inflammability limits; to the nature of the fire; to the high taxes of emission of pollutants, during the combustion process. Any that are these limits, the aerodynamic study of the chambers is relevant in the measure that the formation of areas where there is recirculation, for instance, has strong influence in the time of residence and in the efficiency of the combustion. In spite of the progress in the development of mathematical models more and more sophisticated, to simulate the behavior of combustion in gas turbines, there isn't a specific application for the combustion chambers of the annular type, only of the types tubular or cylindrical for great gas turbines. The present work accomplishes a comparative analysis among the different models of flows with the purpose of the best models that describe the aerodynamic behavior for combustion chambers of the annular type, using the tool computational CFX v 5.7 ®.

Keywords: Aerodynamics, CFD Calculation, Combustion Chamber, Gas Turbines

PACS: 47.85.Gj

1. Introduction

The uses of micro turbines consist of the generation of electricity and vapour production in thermal systems. They are plants that use fuels of low calorific power with power bettwen 15 to 300 kW, whose efficiency can be bettwen 30 % (for regeneratived cycles) and 80 % (for cogenerated cycles). Its main limitations are: the instability of the fire; the inflammability limits; the nature of the fire; the high taxes of emission of pollutants, during the combustion process.

Its application started in 70′s decade like an alternative energy source because of oil crisis in 1973. This alternative has demostrated to be a good opportunity in the electrical generation in isolated regions or unsupplied of electrical lines. In Brazil, this alternative together other alternative energies, for instance, solar, wind and biomass, is a part of investments accomplished by small and big dealerships of energy, which see in the same a promising way for the complete electrification of the country. Today, Brazil has more than 10 % of electrical energy generated by thermoelectrical system. Year by year, the investiments in this sector has grown.

Besides, some these facilities has looked to assist the programs of environmental regulation in energy, that detach the control of the emission of the pollutant agents of the types NOx, SOx and CO. In Brazil, the government organization known by CONAMA – National Council of Environment - is responsible by approval of enterprises in the electric section that can not harm the environment.

[†] Corresponding author. Mechanical Engineer and Student of Doctorclass at UNIFEI. E-mail: harley.alencar@power.alstom.com

[‡] Academic advisor. Main advisor from UNIFEI. E-mail: marcoantonio@unifei.edu.br

[§] Academic Secondary advisor from UNIFEI. E-mail: patricia.foroni@itelefonica.com.br

The main places in Brazil where the micro turbines has been used in Amazonic Region. The limitant factor is the kind of fuel, in this case, the gas natural. The same has benn expensive yet because of technical limitation to construct the gas pipes along big distances. Although, many researchers has studied alternatives of combustion chambers for microturbines, for instance, the kind of annular chambers. There are analitical models of combustion chambers using biomass like fuel once that this a good alternative that can used in northwest of Brazil, for instance.

In kind of annular chamber, the aspect of construction is different from tubular chamber. While in the tubular chamber, there is only one nozzle of fuel, the annular chamber can have many nozzles of fuel. Some cases of annular chambers have inclinated nozzles to permit a rotative flow inside of chamber. It permitts a good combustion process, because of the increase of residence time.

This characteristic is very important because it involves the aerodynamics and thermodynamics concepts which can be optimized since some known dimensional and physical parameters are respected. Among them, the main parameters are the pressure, the speed of flow and the temperature.

Soon, the aerodynamic study of the chambers is relevant in the measure that it permitts to identify special regions in flow where there is a high pressure or temperature gradients. While the high pressure gradient is a high load loss, the high temperature gradient is a flame with high diffusion, which can prejudice the mechanical resistance of walls of chamber. Other phenomena that can seen is the formation of areas where there are vortex.

Any that are these physical phenomena, they have strong influence in the residence time and in the efficiency of the combustion.

In spite of the progress in the development of mathematical models more and more sophisticated, to simulate the behavior of combustion in gas turbines, there isn't a specific application for the combustion chambers of the annular type, only of the types tubular or cylindrical for great gas turbines. In this last case, good issues can be mentioned such as:

a) Melick et al. (1998) and Wakabayashi et. al.. (2002), which present a special solution against the instability of the fire using a valve to regulate the mixture air / fuel in nozzles for tubular chambers. But, this solution is not promising in the case of the chamber using biomass, due to the blockage possibility for the particles;

b) Yadigaroglu et al. (1998), which presents the influence of the load loss in the emission of the pollutant, that due to their dimensions, they need to have smaller load losses in way to induce small residence time; and

c) Lefebvre (1995) and Kuo (1986), which present some concepts about the simplification of theorical models with separation of the aerodynamic and thermodynamic studies, turning promising the application of numeric methods in the solution of the physical problems in combustion.

With this intention, different numeric methods were applied to analyze each component of the combustion chamber. Among these numeric methods, the calculation CFD (" Computational Fluid Dynamics"), based on Finite Volume Method, has been an inportant tool to analyse the combustion in tubular and annular chambers.

From this way, the present work proposes to do a contribution in the study of annular chamber from the comparative analysis among the different models of flows with the purpose to look the best model that describes the aerodynamic behavior for annular chambers, using the tool computational CFX v 5.7 ®.

2. Modelling

The modelling is divided in six steps: (a) to draw of full geometry in Mechanical DeskTop and a simplifiel geometry in CFX; (b) to generate a tetraedric mesh in CFX; (c) to set the physics domain; (d) to set the boundary conditions and convergence for calculation; (e) solve the problem using adequated models of turbulence flow; and (f) visualization of results.

The tested geometry is showed in next figure, with some technical data.

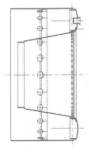 

Models: T - 62T. 32 (THOUSAND. W. 5088)
Manufacturer: Systens Wire
Number of nozzlers: 06 (sloping)
Power: 30 [kW]

Figure 1: The prototype and some technical data.

The developed methodology to solve the physic problem consists in the use of the Finite Volume Method in partial differential equations of mass and momentum conservation, in the general ways:

Mass Conservation – Continuity equation:

$$\frac{\partial \rho}{\partial t} + \nabla \bullet (\rho . v) = 0 \tag{1}$$

Momentum Conservation – Navier Stokes Equation:

$$\frac{\partial (\rho.u)}{\partial t} + (\rho.u) . \nabla \bullet u = -\nabla p + \frac{\partial}{\partial x_j}\left(\mu.\left[\frac{\partial u_i}{\partial x_j} + \frac{\partial u_j}{\partial x_i}\right]\right) + \Psi \tag{2}$$

Where Ψ is the variation of the momentum due to effects of the turbulence.

In a general way, the first member of the left of the continuity equation means the change of the specific mass for a fixed position in the control volume due to the efects of compressivity, while the second member of the left, means the kinetic flow. Besides, in Navier Stokes's equation, the flow is composed by convective and difusive flow, whose perturbation is presented by turbulence.

Considering that the flow is in permanent and incomprehensible regime, the equations (1) and (2) are changed for the following expressions:

$$\frac{\partial (\rho.Uj)}{\partial xj} = 0 \tag{3}$$

$$\frac{\partial (\rho.Ui.Uj)}{\partial xj} = -\frac{\partial p}{\partial xi} + \frac{\partial}{\partial x_j}\left[\mu.\frac{\partial Ui}{\partial xj} - \rho.(Ui.Uj)\right] \tag{4}$$

The solution of the equation (4) is dependent of the model that solves the third term in right member. The main turbulence models: the **Standard K-ε model**; the **RNG K-ε model**; and the **Differential Reynolds Stress Model (DRSM)**. All of them have good results in flows on curved surfaces.

In the next figure, it is shown the geometric model and mesh together the prototype.

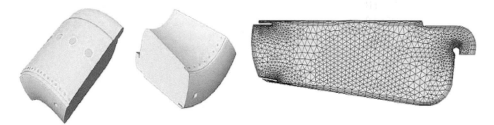

Figure 2: Different positions of model and its tethaedric mesh

This part 6TH of model has the following boundary conditions: (a) Secundary air in inlet with mass flow equal to 0,1475 (kg/s); (b) Primary air in nozzles with mass flow equal to 0,0225 (kg/s); and (c) Mass flow of fuel in nozzles equal to 29,125 (kg/s). Besides, the convergence condition has the goal error equal to 10^{-5} using 100 iterations.

3. Preliminary Results and Conclusions

From the compartion among the different models of turbulence flow applied in combustion chamber of kind annular, the prelimanry results are pressure and speed distribuitions, which are showed in figure 3.

The expected convergence for the three models of turbulence is about 70 iterations, where all of them can identify areas with high vortex around the nozzles, whose position can influence the residence time of combuston process and in the form of dissipation of the fire. As in this work it was not considered the effects of the transfer of heat, due to the combustion, it is intended in future works:

  1) to study the thermodynamic influence from combustion process and to set the distribuition of temperature and the main characteristics of fire;
  2) to compare different combustion models;
  3) to analyze the emission of gases pollutant agents; and
  4) to determine the efficiency of the combustion.

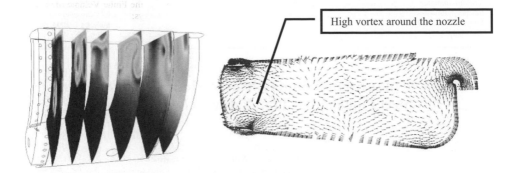

High vortex around the nozzle

Figure 3: Distribuition of speed and the formation of vortex around the position of nozzles.

References

[1] AEA Technologies, 2003, Tutorial of CFX v5 .7®., www.ansys.com/cfx

[2] COHEN H., et al., 1987, Gas Turbine Theory, 3rd Edition, Longman Scientific & Technical, New York, USES.

[3] Croker D. S., Nickolaus D. , Smith C. E., 2002, CFD Modeling of to Gas Turbine Chamber from Compressor Exit to Turbine Inlet, ASME.

[4] GOSSELIN P., DeChamplain S. K., Kretschmer D., 2000, Dimensional Three CFD Analysis of to Gas Turbine Combustor, 36th AIAA / ASME / it EXITS / ASEE Joint Propulsion Conference and Exhibit, pp 11, Huntsville, Alabama

[5] LEFEBVRE THE. H., 1995, The Rolls of Fuel Preparation in Low Emission Combustion, ASME Journal of Engineering goes Gas Turbines and Power, Vol 117, pp. 617-654

[6] LEFEBVRE THE. H., 1987, Gas Turbine Combustion, Mc-GrawHill Book Company, New York, USES

[7] KUO K.K., 1986, Principles of Combustion, John Wiley & Sons Edition, New York, USA.

[8] MELICK T. THE. et al., 1998, Burner Modifications goes Cost Effective Nox Control, Proceedings of the American Power Conference, Vol. 60. II, pp 855 - 860

[9] MUKUNDA, H.S., DASAPPA, S., SHRINIVASA, U., 1993, Open-top Wood Gasifiers, Renewable Energy: Soucers goes Fuels and Eletricity, Ed. T.B. Johansson, p. 699-728, London, UK.

[10] NICKOLAUS D. A., CROKER D. S., SMITH C. E., 2002, Development of Lean Direct Fuel Injector goes Low Emission Aero Gás Turbine, ASME

[11] YADIGAROGLU G. et al., 1998, Numerical and Experimental Study of Swirling Flow in Model Combustor, Heat Mass Transfer Journal, Vol. 41, no. 11, pp. 1485-1497.

[12] WAKABAYASHI T. et al., 2002, Performance of Dry Low Nox Gas Turbine Combustor Designed with to New Fuel Supply Concept, Engineering Goes Gas Turbines and Power Journal, ASME, Vol. 124, pp. 771-775

[13] WILLIANS. F., 1965, Combustion Theory, Addison. Wesley Publishing Company Inc., London, UK.

VSP International
Science Publishers
P.O. Box 346, 3700 AH Zeist
The Netherlands

*Lecture Series on Computer
and Computational Sciences*
Volume 1, 2004, pp. 20-24

Three-dimensional Finite Element Mesh Generation for Steel Pipeline Used for Natural Gas Transport

Y. Y. AL-Obaid [1]

Kuwait Oil Company, Water handling team
North Kuwait, P. O. Box 9758
61008 Ahmadi, Kuwait

Y. F. AL-Obaid[2]

Mechanical Engineering Department
Faculty of Technological Studies, PAAET
P.O. Box 42325, Shuwaikh
70654 Kuwait

Received 20 July, 2004; accepted in revised form 14 August, 2004

Abstract: A three- dimensional mesh generation technique is developed in this paper. Problems arising due to interactive modeling are highlighted in relation to complex geometries. A preprocessor finite element grid generator. FEMGEN, has been modified for mesh generation.

The modified FEMGEN program is linked with a three dimensional finite element program. Pipe sample is chosen to describe the flexibility and capability of the modified FEMGEN program. The appendix I shows typical subroutines to support the computer program discussed in the paper. Discussions and conclusions are given to validate the computer-aided approach.

Keywords: Isoparametric elements, three-dimensional finite element mesh generation, FEMGEN

1. Introduction

Generally a finite element preprocessor performs the task of preparing and checking a data structure and also improves the accuracy and the quality of the data structure itself.

The general problem connected with three-dimensional preprocessing is the fact that both tablet and vectorscope are basically two-dimensional devices hence unsuited for handling three-dimensional geometries. In this paper, an appropriate strategy has been devised to permit input of geometrical and topological data without extensive use of a keyboard to type in nodal related information. At the same time, the graphic display of the data has been organized in such a way as to allow the individual to preceive fully the three-dimensional nature of the structures and their elements.

In the present approach, the mapping is based on a discretization of a complete family of boundary surfaces which are defined by assuming that a set of planar cross sections can adequately describe the solid domain. When cross sections have been discretized into planar meshes, blending functions are used to generate the solid models, by interpolating between the sections. In this way, the problem of three dimensional element topology is considerably reduced. No heavy burden is placed on the analyst to manually compute and input a large number of nodal coordinates and topological data. The analyst hence is not tied to the drawing board the digitizing tablet-video display system is designed to replace.

---

[1] E-mail: yaobaid@kockw.com
[2] E-mail: yobaid@paaet.edu.ku

2. Mesh Generation Process

By combining the dual capabilities of (a) creating a varying number of subdivisions around the zone, and (b) grading the nodes on any side of the zone, the program subdivides a zone into meshes to satisfy most combinations of different corner element sizes.

This mesh generation process is applied in three stages to each zone in turn:

Stage I. Topological Node Generation

The topological structure of the mesh is established within a component of the normalised (ξ, η, ς) solid model.

Each side of the component is subdivided into a number of equal sub-divisions determined from a knowledge of the four target element sizes at the vertices of the side. There are some restrictions on the difference in the number subdivisions possible on opposite of the component.

Stage II.   Mesh Grading

Adjustment are made to the nodal positions, established in the first stage, in order to produce grading of element sizes along one or more of the zone/sides as required.

Stage III. Mesh Mapping

In the mesh mapping, there is a basic change to the Program FEMGEN and this is beginning with the Jacobian Matrix (J) Using 20-noded elements, the domain transformation TD in curvilinear coordinate which maps with that in the cartesian coordinate. The transfinite interpolation takes the form:

$$\text{TFD} = [X,Y,Z]^{T} = \sum_{i=1}^{n} N_{i} \left[(\xi,\eta,\varsigma) \right] (X_{i},Y_{i},Z_{i})$$

(1)

By knowing the shape functions of 20 noded elements (AL-OBAID [1984]), the six faces of the isoparametric element can then be determined by replacing (ξ, η, ς) to -1 or 1.

The generic form of equation 1 is written as:

$$\begin{bmatrix} x(\xi,\eta,\varsigma) \\ \dots \\ y(\xi,\eta,\varsigma) \\ \dots \\ \dots \\ z(\xi,\eta,\varsigma) \end{bmatrix} = \sum_{i=1}^{8} N_{i} \left[(\xi,\eta,\varsigma) \right] \begin{bmatrix} x_{i} \\ \dots \\ y_{i} \\ \dots \\ \dots \\ z_{i} \end{bmatrix}$$

(2)

For 20-noded element, twelve of twenty shape functions are automatically zero on a specific face. The remaining eight shape functions N_{i} are relabeled as $\acute{N}_{1}$, $\acute{N}_{2}...\acute{N}_{8}$. Hence in equation 2 the associated coordinates related to these functions are reordered.

Assuming that the equation of any plane is given as

$$A_{1}x + A_{2}y + A_{3}z + A_{4} = 0$$

(2a)

and if this plane is intersecting an element face already determined by equation 2 and if it is also desired to know the curved element in 3-D space, it is essential to solve numerically the non-linear equation for the locus of points such as

$$X(\xi,\eta,\varsigma), \ Y(\xi,\eta,\varsigma), \ Z(\xi,\eta,\varsigma) \ \text{etc.}$$

The non-linear equation chosen appropriately to suit equation 2a and for evaluation of such locus of points is given by:

$$f(\xi,\eta,\varsigma) = A_1 X(\xi,\eta,\varsigma) + A_2 Y(\xi,\eta,\varsigma) + A_3 Z(\xi,\eta,\varsigma) + A_4 = 0 \qquad (3)$$

Equation (3) can be solved trice. First for ξ as a function of η then η as a function of ς and finally of ξ, ς. If r_Λ is a unit normal to a given plane in equation 2a, a coordinate system can be established by orthogonal unit vectors:

$$n_1 = \frac{P_1 - P_0}{\|P_1 - P_0\|}$$

$$n_2 = r_A \times n$$

where $\quad r_A = x_i + y_j + z_k$

$P_0(X_0,Y_0,Z_0)$ and $P_1(X_1, Y_1, Z_1)$ are points in the plane such that $(P_1 - P_0) \, r_A = 0$

Any curved element can now be plotted in the plane relative to the establishing coordinate system.

After establishing a coordinate system in space, it became essential to create a three-dimensional geometry by interpolating between the cross-sections. Each cross-section is now discretised into planar meshes. In order to generate a solid three -dimensional model, blending functions are used as described by Perucchio et al (1982) first by interpolating between the cross-sections. These blending functions are given below:

$$\sum_{i=1}^{n} f_i^{B}(\xi,\eta,\varsigma)\delta_{ii} \begin{bmatrix} 1=i=j \\ \cdots \\ 0 \quad i=j \end{bmatrix} \qquad (4)$$

where f_i^{B} are blending functions and δ_{ii} Kronecker delta

In some cases it becomes essential to generate cardinal spline curves and sectional meshes at faces of solid isoparametric elements. The above blending functions are chosen to suit the nature of a mapping functions "f" and hence the projector $P\left[f_i^{B}\right]$ can be written as:

$$P\left[f_i^{B}\right]=\sum_{i=0}^{n} N_i^{B}(\zeta) \ f(\xi,\eta,\zeta) \qquad (5)$$

where N_i^{B} is known.

As suggested by Perucchio et al (1982) where interior generalised coordinate surfaces are specified, cubic spine blending functions are adopted. The interpolators are defined as:

$$P\left[f_i^B\left(\xi,\eta,\zeta\right)\ f\left(\xi,\eta,\zeta\right)\right]=\sum_{j=0}^{1}\ \sum_{k=0}^{1}\Psi_i\left(\xi\right)\Psi_k\left(\eta\right)\ f\left(\xi_j,\eta_k,\zeta_i\right)$$

where Ψ_i, Ψ_k are the Lagrange polynomial interpolators defined by Perucchio et al (1982).

It is necessary to refer to some additional features which have been included, and these are:

1. Screen Wipe (SWIPT)

2. The clearance of the display log and the requirement of the picture to be scaled and   then positioned centrally on the screen.

3. Colour and Hard Copy Production

The plot has a modified version of Hard colour Copy View (HCPV) which has instructions in the file for later creation of a hard copy of exactly what was on the screen at the time of issuing the command HCPV. Command copies the picture described by the display lot into an off-file in the form suitable for plotting by a hard plotter. Colour is also used for the resolution of picture parts which represent objects of different types. All screen colour is transmitted through into the plotter file for hard copy production with conventional pen plotters. Some of the applications of the modified program are given by the author (AL-OBAID [1984,1986, 2003].

3. Example

Figure 1 shows Natural gas pipeline transport. Figure 2 shows a three-dimensional finite element mesh generations for Natural gas pipeline transport.

Fig. 1. Natural gas pipeline transport.

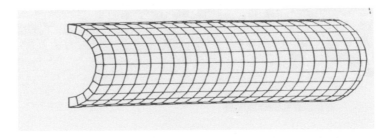

Fig. 2. Three-dimensional finite element mesh generation for Natural gas pipeline

References

[1] Rutherford Engineering Laboratory, U.K., 1981, "Finite Element Mesh Generation" User Manual, Version October.

[2] Perucchio, R et al, 1982 "Interactive Computer Graphic Pre-Processing for Three-dimensional Finite Element Analysis", Int. J. For Num. Met. Eng., Vol. 18., P. 909-920

[3] Al.-Obaid, Y.F., AL-Hassani " S.T.S., Bangash, Y, 1984, "Three-Dimensional Dynamic Finite Element for Indentation problems as applied to shot-peening Mechanics". 2nd international conference on shot-peening, Chicago, U.S.A.

[4] AL-Obaid, Y.F., 1986, "Finite Element Approach to Shot-peening Mechanics", International Conference on Steel Structure, Yugoslavia.

[5] AL-Obaid, Y.F., 2003, "Three-dimensional bond analysis of octagonal prestressed concrete slabl", The Journal of Kuwait University Science and Engineering, Vol. 30,No. 1.

VSP International
Science Publishers
P.O. Box 346, 3700 AH Zeist
The Netherlands

*Lecture Series on Computer
and Computational Sciences*
Volume 1, 2004, pp. 25-28

An Adaptive Version of a Fourth Order Iterative Method
for Quadratic Equations

Sergio Amat[1]

Departamento de Matemática Aplicada y Estadística. Universidad Politécnica de Cartagena
(Spain).

Sonia Busquier[2]

Departamento de Matemática Aplicada y Estadística. Universidad Politécnica de Cartagena
(Spain).

J.Manuel Gutiérrez[3]

Departamento de Matemáticas y Computación. Universidad de La Rioja (Spain).

Received 6 April, 2004; accepted in revised form 24 June, 2004

Abstract: A fourth order iterative method for quadratic equations is considered. An adaptive implementation with multiresolution scheme is proposed. A semilocal convergence theorem is presented. Finally, some numerical results are studied.

Keywords: nonlinear quadratic equations, fourth order, wavelets

Mathematics Subject Classification: 65J15, 65B05, 41A05, 41A10, 65D05, 65D17

Determining the zeros of a nonlinear equation is a classical problem. These roots cannot in general be expressed in closed form. A powerful tool to study these equations is the use of iterative processes, starting from an initial guess x_0 successive approaches (until some predetermined convergence criterion is satisfied) x_i are computed , $i = 1, 2, \ldots$, with the help of certain iteration function $\Phi : X \rightarrow X$,

$$x_{n+1} := \Phi(x_n), \quad n = 0, 1, 2 \ldots \tag{1}$$

In general, an iterative method is of p-th order if the solution x^* of $F(x) = 0$ satisfies $x^* = \Phi(x^*)$, $\Phi'(x^*) = \cdots = \Phi^{(p-1)}(x^*) = 0$ and $\Phi^{(p)}(x^*) \neq 0$. For such a method, the error $\|x^* - x_{n+1}\|$ is proportional to $\|x^* - x_n\|^p$ as $n \rightarrow \infty$. It can be shown that the number of significant digits is multiplied by the order of convergence (approximately) by proceeding from x_n to x_{n+1}.

Newton's method and similar second order methods are the most used. Higher order methods require more computational cost than other simpler methods, which makes them disadvantageous to be used in general, but, in some cases, it pays to be a little more elaborated.

In this paper, we are interesting in the solution of *quadratic equations*,

$$F''(x) = B,$$

where B is a bilinear constant form. This is an example where third order methods are a good alternative to Newton's type methods. Some particular cases of these type of equations, that appear in many applications, are Riccati's equations.

---

[1]Research supported in part by PI-53/00809/FS/01. E-mail: sergio.amat@upct.es

[2]E-mail: sonia.busquier@upct.es

[3]E-mail: jmguti@dmc.unirioja.es

To approximate a solution of the nonlinear equation

$$F(x) = 0, \qquad (2)$$

$F : X \to Y$, X, Y Banach spaces, the classical Chebyshev ($\beta = 0$), Halley ($\beta = \frac{1}{2}$) and Super-Halley ($\beta = 1$) methods can be written as

$$x_{n+1} = x_n - \left(I + \frac{1}{2} L_F(x_n) \left[I - \beta L_F(x_n) \right]^{-1} \right) F'(x_n)^{-1} F(x_n), \qquad (3)$$

where

$$L_F(x_n) = F'(x_n)^{-1} F''(x_n) F'(x_n)^{-1} F(x_n),$$
$$\beta \in [0, 1].$$

On the other hand, the third order methods can be written as

$$x_{n+1} = x_n - \left(I + \frac{1}{2} L_F(x_n) + O(L_F(x_n)^2) \right) F'(x_n)^{-1} F(x_n).$$

In particular, each iteration of a third order method is between two C-methods,

$$x_{n+1} = x_n - \left(I + \frac{1}{2} L_F(x_n) + C L_F(x_n)^2 \right) F'(x_n)^{-1} F(x_n). \qquad (4)$$

In this paper, we find and analyze a C-method with fourth order of convergence for quadratic equations. The advantage of this method is that the matrix of the different associated linear systems in each iteration is the same. Finally, using a wavelet transfrom we obtain an adaptive version of the scheme with better computational properties.

In general, for the C-methods $\Phi'''(x^*) \neq 0$, but taking $C = C(x) = \frac{1}{2}(1 - \frac{L_{F'}(x)}{3})$, we can improve the accuracy since in this case

$$\Phi'''(x^*) = 0.$$

For quadratic equations $L_{F'}(x) = 0$, and $C = C(x)$ will be the constant $C = \frac{1}{2}$.
On the other hand, the $\frac{1}{2}$-method

$$x_{n+1} = x_n - \left(I + \frac{1}{2} L_F(x_n) + \frac{1}{2} L_F(x_n)^2 \right) F'(x_n)^{-1} F(x_n), \qquad (5)$$

can be written as,

$$y_n = x_n - F'(x_n)^{-1} F(x_n),$$
$$z_n = y_n + \frac{1}{2} L_F(x_n)(y_n - x_n),$$
$$x_{n+1} = z_n + L_F(x_n)(z_n - y_n).$$

We are interesting in quadratic equations. In this case, $F''(x)$ is a constant bilinear operator that we denote by B.

Using Taylor expansions, the method becomes

$$y_n = x_n - F'(x_n)^{-1} F(x_n),$$
$$z_n = y_n - F'(x_n)^{-1} F(y_n),$$
$$x_{n+1} = z_n - F'(x_n)^{-1} (2F(z_n) - \frac{1}{2} B(z_n - y_n)(z_n - x_n)),$$

equivalently,

$$
\begin{aligned}
F'(x_n)(y_n - x_n) &= -F(x_n), \\
F'(x_n)(z_n - y_n) &= -F(y_n), \\
F'(x_n)(x_{n+1} - z_n) &= -2F(z_n) + \frac{1}{2}B(z_n - y_n)(z_n - x_n).
\end{aligned}
\tag{6}
$$

Remark 1 *Notice you, that for these equations the classical third order iterative methods can be written as*

$$
\begin{aligned}
y_n &= x_n - F'(x_n)^{-1}F(x_n), \\
z_n &= x_n + \theta(y_n - x_n), \ 0 \le \theta \le 1, \\
x_{n+1} &= y_n - F'(y_n)^{-1}F(y_n),
\end{aligned}
\tag{7}
$$

($\theta = 0$ Chebyshev, $\theta = \frac{1}{2}$ Halley and $\theta = 1$ Super-Halley).
Then, only for Chebyshev's method the matrix of the different linear systems is the same.

The evaluations of F in the method will be computed as follows.

Lemma 1 *Let F a quadratic operator, x_n, y_n and z_n the sequences defined in (6). Then*

$$
F(y_n) = \frac{1}{2}B(y_n - x_n)^2,
\tag{8}
$$

$$
F(z_n) = B(z_n - y_n)(y_n - x_n) + \frac{1}{2}B(z_n - y_n)^2,
\tag{9}
$$

$$
\begin{aligned}
F(x_{n+1}) &= -\frac{1}{2}B(z_n - y_n)(y_n - x_n) + \frac{1}{2}B(x_{n+1} - z_n)^2 \\
&\quad + \frac{1}{2}B(z_n - x_n)(x_{n+1} - z_n) + \frac{1}{2}B(z_n - y_n)(x_{n+1} - z_n).
\end{aligned}
\tag{10}
$$

The adaptive version of the method is performed using wavelet transforms.

Given a matrix A, its standard form, denoting by M the direct wavelet algorithm and by M^{-1} its inverse algorithm, is the matrix

$$
A^s = MAM^{-1}.
\tag{11}
$$

Equation (11) is equivalent to applying the direct algorithm to the columns of A, and the transpose of the inverse algorithm to the rows of the resulting matrix.

The standard form satisfies

$$
M(Af) = A^s M(f) \quad \text{for every vector } f; \quad (A^2)^s = A^s A^s.
$$

Both M and M^{-1} are continuous transformations. Thus if an $N \times N$ matrix A represents a piecewise smooth operator, each transformed row in AM^{-1} represents a piecewise smooth operator. Sparsity is obtained by truncation of each transformed column in MAM^{-1}. The matrix A^s will have $O(N log N)$ significant coefficients, $O(log N)$ for each column. The transformation of the columns leaves the significant coefficients in the typical *finger shape*.

Finally, we are interesting to obtain a standard form for bilinear operators. For simplicity, we consider

$$
B : \mathbb{R}^2 \times \mathbb{R}^2 \to \mathbb{R}^2
$$

the extension to dimension N is self-evident.

Given two vector $x, y \in \mathbb{R}^2$, we have

$$B(x,y) := (x_1, x_2) * \left(\begin{array}{c} B_1 \\ B_2 \end{array} \right) \left(\begin{array}{c} y_1 \\ y_2 \end{array} \right),$$

where $*$ denotes the product by blocks,

$$(x_1, x_2) * \left(\begin{array}{c} B_1 \\ B_2 \end{array} \right) = \left(\begin{array}{cc} b_1^{11} x_1 + b_1^{21} x_2 & b_1^{12} x_1 + b_1^{22} x_2 \\ b_2^{11} x_1 + b_2^{21} x_2 & b_2^{12} x_1 + b_2^{22} x_2 \end{array} \right).$$

Considering

$$H_1 := \left(\begin{array}{cc} b_1^{11} & b_1^{21} \\ b_2^{11} & b_2^{21} \end{array} \right),$$

$$H_2 := \left(\begin{array}{cc} b_1^{12} & b_1^{22} \\ b_2^{12} & b_2^{22} \end{array} \right),$$

it is easy to check that

$$(x_1, x_2) * \left(\begin{array}{c} B_1 \\ B_2 \end{array} \right) = \left(\begin{array}{c} H_1 \\ H_2 \end{array} \right) * \left(\begin{array}{c} x_1 \\ x_2 \end{array} \right).$$

Therefore,

$$
\begin{aligned}
B(x,y) &= M((x_1, x_2) \left(\begin{array}{c} B_1 \\ B_2 \end{array} \right) \left(\begin{array}{c} y_1 \\ y_2 \end{array} \right)) \\
&= M * \left(\begin{array}{c} H_1 \\ H_2 \end{array} \right) * M^{-1} M \left(\begin{array}{c} x_1 \\ x_2 \end{array} \right) M^{-1} M \left(\begin{array}{c} y_1 \\ y_2 \end{array} \right).
\end{aligned}
$$

Thus to compute $MB(x,y)$ we can do

a) $B^s := M * \left(\begin{array}{c} H_1 \\ H_2 \end{array} \right) * M^{-1}$, that we define as the standard form of the bilinear operator B.

This equation is equivalent to applying the direct algorithm to the columns of H_1 and H_2, and the transpose of the inverse algorithm to the rows of the resulting matrix block.

b) $Mx = M \left(\begin{array}{c} x_1 \\ x_2 \end{array} \right)$,

c) $C := (B^s Mx) M^{-1}$,

d) $My = M \left(\begin{array}{c} y_1 \\ y_2 \end{array} \right)$,

e) CMy.

On the other hand, we present a semilocal convergence theorem for the introduced method.

Theorem 1 *Given x_0 such that there exists $F'(x_0)^{-1}$ and condition*

$$||F'(x_0)^{-1} B|| \, ||F'(x_0)^{-1} F(x_0)|| \le \frac{1}{2} \tag{12}$$

holds, then method (6) is well defined and converges to x^*, solution of $F(x) = 0$. Denoting by $\{\tilde{x}_n\}$ the sequence of the adaptive version of the method, if we consider the truncation parameters $\epsilon_k(n) = \epsilon(n)$ sufficiently small, then it will be well defined and there will exist $n_{Tol} \in \mathbb{N}$ such that

$$||x^* - \tilde{x}_n|| \le Tol, \ \forall n \ge n_{Tol}.$$

Finally, we present some numerical examples.

VSP International
Science Publishers
P.O. Box 346, 3700 AH Zeist
The Netherlands

Lecture Series on Computer
and Computational Sciences
Volume 1, 2004, pp. 29-32

A Fully Adaptive PPH Multiresolution Scheme for Image Processing [1]

Sergio Amat[2]

Departamento de Matemática Aplicada y Estadística. Universidad Politécnica de Cartagena (Spain).

Rosa Donat[3]

Departament de Matemàtica Aplicada. Universitat de València (Spain).

Jacques Liandrat[4]

Ecole Généraliste d'Ingénieurs de Marseille (EGIM). Laboratoire d'Analyse Topologie et Probabilites (LATP). (France)

J.Carlos Trillo[5]

Departamento de Matemática Aplicada y Estadística. Universidad Politécnica de Cartagena (Spain).

Received 7 March, 2004; accepted in revised form 10 March, 2004

Abstract: In this paper, a new nonlinear multiresolution transform is presented. It is based on a centered piecewise nonlinear polynomial interpolation fully adapted to discontinuities. Compression properties of the multiresolution are studied.

Keywords: Multiresolution, nonlinear scheme, Harmonic mean

Mathematics Subject Classification: 41A05, 41A10, 65D05, 65D17

In this paper, we present a new nonlinear multiresolution transform for discrete data. Given f^L a data where L stands for a resolution level, a multiresolution representation of f^L is any sequence of type $\{f^0, d^1, \ldots, d^L\}$ where f^k is an approximation of f^L at resolution $k < L$ and d^{k+1} stands for the details required to recover f^{k+1} from f^k. The couple $\{f^k, d^{k+1}\}$ contains the same information as f^{k+1} and therefore the same is true for $\{f^0, d^1, \ldots, d^L\}$ and f^L. Inter-resolution operators are named decimation (from fine to coarse) and prediction (from coarse to fine). The reconstruction and discretization operators act between the continuous level to any discrete level.

Linear multiresolution representations of data, like wavelet decompositions, are multiresolutions involving inter-resolution linear operators. The efficiency of linear multiresolution decompositions is generally limited by the presence of discontinuities or edges. On one hand, the numerically significant detail coefficients d_j^k are mainly those for which the wavelet support is intersected by such discontinuities. On the other hand, to obtain few significant coefficients in the smooth regions of the signal, high order wavelets (with larger supports) are needed.

[1] Research partially supported by european network "Breaking complexity" # HPRN-CT-2002-00286
[2] Research supported in part by PI-53/00809/FS/01. E-mail: sergio.amat@upct.es
[3] E-mail: donat@uv.es
[4] E-mail: liandrat@esm2.imt-mrs.fr
[5] E-mail: jc.trillo@upct.es

A *specific adaptive treatment of singularities*, can be performed in multiresolutions within Harten's setting. The greatest advantage of this general framework lies in its adaptability, where it is indeed possible to consider nonlinear reconstruction operators, which lead to near optimal data-compression rates.

A first step towards non linear adaption near singularities has been proposed by ENO or PPH type multiresolutions. The PPH reconstruction avoids the stencil selection and the stencil decentering that are known to reduce the efficiency of ENO multiresolution. Moreover, it maintains the quality of approximation of the linear scheme in smooth regions and it does not deteriorate too much the accuracy around singularities, except in the intervals containing the singularities.

The aim of this paper is to describe and analyze the nonlinear multiresolution transform that results from considering a modified version of the PPH reconstruction techniques fully adapted to the singularities improving the behaviour of the original PPH multiresolution.

First, one wants to obtain adaptation in the intervals containing the singularities.

Second, introducing the different divided differences $e_{j-\frac{1}{2}} = f[x_{j-1}, x_j]$, $e_{j+\frac{1}{2}} = f[x_j, x_{j+1}]$, $e_{j+\frac{3}{2}} = f[x_{j+1}, x_{j+2}]$, $D_j = f[x_{j-1}, x_j, x_{j+1}]$ and $D_{j+1} = f[x_j, x_{j+1}, x_{j+2}]$, one aims to avoid the accuracy reduction of PPH reconstruction when $D_j D_{j+1} \leq 0$.

a) We consider the set of points $f_{j-1}, f_j, f_{j+1}, f_{j+2}$ and describe the prediction of the mid point $f_{j+\frac{1}{2}}$ assuming (after we will see how to know this information) the discontinuity is in the sub-interval $(x_{j+\frac{1}{2}}, x_{j+1})$.

We first consider the Lagrange interpolatory polynomial P_j of degree three.

It is easy to check that P_j is equivalently defined by

$$\begin{cases} P_j(x_m) &= f_m, \quad j-1 \leq m \leq j \\ a_1 &= \frac{f_{j-1} - 27f_j + 27f_{j+1} - f_{j+2}}{24h} \\ a_3 &= \frac{-f_{j-1} + 3f_j - 3f_{j+1} + f_{j+2}}{6h^3}. \end{cases} \tag{1}$$

Using the divided differences introduced above, the coefficients a_1 and a_3 take the form

$$a_1 = \frac{-e_{j+\frac{3}{2}} + 25e_{j-\frac{1}{2}}}{24} + 13\frac{e_{j+\frac{1}{2}} - e_{j-\frac{1}{2}}}{12}$$

$$a_3 = \frac{e_{j+\frac{3}{2}} - e_{j-\frac{1}{2}}}{6h^2} + 2\frac{-e_{j+\frac{1}{2}} + e_{j-\frac{1}{2}}}{6h^2}.$$

Let us consider $\varepsilon_{j-\frac{1}{2}} := e_{j-\frac{1}{2}}(1+h)$. Introducing the following new second order divided differences, associated to $\varepsilon_{j-\frac{1}{2}}$,

$$E_1 = \frac{\varepsilon_{j-\frac{1}{2}} - e_{j-\frac{1}{2}}}{h}$$

$$E_2 = \frac{e_{j+\frac{1}{2}} - \varepsilon_{j-\frac{1}{2}}}{h}$$

$$E_3 = \frac{e_{j+\frac{3}{2}} - \varepsilon_{j-\frac{1}{2}}}{h},$$

a_1 and a_3 can be rewritten as

$$a_1 = e_{j-\frac{1}{2}} - \left(\frac{1}{12}\frac{E_3 + E_1}{2} - \frac{13}{6}\frac{E_2 + E_1}{2}\right)h$$

$$a_3 = \frac{1}{3h}\left(\frac{E_3 + E_1}{2} - 2\frac{E_2 + E_1}{2}\right).$$

We are now in a similar situation as the original PPH reconstruction and we can find the adaptation to the singularity using harmonic means.

If the singularity is in the sub-interval $(x_j, x_{j+\frac{1}{2}})$, considering $\varepsilon_{j+\frac{3}{2}} := e_{j+\frac{3}{2}}(1+h)$ and

$$E_1 = \frac{\varepsilon_{j+\frac{3}{2}} - e_{j-\frac{1}{2}}}{h}$$

$$E_2 = \frac{\varepsilon_{j+\frac{3}{2}} - e_{j+\frac{1}{2}}}{h}$$

$$E_3 = \frac{e_{j+\frac{3}{2}} - \varepsilon_{j+\frac{3}{2}}}{h},$$

a_1 and a_3 can be rewritten as

$$a_1 = e_{j+\frac{3}{2}} + (\frac{1}{12}\frac{E_3 + E_1}{2} - \frac{13}{6}\frac{E_3 + E_2}{2})h$$

$$a_3 = \frac{1}{3h}(-\frac{E_3 + E_1}{2} + 2\frac{E_3 + E_2}{2}).$$

How to find the location of the discontinuities?

Looking the size of the absolute value of the divided differences $|e_{j-\frac{1}{2}}|$, $|e_{j+\frac{1}{2}}|$ and $|e_{j+\frac{3}{2}}|$, we select the interval where the singularity is.

In the present case, that is, the selected interval becomes (x_j, x_{j+1}), we have to find the position of the jump singularity respect to $x_{j+\frac{1}{2}}$.

Following a similar idea as in the Harten's subcell-resolution technique, we consider the function

$$G(x) = P_1(x) - P_2(x)$$

where $P_1(x)$ and $P_2(x)$ are the Lagrange polynomials for the primitive function associated to the points $\{x_{j-2}, x_{j-1}, x_j\}$ and $\{x_{j+1}, x_{j+2}, x_{j+3}\}$ respectively. Then, if

$$G(x_{j+\frac{1}{2}})G(x_j) \geq 0$$

then the singularity will be in the sub-interval $[x_{j+\frac{1}{2}}, x_{j+1}]$ otherwise in the sub-interval $[x_{j-1}, x_{j+\frac{1}{2}})$. With this strategy the exact location of the singularity can be recovered with an error of $O(h^2)$.

b) Finally, we present a way not to deteriorate the accuracy when the second order divided differences have different signs. A possibility could be to make first a translation of the values D_j, D_{j+1} such that the translated values get the same sign, to perform then the non linear mean and finally to make the inverse translation.

Assume $|D_j| \leq |D_{j+1}|$ (the other case is similar). Let us consider

$$T = -D_j - D_j h + D_{j+1}h^p$$

where p is a natural number big enough. Then $\frac{D_j + D_{j+1}}{2}$ will be replaced by

$$2\frac{(D_j + T)(D_{j+1} + T)}{(D_j + T) + (D_{j+1} + T)} - T$$

We are now able to analyze the properties of the new mean:

1) The term $D_{j+1}h^p$ is the reponsable to do translation when $D_j = 0$. We can consider the number p such that $D_{j+1}h^p = O(h)$.

2) Using the definition of the harmonic mean, we deduce that

$$|2\frac{(D_j + T)(D_{j+1} + T)}{(D_j + T) + (D_{j+1} + T)} - T| \leq \max\{|D_j|, |D_j - D_j h + D_{j+1}h^p|\}$$

In particular, the new mean is adapted, as the original harmonic mean, to the presence of singularities.

3) In smooth regions, the difference between the arithmetic and the new one is $O(h^2)$ and therefore the fourth order of accuracy of the associated reconstruction remains.

References

[1] Amat S., Aràndiga F., Cohen A. and Donat R., (2002). Tensor product multi resolution analysis with error control for compact image representation. *Signal Processing*, **82**(4), 587-608.

[2] Amat S., Donat R., Liandrat J. and Trillo J.C., (2004). Analysis of a fully nonlinear multi resolution scheme for image processing, Preprint U.P. Cartagena **38** 2002, submitted.

[3] Aràndiga F. and Donat R., (2000). Nonlinear Multi-scale Decomposition: The Approach of A.Harten, *Numerical Algorithms*, **23**, 175-216.

[4] Cohen A., Dyn N. and Matei B., (2003). Quasi linear subdivision schemes with applications to ENO interpolation. *Applied and Computational Harmonic Analysis*, **15**, 89-116.

[5] Daubechies I., Runborg O. and Sweldens W., (2004). Normal multi resolution approximation of curves, *Const. Approx.*, Published Online 2003.

[6] Dyn N., (1992). Subdivision schemes in computer aided geometric design, *Advances in Numerical Analysis II., Subdivision algorithms and radial functions, W.A. Light (ed.), Oxford University Press*, 36-104. Prentice-Hall.

[7] Harten A., (1996). Multiresolution representation of data II, *SIAM J. Numer. Anal.*, **33**(3), 1205-1256.

[8] Oswald P., (2004). Smoothness of Nonlinear Median-Interpolation Subdivision, *Adv. Comput. Math.*, **20**(4), 401-423.

[9] Trillo J.C., (2004). Multirresolución no lineal y Aplicaciones, PhD in the University of Valencia, Spain.

VSP International
Science Publishers
P.O. Box 346, 3700 AH Zeist
The Netherlands

*Lecture Series on Computer
and Computational Sciences*
Volume 1, 2004, pp. 33-36

Trigonometrically Fitted Runge-Kutta Methods of Order Five for the Numerical Solution of the Schrödinger Equation

Z.A. Anastassi and T.E. Simos [1] [2] [3] [4]

Department of Computer Science and Technology,
Faculty of Sciences and Technology,
University of Peloponnese,
GR-221 00 Tripolis, Greece

Received 21 September, 2004

Abstract: We are constructing two trigonometrically fitted methods based on a classical Runge-Kutta method. The new methods maintain the algebraic order of the classical one but also have some other significant properties. The most important one is that in the local truncation error of the new methods the powers of the energy are lower and that keeps the error at lower values, especially at high values of energy. The error analysis justifies the actual results when integrating the radial Schrödinger Equation, where the high efficiency of the new methods is shown.

Keywords: Explicit Runge-Kutta methods, exponential fitting, trigonometrical fitting, radial Schrödinger equation, resonance problem, energy.

PACS: 0.260, 95.10.E

1   Introduction

Much research has been done on the numerical integration of the radial Schrödinger equation:

$$y''(x) = \left(\frac{l(l+1)}{x^2} + V(x) - E \right) y(x) \tag{1}$$

where $\frac{l(l+1)}{x^2}$ is the *centrifugal potential*, $V(x)$ is the *potential*, E is the *energy* and $W(x) = \frac{l(l+1)}{x^2} + V(x)$ is the *effective potential*. It is valid that $\lim_{x\to\infty} V(x) = 0$ and therefore $\lim_{x\to\infty} W(x) = 0$.

Many problems in chemistry, physics, physical chemistry, chemical physics, electronics etc., are expressed by equation (1).

In this paper we will study the case of $E > 0$. We divide $[0, \infty]$ into subintervals $[a_i, b_i]$ so that $W(x)$ is a constant with value $\bar{W}_i$. After this the problem (1) can be expressed by the approximation

---

[1]President of the European Society of Computational Methods in Sciences and Engineering (ESCMSE)

[2]Active Member of the European Academy of Sciences and Arts

[3]Corresponding author. Please use the following address for all correspondence: Dr. T.E. Simos, 26 Menelaou Street, Amfithea - Paleon Faliron, GR-175 64 Athens, GREECE, Tel: 0030 210 94 20 091

[4]E-mail: tsimos@mail.ariadne-t.gr

$$y_i'' = (\bar{W} - E)\,y_i, \qquad \text{whose solution is}$$
$$y_i(x) = A_i \exp\left(\sqrt{\bar{W} - E}\,x\right) + B_i \exp\left(-\sqrt{\bar{W} - E}\,x\right), \quad A_i, B_i \in \Re. \tag{2}$$

This form of Schrödinger equation reveals the importance of exponential fitting when constructing new methods. In the next section we will present the most important parts of the theory used.

2 Basic theory

2.1 Explicit Runge-Kutta methods

An s-stage explicit Runge-Kutta method used for the computation of the approximation of $y_{n+1}(x)$, when $y_n(x)$ is known, can be expressed by the following relations:

$$y_{n+1} = y_n + \sum_{i=1}^{s} b_i\,k_i$$

$$k_i = h\,f\left(x_n + c_i h,\ y_n + h\sum_{j=1}^{i-1} a_{ij}\,k_j\right),\ i = 1,\dots,s \tag{3}$$

where in this case $f\left(x, y(x)\right) = (W(x) - E)\,y(x)$.

Actually to solve the second order ODE (1) using first order numerical method (3), (1) becomes:

$$z'(x) = (W(x) - E)\,y(x)$$
$$y'(x) = z(x)$$

while we use two pairs of equations (3): one for y_{n+1} and one for z_{n+1}.

The method shown above can also be presented using the Butcher table below:

$$
\begin{array}{c|ccccc}
0 \\
c_2 & a_{21} \\
c_3 & a_{31} & a_{32} \\
\vdots & \vdots & \vdots \\
c_s & a_{s1} & a_{s2} & \cdots & a_{s,s-1} \\
\hline
 & b_1 & b_2 & \cdots & b_{s-1} & b_s
\end{array}
\tag{4}
$$

Coefficients $c_2, \dots, c_s$ must satisfy the equations:

$$c_i = \sum_{j=1}^{i-1} a_{ij},\ i = 2,\dots,s \tag{5}$$

Definition 1 [1] A Runge-Kutta method has algebraic order p when the method's series expansion agrees with the Taylor series expansion in the p first terms: $y^{(n)}(x) = y_{app.}^{(n)}(x), \quad n = 1, 2, \dots, p$.

A convenient way to obtain a certain algebraic order is to satisfy a number of equations derived from Tree Theory. These equations will be shown during the construction of the new methods.

2.2 Exponentially fitted Runge-Kutta methods

The method (3) is associated with the operator

$$L(x) = u(x+h) - u(x) - h\sum_{i=1}^{s} b_i\, u'\,(x + c_i h, U_i)$$

$$U_i = u(x) + h\sum_{j=1}^{i-1} a_{ij} u'\,(x + c_j h, U_j), \quad i = 1, \ldots, s \tag{6}$$

where u is a continuously differentiable function.

Definition 2 [2] The method (6) is called exponential of order p if the associated linear operator L vanishes for any linear combination of the linearly independent functions $\exp(v_0 x)$, $\exp(v_1 x)$, ..., $\exp(v_p x)$, where $v_i | i = 0(1)p$ are real or complex numbers.

Remark 1 [3] If $v_i = v$ for $i = 0, 1, \ldots, n$, $n \leq p$, then the operator L vanishes for any linear combination of $\exp(vx)$, $x\exp(vx)$, $x^2\exp(vx)$, ..., $x^n\exp(vx)$, $\exp(v_{n+1}x)$, ..., $\exp(v_p x)$.

Remark 2 [3] Every exponentially fitted method corresponds in a unique way to an algebraic method (by setting $v_i = 0$ for all i)

Definition 3 [2] The corresponding algebraic method is called the classical method.

3 Construction of the new trigonometrically fitted Runge-Kutta methods

The first method we construct will integrate exactly the functions:

$\{1, x, x^2, x^3, x^4, \exp(Iwx)\}$ or equivalently
$\{1, x, x^2, x^3, x^4, \cos(wx), \sin(wx)\}$,

where w is a real number and it is called frequency and $I = \sqrt{-1}$.

The second method we construct will integrate exactly the functions:

$\{1, x, x^2, \exp(Iwx), x\exp(Iwx)\}$ or equivalently
$\{1, x, x^2, \cos(wx), \sin(wx), x\cos(wx), x\sin(wx)\}$

4 Numerical results - The resonance problem

The efficiency of the two new constructed methods will be measured through the integration of problem (1) with $l = 0$ at the interval $[0, 15]$ using the well known Woods-Saxon potential

$$V(x) = \frac{u_0}{1+q} + \frac{u_1\, q}{(1+q)^2}, \qquad q = \exp\left(\frac{x - x_0}{a}\right), \quad \text{where} \tag{7}$$

$$u_0 = -50, \quad a = 0.6, \quad x_0 = 7 \quad \text{and} \quad u_1 = -\frac{u_0}{a}$$

and with boundary condition $y(0) = 0$.

The potential $V(x)$ decays more quickly than $\frac{l(l+1)}{x^2}$, so for large x (asymptotic region) the Schrödinger equation (1) becomes

$$y''(x) = \left(\frac{l(l+1)}{x^2} - E\right) y(x) \tag{8}$$

The last equation has two linearly independent solutions $k\,x\,j_l(k\,x)$ and $k\,x\,n_l(k\,x)$, where j_l and n_l are the *spherical Bessel* and *Neumann* functions. When $x \to \infty$ the solution takes the asymptotic form

$$\begin{aligned} y(x) &\approx A\,k\,x\,j_l(k\,x) - B\,k\,x\,n_l(k\,x) \\ &\approx D[sin(k\,x - \pi l/2) + \tan(\delta_l)\,\cos(k\,x - \pi l/2)], \end{aligned} \tag{9}$$

where δ_l is called *scattering phase shift* and it is given by the following expression:

$$\tan(\delta_l) = \frac{y(x_i)\,S(x_{i+1}) - y(x_{i+1})\,S(x_i)}{y(x_{i+1})\,C(x_i) - y(x_i)\,C(x_{i+1})}, \tag{10}$$

where $S(x) = k\,x\,j_l(k\,x)$, $C(x) = k\,x\,n_l(k\,x)$ and $x_i < x_{i+1}$ and both belong to the asymptotic region. Given the energy we approximate the phase shift, the accurate value of which is $\pi/2$ for the above problem. As regards to the frequency w we will use the suggestion of Ixaru and Rizea [4]:

$$\begin{aligned} w &= \sqrt{E - 50}, \quad x \in [0,\,6.5] \\ &\sqrt{E}, \quad x \in [6.5,\,15] \end{aligned}$$

We compare the two new trigonometrically fitted methods to a variety of well known classical Runge-Kutta methods. The new methods and especially the second one that integrates $\{1,\ x,\ x^2,\ \exp(Iwx),\ x\exp(Iwx)\}$ are more efficient than the classical ones. The main reason for this high efficiency is that there are lower powers of energy E in the local truncation error of the new methods than in error of the other methods.

References

[1] Hairer E., Nørsett S.P., Wanner G., *Solving Ordinary Differential Equations I, Nonstiff Problems*, Springer-Verlag, Berlin Heidelberg, 1993

[2] Simos T.E., An exponentially-fiited Runge-Kutta method for the numerical integration of initial-value problems with periodic or oscillating solutions, **115**, 1-8(1998).

[3] Lyche T., Chebyshevian multistep methods for Ordinary Differential Eqations, *Num. Math.* **19** 65-75(1972).

[4] Ixaru L.Gr., Rizea M., A Numerov-like scheme for the numerical solution of the Schrödinger equation in the deep continuum spectrum of energies, *Comp. Phys. Comm.* **19** 23-27(1980).

VSP International
Science Publishers
P.O. Box 346, 3700 AH Zeist
The Netherlands

Lecture Series on Computer
and Computational Sciences
Volume 1, 2004, pp. 37-40

Numerical Flow Analysis Around Structures in Low Speed Fluids and its Application to Environmental Sciences

T. Arima[1], Y. Matsuura[2], S. Oharu[3]

[1]Wako Research Center, Honda R&D Co., Ltd. Saitama, Japan
[2]Faculty of Information Sciences, Hiroshima City University, Hiroshima, Japan
[3]Department of Mathematics, Chuo University, Tokyo, Japan

Received 13 August, 2004; accepted in revised form 31 August, 2004

Abstract: Importance and application of numerical flow analysis to environmental science and technology are outlined. Fluid phenomena in the ocean, rivers, atmosphere and the ground are investigated by means of numerical methods and in turn proposals for the control, restoration and counterplans against the so-called environmental disrupters which destroy natural environment as well as ecological systems in nature. All such environmental disrupters diffuse in and are transported by environmental fluids. Those disrupters sometimes react on some other chemicals to generate more poisonous materials. Environmental fluid dynamics is effective for the evaluation, prediction and restoration of the environmental damage. In this paper a mathematical model of environmental fluid is presented and results of numerical simulations based on the model are exhibited.

Keywords: Environmental fluid, computational fluid dynamics, numerical simulation, environmental restoration technology, three dimensional visualization.

Mathematics Subject Classification: 39A12, 62P12, 65M06, 65M12, 76D05

1 Environmental fluids and environmental restoration technology

Fluid dynamical technologies are increasingly becoming important in the field of environmental science and technology. Evaluation of environmental fluid flows using numerical methods is particularly useful to understand the complex fluid motion and make it possible to control the flow fields from the point of view of environmental restoration. The environmental fluid problems may be classified by three types of applications. The first application is concerned with ultimate use of exergy. This is the most important subject for existing engines that use fossil fuel for combustion. Secondly, new energy sources such as wind and wave power generation should be extensively researched and developed. Thirdly, it is indispensable to develop not only efficient and harmless energy sources but also technologies to restore the environment which has already been polluted by exhaust gases through combustion of fossil fuel. It is also important to develop effective methods for protecting the environmental fluids against pollutants. In this paper the field of studies in evaluation, control and prediction of transport phenomena which arise in a variety of environmental problems is called *environmental fluid dynamics*. Obviously, the environmental fluid dynamics is one of the key theories to invent efficient technologies for the preservation and restoration of the natural environment. Here we focus our attention on dynamical analysis of diffusion and transport processes of pollutants in environmental fluids. Mathematical models of environmental fluids are

---

[2]Corresponding author.@E-mail:matsuura@im.hiroshima-cu.ac.jp

presented and results of simulations based on the models are exhibited. It is then expected that new environmental restoration technology will be extensively developed by means of the computational fluid dynamics.

2 A mathematical model of environmental fluids

As a mathematical model describing the motion of environmental fluids, we employ the compressible Navier-Stokes system. Although it is known (see [1]) that the application of numerical methods for the compressible Navier-Stokes system to low-speed flows does not necessarily provide us with satisfactory results in the numerical computation. This would imply that numerical simulations become inefficient and the associated computational results turn out to be inaccurate.

In this paper we apply the Boussinesq approximation to the compressible Navier-Stokes system and formulate the following system of equations (1-3) as our mathematical model for describing the motion of environmental fluids:

$$\nabla \cdot \mathbf{v} = 0 \tag{1}$$

$$\rho \left[\mathbf{v}_t + (\mathbf{v} \cdot \nabla)\mathbf{v} \right] = -\nabla p + \mu \Delta \mathbf{v} - \rho \beta (T - T_0)\mathbf{g} \tag{2}$$

$$\rho C p \left[T_t + (\mathbf{v} \cdot \nabla)T \right] = \nabla(\kappa \nabla T) + S_c \tag{3}$$

Here the parameters $\mathbf{v}$, ρ, p, μ, β, $\mathbf{g}$, T and T_0 represent the velocity vector, density, pressure, viscosity coefficient, rate of volume expansion, the acceleration of gravity, temperature and its reference temperature, respectively. The third term on the right-hand side of (2) represents the thermal effect on buoyancy. Also, the coefficient κ means the thermal conductivity and S_c stands for the sum of heat sources in the fluid under consideration.

Our main objective here is to obtain numerical data describing the flow field around bodies in an environmental fluid under consideration. For this purpose we impose Dirichlet boundary conditions for $\mathbf{v}$ and T and homogeneous Neumann boundary conditions for p on the inflow boundary. On the outflow boundary we impose homogeneous Neumann conditions for $\mathbf{v}$, T and Dirichlet boundary conditions for p. On the surface of each body standing in the fluid, we impose the non-slip condition for $\mathbf{v}$ and homogeneous Neumann condition for T and employ an inhomogeneous Neumann boundary condition for p which is obtained from equation (2). In this paper a new numerical scheme is proposed such that an fully implicit Euler scheme for the velocity is involved.

3 Numerical Models

Making discretization in time of (1) by use of the Euler implicit method, we obtain the following system of equations:

$$\nabla \cdot \mathbf{v}^{n+1} = 0 \tag{4}$$

$$\frac{\mathbf{v}^{n+1} - \mathbf{v}^n}{\Delta t} = -(\mathbf{v}^{n+1} \cdot \nabla)\mathbf{v}^{n+1} - \frac{1}{\rho}\nabla p^{n+1} + \frac{\mu}{\rho}\Delta \mathbf{v}^{n+1} - \beta(T^{n+1} - T_0)\mathbf{g} \tag{5}$$

$$\frac{T^{n+1} - T^n}{\Delta t} = -(\mathbf{v}^{n+1} \cdot \nabla)T^{n+1} + \frac{1}{\rho C_p}\nabla(\kappa \nabla T^{n+1}) + \frac{S_c}{\rho C_p} \tag{6}$$

In what follows, we regard equations (4)-(6) as the governing equations for the motion of numerical fluids and then take the standpoint that the numerical solvability of this basic model should be investigated. In view of the unilateral positive direction of time, we linearize the nonlinear term on the right-hand side of (5) as $(\mathbf{v}^{n+1} \cdot \nabla)\mathbf{v}^{n+1} \simeq (\mathbf{v}^n \cdot \nabla)\mathbf{v}^{n+1}$ we obtain the following Helmholtz equation for the velocity vector $\mathbf{v}^{n+1}$:

$$\left[1 + \Delta t(\mathbf{v}^n \cdot \nabla) \right] \mathbf{v}^{n+1} - \frac{\mu \Delta t}{\rho}\Delta \mathbf{v}^{n+1} = \mathbf{v}^n - \frac{\Delta t}{\rho}\nabla p^{n+1} - \beta \Delta t(T^{n+1} - T_0)\mathbf{g} \tag{7}$$

Substitution of equation (5) into equation (4) implies Poisson's equation for the pressure field p^{n+1}:

$$\Delta p^{n+1} = -\rho \left[\nabla \cdot \left\{ (\mathbf{v}^{n+1} \cdot \nabla) \mathbf{v}^{n+1} \right\} - \frac{\nabla \cdot \mathbf{v}^n}{\Delta t} \right] - \rho \beta \nabla \cdot T^{n+1} \mathbf{g} \tag{8}$$

Also, the Helmholtz equation for the temperature field is obtained from equation (6) in the following form:

$$\left[1 + \Delta t (\mathbf{v}^n \cdot \nabla) \right] T^{n+1} - \frac{\kappa \Delta t}{\rho C_p} \Delta T^{n+1} = T^n + \frac{\Delta t Sc}{\rho C_p} \tag{9}$$

Elliptic equations (7), (8) and (9) are discretized with respect to the space variables and the resultant system of finite-difference equations are solved by applying an appropriate iteration method. Namely, given velocity field $\mathbf{v}^n$, pressure field p^n and temperature field T^n at the nth time step, the velocity field $\mathbf{v}^{n+1}$, pressure field p^{n+1} and temperature field T^{n+1} are obtained at the $(n+1)$th step through the iteration procedures, respectively. As to the discretization of the physical space Ω, we employ a collocation form such that the scalar and vector quantities on the same grid points. For the discretization of the convective terms, we apply an up-wind scheme of the third order accuracy which is proposed in Kawamura [2]:

$$v \left(\frac{\partial v}{\partial x} \right)_i = v_i \frac{-v_{i+2} + 8(v_{i+1} - v_{i-1}) + v_{i-2}}{12\Delta x} + |v_i| \frac{v_{i+2} - 4v_{i+1} + 6v_i - 4u_{i-1} + v_{i-2}}{4\Delta x} \tag{10}$$

For the other terms we employ central difference schemes of the second order accuracy. It should be mentioned here that the above-mentioned finite-difference scheme is unconditional stable.

4   Results of Numerical Simulations

Computation is started with an adequate initial data and qualitative features are investigated by analyzing the numerical results of the simulation at time steps at which the flow field is well developed and approaches a quasi-stationary state. Figure 1(a) depicts the contours of the pressure on the cross section containing the axes of the two cylinders. In the figure of contours of the pressure it is observed that a vertical sequence of separate regions like cells of negative pressure are formed. This is due to the presence of nonstationary vortices of Karmann-type.

Furthermore, a vertical sequence of regions of positive pressure are observed in the front of the rear cylinder. This phenomenon suggests that the nonstationary vortices generated by the front cylinder interact the regions of stagnation existing in the front of the rear cylinder and deteriorate the stagnation pressure. Figure 1(b) depicts the iso-surfaces of vorticity and illustrates the 3D structure of the pressure field and vorticity distribution.

(a) Pressure contour   (b) Absolute vorticity

Figure 1: Computed pressure contour and vorticity iso-surfaces

On the other hand, concerning the vorticity distribution, regular variation due to the formation of vortex pairs of Karmann type is observed in the top part of the rear cylinder in the same way as in the pressure field. Vorticity distribution is concentrated in the back of the middle part of the rear cylinder, although the vorticity distribution in the bottom part is comparatively diffusive.

(a) streamline   (b) Particle trajectory

Figure 2: Upward flow motion observed behind two circular cylinders

This suggests that the flows running around the bottom parts of the cylinders form longitudinal vortices. It is then inferred that these longitudinal vortices would motivate the upward flows behind the two cylinders. In Figure 2 the stream lines and trajectories of particles in the fluid are depicted. Trajectories of particles are drawn in the following way: We released the particles from the back of each cylinder and traced the trajectories forward and backward in time until the particles reach the boundaries of the computational domain and those of the bodies in the fluid. It is seen from Figure 2(a) that upward flows behind the cylinders are rolling up towards the top. Furthermore, the motion of longitudinal vortices around the bottom sides can be observed as inferred from the iso-surfaces of vorticities. Figure 2(b) is obtained by arranging particles on the same trajectories as in Figure 2(a) at regular time intervals. From this it is seen that the particle distribution represents how long a particle stay in the flow, and that particles are concentrated in the back of the front cylinder.   These results of numerical simulations may have applications to

(a) Oyster raft   (b) Street trees

Figure 3: Structures in the ocean and atmosphere

various environmental problems. The farming part of an oyster raft may be regarded as a regular arrangement of cylinders with bottom ends. Our results suggests that red tide plankton would stay in the upward vortices rolling up along the back of the cylinders provided that the oyster raft is moved or water currents exist against the raft. Another application is that street trees are regarded as a sequence of cylinders with top ends, and that automobile exhaust fumes may be caught in the back of each tree which assimilate those poisonous gases. See Figure2(a) and 2(b). It is then expected that new environmental restoration technology will be significantly improved by applying the results of numerical simulations of the motion of environmental fluids.

References

[1] E. Turkle, "Preconditioned Method for Solving the Incompressible and Low Speed Compressible Equations," *J. Comput. Phys.*, Vol.72, No. 2, pp.277-298, 1987

[2] T. Kawamura and K. Kuwahara, "Computation of High Reynolds Number Flow around a Circular Cylinder with Surface Roughness," AIAA Paper No. 84-0340, 1984.

VSP International
Science Publishers
P.O. Box 346, 3700 AH Zeist
The Netherlands

*Lecture Series on Computer
and Computational Sciences*
Volume 1, 2004, pp. 41-44

Derivation and Optimization of the Material Cost Function for Natural Gas Pipe-Lines

P.I. Arsenos[1], C. Demakos[2], Ch. Tsitouras[3], D.G. Pavlou[1] and M.G. Pavlou[4]

Received 5 September, 2004; accepted in revised form 12 September, 2004

Abstract: Pipe-lines used for natural gas transfer are mainly operating at high pressure. The latter condition is associated with their wall thickness which determines construction cost. Currently, it is recognized that specific construction techniques that are based in reinforcing the pipe-line with rigid rings can result in significant reduction of material cost. This cost is strongly influenced by the mechanical parameters, material strength and material price. The objective of the present study is the derivation and optimization of the cost function [1,2] versus mechanical parameters. In order to obtain the function that is to be optimized, a numerical procedure based on matrix analysis and theory of elasticity will be used.

Keywords: Pressure pipes, beam on elastic foundations, transfer matrices.

Mathematics Subject Classification: 74B05, 74P05, 74G99

1. Formulation of the mechanical problem

A long axisymmetric pipe of radius R and wall thickness t is reinforced by rigid rings of diameter d. The rings are uniformly distributed along the pipe-line and the distance between two rings is L. The pipe is loaded by internal static pressure Po due to natural gas flow. Under these conditions, at every point x along the wall, a radial displacement $w=w(x, Po, t, L)$ is developed. The boundary conditions of the problem are:

$$w(0) = 0, \; w(L) = 0, \; w'(0) = 0, \; w'(L) = 0 \tag{1}$$

2. Relation of the material cost with the construction parameters

In order to develop the basic equation that describes the radial displacement distribution w(x) of the pipe's wall, the equilibrium of a longitudinal strip of length L and width $\alpha\Delta\theta=1$ of the pipe is considered [3]. It is well known by the mechanics that the longitudinal displacement $u(x,y)$ of a point $A(x,y)$ of a beam is given by the simple equation [4]:

$$u(x,y) = yw' \tag{2}$$

where w' is the slope of the centroidal axis after bending. According to the definition of strain it can be written :

$$\varepsilon = \frac{\partial u}{\partial x} = yw'' \tag{3}$$

Taking into account the relation between stress σ and strain ε given by the Hooke's law

$$\sigma = E\varepsilon$$

where E is the elasticity modulus, then the equations (2) and (3) result to:

$$\sigma = Eyw''$$

Using the above expression, the equilibrium equation

[1] TEI of Chalkis, Department of Mechanical Engineering, GR34400 Psahna, Greece
[2] TEI of Piraeus, Department of Civil Engineering, GR122 44 Athens, Greece. E-Mail: cdem@teipir.gr
[3] TEI of Chalkis, Department of Applied Sciences, GR34400 Psahna, Greece. E-Mail: tsitoura@teihal.gr
[4] ERGOSE S.A., Works of Greek Federal Organization, Karolou 27, 10437 Athens, Greece

$$M = \int_F y\sigma dF$$

results to

$$M = Ew''J$$

where

$$J = \int_F y^2 dF$$

Considering the equilibrium equations

$$Q = \frac{\partial M}{\partial x}$$

and

$$q = \frac{\partial Q}{\partial x}$$

the following differential equation is obtained:

$$EJw'''' = q \qquad (4)$$

The distributed load q is the superposition of the action $q**$ of the internal pressure Po and the reaction $q*$ of the wall [5], so that:

$$q = q** - q* \qquad (5)$$

where

$$q** = P_0 \qquad (6)$$

and

$$q* = \frac{Et}{R^2} w \qquad (7)$$

Combining equations (4)-(7) the following fundamental differential equation is derived:

$$EJw'''' + \frac{Et}{R^2} w = P_0$$

The general solution of the above equation [6] is:

$$w = B_1 ch\cos + B_2 ch\sin + B_3 sh\cos + B_4 sh\sin - Po/k$$

$$k = Et/R^2, \quad \beta = \sqrt[4]{\frac{k}{4EJ}}$$

$$ch\cos = \frac{e^{\beta x} + e^{-\beta x}}{2}\cos\beta x, \qquad ch\sin = \frac{e^{\beta x} + e^{-\beta x}}{2}\sin\beta x$$

$$sh\cos = \frac{e^{\beta x} - e^{-\beta x}}{2}\cos\beta x, \qquad sh\sin = \frac{e^{\beta x} - e^{-\beta x}}{2}\sin\beta x$$

Taking into account the boundary conditions (1), the state vector of the beam at the ring locations (i.e. $x=0$ and $x=L$) can be obtained by a matrix equation of the following form:

$$\begin{bmatrix} w(L) \\ w'(L) \\ M(L) \\ Q(L) \\ 1 \end{bmatrix} = \begin{bmatrix} A_{11} & A_{12} & . & . & A_{15} \\ A_{21} & A_{22} & . & . & A_{25} \\ . & . & . & . & . \\ A_{41} & A_{42} & . & . & A_{45} \\ 0 & 0 & 0 & 0 & 1 \end{bmatrix} \begin{bmatrix} w(0) \\ w'(0) \\ M(0) \\ Q(0) \\ 1 \end{bmatrix}$$

or

$$
\begin{bmatrix}
A_{11} & A_{12} & . & A_{14} & -1 & 0 & 0 & 0 \\
A_{21} & A_{22} & . & A_{24} & 0 & -1 & 0 & 0 \\
. & . & . & . & 0 & 0 & -1 & 0 \\
A_{41} & A_{42} & . & A_{44} & 0 & 0 & 0 & -1 \\
1 & 0 & 0 & 0 & 0 & 0 & 0 & 0 \\
0 & 1 & 0 & 0 & 0 & 0 & 0 & 0 \\
0 & 0 & 0 & 0 & 1 & 0 & 0 & 0 \\
0 & 0 & 0 & 0 & 0 & 1 & 0 & 0
\end{bmatrix}
\begin{bmatrix}
w(0) \\ w'(0) \\ M(0) \\ Q(0) \\ w(L) \\ w'(L) \\ M(L) \\ Q(L)
\end{bmatrix}
=
\begin{bmatrix}
-a_{15} \\ -a_{25} \\ -a_{35} \\ -a_{45} \\ 0 \\ 0 \\ 0 \\ 0
\end{bmatrix}
$$

The solution of the above linear system results to:

$$
M(0) = M(L) = \frac{12b^4 EJL^2 (12 + b^4 L^4) Po}{k(432 - 72b^4 L^4 + 72b^5 L^5 + b^8 L^8)}
$$

$$
Q(0) = Q(L) = -\frac{24b^4 EJL(36 - 2b^4 L^4 + 3b^5 L^5) Po}{k(432 - 72b^4 L^4 + 72b^5 L^5 + b^8 L^8)}
$$

The thickness of the pipe's wall and the diameter of each ring can be determined by the following well-known equations:

$$
\frac{6M_{max}}{t^2} = \frac{S_y}{N} \quad \text{and} \quad \frac{4Q_{max}}{\pi d^2} = \frac{S_y}{N}
$$

or

$$
t = \sqrt{\frac{6M_{max} N}{S_y}} \quad \text{and} \quad d = \sqrt{\frac{4Q_{max} N}{\pi S_y}} \tag{8}
$$

The cost c of the required material is proportional to the weight. Therefore:

$$
c = \lambda \left(2\pi RLt + 2\pi R \frac{\pi d^2}{4} \right)
$$

The normalized unit cost $c* = c/2\pi R\lambda L$ can be written:

$$
c* = \frac{1}{L} \left(Lt + \frac{\pi d^2}{4} \right)
$$

Taking into account equations (8) the normalized unit cost function $c*$ it can be expressed:

$$
c* = \frac{1}{L} \left(L\sqrt{\frac{6M_{max} N}{S_y}} + \frac{Q_{max} N}{S_y} \right) \tag{9}
$$

In order to minimize the required cost [7,8] for the pipe-line construction, the distance L between the rigid rings can be determined by the following condition:

$$
\frac{\partial c*}{\partial L} = 0 \tag{10}
$$

3. Numerical example

A long pipe-line with mechanical parameters $EJ=9.84\times10^9$ Nmm^2/mm, $k=0.168$ Nmm^2/mm is constructed by steel with yield stress $S_y=240$ Mpa and elasticity modulus $E=2.1\times10^5$ N/mm^2. Within the pipe, natural gas with pressure $Po=3\times10^5$ Pa is flow. For the reinforcement of the pipe-line a uniform distribution of rigid rings with diameter d have been applied. During loading, two sections along the length are critical for the constructions strength: (a) the section located at the rings ($x=0$ or $x=L$) and (b) the section located at the middle between two rings ($x=L/2$). The stressing of the section of the case (b) is predominant as the distance between the rings increased. The bending moment $M(L/2)$ at this section is determined by a matrix equation of the following form:

$$\begin{bmatrix} w(L/2) \\ w'(L/2) \\ M(L/2) \\ Q(L/2) \\ 1 \end{bmatrix} = \begin{bmatrix} B_{11} & B_{12} & . & . & B_{15} \\ B_{21} & B_{22} & . & . & B_{25} \\ . & . & . & . & . \\ B_{41} & B_{42} & . & . & B_{45} \\ 0 & 0 & 0 & 0 & 1 \end{bmatrix} \begin{bmatrix} w(0) \\ w'(0) \\ M(0) \\ Q(0) \\ 1 \end{bmatrix}$$

For the calculation of the normalized unit cost $c*$ given by (9), the following results are used:

$$M_{max} = \max\left[M(0), M(L/2)\right]$$

$$Q_{max} = Q(0)$$

The normalized cost versus the parameter L indicating strong influence by the construction technique. The equation (10) is solved numerically using the commercial program Mathematica, taking for the safety factor the value $N=5$. The solution of the above equation resulted:

$$L_{optimum} \approx 6743 \ mm \ .$$

References

[1] Cook, W.D. and Kress, M. (1991), A multiple criteria decision model with ordinal preference data, *European Journal of Operational Research*, 54, 191-198.

[2] Doumpos, M, Zopounidis, C. and Pardalos, P.M. (2000a), Multicriteria sorting methodology: Application to financial decision problems, *Parallel Algorithms and Applications*, 15/1-2, 113-129.

[3] Wu, T.Y. and Liu, G.R. (2000), Axisymmetric bending solution of shells of revolution by the generalized differential quadrature rule, *International Journal of Pressure Vessels and Piping*, 77, 149-157.

[4] Timoshenko, S.P. and Goodier, J.N. (1970), *Theory of elasticity*, 3rd Ed., McGraw-Hill, New York.

[5] Boresi, A.P. and Sidebottom, O.M. (1985), *Advanced Mechanics of Materials*, 4th Ed., Wiley, New York.

[6] King, A.C., Billingham, J., and Otto, S.R. (2003), *Differential Equations*, Cambridge Univ. Press, Cambridge.

[7] Koehler, G.J. and Erenguc, S.S. (1990), Minimizing misclassifications in linear discriminant analysis, *Decision Sciences*, 21, 63-85.

[8] Stam, A. (1990), Extensions of mathematical programming-based classification rules: A multicriteria approach, *European Journal of Operational Research*, 48, 351-361.

VSP International
Science Publishers
P.O. Box 346, 3700 AH Zeist
The Netherlands

*Lecture Series on Computer
and Computational Sciences*
Volume 1, 2004, pp. 45-48

A Semi-Algorithm to Find Elementary First Order Invariants of Rational Second Order Ordinary Differential Equations

J. Avellar[1], L.G.S. Duarte[2], S.E.S. Duarte[3], L.A.C.P. da Mota[4]

Universidade do Estado do Rio de Janeiro,
Instituto de Física, Departamento de Física Teórica,
R. São Francisco Xavier, 524, Maracanã, CEP 20550–013,
Rio de Janeiro, RJ, Brazil.

Received 2 September, 2004; accepted in revised form 17 September, 2004

Abstract: Here we present a method to find elementary first integrals of rational second order ordinary differential equations (SOODEs) based on a Darboux type procedure [3, 4, 5]. Apart from practical computacional considerations, the method will be capable of telling us (up to a certain polynomial degree) if the SOODE has an elementary first integral and, in positive case, finds it via a single quadrature.

Keywords: Elementary first integrals, semi-algorithm, Darboux, Lie symmetry

PACS: 02.30.Hq

1 Earlier Results

In the paper [1], one can find an important result that, translated to the case of SOODEs of the form

$$y'' = \frac{M(x,y,y')}{N(x,y,y')} = \phi(x,y,y'), \tag{1}$$

where M and N are polynomials in (x,y,y'), can be stated as:

Theorem 1: *If the SOODE (1) has a first order invariant that can be written in terms of elementary functions, then it has one of the form:*

$$I = w_0 + \sum_i^m c_i \ln(w_i), \tag{2}$$

where m is an integer and the $w's$ are algebraic functions[5] of (x,y,y').

---

[1] E-mail: javellar@dft.if.uerj.br - *Fundação de Apoio à Escola Técnica, E.T.E. Juscelino Kubitschek, 21311-280 Rio de Janeiro – RJ, Brazil*
[2] E-mail: lduarte@dft.if.uerj.br
[3] E-mail: lduarte@dft.if.uerj.br
[4] E-mail: damota@dft.if.uerj.br
[5] For a formal definition of algebraic function, see [2].

The integrating factor for a SOODE of the form (1) is defined by:

$$R(\phi - y'') = \frac{dI(x, y, y')}{dx} \tag{3}$$

where $\frac{d}{dx}$ represents the total derivative with respect to x.

Bellow we will present some results and definitions (previously presented on [6]) that we will need. First let us remember that, on the solutions, $dI = I_x\, dx + I_y\, dy + I_{y'}\, dy' = 0$. So, from equation (3), $R(\phi\, dx - dy') = I_x\, dx + I_y\, dy + I_{y'}\, dy' = dI = 0$. Since $y'\, dx = dy$, we have

$$R\left[(\phi + S\, y')\, dx - S\, dy - dy'\right] = dI = 0, \tag{4}$$

adding the null term $S\, y'\, dx - S\, dy$, where S is a function of (x, y, y'). From equation (4), we have: $I_x = R(\phi + Sy')$, $I_y = -RS$, $I_{y'} = -R$ that must satisfy the compatibility conditions. Thus, defining the differential operator $D \equiv \partial_x + y'\partial_y + \phi\,\partial_{y'}$, after a little algebra, that can be shown to be equivalent to:

$$\mathcal{D}[R] = -R(S + \phi_{y'}), \tag{5}$$
$$\mathcal{D}[RS] = -R\phi_y. \tag{6}$$

2 The theoretical foundations for the algorithm

Let us start this section by stating (without presenting the demonstration here in the extended abstract) a corollary to **theorem 1** concerning S and R.

Corollary: *If a SOODE of the form (1) has a first order elementary invariant then the integrating factor R for such an SOODE and the function S defined in the previous section can be written as algebraic functions of (x, y, y').*

Besides that, working on equations (5) and (6), we get:

$$D[S] = S^2 + \phi_{y'}\, S - \phi_y. \tag{7}$$

That equation will be regarded as a first order ordinary differential equation (FOODE) for $S(x)$ over the solutions for the SOODE (1). Concerning eq.(7) we can demonstrate the following theorem:

Theorem 2: *Consider the operator*

$$D_S = ((NS)^2 + (NM_{y'} - MN_{y'})\, S - (NM_y - MN_y))\, \partial_S + N^2\, D. \tag{8}$$

If P is an eigenpolynomial of D_S (i.e., $D_S[P] = \lambda P$) that constains S, then $P = 0$ is a particular solution of eq.(7).

Since the existence of an elementary first order invariant I (first integral) for the SOODE (1) implies that S can be written as an algebraic function, we have the result:

Theorem 3: *If the SOODE (1) has an elementary first integral, then the operator D_S defined above has an eigenpolynomial P that contains S.*

This result provides us a algorithm to find S. In order to obtain the integrating factor R, let us show a relation between the function S and a symmetry of the SOODE (1). Making the following transformation

$$S = -\frac{D[\eta]}{\eta} \tag{9}$$

eq.(7) becomes

$$D^2[\eta] = \phi_{y'} D[\eta] + \phi_y \eta. \tag{10}$$

¿From Lie theory we can see that eq.(10) represents the condition for a SOODE (1) to have a symmetry $[0, \eta]$. So, from (9) we can find a symmetry given by

$$\eta = e^{-\int S \, dx}. \tag{11}$$

3 Finding the integrating factor and the first integral

Looking at eq.(4) we can infer that R is also an integrating factor for the auxiliary FOODE defined by

$$\frac{dz}{dy} = S, \tag{12}$$

where x is regarded as a parameter. Besides, for this FOODE, $[\eta, D[\eta]]$ is a point symmetry. So, R is given by

$$R = \frac{1}{\eta S + D[\eta]}. \tag{13}$$

If in (13) we use η defined by (11) we would get a singular R. However, eq.(10) has another solution independent from η given by

$$\bar\eta = \eta \int \frac{e^{\int \phi_{y'} dx}}{\eta^2} \, dx. \tag{14}$$

Using this $\bar\eta$ in (13) we get R and, once this is done, we can calculate the first integral I by using simple quadratures.

4 Example and conclusions

Consider the SOODE

$$y'' = -\frac{2\, y'^2 x^2 - 2\, y'^2 - 2\, xy'y - y^2 x^4 + 2\, y^2 x^2 - y^2}{2y\,(x^2 - 1)}. \tag{15}$$

Constructing the D_S operator we get that

$$y^2\,(x^2 - 1) - S^2 y^2 - z^2 + 2\, Szy \tag{16}$$

is an eigenpolynomial of it. Then S is given by $S = z/y + \sqrt{x^2 - 1}$ and $\eta, \bar\eta$ are respectively

$$\eta = \frac{\sqrt{x + \sqrt{x^2 - 1}}}{y\, e^{\frac{x\sqrt{x^2-1}}{2}}}, \quad \bar\eta = \frac{\sqrt{x + \sqrt{x^2 - 1}} \int \frac{\sqrt{x-1}\sqrt{x+1}e^x \sqrt{x^2-1}}{x + \sqrt{x^2-1}} dx}{y\, e^{\frac{x\sqrt{x^2-1}}{2}}}. \tag{17}$$

R can be written as

$$R = \frac{y\sqrt{x + \sqrt{x^2 - 1}}}{\sqrt{x^2 - 1}e^{\frac{x\sqrt{x^2-1}}{2}}} \tag{18}$$

leading to the following first integral:

$$I = \frac{\sqrt{x + \sqrt{x^2 - 1}} \left(2zy + \sqrt{x^2 - 1}y^2\right)}{\sqrt{x^2 - 1}e^{\frac{x\sqrt{x^2-1}}{2}}}. \tag{19}$$

Here we presented a semi-algorithm do find an elementary first integral for a rational SOODE. We are working on a full implementation of it in the MAPLE system.

References

[1] M Prelle and M Singer, Elementary first integral of differential equations. *Trans. Amer. Math. Soc.*, **279** 215 (1983).

[2] Davenport J.H., Siret Y. and Tournier E. *Computer Algebra: Systems and Algorithms for Algebraic Computation.* Academic Press, Great Britain (1993).

[3] Y K Man and M A H MacCallum, A Rational Approach to the Prelle-Singer Algorithm. *J. Symbolic Computation*, **11** 1–11 (1996), and refferences therein.

[4] L.G.S. Duarte, S.E.S.Duarte and L.A.C.P. da Mota, *A method to tackle first order ordinary differential equations with Liouvillian functions in the solution*, in *J. Phys. A: Math. Gen.* **35** 3899-3910 (2002).

[5] L.G.S. Duarte, S.E.S.Duarte and L.A.C.P. da Mota, *Analyzing the Structure of the Integrating Factors for First Order Ordinary Differential Equations with Liouvillian Functions in the Solution*, *J. Phys. A: Math. Gen.* **35** 1001-1006 (2002)

[6] L G S Duarte, S E S Duarte, L A C P da Mota and J E F Skea, Solving second order ordinary differential equations by extending the Prelle-Singer method, *J. Phys. A: Math.Gen.*, **34** 3015-3024 (2001).

VSP International
Science Publishers
P.O. Box 346, 3700 AH Zeist
The Netherlands

*Lecture Series on Computer
and Computational Sciences*
Volume 1, 2004, pp. 49-52

Simultaneous Estimation of Heat Source and Boundary Condition in Two-Dimensional Transient Inverse Heat Conduction Problem

A. Azimi,[1,2] S. Kazemzadeh Hannani and B. Farhanieh

School of Mechanical Engineering, Sharif University of Technology, Tehran- Iran
[1] Energy Research Centre, Research Institute of Petroleum Industry, Tehran- Iran

Received 2 August, 2004; accepted in revised form 15 August, 2004

Abstract: In this research, a simultaneous estimation of heat source and boundary conditions using parameter and function-estimation techniques is presented for the solution of two-dimensional transient inverse heat conduction problem (IHCP). The IHCP involves simultaneous unknown time varying heat generation and time-space varying boundary conditions estimation. Two techniques are considered, Levenberg-Marquardt scheme for parameter estimation and adjoint conjugate gradient method for function estimation. To have fewer numbers of unknown coefficients for estimation, polynomials are used in the parameter estimation scheme. The measured transient temperature data needed in the inverse solution are made by exact or inexact (noisy) data. The results of the present study are compared to exact heat source and temperature (heat flux) boundary conditions.

Keywords: Inverse Heat Conduction- Parameter and Function Estimations- Simultaneous Estimation

1. Introduction

The development of engineering softwares for the study of Inverse Heat Conduction Problems (IHCPs) has widely investigated for applications in industrial and research areas. Time-space IHCPs have been used to estimate time or time-space varying unknown heat generation, surface boundary conditions and thermophysical properties using one or more temperatures measured by sensors inside or on the boundaries of the physical domain.

Mathematically, IHCPs, unlike the direct heat conduction problems, belong to a class of "ill-posed" problems which do not satisfy the "well-posed" conditions introduced by Hadamard [1]. In fact, the IHCPs are very sensitive to random errors in the measured temperature data, thus requiring special techniques for their solutions in order to satisfy the stability condition, one of the "well-posed" conditions. Minimization of error is an aim of the inverse Analyses. Inverse analysis is related to analytical design theory. The concept of choosing the best experiments in the inverse analysis and minimizing it is common to choose the best cost function in the analytical design theory. A number of parameter and function-estimation schemes for inverse analysis have been proposed to treat the ill-posed nature of IHCPs [2, 3].

In this research, a simultaneous estimation of unknown heat source and boundary conditions is presented for the solution of two-dimensional transient inverse heat conduction problem (IHCP). The IHCP involves simultaneous unknown time varying heat generation and time-space varying boundary conditions estimation. Two techniques Levenberg-Marquardt scheme for parameter estimation and adjoint conjugate gradient method for function estimation are considered.

2. Mathematical Formulation

The governing equation in non-dimensional form is the two-dimensional transient heat conduction equation with heat source. The governing equation expressed in Cartesian coordinate system is then:

$$\frac{\partial T}{\partial t} = \frac{\partial^2 T}{\partial x^2} + \frac{\partial^2 T}{\partial y^2} + G(t) \qquad t > 0, \ 0 < x < 1, \ 0 < y < 1 \tag{1}$$

---

[2] Corresponding author. Ph.D. Candidate of the Sharif University of Technology and a member of the Energy Research Centre, Research Institute of Petroleum Industry. E-mail: aazimi@mehr.sharif.edu, azimia@ripi.ir

where function $G(t)$ is time varying heat generation in a square solid. The initial and boundary conditions are as follows:

$$
\begin{aligned}
T(x,y,t) = 0 && t>0, \quad 0<y<1, && x=0 \\
T(x,y,t) = F(y,t) && t>0, \quad 0<y<1, && x=1 \\
T(x,y,t) = 0 && t>0, \quad y=0,1, \quad 0<x<1 \\
T(x,y,t) = 0 && t=0, \quad 0<y<1, \quad 0<x<1
\end{aligned}
\tag{2}
$$

$F(y,t)$ is a polynomial function of time and space in y direction. The non-dimensional groups are as:

$$
t = \frac{t^*}{\alpha/L_{ref}^2}, \quad x = \frac{x^*}{L_{ref}}, \quad y = \frac{y^*}{L_{ref}}, \quad T = \frac{T^*}{T_{ref}}, \quad G = \frac{G^*}{kT_{ref}/L_{ref}^2}
\tag{3}
$$

where α and k are thermal diffusivity and thermal conductivity, respectively.

3. Direct Numerical Scheme

Any IHCP algorithm, regardless of its theoretical approach, requires the usage of a suitable solution routine for direct heat conduction calculations. This solution routine may be called upon numerous times by the main IHCP computational routine.

In this present study, the Galerkin method of weighted residuals as a direct numerical method is employed to solve equation (1) with initial and boundary conditions, equation (2). Therefore, equation (1) is multiplied by weighting function and integrated by part and employing Gauss theorem over the domain Ω leads to [4]:

$$
\int_\Omega \left(w\frac{\partial T}{\partial t} + \frac{\partial w}{\partial x}\frac{\partial T}{\partial x} + \frac{\partial w}{\partial y}\frac{\partial T}{\partial y} \right) d\Omega = \int_\Omega wGd\Omega + \int_\Gamma w\left[\frac{\partial T}{\partial x}n_x + \frac{\partial T}{\partial y}n_y \right] d\Gamma
\tag{4}
$$

A four node element is used, along with bilinear interpolation to approximate temperature. The Galerkin weighting function and temperature in the four node element are defined by:

$$
w_i = N_i
\tag{5}
$$

$$
T^e = \sum_{j=1}^{4} N_j T_j
$$

By replacing the above relations into equation (3), the matrix form of this equation is

$$
\mathbf{M}\dot{\mathbf{T}} + \mathbf{K}\mathbf{T} = \mathbf{F}
\tag{6}
$$

where matrices $\mathbf{M}$, $\mathbf{K}$ and $\mathbf{F}$ are determined as follow:

$$
m_{ij} = \int_\Omega N_i N_j d\Omega
$$

$$
k_{ij} = \int_\Omega \left(\frac{\partial N_i}{\partial x}\frac{\partial N_j}{\partial x} + \frac{\partial N_i}{\partial y}\frac{\partial N_j}{\partial y} \right) d\Omega
\tag{7}
$$

$$
f_{ij} = \int_\Omega N_i G d\Omega + \int_\Gamma N_i \left(\frac{\partial T}{\partial x}n_x + \frac{\partial T}{\partial y}n_y \right) d\Gamma
$$

The following relation is used to linearize the temperature with respect to time:

$$
\frac{\mathbf{T}^{n+1} - \mathbf{T}^n}{\Delta t} = (1-\theta)\dot{\mathbf{T}}^n + \theta\dot{\mathbf{T}}^{n+1}
\tag{8}
$$

By replacing equation (8) in equation (6), the matrix form of direct numerical problem is written as:

$$
\hat{\mathbf{K}}^{n+1}\mathbf{T}^{n+1} = \hat{\mathbf{K}}^n\mathbf{T}^n + \hat{\mathbf{F}}^n
\tag{9}
$$

where

$$\hat{\mathbf{K}}^{n+1} = \mathbf{M}^n + \theta \, \Delta t \, \mathbf{K}^n$$
$$\hat{\mathbf{K}}^n = \mathbf{M}^n - (1-\theta)\Delta t \, \mathbf{K}^n \qquad (10)$$
$$\hat{\mathbf{F}}^n = \left[(1-\theta)\mathbf{F}^n + \theta \, \mathbf{F}^{n+1}\right]\Delta t$$

and θ is a constant which depends on the numerical scheme in time.

4. Inverse Analysis Schemes

4.1 Concepts

In inverse analysis, unlike direct solution, there are one or more parameters or functions which must be determined by the sensor values of temperature measured inside the field or on the boundaries [2]:

$$\left[\vec{T}^m\right]^t = \left[T_1^m, T_2^m, \ldots, T_M^m\right] \qquad (11)$$

where $m = 1, 2, \ldots, M$ is the number of sensors. In inverse analysis, error is defined by the difference between measured and computed temperatures:

$$\vec{e} = \vec{T}^m - \vec{T}^c \qquad (12)$$

where the measured temperatures are different from exact temperatures

$$\vec{T}^m = \vec{T}^{exact} + \vec{\varepsilon} \qquad (13)$$

Minimization of this error is the main goal of inverse analysis. One of the minimization strategies is to apply the least squares method

$$S(\vec{P}) = (\vec{T}^m - \vec{T}^c)^t (\vec{T}^m - \vec{T}^c) \qquad (14)$$

or the weighted least squares method:

$$S(\vec{P}) = (\vec{T}^m - \vec{T}^c)^t \, \mathbf{W} (\vec{T}^m - \vec{T}^c) \qquad (15)$$

4.2 Algorithms

For inverse analysis, methods have been introduced based on the concept of the sensitive analysis. In this work, two iterative algorithms are presented, the Levenberg-Marquardt scheme for parameter estimation and the adjoint conjugate gradient scheme for function estimation [3]. The key relation of the Levenberg-Marquardt scheme is the following equation:

$$\vec{P}^{k+1} = \vec{P}^k + \left[(\mathbf{J}^k)^T \mathbf{J}^k + \mu^k \, \mathbf{\psi}^k\right]^{-1} (\mathbf{J}^k)^T (\vec{T}^m - \vec{T}^c) \qquad (16)$$

where $\mathbf{J}$, $\mathbf{\psi}$ and μ are the sensitive matrix, a diagonal matrix and a damping parameter, respectively.
The key relations of the adjoint conjugate gradient scheme are the sensitive problem, the adjoint problem and the gradient equation.
These iterative algorithms need a criterion to stop the procedure of solution. Three stopping criteria based on Discrepancy Principle [3] are as follows:

$$S(\vec{P}) < \delta$$
$$\left\|(\mathbf{J}^k)^T (\vec{T}^m - \vec{T}^c)\right\| \text{ or } \left|\vec{T}^m - \vec{T}^c\right| < \delta \qquad (17)$$
$$\left|\vec{P}^{k+1} - \vec{P}^k\right| < \delta$$

where δ is a sufficient small positive value. Figure 1 and table 1 show a sample result of present computation for simultaneous solution for the case of unknown time varying point heat source and unknown time-space varying temperature profile in the right boundary by parameter estimation of the Levenberg-Marquardt scheme:

- Heat source term at $x = 0.22$ and $y = 0.45$: $G(t) = p_{11} + p_{12}t$ where t is time.
- Temperature boundary condition: $T(y,t) = p_{21} + p_{22} t + p_{23} y + p_{24} t y$.

Sensor locations are

- Heat source term at $x = 0.50$ and $y = 0.45$ ($0 \leq t \leq 3.5$).
- Temperature boundary condition: at $x = 0.95$ and $y = 0$ to 1 (20 sensors).

Table 1: Simultaneous parameter estimation of unknown time varying point heat source and unknown time-space varying temperature profile in the right boundary.

Parameter	p_{11}	p_{12}	p_{21}	p_{22}	p_{23}	p_{24}
Exact	1.0	1.0	1.0	1.0	1.0	1.0
Inverse Analysis	1.000	1.000	1.000	1.000	1.000	1.000

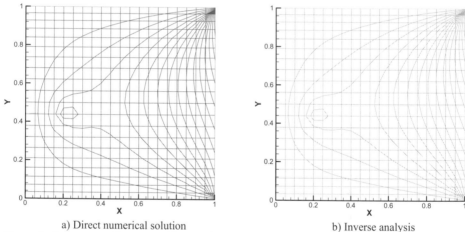

a) Direct numerical solution   b) Inverse analysis

Figure 1: Temperature Contours of inverse heat conduction with unknown time varying point heat source and unknown time-space varying temperature profile in the right boundary.

References

[1] J. Hadamard, *Lectures on Cauchy's Problem in Linear Differential Equations*, Yale University Press, New Haven, CT, 1923.

[2] J.V. Beck and K.J. Arnold: *Parameter Estimation in Engineering and Science*. John Wiley & Sons, New York, 1977.

[3] M.N. Ozisik and H.R.B. Orlande: *Inverse Heat Transfer: Fundamentals and Applications*. Taylor & Francis, New York, 2000.

[4] D.W. Pepper and J.C. Heinrich: *The Finite Element Method: Basic Concept and Applications*. Hemisphere, Washington, 1992.

VSP International
Science Publishers
P.O. Box 346, 3700 AH Zeist
The Netherlands

*Lecture Series on Computer
and Computational Sciences*
Volume 1, 2004, pp. 53-56

Efficient Training of Hidden Markov Models
for Protein Sequence Analysis

P.G. Bagos[1], Th.D. Liakopoulos, S.J. Hamodrakas

Department of Cell Biology and Biophysics,
Faculty of Biology,
University of Athens,
GR-157 01 Panepistimiopolis, Athens, Greece

Received 31 July, 2004; accepted in revised form 17 August, 2004

Abstract: We investigate the effectiveness of a new gradient descent method for Conditional Maximum Likelihood (CML) and Maximum Likelihood (ML) training of Hidden Markov Models (HMMs), which significantly outperforms traditional gradient descent. Instead of using a fixed learning rate for every adjustable parameter of the HMM, we propose the use of independent learning rate adaptation, a strategy that has been proved valuable in Artificial Neural Networks training. We show that this approach, compared to standard gradient descent, performs significantly better. The convergence speed is increased up to five times, while at the same time the training procedure becomes more robust, as tested on applications from molecular biology. We also show that, if the labels of the HMM are well defined, CML training performs better than ML; using this approach, we may obtain better results without any additional computational complexity or the need for parameter tuning.
Keywords: Hidden Markov Models, Gradient descent, Bioinformatics, Membrane proteins

1. Introduction

Hidden Markov Models (HMMs) are probabilistic models suitable for a wide range of pattern recognition applications. Initially developed for speech recognition [1], during the last few years they became very popular in molecular biology for protein modeling and gene finding [2]. The parameters of an HMM could be optimized according to the Maximum Likelihood (ML) criterion, with the use of the efficient Baum-Welch algorithm [3], or using some gradient descent method capable of the same task; the latter offers a number of advantages, including smoothness and on-line training abilities [4]. Using labeled sequences along with the HMM [5], we may train the model according to the ML criterion, by applying some trivial modifications of the Baum-Welch algorithm [5,6] or using a gradient descent method [7]. Alternatively, we may perform Conditional Maximum Likelihood (CML) training using a gradient-based technique, which is shown to perform better in several applications [7]. The main advantage of the Baum-Welch algorithm is its inherent simplicity, along with the fact that it requires no parameter tuning. Furthermore, it achieves significantly faster convergence rates [7]. On the other hand, gradient descent (for both ML and CML) requires a careful search in the parameter space for an appropriate learning rate in order to achieve the best possible performance.

Here, we extend the gradient descent approach for both ML and CML training of HMMs [7]. Inspired from the literature regarding training techniques applied on Neural Networks [8,9], we introduce a new algorithm for gradient descent HMM optimization. We use independent learning rate adaptation for every trainable parameter of the HMM, and we show that it does not only outperform significantly the convergence rate of the standard gradient descent, but also leads to a much more robust training procedure; at the same time it is sufficiently simple, since it requires almost no parameter tuning. We apply the proposed algorithm in problems from Molecular Biology, namely training models to predict the transmembrane segments of both α-helical and β-barrel membrane proteins. Adaptive learning rate optimization provides much faster training for CML, compared to standard gradient descent, and also speeds up the convergence rate of ML training up to five times (even though, in terms of speed, it cannot outperform the Baum-Welch algorithm). We also show that for this class of models, and especially when the labels are well defined, CML training is more

---

[1] Corresponding author. E-mail: pbagos@biol.uoa.gr

effective, and should be preferred. We observe that, although the Baum-Welch algorithm still achieves the fastest convergence rates for ML training, our approach constitutes a potential replacement for efficient HMM training, given its simplicity.

2. Parameter estimation

A HMM is composed of a set of states, a set of observable symbols and a set of transition and emission probabilities [1,2]. Two states k, l are connected by means of the transition probabilities a_{kl}, forming a 1st order Markov chain. Each state k is associated with an emission probability $e_k(x_i)$, which is the probability of a symbol x_i to be emitted by that state [2]. With labeled sequences, each amino acid sequence x is accompanied by a sequence of labels y for each position i in the sequence. Consequently, we need a new probability distribution, the probability $\delta_k(c)$ of a state k having a label c [5]. The total probability P of a sequence x given a model θ is calculated by summing over all possible paths π [2]:

$$P(x \mid \theta) = \sum_{\pi} P(x, \pi \mid \theta) = \sum_{\pi} a_{0\pi_1} \prod_{i=1} e_{\pi_i}(x_i) a_{\pi_i \pi_{i+1}} \tag{1}$$

P is calculated using the well-known forward algorithm [1,2]. Utilizing modified versions of the forward and backward algorithms [5], we can incorporate the concept of labeled data. Thus, we can also use the joint probability of the sequence x and the labeling y given the model [10]:

$$P(x, y \mid \theta) = \sum_{\pi} P(x, y, \pi \mid \theta) = \sum_{\pi \in \Pi_y} P(x, \pi \mid \theta) = \sum_{\pi \in \Pi_y} a_{0\pi_1} \prod_{i=1} e_{\pi_i}(x_i) a_{\pi_i \pi_{i+1}} \tag{2}$$

The Maximum Likelihood estimate for any arbitrary model parameter, is denoted by:

$$\theta^{ML} = \arg \max_{\theta} P(x \mid \theta) \tag{3}$$

Whereas in the CML approach, the goal is to maximize the probability of correct labeling, instead of the probability of the sequences [5,6,10]. This is formulated as:

$$\theta^{CML} = \arg \max_{\theta} P(y \mid x, \theta) = \arg \max_{\theta} \frac{P(x, y \mid \theta)}{P(x \mid \theta)} \tag{4}$$

It is more convenient to switch to logarithms, and denote the logarithms of these likelihoods, ℓ (log-likelihood). ML estimation could be performed either by the Baum-Welch algorithm, or by a gradient descent method, whereas CML estimation is performed mainly by gradient descent. We show that gradient descent optimization could be performed in a much faster and more reliable way, by using an approach based on individual learning rate adaptation. For instance, we calculate the transition probabilities using following algorithm:

for each k
{
  for each l
  {
    if $\dfrac{\partial \ell}{\partial z_{kl}}^{(t)} \cdot \dfrac{\partial \ell}{\partial z_{kl}}^{(t-1)} > 0,$ **then**
    {
      $\eta_{kl}^{(t)} = \min\left(\eta_{kl}^{(t-1)}.a^+, \eta_{max}\right)$
    }
    else if $\dfrac{\partial \ell}{\partial z_{kl}}^{(t)} \cdot \dfrac{\partial \ell}{\partial z_{kl}}^{(t-1)} < 0,$ **then**
    {
      $\eta_{kl}^{(t)} = \max\left(\eta_{kl}^{(t-1)}.a^-, \eta_{min}\right)$
      $\dfrac{\partial \ell}{\partial z_{kl}}^{(t)} = 0$
    }

    $$\alpha_{kl}^{(t+1)} = \frac{\alpha_{kl}^{(t)} \exp\left(-\eta_{kl}^{(t)} \dfrac{\partial \ell}{\partial z_{kl}}^{(t)}\right)}{\sum_{l'} \alpha_{kl'}^{(t)} \exp\left(-\eta_{kl'}^{(t)} \dfrac{\partial \ell}{\partial z_{kl'}}^{(t)}\right)}$$
  }
}

For the transition probabilities above, we will use the partial derivative of the likelihood ℓ w.r.t. the transformed variables z_{kl} and not w.r.t. the α_{kl}, in order to ensure that the probabilities will be properly normalized in the range [0,1]. The derivatives of the likelihood are computed, and the transformation is performed using the expected counts for each parameter, as described in [7,10]. If the partial derivative possesses the same sign for two consecutive steps, the learning rate is increased (multiplied by a factor of $a^+ > 1$), whereas if the derivative changes sign, the learning rate is decreased (multiplied by a factor of $a^- < 1$). In the second case, we set the partial derivative equal to zero and thus prevent an update of the model parameter. This ensures that in the next iteration the parameter is modified according to the reduced learning rate, using though the actual gradient. We chose to have the learning rates bound by some minimum and maximum values denoted by the parameters η_{min} and η_{max}. The *sign* operator returns 1 if the argument is positive, -1 if it is negative and 0 otherwise, whereas *min* and *max* operators are the usual minimum and maximum of two arguments. It is completely straightforward to derive the appropriate expressions for the emission probabilities as well

3. Applications to transmembrane proteins

We have applied our algorithm, for predicting the topology of transmembrane proteins, a hard problem in current bioinformatic research [11]. Transmembrane proteins are divided in two distinct structural classes. The first is the α-helical membrane protein class, found in nearly all cell and organelle membranes of eukaryotes the inner membrane of prokaryotes [11], and the second is the β-barrel membrane protein class, found only in the outer membrane of gram-negative bacteria [12]. For α-helical membrane proteins, we used a HMM with 110 states and a training set of 30 non-homologous multi-spanning membrane proteins with structures deposited in the Protein Data Bank (PDB) [13]. The model of β-barrel membrane proteins consists of 61 states [14,15], whereas the training set includes 16 non-homologous β-barrel membrane proteins deposited in PDB [15]. The labels in both cases were defined by observing the 3-dimensional structures of the proteins.

On both models, we performed ML training using the Baum-Welch algorithm, standard gradient descent and individual learning rate adaptation, whereas for CML training we used standard gradient descent and the individual learning rate adaptation approach. For all methods (except Baum-Welch), we used various initial learning rates ranging from 0.001 to 0.1. For the decoding, we use the standard Viterbi algorithm [2] and the N-best algorithm [6]. We evaluated the fraction of correctly predicted residues and the correlation coefficient.

4. Results and Discussion

For both α-helical and β-barrel membrane proteins, the individual learning rate adaptation significantly outperforms standard gradient descent, in terms of speed of convergence, robustness and prediction accuracy. This is the case with either ML or CML training. For ML training, the Baum-Welch algorithm is still faster from gradient descent, even though with the individually learning rate adaptation the difference is rather small. However, when comparing ML and CML training in terms of the prediction accuracy, we observe that better results could be obtained when the models are trained according to the CML criterion. This is related with the quality of the labels used, arising from the crystallographically solved structures. It is well known that CML is very sensitive to data mislabeling [6]; thus, in cases where the labels are well defined, one should expect the accuracy of CML to be better.

On the other hand, the algorithms that we propose for individually learning rate adaptation, are much more reliable compared to standard gradient descent. They converge much faster, need no parameter tuning and yield better prediction accuracies, irrespective of the initial values of the learning rate; thus, they should be preferred for CML training of HMMs. We conclude that, for this kind of problems, CML training should be preferred. With the use of the independent learning rate adaptation, the HMM learning procedure with CML becomes a very easy task; moreover, given that CML performs better than ML training, we can benefit from the advantages of CML (discriminative power, better accuracy) without suffering from its weaknesses (slow convergence, unstable training). These algorithms could be further applied on hybrid methods such as Hidden Neural Networks [10], where work is on progress, in order to allow their application on biological problems.

Acknowledgments

P.B. was supported by a grant from the IRAKLEITOS fellowships program of the Greek Ministry of National Education, supporting basic research in the National and Kapodistrian University of Athens.

References

[1] L. Rabiner, A tutorial on hidden Markov models and selected applications in speech recognition. *Proc IEEE.* **77**(2) 257-286(1989)

[2] R. Durbin, S. Eddy, A. Krogh, and G. Mithison, Biological sequence analysis, probabilistic models of proteins and nucleic acids. *Cambridge University Press* (1998).

[3] L. Baum, An inequality and associated maximization technique in statistical estimation for probabilistic functions of Markov processes. *Inequalities.* **3** 1-8(1972)

[4] P. Baldi, Y. Chauvin, Smooth On-Line Learning Algorithms for Hidden Markov Models. *Neural Comput.* **6**(2) 305-316(1994)

[5] A. Krogh, Hidden Markov models for labeled sequences, *Proceedings of the12th IAPR International Conference on Pattern Recognition* 140-144(1994)

[6] A. Krogh, Two methods for improving performance of an HMM and their application for gene finding. *Proc Int Conf Intell Syst Mol Biol.* **5** 179-86(1997)

[7] P.G. Bagos, T.D. Liakopoulos and S.J Hamodrakas, Maximum Likelihood and Conditional Maximum Likelihood learning algorithms for Hidden Markov Models with labeled data-Application to transmembrane protein topology prediction. In Simos, T.E. (ed): Computational Methods in Sciences and Engineering, Proceedings of the International Conference 2003 (ICCMSE 2003) 47-55 (2003)

[8] C.M. Bishop, Neural Networks for Pattern Recognition. Oxford University Press (1998)

[9] W. Schiffmann, M. Joost and R. Werner, Optimization of the Backpropagation Algorithm for Training Multi-Layer Perceptrons. *Technical report* University of Koblenz, Institute of Physics (1994)

[10] A. Krogh and S.K. Riis, Hidden neural networks. *Neural Comput.* **11**(2) 541-63(1999)

[11] G. Von Heijne, Recent advances in the understanding of membrane protein assembly and function. *Quart. Rev. Biophys.,* **32**(4) 285-307(1999)

[12] Schulz, G.E.: The structure of bacterial outer membrane proteins, Biochim. Biophys. Acta., 1565(2) (2002) 308-17

[13] H.M. Berman, T. Battistuz, T.N. Bhat, W.F. Bluhm, P.E. Bourne, K. Burkhardt, Z. Feng, G.L. Gilliland, L. Iype, S. Jain, P. Fagan, J. Marvin, D. Padilla, V. Ravichandran, B. Schneider, N. Thanki, H. Weissig, J.D. Westbrook and C. Zardecki, The Protein Data Bank. *Acta Crystallogr D Biol Crystallogr.* 58(Pt 6 No 1), 899-907(2002).

[14] P.G. Bagos, T.D. Liakopoulos, I.C. Spyropoulos and S.J Hamodrakas, A Hidden Markov Model capable of predicting and discriminating β-barrel outer membrane proteins. *BMC Bioinformatics* **5**:29 (2004)

[15] P.G. Bagos, T.D. Liakopoulos, I.C. Spyropoulos and S.J Hamodrakas, PRED-TMBB: a web server for predicting the topology of beta-barrel outer membrane proteins. *Nucleic Acids Res.* **32**(Web Server Issue) W400-W404(2004)

VSP International
Science Publishers
P.O. Box 346, 3700 AH Zeist
The Netherlands

*Lecture Series on Computer
and Computational Sciences*
Volume 1, 2004, pp. 57-60

A Bolza's Problem in Hydrothermal Optimization

L. Bayón[1]; J.M. Grau; M.M. Ruiz; P.M. Suárez

Department of Mathematics, E.U.I.T.I.,
University of Oviedo, Spain

Received 2 June, 2004; accepted in revised form 24 June, 2004

Abstract: This paper studies the optimization of hydrothermal systems. We shall use Pontryagin's Minimum Principle as the basis for proving a necessary condition for the stationary functions of the functional, setting out our problem in terms of optimal control in continuous time, with the Bolza-type functional. This theorem allows us to elaborate the optimization algorithm that leads to determination of the optimal solution of the hydrothermal system. Finally, we present a example employing the algorithm developed for this purpose with the "Mathematica" package.

Keywords: Optimal Control, Bolza's Problem, Pontryagin's Principle, Hydrothermal

Mathematics Subject Classification: 49J24

1 Introduction

In this paper we propose Pontryagin's Minimum Principle (PMP) to solve the optimum scheduling problem of hydrothermal systems. Several applications of optimal control theory (OCT) in hydrothermal optimization have been reported in the literature. These range from the initial studies corresponding to El-Hawary and Christensen [1], to more recent works such as [2] or [3]. In a previous study [4], it was proven that the problem of optimization of the fuel costs of a hydrothermal system with m thermal power plants may be reduced to the study of a hydrothermal system made up of one single thermal power plant, called the thermal equivalent. We will call this problem: the $(H_1 - T_1)$ Problem, and in Section 2 we shall see that this problem consists in the minimization of a functional

$$F(z) = \int_0^T L(t, z(t), z'(t))dt + S[z(T)]$$

within the set of piecewise C^1 functions $(\widehat{C}^1)$ that satisfy $z(0) = 0$, $z(T) \leq b$ and the constraints $0 \leq H(t, z(t), z'(t)) \leq P_d(t), \forall t \in [0, T]$. Hence, the problem involves non-holonomic inequality constraints (differential inclusions). Using classic mathematical methods (see for example [5]), we shall focus in the present paper on the development of the applications of optimal control theory (OCT) to the specific problem of hydrothermal optimization.

In Section 3 we shall establish a necessary condition for the stationary functions of the functional and we shall use PMP as the basis for proving this theorem. We shall see that the treatment of the constraints of the problem using this new approach will be very simple. The development enables the construction, in Section 4, of the optimization algorithm that leads to determination of the optimal solution of the hydrothermal system. Finally, in Section 5, we present a example employing the Algorithm developed for this purpose with the "Mathematica" package.

---

[1]Corresponding author. C./ Manuel Llanoza 75, 33208 Gijón, Asturias (Spain). E-mail: bayon@uniovi.es

2   Statement of the Bolza's Problem: Water Cost

The $(H_1 - T_1)$ problem (one Hydraulic plant - Thermal equivalent) consists in minimizing the cost of fuel needed to satisfy a certain power demand during the optimization interval $[0, T]$. Said cost may be represented by

$$\int_0^T \Psi(P(t))dt + S[z(T)] \tag{2.1}$$

where Ψ is the function of thermal cost of the thermal equivalent, $P(t)$ is the power generated by said plant, and $S[z(T)]$ is the cost assigned to the water discharged. Moreover, the following equilibrium equation of active power will have to be fulfilled

$$P(t) + H(t, z(t), z'(t)) = P_d(t), \forall t \in [0, T]$$

where $P_d(t)$ is the power demand and $H(t, z(t), z'(t))$ is the power contributed to the system at the instant t by the hydro-plant, $z(t)$ being the volume that is discharged up to the instant t by the plant, and $z'(t)$ the rate of water discharge of the plant at the instant t. In this paper, we propose to study the problem when the final instant T is given and the final state has an upper boundary: $z(T) \leq b$. The following boundary conditions will have to be fulfilled $z(0) = 0$, $z(T) \leq b$. Taking into account the equilibrium equation, our objective functional in the Bolza's form is

$$F(z) = \int_0^T L(t, z(t), z'(t))dt + S[z(T)] \tag{2.2}$$

with L having the form $L(t, z(t), z'(t)) = \Psi(P_d(t) - H(t, z(t), z'(t)))$ over the set Θ_b

$$\Theta_b = \{z \in \widehat{C}^1[0, T] \mid z(0) = 0, z(T) \leq b, 0 \leq H(t, z(t), z'(t)) \leq P_d(t), \forall t \in [0, T]\}$$

If z satisfies Euler's equation for the functional F we have that, $\forall t \in [0, T]$

$$L_z(t, z(t), z'(t)) - \frac{d}{dt}(L_{z'}(t, z(t), z'(t))) = 0 \tag{2.3}$$

If we divide Euler's equation (2.3) by $L_{z'}(t, z(t), z'(t)) < 0$, $\forall t$, and integrating we have that

$$-L_{z'}(t, z(t), z'(t)) \cdot \exp\left[-\int_0^t \frac{H_z(s, z(s), z'(s))}{H_{z'}(s, z(s), z'(s))}ds\right] = -L_{z'}(0, z(0), z'(0)) = K \in \mathbb{R}^+ \tag{2.4}$$

We shall call relation (2.4) the coordination equation, and the positive constant K will be termed the coordination constant of the extremal. Let us now see the fundamental result (The Main Coordination Theorem), which enables us to characterize the extremals of the problem and which is also the basis for elaborating the optimization algorithm that leads to determination of the optimal solution of the hydrothermal system. We shall use the above coordination equation (2.4) in the development of the proof of the theorem.

3   The Main Coordination Theorem

We shall use PMP as the basis for proving this theorem, setting out our problem in terms of optimal control in continuous time, with the Bolza-type functional. In this paper we generalize a previous study [6] and we present the problem considering the state variable to be $z(t)$ and the control variable $u(t) = H(t, z(t), z'(t))$. Moreover, as $H_{z'} > 0$, the equation $u(t) - H(t, z(t), z'(t)) = 0$ allows the state equation $z' = f(t, z, u)$ to be explicitly defined.

The optimal control problem is thus:

$$\min_{u(t)} \int_0^T L(t, z(t), u(t))dt + S[z(T)] \quad \text{with} \quad \begin{cases} z' = f(t, z, u) \\ z(0) = 0, \quad z(T) \le b \\ u(t) \in \Omega(t) = \{x \mid 0 \le x \le P_d(t)\} \end{cases}$$

with $L(t, z(t), u(t)) = \Psi(P_d(t) - u(t))$. We shall see that with this approach we shall arrive at the coordination equation (2.4). It can be seen that from the relations $u(t) - H(t, z(t), z'(t)) = 0$ and $z' = f(t, z, u)$, we easily obtain $f_z = -\frac{H_z}{H_{z'}}$; $f_u = \frac{1}{H_{z'}}$. We define the following function.

Definition 1. Let us term the coordination function of $q \in \Theta_b$ the function in $[0, T]$, defined as follows

$$\mathbb{Y}_q(t) = -L_{z'}(t, q(t), q'(t)) \cdot \exp\left[-\int_0^t \frac{H_z(s, q(s), q'(s))}{H_{z'}(s, q(s), q'(s))}ds\right]$$

Theorem 1. The Main Coordination Theorem.
If $q \in \widehat{C}^1$ is a solution of problem $(H_1 - T_1)$, then $\exists K$ such that
i) If $0 < H(t, q(t), q'(t)) < P_d(t) \Longrightarrow \mathbb{Y}_q(t) = K$.
ii) If $H(t, q(t), q'(t)) = P_d(t) \Longrightarrow \mathbb{Y}_q(t) \ge K$.
iii) If $H(t, q(t), q'(t)) = 0 \Longrightarrow \mathbb{Y}_q(t) \le K$.
and

$$K \ge \frac{\partial S[q(T)]}{\partial z} \cdot \frac{-\mathbb{Y}_q(T)}{L_{z'}(T, q(T), q'(T))} \tag{3.1}$$

4 Construction of the Optimal Solution

¿From the computational point of view, the construction of the optimal solution can be performed with the next procedure:
i) For each K we construct q_K, where q_K satisfies the conditions i), ii) and iii) of theorem 1 and the initial condition $q_K(0) = 0$.
In general, the construction of q'_K cannot be carried out all at once over all the interval $[0, T]$. The construction must necessarily be carried out by constructing and successively concatenating the extremal arcs $(0 < H(t, q_K(t), q'_K(t)) < P_d(t))$ and boundary arcs $(H(t, q_K(t), q'_K(t)) = P_d(t)$ or $H(t, q_K(t), q'_K(t)) = 0)$ until completing the interval $[0, T]$. This is relatively simple to implement, with the use of a discretized version of the equations.
ii) Varying the coordination constant K, we would search for the extremal that fulfils the second boundary condition $z(T) \le b$ and (3.1).
Firstly, we search for the value of K whose associated extremal satisfies $q_K(T) = b$. The procedure is similar to the shooting method used to resolve second-order differential equations with boundary conditions. Effectively, we may consider the function $\varphi(K) := q_K(T)$ and calculate the root of $\varphi(K) - b = 0$, which may be realized approximately using elemental procedures like the secant method.
If the relation (3.1) is fulfilled then $q_K(t)$ is the optimal solution and all the available water, b, is consumed. If the encountered K does not verify (3.1), the value of K that fulfills the equality in (3.1) is the optimal solution, and the optimal final volume in this case is $q_K(T) < b$.

5 Application to a Hydrothermal Problem

Let us now see a hydrothermal problem whose solution may be constructed in a simple way taking into account the above theorem 1. A program that resolves the optimization problem was elaborated using the Mathematica package and was then applied to one example of hydrothermal system made up of 8 thermal plants and one hydraulic plant of variable head. We consider the functional (2.1).

For the fuel cost model of the equivalent thermal plant Ψ, we use the quadratic model

$$\Psi(P(t)) = \alpha_{eq} + \beta_{eq}P(t) + \gamma_{eq}P(t)^2$$

We use a variable head model and the hydro-plant's active power generation P_h is function of $z(t)$ and $z'(t)$. Hence the function P_h is defined as

$$P_h(t, z(t), z'(t)) := A(t) \cdot z'(t) - B \cdot z(t) \cdot z'(t)$$

We consider that the transmission losses for the hydro-plant are expressed by Kirchmayer's model, where b_{ll} is the loss coefficient. So, the function of effective hydraulic generation is

$$H(t, z(t), z'(t)) := P_h(t, z(t), z'(t)) - b_{ll}P_h^2(t, z(t), z'(t))$$

Furthermore, we shall consider a linear model for the associated water cost

$$S[z(T)] = \nu \cdot z(T)$$

where ν is a water conversion factor, which accounts for the unit conversion from (m^3) to ($\$$). In this example we present two cases: (a) $\nu = 0.00375(\$/m^3)$ and (b) $\nu = 0.00475(\$/m^3)$. The optimal power for the hydro-plant, $P_h(t)$, for both cases is shown in the next figure.

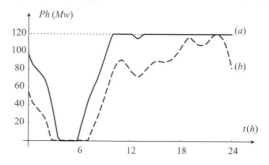

References

[1] El-Hawary, M. E., Christensen, G. S., 1979. *Optimal economic operation of electric power systems*. New York, Academic Press.

[2] Wong, K. H., Teo, K. L. and Jennings, L. S., 1993. A class of nonsmooth discrete-time constrained optimal control problems with application to hydrothermal power systems. *Cybernet. Systems* 24 (4), 341–354.

[3] Mousavi, H. and Ramamurthy, A.S., 2002. Multi-reservoir design using Pontryagin principle. *Advances in Water Resources*, 25 (6), 677-687.

[4] Bayón, L., Grau, J., Ruiz, M.M. and Suárez, P.M., 2003. New Developments on Equivalent Thermal in Hydrothermal Optimization. An Algorithm of Approximation. *Proceedings IC-CMSE 2003*, Greece 2003, 1, 83-86.

[5] Clarke, F.H., 1983. *Optimization and nonsmooth analysis*. New York, John Wiley & Sons.

[6] Bayón, L., Grau J. and Suárez, P.M., 2003. A New Algorithm for the Optimization of a Simple Hydrothermal Problem. *Commun. Nonlinear Sci. Numer. Simul.*, 9(2), 197-207.

VSP International
Science Publishers
P.O. Box 346, 3700 AH Zeist
The Netherlands

Lecture Series on Computer
and Computational Sciences
Volume 1, 2004, pp. 61-64

Fuzzy Control of Inverted Pendulum and
Concept of Stability Based on Java

Yasar Becerikli[1]*, B. Koray Celik[2]

Kocaeli University, Computer Engineering Department, Izmit, Turkey
[1] becer@kou.edu.tr [2] bkoraycelik@yahoo.co.uk

Received 30 April, 2004; accepted in revised form 24 June, 2004

Abstract: In this paper, a fuzzy controller for inverted pendulum system is presented in two stages. These stages: investigation of fuzzy control system modeling methods and solution of "Inverted Pendulum Problem" by using Java programming with Applets for internet based control education. In the first stage, fuzzy modeling and fuzzy control system, Java programming language, classes and multithreading were introduced. In the second stage specifically, simulation of inverted pendulum problem was developed with Java Applets and the simulation results were given. Also some stability concept is introduced.

Keywords: Fuzzy control, Java, Stability, web based simulation
Mathematics Subject Classification: 07.05.Mh, 07.05.Dz, 07.05.

1. Introduction

As we move into the information area, human knowledge becomes increasingly important. So a theory is necessary to formulate human knowledge and heuristic in a systematic manner and put into engineering systems, together with other information such as mathematical models and sensory measurements. This aspect is a justification for fuzzy systems in the literature and characterizes the unique feature of fuzzy systems theory. For many practical systems, important information comes from two sources: one source is human experts who describe their knowledge about the system in natural languages; the other is mathematical models that are derived according to physical laws and sensory measurements.[2] Therefore, we are faced with an important task of combining these two types of information into systems design. To manage this combination, we should answer the question that is how to transform a human knowledge and heuristic base into a mathematical model. Essentially, a fuzzy system performs this transformation.[1]

Fuzzy systems are knowledge-based or rule-based systems that contain descriptive IF-THEN rules that are created from human knowledge and heuristic. Also fuzzy systems are multi-input-single-output mappings from a real-valued vector to a real-valued scalar but for large scale nonlinear systems the multi-output mapping can be decomposed into a collection of single-output mapping as shown in Fig.1. [5]. An important contribution of fuzzy systems theory is that it provides a systematic procedure for transforming a knowledge base into a nonlinear mapping. So we can use this transformation in engineering systems (control) in same manner as we use mathematical models and sensory measurements. Consequently, by means of the fuzzy systems, we can perform analysis and design of engineering systems in a mathematically rigorous manner [3]. Fuzzy systems have been applied to a wide variety of fields ranging from control, signal processing, communication, medicine, expert systems to business, etc. However, most significant applications have concentrated on control problems. Fuzzy systems that are shown in Fig.2 can be used either as closed-loop controllers and open-loop controllers. As shown in Fig.3, when the fuzzy system is used as an open-loop controller, the system usually sets up control parameters and then the system operates according to these parameters. When it is used as a closed-loop controller as shown in Fig.4, the fuzzy system takes the outputs of the controlled system and applied the control action on the controlled system continuously. In this figures, controlled system can be considered as an application process.[3,4,5]

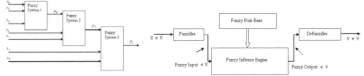

Fig. 1 Decomposing the multi-output mapping into a collection of a single-output mapping

Fig. 2 General configuration of fuzzy system

---

* Corresponding author

Fig. 3 Open-loop controller.   **Fig. 4** Closed-loop controller

The goal of this text is to show how transformation of knowledge base into a nonlinear mapping is done, how the analysis and design are performed on the control systems. As a nonlinear system, inverted pendulum system is often used as benchmark to achieve the goal of verifying the performance and effectiveness of a control method because of the simple structure. Recently, a lot of researches on control of inverted pendulum system by using fuzzy control systems contain fuzzy inference have been done.

Margaliot [6] showed a new approach to determining the structure of fuzzy controller for inverted pendulum by fuzzy Lyapunov synthesis. Yamakawa [7] demonstrated a high-speed fuzzy controller hardware system and used only seven fuzzy rules to control the angle of inverted pendulum system in 1989.Although stabilization control of an inverted pendulum system should also include the position control of the cart besides the angular control of the pendulum because of limit length of the rail, the above stated approaches only took into consideration the angular control of the pendulum. Yubazaki [9,10] built a new fuzzy controller for inverted pendulum system. The fuzzy controller has four input items, each with a dynamic importance degree.[8]

In this paper, a fuzzy controller for inverted pendulum system that needs two input items, which one is angle between pendulum and the vertical position, and the other is derivation of angle (angular velocity) of the pendulum presented simply for educational purpose. The fuzzy controller takes the angle and angular velocity of pendulum from the inverted pendulum system, aggregates inputs with defined IF-THEN rules and drives obtained force as an output item by means of inference methods.

However, recently, obtaining the information resource quickly becomes increasingly important. So, by using internet technologies (Java Applets), designing a simulation program for fuzzy controller of inverted pendulum system is unavoidable. The reason of choosing Java Applet technology arise from supporting all features necessary for extending the Web in ways previously impossible and Java bases on object-oriented technology that has evolved in diverse segments of the computer sciences as a means of managing the complexity inherent in many different kinds of systems.[18]

2. Inverted Pendulum System

The inverted pendulum system defined here as shown in Fig.5, which is formed from a cart, a pendulum and a rail for defining position of cart. The Pendulum is hinged on the center of the top surface of the cart and can rotate around the pivot in the same vertical plane with the rail. The cart can move right or left on the rail freely. Given that no friction exists in the system between cart and rail; between cart and pendulum.[5]

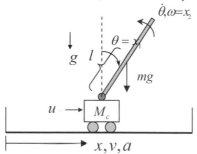

Fig. 5 Configuration of inverted pendulum system

The dynamic equation of the inverted pendulum system can be expressed as [2,11]

$$\dot{x}_1 = x_2$$

$$\dot{x}_2 = \ddot{\theta} = \dot{\omega} = \alpha = \frac{g\sin\theta - \dfrac{m_p l \dot{\theta}^2 \cos\theta \sin\theta}{m_c + m_p} + \dfrac{\cos\theta}{m_c + m_p}u}{l\left(\dfrac{4}{3} - \dfrac{m_p \cos\theta^2}{m_c + m_p}\right)}$$

Here, the parameters, m_c and m_p are, respectively, the mass of the cart and the mass of the pendulum in the unit (kg), and $g = 9.8 m/s^2$ is the gravity acceleration. The parameter l is the half length of the pendulum in unit (m). The variable u means the control force in the unit (N) applied horizontally to the cart. The variables θ, ω, α represent the angle between pendulum and upright position, the angular velocity and the angular acceleration of the pendulum, respectively. Also clockwise direction is positive. The variables x, v, a denote the position of the cart from the rail origin, its velocity, its acceleration and right direction is positive. The variables θ, ω, x, v are the four state variables to describe the dynamic system but, we use the variables θ and $\omega(\dot{\theta})$ to control the inverted pendulum system.

3. Inverted Pendulum Fuzzy Control System

The inverted pendulum fuzzy control system is based on the closed-loop fuzzy system shown in Fig.4 and designed as shown in Fig.6. Then stages of control methods design constructed step by step.

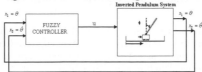

Fig. 6 Configuration of Inverted Pendulum Fuzzy Control System

3.1. Constructing the membership functions

To control the system state variable vector is chosen as $\begin{pmatrix} \theta \\ \dot{\theta} \end{pmatrix} = \begin{pmatrix} x_1 \\ x_2 \end{pmatrix}$ and then specifically, we assume the universe of discourse fro the two state variables to be $-40° \le x_1 \le 40°$ and $-8dps \le x_2 \le 8dps$ (degree per second). First we construct seven membership functions for x_1 on its universe, that is, for the values positive very big (PVB),positive big (PB), positive (P), zero (ZO), negative (N), negative big (NB),negative very big (NVB) as shown in Fig.7.We then construct five membership functions for x_2 on its universe, that is, for the values positive big (PB), positive (P), zero (ZO), negative (N), negative big (NB),negative very big (NVB) as shown in Fig.8. To partition the control space as output we construct nine membership functions for u on its universe which is $-32N \le u \le 32N$ as shown in Fig.9.

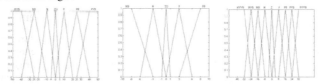

Fig. 7 Membership functions for state variable x₁   **Fig. 8** Membership functions for state variable x_2 **Fig. 9** Membership functions for output variable u

3.2. Constructing rule base

After the constructing membership functions stage, we construct a Fuzzy Associative Memory (FAM) Table for rule base.
According to FAM table our rule base can be expressed as
R1: IF (x_1 = PVB) and (x_2 =PB) THEN u =PVVB
⋮

R35: IF (x_1 = NVB) and (x_2 =NB) THEN u =NVVB
As it seems, the rule base is formed by thirty five rules to control the inverted pendulum system.
Then we determine the fuzzification method with Singleton Fuzzifier, inference system with Mamdani Method, and defuzzification method with Center Average for this fuzzy control system.

4. Java Programming Language

Java is a programming language, a runtime system, a set of development tools, and an application programming interface (API) [18].
A software developer writes programs in the Java language that use predefined software packages of the Java API. The developer compiles his or her programs using the Java compiler. This results in what is known as **compiled bytecode.** Bytecode is in a form that can be executed on the Java virtual machine, the core of the Java runtime system. The Java runtime system consists of the virtual machine plus additional software, such as dynamic link libraries, that are needed to implement the Java API on operating system and hardware. The detailed information can be found from [17], [18].

5. Simulation Results

As shown in Fig.10, inverted pendulum fuzzy control system Java Applets was developed. This Java Applet can be used for educational goals.

The mass of the pendulum is m_p =0.1 kg; the half length of the pendulum is l =1 m; and the mass of the cart is m_c =1.0 kg. The angle and the angular velocity of the pendulum is limited to [− 40°,+40°],[−8dps,+8dps] respectively. Also In the program for angle and angular velocity initial conditions are determined as $x_1 = -30°$, $x_2 = 2dps$.

In the first test, we selected the mass of the cart 2 N and we compared the results as shown in Fig.11. As it seems, the increasing of the mass of the cart makes inverted pendulum system difficult to control. In the second test, we selected the pole length 0.5 m and we compared the results as shown in Fig.12.Here, the increasing of the pole length makes inverted pendulum system difficult to control. According to the results of two tests, we can control the inverted pendulum, by defining short pole length values and less weight cart mass values. Also, as shown in Fig.13 we can learn the approximate stability of inverted pendulum system although it is difficult to determine the stability. According to Fig.13 we can say the inverted pendulum is asymptotically stable.[16]

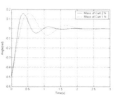

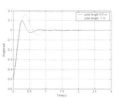

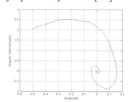

Fig. 10. The Java Application for inverted pendulum fuzzy control system

Fig. 11 The simulation result for changing of cart mass value

Fig.12 The simulation result for changing of pole length

Fig.13 State Space trajectory of inverted Pendulum System

6. Conclusions

Recently, the interest in fuzzy systems becomes increasingly important. Therefore, the solution of "the inverted pendulum problem" has been chosen and has presented by using Java Applets in the Web for educational goals. The simulation results show that the system with short pole and light cart can be controlled easily and quickly.

Moreover, for next projects, we can generalize the fuzzy system based on Java application, so the corresponding user can change the rule base and the structure of the system for purpose of learning the finer points of the inverted pendulum fuzzy system design. Also we can develop a fuzzy laboratory on the Net with Java Applets to provide opportunities to learn fuzzy logic.

References

[1]   Çelik, B.K., *Fuzzy Control Systems and Fuzzy Logic Based Inverted Pendulum Control System*, Term Project, Computer Engineering Department Kocaeli University, January 2004.

[2]   Wang, L., *A Course in Fuzzy Systems and Control,* Prentice-Hall Inc., 1997.

[3]   Ross, J.T., *Fuzzy Logic with Engineering Applications*, McGraw-Hill, Inc., 1995.

[4]   Klir, G.J., B., Youan, *Fuzzy Sets and Fuzzy Logic Theory and Applications*, Prentice Hall Inc., 1995.

[5]   Z., Yeh and K., Li, "A Systematic Approach for Designing Multistage Fuzzy Control Systems, *Fuzzy Sets and Systems,* Elsevier, 7 May 2003.

[6]   M., Margaliot,G., Langholz, "Adaptive Fuzzy Controller Design via Fuzzy Lyapunov Synthesis ".*Proceedings of FUZZ-IEEE'98*, 1998. p. 354–

[7]   T., Yamakawa, "Stabilization of an Inverted Pendulum by a High-Speed Fuzzy Logic Controller Hardware System. *Fuzzy Sets and Systems*, Elsevier, 1989; 32:161–80.

[8]   S., Kawaji , T., Maeda. "Fuzzy Servo Control System For an Inverted Pendulum. *Proceedings of IFES'91*, vol. 2, 1991. p. 812–23.

[9]   N., Yubazaki, J., Yi , M., Otani, K., Hirota "SIRMs Dynamically Connected Fuzzy Inference Model and its Applications". *Proceedings of IFSA'97*,vol. 3, 1997. p. 410–5.

[10]   J.,Yi, N., Yubazaki, K., Hirota, "Upswing and stabilization control of inverted pendulum system based on the SIRMs dynamically connected fuzzy inference model", *Fuzzy Sets and Systems* , Elsevier, March 2000.

[11]   M., Margaliot, and G., Langholz, "A New Approach to Fuzzy Modeling and Control of   Discrete-Time Systems" *IEEE Transactions on Fuzzy Systems*, Vol 11, August 2003.

[12]   J., Yi , And N., Yubazaki , "Stabilization fuzzy control of inverted pendulum systems", *Artificial Intelligence in Engineering IEEE Transactions on Systems*, Vol 14.,April 2000.

[13]   L.A., Zadeh, "From circuit theory to systems theory", *Proc. Institution of Radio Engineers*, 50, pp. 856-865, 1962.

[14]   L.A., Zadeh,. "Fuzzy Algortihms",*Information and Control,*12,no.2,pp.94-102,1968.

[15]   L.A., Zadeh, "Outline of a New Approach to the Analysis of Complex Systems and Decision Processes", *IEEE Transactions on systems ,Man and Cybernetics,*Vol.3,1973

[16]   Slotine,J.-J.E.,W.Li,*Applied Nonlinear Control*, Prentice-Hall,Inc.,1991

[17]   Rao, V.B., *C++ Neural Networks and Fuzzy Logic*, IDG Boks Worldwide,Inc.,1995.

[18]   Jaworski J.,*Java Developer's Guide* ,Sams.net, 1996

VSP International
Science Publishers
P.O. Box 346, 3700 AH Zeist
The Netherlands

Lecture Series on Computer
and Computational Sciences
Volume 1, 2004, pp. 65-69

The Need for High-Resolution Signal Processing in Magnetic Resonant Spectroscopy for Brain Tumor Diagnostics

Karen Belkić[1]

Karolinska Institute,
Department of Oncology/Pathology, Medical Radiation Physics
P. O. Box 260, S-171 76 Stockholm, Sweden

and

University of Southern California School of Medicine
Institute for Prevention Research

Received 23 August, 2004; accepted in revised form 28 August, 2004

Abstract: In the very recent period Magnetic Resonance Spectroscopy (MRS) and Spectroscopic Imaging (MRSI) have become key diagnostic modalities for neuro-oncology. MRS and MRSI are now applied extensively for initial detection of brain tumors, for histopathologic classification, tumor localization and grading, as well as for assessment of response to therapy and for follow-up surveillance, striving, in particular, for earlier identification of recurrence. We have performed a systematic review of this very recent body of knowledge, which clearly demonstrates that MRS and MRSI have made tremendous advances in brain tumor diagnostics. Notwithstanding these achievements, important shortcomings still exist in the present applications of MRS and MRSI in neuro-oncology. Many of these are directly related to the reliance upon the conventional Fourier-based framework for data analysis. The Fast Padé Transform (FPT) relative to the Fast Fourier Transform (FFT) offers distinct advantages that could improve the diagnostic performance of MRS and MRSI for brain tumor diagnostics.

Keywords: Magnetic resonance spectroscopy, Brain tumors, Fast Fourier Transform, Fast Padé Transform

Mathematics Subject Classification: 32.30.Dx: 32.30.RJ:33.25. +k

PACS: .30.Dx: 32.30.RJ;33.25. +k

The area of neuro-oncology has undergone a dramatic transformation in the most recent period. Non-invasive molecular imaging via Magnetic Resonance Spectroscopy (MRS) and Spectroscopic Imaging (MRSI) have now become key modalities for nearly all aspects of brain tumor diagnostics. In fact, in no other area of oncology have MRS and MRSI been so widely incorporated into clinical practice. There is a literal "explosion" of information on MRS and MRSI in neuro-oncology in the last two years. We have performed a systematic review of this very recent body of knowledge, which, indeed, has represented an important advance for the detection and characterisation of tumors of the brain.

From this review, we see, however, that there are still major shortcomings of the present applications of MRS and MRSI in neuro-oncology. First of all, very few of the currently assessed metabolite concentrations or ratios unequivocally distinguish intra-cerebral tumors from normal brain tissue. Moreover, changes in each of the metabolite concentrations and ratios are non-specific for cancer of the brain. In other words, non-neoplastic processes such as infection, stroke, demyelinating disorders, *inter alia*, frequently show spectral changes that are identical to those seen in brain tumors. Histopathological characterization and tumor grading, both of crucial importance for clinical decision-making, have been greatly aided by MRS and MRSI. However, there are numerous contradictory

---

[1] Corresponding author. Dr. Karen Belkic E-mail: Karen.Belkic@radfys.ki.se.

findings in the literature. Simply stated, improvements in the diagnostic accuracy of MRS and MRSI for neuro-oncology are still sorely needed.

Many of the current problems and dilemmas of MRS and MRSI in neuro-oncology are directly related to the reliance upon the conventional Fourier-based framework for data analysis. Among the most critical are limitations in resolution and signal-to-noise ratio that compromise detection of small brain tumors. Using the FFT, a spectrum is given as a single polynomial, $F(\omega_k)$ with pre-assigned angular frequencies, ω_k whose minimal separation $\omega_{min}= 2\pi\,k/T$ is determined by the given epoch (or the total acquisition time) T. The FFT spectrum is defined *only* on the Fourier grid points $\omega_k = \pm\,k\,/T$ where ($k = 0,1,2,3...N-1$) and N is the signal length. The main strategy applied in attempts to improve resolution has been to increase T, and thereby to decrease ω_{min}. However, in clinical practice, at larger T the signal becomes heavily corrupted with background noise[2]. The reason is that envelopes of time signals, such as those observed in MRS, decay exponentially, so that the larger signal intensities are found early in the recording. It is therefore advantageous to encode the time signal as rapidly as possible, i.e. to avoid long acquisition times at which mainly noise will be recorded. In other words, there are two mutually exclusive requirements and as a result, within the FFT, attempts to improve resolution lead to a worsening of the signal-to-noise ratio (S/N).

In brain tumor diagnostics a number informative metabolites (certain lipids, glutamine/glutamate, myoinositol) decay rapidly and can only be detected at short echo-times (TE). Moreover, protocols from different centers use different TE, which will dramatically affect the assessment of even the longer-lived metabolites and their ratios. This is one of the problems that has precluded comparisons among various centers, a key prerequisite for the establishment of norms. Moreover, the use of metabolite ratios e.g. choline to N-Acetyl Aspartate or to creatine, is problematic because the denominator often varies for reasons unrelated to the oncologic process of interest.

Another reason why data compatibility among various centers has been hampered by the reliance on FFT is because of its requirements for fitting. This can lead both to spurious peaks (over-fitting) and true metabolites being undetected (under-fitting). This is not only unacceptable to diagnosticians, but renders inter-study comparisons tenuous, at best, unless the same basis set is used in e.g. the LCModel[3] to predetermine the number of metabolites. Such fitting is often based upon prior-knowledge/measurement, e.g. LCModel, prior to analysis of the actual *in vivo* spectrum [1]. Although claims have been made that fitting procedures can be automatic and objective [2], their major pitfalls and inherent subjectivity have been highlighted [1]. As recently pointed out, if one does not include an *in vitro* metabolite in the basis set, then one is going to have a very bad fit precisely at the frequency location where the missing metabolite is expected to occur in the studied *in vivo* spectrum and this leads to subjectivity [3]. The FFT is a non-parametric estimator, which provides only the shape of spectral structures, but not quantification. The peak parameters are extracted by fitting the obtained structures to a sum of Lorentzians or Gaussians [4]. Thus, much information that is contained in the signal is not obtained, such as the actual position, width, height and phase of each metabolite.

These problems are particularly pronounced with respect to overlapping metabolites. Various fitting procedures such as AMARES[4], MRUI[5], as well as the LCModel can be applied, none of which are generated with certainty, especially for the number of metabolites and peak heights.   Several overlapping resonances are critically important in brain tumor diagnostics, these include lipids, lactate, and alanine; glutamine/glutamate with NAA and with myoinositol.

Reliance upon fitting and quantification issues have represented two inter-connected and crucially important issued for MRS and MRSI in neuro-oncology. A salient example in brain tumor diagnostics is the inclusion or non-inclusion of glutathione in various studies, which has lead to marked differences in the estimation of other metabolites such as glutamine/glutamate.

Furthermore, the current Fourier-based applications of MRS and MRSI in brain tumor diagnostics yield a relatively small number of metabolites (low molecular weight, high concentration) that are observable on clinical scanners. In other words, we are provided with an extremely limited view of normal and abnormal brain chemistry.

_____

[2] Within the FFT a strategy known as zero filling is used as an attempt at quasi-interpolation.
[3] LCModel denotes Linear Combination of Model *in vitro* Spectra [2]
[4] AMARES is the acronym for Advanced Method for Accurate, robust and Efficient Spectral Fitting
[5] MRUI is the acronym for Magnetic Resonance User Interface

It has been conclusively shown in a series of recent papers [1, 3-11], that the Fast Padé transform (FPT), overcomes many of the limitations of the FFT relevant to *in vivo* MRS and MRSI. The FPT is a non-linear rational approximation of two polynomials, $A(z^{-1})/B(z^{-1})$ to the Taylor power series expansion of the raw time signal in the variable z^{-1}, where $z = exp\,(-i\omega\tau)$, τ is the sampling time and ω is the angular frequency. As a high resolution, parametric method, the FPT dramatically improves S/N, and fulfils the most stringent requirements for tumor diagnostics: no post-processing fitting, provides precise numerical results for all the peak parameters for each resonance, and specifies the *exact number* of metabolites (including those that overlap) from the encoded data.

The figure illustrates the advantages of the FPT over the FFT at short signal lengths. Here the magnitude spectrum is presented of signals encoded from an *in vivo* MRS recording at 4T of the brain of a healthy volunteer. These data of full signal length N=2048 encoded by the group at the Center for Magnetic Resonance Research, University of Minnesota, Minneapolis, USA [12] have been kindly made available to us. We present the FFT (left column) and FPT (right column) magnitude spectra at three signal lengths (N/32 = 64, N/16 =128, N/8 = 256). The most dramatic difference between the Fourier and Padé spectra occurs at the largest truncation level (N/32 = 64), top panel. Here, practically no distinct metabolite is discerned by the FFT. On the other hand, with the FPT the NAA peak at 2.0 ppm and creatine near 3.0 ppm are already clearly visible. In the middle panel (N/16 = 128) over 70% of the NAA concentration is predicted by FPT, whereas this is still below 30% for FFT. On the right bottom panel for FPT (N/8=256), the NAA peak at 2.0 is near its full height and the two peaks at 3.0 and 3.3 (creatine and choline, respectively) are now clearly delineated with their ratios being close to that at full signal length. In contrast, creatine, choline and NAA are still all grossly underestimated by the FFT (bottom left panel). Most importantly, the FPT is observed throughout the figure to produce no spurious metabolite or other spectral artifacts in the process of converging in a strikingly steady fashion as a function of increased signal length. Moreover, the rate of convergence of the FPT is found to be much faster than in the FFT.

Clearly, the area of brain tumor diagnostics is one in which the most delicate of clinical decisions are made. Such decisions require maximal information of the highest possible reliability. It is precisely here that the advances in computational methods are uniquely contextualized -- for the promise they hold in helping the fight against one of the most difficult cancers known to humankind. The next, and urgently needed step, is to more widely apply the FPT to *in vivo* MRS and MRSI signals from patients with brain tumors with the aim of tackling, in actual practice, the diagnostic dilemmas still plaguing neuro-oncology.

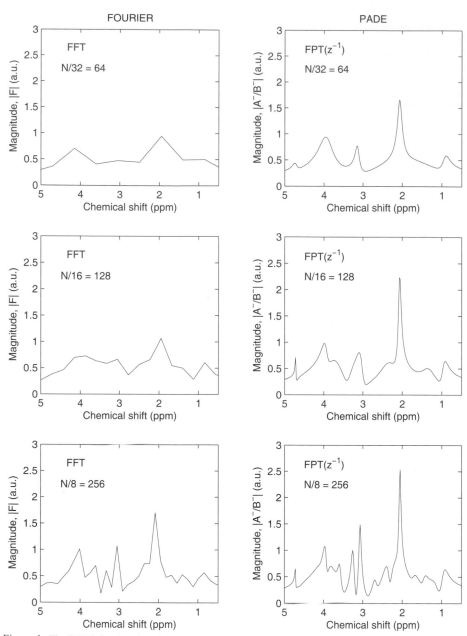

Figure 1: The FFT (left column) and FPT (right column) magnitude spectra at three signal lengths (N/32 = 64, N/16 =128, N/8 = 256) from an *in vivo* MRS recording at 4T of the brain of a healthy volunteer, from Ref. [12].

Acknowledgments

This work was supported by the following four Swedish funds: Stockholm's County Council through FoUU at the Karolinska Hospital, the Swedish Scientific Research Council (Vetenskapsradet), the Cancer Foreningen and Gustav the Fifth Jubileum Fund.

The author would like to thank Dzevad Belkic, Professor of Mathematical Radiation Physics, Karolinska Institute, for critical review of this abstract and fruitful discussions.

References

[1] Dz. Belkić, Non-Fourier based reconstruction techniques, *MAGMA* **15** (Suppl.1), 36-37 (2002).

[2] S.W. Provencher, Estimation of metabolite concentrations from localized in vivo proton NMR spectra, *Magn. Reson. Med.* **30**, 672-679 (1993).

[3] Dz. Belkić, Padé-based Magnetic Resonance Spectroscopy (MRS) *J. Com. Meth. Sci. Eng.* **3**, 563-733 (2003).

[4] Dz. Belkić, Exact analytical expressions for any Lorentzian spectrum in the Fast Padé transform (FPT), *J. Com. Meth. Sci. Eng.* **3**, 109-186 (2003).

[5] Dz. Belkić, Fast Padé transform for magnetic resonance imaging and computerized tomography, *Nucl. Instr. Meth. Phys. Res. A.* **471**, 165-169 (2001).

[6] Dz. Belkić, Strikingly stable convergence of the fast Padé transform: Applications to Magnetic Resonance Spectroscopy. *J. Com. Meth. Sci. Eng.* **3**, 299-382 (2003).

[7] Dz. Belkić, Strikingly stable convergence of the Fast Padé transform (FPT) for high-resolution parametric and non-parametric signal processing of Lorentzian and non-Lorentzian spectra, *Nucl. Instr. Meth. Phys. Res. A.* **525**, 366-371 (2004).

[8] Dz. Belkić, Analytical continuation by numerical means in spectral analysis using the fast Padé transform (FPT), *Nucl. Instr. Meth. Phys. Res. A.* **525**, 372-378 (2004).

[9] Dz. Belkić, Error analysis through residual frequency spectra in the fast Padé transform (FPT), *Nucl. Instr. Meth. Phys. Res. A.* **525**, 379-386 (2004).

[10] K. Belkić, The need for quantitative biomedical spectroscopic imaging through magnetic resonance in oncology beyond the conventional Fourier-based signal processing, *J. Com. Meth. Sci. Eng.* **3**, 535-561 (2003).

[11] K. Belkić, Magnetic resonance spectroscopic imaging in breast cancer detection: possibilities beyond the conventional theoretical framework for data analysis, *Nucl. Instr. Meth. Phys. Res. A.* In press, **525**, 313-321 (2004).

[12] I. Tkáč, P. Andersen, G. Adriany, et al., In vivo 1H NMR spectroscopy of the human brain at 7 T, *Magn. Reson. Med.* **46**, 451-456 (2001).

VSP International
Science Publishers
P.O. Box 346, 3700 AH Zeist
The Netherlands

*Lecture Series on Computer
and Computational Sciences*
Volume 1, 2004, pp. 70-74

A Study of the Mechanical Behavior of an Injected Sand by a Numerical Homogenization Approach

F. Bouchelaghem[1], A. Ben Hamida, H. Dumontet

Laboratoire de Modelisation, Materiaux et Structures,
CNRS UMR 7143,
Universite Paris VI,
4 place Jussieu, 75252 Paris Cedex 05, France

Received 31 July, 2004; accepted in revised form 9 August, 2004

Abstract: This paper deals with the mechanical behavior of a sand grouted with a cement-based grout. The injected sand is viewed as a composite material comprising two distinct phases : the initial granular soil and the solidified grout matrix which cements the grains together. As a result, a multi-scale modelling approach, the Homogenization Method of Periodic Structures, has been employed in order to characterize the macroscopic behavior of the treated medium. The damage behavior of the grout material is investigated by relying on experimental tests found in the literature. The numerical resolution of the cellular problems is performed by the Finite Element Method. The computation of the homogenized coefficients and the resulting macroscopic relationship obtained between the equivalent strains and stresses allow to display the strong nonlinearity of the mechanical behavior induced by the heterogeneous microstructure and the grout damage.

Keywords: injected granular material; effective behavior; homogenization of periodic structures; finite element method

1 Introduction

Permeation grouting consists in injecting under pressure a fluid grout within the voids of the initial soil, and the progressive solidification of the grout will confer improved mechanical or hydraulic characteristics to the treated medium.

Despite the relevance of soil grouting among the usual ground improvement techniques, the mechanical behavior of injected structures is often under-estimated, owing to the uncertainties in the treatment and the lack of models describing the mechanical behavior of an injected soil. Moreover, the existing models are macroscopic ones, relying on classical triaxial tests to identify failure criteria for the injected soils in the large strains domain, and thus discarding the multiphase nature of the material and the stress-strain relation before and after failure [3, 6]. In order to fully characterize the behavior of an injected sand on a large range of strains and stresses, ranging from the small strains domain to the failure state, and to gain a real insight in the micromechanical parameters governing the effective behavior of an injected sand, a micromechanical model is proposed, whereby the injected soil is viewed as a composite material. Among the multi-scale methods, the homogenization approach of periodic structures has been retained, because it provides a mathematically rigorous definition of the cellular problem, and gives the possibility to attain, through

---

[1]Corresponding author. E-mail: fatiha.bouchelaghem@cnrs-orleans.fr

a localization procedure, the microscopic strains and stresses required for any micromechanical study. The numerical homogenization approach used allows to account for the behavior of the individual constituents, gives a realistic representation of the solidified grout distribution within the void space, and offers a useful tool to assess the material deterioration caused by cement grout damage during loading.

2   Modelling approach : homogenization of periodic media

The homogenization methods consist in replacing a strongly heterogeneous material by an equivalent homogeneous material. Those methods rely on the choice of a Representative Elementary Volume (REV) Ω, and the resolution over this domain of a cellular problem. By using a strain approach, an average strain $\mathbf{E}$ is imposed to the volume Ω. The homogeneous equivalent behavior is then defined by the relationship between the average stress within Ω and the imposed strain. For elastic constituents, the cellular problem to be solved is written in the following form [2] :

$$
\begin{aligned}
\mathrm{div}\sigma\,(y) &= 0 \text{ in } \Omega = \Omega_{\mathrm{g}} \cup \Omega_{\mathrm{m}} && (1)\\
\sigma\,(y) &= \mathbf{a}^{\alpha}\,(y)\,\mathbf{e}\,(y) \text{ in } \Omega_{\alpha},\ \alpha = \mathrm{g} \text{ or } \mathrm{m}\\
\mathbf{e}\,(y) &= \frac{1}{2}\left(\mathrm{grad}\mathbf{u}\,(y) + \mathrm{grad}\mathbf{u}\,(y)^{T}\right) \text{ in } \Omega\\
\langle \mathbf{e}\,(y)\rangle_{\Omega} &= \mathbf{E}
\end{aligned}
$$

where $\mathbf{u}\,(y)$, $\mathbf{e}\,(y)$ and $\sigma\,(y)$ represent the microscopic displacement field, the microscopic strain field and the microscopic stress field within the representative volume Ω, $\mathbf{a}^{\alpha}\,(y)$ is the stiffness tensor of the $\alpha-$constituent, and $\langle f\rangle_{\Omega}$ designates the average of the quantity f taken over the domain Ω.

The stiffness tensor of the equivalent medium is then defined by the following homogenized law of behavior :

$$
\mathbf{\Sigma} = \langle \sigma\,(y)\rangle_{\Omega} = \mathbf{a}^{*}\mathbf{E} = \mathbf{a}^{*}\,\langle \mathbf{e}\,(y)\rangle_{\Omega}
$$

which relates the volume average of the microscopic stress $\sigma\,(y)$ with the macroscopic strain $\mathbf{E}$, and allows thus to connect the mechanical behavior of the effective homogeneous medium with the mechanical behavior on the REV.

Observation of in-situ and laboratory samples of injected sand using Scanning Electron Microscopy [8] has motivated the representation of a grouted sand by the microstructure illustrated in Figure 1 for the case of a completely grouted sand. Microstructural observations attest that the grout is essentially located at the contact zones between the grains, and may fill partially or completely the initial intergranular voids. As a result, in all the microstructures retained the intergranular contacts are ensured by the grout matrix, and the grout distribution may vary depending on the final residual porosity value attained after grouting is complete. The numerical results obtained in the linear elastic case with a similar microstructure compare well with laboratory measurements made in the small strains domain [1], giving thus confidence in the microstructural representation chosen for the injected soil and its applicability when the grout behavior is nonlinear.

The effective behavior is computed using the Finite Element Method and the software Zebulon [4]. The following homogeneous boundary condition is imposed on the edge $\partial\Omega$ of the REV Ω :

$$
\mathbf{u} = \mathbf{E}y \text{ on } \partial\Omega \qquad (2)
$$

The mechanical problem 1 defined over the REV is solved using the previous boundary condition 2, which gives the microscopic fields σ and ε. The volume average $\mathbf{\Sigma}$ of the stress tensor σ can

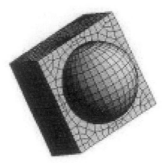

Figure 1: Microstructure used for a completely injected soil.

then be computed, and the exploitation of the linearity of the problem to be solved allows us to compute the matrix of homogenized coefficients by imposing a deformation tensor of norm unity [7]. In the nonlinear case, a similar procedure is followed except that the macroscopic strain $\mathbf{E}$ is increased in an incremental manner, and the linearization of the behavior over an increment of strain $\Delta\mathbf{E}$ gives the relationship between the macroscopic stress $\mathbf{\Sigma}$ and the macroscopic strain $\mathbf{E}$.

3   Application to an injected sand

In a first approach, the effective behavior of an injected sand has been obtained by assuming the behavior of the constituents to be linear elastic [1]. However, experimental tests made on pure cement grout show that under increasing uniaxial or triaxial loading in compression, microcracking initiates and develops within the grout matrix [3]. As a consequence, isotropic damage elasticity is employed to define the cement grout behavior. The damage state is characterized by a scalar damage variable D, the damage law of evolution is given by a thermodynamic potential [5], and the yield for damage initiation is governed by a given value of the elastic energy density identified on experimental tests [7].

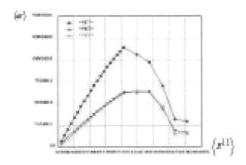

Figure 2: Evolutions of the macroscopic stresses Σ_{11} and $\Sigma_{22} = \Sigma_{33}$ with the imposed macroscopic strain $\mathbf{E}$.

The effective behavior obtained between the macroscopic stresses and the macroscopic strain $\mathbf{E}$ is illustrated in Figure 2 showing the evolutions of the macroscopic uniaxial stress Σ_{11} and the lateral stresses $\Sigma_{22} = \Sigma_{33}$ with an increasing uniaxial stress $\mathbf{E} = \mathbf{E}_{11}$. The computed curves, with their maximum stress state followed by a steep decrease are qualitatively similar to experimental curves obtained with injected sands grouted with cement-based grouts [3], which confirms that grout damage plays an essential role in the overall behavior of the treated medium.

Figure 3 shows the localization of the stresses, the strains and the damage variable D under uniaxial and shear loading tests. The maximum values of compression and shear stress are localized in the cement matrix.

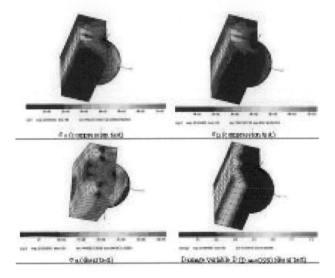

Figure 3: Localization of the stresses and damage variable.

4 Conclusion

With the multi-scale model proposed, the effective behavior of an injected soil has been obtained depending on micromechanical characteristics such as the evolution of the mechanical properties of the individual constituents and the grout distribution within the void space. The stress-strain relationships and the evolutions of the effective stiffness moduli computed at the macroscopic level are similar to experimental results reported in the literature [3]. The study of the localization of the stress and damage variables, which is essential to understand the initiation of damage, is straightforward with the approach used, and constitutes the first step towards the identification of macroscopic failure criteria on the basis of a realistic description of the degradation mechanims within the microstructure.

References

[1] A. Benhamida, F. Bouchelaghem, H. Dumontet: Effective Properties of a Cemented or an Injected Granular Material, submitted in *Int. J. for Num. and Analytical Methods in Geome-*

chanics, 2004.

[2] M. Bornert, T. Bertheau, P. Gilormini: *Homogénéisation en Mécanique des Matériaux*, Hermès, 2001.

[3] C. Dano: *Comportement Mécanique des Sols Injectés*. Doctoral Dissertation, Ecole Centrale de Nantes, 2001.

[4] F. Feyel: *Application du Calcul Parallèle aux Modèles à Grand Nombre de Variables Internes*. Doctoral Dissertation, Ecole Nationale Supérieure des Mines de Paris, France, 1998.

[5] J. Lemaitre: *A Course on Damage Mechanics*, 2nd edition, Springer-Verlag, Berlin Heidelberg, 1996.

[6] E. Nicolini, R. Nova: Modelling of a Tunnel Excavation in a Non-cohesive Soil Improved with Cement Mix Injections, *Computers & Geotechnics* **27** 249-272(2000).

[7] K. D. Pham: *Une approche Multi-échelles du Comportement d'un Sable Injecté*. Rapport de D.E.A., University Pierre et Marie Curie, 2004.

[8] E. Ribay-Delfosse: *Etude du Comportement au Fluage et sous Chargement Cyclique du Sable Injecté*. Doctoral Dissertation, University of Cergy-Pontoise, France, 2001.

VSP International
Science Publishers
P.O. Box 346, 3700 AH Zeist
The Netherlands

*Lecture Series on Computer
and Computational Sciences*
Volume 1, 2004, pp. 75-78

Densification and Dynamic Canonical Descent:
An Optimization Algorithm

K. Bousson [1] and S. D. Correia

Avionics and Control Laboratory,
Department of Aerospace Sciences,
University of Beira Interior,
6201-001 Covilhã, Portugal

Instituto Politécnico de Portalegre,
Escola Superior de Tecnologia e Gestão,
Lugar da Abadessa, Apartado 148,
7301 Portalegre, Portugal

Received 6 August, 2004; accepted in revised form 22 August, 2004

Abstract: Stochastic methods have gained some popularity in global optimization in that most of them do not assume the cost functions to be differentiable, have capabilities to avoid being trapped by local optima, and may converge even faster than gradient-based optimization methods on some problems. The present paper proposes an optimization method which reduces the search space by means of densification curves, coupled with the Dynamic Canonical Descent Algorithm. The performances of the new method are shown on several known problems classically used for testing optimization algorithms, and proved to outperform famous algorithms such as Simulated Annealing and Genetic Algorithms.

Keywords: Algorithms, Optimization and Optimal Control

Mathematics Subject Classification: Multivariable systems, Nonlinear systems, Derivative-free methods.

PACS: 93C35, 93C10, 90C56.

1. Introduction

Global optimization provides a framework for dealing with a wide spectrum of real world problems in science and engineering. Practically any engineering problem may be expressed in terms of a global optimization problem even though it may seem at first sight to have nothing to do with optimization. Because of that, an expressive number of methods have been suggested to find a tradeoff between time and space complexities and the capabilities to locate the global optimum up to a certain precision. Most of these methods can be cast into three main classes: gradient-based, enumerative, and stochastic algorithms.

The gradient-based schemes [1], although thoroughly used, have the limitations to often yield local optima instead of actual global optima, and to be based on differentiability assumptions underlying the cost function. Therefore, they are limited in use with respect to all the possible optimization problems that may involve non-differentiable cost functions and for which global optimization may be required. Enumerative algorithms are devoted to scan the search space with the hope to find out the global optima. Although such algorithms are obviously guaranteed to find global optima, they are inefficient for real-world problems for which the search space may be too large for being scanned in a reasonable amount of time.

Stochastic algorithms have been proposed for overcoming the limitations of pure enumerative searches. Two families of such algorithms that have gained some popularity are genetic algorithms (GA) and simulated annealing (SA) which do not assume differentiability of the cost function of interest.

Considering GA [2], there has not been a clear way of choosing the size of initial population. The way to maintain diversity, to enhance exploration and exploitation, to escape from local minima and to fine tune local search still remains unsolved.

---

[1] Corresponding author. E-mail: bousson@ubistb.ubi.pt

The disadvantages of SA [3] are such that it cannot focus on the regions which are susceptible of containing the global optimizer, and that it may require several function evaluations even for not so complicated optimization problems.

The present paper proposes a fast and efficient derivative-free global optimization algorithm. Based on the Dynamic Canonical Descent (DCD), initially proposed by Bousson [4], where the number of variables of the original problem is reduced by the use of densification curves.

2. Optimization Algorithm

The basis of the DCD algorithm is rather to find at each iteration, along a choosing individual coordinate direction, a point at which the value of the cost function is less than its current value, applying a cyclic choice of the coordinate directions. By construction, the method is endowed with a high capability to jump away from most of the local minima, thus enabling to find at least local minima relatively close to global minima, or even global minima.

Let us assume that the current best value of function f is achieved at a point $x = (x_1,...,x_n) \in X$ and that one shall find out a better point along the i-th coordinate direction. In means one has to find a $\xi \in X_i$ such that:

$$f(x_1,...,x_{i-1},\xi,x_{i+1},...,x_2) < f(x_1,...,x_{i-1},x_i,x_{i+1},...,x_2) \tag{1}$$

Methods borrowed from domain densification [5,6] may be used to improve the convergence rate of Dynamic Canonical Descent, since the densification procedure enables to approximate any multidimensional function by a monodimensional function.

Let $B = \prod_{i=1}^{r}[a_i,b_i] \subset \Re^n$ be a box (or hyper-rectangle) and $J = [a, b]$ be a real interval. Then a function $h : J \to B$ is said to be an α-dense curve in B, if:

$$\forall x \in B, \quad \exists t \in J : \quad \|h(t) - x\| \le \alpha. \tag{2}$$

where $\|.\|$ denotes the euclidean norm in $\Re^n$.

There are many methods for constructing such curves, which are dense up to a given rate α in a determined box. We propose here a new densification curve that allows to densify the space with fewer points. Let us consider the curve $(h_1(\theta), h_2(\theta),..., h_n(\theta)) : J = [0,1] \to B = \prod_{i=1}^{n}[a_i, b_i]$ defined by:

$$h_1(\theta) = \frac{(a_1 - b_1)}{2}\cos(\alpha_1\theta) + \frac{(a_1 + b_1)}{2}$$

$$h_2(\theta) = \frac{(a_2 - b_2)}{2}\cos(\alpha_2\theta) + \frac{(a_2 + b_2)}{2}$$

$$...$$

$$h_n(\theta) = \frac{(a_n - b_n)}{2}\cos(\alpha_n\theta) + \frac{(a_n + b_n)}{2}$$

with

$$\alpha_n = \mu^{n-1}2^n\pi$$

We can prove that the curve is $\dfrac{\sqrt{n-1}}{4\mu} M\pi$ -dense in $\prod_{i=1}^{n}[a_i, b_i]$, where μ is a free parameter to configure the densification degree and $M = \max\limits_{i=1,...,n}\{(b_i - a_i)\}$. We can also prove that the minimum increment $\Delta\theta$ on $J = [0, 1]$ that guarantees the α-densification of box B can be given by the expression (3).

$$\Delta\theta = \left(\frac{\alpha}{M\pi}\right)\frac{2^{n-1}}{\sqrt{n}\left(\sqrt{n-1}\right)^{n-1}} \tag{3}$$

The problem found in the present theory is that for a high number of variables the increment $\Delta\theta$ becomes too small, or in other words, the number of points necessary to explore box B along the densifying curve becomes too high, slowing down the optimization method or even preventing the procedure to converge to any minimizer. To overcome this limitation we propose to reduce a multivariable function to a new multivariable function but with much less variables than the original function. In this way we want to find a new function g in such a way that:

$$f(x_1,...,x_n) \approx f(h_{11}(\theta_1), h_{12}(\theta_1),..., h_{1p}(\theta_1), h_{21}(\theta_2), h_{22}(\theta_2),... \\ ..., h_{2p}(\theta_2), h_{m1}(\theta_m), h_{m2}(\theta_m),..., h_{mp}(\theta_m)) = g(\theta_1,...,\theta_m) \tag{4}$$

with $2 \le m << \dfrac{n}{2}$,

where n corresponds to the original function dimension, m to the new function dimension and p the dimensional coefficient reduction defined by expression (5).

$$p = \frac{n}{m}, \qquad p \in N^+, \qquad p >> 1 \tag{5}$$

The densification of each sub-space can be done with the curve presented earlier in this paper. If we want to densify the domain (or the box) with a densification degree α considering m subspaces with densification degrees $\alpha_1,..., \alpha_m$, then these must satisfy the constraint $\sum_{i=1}^{m}\alpha_i^2 \le \alpha^2$.

3. Results and Discussion

The proposed algorithm was applied to several test problems (Eq. 6 and 7) and compared with other famous algorithms like Simulated Annealing and Genetic Algorithms.

Example 1: Axis Parallel Hyper-Ellipsoid.

$$f_1(X) = \sum_{i=1}^{n} i.x_i^2 \tag{6}$$

with $-5.12 < x_i < 5.12$, $i = 1,...,n$ and $\min(f_1(X)) = f_1(X^*) = 0$ for $x_i^* = 0$.

Example 2: Griewank's Function.

$$f_3(X) = \sum_{i=1}^{n}\frac{x_i^2}{4000} - \prod_{i=1}^{n}\cos(\frac{x_i}{\sqrt{i}}) + 1 \tag{7}$$

with $-600 < x_i < 600$, $i = 1,...,n$ and $\min(f_3(X)) = f_3(X^*) = 0$ for $x_i^* = 0$.

It can be shown from figures 1 and 2 that the proposed algorithm archives better results in the sense that that the solutions of the proposed problems where always found with less function evaluations than competitive algorithms.

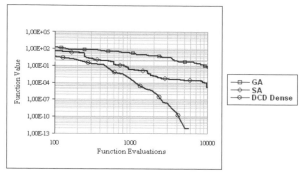

Fig. 1 – Axis Parallel Hyper-Ellipsoid Convergence Result

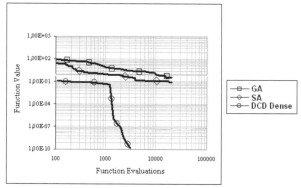

Fig. 2 – Griewank's Function Convergence Result

References

[1] D. P. Bertsekas, *Nonlinear Programming*, second edition, Athena Scientific (1999).

[2] D. E. Goldberg, *Genetic Algorithms in Search, Optimization, and Machine Learning,* Addison-Wesley Publishing Company, Incorporated, Reading, Massachusetts (1989).

[3] S. Kirkpatrick et all, *Optimisation by Simulated Annealing*, Science, vol. 220, n.º 4598 (1983).

[4] K. Bousson, *Efficient Global optimization Based on Dynamic Canonical Descent.* Systems Science, Vol. 26, No. 4, pp. 61-78 (2000).

[5] Y. Cherruault, *α-Dense Curves and Global Optimisation.* Kybernetes, vol. 32, nº 3, pp. 369-375 (2003).

[6] G. Mora and Y. Cherruault, *Characterization and Generation of α-Dense Curves*, Computers Math. Applications, vol. 33, n.º 9, pp. 83-91 (1997).

VSP International
Science Publishers
P.O. Box 346, 3700 AH Zeist
The Netherlands

*Lecture Series on Computer
and Computational Sciences*
Volume 1, 2004, pp. 79-82

New Kernel Definition for Robust to Noise Morphological Associative Memories

Yiannis.S. Boutalis[1]

Department of Electrical and Computer Engineering,
Democritus University of Thrace,
GR-67100 Xanthi, Hellas (Greece)

Received 18 August, 2004; accepted in revised form 30 August, 2004

Abstract: Morphological Associative Memories (MAM's) have been used in two different ways: a direct approach which is suitable for input patterns containing either dilative or erosive noise and an indirect one for arbitrarily corrupted input patterns which is based on kernel vectors. Although kernel vectors represent an elegant representation for the retrieval of arbitrarily corrupted patterns they severely suffer from noise, especially when this noise affects one or more points of the input pattern corresponding to respective kernel points. In this paper, a theoretical proof of this observation is established and a new kernel definition is proposed which is more robust to noise than the traditional kernel vectors. The new kernels are not binary but they contain elements with values in the interval [0, 1], where each selected kernel value is related to the frequency and the position of the respective elements of the training patterns. The performance of the new method is also demonstrated by simulation examples.

Keywords: Neural networks, Associative memory, Kernel vectors, Noise Robustness

1. Introduction

The concept of morphological neural networks (MNNs) grew out of the theory of image algebra developed by Ritter [1]. A sub-algebra of an image algebra, called "lattice", can be viewed as the mathematical background not only for morphological image processing but also for morphological neural networks [2]. Unlike Hopfield network [3,4], MNNs provide the result in one pass through the network, without any significant amount of training. A number of researchers devised MNNs for very specialized applications like those appearing, for example, in [5-8].

Artificial neural network models are specified by the network topology, unit characteristics, and training or learning rules. The underlying algebraic system used in these models is the set of real numbers R together with the operations of addition and multiplication and the laws governing these operations. This algebraic system, known as ring, is commonly denoted by (R,+,x). The basic computations occurring in morphological networks are based on the algebraic lattice structure $(R,+,\wedge,\vee)$, where $\wedge,\vee$ denote the binary operations for minimum and maximum, respectively. The algebra of matrices over R±∞, which has found widespread applications in various engineering sciences provides for an elegant way to express the total input effect on a morphological neural network layer [9]. The basic axiomatic operations of lattice algebra are:

$$
\begin{aligned}
(\alpha \vee -\infty) &= (-\infty \vee \alpha) = \alpha \quad \forall \alpha \in R_{-\infty} \\
\alpha \wedge \infty &= \infty \wedge \alpha = \alpha \quad \forall \alpha \in R_{\infty} \\
\alpha + 0 &= \alpha \quad \forall \alpha \in R \\
0 + \alpha &= \alpha \quad \forall \alpha \in R
\end{aligned}
\tag{1}
$$

---

[1] Corresponding author.. E-mail: ybout@ee.duth.gr

An *associative memory* (AM) is an input-output system that describes a relation $R \subseteq R^m \times R^n$. If $(x, y) \in R$, i.e. if the input x produces the output y, then the associative memory is said to *store* or *record* the *memory association* (x, y). NN models serving as associative memories are generally capable of retrieving a complete output pattern y even if the input pattern x is corrupted or incomplete. The purpose of *auto-associative* memories is the retrieval of x from corrupted or incomplete versions $\tilde{x}$. If an artificial associative memory stores associations (x, y), where x cannot be viewed as a corrupted or incomplete version of y, then we speak about *hetero-associative* memory.

The earliest neural network approach to associative memories is the *linear associative memory* or *correlation memory* [10]. Association of patterns $x \in R^n$ and $y \in R^m$ is achieved by means of a matrix-vector product $y = W \cdot x$. If we suppose that the goal is to store k vector pairs $(x^1, y^1),...,(x^k, y^k)$, where $x^\xi \in R^n$ and $y^\xi \in R^m$ for all $\xi = 1,...,k$, then W is an $m \times n$ matrix given by the following outer product rule:

$$\sum_{\xi=1}^{k} y^\xi \cdot (x^\xi)' \tag{2}$$

If X denotes the matrix whose column are the input patterns $x^1,...,x^k$ and if Y denotes the matrix whose columns are the output patterns $y^1,...,y^k$, then this equation can be written in the simple form $Y \cdot X$. If the input patterns $x^1,...,x^k$ are orthonormal, then

$$W \cdot x^i = ((y^1 \cdot x^1)' + ... + (y^k \cdot x^k)') \cdot x^i = y^i \tag{3}$$

Thus, we have perfect recall of the output patterns $y^1,...,y^k$. In case the patterns are not orthonormal the capacity of the memory is extremely limited and its ability to retrieve associations is further reduced when input patterns are noisy [11].

There are two basic approaches to record k vector pairs $(x^1, y^1),...,(x^k, y^k)$ using a *morphological associative memory* (MAM) [9]. The first approach consists of structuring an $m \times n$ matrix W_{XY} with elements computed by

$$w_{ij} = \bigwedge_{r=1}^{p} (y_i^r - x_j^r), \quad i = 1,...,m \quad j = 1,...,n \tag{4}$$

The dual approach consists of constructing an $m \times n$ matrix M_{XY} with elements computed by

$$m_{ij} = \bigvee_{r=1}^{p} (y_i^r - x_j^r), \quad i = 1,...,m \quad j = 1,...,n \tag{5}$$

If the matrix W_{XY} receives a vector x^r as input, the product $y^r = W_{XY} \otimes x^r$ is formed. The product is called *max product* and each element of the resulting vector y^r is computed by the formula

$$y_i^r = \bigvee_{j=1}^{n} (w_{ij} + x_j^r), i = 1,...,m \tag{6}$$

Likewise, if the matrix M_{XY} receives x^r as input the so-called *min product* $y^r = M_{XY} \oplus x^r$ is formed, where each element of the resulting vector y^r is computed by the formula.

$$y_i^r = \bigwedge_{j=1}^{n} (m_{ij} + x_j^r), i = 1,...,m \tag{7}$$

Both products result in the retrieval of the vector y^r of pair (x^r, y^r). Matrices W_{XY} and M_{XY}, computed by (4) and (5) respectively, constitute the memory of the MAM. The only required training of the network is simply the computation of either of the two matrices W or M. The difference between the two memories arises when noisy patterns appear at their input. Memory M_{XY} is able to retrieve y^r in case a noisy vector $\tilde{x}^r$ corrupted by dilative noise appears at its input while memory W_{XY} is able to retrieve y^r in case an eroded version of the input pattern appears at its input. In case the input pattern $\tilde{x}^r$ is corrupted by both erosive and dilative noise none of the memories is able to recall y^r.

To overcome this problem Ritter and his coworkers [12], [13] proposed the idea of two-step MAMs and the production of the so-called kernel vectors, as an elegant representation of the associations (x^i, y^i). These vectors, in conjunction with matrices M and W, are suitable to recall y^r when an arbitrarily corrupted (both eroded and dilated) input pattern $\tilde{x}^r$ appears at the input of the MAM. A binary vector z is said to be a kernel of the association (x, y) between the binary vectors x and y if it is a subset of x and satisfies the following conditions.

$$M_{zz} \oplus x = z \tag{8}$$

$$W_{ZY} \otimes z = y \tag{9}$$

where matrices M_{ZZ} and W_{ZY} are computed according to (4), (5). In case of autoassociation (8) and (9) are written as $M_{zz} \oplus x = z$ and $W_{ZX} \otimes z = x$ respectively. When an arbitrarily corrupted input pattern $\tilde{x}$ appears at the input of a morphological autoassociative network, the original uncorrupted input pattern x is recalled by the following equation:

$$W_{ZX} \otimes \left(M_{ZZ} \oplus \tilde{x} \right) = x \tag{10}$$

2. Scope and organization of the paper

There are two major issues related to the kernel method. The first is the very selection of representative kernel vectors. Equations (8), (9) provide the necessary and sufficient conditions for a vector z to be a kernel vector but they do not provide with a fast method for selecting such vectors. An apparent procedure is the random selection of kernel vectors and their acceptance or rejection on the basis of (8) and (9). This procedure is quite time consuming, especially when the pattern vectors are of large dimension. Additional conditions for kernels and selection procedures are proposed in [13-15].

The second major issue, which is of this paper concern, is related to the behavior of kernel method in the presence of noisy input patterns. Although this issue has not received the required attention in the relevant literature, the kernel method is not entirely robust to noise. It can be experimentally verified that, although kernel vectors are quite satisfactory in retrieving associations based on noisy (both dilated and eroded) data, in some cases, they entirely fail even in the presence of very small noise percentages. This happens when one or more nonzero element of the input pattern, which correspond to respective nonzero elements of the kernel vector are "hit" by noise, which converts them to zero-valued elements. In this case, the retrieval equation (10) recalls nothing.

In this paper, a theoretical proof of this observation is established and the mechanism of producing this failure is explained. Next a new kernel definition method is proposed which overcomes this problem and is more robust to noise than the traditional kernel vectors. The new kernels are not anymore binary but they contain elements of variable values in the range [0, 1]. The values of the elements of each selected kernel are related to the frequency and the position of the respective elements of the corresponding input training patterns. The performance of the new method is demonstrated by simulations on character recognition examples.

The paper is organized as follows. First a complete introduction to the morphological associative networks is given. Then, the kernel method is reported and its weaknesses are demonstrated. In the sequence, the theoretical justification of the failure of the method in the presence of the above described special form of noise and the theoretical basis for the proposed kernel definition method is established. The paper is supported by an experimental section which demonstrates the performance of the method on a character recognition experiment.

References

[1] G.X.Ritter, J.N.Wilson and J.L.Davidson, Image Algebra: an overview, *Computer Vision Graphics Image Processing* **49** (3), 297–331 (1990).

[2] J.L.Davidson, G.X.Ritter, A theory of morphological neural networks, in: *Digital Optical Computing II*, **1215**, *Proc. SPIE*, 378–388(July 1990).

[3] J.J.Hopfield, Neural networks and physical systems with emergent collective computational abilities, **79**, *Proc. of the National Academy of Sciences, U.S.A.*, 2554–2558 (April 1982).

[4] J.J.Hopfield, Neurons with graded response have collective computational properties like those of two-state neurons, **81**, *Proc. of the National Academy of Sciences, U.S.A.*, 3088–3092 (May 1984).

[5] J.L.Davidson, Simulated annealing and morphological neural networks, in: *Image Algebra and Morphological Image Processing III*, **1769**, *Proc. SPIE, San Diego*, 119–127(1992).

[6] J.L.Davidson, F.Hummer, Morphology neural networks: an introduction with applications, *IEEE Systems Signal Processing,* **12** (2), 177–210(1993).

[7] C.P.Suarez–Araujo, Novel neural network models for computing homotethic invariances: an image algebra notation, *J. Math. Imaging Vision* **7** (1), 69–83(1997).

[8] Y.Won, P.D.Gader, Morphological shared weight neural network for pattern classification and automatic target detection, *Proc. of the IEEE International Conference on Neural Networks*, Perth, Australia, 2134–2138(November 1995).

[9] G.X.Ritter, P.Sussner, J.L.Diaz de Leon, Morphological associative memories, *IEEE Trans. Neural Networks* **9** (2), 281–293(1998).

[10] T.Kohonen, Correlation matrix memory, *IEEE Trans. Comput.* **C–21,** 353–359(1972).

[11] Toshiyuki Tanaka, Shinsuke Kakiya, and Yoshiyuki Kabashima Capacity Analysis of Bidirectional Associative Memory *Proc. Seventh Int. Conf. Neural Information Processing*, Taejon, Korea, **2**, 779-784 (Nov. 2000).

[12] G.X.Ritter, J.L.Diaz de Leon and P.Sussner, Morphological bidirectional associative memories, *Neural Networks* **12**, 851-867 (1999).

[13] Peter Sussner, Observations on morphological associative memories and the kernel method, *Neurocomputing* **31**, 167-183 (2000).

[14] T. Ida, S. Ueda, M. Kashima, T. Fuchida and S. Murashima, On a method to decide kernel patterns of morphological associative memory, *Systems and Computers in Japan* **33**(7), 85-94 (2002).

[15] M. Hattori, A. Fukui and H. Ito, A fast method of constructing kernel patterns for morphological associative memory, *Proceedings of 9th International Conference on Neural Information Processing*, 1058-1063 (Singapore 2002).

VSP International
Science Publishers
P.O. Box 346, 3700 AH Zeist
The Netherlands

*Lecture Series on Computer
and Computational Sciences*
Volume 1, 2004, pp. 83-86

A numerical solution of the Boussinesq equation using the Adomian method

A. G. Bratsos [1], D. P. Papadopoulos [2] and Ch. Skokos [3]

[1,2]Department of Mathematics,
Technological Educational Institution (T.E.I.) of Athens,
GR-122 10 Egaleo, Athens, Greece.
[3]Research Center for Astronomy and Applied Mathematics
Soranou Efesiou 4, GR-11527, Athens, Greece

Received 31 July, 2004; accepted in revised form 18 August, 2004

Abstract: The Adomian decomposition method is applied to a known from the bibliography finite-difference scheme concerning the numerical solution of the Boussinesq equation. The method is examined numerically for both the single and the double-soliton waves.

Keywords: Soliton, Boussinesq, Adomian decomposition

Mathematics Subject Classification: 35Q53, 35L70

PACS: 02.60.Lj

1   Introduction

In the past few years interest has increased in the solution of partial differential equations governing non-linear waves in dispersive media. As a result several texts and numerous research papers have been devoted to the subject. In parallel with the mathematical treatment a considerable literature has grown dealing with the numerical solution of such problems. Among them a great interest has been developed for equations, which possess special solutions in the form of pulses, which retain their shapes and velocities after interaction amongst themselves. Such solutions are called *solitons*. Solitons are of great importance in many physical areas, as for example, in dislocation theory of crystals, plasma and fluid dynamics, magnetohydrodynamics, laser and fiber optics etc., as well as in the study of the water waves. The development of analytical solutions of soliton type equations has been with us for many years (see for example Ablowitz and Segur [1] who implemented the inverse scattering transform method, Hirota [9] who constructed the N soliton solutions using the bilinear form, as well as Whitman [17] etc.). A part of these equations is going to be examined at the present paper.

The archetypal equation introduced by Korteweg & de Vries (KdV) [11], which describes long gravity waves moving over stationary water is written as

$$\frac{\partial u}{\partial t} + 6u\frac{\partial u}{\partial x} + \frac{\partial^3 u}{\partial x^3} = 0, \ x \in \Re \tag{1}$$

---

[1]Corresponding author. E-mail: bratsos@teiath.gr

Once the general method of solution of the KdV equation was obtained (see Gardner [8]), many other equations and mathematical approaches followed. In the particular field of water waves, two families of evolution equations occur: one is the KdV family of equations and the other is based on the nonlinear Schrödinger (NLS) equation

$$i\frac{\partial u}{\partial t} + \frac{\partial^2 u}{\partial x^2} + u|u|^2 = 0.$$

The Boussinesq (BS) nonlinear equation, which belongs to the KdV family of equations, describes shallow water waves propagating in both directions, is given by

$$\frac{\partial^2 u}{\partial t^2} = \frac{\partial^2 u}{\partial x^2} + q\frac{\partial^4 u}{\partial x^4} + \frac{\partial^2 \left(u^2\right)}{\partial x^2} ; \quad L_0 < x < L_1, \ t > 0. \tag{2}$$

where $u = u(x,t)$ is a sufficiently often differentiable function and $|q| = 1$ is a real parameter. Taking $q = -1$ gives the *Good Boussinesq* or *well-posed* equation (GB), while taking $q = 1$ gives the *Bad Boussinesq* or *ill-posed* equation (BB). Besides to Hirota [9] bilinear formalism, Nimmo and Freeman [13] introduced an alternative formulation of the N-soliton solutions in terms of some function of the Wronskian determinant of N functions, Kaptsov [10] implemented Hirota's method to construct a new set of exact solutions of BS equation, while recently, among others, numerical solutions of the BS equation have been given by Bratsos [3] using the method of lines and Wazwaz [15] using the Adomian decomposition method.

The initial displacements associated with BS equation will be assumed to have the form,

$$u(x,0) = g(x) \text{ and } \frac{\partial u(x,0)}{\partial t} = \hat{g}(x) ; \ L_0 \le x \le L_1. \tag{3}$$

1.1 The single-soliton solution

Following Manoranjan et *al* [12] the theoretical solution of BS equation is given by

$$u(x,t) = q_1 \left\{ A \operatorname{sech}^2 \left[\sqrt{\frac{A}{6}} (x - ct + x_0) \right] + \left(b - \frac{q_1}{2} \right) \right\}. \tag{4}$$

where A is the amplitude of the pulse, b is an arbitrary parameter, x_0 is the initial position of the pulse and $c = \pm \left[2q_1 (b + A/3) \right]^{1/2}$, where $q_1 = 1$ for the BB and $q_1 = -1$ for the GB equation.

1.2 The double-soliton solution

Similarly following Hirota [9] and Manoranjan et *al* [12], who obtained a double-soliton solution for GB, the double-soliton solution for both GB and BB can be written as,

$$u(x,t) = 6q_1 \frac{\partial^2}{\partial x^2} \left[\log_e f(x,t) \right], \tag{5}$$

where $f(x,t) = 1 + \exp(\eta_1) + \exp(\eta_2) + \alpha \exp(\eta_1 + \eta_2)$ with $\eta_i = P_i \left[x - \epsilon_i \left(1 + q_1 P_i^2 \right)^{1/2} t - q_1 x_i^0 \right]$ for $i = 1, 2$ and $\epsilon_i = +1$ or -1 showing the direction in which the two solitons are traveling,

$$\alpha = \frac{\left(\epsilon_1 v_1 - \epsilon_2 v_2 \right)^2 + 3q_1 \left(P_1 - P_2 \right)^2}{\left(\epsilon_1 v_1 - \epsilon_2 v_2 \right)^2 + 3q_1 \left(P_1 + P_2 \right)^2}.$$

in which $v_i = \left(1 + q_1 P_i^2 \right)^{1/2}$ and $P_i^2 = \frac{2}{3} A_i$; $i = 1, 2$, where A_i is the amplitude and x_i^0 the initial position of the i-th soliton (see also Bratsos [4]).

2 The Adomian method

Consider Eq. (2) written in an operator form as

$$\mathrm{L}\,u = \frac{\partial^2 u}{\partial x^2} + q\frac{\partial^4 u}{\partial x^4} + \frac{\partial^2 \left(u^2\right)}{\partial x^2}\;;\;\; L_0 < x < L_1,\; t > 0, \tag{6}$$

where

$$\mathrm{L} = \frac{\partial^2}{\partial t^2}, \tag{7}$$

is a twice integrable differential operator with

$$\mathrm{L}^{-1}(.) = \int_0^t \int_0^t (.)\,dt\,dt. \tag{8}$$

Following the Adomian decomposition method (see Adomian [2]) the unknown solution function u is assumed to be given by a series of the form

$$u\left(x,t\right) = \sum_{n=0}^{\infty} u_n\left(x,t\right), \tag{9}$$

where the components $u_n\left(x,t\right)$ are going to be determined recurrently, while the nonlinear term $F\left(u\right) = \left(u^2\right)_{xx}$ in Eq. (2) is decomposed into an infinite series of polynomials of the form

$$F\left(u\right) = \sum_{n=0}^{\infty} A_n, \tag{10}$$

with A_n the so-called Adomian polynomials of $u_0, u_1, ..., u_n$ defined by

$$A_n = \frac{1}{n!}\frac{d^n}{d\lambda^n}\left[F\left(\sum_{i=0}^{\infty}\lambda^i u_i\right)\right]_{\lambda=0} \quad \text{for } n = 0,1,2,... \tag{11}$$

and constructed for all classes of nonlinearity according to algorithms given either by Adomian [2] or alternatively by Wazwaz [14], [16].

3 Numerical experiments

The Adomian decomposition method for solving the BS equation was tested numerically to the problems proposed by Bratsos [3], [4] with boundary lines $L_0 = -80$ and $L_1 = 100$, initial conditions defined by Eq. (3) and theoretical solutions given by Eq. (4) for the single and Eq. (5) for the double-soliton waves.

Acknowledgment

    This research was co-funded by 75% from E.E. and 25% from the Greek Government under the framework of the Education and Initial Vocational Training Program - Archimedes, Technological Educational Institution (T.E.I.) Athens project 'Computational Methods for Applied Technological Problems'. Ch. Skokos was partially supported by the Research Committee of the Academy of Athens.

References

[1] M. J. Ablowitz and H. Segur, *Solitons and the Inverse Scattering Transform*, SIAM Studies in Applied Mathematics 4, Society for Industrial and Applied Mathematics, Philadelphia, 1981.

[2] G. Adomian, *Solving frontier problems of physics: the decomposition method*, Boston: Kluwer Academic Publishers, 1994.

[3] A. G. Bratsos, The solution of the Boussinesq equation using the method of lines, Comput. Methods Appl. Mech. Engrg. **157** 33-44(1998).

[4] A. G. Bratsos, A parametric scheme for the numerical solution of the Boussinesq equation, Korean J. Comput. Appl. Math. **8** no 1, 45-57(2001).

[5] P. Daripa and W. Hua, A numerical study of an ill-posed Boussinesq equation arising in water waves and nonlinear lattices: filtering and regularization techniques, Appl. Math. Comput. **101** no. 2-3, 159-207(1999).

[6] Z. S. Feng, Traveling solitary wave solutions to the generalized Boussinesq equation, Wave Motion Vol. **37** no 1, 17-23(2003).

[7] R. S. Johnson, A two-dimensional Boussinesq equation for water waves and some of its solutions, J. Fluid Mech. **323** 65-78(1996).

[8] C. S. Gardner, J. M. Greene, M. D. Kruskal, R. M. Miura, *A method for solving the Korteweg-de Vries equation*, Phys. Rev. Lett. 19 (1967), 1095-1097.

[9] R. Hirota, Exact N-soliton solutions of the wave of long waves in shallow-water and in nonlinear lattices, J. Math. Phys. 14 (1973), 810-814.

[10] O. V. Kaptsov, Construction of exact solutions of the Boussinesq equation, J. Appl. Mech. and Tech. Phys. **39** no 3, 389-392(1998).

[11] D. J. Korteweg & G. de Vries, On the change of form of long-waves advancing in a rectangular canal, and on a new type of long stationary waves, Phil. Mag. **39** no 5, 422-443(1895).

[12] V. S. Manoranjan, A. R. Mitchell and J. Ll. Morris, Numerical solutions of the Good Boussinesq equation, SIAM J. Sci. Stat. Comput. **5** 946-957(1984).

[13] J. J. C. Nimmo and N. C. Freeman, A method of obtaining the N-soliton solutions of the Boussinesq equation in terms of a Wronskian, Phys. Lett. **95**A 4-6(1983).

[14] A. M. Wazwaz, A new algorithm for calculating Adomian polynomials for nonlinear operators, Appl. Math. and Comput. **111** 53-69(2000).

[15] A. M. Wazwaz, Constructions of soliton solutions and periodic solutions of the Boussinesq equation by the modified decomposition method, Chaos solitons & Fractals Vol. **12** no 8, 1549-1556(2001).

[16] A. M. Wazwaz, *Partial Differential Equations: methods and Applications*, Balkema, The Netherlands, 2002.

[17] G. B. Whitham, *Linear and Nonlinear Waves*, New York:Wiley-Interscience, 1974.

VSP International
Science Publishers
P.O. Box 346, 3700 AH Zeist
The Netherlands

*Lecture Series on Computer
and Computational Sciences*
Volume 1, 2004, pp. 87-90

Phase-Fitted Numerov type methods

A. G. Bratsos[1], I. Th. Famelis[2] and Ch. Tsitouras[3]

[1]TEI of Athens, Department of Mathematics, GR12210 Aegaleo, Greece

[2]Nat. Tech. Univ. of Athens, Dept. of Mathematics, GR15780 Zografou Campus, Greece

[3]TEI of Chalkis, Department of Applied Sciences, GR34400, Psahna, Greece

Received 7 September, 2004; accepted in revised form 10 September, 2004

Abstract: We present a hybrid two step method for the solution of second order initial value problem. It costs only five function evaluations per step and attains sixth algebraic order. The coefficients depend on the frequency of the problem and satisfy zero dissipation and phase-fitted properties. So the method is well suited for facing problems with oscillatory solutions. Numerical tests justify our effort.

Keywords: Initial Value Problem, Second Order, Oscillatory solutions.

Mathematics Subject Classification: 65L05, 65L06

1 Introduction.

We are interested in solving the initial value problem of second order

$$y'' = f(x, y), \ y(x_0) = y_0, \ y'(x_0) = y'_0. \tag{1}$$

In this paper we investigate the class of the above problems with oscillatory solutions. Our result are methods which can be applied to many problems in celestial mechanics, quantum mechanical scattering theory, in theoretical physics and chemistry and in electronics.

Hybrid two step methods are used for about twenty years for solving (1), [1, 2]. Their construction is usually based on interpolatory nodes. These nodes carry a lot of information which is useless even for conventional methods. So, an alternative implementation of such methods was introduced in [5, 8], and studied theoretically by Coleman [3] through B-series. For the new method we propose here, we only need y''_n and four extra function evaluations f_a, f_b, f_c and f_d, summing to five function evaluations per step. The new method has the form:

$$
\begin{aligned}
f_n &= f(y_n) \\
f_a &= f((1-c_1)y_n + c_1 y_{n-1} + h^2(d_{11}f_{n-1} + d_{12}f_n)) \\
f_b &= f((1-c_2)y_n + c_2 y_{n-1} + h^2(d_{21}f_{n-1} + d_{22}f_n + a_{21}f_a)) \\
f_c &= f((1-c_3)y_n + c_3 y_{n-1} + h^2(d_{31}f_{n-1} + d_{32}f_n + a_{31}f_a + a_{32}f_b)) \\
f_d &= f((1-c_4)y_n + c_4 y_{n-1} + h^2(d_{41}f_{n-1} + d_{42}f_n + a_{41}f_a + a_{42}f_b + a_{43}f_c)) \\
y_{n+1} &= 2y_n - y_{n-1} + h^2(w_1 f_{n-1} + w_2 f_n + b_1 f_a + b_2 f_b + b_3 f_c + b_4 f_d).
\end{aligned}
\tag{2}
$$

---

[1]E-Mail: bratsos@teiath.gr
[2]E-Mail: ifamelis@math.ntua.gr
[3]Corresponding author, E-Mail: tsitoura@teihal.gr

2 Algebraic order of the new method.

When solving (1) numerically we have to pay attention in the algebraic order of the method used, since this is the main factor of achieving higher accuracy with lower computational cost. Thus this is the main factor of increasing the efficiency of our effort. Using the notation of Nyström methods we consider the following matrices.

$$
A = \begin{bmatrix}
0 & 0 & 0 & 0 & 0 & 0 \\
0 & 0 & 0 & 0 & 0 & 0 \\
d_{11} & d_{12} & 0 & 0 & 0 & 0 \\
d_{21} & d_{22} & a_{21} & 0 & 0 & 0 \\
d_{31} & d_{32} & a_{31} & a_{32} & 0 & 0 \\
d_{41} & d_{42} & a_{41} & a_{42} & a_{43} & 0
\end{bmatrix},
$$

$$
b = \begin{bmatrix} w_1 & w_2 & b_1 & b_2 & b_3 & b_4 \end{bmatrix},
$$

and

$$
c = \begin{bmatrix} 1 & 0 & c_1 & c_2 & c_3 & c_4 \end{bmatrix}^T.
$$

Now the method can be formulated in a table like the Butcher tableau,

$$
\begin{array}{c|c}
c & A \\
\hline
 & b
\end{array}.
$$

Under the simplifying assumption

$$
Ae = \frac{1}{2} \left(c^2 - c \right), \tag{3}
$$

with $e = \begin{bmatrix} 1 & 1 & 1 & 1 & 1 & 1 \end{bmatrix}^T$ and $c^i = \begin{bmatrix} 1 & 0 & c_1^i & c_2^i & c_3^i & c_4^i \end{bmatrix}$, we get the following sixth order conditions [5]:

$$
b \cdot e = 1, \; b \cdot c = 0, \; b \cdot c^2 = \frac{1}{6}, \; b \cdot c^3 = 0, \; b \cdot A \cdot c = 0, \; b \cdot c^4 = \frac{1}{15}, \; b \cdot A \cdot c^2 = \frac{1}{180}
$$

$$
b \cdot c \cdot A \cdot c = -\frac{1}{60}, \; b \cdot c^5 = 0, \; b \cdot c^2 \cdot A \cdot c = 0, \; b \cdot c \cdot A \cdot c^2 = -\frac{1}{72}, \; b \cdot A \cdot c^3 = 0, \; b \cdot A^2 \cdot c = 0.
$$

Our methods have 24 parameters. Seventeen equations are required assuming order conditions and satisfaction of (3). This leaves seven coefficients as free parameters.

3 Periodic problems.

Following Lambert and Watson [4] and in order to study the periodic properties of methods posed for solving (1), it is constructive to consider the scalar test problem

$$
y' = -\omega^2 y, \quad \omega \in \Re. \tag{4}
$$

When applying an explicit two step hybrid method of the form (2) to the problem (4) we obtain a difference equation of the form

$$
y_{n+1} + S \left(v^2 \right) y_n + P \left(v^2 \right) y_{n-1} = 0, \tag{5}
$$

where $y_n \approx y (nh)$ the computed approximations at $n = 1, 2, \ldots$, $v = \omega h$, h the step size used, and $S \left(v^2 \right), P \left(v^2 \right)$ polynomials in v^2.

Zero dissipation property is fulfilled by requiring $P\left(v^2\right) \equiv 1$, and helps a numerical method that solves (4) to stay in its cyclic orbit. We observe that

$$P\left(v^2\right) = 1 + v^7 b \cdot A^3 \cdot c - v^9 b \cdot A^4 \cdot c.$$

This means that we have to solve $b \cdot A^3 \cdot c = 0$ and $b \cdot A^4 \cdot c = 0$, demanding another two coefficients and leaving only five free parameters.

The phase-lag of the method is the angle difference between numerical and theoretical cyclic solution of (4). Since the solution of (4) is $y(x) = e^{i\omega x}$, we may write equation (5) as

$$e^{2iv} + S\left(v^2\right) \cdot e^{iv} + 1 = O.$$

This equation is satisfied if $S\left(v^2\right) = 2 \cdot \cos\left(v\right)$. Since $S\left(v^2\right) = 2 - v^2 b \cdot \left(I + v^2 A\right)^{-1} \cdot (e - c)$ we spend another coefficient for that last requirement. Finally we have the four nodes c_1, c_2, c_3 and c_4 free to choose. Taking them symmetrically we get the following method.

$$c_1 = \frac{2}{3},\ c_2 = \frac{1}{3},\ c_3 = -\frac{1}{3},\ c_4 = -\frac{2}{3},\ w_1 = 0,\ w_2 = \frac{19}{40},\ b_1 = b_4 = \frac{13}{80},\ b_2 = b3 = \frac{1}{10},$$

$$d_{11} = 0,\ d_{12} = \frac{1}{9},\ a_{32} = 0,\ a_{42} = a_{43} = \frac{2}{13},\ d_{42} = \frac{29}{117} + \frac{1920}{13\,v^8} - \frac{960}{13\,v^6} + \frac{80}{13\,v^4} + \frac{8}{39\,v^2} - \frac{1920\cos(v)}{13\,v^8},$$

$$d_{22} = \frac{41}{144} + \frac{120}{v^8} - \frac{60}{v^6} + \frac{5}{v^4} - \frac{1}{6\,v^2} - \frac{120\cos(v)}{v^8},\ a_{21} = -\frac{7}{16} - \frac{360}{v^8} + \frac{180}{v^6} - \frac{15}{v^4} + \frac{1}{2\,v^2} + \frac{360\cos(v)}{v^8},$$

$$d_{31} = -\frac{1}{24} - \frac{720}{v^8} + \frac{360}{v^6} - \frac{30}{v^4} + v^{-2} + \frac{720\cos(v)}{v^8},\ d_{32} = \frac{25}{144} + \frac{360}{v^8} + \frac{180}{v^6} - \frac{15}{v^4} + \frac{1}{2\,v^2} + \frac{360\cos(v)}{v^8},$$

$$a_{31} = \frac{7}{16} + \frac{1080}{v^8} - \frac{540}{v^6} + \frac{45}{v^4} - \frac{3}{2\,v^2} - \frac{1080\cos(v)}{v^8},\ d_{41} = \frac{3840}{13\,v^8} - \frac{1920}{13\,v^6} + \frac{160}{13\,v^4} + \frac{16}{39\,v^2} - \frac{3840\cos(v)}{13\,v^8},$$

$$d_{21} = \frac{1}{24} + \frac{240}{v^8} - \frac{120}{v^6} + \frac{10}{v^4} - \frac{1}{3\,v^2} - \frac{240\cos(v)}{v^8},\ a_{41} = \frac{-5760}{13\,v^8} + \frac{2880}{13\,v^6} - \frac{240}{13\,v^4} + \frac{8}{13\,v^2} + \frac{5760\cos(v)}{13\,v^8}.$$

4 Numerical Tests.

Two problems are chosen for our comparisons that are well known in the relevant literature.

4.1 Bessel equation

First we considered the following problem

$$y'' = \left(-100 + \frac{1}{4x^2}\right) y,\ y\left(1\right) = J_0\left(10x\right),\ y'\left(1\right) = -0.5576953439142885,$$

whose theoretical solution is $y(x) = \sqrt{x} J_0\left(10x\right)$. We solved the above equation in order to find the 100th root of the solution which is equal to 32.59406213134967.

4.2 Inhomogeneous equation

Our second test problem was an inhomogeneous problem:

$$y'' = -100y(x) + 99\sin(x),\ y(0) = 1,\ y'(0) = 11$$

with analytical solution $y(t) = \cos(10x) + \sin(10x) + \sin(x)$. We integrated that problem in the interval $x \in [0, 10\pi]$ as in [7, 9]. Both problems were tested for the same computational cost using

Table 1: Accurate digits for Besell equation

stages	800	1200	1600	2000	2400	2800	3200	3600	4000	4400
[8]	1.4	2.4	3.2	3.8	4.2	4.6	5.0	5.3	5.6	5.8
NEW	4.3	5.7	6.6	7.4	8.0	8.6	9.1	9.6	9.7	9.7

Table 2: Accurate digits for Inhomogeneous equation

stages	800	1200	1600	2000	2400	2800	3200	3600	4000	4400
[8]	0.6	1.7	2.5	3.1	3.6	4.0	4.4	4.7	5.0	5.2
NEW	10.3	13.0	13.0	12.9	12.5	12.6	12.6	12.6	12.7	12.8

$\omega = 10$. We recorded the end point global error achieved by our previous method given in [8] and the new method in Tables 1 and 2.

We observe an extraordinary improvement. Other constant step methods can been found in the literature that are special tuned for oscillatory problems [1, 2, 5, 6, 7, 9], but the gain for these methods is usually only one digit.

Acknowledgment

The present work was financed by the program Archimedes for supporting research for TEI.

References

[1] M. M. Chawla and P. S. Rao, Numerov type method with minimal phase lag for the integration of second order periodic initial value problems II. Explicit method, *J. Comput. Appl. Math.* **15** (1986) 329-337.

[2] M. M. Chawla and P. S. Rao, An explicit sixth-order method with phase-lag of order eight for $y'' = f(t,y)$, *J. Comput. Appl. Math.* **17** (1987) 365-368.

[3] J. P. Coleman, Order conditions for a class of two-step methods for $y'' = f(t,y)$, *IMA J Numer. Anal.* **23** (2004) 197-220.

[4] J. D. Lambert and I. A. Watson, Symmetric multistep methods for periodic initial value problems, *J. Inst. Math. Appl.* 18 (1976) 189-202.

[5] G. Papageorgiou, Ch. Tsitouras and I. Th. Famelis, Explicit Numerov type methods for second order IVPs with oscillating solutions, *Int. J. Mod. Phys. C* **12**(2001) 657-666.

[6] S. N. Papakostas and Ch. Tsitouras, High algebraic order, high phase-lag order Runge-Kutta and Nyström pairs, *SIAM J. Sci. Comput.* **21** (1999) 747-763.

[7] T. E. Simos, I. Th. Famelis and Ch. Tsitouras, Zero dissipative, explicit Numerov type methods for second order IVPs with oscillating solutions, *Numer. Algorithms* **34** (2003) 27-40.

[8] Ch. Tsitouras, Explicit Numerov type methods with reduced number of stages, *Comput. & Maths with Appl.,* **45** (2003) 37-42.

[9] Ch. Tsitouras and T. E. Simos, Explicit high order methods for the numerical integration of periodic initial value problems, *Appl. Math.& Comput.* 95 (1998) 15-26.

VSP International
Science Publishers
P.O. Box 346, 3700 AH Zeist
The Netherlands

*Lecture Series on Computer
and Computational Sciences*
Volume 1, 2004, pp. 91-94

Determination of elastic buckling loads for lateral torsional buckling of beams including contact

P. Buffel[1], G. Lagae, R. Van Impe, W. Vanlaere

Laboratory for Research on Structural Models,
Department of Structural Engineering,
Ghent University,
B-9000 Ghent, Belgium

Received 28 July, 2004; accepted in revised form 28 August, 2004

Abstract: Elastic buckling loads can be calculated with finite element codes by an eigen-value extraction. Such an analysis is not possible if the model contains non linear constraints such as contacts that can open and close. This paper presents a method to obtain in such cases the buckling load for lateral torsional buckling of beams. Only a limited series of simulations including geometrically non linear behaviour must be made of the excentrically loaded beam. The load displacement curves exhibit a geometrical stiffening part and there is a point of contraflexure. It is demonstrated that the loads corresponding to this point converge to the buckling load when the in plane deformations prior to buckling are taken into account.

Keywords: Buckling, Elastic critical load, Lateral torsional buckling, Large displacement analysis

1   Introduction

Beams loaded in the plane of their web can exhibit lateral torsional buckling. The elastic buckling load is usually calculated by an eigenvalue extraction for a perfect model of the beam. In some cases parts of the beam can come into contact or loose contact. It is then cumbersome to use an eigenvalue extraction as it is not known in advance which parts will effectively be in contact during buckling.

But it is always possible to analyse the behaviour of the imperfect beam including contact resulting in load-displacement curves. It will be demonstrated how to derive the buckling load from a series of such curves. This is first illustrated for a simple system with one degree of freedom. It is shown that the same method is valid in the case of lateral torsional buckling of a cantilever beam. The method is then applied to the same model after adding a contact condition.

2   System with one degree of freedom

The system of figure 1 consists of two rigid bars of length ℓ connected with a rotational spring with stiffness K and is loaded with a compression force P. The exact relation between the displacement v and the force P [1]

$$P = P_{cr} \frac{\arcsin\left(\frac{v_0+v}{\ell}\right) - \arcsin\left(\frac{v_0}{\ell}\right)}{\frac{v_0+v}{\ell}} \tag{1}$$

---

[1]Corresponding author. E-mail: Peter.Buffel@UGent.be

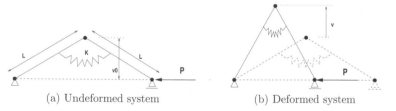

(a) Undeformed system   (b) Deformed system

Figure 1: System with one degree of freedom.

contains the buckling load $P_{cr} = 2K/\ell$ which is the elastic critical load for the system without imperfection (i.e. $v_0 = 0$). The linearised version of this equation can be transformed into $v + v_0 = v_0/(1 - P/P_{cr})$ and is plotted on figure 2-(a) for various values of the dimensionless imperfection v_0/ℓ. The displacements can grow beyond the length ℓ of each bar, a physically impossible

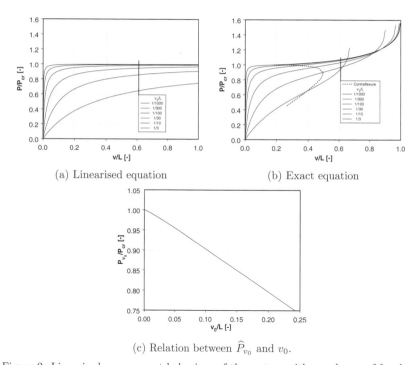

(a) Linearised equation   (b) Exact equation

(c) Relation between $\widehat{P}_{v_0}$ and v_0.

Figure 2: Linearised versus exact behaviour of the system with one degree of freedom.

situation. When the exact relation is plotted (figure 2-(b)), the displacements are bounded to $\ell - v_0$ and the applied load can grow above P_{cr}. Starting from the origin, the inclination of the curves decreases until the point of contraflexure at a load $\widehat{P}_{v_0}$. This load is a nearly linear function of the magnitude of the imperfection (figure 2-(c)). The values of $\widehat{P}_{v_0}$ always remain a lower bound of P_{cr} and converge for $v_0 = 0$ to $\widehat{P}_{lim}$ that is equal to P_{cr}. Apparently the method is only applicable when exact and not approximated geometrically non linear behaviour is accounted for.

3 Lateral torsional buckling of a cantilever beam

A prismatic cantilever beam with an I-section is loaded at its tip with a vertical concentrated force positioned at the upper flange with an excentricity e to the web (figure 3-(a)). The behaviour of the beam is simulated with the finite element package ABAQUS using a model of shell elements (figure 3-(b)). The material is linear elastic and geometrical non linear behaviour is accounted for in the analyses. Further details on the simulations can be found in [2].

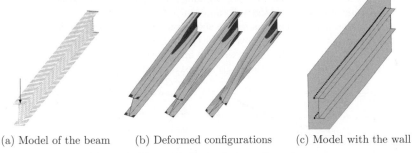

(a) Model of the beam   (b) Deformed configurations   (c) Model with the wall

Figure 3: Excentrically loaded cantilever beam.

The load-displacement curves of figure 4-(a) also show the presence of a type of geometrical stiffening for small as well as for large excentricities. The loads corresponding to the point of

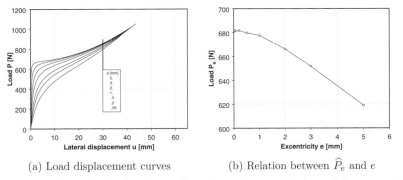

(a) Load displacement curves   (b) Relation between $\widehat{P}_e$ and e

Figure 4: Simulation results for the cantilever beam for several values of the excentricity.

contraflexure $\widehat{P}_e$ can be plotted in function of the excentricity e (figure 4). The curve is very smooth and the loads converge to $\widehat{P}_{lim} = 681.8$N. A buckling load of $P_{cr} = 677.8$N was calculated with ABAQUS for the same finite element model with $e = 0$. The slight difference between $\widehat{P}_{lim}$ and P_{cr} can be explained as follows. The value of P_{cr} was determined by neglecting the in plane deformations that correspond with it, whereas $\widehat{P}_{lim}$ automatically takes the occuring in plane deformations into account.

In ABAQUS the eigenvalue analysis can take into account the effects of a preload. This preload can be increased until the eigenvalue is 0. The preload then equals the buckling load $\widetilde{P}_{cr}$ that takes into account the deformations prior to buckling. This methodology results for the model of the beam in $\widetilde{P}_{cr} = 682.1$N. This value corresponds extremely well to the prediction $\widehat{P}_{lim}$ based on the points of contraflexure. This example clearly indicates that in the case of lateral torsional buckling the loads $\widehat{P}_e$ converge to the buckling load $\widetilde{P}_{cr}$ obtained by taking into account the in plane deformations prior to buckling.

4   Model including a contact condition

The model of the previous section can be extended with a vertical flat rigid wall (figure 3-(c)). This wall is initially in contact with the border of the flanges over the full length of the beam. During the simulations, the beam is allowed to slide without friction and with possible loss of contact so that only displacements away from the wall can occur.

The resulting load-displacement curves are shown on figure 5-(a) and the loads corresponding to the point of contraflexure $\widehat{P}_e$ now converge to $\widehat{P}_{lim} = 809.3$N (figure 5-(b)). This load is larger than for the case without the wall because the wall inhibits the fully free deformation of the beam. An advantage of the method is that the $\widehat{P}_{lim}$-e curves are rather smooth. In order to get a usefull

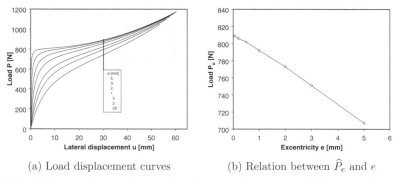

(a) Load displacement curves   (b) Relation between $\widehat{P}_e$ and e

Figure 5: Simulation results for the cantilever beam with the rigid wall for several values of the excentricity.

estimation of $\widetilde{P}_{cr}$ it is not always necessary to do simulations with extremely small excentricities, as such simulations do not always run stable or do not give accurate results. It is clear that the simulations may be terminated shortly after the point of contraflexure is reached.

5   Conclusion

The load-displacement curve of an excentrically loaded elastic beam exhibits geometrical stiffening and has a point of contraflexure at a load $\widehat{P}_e$. With decreasing magnitude of the excentricity, the loads $\widehat{P}_e$ converge to the elastic buckling load $\widetilde{P}_{cr}$ taking into account the in plane deformation of the beam prior to buckling. It was shown that the method is also applicable when the model contains contact conditions.

References

[1] D. Vandepitte, *Berekening van constructies*, **I** 87-88,
http://www.berekeningvanconstructies.be/

[2] P. Buffel, *Numerical and experimental investigation of lateral torsional buckling of web tapered cantilever beams loaded at their tip.* Doctoral Dissertation, Ghent University, Belgium, 2004 (in Dutch).

VSP International
Science Publishers
P.O. Box 346, 3700 AH Zeist
The Netherlands

Lecture Series on Computer
and Computational Sciences
Volume 1, 2004, pp. 95-97

Probability to Break Certain Digital Signature Schemes

P. Caballero-Gil [1]

Department of Statistics, Operations Research and Computing,
Faculty of Maths,
University of La Laguna,
38271 Tenerife, Spain

Received 27 July, 2004; accepted in revised form 10 August, 2004

Abstract: A theoretical analysis of success probabilities of sequential meet-in-the-middle attacks on certain digital signature schemes is here provided. Two variations of such attacks with known used signatures are considered, and several practical hints for the design of both optimal attack strategies and of secure digital signature schemes are pointed out.

Keywords: Hash functions, digital signature, meet-in-the-middle attack

1   Introduction

One of the most used digital signature schemes is based on a cryptographic hash function h and a public-key cryptosystem (E_{KRa}, D_{KUa}) because it produces a fixed length digital signature that depends on the whole message and ensures the authenticity of the signer [4]. With such schemes a sender A signs the message $x \in M$ generally in two steps. Firstly, the message x is hashed into a short value $h(x)$, the digest [1], [5]. Secondly, the encryption of the digest $h(x)$ with a secret key KR_a is obtained. After the reception of the message with its added signature, the receiver B verifies through the sender's public key KU_a whether the digest of the message and the deciphering of the signature y are equal. So, the ciphered hash value uniquely identifies the signer. Such a method of producing signatures is particularly convenient when the messages are long but may be forged with a special attack known as meet-in-the-middle that takes less time and memory than an exhaustive attack. Such an attack is closely related to the famous "birthday paradox" that comes from the computation of the probability of finding at least two people with the same birthday.

The so-called meet-in-the-middle attack is applicable to any scheme that uses some sort of block chaining in its structure. A general meet-in-the-middle attack was formulated as a parallel collision search in [6]. The version of the meet-in-the-middle attack for digital signature schemes consists of finding a pair (x, y) such that $h(x) = D_{KUa}(y)$ [2]. Depending on the way of testing messages, parallel and sequential attacks can be distinguished and each one of these attacks may be based on known or unknown signatures. In [3] the success probabilities of parallel meet-in-the-middle attacks on both known and unknown digital signatures were analyzed. In this work several theoretical results about success probabilities of sequential attacks against known signatures are given, and some hints for practical design of digital signature schemes are deduced.

---

[1]Corresponding author. E-mail: pcaballe@ull.es
Research partially supported by Spanish SEG2004-04352-C04-03 Project.

Table 1: Conditions of increase of P

r	s	c	t	m
Constant	Increases $(rs \leq c)$	Constant	Increases $(rt \leq 1)$	
Constant	Decreases$(rs \geq c)$	Constant	Decreases $(rt \geq 1)$	
Constant	Constant	Decreases $(rs \leq c)$	Constant	Decreases
Constant	Constant	Increases $(rs \geq c)$	Constant	Decreases
Decreases	Constant	Constant	Constant	Constant

2   Success Probabilities of Sequential Attacks

A sequential attacker on known signatures proceeds according to the following six steps:

1.chooses a set of known signatures S

2.computes $D_{KUa}(y) \forall y \in S$

3.sorts them out

4.chooses at random a message $x \in M$

5.calculates $h(x)$

6.checks whether it satisfies the relation $h(x) = D_{KUa}(y)$ for some $y \in S$.

The time required by this attack depends on the cardinality of S, $s = |S|$ and the number of tested messages until success, denoted by X. Random choice of messages may be made with or without replacement, which produces two versions of the attack. The memory size needed in the first case is s blocks, where the block length depends on the used cryptosystem. In the second version of the attack, the amount of memory also depends on X. In the following, success probability of the attack is evaluated. If c denotes the cardinality of the range of h, then $s \leq c$ because the cipher should be injective. In case the attack proceeds with replacement, X follows a geometric distribution $Geo(\frac{s}{c})$ and the success probability is

$P = Prob\{X = r\} = \frac{s}{c}(1 - \frac{s}{c})^{r-1}, r = 1, 2, ...$

Its derivatives are

$\frac{\partial P}{\partial s} = (1 - \frac{s}{c})^{r-2}\frac{1}{c}(1 - \frac{rs}{c})$ and

$\frac{\partial P}{\partial c} = (1 - \frac{s}{c})^{r-2}\frac{s}{c^2}(\frac{rs}{c} - 1)$.

So, provided that either $rs \leq c$ or $rs \geq c$ holds, different properties of P shown in Table 1 are fulfilled.

Denoting by $t = s/c$, the success probability may be rewritten as

$P = Prob\{X = r\} = t(1 - t)^{r-1}$.

Since $t < 1$ and $\frac{\partial P}{\partial t} = (1 - t)^{r-2}(1 - rt)$, provided that either $rt \leq 1$ or $rt \geq 1$ holds, a different relationship between P and t is described in Table 1.

Let $m = |M|$, then, according to hash functions design recommendations, c should divide m. So, if the attack proceeds without replacement, then success probability in every trial varies

$P = Prob\{X = r\} = \frac{mt(m-mt)_{r-1}}{(m-r+1)(m)_{r-1}}, r = 1, 2, ..., (1 - t)m + 1$.

In such a case, the relationship among P and parameters m, t and r is shown in Table 1. ¿From both expressions of success probabilities two practical observations may be made. Firstly, successful attacks are more likely for small values of r, which implies for the attacker it is better to initiate a new attack than to continue with a large unsuccessful one. Secondly, system designers should choose the parameters t and m respectively small and large enough in order to avoid the success of these attacks.

3   Conclusions

In this work success probabilities of sequential meet-in-the-middle attacks against digital signature schemes based on a hash function and a public key cryptosystem have been evaluated. Attacks where a set of known used signatures is used have been studied and classified into "with replacement" and "without replacement". In both cases, different expressions for success probabilities have been obtained. The conclusions of their analysis may be useful when choosing parameters of the schemes. Two interesting problems that deserve further research are the study of similar attacks against unknown signatures, and the asymptotical analysis of the computed probabilities.

References

[1] S. Bakhtiari, R. Safavi-Naini and J. Pieprzyk, "Cryptographic Hash Functions: A Survey", Technical Report 95-09, Department of Computer Science, University of Wollongong, 1995.

[2] K. Nishimura and M. Sibuya, "Probability to Meet in the Middle", Journal of Cryptology 2, pp. 13-22, 1990.

[3] K. Ohta and K. Koyama, "Meet-in-the-middle Attack on Digital Signature Schemes", Advances in Cryptology-ASIACRYPT'90, LNCS 453, pp. 140-154, Springer-Verlag, 1990.

[4] B. Preneel, "The State of Cryptographic Hash Functions", LNCS 1561, pp. 158-182, 1999.

[5] D.R. Stinson, "Some Observations on the Theory of Cryptographic Hash Functions", 2002.

[6] P.C. Van Oorschot and M.J. Wiener, "Parallel Collision Search with Cryptanalytic Applications", Journal of Cryptology 12 (1), pp. 1-28, 1999.

[7] D. Wagner, "A Generalized Birthday Problem", 2002.

VSP International
Science Publishers
P.O. Box 346, 3700 AH Zeist
The Netherlands

Lecture Series on Computer
and Computational Sciences
Volume 1, 2004, pp. 98-101

A Rational Approach to Cryptographic Protocols

P. Caballero-Gil[1], C. Hernández-Goya and C. Bruno-Castañeda

Department of Statistics, Operations Research and Computing,
Faculty of Maths, University of La Laguna,
38271 Tenerife, Spain

Received 27 July, 2004; accepted in revised form 10 August, 2004

Abstract: This work analyzes several two-party cryptographic protocols from a rational point of view using a game-theoretical approach, which is used to represent not only the protocols but also possible misbehaviours of parties. With these tools, protocols are studied in a malicious model in order to find equilibrium conditions that allow to protect honest parties against all possible strategies of adversaries.

1 Introduction

Cryptology may be seen as a continuous struggle between cryptographers and cryptanalysts. On the other hand, Game Theory is devoted to the study of decision making in conflictive situations. According to these definitions, Game Theory and Cryptology seem to have certain common scenarios, so it is natural that tools from one area may be applied in the other. In fact, the main objective of this work is to model several two-party cryptographic protocols as two-person games. In both basic notions of two-person games and two-party cryptographic protocols the following assumptions are usual:

1)Each party (player or protocol participant) has available two or more well-specified choices for making moves or actions.

2)Every possible combination of choices available to the parties leads to a well-defined outcome (win, loss or draw, or protocol's success or fail) that terminates the algorithm (game or protocol).

3)The payoff of each party depends both on his/her choice and the choice of the other party, and is associated with each possible outcome.

4)Each party has perfect knowledge of the rules of the algorithm as well as the payoffs of the other party.

5)Both parties' strategies are rational, that is to say, each party uses a possibly randomized method for his/her choice in order to maximize his/her (expected) payoff.

We should single out [1] as the main starting point of this work since there the concept of rationality applied to exchange was introduced. This paper is part of a work in progress, and is organized as follows. First we introduce briefly the game theoretic notions used throughout the paper. Then we provide background on two-party cryptographic protocols and the particular classification we use for our analysis. In the rest of the paper a theoretic game model will be used to describe and analyze respectively symmetric and asymmetric protocols. Finally, conclusions of the work and comments on further investigation close the paper.

---

[1]Corresponding author. E-mail: pcaballe@ull.es
Research partially supported by Spanish SEG2004-04352-C04-03 Project.

2   Game Theoretic Concepts

If a group P of parties or players i agree to obey certain rules and to act individually or in coalition, the results of their joint action lead to certain situations called outcomes. A game G defines the set of rules that specify a sequence of actions $a \epsilon Q$ allowed to specific parties or to a random experiment. The rules of the game specify what amount of information about all the previous actions and the alternatives chosen can be given to each party before making an specific choice. The game also specify a termination when some specific sequences of choices are made and no more actions are allowed. Each termination produces an outcome in the form of scores or incomes y^+, and payments or expenses y^- for each party. It is assumed that each party i has a preference relation $\leq_i$ over the outcomes reflected in his/her scores and payments.

If q is a finite action sequence, $q.a$ denotes the finite action sequence that consists of q followed by a, whereas $p(q)$ denotes a function that assigns q to a party i so that i is the following party in choosing an action after q. An action sequence q is said to be terminal if it is infinite or if there is no a such that $q.a \epsilon Q$. The set Z of terminal action sequences represents the possible outcomes of the game. For every non-terminal action sequence q, $A_i(q)$ denotes the set $\{a : q.a \epsilon Q\}$ of available actions after q for party i. The real-valued function $y(q) = (y_i(q))_{i \epsilon P}$ that assigns the payoffs for every party i after every terminal action sequence $q \epsilon Z$ is called outcome or payoff function. These payoff values may be negative, in which case they are interpreted as losses. Also these payoffs may verify that $\sum_{i \epsilon P} y_i(q) = 0$ for any $q \in Z$, in which case the game is called zero-sum. The preference relations of the parties are often represented in terms of their payoffs in such a way that for any $q, q' \epsilon Z$ and $i \epsilon P$, $q \leq_i q'$ iff $y_i(q) \leq y_i(q')$. On the other hand, the so called utility function u_i is just a mathematical representation of i's preferences.

A strategy of party $i \epsilon P$ is a function $s_i \epsilon S_i$ that assigns an action in $A_i(q)$ to every non-terminal action sequence $q \epsilon Q \backslash Z$ such that $p(q) = i$. A strategy profile is a vector $(s_i)_{i \epsilon P}$ of strategies, where each s_i is a member of S_i. When we write $(s_j, (s_i)_{i \epsilon P \backslash j})$ we emphasize that the strategy profile specifies strategy s_j for party j. Finally, let $o((s_i)_{i \epsilon P})$ denote the resulting outcome when the parties follow the strategies in the strategy profile $(s_i)_{i \epsilon P}$. On the one hand, a strategy profile $(s_i^*)_{i \epsilon P}$ is called dominant iff for any other strategy s_i we have that $o(s_i) \leq o(s_i^*)$. On the other hand, a strategy profile $(s^*_i)_{i \epsilon P}$ is called a Nash equilibrium iff for every party $j \epsilon P$ we have that $o(s_j, (s_i^*)_{i \epsilon P \backslash j}) \leq_j o(s_j^*, (s_i^*)_{i \epsilon P \backslash j})$. This means that if every party i other than j follows strategy s_i^*, then party j is also motivated to follow strategy s_j^*. So, the main difference between dominant strategies and Nash equilibrium is that in the former the choices are made independently whereas in Nash equilibrium depends on the other's possible strategies.

3   Cryptographic Protocols Concepts

A two-party cryptographic protocol may be defined as the specification of an agreed set of rules on the computations and communications that need to be performed by two entities, A (Alice) and B (Bob), over a communication network, in order to accomplish some mutually desirable goal, which is usually something more than simple secrecy. Several essential properties of cryptographic protocols are the following:

1. Correctness, which guarantees that every honest party should get his/her agreed output.

2. Privacy, which includes the protection of every party' secrets.

3. Fairness, which means that in case a dishonest party exists, then neither he/she may gain anything valuable, nor honest party may lose anything valuable.

In the game-theoretic model two new properties regarding dishonest behaviours can be defined

1. Exclusivity, which implies that one or both parties cannot receive their agreed output.

2. Voyeurism, which is the contrary of privacy because it implies that one or both parties may discover the other's secret.

A two-party cryptographic protocol may be seen as a repeated game formed by a sequence of iterations of the following two phases:

1)Send: Party A (B) sends to B (A) a message M generated depending on her (his) state.

2)Receive: Party A (B) receives from B (A) a message M and makes a state transition.

For the sake of simplicity, in this paper we consider also as a delivery the non-intentional loss of control over message M, so we denote by $rcv_A(M)$ $(rcv_B(M))$ both the cases when party $B(A)$ sends message M to $A(B)$, and when $A(B)$ is able to receive it.

In order to formalize the notion of cryptographic protocols in terms of functions, let's denote by f a two-argument finite function, $f : X_A \times X_B \rightarrow Y_A \times Y_B$ where X_i and Y_i, $i\epsilon\{A, B\}$, denote respectively input and output sets for party i. Intuitively, a two-party cryptographic protocol may be generally described through a function f whose output is defined by the expression $f(M_A, M_B) = (f_A(M_A, M_B), f_B(M_A, M_B))$, where it is understood that party i receives the output of f_i on inputs M_A and M_B.

When designing a two-party cryptographic protocol one of two possible models should be considered:

- Semi-honest model: When it is assumed that both parties follow the protocol properly but adversaries may keep a record of all the information received during the execution and use it to make a later attack.

- Malicious model: Where it is assumed that parties may deviate from the protocol. In this case, during the interaction, each party acts non cooperatively and has different choices which may determine the output of the protocol.

The payoff $y_i(q)$ of a party i, assigned after a terminal action sequence q may defined as $y_i(q) = y^+{}_i(q) - y^-{}_i(q)$, where $y^+{}_i(q)$ and $y^-{}_i(q)$ represent respectively the incomes and expenses of i after q. These incomes and expenses functions will be defined in terms of utilities according to the concrete definitions of each protocol. Here we denote the utility that a secret M_j is worth to party i by $u_{ij} = u_i(M_j)$, value which may be difficult to quantify in practical situations.

A two-party cryptographic protocol is said to be closed when if a party gains something, then the other party must lose something. This property may be expressed in terms of the incomes and expenses functions in the following way: $\forall q\epsilon Z, y^+{}_i(q) > 0 \Rightarrow y^-{}_j(q) > 0$. In this work we assume the closeness of the protocols since in the definition of the payoff function we always consider both the wish of one party to know the other's secret and the wish of the other party to prevent that from happening.

According to the aforementioned functional definition of a two-party cryptographic protocol f, at the end of the execution, party i should receive the output of f_i on secrets M_A and M_B. Depending on whether $f_A = f_B$ we may distinguish between symmetric and asymmetric protocols. From the first group we will study the protocols of Fair Exchange, Secure Two-Party Computation and Coin Flipping. On the other hand, representative protocols of the group of asymmetric protocols are Oblivious Transfer, Bit Commitment and Zero Knowledge Proof. This classification is important for the proposed game theoretic model because it implies the translation to a symmetric game where possible payoffs and outputs of both parties coincide, or to asymmetric games where that does not occur.

For every analyzed protocol we give formal definitions of income, expense and payoff functions for each party in every possible combination of behaviours and misbehaviours of parties, and obtain necessary hypothesis to guarantee the existence of a honest strategy profile being a Nash equilibrium. Also we rank properties of exclusivity, voyeurism, correctness and privacy.

Although we consider the possibility of misbehaviours by both parties, in this paper we analyze specially the case when exactly one of them is dishonest. Concretely, we deal with the idea of modeling cryptographic protocols design as the search of an equilibrium in order to defend honest parties against all possible strategies of malicious parties. So, our main objective is to illustrate the close connection between protocols and games and to use game theoretic techniques for the definition and analysis of cryptographic protocols.

Since this work is in progress, many questions are still open. On the one hand, one direction for further investigation involves the use of a similar game-theoretic approach to describe and analyze multiparty cryptographic protocols. On the other hand, the relationship between properties defined here and other properties described by different authors, and a possible extension of this study to stronger and weaker definitions of essential properties may also deserve further study.

References

[1] Buttyan, L., Ph.D.Thesis, Building Blocks for Secure Services: Authenticated Key Transport and Rational Exchange Protocols, Laboratory of Computer Communications and Applications, Swiss Federal Institute of Technology – Lausanne (EPFL), 2002.

VSP International
Science Publishers
P.O. Box 346, 3700 AH Zeist
The Netherlands

*Lecture Series on Computer
and Computational Sciences*
Volume 1, 2004, pp. 102-105

An Improved Dynamic Programming Algorithm for Optimal Manpower Planning[1]

X. Cai[†,♯2], Y.J. Li[‡2], F.S. Tu[♯]

† Department of System Engineering & Engineering Management
The Chinese University of HongKong, Shatin N. T., Hong Kong

‡ School of Economics and Management
Tsinghua University, Beijing 100084 P. R. China

♯ Department of Automation
Nankai University, Tianjin 300071, P. R. China

Received 4 August, 2004; accepted in revised form 22 August, 2004

Abstract: We consider a manpower planning problem (MPP) over a long planning horizon. Dynamic demands for manpower must be satisfied by allocating enough number of employees, under the objective to minimize the overall cost including salary, recruitment cost, and dismissal cost. We first formulate the problem as a multi-period decision model. We then reveal several properties of the optimal solution and develop an improved dynamic programming algorithm with polynomial computational complexity.

Keywords: Manpower planning, dynamic program, computational complexity

Mathematics Subject Classification: 93B40, 37N40

1   Introduction

With the rapid development of economy, manpower planning has become an important problem in today's business world, especially in labor-intensive corporations, where the workforce plays a prominent role in determining the effectiveness and cost of the organization. As such, studies of optimal manpower planning have received extensive attention in the last two decades; see, e.g., [1], [2], and [3]. Considering the dynamic fluctuations of manpower demands, it is natural for an organization to determine the optimal size of its workforce by making proper and dynamic decisions on recruitment and dismissal over different periods of time. Such models, however, have not received much attention in the literature, due to properly the inherent complexity in deriving the optimal dynamic solutions. It is often that such a problem becomes a dynamic optimization model, which requires very sophisticated computational algorithms to search for the optimal or near-optimal solutions.

In this article we will first model the manpower planning problem with dynamic demands as a multi-period decision process with constraints. We will then propose an improved dynamic

---

[1]This research was supported in part by NSFC Research Funds No. 60074018 and No. 70329001. The second author would also like to acknowledge the support of the China Postdoctoral Science Foundation No. 2003034020 and the Tsinghua-Zhongda Postdoctoral Science Foundation (2003).
[2]Corresponding authors. E-mail: xqcai@se.cuhk.edu.hk (X. Cai), yongjianli@em.tsinghua.edu.cn (Y.J. Li)

programming algorithm to derive the optimal solution. Our approach will be devised based on analysis on the properties of the optimal decisions. Our approach is not only computationally efficient, but also capable to reveal some useful insights on the desirable solutions with respect to the data, and therefore allowing management to have better understanding of the impacts and benefits of the solutions.

2 The Model

We consider the following manpower planning problem for an organization. Suppose that based on forecast of its business, the number of employees required in each time period has been specified in advance. The organization can decide on the number of the employees to be recruited or dismissed at the end of every period, subject to the constraint that the workforce available in the coming period would meet the demand for manpower. The objective is to find a series of optimal decisions on recruitment/dismissal in all time periods over the planning horizon, so that the total manpower-related cost is minimized.

Notation:

T The number of time periods being considered.

α The salary per employee in each time period.

β^+/β^- The recruitment/dismissal cost when an employee is recruited/dismissed.

$X[t]$ The number of employees available in period t, which is called the state in the period t.

$u[t]$ The number of employees being recruited $(u[t] > 0)$/dismissed$(u[t] < 0)$ at the end of period t.

D_t The manpower demand in period t, where $t = 0, \cdots, T$. (Assume $D_0 = 0$).

The problem under consideration can be formulated as

$$\textbf{MP1:}\quad J = \min_{u[t]} \left\{ \alpha X[T] + \sum_{t=0}^{T-1} \left[\alpha X[t] + \beta^+ \max\{u[t], 0\} + \beta^- \max\{-u[t], 0\} \right] \right\} \tag{1}$$

s. t.

$$X[t+1] \;=\; X[t] + u[t], \quad t = 0, 1, \cdots, T-1 \tag{2}$$
$$X[t] \;\geq\; D_t, \quad t = 1, 2, \cdots, T \tag{3}$$
$$X[0] \;=\; X_0 \tag{4}$$

The constraint (2) shows the dynamics of the workforce available in the organization. The constraint (3) indicates that the demand must be met by the available employees in each period. The constraint (4) specifies the value of initial state, where X_0 is a given constant.

3 A Standard Dynamic Programming Approach

One can see that MP1 is a dynamic optimization problem. For each pair (t, x), let $f_t(x)$ be the minimum total cost from the periods t to T, subject to the constraint that there are x employees in period t. For a feasible decision $u[t] = u$, the number of employees in period $t+1$ is $x_{t+1} = x + u$. Let $D_{max} = \max\{D_t, t = 1, 2, \ldots, T\}$, we can have the following dynamic program:

$$\textbf{DP1:}\quad f_T(x) \;=\; \alpha x, \quad D_T \leq x \leq D_{max} \tag{5}$$

$$f_t(x) \;=\; \alpha x + \min_{x_{t+1}} \Big\{ \beta^+ \max\{x_{t+1} - x, 0\} + \beta^- \max\{x - x_{t+1}, 0\}$$

$$+ f_{t+1}(x_{t+1}) | D_{t+1} \leq x_{t+1} \leq D_{max} \Big\}, \quad D_t \leq x \leq D_{max}, 1 \leq t < T \tag{6}$$

The optimal solution is obtained when $f_0(X_0)$ is computed. We can show that the time complexity of the algorithm DP1 is bounded above by $O(TD_{max}^2)$. This time complexity is too high, in particular when we have a large D_{max} or a large T. In the following we present our improved algorithm, which requires substantially less computational time.

4 Improved Dynamic Programming Approach for MP1

In the algorithm DP1, for any given state x ($D_t \leq x \leq D_{max}$) we compute the value $f_t(x)$ over all possible $D_{max} - D_{t+1}$. This is actually not necessary, if we can utilize some properties of the problem MP1. Some analysis is given below.

Let us look at the case ($D_t \leq D_{t+1}$) as illustrated in Fig. 1. We can first define $\overline{d}_t = \max\{D_{t+1}, D_{t+2}, \cdots, D_{t+L}\}$, where $L = \lceil \frac{\beta^+ + \beta^-}{\alpha} \rceil$ ($\lceil x \rceil$ is defined as the smallest integer greater than or equal to x), and then divide the possible $X[t] = x$ into two values $\overline{d}_t$ and D_{t+1}. Further, since $D_t \leq x \leq D_{max}$, the entire interval of the value x can be divided into three parts: $[D_t, D_{t+1}]$, $(D_{t+1}, \overline{d}_t)$ and $[D_{t+1}, \overline{d}_t]$. Thus, we can get the states x_{t+1} of equation (6) as follows:

When $D_t \leq x \leq D_{t+1}$, $x_{t+1} = D_{t+1}$;
When $D_{t+1} < x \leq \overline{d}_t$, $x_{t+1} = x$;
When $\overline{d}_t < x \leq D_{max}$, $x_{t+1} = \overline{d}_t$.

This allows us to construct the following algorithm.

Algorithm: IDP1
Step 1. $x_T = x$ and $f_T(x) = \alpha x$ for $D_T \leq x \leq D_{max}$.
Step 2. For $t = T - 1, T - 2, \ldots, 1$, do loop

Let $L_e = \lceil \frac{\beta^-}{\alpha} \rceil$, we compute $\overline{d}_t$ and $\overline{d}_e(t) = \max\{D_T, D_{T-1}, \cdots, D_{\max\{T-L_e,t\}}\}$.
For $x = D_t, D_t + 1, \ldots, D_{max}$ (when $t = 0$, $x = X_0$), do loop
  - If $(D_{t+1} < x \leq \overline{d}_t)$ or $(t \geq T - L_e$ and $x \leq \overline{d}_e(t))$, then $x_{t+1} = x$ and $f_t(x) = \alpha x + f_{t+1}(x)$.
  - If $x > \overline{d}_t$, then $x_{t+1} = \overline{d}_t$ and $f_t(x) = \alpha x + \beta^-(x - \overline{d}_t) + f_{t+1}(\overline{d}_t)$.
  - If $x \leq D_{t+1}$, then $x_{t+1} = D_{t+1}$ and $f_t(x) = \alpha x + \beta^+(D_{t+1} - x) + f_{t+1}(D_{t+1})$.

Then the optimal value is $f_0(X_0)$, and the optimal state trajectory over the entire planning horizon can be obtained by performing a search from periods 0 to T, since we have computed the neighboring state x_{t+1} for each feasible state x in period t.

Theorem 1. *The algorithm IDP1 can optimally solve the problem MP1 in $O(TD_{max})$ time.*

In the algorithm IDP1, we need to compute $f_t(x)$ for every feasible state x in period t. However, we can see further that the optimal state trajectory over the entire planning horizon can be constructed through computing only $f_t(x)$ for $x = D_t$ in each period t.

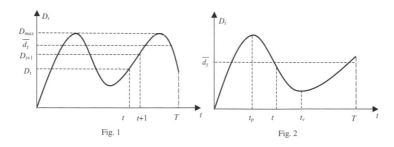

Fig. 1   Fig. 2

Let $x_t = D_t$. Then for the case $D_t \leq D_{t+1}$, we can compute $f_t(D_t)$ since we have already computed $f_{t+1}(D_{t+1})$ in period $t + 1$. However, if $D_t > D_{t+1}$, we need to compute $f_{t+1}(x_{t+1})$

first before computing $f_t(D_t)$ since we don't know $f_{t+1}(x_{t+1})$ in the case $x_{t+1} \neq D_{t+1}$. The value of $f_{t+1}(x_{t+1})$ can be obtained by a simple method as illustrated in Fig. 2, where t_v is a period at the valley of the demand trajectory and t_p is a period before t_v corresponding to a peak of the demand trajectory. Then, for a period t, $t_p \leq t < t_v$, if $D_t < \bar{d}_t$, then $x_{t+1} = D_t$ and $f_t(D_t) = \alpha D_t + f_{t+1}(D_t)$ according to the algorithm IDP1. Moreover, by the algorithm IDP1, we have $f_{t+1}(D_t) = \alpha D_t + f_{t+2}(D_t)$ if $D_t \geq D_{t+2}$ since $D_t \leq \bar{d}_t \leq \bar{d}_{t+1}$ from the definition of $\bar{d}_t$. This is repeated until we reach a period t_l when $D_{t_l-1} \leq D_t < D_{t_l}$. Consequently, we have $f_t(D_t) = \alpha(t_l - t + 1)D_t + \beta^+(D_{t_l} - D_t) + f_{t_l}(D_{t_l})$. Let $X^{(t_0)}[t]$ be the optimal states from t_0 to T under the initial value $X^{(t_0)}[t_0] = D_{t_0}$. Then, the optimal states $X^{(t)}[t'] = D_t$ for $t \leq t' \leq t_l - 1$ and $X^{(t)}[t'] = X^{(t_l)}[t'](t_l \leq t' \leq T)$. Similarly, for the case $D_t \geq \bar{d}_t$, we can prove that $f_t(D_t) = \alpha D_t + \alpha(L-1)\bar{d}_t + \beta^-(D_t - \bar{d}_t) + f_{t+L}(D_{t+L})$ and the optimal states $X^{(t)}[t'] = \bar{d}_t$ for $t + 1 \leq t' \leq t + L - 1$ and $X^{(t)}[t'] = X^{(t+L)}[t'](t + L \leq t' \leq T)$.

The analysis above allows us to further improve IDP1 to obtain the following algorithm.

Algorithm: IDP2

Step 1. Compute L and L_e. Let $f_T(D_T) = \alpha D_T$ and $X^{(T)}[T] = D_T$.

Step 2. For $t = T - 1, T - 2, \dots, 0$, do loop

Compute $\bar{d}_t$ and let $X^{(t)}[t] = D_t$ (When $t = 0$, we let $D_0 = X_0$ in the following computation).

- If $D_t \leq D_{t+1}$, then we get $X^{(t)}[t'] = \begin{cases} D_{t+1} & , & \text{if } t' = t + 1 \\ X^{(t+1)}[t'] & , & \text{if } t + 1 < t' \leq T \end{cases}$.

- If $D_t > D_{t+1}$ and $D_t < \bar{d}_t$, then we search the first period t_l after the period t that satisfies
$D_{t_l-1} \leq D_t < D_{t_l}$, and let $X^{(t)}[t'] = \begin{cases} D_t & , & \text{if } t + 1 \leq t' < t_l \\ X^{(t_l)}[t'] & , & \text{if } t_l \leq t' \leq T \end{cases}$.

- If $D_t > D_{t+1}$ and $D_t \geq \bar{d}_t$, then we get $X^{(t)}[t'] = \begin{cases} \bar{d}_t & , & \text{if } t + 1 \leq t' < t + L \\ X^{(t+L)}[t'] & , & \text{if } t + L \leq t' \leq T \end{cases}$.

- If $t \geq T - L_e$ and $D_t \geq D_{t+1}$, then $X^{(t)}[t'] = D_t$ for $t \leq t' \leq T$.

Step 3. The optimal states over the entire planning horizon are $X^{(0)}[t]$ ($0 \leq t \leq T$). Correspondingly, we can get the optimal value $f_0(X_0)$.

Theorem 2. *The algorithm IDP2 can generate an optimal solution of MP1 in $O(T)$ time.*

5 Concluding Remarks

We have modelled a dynamic manpower planning problem as a multi-period decision process, and developed an improved dynamic programming algorithm to compute its optimal solution. Our proposed algorithm is efficient, requiring only a time complexity $O(T)$, which is independent of the magnitude of the manpower demands. The reduction of computing time of our algorithm has been achieved by analysis of the optimal properties of the problem.

Topics for further investigation include consideration of multiple skills, training requirements and costs, promotion, leave, etc.

References

[1] D.J. Bartholomew, A.F. Forbes and S.I. Mclean, *Statistical techniques for manpower planning* (2nd). Wiley, New York, 1991.

[2] A. M. Bowey, Coorporate manpower planning, *Management Decision*, 15(1977) 421-469

[3] H.K. Alfares, Survey, Categorization, and Comparison of Recent Tour Scheduling Literature, *Annals of Opereraion Research*, 127(2004) 145-175

[4] P.P. Rao, A dynamic programming approach to determine optimal manpower recruitment policies, *Journal of Operations Research Society*, 41(1990) 983-988.

VSP International
Science Publishers
P.O. Box 346, 3700 AH Zeist
The Netherlands

*Lecture Series on Computer
and Computational Sciences*
Volume 1, 2004, pp. 106-109

Parallel Computation of Two Dimensional Electric Field Distribution Using PETSC

O. Ceylan[1] Ö. Kalenderli[2]

[1]Computational Science and Engineering Program
Informatics Institute
Istanbul Technical University
34469, Maslak, Istanbul, Turkey

[2] Department of Electrical Engineering
Istanbul Technical University
34469, Maslak, Istanbul, Turkey

Received 4 August, 2004; accepted in revised form 25 August, 2004

Abstract: Parallel programming is an efficient way for solving large scaled problems. In this study electric field distribution of an example system is approximately computed using finite difference method by PETSC. For different size of grid values and processor numbers results are given and compared. With the help of PETSC features, such as parallel preconditioners and Krylov subspace methods different solution techniques are applied and their performances are compared.

Keywords: Electric field, finite difference method, parallel computation, PETSC

Mathematics Subject Classification: 65y05, 65Y05

PACS: 0270Bf

1 Introduction

Parallel computation is an efficient way when one deals with large scaled problems. In order to use this way electric field problems can be given as an example large scaled problems. In the numerical computation of these fields, for example, they can be modelled in two dimensional cartesian coordinates with equal or non-equal sized grids. There are several methods used for numerical computation of electric fields depending on solution of Laplace and Poisson equations. One of the methods is finite difference method (FDM). By using finite difference method which is an approximate method for solving partial differential equations, good results can be handled if grid sizes are chosen very big.

Finite difference method is widely used in scientific problems. These include linear and non-linear, time independent and dependent problems. This method can be applied to problems with different boundary shapes, different kinds of boundary conditions,and for a region containing a number of different materials [1]. Some finite difference applications can be found in [2-6].

---

[1]Corresponding author: E-mail: oguzhan@be.itu.edu.tr, Phone: +902122857077, Fax: +90 212 2857073
[2]Corresponding author: E-mail: ozcan@elk.itu.edu.tr, Phone: +902122856759, Fax: +90 212 2856700

In this study PETSC (the Portable, Extensible Toolkit for Scientific Computation) is used. PETSC is a suite of data structures and routines for the scalable (parallel) solution of scientific applications modeled by partial differential equations.

It employs the MPI (Message Passing Interface) standard for all message-passing communication. PETSC software is a freely available tool and provides a rich environment for writing parallel codes. The modules in PETSC are appealing as it integrates a hierarchy of components, including data structures, Krylov solvers and preconditioners that can be imbedded into various application codes [7, 8].

The program was run on a SunFire 12K high-end server. This server gives researcher shared memory environment. It has 16 -900 MHz UltraSPARC III Cu- CPU with 32 GB memory and 16-1200 MHz UltraSPARCIII Cu-CPU with 32 GB memory.

2   Electric Field Computation

Steady state distribution of electrical potential V(x,y) on a two dimensional plane can be given by Laplace equation;

$$\frac{\partial^2 V(x,y)}{\partial x^2} + \frac{\partial^2 V(x,y)}{\partial y^2} = 0 \tag{1}$$

Equation (1) can be discretized, on two dimensional cartesian coordinates with equal sized grids, by using first order central differences benefiting from Taylor series as follows;

$$\frac{\partial^2 V(x,y)}{\partial x^2} \simeq \frac{V(x+h,y) - 2V(x,y) + V(x-h,y)}{h^2} \tag{2}$$

$$\frac{\partial^2 V(x,y)}{\partial y^2} \simeq \frac{V(x,y+h) - 2V(x,y) + V(x,y-h)}{h^2} \tag{3}$$

By summing Equation (2) and Equation (3);

$$V(x+h,y) + V(x-h,y) + V(x,y+h) + V(x,y-h) - 4V(x,y) = 0 \tag{4}$$

and leaving V(x,y) alone;

$$V(x,y) = \frac{1}{4}\left[V(x+h,y) + V(x-h,y) + V(x,y+h) + V(x,y-h)\right] \tag{5}$$

By using Equation (5) for all grid points one can write grid number times equations for this system. This yields a linear system and can be solved by numerous numerical methods (direct methods and iterative methods). On the other hand as given in Figure 1 for 2d plane, one grid point has four neighbour points. Hence the system is in a sparse structure and the constructed matrix is a five banded matrix.

3   Numerical Example

An example system of [9] is used for electric field computing. Boundary conditions are given in Figure 2. While writing the code using PETSC, petscles header file must be included, so that SLES (Scalable Linear Equation Solvers) solvers can be used. With MatCreate command laplace matrix A was created and in a loop, it's appropriate spaces were filled with contigious chunks of rows across the processors. Considering boundary conditions vector b was created. System Ax = b was solved by Krylov solvers with the help of PETSC features.

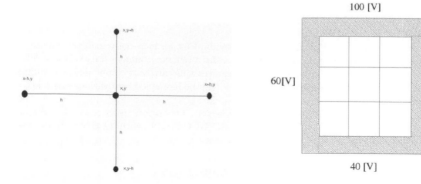

Figure 1: One point and its neighbour points

Figure 2: 3x3 grid and boundary values for the example system

Table 1: Wall clock time (s) for different mesh sizes and number of processors

Mesh Size	Processor Number					
	1	2	4	8	12	16
100×100	3.429	1.852	1.757	1.325	1.510	1.565
200×200	33.715	20.044	10.918	6.08	4.406	3.860
300×300	117.23	89.917	40.05	23.155	15.224	14.104
400×400	409.909	212.011	109.869	65.322	42.352	30.131
500×500	910.686	460.202	229.922	139.284	101.271	58.748
600×600	1616.726	788.18	316.280	255.186	196.386	146.062

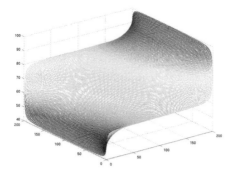

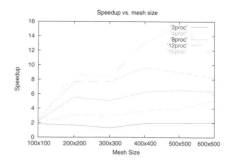

Figure 3: Two dimensional electric field distribution for 200x200 mesh size

Figure 4: Speedup vs. mesh size graphic for different number of processors

An example result for 200*200 grid is shown in Figure 3. Wall clock time (s) for the solution of the electric field distribution problem for different sizes of grids and processors is shown in Table

1. The speedup vs. mesh size graphic is plotted in Figure 4.

4   Conclusion

In this study we are focused on an important engineering problem, the computation of electric field using finite difference methods. We computed electric field in parallel by using a freely available tool PETSC. As shown in Figure 4 the speedup increases with the increase of the processor numbers. From Table 1 it is easily seen that for a specific number of grids the solution time is decreasing when the processor number increases. For small sized experiments this decrease is not as much as big sized experiments. When problem size gets bigger in size, solution by finite difference method is more accurate but also the computation time and computation cost increases.

References

[1] P. B. Zhou, *Numerical Analysis of Electromagnetic Fields*, Springer-Verlag, 1993.

[2] M. V. Schneisder, *Computation of Impedance and Attenuation of TEM-Lines by Finite Difference Methods*, IEEE Transactions on Microwave Theory and Techniques, Vol. 13, pp. 792-800, November, 1965.

[3] P. Basappa, V. Lakdawala, G. Gerdin, *Computation of the Electric Field Around a Wet Polluted Insulator by Analytic and Finite Difference Techniques*, Conference on Electric Insulation and Dielectric Phenomena, 2000.

[4] E. A. Erdelyi, E. F. Fuchs, *Fields in Electric Devices Containing Soft Nonlinear Magnetic Materials Techniques*, IEEE Transactions on Magnetics, Vol. 4, pp. 1103-1108, 1974.

[5] H. I. Saleheen, K. T. Ng, *Parallel Finite Difference Solution of General Inhomogeneous Anistropic Bio-Electrostatic Problems*, IEEE-EMBC and CMBBC, pp. 247-248, 1995.

[6] J. Duncan, *The Accuracy of Finite-Difference Solutions of Laplace's Equation*, IEEE Transactions on Microwave Theory and Techniques, Vol 15, No 10, pp. 575-582, 1967.

[7] S. Balay, K. Buschelman, W. D. Gropp, D. Kaushik, M. Knepley, L. C. McInnes, B. F. Smith, H. Zhang, *PETSC users manual*, Technical Report ANL-95/11 Revision 2.1.3, Argonne National Laboratory, 2002.

[8] S. Balay, W. D. Gropp, L. C. McInnes, B. F. Smith *PETSC web page*, http://www.mcs.anl.gov/petsc.

[9] D. Trybus, Z. Kucerovsky, A. Ieta, T.E Doyle, *Distributed Electric Field Approximation*, Proceedings of the 16th Annual International Symposium on High Performance Computing Systems and Applications, 2002.

VSP International
Science Publishers
P.O. Box 346, 3700 AH Zeist
The Netherlands

*Lecture Series on Computer
and Computational Sciences*
Volume 1, 2004, pp. 110-114

Hydrogen Bonding In Aqueous Mixtures Containing Aprotic Solvents: A Computer Simulation Study

M. Chalaris[1], J. Samios[2]

Laboratory of Physical Chemistry,
Department of Chemistry,
National and Kapodistrian University of Athens,
Panepistimiopolis 15771 Athens, Greece

Abstract: Molecular dynamics simulations have been performed to investigate the thermodynamic and structural properties as well as the self diffusion coefficients and hydrogen bonding in aqueous (binary) mixtures containing the aprotic solvents acetone, acetonitrile, dimethyl sulfoxide (DMSO) and dimethylformamide (DMF). The self diffusion coefficients of both components in the mixtures were calculated and compared. The results show a non-linear dependence with composition. In each case, this property also reveal a peculiar minimum at mole fraction in the range x_{AS}=0.35 -0.40 (AS=aprotic solvent). The diffusion coefficients of water–acetone are found to be anomalous and, the diffusion coefficients of water show a different behavior in this mixture compared with that in the other two mixtures. The intermolecular structure of each mixture was investigated and a number of interesting structural effects have been noted. In the case of acetone and acetonitrile water mixtures, there is a loss of tetrahedral water coordination in the systems In the case of DMSO and DMF water mixtures, the analysis of several site – site pair distribution functions reveals that the average tetrahedral coordination of water is preserved for mole fractions of DMSO or DMF, up to 0.4 . Finally, the hydrogen bonding statistics are obtained and compared with available experimental results.

Keywords: Statistical Mechanics, MD simulation; aprotic solvents; diffusion coefficients; hydrogen bonding

Mathematics SubjectClassification: 82B05, 82D15, 82C70

PACS: Here must be added the AMS-MOS or PACS Numbers

1. Introduction

Binary mixtures in which water is one of the components are interesting molecular liquid systems due to their importance in many fields of solution chemistry. Following the literature, we can notice that considerable effort has been made so far to clarify the non-ideal behaviour of the physicochemical properties of such systems. In particular, over the last two decades the problem of the microstructure (local order) in aqueous solutions of organic molecules has attracted much experimental and theoretical interest. This intensive scientific interest is due to the fact that many questions regarding the properties of such associated hydrogen-bonded solutions have not yet been definitively answered. Notice that hydrogen bonding is known to be one of the most important weak interactions between the molecules leading to the formation of well-defined molecular aggregates. It should be emphasized that enhanced intermolecular structure is presumably responsible for the non-ideal mixing behaviour in many physicochemical properties of such solutions.

Acetonitrile(AN), Acetone(Ac), Dimetylsulfoxide(DMSO) and Dimethylformamide(DMF) belong to a class of aprotic (AS) polar organic solvents, which are miscible with water in all proportions. Furthermore, the AS-water mixtures are particularly interesting systems because of their potential applicability in many fields such as organic chemistry, liquid chromatography and solvent extraction.

The present work describes the use of the Molecular Dynamics (MD) simulation technique to study the properties of these mixtures at ambient conditions over a wide range of mole fractions ($0<X_{AS}<1$). The

---

[1] National Fire Academy, GR-14510, 10 Matsa Str, K. Kifisia, GREECE . E-mail: mhalaris@cc.uoa.gr
[2] Corresponding Author: E-mail: isamios@cc.uoa.gr

main purpose here is to understand the various thermodynamic, dynamic and structural properties of the systems and to explore the intermolecular forces responsible for the formation of Hydrogen bonds among the molecules of each system.

2. Model and Simulation Methods

Molecular dynamics simulations in the microcanonical (NVE) statistical mechanical ensemble were carried out to study the mixtures. Systems of 500 rigid molecules were simulated using periodic boundary conditions at an average temperature of 298 K. The temperature was maintained using the well-known Nose-Hoover thermostat. The effective pair potentials used were the following: the potential of Ferrario et al. for acetone [1], Bohm et al. for acetonitrile[2] , MCS for DMSO[3], CS2 for DMF[4] and SPC for water[5]. The selection of the mole fractions of the mixtures studied in the present simulation was mainly based on the available experimental data, concerning the thermodynamic properties of the system. The intermolecular interaction AS – AS, H_2O - H_2O, AS - H_2O were of the type site – site Lennard – Jones (LJ) with electrostatic terms. Also, the Lorentz-Berthelot mixing rules were used to calculate cross interaction parameters. The direct Ewald method was employed to treat the long-range coulombic interactions. The translational and rotational equations of motions are solved using Leapfrog algorithms. The dimensions of the simulation cells have been selected to allow the derivation of the experimental densities at T=298 K and P =1 atm. The integration time step was $1 \cdot 10^{-15}$ ps. After equilibration, each MD run was extended to additional configurations for about 1 ns.

3. Results and Discussion

3.1 Thermodynamic and transport properties

Table 1 shows the simulation results obtained for each mixture. The simulation results of internal energy and self-diffusion coefficients agree considerably well with the experimental values over a wide density range of the mixtures [6]. The dependence of the self-diffusion coefficients D_x (x= Ac, AN, DMSO, DMF, H_2O) of the molecules with concentration is also presented in Table 1. The calculations have been performed as described elsewhere [7] using the equations

$$D_v = \frac{1}{6}\left[\frac{dJ_v(t)}{dt}\right]_{t\to\infty} \quad v=1,2 \tag{1}$$

$$J_v(t) = \sum_{i=1}^{N_v} \frac{\left[\bar{r}_i^v(t) - \bar{r}_i^v(0)\right]^2}{N_v}, \quad v=1,2 \tag{2}$$

where J(t) is the mean squared displacement of each particle as a function of time t and averaged over all particles of the same component in the mixture for v=1or 2.

Table 1: Simulation results obtained for each water-aprotic solvent (Ac/AN/DMSO/DMF) mixture at molar volumes $V_m[cm^{-1}]$ corresponding to the mole fractions studied here. The equilibrium properties are: potential energy $U_p[KJ/mol]$, pressure P[Kbar], and the self-diffusion coefficients of both species $D[10^{-9} m^2s^{-1}]$.

Acetone						Acetonitrile					
X_{Ac}	V_m	P	$-U_p$	D_w	D_{AC}	X_{AN}	V_m	P	$-U_p$	D_w	D_{AN}
0.00	19.46	0.3	41.75	3.7	-	0.05	19.465	-0.03	38.70	4.47	2.61
0.25	30.67	0.085	37.55	2.69	2.78	0.12	21.545	-0.04	37.85	4.24	3.04
0.50	44.51	-0.006	32.08	3.54	2.90	0.30	28.021	-0.004	35.84	3.81	3.81
0.75	59.05	-0.002	30.09	2.28	3.05	0.50	35.046	0.035	33.55	3.41	3.44
1.00	73.87	0.52	29.3	-	3.10	0.80	45.856	-0.004	45.11	3.71	3.78
DMSO						DMF					
X_{DMSO}	V_m	P	$-U_p$	D_w	D_{DMSO}	X_{DMF}	V_m	P	$-U_p$	D_w	D_{DMF}
0.055	20.871	0.014	44.31	2.59	1.33	0.07	21.93	-0.03	43.49	2.38	1.05
0.19	27.410	-0.22	47.46	0.89	O.52	0.20	29.096	0.46	44.57	1.32	0.72
0.30	33.125	-0.39	49.33	0.55	0.40	0.30	34.79	0.57	44.75	1.03	0.68
0.35	35.734	-0.55	50.20	0.33	0.42	0.39	40.03	0.37	45.592	0.71	0.52
0.67	52.819	-0.60	50.55	0.36	0.48	0.70	58.766	0.12	45.20	0.90	0.80

We have also studied the concentration dependence of the water self-diffusion coefficients in each mixture and we found that a model diffusion function reproduces successfully our MD results. This model diffusion function is given below by equation 3

$$D_W = \sum_{i=0}^{4} C_i \cdot x_{AS}^i \qquad (3)$$

where x_{AS} is the mole fraction of the aprotic solvent and C_i the estimated coefficients. The obtained values of the parameters C_i are given below.

In the case of the mixture AN/H_2O, we have obtained the following equation:

$$D_W = 4.68168 - 4.7467 \, X_{AN} + 11.3328 \, X_{AN}^2 - 22.5318 \, X_{AN}^3 + 17.3548 \, X_{AN}^4$$

For the mixture $DMSO/H_2O$:

$$D_W = 4.26672 - 39.5436 \, X_{DMSO} + 168.904 \, X_{DMSO}^2 - 329.543 \, X_{DMSO}^3 + 228.157 \, X_{DMSO}^4$$

For the mixture DMF/H_2O:

$$D_W = 3.2556 - 13.1442 \, X_{DMF} + 21.7936 \, X_{DMF}^2 - 12.0063 \, X_{DMF}^3 + 1.21349 \, X_{DMF}^4$$

Finally, in the case of Ac/H_2O mixture, the model function used seems to be inadequate to predict successfully the simulated self – diffusion coefficients of water.

3.2. Structural properties

It is well known that a complete description of the intermolecular structure of a liquid system, requires the evaluation of all static site-site pair correlation functions $g_{ij}(r)$ (PCFs). The amplitudes of the first molecular coordination shell peaks for the radial distributions, illustrated in Fig 1, indicate that nearest water molecules are hydrogen bonded to the oxygen in Ac, DMSO and DMF molecules and to the nitrogen in AN molecule, respectively.

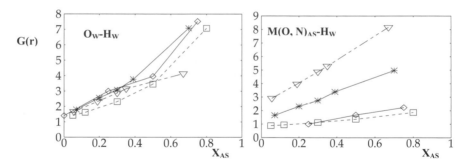

Figure 1: <u>Left</u>: Variation of the first peak height of g(R)'s for $O_w - H_w$ in Ac (◊), AN (□), DMSO (▽) and DMF (*) - water mixtures, respectively, as a function of the molar fraction of aprotic solvent.
<u>Right</u>: Variation of the first peak height of g(R)'s for O_M – Hw in Ac (◊), DMSO (▽) and DMF (*) - water mixtures and for N_{AN} – Hw in AN-water mixtures as a function of the molar fraction of aprotic solvent

In the following we may consider the dependence of the radial distribution functions of water molecules with concentration of each aprotic solvent. From the analysis of the g(R)'s it is evident that the nearest neighbor tetrahedral structure of water in the mixtures with DMSO and DMF is preserved at concentrations up to 0.4 molar fraction of aprotic solvent. The g(R)'s of the AN-water and Ac – water systems show that the tetrahedral rearrangement of the hydrogen bond donors to the water molecule is destroyed. The average DMSO–water and DMF-water H-bond distance ($\approx 1.6 \, \overset{\circ}{A}$) is somewhat smaller than the average water–water H-bond length ($\approx 1.8 \, \overset{\circ}{A}$). It means that the water may form stronger H – bonds with DMSO and DMF than with water.

3.3. Hydrogen bonding

To extend the analysis of the effect of the aprotic solvents on the water structure, we use molecular dynamics to look at the hydrogen bond statistics as a function of the mole fraction. For Ac, DMSO and DMF – water mixtures we have employed the same geometric criterion as used in a previous work [8]. For the above three solvent-water mixtures it is found that, two molecules are regarded as being hydrogen bonded if their separations satisfy that $R_{OO} \leq 3.20 \, \overset{\circ}{A}$, $R_{OH(w)} \leq 2.40 \, \overset{\circ}{A}$ and the angle $H_wO...O \leq 30°$, where H_w is the hydroxyl proton. In the case of the AN–water mixtures, according to the geometrical criterion an acetonitrile molecule is hydrogen bonded to a water molecule, if the distances $R_{N-H} \leq 2.6 \text{Å}$, $R_{N-O} \leq 3.3\text{-}3.5 \text{Å}$ and the angle O...N...H $\leq 45^0$ are fulfilled. The corresponding values for the water-water hydrogen bond criterion are: $R_{O-O} \leq 3.60 \text{Å}$, $R_{O-H} \leq 2.60 \text{Å}$, and for the angle H-O...O $\leq 30^0$. We also computed the average number of hydrogen bonds (N_{MD}) for the different kinds in each mole fraction of the systems. The results are presented in the following figure 2 for all the types of hydrogen bonds. We observe that the number of hydrogen bonds donated by water is almost constant with composition. The number of hydrogen bonds of water reveals that it is independed at very low water concentrations. Moreover, the hydrogen bonding network of the water molecules in AS/water mixtures are studied. Especially, at low concentrations of Ac and AN in the mixtures, the Ac and AN molecules have an identical effect on water hydrogen bonding, but differences between Ac and AN mixtures arise upon further dilution of water. These differences appear to be primarily related to the strength of hydrogen bonding between water and its AS

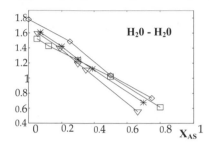 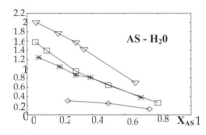

Figure 2: In the left side is the average number of water – water hydrogen bonds and in the right side is the average number of AS – water hydrogen bonds, as function of the molar fraction of aprotic solvent. The symbols in both parts are as in figure 1.

4. Conclusions

Molecular dynamics results on aqueous mixtures have been tested against available structural, thermodynamic and diffusion data at room temperature and at a wide density range. The non-ideal behaviour of the diffusion properties of AS–water mixtures has been analyzed and we found that a model function predicts successfully the simulated resultsof the water at these states. Moreover, we found that, there is a loss of the tetrahedral water structure on mixing and furthermore water molecules have a strong tendency to aggregate in the AN and Ac mixtures. Especially for DMSO and DMF mixtures the results indicate that the tetrahedral structure of water is destroyed as the DMSO, DMF concentration increases and breaks more quickly in the DMSO, DMF- rich region. The shapes of the PCFs $M(O, N)_{AS}\text{-}H_W$ and $O_W\text{-}H_W$ reveal the existence of hydrogen bonds.

References

[1] M. Ferrario, M. Haughney, I.R. McDonald and M. Klein: *Molecular-dynamics simulation of aqueous mixtures: Methanol, acetone, and ammonia*, Journal of Chemical Physics **93** 5156-5166(1990).

[2] H. Kovacs and A. Laaksonen: *Molecular Dynamics Simulation and NMR Study of Water-Acetonitrile Mixtures,* Journal of American Chemical Society **113** 5596-5605(1991)

[3] M.E. Chalaris: *Statistical mechanical studies via molecular dynamics simulation techniques of liquid systems consisted of protic (MeOH, H_2O) and aprotic (DMSO, DMF) solvents*: Doctoral Dissertation, National and Kapodistrian University of Athens, Greece, 1999 (in Greek).

[4] M. Chalaris and J. Samios: *Systematic molecular dynamic studies of liquid N,N-dimethylformamide using optimized rigid force field: Investigation of the thermodynamic, structural, transport and dynamic properties*, Journal of Chemical Physics **112** 8581-8594(2000).

[5] W.L. Jorgensen, J. Chandrasekhar, J.D. Madura, R.W. Impey and M.L. Klein: *Comparison of simple potential functions for simulating liquid water*, Journal of Chemical Physics **79** 926-935(1983).

[6] M.E. Chalaris and J. Samios: *Computer simulation studies of the liquid mixtures water-dimethylsulfoxide using different effective potential models: Thermodynamic and transport properties*, Journal of Molecular Liquids **98-99** 399-409(2002).

[7] M.E. Chalaris, A. Koufou and J. Samios: *Molecular Dynamics Simulations of the liquid mixtures N,N-dimethylformamide-water using available potential models*, Journal of Molecular Liquids **101** 69-79(2002).

[8] H.J.V. Tyrell and K.R. Harris, Diffusion in Liquids, Butterworth & Co, London, 1984.

[9] R. L. Mancera, M. Chalaris, K. Refson and J. Samios: *Molecular dynamics simulation of dilute aqueous DMSO solutions. A temperature-dependence study of the hydrophobic and hydrophilic behaviour around DMSO*, Phys. Chem. Chem. Phys **6** 94-102(2004).

VSP International
Science Publishers
P.O. Box 346, 3700 AH Zeist
The Netherlands

Lecture Series on Computer
and Computational Sciences
Volume 1, 2004, pp. 115-117

Quantum Theory and Simulation Aspects of the Design of Molecules for Nonlinear Optics Applications

Benoît Champagne[1]

Laboratoire de Chimie Théorique Appliquée,
Facultés Universitaires Notre-Dame de la Paix,
Rue de Bruxelles, 61, B-5000 Namur, Belgium

Received 23 July, 2004; accepted in revised form 6 August, 2004

Abstract: The paper surveys the evolution of the molecular structures for application in nonlinear optics by emphasizing the aspects related to the molecular design by quantum chemical methods. In particular, the effects of chirality, charge, and spin multiplicity on the nonlinear optical responses are addressed whereas recent methodological developments are discussed.

Keywords: nonlinear optical properties, donor/acceptor compounds, chiral systems, charge and spin multiplicity effects, quantum chemistry simulations.

PACS: 31.15.-p, 31.25.-v, 33.15.Kr, 33.55.Fi

1. Introduction

Since 30 years, nonlinear optical (NLO) materials have been attracting a lot of interest because of foreseeable applications in optoelectronic technology. In particular, organic compounds are attractive due to their large and fast responses combined with the flexibility of organic synthesis that fastens the design of new chromophores. So far, most of the achievements have been made for second-order NLO responses because the available laser intensities enable easily to display the second-order NLO effects.

2. Donor/Acceptor compounds

The first generation of second-order NLO organic chromophores consists in one-dimensional (1-D) π-conjugated systems end-capped with donor (D) and acceptor (A) moieties. Experimental and theoretical works have shown during the last decades how to choose the D and A and the π-segments for optimizing the second-order NLO responses [1]. Later, two-dimensional (2-D) octupolar molecules such as 1,3,5-triamino-2,4,6-trinitrobenzene have been shown, on the one hand, to exhibit second-order NLO responses that can be as large as for their dipolar analogs and, on the other hand, to correct some of the limitations (*e.g.* transparency-efficiency tradeoff, centro-symmetric arrangement) of the 1-D dipolar structures [2]. Improvements with respect to the dipolar structures, of which several will be detailed in the talk, have also been highlighted in the case of i) 2-D Λ-shape molecules where two D/A pairs intersect at the donor or acceptor group forming DAD- or ADA-like structures [3-4] and of ii) three-dimensional (3-D) tetrahedral D/A compounds [5].

3. Chiral systems

---

[1] Corresponding author. benoit.champagne@fundp.ac.be

An alternative strategy for optimizing the microscopic – as well as macroscopic – second-order NLO response is to use non-racemic chiral structures which are intrinsically non-centrosymmetric. For instance, in *N*-(4-nitrophenyl)-(L)-prolinol (NPP), chirality – as well as hydrogen bonds – prevents centro-symmetric packing and ensures an NLO-efficient molecular stacking. Chirality is also at the origin of second-order – as well as third-order – NLO responses of isotropic media encompassing nonlinear circular dichroism, sum-frequency generation (SFG), and electric-field-induced SFG [6]. In absence of chirality or for racemic mixtures, these phenomena disappear. Moreover, certain second harmonic generation $\chi^{(2)}$ tensor components of organic thin films are due to chirality, and are either associated with a pure electric phenomenon or with mixed electric-magnetic effects [7]. Helicenes – as well as other helical structures – constitute a special class of chiral compounds because, contrary to NPP, in helicenes, the π-conjugated segment is the chiral helix itself, which opens the way for enhanced nonlinear responses. The talk will describe recent results on such systems [8] with the aim of deriving structure/property relationships, *i.e.* finding the key substituent positions and helix size to maximize the second-order NLO response.

4. Charge and spin multiplicity effects

Another stream for increasing the NLO responses consists in charging the molecules [1, 9] or forming open-shell systems [10]. This later solution also provides the possibility of fabricating multifunctional compounds when combining the nonlinear electric field effects with the control of the molecular structure and properties (including spin multiplicity) by applying an external magnetic field. In polyene chains bearing a charged soliton, the charged defect can easily move under the effect of the external field and therefore be associated with substantial NLO responses. In such a case, the choice of the dopant (having the opposite charge) turns out to be of paramount importance for optimizing the response.

5. Methodological aspects

Simulation of the NLO responses of organic conjugated compounds is certainly a difficult task because many aspects need to be tackled with the same acuity. The choice – and more specifically, the validation – of a suitable basis set for studying large conjugated compounds is a complex task, closely linked to the level of theory employed in the simulation. Indeed, whereas most of such studies deal with small systems [11], in many cases, electron correlation plays a critical role and can change the conclusions on the structure/property relationships. Concerning π-conjugated open-shell systems, although model systems have recently been studied, there remain questions about the adequacy of restricted and unrestricted schemes. Moreover, in this regards, the elaboration of suitable exchange-correlation functionals would turn out to be a big step forward for studying large systems [12]. Indeed, in many cases correlated wave-function methods remain difficult to apply to large π-conjugated systems whereas density functional theory approaches require similar computational requirements to the Hartree-Fock scheme but have been shown to present severe drawbacks when using conventional exchange-correlation functionals. Some solutions have recently been proposed to cure this incorrect behavior. In addition to the electronic contributions, there exist vibrational contributions: a pure vibrational contribution dominated by finite field nuclear-relaxation terms [13] and a zero-point vibrational-averaging correction [14]. *Ab initio* calculations have shown that, as a function of the chemical composition and NLO process, the vibrational contribution can be either small and negligible or large and dominant. Some simulation aspects on electron correlation effects and vibrational contributions will be tackled in the talk and illustrated with examples on prototypical π-conjugated systems.

Acknowledgments

The author thanks the Belgium National Fund for Scientific Research for his Senior Research Associate position.

References

[1]   D.R. Kanis, M.A. Ratner, and T.J. Marks, Chem. Rev. **94**, 195 (1994); T. Verbiest, S. Houbrechts, M. Kauranen, K. Clays, and A. Persoons, J. Mater. Chem. **7**, 2175 (1997); Nonlinear Optical Materials, Vol. 9 of Handbook of Advanced Electronic and Photonic

Materials and Devices, edited by H.S. Nalwa (Academic Press, San Diego, 2001) and references therein.

[2]   J. Zyss and I. Ledoux, Chem. Rev. **94**, 77 (1994).

[3]   C.R. Moylan, S. Ermer, S.M. Lovejoy, I.-H. McComb, D.S. Leung, R. Wortmann, P. Krdmer, and R.J. Twieg, J. Am. Chem. Soc. **118**, 12950 (1996).

[4]   M. Yang and B. Champagne, J. Phys. Chem. A **107**, 3942 (2003).

[5]   M. Cho, S.-Y. Am, H. Lee, I. Ledoux, and J. Zyss, J. Chem. Phys. **116**, 9165 (2002).

[6]   F. Hache, H. Mesnil, and M.C. Schanne-Klein, Phys. Rev. B **60**, 6405 (1999); M.A. Belkin and Y.R. Shen, Phys. Rev. Lett. **91**, 213907 (2003); P. Fischer, A.D. Buckingham, K. Beckwitt, D.S. Wiersma, and F.W. Wise, Phys. Rev. Lett. **91**, 173901 (2003).

[7]   T. Verbiest, M. Kauranen, and A. Persoons, J. Mater. Chem. **9**, 2005 (1999).

[8]   O. Quinet and B. Champagne, Int. J. Quantum Chem. **85**, 463 (2001); E. Botek, B. Champagne, M. Türki, and J.M. André, J. Chem. Phys. **120**, 2042 (2004).

[9]   M. Nakano, I. Shigemoto, S. Yamada, and K. Yamaguchi, J. Chem. Phys. **103**, 4175 (1995). B. Champagne, M. Spassova, J.B. Jadin, and B. Kirtman, J. Chem. Phys. **116**, 3935 (2002); L.N. Oliveira, O.A.V. Amaral, M.A. Castro, and T.L. Fonseca, Chem. Phys. **289**, 221 (2003).

[10]   I. Ratera, S. Marcen, S. Montant, D. Ruiz-Molina, C. Rovira, J. Veciana, J.F. Létard, and E. Freysz, Chem. Phys. Lett. **363**, 245 (2002); M. Nakano, T. Nitta, K. Yamaguchi, B. Champagne, and E. Botek, J. Phys. Chem. A **108**, 4105 (2004).

[11]   G. Maroulis, J. Chem. Phys. **111**, 583 (1999); G. Maroulis, J. Chem. Phys. **118**, 2673 (2003).

[12]   S.J.A. van Gisbergen, P.R.T. Schipper, O.V. Gritsenko, E.J. Baerends, J.G. Snijders, B. Champagne, and B. Kirtman, Phys. Rev. Lett. **83**, 694 (1999).

[13]   J.M. Luis, M. Duran, B. Champagne, and B; Kirtman, J. Chem. Phys. **113**, 5203 (2000).

[14]   O. Quinet, B. Kirtman, and B. Champagne, J. Chem. Phys. **118**, 505 (2003).

VSP International
Science Publishers
P.O. Box 346, 3700 AH Zeist
The Netherlands

*Lecture Series on Computer
and Computational Sciences*
Volume 1, 2004, pp. 118-121

Electronic and Structural Properties of the B1-Aln/Tin Superlattices

D. Chen, X. L. Ma [1]

Shenyang National Laboratory for Materials Science,
Institute of Metal Research, Chinese Academy of Sciences,
Shenyang 110016, China

Received 31 July, 2004; accepted in revised form 10 August, 2004

Abstract: Interfacial structures in the B1-AlN/TiN(001) superlattices have been studied by the first-principle plane wave pseudopotential method based on density functional theory. The theoretical calculations show that the preferred bonding configuration across the interface is cation-anion bonding. It is proposed that the bond directionality and charge redistribution in the B1-AlN/TiN(001) superlattice would play an important role in superlattices hardness.

Keywords: the B1-AlN/TiN(001) superlattices, bonding configuration, electronic property

Mathematics SubjectClassification: PACS

PACS: 73.20.–r, 31.15.Ew, 68.35.Ct

1. Introduction

Recently, due to the excellent thermal stability and relatively high peak hardness [1,2], the rock salt AlN (B1-AlN) has been epitaxially grown on the cubic crystal structure of TiN substrate by experiment [3,4]. It is known that the interfacial properties would play an important role in understanding B1-AlN/TiN applications. In present work, we have employed the first-principle calculation to investigate the electronic and structural properties of B1-AlN/TiN(001) superlattices.

2. Calculation method

To lend confidence to carry out calculations of the structural properties and electronic structure for the B1-AlN/TiN(001) superlattice, the lattice constants and modulus for the bulk form of B1-AlN and TiN were firstly calculated. We have used the first-principle plane wave pseudopotential (PWPP) method based on density functional theory implemented in CASTEP code [5]. With a cutoff energy of the plane-waves of 400 eV (29.4 Ry), the ultrasoft pseudopotentials and the gradient-corrected form of the exchange-correlation functional were taken. After the final self-consistency cycle, the remaining forces on the atoms were less than 0.03 eV/Å, and the remaining stress was less than 0.05 Gpa.

Table 1 presents the equilibrium lattice constants and bulk modulus for B1-AlN and TiN according to the experiment and calculations based on various methods. The calculated results in the present study agree well with the experimental data and other theoretical results.

3. Results and Discussions

A. Bonding configuration and interfacial energy

Both B1-AlN and TiN have a NaCl-type structure. According to the experimental observations, the pseudomorphic cubic B1-AlN(001) was epitaxially stabilized on TiN(001), and the coherent interface could be formed in the B1-AlN/TiN(001) superlattice [3,4]. Then, referring to our work in the TiN/MgO(001) interface system [11], the interface structures of the B1-AlN/TiN(001) superlattice have

---

[1] Corresponding author. E-mail: xlma@imr.ac.cn

been studied for three possible bonding configurations: cation-anion counterpart (mode I), cation-cation counterpart (mode II), and ion-center gap site (the mid of two nearest-neighbor atoms along [110]) (mode III) (seeing Fig.3 in Ref. [11]). The supercell of multilayer geometry was chosen with three-dimensional periodicity and two identical interfaces per unit cell. The repeated supercell geometry was denoted by the notation $n+n$, where n is the number of atomic layers in the B1-AlN or TiN geometry.

Table 1: Lattice constants and bulk modulus for B1-AlN and TiN according to the experiment and calculations based on various methods.

	a (Å)	B_0 (GPa)	Method	Ref.
B1-AlN	4.043	221	Expt.	6
	3.956	215	LDF[a]	6
	3.995	270	FP-LMTO[b]	7
	3.982	247	PWPP	This study
TiN	4.240	288	Expt.	8
	4.250	282	T. M. PP[c]	9
	4.159	310	AEPP[d]	10
	4.233	281	PWPP	This study

[a]Local density function
[b]Full-potential linear muffin-tin-orbital calculation
[c]Troullier and Martins pseudopotential
[d]All-Electron pseudopotential

For each relaxed configuration of the B1-AlN/TiN superlattices, the interface energy (E_{inter}) of superlattice can be obtained (referring to Equation (1) in Ref. [11]). Table 2 listed the interface spacing range and energy for different modes of 5+5 superlattice. It can be clearly seen that the interfacial energy for mode I is much lower than that the others, and the calculated interface spacing range of mode I well agreed with the interface spacing value obtained by experiment [3]. These results all show that the preferred bonding configuration at the B1-AlN/TiN(001) interface corresponds to the Al-N, N-Ti counterpart.

Table 2: Interface spacing range and energy for different modes of 5+5 B1-AlN/TiN supperlattice.

Configuration	Equilibrium spacing range (Å)	Interface energy (J/m²)	Method	Relaxation
	2.000	—	Expt. [3]	—
Mode I	2.005~2.167	0.775	This study	Yes
Mode II	4.216~4.495	3.271	This study	Yes
Mode III	3.194~3.436	3.193	This study	Yes

According to this mode, the supercells of the B1-AlN(n)/TiN(n)(n=3 and 7) superlattices were calculated to check the convergence of the interface energy. The results of the B-AlN(n)/TiN(n) (n=3, 5 and 7) superlattice give 0.738, 0.775 and 0.814 J/m² respectively, indicating the PWPP calculations are converged to about 0.04 J/m². No significant difference in interface energy was observed.

B. Electronic structure

To study the bonding configuration by the electronic properties of mode I, the projected density of states (PDOS) of the first layer (L1) of B1-AlN and TiN in 5+5 superlattice can be calculated. At the B1-AlN/TiN(001) interface, the N 2p state (L1 of B1-AlN) and Ti 3d state (L1 of TiN), the Al 3p (3s) state (L1 of B1-AlN) and N 2s (2p) (L1 of TiN) could be hybridized each other, and would form the chemical interaction between the interfacial atoms.

As we know, the interface spacing (distance between Al-N and N-Ti atoms) for mode II and III are larger than that of mode I (Table 2). So, the interfacial interactions between Al-N and N-Ti atoms in mode II and III are already weaker than in mode I. Otherwise, Figure 1 gives the comparison of Al and Ti projected density of states because there are little changes in the PDOS of N atoms for the three different modes. Comparing with mode I, one can notice an upward shift of the centers of gravity of Al 3s and 3p states with respect to the Fermi level in mode II, indicating a weaker interaction with N.

Similar to mode II, the interaction further decreases because the Al-Ti atom distance in the interface of mode III is longer than that in mode II. Also, antibonding Al 3s, 3p and Ti 3p states above the Fermi level increase obviously relative to mode I. Therefore, the interfacial interactions in mode II and III are much weaker than in mode I, the configurations of mode II and mode III are not preferred and the mode I is the favorable configuration.

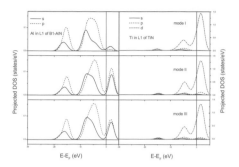

Figure 1: Projected density of states of Al and Ti atoms (solid lines indicate s states, dash lines p states and dash-dot lines d states) in layer 1 of B1-AlN and TiN for different modes of 5+5 superlattice

C. Discussion

For the NaCl structure materials, the shear modulus G, which defines the resistance to reversible deformation upon shape change, is a better predictor of hardness. It is correlated to various physical properties (ionicity, band gap, cohesive energy, etc.) and can thus be studied indirectly. Eberhart proposed the shear modulus is closely correlated to the crystallographic structures, bond directionality and charge redistribution [12].

For the superlattice, we define the bond directionality as the ratio of the bond orders (BOs) on planes parallel to interface to the bond orders on planes perpendicular to interface. The larger the ratio, the greater the bond directionality; the more shear intensity on slip plane. After calculations, the BO_{Al-N} (bond order between Al and N) is 0.90 in bulk B1-AlN, whereas the BO_{Ti-N} is 0.66 in bulk TiN. In the superlattice, however, the bonding configurations and the covalent characters of B1-AlN and TiN layers have changed due to the presence of interface. For the 5+5 B1-AlN/TiN(001) superlattice of mode I, the average BO_{Al-N} and BO_{Ti-N} on (002) plane are 1.49 and 1.28; and the average BO_{Al-N} and the BO_{Ti-N} on (002) or (200) planes are 0.18 and 0.09, respectively. The bond orders on (002) are more bonding than those on (020) or (200). Then, the bond directionality is at least several times larger than that in bulk materials in same slip system. That is to say, the bond directionality can reflect the effect of interface region on the superlattice hardness, and the great bond directionality would correspond the hardness increment of the B1-AlN/TiN(001) superlattice.

In addition, if B and A denote two adjacent atoms, and B is the most electronegative element in the material, the hardness of a given material is adjusted by $1+Z_A/Z_B$, where Z_B and Z_A are the effective nuclear charges for atoms B and A [13]. From our calculated results, in the AlN/TiN(001) superlattice, the Al ionicity of B1-AlN layer adjacent to interface is +1.25 less than +1.31 in bulk B1-AlN, where N ionicity is –1.18 larger than -1.31 in bulk B1-AlN. Then, compare to the B1-AlN, the effective nuclear charge Z_{Al} increases and Z_N decreases because the charges change and redistribute in the layer neighbor interface. However, the ionicities of TiN layer adjacent to interface are almost identical to those in bulk TiN (Ti +0.79, N –0.79). Therefore, the larger $1+Z_{Al}/Z_N$ would make a contribution to the B1-AlN layers hardness of the superlattice.

Generally, it is considered that the formation of the cubic phase of AlN leads to high hardness of the AlN/TiN superlattice [1,2]. In our views, the formation of the multilayer structure of cubic AlN and TiN would result in a difference in covalent and ionicity characters of different layers. The bond directionality and charge redistribution in the B1-AlN/TiN(001) superlattice would also play an important role in superlattices hardness.

4. Conclusions

The electronic and structural properties of the B1-AlN/TiN(001) superlattices are studied using the first principles plane wave pseudopotential method. The following results are obtained.

(1) For the interface of the B1-AlN/TiN(001) superlattice, the favorable bonding configuration is cation-anion bonding (Al-N, N-Ti counterpart).

(2) The characteristic projected densities of states explain that the cation-anion bonding configuration is also preferred by the electronic structure of the superlattice.

(3) Compared with bulk B1-AlN and TiN, the covalent and ionicity characters in the B1-AlN/TiN(001) superlattice have changed due to the presence of interface.

(4) The bond directionality and charge redistribution in the B1-AlN/TiN(001) superlattice would also play an important role in superlattices hardness.

Acknowledgments

Thanks are given to the Hundred Talents Project of Chinese Academy of Sciences and National Outstanding Young Scientist Foundation for Grant No. 50325101 to X. L. Ma. The authors are grateful to Dr. J. Y. Wang, High-Performance Ceramic Division at this laboratory for computer resources and helpful discussions. We also gratefully acknowledge financial support from the Special Funds for the Major State Basic Research Projects of China (Grant No. 2002CB613503).

References

[1] M.Setoyama, M.Irie, H.Ohara, M.Tsujioka, T.Nomura and N.Kitagawa, Thermal stability of TiN/AlN superlattices, *Thin Solid Films* 341 126-131(1999).

[2] M. S. Wong, G. Y. Hsiao and S. Y. Yang, Preparation and characterization of AlN/ZrN and AlN/TiN nanolaminate coatings, *Surface and Coatings Technology* 133-134 160-165 (2000).

[3] A. Madan, I. W. Kim, S. C. Cheng, P. Yashar, V. P. Dravid, and S. A. Barnett, Stabilization of cubic AlN in epitaxial AlN/TiN superlattices, *Physical Review Letters* 78 1743-1746(1997).

[4] V. Pankov, M. Evstigneev, and R. H. Prince, Enhanced stability of rocksalt-type AlN phase in AlN/TiN superlatttices synthesized by room-temperature pulsed laser deposition, *Journal Applied Physics* 92 4255-4260(2002).

[5] M. D. Segall, P. J. D. Lindan, M. J. Probert, C. J. Pickard, P. J. Hasnip, S. J> Clark and M. C. Payne, First-principles simulation: ideas, illustrations and the CASTEP code, *Journal of Physics: Condensed Matter* 14 2717-2744(2002).

[6] Q.Xia, H.Xia, and A.L.Ruoff, Pressure-induced rocksalt phase of aluminum nitride: A metastable structure at ambient condition, *Journal Applied Physics* 73 8198-8200(1993).

[7] N. E. Christensen, and I. Gorczyca, Calculated structural phase transitions of aluminum nitride under pressure, *Physical Review B* 47 4307-4314(1993).

[8] W.Spengler, R.Kaiser, A.N.Christensen and G.M ü ller-Vogt, Raman scattering, superconductivity, and phonon density of states of stoichiometric and nonstoichiometric TiN, *Physical Review B* 17 1095-1101(1978).

[9] K.Kobayashi, First-principles study of the electronic properties of transition metal nitride surfaces, *Surface Science* 493 665-670(2001).

[10] A.Šimůnek and J.Vackář, Correlation between core-level shift and bulk modulus in transition-metal carbides and nitrides, *Physical Review B* 64 235115 (2001).

[11] D.Chen, X.L.Ma, Y.M.Wang and L.Chen, Electronic properties and bonding configuration at the TiN/MgO(001) interface, *Physical Review B* 69 155401(2004).

[12] M.E.Eberhart, The metallic bond: elastic properties, *Acta Materialia* 44 2495-2504(1995).

[13] D.G.Clerc, H.M.Ledbetter, Mechanical hardness: a semiempirical theory based on screened electrostatics and elastic shear, *Journal of Physics and Chemistry of Solids* 59 1071-1095(1998).

VSP International
Science Publishers
P.O. Box 346, 3700 AH Zeist
The Netherlands

*Lecture Series on Computer
and Computational Sciences*
Volume 1, 2004, pp. 122-126

A Self-Adaptive Chaotic Particle Swarm Algorithm for Short-Term Hydro System Scheduling in Deregulate Environment

Jiang Chuanwen [1]
Department of Electrical Engineering,
Shanghai Jiaotong University,
Shanghai, 200030, China

Etorre Bompard
Department of Electrical and Computer Engineering,
Polectenico Di Torino
Torino, Italy

Received 14 July, 2004; accepted in revised form 14 August, 2004

Abstract: This paper proposed a short-term hydro dispatch model based on the rule of making the benefit the most. For the optimal dispatch model is a large-scale nonlinear planning problem with multi-constraint and multi-variable, this paper proposes a novel self-adaptive chaotic particle swarm optimization algorithm to solve the short-term generation scheduling of hydro system in deregulate environment. Since chaotic mapping enjoys certainty, ergodicity and stochastic property, the proposed approach introduces chaos mapping and an adaptive scaling term into the particle swarm optimization algorithm, which aim to increase its convergence rate and results' precision. The new method has been examined and test on a practical hydro system, the results are promising and show the effectiveness and robustness of the proposed approach, compared to the traditional particle swarm optimization algorithm.

Keywords: particle swarm optimization, Chaos, generation scheduling of hydro system

AMS (MOS): 90C30, 90C59

I. Introduction

The purpose of short-term hydro system scheduling is to find the optimum hourly generation of hydro units by utilizing the limited water resource in a schedule horizon so as to optimize the total benefit of hydro generated energy, while satisfying various constraints. Traditional short-term hydro system dispatch, whose object is to make the mount of generation the maximal, With the development of the market economy, this model is no longer suitable for the deregulate environment. This paper proposed a short-term hydro system optimal dispatch model based on the rule of making the benefit the most, and made analysis of the benefit function of hydropower plant based on this model. For the short-term hydro system optimal dispatch model is a large-scale nonlinear planning problem with multi-constraint and multi-variable, the entire procedure of optimization is very complex and has been a focus of many researchers for a long time. Lots of optimal technologies have been used to solve this problem [1-6]. Generally, there are some defects in those methods to some extent, such as dimensionality difficulty, the un-stability of convergence and the complication of algorithm. Considering the characteristics of short-term hydro system scheduling in deregulate environment, this paper proposes a novel self-adaptive chaotic particle swarm optimization algorithm to solve the short-term hydro system optimal dispatch scheduling in deregulate environment. Since chaotic mapping enjoys certainty, ergodicity and stochastic property, the proposed approach introduces chaos mapping and an adaptive scaling term into the particle swarm optimization algorithm, which aim to increase its convergence rate and results' precision. The new method has been examined and test on a practical hydro system, the results are promising and show the effectiveness and robustness of the proposed approach, compared to the traditional particle swarm optimization algorithm.

---

[1] Corresponding author. Tel./fax:86-21-54747501. E-mail: **jiangcw_sjtu@yahoo.com.cn (C. Jiang)**

II.The Model of Hydropower Plant Optimal Dispatching

2.1 Objective Function
In the mode of electric power market, the optimal dispatching of hydropower plant is to make the benefit of generation in certain period the biggest under the condition of security and reliability. Its objective function is:

$$F = \max \sum_{i=1}^{T} B_i[e_i(u_i)] \tag{1}$$

Where:

i represents the order number of time period, T represents the total number of time periods; B_i

represents the benefit function of the period, e_i represents the amount of generation in period , u_i

represents the decision variable in period i.
In the environment of market, because the price is different under different period, and the demand of power in one period is a stochastic variable, we use the method of stochastic analysis to do research on the benefit function of the hydropower substation.

Suppose that the rating capacity of the a hydropower station is N_g, and there are t_1 periods when the

load is high in a day and the price is c_1, t_2 is the number of the periods when the load is normal and

the price is c_2, $t_3 = 24 - t_1 - t_2$ is the he number of the periods when the load is low and the price is

c_3 ($c_1 > c_2 > c_3$). Also, suppose that the allowed amount of generation of hydropower station in peak

load periods e_1 is a stochastic variable, the probability function of which is $p_1(x)(0 \le x \le t_1 N_g)$.

the allowed amount of generation of hydropower station in normal load periods e_2 is a stochastic

variable, the probability function of which is $p_2(x)(0 \le x \le t_2 N_g)$. The allowed amount of

generation of hydropower station in low load periods e_3 is a stochastic variable, the probability

function of which is $p_3(x)(0 \le x \le t_3 N_g)$.

When the planned amount of generation in low load periods is e , but the allowed amount of generation

is e_3, so the benefit of the station is $c_3 \min(e, e_3)$, the expectation value of benefit is:

$$B_3(e) = c_3 \int_0^{t_3 N_g} \min(e, x)P_3(x)dx \tag{2}$$

When the planned amount generation in normal load periods and low load periods is e , and the allowed amount of generation in normal load periods is e_2 , expectation value of benefit is:

$$B_2(e) = \int_0^{t_2 N_g} [c_2 \min(e, x)]P_2(x)dx + B_3(e - \min(e, x)) \tag{3}$$

Similarly, when the planned amount of generation in a day is e , the expectation value of the available benefit is:

$$B_1(e) = \int_0^{t_1 N_g} [c_1 \min(e, x)]p_1(x)dx + B_2(e - \min(e, x)) \tag{4}$$

Which is the expectation function of the benefit of the hydropower plant.
2.2 Constrained Conditions
1) Water dynamic balance equation with travel time:

$$V_{i+1} = V_i + q_i - Q_{xi} - \Delta q_i \tag{5}$$

Where:
V_i, V_{i+1} represents separately the amount of water in reservoir in starting point and ending point of

period i. q_i represents the average amount of natural water flowing into the reservoir in period i. Q_{xi}

represents the amount of water flowing downwards of the station. Δq_i represents the loss of water.
2) Reservoir storage volumes limits:

$$V_{min,i} \le V_i \le V_{max,i} \tag{6}$$

3) Hydro plant power generation limits:

$$N_{min,i} \le N_i \le N_{max,i} \tag{7}$$

4) Hydro plant discharge limits:

$$Q_{min,f} \le Q_{fi} \le Q_{max,f} \tag{8}$$

III. A Self-Adaptive Chaotic Particle Swarm Optimization Algorithm

Like evolutionary algorithms, PSO technique conducts search using a population of particles, corresponding to individuals. Each particle represents a candidate solution to the problem at hand. The procedure of the self-adaptive chaotic PSO for short-term hydro system scheduling can be described as follows.

Step1 Initialization: Q_i are regarded as control variables. Set $t=0$. Let $X_j = \{Q_i^j\}$ be a particle, generate randomly n particles $\{X_j(0), j = 1, \cdots n\}$ (set n to 50 in this paper). All particles are set between the lower and upper limits. Similarly, generate randomly initial velocities of all particles, $\{\vec{V}_j(0), j = 1, \cdots n\}$, where $\vec{V}_j(0) = \{\vec{v}_{j,1}(0), \cdots, \vec{v}_{j,m}(0)\}$. $\vec{v}_{j,k}(0)$ is generated by randomly selecting a value with uniform probability over the kth dimension $\left[-\vec{v}_k^{max}, \vec{v}_k^{max}\right]$. The constrained optimization problem is converted into unconstrained optimization problem using penalty factor.

$$J' = -f + \phi_1 \sum_{i \in T} (N_i - N_{min,i})^2 + \phi_2 \sum_{i \in T} (V_i - V_{min,i})^2 \tag{9}$$

Where, T represents the total number of time periods; ϕ_1, ϕ_2 are penalty factors.
Each particle in the initial population is evaluated using the equation (15).

$$J = J'_{MAX} - J' + \beta(J'_{MAX} - J'_{MIN}) \tag{10}$$

Where, J'_{MAX}, J'_{MIN} are, respectively, the maximum and the minimum of the objective functions for present particles; β is the control parameter and its value choice ranges from 0.01 to 0.1.
For each particle, set $X_j^*(0) = X_j(0)$ and $J_j^* = J_j, j = 1, \cdots n$.
Let $J^{**} = \min\{J_1^*, \cdots J_n^*\}$. Set the particle associated with J^{**} as the global best, $X^{**}(0)$.

Step2 Velocity and Position updating: Let $t=t+1$. Using the global best and individual best of each particle, the ith particle velocity and position in the jth dimension is updated using the equation (11)-(14).

$$v_{i,j}(t) = w(t) \times v_{i,j}(t-1) + c_1 \times r_1 \times (x_{j,k}^*(t-1) - x_{j,k}(t-1)) + c_2 \times r_2 \times (x_{j,k}^{**}(t-1) - x_{j,k}(t-1)) \tag{11}$$

$$x_{j,k}(t) = x_{j,k}(t-1) + v_{j,k}(t) \tag{12}$$

$$(j = 1,2, \cdots n \quad k = 1,2, \cdots m)$$

Where:

$$r_i(t) = 4.0 \times r_i(t-1) \times (1 - r_i(t-1)) \qquad i = 1,2 \tag{13}$$

$$w(t) = w(0) \exp(-\rho \frac{k}{N_{max}}) \tag{14}$$

Step3 Individual and global best updating: Each particle is evaluated according to its updated position.
If $J_j < J_j^*, j = 1, \cdots, n$, then $X_j^*(t) = X_j(t)$ *and* $J_j^* = J_j$. Else go to *Step3*
Search for the minimum value J_{min} among J_j^*.
If $J_{min} < J^{**}$ then $X^{**}(t) = X_{min}(t)$ *and* $J^{**} = J_{min}$. Else go to *Step3*.
Step4 Stopping criteria: If one of the stopping criteria is satisfied then stop; else go to *Step2*.

IV. Numerical Examples

The practical hydro system lies in Enshi in Hubei province of China. The area is adopting the primary power market(zonal time price). The capacity of the reservoir of the plant is 1.205 million cube meter and the total rating capacity of the generators is 45MW(3*15MW). For the chaotic particle swarm optimization, the initial values are chosen as: the weights c_1, c_2 are 2 and $w(0) = 3, r_i(0) = 0.48$.

The optimal results of the day dispatch are shown in the table 1. The figure 1 shows the iteration performances of the two algorithms. It explicitly tells us that the new algorithm achieves a better performance regarding efficiency and convergence rate.

Table 1. The computing result of the day dispatch during the period from 1st, April to 10th, April in 2000

D	$Z_{d,i}$	Q_d	$Q_{d,q}$	$P_{d,aver}$	E_d	$E_{d,p}$	$E_{d,n}$	$E_{d,l}$	B_d	$Z_{d,i+1}$
4.1	505.49	70	73.1	34.57	81.5	30	42.5	8	39.76	505.46
4.2	505.46	56.9	64.89	31.42	76	30	46	0	35.87	505.68
4.3	505.68	49	62.55	28.12	67.5	31.5	36	0	32.07	505.93
4.4	504.93	43.5	54.32	25.62	62.5	31.5	31	0	29.57	504.47
4.5	504.47	140	91.13	43.66	100.64	31.5	45	24.14	41.59	505.65
4.6	505.65	242	106.5	45	106	31.5	45	29.5	45.05	508

Where:

D represents the day in the given period; $Z_{d,i}$, $Z_{d,i+1}$ represent separately the start position and the end position of the water in the reservoir, the unit of which is meter; Q_d represents the natural water flowing in, the unit of which is cube meter; $Q_{d,q}$ represents the water flowing out, the unit of which is cube meter; $P_{d,aver}$ represents the average out force of generation, the unit of which is MW; E_d represents the total amount of generation in a day, the amount of which is MWh; $E_{d,p}, E_{d,n}, E_{d,l}$ represents separately the amount of generation in peak load period, normal load period and low load period, the unit of which is MWh; B_d represents the benefit in a day, the unit of which is ten thousand RMB.

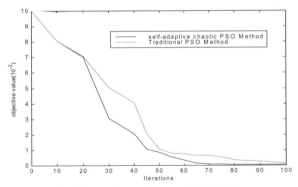

Fig.1 The iteration performances of both algorithms

V. Conclusions

This paper proposed a hydropower plant dispatch model based on the rule of making the benefit the most. For the short-term hydro system optimal dispatch model is a large-scale nonlinear planning problem with multi-constraint and multi-variable, this paper proposes a novel self-adaptive chaotic particle swarm optimization algorithm to solve the short-term generation scheduling of hydro system in deregulate environment. Since chaotic mapping enjoys certainty, ergodicity and stochastic property, the proposed approach introduces chaos mapping and an adaptive scaling term into the particle swarm optimization algorithm, which aim to increase its convergence rate and results' precision. The new method has been examined and test on a practical hydro system, results show that the proposed approach is feasible and effective for the large-scale constrained nonlinear optimization problem.

References

[1] L. Christiano, R. Luiz, A multi-objective approach to the short-term scheduling of a hydroelectric power system, *IEEE Trans. PAS* 10 (4) (1995) 1750–1754.

[2] M. Papageorgiou, Optimal multi reservoir network control by the discrete maximum principle, *Water Resources Res.* 21 (2) (1985) 1824–1830.

[3] M.R. Piekutowski, T. Litwinowicz, R.J. Frowd, Optimal short-term scheduling for a large-scale cascaded hydro system, *IEEE Trans. PAS* 9 (2) (1994) 805–811.

[4] Q. Xia, N. Xiang, S. Wang, B. Zhang, M. Huang, Optimal daily scheduling of cascaded plants using a new algorithm of non-linear minimum cost network flow concept, *IEEE Trans. PWRS* 3 (3) (1988) 929–935.

[5] S.A. Soliman, G.S. Christensen, Application of functional analysis to optimization of variable head multi reservoir power system for long term regulation, *Water Resources Res.* 22 (6) (1986) 852–858.

[6] H. Habibollahzadeh, J.A. Bubenko, Application of decomposition techniques to short term operation planning of hydro-thermal power system, *IEEE Trans. PWRS* 1 (1) (1986) 41–47.

[7] S.O. Orero, M.R. Irving, A genetic algorithm modeling framework and solution technique for short term optimal hydrothermal scheduling, *IEEE Trans. PAS* 13 (2) (1998) 501–518.

[8] Xiaohui Yuan, Yanbin Yuan, Yongchuan Zhang. A hybrid chaotic genetic algorithm for short-term hydro system scheduling. *Mathematics and Computers in Simulation.* 59 (2002) 319-327

[9] R. Naresh, J. Sharma, Hydro system scheduling using ANN approach, IEEE Trans. PAS 15 (1) (2000) 388-395.

[10] Kennedy J, Eberhart R. Particle Swarm Optimization. *In Proceedings of IEEE International Conference on Neural Networks*, 1995(4): 1942~1948

[11] M.A.Abido. Optimal power flow using particle swarm optimization. *Electrical Power and Energy Systems,* 2002,(24):563-571

[12] Kenndey J. The particle swarm: social adaptation of knowledge. *In:Proceedings 1997 IEEE International conference on evolutionary computation.* Piscataway: IEEE Service center. 1997:303-308

[13] Ioan Cristian Trelea. The particle swarm optimization algorithm: convergence analysis and parameter selection. *Information Processing Letters*, 2003,(85):317-325

[14] Shigenori Naka, Takamu Genji, Toshiki Yura, Yoshikazu Fukuyama. A Hybrid Particle Swarm Optimization for Distribution State Estimation. *IEEE Transactions on Power Systems,* 2003(18): 60~68

VSP International
Science Publishers
P.O. Box 346, 3700 AH Zeist
The Netherlands

*Lecture Series on Computer
and Computational Sciences*
Volume 1, 2004, pp. 127-130

Model Selection for Support Vector Classifiers via Direct Search Simplex Method

Gilles Cohen[1]

Medical Informatics Service,
University Hospital of Geneva,
1211 Geneva, Switzerland

Patrick Ruch[2]

Theoretical Computer Science Lab,
Swiss Federal Institute of Technology
1015 Lausanne, Switzerland

Received 31 July, 2004; accepted in revised form 28 August, 2004

Abstract: This paper addresses the problem of tuning hyperparameters in support vector machine modeling. A Direct Search Simplex (DSS) method, which seeks to evolve hyperparameter values using an empirical error estimate as steering criterion, is proposed and experimentally evaluated on real-world datasets. DSS is a robust hill climbing scheeme, a popular derivative-free optimization method, suitable for low-dimensional optimization problems for which the computation of the derivatives is impossible or difficult. This method produces satisfactory results.

Keywords: Model Selection, Support Vector Machine, Direct Search Method, Optimization, Machine Learning.

1   Introduction

Support vector machines (SVM) are a powerful machine learning method for classification problems. However, to obtain good generalization performance, a necessary condition is to choose sufficiently good model hyperparameter set (i.e regularization parameter (C) and kernel parameters) dependly of the data. The choice of SVM model parameters can have a profound affect on the resulting model's generalization performance. Most approaches use trial and error procedures to tune SVM hyperparameters while trying to minimize the training and test errors. Such an approach may not really obtain the best performance while consuming an enormous amount of time. Recently others approaches as to parameters tuning have been proposed [1]. These methods use a gradient descent search to optimize a validation error, a leave-one-out (LOO) error or an upper bound on the generalization error [4]. However, gradient descent oriented methods may require restrictive assumptions regarding, e.g., continuity or differentability. Typically the criteria, such as LOO error, is not differentiable, so these approaches based on gradient descent are not generally applicable in using cross-validation type of criterion.

---

[1] Corresponding author. E-mail: Gilles.Cohen@sim.hcuge.ch
[2] Patrick.Ruch@sim.hcuge.ch

In the present work we propose a Direct Simplex Search methodology to tune SVM hyperparameters and illustrate its effectiveness in classification tasks. The main advantages of a DSS strategy lie in the suitability for problems for which it is impossible or difficult to obtain information about the derivatives. The paper is organized as follows. In Section 2 the Direct Search Simplex algorithm is described. In Section 3 DSS for SVM model selection for classification are described. Experiments conducted to assess this approach as well as results are described in Section 4. Finally, Section 5 draws a general conclusion.

2   Direct Search Simplex Algorithm

Direct search methods belong to a class of optimization methods that do not compute derivatives. The direct search method we used is the Nelder-Mead (NM) Simplex method [5], first published in 1965. The NM method is the most popular direct search method, used in solving a lot of problems, especially in chemistry, chemical engineering, and medicine. Although there are no theoritical results on the convergence of the algorithm, it works very well on a range of practical problem. The Nelder-Mead method is conceptually simple. It performs a search in n dimensional space using heuristic ideas. Its main strength are that it requires no derivatives to be computed and that it doesnt require the objective function to be smooth. The NM method attempts to minimize a scalar-valued nonlinear function of n real variables using only function values, without any derivative information (explicit or implicit). Many of the most well known direct search methods, including the NM method, maintain at each step a nondegenerate simplex, a geometric figure defined by $n + 1$ vertices (real n-vectors). (For example, a simplex in two dimensions is a triangle and a tetrahedron forms a simplex in three dimensions.) Suppose that we are minimizing the function $f(\mathbf{x})$, where $\mathbf{x}$ denotes a real n-vector. The NM method includes four possible operations on the current simplex, each associated with a coefficient: reflection (ρ), expansion (ξ), contraction (γ), and shrinkage (σ). The result of each NM iteration is either: (1) a single new vertex the accepted point which replaces the current worst vertex (the vertex with the largest function value) in the set of vertices for the next iteration; or (2) if a shrink is performed, a set of n new points that, together with the previous best vertex, form the simplex at the next iteration. A single iteration of the NM method is defined according to the following steps.

1. **Order** : Order the $n+1$ vertices to satisfy $f(\mathbf{x}_1) \leq f(\mathbf{x}_2) \leq \cdots \leq f(\mathbf{x}_{n+1})$ using a consistent tie-breaking rule.

2. **Reflect** : Compute the *reflexion point* $\mathbf{x}_r$ from $\mathbf{x}_r = \mathbf{x} + \rho(\mathbf{x} - \mathbf{x}_{n+1})$ where $\hat{\mathbf{x}} = \sum_{i=1}^{n} \mathbf{x}_i/n$ is the centroid of the n best points. Evaluate $f_r = f(\mathbf{x}_r)$. If $f_1 \leq f_r < f_n$, accept the reflected point $\mathbf{x}_r$ and terminate the iteration.

3. **Expand** : if $f_r < f_1$ calculate the *expansion point* $\mathbf{x}_e$: $\mathbf{x}_e = \hat{\mathbf{x}} + \rho\xi(\hat{\mathbf{x}} - \mathbf{x}_{n+1})$ and evaluate $f_e = f(\mathbf{x}_e)$. If $f_2 < f_r$ accept $\mathbf{x}_e$ and stop the iteration; otherwise accept $\mathbf{x}_r$ and stop the iteration.

4. **Contract**: If $f_r \geq f_n$, perform a *contraction* between $\hat{\mathbf{x}}$ and the better of $\mathbf{x}_{n+1}$ and $\mathbf{x}_r$.

5. **Outside** If $f_n \leq f_r \leq f_{n+1}$ perform an *outside contraction*: calculate $\mathbf{x}_c = \hat{\mathbf{x}} - \gamma\rho(\hat{\mathbf{x}} - \mathbf{x}_{n+1})$ and evaluate $f_c = f(\mathbf{x}_c)$. If $f_c \leq f_r$, accept $\mathbf{x}_c$ and terminate the iteration; otherwise go to last step (perform a schrink).

6. **Inside** If $f_r \geq f_{n+1}$ perform an *inside contraction*: calculate $\mathbf{x}_{cc} = \hat{\mathbf{x}} - \gamma(\hat{\mathbf{x}} - \mathbf{x}_{n+1})$ and evaluate $f_{cc} = f(\mathbf{x}_{cc})$. If $f_{cc} < f_{n+1}$, accept $\mathbf{x}_{cc}$ and terminate the iteration; otherwise go to next step (perform a schrink).

7. **Perform a shrink step** Evaluate f at the n points $v_i = \mathbf{x}_1 + \sigma(\mathbf{x}_i - \mathbf{x}_1)$, $i = 1, \ldots, n+1$. The vertices of the simplex at the next iteration consist of $\mathbf{v}_1, \mathbf{v}_2 \ldots, \mathbf{v}_{n+1}$

Since Direct Search method cannot be guaranteed to converge to the global optimum, standard approaches to allievate this problem are to apply the Nelder-Mead algorithm with multiple starting points generated randomly.

3   SVM Model Selection via DSS method

To apply DSS to SVM classification one need to define the model and model quality function to be optimized (model selection criteria).

Support vector machines [6, 2] (SVM) are state of the art learning machines based on the *Structural Risk Minimization principle* (SRM) from statistical learning theory. The SRM principle seeks to minimize an upper bound of the generalization error rather than minimizing the training error (Empirical Risk Minimization (ERM)). This approach results in better generalization than conventionnal techniques generally based on the ERM principle.

For a separable classification task, the idea is to map each data points of the training set into a high dimensional space by some function ϕ, and seeks for a canonical a separating hyperplane $(\mathbf{w}, b)$, with $\mathbf{w}$ the weight vector and b the bias, in this space which maximises the *margin* or distance between the hyperplane and the closest data points belonging to the different classes. When nonlinear decision boundaries is not needed ϕ is an identity function, otherwise ϕ is performed by a non linear function $k(.,.)$, also called a *kernel*, which defines a dot product in the feature space. We can then substitute the dot product $\langle \phi(\mathbf{x}), \phi(\mathbf{x}_i) \rangle$ in feature space with the kernel $k(\mathbf{x}, \mathbf{x}_i)$. Conditions for a function to be a kernel are expressed in a theorem by Mercer [3]. The optimal separating hyperplane can be represented based on kernel funtion:

$$f(\mathbf{x}) = sign\left(\sum_i^n \alpha_i k(\mathbf{x}, \mathbf{x}_i) + b \right) \tag{1}$$

For a separable classification task, a such optimal hyperplane exists but very often, the data points will be almost linearly separable in the sense that only a few of the members of the data points cause it to be non linearly separable. Such data points can be accommodated into the theory with the introduction of slack variables that allow particular vectors to be misclassified. The hyperplane margin is then relaxed by penalising the training points misclassified by the system. Hence, the optimum hyperplane equation can be define as $y_i(\langle \mathbf{w}, \phi(\mathbf{x}_i) \rangle + b) \geq 1 - \xi_i$, $\xi_i \geq 0 \; \forall i$ and the following equation is optimized in order to obtain the maximal margin hyperplane

$$\frac{1}{2}\|\mathbf{w}\|^2 + C\sum_{i=1}^n \xi_i \tag{2}$$

where ξ_i is a positive slack variable that measures the degree of violation of the constraint. The penalty C is a regularisation parameter that controls the trade-off between maximizing the margin and minimizing the training error. This approach is called *soft margin* [3]. To obtain a good performance, some parameters in SVMs have to be selected carefully. These parameters include (1) the regularization parameters C, which determine the tradeoff between minimizing model complexity and the training error and (2) parameter of the kernel function, encoded into a vector $\boldsymbol{\theta} = (\theta_1, \ldots, \theta_n)$, that implicitly defines the non linear mapping to some high-dimensional feature space. These "higher level" parameters are usually refered as metaparameters or hyperparameters. We used the radial-basis function kernel $k(\mathbf{x}, \mathbf{z}) = e^{-\gamma\|\mathbf{x}-\mathbf{z}\|^2}$.

The model selection problem is to select from a candidate set of models the best one so the generalization error is minimized over all possible examples drawn from an unknown distribution

$P(\mathbf{x}, y)$. As the data distributions P in real problems are not known in advance, generalization error is not computable and one needs some reliable estimates of the generalization performance.

To estimate the quality of the model, we use a popular technique Cross-validation (CV). In k- fold cross-validation the original training set S is randomly partitioned into k non-overlapping subsets S_j of approximately equal size. The learning machine is trained on the union of $(k-1)$ subsets; the remaining k-th subset is used as a test set and measures the associated classification performance. This procedure is cycled over all possible k test sets, and the average test error gives an estimate of the expected generalization error.

4   Results

The experimental goal was to assess a DSS optimization method for tuning SVM hyperparameters. To train our SVM classifiers we use a radial basis kernel. Thus the corresponding hyperparameters set to tune is $\boldsymbol{\theta} : (\sigma, C)$. We use four different real-world dataset. To evaluate our model we use a 10-fold cross-validation.

Table 1: Performance of SVMs for optimum parameters settings using an RBF Gaussian kernel (γ, C) found via DSS method.

Dataset	$log_2(C)$	$log_2(\gamma)$	accuracy %
image	8.1	-4.2	96.7
splice	12.4	-6.8	88.2
banana	10.5	-1.6	94
waveform	1.3	-5.2	92.8

5   Conclusion

We have presented an algorithm based on a DSS method that can reliably find very good hyperparameter settings for SVMs with RBF kernels in a fully automated way. We selected DSS method because of its robustness, simplicity and ease of implementation. To improve our method we plan to merge DSS with Genetic Algorithm by first performing a coarse search of the global minimum by a genetic algorithm and then refining the solution by a DSS approach.

References

[1] O. Chapelle, V. Vapnik, O. Bousquet, and S. Mukherjee. Choosing multiple parameters for support vector machines. *Machine Learning*, 46(1):131–159, 2002.

[2] C. Cortes and V. Vapnik. Support vector networks. *Machine Learning*, 20(3):273–297, September 1995.

[3] N. Cristianini and Taylor J.S. *An Introduction to Support Vector Machines*. Cambridge University Press, 2000.

[4] K. Duan, S.S. Keerthi, and A.N. Poo. Evaluation of simple performance measures for tuning svm hyperparameters. In *Technical Report CD-01-11, Singapore*, 2001.

[5] J.A.Nelder and R.Mead. A simplex method for function minimization. *ComputerJournal*, 7:308–313, 1965.

[6] V. Vapnik. *Statistical Learning Theory*. Wiley, 1998.

VSP International
Science Publishers
P.O. Box 346, 3700 AH Zeist
The Netherlands

*Lecture Series on Computer
and Computational Sciences*
Volume 1, 2004, pp. 131-135

Molecular Dynamics of Cholesterol in Biomembranes

Z. Cournia, G.M. Ullmann[1] and J.C. Smith[2]

Computational Molecular Biophysics,
Interdisciplinary Center for Scientific Computing (IWR),
Im Neuenheimer Feld 368,
University of Heidelberg,
69120 Heidelberg, Germany

Received 31 August, 2004; accepted in revised form 10 September, 2004

1 Introduction

Cholesterol is a major component of the mammalian plasma cell membranes and represents about 50% of the membrane lipids. It has been proposed that evolution has selected cholesterol because it optimizes the physical properties of membranes for biological function [1]. Cholesterol influences strongly the mechanical and thermodynamical properties of the membrane [2, 3, 4, 5] and thus regulates membrane fluidity, permeability and adjusts the lateral mobility of membrane proteins. Using Molecular Dynamics (MD) simulations, it is possible to interpret experimental results of complex membrane systems and gain insight on their interactions at the microscopic level. The field of lipid bilayer simulations is growing rapidly and the level of complexity is increasing with explicit inclusion of membrane proteins [6, 7] and cholesterol [8, 9, 10] in the simulated systems.

The reliability of a molecular mechanics calculation is dependent on both the functional form of the force field and on the numerical values of the parameters implemented in the force field itself. Thus, the first necessary step for a reliable MD simulation is the parametrization procedure. Most "all-atom" empirical force fields used in common MD packages (such as CHARMM [11]), are equipped with parameter sets for modelling and combining the basic building blocks of biomolecules, but often not for exotic molecules such as cholesterol.

In this talk, the development of a force field for cholesterol [12] as well as structural and dynamical aspects of cholesterol in membranes derived from MD simulations are being presented. The results are compared with previous experimental and theoretical studies.

2 Computational Methods

The parameters for cholesterol were obtained using an automated frequency matching method (AFMM) [13, 14]. The method involves careful choice of an initial parameter set, which is then fitted to match vibrational eigenvector and eigenvalue sets derived from quantum chemical calculations. This method is based on a harmonic approximation of the molecular potential energy surface and is thus well suited for modelling physical properties of rigid molecules, as in the case

---

[1] Present Address: Lehrstuhl für Biopolym ere, Strukturbiologie/Bioinformatik Universitätsstr.30, Gebaude B14, 95447 Bayreuth, Germany
[2] Corresponding author. E-mail: biocomputing@iwr.uni-heidelberg.de

of cholesterol. Partial atomic charges were calculated using NWchem 4.1 at the 6-31G(d) basis set level and the CHELPG analysis.

To construct the cholesterol-DPPC system we used coordinates of DPPC (1,2-dipalmitoylphosp hatidylcholine) molecules determined by Sundaralingam[15]. Coordinates for the cholesterol molecule were taken from the crystal structure by Shieh et al.[16]. The MD simulation was performed on a system of 120 DPPC, 80 cholesterol molecules and 400 TIP3P [17] waters. We used the CHARMM 22 force field for DPPC and our derived force field for cholesterol[12]. The simulation was performed at constant pressure (1 atm) and temperature (Nose-Hoover thermostat at 309 K) with periodic boundary conditions for 10 ns.

3   Results

3.1   Parametrization of cholesterol

An initial parameter set was used for minimization and calculation of normal modes (eigenvalues and eigenvectors) with CHARMM. The normal modes obtained were then directly compared with the normal modes calculated from the quantum chemistry methods, used as reference values, employing AFMM [13]. Using an iterative procedure, the parameters were refined to reproduce the reference set normal modes. Parameters for cholesterol were developed using a three step procedure. Initially the AFMM method was used to obtain a first set of parameters. In the second step parameters for the hydroxyl group region were further refined using single point QM energy calculations performed on hexanol. We chose hexanol to model the H-O-C-C rotational energy barrier based on the resemblance of the first cholesterol steroid ring and this molecule. Finally, all remaining parameters were refined using AFMM.

3.2   Structural Analysis of the System

3.2.1   Volume and Surface Area per Lipid

There is no obvious solution of how to calculate the "Surface Area per Lipid" in a mixture of two components such as the DPPC-cholesterol system. A simple approach to the problem is given as follows: The area occupied by a DPPC molecule in the bilayer can be written in terms of volume and thickness as:

$$A_{DPPC}(x) = \frac{2V_{DPPC}(x)}{h(x)} \qquad (1)$$

where V_{DPPC} is the volume of the lipid and h(x) is the average thickness of the membrane which corresponds to the average distance of two phosphorus atoms in opposite layers.

The volume that a DPPC molecule occupies in the membrane could be calculated as follows:

$$V_L = \frac{V_T - N_W \cdot V_W - N_C \cdot V_C}{N_L} \qquad (2)$$

where V_T is the total volume of the system in each frame of the trajectory, $N_L = 120$ the total number of the lipids, $N_W = 1600$ the number of waters, $V_W = 29.9$ Å^3 the volume of one water molecule and $N_C = 80$ the number of cholesterol molecules. The volume of a cholesterol molecule, $V_C = 629.1$ Å^3 was calculated from the crystal structure, namely from the size of the uni t cell, which is 5032.6 Å^3 and contains 8 cholesterol molecules. In this approach we consider V_W and V_C fixed as cholesterol is a fairly rigid body .

The mean volume is 1176.4 ± 5.2 Å^3 (all deviations refer to standard deviations) and the mean surface area per lipid is 47.0 ± 0.2 Å^2. The bilayer thickness was set to 50.0 Å by calculating the distance of the two peaks of the electron density profile.

3.2.2   Deuterium Order Parameters

The most popular quantity to characterize the order of the hydrocarbon chains in lipid bilayers is the deuterium NMR order parameters. Such an order parameter may be defined for every CH_2 group in the chains as:

$$S_{CD}^i = \frac{1}{2}(3\langle cos^2\theta_{CD}^i \rangle - 1) \tag{3}$$

where θ_{CD}^i is the angle between a CD-bond (in the experiment) or a CH-bond (in the simulation) of the ith carbon on the acyl chain and the membrane normal (z-axis). The brackets indicate averaging over the two bonds in each CH_2 group, all the lipids and time.

Our results show good overall agreement with those obtained in the studies of Urbina et al. [18] and Faure et al. [19]. Our order parameter profile is also consistent with all simulation results obtained for similar conditions.

3.2.3   Electron Density Profiles

The first atomic scale picture of the average structure of the lipid bilayer:water interface can be produced as a measurement of the density distributions of different types of atoms along the bilayer normal (z-axis) by neutron and X-ray diffraction studies. The corresponding electron density profile for the bilayer has been provided by our MD simulation and is in good agreement with experimental results obtained by various experimental groups.

The electron density profile is calculated every 1 ps and averaged out in the 1ns trajectory by dividing the simulation cells into 0.5 Å slabs and determining the time-averaged number of electrons in each slab. The peaks show the electron-rich phosphate region of the headgroup. Defining the bilayer thickness as the distance between the peaks in the total electron density, we obtain 50.0 Å which is slightly bigger than the values determined by X-ray diffraction analysis and by Smondyrev et al. [20] and Hofsäss et al. [21].

3.2.4   Cholesterol Tilt Angle

We also measured the distribution of the tilt of cholesterol in the lipid bilayer. The tilt is defined as the angle between the bilayer normal and the vector connecting carbon atoms C_3 and C_{17} in the sterol ring system. The average cholesterol tilt angle in the DPPC membrane is 8.1°. This value is very close to the one measured by Smondyrev and Berkowitz [8], 10.6°, for a DMPC bilayer. However it is much lower than the value measured by the same group for 11% cholesterol in DMPC (22.2°). The study of Murari et al. [22] for 1:1 DPPC:Chol mixture at 24° obtain s an average cholesterol tilt of 16° from quadrupolar splittings.

4   Conclusions

We have presented the development of a force field for cholesterol for the CHARMM simulation package. A 10-ns MD simulation of a hydrated cholesterol-DPPC system was performed to illustrate the effect of cholesterol on the lipid bilayer. The results show an ordering of the hydrocarbon chains with respect to their ordering in the pure bilayer and a decrease of the average area per lipid. The orientation of the cholesterol hydroxyl in the bilayer is found to be in the polar headgroup region and the average tilt angle of the cholesterol's steroid ring to be 8° with respect to the membrane normal. Although structural aspects of cholesterol in membranes have been investigated so far, our knowledge for cholesterol dynamics in biomembranes is quite limited. A combination of quasielastic neutron scattering experiments and MD simulations should help us investigate the dynamical effects of sterols in membranes and examine biologically-relevant structure-function relationships from a dynamical point of view.

Acknowledgment

Z.C. would like to thank Dr.A.C.Vaiana for fruitful discussions and gratefully acknowledges funding from BMBF project 03SHE2HD. QM and MD calculations have been performed on HELICS, IWR - Universität Heidelberg (HBFG funds, hww cooperation).

References

[1] K. Bloch. *Cholesterol, evolution of structure and function.*, pages 1–24. in Biochemistry of Lipids and Membranes, Eds. J. E. Vance and D. E. Vance, Benjamin/Cummins Pub. Co. Inc., New York, 1985.

[2] L. Finegold (eds.). . CRC Press, Boca Barton, FL, 1993.

[3] M.R. Vist and J.H. Davis. Phase equilibria od cholesterol/dipalmitoylphosphatidylcholine mixtures: ^{2}H nuclear magnetic resonance and differential scanning calorimetry. *Biochemistry*, 29:451–464, 1990.

[4] A. Kusumi, M. Tsuda, T. Akino, O. Ohnishi, and Y. Terayama. Protein-phospholipid-cholesterol interaction in the photolysis of invertebrate rhodopsin. *Biochemistry*, 22:1165–1170, 1983.

[5] T.H. Haines. Do sterols reduce proton and sodium leaks through lipid bilayers? *Prog. Lipid Res.*, 40:299–324, 2001.

[6] D. Mihailescu and J.C. Smith. Atomic detail peptide-membrane interactions: Molecular Dynamics of gramicidine S in a DMPC bilayer. *Biophys. J.*, 79:1718, 2000.

[7] L. Forrest, A. Kukol, I. Arkin, A. Tielman, and M. Sansom. Exploring the models of Influenza M2 channel - MD Simulations in a phospholipid bilayer. *Biophys. J.*, 78:79, 2000.

[8] A. Smondyrev and M. L. Berkowitz. MD Simulation of the structure of DMPC bilayers with Cholesterol, Ergosterol, and Lanosterol. *Biophys. J.*, 80:1649, 2001.

[9] S. W. Chiu, E. Jacobsson, and H. L. Scott. Combined MC and MD simulation of hydrated lipid-cholesterol lipid bilayers at low Cholesterol concentration. *Biophys. J.*, 80:1104, 2001.

[10] K. Tu, M. Klein, and D. Tobias. Constant-Pressure MD investigation of Cholesterol effects in a DPPC bilayer. *Biophys. J.*, 75:2147, 1998.

[11] B. R. Brooks, R. Bruccoleri, B. D. Olafson, D. J. States, S. Swaminathan, and M. Karplus. CHARMM: A Program for Macromolecular Energy, Minimization and Dynamics Calculations. *J. Comp. Biol.*, 4:187, 1983.

[12] Zoe Cournia, Andrea Vaiana, Jeremy C. Smith, and G. Matthias Ullmann. Derivation of a molecular mechanics force field for cholesterol. *Pure Appl. Chem.*, 76(1):189, 2004.

[13] A.C. Vaiana, A. Schulz, J. Worfrum, M.Sauer, and J.C. Smith. Molecular mechanics force field parametrization of the fluorescent probe rhodamine 6G using automated frequency matching. *J. Comp. Chem.*, 24:632, 2002.

[14] A.C. Vaiana, Z. Cournia, I.B. Costescu, and J.C. Smith. AFMM: A Molecular Mechanics Force Field Parametrization Program. *Computer Physics Communications*, submitted, 2004.

[15] M. Sundaralingam. Molecular structures and conformations of the phospholipids and shingomyelins. *Ann. N. Y. Acad. Sci.*, 195:324, 1972.

[16] H. Shieh, L.G. Hoard, and C.E. Nordman. The structure of cholesterol. *Acta Cryst.*, B37:1538, 1981.

[17] W.L. Jorgensen, J. Chandrasekhar, J.D. Madura, R.W. Impey, and M.L. Klein. Comparison of simple potential functions for simulating liquid water. *J. Chem. Phys.*, 79:926, 1983.

[18] J.A. Urbina, S. Pekerar, H. Le, J. Patterson, B. Montez, and E. Oldfield. Molecular order and dynamics of phosphatidylcholine bilayer membranes in the presence of cholesterol, ergosterol and lanosterol: a comparative study using ^{2}H-, ^{13}C- and 31P-NMR spectroscopy. *Biochim. Biophys. Acta*, 1238:163, 1995.

[19] C. Faure, J. Transchant, and E.J. Dufourc. Comparative effects of cholesterol and cholesterol sulfate on hydration and ordering of DMPC membranes. *Biophys. J.*, 70:1380, 1996.

[20] A.M. Smondyrev and M.L. Berkowitz. Structure of DPPC/Cholesterol bilayer at low and high cholesterol concentrations: molecular dynamics simulation. *Biophys. J.*, 77:2075, 1999.

[21] C.Hofsaess, E. Lindahl, and O. Edholm. Molecular dynamics simulations of phospholipid bilayers with cholesterol. *Biophys. J.*, 84:2192, 2003.

[22] R. Murari, M. Murari, and W.J. Baumann. Sterol orientations in phosphatidylcholine liposomes as determined by deuterium NMR. *Biochemistry*, 25:1062, 1986.

VSP International
Science Publishers
P.O. Box 346, 3700 AH Zeist
The Netherlands

*Lecture Series on Computer
and Computational Sciences*
Volume 1, 2004, pp. 136-139

Advanced System-Approach Based Methods
for Modeling Biomedical Systems

L. Dedík[1] and M. Ďurišová[2]

[1]Department of Automation and Measurement,
Faculty of Mechanical Engineering,
Slovak University of Technology,
SK-812 31 Bratislava, Slovak Republic

[2]Department of Pharmacokinetics,
Institute of Experimental Pharmacokinetics,
Slovak Academy of Sciences,
SK-841 04 Bratislava, Slovak Republic

Received 5 August, 2004; accepted in revised form 22 August, 2004

Abstract: In the field of bio-medicine, tools of the system approach are commonly employed to develop derivations of analytical solutions of the mathematical models which are most frequently used in the given field, *i.e.* the deterministic linear compartment models, for commonly utilized inputs. In practice, however, usually only analytical solutions of these models obtained for single inputs in the form of the Dirac delta function are fitted to measured time profiles, using non-linear regression methods. In contract to this, in our study we present the utilization of tools of the system approach for building mathematical models that provide either phenomenological or mechanism-based mathematical descriptions of dynamical processes under study. The main advantage of the use of the modeling methods based on the system approach over the classic modeling methods in the field of bio-medicine is the fact that the former methods allow the development mathematical models of various dynamical processes represented by dynamical systems in a methodically, conceptually, and computationally unified way.

Keywords: System approach, Dynamical system, Differential equation, Time-delay

Mathematics Subject Classification: 32G34, 54H20

PACS: 32G34, 54H20

1. Structural versus Non-structural Modeling

The study exemplifies two classes of modeling methods based on the system approach, the first one proposed for selecting mathematical models that provide phenomenological descriptions of linear deterministic dynamical systems (thereafter dynamical systems) and the second one for building mechanism-based mathematical models of the given systems. The methods presented in this study are implemented in the integrated software package named CTDB (Clinical Trials DataBase), a version of which is available from http://www.uef.sav.sk/advanced.htm.

The method for the determination of mathematical models which provide phenomenological descriptions of dynamical processes under study consists from the following basic steps: *1)* The representation of a dynamic process by a dynamical system and the definition of the dynamical system by its transfer function, utilizing the Laplace transform of the cause which starts the process (the

---

[2] Corresponding author: E-mail: exfamadu@savba.sk

system input) and of the measured outcome of the process (the system output); *2)* The calculation of the frequency response of the defined dynamical system, utilizing the method of the Fourier transformation; *3)* The presentation of the frequency response of the dynamical system in the form of the Nyquist diagram; *4)* The selection of an optimal structure of a mathematical model of the dynamical system in the form of a linear differential equation, without or with time-delays, and the determination of point-estimates of model parameters, utilizing the non-iterative Levy's method in the complex domain; *5)* The determination of the time-domain output of the selected model for the actual input of the dynamical system utilizing the Euler method, the refinement of the selected model by the iterative Monte-Carlo and Gauss-Newton method in the time domain, and the determination of interval estimates of model parameters [1-10]. Since the modeling technology outlined above starts with a non-iterative procedure it does not require initial estimates of model parameters, which markedly simplifies and speeds up the modeling procedure. Moreover, this technology enables to use identical model structures for the determination of mathematical models of various processes, *e.g.* the availability of a drug in the blood circulation after an extravascular input into the body [2-5], the behavior of a drug in the body [6], physiological processes [7], the dissolution of a drug under *in vitro* conditions [8], the formation of a metabolite a parent drug in the body [9], or the effect of a drug in the body [10], *e.t.c.*

The second class of the modeling methods presented in this study enable to determine mechanism-based mathematical models of the dynamical processes under study. These models can be determined either by combined modeling in the frequency and time domain [4], or by the method which we can be called the computer controlled sequential simulation in the time domain [11,12]. The latter method consists form the sequential identification of those individual subsystems which play predominant role on the performance of the whole complex dynamical system under study over sequential time intervals. The utilization of this method is illustrated by the following example: Based on data obtained in the standard intravenous glucose tolerance test, which is commonly utilized in the diabetes research [11], a circulatory, mechanism-based mathematical model of the human glucose-insulin control system was developed. This model is capable of quantifying: *1)* the uptake of glucose by body cells after the glucose administration, which is denoted as Effect 1 in Figures 1 and 2; *2)* the time-delayed cessation of the glucose output from liver, leading to the decrease of glucose concentrations in plasma, which is denoted as Effect 2 in Figures 1 and 2.

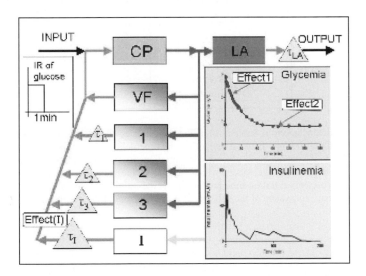

Figure 1. Scheme of the circulatory, mechanism-based mathematical model of the human glucose–insulin control system. *IR* - the glucose infusion. *CP* - the cardiopulmonary subsystem. *VF* - the body subsystem which rapidly returns the exogenous glucose from the arterial to the venous side over the time period up to time τ_1. The subsystems in the remaining backward paths of the model transit the exogenous glucose more slowly than does the subsystem *VF* and thus they return only the basal glucose from the arterial to venous side over the period up to time τ_1. *LA* - the subsystem representing the passage of glucose through the arterial and venous blood in the left arm in which the measurement of the glucose profile is performed. τ_{LA} – time delay of the subsystem *LA*. $\tau_1 < \tau_2 < \tau_3$ -

the time delays of the subsystems 1, 2, and 3 respectively in the backward model paths. *I* - the subsystem that represents the cessation of the glucose output from the liver, as a consequence of the increased concentration of insulin in plasma. τ_I – time delay of the subsystem *I*. Effect(I) is the output of the subsystem *I*. The response of the model to the glucose load is shown in the upper right window.

The presented circulatory, mechanism-based mathematical model of the human glucose–insulin control system is a new alternative to the classic minimal model (MINMOD) [11] and its recent variants which are commonly employed in the diabetes research. In contrast to the latter models which are non-linear, the presented model has the advantage of being a linear model. Moreover, the presented model has the advantage of being capable of quantifying the predominant effects that develop in the body as a consequence of the glucose load, *i.e.* Effect 1 and Effect 2. Finally, the presented model explicitly takes into account the blood circulation in the body, and consequently it is physiologically more appropriate than MINMOD and its recent variants. The developed model exhibits a structure and yields parameters which diabetologists and physiologists might consider acceptable. However, as any model of a biological system, the developed circulatory model requires further analyses in order to establish completely all its characteristics.

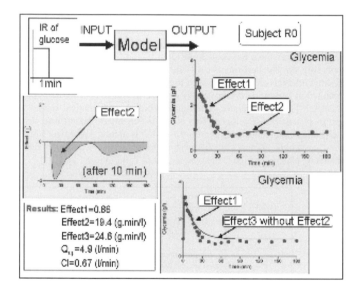

Figure 2. Modeling results of Subject R0. Effect 1 quantifies the uptake of glucose by body cells after the glucose administration. Effect 2 quantifies the time-delayed cessation of the glucose output from liver, leading to the decrease of   glucose concentrations in plasma. Effect 3 quantifies the difference between the response of the model which does not take into the consideration Effect 2 and the measured glucose concentration-time profile. Q_{cp} is the estimate of the plasma blood flow through the cardio-pulmonary subsystem of the subject. Cl is the estimate of the plasma clearance of glucose of the subject.

Acknowledgments

The author wishes to thank the anonymous referees for their careful reading of the manuscript and their fruitful comments and suggestions. This work was partially supported by Grant 1/0521/03 from the Slovak Grant Agency.

References

[1] L. Ljung, *System Identification – Theory for the User*. 2nd ed. PTR Prentice Hall, Upper Saddle River, 1999.

[2] L. Dedík and M. Ďurišová, *System Approach in Technical, Environmental and Bio-medical Studies*. Publishing House of Slovak University of Technology, Bratislava, 1999.

[3] L. Dedík and M. Ďurišová, Frequency response method in pharmacokinetics, *Journal of Pharmacokinetics and Biopharmaceutics* **22** 293-307(1994).

[4] M. Ďurišová, L. Dedík and M. Balan, Building a structured model of a complex pharmacokinetic system with time delays, *Bulletin of Mathematical Biology* **57** 787-808(1995).

[5] L. Dedík and M. Ďurišová, CXT-MAIN: a software package for determination of the analytical form of the pharmacokinetic system weighting function, *Computer Methods Programs in Biomedicine* **51** 183-192(1996).

[6] L. Dedík and M. Ďurišová, CXT - A programme for analysis of linear dynamic systems in the frequency domain, *International Journal of Bio-Medical Computing* **39** 231-241(1995).

[7] L. Dedík and M. Ďurišová, Combination of Modeling in Frequency and Time Domain in Surrogate Endpoint Evaluations, *Proceedings of Third Symposium on Mathematical Modeling MATHMOD IMACS 2000* (Eds.: I. Troch and F. Breitenecker) *University of Technology Vienna* 563-566(2000).

[8] L. Dedík and M. Ďurišová, System-approach methods for modeling and testing similarity of in vitro dissolutions of drug dosage formulations, *Computer Methods Programs in Biomedicine* **69** 49-55(2002).

[9] L. Dedík and M. Ďurišová, System approach to modeling metabolite formation from parent drug: A working example with methotrexate, *Methods and Findings in Experimental and Clinical Pharmacology* **24** 481-486(2002).

[10] L. Dedík, M. Ďurišová, V. Svrček, R. Vojtko, V. Kristová and M. Kriška, Computer-based methods for measurement, recording, and modeling vessel responses in vitro: A pilot study with noradrenaline, *Methods and Findings in Experimental and Clinical Pharmacology* **25** 441-445(2003).

[11] L. Dedík, M. Ďurišová and A. Penesová, Circulatory model for glucose – insulin interaction after intravenous administration of glucose to healthy volunteers, *Klinická Farmakologie a Farmacie* **17** 132-138(2003).

[12] L. Dedík, M. Ďurišová and A. Penesová, Model for evaluation of data for oral glucose tolerance test, *Klinická Farmakologie a Farmacie* **17** 139-144(2003).

VSP International
Science Publishers
P.O. Box 346, 3700 AH Zeist
The Netherlands

*Lecture Series on Computer
and Computational Sciences*
Volume 1, 2004, pp. 140-144

Non-linear Analysis of Sheet Cover (Umbrella) of Reinforced Concrete of 40 m diameter

J. J. del Coz Díaz[1], L. Peñalver Lamarca[1], P. J. García Nieto[2], J. L. Suárez Sierra[1] , F. J. Suárez Domínguez[1] and J. A.Vilán Vilán [3]

(1) Department of Construction Engineering
High Polytechnic School,
University of Oviedo,
Edificio Departamental N° 7 – 33204 – Gijón (Spain)
(2) Department of Mathematics
Faculty of Sciences,
University of Oviedo,
C/ Calvo Sotelo s/n – 33007 – Oviedo (Spain)
(3) Department of Mechanical Engineering
High University School of Industrial Engineering
University of Vigo
Campus de Lagoas-Marcosende - 36200 – Vigo (Spain)

Received 2 August, 2004; accepted in revised form 18 August, 2004

Abstract: The aim of this work is to determine the distribution of strains and stresses throughout a sheet cover known as "umbrella" due to the dead and alive loads taking into account large displacements by the finite element method (FEM) [2-3, 5, 7, 10-11] . The non-linearity is due to large displacements that governs the phenomenon. Finally we compare the analytical calculations with the FEM results and expose the conclusions reached in this study.

Keywords: Finite element modeling, numerical methods, elasticity of concrete, large displacements

PACS: 74S05, 74B10, 74C20

1. Introduction

Following the study carried out by Ildefonso Sánchez del Río y Pisón, who designed one of the most original Civil Engineering constructions, we have analysed an "umbrella" [8] (see figure 1) made of concrete reinforced with steel of 40 m diameter. The problem was originally solved by means of approximate analytical calculations [8] and we have solved it by means of the finite elements method (FEM) [2-3, 5, 7, 10-11] in order to verify the accuracy.

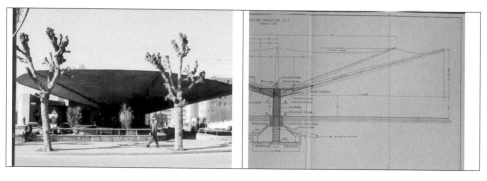

Figure 1: Umbrella (left) and drawing of a section (right).

2. Geometry and finite element modeling

The ribs and cloths are made of concrete reinforced with steel [4]. The cloth is very tight and made of a thin shell (35 millimetres of thickness). It works like "catenaries" 15 meters long at the extreme with a "sagitta" around 0.30 meters.

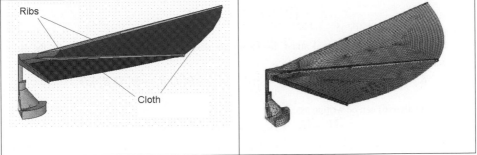

Figure 2: A geometrical quarter of the umbrella (left) and its finite element model (right).

Due to the symmetry only a quarter of the umbrella has been modeled (see figure 2). We have employed a solid tetrahedrical structural element [1] with twenty nodes in the ribs, base and central column, and a shell element [1] with four nodes and six degrees of freedom per node in the cloths. Both membrane and bending capabilities [2, 5] with Allman's functions and large displacements formulation [10-11] have been considered in the shell elements to achieve convergence [6, 9].

3. Mathematical model of the elasticity problem with large displacements

In many problems discussed it has been implicitly assumed that both displacements and strains developed in the structure are small. In practical terms this means that geometry of the elements remains basically unchanged during the loading process and that first-order, infinitesimal linear strain approximations can be used. In case of the umbrella, the thin cloths show an important membrane phenomenon, so that non-linear effects due to large displacements are taken into account [2, 5, 11].

Whether the displacements (or strains) are large or small, equilibrium conditions between internal and external 'forces' have to be satisfied. Thus, using the virtual work principle, we can obtain the following equilibrium equations [3, 11]:

$$\Psi(\vec{a}) = \int_V \overline{B}^T \vec{\sigma} \, dV - \vec{f} = 0 \tag{1}$$

where Ψ once again represents the sum of external and internal generalized forces, and in which $\overline{B}$ is defined from the strain definition as $\vec{\varepsilon}$ [7-8, 10] :

$$d\vec{\varepsilon} = \overline{B} \, d\vec{a} \tag{2}$$

The bar suffix in the matrix $\overline{B}$ has been added for, if displacements are large, the strain depend non-linearly on displacement, and the matrix $\overline{B}$ is now dependent on $\vec{a}$ in the form [11]:

$$\overline{B} = B_0 + B_L(\vec{a}) \tag{3}$$

in which B_0 is the same matrix as in linear infinitesimal strain analysis and only B_L depends on the displacement. In general, B_L will be found to be a *linear function* of such displacements. Clearly the solution of equation (1) will have to be approached interatively. In our case, the Newton-Raphson method [9] is to be adopted to find the relation between $d\vec{a}$ and $d\Psi$. Thus taking appropriate variations of equation (1) with respect to $d\vec{a}$, we have:

$$d\Psi = \int_V d\overline{B}^T \vec{\sigma} \, dV + \int_V \overline{B}^T d\vec{\sigma} \, dV = K_T \, d\vec{a} \tag{4}$$

and using $d\vec{\sigma} = D \, d\vec{\varepsilon}$ and equation (2) we obtain:

$$d\vec{\sigma} = D \, d\vec{\varepsilon} = D\overline{B} \, d\vec{a} \tag{5}$$

and taking into account the equation (3), it is verified that $d\overline{B} = dB_L$. Therefore,

$$d\Psi = \int_V dB_L^T \vec{\sigma}\, dV + \overline{K}\, d\bar{a} = K_\sigma\, d\bar{a} + \overline{K}\, d\bar{a} \tag{6}$$

where

$$\overline{K} = \int_V \overline{B}^T D \overline{B}\, dV = K_0 + K_L \tag{7}$$

in which K_0 represents the usual small displacements stiffness matrix and K_L is due to the large displacement, and are given by [2, 11]:

$$K_0 = \int_V B_0^T D B_0\, dV; \qquad K_L = \int_V \left(B_0^T D B_L + B_L^T D B_L + B_L^T D B_0 \right) dV \tag{8}$$

Finally, equation (6) can be written as:

$$d\Psi = \left(K_0 + K_\sigma + K_L \right) d\bar{a} = K_T\, d\bar{a} \tag{9}$$

with K_T being the total, *tangential stiffness*, matrix [11]. Newton-type iteration [5, 9] can once more be applied precisely to solve the problem.

4. Analysis of the results

In a first load step we have solved the problem assuming that the cloths are perfectly flat and that the ribs are vertically supported because the small curvature in the clothes causes serious geometrical distortions in the mesh. In a second load step we have updated the geometry of the nodes and eliminated the boundary conditions. As can be observed in figure 3 (left and right) the total displacements in the umbrella are about 0.4877 m in the first load step and 0.0048 m (relative displacement) in the second one.

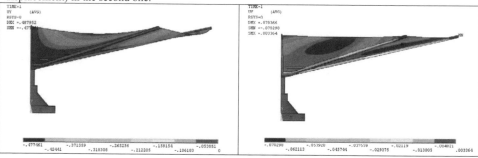

Figure 3: Total displacements in a first load step (left) and a second load step (right).

In figure 4, the von Mises stress [3] in the ribs is shown as well as the normal stress [3] in a section near the central column. In figure 5, we show the average stress in the cloths, principal stresses S1, S2 and S3, and von Mises stress [2].

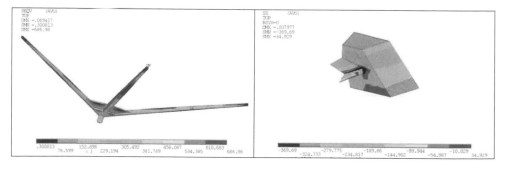

Figure 4: von Mises stress in the ribs (left) and normal stress in a section near the central column (right).

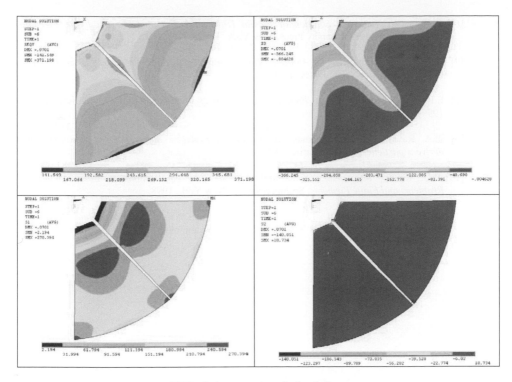

Figure 3: Average stress in the cloths.

5. Summary and conclusions

To summarize, the geometrical modeling of surface elements with very small curvature shows serious distortions and different meshing procedures must be developed for such cases. The method of the finite elements [2-3, 10-11] has been shown as a suitable tool in the modeling and nonlinear analysis of thin elements, although it has not been possible to eliminate the local bending completely in the cloths in spite of using the Allman's functions in the shell elements [1]. Finally, the comparison between the analytical calculations and the numerical procedures (FEM) [5, 9] has shown a very good agreement.

Acknowledgments

The authors express deep gratitude to Construction Department and Department of Mathematics at Oviedo University and Department of Mechanical Engineering at Vigo University for useful assistance. Helpful comments and discussion are gratefully acknowledged. We thanks to Swanson Analysis Inc. for the use of ANSYS University Intermediate program.

References

[1] ANSYS User's Manual: *Procedures, Commands and Elements* Vols. I, II and III. Swanson Analysis Systems, 2004.

[2] K. Bathe, *Finite Element Procedures*, Englewood Cliffs, Prentice-Hall, New York, 1996.

[3] T. Chandrupatla and A. Belegundu, *Introduction to Finite Elements in Engineering*, Englewood Cliffs, Prentice-Hall, New Jersey, 1991.

[4] W.F. Chen, *Plasticity in reinforced concrete*, McGraw-Hill Book Company, New York, 1982.

[5] P.G. Ciarlet, *The Finite Element Method for Elliptic Problems*, North-Holland Publishing Company, Amsterdam, 1978.

[6] C. Johnson, *Numerical Solution of Partial Differential Equations by the Finite Element Method*, Cambridge University Press, New York, 1987.

[7] S. Moaveny, *Finite Element Analysis: Theory and Application with ANSYS*, Prentice-Hall, Upper Saddle River, New Jersey, 1999.

[8] E.P. Popov and T.A. Balan, *Engineering Mechanics of Solids*, Prentice-Hall, New Jersey, 1999.

[9] P.A. Raviart and J.M. Thomas, *Introduction à l'analyse numérique des équations aux dérivées partielles*, Masson, Paris, 1983.

[10] B.D. Reddy, *Introductory Functional Analysis with Applications to Boundary Value Problems and Finite Elements*, Springer-Verlag, New York, 1998.

[11] O.C. Zienkiewicz and R.L. Taylor, *The Finite Element Method: Solid and Fluid Mechanics and Non-linearity*, McGraw-Hill Book Company, London, 1991.

VSP International
Science Publishers
P.O. Box 346, 3700 AH Zeist
The Netherlands

Lecture Series on Computer
and Computational Sciences
Volume 1, 2004, pp. 145-149

Non-linear Analysis and Warping of Tubular Pipe Conveyors

J. J. del Coz Díaz[1], P. J. García Nieto[2], J. A. Vilán Vilán[3], A. Martín Rodríguez[1], F. J. Suárez Domínguez[1] , J. R. Prado Tamargo[1] and J. L. Suárez Sierra[1]

(1) Department of Construction Engineering
High Polytechnic School,
University of Oviedo,
Edificio Departamental N° 7 – 33204 – Gijón (Spain)
(2) Department of Mathematics
Faculty of Sciences,
University of Oviedo,
C/ Calvo Sotelo s/n – 33007 – Oviedo (Spain)
(3) Department of Mechanical Engineering
High University School of Industrial Engineering
University of Vigo
Campus de Lagoas-Marcosende - 36200 – Vigo (Spain)

Abstract: In this paper, an evaluation of distribution of strains and stresses and a warping effect are determined throughout a tubular pipe conveyor due to the geometry and local effects by the finite element method (FEM) [1, 5]. The non-linearity is due to the 'contact problems' that governs the phenomenon. Finally the forces and moments are determined on the different elements of the tubular pipe conveyor, given place to the conclusions that are exposed in this study.

Keywords: Finite element modeling, numerical methods, contact problems

PACS: 74S05, 74M10, 74M15

1. Introduction

The commercial use of the tubular pipe conveyor is relatively recent and many of the design, engineering and maintenance techniques are common to the conventional systems. The tubular pipe conveyor is a natural evolution of the conventional belt conveyors so that conserves all the advantages of these and it eliminates many of its inconveniences, becoming a very attractive option for the design of material transport's systems. However, this type of transport has presented important problems in its practical application, such as the warping [5] of the belt in the curved sections. Consequently it is necessary a more exhaustive theoretical study of this problem. In order to solve this question we have developed a numerical model by means of the finite element method (FEM) [2, 7] taking into account the contact behavior with friction between the belt and the rollers.

2. Geometry

To define the geometry of a tubular pipe conveyor is complex using an analysis program by finite elements (see figure 1). For this reason, an advanced parametric design language (APDL) was used [1]. Firstly, we build each of the sections that describe the belt of the pipe conveyor. Then we introduced the coordinates of the roller stations and joined them with the belt by means of the contact elements with friction, obtaining the graphics shown in figure 2. With this procedure a total of 15 cases were modeled with a 0.3 m of diameter in each section and curves understood between 6° and 90° (angle α in figure 1) with a total belt length ranges from 240 to 390 m.

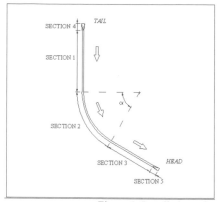
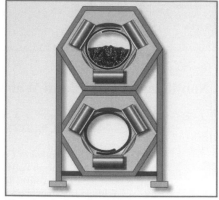

Figure 1: Installation scheme (left) and typical section (right).

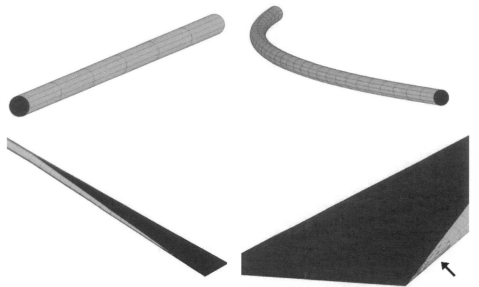

Figure 2: Geometry of the finite element model.

3. Mathematical model

A particularly difficult nonlinear behavior to analyze is the contact between two or more bodies. Contact problems range from frictionless contact in small displacements to contact with friction in general large strain inelastic conditions. Although the formulation of the contact conditions is the same in all these cases, the solution of the nonlinear problems can in some analyses be much more difficult than in other cases. The nonlinearity of the analysis problem is now decided by the contact conditions. The objective is to briefly state the contact conditions in the context of a finite element analysis and present a general approach for solution.

Let us consider N bodies that are in contact at time t. Let S_c^t be the complete area of contact for each body L, $L = 1...N$; then the principle of virtual work for the N bodies at time t gives [2-3, 7]:

$$\sum_{L=1}^{N}\left\{\int_{V^t}\tau_{ij}^t\delta_ie_{ij}dV^t\right\}=\sum_{L=1}^{N}\left\{\int_{V^t}\delta u_i\left(f_i^B\right)^t dV^t+\int_{S_f^t}\delta u_i\left(f_i^S\right)^t dS^t\right\}+\sum_{L=1}^{N}\int_{S_C^t}\delta u_i^c\left(f_i^c\right)^t dS^t \quad (1)$$

where the part given in brackets corresponds to the usual terms:

τ_{ij}^t = Cartesian components of the Cauchy stress tensor (forces per unit areas in the deformed geometry).

$\delta_t e_{ij}$ = strain tensor corresponding to virtual displacements.

δu_i = components of virtual displacement vector imposed on configuration at time t, a function of $x_j^t, j = 1,2,3...$

x_i^t = cartesian coordinates of material point at time t.

V^t = volume at time t.

$\left(f_i^B\right)^t$ = components of externally applied forces per unit volume at time t.

$\left(f_i^S\right)^t$ = components of externally applied surface tractions per unit surface area at time t.

S_f^t = surface at time t on which external tractions are applied.

δu_i^S = δu_i evaluated on the surface S_f^t (the δu_i components are zero and corresponding to the prescribed displacements on the surface S_u^t).

and the last summation sign in equation (1) gives the contribution of the contact forces. The contact force effect is included as a contribution in the externally applied tractions. The components of the contact tractions are denoted as $\left(f_i^c\right)^t$ and act over the areas S_c^t (the actual area of contact for body at time t), and the components of the known externally applied tractions are denoted as $\left(f_i^S\right)^t$ and act over the areas S_f^t. It is possible to assume that the areas S_f^t are not part of the areas S_c^t, although such an assumption is not necessary.

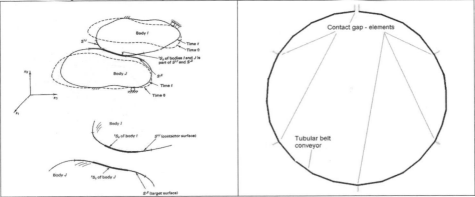

Figure 3: Bodies in contact at time t (left) and contact elements (right).

Figure 3 (left) illustrates schematically the case of two bodies, which are now considered in greater detail. In this paper, the two bodies in contact are denoted as body I and body J. Note that each body is supported such that without contact no rigid body motion is possible. Let $\left(\vec{f}^{IJ}\right)^t$ be the vector of contact surface tractions on body I due to contact with body J, then $\left(\vec{f}^{IJ}\right)^t = -\left(\vec{f}^{JI}\right)^t$. Hence, the virtual work due to the contact tractions in (1) can be written as:

$$\int_{S^{IJ}} \delta u_i^I \left(f_i^{IJ}\right)^t dS^{IJ} + \int_{S^{JI}} \delta u_i^J \left(f_i^{JI}\right)^t dS^{JI} = \int_{S^{IJ}} \delta u_i^{IJ} \left(f_i^{IJ}\right)^t dS^{IJ} \tag{2}$$

where δu_i^I and δu_i^J are the components of the virtual displacements on the contact surfaces of bodies I and J, respectively, and :

$$\delta u_i^{JI} = \delta u_i^I - \delta u_i^J \tag{3}$$

The pair of surfaces S^{IJ} and S^{JI} are termed a 'contact surface pair' and note that these surfaces are not necessarily of equal size. However, the actual area of contact at time t for body I is S_c^t of body I, and for body J it is S_c^t of body J, and in each case this area is part of S^{IJ} and S^{JI}. It is convenient to call S^{IJ} the 'contactor surface' and S^{JI} the 'target surface'. Therefore, the right-hand side of (2) can be interpreted as the virtual work that the contact tractions produce over the virtual relative displacements on the contact surface pair.

These conditions can now be imposed on the principle of the virtual work equation using a Penalty Approach (PA), Lagrange Multiplier Method (LMM) or Augmented Lagrangian Method (ALM). This work uses the technique of PA since it proves to be more efficient from a numerical point of view [4, 6].

4. Analysis of the results

As can be observed in figure 4 (upper and lower) the values of the warping moment in the contact elements nearest to the transition from curved to straight sections are increased with the belt conveyor angle α and with the percentage loading.

In the case studied, this rise in value shows a non-linear oscillatory behavior but increasing in all cases. In our case, the maximum reaction in the roller station is 1,057.7 N (80% percentage loading), precisely in the transition from curved to straight section.

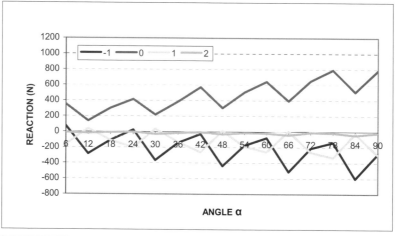

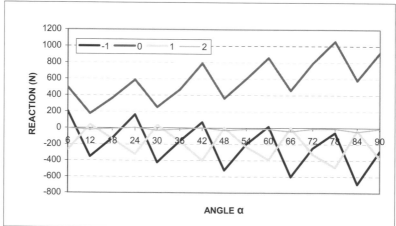

Figure 4: Roller Stations Reactions in the 'curved to straight' transition as a function of the belt conveyor angle α and the percentage loading (50% -upper- and 80% -lower-).

5. Summary and conclusions

In summary, the governing equations to be solved for the two-body contact problem are the usual principle of virtual work equation (1), with the effect of the contact tractions included through externally applied (but unknown) forces, plus the constraint equation. The finite element solution of the governing continuum mechanics equations is obtained by using the discretization procedures for the principle of virtual work, and in addition now also discretizing the contact conditions.

Thanks to the distribution of load inside tubular pipe and the form of contact between material and belt it is possible to get slopes of the tubular pipe conveyor larger than 30°, opposite to $17^\circ\text{-}20^\circ$ in a conventional belt. This characteristic is very outstanding when the instalation is situated in mines or in plants in which it is neccesary to get important slopes. Besides it is possible to design paths with curved sections. In this way we can carry out curved sections greater than 45° so that the number of transference points required are reduced. Nowadays it is possible to carry out curved sections up to 90°.

The method of the finite elements has been shown as a suitable tool in the modeling and nonlinear analysis of tubular pipe conveyors.

Acknowledgments

The authors express deep gratitude to Construction Department and Department of Mathematics at Oviedo University and Department of Mechanical Engineering at Vigo University for useful assistance. Helpful comments and discussion are gratefully acknowledged. We thanks to Swanson Analysis Inc. for the use of ANSYS University Intermediate program.

References

[1] ANSYS User's Manual: *Procedures, Commands and Elements* Vols. I, II and III. Swanson Analysis Systems, 2004.

[2] K. Bathe, *Finite Element Procedures*, Englewood Cliffs, Prentice-Hall, New York, 1996.

[3] T. Chandrupatla and A. Belegundu, *Introduction to Finite Elements in Engineering*, Englewood Cliffs, Prentice-Hall, New Jersey, 1991.

[4] J.J. del Coz Díaz, P.J. García Nieto, F. Rodríguez Mazón and F.J. Suárez Domínguez, Design and finite element analysis of a wet cycle cement rotary kiln, *Finite Elements in Analysis and Design* **39** 17-42(2002).

[5] E.P. Popov and T.A. Balan, *Engineering Mechanics of Solids*, Prentice-Hall, New Jersey, 1999.

[6] J.C. Simo and T.A. Laursen, An augmented Lagrangian treatment of contact problems involving friction, *Computers and Structures* **42** 97-116(1992).

[7] O.C. Zienkiewicz and R.L. Taylor, *The Finite Element Method: Solid and Fluid Mechanics and Non-linearity*, McGraw-Hill Book Company, London, 1991.

VSP International
Science Publishers
P.O. Box 346, 3700 AH Zeist
The Netherlands

*Lecture Series on Computer
and Computational Sciences*
Volume 1, 2004, pp. 150-153

An hybrid linking approach for solving the conservation equations with an Adaptive Mesh Refinement method

S. Delage*,[1], S. Vincent*, J.P Caltagirone* and J.P. Heliot+

*Laboratoire TREFLE,
University of Sciences Bordeaux1,
16 avenue PeyBerland,
33607 Pessac, France.
+CEA-CESTA,
BP n 2,
33114 Le Barp, France.

Received 22 July, 2004; accepted in revised form 15 August, 2004

Abstract: Solving the conservation equations for complex incompressible flows on a single grid turns out to be hardly possible. Numerical diffusion schemes and different scale phenomena compel us to use very fine meshes. Therefore, the obtention of an accurate solution requires expensive calculation time and computer memory. We aim at evaluating the efficiency of an implicite adaptive mesh refinement method so as to partially circumvent these problems.

Keywords: Conservation equations, Implicit, Adaptive Mesh Refinement (AMR), Interpolation, Connections.

Mathematics Subject Classification: $02.70. - c$, $02.60. - x$, $02.06.Cb$.

PACS: 47, $47.11. + j$, $47.27.Eq$, $47.55.Kf$, $02.70. - c$, $02.60. - x$.

1 Introduction

Understanding phenomena which happen on incompressible turbulent and multiphasic flows need an accurate solution. On the one hand, we have to capture different scale phenomena, therefore using a very fine mesh is essential to catch small ones even if they appear locally in space and time. On the other hand, the use of monotone schemes for advection and diffusion problems prevent us from well interpreting results, given that physical diffusion is covered up by numerical one in most real configurations. Thereby, based on the [1]'s ideas for compressible fluids, an original One-Cell Local Multigrid method (OCLM) has been developped by [5] to solve the Naviers-Stokes equations for two-phase flows. The equations are approximated by Finite Volumes on a MAC grid. This method consists in generating fine grids from a coarser one, by means of gradient criterion ie: if a point M on level G_{l-1} verifies the criterion, the control volume around M is refined and level G_l is built (The refined control volume is called AMR cell). An odd cutting is needed to ensure a conform connection between the fine grids of level G_l. When a third level G_{l+1} is generated, it is embedded in level G_l, which is itself embedded in level G_{l-1}. The solution on points of level

---

[1]corresponding author. E-mail: delage@enscpb.fr

G_l generated by a point from level G_{l-1} is treated as follows: For the limit points, the coarse solution is prolongated on level G_l using a classical Q1 interpolation procedure and or the interior points, the conservation equations are solved on level G_l. Then, the discrete solution on level G_l is restricted to level G_{l-1} using a direct injection procedure or a Full Weighting Interface Control Volume (FWICV) [2].

However, this method presents some failure. Firstly, level G_l has to be considered as a serie of independant AMR cells. In fact, the information does not go through cell to cell but from level G_{l-1} to level G_l. Hence, a numerous interpolated points make the solution less accurate.

Secondly, the classical Q_1 interpolation procedure coupled with an implicit solver does not verify the incompressibility condition. Given the fact a penalty method is used to take the interpolation into account, the flow results to be constrained.

Here we propose an improvement of this method to face with these two problems namely a linking method between AMR cells of a single level G_l and an implicit interpolation procedure.

2 Connections

When several adjoining points of level G_{l-1} generate AMR cells on level G_l, there are overlapping. As a matter of fact, the points belonging to an AMR cell limit are overlapped by interior points of another AMR cell to which they are connected. Hence, these limit points can be solved instead of being interpolated. These connections enable a good transmission of information cell to cell and preserve the solution accuracy.

As a result, level G_l has to be considered as a single block and not as a serie of independent AMR cells where the block' interior points are solved and the limit ones are interpolated. Figure 2 presents connected (so solved), solved and interpolated points on level G_l. We can notice that the number of interpolated points is small compared with the solved ones on a single level G_l.

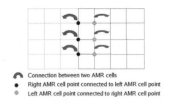

⌢ Connection between two AMR cells
● Right AMR cell point connected to left AMR cell point
◉ Left AMR cell point connected to right AMR cell point

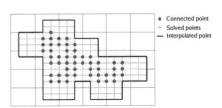

● Connected point
+ Solved points
— Interpolated point

Figure 1: Linking procedure on level G_l.

3 Interpolation and Implicit resolution

An implicit solver can be used to solve a part or the whole of the conservation equations. So, the discretized equations can be put in the following form $A^{n+1}X^{n+1} = B^n$, where B^n is the second member at iteration n, X^{n+1}, the unknown at iteration $n+1$ and A^{n+1} the matrix to inverse.

The lines i of the matrix which correspond to points detected to be interpolated are replaced by the interpolation coefficients. Different classical interpolation procedures namely Q_3, Q_2 and Q_1 have been tested to prolongate the coarse solution on level G_l for the limit points. We have noted oscillations when working with strong gradients and high order level interpolation such as Q_3 or Q_2 (Gibbs phenomena) whereas a Q_1 interpolation remains monotone. So, a Q_1 interpolation will be used for discontinuous solutions whereas a Q_3 interpolation will be implemented for more regular ones.

Once the matrix filled, a new matrix $\tilde{A}^{n+1}$ is obtained with the same dimensions as A^{n+1} and the system to solve becomes $\tilde{A}^{n+1}X^{n+1} = \tilde{B}^n$ (figure 3). A zero is put on the second member B_i^n. Hence, the adaptive mesh refinement is treated on an implicit way which avoids time-lags between solved points and interpolated ones. Thus the flow is not constrained anymore (see also [3]).

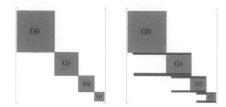

Figure 2: Typical shape of A^{n+1} (left) and $\tilde{A}^{n+1}$ (right).

4 Explicit resolution

The transport or advection equations cannot be solved in an implicit way because discretization schemes used in the implicit solver are either diffusive when they are monotone or dispersive. So we have no choice but to use explicit TVD or WENO schemes to discretize advection terms.[4]

5 Results

The performances of the AMR method are evaluated on the solving of a scalar advection-diffusion equation. A disc of concentration 1, submerged in another fluid is sheared by a rotating velocity field of intensity 2 rad.s^{-1}. The evolution of the concentration field is calculated either on a classical refined mesh ($144 * 144$) or on a coarse one ($16 * 16$) with several AMR levels (here 2) which corresponds to a local refined mesh ($144 * 144$).

The efficiency of AMR on reducing numerical diffusion is first tested. On the warped disc (figure 3), we can note that the higher the grid level, the weaker numerical diffusion is. So our AMR method allows us to control the numerical diffusion induced by the discretization scheme. Now, if the diffusion coefficient of $10^{-10}m^2.s^{-1}$ is raised up to $10^{-4}m^2.s^{-1}$, it is shown on left and centered plots of figure 4 that AMR solution captures the effect of the molecular diffusion.

Concerning the performances of the AMR method, it requires less computer memory than the use of a classical refined mesh (35% to 90% less)(figure 4 right). However, given that it requires expensive calculations in time and has not been optimized with respect to time yet, the rate of profit is about 6% in time. In 3D and for coupled vectorial equations, benefit will be greater.

6 Conclusion

A new hybrid linking technique dedicated to AMR methods has been proposed. Its efficiency has been demonstrated on scalar equation solving concerning the decrease in numerical diffusion, computer memory and time cost.

Vectorial conservation equation solving is being achieved and will be presented in the congress and the full article version.

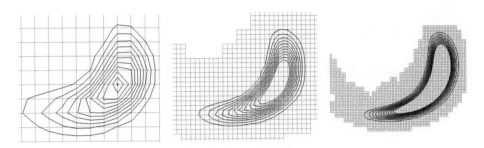

Figure 3: Numerical diffusion on level G_0 (left), G_1 (center) and G_2 (right) - $t = 2s$, timestep 10^{-3}s, diffusion coefficient $10^{-10}m^2.s^{-1}$.

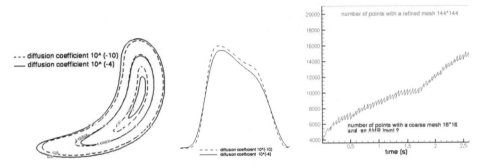

Figure 4: Effect of the molecular diffusion when 2 AMR level are considered in the xz plan (left)and in a z-slice (center) concentration fields - Comparison of the memory performances between the AMR procedure and the regular grids one (right) - $t = 2s$, timestep 10^{-3}s, diffusion coefficient $10^{-4}m^2.s^{-1}$.

References

[1] M.J. BERGER and P. COLLELA, Local adaptive mesh refinement for hyperbolic partial differential equations. *J. Comput. Phys.*, 82, 64-84, 1989.

[2] W. HACKBUSH, Multi-grid methods and applications.*SCM* , vol. 4, Springer, Berlin, 1985.

[3] ROME C. and GLOCKNER S.,An implicit multiblock coupling for the incompressible Navier-Stokes equations. Accepted for publication in*Journal for Numerical Methods in Fluids*.

[4] S. VINCENT and J.P. CALTAGIRONE, Solving incompressible two-phase flows with a coupled TVD interface capturing / local mesh refinement method.*Godunov Methods: Theory and applications* , E.F. Toro, Klumer Academic/Plenum Publishers, New York, 1007-1014, 2001.

[5] S. VINCENT and J.P. CALTAGIRONE, One Cell Local Multigrid method for solving unsteady incompressible multi-phase flows, *J. of Comput. Phys.* **1** 163, 172-215, 2000.

VSP International
Science Publishers
P.O. Box 346, 3700 AH Zeist
The Netherlands

*Lecture Series on Computer
and Computational Sciences*
Volume 1, 2004, pp. 154-158

Performance Analysis of Branch-and-Bound Skeletons

I. Dorta, C. León, C. Rodríguez[1]

Departamento de Estadística, I.O. y Computación,
Universidad de La Laguna, E-38271 La Laguna, Tenerife, Spain
http://nereida.deioc.ull.es

Received 2 August, 2004; accepted in revised form 15 August, 2004

Abstract: This article proposes a study of load balancing for Branch-and-Bound algorithms. Concretely, sequential and parallel generic skeletons to implement this algorithmic technique are presented. To accomplish the work the CALL tool is used. CALL allows to annotate the code by hand at the special points with a complexity function. Also, some preliminary computational results are presented.

Keywords: Algorithmic Skeletons, Branch-and-Bound Technique, Work Load Balancing

Mathematics Subject Classification: 68R99

1 Introduction

The Branch-and-Bound Technique is a general method to solve combinatorial optimization problems. The MaLLBa [2] skeleton provides a set of classes to implement automatically this technique. In this paper, a sequential implementation of the Knapsack Problem is presented, using the MaLLBa pattern. The complexity analysis of an algorithm produces as a result a "complexity function" that provides an approximation of the number of operations to accomplish. The CALL tool allows the logging of the C code that implements the algorithm with that complexity expression. The LLAC [1] tool is used to analyze the obtained results of the execution of the experiments with CALL.

This article is organized as follows. In second section, a sequential implementation of the Knapsack Problem using the skeleton for Branch-and-Bound technique is analyzed using CALL tool, and a parallel skeleton is briefly described. In the third section the preliminary computational results are presented.

2 Using call tool to accomplish performance analysis

The algorithm presented in Figure 1 shows the recursive code used to solve the Knapsack Problem. The number of objects for insertion into the knapsack is stored in the variable N, the variables w and p store the weights and the benefits of each of them, while the variable M represents the capacity. Since it is a maximization problem, the initial value $-\infty$ is assigned to the variable that stores the best solution found until that moment; bestSol (lines 1 to 4). Between lines 8 and 17 the bound function (lowerUpper) is defined. We use the same function to calculate the lower and upper bounds. The lower bound is defined as the maximum benefit that can be obtained from a

---

[1]E-mail: {isadorta, cleon, casiano}@ull.es

```
1    /* main files */
2    number N, M;
3    number w[MAX], p[MAX];
4    number bestSol = -INFINITY;
5
6    #pragma cll code double numvis;
7
8    void lowerUpper(number k, number C, number P, number *L, number *U) {
9       number i, weig, prof;
10      if (C <0) {*L = -INFINITY; *U = -INFINITY; }
11      else {
12         for (i = k, weig = 0, prof = P; weig <= C; i++)
13            {weig += w[i]; prof += p[i];}
14         i--;
15         weig -= w[i]; prof -= p[i];
16         *L = prof;   *U = prof+(p[i]*(C-weig))/w[i];
17   } }
18
19   number knap(number k, number C, number P) {
20      number L, U, next;
21       if (k < N) {
22          lowerUpper(k,C,P,&L,&U);
23          if (bestSol < L) { bestSol = L; }
24          if (bestSol < U) { /* L <= bestSol <= U */
25             next = k+1;
26             knap(next, C - w[k], P + p[k]);
27             knap(next, C, P);
28   #pragma cll code numvis += 2 ;
29          } }
30          return bestSol;
31   }
32
33   int main (int argc, char ** argv) {
34   number sol;
35   readKnap (data);
36   #pragma cll code double numvis = 0.0;
37   #pragma cll kps kps[0] * unknown(numvis) posteriori numvis
38      sol = knap( 0, M, 0); /* next obj., current capacity, profit */
39   #pragma cll end kps
40      printf("\nsol = ", sol);
41   #pragma cll report all
```

Figure 1: Using CALL in the Sequential Implementation

given subproblem, while the upper bound includes the proportional part of the benefit of the last object that could not be inserted into the knapsack. The function, knap (lines 19-31), implements a recursive Branch-and-Bound algorithm. The call to the function that studies the insertion of the object k is accomplished in line 26, while line 27 considers not inserting it. The condition to update the value of the best solution (bestSol) found until that moment is implemented in lines 21 to 23.

We are interested in knowing the behaviour of the algorithm. To be precise, we would like to know the number of search space nodes visited until the best solution is found. To accomplish this study, we note down the code with CALL directives. Line 6 is used to specify the CALL definition of a double type variable to store the number of visited nodes, numvis, whose initial value is zero (line 40). The directive CALL code provides the possibility of inserting source code to the logging program. This code does not modify the original program, it is only used in the CALL sentences.

The string `cll` written after the reserved word `#pragma`, gives the indication to the C compiler that it is a directive for the CALL compiler. A typical CALL experiment named `kps` is created in line 37. This experiment measures the execution time of the sentences between the beginning and ending directives (line 39). The sentence `pragma cll end` followed by the experiment name stops the CALL chronometer and the times are stored. The expression specified after the experiment `kps` (line 37) synthesizes the complexity function, which describes the behaviour of the time. The constants `ksp[0],...,kps[i]` are associated to the complexity function. The CALL syntax requires the use of the experiments name to index the constants associated with the complexity function. These constants will be evaluated with the LLAC tool. Intuitively, we can say the time spent by the Branch-and-Bound algorithm is related to the number of visited nodes in the search space. Thus, our expression has to be written in terms of the variable `numvis` - number of visited nodes. However, the final value of `numvis` will only be known once the problem is solved, i.e., when the call to `knap(0,M,0)` finishes. Therefore, this specified value is not know (`unknown`) in the complexity expression and it will be evaluated at the end (`posteriori`). The number of visited nodes (`numvis`) increases when two new subproblems obtained from each subproblem are studied, i.e., when the problems that include (line 26) and do not include the following object (line 27) are solved. We have to indicate this CALL increase with the corresponding directive, as it is shown in the line 28. Finally, the directive `report` has to be added after the specification of all the experiments. This directive indicates to CALL compiler to generate a file with all the results obtained during the execution. In our case, we add it before the return sentence of the main function (line 41). The word `all` has been used to keep the results of all the experiments defined in the program. Alternatively, a list of identifiers can be used to specify the experiments whose results will be stored. Once the code is annotated, the source code with the CALL directives can be compiled by any C compiler and can be executed exactly as if it had not been modified. The reason for this being that the compiler ignores all the CALL directives considering them as commentaries.

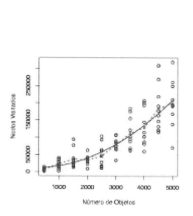

(a) Sequential Execution

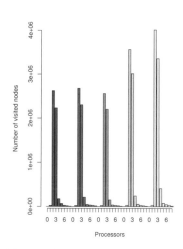

(b) Load Balance for 5 parallel Executions of Size 50,000

Figure 2: LLAC results

The parallel version of Branch-and-Bound algorithm in C follows a Master/Slave scheme. It was implemented using Message Passing. As in the sequential case, we wanted to study the number of visited nodes (`numvis`). This value was distributed between the slaves, because the master did not perform task bounding. Therefore, we will annotate the parallel code exactly the same as the sequential code, where the only necessary change is the line where the number of visited nodes is increased. The CALL directives have been added in the points where the problems are extracted from the local queue of each slave to study them or to send them to another slave to be solved. The Master code has not changed.

3   Preliminary computational results

As the LLAC tool is an extension of R [3], it allows us to analyze the obtained results of the execution of the experiments with CALL. With the same samples different types of representations to analyze the similarity between the expression with which we annotated the program and the obtained results can be accomplished.

The sequential executions of the knapsack problem were run on a AMD-DURON processor at 800 MHz with 256 Mb of memory. The experiment consisted of ten randomly generated knapsack problems with capacities in the range of [500, 5000]. Figure 2 (a), generated with LLAC, shows the obtained results. Each one of the ten round points associated with each size between (500-5000) represents the number of visited nodes to solve that problem. The "x" linked with a dotted line represents the average of visited nodes. The continuous line represents a second degree polynomial. It indicates the average number of visited nodes is approximated to a parable. We conclude that the number of visited nodes was more scattered, when the number of objects were larger. This behaviour could be due to the generator of random problems that we were using.

A very interesting parameter for studying the parallel Branch-and-Bound implementation is the "balanced work load" among the processors that take part in the execution. Figure 2 (b) shows the distribution of visited nodes among eight processors for five executions of a randomly generated knapsack problem of size 50,000. The first two processors were AMD-DURON at 800 MHz and the rest at 500 MHz, all of them with 256 Mb of memory. It can be appreciated that processor zero does not visit any node, because it is the Master. We believe the first slave explores the width part of the search space because it is the fastest, and due to the size of the problem, not much work remains for the rest of the processors.

Acknowledgement

This work was partially supported by the Spanish "Ministerio de Ciencia y Tecnología" and FEDER project TIC02-04498-C05-05.

References

[1] I. Dorta, C. León, C. Rodríguez, G Rodríguez., and A. Rojas. *Complejidad Algorítmica: de la Teoría a la Práctica*. Actas de las IX Jornadas de Enseñanza Universitaria de la Informática, 2003.

[2] I. Dorta, C. León, and C. Rodríguez. *Parallel Branch-and-Bound Skeletons: Message Passing and Shared Memory implementations*. Fifth International Conference on Parallel Processing and Applied Mathematics (PPAM2003) Proceedings Published in the Springer Verlag Series LNCS 3019, pp. 286-291, 2004.

[3] R. Ihaka and R. Gentleman. *R, A language for Data Analysis and Graphics*. Journal of Computational and Graphical Statistics, 5: 3:299–314, 1996.

VSP International
Science Publishers
P.O. Box 346, 3700 AH Zeist
The Netherlands

*Lecture Series on Computer
and Computational Sciences*
Volume 1, 2004, pp. 159-162

Risk Quantitative Analysis Using Fuzzy Sets Theory

M. El-Cheikh, J. Lamb, N. Gorst

Department of Civil Engineering,
School of Engineering
University of Birmingham
B15 2TT, Birmingham, United Kingdom

Received 10 May, 2004; accepted in revised form 24 June, 2004

Abstract: Estimating the likelihood and the impact of risks, and estimating their implications on project's objectives (quantitative risk analysis) requires expert knowledge. In addition, statements made by experts usually contain imprecision. Previous studies have demonstrated the use of the CIRIA document "A Simple Guide to Controlling Risk", Monte Carlo simulation or methods involving fuzzy sets, in order to quantify the uncertainty associated with construction activities' risks. These studies, however, did not address the processing of information for generating a complete quantitative risk analysis method.

This paper presents a new analytical modelling method that simulates and supports the existing risk assessment analysis and Risk Management Systems and based on Fuzzy Sets Theory. The proposed method incorporates a number of new techniques that facilitate: the representation of imprecise risks, the calculation of the actual cost of risks, and the interpretation of the results. A worked example, using the three new techniques, illustrates the use of the proposed method and allows comparison with Monte Carlo simulation. The calculations are shown to be simpler, requiring less computational effort than that needed in Monte Carlo simulation. The research aims to determine the modification factor ξ, which should be used to combine the effects of multiple risk factors on a project's overall risk (and on the component work packages), in order to deliver improved results and to bridge the gap between monitoring and controlling the risk management process. The authors argue that the proposed method is practical and can be easily computerised.

Keywords: Construction project, fuzzy set theory, fuzzy logic, qualitative risk assessment, quantitative risk analysis, risk management.

Mathematics SubjectClassification: 03E72 (Fuzzy set theory) & 03E75 (Applications of set theory)

1. Introduction and Background

Ang & Tang (1975, 1984) state that: "In the analytic treatment, it is assumed that the total uncertainty, Ω, consists of the uncertainty due to inherent randomness, ε, and the uncertainty associated with the error in the prediction, Λ." (cited in Timothy J. Ross and Jonathan L. Lucero, p. 196).

$$\Omega^2 = \varepsilon^2 + \Lambda^2 \tag{1}$$

In her survey, the primary author found that there is no consideration for the total uncertainty, Ω, of the above mentioned concept or the randomness concept. The authors propose an analytical modelling method using both fuzzy sets, which give a sufficient range to list all the possible risks, and fuzzy estimation which minimizes Λ, the error in the prediction, in a way that balances the current quantitative risk systems (current state) and the new proposed method (desired state). This is shown schematically in Figure 1. The primary author argues that completing this loop will be achieved by bridging the gap between monitoring and controlling the risk management process, i.e., linking the feedback records with the inputs as shown in Figure 2.

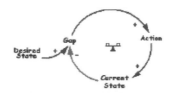

Figure 1: Balancing Loop in the Systems Thinking Concept (After Gene Bellinger, OutSights, Inc.)

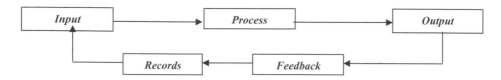

Figure 2: Proposed Method to Bridge the Gap.

2. Logic Alpha-Cut

The alpha-cut analysis, which was introduced by Zadeh in 1965, is one of the methods that are widely used in representing imprecise or uncertain knowledge. Uncertain parameters are considered to be fuzzy numbers with given membership functions. Figure 3 shows a parameter P represented as a triangular fuzzy number with a support of A_0. The wider the support of the membership function is, the grater is the uncertainty (Abebe, Guinot, and Solomatine, 2000).

Li & Vincent stated in 1995: "The fuzzy set that contains all elements with a membership of $\alpha \in [0,1]$ and above is called the α-cut of the membership function. At a resolution level of α, it will have support of A_α. The higher the value of α, the higher the confidence in the parameter". (quoted in Abebe, Guinot, and Solomatine, 2000, p.3, www.hydroinformatics.org.). In this research the impact of each risk associated with a construction activity is an uncertain parameter, which is presented as a triangular fuzzy number with a support of A_0, with A_0 being the duration of the construction activity. As a result, the larger the duration is, the greater the uncertainty is regarding the impact of the risk on the duration activity. Each uncertain parameter is cut by α, which is a measure of the likelihood of each risk; the higher the likelihood (α value) is, the grater is the confidence of the occurrence of that risk.

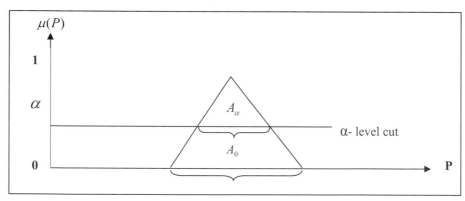

Figure 3: Fuzzy number, its support and α –cut (After Abebe, Guinot, and Solomatine, 2000)

3. Modelling Techniques

Let project A have n activities. The array A(n) contains the project's activities. Each activity A(i), where i=1 to n, has a fuzzy duration represented as a triangular distribution (optimistic, most likely, pessimistic). Each activity A(i) is assumed to be under the influence of simultaneous risks, where each of which has a likelihood and impacts on time, cost, quality and safety. Let us assume A(20) is one of the project's activities. The duration of A(20) in days is represented by the fuzzy number (9, 11, 13) and its daily cost is £1000, so the costs for this activity are therefore £ (9000, 11000, 13000). A possible risk associated with activity A(20) is given in table 1:

Act.	Risk	Likelihood	Impact			
			Time	*Cost*	*Quality*	*Safety*
A(20)	New Technology	Medium(0.4)	Very Low(0.1)	Low(0.2)	Low(0.3)	Medium(0.5)

Table 1: Example about the Risk Matrix

In this study the risk's impacts are initially expressed using linguistic variables that are then converted to numbers to be assigned to the duration. The likelihood of the risk is represented using an α-Cut. Figure 4 shows the trapezoidal intersection area bounded by the triangular distributions and the α-Cut. The shaded area represents the possibility of having all these impacts together with α-Cut likelihood (0.4 in Figure 4). This fuzzy set is then converted into a single output value by a process known as defuzzification. The most common method of defuzzification is the centre of gravity, or centroid method.

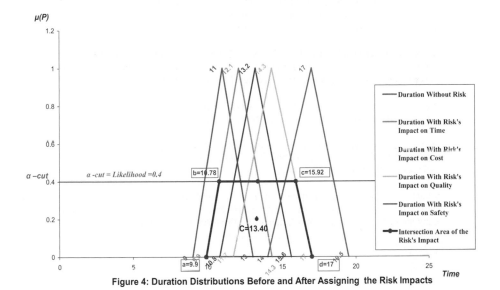

Figure 4: Duration Distributions Before and After Assigning the Risk Impacts

The geometric centroid (C) of a trapezoid (a, b, c, d) can be calculated as follow (Lorterapong and Moselhi, 1996, p.313):

$$C = \frac{\left(-a^2 - b^2 + c^2 + d^2 - a \times b + c \times d \right)}{3 \times \left(-a - b + c + d \right)} \qquad (2)$$

By applying Equation 2 to the trapezoid of Figure 4 the numerical value of C will be:

$$C = \frac{(-9.9^2 - 10.78^2 + 15.92^2 + 17^2 - 9.9 \times 10.78 + 15.92 \times 17)}{3 \times (-9.9 - 10.87 + 15.92 + 17)} = 13.40$$

C represents the most likely duration. Hence, the most likely cost will be:

$$Cost = 13.40 \times 1000 = £13400$$

A key factor in the usefulness of any model used to present project risk is the accuracy of the estimated durations and costs. Further is planned to take this into consideration in order to find an appropriate way to present these outputs as accurately as possible. The model will be expanded to include additional project characteristics, such as critical path, criticality index, possibility measure, and cash flows.

Acknowledgments

The author wishes to thank Professor CDF Rogers, the head of the Infrastructure Engineering and Management Research Centre at the University of Birmingham, for his funding of this work.

References

[1] Abebe, A. J.; Guinot, V.; Solomatine (2000). Fuzzy alpha-cut vs. Monte Carlo techniques in assessing uncertainty in model parameters. (4th International Conference on Hydro informatics). Iowa, USA, July 2000. [Internet]. Available from:
     http://www.hydroinformatics.org/hi/sol/papers/HI2000-AlphaCut.pdf
     [Accessed 10th Nov, 2003].

[2] Gene Bellinger, OutSights, Inc. Systems Thinking, An Operational Perspective of the Universe. 2004: [Internet]. Available from:http://www.systems-thinking.org/systhink/systhink.htm[Accessed 17th April, 2004].

[3] Pasit Lorterapong and Osama Moselhi (1996). Project Network Analysis Using Fuzzy Sets Theory. Construction Engineering and Management, (308) Volume 122, No. 4.

[4] Timothy J. Ross and Jonathan L. Lucero (2002). Fuzzy Logic and Probability applications, Bridging the Gap. ASA- SIAM Series on Statistics and Applied Probability. Pheladelphia, Alexandria, Virginia. SIAM & ASA.

[5] Zadeh, A. Lotfi, (1996). Fuzzy Sets. In Klir, G. and Yuan, B. (ed), Fuzzy Sets, Fuzzy Logic, and Fuzzy System (Issues in Advances In Fuzzy Systems- Applications And Theory), Singapore, NJ and London: World Scientific Publishing Co Pte Ltd.

VSP International
Science Publishers
P.O. Box 346, 3700 AH Zeist
The Netherlands

Lecture Series on Computer
and Computational Sciences
Volume 1, 2004, pp. 163-168

Neutron Diffusion Problem Solutions Using The Method of Fundamental Solutions with Dual Reciprocity Method

C. Erdönmez[a,1], H. Saygın[a,b]

[a]Istanbul Technical University, Institute of Informatics, Computational Science and Engineering
Program, 34469 Maslak, Istanbul, Turkey
[b]Istanbul Technical University, Institute of Energy, 34469 Maslak, Istanbul, Turkey

Received 30 July, 2004; accepted in revised form 20 August, 2004

Abstract: The Method of Fundamental Solutions Method (MFS) is used to solve one group and two group Neutron Diffusion problems. The inhomogeneous terms were modeled with the Dual Reciprocity Method (DRM) by using the well-known thin-plate spline Radial Basis Function (RBF) as the approximating function. A test problem is presented first and the results obtained by using MFS are compared with FEM solutions. Then one group and two group Neutron Diffusion equations in a rectangular domain were solved using MFS. Results for each group problems will be presented in the tables. Results indicate that MFS gives accurate solutions and are consistent with the exact solution.

Keywords: Method of Fundamental Solutions, Dual Reciprocity Method, Thin Plate Spline, Neutron Diffusion Equation

Mathematics Subject Classification: 82D99, 34B99

PACS: 28.20.Gd.

1   Introduction

The MFS is a mesh-free method which uses a fictitious exterior boundary to solve problems. The method was introduced by Kupradze and Aleksidze [1], and was only applicable to homogeneous problems at first. But recently, the method was extended to solve nonhomogeneous problems by Goldberg and Chen [2]. In this paper a test problem of modified Helmholtz type with Dirichlet boundary condition is solved with MFS and the result is presented in graphics. Also the result is compared with the FEM solution by using Matlab (release 6.5) program.

One group and two group Neutron Diffusion Equations are solved using MFS and compared with the exact solutions. Particular solutions are found by using approximation functions. Dual Reciprocity Method (DRM) is used to find particular solutions with the well-known thin-plate spline Radial Basis Function (RBF). The MFS results agree well with both FEM and exact solutions of the Neutron Diffusion Equations.

2   The Method of Fundamental Solutions with Dual Reciprocity

Here the MFS is used to solve a modified Helmholtz problem. The inhomogeneous terms are modeled using the DRM, employing thin-plate spline approximation functions as the RBF for the homogeneous solution, fictitious boundary points first introduced by Bogomolny [3].

---

[1]E-mail: erdonmez@be.itu.edu.tr, Tel: (0532)2513366

The problem which will be investigated in this paper is in the form of;

$$\nabla^2\Phi - \lambda^2\Phi = f(x) \quad , x \in \Omega \tag{1}$$

with the Dirichlet or Neumann boundary conditions,

$$\Phi = 0 \quad , x \in \partial\Omega. \tag{2}$$

$$\frac{\partial\Phi}{\partial n} = 0 \quad , x \in \partial\Omega. \tag{3}$$

Equation (1) can be solved by using MFS [4] by first defining a fictitious surface Q exterior to Ω as shown in figure 1.

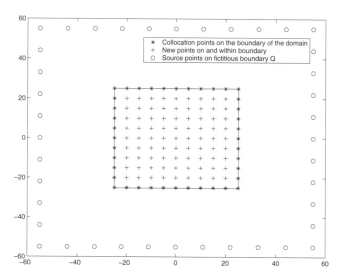

Figure 1: Fictitious boundary definition and geometry for the test problem.

To approximate the source term f(x), DRM [5] is used with the thin-plate spline RBF as $r^2 log(r)$ [4]. Error estimates are computed by Root Mean Square (RMS) [6] formula which can be defined as,

$$RMS = \sqrt{\sum_{i=1}^{N}(\bar{\Phi} - \Phi)^2/N}. \tag{4}$$

3 Solution of the test problem by using MFS

Two dimensional, time independent one group Neutron Diffusion equation with dirichlet boundary condition is selected as a test problem. The definition of the test problem is as follows:

$$\nabla^2\Phi(r) - k^2\Phi(r) = -\frac{S(r)}{D} \quad , r \in \Omega \tag{5}$$

with the Dirichlet boundary condition,

$$\Phi = 0 \quad , r \in \partial\Omega. \tag{6}$$

on a square domain bounded by $-25 \leq x, y \leq 25$. The constant k can be computed by,

$$k = \sqrt{\frac{\Sigma_a}{D}} \tag{7}$$

where Σ_a and D denote cross-section and group diffusion constant respectively.

Table (1) : One group constants and analytical solution

S $(neutron/cm^3)$	Σ_a (cm^{-1})	D (cm)	Φ $(neutron/cm^2 s)$ (analytical solution)
1	1	1	1

It is assumed that there is a constant source of neutron (S) everywhere in the system. The system has reflection condition from each side of the area. Analytically computed flux is constant and can be computed by

$$\Phi = \frac{S}{\Sigma_a}. \tag{8}$$

When the neutron source, cross-section and group diffusion constants are taken as in table (1), analytical flux is computed as $1 \ neutron/cm^2 s$ everywhere in the domain.

Figures (2) and (3) illustrate the numerical solution of the test problem. FEM uses 2753 points over the domain while MFS only uses 121. These results show that MFS is easy to program and gives accurate solutions while using less number of points than FEM over the domain.

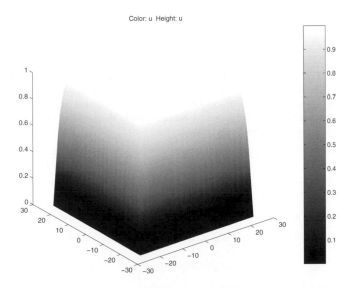

Figure 2: Solution of the test problem by FEM using 2753 points.

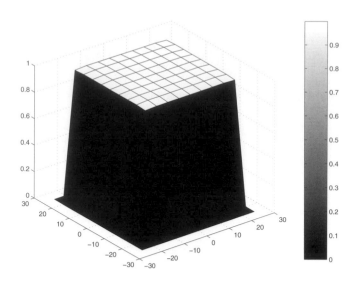

Figure 3: Solution of the test problem by MFS using 121 points.

4 Problem formulation for two dimensional one group and two group Neutron Diffusion equation

4.1 Two Dimensional One Group Neutron Diffusion Equation

Formulation of the time independent one region one group neutron diffusion equation is given in equation (5). By choosing Neumann boundary condition

$$\frac{\partial \Phi}{\partial n} = 0 \quad , r \in \partial\Omega, \tag{9}$$

on a square domain bounded by $-25 \leq x, y \leq 25$ and using constant values defined in table (1), analytical solution is computed 1 $neutron/cm^2s$ everywhere in the domain. When the constant values are put in the equation (5), modified Helmholtz equtaion will be created. Solving the problem with both FEM and MFS gives 1 $neutron/cm^2s$ everywhere in the domain which is the exact solution. Table (2) shows the solutions.

Table (2) : One group constants and solutions compared with FEM and MFS

S $(neutron/cm^3)$	Σ_a (cm^{-1})	D (cm)	Φ $(neutron/cm^2s)$ (analytical solution)	Φ (FEM)	Φ (MFS)
1	1	1	1	1	1

4.2 Two Dimensional Two Group Neutron Diffusion Problem

Problem for time independent one region two group Neutron Diffusion equation is given as,

$$\nabla^2 \Phi^g(r) - (k^g)^2 \Phi^g(r) = -\frac{S^g(r)}{D^g} \quad , r \in \Omega, \ g = 1, 2, \tag{10}$$

with the Neumann boundary conditions

$$\frac{\partial \Phi^g}{\partial n} = 0 \quad , r \in \partial\Omega, \tag{11}$$

on a square domain bounded by $-25 \leq x, y \leq 25$, where k^g is given by

$$k^1 = \sqrt{\frac{\Sigma_{r,1}}{D^1}}, \qquad k^2 = \sqrt{\frac{\Sigma_{a,2}}{D^2}}, \qquad (12)$$

where D^g, $\Sigma_{r,1}$, $\Sigma_{a,2}$ and S^g denote that g'th group diffusion constant, 1'st group subtraction effect cross-section, 2'nd group absorb effect cross-section and g'th group source term respectively.

Diffusion equations and group flux are given by $\Phi^1 = \frac{S^1}{\Sigma_{r,1}}$, $\Phi^2 = \frac{\Sigma_{s,2\leftarrow1}\Phi^1 + S^2}{\Sigma_{a,2}}$. The constant values are figured in table (3).

Table (3) : Two group neutron diffusion constants

Group	S	$\Sigma_{r,1}$	$\Sigma_{a,2}$	$\Sigma_{s,2\leftarrow1}$	D
	$(neutron/cm^3s)$	(cm^{-1})	(cm^{-1})	(cm^{-1})	(cm)
1	1	0.054577	-	-	2.68
2	1	-	0.014496	0.040792	1.5788

If equation (10) is rearranged according to the given values in table (3) one can find out these two equations for g=1,2 respectively,

$$\nabla^2\Phi^1(r) - \frac{0.054577}{2.68}\Phi^1(r) = -\frac{1}{2.68}, \qquad (13)$$

$$\nabla^2\Phi^2(r) - \frac{0.014496}{1.5788}\Phi^2(r) = -\frac{1. + 0.040792 * \Phi^1}{1.5788}. \qquad (14)$$

Equation (13) is a modified Helmholtz equation and can be solved using MFS as in one group solution. Solution of equation (13) which is Φ^1, is used as a parameter to solve equation (14) and Φ^2 is computed by solving this equation. The results of the two group Neutron Diffusion equation is given in table (4) below.

Table (4) : Two group neutron diffusion solutions Φ^g; FEM, MFS and exact solution.

Group	Analytical Solution $(neutron/cm^2s)$	FEM (188 points) RMS error	MFS (100 points) RMS error
1	18.322737	18.322737	18.322737
		3.570636E-012	2.131628E-015
2	120.545052	120.545052	120.545052
		1.176279E-011	1.069219E-013

5   Conclusion

The solutions of the one group and two group Neutron Diffusion equations found by using MFS are compared both with the FEM solution obtained by Matlab program and analytical solution. The results are presented in tables (2) and (4). It has been obtained that as a meshless method MFS gives much more accurate results than FEM, needs little effort to program, requires small amount of points within the problem domain and its discretization is not as difficult as in FEM. For the further studies, MFS should be applied to domains including different materials and also be applied to three dimensional problems.

References

[1] Kupradze V.D., Aleksidze M.A., A method for the approximate solution of limiting problems in mathematical physics. USSR Comp Math and Math Phys 1964;4:199-205.

[2] Golberg MA, Chen CS. The method of fundamental solutions for potential, Helmholtz and diffusion problems. In: Golberg MA, editor. Boundary integral methods: numerical and mathematical aspects, Boston/Southampton: WIT Press, 1998. p. 103-76.

[3] Bogomolny A. Fundamental solutions method for elliptic boundary value problems, SIAM Journal 1985;22:644-69.

[4] Golberg M.A., Chen C.S., Discrete Projection Methods for Integral Equations, Computational Mechanics Publications, 2001

[5] Partridge P.W., Brebbia C.A., Wrobel L.C., The Dual Reciprocity Boundary Element Method, Computational Mechanics Publications Southampton Boston, 1992

[6] Hoschek, J., Fundamentals of Computer Aided Geometric Design, 1993

VSP International
Science Publishers
P.O. Box 346, 3700 AH Zeist
The Netherlands

*Lecture Series on Computer
and Computational Sciences*
Volume 1, 2004, pp. 169-172

A Numerical Method for Cyclic Loading Analysis of Beams

M. R. Eslami[1] and H. Mahbadi [2]

ME. Dept., Amirkabir Univ. of Tech., Hafez Ave., Tehran, Iran, 1594.

Received 23 June, 2004; accepted in revised form 8 August, 2004

Abstract: The present article, proposes an effective numerical method to evaluate the cyclic loading analysis of the beams, under the mechanical and thermal loads. It is assumed that the beams material follows the nonlinear hardening curve. The method is used for different kinematic hardening models such as Prager, Frederick-Armstrong and Chaboche kinematic hardening models.

1 Introduction

Naderan and et al. [1] proposed a numerical method to evaluate the residual stresses and strains of the beams and thick cylindrical vessels. Mahbadi and Eslami [2] used this method to predict the cyclic loading results of the structures. But since the isotropic hardening model does not consider the buashinger effect during the unloading of the structures, the cyclic loading results are not well predicted based on this method. Tai-Ran Hsu [3] divided the nonlinear part of the stress-strain curve into a number of small linear segments, called the piecewise linear, and proposed a finite element algorithm for the elasto-plastic analysis of structures. The method is proposed for the Prager [4] kinematic hardening model.

In this article a new effective numerical method is proposed to evaluate the cyclic loading analysis of the beams of the strain hardening materials. The material hardening curve is considered to be nonlinear. The beams under different mechanical and thermal loads are analyzed using this method. The method can be used with different types of the kinematic hardening models such as Prager [4] and Frederick-Armstrong[5] and [6] kinematic hardening models.

2 Mathematical Formulation

Consider a beam of isotropic material under axial load P, bending moment M, and transverse temperature distribution. Axial total strain in the beam is:

$$\epsilon_x = \frac{\sigma_x}{E} + \alpha T + \epsilon_x^p + \epsilon_x^{Res} \tag{1}$$

where σ_x is the axial stress, ϵ_x^p is the axial plastic strain, ϵ_x^{Res} is the axial residual strain which is obtained from the previous load cycle, T is temperature distribution across the beam thickness, E is the modulus of elasticity, and α is the linear coefficient of thermal expansion. The dimensionless quantities are defined as:

$$S = \frac{\sigma_x}{\sigma_0}, \qquad \tau = \frac{\alpha T}{\epsilon_0}, \qquad e_x = \frac{\epsilon_x}{\epsilon_0}, \qquad e_x^p = \frac{\epsilon_x^p}{\epsilon_0}, \qquad e_x^{Res} = \frac{\epsilon_x^{Res}}{\epsilon_0}, \qquad \eta = \frac{y}{h} \tag{2}$$

---

[1]Corresponding author. Professor. Email: eslami@aut.ac.ir
[2]Assistant Professor of ME. Dept., Azad Univ., Central Tehran Branch, Tehran, Iran. E-mail: h_mahbadi@yahoo.com

where $2h$ is the height of beam cross section, σ_0 is the initial yield stress, and ϵ_0 is the initial yield strain. The boundary conditions are as follows:

$$\int_{-1}^{1} S d\eta = \frac{P}{\sigma_0 ch} = P^* \quad and \quad \int_{-1}^{1} S\eta d\eta = \frac{M}{\sigma_0 b^2 h} = M^* \tag{3}$$

where b is the beam width. Calling the mechanical strain by $e_x^M = e_x - \tau$, Normalizing Eq. (1) and solving the compatibility equation using the boundary conditions (3), the mechanical strain may be found as:

$$e_x^M = \frac{P^*}{2} + \frac{3}{2}M^* - \tau + \frac{1}{2}\int_{-1}^{1} \tau d\eta + \frac{3}{2}\int_{-1}^{1} \eta\tau d\eta + \frac{1}{2}\int_{-1}^{1} e_x^p d\eta$$
$$+ \frac{3}{2}\int_{-1}^{1} e_x^p \eta d\eta + \frac{1}{2}\int_{-1}^{1} e_x^{Res} d\eta + \frac{3}{2}\int_{-1}^{1} e_x^{Res} \eta d\eta \tag{4}$$

3 Hardening Model

Different kinematic hardening models are available for the plastic analysis of structures. The Prager kinematic hardening model (1956) is proposed for the linear strain hardening materials. The Frederick and Armstrong (1966) modified the Prager model so that the transformation of yield surface in the stress space is different during loading and unloading. Chaboche (1986, 1991) proposed a model which was composed from three Frederick-Armstrong models. Based on this assumption better approxiamation can be found for uniaxial stress-strain curve. The von Mises yield criterion for the kinematic hardening model is in the following form:

$$f(\sigma_{ij} - \alpha_{ij}) = [\frac{3}{2}(s_{ij} - a_{ij}) \cdot (s_{ij} - a_{ij})]^{\frac{1}{2}} = \sigma_0 \tag{5}$$

where, σ_{ij} and α_{ij} are the stress and back stress tensors, and s_{ij} and a_{ij} are the stress and back stress deviatoric tensors in the stress space. The flow rule is:

$$d\epsilon_{ij}^p = \frac{1}{H} < \frac{\partial f}{\partial \sigma_{ij}} \cdot d\sigma_{ij} > \frac{\partial f}{\partial \sigma_{ij}} \tag{6}$$

where H is the plastic modulus. The kinematic hardening models are:
i) Prager kinematic hardening model:

$$da_{ij} = Cd\epsilon_{ij}^p \tag{7}$$

where C is a positive non-constant multiplier and may be found from the uniaxial stress-strain curve.
ii) Frederick-Armstrong kinematic hardening model:

$$da_{ij} = \frac{2}{3}Cd\epsilon_{ij}^p - \gamma a_{ij} |d\epsilon_p| \tag{8}$$

where C and γ are two material constants in the Frederick-Armstrong kinematic hardening model and they will be found from the uniaxial strain controlled stable hystersis curve.
iii) Chaboche kinematic hardening model:

$$da_{ij} = \sum_{i=1}^{3}(\frac{2}{3}C_i d\epsilon_{ij}^p - \gamma_i a_{ij}^i |d\epsilon_p|), \quad i = 1 \ldots 3 \tag{9}$$

As it can be seen from Eq. (9), back stress tensor in the Chaboche model is composed from three Frederick-Armstrong kinematic hardening rules. In this equation a^i are the back stress tensors for the Frederick-Armstrong kinematic hardening models, C_i and γ_i are the matreial constants for these models which are found from the experimental tests and best fit the uniaxial stress-strain curve.

4   Numerical Method

The method of analysis used in this paper is based on a numerical iterative method. Due to the complexities incurred by the assumed kinematic hardening theories, the proposed method is quite capable and unique to handle the cyclic loading calculations. This method may be used for the analysis of uniaxial as well as multiaxial state of loadings.

Solution procedure is as follow: 1 - The elastic analysis of structure is obtained and the critical load to bring the structure to plastic yielding is calculated. 2 - The difference of the final and critical load is divided into n steps. 3 - The accumulated plastic strain for the first increment of load in the plastic region is set to zero. 4 - A value is assumed as the first guess for the current value of equivalent plastic stain ($\Delta\epsilon_p$). 5 - Calculate the current value of the plastic strain ($\Delta\epsilon_{ij}^p$), using the flow rule. 6 - Calculate the total plastic strain (ϵ_{ij}^p). 7 - Calculate the value of C in Prager kinematic model using the slope of the experimental stress-strain curve. 8 - Calculate the current value of back stress tensor (Δa_{ij}). 9 - Calculate the total value of the back stress tensor (a_{ij}). 10 - Calculate the stress components from the solution of the differential equations using the current value of the assumed plastic strain components. 11 - Calculate the new values of the current plastic components with stress components found from step 10, using the flow rule. 12 - Calculate a new value for the current increment of the equivalent plastic strain. 13 - Repeat the procedure from step 5 through 12, until the current value of the increment of equivalent plastic strain converges. 14 - Add the converged value of the increment of current equivalent plastic strain to the accumulated plastic strains. 15 - The load is increased one step and the procedure is repeated from step 4 through 15, until the final value of load is reached.

For unloading, we may define two different coordinate axes showing the loading and unloading behavior of the structure. A coordinate system is initially fixed at the origin, showing the loading curve. At the end of the loading curve, we may fix the second coordinate system with opposite coordinate directions showing the unloading curve. In the second coordinate system the effective stress, as the result of unloading, is always positive and is increasing. The final residual stresses are obtained by adding up the stresses in the second coordinate system to those in the first system, where proper change of sign are considered (Jiang [7]).

If Frederick-Armstrong model is used to evaluate the plastic behavior of the beam, items (7) and (8), which are used to calculate the value of the back stress tensor, will be changed.

5   Results and Discussion

Figure (1) shows the stress-strain cycle of a beam obtained using the proposed numerical method, with the ANSYS computer program. The beam data are: modules of elasticity $E = 193\ Gpa$, yield stress $\sigma_0 = 193\ Mpa$, height $h = 50\ mm$, width $b = 25\ mm$, $m = 896\ MPa$, $n = 0.5$. The Prager kinematic hardening rule is used and linear temperature distribution across the beam height with fixed ends (where the axial displacements at ends are zero) is cycled. The temperature at top of the beam is cycled between ($-220\ to\ 220^oC$), and at the same time temperature at bottom of the beam is cycled between ($220\ to\ -220^oC$). Now, the Frederick-Armstrong kinematic hardening theory is used to evaluate the cyclic behavior of the beam. The parameters C_f and γ are assumed as $43.3\ Mpa$ and 280, respectively. In the second example, consider a beam of rectangular cross section made of $CS1026$ steel under axial deformation, with the following data: modules of elasticity $E = 173.2\ Gpa$, yield stress $\sigma_0 = 241\ Mpa$. In Fig. (2) the beam is under cyclic bending moment of $-4\ KN$ to $5\ KN$, where ratcheting behavior is observed.

Different kinds of load and deformation contrlled cyclic loading with the named kinematic hardening models are solved using this method. Howevere the results are not showed hear.

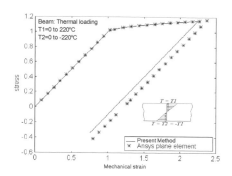 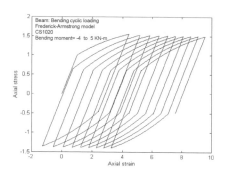

Figure 1: Verification of iterative numerical method for beam with ANSYS (left). Beam under cyclic bending moment load based on the frederick-Armstrong model (right).

6 Conclusion

A new effective iterative method for cyclic loading analysis of the beams is proposed in this article. This method may be used for the analysis of uniaxial as well as multiaxial state of loadings. Different beam problems considering the Prager, Frederick-Armstrong and Chaboche kinematic hardening model are solved using this method.

References

[1] Naderan-Tahan, K., Eslami, M.R. and Mahbadi, H., A Cyclic Loading Analysis of Structures of Strain Hardening Material, Proc. ISME2000 Conf., Sharif Univ. of Tech., Tehran, 2000, pp. 67-74.

[2] Mahbadi, H., and Eslami, M.R., "Load and Deformation Controlled Cyclic Loading of Beams, Based on the Isotropic Hardening Model", Accepted for publication. Trans. ISME.

[3] Tai-Ran Hsu, The Finite Element Method in Thermomechanic, Prentice-Hall Inc. Boston, 1984.

[4] Prager, W., A New Method of Analyzing Stresses and Strains Work-Hardening Plastic Solids, Journal of Applied Mechanics, December 1956, pp. 493-496.

[5] Armstrong, P.J. and Frederick, C.O., A Mathematical Representation of the Multiaxial Bauschinger Effect. CEGB Report No. RD/B/N 731, 1966.

[6] Chaboche, J.L., Dang-Van, K. and Cordier, G. Modelization of the Strain Memory Effect on the Cyclic Hardening of 316 Stainless Steel. Proceedings of the International Conference on SMiRT, 1979, Div. L, Berlin, Germany.

[7] Jiang, W., "New Kinematic Hardening Model", Journal of Engineering Mechanics", Vol. 120. No. 10, October 1994, pp. 2000-2020.

VSP International
Science Publishers
P.O. Box 346, 3700 AH Zeist
The Netherlands

*Lecture Series on Computer
and Computational Sciences*
Volume 1, 2004, pp. 173-176

Higher Order Elements for the Analysis of the Generalazied Thermoelasticity of Disk Based on the Lord Shulman Model

M. R. Eslami[1] and A. Bagri [2]

Department of Mechanical Engineering
Amirkabir University of technology,
Tehran, Iran

Received 17 July, 2004; accepted in revised form 14 August, 2004

Abstract: The generalized coupled thermoelasticity model of a disk based on the Lord-Shulman theory is presented in this paper. A transfinite element method using the Laplace transform is proposed to solve the coupled equations for an axisymmetrically loaded disk in the transformed domain. The dimensionless temperature and displacement in the transformed domain are inverted to obtain the actual physical quantities using the numerical inversion of the Laplace transform method. Results are compared for various orders of the elements.

1 Introduction

Some modified dynamic thermoelastic models are proposed to analyze the problems with second sound effects, such as Lord-Shulman (LS) [1], Green-Lindsay (GL) [2], and the Green-Naghdi (GN) [3] theories. These non-classical theories are referred to as the generalized thermoelasticity theories, or thermoelasticity theories with finite thermal wave speed.

Except for the particular problems, the analytical solution of these theories are difficult to obtain. Therefore, the numerical solutions are commonly employed to analyze the problems under the assumption of these theories. Chen and Lin [4] proposed a hybrid numerical method based on the Laplace transform and control volume method to analyze the transient coupled thermoelastic problems with relaxation times involving a nonlinear radiation boundary condition. Hosseini Tehrani and Eslami [5] considered the boundary element formulation for the analysis of coupled thermoelastic problems in a finite domain and studied the coupling coefficient and relaxation times effects on thermal and elastic waves propagations.

In this paper, a transfinite element method using the Laplace transform is used to solve the coupled equations for a axisymmetrically loaded disk in the transformed domain. Elements with various orders are employed to investigate the effects of the number of nodes in an element. Finally, the temperature and displacement are inverted to obtain the actual physical quantities, using the numerical inversion of the Laplace transform method proposed by Honig and Hirdes [6]. The results are qualitatively validated with the known date in the literature.

[1]Professor of ME. Dept. Amirkabir Univ. of Tech., Tehran, Iran. E-mail: eslami@aut.ac.ir
[2]Ph.D Student ME. Dept. Amirkabir Univ. of Tech., Tehran, Iran. E-mail: bagri_a@aut.ac.ir

2   Governing Equations

In the absence of heat source and body forces and for isotropic materials, the nondimensionalized form of the generalized coupled thermoelastic equations of the axisymmetrically loaded circular disk based on the Lord-Shulman theory in terms of the displacement and temperature may be written as [7]

$$
\left\{ \frac{\partial^2}{\partial r^2} + \frac{1}{r}\frac{\partial}{\partial r} - \frac{1}{r^2} - \frac{\partial^2}{\partial t^2} \right\} u - \frac{\partial T}{\partial r} = 0 \tag{1}
$$

$$
\left\{ \frac{\partial^2}{\partial r^2} + \frac{1}{r}\frac{\partial}{\partial r} - \frac{\partial}{\partial t}\left(1 + t_0\frac{\partial}{\partial t}\right) \right\} T - C \left\{ t_0 \left[\frac{\partial^3}{\partial r \partial t^2} + \frac{1}{r}\frac{\partial^2}{\partial t^2} \right] + \frac{\partial^2}{\partial r \partial t} + \frac{1}{r}\frac{\partial}{\partial t} \right\} u = 0 \tag{2}
$$

Here, $C = T_0 \bar{\beta}^2 / \left[\rho c_e \left(\bar{\lambda} + 2\mu \right) \right]$ is the coupling coefficient. For the plane stress condition $\bar{\lambda} = \frac{2\mu}{\lambda+2\mu}\lambda$ and $\bar{\beta} = \frac{2\mu}{\lambda+2\mu}\beta$. In the preceding equations ρ, u, T_0, T, $\bar{\beta}$, c_e and t_0 are the density, radial displacement, reference temperature, temperature change, stress-temperature moduli, thermal conductivity, specific heat and relaxation time (proposed by Lord and Shulman), respectively, while λ and μ are the Lamé constants. The dimensionless thermal and mechanical boundary conditions are

$$
q_{in} = -\frac{\partial T}{\partial r} \qquad ; \qquad u = 0 \qquad\qquad at\ \ r = a
$$

$$
T = 0 \qquad ; \qquad \sigma_{rr} = \frac{\partial u}{\partial r} + \frac{\bar{\lambda}}{\bar{\lambda}+2\mu}\frac{u}{r} - T = 0 \quad at\ \ r = b \tag{3}
$$

where σ_{rr}, a and b are the radial stress, dimensionless inner and outer radii, respectively.

3   Transfinite Element Formulation

In order to derive the transfinite element formulation, the Laplace transformation is used to transform the equations into the Laplace domain. Applying the Galerkin finite element method to the governing equations (1) and (2) for the base element (e), yields

$$
\int_0^L \left\{ -\left[\left\{ \frac{1}{(R+r_i)}\frac{\partial}{\partial R} - \frac{1}{(R+r_i)^2} - s^2 \right\} u - \frac{\partial T}{\partial R} \right] N_m(R+r_i) \right.
$$
$$
\left. + \frac{\partial\left(N_m(R+r_i)\right)}{\partial R}\frac{\partial u}{\partial R} \right\} dR = N_m(R+r_i)\frac{\partial u}{\partial R}\bigg|_0^L \tag{4}
$$

$$
\int_0^L \left\{ -\left[\left\{ \frac{1}{(R+r_i)}\frac{\partial}{\partial R} - s(1+t_0 s) \right\} T - C\left(t_0 s^2 + s\right)\left(\frac{\partial}{\partial R} + \frac{1}{R+r_i} \right) u \right] \right.
$$
$$
\left. \times N_m(R+r_i) + \frac{\partial\left(N_m(R+r_i)\right)}{\partial R}\frac{\partial T}{\partial R} \right\} dR = N_m(R+r_i)\frac{\partial T}{\partial R}\bigg|_0^L \tag{5}
$$

where $u = \sum_{m=1}^n N_m U_m$ and $T = \sum_{m=1}^n N_m T_m$. In the preceding equations s, N_m, $R = r - r_i$, r_i, L, U_m, T_m are the Laplace parameter, shape function, local coordinates, the radius of the i-th node of the base element, the length of element in the radial direction, nodal displacement, and the nodal temperature respectively. The terms on the righthand sides of Eqs. (4) and (5) cancel each other between any two adjacent elements, except the nodes located on the boundaries

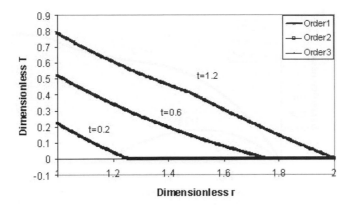

Figure 1: distribution of the dimensionless temperature along the radius of the disk at several times for three types of elements

of the solution domain. These boundary conditions are

$$a\left.\frac{\partial T}{\partial R}\right|_1 - aq_{in} \quad ; \quad U_1 = 0$$

$$T_M = 0 \quad ; \quad b\left.\frac{\partial u}{\partial R}\right|_M = -\frac{\bar{\lambda}}{\bar{\lambda} + 2\mu}U_M + bT_M \qquad (6)$$

The subscript 1 and M are referred to the first and last nodes of the solution domain, respectively.

4 Numerical Results and Discussions

To investigate the accuracy of the method, a numerical example is considered. The material of the disk is assumed to be aluminum. The dimensionless inside and outside radii are $a = 1$ and $b = 2$. The dimensionless input heat flux is defined as the Heaviside unit step function. Since the applied boundary conditions are assumed to be axisymmetric, the radius of the disk is divided into 100 elements. Three types of shape functions, linear, second order, and third order polynomials are used for the finite element model of the problem. Results for each of these orders are plotted and are compared.

Figures (1) and (2) show the wave propagations of temperature and radial displacement along the radial direction. The numerical values of the coupling parameter and the dimensionless relaxation time are assumed to be 0.01 and 0.64, respectively. The wave propagations are shown at several times. Two wave fronts for elastic and temperature waves are detected from the figures, as expected from the LS model. It is seen from the figures that the results of the three types of shape functions for the assumed number of elements coincide. For smaller number of elements, the difference between the results obtained for different shape functions increase noticibly. For the assumed number of elements, the curves for radial displacement and temperature distribution are checked againts the known data in the literature, where very close agreement is observed. Figure (1) clearly shows the temperature wave front (the second sound effect), which is propagating along the radius of the disk.

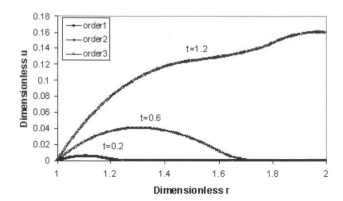

Figure 2: distribution of the dimensionless displacement along the radius of the disk at several times for three types of elements

5   Conclusions

The generalized coupled thermoelasticity of a hollow disk based on the LS theory is studied in this paper. A suitable transfinite element method is applied to provide an appropriate results in time domain. Distributions of temperature and displacement at several times and along the radius of the disk are obtained and shown in the figures. It is seen that the Galerkin finite element method rapidly converge for different order of shape functions, when the number of elements are suitably increased.

References

[1] Lord, H.W. and Shulman, Y., A Generalized Dynamical Theory of Thermoelasticity, J. Mech. Phys. Solids, Vol. 15, pp. 299-309, 1967.

[2] Green, A.E. and Lindsay, K.A., Thermoelasticity, J. Elasticity, Vol. 2, No. 1, pp. 1-7, Mar 1972.

[3] Green, A.E., and Naghdi, P.M., Thermoelasticity Without Energy Dissipation, J. Elasticity, Vol. 31, pp. 189-208, 1993.

[4] Chen, H., and Lin, H., Study of Transient Coupled Thermoelastic Problems with Relaxation Times, Trans. ASME, Vol. 62, pp. 208-215, 1995.

[5] Hosseini Tehrani, P., and Eslami, M.R., Boundary Element Analysis of Coupled Thermoelasticity with Relaxation Time in Finite Domain, AIAA J., Vol. 38, No. 3, pp. 534-541, 2000.

[6] Honig, G., and Hirdes, U., A Method For The Numerical Inversion of Laplace Transforms, J. Comp. App. Math., Vol. 10, pp. 113-132, 1984.

[7] Bagri, A., and Eslami, M.R., Generalized Coupled Thermoelasticity of Disks Based on The Lord-Shulman Model, to be published in Journal of Thermal Stresses, Vol. 27, 2004.

VSP International
Science Publishers
P.O. Box 346, 3700 AH Zeist
The Netherlands

*Lecture Series on Computer
and Computational Sciences*
Volume 1, 2004, pp. 177-180

Numerical Simulation of Two-Layered, Unsteady, Stratified Flow

E. D. Farsirotou[†], J. V. Soulis[2], V. D. Dermissis[3]

[1] Filippou Ioannou 59, 38222 Volos, Greece.
e-mail: efars@uth.gr tel.: 2421026552
(Corresponding author)

[2]Fluid Mechanics/Hydraulics Division
Department of Civil Engineering
Democrition University of Thrace
Xanthi 67100, GREECE
e-mail: jvsoulis@med.auth.gr tel.: 2541079617, fax: 2310994838

[3]Hydraulics and Environmental Engineering Division
Department of Civil Engineering
Aristotle University of Thessaloniki
Thessaloniki 54006, GREECE
e-mail: vdermiss@civil.auth.gr tel.: 2310995691, fax: 2310995664

Received 26 August, 2004; accepted in revised form 4 September, 2004

Abstract: The environmental protection of the rivers estuaries or channels near coastal areas from seawater entrance is of paramount importance. Stratified flows between two distinct layers of fluids, of slightly different density, are correlated with a variety of practical problems such as water quality, the length of saline and warm arrested wedges. The current numerical research work is based on a laboratory experimental investigation (previously conducted) of a two-layered system. Main scope was the correlation of the mean interfacial friction factor with the flow and salinity parameters. The hydraulic flow conditions of a two-layered system comprised of fresh and salt water are numerically simulated under unsteady flow conditions. Numerical predictions and measurements of the vertical distribution of the relative density under different freshwater discharges and salt water densities are satisfactorily compared.

Keywords: Stratified flow, salt water-freshwater interface, density variation, computational fluid dynamics.

Mathematics SubjectClassification: 35Q30

1. Introduction

As rivers estuaries are sensitive to a reduction in the supply of freshwater many studies have been conducted in order to succeed a good water quality management and to prevent from salinity hazards. Coates M. J. et al [2] have experimentally simulated the motion of the trapped salt wedge within the estuary and Burrows R. et al [1] studied experimentally and numerically the mixing and entrainment between fresh and salt water. The hydraulic flow conditions of a two-layered system comprised of fresh and salt water are to be numerically simulated with measurements under unsteady flow conditions.

2. Experimental set up

An experimental apparatus whose main part is a closed rectangular duct was designed and constructed at the Hydraulics Laboratory, of Civil Engineering Department, Aristotle University of Thessaloniki [3] and [4] in order to investigate the hydraulics of a two-layered system, fresh and salt water, under quasisteady conditions with freshwater flowing over a quasi-stagnant salt water pool. The testing laboratory flume was 15.0 m long, 0.5 m wide and 0.2 m deep. The blocking of salt water was achieved

---

[†] Corresponding author. Dr Civil Engineer. E-mail: efars@uth.gr

using two piers at the two edges of the duct. The depth of the two-layered system was stable. During the experiments the inlet velocity of the freshwater, the original depth of the upper layer and the density difference between the two layers were varied. The original depth of the upper layer (freshwater) was defined from the piers height. A sketch of the closed duct is given in Figure 1, where the section, named B′, of the measured freshwater depth is also shown. All experimental measurements were conducted under steady conditions, during the course of the first 2-3 minutes, in order to avoid any variation in the interface slope.

3. Numerical formulation

The Computational Fluid Dynamics computer software "Fluent" [5] was used to simulate the shape of salt water-freshwater interface and the entrainment through the interface in the experimental closed duct. A three-dimensional, unsteady, turbulent flow simulation, based on the k-ω turbulence model, is conducted to investigate the flow parameters and the density variation of the aforementioned two-layered system with fresh and salt water. The numerical code solves the governing Navier-Stokes equations,

$$\frac{\partial \rho}{\partial t} + \frac{\partial (\rho u_j)}{\partial x_j} = S_m \tag{1}$$

$$\frac{\partial (\rho u_j)}{\partial t} + \frac{\partial (\rho u_j u_j)}{\partial x_j} = -\frac{\partial p}{\partial x_i} + \frac{\partial \tau_{ij}}{\partial x_j} + \rho g_i + F_i \tag{2}$$

and the convection/diffusion equation for a scalar quantity ϕ equal to the density difference between the two liquids,

$$\frac{\partial (\rho \phi)}{\partial t} + \frac{\partial (\rho u_j \phi)}{\partial x_j} - D_j \frac{\partial^2 \phi}{\partial x_i^2} = S_\phi \tag{3}$$

where ρ (kg/m^3) is the density, t (sec) is the time, u_i (m/s) are the velocity components along the x_i (m) axes, S_m is the added or subtracted mass of fluid in the flow field, p (N/m^2) is the static pressure, τ_{ij} (N/m^2) is the shear stress tensor, ρg_i (N/m^3) and F_i (N/m^3) are gravity and externally acting forces, respectively, D_i is the diffusion coefficient and S_ϕ is the source term. The channel was carrying a uniform inflow discharge of fresh water (laboratory measurements) and a uniform velocity is set as inlet boundary condition. The salt water is retained in the region enclosed between the two piers and this situation is simulated as the initial flow condition for the numerical solution. The interface between the two fluids is expected to vary.

4. Results

Figure 2 and 3 shows characteristic comparisons between numerical predictions and measurements of the vertical distribution (h is the distance along the vertical direction) for the relative density $\Delta\rho/\rho$ (where $\Delta\rho$ is the density difference between the two layers, ρ is the density of fresh water). In these simulations the relative density of the two layers is equal to 25°/oo and the inflow discharges of freshwater were set equal to Q=0.002829m^3/s and Q=0.006964m^3/s, respectively. The initial depth of the salt water is set equal to 0.044 m. The expected intense entrainment through the interface is, numerically, well simulated. The vertically distribution of the relative density along the depth of the duct is adequately predicted.

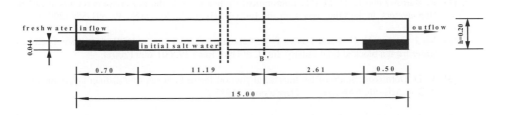

Figure 1: Sketch of the experimental closed duct. All dimensions in meters.

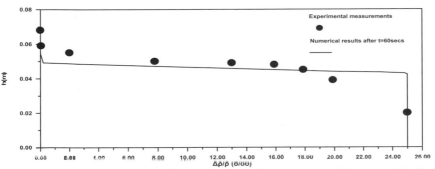

Figure 2: Comparison between numerical predictions and measurements of the vertical distribution of the relative density $\Delta\rho/\rho$ under Q=0.002829m^3/s.

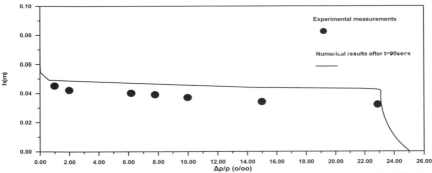

Figure 3: Comparison between numerical predictions and measurements of the vertical distribution of the relative density $\Delta\rho/\rho$ under Q=0.006964m^3/s.

The shape of the salt water-freshwater interface and the width of the interfacial zone are satisfactorily simulated. The calibrated numerical code can be used for a variety of two-layered stratified flows.

Acknowledgments

The author wishes to thank the anonymous referees for their careful reading of the manuscript and their fruitful comments and suggestions.

References

[1] R. Burrows and K. H. M. Ali, Entrainment studies towards the preservation (containment) of 'freshwaters' in the saline environment, Int. J. of Hydraulic Research, 39, 591-599 (2001).

[2] M. J. Coates, Y. Guo and Davies P.A., Laboratory model studies of flushing of trapped salt water from a blocked tidal estuary, Int. J. of Hyd. Res., 39, 601-609 (2001).

[3] V. D. Dermissis, Study of the interfacial friction factor between two distinct layers of fluids, PhD in Aristotle University of Thessaloniki (1977).

[4] V. D. Dermissis and E. Partheniades, Interfacial resistance in stratified flows, Int. J. of Waterway, Port, Coastal and Ocean Engineering, 110(2), 231-250 (1984).

[5] Fluent Inc., Fluent (version 6.0) User's Guide, Lebanon, USA (2001).

VSP International
Science Publishers
P.O. Box 346, 3700 AH Zeist
The Netherlands

*Lecture Series on Computer
and Computational Sciences*
Volume 1, 2004, pp. 181-184

Clustering Chemical Data Bases Through Projection of MOS Similarity Measures on Multidimensional Spaces

Gonzalo Cerruela García, Irene Luque Ruiz[1], and Miguel Ángel Gómez-Nieto

Department of Computing and Numerical Analysis. University of Córdoba
Campus Universitario de Rabanales, Building C2, Plant 3, E-14071 Córdoba (Spain)

Received 24 May, 2004; accepted in revised form, 24 June 2004

Abstract: In this paper we propose a new method for the clustering of chemical databases based on the representation of the measures of structural similarity on multidimensional spaces. The proposed method allows the tuning of the clustering process by means of the selection of the number of spaces, the normal vectors and the sensibility of the projection process. The structural similarity of each element with regard to the elements of the database is projected on to the defined spaces generating clusters that represent the pattern and diversity of the database and whose size and characteristics can be easily adjusted.

Keywords: Clustering, Chemical databases, Similarity measures

Mathematics SubjectClassification: 62H30, 92E10, 68P20

1. Introduction

The clustering method proposed is based on the calculation of the measures of structural similarity among the elements of the chemical database. Given two molecular graphs G_A and G_B representing the structure of chemical compounds A and B respectively, it is possible to obtain a set M of all the maximum common substructures among the G_A and G_B graphs [1]. Each element M_i of the M set is composed of a set of nodes and edges present in the G_A and G_B graphs (atoms and bonds present in A and B), for which the M set represents the maximum overlapping or the maximum matching among the molecular graphs G_A and G_B. Knowing the sizes (number of nodes and edges) n_A, n_B, e_A, e_B of the G_A and G_B graphs, and knowing the sizes of the matching vectors M_i, we can obtain a series of similarity measures $ns_{ij}-m+1$ between the G_A and G_B graphs, where m is the number of elements of the M set.

For a database of chemical compounds with db elements, a symmetrical S matrix can be obtained with information of the maximum overlapping (MOS) among each couple of elements (i, j) of the database, and therefore $(db^2-db)/2$ measures of structural similarity S_{ij}, which can be obtained in one or several of the commonly accepted indexes of similarity [2].

In this paper we propose a clustering method based on the MOS measures and the projection of these measures onto different spaces that represent intervals of similarity in those that can represent the similarity of a molecule with the elements of a chemical database.

2. Description of the clustering method

Given a chemical database with db elements, for each pair of elements (i, j) a value s_{ij} can be obtained that stores the similarity among the elements i and j calculated starting from the Maximum Overlapping Substructure [1].

Thus, the clustering process requires a preprocessing stage in which the s_{ij} values for each couple of elements of the database are calculated. In our paper we have used the cosine similarity index for this calculation.

---

[1] Corresponding author.. E-mail: ma1lurui@uco.es

$$s_{ij} = \frac{\overline{G_i \cap G_j}}{\overline{G_i \cup G_j}} = \frac{n_c^m + e_c^m}{n_i + n_j + e_i + e_j} \tag{1}$$

where: $\overline{G}$ represents the size of the molecular graphs, n is the number of nodes, e is the number of edges, n_c^m and e_c^m are the common number of nodes and edges belonging to the maximum common substructure between the G_i and G_j graphs.

When this process is carried out for the *db* elements of the database, a symmetrical similarity matrix S is obtained, in which the element $S(i,j) = S(j,i)$ stores the s_{ij} similarity value between the G_i and G_j graphs.

As in most of the clustering methods, the preprocessing stage is the most expensive computationally — $O(n^2)$—. Having obtained the information corresponding to the similarity values among all the database elements, the clustering process is carried out in four steps.

1. The number and range of the intervals of similarity values in which the database will be projected are selected.
2. The projection process of the S matrix of similarity values on the defined n-dimensional space of similarity is carried out.
3. The arrays of similarity of each element of the database are normalized in the n-dimensional space of similarities.
4. The classes or clusters are defined and the database elements are assigned to them.

The intervals of similarity will determine the granularity of the clustering process. As the number of intervals of similarity increase (and therefore smaller in size) the granularity of the clustering process will be greater, and vice versa. As the granularity increases the number of created clusters also increases and therefore the clusters population diminishes.

In the proposed clustering method the number and size of the intervals of similarity is dynamic and it can be re-adjusted conveniently in function of the results observed. So, an I array is defined corresponding to the intervals of similarity where:

1. Each element $I(i)$ defines an interval of similarity $[x, y[$, where $y > x$.
2. The first element $I(1)$ is an interval defined as $[0, y[$.
3. The last element $I(n)$ is an interval defined as $[x, 1]$.
4. The defined intervals of similarity in the I array are disjoint intervals, so that $x_k > y_{k-1}$.

Having selected the projection spaces, for each row of the S matrix (each element of the database) an array in the n-dimensional space of similarity is obtained by means of the following expression:

$$b_i = \left[\frac{\sum\limits_{j}^{I(1)} S(i,j)}{M_i^{I(1)}}, \frac{\sum\limits_{j}^{I(2)} S(i,j)}{M_i^{I(2)}}, \frac{\sum\limits_{j}^{I(3)} S(i,j)}{M_i^{I(3)}}, \dots\dots, \frac{\sum\limits_{j}^{I(db)} S(i,j)}{M_i^{I(db)}} \right] \tag{2}$$

where: $\overset{I(k)}{S(i,j)}$ represents the MOS similarity value among the elements i and j into the $I(k)$ interval, and $\overset{I(k)}{M_i}$ represents the total number of matching of the element i whose value of similarity is in the $I(k)$ interval.

Once the (b_i) arrays representing each element of the database in the n-dimensional space of similarity are obtained, the database can be represented by means of a B matrix of size $db \times k$, where k is the number of dimensions of the n-dimensional space of similarity. This B matrix can be normalized in the interval $[0,1]$ which bears the normalization of the space of similarity, in the following way:

$$\forall k, \overline{B(i,j)} = \frac{B(i,j) - \min(B(k,j))}{\max(B(k,j)) - \min(B(k,j))} \tag{3}$$

Next is carried out a grid of the n-dimensional space of similarity. This grid consists of the construction of a set of n-dimensional cells or bins whose size can be equal or different in each dimension.

The grid size, together with the number of dimensions determines the maximum number of classes or clusters in which the database elements can be classified, which is given by the expression: $(1 / grid)^k$, where k is the number of dimensions of the space of representation.

Table 1: Some results of the clustering for different grid sizes and projection spaces for a database of 500 elements. C_T: number of classes, CE: cluster entropy, ENC effective number of clusters, %ENC: percentage of effective number of clusters, %S: percentage of singletons, %D: percentage of doubletons, AP: average of the clusters population

Projection	Grid	C_T	CE	ENC	%ENC	%S	%D	AP
	0.20	19	3.1	8.6	45.3	26.3	15.8	26.3
[0.00 – 0.50 – 1.00]	0.10	46	4.6	25.1	54.6	26.1	13.0	10.9
	0.05	112	6.2	71.6	63.9	35.7	13.4	4.5
	0.20	46	4.9	29.2	63.5	26.1	6.5	10.9
[0.00 – 0.33 – 0.66 – 1.00]	0.10	144	6.7	106.3	73.8	34.7	15.3	3.5
	0.05	314	8.1	265.9	84.7	66.6	18.2	1.6
	0.20	57	4.3	20.4	35.8	31.6	24.6	8.8
[0.00 – 0.25 – 0.50 – 0.75 – 1.00]	0.10	134	6.2	72.7	54.3	44.8	20.9	3.7
	0.05	271	7.7	211.2	77.9	62.7	17.0	1.8

Once the grid is built, each element of the database is assigned to one of the generated cells and we proceed to the obtaining of a series of measures used to analyse the usefulness of the clustering process (see Table 1).

1. The number of classes and its populations.
2. The number and percentage of singletons and doubletons.
3. The entropy of the clustering process (*CE*), the number of effective cluster (*ENC*) and its percentage *(% ENC)* [3], calculated from the following expressions:

$$CE = -\sum_{i=1}^{q} f_i \log_2 f_i \qquad (4)$$

where: q is the clusters number and $f_i = n_i/db$ is the population frequency in each cluster, calculated as the ratio among the population of each cluster (n_i) and the number of elements of the database (db). Knowing *CE*, the effective number of clusters can be calculated with the expression:

$$ENC = 2^{CE}, \% ENC = \frac{ENC}{q} \qquad (5)$$

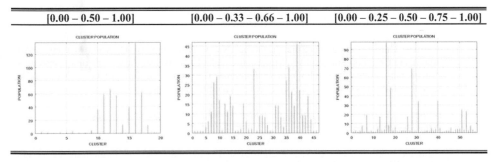

| [0.00 – 0.50 – 1.00] | [0.00 – 0.33 – 0.66 – 1.00] | [0.00 – 0.25 – 0.50 – 0.75 – 1.00] |

Figure 1: Representation of clusters populations for grid equal to 0.2 and different projection spaces

Discussion

The tests of the method of proposed clustering have been carried out on several public domain databases with very different characteristics as to size and diversity of the chemical compounds that store [4]. A summary of the obtained results is shown in Table 1 and Figure 1 with a database of 500 elements.

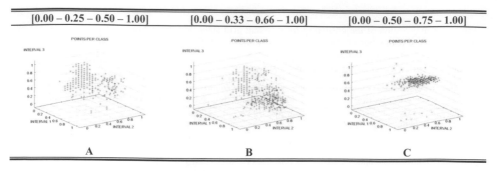

Figure 2: Representation of clusters distribution for grid equal to 0.05 and different similarity intervals in 3D space

As we can appreciate for the same number of projection spaces, as the size of the cells (from 0.2 at 0.05) diminishes, the number of generated classes increases, diminishing the population of the classes and increasing the singletons and doubletons percentage.

As Table 1 and Figures 1 show, the characteristics of the clustering depend on the number of projection spaces, the grid size and the intervals of similarity considered for each projection dimension.

The projection in 3D generates an increase in the number of classes with regard to 2D for all the grid sizes for the intervals of similarity considered in the projection dimensions that are shown in Table 1, while the projection in 4D produces a decrease in the number of classes with regard to 3D for small grids.

The consideration of other intervals of similarity has proven that it causes changes in its behaviour, which allows the refinement of the clustering process based on the characteristics of the elements of the database.

Figure 2 shows the behaviour of the clustering for a projection in 3D for different intervals of similarity. It can be observed that when we displace the intervals toward greater values of similarity (from A to C), a cluster of the classes in well differentiated groups takes place.

The tuning of parameters as intervals of similarity, number of projection spaces and grid size allow for adjusting the clustering method to adapt to different objectives, as: a) search of patterns, screening in databases, or to characterize diversity in databases, as can be appreciated in Figure 2, in which for a grid of 0.05 clusters with a very different behaviour are observed.

References

[1] Cerruela García, G., Luque Ruiz, I., Gómez-Nieto, M.A. A New Algorithm to Obtain All Maximum Common Subgraphs in Molecular Graphs Using Binary Arithmetic and Constraints Satisfaction Model. Proceedings of International Conference of Computational Methods in Sciences and Engineering, Kastoria, Greece, September 2003. pp. 135-138.

[2] Willet, P., Barnard, J.M., Downs, G. Chemical Similarity Searching. *J. Chem. Inf. Comput. Sci.* 1998, 38(6), 983-996.

[3] Taraviras, S.L., Ivanciuc, O., Carbol-Bass, D. Identification of Groupings of Graph Theoretical Molecular Descriptors Using a Hybrid Cluster Analysis Approach. *J. Chem. Inf. Comput. Sci.* 2000, 40(5), 1128-1146.

[4] SPECS and BioSPECS B.V. http://www.specs.net

VSP International
Science Publishers
P.O. Box 346, 3700 AH Zeist
The Netherlands

*Lecture Series on Computer
and Computational Sciences*
Volume 1, 2004, pp. 185-189

A Potential – Based Cellular Automaton Model for Earthquake Simulation

I.G. Georgoudas, G.Ch. Sirakoulis[1], I. Andreadis

Laboratory of Electronics,
Department of Electrical and Computer Engineering,
Democritus University of Thrace,
GR-671 00 Xanthi, Greece

Received 28 July, 2004; accepted in revised form, 18 August 2004

Abstract: In this paper a potential - based model for earthquake simulation that is a two-dimensional cellular automaton dynamic system constituted of cells-charges, is presented. The proposed model is constructed in order to simulate earthquake activity in correspondence to the quasi-static two-dimensional version of the Burridge-Knopoff spring-block model of earthquakes, as well as, to the Olami-Feder-Christensen (OFC) earthquake model. The simulation results are found in good quantitative and qualitative agreement with the Gutenberg–Richter (GR) scaling relation predictions. Numerical results for various cascade (earthquake) sizes and different critical states are presented. Furthermore, the parameter of different neighbourhood in the proposed model is also explored.

Keywords: Cellular Automata, Earthquake modelling, critical state, earthquake size, neighbourhood.

Mathematics Subject Classification: 37B15, 68Q80, 86A15, 86A17, 82C27.
PACS: 91.30.Px, 05.45.-a, 02.90.+p, 05.65.+b.

1. Introduction

Seismologists, and more recently computational and condensed-matter physicists, have made extensive use of computer modelling to investigate the physics of earthquakes. Numerical modelling of earthquake processes has become an important proving ground for ideas that have no other experimental arena. Earthquakes primarily occur at the boundaries of rigid tectonic plates. As the plates drift, nearly all relative motion occurs across a narrow network of cracks or faults that form at the interface. Friction across the rough fault faces prevents steady sliding, instead forcing the nearby ground to deform and accumulate strain. The resulting increase in stress will eventually force the rock to fracture along a weakened fault plane. In this manner, strains accumulated at rates of a few centimetres per year over decades and centuries release in sudden shifts of several meters. The energy once stored as elastic strain is released in the form of heat, damage to the rock, and destructive seismic waves [1]. This process is innately nonlinear, involving the interplay of viscoelastic and frictional forces, under the influence of heterogeneities of the crust, pore fluids in the rock, and innumerable other complications. There is no successful strategy for predicting when and where such ruptures will occur or how big they will grow to be [1]. Emerging from these seemingly insurmountable complexities is a surprisingly simple and exacting pattern: the Gutenberg–Richter (GR) scaling relation.

More specifically, in 1956, Gutenberg and Richter [2] determine that the frequency F of earthquakes of magnitude $M > m$ is given by:

$$\log F\,(M > m) = a - b \cdot m \tag{1}$$

The parameter b presents a wide range of values that depends on the fault. For small earthquakes it was found that $b \in [0.80, 1.06]$ while for large ones $b \in [1.23, 1.54]$, [3]. This equation is known as the Gutenberg-Richter law. By the other hand, the energy E released during an earthquake is believed to have the following behaviour:

$$\log E\,(M > m) = c + d \cdot m \tag{2}$$

[1]Corresponding author, email: gsirak@ee.duth.gr

where $d = 1$ and $d = 3/2$ for small and large earthquakes respectively [3]. By combining these equations:

$$F(E > e) \propto E^{-\beta} \tag{3}$$

As a result, relatively simple models will properly capture the essential physics responsible for these behaviours [1]. The use of minimalistic models began when Burridge and Knopoff published results indicating GR-like power-law behaviour from a simple chain of blocks and springs being pulled across a rough surface. In such a model, springs connect blocks representing contiguous sections of a fault to provide a linear elastic coupling [4]. Seismologists became interested in cellular automata as possible analogues of earthquake fault dynamics when Bak and Tang demonstrated that even highly simplified, nearest neighbour automata produce power-law event size distributions [5]. The sand-pile automaton of Bak et al. [6] was employed to demonstrate the emergence of power-law statistics as the result of the combined action of a large number of simple elements which interact only with nearby cells via a pre-specified interaction rule. Bak et al. [6] termed such behaviour Self-Organised Critical behaviour [6]. Carlson and Langer proposed a 1D dynamical version of the Burridge and Knopoff model (BK model) [7], while Olami, Feder and Christensen introduced a generalized, continuous, non-conservative cellular automaton for modelling earthquakes, known as OFC earthquake model [8].

Here, it is described a simple cellular automaton model with continuous states and discrete time, constituted of cells-charges that aim to simulate earthquake activity with the usage of potentialities. The produced simulation results are found in good quantitative and qualitative agreement with the Gutenberg–Richter (GR) scaling relation predictions, while numerical results for various cascade (earthquake) sizes and different critical states are presented and found in good agreement with others CA earthquake models. It should be mentioned that, in the literature, the BK model is treated as an example of weakly driven dissipative system with many meta-stable states. In most of these studies a rough purely velocity-weakening friction law is used. But any phenomenological friction law has to be velocity strengthening for large velocities. At the presented model the use of potentialities is able to produce a more realistic friction for small velocities. Additionally, the physics of the model using potential dynamics reveal a more natural structure than that described at literature [4], [8]. Finally, the parameter of different neighbourhood, meaning the usage of Moore neighbourhood, instead of von Neumann neighbourhood, is also examined resulting in the corresponding measurements.

2. CA model description

Potential - based model for earthquake simulation is a two-dimensional dynamic system constituted of cells-charges. Its aim is the simulation of seismic activity with the use of potentialities. It is assumed that the system balances through the exercitation of electrostatic Coulomb-forces among charges, without the existence of any other form of interconnection in-between. Such kind of forces is also responsible for this level to be bonded with a rigid but moving plane below. Cells are being recomposed by the alteration of this plane's potential.

The major characteristic of this model is that the whole study of the seismic activity stands on a potential-based analysis for each cell and not on the electrostatic forces that are being developed among them. The equivalence of the potential-based study with the force study is being ensured under the condition that the system is conservative. Obviously, the fact that in our system exist only conservative forces (Coulomb-forces among the charges) very well satisfies the above-mentioned condition. Furthermore, this is confirmed by the results show that the system presents self - organized criticality [9]. The hence coming advantage of this method is that vector analysis that would require additional compromises bringing on a decay of the results received is being overcome. On the other hand the manipulation of magnitude-only sizes provides calculate simplification as well as increased level of reliability.

As far as it concerns the dynamics of the potential-based model, this is ensured by the existence of simple update rules that takes place in discrete steps. In fact, there exists a model of cellular automaton, CA. Specifically, if at time moment $t = 0$ the potential $V_{i,j}$ of a cell placed at (i,j) exceeds the threshold value V_{th} of the level below, the balance is disturbed, and the cell is being moved. The moving cell is transferred at a point of lower energy, hence at a state of lower but nonzero potential.

The removal of the cell reorders the values of the potential at its nearest neighbours. The model study has been made taking under consideration four (Von Neumann neighbourhood) or eight active neighbours (Moore neighbourhood). In both cases, the moving cells, wherever placed in the lattice, interact with a constant number of neighbours contributing in the systems homogeneity.

Potential values conversion results in new cells to become unstable driving to the appearance of the well known as cascade phenomenon, which is the earthquake's equivalent of the proposed model [1].

The updated potential values following the local displacement of the (i,j) cell-charge is proven to be given by the relationship below:

$$V_{i\pm1,j} \rightarrow V_{i\pm1,j} + (a*r)V_{i,j}$$
$$V_{i,j\pm1} \rightarrow V_{i,j\pm1} + (a*r)V_{i,j}$$
$$V_{i\pm1,j\pm1} \rightarrow V_{i\pm,j\pm1} + (a*r)V_{i,j} \,(Moore)$$
$$V_{i,j} \rightarrow 1$$

(4)

This model needs only to store one real-value field, which is the potential. Being also accompanied by the merit that this is a magnitude-only size and hence it can be fully described by a unique value, further assumptions are avoided.

Another crucial factor, cell displacement, $dx_{i,j}$, is present in the model. Not only it is taken into account by the presence at the update rule of parameter a, which is inverse proportional to cell displacement $dx_{i,j}$, but also by the definition of an upper displacement limit. In other words, a maximum value for cell movement is introduced, away from which the return to an equilibrium state becomes impossible (charge would tend to infinite hence its potential value to zero). More specifically:

$$x_{i,j} \leq x_{th'} \Rightarrow \frac{k}{x_{i,j}} \geq \frac{k}{x_{th'}} \Rightarrow V_{i,j} \geq V_{th'} \quad \text{where } k > 0, cons\tan t$$

(5)

It is obvious that the displacement condition implies an additional one for the respective potential value. Consequently, the potential at each place (i,j) should be constantly larger than the value of the lower potential limit, $V_{th'}$. The dynamics of the proposed model is defined by the use of the parallel updating of the lattice [1]. All cells are being tested and change their potential values until none of them has a value greater than V_{th}. As soon as the whole procedure is completed, the earthquake phenomenon, which has been generated by the presence of value $V_{i,j} > V_{th}$, has been fully simulated.

The successive discrete step, which also implies the recurrence of an earthquake process, initiates with the application of the so-called increment method. According to this method, the cell with its value closer to the limit V_{th} is being found. This value minimizes the difference $(V_{th}-V_{i,j})$, which is then being added to the potential value of each cell.

Finally, as far as it concerns the boundary conditions the use of closed (zero) boundary conditions has been preferred. This decision relies on the fact that the model should present strong forms of in-homogeneity at the boundaries, since in-homogeneity is an attribute of earthquake faults at the surface.

3. Measurements and simulation results

There have been made two kinds of measurements so far. They are defined as follows [3]:

- **Critical state:**

It presents the state of the system after a large number of earthquake simulations, which practically takes place after the successive addition of the quantity $(V_{th}-V_{i,j})_{min}$ at the value of the potential of each cell-charge.

- **Cascade (earthquake) size:**

It is defined as the total number of cells that participate at a single earthquake procedure, which stands as long as the condition $V_{i,j} > V_{th}$ is true. This magnitude can be treated as a measure of the total energy released during the evolution of the earthquake or even to be interpreted as a measure of the earthquake's magnitude.

The results received reveal for the model the validity of the Gutenberg-Richter scaling law.

Below follow results for a variety of single parameters, given in forms of contours as far as it concerns critical state measurements or plots regarding measurements that prove the validity of the Gutenberg-Richter law.

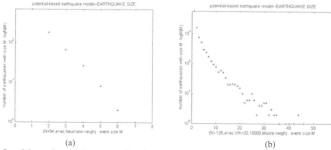

(a) (b)

Figure 1: (a) Semi-log plot of the cascade size vs. the number of earthquakes for N=64, V_{th}=2 (Von Neumann neighborhood-4 active neighbours) and 1000 events. (b) Semi-log plot of the cascade size vs. the number of earthquakes for N=128, V_{th}=30 (Moore neighborhood-8 active neighbours) and 10000 events.

At the above given plots, the dependence of the number of seismic procedures with a certain 'magnitude' M, named N(M), with this 'magnitude' M is depicted by a semi-logarithmic scale. It is obvious that these two sizes [log N(M), M] hold a linear relationship with a constant slope –p. Consequently, this is a result-based confirmation of the existence of the fundamental Gutenberg-Richter law in this model (Figure 1). The corresponding results, which include an increased number of active neighbours (Moore neighbourhood, including eight active neighbours) display also the maintenance of this substantial relationship [Figure 1 (b)]. It should be noted that the larger the number of interacted neighbours and of the lattice size the more accurate the results are. The drawback of such an approach is that the accumulation of neighbours widens the range of 'magnitudes' without though severely destroying the linear dependence. In any case, this is successfully confronted by increasing the value of the potential limit, V_{th}.

The following contours depict the form of the critical state for different values of the system's parameters. It is of major concern the size of the lattice together with the threshold value V_{th} as well as the number of the active neighbourhoods.

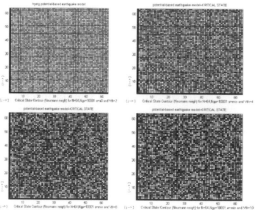

Figure 2: Critical state contour for N=64 and V_{th}=2, V_{th}=4, V_{th}=8, V_{th}=10, (from up left to right down) (Von Neumann neighbourhood - 4 active neighbours), in correspondence, and 10000 events.

The figures above disclose the complicated structure of the final state. The main feature is the reduction of the cells that reach their maximum value (which is depicted by bright colours) while the threshold of the potential value V_{th} increases. A rational explanation is based on the fact that the growth of the value V_{th} attenuates the probability of a potential value close to the threshold one to appear at the active neighbourhood. The shape is very similar to that observed in discrete sand-pile automaton, thus proving the existence of self-organised criticality in the model [6].

Similar results are obtained following the application of the Moore neighbourhood with eight active neighbour-cells (Figure 3).

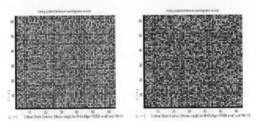

Figure 3: Critical state contour for N=64 and V_{th}=4, V_{th}=10, (from left to right) (Moore neighbourhood - 8 active neighbours), in correspondence, and 10000 events.

4. Conclusions

The use of simple models, amenable to making the best use of today's computing power as well as yielding to analytical treatment, is without a doubt a powerful approach to difficult nonlinear problems. The presented potential-based CA model was constructed in order to simulate earthquake activity in correspondence to the quasi-static two-dimensional version of the BK spring-block model of earthquakes, as well as, to the OFC earthquake model. Numerical results for various earthquake sizes and different critical states were presented and the parameter of different neighbourhood in the proposed model was also explored. The simulation results were found in good quantitative and qualitative agreement with the GR scaling relation predictions. Consequently, complex real systems, like earthquakes, can be studied with simple cellular automata models.

References

[1] E.F. Preston, J.S. SaMartins, J.B. Rundle, M. Anghel, and W. Klein, Models of Earthquake Faults with Long Range Stress Transfer, *IEEE Computing in Science and Engineering*, 2 34-41(2000).

[2] B. Gutenberg, and C.F. Richter, Magnitude and Energy of Earthquakes, *Ann. Geophys.*, 9 1-15(1956).

[3] G. Hernandez, Parallel and distributed simulations and visualizations of the Olami-Feder-Christensen earthquake model, *Physica A*, 313 301-311(2002).

[4] R. Burridge, and L. Knopoff, Model and theoretical seismicity, *Bulletin of Seismological Society of America*, 57 3 341-371(1967).

[5] P. Bak, and C. Tang, Earthquakes as a Self-organised Critical Phenomenon, *J. Geophys. Res.* 94 B11 15635–15637(1989).

[6] P. Bak, C. Tang, and K. Wiesenfield, Self-organised Criticality: an Explanation of 1/f Noise, *Phys. Rev. Lett.*, 59 381–384(1987).

[7] J.M. Carlson, and J.S. Langer, Mechanical model of an earthquake fault, *Phys. Rev. A*, 40 6470–6484(1989).

[8] Z. Olami, H.J.S. Feder, and K. Christensen, "Self-Organized Criticality in a Continuous, Nonconservative Cellular Automaton Modelling Earthquakes", *Phys. Rev. Lett.*, 68 8 1244-1247(1992).

[9] T. Hwa, and M. Kardar, Dissipative transport in open systems: An investigation of self-organized criticality, *Phys. Rev. Lett.*, 62 1813-1816(1989).

VSP International
Science Publishers
P.O. Box 346, 3700 AH Zeist
The Netherlands

*Lecture Series on Computer
and Computational Sciences*
Volume 1, 2004, pp. 190-193

On Transport-Chemistry Interaction in Laminar Premixed Methane-Air Flames

D.A. Goussis[†,1] G. Skevis[†] and E. Mastorakos[#]

[†]Institute of Chemical Engineering and High Temperature Processes, FORTH/ICE-HT, 26500 Rio,
Patra, Greece
[#]Engineering Department, University of Cambridge, Cambridge CB2 1PZ, UK

Received 6 August, 2004; accepted in revised form, 18 August 2004

Abstract: A laminar premixed CH_4-Air flame is analyzed with tools from the Computational Singular Perturbation method. It is shown that in the flame zone, transport has no influence in the directions in the phase space, along which act the fastest chemical time scales. Diffusion starts becoming important in the directions along which act the intermediate chemical time scales, while along the slowest directions both diffusion and convection dominate. It is also demonstrated that the stoichiometry of the mixture has no significant effect in the dynamics along the fastest chemical directions, influencing the slow ones, along which chemistry interacts with transport. The analysis presented here is based on numerical CSP data. The ease in their interpretation makes CSP a well suited tool for the study of complex and unexplored chemical mechanisms.

Keywords: Singular Perturbations, Computational Methods, Transport-Chemistry Interaction.

1. Introduction

A thorough understanding of the coupling between chemistry and molecular transport processes is essential for the accurate description of laminar and turbulent flame structures and limiting phenomena such as ignition and extinction. Traditional flame analysis methods include reaction path and sensitivity analyses [e.g., 1, 2] and asymptotic expansion methods [e.g., 3]. Here, the algorithmic *local* Computational Singular Perturbation (CSP) analysis [4-5] will be employed for the analysis of laminar flames. CSP has been applied to a wide range of combustion problems [e.g., 6-8] and is based on the realization that the solution of the species equation follows a trajectory, which lies on a manifold in the mass fraction space [4-6]. This manifold is created by the equilibration of the "fast" part of chemistry, due to the action of fast chemical time scales acting perpendicular to the manifold; directions along which the effect of diffusion and convection time scales is negligible [7]. The "slow" part of chemistry and the other physical processes present, such as convection and diffusion, are responsible for moving the solution along the manifold. CSP will be employed here to study the transport-chemistry interaction in laminar premixed CH_4-Air flames. The aim of this study is to investigate the extent to which transport couples with chemistry and the physical mechanism by which such a coupling is possible.

2. The CSP Tools

Given a kinetic mechanism consisting of N species, E elements and K elementary reactions (forward and backward reactions counted separately), the species N-dim. system of equations is:

$$\frac{d\mathbf{y}}{dt} = \mathbf{L}_c(\mathbf{y}) + \mathbf{L}_d(\mathbf{y}) + \frac{1}{\rho}\mathbf{W}\left[\mathbf{S}_1 R^1(\mathbf{y}) + \ldots + \mathbf{S}_K R^K(\mathbf{y})\right] \tag{1}$$

where, $\mathbf{y}$ is the vector of the species mass fractions, $\mathbf{L}_c$ and $\mathbf{L}_d$ are the convective and diffusive spatial vector differential operators, ρ is the mixture density, $\mathbf{W}$ is a diagonal matrix with the species'

---

[1] Corresponding author. E-mail: dagoussi@iceht.forth.gr

molecular weights as entries, $\mathbf{S}_k$ is the stoichiometric vector of the k-th elementary reaction and R^k is the k-th reaction rate. Equation (1) can be cast as:

$$\frac{d\mathbf{y}}{dt} = \mathbf{a}_1 h^1 + \ldots + \mathbf{a}_N h^N \tag{2}$$

where

$$h^i = \mathbf{b}^i \cdot \mathbf{L}_c(\mathbf{y}) + \mathbf{b}^i \cdot \mathbf{L}_d(\mathbf{y}) + \frac{1}{\rho} f^i \tag{3}$$

$$\mathbf{b}^i \cdot \mathbf{L}_c(\mathbf{y}) = \sum_{n=1,N} b_n^i L_c^n, \quad \mathbf{b}^i \cdot \mathbf{L}_d(\mathbf{y}) = \sum_{n=1,N} b_n^i L_d^n, \quad f^i = \sum_{k=1,K} \left(\mathbf{b}^i \cdot \mathbf{W} \mathbf{S}_k\right) R^k \tag{4a-c}$$

$\mathbf{a}_i$ and $\mathbf{b}^i$ are the N-dim. CSP basis vectors ($\mathbf{b}^i \mathbf{a}_k = \delta_k^i$), h^i are the related amplitudes, b_n^i is the n-th component of $\mathbf{b}^i$ and L_c^n and L_d^n are the n-th components of $\mathbf{L}_c$ and $\mathbf{L}_d$ [5, 9]. The signs of $\mathbf{b}^i$ and $\mathbf{a}_i$ are adjusted so that $f^i > 0$ for i=1,N-E, while $f^i \equiv 0$ for i=(N-E+1),N due to the conservation of elements in each elementary reaction. The CSP-modes $\mathbf{a}_i h^i$ are ordered in eq. (2) according to the magnitude of the chemical time scales they relate to; i.e. the i=1 term relates to the fastest scale $\tau_{chem,1}$, etc. Here, the CSP basis vectors $\mathbf{a}_i$ and $\mathbf{b}^i$ are approximated by their leading order term: the right and left, respectively, eigenvectors of the Jacobian of the source term in eq. (1) [5, 7]. The chemical time scales are defined as $\tau_{chem,i} = 1/|\lambda_i|$, where λ_i is the i-th eigenvalue acting along the direction $\mathbf{a}_i$. Equation (1) can also be cast as:

$$\frac{d\mathbf{y}}{dt} = \frac{1}{\rho} \mathbf{W} \left[\tilde{\mathbf{a}}_1 \tilde{h}^1 + \ldots + \tilde{\mathbf{a}}_N \tilde{h}^N \right] \tag{5}$$

where $\tilde{\mathbf{a}}_i = \mathbf{W}^{-1} \mathbf{a}_i$ and $\tilde{h}^i = \rho h^i$, so that K elementary reactions, convection and diffusion in eq. (1) can all be replaced by N non-physical reactions, the stoichiometric vector and rate of which are $\tilde{\mathbf{a}}_i$ and $\tilde{h}^i$ (*i-th CSP reaction and rate*). Inspecting the elements in $\tilde{\mathbf{a}}_i$ and the major contributors in the expression for $\tilde{h}^i$, valuable information can be extracted regarding the most important chemical paths, which either are in equilibrium or drive the system.

3. Chemistry-Transport Interaction

Steady, freely propagating, one-dimensional, laminar, premixed atmospheric CH_4-Air flames were simulated with the RUN-1DL code [10], using the detailed GRI 3.0 mechanism [11], which incorporates 53 species (N=53), 325 reversible reactions (K=650) and 5 elements (E=5). In the flames reported here the inlet mixture temperature was considered at $T_0 - 300K$, while the exit product temperature was computed $T_\infty = 1670K$ for $\phi=0.6$, $T_\infty = 2231K$ for $\phi=1.0$ and $T_\infty = 1909K$ for $\phi=1.5$. Since the problem considered here is steady, all *CSP rates* $\tilde{h}^i$ are equilibrated:

$$\tilde{h}^i = \rho \mathbf{b}^i \mathbf{L}_c(\mathbf{y}) + \rho \mathbf{b}^i \mathbf{L}_d(\mathbf{y}) + f^i = 0 \qquad i=1,N-E \tag{6a}$$

$$\tilde{h}^i = \rho \mathbf{b}^i \mathbf{L}_c(\mathbf{y}) + \rho \mathbf{b}^i \mathbf{L}_d(\mathbf{y}) = 0 \qquad i=(N-E+1),N \tag{6b}$$

The N-E eqs. (6a) are projections of the N-dim. system of eqs. (1) on the N-E directions in the phase space, along which chemistry is active and is allowed to interact with transport. The E eqs. (6b) govern the conserved scalars and by definition chemical activity is absent.

The equilibration of each *CSP rate* $\tilde{h}^i$ is associated with a time scale, the magnitude of which depends on both transport and chemistry. In particular, the time scales characterizing the fastest *CSP rates* are dominated by the fast components of chemistry only. The time scales of the next *CSP rates* are influenced by both chemistry and transport; the effect of the latter becoming more important as the time scales become slower, dominating completely those of the E *CSP rates*, eqs. (6b). In order to relate the above with the different processes contributing in the cancellations in $\tilde{h}^i = 0$, we introduce the quantities:

$$I_{src}^i = \sum_{k=1,K} \left| I_{src,k}^i \right| \qquad I_{con}^i = \sum_{n=1,N} \left| I_{con,n}^i \right| \qquad I_{dif}^i = \sum_{n=1,N} \left| I_{dif,n}^i \right| \tag{7a-c}$$

which, since $\left|I^i_{src}\right| + \left|I^i_{con}\right| + \left|I^i_{dif}\right| = 1$, are indicative of the chemical, convective and diffusive activity levels taking place in each *CSP rate* $\tilde{h}^i$.

The quantities I^i_{src}, I^i_{con} and I^i_{dif} for all i=1,N-E=48 modes that allow transport-chemistry interaction were computed at a point where temperature attains an average value, $T_a = (T_0 - T_\infty)/2$, for the three stoichiometries ϕ=0.6 (T_a=1000K), ϕ=1.0 (T_a=1286K) and ϕ=1.5 (T_a=1181K) and are displayed in Fig. 1. It is shown that chemistry dominates the first (chemically fastest) *CSP rates*, while transport dominates the last (chemically slowest) ones. These findings, simply indicative of what happens in the flame zone, allow us to conclude that:

- Along the "fast" chemical directions, chemistry dominates and transport (mainly diffusion) plays a minor role, being little more effective in the stoichiometric flame than in the lean/rich ones.
- Along the "intermediate" chemical directions the action of diffusion and convection becomes noticeable, with that of diffusion being still more pronounced.
- Along the "slow" chemical directions both diffusion and convection attain equal importance to chemistry, eventually dominating the "slowest" directions, their action being more effective in the lean/rich flames than in the stoichiometric one.

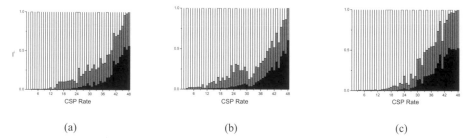

(a)   (b)   (c)

Figure 1:   Relative measure of convection I^i_{con} (black), diffusion I^i_{dif} (grey) and chemical I^i_{src} (white) activities in the flame zone of (a) lean (ϕ=0.6, T=1000K), (b) stoichiometric (ϕ=1.0, T=1286K) and (c) rich (ϕ=1.5, T=1181K) flames.

The rising influence of transport along the chemically "slow" directions is the result of the time scales characteristic of transport along these directions in the phase space, catching up with the chemical ones. A demonstration of this coupling is provided by the results shown in Fig. 2, were the chemical time scales for the lean (ϕ=0.6, T=1000K), stoichiometric (ϕ=1.0, T=1286K) and rich (ϕ=1.5, T=1181K) flames are displayed. It is shown that while the mixture composition has no effect on the fastest chemical time scales, the slow ones tend to become slower in lean/rich mixtures than in stoichiometric ones. These findings are in agreement with the results shown in Figs 1a-c, according to which the increased activity of transport along the chemically "slow" directions is more intense in the lean/rich flames than in the stoichiometric.

4. Conclusions

It became clear that the dynamics of flame structure inside the flame and of limiting phenomena such as ignition and extinction cannot be discussed only in kinetic terms, the convective/diffusive processes and their interaction with chemistry being equally important. The significance of such conclusions has already been recognised by researchers trying to develop reduced kinetics models of high accuracy, which include the effects of diffusion [e.g., 12]. Moreover, it was shown that the dynamics of the fastest part of chemistry is unaffected by the stoichiometry of the inlet mixture because the fast chemistry is fully uncoupled from transport. This feature has also been observed in flames of different geometries [13]. The above methodology can also be used to quantify in a fully algorithmic manner other important chemical processes, such as the role of transport-chemistry interaction for the thermal and prompt NO formation.

Acknowledgments

The financial support of the EU, through the FLAMESEEK project (ENK5-CT-2000-00115), is gratefully acknowledged.

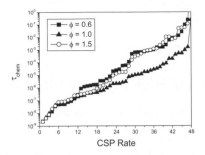

Figure 2: Chemical time scales in the flame zone of lean (ϕ=0.6, T=1000K), stoichiometric (ϕ=1.0, T=1286K) and rich (ϕ=1.5, T=1181K) flames.

References

[1] J. Warnatz, U. Maas and R.W. Dibble, *Combustion*. Springer Verlag, Berlin, 2000.

[2] A. Tomlin, T. Turanyi and M. Pilling, in M. Pilling (Ed.) *Oxidation Kinetics and Autoignition of Hydrocarbons*. Elsevier, New York, 1997, p. 293.

[3] N. Peters and F.A. Williams, *Combust. Flame* 68 (1987) 185-207.

[4] S.H. Lam and D.A.Goussis, *Proc. Combust. Inst.* 22 (1988) 931-941.

[5] S.H. Lam and D.A.Goussis, *Int. J. Chem. Kinet.* 26 (1994) 461-486.

[6] C.G. Fotache, H. Wang and C.K. Law, *Combust. Flame* 117 (2003) 777-794.

[7] M.Valorani, H.N. Najm and D.A.Goussis, *Combust. Flame* 134 (2003) 35-53.

[8] T. Lovas, P. Amneus, F. Mauss and E. Mastorakos, *Proc. Combust. Inst.* 29 (2002) 1387-1393.

[9] M. Hadjinicolaou and D.A. Goussis, *SIAM J. Sci. Comput.* 20 (1999) 781-810.

[10] B. Rogg, in N. Peters and B. Rogg (Eds.) *Reduced Kinetic Mechanisms for Applications in Combustion Systems*, Springer, Berlin, 1993, p.350.

[11] G.P. Smith, D.M.Golden, M. Frenklach, N.W. Moriarty, B. Eiteneer, M. Goldenberg, C.T Bowman, R.K.Hanson, S. Song, W.C. Gardiner Jr., V.V. Lissianski and Z. Qin, *The GRI-3.0 Detailed Mechanism*, http://www.me.berkeley.edu/gri_mech/.

[12] H. Bongers, J.A. van Oijen and L.P.H. de Goey, *Proc. Combust. Inst.* 29 (2002) 1371-1378.

[13] D. Diamantis, E. Mastorakos and D.A. Goussis, *Combust. Theory and Modeling*, 6 (2002) 383-411.

VSP International
Science Publishers
P.O. Box 346, 3700 AH Zeist
The Netherlands

*Lecture Series on Computer
and Computational Sciences*
Volume 1, 2004, pp. 194-197

Ab inito Calculations of High Pressure Effect on Fe$_3$Pt by CASTEP

Yousong Gu, Yue Zhang and Zhen Ji

Department of Material Physics and Chemistry,
University of Science and Technology in Beijing,
Beijing 100083, People's Republic of China

Received 2 August, 2004; accepted in revised form, 18 August 2004

Abstract: The electronic and magnetic properties of Fe$_3$Pt under high pressures had been studied by the CASTEP codes. Both cubic and tetragonal phases have been investigated under high pressures up to 8GPa, and the magnetic and non-magnetic states were compared. In the case of cubic, ferromagnetic phases, the linear compression rate is 1.39×10^{-3}/GPa. As pressure increases, s band population increases while p band population decreases, and d band population increases. The magnetic moment of Fe atoms decreases while that of Pt atoms increases. The cell moment increase and the saturated magnetization decreases. Calculations performed on tetragonal phases show that the cell volumes and magnetizations are smaller than those of the corresponding cubic phases; but the total energies are almost the same. Two phases with different cell volume and magnetization could exist in Fe$_3$Pt. The results on non magnetic phase showed a small cell volume and high total energy, indicating that the ferromagnetic state is the ground state of Fe$_3$Pt. Geometry optimization of non magnetic tetragonal phase yield the same values for a and c, such reduced to the cubic phase. It shows that the cubic to tetragonal martensitic transformation could not exist.

Keywords: Ab inito Calculations, High Pressure Effect, Magnetic Properties, Fe$_3$Pt

Mathematics SubjectClassification: 03.65.Db, 71.15.Mb, 71.20.Be, 75.50.Bb

PACS: 03.65.Db, 71.15.Mb, 71.20.Be, 75.50.Bb

1. Introduction

We employ a density functional approach, the CASTEP code, in studying the electronic and magnetic properties of Fe$_3$Pt under high pressure. This method adopted plane wave basis sets and pseudo potential to speed up the calculation process, and made the calculations of interesting properties of material possible on desktop PCs [1].

Fe$_3$Pt exhibited a lot of interesting properties, such as giant magnetostriction and cubic to tetragonal martensitic transformation [2], Invar behavior in disorder fcc phase [3] and Anomalous magnetic moments found in Fe-Pt alloys under high pressure [4]. Spin fluctuations in Fe$_3$Pt Invar have been studied by local density functional calculations [5] and the electron momentum density distribution have been studied [6].

In this paper, we presented our calculated results of Fe$_3$Pt under high pressure by the CASTEP code. We will look at the structure, electronic and magnetic properties of Fe$_3$Pt as functions of pressure. Our calculations will give some insight on Fe$_3$Pt and its behavior under high pressure.

2. Calculation Methods

The properties were calculated after performing geometry optimization at ultra fine convergence criteria, i.e, three out of four criteria: energy change less than 5.0×10^{-6} eV/atom, force smaller than 0.01 eV/Å, stress less than 0.02GPa and displacement shorter than 5.0×10^{-4} Å. BFGS algorithm was selected for geometrical optimization. GGA functional was used to describe electron densities, PBE type of exchange functions were implemented. The spins were treated as polarized in order to deal with magnetic properties. In the self-consistent electronic minimization, ultrasoft type pseudopotentials were used, and energy cut offs is set to 330eV. K point separations were chosen to be 0.03/Å. Density mixing was used to stabilize the process, and the tolerance is 5.0×10^{-7} eV/atom. Band structure and density of states were calculated, as well as partial DOS and band populations. With these results, we can deliberate on the high pressure behavior of Fe₃Pt.

Calculations were performed on a series of Fe₃Pt under pressure of 0GPa up to 8GPa. Tetragonal phase were also calculated, since there are cubic to tetragonal martensitic transformation [2]. The non magnetic or paramagnetic phases were also calculated for comparison.

3. Results and discussion

In the case of cubic, ferromagnetic phases, geometry optimization shows that as pressure increases all the way up to 8GPa, the lattice constant decreased from 3.720, to 3.710, 3.699, 3.689 and 3.679Å, under the pressure of 2GPa, 4GPa, 6GPa and 8GPa, respectively. We get a linear compression coefficient of 1.39×10^{-3}/GPa, which is quite small. The total energy, band populations, atomic and cell magnetic moments and saturated magnetization were evaluated.

Figure 1 shows the band population of Fe and Pt atoms as functions of pressure. We can seen that as the pressure increases, the s band populations increase and p band populations decrease for both Fe and Pt. The 3d band population of Fe shows slight increase, while 5d band population of Pt remains almost the same.

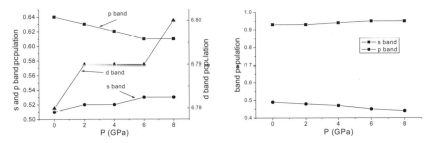

Figure 1: The band populations of Fe (left) and Pt (right) in Fe₃Pt as functions of pressure.

Figure 2 shows the magnetic moments of Fe and Pt atoms and the cell as a whole, as well as the saturation magnetization as functions of pressure. Due to the calculation accuracy, the change in atomic moments is small, but the tendency is clear, the magnetic moment of Fe atoms decrease, while that of Pt increase. The cell moment decreases while the saturated magnetization increase due to volume compression.

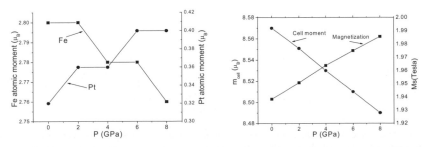

Figure 2: The atomic magnetic moments of Fe and Pt (left) and saturation magnetization of Fe₃Pt as functions of pressure.

In the case of tetragonal crystal structure, we performed the same calculation and obtained similar results. Table 1 lists the calculated results. After geometry optimization, the lattice constant *a* is larger and *c* is smaller than that of the cubic phase, and the cell volume is smaller than that of cubic phase.

The total energy is the same as the cubic phase within the calculation accuracy. Therefore, the present of tetragonal phase is possible.

Table 1 calculated electronic and magnetic properties of tetragonal Fe_3Pt as functions of pressure.

P (GPa)	0		2		4		6		8	
a(Å)	3.724		3.714		3.7034		3.694		3.684	
c(Å)	3.705		3.695		3.684		3.674		3.665	
V(Å³)	51.372		50.975		50.524		50.127		49.736	
E(eV)	3320.99		3320.35		3319.72		3319.10		3318.47	
$m_{cell}(\mu_B)$	8.485		8.470		8.454		8.440		8.425	
M(T)	1.925		1.936		1.950		1.962		1.974	
Atoms	Fe	Pt	Fe	Pt	Fe	Pt	Fe	Pt	Fe	Pt
s	0.51	0.93	0.52	0.93	0.52	0.94	0.53	0.95	0.54	0.95
p	0.63	0.49	0.62	0.48	0.62	0.46	0.61	0.45	0.60	0.44
d	6.79	8.80	6.79	8.79	6.80	8.79	6.80	8.79	6.80	8.79
net spin	1.39	0.08	1.38	0.08	1.38	0.09	1.38	0.09	1.37	0.09

The band populations behavior in a similar fashion as pressure increases, and so are the atomic and cell moments, as well as the saturated magnetization, as shown in Figure 3. However, the magnetic moments of Fe, Pt and the cells are smaller that these of cubic phase and the saturated magnetizations are also smaller.

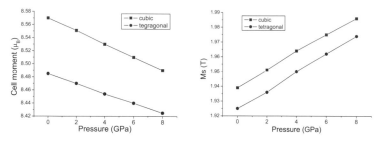

Figure 3 The cell magnetic moment (left) and saturated magnetization (right) of cubic and tetragonal phases.

The existence of two phases with different volume and saturated magnetization but almost identical total energy may be the causes of phenomena and properties observed in Fe_3Pt. Further studies of this subject may yield interesting results.

The non magnetic, cubic phases were also studies. After geometry optimization, it is found that the lattice sizes are smaller than that of the magnetic phase, from 3.720, 3.710, 3.699, 3.689, 3.679Å to 3.600, 3.592, 3.584, 3.570, 3.568Å, respectively. The band populations change with pressure in a similar fashion, but the values are shifted due to reduced cell volumes, as shown in Figure 4. The total energy is increased from 3320.99, 3320.35, 3319.72, 3319.10 and 3318.47eV to 3320.08, 3319.51, 3318.94, 3318.41 and 3317.80eV, respectively. It is unfavorable in energy even under press as high as 8GPa. This is to say that the Fe_3Pt favors ferromagnetic states.

We also performed calculations on non magnetic tetragonal phases. However, after geometry optimization, the tetragonal phases reduced to cubic phases since the optimized lattice constants a and c were identical within calculation limit. This is to say that in non magnetic states, cubic to tetragonal Martensitic transformation could not occur.

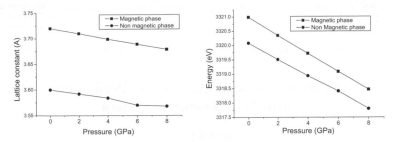

Figure 4 The lattice constants of magnetic and non magnetic phase (left) and the total energy (right) as functions of pressure.

4. Conclusion

Density functional calculations have been performed on Fe_3Pt under various high pressures, in order to study the pressure effect on structure, electronic and magnetic properties. In case of cubic, magnetic phases, the linear compression rate is $1.39 \times 10^{-3}/GPa$. As pressure increases, s band population increases while p band population decreases, and d band population increases. The magnetic moment of Fe atoms decreases while that of Pt atoms increases. The cell moment increase while the saturated magnetization decreases. Calculations performed on tetragonal phases show that the cell volumes and magnetizations are smaller than that of the corresponding cubic phases; the total energy is almost the same. Two phases with different cell volume and magnetization could exist. The results on non magnetic phase showed a small cell volume and high total energy, indicating that the ferromagnetic state is the ground state of Fe_3Pt. Geometry optimization of non magnetic tetragonal phase yield the same values for a and c, such reduced to the cubic phase. It shows that the cubic to tetragonal martensitic could not exist.

Acknowledgments

This work was supported by the National Science Fund for Distinguished Young Scholars and the National Natural Science Foundation of China (No. 50325209, 50172006, 50232030).

References

[1] M.D. Segall, P.J. Lindan, M.J. Probet C. J. Pickard, P. J. Hasnip, S. J. Clark, and M. C. Payne, First-principles simulation: ideas, illustrations and the CASTEP code, *J. Phys.: Condens. Mater.* **14** 2717-2743 (2002)

[2] T. Kakeshita, T, Kakeuchi, T. Fukuda, M. Tsujiguchi, T. Saburi, R. Oshima and S. Muto, Giant magnetostriction in an ordered Fe_3Pt single crystal exhibiting a Martensitic transformation, *Appl. Phys. Lett.* **77** 1502-1504 (2000)

[3] R. Hayn and V. Drchal, Invar behavior of disordered fcc-Fe_xPt_{1-x} alloys, *Phys. Rev.* **B 58** 4341-4344 (1998)

[4] M. Matsushita, T. Nishimura, S. Endo, M. Ishizuka, K. Kindo and F. Ono, Anomalous magnetic moments in Fe-Pt and Fe-Pd Invar alloys under high pressure. *J. Phys.: Condens. Mater.* **14** 10753-10757 (2002)

[5] M. Uhl, L.M. Sandratskii and J. Kubler, Spin fluctuations in γ-Fe and in Fe_3Pt Invar from local-density-fuctional calculations, *Phys. Rev.* **B 50** 291-301 (1994)

[6] J.W. Taylor, J.A. Duffy, A.M. Bebb, J.E. McCarthy, M.R. Lees, M.J. Cooper and D.N. Timms, Spin-polarized electron momentum density distribution in the Invar system Fe_3Pt, *Phys. Rev.* **B 65** 224408 (2002)

VSP International
Science Publishers
P.O. Box 346, 3700 AH Zeist
The Netherlands

*Lecture Series on Computer
and Computational Sciences*
Volume 1, 2004, pp. 198-199

Building Financial Time Series Predictions
with Neural Genetic System

Serge Hayward[1]

Department of Finance
Ecole Supérieure de Commerce de Dijon,
29, rue Sambin, 21000, Dijon, France

Received 2 August, 2004; accepted in revised form, 18 August 2004

Keywords: Artificial Neural Network; Genetic Algorithm; Performance Surface; Evaluation Criteria;
Summary Statistics; Economic Profitability; Stock Trading Strategies

Problems with applications of computational methods in finance are often due to the lack of common methodology and statistical foundations of its numerous techniques. At the same time, relationships between summary statistics used for predictions' evaluation and profitability of investment decisions based on these predictions are not straightforward in nature. The importance of the latter is particularly evident for applications of an evolutionary / artificial neural network (E/ANN) under supervised learning, where the process of network training is based on a chosen statistical criterion, but when economic performance is generally sought. This paper is a step towards the econometric foundation of computational methods in finance.

For our experiment we develop the dual network structure, where forecasting network feeds into the action network. The model is evolutionary in the sense it considers a population of networks (individual agents facing identical problems/instances) that generate different solutions, which are assessed and selected on the basis of their fitness.

Considering traditional performance measures as common errors, accuracy (directional), correlation (the desired and ANN output), improvement over 'efficient prediction', their relationships with stock trading strategies' profitability were found to be of complicated and non-conclusive nature. Only the degree of improvement over 'efficient prediction' shows some robust links with returns' measures.

The experiment establishes that training ANN with the performance surface optimised with genetic algorithm (GA) for directional accuracy (DA), discounting least recent values or minimizing number of large errors generally improves strategies' profitability. The simulation shows that among three optimisation of performance surface considered, strategies trained on learning the sign of the desired output were generally superior to those, trained to reduce the number of large errors or focusing learning on recent values. The results also demonstrate that DA (alone or always) does not guarantee profitability of a trading strategy trained with that criterion.

Addressing the issue of ANN topology dependency, simulations reveal optimal for financial applications network settings. Optimality of discovered ANN topologies' is explained through their links with the ARMA processes, thus presenting identified structures as nonlinear generalizations of such processes. Optimal settings examination demonstrates the weak relationships between statistical and economic criteria. Model discovery and performance surface optimisation with GA demonstrate profitability improvement with an inconclusive effect on statistical criteria.

A trading strategy choice (with regards to the time horizons) is a function of market conditions, which themselves are a function of strategies used by agents, populated this market. In these settings market

---

[1] Corresponding author. E-mail: shayward@escdijon.com

conditions (or strategies used by the dominant type of traders) determine the optimal memory length. This approach is an effort towards considering a market environment endogenously in financial data mining.

To model the turmoil in an economic system with frequent shocks, short memory horizons are considered optimal, as older data is not necessarily informative for the current/future state modelling/forecasting. In thinner markets there is higher likelihood that short memory horizons agents take the dominant position, influencing price movements. Thin markets' dominance by a particular traders' type facilitates a better environment for learning with agent-based methods of profitable strategies, used by dominant traders.

The research demonstrates that the performance surface set-up is a crucial factor in search of a profitable prediction with an agent-based model. Evaluation criteria used to assess predictive power of stock trading strategies, generated with agent-based models, might significantly differ from criteria leading to their profit maximization. The choice of evaluation criteria combining statistical and economic qualities is viewed as essential for an adequate analysis of social dynamics.

Fine-tuning the ANN settings is considered to be an important stage in the computational model set-up for results' improvement and mechanism understanding. GA is proposed to be used for model discovery, making technical decisions less arbitrary and adding additional explanatory power to the analysis of social dynamics with agent-based methods and tools.

VSP International
Science Publishers
P.O. Box 346, 3700 AH Zeist
The Netherlands

*Lecture Series on Computer
and Computational Sciences*
Volume 1, 2004, pp. 200-203

Numerical Solutions to the Reheater Panel Overheating of a Practical Power Plant

Boshu He [A, 1], Meiqian Chen [A], Laiyu Zhu [B], Jianmin Wang [B], Shumin Liu [A], Lijuan Fan [A], Qiumei Yu [A]

[A] School of Mechanical, Electronical and Control Engineering, Beijing Jiaotong University, Beijing 100044, China
[B] Dagang Power Plant, Tianjin 300272, China

Received 17 June, 2004; accepted in revised form 12 August, 2004

Abstract: The commercially available CFD package, FLUENT was utilized to numerically solve the metal surface overheating issues of the reheater pendants that exist in the full-scale No.3 boiler of Dagang Power Station, Tianjing, China. Some factors that may affect the velocity and temperature distributions at the final reheater inlet section (final superheater outlet) had been taken into account when the designated coal was burned, such as the quantity and fashion of counter-flow in the operation, the pressure difference in the air box, and the downward inclination of the secondary air injection. The basic conclusion is that some corresponding measures must be taken to rebuild the flow constructions to effectively avoid the boiler reheater pendant metal overheating. To obtain background detail, eight reformation cases were arranged on the main field, influencing reasons to diagnose this boiler numerically. Compared to the base case, all of the reformation cases had some emendatory effects to the flow and temperature distributions. The most outstanding among the reformations was Case I, where the secondary air (OFA and the upper secondary air of primary air burner D) was operated with counter-flow with a downward angle, and the pressure difference in the air box was increased. It can then be concluded that Case I can more efficiently modify the velocity and temperature deviations in the overheating regions, to ensure the furnace will operate within stable and safe conditions. Undoubtedly, these conclusions are of value to the other units of this power plant, and to other power plant furnaces throughout China that have similar construction and capacity.

Keywords: FLUENT; numerical retrofitting; tangentially fired furnace; residual swirling; velocity deviation; temperature deviation; secondary air counter-flow

Mathematics Subject Classification: 62P30, 76F70

1. Introduction

The No.3 boiler of Dagang Power Station was a tangentially fired oil furnace designed and manufactured by the FRANCO TOSI INDUSTRIALE S.p.A Co., Italy. There exists panel overheating of the reheater pendants after the fuel was switched to pulverized Datong coal in 2001. This overheating of the tube surface of reheater pendants greatly effects the operation of the boiler, in both safety and economical aspects. After the retrofitting with the introduction of counter-flow of the over fire air (OFA), the overheating had been preliminarily controlled with less frequency.

It has been recognized that residual swirling widely exists in tangentially fired pulverized coal furnaces, and this results in velocity and temperature deviations. These effects would be more severe with the increased capacity of the boiler [1-5]. At the same time, this kind of furnace is common throughout China, due to the many advantages it possesses as had been previously mentioned [3]. The overheating of the tube surface in the No. 3 boiler of Dagang Power Station, Tianjin, China, originates from the velocity and temperature deviations due to the thermo-mechanism aspect. Even though the overheating was greatly decreased with the introduction of OFA counter-flow, it did appear frequently with the BCD grinders operation, and the thermo-mechanism was not very clear. For example, how to obtain the most reasonable distribution of velocity and temperature fields for the designated coal, by the different

---

[1] Corresponding author. E-mail: hebs@center.njtu.edu.cn, boshu.he@wku.edu

conditions of the flow rates or the optimal times to introduce the counter-flow into the furnace. In what directions will the overheating occur when the coal is switched from the fuel that had been designated for use. These rules, of course, could be taken as controlling factors through the actual operation tests, with higher costs and tremendous risks.

The above influences can be economically simulated in a faster and a more effective way with the development of computerized numerical techniques. Over the last 20 years, the numerical techniques have gained the reputation of being effective tools in identifying and solving problems related to pulverized coal combustion [3, 6]. The numerical calculations generally require the application of computational fluid dynamics (CFD), which is not an easy task, especially for three-dimensional engineering modeling. For this reason, the research undertaken herein has been performed using the commercial CFD package from Fluent Inc [7]. It was employed to numerically retrofit the furnace by the introduction of 9 Cases, including one baseline case (Case A), to find the solution to the reheater pendant overheating problem. The results obtained would undoubtedly be instrumental for this boiler to operate in a more economical way, and be referential for other units of this station and other power stations throughout China.

2. Models and numerical methods

The simulation of the operational furnace is complex and exacting. Many mathematical models and pre-digestions will be introduced. These models include Turbulent Flow, Coal Devolatilization, Coal Combustion, and Heat Radiation Transfer. The CFD software from Fluent was used to numerically retrofit the operation of the No.3 full-scale boiler of Dagang Power Station. The gas-phase turbulent reacting flow was predicted by solving the steady-state conservation equations for continuity, momentums, pressure, enthalpy, mixture fractions, turbulence energy, and dissipation. The gas and particle phases are coupled through the particle source terms.

The gas-phase combustion was modeled using the fast chemistry mixture fraction approach. The mixture fraction is defined as the local mass fraction of burned and unburned fuel stream elements (C, H, etc.) and its values are then used to compute the individual species molar fractions, density, and temperature with equilibrium chemistry and pdf approaches [6, 7]. Moreover, a group of coal particle injectors was used that enabled to track the coal particles during the solution procedure. Some models nested with the software are selected in the simulation work.

3. Retrofitting cases, results and analysis

The overheating of the metal surfaces of the reheater pendants always occurs with the combination of BCD grinders in operation, and it is rare with the other grinders' combinations such as ABC, ACD, ABD, etc. This indicates that the BCD combination should be thoroughly analyzed. There are 9-sets of CFD retrofitting cases established for this furnace, with the BCD combination, which are labeled from A to I. Each is identified by the counter-flow second air jets, the pressure difference in the air box, and the axis tilt-angle of the counter-flow jets. Where case A is set as the base case with no counter-flow jets, the designed air box pressure difference, horizontal jets of secondary air, and the details for the other cases can be found in Table 1.

The CFD based solutions have been calculated for Cases B to I where the counter-flow jets are put into operation and the results will be compared with those from the base case A. The overheating occurs frequently in the L2 vertical plane, which is the outlet of the final superheater, or the inlet of the final reheater and is shown in Figure 1 where the *a*, *b* and *c* are the positions partitioning the horizontal flue into 4 equal segments in the vertical plane. The most frequent position that the overheating occurs is the place around position *a*. Therefore, the focus will be the distributions of the velocities and temperatures at the position *a* in this CFD based retrofitting analysis.

Table 1: Retrofitting Cases and explanations

Cases	Counter-flow combinations	General situation	
A	void	Secondary air jets with horizontal injection	Furnace air box with designed pressure difference: 1000 Pa Primary air velocity: 28 m/s Secondary air velocity: 50 m/s
B	OFA		
C	OFA & 4.2		
D	OFA	Secondary air jets with 20 degree downward injection	
E	OFA & 4.2		
F	OFA	Secondary air jets with horizontal injection	Furnace air box with increased pressure difference: 2000 Pa Primary air velocity: 28 m/s Secondary air velocity: 55 m/s
G	OFA & 4.2		
H	OFA	Secondary air jets with 20 degree downward injection	
I	OFA & 4.2		

The velocity magnitudes are compared in Figure 2 in the positions of L2-*a* for all of the 9 cases. The *x* component velocity takes a very similar profile to the velocity magnitude, since this component is the main one at these positions for all cases. From this Figure one can observe that the characteristics of the velocity distributions have similar profiles for all of the retrofitting cases, i.e. the velocities near the right wall reveal higher magnitudes than those near the left wall for the position of L2-*a*, and the relatively low velocity magnitudes occur near the left wall. Among all of the cases, the greatest velocity difference occurs in case A, as indicated by Figure 2, a distinct high velocity value is present near the right side while a relatively low value is present close to the central and the left regions. The velocity differences between the high and low values will inevitably result in the homologous high and low heat transfer coefficients and will result in the heat absorption differences by the water vapor. Therefore, case A is indeed the most dangerous one, as had been substantiated by the actual operation, and effective measures must be taken to rebuild the velocity fields to the levels in cases through B to I. Compared to case A, the velocity differences between the right and left side walls becomes obviously less in some extent for the other cases where some retrofitting techniques have been put into operation (the details can be found in Table 1) and the least amount of difference is in case I.

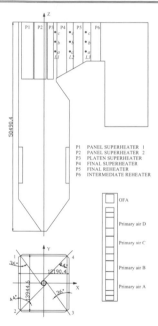

Figure 1: Geometric diagram, burner and injection arrangements for the numerical retrofitting

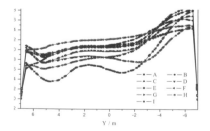

Figure 2: Velocity distributions at line L2-*a* for the retrofitting cases

Figure 3: Temperature distributions at line L2-*a* for the retrofitting cases

The temperature magnitudes are compared in Figure 3, in the positions of L2-*a* for all of the 9 cases. It can be concluded that the temperature levels for all of the retrofitting cases appears to have substantial decrease at this location as compared to base case A. The smallest temperature deviation is fortunately also found to be in case I.

Both the velocity and temperature distributions at the L2-*a* location for all of the numerical cases, which have been shown in Figures 2 and 3, indicate that case I can best modify those distributions for base case A than any of the other cases. Among all of the other numerical cases, the least variations for the velocity and temperature distributions can be obtained by case I. This means that case I may effectively eliminate the residual swirling that originates from the Concentric Firing System (CFS) of this tangentially fired furnace. Case I has been recommended and accepted to be in the future operation by the power station owner.

4. Conclusions

The numerical simulations, based upon one base case and 8 retrofitting cases for this boiler with the designated coal, indicate:

- There exist large deviations for velocity and temperature between the two side walls near the reheater pendants at the base case A where there is no reformation. Some particular and effective measures must be incorporated to rebuild the velocity and temperature fields in order to guarantee the furnace will be operating in a both steady and safe manner.
- Whether the unreasonable velocity field or the temperature profile can be rectified to some extent, by introduction of the counter-flow secondary air, and the final effect is influenced by the amount and the direction of the counter-flows, and the pressure difference in the air box.
- Among all of the 8 retrofitting cases, it is case I, with the introduction of counter-flow jets with a 20 degree downward tilt and doubling the air box pressure difference, that the fields can be effectively modified and result with minimized deviations of velocity and temperature close to the reheater pendants. This case can then most probably prevent the metal surface of the reheaters from overheating.

Acknowledgments

The authors gratefully acknowledge financial supports from the Dagang Power Station Project foundation and the BJTU (Beijing Jiaotong University) paper foundation

References

[1] B.S.He, On the separated vortices around the panels and the evolution of the swirl flow in tangentially fired furnaces: Ph.D. Dissertation. Xi'an: Xi'an Jiaotong University, 1999. (in Chinese)

[2] B.S. He, S.F. Ding, Y.F. Diao, J.Y. Xu, C.H. Chen, Numerical Study on the Contrary Modes of Air Jets in a Large Unity Furnace, Proc. Chinese Society for Electrical Engineering, 21(2001), 60-64 (in Chinese).

[3] B.S. He, M.Q. Chen, Q.M. Yu, S.M. Liu, L.J. Fan, S.G. Sun, J.Y. Xu, W.P. Pan, Numerical Study of the Optimum Counter-flow Mode of Air Jets in a Large Utility Furnace, Computers and Fluids, 33 (2004), 1201–1223.

[4] C.G. Yin, S. Caillat, J.L. Harion, B. Baudoin, E. Perez, Investigation of the flow, combustion, heat-transfer and emission from a 609 MW utility tangentially fired pulverized-coal boiler, Fuel, 81(2002), 997-1006.

[5] C.G. Yin, L. Rosendahl, T.J. Condra, Further study of the gas temperature deviation in large-scale tangentially coal-fired boilers, Fuel, 82(2003), 1127-1137.

[6] A.M. Eaton, L.D. Smoot, S.C. Hill, C.N. Eatough, Components, formulations, solutions, evaluation, and application of comprehensive combustion models, Prog Energy Combust Sci 25(1999), 387-436.

[7] FLUENT6.1-User's guide. Fluent Inc., 2003.

VSP International
Science Publishers
P.O. Box 346, 3700 AH Zeist
The Netherlands

*Lecture Series on Computer
and Computational Sciences*
Volume 1, 2004, pp. 204-206

Direct Determination of Pair Potential Energy Function from Extended Law of Corresponding States and Calculation of Thermophysical Properties for CO_2-N_2

T. Hoseinnejad[†] and H. Behnejad

Department of Chemistry, Tehran University, Tehran, Iran

Received 23 June, 2004; accepted in revised form, 15 August 2004

Abstract: The isotropic reduced intermolecular potential energy function for CO_2-N_2 has been determined using a direct inversion of the experimentally reduced viscosity collision integrals obtained from the corresponding states correlation then we used these results in conjunction with the second virial coefficient to calculate the outer branch of the potential well. The results are then fitted to obtain a best MSV potential model. Our obtained interaction potential function has been used to predict thermophysical data in a wide temperature and composition range which provides a reasonable agreement with the experimental data.

Keywords: intermolecular potential energy for CO_2-N_2, direct inversion method, collision integrals, corresponding states principle, thermophysical properties

Mathematics subject Classification: Thermodynamics of mixtures

PACS: 80A10

1. Inversion Procedure

The prediction and interpretation of most phenomena involving atoms and molecules depend on knowledge of intermolecular pair potential function. In the early 1970s most information concerning intermolecular forces was inferred from study of the thermophysical properties but the functions resulting from this extremely difficult method do not appear to be unique.

The basic purpose of inversion method is to obtain the potential energy by considering the experimental data instead of fitting the data to a constrained potential model having a few parameters[1]. The direct inversion procedure for the viscosity is based on the idea that at a given T^* the values of reduced orientation averaged viscosity collision integral, $\left\langle \Omega^{(2,2)*} \right\rangle$, is determined by the potential over only a small range of separation around a value r^*[2]. $\left\langle \Omega^{(2,2)*} \right\rangle$ for CO_2-N_2 has been computed on the assumption of fixed relative orientation per collision from the extended law of corresponding states[3]. We could define an inversion function, $G(T^*)$, so that

$$\left\langle V^* \right\rangle = T^* G(T^*) \tag{1}$$

$$\left\langle \Omega^{(2,2)*} \right\rangle = r^{*2} \tag{2}$$

where $G(T^*)$ is almost a universal function of T^*. Inserting the initial estimate values of $G(T^*)$ in relation (1) gives the reduced potential energy $\left\langle V^* \right\rangle$. The corresponding values of r^* may be obtained from relation (2) using experimental viscosity collision integrals.

This iteration process is repeated until a pre-determined minimum deviation is reached. In this work, convergence is obtained after three iterations. The average absolute deviation between the corresponding states values of $\left\langle \Omega^{(2,2)*} \right\rangle$ and its calculated values obtained from inversion process is

[†] Corresponding author. E-mail: hnezhad@khayam.ut.ac.ir

about 0.3%. Three successive numerical integrations are required to obtain $\langle\Omega^{(2,2)*}\rangle$ that is made by using Gatland version of the computer program developed by O'Hara and Smith[4].

2. Results and discussion

Knowing the inner branch of the potential well from viscosity we have inverted the second virial coefficient data [5] according to the following formulas for CO_2-N_2 at $T < \varepsilon/k$ to yield $\langle V^*\rangle$ as a function of r^* which results are shown in figure 1 as outer branch of the potential well.

$$V = kT - \varepsilon \qquad (3)$$
$$r_R^3 - r_I^3 = -(B - b_0) N(T^*) \qquad (4)$$

Our inverted reduced potential energy for CO_2-N_2 can be used to compute collision integrals and their ratios occur frequently enough in the expression for calculation of transport coefficients. The numerical values of these quantities are given in table 1.

Our obtained values for $\langle\Omega^{(2,2)*}\rangle$, $\langle\Omega^{(1,1)*}\rangle$ and its derivatives has been applied to calculate thermophysical properties [6] such as diffusion, viscosity, thermal conductivity and thermal diffusion factor of CO_2-N_2 in a wide temperature range from $T^*=1$ to the onset of ionization and three molar compositions: 25%-75%, 50%-50%, 75%-25%. The results have been compared in a deviation plot with accurate experimental data which demonstrates deviation within experimental uncertainty.

So we conclude that our isotropic potential energy directly determined from the corresponding states viscosity allows the collision integrals needed to compute other transport properties to be obtained with more accuracy than is possible by a corresponding states analysis.

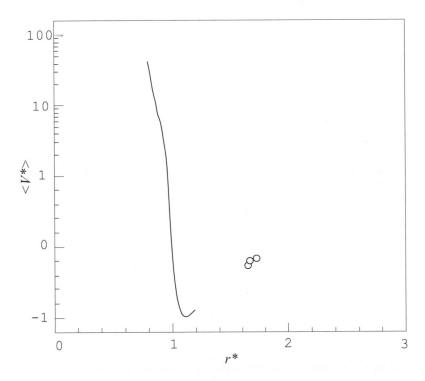

Figure 1: isotropic reduced intermolecular potential energy by invert of corresponding states viscosity collision integrals (———) and the second virial coefficient (O) for CO_2-N_2. Note that the scale changes from linear to logarithmic at $V/\varepsilon =1.0$

Table1: Dimensionless collision integrals and the related ratios for CO_2-N_2.

$\log_{10} T^*$	$\langle \Omega^{(1,1)*} \rangle$	A^*	B^*	C^*	E^*	F^*
0.0	1.5863	1.1209	1.2067	0.8473	0.8784	0.9235
0.1	1.4221	1.1134	1.1761	0.8564	0.8853	0.9284
0.2	1.2861	1.1072	1.1507	0.8692	0.8969	0.9367
0.3	1.1767	1.1036	1.1317	0.8832	0.9102	0.9475
0.4	1.0897	1.1026	1.1184	0.8968	0.9227	0.9594
0.5	1.0199	1.1035	1.1093	0.9089	0.9332	0.9711
0.6	0.9628	1.1055	1.1029	0.9190	0.9413	0.9819
0.7	0.9148	1.1077	1.0984	0.9271	0.9475	0.9909
0.8	0.8738	1.1100	1.0962	0.9335	0.9526	0.9976
0.9	0.8384	1.1132	1.0980	0.9379	0.9575	1.0019
1.0	0.8078	1.1187	1.1048	0.9401	0.9618	1.0044
1.1	0.7809	1.1271	1.1148	0.9400	0.9640	1.0068
1.2	0.7551	1.1366	1.1227	0.9380	0.9622	1.0110
1.3	0.7274	1.1437	1.1227	0.9359	0.9561	1.0172
1.4	0.6960	1.1443	1.1119	0.9355	0.9482	1.0238
1.5	0.6616	1.1364	1.0919	0.9383	0.9421	1.0286
1.6	0.6264	1.1205	1.0673	0.9447	0.9401	1.0303
1.7	0.5934	1.0996	1.0431	0.9537	0.9433	1.0288
1.8	0.5650	1.0771	1.0228	0.964	0.9507	1.0254
1.9	0.5423	1.0563	1.0079	0.9741	0.9607	1.0213
2.0	0.5256	1.0388	0.9981	0.983	0.9711	1.0173

Acknowledgement

Authors are very indebted to Research Committee of the University of Tehran due to its authorities for financial support during the tenure of which work was completed.

References

[1] P. Clancy, D.W. Gough, G.P. Matthews and E.B. Smith , Simplified methods for the inversion of thermophysical data, *Molecular Physics* **30** 1397-1407(1975).

[2] H. Behnejad, A. Maghari, M. Najafi, The extended law of corresponding states and the intermolecular potentials for He-He and Ne-Ne, *Journal of ComputationalChemistry* **16** 441-444(1995).

[3] A. Boushehri, J. Bzowski, J. Kestin, E.A. Mason, Equilibrium and transport properties of eleven polyatomic gases at low density, *journal of physical chemical reference data* **16** 445-466 (1987) & Erratum **17** (1988)255.

[4] H. O'Hara, F. J. Smith, *Journal of Computational Physics* **5** 328 (1970).

[5] J.H. Dymond and E.B. Smith: The Virial Coefficients of Pure Gases and Mixtures ,Oxford University Press, London, 1980.

[6] J.O. Hirschfelder, C.F. Curtiss and R.B. Bird: Molecular Theory of Gases and Liquids, John Wiley & Sons, New York ,1964.

VSP International
Science Publishers
P.O. Box 346, 3700 AH Zeist
The Netherlands

*Lecture Series on Computer
and Computational Sciences*
Volume 1, 2004, pp. 207-210

Variable Reduction Scheme for Finite-Difference Modeling of Elastic Problems of Solid Mechanics

M Zubaer Hossain[†]

Department of Mechanical Engineering,
Bangladesh University of Engineering and Technology,
Dhaka-1000, Bangladesh

Received 2 August, 2004; accepted in revised form, 18 August 2004

Abstract: The paper presents a new variable reduction scheme for finite-difference modeling of mixed-boundary-value elastic problems of solid mechanics. The application of the scheme is demonstrated for both the cases of two- and three- dimensional problems. In this approach, the elastic problems are formulated in terms of a single potential function, defined in terms of the displacement components. Based on the scheme, an efficient finite-difference method of solution has been developed for the mixed-boundary-value stress problems. Compared to the conventional models, the present method is capable of providing numerical solution of higher accuracy with a tremendous saving in the computation effort.

Keywords: Variable reduction scheme, stress analysis, potential function, finite-difference-technique

1. Introduction

Stress-analysis of material bodies has now become a classical subject in the field of engineering. But, somehow, the subject is still suffering from a lot of shortcomings and thus it is constantly coming up in the literature [1-5]. Earlier mathematical models of elasticity were very deficient in handling the practical problems. The difficulties involved in trying to solve practical stress problems using the existing models are clearly pointed out in our previous reports [6-9] and also by Durelli [2]. Analysis of stresses in a material body is basically a three-dimensional problem. However, still now, in most of the cases, the problems are approximated to either two- or one-dimensional ones. This is mainly because of the lack of useful method of solution by which three-dimensional problems can be formulated efficiently. Elastic problems of solid mechanics are usually formulated either in terms of deformation parameter or stress parameter. The stress formulation is not suitable for solving mixed-boundary-value problems. The displacement formulation, on the other hand, involves finding three displacement functions simultaneously from three-second order partial differential equations of equilibrium. But solving for three functions simultaneously under variously mixed conditions on the surfaces is well neigh impossible.

Stress analysis of structural problems is mainly handled by numerical techniques. Recently, extensive researches are being carried out with finite-element methods (FEM). Although the adaptation of FEM relieved us from the major inability of solving 3-D problems as well as managing the odd boundary shapes, we are aware of the lack of sophistication and doubtful quality of FEM solutions, especially for the stresses at the surfaces [11]. This is primarily because of the manifold increase of the computational work and a lot of loss in sophistication in satisfying the boundary conditions, especially in the regions of transition, where the boundary conditions change from one type to the other. On the other hand, the superiority of finite-difference method in predicting the state of stresses along the boundary has been verified repeatedly in our previous researches [6-9]. And also by Dow et. al [4].

The present paper describes the development of an efficient finite-difference method of solutions for the elastic problems of solid mechanics, where the use of a new scheme of reduction of unknowns is emphasized. The scheme reduces both the two and three-dimensional problems to finding a single

---

[†] Corresponding author. Lecturer, Department of Mechanical Engineering, BUET, E-mail: zubaexy@me.buet.ac.bd

potential function, instead of the two and three displacement components at the nodal points, respectively, and hence a tremendous saving in computation work is achieved through the present method.

2. Scheme of Reduction of Variables

The solution of three-dimensional elastic problems can be obtained by solving the three equilibrium equations in terms of the three displacement components, u_x, u_y and u_z. They are [10,12]

$$G'\frac{\partial^2 u_x}{\partial x^2} + G\left[\frac{\partial^2 u_x}{\partial y^2} + \frac{\partial^2 u_x}{\partial z^2}\right] + G''\left[\frac{\partial^2 u_y}{\partial x \partial y} + \frac{\partial^2 u_z}{\partial x \partial z}\right] = 0 \tag{1a}$$

$$G'\frac{\partial^2 u_y}{\partial y^2} + G\left[\frac{\partial^2 u_y}{\partial x^2} + \frac{\partial^2 u_y}{\partial z^2}\right] + G''\left[\frac{\partial^2 u_x}{\partial x \partial y} + \frac{\partial^2 u_z}{\partial y \partial z}\right] = 0 \tag{1b}$$

$$G'\frac{\partial^2 u_z}{\partial z^2} + G\left[\frac{\partial^2 u_z}{\partial y^2} + \frac{\partial^2 u_z}{\partial x^2}\right] + G''\left[\frac{\partial^2 u_x}{\partial z \partial x} + \frac{\partial^2 u_y}{\partial y \partial z}\right] = 0 \tag{1c}$$

where, $\quad G = \dfrac{E}{2(1+\mu)}, \quad G' = \dfrac{E(1-\mu)}{(1+\mu)(1-2\mu)} \quad$ and $\quad G'' = \dfrac{E}{2(1+\mu)(1-2\mu)}$.

In the present approach, the problem is reduced to the determination of a single variable instead of evaluating three functions, u_x, u_y, u_z, simultaneously, from the equilibrium equations (1). A potential function of space variables ψ (x,y,z) is thus defined in terms of three displacement components as follows:

$$u_x = \alpha_1\frac{\partial^2\psi}{\partial x^2} + \alpha_2\frac{\partial^2\psi}{\partial y^2} + \alpha_3\frac{\partial^2\psi}{\partial z^2} + \alpha_4\frac{\partial^2\psi}{\partial x\partial y} + \alpha_5\frac{\partial^2\psi}{\partial y\partial z} + \alpha_6\frac{\partial^2\psi}{\partial x\partial z} \tag{2a}$$

$$u_y = \alpha_7\frac{\partial^2\psi}{\partial x^2} + \alpha_8\frac{\partial^2\psi}{\partial y^2} + \alpha_9\frac{\partial^2\psi}{\partial z^2} + \alpha_{10}\frac{\partial^2\psi}{\partial x\partial y} + \alpha_{11}\frac{\partial^2\psi}{\partial y\partial z} + \alpha_{12}\frac{\partial^2\psi}{\partial x\partial z} \tag{2b}$$

$$u_z = \alpha_{13}\frac{\partial^2\psi}{\partial x^2} + \alpha_{14}\frac{\partial^2\psi}{\partial y^2} + \alpha_{15}\frac{\partial^2\psi}{\partial z^2} + \alpha_{16}\frac{\partial^2\psi}{\partial x\partial y} + \alpha_{17}\frac{\partial^2\psi}{\partial y\partial z} + \alpha_{18}\frac{\partial^2\psi}{\partial x\partial z} \tag{2c}$$

Here the α's are unknown material constants.

Now combining Eqs. (1) and (2), the values of the unknown α's are determined in such a way that two out of three equilibrium equations are automatically satisfied. Solving the set of the resulting homogeneous equations for α's, the three displacement components can be explicitly expressed as follows [12]:

$$u_x = \frac{\partial^2\psi}{\partial x\,\partial y} \tag{3a}$$

$$u_y = -\left[2(1-\mu)\frac{\partial^2\psi}{\partial x^2} + (1-2\mu)\frac{\partial^2\psi}{\partial y^2} + 2(1-\mu)\frac{\partial^2\psi}{\partial z^2}\right] \tag{3b}$$

$$u_z = \frac{\partial^2\psi}{\partial y\,\partial z} \tag{3c}$$

When the displacement components of Eq. (1) are replaced by those given by Eq. (3), Eqs. (1a) and (1c) are found to be satisfied automatically, and the only equilibrium condition [Eq. (1b)] that ψ has to satisfy becomes

$$\frac{\partial^4\psi}{\partial x^4} + \frac{\partial^4\psi}{\partial y^4} + \frac{\partial^4\psi}{\partial z^4} + 2\frac{\partial^4\psi}{\partial x^2\partial y^2} + 2\frac{\partial^4\psi}{\partial y^2\partial z^2} + 2\frac{\partial^4\psi}{\partial z^2\partial x^2} = 0 \tag{4}$$

Therefore, the three dimensional problem is reduced to the evaluation of the single potential function $\psi(x,y,z)$ from the single governing differential equation of equilibrium [Eq. (4)], satisfying the associated boundary conditions on the surfaces of the body. Similarly in the case of a two-dimensional problem, the potential function of space variables $\psi(x,y)$ is defined in terms of the two displacement components as

$$u_x = \alpha_1 \frac{\partial^2 \psi}{\partial x^2} + \alpha_2 \frac{\partial^2 \psi}{\partial x \partial y} + \alpha_3 \frac{\partial^2 \psi}{\partial y^2} \tag{5a}$$

$$u_y = \alpha_4 \frac{\partial^2 \psi}{\partial x^2} + \alpha_5 \frac{\partial^2 \psi}{\partial x \partial y} + \alpha_6 \frac{\partial^2 \psi}{\partial y^2} \tag{5b}$$

Substituting Eq. (5) into the two equilibrium equations expressed in terms of u_x, and u_y [10], and following the procedure explained above for the case of 3D problems, the governing differential equation for the 2D problem in terms of the function $\psi(x,y)$ becomes

$$\frac{\partial^4 \psi}{\partial x^4} + 2 \frac{\partial^4 \psi}{\partial x^2 \partial y^2} + \frac{\partial^4 \psi}{\partial y^4} = 0 \tag{6}$$

where the two-dimensional potential function $\psi(x,y)$ is defined explicitly by the following equations [7-8]

$$u_x(x,y) = \frac{\partial^2 \psi}{\partial x \partial y}$$

$$u_y(x,y) = -\frac{1}{1+\mu}\left[2\frac{\partial^2 \psi}{\partial x^2} + (1-\mu)\frac{\partial^2 \psi}{\partial y^2}\right] \tag{7}$$

The displacement boundary conditions of 3D and 2D problems have been expressed by the Eqs. (3) and (7), respectively. Now the boundary conditions in terms of stress components can readily be expressed as a function of ψ where the displacement components in the stress-displacement relations are replaced by ψ. The corresponding central difference stencils of the governing differential equations (4) and (6) are illustrated in Figs. 1 and 2, respectively.

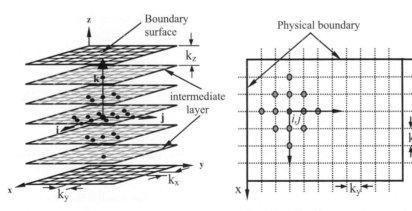

Fig. 1 Stencil of the governing equation (4) at internal nodal point of a 3D body

Fig. 2 Stencil of the governing equation (6) at internal nodal point of a 2D body

3. Reduction of Computational Effort

For a three-dimensional problem, the usual computational methods involve finding three parameters (like, u_x, u_y, u_z) at each nodal point. Here, in the present approach, we have to find only one parameter, ψ, at each nodal point. So, the reduction of the unknowns between the usual and the present approaches is from 3n to n, respectively, where n is the number of nodal points considered.

Therefore, the percentage saving in the number of division is:

$$\left[\left\{3n + \sum_{i=1}^{3n-1} i\,(3+i)\right\} - \left\{n + \sum_{i=1}^{n-1} i\,(3+i)\right\}\right] \Big/ \left\{3n + \sum_{i=1}^{3n-1} i\,(3+i)\right\} \approx 96\,\%$$

Similarly the percentage saving in the number of subtractions can be shown to be approximately 96%. Hence, the present finite-difference method of solution would save a significant amount of computational effort for the solution of a problem, in comparison with that required for the usual computational approaches.

4. Conclusion

The philosophy of the present variable reduction scheme is that both the 2D and 3D problems are formulated in terms of a single potential function, defined in terms of the respective displacement components. It has been shown that the present numerical approach would save the computational effort significantly in comparison to that taken by the usual methods. Because of the drastic reduction in the number of variables to be solved, both the accuracy as well as the saving in the computational time will be improved significantly. The saving in computational effort through the use of the present variable reduction scheme along with the finite-difference can also be extended to other fields of science and engineering.

References

[1]   C.O. Horgan, J.K. Knowels, Recent Developments Concerning Saint Venant's Principle, *Advances in Applied Mechanics*, 23, pp. 179-269, 1983.

[2]   A.J. Durelli and B. Ranganayakamma, Parametric solution of stresses in beams, *Journal of Engineering Mechanics*, 115 (2), pp. 401-415, 1989.

[3]   Durelli, A.J., and Ranganayakamma, B., On the use of Photoelasticity and some numerical methods, *Photomechanics and Speckle Metrology*, SPIE 814, pp. 1-8, 1987.

[4]   J.O. Dow, M.S. Jones and S.A. Harwood, A new approach to boundary modeling for finite difference applications in solid mechanics, *International Journal for Numerical Methods in Engineering*, 30, pp. 99-113, 1990.

[5]   Hardy, S.J., and Pipelzadeh, M.K., Static Analysis of Short Beams, *Journal of Strain Analysis*, 26 (1), pp. 15-29, 1991.

[6]   S.R. Ahmed, M.R. Khan, K.M.S. Islam and M.W. Uddin, Analysis of Stresses in Deep Beams Using Displacement Potential Function, *Journal of Institution of Engineers (India)*, 77, pp. 141-147, 1996a.

[7]   S.R. Ahmed, A.B.M. Idris and M.W. Uddin, Numerical Solution of Both Ends Fixed Deep Beams, *Computers & Structures*, 61 (1), pp. 21-29, 1996b.

[8]   S.R. Ahmed, M.R. Khan, K.M.S. Islam and M.W. Uddin, Investigation of stresses at the fixed end of deep cantilever beams, *Computers & Structures*, 69, pp. 329-338, 1998.

[9]   M.A.S. Akanda, S.R. Ahmed and M.W. Uddin, Stress Analysis of Gear Teeth using Displacement Potential Function and Finite-Differences, *International Journal for Numerical Methods in Engineering*, 53 (7), pp. 1629-1649, 2001.

[10]   S.P. Timoshenko and J.N. Goodier, *Theory of Elasticity*, 3rd Ed., McGraw-Hill Book Company, New York, 1979.

[11]   Richards, T.H. and Daniels, M.J., Enhancing Finite Element Surface Stress Predictions: A Semi-Analytic Technique for Axisymmetric Solids, *Journal of Strain Analysis*, Vol. 22, No. 2, pp. 75-86, 1987.

[12]   Hossain, M. Z., A New Approach to Numerical Solution for Three-Dimensional Mixed-Boundary-Value Elastic Problems, *M. Sc. Thesis*, Bangladesh University of Engineering and Technology (BUET), 2004.

VSP International
Science Publishers
P.O. Box 346, 3700 AH Zeist
The Netherlands

*Lecture Series on Computer
and Computational Sciences*
Volume 1, 2004, pp. 211-214

Generalized Kinematics for Energy Prediction of N-link Revolute Joint Robot Manipulator

M Zubaer Hossain[†]

Department of Mechanical Engineering,
Bangladesh University of Engineering and Technology,
Dhaka-1000, Bangladesh

Received 2 August, 2004; accepted in revised form, 18 August 2004

Abstract: This paper presents a mathematical formulation to predict the kinetic and potential energy of a robot manipulator comprising n numbered links. While doing kinematics analysis of robot manipulators utmost importance is given to evaluate the energy required for making the robot move from one position to another for accomplishing any task to be done. Forward kinematics solutions give the positional vectors for computing the position of a robot manipulator after corresponding positional changes due to rotation and translation. Here, using mechanics considerations a general mathematics has been presented to evaluate the energy of a n-link robot manipulator at any certain position from their positional vectors which is found to be useful for dynamic analysis of automated systems as well. The analysis has taken into consideration the mass of each link, their centroidal positions, angular speeds and dimensional parameters. Compared to the conventional methods it ensures easiness to computational mathematics for predicting both the kinetic and potential energy of a robot manipulator and generalizes the computational scheme to be incorporated while designing any automated system of n-links.

Keywords: kinematics, kinetic energy, potential energy, n-link, robot manipulator

1. Introduction

Robot systems, nothing but a series of links, are of course imitates natural systems to produce artificial mechanical objects that eventually are intended to act like humanoids. It is highly evident to emulate and simulate different natural mechanical systems into programmable and reproducible artificial systems. Before starting emulation completely energy consideration must come first that would eventually make the system run and do things we desire to accomplish. But no general mathematical formulation is found to be available in the literature, which can be employed to predict the instantaneous energy of a robot manipulator having n numbered links. In this paper, using simple mechanics a general mathematics has been presented sequentially that enables the evaluation of energy of a n-link robot manipulator at any instant. While deriving the formulation all the bodily as well as dynamic parameters have been included.

Unlike human joints robot joints have one degree of freedom and are either revolute or prismatic. Two different types of forces are responsible for link motion, namely, conservative and non-conservative [2]. Conservative forces such as gravitational force does not consume kinetic energy therefore these forces have a potential energy, as taken into consideration in this paper as V, whereas energy consuming forces such as friction and damping do not have any potential energy which are usually taken aside [1,3]. Forces and torques are appraised in the contribution of causing resultant linear and rotational motion [4].

In this approach, links are assumed not to have their mass center at their geometric centers, for which handling different shaped two-dimensional links has been made possible. To calculate the potential

---

[†] Corresponding author. Lecturer, Department of Mechanical Engineering, BUET, E-mail: zubaexy@me.buet.ac.bd

energy of i-th link of open-chain kinematics height, hi of its mass center w. r. t. world reference frame is considered whereas kinetic energy for each link is apportioned to the absolute transnational velocity which in fact the resultant of relative velocity w. r. t. the neighboring reference joint and the motion of the joint itself and the angular velocity w. r. t. the mass center of the link [4].

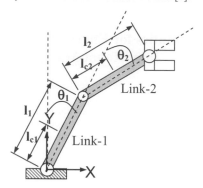

Figure 1: Schematic diagram of a 2-link robot manipulator
with the world reference coordinate (x, y)

2. Formulation of the General Kinematics

Total energy of the links that are individually set to some transnational motion w. r. t. its center and angular motion w. r. t. its local coordinate frame can be expressed as the summation of potential and kinetic energy at any instant. Hence, total energy $= V+T$; where, V is the overall potential energy and T is the overall kinetic energy of the manipulator. Mathematically,

$$T = \frac{1}{2}\sum_{i=1}^{n} m_i v_i^2 + \frac{1}{2}\sum_{i=1}^{n} I_i \Omega_i^2 \tag{1}$$

$$V = \sum_{i=1}^{n} V_i = \sum_{i=1}^{n} m_i g h_i \tag{2}$$

where, I_i = mass moment of inertia, m_i = mass of the links, Ω_i = absolute angular velocity, v_i = absolute linear velocity. Now, with reference to the Fig. 1, the velocity vectors of the links w. r. t. the world deference coordinate frame can be expressed as follows:

$$\bar{v}_1 = v_{1x} i + v_{1y} j = \left(\omega_1 l_{c1} \cos\theta_1\right) i - \left(\omega_1 l_{c1} \sin\theta_1\right) j$$

$$\Rightarrow \begin{bmatrix} v_{1x} \\ v_{1y} \end{bmatrix} = \begin{bmatrix} l_{c1} \cos\theta_1 & 0 \\ -l_{c1} \sin\theta_1 & 0 \end{bmatrix}\begin{bmatrix} \omega_1 \\ \omega_2 \end{bmatrix}$$

$$\text{or, } \bar{v}_1^2 = \omega_1^2 l_{c1}^2 \tag{3a}$$

Where, θ_i = relative angle between i-th & (i-1)-th link,
ω_i = Angular velocity of i-th link w. r. t. its center of gravity;
l_i, l_{ci} = length & centroidal distance of i-th link.

Similarly, for n-th link general expression for the velocity vector would be:

$$\bar{v}_n = \left[\sum_{i=1}^{n-1}\omega_i l_i \cos\left(\sum_{j=1}^{i}\theta_j\right) + \omega_n l_{cn} \cos\left(\sum_{j=1}^{n}\theta_j\right)\right] i - \left[\sum_{i=1}^{n-1}\omega_i l_i \sin\left(\sum_{j=1}^{i}\theta_j\right) + \omega_n l_{cn} \sin\left(\sum_{j=1}^{n}\theta_j\right)\right] j$$

i.e.

$$\begin{bmatrix} v_{nx} \\ v_{ny} \end{bmatrix} = \begin{bmatrix} l_1 \cos\theta_1 & l_{c2}\cos(\theta_1+\theta_2) & \ldots & l_{cn}\cos\left(\sum_{i=1}^{n}\theta_i\right) \\ -l_1\sin\theta_1 & -l_{c2}\sin(\theta_1+\theta_2) & \ldots & -l_{cn}\sin\left(\sum_{i=1}^{n}\theta_i\right) \end{bmatrix}\begin{bmatrix}\omega_1 \\ \omega_2 \\ \ldots \\ \omega_n\end{bmatrix}$$

Hence, $\bar{v}_n^{\ 2} = \left(\sum_{i=1}^{n-1}\omega_i^{\ 2}l_i^{\ 2} + \omega_n^{\ 2}l_{cn}^{\ 2}\right) + 2\sum_{k=1}^{n-2}l_k\omega_k\sum_{i=k+1}^{n}\omega_i l_i\cos\left(\sum_{j=k+1}^{n}\theta_j\right) + 2\omega_{n-1}l_{n-1}\omega_n l_{cn}\cos\theta_n$

Substituting the above relations, Eq. (3), in the first part of Eq. (1), we get the general kinetic energy expression for the translation velocity of the link assembly as follows:

$$T_v = \frac{1}{2}\left[\left(m_1 l_{c1}^{\ 2} + l_1^{\ 2}\sum_{i=2}^{n}m_i\right)\omega_1^{\ 2} + \ldots + m_n l_{cn}^{\ 2}\omega_n^{\ 2}\right] + \omega_1\omega_2\left(m_2 l_1 l_{c2}\cos\theta_2 + l_1 l_2\cos\theta_2\sum_{i=3}^{n}m_i\right)$$

$$\ldots + \omega_1\omega_n m_n l_1 l_{cn}\cos\left(\sum_{i=2}^{n}\theta_i\right) + \ldots + \omega_2\omega_3\left(m_3 l_2 l_{c3}\cos\theta_3 + l_2 l_3\cos\theta_3\sum_{i=4}^{n}m_i\right) +$$

$$\ldots + \omega_2\omega_n m_n l_2 l_{cn}\cos\left(\sum_{i=3}^{n}\theta_i\right) + \ldots + \omega_{n-1}\omega_n m_n l_{n-1} l_{cn}\cos\theta_n$$

Furthermore, kinetic energy for the rotational velocity can be written as follows

$$T_w = \frac{1}{2}\left[\omega_1^{\ 2}\sum_{i=1}^{n}I_i + \omega_2^{\ 2}\sum_{i=2}^{n}I_i + \ldots + \omega_n^{\ 2}I_n\right] + \left[\omega_1\omega_2\sum_{i=2}^{n}I_i + \omega_1\omega_3\sum_{i=3}^{n}I_i + \ldots + \omega_1\omega_n I_n\right]$$

$$+ \left[\omega_2\omega_3\sum_{i=3}^{n}I_i + \omega_2\omega_4\sum_{i=4}^{n}I_i + \ldots + \omega_2\omega_n I_n\right] + \ldots + \omega_{n-1}\omega_n I_n$$

Hence, total kinetic energy would be: $T = T_v + T_w = \begin{bmatrix}\omega_1 \\ \omega_2 \\ \ldots \\ \omega_n\end{bmatrix}^T \begin{bmatrix} T_{11} & T_{12} & \ldots & T_{1n} \\ T_{11} & \ldots & \ldots & \ldots \\ \ldots & \ldots & \ldots & \ldots \\ T_{n1} & \ldots & \ldots & T_{nn}\end{bmatrix}\begin{bmatrix}\omega_1 \\ \omega_2 \\ \ldots \\ \omega_n\end{bmatrix}$

where, $T_{ii} = \frac{1}{2}\left[m_i l_{ci}^{\ 2} + l_i^{\ 2}\sum_{j=i+1}^{n}m_j + \sum_{k=1}^{n}I_k\right]$; $i = 1,2,3,\ldots\ldots,n$

$T_{ij} = m_j l_i l_{cj} + l_i l_j \sum_{kk=j+1}^{n}m_{kk}\cos\left(\sum_{k=i+1}^{j}\theta_k\right)$; $j > i$, $i = 1,2,3,\ldots\ldots,n$; $j = 1,2,3,\ldots\ldots,n$

and $T_{ij} = \sum_{k=i}^{n}I_k$; $i = 1,2,3,\ldots\ldots,n$; $j = 1,2,3,\ldots\ldots,n$; $j < i$

Potential energy, on the other hand, can be expressed as a summation of the potential energy of the individual links w. r. t. the world reference coordinate frame, i.e.,

$$V = \sum_{i=1}^{n}m_i g h_i = m_1 g l_{c1}\cos\theta_1 + m_2 g[l_1\cos\theta_1 + l_{c2}\cos(\theta_1+\theta_2)] +$$

$$\ldots\ldots\ldots + m_n g\left[l_1\cos\theta_1 + l_2\cos(\theta_1+\theta_2) + \ldots\ldots + l_{cn}\cos\left(\sum_{i=1}^{n}\theta_i\right)\right]$$

Hence, total potential energy, $V = \begin{bmatrix} V_{11} \\ V_{21} \\ .... \\ V_{n1} \end{bmatrix} \begin{bmatrix} \cos\theta_1 \\ \cos(\theta_1+\theta_2) \\ .... \\ \cos\left(\sum_{i=1}^{n}\theta_i\right) \end{bmatrix}^{T}$ (5)

where, $V_{i1} = m_i gl_{ci} + \sum_{k=i+1}^{n} m_k gl_i \; ; i = 1,2,3,........,n$

Total energy for n-link manipulator at any certain instant of the planar motion can therefore be expressed combining Eq. (4) and (5) as follows:

$$\text{Total energy} = T + V = \begin{bmatrix} \omega_1 \\ \omega_2 \\ ... \\ \omega_n \end{bmatrix}^{T} \begin{bmatrix} T_{11} & T_{12} & ... & T_{1n} \\ T_{11} & ..... & .. & ... \\ .... & ... & .. & ... \\ T_{n1} & ... & ... & T_{nn} \end{bmatrix} \begin{bmatrix} \omega_1 \\ \omega_2 \\ ... \\ \omega_n \end{bmatrix} + \begin{bmatrix} V_{11} \\ V_{21} \\ .... \\ V_{n1} \end{bmatrix} \begin{bmatrix} \cos\theta_1 \\ \cos(\theta_1+\theta_2) \\ .... \\ \cos\left(\sum_{i=1}^{n}\theta_i\right) \end{bmatrix}^{T}$$

3. Conclusion

Since energy of the total manipulator depends on the link masses, their length, angular velocity, position, and inertia [5], it is indispensable to investigate the potential and kinetic energy at any instant of the mechanism for any automated system of links. As link change in link parameters and given motion cause change in energy requirement for making the mechanism move, it is advisable to design a system which would be able to minimize or optimize the energy requirement for the task to be completed. The present formulation can also be extended to evaluate the frictional loss for any numbered link robot manipulator.

References

[1] Kram, R., Wong, B. and Full, R. J. (1997), Three-dimensional kinematics and limb kinetic energy of running cockroaches, *Journal of Experimental Biology*, 200, pp. 1919-1929.

[2] Kown, S. J., Youm, Y. and Chung, W. K. (1994), General Algorithm For Automatic Generation of The Workspace For N-Link Redundant Manipulators, *ASME Journal of Mechanical Design*, Vol. 116, No. 3, pp 967-969

[3] Lee, D., Youm, Y. and Chung, W. K. (1996), Mobility Analysis of Spatial 4- and 5-link Mechanisms of the RS Class, *Mechanism and Machine Theory*, Vol. 31, No. 5, pp 673-690

[4] Beer, F. and Johnston, R. (2002), *Vector Mechanics for Engineers: Statics and Dynamics*, McGraw-Hill Education.

[5] Silva, F. M. and Machado, J. A. T. (1999), *Energy Analysis during Biped Walking*, IEEE International Conference on Robotics and Automation, Michigan, USA.

VSP International
Science Publishers
P.O. Box 346, 3700 AH Zeist
The Netherlands

*Lecture Series on Computer
and Computational Sciences*
Volume 1, 2004, pp. 215-218

Stationary Solutions of the Gross-Pitaevskii Equation for Neutral Atoms in Harmonic Trap

Wei Hua, Xueshen Liu[1] and Peizhu Ding

Institute of Atomic and Molecular Physics,
Jilin University,
Changchun 130012, P. R. China

Received 30 June, 2004; accepted in revised form, 8 August 2004

Abstract: We solve the time-independent Gross-Pitaevskii equation that describes the dilute Bose-condensed atoms in harmonic trap at $T=0$ by symplectic shooting method (SSM). Both the repulsive nonlinearity and the attractive nonlinearity cases are studied, and the bounded state eigenvalues as well as the corresponding wave functions are presented.

Keywords: Bose-Einstein Condensation, Gross-Pitaevskii equation, stationary solution

PACS: 02.70.-c, 03.65.Ta

1. Gross-Pitaevskii equation and the numerical method

The Gross-Pitaevskii equation for the neutral atoms in 3D spherically harmonic potential trap is [1-2]

$$\left[-\frac{\hbar^2}{2m}\nabla^2 + V(r) + NU_0|\psi(r)|^2 \right]\psi(r) = \mu\psi(r) \tag{1}$$

and the normalization condition is

$$\int|\psi(\bar{r})|^2 d^3r = 1$$

where $V(r) = \frac{1}{2}m\omega_t^2 r^2$ is the spherical harmonic trap, m is the mass of a single atom, ω_t is the angular frequency of the trap, N is the number of atoms in the condensate. $\psi(r)$ is the 'wave function' of the condensate, and μ is the corresponding eigenvalues. $U_0 = \frac{4\pi \hbar^2 a_s}{m}$ represents the interaction between two atoms, and a_s is the scattering length.

If the following harmonic oscillator units are used

$$r = \left[\frac{\hbar}{2m\omega_t}\right]^{\frac{1}{2}}x, \ \beta = \frac{\mu}{\hbar\omega_t}, \ \text{and} \ \psi(r) = \frac{1}{\sqrt{4\pi}\left[\dfrac{\hbar}{2m\omega_t}\right]^{\frac{3}{4}}}\frac{\Phi(x)}{x},$$

we can obtain the dimensionless form

---

[1] Corresponding author. E-mail: yashuliu@email.jlu.edu.cn

$$\left[-\frac{d^2}{dx^2} + \frac{x^2}{4} + \alpha \frac{\Phi(x)^2}{x^2} - \beta \right] \Phi(x) = 0 \tag{2}$$

the corresponding normalization condition is

$$4\pi \int_0^\infty \psi(r)^2 r^2 dr = \int_0^\infty \Phi(x)^2 dx = 1. \tag{3}$$

In Eq. (2), $\alpha = 2Na_s / \left[\dfrac{\hbar}{2m\omega_t} \right]^{\frac{1}{2}}$ is the nonlinear coefficient. We rewrite Eq. (2) into two 1-order

ODEs

$$\frac{d\Phi(x)}{dx} = \Psi(x) = -f(\Psi) \tag{4}$$

$$\frac{d\Psi(x)}{dx} = \left[\frac{x^2}{4} - \beta + \alpha \frac{\Phi(x)^2}{x^2} \right] \Phi(x) = g(\Phi, x) \tag{5}$$

If we take x as the 'time' variable, $\Phi(x)$ the general position, and $\Psi(x)$ the general velocity, it is easy to see that the system (4), (5) is of the form of a Hamiltonian equation that is a separable Hamiltonian system. The Hamiltonian function can be written as follows

$$H = \frac{\Psi(x)^2}{2} + \frac{1}{2}\left[-\frac{x^2}{4} + \beta \right] \Phi(x)^2 - \frac{1}{4}\alpha \frac{\Phi(x)^4}{x^2} \tag{6}$$

So, the reliable method for this kind of problem is the structure-preserving method. We adopt the 4-order explicit symplectic algorithm to solve this equation [3-4]. We use the symplectic scheme-shooting method (SSSM) that combines the structure-preserving method and the shooting method that is accompanied by half-interval method to this problem. We adopted the well-known boundary conditions [5]

$$x \sim 0, \quad \Phi(x) = \varepsilon \Phi'(x), \quad \left(\varepsilon = 10^{-6} \right) \tag{7}$$

$$x \sim \infty, \quad \frac{\Phi(x)_{num}}{\Phi'(x)_{num}} = \frac{\Phi(x)_{asym}}{\Phi'(x)_{asym}} = \left[-\frac{x}{2} + \left[\beta - \frac{1}{2} \right] \frac{1}{x} \right]^{-1} \tag{8}$$

Taking into consideration the normalization of the NLSE, the criterion for our SSSM at x_{max} is

$$\left| \Phi(x)_{num} - \Phi'(x)_{num} \left[-\frac{1}{2} + \left[\beta - \frac{1}{2} \right] \frac{1}{x} \right]^{-1} \right| < \varepsilon_1 \quad \left(\varepsilon_1 = 10^{-10} \right) \tag{9}$$

$$\left| \int |\Phi(x)|^2 dx - 1 \right| < \varepsilon_2 \quad \left(\varepsilon_2 = 10^{-5} \right) \tag{10}$$

where $\varepsilon_1, \varepsilon_2$ is the error in our numerical computation.

2. Numerical Results

We solve the time-independent NLSE with an external potential by the general method presented in literature [6]. Given a nonlinear coefficient α, we take $\Phi'(0)$ and β as variables, and then the SSM is used.

2.1 The repulsive nonlinearity

In this case, $\alpha > 0$, the solution of the condensation is stable. Let α range in 0.1~ 50, we work out the ground state wavefunction and the eigenvalues of the ground state and the first excited state, which are displayed respectively in Figure 1 and Figure 2, our computation results are consistent with Ref. [6].

2.2 The attractive nonlinearity

For attractive nonlinearity, we restudied the similar case in literature [5] that deals with the exact stationary NLSE. This literature presents another way of normalization that saves much computation. The author pointed out that n is given out as the nonlinear coefficient in literature [2]. For the attractive nonlinearity case we have $-n = \alpha$. We list part of their results in the first three columns of Table 1 and our results in the last three columns. We set $x_{max} = 7$, to be the same with literature [5]. From Table 1 we can see that the boundary values $\Phi'(0)$ of the two methods are not the same. It is a problem, and we believe that it is not the end of the story. On the whole, the results of the two methods coincide very well except one point ($n = -1.6237$). Furthermore, we give out the ground state wave functions normalized to unity in Figure 3 that is not presented in literature [5]. In Figure.3 we can see clearly the shape of the condensates subject to different negative scattering lengths.

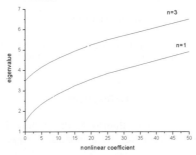

Figure 1: The ground state wave function of the condensate. The nonlinear coefficient is taken to be 0.1, 1, 3, 5, 10, 15, 25, and 50 down the vertical axis. Horizontal axis is x, and vertical axis is $\Phi(x)/x$.

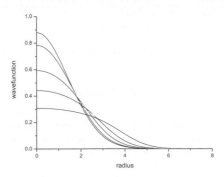

Figure 2: The eigenvalue of the gound state (n=1) as well as the first excited state (n=3) with the nonlinear coefficient. n is the principle quantum number.

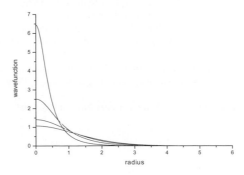

Figure 3: The ground state wave function of the condensate. β is taken to be -2.0, -1.0, 0.0, 0.4, 0.8, 1.2 down the vertical axis. Horizontal axisi s x and vertical axis is $\Phi(x)/x$.

Table 1: The first three columns are the results of literature [5], and the last three columns are our computation results.

β	$\Phi'(0)$	$-n$	α	$\Phi'(0)$	β
1.5	0	0	0	0.89325	1.50000
1.4	0.5448721	-0.3310	-0.3310	0.94703	1.40000
1.2	0.9939222	-0.8597	-0.8597	1.07195	1.20000
1.0	1.3567267	-1.2282	-1.2282	1.22406	1.00009
0.8	1.7022822	-1.4607	-1.4607	1.40844	0.80003
0.6	2.0495486	-1.5839	-1.5839	1.62844	0.60006
0.4	2.4045809	-1.6254	-1.6254	1.88594	0.40009
0.2	2.5851166	-1.6237	-1.6237	2.03008	0.29906
0.0	3.1340461	-1.5632	-1.5632	2.50742	-0.00050
-1.0	4.8924036	-1.2234	-1.2234	4.42383	-1.00041
-2.0	6.3914678	-0.9843	-0.9843	6.44258	-2.00026

3. Conclusion

In this paper, we have considered the time-independent Gross-Pitaevskii equation, which describes the stationary states for neutral atoms in a harmonic trap. Since it is a Hamiltonian system, and has symplectic structure, so structure-preserving method is the efficient method to solve this kind of problem. By the general method presented in literature and the well-known boundary condition, we apply SSSM accompanied by the half-interval method to this Hamiltonian system. Both the repulsive nonlinearity and the attractive nonlinearity cases are considered. It can be assured that our method is reliable and efficient from the computation results.

Acknowledgments

This work was supported by The National Natural Science Foundation of China (10171039), The Special Funds for Major State Basic Research Projects (G1999032804) and The Young Teacher Foundation of Jilin University.

References

[1] P. A. Ruprecht, M. J. Holland, K. Burnett, Mark Edwards. Time-dependent solution of the nonlinear Schrödinger equation for Bose-condensed trapped neutral atoms. *phys. Rev., A*51(1995), 4704-4711.

[2] M. Edwards, K. Burnett. Numerical solution of the nonlinear Schrödinger equation for small samples of trapped neutral atoms. *Phys. Rev., A*51 (1995) 1382-1386.

[3] M. Z. Qin, D. L. Wang, M. Q. Zhang, Explicit Symplectic Difference Schemes for Separable Hamiltonian System, *J. Comp., Math.*, (1991) 9(3), 211-221.

[4] X. S. Liu, L.W. Su, and P. Z. Ding, Symplectic algorithm for use in computing the time-independent Schrödinger equation, *International Journal of Quantum Chemistry*, 87(1), 2002,1-11.

[5] A.Gammal. T. Frederico, Lauro Tomio. Improved numerical approach for the time-independent Gross-Pitaevskii nonlinear Schrödinger equation. *Phys. Rev. E* 60, (1999) 2421-2424.

[6] K. Yan, W. Tan, *Chin. Phys. Soc.* Vol.48 (1999), No.7, 1185-1991.

VSP International
Science Publishers
P.O. Box 346, 3700 AH Zeist
The Netherlands

*Lecture Series on Computer
and Computational Sciences*
Volume 1, 2004, pp. 219-222

Formalism of Second-Kind Phase Transitions in the Dynamic Failure Phenomenon

R.I. Il'kaev, V.T. Punin, A.Ya. Uchaev[1], S.A. Novikov, N.I. Sel'chenkova

Russian Federal Nuclear Center – VNIIEF,
607188, Nizhni Novgorod region,
Sarov, Mira prospect 37, Russia

Received 7 July, 2004; accepted in revised form, 8 August 2004

Abstract: The paper contemplates the possibility for applying the apparatus of critical phenomena theory and second-kind phase transitions theory to description of the failure process at the final stage in the dynamic longevity range ($t \sim 10^{-6} \div 10^{-11}$ s). As a result of a large number of experimental and calculation-and-theoretical studies it was shown that the body resists the external action in the longevity dynamic range through originating dissipative structure – failure centers cascade. The failure centers cascade is a fractal cluster when distribution of failure centers by sizes is determined by the degree law $N(D) \sim D^{-\alpha}$; $N(D)$ – failure centers number of dimensions D, $\alpha > 1$ (meso-II). Such a dependence means that the relation of the number of clusters of one size to the number of clusters of another size depends not only on their dimension but also on the relation of sizes. The same degree relations describe distributions (by sizes) of glide lines and strips occurring near the site of failure centers formation characterizing the turbulent mixing of crystal lattice (meso-I) and distribution of "roughness of mountain relief" of the failure centers surface (nanolevel) by sizes. There were studied physics preconditions of application of percolation models for description of metals failure process in the dynamic longevity range . In the loaded state the failure centers density ρ increases, and when reaching the critical density ρ_c there originates connectivity in the system of failure centers, changing the body connectivity, i.e. macro-failure occurs. At the final stage of the dynamic failure the process is controlled by concentration criteria, when failure centers dimension and the average distance between them are connected by the definite relation. The critical phenomena are conditioned by the properties of the whole complex of system particles, but not by each particle individual properties.The foregoing determines the universal properties of metals behavior in the dynamic failure phenomenon. The unique mechanism of the process of dynamic failure – the loss of connectivity of the system through clusterization of failure centers cascade (equal dimensionality of order parameter, unique class of versatility) and equal space dimensionality, where the process occurs, determine the possibility for prediction of behavior of unstudied metals in the extreme conditions and for "constructing" of new materials resistant to definite types of exposure by the means of computer. Application of apparatus of critical phenomena theory and theory of second-kind phase transitions for the processes of dynamic failure at the final stage allowed determination of universal properties of metals behavior in the phenomenon of dynamic failure conditioned by self-arrangement and instability in dissipative structures.

Keywords: apparatus of critical phenomena theory, second-kind phase transitions theory, dissipative structure, failure centers cascade, processes of dynamic failure

PACS: Here must be added the AMS-MOS or PACS Numbers

The paper studies the possibility for applying an apparatus of critical phenomena theory and second-kind phase transition theory to description of failure process at the final stage in the dynamic longevity range ($t \sim 10^{-6} \div 10^{-11}$ s).

Critical phenomena are cooperative phenomena. They are conditioned by the properties of the whole integrity of the system particles but not by individual properties of each particle. The radius of correlation characterizes the distance at which the structure elements influence each other, and, thus, they are dependent.

The correlation radius characterizes the distance at which the structure elements influence each other and are dependent. For all fractal systems this correlation radius is described by the power law [1].

---

[1] Corresponding author. Head of Department RFNC-VNIIEF. E-mail: uchaev@expd.vniief.ru

As a result of a large volume of calculation-and-theoretical and experimental studies [2-4] it was shown that the originating dissipative structure - failure centers cascade puts up the body resistance to the external action in the dynamic longevity range.

The cascade of failure centers is a fractal cluster when distribution of failure centers by sizes is determined by the degree law $N(D) \sim D^{-\alpha}$; $N(D)$ – number of failure centers of dimension D, $\alpha > 1$ (mesa-level-II). Such a dependence means that the relation of the number o clusters of one size to the number of clusters of another size depend not on their size, but on dimensions relation.

The same power relations describe the distributions by sizes of strips and slip bands emerging near failure centers formation characterizing turbulent mixing of crystal lattice (meso-1) and distribution by sizes "mountain relief roughness" of failure centers surface (nanolevel) [2-4].

In contrast to the theory of temperature phase transitions where the transition between two phases occurs at a critical temperature, the percolation transition is a geometry phase transition. The percolation threshold separates two phases: in one phase there exist final clusters, in another one – a single infinite cluster.

Let us view, for example, the magnetic phase transition. At low temperatures some materials has non-zero spontaneous magnetization. As temperature increases the spontaneous magnetization continuously decreases and disappears at critical temperature. In the percolation theory concentration of occupied sites plays the same part as the temperature in temperature phase transitions. The probability that the site belongs to the infinite cluster is similar to the order parameter in the temperature phase transition theory. Many important cluster characteristics (correlation length, average site number) near the transition are described by power function with different critical indices (see expression 1.2) [5]. Thus the infinite cluster power is described by expression of the form

$$P(x) \sim |x-x_c|^{\alpha}, \tag{1}$$

where x_c – percolation threshold at which the infinite cluster emerges. The average number of final cluster sites (receptivity analogue) at $x - x_c \to 0$ behaves as

$$S(x) \sim |x-x_c|^{-\gamma}, \tag{2}$$

where γ - critical index [5, 6].
Universal indices [5, 6] do not depend on the lattice type, percolation type but only on dimensionality of the space of the problem.

The results of studies given in papers [7, 8] show that with increasing lattice scale L, the mass of percolation cluster (M) (roughness of mountain relief of fracture surface was contemplated) grows as $M(L) \sim L^D$, where D – fractal dimensionality.

Fig. 1, a, b presents samples fracture surfaces under the action of high-current beams of relativistic electrons at the energy input rate $dE/dt \sim 10^5 \div 10^{11}$ J/(g·s), in the temperature range $T_0 \sim 4K \div 0,8 \, T_{melt}$, in the longevity range $t \sim 10^{-6} \div 10^{-10}$ s.

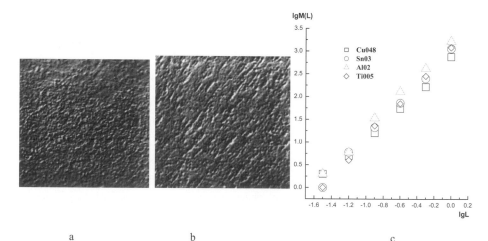

a	b	c

Figure 1: View of fracture surface: a – Cu Δ = 0,48 mm; b – Ti Δ = 0,05 mm; c- masses of percolation clusters of fracture surfaces as some metals lattice scale increases, where Δ - sample thickness.

Fig. 1, c gives the mass of percolation cluster of fracture surface as some metals lattice scale increases under the action of high-current beams of relativistic electrons.

The results of processing show that the fractal dimensionality D obtained at processing of some metals fracture surfaces (see fig.2) is close to the data given in the modern literature [7,8] obtained by calculations D_p = 1.89. Fractal dimensionality (see fig.1) of the mass of percolation clusters of the fracture surface has the value of D ~ 1,8 as some metals lattice scale increases.

Numerical simulation of percolation clusters at the lattice 50x50 with different value of x (x > x_c и x < x_c) was performed, where x_c – percolation threshold. The mass M(L) of clusters given in fig.3 for different values of L of the lattice scale was determined. Fig.2 presents a percolation cluster of 50x50 square lattice for the occupation probability p = 0,3 (fig. 2, a), p = 0,593 (fig. 2, b).

Fig. 2, c presents the "mass" of percolation cluster at the square lattice for the occupation probability of 0,3; 0,593.

Fractal dimensionality D has a value of (for x > x_c) D = 1.6 for the occupation probability p = 0,3; D = 1.79 for occupation probability p = 0,593. Values of D obtained at processing of computer clusters and fracture surfaces are close.

I.e., the results of studies presented in fig.1, 2 prove that fracture surface roughness for a number of studied structural materials under the action of high-current beams of relativistic electrons is a percolation cluster.

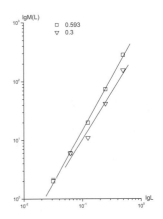

Figure 2: Percolation cluster of 50x50 square lattice for p occupation probability: a) p = 0.3 b) p = 0.593; c) growth of mass of percolation cluster for the occupation probability p=0,3; 0,593.

I.e. application of the apparatus of the critical phenomena theory and the theory of second-kind phase transitions for the dynamic failure process at the final stage allowed determining of universal properties of metals behavior in the phenomenon of dynamic failure conditioned by self-arrangement and instability in dissipative structures.

The unique mechanism of the process of dynamic failure of a number of studied metals – the loss of connectivity of the system through clusterization of the failure centers cascade (equal dimensionality of the order parameter, the unique universality class) and the equal space dimensionality where the process occurs determine the possibility for forecasting of unstudied metals behavior under extreme conditions as well as for "constructing" by computer method new materials resistant to certain kinds of effect.

The above said defines the universal properties of metals behavior in the dynamic failure phenomenon.

References

[1] M. Shreder. Fractals, chaos, power laws. Izhevsk: 2001, RKhD, 528 p.

[2] E.K. Bon'ushkin, N.I. Zavada, S.A. Novikov, A.Ya. Uchaev Kinetics of dynamic metal failure in the mode of pulse volume heating-up. Scientific edition. Edited by R.I. Il'kaev// Sarov, 1998, 275p.

[3] R.I. Il'kaev, A.Ya. Uchaev, S.A. Novikov, N.I. Zavada, L.A. Platonova, N.I. Sel'chenkova. Universal metals properties in the dynamic failure phenomenon. // Report of Academy for Science (RAS), 2002, v. 384, № 3, P. 328-333.

[4] R.I. Il'kaev, V.T. Punin, A.Ya. Uchaev, S.A. Novikov, E.V. Kosheleva, L.A. Platonova, N.I. Sel'chekova, N.A. Yukina. Time regularities of the process of dynamic metals failure conditioned by hierarchic properties of dissipative structures – failure centers cascade // RAS, 2003, vol. 393, No.3.

[5] Yu.Yu. Tarasevich. Percolation: theory, applications, algorithms:Education book. M.: Editorial URSS, 2002. – 112 p., illustr.

[6] I.M. Sokolov. Dimensionalities and other geometric critical indexes in the percolation theory. // UFN, 1986, vol. 150, №2. P. 221-255.

[7] Feder E. Fractals // M.: Mir. 1991. 264 p.

[8] Stauffer, D. "Scaling Theory of Percolation Clusters," Phys. Reports, Vol. 54, No. 1, 1-74 (1979).

VSP International
Science Publishers
P.O. Box 346, 3700 AH Zeist
The Netherlands

*Lecture Series on Computer
and Computational Sciences*
Volume 1, 2004, pp. 223-225

Theoretical Investigation of Conductance
Properties of Molecular Wires

S. Jalili[1,2,*] and F. Moradi[1]

[1]Department of chemistry, K. N. Toosi University of Technology, P. O. Box 16315-1618, Tehran, Iran
[2]Computational Physical Sciences Research Laboratory, Department of Nano-Science, Institute for
Studies in Theoretical Physics and Mathematics (IPM), P.O. Box 19395-5531, Tehran, Iran

Received 2 August, 2004; accepted in revised form 25 August, 2004

Abstract: Quantum-mechanical based methods, such as Density Functional Theory and *ab initio*, in
conjunction with Non-equilibrium Green's Function method are used to obtain the current-voltage (I-V)
characteristic curves of a molecular nano-wire bridging two metallic electrodes. By using the improved
chemical software's, the Hamiltonians of the three main parts of system, i.e. the right lead, the device, the left
lead and conductance properties of a molecular wire were calculated.

Keywords: Conductance properties, Molecular wire, *ab initio*, Non-equilibrium Green's Function, Density
Functional Theory.

PACS: 77.63.-b, 73.40.-c, 85.65.+h

A molecular nano-wire normally refers to a system composed of a molecule bridging two electron
reservoirs. The emerging field of Molecular Electronics is concerned with constructing information
processing devices by coupling single molecules, with electronic functionalities, together and connecting
the resulting nano-wire to external electrodes. The design of such a system poses several theoretical,
computational and experimental challenges. Ability of electronic transport could be theoretically
observed in almost all of the aromatic systems, such as benzene, thiophene and etc. It seems that
bithiolate derivatives of these molecules connected to suitable metallic contacts, like gold leads, is an
ideal configuration to study conductance properties of them.

Several research groups have studied phenyl bithiol molecule as a molecular wire which bridging two
golden contacts, some of the recent researchers -and also us- are interested to study a thiophene bithiol
molecule (TBT) as a molecular wire using computational approaches such Quantum-Mechanical based
methods.

In this work we have assumed that the terminal S atoms of the TBT molecule are connected to the
triangular hollow sites in the two gold electrodes. Recent experiments have suggested that the binding to a
hollow site is energetically more favorable, while others suggest this to be true for a single-atom
connection. Consequently, the geometry of a molecule-metal contact is, as yet, not well understood. The
TBT molecule contact-geometry and the associated bond lengths of the TBT molecule were determined
from Density Functional Theory (DFT) calculations.

By dividing the main system to three parts, that is, the left lead, the device (TBT and some of the surface
atoms from leads) and the right lead, we calculated the Hamiltonian of each part independently and then
added them to obtain the total Hamiltonian of the system. The obtained Hamiltonian should be added to
the self-consistent potential to get the Fock matrix of the whole system. This matrix includes the effects
of external fields, kinetics energy of electrons, electron-electron interactions, and electron-nuclear
interactions. This matrix calculated directly by a self-consistent procedure using the DFT approach with
Beck-3 exchange and Perdew-Wang 91 correlation and LANL2DZ basis set which incorporates the
relativistic core pseudo potentials.

Consequently we could describe the Green's function of the system according to a Non-equilibrium Green's
Function (NEGF) formalism,

---

* Corresponding author. E-mail address: sjalili@kntu.ac.ir

$$G = (ES - F)^{-1} = \begin{bmatrix} ES_{dd} - F_{dd} & \tau \\ \tau^+ & D \end{bmatrix}^{-1} \qquad (1)$$

The first left block of this matrix is the device Green's Function (G_{dd}) in which S_{dd} and F_{dd} are the overlap and the Fock matrices of the device respectively. $\tau D^{-1}\tau^+$ is the self-energy term, which describes the effects of contacts on the device. Due to the fact that many of the elements of coupling matrix (τ) are zero, we used the only surface term of D^{-1} that is the surface Green's function g, and thus the reduced device Green's function could be written as:

$$G_{dd}(E) = [ES_{dd} - F_{dd} - \Sigma_1(E) - \Sigma_2(E)]^{-1} \qquad (2)$$

where the self-energies Σ_1 and Σ_2 are non-Hermitian matrices that are related to the non-zero part of the corresponding τ and g through

$$\Sigma_{1,2} = \tau_{1,2} g_{1,2} \tau_{1,2}^+ \qquad (3)$$

Using the anti-Hermitian components of the self-energies

$$\Gamma_{1,2} = i[\Sigma_{1,2} - \Sigma_{1,2}^+] \qquad (4)$$

and the device Green's function we calculated the Transmission Function

$$T(E) = trace[\Gamma_1 G \Gamma_2 G^+] \qquad (5)$$

in order to obtain the I-V curve of the TBT molecular wire the following equation was used:

$$I = \frac{2e}{h} \int_{-\infty}^{\infty} dE T(E)[f_1(E) - f_2(E)] \qquad (6)$$

where $f_{1,2}(E)$ are the Fermi functions with electrochemical potentials $\mu_{1,2}$ (which are supposed to be equal to each other and to gold's Fermi energy, ie, –5.31 eV at equilibrium)

$$f_{1,2}(E) = [1 + \exp(\frac{E - \mu_{1,2}}{K_B T})]^{-1} \qquad (7)$$

The obtained I-V curve (Fig.1) represents a resonant mechanism for TBT as well as the results have been noticed for other aromatic rings such as phenyl bithiol. This representation, beside others could be used to prove the theory of non-ohmic current-voltage relation in these systems. In this work, other properties such as DOS, Transmission Function, voltage drop through TBT molecule and etc will be discussed.

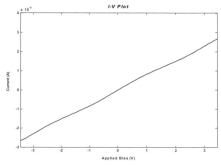

Figure 1: I-V curve of thiophene bithiol connected to two gold lead

References

[1] S. Datta, *Electronic Transport in Mesoscopic Systems*, Cambridge University Press, Cambridge, 1995.

[2] P. Damle, A. W. Ghosh, S. Datta, *Chem. Phys* 281 (2002) 171.

[3] M. Di. Ventra, N. D. Lang, S. T. Pantelides, *Chem. Phys* 281 (2002) 189.

[4] W. Tian, S. Datta, S. Hong, R. Reienberger, J. I.Henderson, C. P.Kubiak, *J. Chem. Phys* 109(7) (1998) 2874.

[5] E. Emberly, G. Kirczenow, *Phys. Rev. B* 58 (1998) 10911.

[6] Y. Xue, S. Datta, M. Ratner, *J. Chem. Phys* 115 (2001) 4292.

[7] C. Majumder, H. Mizuseki, Y. Kawazaoe, *J. Chem. Phys* 118(21) (2003) 9809.

[8] N. B. Larsen, H. Biebuyck, E. Delamarche, B. Michel, *J. Am. Chem. Soc* 119 (1997) 3107.

[9] N. Camillone, C. E. D. Chidsey, G.Y. Liu, G. Scoles, *J. Phys. Chem* 98 (1993) 3503.

[10] P. Hay, W. Wadt, *J. Chem. Phys* 82 (1985) 270.

[11] P. Hay, W. Wadt, *J. Chem. Phys* 82 (1985) 284.

[12] A. D. Becke, *J. Chem.Phys* 98 (1993) 5648.

[13] J. P. Perdew, in: P. Ziesche, H. Eschrig (Eds.), *Electronic Structure of Solids*, Akademie Verlag, Berlin, 1991, pp. 11–20.

[14] L. A. Bumm and et al. *Science* 271 (1996) 1705.

[15] M. A. Reed and et al. *Science* 278 (1997) 252.

VSP International
Science Publishers
P.O. Box 346, 3700 AH Zeist
The Netherlands

Lecture Series on Computer
and Computational Sciences
Volume 1, 2004, pp. 226-229

Information Leakage in a Quantum Computing Prototype System: Stochastic Noise in a Microcavity

L.A.A. Nikolopoulos[1] and A. Maras

Department of Communication Sciences and Technology,
Faculty of Sciences and Technology,
University of Peloponnese,
GR-221 00 Tripolis, Greece

Received 1 July, 2004; accepted in revised form 15 July, 2004

Abstract: One serious problem in the physical implementation of quantum computing is that of maintaining quantum noise. Noise (decoherence) of a quantum system, due to its interaction with the surrounding environment, is the main source of spoiling in the execution of a quantum algorithm. In this work we investigate quantum noise through methods from classical and quantum information theory. Decoherence is seen as the noise effect of a channel operation on a single quantum bit (qubit) and its properties are investigated through the quantum entropy evolution and fidelity transmission. One of the fundamental physical systems that appears promising for quantum computation is that of an atom and electromagnetic (E/M) field in a cavity [1].

Keywords: Quantum information, quantum computation, quantum communication, entanglement, quantum entropy, decoherence, density matrix

PACS: 03.67.Pp, 03.67.Kx, 5.40.-a

1 Introduction

Quantum computation is a rapidly growing field of modern science [2] which has attracted a large number of scientists from areas other than physics, such as mathematics, computing, information and communication theory. Quantum computer is nothing else than a physical system whose evolution can be manipulated in such way so that to perform specific computational tasks. Current technology is faced with the implementation of practical quantum computation algorithms, which when compared with the classical counterpart algorithms (i.e. prime factorization, discrete Fourier transform), achieve an amazing computational performance [3, 4]. One of the main obstacles of quantum computational systems and algorithms is the inevitable imprecision due to interaction of the fundamental quantum system with its environment. Decoherence appears in any quantum computation resulting in a loss of information when a quantum signal is sent in time and/or space through noisy quantum channels. Theoretical and experimental methods have to be developed for the control of such problems which either have pure quantum mechanical origin or they have their analogue in classical information field.

Noise (decoherence) in a quantum physical system

An important topic of research in an open quantum system is the dynamics, induced by the surrounding environment. Noise in such quantum systems, as is the system that we are going to consider, can be

---

[1]Corresponding author. E-mail: nlambros@uop.gr

described by a master equation for the density operator of the system. Assuming that the total hamiltonian (system + environment) is given by the the the hamiltonian $H = H_S + H_E + H_{SE}$, the time evolution of the total density operator is given as the solution of the Liouville equation, where the initial density operator is considered to be in a factorized form as $\rho(0) = \rho_s(0) \otimes \rho_e(0)$. Here H_{SE} denotes the system-environment interaction and ρ_s, ρ_e the density operators of the system and the environment, respectively.

Tracing out the environmental degrees of freedom together with a Markov and a Born approximation [5] results in the following reduced density matrix equation for the system alone:

$$\frac{\partial}{\partial t}\rho_s(t) = -i[H_S, \rho_s] + \sum_{i=a,c} \left[L_i \rho_s L_i^\dagger - \frac{1}{2}\{L_i^\dagger L_i, \rho_s\} \right], \quad i = a, c. \tag{1}$$

The first term of the right-hand-side describes the coherent evolution of the system (noiseless channel) while the second and third terms ($i = a, c$) describe the noise inserted into the system. The Lindblad operators $L_i = \sqrt{k_i}\sigma_i, i = a, c$ [6] describe the decoherence of the system due to its interaction with the environment. Decoherence time, in the first approximation, is given by the quantity $1/k_i, i = a, c$. Operators σ_a, σ_c represent the destruction operators in the Hilbert space of the system.

Noise in quantum computation

Assume that the initial state $|\psi_s(0)\rangle$ of the system is the input (qubit) of some arbitrary quantum computing operation. The system's evolution, in absence of noise, denote the desired quantum operation (i.e. AND, NOT, XOR, CNOT). The final state of the system $|\psi_s(t_f)\rangle$ represents the value of the qubit after the operation. Noise (decoherence) is now inserted due to the fact that no quantum system is completely isolated from the environment. This affects the final state of the system (final value of the qubit) in a unpredictable way. Let's now consider the case that we are going to examine.

- Evolution of a pure initial state to a final (mixed or not) state.

Assume that the initial state of the system $|\psi_s(0)\rangle - |\psi_a(0)\rangle \otimes |\psi_c(0)\rangle$ is pure and the atom and the cavity field are uncorrelated. As the system evolves according (Eq. 1), in the presence of the environment, it becomes entangled (quantum correlated) in such way that the state of the system is not possible to be written in a factorized form (uncorrelated case). One interesting question is under what conditions in terms of the input (initial) states of the field and the atom the degree of the entanglement can be maximum.

Noise examined with quantum information concepts

From the quantum information point of view, noise effects may be investigated through the concepts of quantum entropy and the density matrix of the state. More over, fidelity of entanglement as a measure of the pure quantum correlation (non-classical) between the two parts (qubits) of the system is considered. The (Von Neumann) quantum entropy S of the bipartite system is defined as:

$$S(\rho_s) = -Tr(\rho_s \log_2 \rho_s) = -\sum_i r_i log_2 r_i, \tag{2}$$

with r_i being the eigenvalues of the density matrix of the system. The classical counterpart is the Shannon entropy [7] ($S_c = -\sum p_i \log_2 p_i$). It has been shown that in general $S \leq S_c$.

Evolution of the state of the system $\rho_s(t)$ may be modeled as a quantum channel. The initial state of the system $\rho_s(0)$, is the state 'sent' by the source and the final state $\rho_s(t_f)$ is the information 'received' by the receiver. Environmental noise (quantum decoherence) again may affect the final state of the system (receiver readout) in an unpredictable way, thus leading to loss of information.

Here we are going to examine the evolution of the linear entropy (S_l) and the fidelity of entanglement (F_s), defined by:

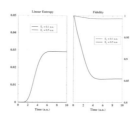

Figure 1: 1 Fidelity transmission and linear entropy for two different constant values of the external E/M field. We have used $\kappa = 2$ and $\gamma = 0.2$ in a.u.

$$S_l = 1 - Tr_s[\rho_s^2(t)] \qquad F_s = \langle \psi_s(t)|\rho_s(0)|\psi_s(t)\rangle = Tr_s[\rho_s(t)\rho_s(0)]. \tag{3}$$

2 Atom and photon in a E/M cavity: quantum computation in a noisy environment

The system is considered to be bipartite, constituted by a typical two-level atomic system (TLS), with hamiltonian $h_a = \omega_0 \sigma_z$, inside an E/M cavity modeled by the hamiltonian $h_c = \omega_c a^\dagger a$. Their interaction is given by the operator $h_{ac} = ig(\sigma_+ a + a^\dagger \sigma_-)$. In addition the system is driven by an external E/M field of frequency ω_L with hamiltonian $h_f = \mathcal{E}_L(t)(a^\dagger e^{i\omega_L t} + a e^{-i\omega_L t})$. The atom-cavity field hamiltonian is given by $H = h_a + h_c + h_{ac} + h_l$. In the interaction picture and by making the rotating wave approximation, we end up with the following transformed form for the system hamiltonian:

$$H_S \to H_s = \Delta_0 \sigma_+ \sigma_- + \Delta_c a^\dagger a + ig(a^\dagger \sigma_- + \sigma_+ a) + \mathcal{E}_L(t)(a^\dagger + a), \tag{4}$$

with $\Delta_a = \omega_L - \omega_0$ and $\Delta_c = \omega_L - \omega_c$ the detunings.

Environment, for this bipartite system, is the continuum of E/M field modes surrounding the atom and the cavity walls. Noise is inserted through the purely quantum mechanic phenomenon of spontaneous emission of the excited state of any atomic system. The excited state of the TLS decays into the ground state with a rate γ. Moreover, coupling of the intracavity E/M field with the walls (non-ideal mirrors) of the cavity cause photons to leak out the cavity mirrors with a rate κ. Noise for the atomic system observables modeled with the Pauli destruction spin matrix while noise for the cavity E/M mode, is modeled by the destruction cavity mode operator, namely,

$$L_a = \sqrt{k_a} \; \mathbf{1}(N_c) \otimes \sigma_a(2) = \gamma \; \sigma_-, \qquad L_c = \sqrt{k_c} \; \sigma_c(N_c) \otimes \mathbf{1}(2) = \sqrt{2\kappa} \; \alpha_c.$$

The number N_c is the number of photons for the cavity mode, here kept equal to five $N_c = 5$. The decoherence characteristic times are of the order $\tau_i = 1/k_i$, $i = a, c$. Let's now assume the initial state of the intracavity field to be the vacuum state $|0_c\rangle$, and the initial state of the atomic system to be in the ground state $|0_a\rangle$. In other words, if we consider that the initial qubit is of the form, $\psi_a(0) = \alpha|0_a\rangle + \beta|1_a\rangle$, then $\alpha = 1, \beta = 0$.

The initial state of the bipartite system (atom+intracavity field) is given by $\psi_s(0) = |\psi_a(0)\rangle \otimes |0_c\rangle = |0_a 0_c\rangle$. The initial entropy of the system is $S(0) = 0$. We have complete information for both of the subsystems independently. The two systems are initially uncorrelated.

The system after sufficient time has been evolved in its steady state ψ_{ss} under the influence of the external E/M field of strength $\mathcal{E}_L$ and the noise induced by the environment, characterized by the parameters κ, γ. Evolution is governed by the differential equation for the density matrix elements (Eq. 1). At that time, the two systems are correlated in a non-classical way and their joint (not-separable) state is called the entangled state. Of crucial importance is the entropy that this entangled state has obtained since it represents the amount of information that we can extract from that state. In addition the fidelity of the transmission gives us a quantitative measure of how well the information has been sent from the 'source' to the 'receiver' or equivalently from the quantum computation point of view, the probability that no error has occurred during the execution of the quantum logical algorithm (gate).

In the present case we assume $\kappa = 2$, $\gamma = 0.2$ (characterize the coupling of the noisy environment with the channel). The coupling between the atom and the cavity E/M field has been set to $g = 1$ (characterizes the transmission channel). We also assume that the detunings are zero ($\Delta_c = \Delta_a = 0$). Thus, we have resonant conditions.

For the case of weak external field $\mathcal{E}_L = 0.1$ a.u. we see that the linear entropy S_l remains very close to zero while the fidelity remains very close to one. This suggests that a small amount of information has been lost to the environment while at the same time the system has been entangled. We have almost complete information both for the joint system and for the subsystems independently. The two systems are almost uncorrelated. When we increase the external field to the value of $\mathcal{E}_L = 0.5$ a.u. we see that entropy is increased in its steady value but still stays very small ($S_l \rightarrow 0.029$) while fidelity decreases more rapidly with the field strength ($F_s \rightarrow 0.854$). This is consequence of the fact that while the information loss for the joint system is small the available information for the the two subsystems separately has been decreased. The degree of the entanglement (degree of quantum correlation) between the two subsystems has been increased. Results for different initial states and other envriromental parameters have also been obtained.

References

[1] B. B. Blinov, D. L. Moehring, L.-M. Duan, and C. Monroe, Observation of entanglement between a single trapped atom and a single photon. *Nature* **428** (2004), 153–157.

[2] C. Monroe, Quantum information processing with atoms and photons. *Nature*,**416** (2002), 238–246.

[3] P. Shor. Scheme for reducing decoherence in quantum computer memory. *Phys. Rev. A*, 52:R2493, 1995.

[4] Lov K. Grover. Quantum mechanics helps in searching for a needle in a haystack. *Phys. Rev. Lett.*, 78:4709–4712, 1997.

[5] N.G. Van Kampen. *Stochastic Processes in Physics and Chemistry*. Elsevier Science B.V. Amsterdam, 1992.

[6] G. Lindblad. On the generators of quantum dynamical semigroups. *Commun. Math. Phys.*, 48:119–130, 1976.

[7] C.E. Shannon. A mathematical theory of communication. *Bell, Syst. Tech. J.*, 27:379–423,623–656, 1948.

VSP International
Science Publishers
P.O. Box 346, 3700 AH Zeist
The Netherlands

*Lecture Series on Computer
and Computational Sciences*
Volume 1, 2004, pp. 230-233

Adaptive Stiff Solvers at Low Accuracy and Complexity

Alessandra Jannelli and Riccardo Fazio[1]

Department of Mathematics, University of Messina
Salita Sperone 31, 98166 Messina, Italy

Received 12 July, 2004; accepted in revised form 18 August, 2004

Abstract: We consider here adaptive stiff solvers at low accuracy and complexity for systems of ordinary differential equations. For the adaptive algorithm we propose to use a novel monitor function given by the comparison between a measure of the local variability of the solution times the used step size and the order of magnitude of the solution. The considered stiff solvers are: a special second order Rosenbrock method with low complexity, and the classical BDF method of the same order. We use a reduced model for the production of ozone in the lower troposphere as a test problem for the proposed adaptive strategy.

Keywords: Stiff ordinary differential equations, linearly-implicit and implicit numerical methods, adaptive step size.

Mathematics Subject Classification: 65L05, 65L07.

1 Introduction.

The main concern of this work is to study the most promising adaptive solvers at low accuracy and complexity that can be used for the numerical integration of stiff systems of ordinary differential equations (ODEs) written here, without loss of generality, in autonomous form:

$$\frac{d\mathbf{c}}{dt} = \mathbf{R}(\mathbf{c}) \tag{1}$$

where $\mathbf{c} \in R^n$ and $\mathbf{R}(\mathbf{c}) : R^n \to R^n$.

Adaptive solvers can be used to automatically adjust the step size according to user specified criteria. Accepted strategies for variable step-size selection are based mainly on monitoring of the local truncation error: Milne's device in the implementation of predictor-corrector methods; embedded Runge-Kutta methods as developed by Sarafyan [4], Fehlberg [3], Verner [6] and Dormand and Price [2]; Richardson local extrapolation proposed by Bulirsch and Stoer [1].

The simple and inexpensive approach to the adaptive step size selection, proposed here, is to require that the change in the solution is monitored in order to define a suitable local step size. This results into a simpler algorithm than the classical ones.

2 A simple adaptive step-size strategy.

In this section, we present a simple adaptive procedure for determining the local integration step-size according with user-specified criteria. First of all, the user has to choose bounds on the

---

[1]Corresponding author. E-mail: rfazio@dipmat.unime.it or rfazio@na-net.ornl.gov

allowable tolerance, say $0 < \eta_{min} < \eta_{max}$, for a monitor function η^n to be defined below. Moreover, a range for the permissible step size is also required to the user ($\Delta t_{min} \leq \Delta t^n \leq \Delta t_{max}$). Large enough tolerance intervals for Δt^n and η^n should be used, otherwise the adaptive procedure might get caught in a loop, trying repeatedly to modify the step size at the same point in order to meet the bounds that are too restrictive for the given problem. However, in general, the step size should not be too small because the number of steps will be large, leading to increased round-off error and computational inefficiency. On the other hand, the interval size should not be too large also because truncation error will be large in this case.

We consider first the simple scalar case. Given a step size Δt^n and an initial value c^n at time t^n, the method computes an approximation c^{n+1} at time $t^{n+1} = t^n + \Delta t^n$, so that we can define the monitoring function

$$\eta^n = \frac{|c^{n+1} - c^n|}{|c^n| + \epsilon_M},$$

where $\epsilon_M > 0$ is of the order of the machine precision. Now, we can require that the step size is modified as needed in order to keep η^n between the tolerance bounds. The basic guidelines for setting the step size are given by the following algorithm:

1. Given a step size Δt^n and an initial value c^n at time t^n, the method computes a value c^{n+1} and, consequently, the monitoring function η^n.

2. If $\eta_{min} \leq \eta^n \leq \eta_{max}$, then t^n is replaced by $t^n + \Delta t^n$; the step size Δt^n is not changed and the next step is taken by repeating Step 1 with initial value c^n replaced by c^{n+1}.

3. If $\eta^n < \eta_{min}$, then t^n is replaced by $t^n + \Delta t^n$ and Δt^n is replaced by $10\Delta t^n$; the next integration step, subject to the check at Step 5, is taken by repeating Step 1 with initial value c^n replaced by c^{n+1}.

4. If $\eta^n > \eta_{max}$, then t^n remains unchanged; Δt^n is replaced by $\Delta t^n/2$ and the next integration step, subject to the check at Step 5, is taken by repeating Step 1 with the same initial value c^n.

5. If $\Delta t_{min} \leq \Delta t^n \leq \Delta t_{max}$, return to Step 1; otherwise Δt^n is replaced by Δt_{max} if $\Delta t^n > \Delta t_{max}$ and by Δt_{min} if $\Delta t^n < \Delta t_{min}$, then proceed with Step 1.

Recall the definition of the monitoring function η^n:

$$\eta^n = \frac{|c^{n+1} - c^n|}{|c^n| + \epsilon_M} = \frac{|c^{n+1} - c^n|}{\Delta t^n} \frac{\Delta t^n}{|c^n| + \epsilon_M} \simeq \left|\frac{dc}{dt}(t^n)\right| \frac{\Delta t^n}{|c^n| + \epsilon_M},$$

and note that this can be considered as a measure of the suitability of the used step size. In fact, we can consider $\left|\frac{dc}{dt}(t^n)\right|$ a measure of the increase or decrease of the solution, Δt^n as the grid resolution, and $|c^n| + \epsilon_M$ the order of magnitude of the solution. So that, in the above formula the product of the derivative times grid resolution is compared with the order of magnitude of solution. When the numerical solution increases or decreases to much, that is the monitor function exceed the upper bound, our algorithm choices to reduce the time step. On the other hand, if the solution slowly varies with respect to the grid resolution over the order of magnitude of the solution, then if the value of η^n is within the chosen range, the step size is unchanged; otherwise the step size is magnified by one order of magnitude.

It is evident that this adaptive approach can be used with a single method. In the vectorial case, a norm has to be considered instead of the absolute values, for instance the two norm $||\cdot||_2$ or the infinity norm $||\cdot||_\infty$.

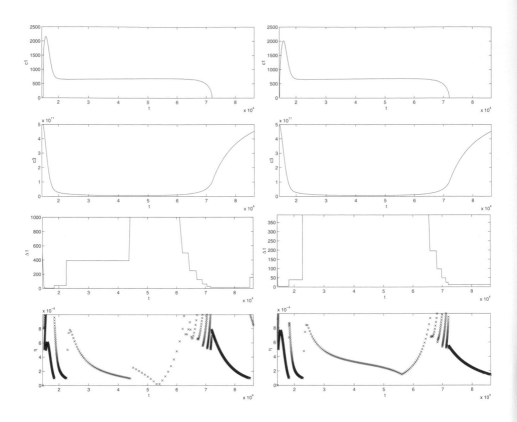

Figure 1: Results obtained by: ROS2 on the left and BDF2 on the right. Top: c_1; middle-up: c_3; middle-down: step-size selection; and bottom: relative error.

3 A test case: ozone production in the lower troposphere.

Ozone is an unsafe gas for human beings and animals even during short term exposures and can damage crops when over long periods its levels are too high. Ozone is formed in many different reactions.

A reduced model used for the production of ozone in the lower troposphere is given below, see [6],

$$
\begin{aligned}
c_1' &= \mu_1 c_3 - \mu_2 c_1 \\
c_2' &= \mu_1 c_3 - \mu_3 c_2 c_4 + s_2 \\
c_3' &= \mu_3 c_2 c_4 - \mu_1 c_3 \\
c_4' &= \mu_2 c_1 - \mu_3 c_2 c_4
\end{aligned}
\tag{2}
$$

with the initial conditions

$$\mathbf{c}(0) = [0, 1.3 \; 10^8, 5 \; 10^{11}, 8 \; 10^{11}]^T \ .$$

Here $'$ is the derivative with respect the independent variable the concentrations c_i for $i = 1, 2, 3, 4$ are given in molecules for cm^3 and time in seconds. The involved parameters are given by

$$\mu_1 = \begin{cases} 10^{-40} & \text{during the night} \\ 10^{-5} \; e^{7 \; \text{sec}(t)} & \text{during the day} \end{cases} \ , \quad \mu_2 = 10^5 \ , \quad \mu_3 = 10^{-16} \ , \quad s_2 = 10^6$$

$$\text{where} \quad \text{sec}(t) = \left(\sin \left(\frac{\pi}{16}(t_h - 4) \right) \right)^{0.2} \quad \text{and} \quad t_h = th - 24 \lfloor th/24 \rfloor \ , \quad th = t/3600 \ ,$$

here $\lfloor \cdot \rfloor$ stands for the floor function.

The figure 1 shows two components of the numerical solution, step-size selection and monitor function obtained with a low complexity second order Rosenbrock (ROS2) method and the BDF2 (Backward Difference Formulas) method of the same order. We set $\eta_{max} = 10^{-3}$ and $\eta_{min} = \eta_{max}/10$, $\Delta t_{min} = 1$ and $\Delta t_{max} = 1000$. The ROS2 method calculates the solution in 0.7 second after 1983 steps using a maximum time step equal to 1000. Instead, the BDF2 obtains the numerical solution in 18 seconds with 2429 steps and a maximum time step equal to 390.

Acknowledgment

Work supported by a grant from the Messina University and partially by the Italian "MIUR".

References

[1] R. Bulirsch and J. Stoer, Numerical treatment of ordinary differential equations by extrapolation methods, *Num. Math.* **8** 1–13(1966).

[2] J.R. Dormand and P.J. Price, A family of embedded Runge-Kutta formulae, *J. Comp. Appl. Math.* **6** 19–26(1980).

[3] E. Fehlberg, Classical fifth-, sixth-, seventh- and eighth order formulas with step size control, *Computing* **4** 93–106(1969).

[4] D. Sarafyan, *Error estimation for Runge-Kutta methods through pseudoiterative formulas.* Techn. Rep. No 14, Lousiana State Univ., New Orleans, 1966.

[5] J. M. VERNER. Explicit Runge-Kutta methods with estimates of the local truncation error. *SIAM J. Num. Anal.*, **15** (1978) 772–790.

[6] J. G. Verwer, W. H. Hundsdorfer and J. G. Blom, Numerical time integration for air pollution models, *Sur. Math. Ind.* **2** 107–174(2002).

VSP International
Science Publishers
P.O. Box 346, 3700 AH Zeist
The Netherlands

Lecture Series on Computer
and Computational Sciences
Volume 1, 2004, pp. 234-238

Refining Existing Numerical Integration Methods

P. Johnson, K. Busawon and S. Danaher

School of Engineering and Technology,
Northumbria University,
Newcastle-Upon-Tyne, NE1 8ST, U.K.

Received 12 July, 2004; accepted in revised form 13 August, 2004

Abstract: In this paper we present a refinement to the single-step numerical methods of solving ordinary differential equations (ODEs). We propose an adaptive algorithm for generating the step length between successive numerical approximations to the solution of an ODE. The algorithm is designed such that more accurate approximations to the solution are made when the solution gradient is high, whilst accuracy is dropped in favour of economising computation effort when the gradient is low. It is demonstrated how this adaptive algorithm can be incorporated into the forward Euler method as well as all other single-step methods. An example is studied and simulations carried out comparing the single-step Euler, Heun and Runge-Kutta methods with their adaptive method counterparts. Analysis of the simulations is carried for each case. Through the simulations and analysis it is demonstrated that the proposed algorithm does indeed refine the existing single-step methods.

Keywords: numerical integration, adaptive step, differential equation

Mathematics Subject Classification: 74H15, 65D30

1   Introduction

Consider the ODE

$$\frac{dx(t)}{dt} = f(x(t), t); \quad x(t_0) = x_0 \tag{1}$$

where $x \in \mathcal{R}$. We assume that the relevant hypotheses for the existence and the unicity of the solution of (1) are satisfied. In addition, there exists a constant $M > 0$ such that $|f(x(t), t)| \leq M$ for all $x \in \mathcal{R}$ and $t \geq 0$. Throughout the paper, the set of real numbers is denoted by $\mathcal{R}$ and the set of natural numbers by $\mathcal{N}$.

We are concerned here with finding the numerical solution of (1) over a finite time interval $[a, b]$. The simplest of all of the single-step numerical integration methods is that of the forward Euler. For this reason we will examine the problem associated with single-step numerical integration techniques using the forward Euler method as an example. First of all, recall that the forward Euler method consists of generating two sequences:

$$t_{n+1} = t_n + T; \quad n = 0, 1, 2, ..., N \tag{2}$$

$$x_{n+1} = x_n + Tf(x_n, t_n) \tag{3}$$

where $N \in \mathcal{N}$, $t_0 = a$, $t_N = b$, $T = \dfrac{b-a}{N} > 0$ is the constant integration step and x_n is an approximation of $x(t_n)$. Note that equation (2) defines a monotone increasing arithmetic sequence $\{t_n\}_{n=0}^N$, which we call the *integration sequence*, and hence gives rise to uniformly spaced integration instances. It is clear that with a small T one may make good approximations to a solution of (1) though at the expense of computational effort, whereas a large T will economise effort at the expense of accuracy. Hence, one is forced to make trade-off between accuracy and computation if a solution demands accuracy and economy in different sub-intervals of $[a, b]$. We show, by the use of a variable integration step, that the error between the numerical and the analytical solution can be significantly reduced without the extra computational effort that the single-step integration methods would require.

2   An adaptive numerical algorithm

We now present a refinement of the forward Euler method. Again, for reasons of ease of illustration, Euler's method is examined as it is the simplest of all single-step methods. Consider the ODE described by (1). As before, we are concerned with finding the numerical solution of (1) over a finite interval $[a, b]$. We propose the following numerical integration algorithm defined by the sequences:

$$t_{n+1} = t_n + \frac{T_0}{\gamma f^2(x_n, t_n) + 1}; \quad n = 0, 1, 2, ..., P \tag{4}$$

$$x_{n+1} = x_n + (t_{n+1} - t_n) \, f(x_n, t_n) \tag{5}$$

where $P \in \mathcal{N}$, $t_0 = a$, $t_P \in [b, b + T_0[$ and both $T_0 > 0$ and $\gamma \geq 0$ are real constants. First of all, note that, by comparing (2)-(3) with (4)-(5), if $T_0 = T$ and $\gamma = 0$ or $f(x_n, t_n) = 0$, then the Euler algorithm and the above proposed algorithm are identical. Otherwise, the integration step

$$\beta_n = \frac{T_0}{\gamma f^2(x_n, t_n) + 1}$$

is inversely proportional to the approximation of the square of the gradient $\dot{x} = f(x(t), t)$, of the solution $x(t)$. Consequently, the integration steps will be relatively small during regions where $x(t)$ contains rapid changes. As a result, one can ensure that more accurate approximations of $x(t)$ will be obtained in intervals of $x(t)$ that exhibit rapid change. On the other hand, if approximations to $x(t)$ are being made in an interval of $x(t)$ which exhibits little or slow change (i.e. $f(x(t), t) \approx 0$) then β_n will increase. Consequently, the integration steps will grow and relatively few computations will be made and stored in such a region. It is clear then, that this method is adaptive in nature, rather than single-step, due to it's variable integration step size. The constant γ is introduced to add a weight factor to the gradient. In other words, it allows β_n to have a tunable gradient sensitivity. Furthermore, since $|f(x(t), t)| \leq M$, we can see that $\frac{T_0}{\gamma M^2 + 1} \leq \beta_n \leq T_0$. Also, since $\beta_n > 0$, the sequence $\{t_n\}_{n=0}^P$ is monotone increasing and hence divergent. This means that $\{t_n\}_{n=0}^P$ is indeed a valid integration sequence. It is important to realise that the divergence property of $\{t_n\}_{n=0}^P$ is necessary so as to ensure that t_n covers the whole interval $[a, b]$.

3   Comparison with Euler method

The adaptive numerical integration method that is the extension to the Euler method outlined above, is compared with the forward Euler method for a particular function $f(x(t), t)$ in (1) over a fixed interval $[t_0, t_P] = [t_0, t_N]$. In the example we studied two cases (more cases are studied in the full paper): Case 1 $T = T_0 = 0.10$ and $\gamma = 1$ - Case 2 $T = T_0 = 0.05$ and $\gamma = 1$. To compare the two techniques objectively we adopt two different approaches. Firstly we use the square error, computed by $(\Delta x)_n^2 = (x(t_n) - x_n)^2$ over the interval $[t_0, t_P]$, to measure the relative mean square error $\Lambda_{[t_0,t_P]} = \dfrac{r_{adaptive}}{r_{euler}}$ where r denotes the mean square error for a particular method. In the second approach we define a variable $S_{[t_0,t_P]}$, referred to as the S factor, which represents the computation effort with respect to the error over the time interval $[t_0, t_P]$, and is defined as follows:

$$S_{[t_0,t_P]} = \frac{R_2}{R_1} \times \left(\frac{r_1}{r_2}\right)^2$$

where

$$R_q = \frac{1}{(P_q + 1)^2} \sum_{n=0}^{P_q} (\Delta x)_n^2 \quad \text{for } q = 1, 2$$

is the mean square error per point for the adaptive numerical solution in case q and $P_q + 1$ is the number of points in the solution for case q. Also

$$r_q = \frac{1}{(P_q + 1)} \sum_{n=0}^{P_q} (\Delta x)_n^2$$

is the mean square error for the adaptive numerical solution in case q. It can be shown that if $S_{[t_0,t_P]} > 1$, then we may say that the adaptive method has improved upon the method of Euler. The improvement being, that the number of points required to reduce the error between numerical and real solution by a given factor using the adaptive method, is less than that which would be required by the Euler method. The following example was used in the comparison:

$$\frac{dx(t)}{dt} = -tx; \quad t_0 = 0; \quad x(0) = 1$$

which has the solution $x(t) = e^{-\frac{t^2}{2}}$ and we set $t_P = t_N = 4$. The results of the Euler comparison are summarised in the table and figure below.

Example	T_0	γ	$\Lambda_{[0,4]}$	$S_{[0,4]}$
Case 1	0.1	1	0.9502	-
Case 2	0.05	1	0.9574	2.1168

The above data illustrates several points. In both cases the mean square error for the adaptive method never exceeds that of the Euler method i.e. $\Lambda_{[t_0,4]} < 1$. Also, one can see that for case 2 $S_{[t_0,4]} > 1$ which demonstrates the low cost at which this additional accuracy comes.

4   Further extensions

It has been shown in the above how one can refine/extend the Euler method for numerically solving (1). The extension is to replace the static integration step size in the Euler algorithm with an adaptive step size. It stands to reason that the same extension may be made to, not just Euler's method, but to any single-step method. Every single-step method employs the same algorithm for generating it's integration sequence. The algorithm is of the form (2) and can be written generally as $t_{n+1} = t_n + h$; $n = 0, 1, 2, ..., Q$ where h is a real positive constant. As with the extension of Euler's method, we propose that one can replace this algorithm with that of (4) i.e. $t_{n+1} = t_n + \beta_n$; $n = 0, 1, 2, ..., P$. In making this substitution we can justifiably expect that in order to reduce the

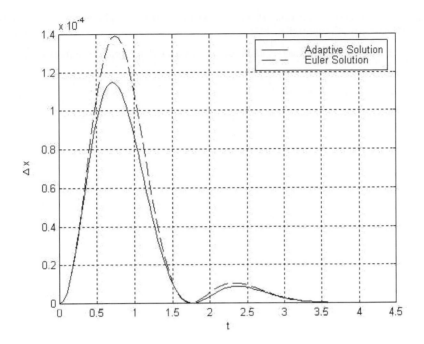

Figure 1: Square errors between the adaptive/Euler solutions for case 2.

error of a particular technique by a given amount, less points will have to be computed/stored than we would otherwise have done with the single-step method equivalent. As a demonstration of this assertion a 4th order Runge-Kutta solution to the problem examined in the above example was generated using the parameters of case 1. The simulation yielded the result $\Lambda_{[0,4]} = 0.5494$.

5 Conclusions

In this paper, we have proposed and implemented an adaptive numerical technique for solving ODEs of the form (1). It has been demonstrated how this method can be considered as a refinement of the all the single-step methods of numerical integration. It has also been shown how the adaptive nature of the proposed algorithm addresses a problem associated with single-step methods. A comparison between the proposed adaptive algorithm and the single-step methods of Euler and Runge-Kutta has been presented via a set of simulations.

Acknowledgement: The authors wish to thank the EPSRC for funding the research.

References

[1] Lambert, J.D. *Numerical Methods for Ordinary Differential Systems: The Initial Value Problem*. John Wiley & Sons, 1991.

[2] Gear, C.W. *Numerical Initial Value Problems in Ordinary Differential Equations.* Prentice-Hall Series in Automatic Computation, 1971.

[3] Jeffreys, H. Swirles, B. *Methods of Mathematical Physics 3rd Edition.* Cambridge University Press, 1956.

VSP International
Science Publishers
P.O. Box 346, 3700 AH Zeist
The Netherlands

*Lecture Series on Computer
and Computational Sciences*
Volume 1, 2004, pp. 239-243

Visualization of Signal Transduction Pathways[§]

Byoung-Hyun Ju and Kyungsook Han[1]

School of Computer Science and Engineering, Inha University, Inchon 402-751, Korea

Received 31 July, 2004; accepted in revised form, 18 August 2004

Abstract: The automatic generation of signal transduction pathways is challenging because it often yields complicated, non-planar diagrams with a large number of intersections. Most signal transduction pathways available in public databases are static images and thus cannot be refined or changed to reflect updated data. We have developed an algorithm for visualizing signal transduction pathways dynamically as three-dimensional layered digraphs. Experimental results show that the algorithm generates clear and aesthetically pleasing representations of large-scale signal-transduction pathways.

Keywords: signal transduction pathways, visualization algorithm

1 Introduction

Advances in biological technology have produced a rapidly expanding volume of molecular interaction data. Consequently, the visualization of biological networks is becoming an important challenge for analyzing interaction data. There are several types of biological networks, such as signal transduction pathways, protein interaction networks, metabolic pathways, and gene regulatory networks. Different types of network represent different biological relationships, and are visualized in different formats in order to convey their biological meaning clearly. The primary focus of this paper is the representation of signal transduction pathways.

A *signal transduction pathway* is a set of chemical reactions in a cell that occurs when a molecule, such as a hormone, attaches to a receptor on the cell membrane. The pathway is a process by which molecules inside the cell can be altered by molecules on the outside [1]. A large amount of data on signal transduction pathways is available in databases, including diagrams of signal transduction pathways [2, 3, 4]. However, most of these are static images that cannot be refined or changed to reflect updated data. It is increasingly important to visualize signal transduction pathway data from databases dynamically.

From the standpoint of creating diagrams, signal transduction pathways are different from protein-protein interaction networks [9] because (1) protein-protein interaction networks are non-directional graphs, whereas signal transduction pathways are directional graphs (digraphs for short) in which a node represents a molecule and an edge between two nodes represents a biological relation between them, (2) protein-protein interaction networks have proteins and edges of uniform types, whereas signal transduction pathways contain nodes and edges of various types, and (3) signal transduction pathways impose more restrictions on edge flows and node positions than protein-protein interaction networks.

Fig. 1A shows the mammalian mitogen-activated protein kinase (MAPK) signaling pathway (http://kinase.uhnres.utoronto.ca/pages/maps.html), represented by a force-directed layout algorithm. This is the most general graph layout algorithm, often used for visualizing protein interaction networks. The drawing shown in Fig. 1A follows several drawing rules faithfully, including no edge crossing and no overlapping among edges and vertices. However, Fig. 1A does not convey its meaning as clearly as Fig. 1B, which represents the same pathway but visualized as a layered graph.

Our experience is that signal transduction pathways convey their meaning best when they are represented as layered digraphs with uniform edge flows, typically either downward or to the right. The Sugiyama algorithm [7, 8] or its variants is most widely used for drawing layered graphs, but the algorithm produces a diagram with various types of edge flow even for an acyclic graph. In the present

[§] This study was supported by a grant of the Ministry of Health & Welfare of Korea under grant 03-PJ1-PG3-20700-0040.
[1] To whom correspondence should be addressed. Email: khan@inha.ac.kr

paper we introduce a new algorithm for automatically representing signal transduction pathways as layered digraphs with uniform edge flows and no edge crossing.

(A)

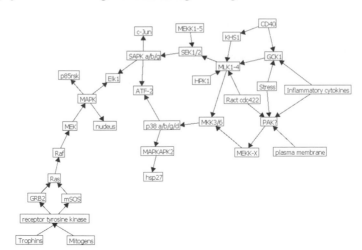

(B)

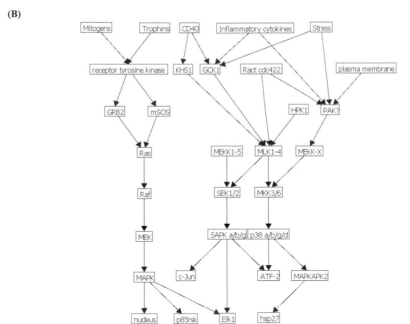

Fig. 1. (A) The mammalian mitogen-activated protein kinase (MAPK) pathway, represented by a force-directed layout algorithm. (B) The same pathway represented as a layered graph by our algorithm.

2. Layout Algorithm

To discuss the layout algorithm, we need to define a few terms. Suppose that $G=(V, E)$ is an acyclic digraph. A layering of G is a partition of V into subsets $L_1, L_2, \ldots, L_h$, such that if $(u, v) \in E$, where $u \in$

L_i and $v \in L_j$, then $i < j$. The *height* of a layered digraph is the number h of layers, and the *width* is the number of nodes in the largest layer. The *span* of an edge (u, v) with $u \in L_i$ and $v \in L_j$ is $j - i$.

We visualize signal transduction pathways as layered digraphs. The visualization algorithm is composed of 4 steps from the top level: (1) layer assignment, (2) dummy node creation, (3) crossing reduction, and (4) z-coordinate adjustment.

Step 1: Layer Assignment
This step assigns a y-coordinate to every node by assigning it to a layer. Nodes in the same layer have the same y-coordinate value. It first places all the nodes with no parent in layer L_1, and then each remaining node n in layer L_p+1, where L_p is the layer of n's parent node. When the layer of a node is already determined, the node is assigned the larger layer value. The initial layering is updated later since the digraph may be too wide and the edge may have a span greater than one. The number of edges whose span > 1 should be minimized because they slow down subsequent steps (steps 2-4) of the algorithm [5]. We place the source node of an edge whose span > 1 in higher layers so that the span of the edge becomes one.

Step 2: Dummy Node Creation
When every edge is represented by a straight line segment it may cross other edges and/or nodes. In order to avoid this we add a dummy node along an edge and bend the edge at the dummy node. A dummy node is inserted into a layer through which an edge with span > 1 passes (see Algorithm 4). Dummy nodes are treated as general nodes when computing their location but are not shown in the final diagrams.

Step 3: Reducing Crossing
The problem of minimizing edge crossings in a layered digraph is NP-complete, even if there are only two layers [6]. We use the barycenter method to order nodes in each layer [7, 8]. In the barycenter method, the x-coordinate of each node is chosen as the barycenter (average) of the x-coordinates of its neighbors.

Step 4: Z-Coordinate Adjustment
We visualize simple signal transduction pathways as two-dimensional (2D) diagrams. However, we visualize complicated signal transduction pathways as three-dimensional (3D) diagrams for two reasons. First, the lengths of the edges can be more easily made uniform in 3D diagrams than in 2D diagrams. Second, most complicated signal transduction pathways contain a large number of crossings that cannot be removed in a two-dimensional diagram. When there is no edge crossing in the graph after step 3, the graph is sufficiently clear as a 2D diagram (that is, all nodes have the same z-coordinate value). For graphs with edge crossing even after step 3 (for example, graphs A1 and B1 in Fig. 2), we divide the nodes involved in the edge crossing into groups and adjust the z-coordinate values of the groups so that the edge crossing is removed (graphs A2 and B2 in Fig. 2). When varying the z-coordinate values, the distance between layers is preserved and the maximum angle between the groups is 90°.

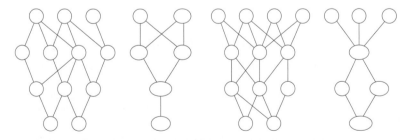

Fig. 2. (A1) front view of a graph. (A2) side view of graph A1 with 2 groups of nodes in layers 1 and 2. (B1) front view of a graph. (B2) side view of graph B1 with 3 groups of nodes in layer 1 and 2 groups of nodes in layer 3.

Step 4 first computes the number of groups by calculating the maximum number of edge crossings from target (child) layer to source (parent) layer (Algorithm 1) and then partitions the nodes into groups if there is an edge crossing (Algorithm 2). The total number of groups required in a given layer is one more than the maximum number of crossings in that layer.

Algorithm 1 getGroupCnt ()

groupCnt=0;
for each group g in targetLayer
    for each node n in g
        for each upward edge e of n {
            if(leftCrossing(e) > groupCnt) groupCnt = leftCrossing(e);
            if(rightCrossing(e) > groupCnt) groupCnt = rightCrossing(e);
        }
groupCnt++;

Algorithm 2 AssignGroupCnt ()

for each upward node n_{ij} connected to group g_i of targetLayer {
    Assign node n_{i1} to group g1;   //n_{ij}: j-th upward node connected to group g_i
    for each node n_{ij} {
    Assign node n_{ij} to group g_{x+y}, where
    x = group number to which (j-1)th node is assigned.
    y = 1 if y is 1 if there is an edge crossing between (j-1)th node and j-th node.
    y = 0 otherwise.
    if (x+y > total number of groups)
        The group number of the j-th node is 1;
    }
}

The leftCrossing and rightCrossing in Algorithm 1 are the number of edge crossings associated with the left and right nodes, computed from equations 1 and 2, respectively.

$$leftCros\sin g(e) = \sum_{\alpha=k+1}^{p} f(\alpha,l), where$$

$$f(\alpha,l) = \begin{cases} 0: \sum_{\beta=1}^{l-1} m_{\alpha\beta} = 0 \\ 1: otherwise \end{cases} \qquad (1)$$

$$rightCros\sin g(e) = \sum_{\alpha=1}^{k-1} f(\alpha,l), where$$

$$f(\alpha,l) = \begin{cases} 0: \sum_{\beta=l+1}^{q} m_{\alpha\beta} = 0 \\ 1: otherwise \end{cases} \qquad (2)$$

3 Results and Discussion

We implemented the algorithms in Microsoft C#. The program runs on any PC with Windows 2000/XP/Me/98/NT 4.0 as its operating system. Fig. 3 shows a T cell signal transduction pathway, visualized with algorithm. The pathway contains exclusively straight-line segments and edges of span=1, but the diagram contains no crossings. The program accepts the input signal transduction data in several formats:

- **pnm**: a pair of names for the source node and target node, separated by a tab, in each line.
- **pid**: a pair of node indices for the source node and target node, separated by a tab, in each line. Additional information is provided by pid_Label and pid_Pos files. Pid_label contains node labels and pid_Pos contains the x-, y-, and z-coordinates of each node.
- **stp**: consists of 2 sections. The node section contains node ID, x-, y-, and z-coordinates of the node, node type and node label. The edge section contains the source node ID, target node ID, and edge type.

- **mdb**: data from a Microsoft Access database, which can provide all the information supported by the pnm and stp formats.

Most diagrams of signal transduction pathways in databases are static images and thus cannot be refined, or changed later, to reflect new data. We have developed an algorithm for automatically representing signal transduction pathways from databases or text files. Unique features of the algorithm include (1) it does not place all sink nodes (nodes with no parent) in layer 1 but moves them to a lower layer so that edge spans can be minimized; (2) there are no curved lines, and edge bends occur only at dummy nodes; (3) the number of edge bends is minimized; (4) edge crossings are removed by adjusting the z-coordinates of nodes. We are currently extending the program to overlay various types of additional information onto the signal transduction pathways.

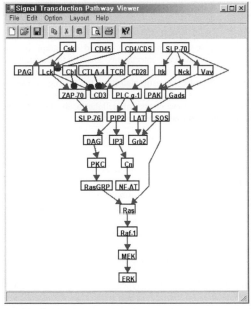

Fig. 3. T Cell signal transduction pathway visualized by our program. Red arrows indicate that the source nodes activate the sink nodes, while blue lines ending with filled circles indicate that the source nodes inhibit the sink nodes. The data were obtained from http://stke.sciencemag.org.

References

1. Hirsch, E.D., Kett, J.F., Trefilet J. (ed.): The New Dictionary of Cultural Literacy. Boston (2002)
2. Kanehisa, M., Goto, S., Kawashima, S., Nakaya, A.: The KEGG databases at GenomeNet. Nucleic Acids Research 30 (2002) 42-46
3. Hippron Physiomics, Dynamic Signaling Maps. http://www.hippron.com/products.htm
4. BioCarta. http://www.biocarta.com
5. Gansner, E.R., Koutsofios, E., North, S.C., Vo, K.-P.: A technique for drawing directed graphs. IEEE Transactions on Software Engineering 19 (1993) 214-230
6. Garey, M.R., Johnson, D.S.: Crossing Number is NP-Complete. SIAM J. Algebraic Discrete Methods 4 (1983) 312-316
7. Sugiyama, K., Tagawa, S., Toda, M.: Method for visual understanding of hierarchical system structures, IEEE Transaction on Systems, Man, and Cybernetics SMC-11 (1981) 109-125
8. Sugiyama, K.: Graph Drawing and Applications for Software and Knowledge Engineering. Singapore (2002)
9. Han, K., Ju, B.-H.: A fast layout algorithm for protein interaction networks. Bioinformatics 19 (2003) 1882-1887

VSP International
Science Publishers
P.O. Box 346, 3700 AH Zeist
The Netherlands

*Lecture Series on Computer
and Computational Sciences*
Volume 1, 2004, pp. 244-247

A Heuristic Algorithm for Finding Cliques and Quasi-Cliques in Molecular Interaction Networks[¶]

Byong-Hyon Ju and Kyungsook Han[*]
School of Computer Science and Engineering, Inha University, Inchon 402-751, Korea

Received 31 July, 2004; accepted in revised form, 18 August 2004

Abstract: Molecular interactions are typically represented as a graph in which nodes represent molecules and edges represent molecular interactions. Identifying hidden topological structures of molecular interaction networks often unveil biologically relevant functional groups and structural complexes. We have developed a heuristic algorithm for finding cliques and quasi-cliques in protein interaction networks. As highly connected subgraphs, the identified cliques and quasi-cliques can be used to predict the function and sub-cellular localization of uncharacterized proteins as well as to abstract complex molecular interaction networks.

Keywords: graph layout, molecular interaction networks, clique

Molecules in a highly connected subgraph of a molecular interaction network usually share a common function [1]. Therefore, finding highly connected subgraphs such as cliques and quasi-cliques in a protein interaction network provides useful information to predict the function of uncharacterized proteins in the highly connected subgraphs. However, finding a clique with a maximum size in a graph is a NP-hard problem [2]. There are several heuristic algorithms for the maximum clique problem, but most of them focus on finding a complete subgraph (that is, clique) and cannot be used to find near cliques or what we call quasi-cliques.

We have developed an efficient, heuristic algorithm that identifies all *edge-disjoint* cliques (i.e., cliques that do not share an edge), as outlined in Algorithms 1 and 2 [3]. After we find all edge-disjoint cliques, we find quasi-cliques of the following types, which are biologically meaningful clusters (see Fig. 1):

A. When a protein outside a clique interacts with two or more proteins in the clique, the node and the clique forms a quasi-clique.
B. When a clique shares a protein with cliques, the cliques forms a quasi-clique.
C. When two or more cliques interact with a common protein outside them via two or more interactions, the cliques and the protein forms a quasi-clique.

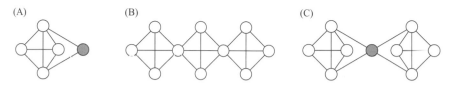

Fig. 1. Quasi-cliques of type A, B, and C.

Quasi-cliques are found as follows: 1) assign every node of a clique the index of the clique containing the node, 2) when a node of a clique has a preassigned clique index, assign the preassigned index to all nodes of the clique (merging two cliques into a quasi-clique of type B), 3) when a node outside a clique interacts with two or more proteins in a clique, assign the clique index to the node (quasi-clique of type A), 4) when two or more cliques interact with a common protein outside them via two or more interactions, merge the cliques and the protein into a quasi-clique (of type C). A quasi-

[¶] This work was supported by the advanced backbone IT technology development program of the Ministry of Information and Communication of Korea under grant IMT2000-C3-4.
[*] To whom correspondence should be addressed. Email: khan@inha.ac.kr

clique is formed by selecting nodes with same cIdx (cliques index) value from those nodes with cIdx > 0 (Algorithms 3 and 4).

Algorithm 1 FindClique
```
 1:   for all v ∈ V do
 2:      for all neighbor u of v do
 3:         Lst ← Ø
 4:         if not isClique(u, v) then
 5:            Lst[0].node ← v
 6:            Lst[0].idx ← v(u)+1          {v(u): index of u in v}
 7:            Lst[1].node ← u
 8:            Lst[1].idx ← u(v)+1
 9:            repeat
10:               if not isClique(Lst[0].node, Lst[0].node[Lst[0].idx] then
11:                  ChkClique(1, Lst[0].node[Lst[0].idx])
12:               end if
13:                  Lst[0].idx ← Lst[0].idx+1
14:            until (Lst[0].idx ≥ |Lst[0].node|)
15:            if (|Lst| ≥ 3) then
16:               DeclareClique(Lst)
17:            end if
18:         end if
19:      end for
20:   end for
```

Algorithm 2 ChkClique(N, NVal)
```
 1:   if (N = |Lst|) then
 2:      Lst[|Lst|].node ← NVal
 3:      Lst[|Lst|].idx ← 0
 4:   else
 5:      if (Lst[N].idx < |Lst[N].node|) then
 6:         if (NVal > Lst[N].node[Lst[N].idx]) then
 7:            repeat
 8:               Lst[N].idx ← Lst[N].idx+1
 9:            until (NVal ≤ Lst[N].node[Lst[N].idx])
10:            if (NVal = Lst[N].node[Lst[N].idx]) then
11:               ChkClique(N+1, NVal)
12:            end if
13:         end if
14:      end if
15:   end if
```

Algorithm 3 AssignQuasiCliqueIdx
```
 1:   for all N ∈ G do
 2:      N.cIdx=0
 3:   end for
 4:   curCIdx=1
 5:   for all N ∈ G do
 6:      if isClique(N) then
 7:         for all E ∈ N do
 8:            if (E.target.cIdx > 0) then
 9:               for all tmpN ∈ G do
10:                  if (tmpN. cIdx = E.target.cIdx) then tmpN. cIdx = curCIdx
11:               end for
12:            else E.target.cIdx = curCIdx
13:         end for
14:         N.cIdx = curCIdx
15:         curCIdx ++
16:      end if
17:   end for
```

Algorithm 4 ExtendQuasiClique
1:     **for all** $N \in G$ **do**
2:         **if** ($N.cIdx = 0$) **then**
3:             $qCliqueCnts \leftarrow \emptyset$
4:             **for all** $E \in N$ **do**
5:                 **if** ($E.target.cIdx > 0$) **then** $qCliqueCnts[E.target.cIdx]$++
6:             **end for**
7:             $qCValue$=0
8:             **for all** $c \in qCliqueCnts$ **do**
9:                 **if** (c>1) **then**
10:                     **if** ($qCValue > 0$) **then**
11:                         **for all** $tmpN \in G$ **do**
12:                             **if** ($tmpN.cIdx = qCValue$) **then** $tmpN.cIdx = qCValue$
13:                         **end for**
14:                     **else** $qCValue$=c
15:                 **end if**
16:             **end for**
17:             $N.cIdx = qCValue$
18:         **end if**
19:     **end for**

(A)

(B)

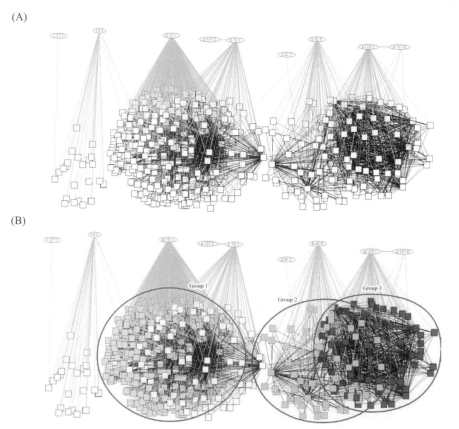

Fig. 2. (A) Part of human protein interaction networks, visualized as a 2-layer graph. Oval nodes at the top layer and square nodes at the bottom layer represent functions and proteins, respectively. The relation of proteins and their functions are represented by cyan edges and protein-protein interactions are represented by black edges. (B) Three quasi-cliques (shown in yellow, green and magenta) found in

the protein interaction network. Proteins in group 1, 2, and 3 share a function b.10.1, b.42.5, and a.123.1, respectively.

Fig. 2A is a part of human protein interaction networks [4], visualized as a 2-layer graph. Oval nodes at the top layer and square nodes at the bottom layer represent functions and proteins, respectively. The relation of proteins and their functions are represented by cyan edges and protein-protein interactions are represented by black edges. Fig. 2B shows 3 quasi-cliques found in the protein interaction network. The proteins in each quasi-clique indeed share a protein function. Fig. 3 shows a network of yeast protein interactions with 6 quasi-cliques. The interaction data and functional catalogues were extracted from the MIPS database (ftp://ftpmips.gsf.de/yeast/PPI/PPI_120803.tab; ftp://ftpmips.gsf.de/yeast/catalogues/funcat/funcat-2.0_data_18032004). Proteins in each quasi-clique share at least one function with other proteins within the quasi-clique. We are currently using the algorithms to predict the function of proteins with unknown function.

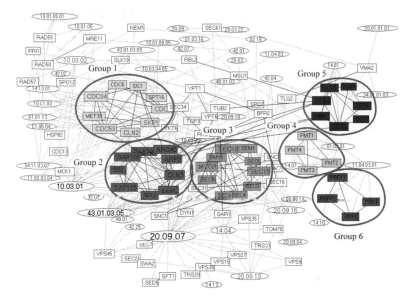

Fig. 3. Six quasi-cliques found in yeast protein interaction networks. Proteins in each quasi-clique share at least one function with other proteins within the quasi-clique.

References

1. Bu, D., Zhao, Y., Cai, L., Xue, H., Zhu, Z., Lu, H., Zhang, J., Sun, S., Ling, L., Zhang, N., Li, G., Chen R.: Topological structure analysis of the protein-protein interaction network in budding yeast. Nucleic Acids Research 31 (2003) 2443-2450
2. Battiti, R., Protasi, M.: Reactive Local Search for the Maximum Clique Problem. Algorithmica 29 (2001) 610-637
3. Ju, B.-H., Han, K.: Complexity management in visualizing protein interaction networks. Bioinformatics 19 (2003) i177-i179
4. Han, K., Park, B., Kim, H., Hong, J., Park, J.: HPID: The Human Protein Interaction Database. Bioinformatics 20 (2004) in press

VSP International
Science Publishers
P.O. Box 346, 3700 AH Zeist
The Netherlands

Lecture Series on Computer
and Computational Sciences
Volume 1, 2004, pp. 248-252

Construction of Assymptotically Symplectic Methods for the numerical solution of the Schrödinger Equation - application to 5th and 7th order.[1]

Z. Kalogiratou

Department of International Trade,
Technological Educational Institute of Western Macedonia at Kastoria,
P.O. Box 30, GR-521 00, Kastoria, Greece

Th. Monovasilis

Department of Computer Science and Technology,
Faculty of Sciences and Technology,
University of Peloponnese,
GR-221 00 Tripolis, Greece
Department of International Trade,
Technological Educational Institute of Western Macedonia at Kastoria,
P.O. Box 30, GR-521 00, Kastoria, Greece

T.E. Simos[2]

Department of Computer Science and Technology,
Faculty of Sciences and Technology,
University of Peloponnese,
GR-221 00 Tripolis, Greece

Received 30 August, 2004; accepted in revised form 10 September, 2004

Abstract: The solution of the one-dimensional time-independent Schrödinger equation is considered by symplectic integrators. The Schrödinger equation is first transformed into a Hamiltonian canonical equation. The concept of assymptotic symplecticness is introduced. A family of assymptotically symplectic methods of order p is developed, and applied in the construction of methods of 5th and 7th order. Numerical results are obtained for the one-dimensional harmonic oscillator and the hydrogen atom.

Keywords: Assymptotically symplectic methods, Schrödinger equation

1 Introduction

The time-independent Schrödinger equation is one of the basic equations of quantum mechanics. Its solutions are required in the studies of atomic and molecular structure and spectra, molecular

---
[1]Funding by research project 71239 of Prefecture of Western Macedonia and the E.U. is gratefully acknowledged.
[2]Corresponding author. Active Member of the European Academy of Sciences and Arts. E-mail: simos-editor@uop.gr, tsimos@mail.ariadne-t.gr

dynamics and quantum chemistry. In the literature many numerical methods have been developed to solve the time-independent Schrödinger equation (see [4]). One of the new developed category of numerical methods for the solution of differential equations are symplectic integrators (see [3,6]), among their properties is the energy preservation, which is an important property in quantum mechanics. Liu et. al [2] applied symplectic numerical methods as developed by Yoshida [6] to solve the one dimensional time-independent Schrödinger equation. In this paper we introduce the concept of assymptotic symplecticness and develope a family of assymptocally symplectic methods of order p. As examples we construct methods of 5th and 7th order. Our new methods are tested on the computation of the eigenvalues of the one-dimensional harmonic oscillator and the hydrogen atom and compared with Yoshida's 6th order method [6]. The new methods are proven to be more accurate and computationally efficient.

In section 2 we transform the Schrödinger equation into a hamiltonian canonical equation. In section 3, give the neccessary results concerning symplectic methods. In section 4 the family of assymptotically symplectic methods is developed.

2   The time-independent Schrödinger equation

The one-dimensional time-independent Schrödinger equation may be written in the form

$$-\frac{1}{2}\frac{d^2\psi}{dx^2} + V(x)\psi = E\psi \tag{1}$$

where E is the energy eigenvalue, $V(x)$ the potential, and $\psi(x)$ the wave function. Equation (1) can be rewritten in the form

$$\frac{d^2\psi}{dx^2} = -B(x)\psi$$

where $B(x) = 2(E - V(x))$, or

$$\begin{aligned} \phi' &= -B(x)\psi \\ \psi' &= \phi \end{aligned} \tag{2}$$

3   Numerical methods

3.1   Symplectic numerical schemes

The key point of applying symplectic methods into numerical calculations is to maintain sympectic structure in the discrete scheme. Given an interval $[a, b]$ and a partition with N points

$$x_0 = a, \ x_n = x_0 + nh, \quad n = 1, 2, \ldots, N,$$

a discrete scheme proceeds as follows

$$\begin{pmatrix} \phi_{n+1} \\ \psi_{n+1} \end{pmatrix} = M_{n+1} \begin{pmatrix} \phi_n \\ \psi_n \end{pmatrix}, \quad M_{n+1} = \begin{pmatrix} a_{n+1} & b_{n+1} \\ c_{n+1} & d_{n+1} \end{pmatrix} \tag{3}$$

where

$$\begin{pmatrix} \phi_0 \\ \psi_0 \end{pmatrix} = \begin{pmatrix} \phi(x_0) \\ \psi(x_0) \end{pmatrix}$$

the initial conditions.

The n-step approximation to the solution can be written as

$$
\begin{pmatrix} \phi_n \\ \psi_n \end{pmatrix} = \begin{pmatrix} a_n & b_n \\ c_n & d_n \end{pmatrix} \begin{pmatrix} a_{n-1} & b_{n-1} \\ c_{n-1} & d_{n-1} \end{pmatrix} \cdots \begin{pmatrix} a_1 & b_1 \\ c_1 & d_1 \end{pmatrix} \begin{pmatrix} \phi_0 \\ \psi_0 \end{pmatrix}
$$

$$
= M_n M_{n-1} \cdots M_1 \begin{pmatrix} \phi_0 \\ \psi_0 \end{pmatrix}
$$

Defining

$$
T = M_n M_{n-1} \cdots M_1 = \begin{pmatrix} A_n & B_n \\ C_n & D_n \end{pmatrix}
$$

we can write the discrete transformation

$$
\begin{pmatrix} \phi_n \\ \psi_n \end{pmatrix} = \begin{pmatrix} A_n & B_n \\ C_n & D_n \end{pmatrix} \begin{pmatrix} \phi_0 \\ \psi_0 \end{pmatrix}
$$

which we want to be a symplectic transformation.

A transformation matrix A is symplectic if $A^T J A = J$ where

$$
J = \begin{pmatrix} 0 & 1 \\ -1 & 0 \end{pmatrix}
$$

The following results can be easily verified:

- A 2×2 matrix is symplectic if its determinant equals 1.

- Given A and B symplectic matrices then their product $A B$ is also symplectic.

- The discrete scheme (3) is symplectic if each matrix M_i is symplectic.

Many authors (see [3,6,10]) developed symplectic integrators of the following form

$$
\begin{aligned}
p_1 &= \phi_n - c_1 h B \psi_n, \\
q_1 &= \psi_n + d_1 h p_1, \\
p_2 &= p_1 - c_2 h B q_1, \\
q_2 &= q_1 + d_2 h p_2, \\
&\vdots \qquad \vdots \\
p_{k-1} &= p_{k-2} - c_{k-1} h B q_{k-2}, \\
q_{k-1} &= q_{k-2} + d_{k-1} h p_{k-1}, \\
\phi_{n+1} &= p_{k-1} - c_k h B q_{k-1}, \\
\psi_{n+1} &= q_{k-1} + d_k h \phi_{n+1}
\end{aligned}
$$

this is the k-stage method.

This form is equivalent with (3), in order to demonstrate we write for each stage

$$
\begin{pmatrix} 1 & 0 \\ d_1 h & 1 \end{pmatrix} \begin{pmatrix} p_1 \\ q_1 \end{pmatrix} = \begin{pmatrix} 1 & -c_1 h B \\ 0 & 1 \end{pmatrix} \begin{pmatrix} \phi_n \\ \psi_n \end{pmatrix}
$$

$$\begin{pmatrix} 1 & 0 \\ d_2h & 1 \end{pmatrix} \begin{pmatrix} p_2 \\ q_2 \end{pmatrix} = \begin{pmatrix} 1 & -c_2hB \\ 0 & 1 \end{pmatrix} \begin{pmatrix} p_1 \\ q_1 \end{pmatrix}$$

$$\vdots \qquad \vdots$$

$$\begin{pmatrix} 1 & 0 \\ d_{k-1}h & 1 \end{pmatrix} \begin{pmatrix} p_{k-1} \\ q_{k-1} \end{pmatrix} = \begin{pmatrix} 1 & -c_{k-1}hB \\ 0 & 1 \end{pmatrix} \begin{pmatrix} p_{k-2} \\ q_{k-2} \end{pmatrix}$$

$$\begin{pmatrix} 1 & 0 \\ d_kh & 1 \end{pmatrix} \begin{pmatrix} \phi_{n+1} \\ \psi_{n+1} \end{pmatrix} = \begin{pmatrix} 1 & -c_khB \\ 0 & 1 \end{pmatrix} \begin{pmatrix} p_{k-1} \\ q_{k-1} \end{pmatrix}$$

for the i stage

$$\begin{pmatrix} p_i \\ q_i \end{pmatrix} = \begin{pmatrix} 1 & -c_ihB \\ -d_ih & 1+c_id_ih^2B \end{pmatrix} \begin{pmatrix} p_{i-1} \\ q_{i-1} \end{pmatrix}$$

we define

$$S_i(B) = \begin{pmatrix} 1 & -c_ihB \\ -d_ih & 1+c_id_ih^2B \end{pmatrix}$$

then

$$\begin{pmatrix} \phi_{n+1} \\ \psi_{n+1} \end{pmatrix} = S_k(B) \cdots S_2(B)S_1(B) \begin{pmatrix} \phi_n \\ \psi_n \end{pmatrix}$$

The last is of the form (3), where the transformation matrix is

$$M_{n+1} = S_k(B) \cdots S_2(B)S_1(B) = \begin{pmatrix} \alpha_{k-1}(B) & \beta_k(B) \\ \gamma_{k-1}(B) & \delta_k(B) \end{pmatrix}$$

where $\alpha_{k-1}(B)$, $\gamma_{k-1}(B)$ are polynomials in B of degree $k-1$, and $\beta_k(B)$ and $\delta_k(B)$ are polynomials in B of degree k.

3.2 Assymptotically symplectic schemes

We give the following definitions of assymptotically symplectic matrix and method

Definition 1 *We say that a matrix A is assymptotically symplectic of order p if*

$$A^T J A = J + O(h^{p+1})$$

Definition 2 *We say that a discrete scheme of the form (3) is assymptotically symplectic of order p if the transformation matrix T is assymptotically symplectic.*

The three results conserning symplectic matrices above also hold for assymptotically symplectic matrices. Therefore in order to prove that a discrete scheme is assymptotically symplectic it suffices to show that the associated matrix M_n is assymptotically symplectic.

4   Numerical results

We consider the one-dimesional eigenvalue problem with boundary conditions

$$\psi(a) = 0, \quad \psi(b) = 0$$

We use the shooting scheme in the implementation of the above methods. The shooting method converts the boundary value problem into an initial value problem where the boundary value at the end point b is tranformed into an initial value $y'(a)$, the results are independent of $y'(a)$ if $y'(a) \neq 0$. The eigenvalue E is a parameter in the computation, the value of E that makes $y(b) = 0$ is the eigenvalue computed.

4.1   The Harmonic Oscillator

The potential of the one dimensional harmonic oscillator is

$$V(x) = \frac{1}{2}kx^2$$

we consider $k = 1$. The exact eigenvalues are given by

$$E_n = n + \frac{1}{2}, \quad n = 0, 1, 2, \ldots$$

4.2   The Hydrogen Atom

The potential of the hydrogen atom is

$$V(x) = -\frac{1}{x} + \frac{l(l+1)}{x^2}, \quad 0 \leq r < +\infty, \quad l = 0, 1, 2, \ldots$$

The exact eigenvalues are given by

$$E_n = -\frac{1}{2n^2}, \quad n = 1, 2, \ldots$$

In order to compute the eigenvalues for the case $l = 0$ by the shooting method we used the interval $[0, b]$ with boundary conditions $y(0) = 0$ and $y(b) = 0$.

References

[1] Arnold V., *Mathematical methods of Classical Mechanics*, Springer-Verlag, 1978.

[2] Liu, X.S., Liu X.Y, Zhou, Z.Y., Ding, P.Z., Pan, S.F., Numerical Solution of the One-Dimensional Time-Independent Schrödinger equation by using symplectic schemes, *International Journal of Quantum Chemistry* 79(2000),343-349.

[3] Ruth R.D., A canonical integration technique, *IEEE Transactions on Nuclear Science*, NS 30 (1983),2669-2671.

[4] Simos T.E., Numerical methods for 1D, 2D and 3D differential equations arising in chemical problems, *Chemical Modelling: Application and Theory*, The Royal Society of Chemistry, 2(2002),170-270.

[5] Sanz-Serna, J.M., Calvo, M.P., *Numerical Hamiltonian Problem*, Chapman and Hall, London, 1994.

[6] Yoshida H., Construction of higher order symplectic integrators, *Physics Letters A* 150(1990),262-268

VSP International
Science Publishers
P.O. Box 346, 3700 AH Zeist
The Netherlands

*Lecture Series on Computer
and Computational Sciences*
Volume 1, 2004, pp. 253-256

CoMFA and CoMSIA 3D-Quantitative Structure-Activity Relationships of New Flexible Antileishmanial Ether Phospholipids

Agnes Kapou, Nikos Avlonitis, Anastasia Detsi, Maria Koufaki, Theodora Calogeropoulou, Nikolas P. Benetis and Thomas Mavromoustakos[1]

National Hellenic Research Foundation, Institute of Organic and Pharmaceutical Chemistry
Vas. Constantinou 48, 11635 Athens, Greece

Received 31 July, 2004; accepted in revised form, 12 August 2004

Abstract: Modern 2D-NMR spectroscopy and Molecular Modeling are valuable tools in 3D-QSAR applications of flexible molecules. CoMFA and CoMSIA have been applied to a novel synthetic class of flexible ring substituted ether phospholipids. A correlation was observed, relating the steric, electrostatic, and hydrophobic properties of the new compounds with the *in vitro* antileishmanial activity against two promastigote Leishmania strains, L. infantum and L. donovani. The obtained 3D-maps indicate the necessary structural features for biological activity and the pharmacophores. These results will assist in the design of new analogs possessing improved biological profile.

Keywords: CoMFA, CoMSIA, Flexible Ether Phospholipids, Leishmaniasis, Molecular Modeling, 2D-NMR

1. Introduction

QSAR studies of rigid antileishmanial compounds such as chalcones have been reported earlier [1]. The real challenge in our approach was to apply 3D-QSAR methodologies such as CoMFA and CoMSIA to flexible ether phospholipids. The problem of flexibility has been addressed before [2], but no 3D-QSAR study has ever confronted a set of molecules with long alkyl chains successfully. The 3D-QSAR models derived by this study, describing quantitatively the spatial and electrostatic requirements of the pharmacophores, will be useful in guiding the design of molecules with improved activity.

Figure 1: The conformation of **I**, concluded by NMR experiments and molecular modeling [3].

Although the molecular target of antileishmanial ether phospholipids is not known yet and the compounds are extremely flexible, there was adequate information supporting the proposed conformation of the active adamantylidene-substituted compound **I** (Fig. 1), derived by 2D-NMR and

[1] Corresponding author. Laboratory for Molecular Analysis, Institute of Organic and Pharmaceutical Chemistry, National Hellenic Research Foundation, 48 Vas. Constantinou Av. 116 35, Athens Greece. E-mail: tmavro@eie.gr

Molecular Modeling [3]. In fact, structural information about an unknown molecular target can be derived from CoMFA and CoMSIA. **I** has served as a template to build the 3D-conformation of the rest of the compounds of the present study. CoMFA and CoMSIA, additionally to their predictive power, have constituted a means of confirmation of the already proposed conformation of **I**.

2. Materials and Methods

The chemical structure of the ether phospholipids, possessing N,N,N-trimethyl-ammonium, N-methyl-piperidino, or N-methyl-morpholino head groups, is shown in Figure 2. The hexadecyl chain in the control compound, miltefosine was replaced by cyclohexylidenealkyl or adamantylidenealkyl groups.

The biological activity of the test molecules was evaluated by examination of the *in vitro* activity against of promastigote cultures of L. infantum and L. donovani, using MTT as a marker of cell viability. The antileishmanial activity and the synthesis of new active ring-substituted ether phospholipids were previously described in [3]. The conformation of **I** in solution (Fig. 1) was determined by 2D-NOESY NMR experiments and Conformational Search studies [3].

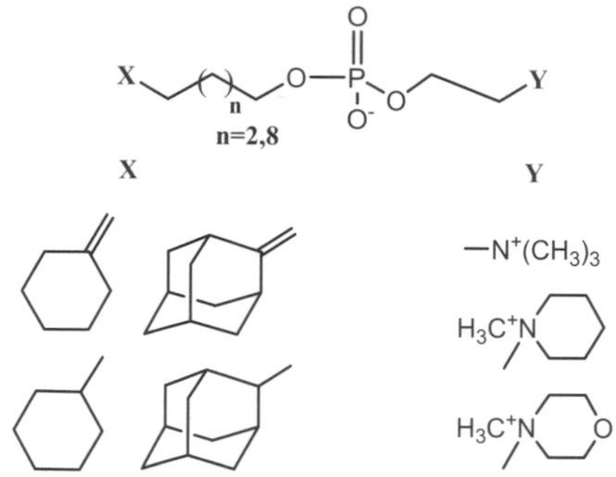

Figure 2: The chemical structure of the ring substituted ether phospholipids.

CoMFA (Comparative Molecular Field Analysis) is a standard 3D-QSAR approach and is usually preferable over classical QSAR when conformational data are available. In CoMFA, the physicochemical properties of each molecule are described as a combination of the *electrostatic* and *steric* interactions of the molecule with *probe atoms* regularly placed in the space surrounding the molecule.

CoMSIA (Comparative Molecular Similarity Indices Analysis) offers a significantly more detailed analysis of the pharmacophore as it involves additional *descriptors* besides electrostatic and steric interactions, such as *hydrophobicity* and *hydrogen–bonding*. These are the principal factors that determine the non-covalent attractive forces and the compatibility between the active site of the "receptor" and the ligand.

CoMFA uses the Coulomb potential to calculate the electrostatic field and the Lennard-Jones potential to calculate the steric field. Both these functions are very steep near the van der Waals surface area, while the distance dependence between the probe atom and the atoms of the molecule in CoMSIA is a Gaussian function. By this way the similarity indices can be calculated at all grid points, both inside and outside the molecular surface while any field calculations inside the molecule are not involved in CoMFA. One consequence of this difference is that in CoMSIA one does not need to use any cut-off values, as it is the case in CoMFA, a fact that makes the *electrostatic field* to be dependent on the orientation of the aligned molecules in the lattice.

In both CoMFA and CoMSIA, first a detailed numerical mapping of certain *potential fields* is constructed, by a *molecular mechanics* (force field) approach. The values of these interactions are subjected to statistical analysis, usually by PLS (Partial Least Squares). The resulting pharmacophore CoMFA model, which is validated by this powerful statistical technique, comprises regions in which *positive* or *negative charge* is preferred and regions in which *more* or *less volume* can be afforded. The CoMSIA model additionally comprises regions where *hydrophobic* or *hydrophilic* groups are preferred and regions where *donors* or *acceptors* can be afforded. In order to achieve the best comparison between the different *test molecules*, a careful *superposition* (alignment) of the rigid parts is required before the statistical analysis.

In the present case, the derived pharmacophore models will be used to predict the expected activity of any proposed substituted lipid molecules, which possess analogy in the polar head group with the template molecule, but comprise different alkyl chains.

3. Results an Discussion

The CoMFA and CoMSIA contour maps and the statistical results of the PLS analyses of the L. infantum and L. donovani biological data are shown in Figure 3 and 4 respectively. The predicted versus the actual values of the biological activity are shown on the corresponding diagrams. In all models there is a clear clustering of the predicted activity of the compounds on the diagrams, analogous to that of the actual values.

A very good correlation is observed in the case of L. infantum CoMFA. i) The most active compounds comprise long alkyl chains and choline, and occupy the upper right area of the diagram; ii) Compounds that possess moderate activity comprise long alkyl chains and mainly non-choline head-groups, and occupy the central part of the diagram; iii) The inactive compounds consist of short alkyl chains and any of the head groups, and occupy the lower left part of the diagram. The active analogs are well accommodated in the region defined by the green contours. The preference for choline over piperidine and morpholine is indicated by the yellow and blue contours in the region of the head group.

In the case of L. donovani CoMFA the correlation is worse. This was expected because the structural features essential for activity are not that well differentiated as in the case of L. infantum, in which the length of the alkyl chain plays the most important role. The negative impact of the use of a head group other than choline is overestimated (yellow contours). However, the model distinguishes between active and inactive analogs.

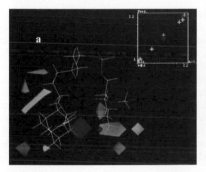 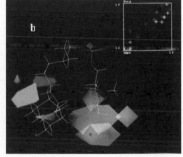

Fig 3: The CoMFA contour maps resulting from the PLS analysis of a. the L. infantum and b. the L. donovani biological data. A quite active analog, comprising a long alkyl chain, an N-methyl-morpholino head group and an adamantylidene group, is shown. Green: bulk is favored, yellow: bulk is disfavored. Blue: positive charge is favored, red: negative charge is favored. a. *Cross-validated*: *PRESS*: 0.302, q^2: 0.805, Optimal no. of components: 6; *without validation*: s: 0.044, r^2: 0.996. b. *Cross-validated*: *PRESS*: 0.511, q^2: 0.449, Optimal no. of components: 3; *without validation*: s: 0.278, r^2: 0.837.

The L. infantum CoMSIA model shows a poorer predictive power compared to the corresponding CoMFA model. The model overestimates the structural differences among the inactive short-chained compounds and the negative impact of bulk in the area of the head group. However, the contours that determine the regions, in which bulky and hydrophobic groups of the alkyl chain are favored or disfavored, coincide, indicating that the model distinguishes between active and inactive compounds, in relation to the length of the alkyl chain.

CoMSIA, on the other hand, gives better results than CoMFA for L. donovani, with improved correlation of the predicted and actual values. This is due to the fact that the model recognizes the major characteristic of this data set, that is, both short and long-chained compounds can be active. The distribution of white contours, which coincide with the green contours in the region of the alkyl chain, justifies the high activity of both long- and short-chained compounds. Moreover, the cyan contour in the region of the head group, coinciding with the yellow contour, explains the preference for choline over piperidine and morpholine. However, the negative impact of bulk and hydrophilicity in the area is overestimated.

In none of the CoMSIA models are the related to hydrogen-bonding fields shown, since this kind of interactions do not seem to be important for activity.

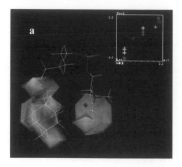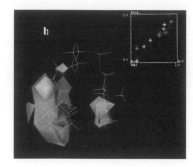

Fig 4: The CoMSIA contour maps resulting from the PLS analysis of a. the L. Infantum and b. the L. Donovani biological data. Colors for steric and electrostatics contours as for CoMFA. White: Hydrophobicity is favored, cyan: Hydrophobicity is disfavored. a. *Cross-validated*: *PRESS*: 0.306, q^2: 0.733, Optimal no. of components: 3; *without validation*: s: 0.168, r^2: 0.919. b. *Cross-validated*: *PRESS*: 0.503, q^2: 0. 0.466, Optimal no. of components: 3; *without validation*: s: 0.247, r^2: 0.871.

4. Conclusions

The present systematic procedure correlating the antileishmanial activity against the two strains of Leishmania, L. infantum and L. donovani, of the newly synthesized, ring-substituted ether phospholipids to the steric, electrostatic, and hydrophobic fields resulted in models predicting the ligand requirements for an active pharmacophore. Better correlation was observed for L. infantum in the CoMFA model, while for L. donovani the CoMSIA model related better structure and activity.

In all models there is a clear clustering of the compounds on the diagrams from the very active to the inactive ones, reflecting the good accordance of predicted and actual values of biological activity.

The proposed models will be further refined with the inclusion of biological data concerning the activity against the intracellular form of the parasite, the cytotoxicity and the hemolytic activity.

Hopefully, this will initiate synthetic strategies for new analogs with an improved biological profile. Furthermore, these results verify the proposed conformation of **I** and the invaluable contribution of NMR and molecular modeling in modern 3D-QSAR studies.

References

[1] Mei Liu, Prapon Wilairat, Simon L. Croft, Agnes Lay-Choo Tan and Mei-Lin Go, Structure-Activity Relationships of Antileishmanial and Antimalarial Chalcones, *Bioorg. Med. Chem.* **11** 2729-2738 (2003).

[2] M. C. Nicklaus, G. W. A. Milne, T. R. Burke, Jr., QSAR of Conformationally Flexible Molecules: Comparative Molecular Field Analysis of Protein-Tyrosine Kinase Inhibitors, *J. Comput.-Aided Mol. Des.*, **6** 487-504 (1992).

[3] Nikos Avlonitis, Eleni Lekka, Athanasia Detsi, Maria Koufaki, Theodora Calogeropoulou, Efi Scoulika, Eleni Siapi, Ioanna Kyrikou, Thomas Mavromoustakos, Andrew Tsotiris, Simona Golic Grdadolnic, and Alexandros Makriyannis, *Antileishmanial Ring-Substituted Ether Phospholipids*, *J. Med. Chem.* **46** 755-767 (2002).

VSP International
Science Publishers
P.O. Box 346, 3700 AH Zeist
The Netherlands

*Lecture Series on Computer
and Computational Sciences*
Volume 1, 2004, pp. 257-260

Sample Size Criteria for Estimating the Prevalence of a Disease

Athanassios Katsis[1]

Department of Social and Education Policy.
University of Peloponnese
20100, Korinthos, Greece

Hector E. Nistazakis

Department of Statistics and Actuarial Science.
University of the Aegean
Karlovasi, 83200, Samos, Greece

Received 8 August, 2004; accepted in revised form 18 August, 2004

Abstract: We investigate the problem of deriving the optimal sample size in a medical setup when misclassification is present. The methodology adopts the average coverage criterion for the binomial parameter utilizing efficient numerical techniques. The approach follows the Bayesian point of view.

Keywords: Sample size, Average Coverage Criterion, Misclassification, Bayesian point of view.

Mathematics Subject Classification: 62K05; 62F15

1 Introduction

In applied statistical analysis, the problem of obtaining the optimal sample size to estimate the binomial parameter when no misclassification exists has been successfully dealt with. Unfortunately, the occurrence of misclassification is a frequent phenomenon especially in a medical setup. Regardless of the classification procedure (diagnostic test and/or physician's examination), a degree of fallible classification usually exists. Moreover, the Bayesian approach seems like a natural choice since any prior beliefs about the unknown parameters may affect the sample size. Thus, in order to calculate the necessary sample size to estimate the prevalence of a disease, the researcher has to account for the existing misclassification as well as for any other prior information.

The literature on sample size specification adjusting for misclassification is rather narrow. From the Bayesian side, Erkanli, Soyer and Angold (1998) discussed two-phase designs using a fallible and an error free device while in the medical context Rahme, Joseph and Gyorkos (2000) derived the required sample size in order to estimate the prevalence of a specified disease when all the prior parameters are independent. Furthermore, Katsis and Toman (2004) proposed a double sampling scheme combining information from a fallible and an error-free device.

---

[1] Corresponding author. E-mail: katsis@uop.gr

In this article, we apply a Bayesian sample size criterion to address the problem of misclassification in a medical context. The dependence among the prior parameters is examined and a practical algorithm for numerical implementation is offered.

2   Bayesian criteria

One of the most widely used criteria in sample size calculations is the the average coverage criterion (ACC). In this framework, we require that n be the smallest sample size such that the average coverage of the posterior credible interval $R(x)$ is at least $1 - \alpha$. The expectation is taken over the marginal distribution of the data x. In particular, we are seeking the smallest value of n that satisfies the following inequality:

$$\int_X \{\int_{R(x)} f(\theta|x)d\theta)\}m(x)dx \geq 1 - \alpha$$

where

$$m(x) = \int l(x|\theta)f(\theta)d(\theta).$$

The interval $R(x)$ is of fixed length and may be either symmetric around the posterior mean or have the highest posterior density. If there is more than one prior parameter, we integrate out the other parameters, obtaining the marginal posterior density of the parameter of interest. Further computational details are provided in Adcock (1997).

Turning our attention to the medical context, let θ denote the prevalence of a disease. Assuming the existence of a diagnostic test, let p denote the probability of a positive test. The following probabilities will be used to describe the case of misclassification in our experiment:

$s = $ P(negative test | person with the disease)

$c = $ P(positive test | person without the disease) It can be shown that

$$p = \theta(1 - s) + (1 - \theta)c$$

If x people are tested positive in a sample of size n, then the likelihood function is given by

$$l(x|\theta, s, c) = \binom{n}{x}\{\theta(1 - s) + (1 - \theta)c\}^x\{\theta s + (1 - \theta)(1 - c)\}^{n-x} \tag{1}$$

Regarding the prior information for the parameters, it is realistic to assume that the frequency of the disease is independent of the accuracy of the diagnostic test. Thus, θ is independently distributed of both s and c. However, any information about the results of the test among the healthy, say, part of the population will undoubtedly provide a rough idea how the test fares among the people with the disease. The concept of statistical dependence between error probabilities in various settings has been discussed in, among others, Galen and Gambino (1975) and Gunel (1984). Thus, the joint prior distribution of θ, s and c may be thought of as a product of a Beta distribution (for θ) with parameters a and b and a Dirichlet distribution (for s and c) with parameters $(\lambda_1, \lambda_2, \lambda_3)$. Therefore we have that

$$f(\theta, s, c) \propto \theta^{a-1}(1 - \theta)^{b-1}s^{\lambda_1 - 1}c^{\lambda_2 - 1}(1 - s - c)^{\lambda_3 - 1} \tag{2}$$

where a, b, λ_i, $(i = 1, 2, 3)$ are non-negative numbers and also that $s + c < 1$. The latter condition is true for almost all practical purposes.

Combining (1) and (2) we obtain the following joint posterior distribution:

$$f(\theta, s, c|x) \quad \propto \quad \{\theta(1-s) + (1-\theta)c\}^x \{\theta s + (1-\theta)(1-c)\}^{n-x} \qquad (3)$$
$$\theta^{a-1}(1-\theta)^{b-1}s^{\lambda_1-1}c^{\lambda_2-1}(1-s-c)^{\lambda_3-1}$$

The marginal posterior distribution of θ is given by

$$f(\theta|x) = \int_0^1 \int_0^1 f(\theta, s, c|x)dsdc \qquad (4)$$

Therefore according to the ACC criterion for a symmetric interval of length w, we seek the minimum value of n such that the following condition is satisfied:

$$\int_X \{\int_{\hat{\theta}-\frac{w}{2}}^{\hat{\theta}+\frac{w}{2}} f(\theta|x)d\theta)\}m(x)dx \geq 1-\alpha \qquad (5)$$

where $\hat{\theta}$ is an estimate of the posterior mean. In general there is no closed form solution to (5), thus resulting to alternative methodology.

3 Numerical techniques

In this section we shall apply the above methodology in specific cases. Since it is very difficult to obtain an analytic expression for (4), we shall employ the non-iterative Monte Carlo method discussed by Ross (1996). We generate k random sample points from the joint prior distribution described in (2). The following weight function $\tau_i(x)$ ($i = 1, \ldots, k$) is defined for each triplet (θ_i, s_i, c_i) of the prior distribution:

$$\tau_i(x) = \frac{l(x|\theta_i, s_i, c_i)}{\sum_{i=1}^k l(x|\theta_i, s_i, c_i)}$$

We then approximate the posterior mean of θ by the following:

$$\hat{\theta} \approx \frac{\sum_{i=1}^k \theta_i \tau_i(x)}{\sum_{i=1}^k \tau_i(x)}$$

The average coverage, $cover(w)$, is estimated by the weighted average of the coverage for each x, which in turn is estimated by the sum $\sum \theta_i \tau_i(x)$. This summation is taken over all the θ_i's that range between $\hat{\theta} - \frac{w}{2}$ and $\hat{\theta} + \frac{w}{2}$. Thus we have that:

$$cover(w) \approx \sum_{x=0}^n \{\sum \theta_i \tau_i(x)\}m(x) \qquad (6)$$

where the marginal probability function $m(x)$ is estimated by

$$m(x) \approx \sum_{i=1}^k \frac{\tau_i(x)}{k}$$

The optimal sample size is derived by setting (6) equal to $1-\alpha$. The factors affecting the optimal sample size are the variance and the mean prior value of the parameter.

Epitomizing, we have illustrated a fully-fledged Bayesian approach to the problem of sample size determination in the planning stages of an experiment when the binomial proportion is subject to misclassification. We have utilized the average coverage criterion accounting for the fact that the data are unknown beforehand.

References

[1] C. A. Adcock, Sample size determination: a review, *Statistician* **46** 261-283(1997).

[2] A. Erkanli, R. Soyer and A. Angold, Optimal Bayesian Two-Phase designs for Prevalence Estimation, *Journal of Statistical Planning and Inference*, **66** 175-191(1998)

[3] R. S. Galen and S. R. Gambino, *Beyond Normality: The Predictive Value and Efficiency of Medical Diagnoses*, John Wiley & Sons, 1975.

[4] E. Gunel, A Bayesian analysis of the multinomial model for a dichotomous response with nonrespondents, *Communications in Statistics, Theory and Methods*, **13** 737-751(1984)

[5] A. Katsis and B. Toman, A Bayesian double sampling scheme for classifying binomial data, *The Mathematical Scientist*, **81**, 49-53(2004)

[6] E. Rahme, L. Joseph and T. W. Gyorkos, Bayesian sample size determination for estimating binomial parameters from data subject to misclassification, *Applied Statistics*, **49**, 119-1287(2000)

[7] S. M. Ross, Bayesians should not resample a prior sample to learn about the posterior, *American Statistician*, **50**, 116(1996)

VSP International
Science Publishers
P.O. Box 346, 3700 AH Zeist
The Netherlands

*Lecture Series on Computer
and Computational Sciences*
Volume 1, 2004, pp. 261-264

The Influence of Contact Interaction Between the Debonded Surfaces on Debonding Arrest in Graphite/Aluminum Composite

G. Papakaliatakis[†]

Democritus University of Thrace, GR-671 00 Xanthi, Greece

Received 8 September, 2004; accepted in revised form, 22 September 2004

Abstract: The influence of the contact and friction interaction taking place between the debonded surfaces on the arrest of further debonding extension in Graphite/Aluminum single-fiber composite materials was studied. In this specimen, a broken Graphite fiber, perpendicular to the fiber axis and a small debonding was considered. The composite is modeled as a two-material cylinder subjected to uniform displacement and consisting of an inner cylinder simulating the fiber and a surrounding shell simulating the metallic matrix. The specimen is subjected to a monotonic increasing uniform displacement perpendicular to the plane of the originally cracked fiber. An elastic-plastic finite element analysis was performed based on the finite element code ABAQUS to study the fracture behavior. A very detailed analysis of the stress field in the vicinity of the crack tip was undertaken. The results of stress analysis were coupled with the strain energy density theory to predict the crack growth behavior including crack initiation, crack growth extension/direction and final termination along the constituents interface. The influence of contact and friction interaction taking place between the debonded surfaces is also considered into the numerical study. It is obtained that the presence of such surfaces being in contact can lead to considerable reduction of the calculated stress fields near the crack front. It is shown that the followed analysis combined with the strain energy density criterion can explain the arrest of further debonding extension observed during experimental testing. The total debonding extension length under the critical applied displacement for fully fractured fiber was calculated.

Keywords: Metal matrix, graphite fiber, composite material, stain energy density criterion, contact, friction, fracture

Mathematics SubjectClassification:

1. Introduction

Metal matrix composite (MMC) materials have been under development for more than 40 years. The major advantages of these materials consist of their high strength and stiffness to weight ratios, high toughness and impact properties, high surface durability and high thermal conductivity. Also, graphite fibers are of great interest as reinforcements in MMCs because of their high strength and stiffness, low density, and potential for large–scale production at low cost. Aluminum alloys reinforced with graphite fibers appear to offer the most advantages as MMCs and consequently have received the most attention. The main problem has been reducing their cost. Thus this composite would be competitive for many applications in aircraft, missiles, electrical machinery, rocket propulsion systems, launch vehicle structures and spacecraft. Aircraft applications are in skins, struts, spars, wing boxes and helicopter blades.

A review of the various graphite-reinforced MMCs is given by Kendall, E.G.[1]. Many experimental studies have been conducted to investigate the mechanisms of crack propagation in MMCs [2-3]. In parallel to experimental investigations of their fracture properties, large efforts have been focused on modeling the fracture behavior of composites analytically. A number of models that predict the response of unidirectional MMCs have been reported [4-5]. Several numerical studies have been reported investigating the fracture behavior of fiber-reinforced MMCs using finite element methods [6-9]. A computational study of fiber crack initiation, crack growth behavior and debonding development

---

[†] E-mail:gpapakal@civil.duth.gr

of a modeled SiC/6061-Al composite with the stain energy density criterion is given by Papakaliatakis G. and Karalekas D.[10,11]

In the present investigation, an elastic-plastic finite element analysis has been carried out to study the fracture behavior of a graphite/Al single-fiber composite specimen containing a small crack parallel to the fiber-matrix interface. The specimen is subjected to monotonic tensile loading with the tensile axis perpendicular to the plane of the originally fully cracked reinforcing fiber. The composite is modeled as a two-material cylinder subjected to uniform displacement. For an existing initial interfacial debonding the calculation of the stress and displacement field near to the debonding crack tip was performed with the ABAQUS computer program and take into account the contact and the friction of the debonding faces. The results of numerical stress analysis were coupled with the strain energy density theory to predict the further debonding growth behavior and explain the arrest of further debonding extension observed during experimental testing. The total debonding extension length under the critical applied displacement for fully fractured fiber and for various values of friction coefficient was calculated.

2. Numerical Analysis

The material used in the present investigation was a Graphite/6061-Al composite, which consisted of an aluminum matrix reinforced with Graphite fibers. The fiber volume fraction was considered to be 44%.

A single-fiber composite (SFC) specimen was considered. Under monotonic loading in the axial direction, the fiber will fracture. Such SFC specimens have been used in the past to conduct micromechanics studies in metal matrix composites. The composite was modeled as a two-material cylinder consisting of an inner cylinder simulating the fiber and a surrounding shell simulating the matrix. Fig. 1 shows the micromechanical model used in this study. The fiber is considered linear elastic up to fracture, while the aluminum matrix exhibits elastoplastic behavior. It was assumed that the two components are perfectly bonded at the interface between the reinforcement and the matrix and, thus, all displacement components were continuous across the interface. In the analysis, it is also assumed that the composite cylinder is infinitely long subjected to a uniform displacement load along its upper and lower faces. The fracture behavior of the modeled composite, under applied displacement, u, increased incrementally, was studied for the following three cases: a) a fractured fiber containing a small central crack perpendicular to fiber long axis, b) a fully fractured fiber and c) a totally broken fiber, with an existing initial interfacial debonding.

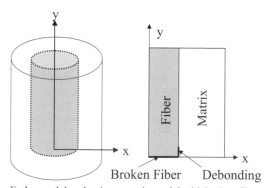

Fig. 1. Composite cylinder model and axisymmetric model with broken fiber and initial debonding.

A nonlinear finite deformation analysis was performed based on the finite code ABAQUS. For the finite element analysis, due to symmetry, one quadrant of the cylindrical element of the composite was modeled. Axisymmetric quadrilateral four-mode elements were used. The element size used was sufficiently small to provide satisfactory results for most areas of the model. However, in order to obtain sufficiently accurate data for the area near the crack tip a much finer mesh was constructed. The influence of contact interaction taking place between the debonded surfaces with a friction coefisient is also considered into the numerical study.

2. Prediction of Crack Growth

Crack growth consists of three stages: crack initiation, subcritical or slow growth and unstable crack propagation. These stages of crack growth can be addressed in a unified manner by the Strain Energy Density (SED) Criterion.

If a small central crack perpendicular to the fiber axis, is created under a critical value of applied displacement, the results of numerical stress analysis combined with the strain energy density theory lead to the conclusion that the crack propagates into the fiber, along the original crack plane and the complete fracture of the fibre befalls.

For the case of a totally broken fiber, the numerically obtained results for the variation of dW/dV versus distance, r, from the crack tip along the direction of crack extension and the fiber/matrix interface are plotted. The application of the stain energy criterion leading to the conclusion that the crack will propagate along the fiber/matrix interface and parallel to the applied displacement direction. This prediction that the crack growth will take place and continue at the interface region close to fibre side is in accordance with the experimental results.

For the case of a totally broken fiber, with an existing initial interfacial debonding, the calculation of the stress and displacement field near to the debonding crack tip was performed with the ABAQUS computer program and take into account the contact and the friction of the debonding faces. The variation of dW/dV versus distance r, from the debonding crack tip along the direction of crack extension for several initial interfacial debonding are plotted and the critical applied displacement for crack extension are calculated. The results showed that, as the existing interfacial debonding length increased, the corresponding critical applied displacement for debonding extension decreased. Also the corresponding critical applied displacement for debonding extension decreased more, as the friction coefficient increased. For the critical applied displacement that lead to the fully fracture of the fiber, until a limiting value for the debonding extension length is reached. When that debonding extension length is reached, a growth arrest of the interface crack is observed. The limiting value for the debonding extension corresponding to several friction coefficients was calculated.

References

[1] E.G. Kentall, Development of Metal_Matrix Composites Reinforced with High-Modulus Graphite Fibers, *Composite materials, volume 4 Metallic Matrix Composites K.G. Kreider, Ed., Academic Press, New York*, 1974.

[2] E.E. Gdoutos, D. Karalekas and I.M. Daniel, Thermal stress analysis of a Silicon Carbide/Aluminum composite, *Experimental Mechanics*, 31(3) (1991) 202-208.

[3] C.J. Lissenden, C.T. Herakovich and M-J. Pindera, Inelastic deformation of metal matrix composites, NASA Contractor Report 191522 (1993).

[4] D. Karalekas, E.E. Gdoutos, and I.M. Daniel, Micromechanical analysis of nonlinear thermal deformation of filamentary metal matrix composites, *Computational Mechanics*, 9(1) (1991) 17-26.

[5] J.G. Boyd, F. Constanzo and D.H. Allen, A micromechanics approach for constructing locally averaged damaged dependent constitutive equations in inelastic composites, *Int. J. Damage Mech.*, 2 (1993) 209-228.

[6] C.J. Lissenden and C.T. Herkovich, Numerical modelling of damage development and viscoplasticity in metal matrix composites, *Computer Methods in Applied Mechanics and Engineering*, 126 (1995) 289-303.

[7] Z. Kassan, R.J. Zhang and Z. Wang, Finite element simulation to investigate interaction between crack and particulate reinforcements in metal-matrix composites, *Materials Science & Engineering A*, 203 (1995) 286-299.

[8] G. Rauchs, P.F. Thomasson and P.J. Withers, Finite element modelling of fractional bridging during fatigue crack growth in fibre-reinforced metal matrix composites, *Computational Materials Science*, 25 (2002) 166-173.

[9] J.E. Davis and J. Qu, Numerical analysis of fiber fragmentation in a SiC/al single-fiber composite specimen, *Composites Science and Technology*, 60 (2000) 2297-2307.

[10] G.Papakaliatakis. and D. Karalekas., Study of debonding development in fibrous metal matrix composites, *in Book of Abstracts and in CD-ROM, (7 pp), Proceedings of the 11th European Conference on Composite Materials (ECCM 11),* Rhodes, Greece, May 31 – June 3, 2004.

[11] G.Papakaliatakis. and D.Karalekas. Computational study of crack growth in SiC/Al composites, *Mathematical and Computer Modelling*, Elsevier Ltd, 2004

VSP International
Science Publishers
P.O. Box 346, 3700 AH Zeist
The Netherlands

*Lecture Series on Computer
and Computational Sciences*
Volume 1, 2004, pp. 265-268

Accurate real-time simulation of semi-deformable tubes

A.N.F. Klimowicz and M.D. Mihajlović[1]

Department of Computer Science,
The University of Manchester,
Manchester, United Kingdom

Received 2 August, 2004; accepted in revised form 20 August, 2004

Abstract: Numerical modelling and simulation of deformable objets is one of the focal issues in mechanical engineering and computational science. In the cases when the computational results need to be rendered haptically and graphically in real time (for example, in a medical training simulator), a suitable trade-off has to be found between accuracy and effiiency. The most commonly used numerical techniques for modelling deformable objects fall mainly into the following two classes: interactive methods, that are fast, but not always accurate, and physically based methods, that are accurate, but generally not interactive. In this paper we present a numerical method for the simulation of semi-deformable tubes based on oriented splines. The numerical results obtained with this method demonstrate its accuracy and interactivness.

Keywords: linear elastic tubes, oriented splines, Lagrange equations, Newmark method, Cholesky factorisation.

Mathematics Subject Classification: 74H15, 74K25, 65M60, 65F05, 65D07.

1 Introduction

Accurate and interactive numerical modelling of deformable tubes represents both an important and a challenging problem in mechanical engineering. For example, a medical training simulator must be capable of responding in real time to the operator constraints both in graphics and in haptics. The conventional elastic tube modelling techniques can be classified into the following two categories: interactive, or real-time models (such as particle systems, material simple splines, etc.), and physically based models (typically using the finite element method (FEM)). The methods from the first category are fast enough to allow real time simulation, but lack physical accuracy, while the FEM is physically accurate, but time-consuming.

As a trade-off between the previously mentioned methods, in this paper we introduce a numerical method for tube dynamics simulation based on oriented splines. The key idea of this method is to use two conventional splines in the context of the Lagrange equations. We perform a semi-discretisation of a linear visco-elastic tube in space using oriented splines. This gives a system of ordinary differential equations for the time variable.

In Section 2 we describe in more detail the physical model. Section 3 covers the corresponding implementation, while Section 4 presents numerical results, in terms of accuracy, execution time, and memory cost, which validate the effectivness of our approach.

---

[1]Corresponding author. E-mail: milan@cs.man.ac.uk

2 Tube physics modelling

In order to describe our approach, we first define the tube geometry. We are modelling a semi-deformable 3D-tube, which can be generated by extruding a non-deformable 2D-ring along a deformable oriented spline curve in space (see Fig. 1).

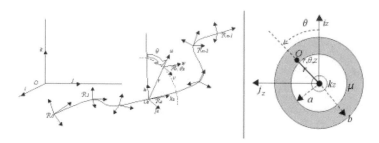

Figure 1: A deformable oriented spline curve in space (left) and a non-deformable 2D ring (right).

Consider the nodes $(\xi, \eta) \in (\mathbb{R}^p)^n \times (\mathbb{R}^q)^n$ of the oriented spline $(\varphi(z), \psi(z)) \in (\mathcal{M}_{p,3}(\mathbb{R}[z]))^n \times (\mathcal{M}_{q,3}(\mathbb{R}[z]))^n$ as defined in [1]. The mappings $\xi \to \left[\overrightarrow{OO_z}\right]_{\mathcal{B}} = \varphi(z)^T \xi$ and $\eta \to \left[\overrightarrow{\Theta\Theta_z}\right] = \psi(z)^T \eta$ define the translation vector $\overrightarrow{OO_z}$, and the rotation vector $\overrightarrow{\Theta\Theta_z}$ from a Galilean referential $\mathcal{R} = (O; \mathcal{B})$ to the solid $\mathcal{R}_z = (O_z, \mathcal{B}_z)$ which parametric abscissa is $z \in [0, n-1]$, as depicted in Fig. 1. The corresponding homogeneous transition matrices are:

$$[\mathcal{R}_z]_{\mathcal{R}} = \begin{pmatrix} [\mathcal{B}_z]_{\mathcal{B}} & \left[\overrightarrow{OO_z}\right]_{\mathcal{B}} \\ 0\ 0\ 0 & 1 \end{pmatrix} \tag{1}$$

The rotation matrix from $\mathcal{B}$ to $\mathcal{B}_z$ is given by the Rodrigues formula [1]:

$$[\mathcal{B}_z]_{\mathcal{B}} = [\cos \omega_z]_3 + \sin \omega_z \cdot \hat{\mathbf{u}}_z + (1 - \cos \omega_z) \cdot \mathbf{u}_z \mathbf{u}_z^T \tag{2}$$

with $\omega_z = \|\overrightarrow{\Theta\Theta_z}\|$ and $\mathbf{u}_z = [\overrightarrow{\Theta\Theta_z}]/\omega_z$ if $\omega_z \neq 0$ and 0 otherwise ($\hat{\mathbf{u}}_z$ is the pre cross-product matrix of $\mathbf{u}_z$). Then considering the polar coordinates $(r, \theta) \in [a, b] \times [0, 2\pi]$ with $0 < a < b$ in $\mathcal{R}_z$ (see Fig. 1) gives the full geometry of the system $[\mathcal{R}_{r,\theta,z}]_{\mathcal{R}} = [\mathcal{R}_z]_{\mathcal{R}} [\mathcal{R}_{r,\theta,z}]_{\mathcal{R}_z}$ from the state vectors (ξ, η).

The tube is given a uniform parametric mass distribution $\hat{\mu} > 0$ (in $kg \cdot m^{-2} \cdot u^{-1}$, where u is the parametric unit), thus yielding the following symmetric positive definite (SPD) quadratic form as kinetic energy [1]:

$$T(\dot{\xi}, \dot{\eta}) = \frac{1}{2} \dot{\xi}^T (\hat{\mu}\alpha\hat{\mathbf{A}})\dot{\xi} + \frac{1}{2}\dot{\eta}^T (\hat{\mu}\beta\hat{\mathbf{B}})\dot{\eta} \tag{3}$$

The linear inertia matrix $\hat{\mu}\alpha\hat{\mathbf{A}}$ and the angular inertia matrix $\hat{\mu}\beta\hat{\mathbf{B}}$ in (3) are time-independent, with $\alpha = \pi(b^2 - a^2)$, $\beta = \alpha(b^2 + a^2)/4$, $\hat{\mathbf{A}} = \int_0^{n-1} \varphi(z)\varphi(z)^T dz$ and $\hat{\mathbf{B}} = \int_0^{n-1} \psi(z)\mathrm{diag}(1,1,2)\psi(z)^T dz$. The deformable oriented spline curve is given a uniform parametric linear stiffness $\hat{k} > 0$ (in

$N \cdot m^{-1} \cdot u^{-1}$) and a uniform parametric angular stiffness $\hat{\ell} > 0$ (in $N \cdot m \cdot rad^{-1} \cdot u^{-1}$), thus yielding the following SPD quadratic form as deformation energy [1]:

$$U(\Delta \xi, \Delta \eta) = \frac{1}{2} \Delta \xi^T (\hat{k}\hat{\mathbf{D}}) \Delta \xi + \frac{1}{2} \Delta \eta^T (\hat{\ell}\hat{\mathbf{E}}) \Delta \eta. \tag{4}$$

The linear stiffness matrix $\hat{k}\hat{\mathbf{D}}$ and the angular stiffness matrix $\hat{\ell}\hat{\mathbf{E}}$ are time-independent, with $\hat{\mathbf{D}} = \int_0^{n-1} \varphi'(z)\varphi'(z)^T dz$ and $\hat{\mathbf{E}} = \int_0^{n-1} \psi'(z)\psi'(z)^T dz$. The semi-deformable tube is given a uniform parametric linear viscosity $\hat{\nu} > 0$ (in $N \cdot m^{-3} \cdot u^{-1} \cdot s$), thus yielding the following generalised internal friction [1]: $\mathbf{S}^* = -\hat{\nu}\alpha \hat{\mathbf{A}}\dot{\xi}$ and $\mathbf{M}^* = -\hat{\nu}\beta \hat{\mathbf{B}}\dot{\eta}$. Hence, the linear and angular damping matrices are $\hat{\nu}\alpha \hat{\mathbf{A}}$ and $\hat{\nu}\beta \hat{\mathbf{B}}$ respectively. We assume that the only external force $\mathbf{S}$ and torque $\mathbf{M}$ acting upon the system come from the boundary conditions (other actions, such as the gravity, can be easily incorporated into the model as well). The state vectors $(\Delta \xi, \Delta \eta)$ are the Lagrange coordinates that satisfy the state equations [1]:

$$\hat{\mu}\alpha \hat{\mathbf{A}}(\ddot{\Delta \xi}) + \hat{\nu}\alpha \hat{\mathbf{A}}(\dot{\Delta \xi}) + \hat{k}\hat{\mathbf{D}}\Delta \xi = \mathbf{S} \tag{5}$$

$$\hat{\mu}\beta \hat{\mathbf{B}}(\ddot{\Delta \eta}) + \hat{\nu}\beta \hat{\mathbf{B}}(\dot{\Delta \eta}) + \hat{\ell}\hat{\mathbf{E}}\Delta \eta = \mathbf{M} \tag{6}$$

The equations (5)-(6) represent an uncoupled system of second-order ordinary differential equations in which all the coefficient matrices are SPD.

3 Implementation details

In our implementation we adopt for both $\varphi(z)$ and $\psi(z)$ the Hermitian interpolation splines at order two [1]. The nodes are defined as $\xi_i = (\vec{\xi}_i^{(k)})$, $\eta_i = (\vec{\eta}_i^{(k)})$, $i = 0, \ldots, n-1$, $k = 0, 1, 2$, where $\vec{\xi}_i^{(k)}$ and $\vec{\eta}_i^{(k)}$ are the 3×1 matrices of the k-th derivative of the translation and rotation vectors, respectively, into the basis $\mathcal{B}$ for the node i. Each node i has 18 degrees of freedom (3 × 3 scalar parameters for both the translation and the rotation). The choice of Hermitian splines at order two is motivated by the need for smooth and flexible shapes. The C^2 continuity provides a continuous curvature at every node, while the spline flexibility guarantees an accurate and stable dynamics. The order two for the Hermitian splines is a good tradeoff between the smoothness requirements for the graphics (adaptive tesselation, texturing, etc.) and the computational efficiency (keeping the number of degrees of freedom per node relatively low).

The assembly of the system(5)-(6) involves the offline computation of the coefficient matrices $\hat{\mathbf{A}}$, $\hat{\mathbf{B}}$, $\hat{\mathbf{D}}$, $\hat{\mathbf{E}}$. The assembly procedure is that of the FEM, since each spline segment is a tube element. Basis integrals are computed numerically using the Gauss-Legendre quadrature rule [2]. The numerical quadrature order is selected optimally (which guarantees machine-tolerance accuracy with minimal computational effort), and we use the Horner algorithm when evaluating polynomials and their first derivatives. This enables us to compute the matrices $\hat{\mathbf{A}}$ and $\hat{\mathbf{D}}$ as well as $\hat{\mathbf{B}}$ and $\hat{\mathbf{E}}$ at the same pass.

Numerial solution of the ordinary differential equations system (5)-(6) is done by the simultaneous application of the Newmark method on the equations (5) and (6) with the appropriate boundary conditions [3]. Being an implicit scheme, the Newmark method requires solution of two linear algebraic systems:

$$\check{\mathbf{A}}\ddot{\Delta \xi} = \check{\mathbf{S}} \quad \text{and} \quad \check{\mathbf{B}}\ddot{\Delta \eta} = \check{\mathbf{M}} \tag{7}$$

with the following SPD coefficient matrices [1]:

$$\check{\mathbf{A}} = (\hat{\mu} + \beta_1 \, dt \, \hat{\nu})\alpha \hat{\mathbf{A}} + (\beta_2 (dt)^2 / 2)\hat{k}\hat{\mathbf{D}}, \qquad \check{\mathbf{B}} = (\hat{\mu} + \beta_1 \, dt \, \hat{\nu})\beta \hat{\mathbf{B}} + (\beta_2 (dt)^2 / 2)\hat{\ell}\hat{\mathbf{E}} \tag{8}$$

The Newmark method is unconditionally stable if we select $\beta_1 = \beta_2 = 0.5$ in (7). In (8) we take $dt > 0$ (with $dt = 0$ only at the first step). The right-hand side vectors $\check{\mathbf{S}}$ and $\check{\mathbf{M}}$ are the explicit functions of the previous kinematic state and the new load state [1]. The linear systems (7) are solved by a direct method (using the Bunch-Kaufman version of the Cholesky factorisation [2]). The systems could also have been solved by a preconditioned iterative algorithm, such as the CG method preonditioned by the algebraic multigrid. However, relatively small size of the linear systems (see Section 4) makes direct methods the preferable choice.

4   Numerical results

In this section we demonstrate the accuracy and the efficiency of our model applied to the example of a straight tube initially deformed into a stressed position. In such situation the tube returns to its reference configuration. We measure the execution time per integration step and the maximal relative errors in the positions as functions of the problem size. The measurements are taken during the transition period, well away of the equilibrium (after 1000 time steps, while the equilibrium occurs after approximately 6000 time steps of $dt = 5ms$). The solution for comparison is taken from the computation with $N_{\text{elt}} = 128$ elements. The program was compiled with MS Visual C++ 6.0 and it was run under Windows XP on a PC with a single Intel Pentium 4 CPU at 1.8GHz and 512MB of RAM. The results are summarised in Table 1. In the table N_{elt} denotes the number of elements, N_{dof} is the total number of degrees of freedom. T is the CPU clock time per integration step (in ms), and M is the memory used to store the data (in KB). For this experiment we adopted the values $a = 0.32m$, $b = 0.4m$ $L = 12m$, $\hat{k} = 1.2\,N\,m^{-1}\,u^{-1}$, $\hat{\mu} = 0.8\,kg\,m^{-2}\,u^{-1}$, $\hat{\ell} = 1.0N\,m\,rad^{-1}\,u^{-1}$, and $\hat{\nu} = 0.5N\,m^{-3}\,u^{-1}\,s$.

Table 1: The execution time T (in ms per time step), memory requirements M (in kB) and the relative errors ε_{lin} and ε_{ang} after 1000 time steps as functions of the problem size

N_{elt}	N_{dof}	T	M	ε_{lin}	ε_{ang}
1	36	0.065	17	4.17	0.13
4	90	0.366	37	1.50	0.15
16	306	1.497	118	0.55	0.14
64	1170	6.640	442	0.06	0.07

From the results it can be seen that there is a steady pattern in error decay when the problem size is increased. At the same time, the computation time per integration step is sufficiently small to allow the simulation in real time (we need $T_s < 0.05s$ to allow the interactive graphic output at $20Hz$ and $T_s < dt = 0.005s$ for the interactive haptic output at 200Hz).

References

[1] A.N.F. Klimowicz: *Accurate real-time deformable tubes*, MSc Thesis, The University of Manchester, 2003, (http://www.cs.man.ac.uk/cnc/students/aklimowicz/home.html).

[2] Numerical Algorithms Group: NAG manual, Fortran Library Mark 20, 2002.

[3] O.C. Zienkiewicz, R.L. Taylor: *The Finite Element Method*, 5th ed., Butterworth-Heinemann, 2000.

VSP International
Science Publishers
P.O. Box 346, 3700 AH Zeist
The Netherlands

Lecture Series on Computer
and Computational Sciences
Volume 1, 2004, pp. 269-273

A posteriori error estimation of goal-oriented quantities for elliptic type BVPs

S. Korotov[1] and P. Turchyn

Department of Mathematical Information Technology
University of Jyväskylä
P.O. Box 35 (Agora), FIN-40014 Jyväskylä, Finland

Received 2 September, 2004; accepted in revised form 10 September, 2004

Abstract:
In engineering practice the process or object under analysis is usually modelled by means of a selected mathematical model, whose approximate solution is computed with help of a certain computer code. This approximate solution necessarily includes various errors related to the approximation itself, special features of the particular method used, round-off errors, etc. Therefore, it inevitably rises the question about the *reliability* of the computed approximations. In the present paper we give a general description of the new effective computational technology designed for a control of the accuracy of approximate solutions in terms of *goal-oriented quantities*. Such quantities are to be chosen by a user depending on solution properties that present a special interest. The technology proposed leads to effective computer codes aimed to control errors of approximate solutions obtained by the finite element method which presents nowadays the main computational tool in industrial software.

Keywords: a posteriori error estimation, goal-oriented quantity, finite element method, differential equation of elliptic type, superconvergence.

Mathematics Subject Classification: 65N15, 65N30, 65N50

1 Model Problem

We consider a model *boundary value problem (BVP) of elliptic type* as follows: Find a function $u = u(x_1, ..., x_d)$ such that

$$-\sum_{i,j=1}^{d} \frac{\partial}{\partial x_i} \left(a_{ij} \frac{\partial u}{\partial x_j} \right) = f \quad \text{in} \quad \Omega, \tag{1.1}$$

$$u = u_0 \quad \text{on} \quad \Gamma_1, \tag{1.2a}$$

$$\sum_{i,j=1}^{d} a_{ij} \frac{\partial u}{\partial x_j} n_i = g \quad \text{on} \quad \Gamma_2, \tag{1.2b}$$

where $\Omega \subset \mathbf{R}^d$ is a bounded connected domain with Lipschitz continuous boundary $\Gamma = \Gamma_1 \cup \Gamma_2$, $a_{ij} = a_{ij}(x_1, ..., x_d) \in L_\infty(\Omega)$, $i, j = 1, ..., d$, are the coefficients of the problem, $f = f(x_1, ..., x_d) \in L_2(\Omega)$, $u_0 = u_0(x_1, ..., x_d) \in H_1(\Omega)$ and $g = g(x_1, ..., x_d) \in L_2(\Gamma_2)$. The symbol n_i denotes the i-th component of the outward unit normal vector $\mathbf{n} = \mathbf{n}(x_1, ..., x_d)$ to the boundary Γ.

---

[1]Corresponding author, e-mail: korotov@mit.jyu.fi

1.1   Finite Element Solution

Let V_h be a finite-dimensional space constructed by means of a selected set of finite element trial functions defined on commonly-used finite element mesh $\mathcal{T}_h$ over Ω. We notice that space V_h is chosen so that its functions w_h vanish on Γ_1. The finite element approximation for the *primal problem* (1.1)–(1.2) is defined then as a function

$$u_h = u_h(x_1, ..., x_d) \in V_h + u_0,$$

such that

$$\int_\Omega \sum_{i,j=1}^d a_{ij} \frac{\partial u_h}{\partial x_j} \frac{\partial w_h}{\partial x_i}\, dx = \int_\Omega f w_h\, dx + \int_{\Gamma_2} g w_h\, ds \quad \forall w_h \in V_h. \tag{1.3}$$

1.2   Goal-Oriented Quantity

Engineers are often interested not only in the overall error $e = u - u_h$, but also in its local behaviour, e.g., in a certain subdomain $\omega \subset \Omega$. One way to get an information about the local behaviour of the error is to measure it in terms of specially selected *goal-oriented quantities*. The most typical quantity is presented by the integral

$$\int_\Omega \varphi(u - u_h)\, dx, \tag{1.4}$$

where $\varphi = \varphi(x_1, ..., x_d)$ is a selected weight-function such that supp $\varphi \subseteq \omega$.

2   Error Estimation of Goal-Oriented Quantity

To present the technology for estimation of the quantity given by (1.4), we describe two auxiliary problems, that must be previously solved. They consist of finding an approximate solution of the so-called *adjoint problem* and making a certain post-processing of it, and also making a post-processing of the finite element approximation u_h.

2.1   Adjoint problem and its finite element solution

Let V_τ be another finite-dimensional space constructed by means of a selected set of finite element trial functions on another standard finite element mesh $\mathcal{T}_\tau$ over Ω. We notice that space V_τ is chosen so that its functions w_τ vanish on Γ_1, and also that $\mathcal{T}_\tau$ need not to coincide with $\mathcal{T}_h$.

Consider the adjoint finite-dimensional problem as follows: Find a function

$$v_\tau = v_\tau(x_1, ..., x_d) \in V_\tau,$$

such that

$$\int_\Omega \sum_{i,j=1}^d a_{ji} \frac{\partial v_\tau}{\partial x_j} \frac{\partial w_\tau}{\partial x_i}\, dx = \int_\Omega \varphi w_\tau\, dx \quad \forall w_\tau \in V_\tau. \tag{2.1}$$

2.2 Gradient averaging procedures

On $\mathcal{T}_h$, we define the *gradient averaging transformation* $\mathbf{G_h}$ mapping the *gradient of the finite element approximation* u_h

$$\nabla \mathbf{u_h} = \left[\frac{\partial u_h}{\partial x_1}, ..., \frac{\partial u_h}{\partial x_d}\right]^T, \tag{2.2}$$

which is constant over each element of the finite element mesh, into a vector-valued continuous piecewise affine function

$$\mathbf{G_h}(\nabla \mathbf{u_h}) = [G_h^1(\nabla u_h), ..., G_h^d(\nabla u_h)]^T, \tag{2.3}$$

by setting each its nodal value as the mean (or weighted mean) value of $\nabla \mathbf{u_h}$ on all elements of the patch $\mathbf{P}(x_*)$ associated with corresponding node x_* in the mesh $\mathcal{T}_h$.

Similarly, on $\mathcal{T}_\tau$, we define the *gradient averaging transformation* $\mathbf{G}_\tau$ mapping the *gradient of the finite element approximation* v_τ

$$\nabla \mathbf{v}_\tau = \left[\frac{\partial v_\tau}{\partial x_1}, ..., \frac{\partial v_\tau}{\partial x_d}\right]^T \tag{2.4}$$

into a vector-valued continuous piecewise affine function

$$\mathbf{G}_\tau(\nabla \mathbf{v}_\tau) = [G_\tau^1(\nabla v_\tau), ..., G_\tau^d(\nabla v_\tau)]^T. \tag{2.5}$$

Such averaging transformations are widely used in the finite element calculations (see, e.g., [2], [6], [7], [11], [13], [14]). Usually they lead to computationally inexpensive algorithms.

2.3 The estimator

Our error estimation method is based upon a new estimator that has been derived and justified first in [8] (see also [10]). It estimates the quantity (1.4) by the quantity $E(u_h, v_\tau)$ given by the following formula:

$$E(u_h, v_\tau) := E_0(u_h, v_\tau) + E_1(u_h, v_\tau), \tag{2.6}$$

where

$$E_0(u_h, v_\tau) = \int_\Omega f v_\tau \, dx + \int_{\Gamma_2} g v_\tau \, ds - \int_\Omega \sum_{i,j=1}^d a_{ij} \frac{\partial u_h}{\partial x_j} \frac{\partial v_\tau}{\partial x_i} \, dx, \tag{2.7}$$

and

$$E_1(u_h, v_\tau) = \int_\Omega \sum_{i,j=1}^d a_{ij} \left(\frac{\partial u_h}{\partial x_j} - G_h^j(\nabla u_h)\right) \left(\frac{\partial v_\tau}{\partial x_i} - G_\tau^i(\nabla v_\tau)\right) \, dx. \tag{2.8}$$

The functional $E(u_h, v_\tau)$ is directly computable once the approximations u_h and v_τ are found.

3 Comments

1. Our approach is different from the techniques proposed in [1, 3, 4, 5, 12] where it is always assumed that the primal and adjoint problems are solved on coinciding meshes. Using our technique one can obtain very good estimates also for the case when the number of nodes in the mesh used for the adjoint problem is considerably smaller than the number of nodes in the mesh used for the primal problem, see [8, 10].

2. The technology proposed can be applied to another linear elliptic equations, e.g. to problems in the linear elasticity [9], provided that the averaged gradients of their solutions demonstrate certain superconvergenve phenomena.

3. The effectivity of the proposed technique, strongly increases when one is interested not in a single solution of the primal problem for a concrete data, but analyzes a series of approximate solutions for a certain set of boundary conditions and various right-hand sides (which is typical in the engineering design when it is necessary to model the behavior of a construction for various working regimes). In this case, the adjoint problem must be solved *only once* for each "quantity", and its solution can be further used in testing the accuracy of approximate solutions of various primal problems.

References

[1] M. Ainsworth, J. T. Oden. *A posteriori error estimation in finite element analysis.* John Wiley & Sons, Inc., 2000.

[2] I. Babuška, T. Strouboulis. *The Finite Element Method and its Reliability.* Oxford University Press Inc., New York, 2001.

[3] W. Bangerth, R. Rannacher. *Adaptive finite element methods for differential equations.* Lectures in Mathematics ETH Zürich. Birkhäuser Verlag, Basel, 2003.

[4] R. Becker, R. Rannacher. *Weighted a posteriori error control in finite element methods.* Preprint 96 – 1, SFB 359, Heidelberg, 1996.

[5] R. Becker, R. Rannacher. A feed-back approach to error control in finite element methods: Basic approach and examples. *East-West J. Numer. Math.*, 4, 237–264, 1996.

[6] J. Brandts, M. Křížek. Gradient superconvergence on uniform simplicial partitions of polytopes. *IMA J. Numer. Anal.*, 23, 489–505, 2003.

[7] I. Hlaváček, M. Křížek. On a superconvergent finite element scheme for elliptic systems. I. Dirichlet boundary conditions. *Apl. Mat.* 32, 131–154, 1987.

[8] S. Korotov, P. Neittaanmäki, S. Repin. A posteriori error estimation of goal-oriented quantities by the superconvergence patch recovery. *J. Numer. Math.* 11, 33–59, 2003.

[9] S. Korotov, P. Neittaanmäki, S. Repin. A posteriori error estimation in terms of linear functionals for the linear elasticity problems. *J. Numer. Math.* 1–16 (submitted).

[10] S. Korotov, P. Turchyn. A posteriori error estimation of "quantities of interest" on tetrahedral meshes, *in Proc. of the European Congress on Computational Methods in Applied Sciences and Engineering (ECCOMAS-2004), Jyväskylä, Finland (eds. P. Neittaanmäki et al.)*, 1–20, 2004.

[11] M. Křížek, P. Neittaanmäki. Superconvergence phenomenon in the finite element method arising from averaging gradients. *Numer. Math.* 45, 105–116, 1984.

[12] J. T. Oden, S. Prudhomme. Goal-oriented error estimation and adaptivity for the finite element method. *Comput. Math. Appl.* 41, 735–756, 2001.

[13] R. Verfürth. *A review of a posteriori error estimation and adaptive mesh-refinement techniques*, Wiley-Teubner, 1996.

[14] O. C. Zienkeiewicz, J. Z. Zhu. A simple error estimator and adaptive procedure for practical engineering analysis. *Internat. J. Numer. Methods Engrg.* 24, 337–357, 1987.

VSP International
Science Publishers
P.O. Box 346, 3700 AH Zeist
The Netherlands

*Lecture Series on Computer
and Computational Sciences*
Volume 1, 2004, pp. 274-278

Direct solver for an inverse problem of the type of the transmission

H. Koshigoe[1]

Department of Urban Environment Systems,
Faculty of Engineering,
Chiba University,
1-33 Yayoi, Inage, Chiba, Japan

Received 31 July, 2004; accepted in revised form 20 August, 2004

Abstract: The purpose of this paper is to control a temperature of a material Ω_0 when the domain Ω_0 is occupied by a dangerous material and one can not directly control its temperature. To do so, we set two materials $\Omega_i (i = 1, 2)$ outside Ω_0 in order to control the temperature of Ω_0. From this point of view, an inverse problem which decide the boundary value of $\Omega_i (i = 1, 2)$ from the information of the interior Ω_0 is derived. In this paper the direct solver for solving the inverse problem will be shown and itstechnique is based on the domain decomposition and succcessive elimination of lines which we call SEL.

Keywords: transmission, inverse, control, direct, SEL

Mathematics Subject Classification: 65F05, 65F22

1 Introduction

In this paper we establish a mathematical formula of finite difference solutions for the inverse problem that determines the boundary value for the transmission problem(cf.[4]) and show a numerical result. Let $\Omega = \Omega_0 \cup \Omega_1 \cup \Omega_2 \subset R^2$ which denotes a cross section of a column in R^3 (cf. Figure 1) and ϵ_i be a positive constant (i=0,1,2). Transmission problem which we consider here are three Poisson equations in each domain Ω_i $(i = 0, 1, 2)$ and the transmission conditions on the interface γ_1, γ_2, and Dirichlet bounadary conditions. We now present an inverse problem as follows.
Problem I. Let $f \in L^2(\Omega)$ and $g_i \in H^{1/2}(\Gamma \cap \overline{\Omega_i})$. For a given $u_0 \in H^2(\Omega_0)$ such that $-\epsilon_0 \Delta u_0 = f$ in Ω_0, find the boundary value $\phi_i \in H^{1/2}(\Gamma_i)$ $(i = 1, 2)$ and $u_i \in H^1(\Omega_i)$ $(i = 1, 2)$ satisfying

$$
\begin{aligned}
-\epsilon_0 \Delta u_0 &= f_0 \quad \text{in } \Omega_0 \\
-\epsilon_i \Delta u_i &= f_i \quad \text{in } \Omega_i \quad (i = 1, 2), \\
u_i &= u_0 \quad \text{on } \gamma_i \quad (i = 1, 2), \\
\epsilon_i \frac{\partial u_i}{\partial n} &= \epsilon_0 \frac{\partial u_0}{\partial n} \quad \text{on } \gamma_i \quad (i = 1, 2), \\
u_i &= g_i \quad \text{on } \Gamma \cap \overline{\Omega_i} \quad (i = 0, 1, 2), \\
u_i &= \phi_i \quad \text{on } \Gamma_i \cap \overline{\Omega_i} \quad (i = 1, 2).
\end{aligned}
\tag{1}
$$

---

[1]E-mail: koshigoe@faculty.chiba-u.jp

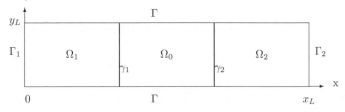

Figure 1 Domain Ω and Boundary Γ

The purpose of this paper is to show the direct method to constract the finite difference solution of the above inverse problem.

2 Distribution formulation

Let χ_i $(i = 0, 1, 2)$ be a characteristic function in Ω_i $(i = 0, 1, 2)$ and set $a = \epsilon_0 \chi_0 + \epsilon_1 \chi_1 + \epsilon_2 \chi_2$. Then the distribution theorem shows that Problem I is equivalent to

Problem II Find $u \in H^1(\Omega)$ and $\phi_i \in H^{1/2}(\Gamma_i)$ $(i = 1, 2)$ such that

$$\text{div}(a\nabla u) = f \quad \text{in } \Omega$$
$$u = g_i \quad \text{on } \Gamma \cap \overline{\Omega_i} \quad (i = 0, 1, 2), \tag{2}$$
$$u = \phi_i \quad \text{on } \Gamma_i \ (i = 1, 2)$$

Moreover we intriduce the function $G_i \in H^2(\Omega_i)$ $(i = 0, 1, 2)$ satisfying

$$G_i = G_0 = u_0 \quad \text{on} \quad \gamma_i$$

respectively, set $u_i = v_i + G_i$ (i=0,1,2) and $v = \epsilon_0 \chi_0 v_0 + \epsilon_1 \chi_1 v_1 + \epsilon_2 \chi_2 v_2$. Then using the fact that $v_i = 0$ on γ_i(i=1,2) and the following equation

$$-\Delta v = -(\epsilon_0 \, \Delta v_0)\chi_0 - (\epsilon_1 \, \Delta v_1)\chi_1 - (\epsilon_2 \, \Delta v_2)\chi_2$$

$$+(\epsilon_1 \frac{\partial v_1}{\partial n} - \epsilon_0 \frac{\partial v_0}{\partial n})\delta_{\gamma_1} + (\epsilon_2 \frac{\partial v_2}{\partial n} - \epsilon_0 \frac{\partial v_0}{\partial n})\delta_{\gamma_2}$$

holds in the sense of distribution(cf.[5]), we get a equivalent problem

Problem III Find $v \in H^1(\Omega)$ and $\phi_i \in H^{1/2}(\Gamma_i)$ $(i = 1, 2)$ such that

$$\begin{aligned}
-\Delta v &= \sigma_1 \delta_{\gamma_1} + \sigma_2 \delta_{\gamma_2} + f \quad \text{in } \Omega \\
v &= 0 \quad \text{on } \gamma_i \ (i = 1, 2) \\
v &= \epsilon_i \cdot g_i \quad \text{on } \Gamma \cap \overline{\Omega_i} \quad (i = 0, 1, 2), \\
v &= \epsilon_i \cdot \phi_i \quad \text{on } \Gamma_i \ (i = 1, 2)
\end{aligned} \tag{3}$$

where δ_{γ_i} denotes the Dirac distribution supported on γ_i(i=1,2), n is the unit normal vector on γ_i directed from Ω_0 to Ω_i (i=1,2) and $\{\sigma_i\}$ is respectly defined by

$$\sigma_1 = -\epsilon_1 \frac{\partial G_1}{\partial n} - \epsilon_0 \frac{\partial G_0}{\partial n}$$

$$\sigma_2 = -\epsilon_2 \frac{\partial G_2}{\partial n} - \epsilon_0 \frac{\partial G_0}{\partial n}.$$

Hereafter we consider Problem III.

3 Direct solver and mathematical formula

The direct solver we show here is based on the domain decomposition and the successive elimination of lines which we call SEL ([1]-[3]).

Step 1:Domain decomposition
We divide the domain Ω into two parts D_1 and D_2 which consit of $\Omega_0 \cup \Omega_1 \cup \gamma_1$ and $\Omega_0 \cup \Omega_2 \cup \gamma_2$ (see below Figure 2).

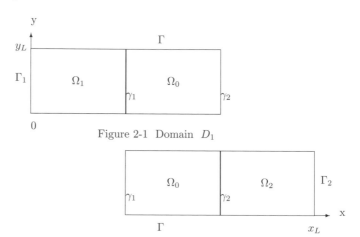

Figure 2-1 Domain D_1

Figure 2-2 Domain D_2

Step 2: Construction of a direct solver
In this subsection we show the mathematical formula of the finite difference solution for the inverse problem: Let Ω be the unit square in R^2 , $h \in R$ be a mesh size such that $h = 1/n$ for an integer n and

$$\Omega_0 = \{(x,y)|\ n_1 h < x < n_2 h,\ 0 < y < 1\ \},$$
$$\Omega_1 = \{(x,y)|\ 0 < x < n_1 h,\ 0 < y < 1\ \},$$
$$\Omega_0 = \{(x,y)|\ n_2 h < x < nh,\ 0 < y < 1\ \},$$

Then we propose the finite difference approximation of Problem III as follows.
Problem F. Find $\{v_{i,j}\}$ $(1 \le i,\ j \le n - 1)$ such that

$$-\left(\frac{v_{i+1,j}\ -\ 2\,v_{ij}\ -\ v_{i-1,j}}{(\Delta x)^2}\ +\frac{v_{i,j+1}\ -\ 2\,v_{ij}\ -\ v_{i,j-1}}{(\Delta y)^2}\right) = f_{i,j} + \frac{\sigma_{m_1,j}}{\Delta x}\,\delta_{i,m_1} + \frac{\sigma_{m_2,i}}{\Delta x}\cdot\delta_{i,m_2} \quad (4)$$

$$v_{n_1,j} = 0,\quad v_{n_2,j} = 0 \quad (j = 1, 2, \cdots, n - 1) \tag{5}$$

where $\delta_{i,j}$ denotes Kronecker' delta and $\Delta x = \Delta y = h$.

Let us introduce the $(n - 1)$ vectors $\{V_i\}$ instead of the $(n - 1)^2$ unknowns $v_{i,j}$. For each i, set $V_i = {}^t[v_{i,1},\ v_{i,2}, \cdots,\ v_{i,n-1}]$ and also $F_i = (f_{i,j} + \frac{\sigma_{n_1,j}}{\Delta x}\cdot\delta_{i,n_1} + \frac{\sigma_{n_2,j}}{\Delta x}\cdot\delta_{i,n_2})_{j=1,2,\cdots,n-1}$ $(1 \le i \le n - 1)$

Theorem1 (SEL) From the equation (14), it follows that for each k $(1 \le k \le n - 1)$,

$$V_k = P\left(D(n-k,1)PV_0 + D(k,1)PV_n + \sum_{i=1}^{k-1} D(n-k,\ i)\ P\ F_i\ + \sum_{i=k}^{n-1} D(k,\ n-i)\ P\ F_i\right) \tag{6}$$

holds. Here $D(l, i)$ $(1 \leq l, i \leq n - 1)$ is a diadonal matrix and P is an othogonal matrix, a_j $(1 \leq j \leq n - 1)$ is determined by the eigen-values of the matrix A. More concletely set

$$a_j = \operatorname{arccosh}(2 - \cos(j\pi/n)))$$

and $D(l, i)$ is defined by

$$D(l, i) = \operatorname{diag} \left[\frac{\sinh(l\, a_j)\, \sinh(i\, a_j)}{\sinh(n\, a_j)\, \sinh a_j} \right]_{1 \leq j \leq n-1}.$$

And the components $p_{i,j}$ of the othogonal matrix P are of these typeG

$$p_{l,j} = \sqrt{\frac{2}{n}}\, \sin\left(\frac{l\, j\, \pi}{n}\right) \quad (1 \leq l, j \leq n - 1).$$

Using SEL in the domain D_1 and D_2 coupled with (15), we have mathematical formulas to calculate $V_0 = \{\phi_1(jh)\}$. and $V_n = \{\phi_2(jh)\}$.

Theorem 2   The finite difference solutions of Problem III are calculated directly as follows.

$$PV_0 = - D(n_2 - n_1, 1)^{-1} \left(\sum_{i=1}^{n_1-1} D(n_2 - n_1, i)PF_i + \sum_{i=n_1}^{n_2-1} D(n_1, n_2 - i)PF_i \right)$$

$$PV_n = - D(n_2 - n_1, 1)^{-1} \left(\sum_{i=1}^{n-n_2-1} D(n_2 \quad n_1, i)PF_{n\ i} + \sum_{i=n-n_2}^{n-n_1-1} D(n - n_2, n - n_1 - i)PF_{n-i} \right)$$

$$(7)$$

4   Numerical experiment

Let $u(x, y) = \sinh(\pi x) \sin(\pi y)$ $((x, y) \in \Omega)$ and $\epsilon_i = 1$ $(i = 0, 1, 2)$. We also denote the approximate solution of Problem III by u_h and use the absolute error as follows.

$$\|u - u_h\|_\infty = \max_{i,j} |u(i\, h, j\, h) - u_{i,j}|.$$

We get the following table by use of Maple 8.

mesh size	$\|u - u_h\|_\infty$
1/16	3.602408×10^{-2}
1/32	9.733142×10^{-3}
1/64	2.480822×10^{-3}

Acknowledgment

The author wishes to thank Prof.K.Kitahara of Kogakuin University and Dr.K.Hirayama of Chiba University for significant discussion.

References

[1] H. Koshigoe, *Direct solver based on FFT and SEL for diffraction problems with distribution.* Computational Science-ICCS2004, Part II,105-112(2004).

[2] H. Koshigoe, *Direct method for solving a transmission problem with discontinuous coefficient and the Dirac Distribution.*Computational Science-ICCS2004, Part III,388-400(2003).

[3] H. Koshigoe and K. Kitahara,*Numerical algorithm for finite difference solutions constructed by fictitious domain and successive eliminations of lines.*Japan SIAM, Vol.10, No.3,211-225(2000)(in Japanese).

[4] J.L. Lions, *Optimal control of systems governed by partial differential equations,*Springer-Verlarg (1971).

[5] S. Mizohata, *The theory of partial differential equations,* Cambridge at the University Press(1973).

VSP International
Science Publishers
P.O. Box 346, 3700 AH Zeist
The Netherlands

Lecture Series on Computer
and Computational Sciences
Volume 1, 2004, pp. 279-282

An Equation of State of Polymeric Melts

M. Kosmas[1]

Chemistry Department ,University of Ioannina, Greece

Received 7 March, 2004; accepted in revised form 10 March, 2004

Abstract: The Gaussian molecular model for the description of systems of macromolecular chains is presented. The long polymeric chains each made of a large number of units interacting between them can obtain many different conformations all of which determine the average macroscopic behaviour. Microscopic parameters in terms of which the macroscopic properties will be analyzed are the number N of units of each chain, proportional to the length and the molecular weight M of the chains and the intensity u of the interactions between the units of the chains. Perturbation theory in the parameter u is used and the series of one loop diagrams is summed for the determination of the partition function and the pressure of the system of n chains occupying a volume V. Comparison with experimental results are included

Keywords: Equation of state, polymer melts

PACS: Equations of state:specific substances 64.30.+t, Statistical mechanics:classical 05.20.-y

The model and results

The basic aim of statistical thermodynamics is to describe the macroscopic properties of matter starting with microscopic models of molecular origin. The writing of a microscopic model is not usually the difficult part of the effort but unfortunately this is not true also for the determination of its average properties which describe the macroscopic observable properties of the system. The reason for this is the large number of different microscopic configurations of the systems capable to reach even the huge Avogadro's number 10^{23} in quantities of few grams. The probability of each configuration depends on the energy E_i according to the Boltzmann Law $P_i = A \exp\{-E_i/kT\}$ so that the macroscopic properties X can be writtens as the averages $X = \sum X_i P_i$ over the corresponding properties of all configurations i. The summation in case of a continuous spectrum of different properties becomes an integration.

The known Gaussian model of polymer chains describes a flexible polymer chain consisting of N units, with a set of consecutive points connected with segments of lengths ℓ. A chain configuration is precisely determined if we fix all position R_i of the ends of the units of the chains as it is shown in the Fig. 1.

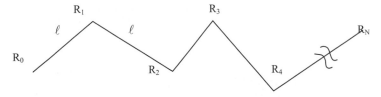

Fig.1. A configuration of a macromolecular chain of N units and segment length ℓ

[1] E-mail: mkosmas@cc.uoi.gr

The probability of each configuration is given by the product of Gaussian functions of all segmental units securing that the segment length is lying about a certain value ℓ .

$$Po\{Ri\} = A \exp\{ - (3/2 \, \ell^{\,2}) [(R_0\text{-}R_1)^2 + (R_1\text{-}R_2)^2 + \cdots (R_{N\text{-}1}\text{-}R_N)^2]\} \tag{1}$$

A is a normalization constant. This ideal probability forces the units to be consecutive, establishing thus the nature of the long chain. It differentiates the case of polymers where the units of the chain are connected and form the chain, from that of a liquid where the units can belong anywhere in the available volume. Average properties can be determined from this probability if we average by integrating over all possible positions R_i of the ends of the segments of the chain. Known result of this ideal probability is the probability of finding the end at a specific point R given by

$$P(R) = (3/2\pi \, \ell^{\,2})N \exp\{ - (3/2 \, \ell^{\,2}) [(R)^2]\} \tag{2}$$

With the mean end to end square distance of the form $<R^2> = N \ell^{\,2}$. Notice that linear dimension scales as $\sim N^{1/2}$ expressing the random nature of the configurations of the chain. When extra interactions are present they are expressed as proper potentials the simplest of which are the delta function potentials of the structure

$$P = Po \exp\{ -u \sum_{all} \delta (Ri\text{-}Rj) \} \tag{3}$$

Intrachain interactions   Interchain interactions

And describe both intra and inter interactions of the units of the chains. Basic property is the configurational partition function give by the multple integral over all positions of the units of the system.

$$Z= \int\!\!\int \; \Pi \, dRi \, P\{Ri\} \tag{4}$$

From kTlnZ which is the free energy of the system the pressure P can be found from the differentiation with respect to volume as

$$P =kTdlnZ/dV \tag{5}$$

And it is a function of the volme V the number n of the chains and the microscopic parametes u and ℓ .

A way to find a solution to the problem is to expand the exponential term of Eq. and solve the corresponding linearized problems. Each delta function brings in contact two points from the same or different chains producing conformations with one or more loops which are described by diagrams representing chains with loops. The first order perturbation theory produces for example the conformation shown Fig2. Two loops coming from the action of two delta functions produce the two loops for a single chain shown in Fig.3.

Fig.2 one loop   Fig.3. Two loops-One chain

Fig.4. One loop –Many chains

Of importance are the one loop diagrams shown in Fig.4 produced from one, two or more chains and having the simplest possible structure of one loop. We sum the contributions of these diagrams and give expressions valid to a broad range of the interaction parameter u. An analytical expression for the pressure P is found in this way in terms of an integral including the Debye function $D(p) = \{exp(-p)-1+p\}/p^2$. The latter is approximated with $D(p) =1/(2+p)$ and the following analytical expression is given

$$p =(c/10) + x c^2 -r[1+(1+20 xc)^{1/2}(10xc-1)], \qquad (6)$$

where $p=(P/kT)10^{-22}$ is a reduced pressure and the density variable $c=(n/V)10^{-21}$ with n the number of macromolecules in the volume V of the system. The other two parameters are: $x=uN^2 10^{20}$ and $r = (2\ 3^{1/2})\ 10^{-22}/[\pi(<R^2>_o)^{3/2}]$, $<R^2>_o=N\ell^2$ with $<R^2>_o=N\ell^2$ the mean end tot end square distanc eof a polymer chain.

Comparing the results of the theoretical equation with experimental results of Polystyrene we take for samples with three different molecular weights M the graphs of Fig.5 An excellent agreement is achieved only if we change the c/10 term into a ac term of order 105 moe . This can be explained from the real nature of polymer melts where the first linear term represents the real liguid nature with its viscoelastic behavior and much large response to pressure changes.

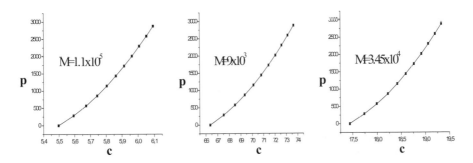

Fig 5. The fitting of experimental data (dots) of the three polystyrenes together with the best lines from the theoretical equation $p =a c + x c^2 -r[1+(1+20 xc)^{1/2}(10 xc-1)]$, taken from Eq.6 with the replacement of the c/10 with a ac term. The three adjustable parametrs a, x and r found from the graphs of the three polymers are given in Table.

M	$\alpha.10^{-4}$	$\alpha/M\cdot 10^1$	$x\cdot 10^{-3}(L)$	$x/M^2 10^6(L)$	$r\cdot 10^4(L^{-1})$	$rM^{3/2}\cdot 10^{-4}(L^{-1})$
$1.1x10^5$	6.378	5.80	12.556	1.038	9.0	3.28
$3.45x10^4$	2.020	5.81	1.253	1.053	50.8	3.26
$9x10^3$	0.508	5.64	0.083	1.025	384	3.28

Table. The adjustable parameters a,x, and r for the three polystyrene samples and their scaling dependence on the molecular weight M.

References

[1] The effect of chain correlations on the size of polymer coils in Binary Polymer Blends, G. Garas, M. Kosmas, J. Chem. Phys. 1996,105,4789
[2] "On the miscibility of chemically identical linear homopolymers of different size" C. Vlahos and M. Kosmas, polymer, 2003,44,503
[3] Comparison of the stability of chemically identical and different homopolymers , M. Kosmas,C.Vlahos, J.Chem.Phys.2003,119,4043

VSP International
Science Publishers
P.O. Box 346, 3700 AH Zeist
The Netherlands

Lecture Series on Computer
and Computational Sciences
Volume 1, 2004, pp. 283-287

A Model and a Numerical Solver for the Flow Generated by an Air-Bubble Curtain in Initially Stagnant Water

C. Koutitas[†] (Dept. of Civil Enineering) M. Gousidou, (Dept. of Mathematics)
Aristotle University of Thessaloniki, Thessaloniki,54006, Greece

Received, June 20th 2004, accepted August 20th 2004.

Abstract : The evolution of an air-bubbles curtain in an initially stagnant water body   and
the generated fully interactive two-phase flow is investigated. The flow is simulated using an Eulerian formulation
and a Finite Differences Navier-Stokes numerical solver for the water phase and a Lagrangian formulation for the
air bubbles phase. Equilibrium conditions are assumed for the bubbles motion (use of kinematic, conditions only).
The air phase is acting on the water by mass displacing, local shearing and regulating the turbulence, while the
water phase is acting on the air bubbles by transporting it by the local deterministic and stochastic velocities. The
numerical solution algorithm is an explicit one and computationally feasible for a large number of material
particles (10^4) on a PENTIUM IV micro computer. The numerical experiments lead to results comparable to
analytic solutions, prove the sensitivity of the model to the eddy viscosity values, and project the influence of the
bubbles diameter to the resulting hydrodynamics.

Keywords: Navier Stokes solver, Particles Technique, 2Phase Flow, Finite Difference
Mathematical Subject Classification :76D25,76T10,82C80.

1. Introduction

The flow domain is a 2DV one , a vertical slice of a water body with predefined geometry, confined by
free surface and perimetric shore and bottom boundaries. A source of air bubbles is located near the
bottom boundary. The air is introduced through the pipe in the form of a «curtain» of small bubbles
moving upwards under the buoyancy force. The discharge of air and the initial bubbles diameter is
defined.   This is a case of two-phase-flow with full interaction between the two phases (air-water).
Contrary to other cases of two phase flows , like water and oil, where the two liquids have finite
geometry and volume, and they are separated by an interface describable by a smooth mathematical
function, in the present case the second phase is composed by billions of almost spherical bubbles of
infinitesimal size (order of 1mm) The air bubbles are driven by their buoyancy, the hydrodynamic
resistance and the fluid velocity generated secondarily by their influence on the stagnant water.
The influence of the air phase on the water phase is triple:
a) the air bubbles displace the water mass
b) the air bubbles exercise on the nearby water a force equal to their buoyancy, counteracted by the
hydrodynamic resistance on them and
c).the air bubbles influence the development of free turbulence.

2. The proposed simulation and the numerical solution

The air phase in the two phase flow has a special physical configuration making very difficult the
application of the classical fluid mechanics equations on that. The proposed simulation is based on the
«Tracer» technique [2],[3],[4],[5], ie the tracking of the trajectories of a large number of material
particles in Lagrangian coordinates, simulating in a statistical manner the "continuum" of the air phase.
In our case the simulation does not refer to a continuum but to the discrete air bubbles.

[†] Corresponding author. Prof. Dept. Civil Eng. E-mail. koutitas@civil.auth.gr

Evidently, the number of the particles used does not coincide with the real number of the bubbles, but each particle represents a certain group of bubbles and the accuracy of the simulation is proportional to the number of "particles" used in the simulation.

The Tracer technique can be applied either for passive or active particles. In the case of air bubbles the particles are active as

a. their volume is considerable and it influences the pre-existing mass continuity by introducing, locally, non-negligible volume of air, as a source term and by consequently mobilising the water phase ,

b. they act dynamically on the surrounding water , by applying each one a force equal and opposite to the hydrodynamic resistance exercised by the water on the bubble. As we assume that there is an equilibrium of buoyancy force and hydrodynamic resistance, the force applied by each particle on the surrounding water is equal to the buoyancy force of the bubbles represented by the moving particle.

c. they regulate the development of free turbulence in the mobilised water masses. The model is synthesized by the following equations

Equations of the water phase

Mass continuity equation, in 2D space x,y

$$\partial p/\partial t + \partial u/\partial x + \partial v/\partial y = VOL * \partial c/\partial t \tag{1}$$

p is the reduced «pressure» , introduced in the classical continuity equation (div $\mathbf{V} = 0$) according to the solution of Navier Stokes equations by the method of «pseudo-compressibility»[1],[7] The r.h.s. is a source term, due to the second phase (air bubbles), proportional to the normalized (per unit area) rate of change of concentration «c» of the air bubbles, responsible for the displacement of water masses.

Forces equilibrium equations along Ox, Oy. They have the form

$$Du/Dt = -\lambda\partial p/\partial x + \partial(N\partial u/\partial x)/\partial x + \partial(N\partial u/\partial y)/\partial y + F_x \tag{2}$$

$$Dv/Dt = -\lambda\partial p/\partial y + \partial(N\partial v/\partial x)/\partial x + \partial(N\partial v/\partial y)/\partial y + F_y \tag{3}$$

N is the eddy viscosity term, estimated via a selected closure technique, F_y is the distributed mass force , resultant of the buoyancy forces on the air bubbles, also normalized (per unit area) , and λ is a fictitious compression waves celerity (due to the assumed pseudo-compressibility) , $O(\lambda) >> O(u)$.

The eddy viscosity N is computed via the following equation

$$N = k Dx^3 |\text{grad } \omega| \text{ , where } \omega \text{ is the local vorticity, and } O(k)=0.1 \tag{4}$$

Equations for the air bubbles phase

The air bubbles are simulated by a large number of particles, introduced at the defined location (near the bottom) at a fixed rate. Each particle represents, according to the actual air discharge and the assumed diameter of the bubbles, a number (BPP) of bubbles, or a certain amount of air volume (VOL), it has a defined buoyancy velocity w_b ,characteristic of the bubbles contained in the particle, and it exercises on the water a force equal to the resultant buoyancy of the contained bubbles .

The full dynamic analysis of the motion of a particle, would involve the solution of the Newton equations relating the forces acting on each bubble to the bubble acceleration, in the form [6]

$$- F_{xk} = d^2x_k/dt^2 \ m_k \tag{5}$$

$$- F_{yk} + B = d^2y_k/dt^2 \ m_k \tag{6}$$

where x_k, y_k the coordinates of the kth particle m_k its mass and F_{xk}, F_{yk} the forces acting on it along the x,y directions and B the buoyancy force on the particle.

The F_x , F_y force components are related only to the hydrodynamic resistance on the moving bubbles .

$$F_{xk} = 1/2C_d \rho \pi R^2 (dx_k/dt - u) |dx_k/dt - u| \tag{7}$$

$$F_{yk} = 1/2C_d \, \rho \, \pi \, R^2 \, (dy_k/dt - v) \, |dy_k/dt - v| \qquad (8)$$

where C_d is the drag coefficient, R the bubble diameter and ρ the fluid density.
The diffusion is simulated according to Einstein by random walks with velocity, which is a stochastic magnitude uniformly distributed between +U, -U, where U is related to the local N value through the well known relation ,

$$U = \sqrt{(6N/Dt)} \qquad (9)$$

Assuming that the bubble mass and inertia are small and that the bubble follows the water motion (deterministic and stochastic) with the exception of the vertical direction along which the air bubble is moving at a relative velocity equal to w_b , it is not necessary to solve the full dynamic equations and only the kinematic conditions define the position of each bubble (or particle).
The numerical solution of the equations for the water phase in the fixed (Eulerian) frame of reference is done by an explicit 2nd order F.D. scheme on a staggered (Arakawa "C" grid) [5]. The velocity components are computed on the mesh sides and the pressure on the mesh center .
During each time step the solution is iterated until convergence is reached for the instant pressure distribution[7] .The air phase is treated separately in a Lagrangian frame of reference.
During each time step, the number of particles in the flow domain is increased by a certain amount, and the "newcomers" are placed at the location of the «air source».During each time step the particles move according to the following procedure, a) they move along the x direction, advected by the existing (generated by them in a secondary manner) local water velocity component «u», and they also make a random walk simulating their diffusion process in accordance with the local N value, b) they move along the y direction in the same way (with the local water «v» velocity component and with the random velocity related to the diffusion coefficient N as for x), but in excess to that motion they also move upwards with the buoyancy velocity w_b, computed by the Stoke' s law.
At a new time level (n+1) the coordinates of a particle «k» are computed from the previous ones (time level n) by the following relations

$$x^{n+1}_k = x^n_k + u_k \, Dt + U(1-2rnd(1)) \, Dt \qquad (10)$$

$$y^{n+1}_k = y^n_k + (v_k + w_b) \, Dt + U(1-2rnd(1)) \, Dt \qquad (11)$$

where rnd(1) is a random number uniformly distributed between 0 and 1, and the u_k, v_k deterministic velocity components are estimated by interpolation of the nearest computed mesh values.
At each time step, the volume concentration of air bubbles inside a cell of the discretisation grid is computed by the summation of the number of particles inside the cell, times the volumetric parameter (VOL), relating each particle to the real volume of air that it represents.
Each time step the force exercised along the vertical (y) direction by the particles inside a cell is computed by the summation of the buoyancy force exercised by each particle over the number of the particles in the cell .
The force that a particle «k» exercises on the water, assuming the validity of the Stoke's formula is given by

$$SHR = BPP * (3\pi\mu D \, w_b) \qquad (12)$$

where D is the bubble diameter, w_b its limit buoyancy velocity , BPP is the parameter giving the Bubbles Per Particle and μ the dynamic viscosity.
The sum of those forces inside a control volume (a cell) produces the distributed (mass) force component on the water to be used in the equilibrium equations

$$F_{y \, i,j} = \Sigma F_{yk} \quad \text{(for k inside the cell with coordinates i, j) or} \qquad (13)$$

$$F_{y \, i,j} = SHR \, C_{i,j} \quad (C_{i,j} \text{ is the concentration, or the number of particles in i,j cell)} \quad (14)$$

Particles reaching the borders of the flow domain are not included in the subsequent computations as those reaching the surface are diffused in the atmosphere and those reaching the lateral boundaries are not returning in the flow domain

3. Applications . Current patterns and bubbles concentrations

A computer code in FORTRAN is synthesized applied and tested against analytic solutions, for the case of an orthogonal flow domain with open perimetric boundaries, a closed bottom boundary and a bubble source located near the bottom. The assumed water depth is 20m.

The flow parameters used are defined in the above paragraph, for various D and Q (bubble diameters and air flow rates).

An illustration of the transient evolution of the bubble curtain for D=0.0005m and Q=0.1m2/sec is contained in Figure 1.

An operational application aims to the investigation of the importance of the bubbles diameter to the resulting current pattern. The same air discharge, 10 lit/sec, is supplied through various bubbles diameters D= 0.0005, 0.001,and 0.002 m. The produced current patterns indicate that the finer the segmentation of the air to bubbles, the stronger are the currents and mixing in the area .

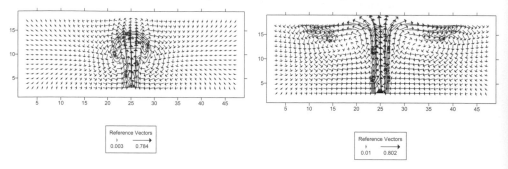

Figure 1 :. Evolution of the air curtain for t=35sec and 70 sec. Q=0.01, D=.0005m

Finally the model results are compared to existing analytical solutions for buoyant vertical plumes. The axial concentrations on the water surface relative to those magnitudes at the source are computed for four air supply rates Q= 0.0033, 0.01, 0.033 and 0.1 m^2/sec) and compared to the analytic solutions. A satisfactory comparison between analytic and numerical solutions is found.

4. Conclusions

A simple and efficient numerical solver is constructed and applied for the Eulerian solution of the water phase motion and the Lagrangian (grided) solution of the air bubbles phase in the two phase flow of an air curtain in an initially stagnant fluid.

The comparison of the numerical solution to integral values of the analytical solution documents the efficiency of the method.

The numerical experiments reveal the sensitivity of the results to the assumed eddy viscosity levels and to the bubble diameter.

A comparison of the present results to the ones obtained by the full dynamic treatment of the air particles , and to a more sophisticate turbulence closure is under way.

References

[1] Chorin A., Numerical Solution of the Navier Stokes Equations, Math. Comp.,22, pp 745-762, 1968

[2] Druzhinin O., Elghobashi S., Direct Numerical Simulation of Bubble Laden Turbulent Flows Using the 2-Fluid Formulation, Phys. Of Fluids, 10(3),685,1998

[3] Gousidou M., Koutitas C., A Computational Model for the Forced Resuspension of Seabed Sediment Caused by Trawling, 5th Int. Conf. on Ecohydraulics, Madrid, Sept. 2004

[4] Koutitas C., Gousidou M., Mathematical Modelling of Oil Slick Pollution in Kavala bay, Proc, Int. ASME Conference on Modelling and Simulation Athens 1984

[5] Koutitas C., Mathematical Models in Coastal Engineering, Pentech Press 1985, pp146

[6] Michaelidis E., Review - The Transient Equation of Motion for Particles, Bubbles and Droplets, J. of Fluids Engineering, Vol.119, pp233,247, 1997

[7] Shen J., Pseudo-Compressibility Methods for the Unsteady Incompressible Navier Stokes Equations, www.math.purdue.edu/~shen/pub, 2001

VSP International
Science Publishers
P.O. Box 346, 3700 AH Zeist
The Netherlands

*Lecture Series on Computer
and Computational Sciences*
Volume 1, 2004, pp. 288-293

Advanced magnetic resonance–based computational modelling in medicine: technology contributing to improved clinical outcomes

A.P. Kurmis[1,2], Slavotinek JP[3], Reynolds KJ[2], Hearn TC[4].

[1] Department of Orthopaedics, Division of Surgery, Repatriation General Hospital, Daw Park, SA 5041, Australia.
[2] School of Informatics and Engineering, Faculty of Science and Engineering, Flinders University, Bedford Park, SA 5042, Australia.
[3] Department of Medical Imaging, Flinders Medical Centre, Bedford Park 5042, SA, Australia.
[4] Orthopaedic Research Unit, Repatriation General Hospital / Flinders Medical Research Institute, Daw Park, SA 5041, Australia.

Received 30 July, 2004; accepted in revised form 19 August, 2004.

Keywords: three-dimensional imaging; 3DMR; image reconstruction; 3D medical imaging.

The integration of computers and computer applications into everyday living situations continues to shape our lives at a rapid pace. The unprecedented rate of technological growth is nowhere more apparent than in the field of clinical medicine. With ever-increasing expectations of quality of life and health, society demands that this profession keep abreast with technology, providing an ever-present pressure to find faster, more accurate and more cost-effective ways of preventing, identifying, managing and treating all manner of injuries, ailments and afflictions. Within this holistic patient care framework, the importance of accurate diagnosis of injury and disease, and effective patient management planning cannot be overstated.

Sectional depiction of *in vivo* anatomy has been available through the field of radiology for decades, using either computed tomography (CT) or more recently magnetic resonance (MR) imaging. The accuracy and reliability of both imaging tools in this form has been well established, documented and accepted. The subsequent capability to progress planar (sectional) images to three-dimensional (3-D) displays has been reported in the medical literature as far back as the late 1970s [1]. In those formative stages, many regarded 3-D image displays as simply 'novel' transformations or manipulations of existing datasets, pursued largely for publicity purposes [1]. They were viewed as being of limited value to clinical practice, representing an unjustifiable time and monetary expense with very little perceivable benefit. The accuracy and reliability of these preliminary works was openly questioned and the early drive for routine clinical acceptance effectively stalled.

Contemporary advances in software and hardware processing capabilities have revitalised interest in the potential application of 3-D computer displays reconstructed from parallel imaging sections and such technology has already been incorporated into many facets of modern medicine. While the use of 3-D reconstructed CT images is now almost universally practiced and may be considered commonplace [2], owing to the ease with which data can be accessed and interactively segmented, the level of technical progression of comparative MR-based applications is substantially less. Reflecting the relative newness of the basic sectional technology, advancement in the area of 3-D MR imaging remains a considerably under-explored diagnostic option, historically limited by difficulties in data access (encrypted proprietary image formats), tissue separation (contrast resolution) and registration, file management and handling (image size) and high associated cost. Ongoing developments in post-scanner software however have allowed much wider access to 3-D reconstructed MR imaging. While driven principally from within radiological circles, many other traditional allied disciplines including orthopaedics, rheumatology, oncology, neurology, general surgery, anatomy, histology and tissue pathology have expressed vested interest in the development, advancement and more widespread availability of such a clinical tool.

---

[1] Corresponding author: Dr. Andrew P. Kurmis - Orthopaedic Research Fellow. E-mail: andrew.kurmis@flinders.edu.au

While successful use has already been reported in 3-D reconstructed MR-based assessment of the human heart and brain, the extension of this technology to musculoskeletal applications is limited. Despite an increasing global interest, reflected by the number of articles being published in the international literature suggesting the potential extension of 3-D MR techniques to musculoskeletal medicine, there is little evidence of substantive work to underpin the validation of many of the proposed software-based models. Owing to inherent differences in image acquisition and data constitution, the successes of 3-D CT cannot be simply extrapolated to MR imaging. Thus the impressive results demonstrated in some 3-D MR applications, those without evidence of technique validation, may be considered unsupported and perhaps unreliable.

In its conventional planar (two-dimensional) format, MR imaging provides an accurate and reliable form of non-invasive assessment of *in vivo* anatomical and selected physiological processes, and is now widely regarded as the 'gold standard' method for soft-tissue evaluation. MR-based approaches are free from the morbidity and post-procedural debilitation associated with exploratory diagnostic surgical methods, and do not utilise an ionising or cavitating energy source (as is the case with other contemporary imaging modalities). Support for the latter consideration inadvertently gained much patient-driven momentum during the second half of the 1990's as public awareness campaigns highlighting the potentially carcinogenic risks associated with ionising radiation sources (in various forms) and notably influenced the mindset of the general lay population [1] with patients today increasingly mindful of the need to avoid unnecessary medical radiation exposure.

More recent advances in MR-based 3-D reconstructive imaging continue to suggest a potential alternative to conventional diagnostic approaches. Preliminary research in this area has suggested that 3-D applications of MR imaging may represent the progression of non-invasive *in vivo* assessment to a new plateau of clinical efficacy. It has been postulated that 3-D MR has the potential to offer significant clinical benefit in influencing prospective patient management, principally through facilitating improvement in both the standard of general diagnostics and surgical planning [1] by presenting anatomical and pathological information in a conceptually simpler form. Three-dimensional displays allow direct visual appreciation of spatial relationships, minimising human errors in clinical perception resultant from inaccurate mental registration of structural relationships.

Using a combination of synthetic materials, ovine tissues and human volunteer subjects, the primary research objective was to develop, test and validate an inter-active 3-D model of the knee, generated from sectional MR images using a commercially available reconstructive software package, as a precursor to future medical applications. A prospective, multi-component format was employed, incorporating a series of independent validation studies, to explore the potential efficacy of 3-D MR-based reconstructive imaging, in a controlled pre-clinical setting, motivated by endpoint musculoskeletal considerations. The project series was designed to allow scientifically rigorous validation of the model to support an evidence-based progression to active targeted clinical applications. Now-published data progressively explored the:

- level of current orthopaedic utilisation of 3-D reconstructed imaging in musculoskeletal evaluation [3];
- effect of base scan plane orientation on the perceived quality of resultant reconstructed images [4];
- level of intra- and inter-observer repeatability in generating meaningful image software outputs using generated 3-D reconstructed MR images [5];
- accuracy of the technique in generating direct anatomical (image) linear measurement in 3-D space [6];
- effect of window level selection on resultant 3-D reconstructed image quality and its effect on direct linear measurement [7];
- accuracy of 3-D reconstructed MR imaging in generating direct volumetric assessment of discrete image components [8];
- time efficacy of 3-D reconstructed MR image generation, contrasting this to existing conventional methods [9].

Having established a preliminary level of scientific validation, as a precursor to planned targeted clinical applications, the last component of the research sought to determine the potential usefulness of the new 3-D reconstructed MR imaging technique in a pair of simulated clinical settings – volumetric assessment of isolated meniscal cartilages [10] and pre-surgical evaluation of depressed-fragment tibial plateau injury [2]. The results of these individual, yet integrated, study components are summarised in the following paragraphs.

A representative sample of Australian primary orthopaedic surgical care facilities was surveyed to gauge the level of use of both 3-D reconstructed CT and MR applications and to establish the willingness of senior orthopaedic staff and departments to explore the development and possible clinical application of 3-D MR technology in the local setting [3]. A 92.9% response rate was achieved. 90.9% of respondents indicated that they were currently employing 3-D CT as part of their standard clinical patient management. Conversely, only 7.1% of respondents indicated that their facility was actively employing 3-D MR for the same purposes, and the sites that were using this technology did so only in rare circumstances. This disparity was somewhat surprising given the similarities associated with the software processing techniques required to generate 3-D images (most modern reconstructive packages will accept data from either CT or MR input sources.) Encouragingly, all of the survey respondents indicated an enthusiasm to explore advances in the development and validation of a 3-D MR model and could foresee potential clinical benefit for targeted or generalised orthopaedic assessment through access to such technology.

As the first stage of the active validation of the 3-D MR model, using a prospective human volunteer cohort, a preliminary study was undertaken to determine the influence of orthogonal acquisition scan plane orientation on the clinically perceived accuracy of resultant 3-D reconstructions [4]. The premise for this work was based on suggestions in the CT literature that anisotropic voxel dimensions would have a significant negative impact on resultant image quality. Computer reconstructions were generated for each volunteer knee from the standardised axial, coronal and sagittal planes. These images were shown to a series of orthopaedic and radiology specialists, blinded to the original plane of acquisition orientation, who were asked to nominate in rank-order their image preference in the assessment of each patient. The study did not demonstrate a difference in sequence orientation preference (P > 0.70), suggesting a consistency in the quality of image generation, independent of scan capture orientation.

Before the measurement accuracy of the model could be determined, it was considered essential to establish the degree of measurement repeatability that could be achieved both between observers and in repeat image evaluation performed by the same individual [5]. To achieve this, a series of previously generated 3-D MR-based images was selected from which a sequence of clinically relevant measurement parameters was identified. A range of measurement magnitudes was purposefully chosen to encompass a spectrum of values likely to be representative of true clinical application. To ameliorate the potentially biasing influence of observer experience, 30 prescribed measurements within the interactive 3-D displays were selected. To establish inter-observer variability, multiple observers from various professional and technical backgrounds (including a lay observer) were asked to independently record magnitudes for each of the 30 measures using the software's inbuilt measurement tools. To determine intra-observer variability, a single observer was asked to record the same 30 measurements on multiple occasions. All measurements for both analyses were recorded in isolation without access to previous measurement results. Statistical analysis suggested high degrees of both intra- and inter-observer agreement (Pearson Correlation Coefficient values of 0.76 and 0.87 respectively).

With an established degree of confidence in the reliability of image feature measurement, linear and volumetric measurement accuracy's were independently assessed using direct 'gold standard' comparisons to true physical parameters. Free-plane linear measurement accuracy was determined using an *ex vivo* knee model and routine MR section acquisitions [6]. Measurements recorded from 3-D reconstructions for a range of clinically relevant structures were statistically contrasted to direct surgical measures, demonstrating excellent correlation, independent of the magnitude of the true measure. In a disarticulated ovine model, the 3-D MR technique revealed an average error of measurement of 0.2 mm across 32 diagnostically-relevant structural features. An independent study, using the same base dataset, showed that time-of-capture window level setting variation played no significant role in influencing resultant linear measurement accuracy from the 3-D displays [7].

Volume measurement accuracy was assessed in three stages. Initially a series of precise geometric solid phantoms were imaged and software-based volume measures obtained for direct comparison to the known physical size [8]. A number of progressive acquisition slice thicknesses, within the current clinical range (1.0 – 5.0 mm) were employed to determine the influence of this dependent variable. Mid-range slices (3.0 mm) level yielded a percentage measurement error (PME) of volume of less than 5.0 %. In the second stage of volumetric analysis, a series of surface irregularities were then introduced into the longitudinal axis of the largest phantom, simulating surface degradation (i.e. as observed in age-related joint cartilage change) and repeat MR scanning performed [8]. With negative lesions volumes measuring between 60.0 and 215.0 mm^3, the findings of this study suggested that lesion volumes of such size could be accurately identified using conventional slice thicknesses, although sub-millimetre acquisitions would be required for high clinical volumetric precision. In the third stage of volumetric analysis, a pair of detailed synthetic meniscal cartilage phantoms was fashioned from a high-density acrylic resin [10]. Using an edge-delineation-optimised scanning sequence as determined by an independent study [11], the phantoms were imaged using various MR slice thicknesses and a 3-D image was generated in each instance (Figure 1). The image-derived volumes were compared to the known mass/density-derived volume measurements. The technique revealed mean volumetric PMEs of 1.1 % and 3.0 % using 1.0 and 2.0 mm slice thicknesses respectively.

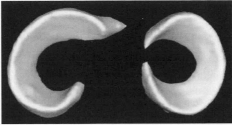

Figure 1: Software-generated reconstructions of the two meniscal (phantoms) [8].

One of the longstanding criticisms of 3-D reconstructive imaging technology has been the suggestion of image generation times disproportionately great for the perceived clinical benefit. Having provided preliminary evidence to support the measurement accuracy of the technique, an independent study was performed to determine the total time taken for 3-D MR-based image generation to a clinical standard, and contrast this to the time taken to generate a routine sectional acquisition series [9]. As a precursor to planned future work in tibial plateau injury assessment and being procedurally-representative of common musculoskeletal assessment, the adult knee was selected as the examination of interest. Mean routine knee MR scan time, using the five sequence institutional protocol, was contrasted to the mean base-acquisition plus 3-D image reconstruction time (using a single targeted sequence). Overall, employing the 3-D approach equated to an average examination time saving of 42 seconds. While perhaps seemingly small in the overall time associated with image generation, of far greater clinical relevance it was noted that employing the 3-D technique resulted in an average saving of 23 minutes and 33 seconds of active MR scanner time, while still generating diagnostic quality images relevant to bony orthopaedic pre-surgical assessment. This finding has significant potential impact regarding increasing patient throughput and associated per MR examination cost-efficacy.

As the final stage of pre-clinical investigation, the progressively refined 3-D MR model was prospectively targeted to the presentation of simulated orthopaedic injury. After departmental discussion, it was decided to test the model in the non-invasive evaluation of depressed-fragment tibial plateau injury in the human specimen [2]. The rationale for this choice was multi-factorial, most notably influenced by the prevalence with which the injury was seen in the adult population (i.e. post motor vehicle accident), the difficulties associated with pre-surgical assessment and management of the injury based on conventional imaging means, and the desire to generate a diagnostic tool that provided not only relevant bony contour information but also exploited the soft-tissue contrast sensitivity of MR imaging to allow concurrent assessment or exclusion of associated internal derangement. Following institutional ethics committee approval, a pair of separated oblique-planar tibial plateau fractures was surgically induced in an ovine tibial model. An MR-inert perspex spacer was inserted into the plateau to allow the maintenance of a precise fracture gap, and to facilitate stabilised infero-medial and infero-lateral fragment depression. With controlled degrees of progressive fragment displacement, the plateau was imaged under T2-weighted TSE conditions (Figure 2).

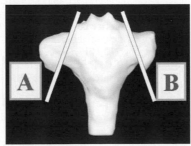

Figure 2: Software-generated reconstruction of the proximal tibia and showing bilateral plateau-relative depression of induced fracture fragments. Note: schematic representation of the perspex spacers are also shown [2].

Subsequent 3-D reconstructions were generated for each displacement level and software-derived measures of fragment volume, plateau-relative depression and fracture gap width were determined and statistically contrasted to the known 'gold standard' physical measurements. Multi-image averaging demonstrated mean PMEs of 2.35%, 2.71% and 3.98% for fragment volume, fragment depression and fracture gap width respectively. In linear displacement measurement of fragment depression and fracture gap width, the PMEs corresponded to true measurement errors of just 0.31 and 0.12 mm respectively. In consultation with clinical specialists, it was concluded that such degrees of accuracy were likely to have already exceeded the level required for surgical confidence.

One of the most encouraging outcomes of the research was the overwhelming clinical and technical interest in the development and potential future application of the model. Negotiations, at various stages of progression, have already taken place with several fields including orthopaedics, radiology, rheumatology, biomedical engineering, computer and software engineering. Other areas such as neurology, oncology and prosthetics have also indicated an interest in exploring the application of the model in their respective domains. While ongoing validation work continues to contribute to the refinement of the model, several large clinical trials are in the process of development and planning, with three currently being considered by key national research and funding bodies.

Extending beyond direct clinical application, the technology has also shown considerable potential benefit in areas including aiding custom orthopaedic prosthesis design and modelling; general teaching—both as a spatially appreciable anatomical guide and as a lead in to inter-active and remote virtual surgery—and as a conceptually simpler form of presentation for the conveyance of diagnoses and clinical management plans to other specialists, allied health professionals, and to patients and their families.

In summary, preliminary investigations performed in a controlled pre-clinical setting suggest that, using the described MR-based model, 3-D reconstructed images provide an accurate reflection of true anatomical structural presentation. The reported findings provide evidence to suggest a role for such applications in contemporary medical practice, with the potential for considerable clinical benefit, positively influencing both pure diagnosis and patient management while also providing time, cost, and resource efficiency. Independent components of this research indicate that the generation, manipulation and interpretation of the 3-D images may be done with a high level of accuracy, reliability and repeatability. These data, and the early clinical impressions, provide a strong indication to warrant the design and implementation of further prospective studies, in targeted *in vivo* patient populations, to further refine the existing model and to explore avenues for further extension and application. The enthusiasm generated in the development and testing of the model provides many avenues for clinical exploration and suggests an exciting future for the application of this technology. Ultimately, ongoing technical advances in the domain of advanced computational modelling in medicine can only serve to improve the standard of holistic patient care and the quality of clinical endpoint outcomes.

Acknowledgments

The authors would like to thank the staff of both the Orthopaedic Research Unit (Division of Surgery, Repatriation General Hospital, South Australia, Australia) and the School of Informatics and Engineering (Division of Engineering, Flinders University, South Australia, Australia) for their support in the successful completion of the work leading to the development of this presentation.

References
[1.] A.P. Kurmis. The developing role of knee MRI in musculo-skeletal radiology: the progression to 3-D imaging. *The Radiographer* **48** (1): 19-26, (2001).

[2.] A.P. Kurmis and J.P. Slavotinek. Reconstructed three-dimensional MR images: application to simulated tibial plateau depression fractures. *Radiography* **10** (2): 95-101, (2004).

[3.] A.P. Kurmis. Orthopaedic utilisation of three-dimensional image displays reconstructed from magnetic resonance imaging. *The Radiographer* **49** (2): 67-71, (2002).

[4.] A.P. Kurmis, J.P. Slavotinek and T.C. Hearn. The effect of scan acquisition plane on the perceived quality of T1 weighted three-dimensional MR reconstructions of bony knee anatomy. *Radiography* **10** (1): 53-9, (2004).

[5.] A.P. Kurmis, T.C. Hearn, K. Grimmer and K.J. Reynolds. Dimensional measurement of structural features of the ovine knee using three-dimensional reconstructed imaging: intra- and inter-observer repeatability. *Radiography* 2004; *In press*.

[6.] A.P. Kurmis, T.C. Hearn, J.R. Field, K. Grimmer and K.J. Reynolds. The accuracy of three-dimensional reconstructions of the ovine knee: dissectional validation. *Computerized Medical Imaging and Graphics* **26**: 171-5, (2002).

[7.]   A.P. Kurmis, T.C. Hearn and K.J. Reynolds. The effect of base image window level alteration on the dimensional measurement accuracy of resultant three-dimensional image displays. *Radiography* **9** (3): 211-8, (2003).

[8.]   A.P. Kurmis, J.P. Slavotinek and K.J. Reynolds. The influence of slice thickness on the volume measurement accuracy of 3-D MR reconstructions of acrylic phantoms: a precursor to knee imaging. *Radiography* 2004; *In press*.

[9.]   A.P. Kurmis. Time generation efficacy of three-dimensional image reconstructions of the human knee. *The Radiographer* **50** (3): 85-9, (2003).

[10.] A.P. Kurmis, J.P. Slavotinek and K.J. Reynolds. Three-dimensional reconstructed MRI of an acrylic meniscal cartilage phantom: The effect of acquisition slice thickness upon accuracy of volume measurement. *The Radiographer* **51** (2): 77-80, (2004).

[11.] A.P. Kurmis and J.P. Slavotinek. MR pulse sequence selection for optimal display of acrylic polymer phantoms. *Radiography* **10** (3): 177-82, (2004).

VSP International
Science Publishers
P.O. Box 346, 3700 AH Zeist
The Netherlands

Lecture Series on Computer
and Computational Sciences
Volume 1, 2004, pp. 294-297

Molecular Dynamics Simulations for
Selection of Kinetic Hydrate Inhibitors

Bjørn Kvamme[1], Tatyana Kuznetsova and Kjetil Aasoldsen

Department of Physics
University of Bergen
Allégt. 55, 5007 Bergen, Norway

Received 20 August, 2004; accepted in revised form 30 August, 2004

Abstract: Natural gas hydrates are ice-like structures composed of water and gas molecules that have long been a problem in petroleum industry. Heavy cost of alcohol and glycol injection has spurred an interest in called 'kinetic inhibitors' able to slow down the hydrate formation rather than prevent it. Since it is not possible to compare directly the macroscopic effects of different inhibitors on the kinetics of hydrate formation in computer experiments, a scheme capable of culling the list of candidates for experimental testing was proposed earlier [Kvamme, B. *et al*, 1997, Molec. Phys., 90, 979]. Molecular dynamics simulations were implemented to test several kinetic inhibitors in a multiphase water-hydrate system with rigid hydrate interface. In addition, a long-scale run was implemented for a system where the hydrate was free to melt and reform. Our conclusion that PVCap will outperform PVP as a kinetic hydrate inhibitor is supported by experimental data. We demonstrate that numerical experiments can be a valuable tool for selecting kinetic inhibitors as well as provide insight into mechanisms of kinetic inhibition and hydrate melting and reformation.

Keywords: molecular modeling; clathrate hydrates; kinetic inhibitors; water; multiphase system; interfacial tension.

PACS: 61.25.Em; 64.70.Dv; 68.08.De

1. Introduction

Testing of potential kinetic inhibitors involves expensive manipulation of hydrates at high pressures. If numerical simulations can be used to cull the experimental candidates, molecular dynamics may provide valuable insights into the mechanisms of hydrate growth inhibition, melting and reformation. While it's not feasible at present to compare directly the effects of inhibitors on the macroscopic kinetics of hydrate formation, a scheme enabling one to evaluate the relative performance of kinetic inhibitors in a multiphase water-hydrate system was devised and tested [1]. In this work, monomers of potential kinetic inhibitor PVP (Poly(N-vinylpyrrolidine)) was studied in a composite system of liquid water and hydrate crystal. In our previous paper [1], it was demonstrated that the PVP monomer had a favorable attachment to the hydrate surface, and that the attachment was also favorable in terms of free energy differences.

2. Simulation details

Simulation set II involved a block of structure I [2] methane hydrate made of 20 unit hydrate cells (982 SPC/E water and 144 OPLS one-site methane molecules). The hydrate was first brought into contact with a liquid slab of 862 SPC/E water molecules and two inhibitor molecules (monomers of PVP). The resulting interfacial system used Periodic Boundary Conditions (PBC) in all three directions. All the molecules were free to engage in translational and rotational motion.

Molecular Dynamics used the MDynaMix package of Lyubartsev and Laaksonen [3], with explicit reversible integrator for NPT-dynamics of Martyna *et al* [4], modified by us to implement implicit

---

[1] Corresponding author. E-mail: bjorn.kvamme@ift.uib.no

quaternion treatment of rigid molecules with Nosé-Hoover thermostat for temperature and pressure [5,6,7]). The time step was set to 0.5 fs. All systems were kept at constant temperature of 261 K and pressure of 7 MPa by means of Nosé-Hoover thermo- and barostat. Linux-based Message Passing Interface (MPI) was used to implement parallel computation on either a cluster or standalone dual-processor PCs. The number of processors ranged from 2 to 8. Possible hydrogen bonding was identified by means of distance-angle criterion of the VMD package [8].

3. Results and discussion

The inhibitor-hydrate-liquid water system was run for approximately 2.5 nanoseconds. The complexity of hydrate structure far surpasses that of ice, due to both the inclusion of guest molecules and two different water structures encaging them (large and small cages) and making up the structure I hydrates. We found that the best results were yielded by combining the analysis of mass-density profiles with the visual inspection of the hydrogen bond network (Figures 1 through 3).

We followed the general approach of Davidchak and Laird [9] and applied finite impulse response (FIR) filters [9,10] to produce a smooth coarse-grained profile from the oscillating *fine-scale* profiles. The shape of the filter was assumed to be Gaussian. Two filters, each with its own N and ε, were used, the first one aimed to average over the oscillations due to hydrate's layered structure but retain the essential features of the interfacial region. The values of $N=68$ and $\varepsilon=42.5$ appeared to satisfy both conditions for our system and were used to generate a coarse-grained profile. The second filter was used to smooth out only the momentary fluctuations in the density profiles while maintaining its oscillatory nature. This filter was much narrower, with $\varepsilon=2.4$.

The characteristic structure of a methane hydrate is well represented in Figure 1 showing the interfacial system approximately 7 ps after the start of the production run. In Figure 1b, one can clearly see both the alternating high- and low-density peaks of encaged methane guests (full magenta line) and the corresponding spacing of oxygen peaks (full cyan line). The low-density methane peaks coincide with the valleys of oxygen density, the high-density peaks, with the peaks in the oxygen profiles. Another characteristic feature of the hydrate structure is provided by the regular pattern of four hydrogen peaks (full blue line). Two of those hydrogen peaks coincide with the methane peaks, the other two (of equal height) are located just off the oxygen peaks. This highly structured pattern of hydrogen density is totally absent in the liquid water phase. The liquid phase also includes methane molecules (released by hydrate dissolution), as well as two PVP molecules (full red line). The dashed lines in Figure 1b are coarse-grained profiles. The picture they present is that of the system with a smooth interface between uniform hydrate bulk and liquid water. The 10-90 interface widths of all filtered density profiles amounted to about 8-10Å.

The absence of methane in a system comprised by just hydrate and pure water (our starting setup) should give rise to a chemical-potential driving force for hydrate dissolution. On the other hand, while it was left undisturbed by the inhibitor, the leftmost water-hydrate interface remained stable, so this force does not play a significant part on the nanosecond time scale, a fact that can be attributed to the low solubility of methane in water.

At the start of the production run, both PVP molecules were positioned closer to the r.h.s. interface at the distance of about 15-18 Å. One of the PVP molecules started moving towards the rightmost interface, triggering a rapid dissolution the of hydrate-liquid interface but never coming into direct contact with the hydrate itself; the closest distance of approach was around 12 Å (Figure 3). The analysis of this PVP-triggered process of hydrate dissociation made us to conclude that the most persisting hydrate formation at the melting hydrate-water interface appeared to be the "half-cage" structures (guest molecules enclosed by a hydrogen-bonded cage of water molecules typical for hydrate on the one side, and halfway surrounded by a liquid-structured environment on the other side). The hydrate melting itself proceeded in a succession of two steps: relatively slow dissociation of the half-cages followed by rapid dissolution of the exposed intact layer of hydrate crystal. Methane released by dissociating hydrate will eventually form a bubble (cluster) whose small curvature will increase the water-hydrocarbon interfacial tension and make it even more attractive for surfactants such as PVP (Figure 3a). As one can see in Figures 1 through 3, the PVP unit showed a strong preference for the water-hydrocarbon interface over the water-hydrate one.

The second water-hydrate interface remained stable for over 1.5 nanoseconds until the approach of the second PVP once again triggered the fast hydrate dissolution (Figure 3). Density profiles in Figures 2 and 3 show that just like the first one, the second PVP molecule attached itself to a small cluster of released methane drifting away from the hydrate interface and towards the larger methane bubble. This withdrawal of the PVP molecule appeared to stabilize the remaining hydrate crystal structure (strengthening of methane density peaks in Figures 2 and 3). Moreover, the liquid phase on the both sides of the hydrate crystal remainder has started to restructure itself into a typical hydrate formation (note the appearance of the fourth methane peak and the distinct cage-like structure of hydrogen bonds around the methane molecules in Figure 3). Note the distinct pattern of hydrogen density peaks starting to emerge. This rapid restructuring bears all the hallmarks of experimental phenomena often found in hydrate literature [11], the so-called "hydrate memory" effect, *i.e.* the fact that when hydrates have been produced from a mixture of water and guest molecules, and then destroyed by changing the conditions, the secondary formation of hydrates will proceed at a much faster rate than the primary one. This observed reformation of hydrates certainly lends weight to our belief that computer simulations can reproduce complex processes occurring in the systems involving hydrates in contact with liquid phases.

The marked preference for water-hydrocarbon interface exhibited by a promising kinetic inhibitor like PVP highlights an important potential problem of industrial inhibitor application. Inhibitors with large specific affinity for the water-hydrate interface are needed, with the affinity provided either by their polar heads or by folding of their carbon backbone over the hydrate surface.

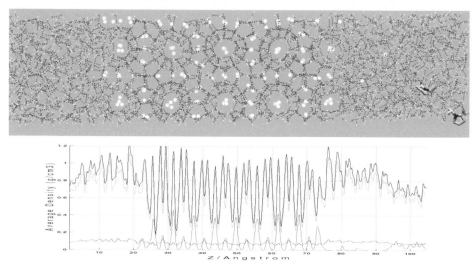

Figure 1: Snapshot and density profile of the PVP-water-hydrate system 7 ps after the start of the run.

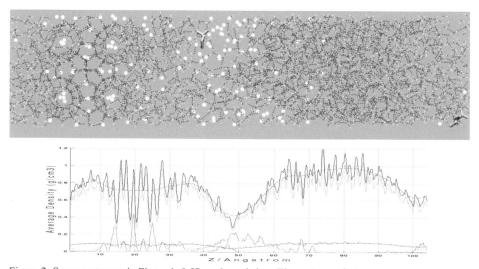

Figure 2: Same system as in Figure 1, 0.57 ps elapsed since Figure 1 snapshot.

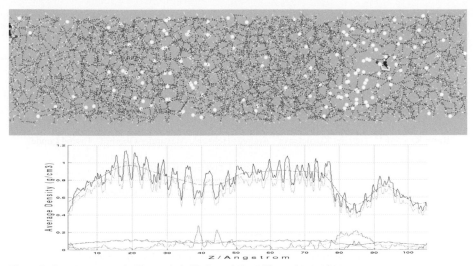

Figure 3: Same system as in Figure 1, 1.45 elapsed since Figure 1 snapshot.

Acknowledgments

One of the authors (TK) is grateful to the Norwegian Research Council for the grant of a postdoctoral scholarship.

References

[1]   Kvamme, B., Huseby, G., and Førrisdahl, O.K. Molecular dynamics simulations of PVP kinetic inhibitor in liquid water and hydrate/liquid water systems. *Molec. Phys.* 1997, **90**, 979-991.

[2]   von Stackelberg, M., and Müller, H.R. Feste Gashydrate II: Structur und Raumchemie. *Z. Elektrochem.* 1954, **58**, 25-39.

[3]   Lyubartsev, A.P., and Laaksonen, A. M.DynaMix - a scalable portable parallel MD simulation package for arbitrary molecular mixtures. *Computer Physics Communications* 2000, **128**, 565-589.

[4]   Martyna, G.J., Tobias, D.J., and Klein M.L. Constant-Pressure Molecular-Dynamics Algorithms. *J. Chem. Phys.* 1994, **101**, 4177-4189.

[5]   Nosé, S. Constant Temperature Molecular-Dynamics Methods. *Progr. Theor. Phys. Suppl.* 1991, **103**, 1-46.

[6]   Hoover, W.G. Canonical Dynamics - Equilibrium Phase-Space Distributions. *Phys Rev A* 1985, **31**, 1695-1697.

[7]   Fincham, D. Leapfrog rotational algorithms. *Molecular Simulation* 1992, **8**, 165-178.

[8]   Humphrey, W., Dalke A., and Schulten, K. VMD: Visual molecular dynamics. *J. Molec. Graphics* 1996, **14**, 33-38.

[9]   Davidchack, R.L., and Laird, B.B. Simulation of the hard-sphere crystal-melt interface. *J. Chem. Phys.* 1998, **108**, 9452-9462.

[10]   Bryk, T., and Haymet, A.D.J. Ice 1h/water interface of the SPC/E model: Molecular dynamics simulations of the equilibrium basal and prism interfaces. *J. Chem. Phys.* 2002, **117**, 10258-10268.

[11]   Sloan, E.D. *Clathrate Hydrates of Natural Gases*, 2nd ed. Marcel Dekker, New York, 1998.

VSP International
Science Publishers
P.O. Box 346, 3700 AH Zeist
The Netherlands

*Lecture Series on Computer
and Computational Sciences*
Volume 1, 2004, pp. 298-301

Measurements and modelling of interfacial tension for water+carbon dioxide systems at elevated pressures

Bjørn Kvamme[1a], Tatyana Kuznetsova[a], Andreas Hebach[b], Alexander Oberhof[b], Eivind Lunde[a]

[a]Department of Physics, University of Bergen
Allègt. 55, 5007 Bergen, Norway
[b]Institute for Technical Chemistry-CPV – Forschungszentrum Karlsruhe GmbH
Postfach 3640, 76021 Karlsruhe, Germany.

Received 4 September, 2004; accepted in revised form 15 September, 2004

Abstract: A novel apparatus, PeDro, for high-pressure measurement of interfacial tension using the quasi-static procedure of pendant drop method is presented. Optimized experimental set-up allowed for accurate measurements in the two-component system of water and compressed carbon dioxide. Experimental error of measurement was smaller than 2%. Measurements were conducted in the range of (278 to 335) K and (0.1 to 20) MPa. The interfacial tension showed a pronounced dependence on pressure and temperature. Molecular dynamics (MD) simulations were performed for liquid–liquid and liquid–vapor interfaces between water and carbon dioxide at elevated pressures. Constant-volume production runs of 0.3 nanoseconds were used to estimate the water–CO_2 interfacial tension at 323 K for a system comprising 512 SPC/E water molecules and 200 three-site CO_2 molecules. Interfacial tension was obtained from the difference between pressure components normal and tangential to the interface. The results showed a good agreement with experimental data; our model system also reproduced the pressure–temperature relationship of the interfacial tension.

Keywords: carbon dioxide; water; interfacial tension; pendant drop; molecular modeling; high pressure

PACS: 61.25.Em; 64.70.Dv; 68.08.De

1. Introduction

Carbon dioxide (CO_2) in its liquid or supercritical state is a promising fluid medium for reactions, separation and cleaning processes. In multiphase systems, interfacial phenomena play an important role in such applications as mass transfer limitations, mass transfer area, wetting behaviour and formation of micro-emulsions. These properties depend heavily on the density of the phases involved. The density of highly compressible fluids like CO_2 varies rapidly with temperature and pressure, affecting the interfacial properties considerably. One objective of this study was to develop an accurate measuring procedure for this system using the pendant drop method. Another objective was to study the behaviour of the interfacial tension with pressure and temperature in a range of (0.1 to 20) MPa and (278 to 335) K. Moreover, a system consisting of water in contact with carbon dioxide, a substance often considered a third type of condensed phase, different from lipophilic and hydrophilic phases, is a worthy target of computer simulation study in its own right. In the last two decades, the demand from industrial separation processes has led to rapid development and frequent application of the theory of inhomogeneous fluids. But certain theoretical progress notwithstanding, determination of the interfacial phenomena at the liquid-liquid water-carbon dioxide interface appears to remain in the realm of experiments and computer simulations.

2. Experimental

Method and apparatus

The interfacial tension data were obtained by using the pendant drop method. The method and apparatus used are described in detail elsewhere [1]. Briefly, the system consists of a high-pressure circuit with a gear pump, a viewing cell, and a sample loop (see Figure 1). The whole system is installed in a climatic chamber, which allows precise and accurate temperature adjustment to within

---

[1] Corresponding author. E-mail: bjorn.kvamme@ift.uib.no

±0.2 K. The pressure stability is ±0.02 MPa. Isotherms of the interfacial tension are measured using the quasi-static method. This method ensures a time regime where the influence of diffusion and drop ageing does not affect the IFT.

Accuracy of the measurements

To determine the accuracy of the equipment, measurements with water and CO_2 were compared with recommended values of the ASME at atmospheric pressure. According to this comparison, the experimental accuracy of the PeDro apparatus is greater than 98%. At higher pressures, especially in the presence of a highly compressible phase, the accuracy in density determination of the phases becomes particularly important. Therefore we measured the saturated densities of CO_2 and water. Within the limits of the accuracy (better than 99.85 %) the carbon dioxide density equals the pure density, but that of the CO_2-saturated water changes within 1.9 % [2].

Experimental results

Several isotherms were measured in the temperature range of (278 to 333) K at pressures from atmospheric pressure up to 20 MPa. To prove that there was no interference with drop ageing processes, three different measuring methods were tested: firstly with increasing pressure, secondly with decreasing pressure, and thirdly with randomly selected data points. All values are identical within the limits of experimental accuracy. Table 1 presents some IFT results (together with their simulational counterparts) at 323K. As pressure increases, a decrease in the interfacial tension is observed. At lower pressures IFT decreases with rising temperature; at higher pressures the opposite effect is found. At higher pressure and temperature, IFT-values reach a plateau. Here, the density difference varies to a smaller extent and the IFT values become comparable to those between water and non-polar organic solvents, such as benzene or hexane under ambient conditions.

The isotherms below 304 K, the critical temperature of CO_2, are separated into two parts, which are interrupted at the vapour-liquid coexistence line of CO_2. Both measurements and simulations display a jump in IFT at this pressure point of the isotherm. Below this pressure the isotherm reflects the interfacial tension of water with gaseous CO_2, which decreases almost linearly with pressure. The interfacial tension between water and liquid CO_2 is measured above this pressure. Isotherms measured above the critical temperature of CO_2 continuously decrease with rising pressure, showing a hyperbola-like behaviour.

3. Simulations

Details

The molecular dynamics used constant-temperature, constant-pressure algorithm from MDynaMix package of Lyubartsev and Laaksonen [3]. The starting interfacial system was constructed from two slabs of bulk water (SPC/E [4], either 256 or 512 molecules, and carbon dioxide [5], 100 to 200 molecules), thermalized initially at 323 K and set side by side, with the periodic boundary conditions applied. The resulting systems, ranging from 22x22x66 Å to 22x22x256 Å in size were subsequently equilibrated for several tens of picoseconds before the average collection began.

Interfacial tension estimation

Interfacial tension γ was obtained by a common method using pressure tensor in case of two interfaces normal to the z–axis:

$$\gamma = 0.5 * \overline{h_z} \left\{ P_z - \frac{1}{2} [P_x + P_y] \right\},$$

where P_i is average system pressure in the i direction, $\overline{h_z}$ is box length in the direction normal to the interface.

Simulational results and discussion

Figures 2 and 3 present density profiles characteristic for the systems containing, respectively, liquid and gaseous CO_2 as one of the phases. The fine-scale profiles of x-y averaged mass densities were generated by partitioning the simulation box into discrete bins in the z direction. Finite impulse response (FIR) filters [6] was applied to produce a smooth coarse-grained profile from the oscillating *fine-scale* profiles. The gaseous-CO_2 profile of Figure 2 shows a significant increase in CO_2 density at the interface relative to its bulk value, reflecting the expected "sticking" effect.

Table 1 shows that except in the lowest-pressure and density region, the simulations (see under "sim IFT") appear to reproduce the experimental data with a fair degree of accuracy, especially given the fact that we used force fields designed mostly for bulk properties at ambient conditions. Still, the simulated data using the standard Lorentz-Berthelot mixing rules appeared to mimic the pressure dependence of the experimental results. To improve the fit of modeling results further, we attempted to modify the well depth in the Lennard-Jones part of water-CO_2 cross-interactions by factors ranging from 0.8 to 1.2; the factor of 1.2 yielded the best results. These are listed in Table 1 under the "Cross-sim IFT" column header. As one can see from Table 1, while the non-Lorentx-Berthlot corrections brought the modeling results closer to the experimental ones for the high- and mid-range CO_2 densities, they slightly worsened the fit in the low-density region. This low-pressure region remains problematic for computer simulations, with the possible solution given by using exp-6 potentials derived from latest *ab initio* studies for both water and CO_2.

Table 1: Experimental and modeling results for water+CO_2 systems at 323K
and various elevated pressures

p	T	density sat CO2	density sat water	Cross-sim IFT	sim IFT	IFT
MPa	K	kg/m^3	kg/m^3	mN/m	mN/m	mN/m
1.1	322.8	18.8484	988.52	51.8	53.2	63.7
2.1	322.8	37.6033	990.71	48.0	50.1	57.1
4.18	322.9	83.397	995.07	48.6	46.7	47.5
6.22	322.9	143.052	997.94	43.8	44.5	44.1
8.26	322.8	235.565	1000.51	41.3	40.7	38.4
10.28	322.8	424.325	1002.38	37.0	36.6	32.5
12.33	322.9	608.399	1003.75	35.7	35.0	31.2
14.4	322.9	686.864	1005.11	33.7	35.4	30.6
18.43	322.8	765.667	1007.07	34.3	35.3	29.8
20.44	322.8	791.304	1008.15	33.7	34.2	29.4
22.45	322.8	812.725	1009.13	31.9	34.7	29.1

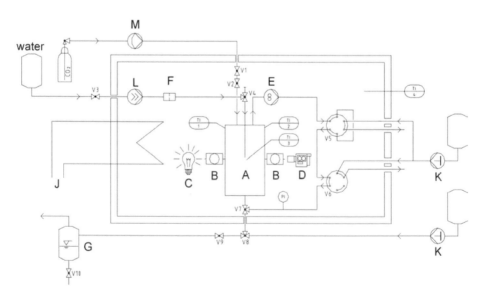

Figure 1: High pressure pendant drop device PeDro, A viewing cell, B inspection window, C light source, D CCD camera, E gear pump, F filter, G phase separator, J thermostat, K piston pump, L syringe pump, M compressor.

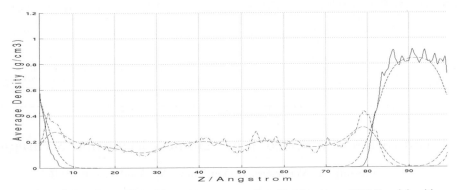

Figure 2: Density profiles of the water-gaseous carbon dioxide system at 323 K and densities corresponding to 8.3 MPa. Full line: fine-scale water density; dashed-dot line: fine-scale CO_2 density (profiles averaged over 50 ps). Dashed lines: coarse-grain profiles obtained by applying FIR.

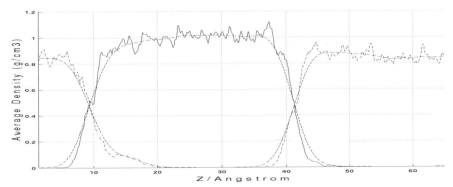

Figure 3: Density profiles of the water-liquid carbon dioxide system at 323 K and densities corresponding to 4.5 MPa. Same conventions are used as in Figure 2.

Acknowledgments

Authors AH and AO are grateful to the German Federal Ministry of Education and Research for its financial support. AH and TK are recipients of a grant form Statoil Co (Norway). One of the authors (TK) is grateful to the Norwegian Research Council for the grant of a postdoctoral scholarship.

References

[1]   Hebach, A., Oberhof, A., Dahmen, N., Kögel, A., Ederer, N., and Dinjus, E. *J. Chem. Eng. Data* 2002, **47**, 1540-1546.

[2]   Hebach, A., Oberhof, A., Dahmen, N., and Dinjus, E. *J. Chem. Eng. Data* 2004, **49**, 950-953.

[3]   Lyubartsev, A.P., and Laaksonen, A. M.DynaMix - a scalable portable parallel MD simulation package for arbitrary molecular mixtures. *Computer Physics Communications* 2000, **128**, 565-589.

[4]   Berendsen, H.J.C., Grigera, J.R., and Straatsma, T.P. The Missing Term in Effective Pair Potentials. *J. Phys. Chem.* 1987, **91**, 6269-6271.

[5]   Harris, J.G., and Yung, K.H. Carbon dioxides liquid-vapor coexistence curve and critical properties as predicted by a simple molecular-model. *J. Phys. Chem.* 1995, **99**, 12021-12024.

[6]   Davidchack, R.L., and Laird, B.B. Simulation of the hard-sphere crystal-melt interface. *J. Chem. Phys.* 1998, **108**, 9452-9462.

VSP International
Science Publishers
P.O. Box 346, 3700 AH Zeist
The Netherlands

*Lecture Series on Computer
and Computational Sciences*
Volume 1, 2004, pp. 302-305

Optimization of a Space Truss Dome - Evolution Strategies

E. Lamkanfi[1], L. Vermaere, R. Van Impe, W. Vanlaere, P. Buffel

Laboratory for Research on Structural Models,
Department of Structural Engineering,
Faculty of Applied Sciences,
Ghent University,
B-9000 Ghent, Belgium

Received 4 August, 2004; accepted in revised form 21 August, 2004

Abstract: The objective of this contribution is to give a framework for the general optimization of 3D trusses using evolution strategies. These stochastic methods have proven their efficiency in the past for several problems. They are very robust and converge in most cases to the true optimum. An important part of the optimization process is the verification of the constraints. This will be explained briefly for the case of truss structures which are often used in civil structures such as bridges and roof structures. The whole optimization technique is implemented in the Space Truss Evolution Strategy (STES) program written in JAVA and in which the cross-sectional areas of the truss members, the coordinates of the nodes and the topology of trusses can be optimized simultaneously. However, in this paper each part of the general optimization is tested seperately and the results are compared with those found in the literature. Finally an example is presented where a one-layered dome is submitted to the evolutionary optimization process. For this particular structural engineering problem it is imperative to verify the global buckling of the dome, since this failure phenomenon tends to be dominant for this kind of structures.

Keywords: Mixed discrete-continu-binary structural optimization, Evolution strategies, Stochastic methods, Truss structures, One-layered dome, Bifurcation

1 Introduction

Three dimensional space trusses are commonly used in structural engineering. In an optimal design the weight of the truss - that must be capable of bearing the loads - has to be as low as possible. In this paper, evolution strategies are used and implemented in a program written in JAVA. This program consists of two interacting parts, the strategy itself and a truss calculation module, in order to check the load bearing capacity of the contruction.

Evolution strategies (ES) are stochastic optimization methods based on the theory of Charles Darwin. These strategies, developed by Rechenberg [1] and Schwefel [2] in the seventies, were in the beginning commonly applied for continuous optimization problems. In the nineties Thierauf and Cai [3] proposed a modified ES algorithm to handle also discrete problems. These stochastic methods differ from the conventional optimization algorithms by using randomized operators such as the mutation and the recombination operator. More detailed information about these strategies can be found in the very extensive literature. Examples of comprehensive works are those of Schwefel [4] and Bäck [5].

---

[1] Corresponding author. E-mail: ebrahim.lamkanfi@lid.kviv.be

In the following section more attention will be payed to the structural part of the optimization process. This is necessary because during the optimization process there has to be information available whether a certain truss structure is 'vital' enough to survive a generation and to make part of the population of the next generation.

2   Truss structures

Besides the understanding of the evolution strategy it is also necessary to have some notions about trusses. During the optimization process, every truss structure has to fulfill some constraints in order to survive a generation. This requires a truss calculation that leads to the values of the nodal displacements and the bar forces. More adavanced constraints such as the mechanism verification and the calculation of the global bifurcation of the truss should also be implemented.

A general definition of a 2D or a 3D truss structure can be described as follows. A truss structure is composed of a number of bars, which are connected with hinges. Because of this type of connections, the bars can only be subjected to tensile or compression forces. An example of a 2D truss is given in Figure 1(A). Notice that the loads P have to work in the nodes of the structure because of the no-bending moments characteristic of the bars. For the calculations of the node displacements and the tensile or compression forces in the bars the *displacement-method* [6] is used in the program.

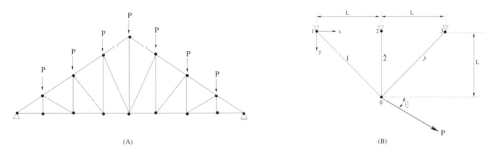

(A)   (B)

Figure 1: An example of a 2D truss (A) and the three-bar truss (B).

3   Comparison of the results with the literature

3.1   Sizing optimization

The example that is handled here is the three-bar truss investigated in the work of Ramaswamy [7] and shown in Figure 1 (B). Two different cross-sectional areas[2] are used : $A_1 = A_3$ and A_2. The optimum values for the cross-sectional areas $(A_1, A_2, A_3) = (5.8, 5.2, 5.8)$ $in.^2$ are exactly the ones that are found by STES.

3.2   Shape optimization

For the optimization of the coordinates the example of a truss structure with two supports is chosen. These supports are taken in the lower left and the lower right nodes. The initial structure used in the evolution process is shown in Figure 2 and is indicated by *Generation 0*. The vertical loads are applied in the nodes between the two supports. In the literature [8] can be found that

---

[2]With A_i the cross-sectional area of bar i.

the optimal structure in this case is an arc-form. This result, shown in Figure 2, is also found by STES.

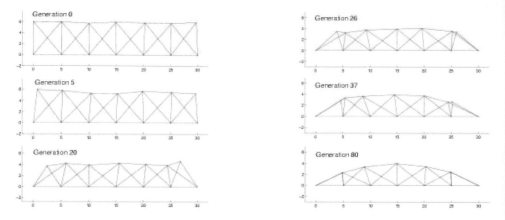

Figure 2: Evolution of the truss structure during the optimization process.

3.3 Topology optimization

A topology example frequently examined in the literature is the 12-bar truss given in Figure 3(A). Besides the optimal structure given in [9] (Figure 3(C)) STES also finds another optimal structure given in Figure 3(B). The latter structure is slightly better when the weight of the structure is taken as the objective function.

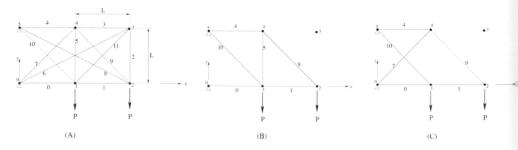

Figure 3: 12-bar truss.

4   A one-layered dome structure

The initial structure of the optimized dome, with a mass of 51.2 ton, is given in Figure 4(A). In this example the cross-sectional areas as well as the coordinates of the nodes are submitted to the evolution algorithm. To prevent the overall buckling phenomenon, a module is programmed in STES that verifies for each dome structure during the optimization process whether it will buckle or not. In Figure 4(B) the optimal structure, with a mass of 12.4 ton, given by STES is showed.

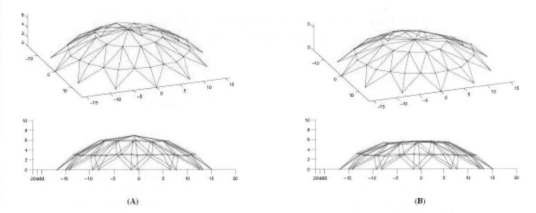

Figure 4: Intial dome structure (A) and the optimized dome structure (B).

5   Conclusion

In this paper a program based on evolution strategies has been presented for the optimization of truss structures. This method is implemented in the program STES. The performance of this program is tested with some examples from the literature. As example a one-layered dome is optimized where the global buckling has been taken into account.

References

[1] I. Rechenberg, *Evolution strategy : optimization of technical systems according to the principles of biological evolution*, Frommann-Holzboog, Stuttgart, 1973.

[2] H.-P. Schwefel, *Numerical optimization for computer models*, Wiley & Sons, Chichester, UK, 1981.

[3] G. Thierauf and J. Cai, *A two level parallel evolution strategy for solving mixed-discrete structural optimization problems*, The 21th ASME Design Automation Conference, **2** 17-221(1995).

[4] H.-P. Schwefel, *Evolution and optimum seeking*, John Wiley & Sons, 1994.

[5] T. Bäck, *Evolutionary Algorithms in Theory and Practice*, Oxford University Press, 1995.

[6] D. Vandepitte, *Berekening van constructies, Boekdeel III, Hoofdstuk 44*, E. Story-Scientia Gent, 1980 (in Dutch).

[7] G. S. Ramaswamy, G. R. Suresh and M. Eekhout, *Analysis, Design, and Construction of Steel Space Frames*, American Society of Civil Engineers, 2002.

[8] B. Kost, *Evolutionary shape optimization with self-adapting mutation distribution based on the Cholesky decomposition*, Computer Aided Optimum Design of Structures VIII, **67**, 2003.

[9] B. Kost and B. Baumann, *Topology Optimization of Trusses - Random Cost Method versus Evolutionary Algorithms*, Computational Optimization and Applications Archive, **14**, 203-218(1999).

VSP International
Science Publishers
P.O. Box 346, 3700 AH Zeist
The Netherlands

*Lecture Series on Computer
and Computational Sciences*
Volume 1, 2004, pp. 306-309

Thermodynamic Property and Relative Stability of 76 Polychlorinated Naphthalene by Density Functional Theory

Wang Zunyao[1,2†], Zhai Zhicai[2], Wang Liansheng[1]

[1]School of the Environment, Nanjing University, 210093, P R China
[2]Yancheng Institute of Technology, Jiangsu Yancheng, 224003, P R China

Received 7 Jun, 2004; accepted in revised form 10 Aug, 2004

Abstract: The thermodynamic functions, including total energy(TE), enthalpy($H°$), entropy($S°$), free energy($G°$), zero point vibrational energy(ZPE), constant volume molar heat capacity($C_V°$) and thermal energy correction(E_{th}) for 76 polychlorinated naphthalenes(PCNs), were predicted by fully optimized calculation at the B3LYP/6-31G* level. The standard heat energy of formation($\Delta H_f°$) and standard formation free energy($\Delta G_f°$) of PCN congeners were obtained by designing isodemic reactions. In addition, the dependences of these thermodynamic parameters on the number and the position of chlorine substitution were discussed, in which the obvious increase of TE, $H°$, $G°$, $\Delta H_f°$ and $\Delta G_f°$ occur once chlorine substitutions are simultaneously at position 1 and 8(or position 4 and 5). And with increasing number of chlorine atoms in PCN molecule, the values of $S°$ and $C_V°$ increase and E_{th} and ZPE decrease. Based on the magnitude of the relative free energy($R\Delta G_f°$), the relative stability order of PCN isomers in each congener was theoretically determined. Compared with the four semiempirical methods, AM1, MINDO/3, MNDO and PM3, the magnitude of standard formation heat energy($\Delta H_f°$) of PCNs obtained from B3LYP/6-31G* is the highest, and the differences resulted from these methods become greater with increasing number of chlorine substitutions.

Keywords: Dioxin, Polychlorinated naphthalene(PCN), Density functional theory(DFT) method, Thermodynamic property, Isodemic reaction, Relative stability

1. Introduction

Polychlorinated naphthalenes(PCNs) are environmentally persistent compounds, and PCN congeners have been detected and quantified in several matrices including sediments[1,2], water[3], air[4,5] and biota[6–8]. Therefore, knowledge of the thermodynamic property of PCN is of vital importance. Due to limits of PCN samples and analytical complexity, to obtain overall physiochemical property of the toxic material in the environment is somewhat difficult. Fortunately, this can be achieved by computation models in quantum chemistry, which provide statistical thermodynamic data of chemical compounds. As reported, some researches have been carried out for PCNs by scientists before using mostly semiempirical methods[9–11]. Because of the well-known relatively low precision of the semiempirical methods, such as AM1 and PM3 etc., it is necessary to use high precision empirical method to calculate the thermodynamic data of PCN congeners often needed for predicting their environmental fate and behavior.

The objective of the present study is to observe the thermodynamic property and relative stability of 76 PCNs by fully optimized calculation at the B3LYP/6-31G* level[12-14].

2. Computational Method

In the present work, B3LPY/6-31G* level in Guassian98 program was employed for calculation of PCNs. In condition of temperature of 298.15 K and pressure of 1 atm, the relationship of $S°$, $H°$, $G°$

---

† Corresponding author: Zhai Zhicai, Tel: + 86-0515-8398092; Fax: +86-025-3707304. Email address: njuzzc0125@yahoo.com.cn (Zhai Zhicai).

and *TE* can be expressed as follows:

$$G° = H° - TS° \tag{1}$$
$$H° = TE + ZPE + E_{th} + RT \tag{2}$$

Where *R* is gas constant; and *T* is temperature. All these computations, however, were performed on a Pentium IV 2.4G computer.

2. Results and Discussion

2.1 Correlation of *TE*, *H°*, *G°* and the position of chlorine substitute

In this work, *TE*, *H°*, *G°* of 76 PCNs were all obtained from computation at B3LYP/6-31G* level. Based on observation of other PCN congeners, it is proposed that the *TE* of PCNs has a close correlation with the number and position of chlorine substitution in their structures.

If the numbers of chlorine atoms at positions 1, 4, 5, 8(α positions) and 2, 3, 6, 7(β positions) are defined as N_α and N_β; the number of chlorine substitution at position 1,8(or position 4, 5) is $N_{1,8}$; and the number of chlorine at ortho, meta and para positions are symbolized as N_o, N_m and N_p, respectively, the correlation of *TE* and the number of *N* for PCN can be described as follows using the least square method:

$$TE = -385.89436 - 459.59431N_\alpha - 459.59525N_\beta + 0.02000N_{1,8} + 0.00599N_o + 0.00140N_m + 0.00197N_p$$
(hartree)
(3)

$r^2 = 1.0000 \quad SE = 0.001$

Where *r* is correlation coefficient, and *SE* is standard error.

Similarly, the correlation of *H°*, *G°* and *N* can be summarized in the following equations:

$$H° = -385.73900 - 459.60300N_\alpha - 459.60400N_\beta + 0.02000N_{1,8} + 0.00584N_o + 0.00133N_m + 0.00202N_p$$
(hartree)
(4)
$r^2 = 1.0000 \quad SE = 0.001$

$$G° = -385.77700 - 459.60600 N_\alpha - 459.60700N_\beta + 0.01636N_{1,8} + 0.00639 N_o + 0.00163N_m + 0.00203N_p$$
(hartree)
(5)
$r^2 = 1.0000 \quad SE = 0.001$

2.2 Correlation of *S°*, E_{th}, *ZPE* and $C_V°$ to the substituted position of chlorine atom

Using the least square method, the correlation of *S°*, E_{th}, *ZPE* and $C_V°$ to the chlorine substitution number *N* can also be expressed in the following Eqs. (6) to (9).

$$S° = 341.57 + 31.38 N_\alpha + 32.05N_\beta + 1.78N_{1,8} - 4.85N_o - 2.63N_m - 0.11N_p \quad (J/mol·K) \tag{6}$$
$r^2 = 0.9690 \quad SE = 7.800$
$$E_{th} = 406.23 - 21.88 N_\alpha - 22.10N_\beta - 1.29N_{1,8} - 0.39N_o - 0.18N_m + 0.14N_p \quad (J/mol) \tag{7}$$
$r^2 = 0.9996 \quad SE = 0.717$
$$ZPE = 388.50 - 25.17N_\alpha - 25.58N_\beta - 1.20N_{1,8} - 0.02N_o - 0.13N_m - 0.15N_p \quad (kJ/mol) \tag{8}$$
$r^2 = 1.0000 \quad SE = 0.150$
$$C_V° = 120.78 + 15.87N_\alpha + 16.61N_\beta - 0.45N_{1,8} - 1.40N_o - 0.28N_m + 1.13N_p \quad (J/mol·K) \tag{9}$$
$r^2 = 0.9908 \quad SE = 2.338$

2.3 Calculated $\Delta H_f°$ and $\Delta G_f°$ value for PCNs

To get more precise thermodynamic information of PCNs therefore, a useful theoretical model was suggested in the present work. In our study, the following isodemic reaction was designed for calculating $\Delta H_f°$ of PCNs:

Naphthalene + n Chlorobenzene = Polychlorinated naphthalene + n Benzene (10)

Based on the calculated result of $\Delta H_f°$ for PCNs, $\Delta G_f°$ was obtained from the Eq. $\Delta G_f = \Delta H_f - T\Delta S_f$. Using the same method as mentioned above, the correlation expressions of $\Delta H_f°$ and $\Delta G_f°$ to the number and position of chlorine substitution are demonstrated in Eqs. (11) and (12), which all clearly represent the influence of the number of chlorine position and spatial chlorine substitution position on the value of $\Delta H_f°$ and $\Delta G_f°$.

$$\Delta H_f^\circ = 224.35 - 12.53 N_\alpha - 17.03 N_\beta + 39.36 N_{1,8} + 11.11\ N_o - 0.76 N_m + 1.60 N_p \quad \text{(kJ/mol)} \qquad (11)$$
$$r^2 = 0.8393 \quad SE = 10.510$$
$$\Delta G_f^\circ = 295.21 - 8.11\ N_\alpha - 12.82 N_\beta + 38.83 N_{1,8} + 12.56\ N_o + 0.02 N_m + 1.64 N_p \quad \text{(kJ/mol)} \qquad (12)$$
$$r^2 = 0.8638 \quad SE = 10.871$$

2.4 Comparison of the calculated ΔH_f° of PCNs with semiempirical methods

In this study, the ΔH_f° of 76 PCNs were calculated by B3LYP/6-31G* and compared to that obtained from these semiempirical methods, AM1, MINDO/3, MNDO and PM3. The results are illustrated in Figure 1. From Figure 1, it can be seen that the ΔH_f° values of PCNs from B3LYP/6-31G* are larger than those from semiempirical methods, especially the differences become larger with increasing number of chlorine substituents. While for the ΔH_f° of the adjacent isomers of a same congener, the difference from B3LYP/6-31G* is distinct than that from semiempirical methods. As for the ΔH_f° from the three semiempirical methods in the same PCN molecule, the ΔH_f° value resulted from MINDO/3 is the largest, while the ΔH_f° data calculated from AM1, MNDO and PM3 are very close.

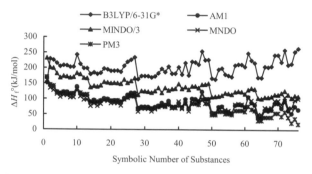

Figure 1: The differences of ΔH_f° obtained from B3LYP/6-31G* and semiempirical methods.

2.5 Relative stability of PCN isomers

Usually the isomer with low free energy is stable, while that of higher free energy is unstable. If based on the lowest ΔG_f° in each PCN congener, the relative formation free energy $R\Delta G_f^\circ$ of PCNs can be obtained. For comparison of these parameters, the most stable and unstable isomers in the 7 categories of PCN congeners are summarized and listed in Table 1.

Table 1: The most stable and unstable isomer in different PCN congener

Substance	The most stable isomer	The most unstable isomer
Mono-	2-	1-
Di-	2,7- 2,6-	1,8-
Tri-	1,3,6-	1,2,8-
Tetra-	1,3,5,7-	1,4,5,8-
Penta-	1,2,4,6,7-	1,2,4,5,8-
Hexa-	1,2,3,5,6,7-	1,2,3,4,5,8-
Hepta-	1,2,3,4,5,6,7-	1,2,3,4,5,6,8-

From Table 1, it is obvious that the isomers with chlorine substituted simultaneously at positions 1 and 8, or 4 and 5, are not stable. On the contrary, the isomers with meta chlorines and no chlorine substituted at 1 and 8, or 4 and 5 positions simultaneously, are relatively stable, which is consistent with conclusion from Eqs. (5) and (12).

In conclusion, all the thermodynamic properties of PCNs were correlated with the number and position of chlorine substitution, for which the obvious increase of *TE*, H°, G°, ΔH_f° and ΔG_f° occur once chlorine substitutions are simultaneously at position 1 and 8(or position 4 and 5). With addition of each chlorine atom in PCN molecule, the values of S° and C_V° increase by 32 J/mol·K and 16 J/mol·K, and E_{th} and *ZPE* decrease by about 22 J/mol and 25.5 kJ/mol, respectively. As for the PCN compounds of a same congener, the isomers with chlorine substituting at positions 1 and 8, or 4 and 5 simultaneously, are not stable. On the contrary, the isomers with meta chlorines and no chlorine substituted at 1 and 8, or 4 and 5 positions simultaneously, are relatively stable. Also, the magnitude of ΔH_f° of PCNs obtained from B3LYP/6-31G* is larger than that from the 4 semiempirical methods, AM1, MINDO/3, MNDO and PM3.

Acknowledgments

The present work was financially supported by the State Science Foundation of China (Grant No. 20177008), National Basic Research Program of China (2003CB415002), and the China Post Doctoral Research Fund (Grant No. 2003033486).

References

[1] K. Kannan, T. Imagawa, A. Blankenship, J.P. Giesy, Isomer-specific analysis and toxic evaluation of polychlorinated naphthalenes in soil, sediment, and biota collected near the site of a former Chlor-Alkali plant, *Environ. Sci. Technol.* 32 2507-2514(1998).

[2] U. Jarnberg, L. Asplund, C. deWit, A. Egeback, U. Wideqvist, E. Jakobsson, Distribution of polychlorinated naphthalene congeners in environmental and source-related samples, *Arch. Environ. Contam. Toxicol.* 32 232-245(1997).

[3] I. Marti, F. Ventura, Polychlorinated naphthalenes in groundwater samples from the Llobregat aquifer (Spain), *Chromatogr. A* 786 135-144(1997).

[4] G. Dorr, M. Hippelein, O. Hutzinger, Baseline contamination assessment for a new resource recovery facility in Germany, Part V: Analysis and seasonal/regional variability of ambient air concentrations of polychlorinated naphthalenes(PCN), *Chemosphere* 33 1563-1568(1996).

[5] T. Harner and T. Bidleman, Polychlorinated naphthalenes in urban air, *Atmos. Environ.* 31 4009-4016(1997).

[6] J. Falandysz and C. Rappe, Spatial distribution in plankton and bioaccumulation features of polychlorinated naphthalenes in a pelagic food chain in southern part of the Baltic proper, *Environ. Sci. Technol.* 30 3362-3370(1996).

[7] J. Falandysz, B. Strandberg, L. Strandberg, P. Bergqvist, C. Rappe, Concentrations and biomagnification of polychlorinated naphthalenes in black cormorants Phalacrocorax carbo sinensis from the Gulf of Gdansk, *Baltic Sea. Sci. Total Environ.* 204 97-106(1997).

[8] A. Lunden and K. Noren, Polychlorinated naphthalenes and other organochlorine contaminants in Swedish human milk, *Arch. Environ. Contam. Toxicol.* 34 414-423(1998).

[9] J. Olivero, T. Gracia, P. Payares, R. Vivas, D. Diaz, E. Daza, P.Geerlings, Molecular structure and gas chromatographic retention behavior of the components of Ylang-Ylang oil, *Pharm. Sci.* 86 625-630(1997).

[10] F. Iino, K. Tsuchiya, T. Imagawa, B. K. Gullett, An isomer prediction model for PCNs, PCDD/Fs, and PCBs from municipal waste incinerators, *Environ. Sci. Technol.* 35 3175-3181(2001).

[11] R. Weber, F. Iino, T. Imagawa, M. Takeuchi, T. Sakurai, M. Sadakata, Formation of PCDF, PCDD, PCB, and PCN in de novo synthesis from PAH: mechanistic aspect and correlation to fluidized bed incinerators, *Chemosphere* 44 1429-1438(2001).

[12] A. D. Becke, Density-functional thermochemistry, III The role of exact exchange, *J. Chem. Phys.* 98 5648-5652(1993).

[13] C, Lee, W, Yang, R. G. Parr, Development of the Colle-Salvetti correlation-energy formula into a functional of the electron density, *Phys. Rev. B* 37 785-789(1988).

[14] W. J. Hehre, L. Radom, P. V. R. Schleyer, J. A. Pole, *Ab initio molecular orbital theory*. Wiley, New York, 1986.

VSP International
Science Publishers
P.O. Box 346, 3700 AH Zeist
The Netherlands

*Lecture Series on Computer
and Computational Sciences*
Volume 1, 2004, pp. 310-312

On A Seven-Dimensional Representation
of RNA secondary Structures

*Bo Liao[1,2], Tianming Wang[1], Kequan Ding[2]

[1]Department of Applied Mathematics, Dalian University
of Technology Dalian 116024,China
[2]Department of Applied Mathematics, Graduate School of
the Chinese Academy of Sciences
Beijing 100039,China

Received 25 May, 2004; accepted in revised form 24 June, 2004

Abstract: In this paper, we proposed a 7-D representation of RNA secondary structures. The use of the 7-D representation is illustrated by constructing structure invariants. Comparisons with the similarity/dissimilarity results based on 7-D representation for a set of RNA3 secondary structures at the 3'-terminus of different viruses, are considered to illustrate the use of our structure invariants based on the entries in derived sequence matrices restricted to a selected width of a band along the main diagonal.

Keywords: RNA secondary structure; Similarity; virus; 7D representation

Mathematics Subject Classification: 92D20

We illustrate the seven-dimensional characterization of RNA secondary structure at the 3'-terminus of RNA3. In 7-D space points, vectors and directions have seven components, and we will assign the following basic elementary directions to the four free bases and two base pairs.

free base   A $(1,0,0,0,0,0,0)$, U $(0,1,0,0,0,0,0)$,C $(0,0,1,0,0,0,0)$,G $(0,0,0,1,0,0,0)$ base pair   A-U $(0,0,0,0,1,0,0)$,C-G $(0,0,0,0,0,1,0)$,G-U $(0,0,0,0,0,0,1)$

Because the seven directions of 7-D space fully equivalent, the above selection is equivalent to any other permutation of labels and directions, hence this selection should not be viewed as introducing any arbitrary decision to influence numerical analysis that follows. We will reduce a RNA secondary structure into a series of nodes $P_0, P_1, P_2, \cdots, P_N$, whose coordinates $x_n, y_n, z_n, s_n, v_n, w_n$ $(n = 0,1,2,\cdots, N$, where N is the length of the RNA secondary structure being studied)satisfy

$$\begin{cases} x_n = A_n \\ y_n = U_n \\ z_n = C_n \\ s_n = G_n \\ v_n = A'_n + U'_n \\ w_n = C'_n + G'_n \\ t_n = G''_n + U''_n \end{cases}$$

Where $A_n, C_n, G_n, U_n, A'_n, U'_n, C'_n, G'_n, U''_n$ and G''_n are the cumulative occurrence numbers of $A, C, G, U, A', U', C', G', U''$ and G'', respectively, in the subsequence from the 1st base to the n-th base in the sequence. We define

---

* Corresponding Email Address: dragonbw@163.com, wangtm@dlut.edu.cn, kqding@yahoo.com

$$A_0 = C_0 = G_0 = U_0 = A'_0 = U'_0 = C'_0 = G'_0 = G''= U''= 0$$

Since we have no graphical representation to associate with a random walk in 7-D space, we have constructed the distance matrix for each such random walk in which any (i,j)entry is the Euclidean distance between corresponding points in 7-D space given by

$$D_{i,j} = \sqrt{\Delta x^2 + \Delta y^2 + \Delta z^2 + \Delta s^2 + \Delta v^2 + \Delta w^2 + +\Delta t^2}$$

We will do mathematical analysis and comparison of distance matrices belonging to different RNA secondary structures. Constructing suitable structural and matrix invariant facilitates such comparisons.

Table 12: Similarity/Dissimilarity between the nine species of Figure 1 based on the Euclidean distances between the end points of the 5-component vectors of the average bandwidth 5 using the full RNA secondary structure

Species	AlMV-3	CiLRV-3	TSV-3	CVV-3	APMV-3	LRMV-3	PDV-3	EMV-3	AVII
AlMV-3	0	0.806230	0.364488	0.104315	0.110627	0.410819	0.254007	0.378070	0.384726
CiLRV-3		0	0.455902	0.730439	0.707109	0.400967	0.558049	0.434756	0.428281
TSV-3			0	0.284045	0.272260	0.070880	0.152847	0.077557	0.074464
CVV-3				0	0.086116	0.330441	0.195951	0.298626	0.304282
APMV-3					0	0.313522	0.152562	0.276765	0.282490
LRMV-3						0	0.174291	0.046235	0.036283
PDV-3							0	0.141350	0.147925
EMV-3								0	0.012117
AVII									0

Table13: The similarity/dissimilarity matrix for the nine species of Figure 1 based on the angle between the end points of the 5-component vectors of the average bandwidth 5 using the full RNA secondary structure

Species	AlMV-3	CiLRV-3	TSV-3	CVV-3	APMV-3	LRMV-3	PDV-3	EMV-3	AVII
AlMV-3	0	0.028949	0.022049	0.012703	0.009230	0.020154	0.009968	0.016279	0.017763
CiLRV-3		0	0.013151	0.021960	0.023912	0.011459	0.022422	0.016625	0.015145
TSV-3			0	0.014356	0.019404	0.007003	0.018937	0.011038	0.010264
CVV-3				0	0.012717	0.011208	0.012660	0.006237	0.007887
APMV-3					0	0.016127	0.002471	0.012522	0.013456
LRMV-3						0	0.015020	0.005735	0.004318
PDV-3							0	0.012050	0.012912
EMV-3								0	0.001730
AVII									0

In Table 12 and Table 13, we show the similarity/dissimilarity table for the nine secondary structure based on cumulative bandwidths of order 5 but using the full structure listed in Figure 1. Interestingly, with the full RNA secondary structure, the most similar are EMV-3 and AVII with the lowest value of 0.001730 followed by AVII and LRMV-3 with a value of 0.004318 and by LRMV-3 and EMV-3 with a value of 0.005735.

Acknowledgments

The author wishes to thank the anonymous referees for their careful reading of the manuscript and their fruitful comments and suggestions.

References

[1] Chantal B.E.M.Reusken, John F.Bol, Structural elements of the 3'-terminal coat protein binding site in alfalfa mosaic virus RNAs, Nucleic Acids Research, 14(1996), 2660-2665.

[2] Milan Randic, Marjan Vracko, Nella Lers, Dejanplavsic, Analysis of similarity/dissimilarity of DNA sequences based on novel 2-D graphical representation, Chemical Physics Letters,371(2003),202-207.

[3] Milan Randic, Marjan Vracko, On the similarity of DNA primary sequences, J.Chem.Inf.Comput.Sci, 40(2000), 599-606.

[4] M.Randic, Condensed representation of DNA primary sequences. J.Chem.Inf.Comput.Sci, 40(2000), 50-56.

[5] M.Randic,Alexandru T.Balanba, On A Four-Dimensional Representation of DNA Primary Sequences. J. Chem. Inf. Comput. Sci, 40(2000), 50-56.

[6] Chunxin Yuan, Bo Liao, Tianming Wang, New 3-D graphical representation of DNA sequences and their numerical characterization, Chemical Physics Letters,379(2003),412-417.

[7] Bo Liao, Tianming Wang, Analysis of similarity/dissimilarity of DNA sequences based on 3-D graphical representation, Chemical Physics Letters, 388(2004)195-200.

[8] Bo Liao, Tian-ming Wang, General combinatorics of RNA hairpins and cloverleaves, J.Chem.Inf.Comput.Sci, 43(4)2003,1138-1142.

[9] V.Bafna, S.Muthukrisnan, R.Ravi, Comparing similarity between RNA strings. Computer Science, Vol 937(1995),1-14

[10] F.corpet, B.Michot, RNAlign program: alignment of RNA sequences using both primary and secondary structures. Computer.Appl.Biosci,10(4)(1995),389-399

[11] S.Y.Le, R.Nussinov, J.V.Mazel, Tree graphs of RNA secondary structures and their comparison, Computer Biomed.Res 22(1989) 461-473

[12] S.Y.Le, J.Onens, R.Nussinov, J.H.Chen, B.shapiro, J.r.Mazel, RNA secondary structures: comparison and determination of frequently recurring substructures by consensus ,Computer Biomed 5(1989)205-210

[13] B.Shapiro An algorithm for comparing multiple RNA secondary structures, Computer. Appl.Biosci, 4(3)(1998), 387-393

[14] B.Shapiro, K.Zhang, Comparing multiple RNA secondary structures using tree comparisons, Computer.Appl.Biosci, 6(4)(1990),309-318

[15] K.Zhang, Computing similarity between RNA secondary structures , Pro.IEEE.Internat. Joint Symp On Intelligence and Sytems Rockviue,Maryland.May, 1998, 126-132

[16] Koper-Zwarthoff, E.C., Brederode, F.Th., Walstra, P., Bol,J.F., Nucleic Acids Research, 7(1979), 1887-1900.

[17] Scott, S.W. and Ge, X., J.Gen.Virol., 76(1995),957-963

[18] Koper-Zwarthoff, E.C., Brederode, F.Th., Walstra, P., Bol,J.F., Nucleic Acids Research\em, 8(1980), 3307-3318.

[19] Cornelissen B.J.C, Janssen H., Zuidema D., Bol J.F., Nucleic Acids Research\em, 12(1984), 2427-2437.

[20] Alrefai R.H., Shicl P.J, Domier L.L., D'Arcy C.J., Berger P.H., Korban S.S., J.Gen.Virol.,75(1994),2847-2850

[21] Scott, S.W. and Ge, X., J.Gen.Virol., 76(1995),1801-1806

[22] Bachman E.J., Scott S.W., Xin G., Bowman Vance V.,Virology , 201(1994),127-131

[23] Houser-Scott F.,Baer M.L., Liem K.F.,Cai J.M., Gehrke L., J.Virol 68(1994), 2194-2205

[24] EMBL/GenBank/DDBJ databases. Accession no.X86352.

VSP International
Science Publishers
P.O. Box 346, 3700 AH Zeist
The Netherlands

Lecture Series on Computer
and Computational Sciences
Volume 1, 2004, pp. 313-316

Frameworks for Intelligent Shopping Support

W.S. Lin[1], N. Cassaigne

Department of Computation,
University of Manchester Institute of Science and Technology,
PO BOX 88 M60 1QD, Manchester, UK

Received in 15 July, 2004; accepted in revised form 5th September, 2004

Abstract: This paper is motivated by the objective to reduce the gap between e-marketing strategies and agent technology by presenting a research framework to design an intelligent shopping agent's knowledge base. Three hypotheses are tested in this paper: (1) e-marketing terms are important in influencing shoppers' decisions to choose an online shop, (2) different shoppers have different shopping styles and these are influenced by e-marketing terms, (3) shoppers behave differently with respect to different types of products. The research framework is two folds, considering on one hand the e-marketing strategies used by vendor, and on the other hand shopper's decision-making style. The experimental results show that shoppers perceive the e-marketing terms identified in this paper to be determinant when shopping online. In addition, the styles of shoppers are significant in indicating shoppers' differences in decision making when choosing an online shop. According to their style, different shoppers perceive marketing terms in a different way. The solution proposed to support intelligent shopping agents to mimic shoppers' decision style when faced to e-marketing strategies focuses on the creation of a knowledge base composed of (1) an ontology of e marketing terms and (2) a number of decision-making rules for each different shopping style.

Keywords: Electronic commerce, e-marketing strategy, intelligent shopping agent, knowledge base development

1. Introduction

The world is changing quickly, with an ever increasing amount of information and knowledge becoming available thanks to technology such as the World Wide Web. At the same time, however, human beings' capability to effectively and efficiently access and assess a huge amount of data to discover relevant information has not increased accordingly. To overcome this human weakness, intelligent shopping agents have been developed in the domain of electronic commerce to carry out searches on behalf of shoppers. Intelligent shopping agents access web sites and compare products based on price and a limited number of keyword search criteria. By doing so, they emulate only a very small part of the shopper's decision-making process [1].

The prospect of executing an online information search presents several challenges and problems as discussed in the literature [2, 3, 4]. Within the context of electronic commerce, intelligent agent technology promises substantial increases in productivity by automating several of the time-consuming stages of searching or buying online. Currently, there are several shopping agents available online, that are mainly used to search for the 'best buy' regarding a particular product [5, 6, 7, 8]. However, maintaining the agents' searching capability at a high level of efficiency remains elusive. Also, this type of intelligent agent does not really know what shoppers want and what their decision style is. To overcome this problem this paper claims that a smarter intelligent agent should categorize shoppers into six decision-making styles in order to understand their information needs and the attraction that e-marketing strategies might have on shoppers' decisions.

This paper presents the conceptual models of an intelligent shopping agent's knowledge base. It propounds an approach to improving the level of agent intelligence by considering the influence of e-marketing terms and online shoppers' decision-making styles. Then, the evaluations results of a user

---

[1] Corresponding author. IEEE member. E-mail: wenshan_lin@yahoo.co.uk.

experiment are presented with the interpretations of results. Finally, the paper concludes with a discussion including suggestions for future work.

2. Research solution and methodology

This paper proposes to apply a soft computing approach, especially knowledge modeling, to acquire essential knowledge about what online shoppers want and what exists online. It is claimed that the e-marketing strategies applied by online vendors to sell or promote products online exert an influence on shoppers' search activities. At the same time, shoppers do not all adopt the same style; the price of products, for example, is of primary concern to some but not to others. Shoppers have different styles of decision-making when shopping online. In other words, this paper aims at improving the intelligence level of internet shopping agents in a user-centric manner. The research objectives we aim to achieve are: (1) to acquire and present a number of e-marketing terms that affect shoppers' decision-making; (2) to model online shoppers' decision-making styles; (3) to validate the research framework with potential customers.

An intelligent shopping assistant's knowledge base is designed using an ontology-driven approach integrated with problem-solving methods. The notion of ontology is used to hierarchically present the e-marketing terms contained in and used by web sites, particularly web storefronts -the first page of commercial web sites- by specifying synonymous and related terms. Conceptually, the e-marketing framework relies on the relatively established literature of traditional marketing and the emerging literature on electronic commerce, seeking to discover the antecedents of online purchase intentions.

The framework of shopper's decision-making style, together with their decision trees, is modeled. Each decision-making style is linked with a number of e-marketing terms in the sense that it is demonstrated that these e-marketing terms have notable value for shoppers whilst shopping online. Intelligent shopping agents' search is supported by rules that are produced based on these shoppers' decision trees to mimic shoppers' behavior according to their decision-making styles.

The research framework was experimentally applied to general online shopping on one hand and online book-shopping on the other hand. The experimental results validate the research contributions claimed in this paper. It also demonstrates the applicability of the research framework to real cases.

In the following sections, the research framework of e-marketing strategies and shoppers' decision-making styles are presented.

3. Research Framework and Evaluation Results

Web stores are developed to represent the e-marketing strategies applied by vendors [9, 10]. These e-marketing strategies are presented as e-marketing terms and displayed in web stores to convey marketing messages to sell and promote products online. Online shoppers' decision-making styles are defined by [11, 12] as an evaluation process regarding the information offered by the web stores. It indicates that there is a connection between the e-marketing strategies applied by vendors, and shoppers' decision-making styles. Therefore, we propose on one hand a framework of e-marketing strategies derived from general marketing strategies and resulting in the creation of an ontology of e-marketing terms including synonyms, relevant terms and contextual rules, and on the other hand a framework of shoppers' decision-making styles derived from the literature of consumer behavior and market segmentation strategies. In the following sections, the two frameworks of e-marketing strategies and shoppers' decision-making styles are further illustrated. Also, the evaluation results based on a user experiment of 124 participants are presented.

3.1 Framework of e-marketing strategies

E-marketing strategies form a new environment for intelligent shopping assistants to work and find information for shoppers. One of the search features of intelligent shopping assistants is to search for product and service information in order to match the profiles of web stores and shoppers'. The framework of e-marketing strategies provides the full integrated definitions of product and service. This framework is translated into ontology and contextual rules for assisting intelligent shopping assistants to take into account shopper's search factors. This framework aggregates and selects other researchers' works about marketing strategies and models a framework of e-marketing strategies that are most suitable for categorising the e-marketing strategies. This paper considers six main e-marketing strategies: place, the web stores; computer-mediated electronic commerce; (3) product; (4) promotion; (5) price and (6) process. For each strategy, there are a number of sub-classes of e-marketing terms. In total 22 e-marketing terms were identified as important in affecting shoppers' decision-making behavior. Figure 1 presents the ontology hierarchy of e-marketing terms. The evaluative results show that: (1) these 22 e-marketing terms are reliable with high internal consistency (alpha value is over 0.7), and (2) they are important in influencing shoppers' decisions: (a) all 22 terms are influential for general shopping, while (b) 'the availability of gift service' is the only e-marketing term not to be perceived as so important for online book-shopping.

Figure 1. A demonstration of ontology hierarchy of e-marketing terms

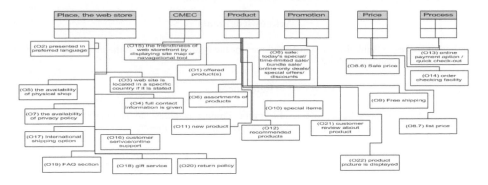

3.2 Framework of shoppers' decision-making styles

In the field of traditional marketing segmentation, shoppers in physical shops are grouped into several categories according to their decision-making styles. Vendors apply marketing segmentation strategy to target different styles of shopper in order to meet shoppers' needs and maximize business profits. In other words, different product information is provided for meeting shoppers' varying needs. Taking this notion into account, strategies of traditional marketing-segmentation are adapted for intelligent shopping agents to match the profiles of web stores and styles of shoppers. Table 1 presents the six styles of online shoppers used in this study. Shoppers' decision making styles, combined with the e-marketing terms that influence them, are translated into decision-making trees and applied to online book shopping. The evaluative results show that while shoppers have a predominant decision-making style that they use for both general online shopping an online book-shopping, shoppers perceive the importance of some e-marketing terms differently depending on the type of shopping. The full length paper will present examples of these decision trees and rules.

Table 1: Shoppers' decision-making styles

Shoppers' decision-making styles	
Decision style	Information needs
General-purpose shopper	Web sites designed to diverse audiences with necessary pieces of information
Security-concerned shopper	Responsive web sites
Value shopper	Best value
Fashionable shopper	Attractive and entertaining web sites
Time-sensitive shopper	Useful web sites with high ordering efficiency and well organised contents
Service-oriented shopper	Web sites with high helpfulness in terms of customer service and functional features

4. Conclusions and Further

This paper demonstrates that e-marketing terms are important in influencing shoppers' decision to choose an online shop. Different shoppers have different shopping styles and these are influenced by e-marketing terms. In other words, shoppers perceive different important level with respect to a number of e-marketing terms and behave differently when buying different types of products. It indicates the possibility to amend shoppers' decision trees for other types of buying. The evaluative results prove that it is substantiated to apply the research framework of decision styles for cases of general online shopping and online book-shopping. For the future, the knowledge base can be used to develop a real intelligent shopping agent. There are also other research opportunities such as applying this research to other type of products, in addition, the ontology of e-marketing terms will be evolved. New terms can be added by online marketers, knowledge engineers or shoppers.

Acknowledgments

The author wishes to thank the anonymous referees for their careful reading of the manuscript and their fruitful comments and suggestions.

References

[1] S. Stanoevska-Slabeva, Internet Electronic Product catalogs: an approach beyond simple keywords and multimedia, *Computer Networks*. **32** 701-715 (2000)

[2] V. Anupam, R. Hull and B. Kumar, *Personalising e-commerce applications with online heuristic decision-making*, Proceedings of 10th World Wide Web Conference, Hong Kong, 2001

[3] J. Domingue, M. Martins, J. Tan, A. Stutt and H. Pertusson, Alice: Assisting online shoppers through ontologies and novel interface metaphors, *Knowledge Engineering and Knowledge Management: Ontologies and the Semantic Web* (Editor: Gómez-Pérez and V. R. Benfamins), Berlin Heidelberg, Springer-Verlag, (2002), 335-351.

[4] F. Menczer, W. N. Street and N. Vishwakarma, *IntelliShopper: A proactive, personal, private shopping assistant*, Proceedings of the First International Joint Conference on Autonomous Agents and Multiagent Systems, Palazzo RE Enzo Bologna, Italy, 2002.

[5] P. Resnick, *Roles for electronic brokers*, [online] Available at: http://ccs.mit.edu/papers/CCSWP179.html, (1994). [Access Date: 10th September, 2003]

[6] C. Dellarocas and M. Klein, *Designing robust, open electronic marketplaces of contract net agents*, Proceedings of 20th International Conference on Information Systems, Charlotte, North Carolina, USA, 1999.

[7] B.N. Grosof, *Building commercial agents: An IBM research perspective.* Proceedings of the second International Conference on the Practical applications of Intelligent Agent and Mulit-agent Technology, London, 1997

[8] H. Tu and J. Hsiang, *An architecture and category knowledge for intelligent information retrieval agents*, Decision Support Systems 28(3) 225-268(2000)

[9] P. McGoldrick, *Retail Marketing*. New York, McGraw-Hill, 2002.

[10] D. Chaffey, R. Mayer, K. Johnston and F. Ellis-Chadwick, *Internet Marketing Strategy, implementation and practice*, Pearson Education, Financial Times, 2000.

[11] Jarvenpaa, S. L. and P. A. Todd, Consumer Reactions to electronic Shopping on the World wide Web, *International Journal of electronic Commerce*, 1(2) 59-88(1997)

[12] M. Solomon, G. Bamossy and S. Askegaard, *Consumer Behaviour: A European Perspective*, New Jersey, USA, Prentice Hall Inc., 1999.

VSP International
Science Publishers
P.O. Box 346, 3700 AH Zeist
The Netherlands

Lecture Series on Computer
and Computational Sciences
Volume 1, 2004, pp. 317-319

Monte Carlo Simulation of Gas Adsorption in Single Walled Nanotubes

George P. Lithoxoos and Jannis Samios[1]

Laboratory of Physical Chemistry,
Department of Chemistry,
University of Athens,
Panepistimiopolis 15771, Athens,
Greece

Received 15 September, 2004; accepted in revised form 23 September, 2004

Abstract: The Grand Canonical Monte Carlo simulation technique was employed to study hydrogen   adsorption in Single Walled Carbon Nanotubes (SWCN) of armchair structure. The results show that SWCN do not satisfy the international standards for hydrogen adsorption.

Keywords: Monte Carlo, Carbon Nanotubes, Silicon Nanotubes, Hydrogen

Mathematics Subject Classification: 8208, 08B80, 82D99

1. Introduction

The present computational work aims to study the possible application of carbon and silicon nanostructured materials like nanotubes for the separation of gaseous mixtures as well as their suitability as storage means for several gases and especially hydrogen. Carbon nanotubes (CNTs) are molecular nanostructures that are consisted of a network of condensed carbon rings and sheets, such as graphite, which are rolled-up into cylinders with diameters up to several nanometers. CNTs are encountered in two forms depending on their production method. Tubes formed by only one single graphite layer are called single wall nanotubes (SWNT). Tubes consisting of multiple coaxial graphite layers are called multi-wall nanotubes (MWNT). These forms of tubes are presented in figure 1.

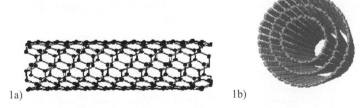

1a)   1b)

Figure 1: Single Walled Carbon Nanotubes (SWCN) (1a) and Multi Walled Carbon Nanotubes (MWCN) (1b).

After the discovering of CNTs [1], many researchers focused their interest on the investigation of possible applications of these new materials thus numerous experimental and theoretical studies have been conducted to study the adsorption of gases, such as NO_2, O_2, NH_3, N_2, CO_2, CH_4, H_2O, H_2, and Ar. The adsorption of gases is an issue of great significance not only from the perspective of basic research,

[1] Corresponding author. E-mail: isamios@cc.uoa.gr

but also from that of technical applications and it could lead to the development of new gas storage technologies with positive effects on everyday life.

Important theoretical and experimental efforts [2] have concentrated on the storage of hydrogen in nanotubes, as well as on the storage of methane, which is the main component of natural gas. Note that molecular hydrogen is the purest energy source and it could offer solution to the environmental pollution problem with use of vehicles of trivial gas emissions, which will operate with electrical motors that will be powered by H_2 fuels cells. Today it is widely accepted that materials based on carbon will form a basic field in the technology of the twenty first century. According to the above statements, the present study is focused on the development of theoretical models in order to investigate the storage of gases (e.g., hydrogen) and the separation of gaseous mixtures. This work also includes the study of model silicon (Si) nanostructures. Preliminary theoretical studies in our group have already shown that Si nanostructures are promising materials for hydrogen storage. The theoretical calculations are based on Statistical Mechanical simulations, such as Molecular Dynamics (MD) and Monte Carlo (MC). Other researchers have used these techniques so far for the simulation of similar systems [3]. The simulation techniques are employed here to study the particle microstructure in the tubes as well as the dynamic properties (e.g. diffusion processes) of stored materials in the tubes. Of great interest is the study of the density profile and of the diffusion of relatively small molecules inside nanotube systems, or pores, within a wide temperature and density range. The results of the theoretical studies are compared with experimental data available for other similar systems and potential "storage tanks" for condensed hydrogen and other gaseous fuels.

2. Model system calculations.

For a theoretical study of SWCN we considered a model consisting of an array of parallel nanotubes. The tubes are of armchair structure. The open ended cylindrical structure of carbon nanotubes enlists them among the promising adsorbents. Hydrogen can be stored either internally or externally in the space between tubes. The amount of molecular hydrogen that is stored in the system is dependent on the attraction induced by the graphite carbon atoms on hydrogen molecules. The interactions between hydrogen molecules are described by a site-site Lennard-Jones potential centered on the position of the two H atoms [4],

$$\upsilon_{LJ}(r) = 4\varepsilon_{H-H}\left[\left(\frac{\sigma_{H-H}}{r}\right)^{12} - \left(\frac{\sigma_{H-H}}{r}\right)^{6}\right]$$

with $\sigma_{HH} = 2.958$ A° and $\varepsilon_{HH}/k_B = 36.7$ K. A quadrupolar moment is taken into account by putting a point charge $q = 0.466|e|$ on both the hydrogen atoms and a charge $-2q$ on the center of mass. The carbon nanotubes we chose are the armchair (11,11) type consisting of 352 carbon atoms each. The simulation box is made up of three rows of three tubes each, making up a total of 9 nanotubes arranged on a two dimensional squared lattice.

First we performed NVT Monte Carlo simulation for the calculation of the chemical potential using Widom's umbrella sampling method. Then by the Grand Canonical Monte Carlo technique we obtained the gravimetric and volumetric densities of the adsorbed hydrogen and also the density profiles of hydrogen inside and outside the tube surface.

Table 1: Gravimetric densities of hydrogen in SWCN for various values of pressure and temperature.

P(Mpa)/T(K)	293.5	274	175
wt%	CN	CN	CN
10	2.68	2.82	3.96
5	1.88	1.98	3.26
1	0.61	0.64	1.1
0.1	0.05	0.053	0.1

The hydrogen density profiles obtained from the simulation are presented in figure 1.

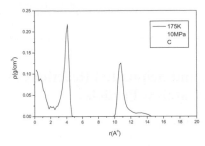
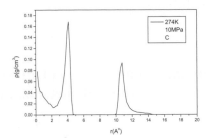

Figure 2: Density profiles of hydrogen in SWCN for 175K and 10Mpa (left) and 274K and 10Mpa (right).

The simulations results were fitted using the Langmuir isotherms with good results. Two of these isotherms are presented in figure 2.

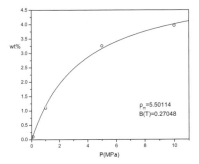
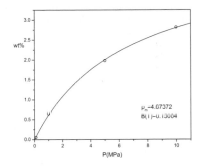

Figure 3: Langmuir isotherms at 175K (left) and 274K (right) of the adsorbed hydrogen in SWCN.

The simulation results of hydrogen adsorption in Single Walled Carbon Nanotubes are in agreement with previous experimental and theoretical studies and show that the international standards for hydrogen adsorption are not yet reached [5]. In contrast the results obtained from simulations of hydrogen adsorption in Silicon Nanotubes are much more encouraging for further investigation [6].

References

[1] S. Iijima, *Nature* 354 56 (1991)

[2] F.L Darkrim, P. Malbrunot and G.P. Tartaglia, *Int. J. Hydrogen Energy*, 27, 193, (2002)

[3] F.L Darkrim., D, Levesque, *J.Chem.Phys.* 109, 12, 4981 (1998)

[4] G. Garberoglio, R. Vallauri, *Physics Letters A,* 316 , 407, (2003)

[5] S. Hynek, W. Fuller, J. Bentley, *Int. J. Hydrogen Energy* 22 6 601 (1997)

[6] G. P. Lithoxoos and J. Samios, Paper in Preparation

VSP International
Science Publishers
P.O. Box 346, 3700 AH Zeist
The Netherlands

Lecture Series on Computer
and Computational Sciences
Volume 1, 2004, pp. 320-323

Dynamic Simulation of Diffusion and Sequential Reactions in Hydrodemetallization Catalyst Particles

F.X. Long and B.S. Gevert[1]

Department of Materials and Surface Chemistry,
Chalmers University of Technology,
SE-412 96 Göteborg, Sweden

Received 17 August, 2004; accepted in revised form 30 August, 2004

Abstract: A framework for dynamic simulation of an individual hydrodemetallization (HDM) particle under two-dimensional diffusion and reactions on common calculating software MATLAB 5.3 is built in this study. From the simulation results it is demonstrated that vanadyl etioporphyrin hydrodemetallization is a diffusion-limited sequential reaction with vanadyl etiochlorin as the reaction intermediate. Moreover, M-shaped profiles of metallic deposition on the catalyst particle in the radial direction are confirmed to be a result of the series reactions.

Keywords: Dynamic simulation; Finite element method; Hydrodemetallization; Intraparticle diffusion.

1. Introduction

Petroleum catalytic hydrodemetallization (HDM) that serves as the first stage of residue hydroprocessing in refineries is a process of industrial importance. It is well known that residual oils are characterized by high contents of large molecular species such as asphaltenes and heteroatoms existing in hydrocarbon derivatives. The molecular size distributions of the organometallic compounds in residua are found in the range of 25-150 Å with the peaks around 50 Å, which is comparable to that of the catalyst pores [1]. Therefore, the restrictive diffusion does happen to HDM catalysts in most industrial applications. Moreover, during the demetallization, vanadium and nickel the most abundant and problematic metals in residua are removed from the stream and deposit on the catalyst surface in metal sulfides. This progressive buildup of the metal deposition will decrease the effective diffusivity and ultimately control the running-length of the process. Therefore, a good understanding and a realistic modeling of the simultaneous diffusion and reactions in HDM catalyst particles will establish a basis for developing improved HDM catalysts and reactors.

The objective of this study is to build a framework for dynamic simulation of an individual HDM particle under two-dimensional diffusion and reactions on common calculating software MATLAB, which is essential to analyzing more complex situations involving catalyst deactivation and catalyst optimization.

2. Mathematical Modeling

The deposited metals on the HDM particles situated in the entrance of the reactor have been well documented by the microprobe to have a characteristic concentration profile resembling the letter M in the radial direction, i.e. the maximum metal deposition is located inside the particle [2]. This profile "record" provides researchers with an unusual advantage to extract information about the diffusion and reactions in HDM catalyst particles, which would otherwise be difficult to obtain. Of course, a successful model should accurately predict this deposition profile.

Unfortunately, according to the classical theory of the pore diffusion controlled reaction, a U-shaped deposition profile is always resulted in if a single irreversible reaction scheme is assumed. Therefore,

---

[1] Corresponding author. E-mail: gevert@chem.chalmers.se.

although the power law rate expression of the first or the second order has been frequently used to the HDM reaction in the literature, the simplified kinetics is obviously not appropriate to the real situation. In their studies using model compound, Agrawal & Wei proposed a sequential reaction mechanism for the HDM reaction [3, 4]:

$$\text{VO-EP} \underset{k2}{\overset{k1}{\rightleftharpoons}} \text{VO-EPH}_2 \xrightarrow{k3} \text{DEPOSITS}$$

where VO-EP is referred to vanadyl etioporphyrin the representative vanadium species in crude oils, VO-EPH$_2$ denotes the dihydrogenated intermediate vanadyl etiochlorin. All steps in the model (k_1-k_3) were assumed to be the first order in metal concentrations. Also, they suggested that the M-shaped deposition profile was resulted from the series reactions. So in this study we take the consecutive kinetic scheme and simulate the diffusion coupled with a two-step sequential reaction to test if the inner maximum will appear as suggested.

A typical particle of commercial HDM catalysts is a cylindrical extrudate with a large ratio of length to diameter. Nevertheless, short cylinders or purposely-cracked particles are not unusually utilized especially in laboratories. In order not to neglect the end effects, we consider a more complex geometry of the particle. Specifically, instead of an infinite cylinder for simplification, a cylinder with a diameter of 0.8 mm and a length of 1 mm is adopted in this work as a representative. Thereby, we will model the diffusion and reactions inside a rectangle occupying a volume Ω in R^2 with the boundary Γ over a time interval $I = [0\ T]$.

Based on the above discussions, the mass balance inside the catalyst particle can be written as:

$$\frac{du_m}{dt} + \nabla \bullet (-De_m \nabla u_m) = f_m \tag{1}$$

where $m = 1, 2, 3$ representing VO-EP, VO-EPH$_2$ and the metal deposition respectively; u_m is the concentration inside the particle; De_m denotes the effective diffusivity while De_3 should be 0 since the third specie is solid; and

$$f_1 = \rho_c(-k_1u_1 + k_2u_2) \tag{2}$$
$$f_2 = \rho_c[k_1u_1 - (k_2 + k_3)u_2] \tag{3}$$
$$f_3 = \rho_c k_3 u_2 \tag{4}$$

A homogeneous Robin boundary condition is chosen to describe the boundary condition for all of the three species. Robin boundary conditions correspond to the concentration flux through the boundary being proportional to the difference of the concentration u_m inside and a given concentration g_m outside Ω:

$$-\frac{\partial u_m}{\partial n} = a_m(u_m - g_m) \tag{5}$$

For VO-EP and VO-EPH$_2$, a large number (10000) is set to a_1 and a_2 to approximate the assumption that the external mass-transfer resistance to the particle is negligible, while a_3 should equal to 0 again because the third specie is solid.

3. Numerical Methods

The partial differential equations (PDE) of Equation (1) are a typical parabolic problem with the initial conditions of $u_m(x,0) = 0$ for $x \in \Omega$ in R^2. Assuming a triangular mesh on Ω, the solution of the equation $u_m(x,t)$ can be approximated in the finite element method basis by

$$u(x,t) = \sum_i U_i(t)\Phi_i(x) \tag{6}$$

Plugging this expansion into the PDE, multiplying with a test function Φ_j, integrating over Ω, and applying Green's formula and the boundary conditions yield:

$$\sum_i \int_\Omega \Phi_j \Phi_i dx \frac{dU_i(t)}{dt} + \sum_i \left(\int_\Omega \nabla\Phi_j \bullet (D_e\nabla\Phi_i) + \int_\Gamma a\Phi_j\Phi_i ds \right) U_i(t) = \int_\Omega f\Phi_j dx + \int_\Gamma ag\Phi_j ds \qquad (7)$$

If we define the following notions:

$$M_{i,j} = \int_\Omega \Phi_j \Phi_i dx \qquad (8)$$

$$D_{i,j} = \int_\Omega (D_e\nabla\Phi_i) \bullet \nabla\Phi_i dx \qquad (9)$$

$$K_{i,j} = \int_\Gamma a\Phi_j\Phi_i ds \qquad (10)$$

$$F_j = \int_\Omega f\Phi_j dx \qquad (11)$$

$$G_j = \int_\Gamma ag\Phi_j ds \qquad (12)$$

Equation (7) can be rewritten as

$$M\frac{dU}{dt} + (D+K)U = (F+G) \qquad (13)$$

which is a large and sparse system of the ordinary differential equations (ODE) with the initial value $U_i(0) = 0$.

Further, in order to solve Equation (13) numerically, we discretize the time in continuous piecewise linear functions as well, and obtain

$$U_i(t) = U_i^{n-1}(x)\Psi_{n-1}(t) + U_i^n(x)\Psi_n(t) \qquad (14)$$

Plugging this expansion into equation (13) and integrating over Ω lead to the following sequence of systems of equations:

$$U^n = (M + \frac{dt}{2}D + \frac{dt}{2}K)^{-1} \cdot \left[(M - \frac{dt}{2}D - \frac{dt}{2}K)U^{n-1} + dtF + dtG \right] \qquad (15)$$

where dt is the tiny time step chosen.

The ODE system is an initial value problem (IVP) and ill conditioned. Therefore, the adaptive time step control involving automatic choice of the time step to satisfy a tolerance on the error is employed in this work. Moreover, since the coefficient matrix F is of function of U that is time dependent, the ODE system is also nonlinear. Thereby, re-evaluating and re-factorizing the matrix F are necessary while the fixed point iteration method is used to solve the nonlinear system. For the systems of the PDE in our case, we first arrange U in the form of $U = [U_{VO-EP} \;\; U_{VO-EPH2} \;\; U_{deposits}]'$ and then the coefficient matrices M, D, K, F and G are changed correspondingly.

The key issues for developing the codes on MATLAB are: (i) Defining the geometry and meshing; (ii) Factorizing the coefficient matrices that are not time dependent; (iii) Time evolving while re-evaluating and re-factorizing the matrix that is time dependent in each time-step; (iv) Solving U at the point of time; (v) Recalculating the time-step by satisfying a tolerance on the error; (vi) Plotting and visualizing.

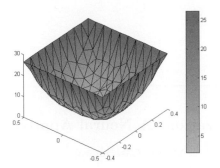 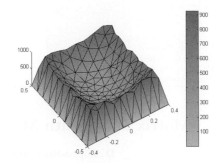

Figure 1: Concentration Profile of Vanadyl Etioporphyrin (VO-EP) after 1 h on Stream.

Figure 2: Concentration Profile of the Metal Deposits after 1 h on Stream.

4. Results and Discussion

After inputting the corresponding parameter values adopted from the experimental results of Run VE-6 [3, 4], the process of simultaneous diffusion and sequential reactions in a catalyst particle was vividly visualized, which was helpful to understand the whole HDM process. At the end, the concentration changes inside the particle were animated and the final concentration profiles of VO-EP and metal deposits were plotted as shown in Figure 1 and Figure 2, respectively. The simulation results indicated that the hydrodemetallization of vanadyl etioporphyrin was a diffusion-limited reaction, since VO-EP was unable to penetrate the catalyst particle deeply. The concentrations of VO-EPH$_2$ with the maximum inside the boundary did not build up during the process, demonstrating that VO-EPH$_2$ was a reaction intermediate. M-shaped deposition profiles in the radial direction were obtained as shown in Figure 2, confirming the suggestion that the M-shaped profile was a result of a series reaction. Furthermore, since dense deposition inside particles could decrease effective diffusivity, catalysts with a large pore size and a high pore volume were demanded. Finally, for short cylindrical particles the diffusion from the ends was significant, so the end effects could not be neglected.

References

[1] M.C. Tsai and Y.W. Chen, Restrictive Diffusion under hydrotreating reactions of heavy residue oils in a trickle bed reactor, *Ind. Eng. Chem. Res.* **32** 1603-1609(1993).

[2] P.W. Tamm, H.F. Harnsberger and A.G. Bridge, Effects of feed metals on catalyst aging in hydroprocessing residuum, *Ind. Eng. Chem. Process Des. Dev.* **20** 262-273(1981).

[3] R. Agrawal and J. Wei, Hydrodemetalation of nickel and vanadium porphyrins. 1. intrinsic kinetics, *Ind. Eng. Chem. Process Des. Dev.* **23** 505-514(1984).

[4] R. Agrawal and J. Wei, Hydrodemetalation of nickel and vanadium porphyrins. 2. intraparticle diffusion, *Ind. Eng. Chem. Process Des. Dev.* **23** 515-522(1984).

VSP International
Science Publishers
P.O. Box 346, 3700 AH Zeist
The Netherlands

Lecture Series on Computer
and Computational Sciences
Volume 1, 2004, pp. 324-327

Matrix Bandwidth Minimization: A Neural Approach

D. López-Rodríguez[1] and E. Mérida-Casermeiro[2]

Department of Applied Mathematics,
E. T. S. I. Informática,
University of Málaga
29071 Málaga, Spain

Received 30 July, 2004; accepted in revised form 28 August, 2004

Abstract: The aim of this paper is to present a neural technique to tackle the bandwidth minimization problem. This neural model, successfully applied to other problems, presents better solutions than those of other classical approaches to this problem, as shown by the simulations performed.

Keywords: Bandwidth Minimization problem. Multivalued neural network. Combinatorial Optimization.

Mathematics Subject Classification: 05C78 (Graph labelling), 05A05 (Combinatorial choice problems).

PACS: 120304

1 Introduction

Bandwidth Minimization Problem (BMP) is one of the most recurrent themes in literature, since it has many applications in science and engineering, not only in linear algebra. The aim of BMP is to find a reordering scheme of rows and columns of a matrix, i.e. a permutation of its rows and columns, such that nonzero elements get as close as possible to the main diagonal.

This problem can be derived from graph theory in terms of minimizing the bandwidth of the adjacency matrix associated to a given graph.

So, BMP has indeed many optimization applications [6], from both linear algebra and graph theory, including circuit design, large hypertext media storage, VLSI design, finite element method for partial differential equations and resolution of large linear systems. In this last case, if matrix size is $N \times N$, Gaussian elimination can be carried out in $O(Nb^2)$ if matrix bandwidth is $b < N$, instead of a much slower $O(N^3)$.

BMP is also interesting due to its NP-completeness, proved in the general case in the middle 70's [10]. In the case of minimizing a graph bandwidth, Garey et al. [4] showed that the problem is NP-complete even if the maximum vertex degree in the graph is 3. So, the intractability of this problem is assured and thus there is need for algorithms to achieve near optimal solutions.

In the late 60's, E. Cuthill and J. McKee [2] introduced the first algorithm, CM, to approach BMP by constructing a level structure of the associated graph. Subsequently, Liu and Sherman [7] showed that reversing Cuthill-McKee's ordering can never increase the bandwidth of the matrix.

---

[1]Corresponding author. E-mail: dlopez@ctima.uma.es
[2]E-mail: merida@ctima.uma.es

Another approach based on the level structure of the associated graph is Gibbs-Poole-Stockmeyer (GPS) algorithm [5], which achieved results comparable to those of CM, with the advantage of being less time-consuming.

A Simulated Annealing procedure was presented by Dueck and Jeffs (1995). Although it finds better results than GPS or RCM in some particular cases, it should be noted that it takes up to 2000 times longer.

Some another approaches consist in a spectral decomposition of the matrix [1], although they are focused on minimizing the 'work-bound' of the matrix, i.e. the cost of a Cholesky factorization, and bandwidth of the permuted matrix can be high. In 2001, Corso and Romani modified this spectral technique to reduce both bandwidth and 'work-bound', but in general their algorithm performed better than RCM only when RCM was used at a preprocessing stage.

Since neural networks have been successfully applied to many other combinatorial optimization problems [8, 9, 11], improving the solutions given by classical algorithms, we expect that the model proposed in this paper achieves better results than other techniques.

2 Definition of the Problem

Bandwidth Minimization Problem (BMP) can be stated from two closely related points of view, one being matrix theory and the other being graph theory.

Let $M = (m_{i,j})$ be a $N \times N$ real symmetric matrix, with zero-diagonal. The bandwidth of M is defined as $w(M) = \max_{i,j=1,\ldots,N}\{|i - j| : m_{i,j} \neq 0\}$.

BMP consists in searching for a permutation of the rows and the columns in M such that its nonzero elements lie as close as possible to the main diagonal, making the bandwidth of the new matrix minimum. Denoting by $\mathcal{S}_N$ the set of permutations of $\{1,\ldots,N\}$ and by $M_\alpha = (m_{\alpha(i),\alpha(j)})$ $(\alpha \in \mathcal{S}_N)$ the permuted matrix, we look for a permutation $\sigma \in \mathcal{S}_N$ such that $w(M_\sigma) = \min_{\alpha \in \mathcal{S}_N} w(M_\alpha)$.

¿From the viewpoint of graph theory, BMP arises when all nodes in a given graph must be optimally labelled. So, let $G = (V, E)$ be an undirected graph with no self-connections, where $V = \{v_i\}$ is the set of vertices and E is the edge set. A *labelling* (or *numbering*) of G is a bijective function $\phi : V \to \{1, 2, \ldots, N\}$ (where N is the cardinality of V) that assigns an index to each vertex. Then, the adjacency matrix of G with respect to ϕ is $A_\phi = (a_{\phi(i),\phi(j)})$, $a_{\phi(i),\phi(j)} = 1$ if and only if the edge connecting $v_{\phi(i)}$ and $v_{\phi(j)}$ exists. In this context, BMP consists in looking for a labelling that minimizes the bandwidth of the associated adjacency matrix.

So, this formulation is a particular case of the former one. Then, we can restrict ourselves to the matricial case.

3 The Neural Model

The proposed neural model (MREM) consists in a group of N neurons whose outputs, denoted by $\{s_1, \ldots, s_N\}$, can take values in a finite set F, usually $F \subset \mathbb{Z}^+$, although F needs not be numerical. Then, the state of the i-th neuron is characterized by s_i. So, the state of the net can be expressed as a vector, called state vector, $\vec{S} = (s_1, \ldots, s_N) \in F^N$. In our case, $F = \{1, 2, \ldots, N\}$ and the only feasible states for the net are given by state vectors $\vec{S}$ associated to permutations in $\mathcal{S}_N \subset F^N$, such that $s_i = \sigma(i)$ for all $i \in F$ and a fixed $\sigma \in \mathcal{S}_N$. Thus, we can identify $\vec{S} \in \mathcal{S}_N$.

In order to solve BMP, an energy function, based entirely on the problem, is associated to state vectors. Given a matrix $M = (m_{i,j})$, this function can be defined in terms of the bandwidth w' of the permuted matrix $M_{\vec{S}} = (m'_{i,j}) = (m_{s_i,s_j})$, and the number $n_{w'}$ of nonzero elements that lie on the diagonals w'-th and $(-w')$-th of $M_{\vec{S}}$.

Table 1: Performance Results.

N	ρ	Best	Av.	RCM	N	ρ	Best	Av.	RCM
10	0.1	1	1.45	1	50	0.1	19	21.76	28
10	0.3	3	4.15	4	50	0.3	33	35.50	40
10	0.5	5	5.44	6	50	0.5	39	41.13	42
10	0.7	6	6.68	8	50	0.7	42	43.97	46
10	0.9	7	7.69	7	50	0.9	45	46.43	46
25	0.1	5	6.92	10	100	0.1	54	58.69	71
25	0.3	12	13.59	17	100	0.3	78	81.46	85
25	0.5	15	17.64	19	100	0.5	86	88.88	92
25	0.7	18	20.00	22	100	0.7	90	92.87	95
25	0.9	22	22.20	23	100	0.9	95	96.03	97

So, the energy function is $E(\vec{S}) = w' + \frac{n_{w'}}{A}$ where A is the number of nonzero elements in M and $n_{w'} = \sum_{k=1}^{N-w'} \left[(1 - \delta_{m'_{k,k+w'},0}) + (1 - \delta_{m'_{k+w',k},0}) \right]$.

The dynamics proposed in this paper attempts to minimize E by updating two neuron outputs, interchanging their values. Only changes that produce a state vector with less energy are allowed, and so, in each step, the energy function will never be increased.

First, the network computes the energy of a randomly generated permutation. At any time, the scheduling selects cyclically two neurons (p and q), and the network computes the increase of energy, resulting of the interchange of their outputs, with respect to the previous saved energy value. If this increase is negative, the next state vector is given by:

$$s_i(t+1) = \begin{cases} s_p(t), & \text{if } i = q \\ s_q(t), & \text{if } i = p \\ s_i(t), & \text{if } i \notin \{p, q\} \end{cases}$$

and the new value of the energy function is saved. This process is repeated until there is no change in the state vector that could decrease the energy. Therefore, the network always achieves a local minimum state in which the bandwidth is minimized.

4 Simulation Results

We have compared the neural algorithm proposed in this paper to the classical RCM algorithm due to its efficiency, as mentioned in the introduction. As RCM is implemented for MATLAB (symrcm command), our algorithm was also implemented and tested under such environment, on a Pentium III processor (451 Mhz).

The test set was formed by random binary symmetric matrices. These matrices were build depending on two parameters, N (size) and ρ (density, meaning that the number of 1's in a matrix is the closest integer to $\rho N(N-1)$). In order to cover a wide range of cases, N was taken from $\{10, 25, 50, 100\}$ and $\rho \in \{0.1, 0.3, 0.5, 0.7, 0.9\}$. Thus, for each N and ρ as above, a total of 100 runs were performed on the test matrix.

Table 1 shows the average and best results of our simulations. So, we can confirm that our neural model improves RCM solutions, giving, in every case, a value for the best bandwidth indeed lower than that of RCM, and achieving, in most cases, better solutions on average, specially when performed over low-density matrices. Note that RCM always produces the same solution, however this neural model achieves better optimal solution by initializing with different state vectors.

In order to improve the knowledge about the performance of the proposed model, a problem obtained from a real situation has been analyzed. The studied matrix is the adjacency matrix of a graph, whose nodes represent some cities and edges representing that it exists a road between them.

For this problem, the number of cities is 13, and the adjacency matrix has its non zero elements {(1,2), (1,6), (1,9) (1,12), (2,3), (2,13), (3,4), (3,13), (4,5) (4,11), (6,7), (6,8) (8,9), (9,10), (11,12)} and their symmetric ones. For this matrix the RCM algorithm obtains a bandwidth of 4, while the proposed model achieves an average of 4.08 and a best bandwidth of 3 (after 100 iterations). Note that, in general, RCM algorithm always obtains the same solution for a problem, while the network achieves different solutions that can improve the best previous result.

References

[1] S. T. Barnard, A. Pothen and H. D. Simon. *A spectral algorithm for envelope reduction of sparse matrices.* Journal Num. Lin. Alg. with Appl. **2**, 317-334, 1995.

[2] E. Cuthill and J. McKee, *Reducing the bandwidth of sparse symmetric matrices.* Proc. ACM National Conference, Association for Computing Machinery, New York, 157-172, 1969.

[3] G.M. Del Corso and F.Romani, *Heuristic Spectral Techniques for the reduction of Bandwidth and Work-bound of Sparse Matrices,* Numerical Algorithms, **28(1-4)**, 117-136, 2001.

[4] M.R. Garey and D.S. Johnson, *Computers and Intractability. A guide to the theory of NP-Completeness.* W. H. Freeman and Company, New York, 1979.

[5] N. E. Gibbs, W. G. Poole and P. K. Stockmeyer. *An algorithm for reducing the bandwidth and profile of a sparse matrix.* SIAM J. Numer. Anal. **13**, 236-250, 1976.

[6] Y. Lai and K. Williams, *A survey of solved problems and applications on bandwidth, edgesum, and profile of graphs,* J. Graph Theory, **31**, 75-94, 1999.

[7] J.W.H. Liu and A.H. Sherman. *Comparative analysis of the Cuthill-McKee and the reverse Cuthill-McKee ordering algorithms for sparse matrices.* SIAM J. Num. Anal. **12**,198-213, 1975.

[8] E. Mérida-Casermeiro, G. Galán-Marín and J. Muñoz-Pérez. *An Efficient Multivalued Hopfield Network for the Traveling Salesman Problem.* Neural Processing Letters, **14**, 203-216, 2001.

[9] E. Mérida-Casermeiro, R. Benítez-Rochel and J. Muñoz-Pérez. *Neural Implementation of Dijkstra's Algorithm.* Lecture Notes in Computer Science, **2686**, 342-349, 2003.

[10] C. Papadimitriou, *The NP-completeness of the Bandwidth Minimization Problem,* Computing, **16**,263-270, 1976.

[11] K. A. Smith, *Neural Networks for Combinatorial Optimization: A review of more than a decade of research.* INFORMS Journal on Computing **11-(1)**, 15-34, 1999.

VSP International
Science Publishers
P.O. Box 346, 3700 AH Zeist
The Netherlands

Lecture Series on Computer
and Computational Sciences
Volume 1, 2004, pp. 328-331

Non-standard Finite Difference Method for Self-adjoint Singular Perturbation Problems

Jean M.-S. Lubuma [1] and **Kailash C. Patidar**

Department of Mathematics and Applied Mathematics
University of Pretoria, Pretoria 0002
South Africa

Received 2 August, 2004; accepted in revised form 15 August, 2004

With a smooth function $f(x)$, we associate the self-adjoint singularly perturbed two point boundary value problem

$$-\varepsilon\left(a(x)y'\right)' + b(x)y = f(x) \quad \text{on} \quad (0,1)$$
$$y(0) = \eta_0, \quad y(1) = \eta_1 \tag{1}$$

where η_0, η_1 are given constants, $0 < \varepsilon << 1$, $a(x)$ and $b(x)$ are sufficiently smooth functions satisfying some positivity and boundedness conditions.

By a standard change of dependent variable, Problem (1) is reduced to the equivalent normal form

$$-\varepsilon V'' + W(x)V = Z(x)$$
$$V(0) = \alpha_0, \quad V(1) = \alpha_1, \tag{2}$$

which we will consider, assuming that $W(x)$ is strictly positive. The main concern with Problem (1) or (2) is its dissipativity nature, which means that the solution has a rapidly varying component that decays exponentially (dissipates) away from a localized breakdown or discontinuity point in the layer region(s) as $\varepsilon \to 0$ [2, 3]. As a result, classical numerical methods fail to provide reliable discrete solution.

Let n denote a positive integer and let the interval $[0, 1]$ be divided into n equal parts through the nodes $x_m = mh$, $m = 0(1)n$ where $h = 1/n$ is the "mesh size".

In this work, we use the non-standard finite difference method introduced by Mickens [4] and defined formally as in [1]:

Definition *A difference equation to determine approximate solutions y_m to the solution $y(x)$ of the problem (1) is called a non-standard finite difference method if at least one of the following conditions is met:*

1. *The classical denominator $\Delta x \equiv h$ of the discrete derivate $D_{\Delta x}y_m$ is replaced by a nonnegative function ϕ such that*

$$\phi(z) = z + O(z^2) \quad \text{as} \quad 0 < z \to 0 \tag{3}$$

[1]Corresponding author, E-mail: jlubuma@scientia.up.ac.za, Tel. +27-12-4202222, Fax. +27-12-4203893. Research supported by the University of Pretoria and the National Research Foundation (South Africa).

2. Nonlinear terms that occur in the differential equation are approximated in a nonlocal way.

Let us first assume that the functions $W(x)$ and $Z(x)$ are constants: $W(x) = W$, $Z(x) = Z$. In this case, the difference scheme

$$-\varepsilon \frac{v_{m+1} - 2v_m + v_{m-1}}{\frac{4}{\rho^2} \sinh^2\left(\frac{\rho h}{2}\right)} + W v_m = Z, \quad \text{where} \quad \rho = \sqrt{W/\varepsilon}, \tag{4}$$

is the so-called exact scheme of Eq. (2) [4]. Thus, this scheme preserves the dissipativity and all other properties of the solution of (2). Coming back to the case when coefficient functions are not constant, Eq. (4) motivates the approximation of Eq. (2) by the non-standard finite difference scheme

$$-\varepsilon \frac{v_{m+1} - 2v_m + v_{m-1}}{\phi_m^2} + W_m v_m = Z_m \tag{5}$$

where $W_m = W(x_m)$, $Z_m = Z(x_m)$, $\rho_m = \sqrt{W_m/\varepsilon}$ and

$$\phi_m \equiv \phi_m(h, \varepsilon) := \frac{2}{\rho_m} \sinh\left(\frac{\rho_m h}{2}\right) = h + O\left(\frac{h^3}{\varepsilon}\right). \tag{6}$$

We show that the non-standard scheme (5) has the following properties:

- It is convergent with error estimate in $O(h^2)$ for a fixed ε

- It replicates a number of physical properties (e.g., dissipativity, etc.) of the singularly perturbed problem. In particular, compared to standard, the non-standard method produces better results in the boundary layer regions as $\varepsilon \to 0$.

To illustrate the power of the non-standard scheme (5), we consider several numerical examples including the following one.

Example Consider problem (1) or equivalently problem (2) with

$$a(x) = 1, \quad b(x) = W(x) = 1 + x(1 - x)$$

$$\begin{aligned} f(x) = Z(x) \;=\; & 1 + x(1-x) + \left[2\sqrt{\varepsilon} - x^2(1-x)\right] \exp[-(1-x)/\sqrt{\varepsilon}\,] \\ & + \left[2\sqrt{\varepsilon} - x(1-x)^2\right] \exp[-x/\sqrt{\varepsilon}\,] \end{aligned}$$

and

$$\eta_0 = \eta_1 = \alpha_0 = \alpha_1 = 0.$$

Its exact solution is given by

$$y(x) = 1 + (x-1) \exp[-x/\sqrt{\varepsilon}\,] - x \, \exp[-(1-x)/\sqrt{\varepsilon}\,].$$

Table 1 demonstrates the above mentioned rate of convergence. Moreover, as ε decreases, the error does not grow which shows the robustness of the non-standard method. In Figure 1, the difference between the exact solution and the approximate solution obtained via the non-standard/standard finite difference method is plotted. One can see clearly that in the boundary layer regions, the standard method performs badly.

Table 1: Numerical Results (Max. Errors)

ε	$n=4$	$n=8$	$n=16$	$n=32$	$n=64$	$n=128$
1.0	0.10E-02	0.26E-03	0.66E-04	0.16E-04	0.41E-05	0.10E-05
1.0E-02	0.47E-02	0.13E-02	0.33E-03	0.83E-04	0.21E-04	0.52E-05
1.0E-03	0.43E-03	0.14E-02	0.58E-03	0.16E-03	0.40E-04	0.10E-04
1.0E-04	0.71E-04	0.70E-03	0.15E-02	0.37E-03	0.11E-03	0.32E-04
1.0E-05	0.75E-11	0.11E-05	0.13E-03	0.28E-03	0.10E-03	0.87E-04
1.0E-06	0.22E-15	0.22E-15	0.85E-09	0.43E-05	0.93E-04	0.48E-04
1.0E-07	0.22E-15	0.22E-15	0.22E-15	0.84E-14	0.14E-07	0.66E-05
1.0E-08	0.11E-15	0.11E-15	0.11E-15	0.11E-15	0.22E-15	0.14E-11
1.0E-09	0.11E-15	0.22E-15	0.22E-15	0.22E-15	0.22E-15	0.22E-15
1.0E-10	0.22E-15	0.22E-15	0.22E-15	0.44E-15	0.44E-15	0.44E-15
1.0E-11	0.22E-15	0.22E-15	0.22E-15	0.22E-15	0.22E-15	0.22E-15
1.0E-12	0.22E-15	0.22E-15	0.22E-15	0.22E-15	0.22E-15	0.22E-15

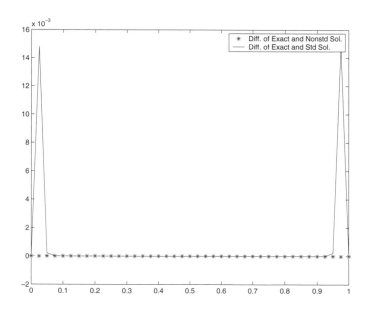

Figure 1: Errors of Standard & Non-standard Solutions for $n=40$, $\varepsilon = 10^{-5}$.

References

[1] R. Anguelov and J.M.-S. Lubuma, Contributions to the mathematics of the nonstandard finite difference method and applications, *Numer. Methods for Partial Differential Equations* **17** (2001), 518-543.

[2] C.M. Bender and S.A. Orszag, *Advanced Mathematical Methods for Scientists and Engineers*, McGraw-Hill, New York, 1978.

[3] J. Kevorkian and J.D. Cole, *Perturbation Methods in Applied Mathematics*, Springer-Verlag, New york, 1981.

[4] R.E. Mickens, *Nonstandard Finite Difference Models of Differential Equations*, World Scientific, Singapore, 1994.

VSP International
Science Publishers
P.O. Box 346, 3700 AH Zeist
The Netherlands

*Lecture Series on Computer
and Computational Sciences*
Volume 1, 2004, pp. 332-336

The Performance of Cfd Methods in Aerodynamic Estimation of Circular and Non-Circular Bodies[†]

S. Mahjoob[i‡], M. Mani[2]

1.   Aerospace Research Institute, Ministry of Science, Research, and Technology, Tehran, Iran.
2.   Aerospace Engineering Department, Amir-Kabir University of Technology, Tehran, Iran.

Abstract: In order to optimize the geometry of the body's cross-section for transportation purposes and also to obtain higher aerodynamic efficiencies, non-circular bodies have gained substantial attention by many researchers. In this work, the aerodynamic characteristics of two bodies having the same cross-sectional areas, but different shapes (one circular and one square with round corners) have been compared. In order to differentiate the non-circularity and the fin effects, the bodies with no fins have been considered. A three-dimensional, compressible, stationary, viscous, turbulent flow has been simulated using the FLUENT CFD code with the standard k-ε model and adaptive grids. The free stream Mach number is 0.83. The results have been compared with the experimental data and the data obtained from two semi-empirical codes, Digital DatCom and Missile DatCom. The results indicate that, although the semi-empirical codes were developed to calculate the aerodynamic parameters of bodies, the accuracy of CFD results is better and more reliable than that of semi-empirical codes. In addition, changing the cross section from circular to square increase the lift coefficient and aerodynamic performance, especially at low angles of attack. By changing the cross section from circular to square, the drag coefficients remain constant at low angles of attack but at higher angles, drag coefficient somehow increase.

Keywords: CFD, Turbulent, Transonic Flow, Aerodynamic, Non-Circular Body Cross Section

1.   Introduction

Nowadays, one of the open discussions in the area of body aerodynamics is increasing the aerodynamic efficiency by optimizing the geometry of its cross-section. In this regard, many researchers' attentions have been towards the bodies with non-circular cross-sections. Even though, this causes some inconvenience in the body's stability and control, with new advances in that area, the aerodynamists suggest non-circular cross-sections without much concerns. Jackson and Sawyer [1] have experimentally investigated bodies with elliptical cross-sections and noticed a considerable increase in aerodynamic efficiency (L/D) for horizontal elliptical cross-sections (compared with circular cross-sections). In family of bodies with non-circular cross-sections, because of the storage and carriage purposes, the ones with square or rectangular cross-sections (with round corners) are used more extensively. In 1987, Sigal and Lapidot [2] performed an extensive experimental investigation, in which three families of bodies with same length and cross-sectional areas were used. Their results showed that C_N was the highest for the horizontal rectangular case and the second highest for the square case. Of course, the problem with the rectangular case is that, when the body rolls and its cross-section changes from horizontal rectangle to vertical, its aerodynamic efficiency drops drastically (even to less than the circular case). Note, their results were consistent in both cases of a body with and without fins. Flow separation effects are highly related to the corners of the square or rectangular cross-sections and cause unfavorable aerodynamic instabilities [3].

In our work, we have compared the aerodynamic characteristics of two bodies with the identical cross-sectional areas, but with different shapes in a transonic regime at different angles of attack. In order to differentiate the non-circularity and the fin effects, we have considered the bodies with no fins. This study has been done using experimental methods, a CFD code and two semi-empirical codes.

2.   Models and Flow Characteristics

The two models with the same length and cross section area have been studied. The bodies have fineness-ratio-3.5 ogive noses. The cross section of one of the models is square with rounded corners and the other model's cross section is circular (Fig. 1). The cross section area of both bodies is $0.001257m^2$. The aerodynamic coefficients have been calculated based on the cross section of the body. The free stream Mach number is 0.83, the static temperature is 266.86 °K, the static pressure is 54400 Pa, and the total pressure is 85000 Pa.

[†] Received 2 August, 2004; accepted in revised form 20 August, 2004

[‡] Corresponding author. E-mail: shmahjoob@ari.ac.ir. Address: Hava-faza alley, Mahestan St., Iranzamin St., Shahrak Gharb, Tehran, Iran, P.O. Box: 14665-834.

3. Experimental Methods

The experimental test has been done in an multi purposed ST-2 wind tunnel [4]. The range of angle of attack is from -2° to 16°. It is noticeable that because of the lack of drier system or filter in the tunnel, the results can have a little error. It happens especially at high speeds (supersonic regimes) that the airflow reaches the two phase flow situation.

4. Computational Methodology

The FLUENT CFD code [5], which uses a cell centered finite volume method and has been proven to work well for different flow regimes around bodies, has been used in this study. The implicit method implemented uses a segregated solution method. Note, all the schemes used here are second order. The SIMPLE algorithm with under-relaxation coefficients is used in the overall discretization of the equations. In order to reduce the dispersion errors (and also to increase the speed of the computations), the multi-grid approach has also been used.

4.1. Governing Equations

The Reynolds averaged governing equations include continuity, momentum, and energy. The flow considered here is three-dimensional, compressible, stationary, single-phase, viscous and turbulent that the standard κ-ε model has been used for turbulence modeling. The fluid is air, for which the viscosity is obtained using Sutherland relation. The turbulence intensity is 1%, the characteristic length scale (the body's diameter of cross-section) is 0.04 m.

4.2. Geometric Modeling

The outer boundaries of the computational domain have been assumed to be 9 radii in the radial direction, 2 body lengths at the front and 3 body lengths behind the body. This size of the domain was shown to be optimal using numerical experimentation. Note, due to the symmetric assumption used, only half of the domain is considered here. Our computational results indicate that this assumption has not affected the results considerably. A structured, body fitted, and non-uniform grid has been used in this study. A 169×21×21 grid was chosen as the basic grid. In order to have more accurate results, adapted grids have been used. The adaption has been done based on the Y^+ and pressure difference parameters.

4.3. Boundary Conditions

Different boundaries of the physical domain have been used in this work. The wall of the body is assumed to be adiabatic. The free stream pressure is assumed at the inlet and outlet of the domain. In the pressure outlet boundary condition, only the static pressure is specified and all other flow quantities are extrapolated from the flow in the interior. The far field pressure condition is also used. This boundary condition is often called a characteristic boundary condition because it uses characteristic information (Reiman invariants) to determine the flow variables at the boundaries. Finally, at the symmetric plane, the symmetric boundary condition has been implemented.

5. Semi-Empirical Methods

One of the most proper methods in aerodynamic studies is using semi-empirical methods. The semi-empirical softwares of Digital DatCom [6] and Missile DatCom [7] have the ability of reasonably analyzing the flying vehicles. These codes can determine the characteristics of the static and dynamic stability and control of airplanes and missiles with an acceptable error.

6. Results

The results of both the circular and the square bodies were calculated by CFD code and two semi-empirical codes of Digital DatCom and Missile DatCom. In the following, the error of each these codes are calculated in comparison with the experimental results of each body [4]. Then, the aerodynamic coefficients of the circular bodies and those of the non-circular bodies, obtained from CFD and semi-empirical codes, are compared.

6.1. CFD and Semi-Empirical Codes Validation Study

Figure 2 shows the drag coefficients of a circular body obtained from wind tunnel test, CFD, Missile DatCom and Digital DatCom codes. According to this figure, it can be observed that the error of the results of CFD is very low. The maximum error percentage of CFD results is 10%. Besides, the error of the results of Digital DatCom code is much lower than the errors of the results of Missile DatCom code. The maximum error percentage of Digital DatCom code is 12.8% but that of Missile DatCom Code is 42.5%.

Figure 3 shows the drag coefficients of a square body obtained from wind tunnel test, CFD, Missile DatCom and Digital DatCom codes. It shows that drag coefficients obtained from CFD are similar to those of experimental method and the results of both experiment and CFD have the same behavior. In addition, Digital DatCom code have much lower errors than those obtained from Missile DatCom. This error is more conspicuous in higher angles of attack. The maximum error percentage of CFD is 20%. The maximum error percentage of Digital DatCom code is 29.6% but that of Missile DatCom Code is 59%. In other words, in both the circular and square bodies, the drag coefficient errors obtained from Digital DatCom code is lower than that of Missile DatCom code. And the results of CFD are more acceptable than those of semi-empirical codes.

In Figure 4, the lift coefficients of a circular body obtained from wind tunnel tests, Missile DatCom and Digital DatCom codes are presented. This figure indicates that the results of CFD are close to those of experiment. Besides, the wind tunnel and CFD results are somewhere between the Digital DatCom results and Missile DatCom results. The error of these two semi-empirical codes is almost similar. However, the results of Missile DatCom have an error percentage range limited between 21% to 33% which is a more acceptable and a more stable condition than those of Digital DatCom. The maximum error percentage of Digital DatCom code is 41.9%. On the whole, in the body alone configuration of both circular and square bodies, the errors of Missile DatCom code in calculating the lift coefficient are lower than those of Digital DatCom Code. The maximum error of CFD is 25%.

In Figure 5, the lift coefficients of a square body obtained from wind tunnel test, Missile DatCom and Digital DatCom codes are presented. According to this figure, it can be seen that the results of CFD are close to those of experiment. Besides, the error of the Missile DatCom results, with a maximum of 19.4%, is much lower than that of Digital DatCom code. The maximum error percentage of Digital DatCom code is 64.8%. In addition, the Missile DatCom results have negligible error especially in the higher angles of attack. The maximum error of CFD results is 20%. In other words, in both the circular and square bodies, the lift coefficient errors obtained from Missile DatCom code is lower than that of Digital DatCom code. And the results of CFD are more acceptable than those of semi-empirical codes.

On the whole, the comparison of the experimental, CFD, and semi-empirical results show that CFD results have better and more reliable accuracy than semi-empirical codes. This is noticeable that these two semi-empirical codes are basically prepared to calculate aerodynamic parameters of bodies.

6.2. The Aerodynamic Effects of the Cross Section Shape

Figure 6 shows the variations of drag coefficients versus the angle of attack for both circular and square bodies obtained from CFD. This figure indicates that at low angles of attack (lower than 4 degrees), the drag coefficients of both circular and square bodies are very close to each other. But by increasing the angle of attack, the square body has a little more drag coefficient. The maximum difference between bodies' drag coefficient is 14% that occurs at the angle of 16 degrees. This is the same result obtained from the experiment. In fact, the experiment data show that at low angles of attack, changing the cross section does not affect drag coefficient very much.

Figure 7 shows the variations of drag coefficients versus the angle of attack for both circular and square bodies obtained from Digital DatCom and Missile DatCom codes. Since the error of the results obtained from Digital DatCom code in calculating the drag coefficients was acceptable, the following analysis is based on the results of this code. Comparing the two curves, the one with circular cross section and the one with square cross section, with each other indicates that for the lower angles of attack, the drag coefficients of the circular body are close to those of the square body. However, the drag coefficients of the square body are slightly higher than those of the circular body and this is exactly the same result obtained from the experiment. But the drag of the square body for higher angles of attack is much lower than the drag of the circular body and the drag results of the square body differ from the drag results of the circular body substantially.

A comparison of the lift coefficients variations versus the angle of attack for circular and square bodies obtained from CFD is presented in Fig. 8. This figure indicates that the square body has more lift coefficient. The difference percentage between the square and circular bodies even reaches 28% that most of the time, higher differences happen at the angles around 4 degrees. In fact, at the angles of attack around 4 degrees, the increase of lift coefficient by changing the cross section from circular to square is more than that of other angles. The experimental results admit this deduction.

Figure 9 shows the variations of lift coefficients versus the angle of attack for both circular and square bodies obtained from Digital DatCom and Missile DatCom codes. The analysis is based on the Missile DatCom code as its results are more acceptable. This figure shows that the lift coefficients for the square body are relatively higher than those of the circular body. For high angles of attack, the results of the circular body are about 24% higher than those obtained from the wind tunnel test, and this made the curve for the circular body approach the curve for the square body. In spite of this, the results of the

square body are still higher than those of the circular body. According to the experimental results, a change of the cross section shape from circular to square, leads to the increase of the lift coefficients.
A comparison of the aerodynamic performance (L/D) variations versus the angle of attack for circular and square bodies obtained from CFD is presented in Fig. 10. This figure indicates that changing the cross section from circular to square increase the aerodynamic performance considerably, specially at low angles of attack (less than 5 degrees). It is noticeable that by increasing the angle of attack from 4 degrees, the increase of aerodynamic performance by changing the cross section is reduced. This is exactly the same result obtained from the experiment.

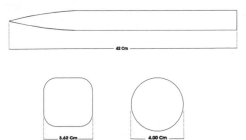

Figure 1: The geometry of models: Body and Studied cross sections, circular and square with rounded corners.

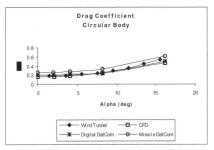

Figure 2: The Comparison of the drag coefficients in the circular body.

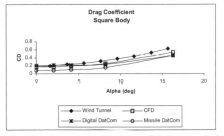

Figure 3: The Comparison of the drag coeff. in the square body.

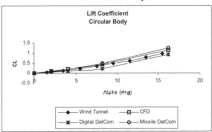

Figure 4: The Comparison of the lift coeff. in the circular body.

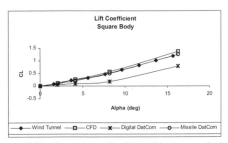

Figure 5: The Comparison of the lift coeff. in the square body.

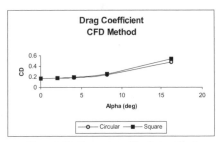

Figure 6: The Comparison of the drag coeffs of circular and square.

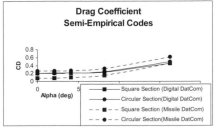

Figure 7: The drag coeff. from Digital and Missile DatCom .

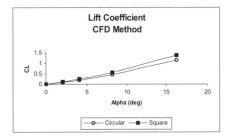

Figure 8: The lift coeff. of circular and square bodies using CFD.

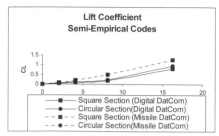

Figure 9: The lift coeff. of bodies obtained using semi-empirical codes.

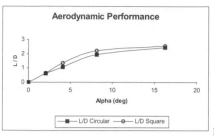

Figure 10: The Aerodynamic Performance of bodies using CFD.

References

[1] Jackson, C.M. and Sawyer, W.C., "Bodies with Non-Circular Cross-Sections and Bank-to-Turn Missiles," Progress in Astronautics and Aeronautics, Vol. 141, pp. 365-389, 1991.

[2] Sigal, A. and Lapidot, E., "The Aerodynamic Characteristics of Configurations Having Bodies with Square, Rectangular, and Circular Cross-Sections at a Mach Number of 0.75," AIAA, Inc., 1987.

[3] Nielsen, J.N., "Problems Associated with the Aerodynamic Design of Missile Shapes," Proceedings of the Second Symposium on Numerical and Physical Aspects of Aerodynamic Flows, CA, Jan. 1983.

[4] Mahjoob, S., Mani, M., Ebnoldin, H., Haghiri, A., "Aerodynamic Study of Bodies with Non-Circular Cross Section Using Experimental Method", Presented in the 8th International and 12th Annual Conference of Iranian Society of Mechanical Engineers, Tehran, Iran, 2004.

[5] "FLUENT 5 User's Guide," FLUENT Incorporated, July 1998.

[6] "The USAF Stability and Control Digital DatCom", McDonnell Douglas Astronautics Company, 1979.

[7] "The USAF Automated Missile DatCom", S.R., Vukelich, S.L., Stoy, M.E., Moore, AFWAL-TR-86-3091, McDonnell Douglas Missile Systems Company, 1991.

VSP International
Science Publishers
P.O. Box 346, 3700 AH Zeist
The Netherlands

*Lecture Series on Computer
and Computational Sciences*
Volume 1, 2004, pp. 337-340

Integrating Data Mining Methods for Modeling Urban Growth Dynamics

P. Manetos [1]

Department of Planning and Regional Development,
University of Thessaly,
GR- 38334 Volos, Greece

Y. N. Photis [2]

Department of Planning and Regional Development,
University of Thessaly,
GR- 38334 Volos, Greece

Received 2 September, 2004; accepted in revised form 13 September, 2004

Abstract: The continuous effort to analyze and model processes and phenomena related to the majority of scientific fields, in conjunction with enhancing computer power has led to the development of Data Mining methods and techniques. Their significance lays on the fact that they approach each problem's solution mainly driven by data structures rather than by theoretical hypotheses. Although, relevant research has increased during the last decade, their use and application in Geography is very recent. In this paper, an integration of Data Mining methods under a consistent methodological framework, namely, the Knowledge Discovery Procedure is proposed in order to model urban growth dynamics. Its application focuses on the greater region of Athens, Greece and the interpretation of results reveals its efficiency in capturing spatial trends and revealing future tendencies.

Keywords: Urban growth, Spatio-temporal modeling, GIS, Fuzzy clustering

Mathematics Subject Classification: 62M30, 91C20, 91D10

1. Introduction

The continuous effort to analyze and model processes and phenomena related to the majority of scientific fields, in conjunction with enhanced computer power has led to the development of Data Mining methods and techniques. Their significance lays on the fact that they approach each problem's solution mainly driven by data structures rather than by theoretical hypotheses. Although, relevant research has increased during the last decade, their use and application in Geography is very recent [1].

The urban planning process presupposes plethora of information and a variety of analytical tools and so, the technological and functional environment in which it will be developed, should be characterized by flexibility and adaptability of a breadth of needs and requirements [2]. In this framework, quantitative and qualitative spatio-temporal demographic changes in urban regions are more than often expressed via urban sprawl and land use patterns modifications. And although large volumes of geographically referenced urban data became available, it is still difficult for planners to exploit them without the assistance of sophisticated tools and methodologies. Furthermore, growth is not random; When a spatial entity is developed then the probability of development for regions sharing a common border is increased [3]. Consequently, there is apparent interest for the conduct of such researches and the quest of suitable methods to support urban models.

Since cities change not only at, but also within their limits in a complex way urban modeling has long been and still is a major scientific challenge. One of several contemporaneous attempts to capture the rich urban environment is UrbanSim. It is a model designed to analyse and define urban growth, by

---

[1] Corresponding author. PhD Candidate. E-mail: pmanetos@prd.uth.gr

[2] Assistant Professor of Spatial Analysis and Locational Planning. E-mail: yphotis@prd.uth.gr

taking into consideration interactions between land use patterns, transport data and public policies [4], mainly using classical statistics. On the other hand, the application of data mining methods has proven to contribute in a variety of problems and especially to the analysis of very large commercial databases. Other artificial intelligence methods such as Neural Networks have also been applied for more than a decade. Example applications can be found in the literature referring to the forecast of the spread of AIDS cases in the state of Ohio and the socio-economic classifications of population based on census data [5]. Moreover, challenges from the application of Fuzzy Logic methods in Geography, and especially in Geodemographics were stated by Openshaw as early as 1989 [6]. The first challenge concerns the descriptive characteristics of spatial units, where, for example, if clustering of neighboring entities is performed by binary logic, there is a possibility those sharing common characteristics, to be assigned to different classes. The second, concerns the geographic phenomenon where while neighboring populations tend to share certain common characteristics in their behavior despite the other differences, are classified in wrong clusters. One reason for these is that the borders of spatial entities are not natural in demographic and economic-social terms [7]. A second issue is that the concepts of space and distance are not easily integrated in data mining tools, because a simple addition of the spatial position of objects (x, y, coordinates), is not treated as an actual position in space, but as two extra variables, misinterpreting the geographical dimension of the spatial objects [8]. However, the primary question remains, whether advanced techniques and capabilities of analysis correspond with the contemporary needs of urban planning and future development of regions.

2. Knowledge Discovery Process (KDP)

Spatio-Temporal Data Mining involves the latest statistical and spatial analysis methods used for the definition of structures and patterns in Spatio-Temporal Databases. These methods exploit the increased volume and the variety of available data, and with the help of a suitable Data Mining process framework, can become an invaluable tool for the identification of cross-correlations between factors that affect urban development.

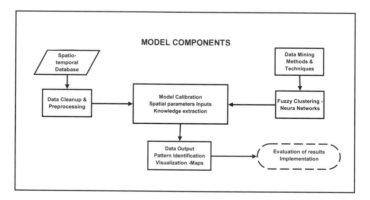

Figure 1: Model components.

The proposed Knowledge Discovery methodological framework (Figure 1) faces both kinds of fuzziness that were mentioned earlier and moreover considers the effect of contiguity to the configuration of the initial clusters in order to forecast their future development. It is applied to the analysis of spatio-temporal patterns in the prefecture of Attica in Greece, unveiling in essence the role of the Athens metropolitan area to its current and future evolution. Fuzzy clustering and Neural Networks are utilized for the model definition. Often, for the observation of a phenomenon spatial entities are clustered based on common features, in order for the changes to be distinct, particularly when the number of spatial entities is augmented. Clustering is characterized by a particular breadth of values in the corresponding variables and constitutes the constant diachronic reference, based on which each entity is classified and described. Grouping both in attribute and in geographical space is considered for more than one variable [9], so the fuzzy c-means (FCM) clustering algorithm was applied. The resulting classes should also be viewed by means of quality criteria. Among the various measures proposed in the literature, the following three were examined for the data used in the model:

partition coefficient, partition entropy and proportion exponent. Data preprocessing proves critical at this point of the procedure, because the main disadvantage of the FCM method is that clusters, which are strung out over long distances will not be identified by an FCM method calculating Euclidean distances. Additionally, several Neural Network validity measures are also examined so as to define how efficiently the model can simulate urban growth. When the model implementation is completed, the final results are registered in a geo-database and visualized in a GIS environment.

3. Model implementation & evaluation: Case Study Attica, Greece

This approach is based on a three-axis theoretical background, which provides the necessary support and flexibility in order to incorporate several concepts. For the determination of geodemographic evolution of urban regions, a spatiotemporal approach was adopted. Observation, interpretation and forecast are accomplished in the flow of time, formulating a thorough perception of its progressive formation and designating in this sense, the importance of time in the decision-making process. The main axis of the proposed methodology comprises the ascertainment that the development of a spatial entity does not solely depend on its own features and characteristics but also on the way neighboring entities develop. Common borders create bidirectional relations of assimilation of characteristics and interaction of activities.

The KDP was applied to the prefecture of Attica. This is the most urbanized region in Greece, since the city of Athens is inhabited by almost 40% of country's population. Data concerning building uses, population, road network characteristics were collected at the municipality level which is the spatial reference unit. Geographical influence factors were added to the initially acquired variables. Following data preprocessing and according to the proposed methodological framework, Fuzzy c-means (FCM) clustering algorithm was applied. For the selection of the optimal number of clusters, reliability criteria of fuzzy clustering were used, according to recent literature [10]. They include the crispiness of clustering, by determining a value proportional with the number of data that are close to the cluster centers. The crispier clustering is the higher the particular value will be.

Visualization is another important component of the Knowledge discovery process, especially in the field in Geography, where maps are the main means of communicating the results of the process. One of the challenges of maps is that the visualization of distinguishable colors between regions implies abrupt changes between them, while the transition is not usually so abrupt. The color depiction of the FL clusters constitutes a modern field of experimentations and research [11], as until today there has not yet been established a common way of depiction. Analysis of the current case study has become richer by the combined use of data mining methods and especially the additional value that they offer.

4. Conclusions & further research

The future development of urban regions traditionally constitutes a challenge for the planning process. The proposed methodological framework is characterized by its simple and simultaneously dynamic structure. Further development is under way, involving the enrichment of the methodology and the integration of the graphical user interface in a GIS environment, so as to aid the user to confront the problem in a unified software environment.

This paper argues that contemporaneous and interdependent spatial phenomena cannot be analyzed and represented efficiently with conventional deterministic techniques. The development of a new integrated spatial toolbox is required, if researchers can acquire the analytical advantage for the understanding of complex systems. The ability of fuzzy clustering to assign each object to more than one clusters, offers a more intuitive knowledge of their future spatial evolution and an increased potential from model implementation and prediction. In equivalence, the added value of NN is not limited exclusively in the simulation of complex and combinational phenomena, but in the designation of the importance and the degree of influence of every problem parameter in the final result.

Concluding, traditional methods cannot exploit the richness and complexity of modern urban regions and their a priori hypotheses prevent a deeper insight in urban growth dynamics. Moreover, analysis and interpretation of results reveals increased capabilities, possibilities and insights in urban growth dynamics, under a consistent and integrated data mining methodological framework.

References

[1] H. J. Miller, J. Han (eds), *Geographic Data Mining & Knowledge Discovery*. Taylor & Francis, New York, 2001

[2] M. Visvalingam, Areal units and the linking of data: some conceptual issues, in L. Worrall *Spatial Analysis and Spatial Policy using Geographic Information Systems*, Belhaven Press, London., 1991

[3] A. Makse, S. Andrade, M. Batty, S. Havlin, E. Stanley, Modeling Urban Growth Patterns with Correlated Percolation, *Phys. Rev. E 58*, 7054-7062. , 1998

[4] P. Waddell, Urban Simulation Project, http://www.urbansim.org/, 2004

[5] B. Hewitson and R. Crane, *Neural Nets: Application in Geography*. Kluwer Academic Publishers, Netherlands, 1994

[6] S. Openshaw, Making Geodemographics more sophisticated, *Journal Market Research Society*, 31 , 1989

[7] Z. Feng and R. Flowerdew, Fuzzy geodemographics: a contribution from fuzzy clustering Methods, in Innovations *in GIS: Innovations in GIS 5*, Taylor & Francis, UK, 1998

[8] S. Openshaw, A. Turner, I. Turton, J. Macgill, and C. Brunsdon, Testing space-time and more complex hyperspace geographical analysis tools in *Innovations in GIS* 7. Taylor & Francis, London, 2000

[9] P. A. Burrough, *Principles of Geographical Information Systems*, Oxford University Press, New York, 1996

[10] MIT, Data Engine: Tutorials and Theory manual, Management Intelligenter Technologien Germany, 1997

[11] K. Nakakiji, Y. Yamamoto, K. Sugiyama, S. Takada. Finding the "Right" Image: Visualizing Relationships among Persons, Images and Impressions", *Designing Effective and Usable Multimedia Systems*, pp. 91-102, Kluwer Academic Publishers, Netherlands, 1998

VSP International
Science Publishers
P.O. Box 346, 3700 AH Zeist
The Netherlands

*Lecture Series on Computer
and Computational Sciences*
Volume 1, 2004, pp. 341-342

Nonequilibrium Molecular Dynamics Simulation of Diatomic Ions in Supercritical Gases in an Electrostatic Field

Nikos D. Margetis and Andreas D. Koutselos

National and Kapodistrian University of Athens,
Department of Chemistry,
Physical Chemistry Laboratory,
Panepistimiopolis, 15771 Athens, Greece.

Received 28 July, 2004; accepted in revised form 18 August, 2004

1. Introduction

The transport of diatomic ions in noble gases at supercritical states[1] or at coexistence of liquid and gas under the action of an electrostatic field[2] has been found experimentally to depend drastically on the density of the medium. In the case of O_2^- in Ar a parametric resonance is observed for the mobility around the critical density.[3] Similarly, the cation mobility in monoatomic and polyatomic molecules such as the noble gases, N_2 and CH_4 is found to vary significantly with the gas density.[2,4]

To reproduce the experimental density and field dependence of the ion transport properties around the critical point of the buffer gases and to probe the interactions and the structure around the ion, we apply a new nonequilibrium molecular dynamics method[5] (NEMD) for the simulation of the ion motion. The method has been developed for the simulation of swarms of ions which can be driven away from equilibrium by the action of external fields but which maintain steady drift due to interaction with buffer gases.

2. Molecular Dynamics Method

The system consists of a few ions not interacting with one another in a gas, which, as being in excess, is assumed to be always at equilibrium. The reproduction of the motion is implemented through consideration of two parallel simulation procedures,[5,6] one for the gas, which remains always in equilibrium, and a second for the ions and their immediate environment. This environment consists of images of neutral atoms which are generated with the aid of the first simulation procedure. The images of atoms are created at the edge of an interaction sphere around an ion, whenever an atom of the first procedure enters such a sphere. The images of atoms belong specifically to a certain ion and last as long as they remain within the interaction sphere of the ion. In the case of moderately dense gases, provision has to be taken for the newly created images of the atoms so that they feel forces from the gas atoms of the first simulation procedure, which lay outside the interaction sphere.[5] This should continue until the images of atoms move deep into the interaction region and away from the edge of the interaction sphere.

3. Ion Transport Properties

The method allows the calculation of molecular distributions, velocity distributions, static and dynamic correlation function, as well as ion transport properties. The latter are defined through analysis of the ion flux, J, in contributions with respect to various causes of motion such as the gradient of ion density, n, the gradient of the temperature, T, or the electrostatic field, E,[7]

$$J = n\boldsymbol{v}_d - \boldsymbol{d}\nabla n - n\boldsymbol{D}_T\nabla \ln T + ...,$$

where v_d, $\mathbf{d}$, $\mathbf{D_T}$, etc. are drift velocity, diffusion coefficient, thermal diffusion coefficient, etc. of the ion, respectively. High order contributions can be considered whenever the ion motion is affected significantly by them. The mobility is defined through $\mu = |v_d|/E$, with the field and v_d in the z-direction. Due to the action of the field, the diffusion coefficient is a diagonal matrix with two independent components, $d_{XX} = d_{YY}$ and d_{ZZ}. The $\mathbf{D_T}$ coefficient acquires a similar structure.

At low field strengths, the various components of the transport coefficients become equal to each other and the above tensorial description becomes obsolete. This situation is common in moderate- to high-density experiments due to experimental limitations. Here, we mainly calculate the mobility through the mean ion velocity, $\langle v_Z \rangle$, and ion diffusion components parallel and perpendicular to the

field through the velocity correlation functions,[8] $d_{AA} = \int_0^\infty M_{AA}(t)\, dt$, where $M_{AA}(t) = \langle (v_A(0) - \langle v_A \rangle)(v_A(t)$

$- \langle v_A \rangle)\rangle$ is the v-correlation function and A can be X, Y or Z.

4. Results And Discussion

The procedure is applied to a model X_2^+–Ar system with 12-6-4 site-site ion-neutral interaction potential and an efficient but accurate at high densities LJ- potential for Ar.[9] We have thus calculated the mobility, ion-diffusion components and mean kinetic energies parallel and perpendicular to the field.

The inferred sensitivity of the results on the interaction potential shows that (effective) two-body interactions can be probed from experimental data[1,3,4] with reasonable accuracy at moderate gas densities. Since at such densities the ions interact mainly with one or two neutrals simultaneously,[5] extension to high densities, where multiple encounters take place, should demand quite high computer power in order the same accuracy to be attained.

Acknowledgments

The authors would like to thank the Institute of Computer Science of the National and Kapodistrian University of Athens. This work was supported by grants from the Research Fund of the National and Kapodistrian University of Athens, No 70/4/6482.

References

[1] A. F. Borghesani, D. Neri, and M. Santini, Phys. Rev. E, **48**, 1379 (1993).
[2] N. Gee, S, S.-S.Huang, T. Wada, and G. R. Freeman, J. Chem. Phys. **77**, 1411 (1982).
[3] A. F. Borghesani, F. Tamburini, Phys. Rev. Lett. **83**, 4546 (1999).
[4] K. F. Volykhin, A. G. Khrapak, and W. F. Schmidt, Sov. Phys. JETP **81**, 901 (1995).
[5] G. Balla and A. D. Koutselos, J. Chem. Phys. **119**, 11374 (2003)
[6] A. D. Koutselos and J. Samios, Pure Appl. Chem. **76**, 223 (2004)
[7] E. A. Mason and E. W. McDaniel, *Transport Properties of Ions in Gases* (Wiley, New York, 1988)
[8] A. D. Koutselos, J. Chem. Phys. **104**, 8442 (1996).
[9] A. Michels, Hub. Wijker, and Hk. Wijker, Physica **15**, 627 (1949).

VSP International
Science Publishers
P.O. Box 346, 3700 AH Zeist
The Netherlands

*Lecture Series on Computer
and Computational Sciences*
Volume 1, 2004, pp. 343-346

Largest Lyapunov Exponent Application to the Pollen Time-Series and Related Meteorological Time-Series

F. Martínez, R. García[1], M. Munuera[2], A. Guillamón & I. Rodríguez[1]

[1]Department of Applied Mathematics and Statistic, Campus Muralla del Mar,
[2]Department of Agriculture Science & Technology, Campus Alfonso XIII
Technical University of Cartagena, Spain.
30203 CARTAGENA (Murcia) – Spain.

Received 2 August, 2004; accepted in revised form 18 August, 2004

Abstract: Pollen time-series can be considered as being generated by biological system with inherently nonlinear dynamics. The purpose of this paper is to characterize pollen and related time-series using the largest Lyapunov exponent. We used the method suggested by Rosenstein et al. for calculating it and we introduce new tools to a previous time-series segmentation and to search the linear region. In our research, they were computed in MATLAB®, over data obtained in Murcia, Spain. The results prove the existence of chaos in the pollen time-series and related meteorological time-series and the utility of largest Lyapunov exponent to distinguish and quantify chaotic events in biological time-series.

Keywords: Nonlinear Dynamic, Environmental chaos, Biosignal processing, chaos characterization, Largest Lyapunov exponent.

1. Introduction

Distinguishing deterministic chaos from noise has become an important problem in many diverse fields, e.g., physiology, economics,.... This is due, in part, to the availability of numerical algorithms for quantifying chaos using experimental time series. In particular, methods exist for calculating correlation dimension (D_2) [3], Kolmogorov entropy [4], and Lyapunov characteristic exponents [2], [5], [6], [7]. Dimension gives an estimate of the system complexity; entropy and characteristic exponents give an estimate of the level of chaos in the dynamical system.

Several algorithms have been proposed for the computation of characteristic invariant measures from an experimental time series. We have developed an algorithm for calculating the largest Lyapunov exponent based on the method proposed by Rosenstein et al.[1]. This method is reliable for small data sets, fast, and easy to implement. Also we have created different algorithms for automatic and manual searching the linear region over the average diverge curve. We have used Matlab to implement the algorithm because the facilities to simulation and display results and to import data from different formats.

One of the most important problems in the time-series analysis is the existence of different dynamics in the same time-series. We use a previous time-series segmentation based on the symbolic dynamic method [8]. The implemented tools have been tested with the total pollen concentration and related meteorological time-series in Murcia, Spain.

From 1 March 1993 to 31 December 1998 a Burkard volumetric seven-day recording spore trap (Hirst 1952) located at about 19 m a.g.l. was operated on the exposed flat roof of the Veterinary Faculty (Murcia University, 110 m a.s.l., 38° 01' N, 01° 10' W, 4 km NW of Murcia city centre). From weekly tapes, daily slides were prepared following the standard methods accepted by the Spanish Aerobiology

---

[1] Corresponding author. E-mail: f.martinez@upct.es, vicalum@upct.es

Network (REA, Red Española de Aerobiología; Domínguez et al. 1991), and subsequently examined by light microscopy. [10], [11] & [12].

2. Methods

For time series produced by dynamical systems, the presence of a positive characteristic exponent indicates chaos. Furthermore, in many applications it is sufficient to calculate only the largest Lyapunov exponent (λ_1).

If we assume that there exists an ergodic measure of the system, then the multiplicative ergodic theorem of Oseledec [9] justifies the use of arbitrary phase space directions when calculating the largest Lyapunov exponent with smooth dynamical systems. We can expect (with probability 1) that two randomly chosen initial conditions will diverge exponentially at a rate given by the Largest Lyapunov exponent. Thus, the largest Lyapunov exponent can be defined using the following equation where $d(t)$ is the average divergence at time t and C is a constant that normalizes the initial separation:

$$d(t) = Ce^{\lambda_1 t} \tag{1}$$

The first step of the algorithm for calculating the largest Lyapunov exponent involves reconstructing the attractor dynamics from a single time series. We use the method of delays [13] & [14]. After reconstructing the dynamics, the algorithm locates the *nearest neighbor* of each point on the trajectory. This approach imposes the additional constraint that nearest neighbours have a temporal separation greater neighbours have a temporal separation greater than the mean period of the time series. This allows us to consider each pair of neighbours as nearby initial conditions for different trajectories. The largest Lyapunov exponent is then estimated as the mean rate of separation of the nearest neighbours.

From the definition of λ_1 given in eq. (1), we assume the *jth* pair of nearest neighbours diverge approximately at a rate given by the largest Lyapunov exponent:

$$d_j(i) \approx C_j e^{\lambda_1 (i\Delta t)} \tag{2}$$

where C_j is the initial separation. By taking the logarithm of both sides of eq. (2), we obtain

$$\ln d_j(i) \approx \ln C_j + \lambda_1 (i\Delta t) \tag{3}$$

Eq. (3) represents a set of approximately parallel lines *(for j=1,2,...,M)*, each with a slope roughly proportional to λ_1. The largest Lyapunov exponent is calculated using a least-squares fit to the "average" line defined by

$$y(i) = \frac{1}{\Delta t} \left\langle \ln d_j(i) \right\rangle \tag{4}$$

where $\langle \ \rangle$ denotes the average over all values of j. After a short transition, there is a long linear region is used to extract the largest Lyapunov exponent.

The search for linear region problem has been solved with some manual and automatic algorithms. We can choose the limits of the linear region as input parameters in the manual algorithm. In the automatic algorithms, there is a initial "marking out" of the search area based in the maximum value of the curve. We have developed different algorithm each one with a different way to estimate the point-slope. We have used numeric methods (Euler, Adams Bashforth methods...) to calculate it. With a tolerance input parameter and the point-slope estimation the automatic algorithms select the most linear region.

The algorithm for calculating the largest Lyapunov exponent is easy to implement and fast because it uses a simple measure of exponential divergence that circumvents the need to approximate the tangent map. The algorithm is also attractive from a practical standpoint because it does not require large data sets and it is robust to variations in embedding dimension, number of data points, reconstruction delay, and noise level.

3. Experimental results

The developed tools have been applied to analyse biological time-series. The analysed time-series correspond with pollen and meteorological parameters related time-series. We have been able to prove the chaotic events existence in all these series and to quantify, in the most, the chaos level with the largest Lyapunov exponent.

The next figures show the analysis process over one of the annual pollen concentration time-series. Fig. 1 shows the previous segmentation and fig.2 shows the curve result of the largest Lyapunov exponent from the selected time-series segment. Table 1 summarises the parameters and numeric results.

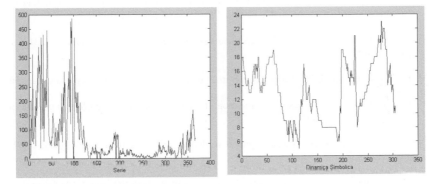

Figure 1: Previous segmentation applied over one of the annual pollen concentration time-series. (The red colour corresponds to the selected segment). On the right, we include the corresponding plot for the symbolic dynamic.

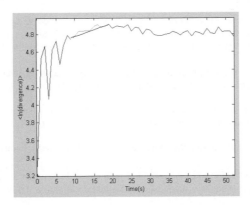

Figure 2: The largest Lyapunov exponent from the selected time-series segment; the black line has a slope equal to the value of λ_1

Table 1: Used parameters and numeric results.

Segment size	Time lag	Embedding dimension	Mean period	Method of search for linear region	Calculated largest exponent
134	3	3	4	Manual	0.0149

4. Conclusions

This study suggests that the largest Lyapunov λ_l is a powerful tool to characterise and to quantify the existence of chaotic events over biological time-series. After the analysis with the developed tools, we have been able to prove the level of chaos in the pollen time-series.

In other hand, the developed algorithm for calculating the largest Lyapunov exponent is robust to changes in the following quantities: embedding dimension, size of data set, reconstruction delay, and noise level. Although we have created different tools to estimate them, we use the mean period, the embedding dimension and the reconstruction delay like input parameters in the algorithm. This lets us change them and study what values of these parameters are better to reconstruct the dynamic system.

References

[1] M. T. Rosenstein, J.J. Collins and C.J. De Luca, A practical method for calculating largest Lyapunov exponents from small data sets. *Physica* D **65** 117-134 (1993).

[2] J.-P. Eckmann and D. Ruelle, Ergodic theory of chaos and strange attractors, *Rev. Mod. Phys.* **57** 617(1985).

[3] P. Grassberger and I. Procaccia, Characterization of strange atrractors, *Phys. Rev. Lett.* **50** 346 (1983).
[4] P. Grassberger and I. Procaccia, Estimation of the Kolmogorov entropy from a chaotic signal, *Phys. Rev. A* **28** 2591 (1983).

[5] J.D. Farmer and J.J. Sidorowich, Optimal shadowing and noise reduction. *Physica D.* **47** 373-392(1991).

[6] M. Sano and Y. Sawada, Measurement of the Lyapunov spectrum from a chaotic time series. *Phys. Rev. Lett.* **55** 1082 (1985).

[7] A.Wolf, J.B. Swift, H.L. Swinney and J.A. Vastano, Determining Lyapunov exponents from a time series, *Physica D* **16** 285(1985).

[8] R. García Valverde: *Método para el análisis de la caoticidad de series temporales basado en el superexponente de Lyapunov. Aplicación a diversas señales de naturaleza biológica. Proyecto fin de carrera.* Polytechnic University of Cartagena, Spain, 2004 (in Spain)

[9] V.I. Oseledec, A multiplicative ergodic theorem. Lyapunov characteristic numbers for dynamical systems. *Trans. Moscow Math. Soc.***19** 197(1968).

[10] E. Domínguez, C.Galán , F. Villamandos. & F. Infante 1991. Manejo y evaluación de los datos obtenidos en los muestreos aerobiológicos, *Monografías REA/EAN* **1**, 1–18.

[11] J.M. Hirst, . 1952. An automatic spore trap. Annals of Applied Biology 39: 257-265.

[12] Munuera, M. 1999. Patrones de variación polínica en la atmósfera de Murcia. Implicaciones alergológicas, prevención y diagnóstico. Facultad de Biología. Universidad de Murcia. Tesis Doctoral. 1999.

[13] F. Takens , Detecting strange attractors in turbulence. Dynamical Systems and Turbulence. *Lecture Notes in Mathematics.* **898**, 366-381, Springer-Verlag. 1981.

[14] N.H. Packard, J.P. Crutchfield , J.D. Farmer and R.S. Shaw Geometry from a time series, *Phys. Rev. Lett.* **45** 712 (1980)

VSP International
Science Publishers
P.O. Box 346, 3700 AH Zeist
The Netherlands

*Lecture Series on Computer
and Computational Sciences*
Volume 1, 2004, pp. 347-350

Using Fractal Dimension to Characterize Pollen Time-Series

F. Martínez, I. Rodríguez, A. Guillamón[1] & M. Munuera[2]

[1]Department of Applied Mathematics and Statistic, Campus Muralla del Mar
[2]Department of Agriculture Science & Technology, Campus Alfonso XIII
Technical University of Cartagena, Spain.
30203 CARTAGENA (Murcia) – Spain.

Received 29 August, 2004; accepted in revised form 8 September, 2004

Abstract: Pollen time-series can be considered as being generated by biological system with inherently nonlinear dynamics. The purpose of this paper is to characterize the several pollinic stations using the fractal dimension. In our research, it was computed in MATLAB®, over data obtained in Murcia, Spain. The fractal dimension is a measure of signal complexity that can characterize different pollinical stations. We used the method suggested by Katz, that it's derived directly from the waveform. The results prove the existence of chaos in the pollen time series and the utility of fractal dimension to distinguish the several periods of pollinic activity.

Keywords: Nonlinear Dynamic, Environmental chaos, Biosignal processing, chaos characterization, Fractal Dimension.

1. Introduction

Recent suggestions that biological and, concretely, ecological and environmental may be a nonlinear process have sparked great interest in the area of nonlinear analysis of these processes giving rise to some studies   [2] & [3]. However, some of these works have not quantified the several nonlinear parameters, which are necessaries to characterize some processes of the mentioned areas.

Several algorithms have been proposed for the computation of characteristic invariant measures from an experimental time series. In this case, we use the fractal dimension. The fractal dimension refers to a non-integer or fractional dimension of a geometric object. This object, called strange attractor, is typically charcerized (to measure its strangeness) by a non-integer (fractal) dimension which is smaller than the number of degrees of freedom. The fractal dimension is calculated for the total pollen concentration in Murcia, Spain, using the method suggested by Katz [1].

From 1 March 1993 to 31 December 1998 a Burkard volumetric seven-day recording spore trap [4] located at about 19 m a.g.l. was operated on the exposed flat roof of the Veterinary Faculty (Murcia University, 110 m a.s.l., 38° 01' N, 01° 10' W, 4 km NW of Murcia city centre). From weekly tapes, daily slides were prepared following the standard methods accepted by the Spanish Aerobiology Network (REA, Red Española de Aerobiología; Domínguez et al. 1991), and subsequently examined by light microscopy. [4] , [9] & [10]. Pollen counts were transformed to pollen concentrations expressed in number of pollen grains per cubic meter.

2. Methods

The fractal dimension (D_f) can be considered as a relative measure of the number of basic building blocks that form a pattern. For this reason, we think that D_f is a measure of signal complexity that can characterize different set of data. The fractal dimension is a geometric invariant measure used for characterizing nonlinear systems, which is of interest to us because of its properties. It provides an

---

[1] Corresponding author. E-mail: f.martinez@upct.es, vicalum@upct.es

alternative technique for assessing signal complexity in the time domain, as opposed to the embedding method based on the reconstruction of the attractor in the multidimensional phase space [5]. The phase space representation of a nonlinear, autonomous and dissipative system can contain one or more attractors with generally fractional dimension. This attractor dimension is invariant, even under different initial conditions. This explains why the fractal dimension of attractors has been used widely for system characterization. However, estimating the fractal dimension of these attractors involves a large computational burden.

This complexity variations and biological signals changes over time, providing a fast computational tool to detect nonstationarities in these signals. There are many algorithms in the literature for estimating the fractal dimension of a waveform, [1], [6], [7]. In our research, the fractal dimension was computed according to the algorithm proposed by Katz [1].

In contrast to other methods, Katz´s D_f calculation is slightly slower, but it is derived directly from the waveform, eliminating the pre-processing step. The D_f of a time sequence analysed $x(j)$, $j=1,2,..., N$, can be defined as:

$$D_f = \frac{\log_{10}(L)}{\log_{10}(d)} \qquad (1)$$

where L is the sum of distances between successive points:

$$L = \sum_{i=1}^{N-1}|x(i+1) - x(i)|$$

and d is the diameter, estimated as the distance between the first point of the sequence and the point of the sequence that provides the farthest distance.

D_f computed in this way method depends on measurement units used. Katz´s approach solves this problem by creating a general unit or yardstick: the average distance between successive points, $\langle L \rangle$. Finally, normalizing distances in equation (1) by this average results in:

$$D_f = \frac{\log_{10}(N-1)}{\log_{10}\left(\frac{d}{L}\right) + \log_{10}(N-1)}$$

3. Results

In this case, the D_f parameter is obtained from the experimental time-series using a sliding window of 180 points, shifted along pollen series with one point of overlap, to make the two important cycles of pollinic activity stand out.This correlation is showed in the Figure No. 1.

It's observed the relationship between the pollen time-series evolution and the computed fractal dimension. This is more prominent in 1998, because of the great pollinic activity.

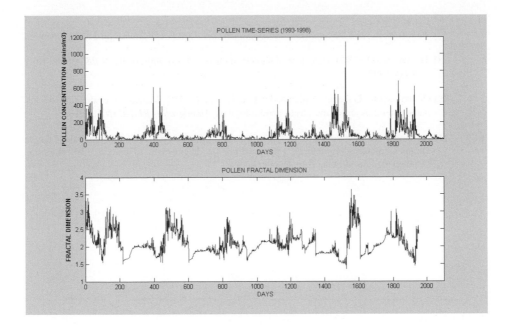

Figure 1: Pollen time series and Fractal Dimension evolution by days. (1993-1998)

4. Conclusions

This study suggests that the D_f is powerful tool to characterize the several pollinic stations, and, in addition the non-integer dimension has practical significance, because it's a way for testing the chaotic behaviour of the dynamical system [8].

In conclusion, the fractal measures expand the distinguishing features in characterizing pollinc biosignals and it gets the possibility of designing an automatic system "on line" for pollinic stations recognition.

References

[1] Katz, M. *Comp Biol. Med.,* vol 18, No.3, pp. 145-156, 1998.

[2] Hunt, J.C.R. Environmental forecasting and turbulence modelling. *Physica D* **133** (1999) 270-295.

[3] Kakonge, J. O. Applications of chaos theory to solving the problems of social and environmental decline in Lesotho. *Journal of Environmental Management* (2002) **65**, 63-78.

[4] Hirst, J. M. 1952. An automatic spore trap. *Annals of Applied Biology* 39: 257-265.

[5] Takens, F., *Lecture Notes in Mathematics* 898, Springer, Berlin, pp. 366-381, 1981.

[6] Higuchi, T. *Physica D*. Vol.31, pp. 277-287 , 1998.

[7] Petrosian, A., *IEEE Symposium on Computer-Based Medical Systems,* pp 212-217, 1995.

[8] Haykin, S & Li, X.B. Detection of signals in chaos. *Proceedings of the IEEE,* Vol 83 No.1, January 1995.

[9] Domínguez E., Galán C., Villamandos F. & Infante F. 1991. Manejo y evaluación de los datos obtenidos en los muestreos aerobiológicos, *Monografías REA/EAN* **1**, 1–18.

[10] Munuera, M. 1999. Patrones de variación polínica en la atmósfera de Murcia. Implicaciones alergológicas, prevención y diagnóstico. Facultad de Biología. Universidad de Murcia. Tesis Doctoral. 1999.

VSP International
Science Publishers
P.O. Box 346, 3700 AH Zeist
The Netherlands

*Lecture Series on Computer
and Computational Sciences*
Volume 1, 2004, pp. 351-353

On the Stability of Exponential Fitting BDF Algorithms: Higher-Order Methods

J. Martín-Vaquero * [1] and **J. Vigo-Aguiar** ** [2]

*ETS Ingenieros industriales. Universidad de Salamanca. Bejar, Spain.
** Facultad de Ciencias. Universidad de Salamanca. Salamanca

Received 30 July, 2004; accepted in revised form 20 August, 2004

Extended Abstract

The exponential fitting is a procedure proposed a few years ago to modify the classical algorithms in a way to make them particularly efficient for solving differential equations with oscillatory solutions or stiff problems.

In Vigo-Aguiar and Ferrndiz [1] we find a particular fast way to construct the coefficients to adapted BDF algorithms with constant steplength to these kind of problems. In this case, the methods integrate without error the following problems:

$$y' - Ay = F(x) \tag{1}$$

where F(x) is a vector $F(x) = [f^1, ..., f^m]$, and f^i belongs to the space generated by the linear combinations of $1, x, x^2, ..., x^k$, $i = 1, ..., m$, where k will be up to five.

To get the methods we start from the formulae

$$\nabla_{P(D)} \sum_{j=0}^{k} \beta_j \nabla^j y_p = h f_{p-s} \tag{2}$$

where $P(x) = x - A$, and $\nabla_{(D-A)} = Id - e^{Ah}E^{-1}$, Id being the identity, $E(f_p) = f_{p+1}$, and ∇ the backward difference operator, that is to say $\nabla^0 f_p = f_p, \nabla^{j+1} f_p = \nabla^j f_p - \nabla^j f_{p-1}$.

Now, we examine the gain to be expected when vectorial first order ODE's are solved by BDF methods whose weights are generated by means of exponential fitting. We give the analytic of the corresponding weights.

So forth, it is easy to prove the following theorem: **Theorem: Let be** $\rho_k(z)$ **the polynomial** $\beta_0^s + \beta_1^s + \beta_2^s z^2 + ... + \beta_k^s z^k$, **then the formulas (2), with order k+1 and k+1 steps, can be written as:**

$$\rho_k(1)y_p = (\sum_{j=1}^{k}(-1)^{j+1}\{\frac{\rho_k^{j)}(1)}{j!} + \{\frac{\rho_k^{j-1)}(1)}{(j-1)!}e^{Ah}\}y_{p-j}) + h f_{p-s} \tag{3}$$

[1]E-mail:jesmarva@usal.es
[2]E-mail:jvigo@usal.es

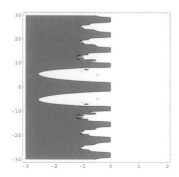

Figure 1: Region of 0-stability (in grey) for the implicit sixth-step method.

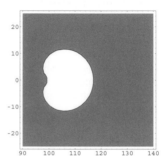

Figure 2: Absolute-stability (in grey) for the implicit sixth-step method, when the parameter of the method $\lambda h = -100$.

The coefficients β_j^s, where s=0 for implicit methods and s=0 for explicit methods, are obtained through the McLaurin series about $\xi = 0$ of the generating function:

$$G_s(\xi, Ah) = (Id - \xi)^{-1-s} \frac{\tilde{\mathrm{P}}(-\ln(Id - \xi))}{L(Id/(Id - \xi))}, \tag{4}$$

where $L(E) = (E - e^{Ah})$, $\tilde{\mathrm{P}}(x) = (x - Ah)$, and Id is the Identity Matrix of dimension $m \times m$.

Exponential fitting algorithms can be constructed in almost cases in this way and, if this is done properly, the accuracy is significantly enhanced.

Then, following the results in Vigo-Aguiar, Martin-Vaquero and Criado [2], we show some plots of zero-stability and absolute-stability of higher-order methods (4, 5 and 6). While explicit methods are not zero-stables, implicit methods get good results:

Finally, using explicit method of order one:

$$\beta_0 y_p = \beta_0 e^{Ah} y_{p-1} + h f_p, \tag{5}$$

with

$$\beta_0 = \frac{-Ah}{Id - e^{Ah}}, \tag{6}$$

and six-steps implicit method, we show results with several stiff oscillatory examples.

References

[1] J. Vigo-Aguiar and J. M. Ferrándiz, *A general procedure for the adaptation of multistep algorithms to the integration of oscillatory problems.* SIAM Journal Numerical Analysis **35**, 4, pp. 1684-1708, (1998).

[2] J. Vigo-Aguiar, J. Martín-Vaquero and R. Criado *On the stability of exponential Fitting BDF Algorithms.* J. Comp. Appl. Math. (accepted).

VSP International
Science Publishers
P.O. Box 346, 3700 AH Zeist
The Netherlands

*Lecture Series on Computer
and Computational Sciences*
Volume 1, 2004, pp. 354-356

Quantum Conductivity of Single Organic Molecules

Hitoshi Maruyama[1], Yoshihiro Asai[1,2], and Koichi Yamashita[1]

[1]Department of Chemical System Engineering,
University of Tokyo
7-3-1 Hongo, Bunkyo, Tokyo 113-8656, Japan
[2]National Institute of Advanced Industrial Science and Technology(AIST)
1-1-1 Umezono, Tsukuba Central 2, Tsukuba, Ibaraki 305-8568, Japan

Received 31 July, 2004; accepted in revised form 24 August, 2004

Abstract: We have developed a new program code for calculating the conductance properties of a single organic molecule connected to two *semi-infinite* leads using the Green's function method. We have applied the code for the study of the conductance of single benzene-dithiolate molecule connected to Au(111) electrodes. The I-V relation is analyzed in terms of the molecular orbitals of the system.

Keywords: Quantum conductance, Green's function method, Molecular Orbitals

1. Introduction

There has been significant progress in exploring the concept of molecular electronics in recent years. We have developed a new program code for calculating the conductance properties of a single organic molecule connected to two *semi-infinite* leads using the Green's function method. This is a surface problem and therefore neither usual quantum chemistry methods for molecules nor band calculation methods with periodic condition for solids can be applied to solve the problem. We addressed this problem by developing Sanvito's method [1] with the tight-binding model for the calculation of surface Green's functions. This method can include correctly the surface effects originating in evanescent states. Using our method, we have studied the conductance of single benzene-dithiolate molecule connected to Au(111) electrodes (see Fig.1).

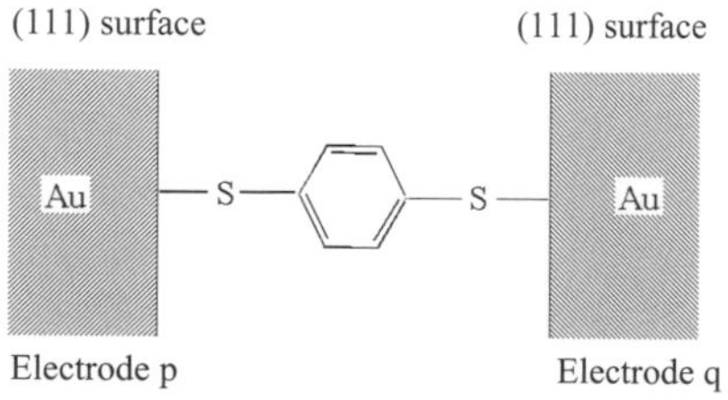

Fig.1 Model system

2. Theory

We evaluate the I-V relation from the expression,

$$I(V) = \frac{2e}{h}\int_{-\infty}^{+\infty} dE\, T(E)\{f_p(E-\mu_p) - f_q(E-\mu_q)\}$$

where T(E) is the transmission probability and f(E) is the Fermi function.

The transmission probability can be evaluated in terms of the Green's function of the system, G_{all} ,

$$T(E) = \mathrm{Tr}[\Gamma_p G_{all} \Gamma_q G_{all}^{\dagger}]$$

$$G_{all} = (E - H_{mol} - \Sigma_p - \Sigma_q)^{-1}$$

$$\Gamma = i[\Sigma - \Sigma^{\dagger}]$$

where H_{mol} is the Hamiltonian of the molecule and $\Sigma_{p,q}$ are the self-energies of the two *semi-infinite* leads. We developed the Sanvito's method [1] with the tight-binding model for the calculation of the system Green's function which can include correctly the surface effects originating in evanescent states.

3. Results

The theoretical I-V characteristics for the system shown in Fig.2 agree nicely with the experimental I-V data of Reed et al [?] Figure 3 shows the analysis of the transmission probability in terms of the molecular orbitals of the system. The difference in conductance properties of two absorbing sites, on-top and hollow, and an application to asymmetric organic molecules will also be discussed.

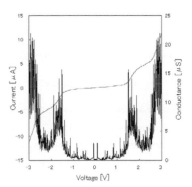

Fig.2 Conductance of the model system of Fig.1.

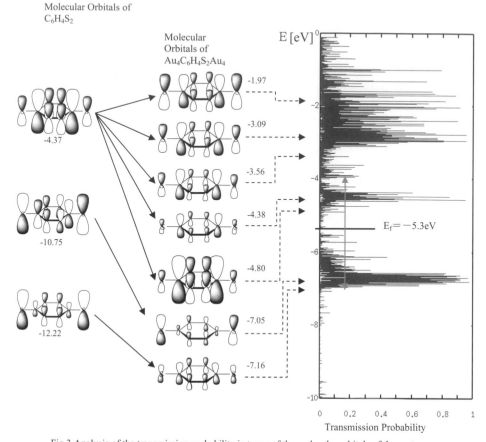

Fig.3 Analysis of the transmission probability in terms of the molecular orbitals of the system

Acknowledgments

This study was supported by a Grant-in-aid for the 21st Century COE Program for Frontiers in Fundamental Chemistry, and by a grant for Scientific Research (B) (No.14340174) from the Ministry of Education, Culture, Sports, Science and Technology of Japan. Some of the numerical calculations were carried out at the Research Center for Computational Science of Okazaki National Research Institute, Japan.

References

[1] S.Sanvito, C.J.Lambert, J.H.Jefferson and A.M. Bratkovsky, *Phys. Rev. B*, **59**, 11936 (1999)

[2] M.A.Reed, C.Zhou, C.J.Muller, T.P.Burgin, J.M.Tour, *Science*, **272**, 252 (1997).

VSP International
Science Publishers
P.O. Box 346, 3700 AH Zeist
The Netherlands

Lecture Series on Computer
and Computational Sciences
Volume 1, 2004, pp. 357-365

Modelling spatial clustering through point processes: a computational point of view

J. Mateu[1] and J.A. Lopez

Department of Mathematics, Campus Riu Sec, University Jaume I,
E-12071 Castellón, Spain

Received 10 August, 2004; accepted in revised form 28 August, 2004

Abstract: Spatial point process models provide a large variety of complex patterns to model particular cluster situations. Usually, the main tools to compare theoretical results with observations in many varied scientific fields such as engineering or cosmology are statistical, so the new theories and observations also initiated an active use of spatial statistics in these fields. However, due to model complexity, spatial statistics often rely on MCMC computational methods, a backbone for using these techniques in practice.
In this paper we introduce point field models, such as the continuum random-cluster process, the area-interaction process and interacting neighbour point processes, that are able to produce clustered patterns of great variety. In particular, we focus on computational aspects for simulation and statistical inference while analyzing the flexibility of these models when used in practical modelling.

Keywords: Area-interaction point processes, Clustered point patterns, Continuum random-cluster processes, Interacting neighbour point processes, Spatial statistics.

Mathematics Subject Classification: 60G12, 60G55.

1 Introduction

Stochastic geometry is the study of random patterns, whether of points, line segments, or objects. In particular, point processes are random point patterns in the plane or in space. The simplest example of these is the *Poisson point process*, which models "completely random" distribution of points in the plane or space. It is a basic building block in stochastic geometry and elsewhere. Its many symmetries typically reduce calculations to computations of area or volume. In stochastic geometry the classic generalization of the Poisson process is the Boolean model; the set-union of (possibly random) geometric figures or grains located one at each point or germ of the underlying Poisson process. Much of the amenability of the Poisson process is inherited by the Boolean model, though nevertheless its analysis can lead to very substantial calculations and theory as discussed by Hall (1988) and Molchanov (1996). It is often convenient to think of Boolean models as being derived from marked Poisson processes, in which the points are supplemented by marks encoding the geometric data (such as disk radius) of the grains; in effect the Poisson process now lives on $R^2 \times T$ where T is the mark space. Poisson cluster processes form a special case, in which the grains are simply finite clusters of points. The famous *Neyman-Scott processes* arise when the clusters are produced by independent sampling. How can we escape the limitations of the independence

---

[1]Corresponding author. E-mail: mateu@mat.uji.es, Fax:+34.964.728429

property of Poisson processes? One possibility is to randomize the intensity of the Poisson process, which produces the *Cox process* (also termed the doubly stochastic Poisson process).

Since the introduction of Markov point processes in spatial statistics, attention has focused on the special case of *pairwise interaction models*. These provide a large variety of complex patterns starting from simple potential functions which are easily interpretable as attractive and/or repulsive forces acting among points. However, these models do not seem to be able to produce clustered patterns in sufficient variety. And this is the reason why other families of Markov point process models, able to produce clustered patterns, are introduced in this paper.

The *area-interaction process* (Baddeley and van Lieshout, 1995), which uses the area of an associated Boolean model, the *quermass-weighted point processes* (Kendall, *et al.* 1999), which use perimeter length or total curvature characteristics, the *continuum random-cluster model* (Häggström, *et al.* 1999) and the *penetrable spheres mixture model* (Häggström, *et al.* 1999) are of interest in spatial statistics in situations where the independence property of the Poisson process needs to be replaced either by attraction or by repulsion between points. They are also highly relevant in statistical physics, where the first and third models provide the most well-known example of a phase transition in a continuous setting.

Ripley and Kelly (1977) introduced Markov processes with fixed finite range interactions. Continuum random-cluster processes, on the other hand, allow interactions with arbitrarily large range. Two points of a continuum random-cluster process are neighbours if they belong to the same connected component. Therefore, two points that are far away from each other interact if there is a chain of other points connecting them. A new class of processes, which is between Ripley-Kelly Markov and connected component Markov processes (Baddeley and Møller, 1989), is that when we allow points to interact if they are close enough (neighbours) or if there is a third point (but not a longer chain of points) connecting them. This new class is named *interacting neighbour processes*. One example of such a process is Geyer's *saturation process* (Geyer, 1999). It resembles the Strauss process, but has an extra parameter which puts an upper bound on the contribution to the density of any single point, and which, therefore, overcomes the normalizing problem in the clustered case. Because of this extra parameter, the saturation process is no longer a Markov process with respect to r-close neighbours but it is a Markov process with respect to r-close neighbours and the neighbours of these. Another example is the *isolated-point-penalization point process* (Hayat and Gubner, 1999) in which the interaction term controls the number of isolated points.

In this paper we introduce the main point field models that are useful as particular models of cluster situations within the context of spatial statistics. We further analyze, through a simulation study, the flexibility of these models when used in practical modelling and focus on inferential and simulation computational tools.

2 Spatial point process methodological setup

Markov point processes are a rich class of stochastic models for spatial patterns, with the virtue of being relatively tractable. They are defined to satisfy one of several spatial counterparts of the Markov conditional independence property. The likelihood takes a simple explicit form, apart from a difficult normalising factor. Indeed typically the likelihood is an exponential family, and the canonical sufficient statistic is often closely related to nonparametric spatial statistics. Typically each process is the equilibrium measure of an associated (space-time) Markov process; thus it is amenable to MCMC simulation and inference. Accordingly there is much current interest in exploring the potential applications of Markov point processes, which include spatial statistics, digital image analysis and geostatistics.

In order that likelihoods may exist, we shall restrict attention to finite simple point processes whose distributions are absolutely continuous with respect to the distribution of the Poisson process.

Such a process may be visualised very easily as a random finite number of points at random locations in a space S. A realisation of the point process X is a finite unordered set of points,

$$\mathbf{x} = \{x_1, ..., x_n\} \tag{1}$$

with $x_i \in S$. The space S in which the points lie is typically a subset of $\Re^d$ but may be any Polish space. Let $\aleph$ be the space of all such realisations.

Suppose that all point process models X here are absolutely continuous with respect to the distribution of the Poisson point process with intensity measure ν on S, where ν is a fixed, nonatomic, finite Borel measure. Then X has a probability density $f : \aleph \to [0, \infty]$ such that

$$P(X \in A) = \exp(-\nu(S)) \sum_{n=0}^{\infty} I_n(f, A) \tag{2}$$

for each $A \in F$, where $I_0(f, A) = 1\{\varnothing \in A\} f(\varnothing)$ and for $n \geq 1$

$$I_n(f, A) = \frac{1}{n!} \int_S ... \int_S 1\{\{x_1, ..., x_n\} \in A\} f(\{x_1, ..., x_n\}) d\nu(x_1)...d\nu(x_n) \tag{3}$$

In the simple case where S is a bounded subset of $\Re^d$ and ν is the restriction to S of the Lebesgue measure,

$$f(\{x_1, ..., x_n\}) dx_1...dx_n \tag{4}$$

is the probability that the process consists of a point near each of the locations $x_1, ..., x_n$ and no other points.

Example 1: Poisson process.

A point process is a *homogeneous planar Poisson process of intensity* λ if: **PP1**: the number, $n(S)$, of events in any planar region S follows a Poisson distribution with mean $\lambda|S|$, where $|S|$ denotes the area of S; **PP2**: Given $n(S) = n$, the n events in S form an independent random sample from the uniform distribution on S; **PP3**: For any two disjoint regions S_1 and S_2, the random variables $n(S_1)$ and $n(S_2)$ are independent.

Let $n(\mathbf{x})$ denote the number of points in a realisation $\mathbf{x} \in \aleph$. If $\alpha, \beta > 0$ are constants,

$$f(\mathbf{x}) = \alpha \beta^{n(\mathbf{x})} \tag{5}$$

defines the density of the Poisson process with intensity measure $\beta\nu(.)$, and the normalising constant α equals

$$\alpha = \exp\{-(\beta - 1)\nu(S)\} \tag{6}$$

Example 2: Inhomogeneous Poisson process.

For a function $\beta : S \to [0, \infty)$,

$$f(\mathbf{x}) = \alpha \prod_{i=1}^{n(\mathbf{x})} \beta(x_i) \tag{7}$$

is the density of the "inhomogeneous" Poisson process with intensity measure $\kappa(B) = \int_B \beta(u) d\nu(u)$ on S and the normalising constant is

$$\alpha = \exp\left\{-\int_S (\beta(u) - 1) d\nu(u)\right\} \tag{8}$$

Example 3: Cox process.

A point process is a *Cox process* if: **CP1**: $\Lambda(\mathbf{x})$ is a non-negative valued stochastic process and **CP2**: conditional on the realisation of $\Lambda(\mathbf{x})$, the point process is an inhomogeneous Poisson process with intensity function $\Lambda(\mathbf{x})$.

Example 4: Poisson cluster process.

A point process is a *Poisson cluster process* if: **PCP1**: *Parents* form a homogeneous Poisson process of intensity ρ; **PCP2**: The number of *offsprings* per parent is a random variable M, realised independently for each parent; **PCP3**: The position of each offspring relative to its parent is a bivariate random variable Y, realised independently for each offspring; **PCP4**: The observed point process consists of the superposition of offsprings from all parents.

2.1 Interpoint interactions

Definition. A finite *Gibbs point process* is a finite simple point process with a density $f(\mathbf{x})$ satisfying the positively condition

$$f(\mathbf{x}) > 0 \Rightarrow f(\mathbf{y}) > 0 \tag{9}$$

for all $\mathbf{y} \subset \mathbf{x}$.

By an application of the Möbius inversion formula or inclusion/exclusion, the density of any finite Gibbs point process can be written in the form

$$f(\mathbf{x}) = \exp\left\{ v_0 + \sum_{i=1}^{n(\mathbf{x})} v_1(x_i) + \sum_{i<j} v_2(x_i, x_j) + \sum_{i<j<k} v_3(x_i, x_j, x_k) + \ldots \right\} \tag{10}$$

where v_0 is constant and $v_k : S^k \to \Re \cup \{-\infty\}$ are symmetric functions, usually called interaction potentials. Thus the loglikelihood of a particular configuration $\mathbf{x}$ is a sum of penalties incurred for the presence of each point $x_i \in \mathbf{x}$, for the interaction between each pair of points $x_i, x_j \in \mathbf{x}$, for the interaction between each triple of points $x_i, x_j, x_k \in \mathbf{x}$, and so on. The sum can be interpreted as the physical potential energy of the configuration.

A **Markov point process** is one in which interpoint interactions occur only between those points which are deemed to be neighbours.

Example 5: Pairwise interaction process

A *pairwise interaction process* on S has a density of the form

$$f(\mathbf{x}) = \alpha \prod_{i=1}^{n(\mathbf{x})} b(x_i) \prod_{i<j} h(x_i, x_j) \tag{11}$$

where $b : S \to \Re^+$ is the activity and $h : S \times S \to \Re^+$ the interaction function, and $\alpha > 0$ is the normalising constant. The terms $b(x_i)$ influence the intensity and location of points, while the terms $h(x_i, x_j)$ introduce dependence or interaction between different points of the process. Typically, α is not known explicitly.

Assume now that the interaction function h has finite range $r > 0$, in the sense that $h(u, v) = 1$ whenever $\|u - v\| > r$. Declare two points $u, v \in S$ to be neighbours, and write $u \sim v$, if they are closer than r units apart:

$$u \sim v \quad \text{iff} \quad \|u - v\| < r \tag{12}$$

Then interactions occur only between neighbours, and the density becomes

$$f(\mathbf{x}) = \alpha \left[\prod_{i=1}^{n(\mathbf{x})} b(x_i) \right] \left[\prod_{x_i \sim x_j} h(x_i, x_j) \right] \tag{13}$$

A Markov point process on S with respect to a symmetric, reflexive realtion $\sim$ is a finite Gibbs point process whose conditional intensity $\lambda(u; \mathbf{x}) = f(\mathbf{x} \cup u)/f(\mathbf{x})$ depends only on u and $\{x_i \in \mathbf{x} : x_i \sim u\}$.

Example 6: Widom-Rowlinson penetrable sphere model

The *Widom-Rowlinson penetrable sphere model*, or *area-interaction process* (Baddeley and van Lieshout, 1995), has density

$$f(\mathbf{x}) = \alpha \beta^{n(\mathbf{x})} \gamma^{-A(\mathbf{x})} \tag{14}$$

where $\beta, \gamma > 0$ are parameters, $\alpha > 0$ is the normalising constant, and $A(\mathbf{x})$ is the area of

$$U_r(\mathbf{x}) = \left(\bigcup_{i=1}^{n(\mathbf{x})} B(x_i; r) \right) \cap S \tag{15}$$

where $B(x_i; r)$ is the disc of radius r centred at x_i. The above density is integrable for all values of $\gamma > 0$. The process produces clustered patterns when $\gamma > 1$, ordered patterns when $\gamma < 1$, and reduces to a Poisson process when $\gamma - 1$. This model has interactions of infinite order.

Example 7: Continuum random-cluster process

The *continuum random-cluster process*, has density

$$f(\mathbf{x}) = \alpha \beta^{n(\mathbf{x})} \gamma^{-c(\mathbf{x})} \tag{16}$$

where $\beta, \gamma > 0$ are parameters, $\alpha > 0$ is the normalising constant, and $c(\mathbf{x})$ denotes the number of connected components. This model exhibits regularity for $\gamma < 1$, and clustering for $\gamma > 1$.

Example 8: Triplets process

The idea of Markov point processes suggests adding the clique of next higher order to get a process that permits positive attraction of pairs of points. Define

$$W(\boldsymbol{x}) = \frac{1}{6} \sum_{i=1}^{n} \sum_{j \neq i} \sum_{k \neq i \neq j} I\left[\|x_i - x_j\| \leq r\right] I\left[\|x_j - x_k\| \leq r\right]$$
$$I\left[\|x_i - x_k\| \leq r\right] \tag{17}$$

the number of triplets of points that are mutual neighbours. Define $t(\boldsymbol{x})$ to be the 3-dimensional vector $(n(\boldsymbol{x}), S(\boldsymbol{x}), W(\boldsymbol{x}))$, where $S(\boldsymbol{x})$ is the same as in the Strauss process. Then, the *triplets process* (Geyer, 1999) has a density depending on three parameters $(\theta = (\beta, \gamma, \eta))$ and takes the form

$$f(\mathbf{x}) = \alpha \beta^{n(\mathbf{x})} \gamma^{S(\mathbf{x})} \eta^{W(\mathbf{x})} \tag{18}$$

The normalizing constant α is finite if and only if $\gamma \leq 1$ and $\eta \leq 1$ or if $\eta < 1$. Thus the canonical parameter space of the family is $\Theta = \{\theta \in \Re^2 : \eta < 1\}$ indicating that this is a good model for clustering.

Example 9: Saturation process

Another simple way of obtaining a model for clustering is by defining for each point $s \in \boldsymbol{x}$

$$m_s(\boldsymbol{x}) = \sum_{u \neq s \in \boldsymbol{x}} I\left[\|s - u\| \leq r\right]. \tag{19}$$

Then $\sum_{s \in \boldsymbol{x}} m_s(\boldsymbol{x})$ is just twice the neighbour pair statistic $S(\boldsymbol{x})$. Now, we put an upper bound $d > 0$ on the influence of any single point and define

$$U(\boldsymbol{x}) = \sum_{s \in \boldsymbol{x}} \min(d, m_s(\boldsymbol{x})). \tag{20}$$

Define $t(\boldsymbol{x})$ to be the 2-dimensional vector $(n(\boldsymbol{x}), U(\boldsymbol{x}))$.

The *saturation process* (Geyer, 1999) has a density depending on the parameters $(\theta = (\beta, \gamma))$ and takes the form

$$f(\mathbf{x}) = \alpha \beta^{n(\mathbf{x})} \gamma^{U(\mathbf{x})} \tag{21}$$

Now the normalizing constant α is finite for all values of γ.

Example 10: Isolated-point-penalization point process

Hayat and Gubner (1999) introduce the *isolated-point-penalization point process* which has the density

$$f(\mathbf{x}) = \alpha \beta^{|\mathbf{x}|} \gamma^{I(\mathbf{x})}$$

with $0 < \gamma < 1$, and $I(\mathbf{x})$ is the number of isolated points with respect to the r-close neighbour relation. Note that two points ξ and η are neighbours if $0 < d(\xi, \eta) \leq r$, where $d(\xi, \eta)$ is the distance between ξ and η.

Example 11: Interacting neighbour point process

A point process X is an *interacting neighbour point* (INP) process with respect to a given neighbour relation $\sim$ if its density has the form

$$f(\mathbf{x}) = \alpha \prod_{x_i \in \mathbf{x}} g(x_i, N_{\mathbf{x}}(x_i)) \tag{22}$$

where α is a normalizing constant and g a measurable function. This class of models was defined by Grabarnik and Särkkä (2001). The conditional intensity can be written as

$$\lambda(\xi; \mathbf{x}) = g(\xi, N_{\mathbf{x} \cup \{\xi\}}(\xi)) \prod_{x_i \in \mathbf{N}_{\mathbf{x}}(\xi)} \frac{g(x_i, N_{\mathbf{x} \cup \{\xi\}}(x_i))}{g(x_i, N_{\mathbf{x}}(x_i))}$$

Thus, the conditional intensity depends not only on ξ and the neighbours of ξ, but also on the neighbours of neighbours of ξ. This shows that there are interacting neighbour processes that are not Markov processes in the sense of Ripley-Kelly (R-K) with respect to the neighbour relation $\sim$. So, we consider another kind of relationship. Two points ξ and η of $\mathbf{x}$ are said to be interated $\sim$-neighbours, and write $\xi \sim_{\mathbf{x}}^2 \eta$, whenever $\xi \sim \eta$ or whenever there is another point $\zeta \in \mathbf{x}$ such that $\zeta \sim \xi$ and $\zeta \sim \eta$. Hayat and Gubner (1996) studied processes with fixed range iterated neighbours and called them 2-step Markov processes. Note that this neighbour relation depends on the realization. Such a neighbour relation is said to be *dynamic* or *realization dependent*. Baddeley and Møller (1989) defined a general class of processes, *nearest neighbour Markov processes*, where the neighbour relation depended on the realization. The iterated relation can be generalized to an m-iterated relation (see Hayat and Gubner, 1996). If m is infinite we obtain the connected component neighbour relation (Baddeley and Møller, 1989). Therefore, the m-step Markov processes as well as the INP processes are special cases of connected component processes.

INP processes include R-K Markov point processes (with respect to (w.r.t.) $\sim$) because a clique w.r.t. the $\sim$-relation is a star clique. But in general, INP processes are not R-K Markov processes w.r.t. the relation $\sim$, but they are R-K Markov with respect to a wider neighbourhood. For example, an INP process with r-close neighbours may not be R-K Markov with r-close neighbours but it is R-K Markov with 2r-close neighbours.

R-K Markov point processes with respect to a basic finite range relation may be unable to model specific features of real data sets. Therefore, iterated or m-iterated (for small m) Markov processes can be an appropriate alternative to the general connected component processes.

Being Markov point processes, homogeneous and inhomogeneous Poisson processes are particular cases of INP processes. Letting $g(\xi, N_{\mathbf{x}}(\xi)) = \beta$ in (22) we obtain the Poisson process with intensity β, and $g(\xi, N_{\mathbf{x}}(\xi)) = \beta(\xi)$ gives the inhomogeneous Poisson process with intensity $\beta(\xi)$. The Strauss process or the area-interaction point process are both Markov with respect to the r-close neighbours, and consequently are particular cases of INP processes. The saturation process above commented is another example of INP processes.

Let us now consider a process of the exponential family form with

$$f(\mathbf{x}) = \alpha \beta^{|\mathbf{x}|} \gamma^{\sum_{x_i \in \mathbf{x}} V(x_i, N_{\mathbf{x}}(x_i))}$$

and $V(x_i, N_{\mathbf{x}}(x_i)) = \max\{0, |N_{\mathbf{x}}(x_i)|(c - |N_{\mathbf{x}}(x_i)|)\}$, where $c > 0$ is an arbitrary constant and $N_{\mathbf{x}}(x_i)$ consists of the points that are within distance r from x_i. In the case $\gamma > 1$, the interaction term $\gamma^{\max\{0,|N_{\mathbf{x}}(x_i)|(c-|N_{\mathbf{x}}(x_i)|)\}}$ obtains its largest value as the number of neighbours equals $c/2$, i.e., the cluster size $c/2 + 1$ is favoured. Thus, we can control the size of clusters by choosing the constant c appropriately. Since for large values of the interaction parameter γ all clusters are of the same size, we call this process a *twin cluster process*. The case $0 < \gamma < 1$ is interesting as well. Then, the interaction term gets its smallest value when the number of neighbours of each point equals $c/2$ and its largest value if points have either no neighbours or c or more neighbours. Hence, instead of being a repulsive model, it is a specific cluster model defining processes with a combination of regular and clustered structures. This is called *bipattern process*.

3 Inference: Pseudolikelihood approach

Given a point pattern $\mathbf{x} = \{x_1, x_2, \ldots x_n\}$, the conditional intensity function of an event at u given the point pattern $\mathbf{x}$ in the remainder of S is defined as

$$\lambda^*(u; \mathbf{x}) = \frac{f(\mathbf{x} \cup u)}{f(\mathbf{x})}.$$

The general log-pseudo-likelihood function for a Gibbs process is given by

$$pl(\mathbf{x}; \theta) = \sum_{i=1}^{n} \log\{\lambda^*(x_i; \mathbf{x} \setminus \{x_i\}; \theta)\} - \int_S \lambda^*(u; \mathbf{x}; \theta) du. \tag{23}$$

The following relation is satisfied for a general Gibbs process

$$\lambda^*(u; \mathbf{x}; \theta) = \beta \exp\left\{-\sum_{j=1}^{n} \Phi(\|u - x_j\|; \theta)\right\} = \beta \lambda_0^*(u; \mathbf{x}; \theta). \tag{24}$$

If we condition on the number of points, the intensity parameter β is estimated by

$$\widehat{\beta} = n/\int_S \lambda_0^*(u;\boldsymbol{x};\theta)du,$$

and the conditioned log-pseudo-likelihood function for a Gibbs process is given by

$$pl_n(\boldsymbol{x};\theta) = \sum_{i=1}^{n} \log\{\lambda_0^*(x_i;\boldsymbol{x}\setminus\{x_i\};\theta)\} - n\log\{\int_S \lambda_0^*(u;\boldsymbol{x};\theta)du\}. \tag{25}$$

Maximization of (23) or (25) yields the maximum pseudo-likelihood estimate.

The evaluation of the integral term in (23) is developed here using a Monte Carlo approximation. The integral term is approximated by numerical quadrature in which a grid of $m = k \times k$ is considered in the region S, thus $S = \cup_{j=1}^m S_j$, and the integral is evaluated by the summation

$$\int_S \lambda^*(u;\boldsymbol{x};\theta)du = \sum_{j=1}^{m} |S_j|\,\lambda^*(u_j;\boldsymbol{x};\theta). \tag{26}$$

4   Aim of the paper and further research

The aim of this paper is to present in a natural way point field models that are useful as particular models of cluster structures based on collection of points, which can be useful in many diverse scientific disciplines. Amongst all the possible models, we focus on some of the point process models presented in this paper that are able to model clustering of points. Thus, we analyze the saturation model, the continuum random cluster model, the area-interaction and the INP processes. We develop the statistical methodology underlying these processes, comparing them and then analyze the computational tricks under several simulation and estimation techniques. We also present a complete simulation study, analyzing the flexibility of these models when used in practical modelling. Aplication to real data analysis and simulation using exact procedures is also discussed.

Acknowledgment

This work has been partially supported by grant BFM2001-3286 from Ministerio de Ciencia y Tecnología.

References

[1] A.J. Baddeley and J. Møller, Nearest-neighbour Markov point processes and random sets, *International Statistical Review* **57** 89-121(1989).

[2] A.J. Baddeley and M.N.M. van Lieshout, Area-interaction point processes, *Annals of the Institute of Statistical Mathematics* **46** 601-619(1995).

[3] C.J. Geyer, Likelihood inference for spatial point processes, In *Stochastic Geometry, Likelihood and Computation*. O.E. Barndorff-Nielsen, W.S. Kendall and M.N.M. van Lieshout (Eds.), Chapman and Hall, London, 1999.

[4] P. Grabarnik and A. Särkkä, Interacting neighbour point processes: some models for clustering, *Journal of Statistical Computation and Simulation*, **68** 103-125(2001).

[5] O. Häggström, M.N.M. van Lieshout and J. Møller, Characterization results and Markov chain Monte Carlo algorithms including exact simulation for some spatial point processes, *Bernoulli* **5** 641-659(1999).

[6] P. Hall, *Introduction to the Theory of Coverage Processes*, Wiley Series in Probability and Mathematical Statistics, New York, John Wiley & Sons, 1988.

[7] M.M. Hayat and J.A. Gubner, A two-step Markov point process, *Technical report ECE-96-2*, Department of Electrical Computer Engineering, University of Wisconsin-Madison, 1996.

[8] M.M. Hayat and J.A. Gubner, Asymptotic distributions for the performance analysis of hypothesis testing of isolated-point-penalization point processes, *IEEE Transactions in Information Theory* **45**, 177-187(1999).

[9] W.S. Kendall, M.N.M. van Lieshout and A.J. Baddeley, Quermass-interaction processes: conditions for stability, *Advances in Applied Probability (SGSA)* **31** 315-342(1999).

[10] I.S. Molchanov, *Statistics of the Boolean Models for Practitioners and Mathematicians*, Chichester, John Wiley & Sons, 1996.

[11] B.D. Ripley and F.P. Kelly, Markov point processes, *Journal of London Mathematical Society* **15** 188-192(1977).

VSP International
Science Publishers
P.O. Box 346, 3700 AH Zeist
The Netherlands

Lecture Series on Computer
and Computational Sciences
Volume 1, 2004, pp. 366-369

Managing Heterogeneity in Time Series Prediction

J. M. Matías[1], W. González-Manteiga[2], J. Taboada[3] and C. Ordóñez[3]

[1]Department of Statistics, University of Vigo, 36200 Vigo
[2]Department of Statistics, University of Santiago de Compostela, Spain
[3]Department of Natural Resources, University of Vigo, Spain

Received 5 August, 2004; accepted in revised form 24 August, 2004

Abstract: In statistical applications it is usual to postulate a single hypothetical distribution for modeling the data generating process. However, this practice it is not realistic when the true distribution can change through the input domain (e.g. time). In this work we propose a technique for modelling the possible heterogeneity of time series. Firstly, we include model selection as part of the likelihood maximisation in the main estimation problem; we use both Yeo-Johnson transformations and the hypernormal distribution and compare the results of each. Then, we generalise the above proposal applying it to each instant of time. In this way we can postulate and estimate a pattern of temporal change in the distribution model using an auto-regressive structure to model its evolution. In this early stage of our research we consider lineal models for this evolution and focus on the problem of the prediction of heteroskedastic time series, -- although there is no reason why the method cannot be applied to more general regression problems.
The proposed technique is evaluated with daily return series for several stock market indices, improving the results obtained when heterogeneity is not assumed.

Keywords: GARCH, heteroskedasticity, Hypernormal densities, model heterogeneity, model selection, power transformations, time series.
Mathematics Subject Classification: 65C60, 62E10, 62G07, 62M10, 62M20, 62P20

1. Introduction

Techniques for predicting time series require the initial postulation of a hypothetical distribution model as a framework in which to estimate parameters such as mean, variance, quantiles, etc. from available data. Since it is not possible to test a priori the veracity of the model, selection will have to be made from a set of alternative models.

For example, in the financial field, the most usual models considered are Gaussian or Student's *t* distributions, or any model estimated non-parametrically. Model selection in this field is critical, given its great impact on the quality of the predictions and their confidence intervals.

The usual selection approach results in a single best model that is supposed to hold valid for the entire time period considered. Although not very realistic for many application fields, the fact that this assumption is made is due to the absence of techniques for managing heterogeneity in the input space of interest.

For time series, for example, it is quite possible that specific days of the year, month or week may be governed by different distribution models (apart from the well-known heteroskedasticity effect). Other fields, such as engineering, the environment, health sciences, etc., also pose the same modelling problem.

However, before dealing with the model heterogeneity problem, it is necessary to solve satisfactorily the (single) model selection problem. In this regard, one approach is to use more flexible and universal distribution models that depend on one parameter that can be estimated in the process of maximising

---

[1] Corresponding author. E-mail: jmmatias@uvigo.es

likelihood and which, to some degree, includes traditional models as particular cases. One well-known technique is based on the Box-Cox [1] family of transformations and subsequent variations [4]. An important advance in this line of research, however, has been the hypernormal distribution [2] -- which includes the Student's t family as a particular case and the Gaussian family as a limit -- and for the moments of which the authors provide analytical expressions. However, despite the great flexibility that the hypernormal family introduces in the modelling process, the problem of selecting the parameters that specify the most suitable model remains unresolved.

One contribution of our research is to include (single) model selection as part of the likelihood maximisation in the main estimation problem; we use both Yeo-Johnson transformations and the hypernormal distribution and compare the results of each.

Nevertheless, our main contribution consists of generalising the above proposal and applying it to each instant of time. In this way we can postulate and estimate a pattern of temporal change in the distribution model using an auto-regressive structure to model its evolution. In this early stage of our research we consider lineal models for this evolution and focus on the problem of the prediction of heteroskedastic time series, -- although there is no reason why the method cannot be applied to more general regression problems.

2. Automatic selection of the distribution model in an ARMA-GARCH model

The Yeo-Johnson transformations [4], which represent an improvement over the Box-Cox transformation family [1], are designed fundamentally to better manage original variable skewness and positive negative values.

Recently, Giacomini and others proposed [2] the so-called hypernormal distribution with density function:

$$f_{\lambda,\varsigma}(x) = \frac{\Gamma(a_{\lambda,\varsigma})}{\Gamma(a_{\lambda,\varsigma} - \frac{1}{2})} \sqrt{\frac{\lambda}{\pi}} (\lambda x^2 + 1)^{-a_{\lambda,\varsigma}} \text{ with } a_{\lambda,\varsigma} = \frac{1 - \lambda^{1+\varsigma}}{2\lambda(1-\lambda)} \text{ for } \lambda \in (0,1) \tag{1}$$

Relevant particular cases for the above family are obtained when $\varsigma - 0$ and $v = 1/\lambda \in (1,\infty) \cap \mathbb{N}$, in which case the Student's t distribution with v degrees of freedom is obtained. The Gaussian distribution is obtained as a limit case when $\lambda \to 0$.

With a view to estimating variance in a heteroskedastic context, it is sufficient to consider that the original variables Y_t with variance h_t are the result of the transformation $Y_t = \sqrt{h_t} Z_t$ where Z_t is a standardised hypernormal random variable and where $f_{\lambda,\varsigma,y}(y) = f_{\lambda,\varsigma,z}(z)/\sqrt{h_t}$ is the density of Y_t. If X_t is a hypernormal random variable with variance $\sigma_{\lambda,\varsigma}^2$ the standardised variable is $Z_t = X_t / \sigma_{\lambda,\varsigma}$ whose density is $f_{\lambda,\varsigma,z}(z) = f_{\lambda,\varsigma,x}(x) \cdot \sigma_{\lambda,\varsigma}$. Hence, the likelihood of the observations $\mathbf{y} = (y_1,...,y_n)$ is:

$$\ln f_{\lambda,\varsigma}(\mathbf{y}) = n \ln \frac{\Gamma(a_{\lambda,\varsigma})}{\Gamma(a_{\lambda,\varsigma} - \frac{1}{2})} - \frac{n}{2} \ln \pi + \frac{n}{2} \ln \lambda + n \ln \sigma_{\lambda,\varsigma} - \frac{n}{2} \ln h_t - a_{\lambda,\varsigma} \sum_{t=1}^{n} \ln(\lambda \frac{\sigma_{\lambda,\varsigma}^2 y_t^2}{h_t} + 1) \tag{2}$$

In the general case of requiring an estimate for the mean μ_t, it is enough to substitute y_t in the above expression for the innovations $\varepsilon_t = y_t - \mu_t$, and to estimate and using the same procedure. Below we will assume $\varsigma = 0$ as this does not reduce appreciably the expressivity of the family.

The family implicit in (2) has great flexibility in modelling heteroskedastic times series whose conditional distribution presents greater kurtosis than the Gaussian family (as is usual for financial series). Nonetheless, we still have to determine a suitable value for the parameter λ. We propose estimating it in the likelihood maximisation process that has to be undertaken to estimate the parameters of the ARMA-GARCH model for time series. This approach can be equally used for any regression problem that relies on maximising likelihood, and thereby avoids the imposition of a rigid model for error distribution.

3. Temporal modeling of models

The modelling process described above is global in that it assumes the existence of a single valid model for the entire time period under consideration. Nonetheless, in many cases such a hypothesis may be somewhat unrealistic.

In theory, any approach that considers the possibility of change in a distribution model should be useful for modelling its temporal patterns and for prediction. Here, we propose an auto-regressive temporal model for the parameter λ from the Yeo-Johnson and hypernormal families. Bearing in mind that kurtosis is a fundamental differentiating feature in the distributions for each family, our temporal model uses the kurtosis in previous instants as predictive variables. Specifically, for return series $\{\varepsilon_t\}$ such as those used in this research to evaluate the different techniques, the temporal model responds to the following expression:

$$\lambda_t = L + A\lambda_{t-1} + B\kappa_{t-1} + C\varepsilon_{t-1}^2 \tag{3}$$

where κ_{t-1} is the kurtosis in the time $t-1$ relative to a recent time window. The general model (3) includes as particular cases those which fix null values for some of its parameters; likewise, this framework can be expanded to a greater number of lags if necessary, as explained in the text.

Logically, as a non-observable parameter, kurtosis should be estimated from the observations in each time window. The estimator used should be robust enough to overcome possible noise. Here we use an estimator based on quantiles:

$$\hat{\kappa} = \frac{F^{-1}(0.975) - F^{-1}(0.025)}{F^{-1}(0.75) - F^{-1}(0.25)} \tag{4}$$

where is the distribution function. This estimator has been evaluated, among other alternatives, in [3]. Other alternatives to the model (3) are discussed in the text.

4. Results

The performance of constant (global) and variable (local) Yeo-Johnson and hypernormal distributions with ARMA-GARCH models have been evaluated with daily return series for the DAX, Dow Jones and Ibex 35 indices for the period 29 September 2000 to 23 June 2004. The Table 1 shows the results of the algorithms on a test set of observations of these return series and compares these with the results for the GARCH models with Gaussian and the best Student's t distribution.

For the series evaluated, the variable (local) distribution models improved on the results of the best fixed (global) models selected using our training sample, with the local hypernormal model always producing the best results by an appreciable margin.

Furthermore, the local techniques permit to obtain and plot the estimated series $\{\hat{\lambda}_t\}$ and interpreting it in terms of the evolution of the conditional distribution model.

5. Conclusions

As far we are aware, no techniques are as yet available that model heterogeneity in the distribution of the variable of interest in a viable and structured way. We have proposed new techniques for modelling evolution over time of conditional distribution in a time series. The results would indicate that the hypernormal family is a potentially useful tool for managing heterogeneity.

Table 1: Results for a selection of stock market indices, featuring the distribution parameter (the mean for the local models) and log-likelihood. Note that small differences in log-likelihood represent significant differences in goodness-of-fit, since this criterion, as well as being logarithmic in scale, reflects the goodness-of-fit of the simultaneous estimates for each the value of the variable and its variance.

	DAX		DOW JONES		IBEX 35	
GARCH Model	**df or λ**	**lnLkhTest**	**df or λ**	**lnLkhTest**	**df or λ**	**lnLkhTest**
Gaussian		992.46		1219.75		1150.98
Student's t_{df}	6	1029.60	7	1230.91	49	1152.60
Yeo-Johnson t_{df} global	-4.0000	1029.17	-2.2854	1226.75	-1.6232	1149.31
Hypernormal global	0.1346	1029.97	0.1194	1232.01	0.0219	1153.44
Yeo-Johnson t_{df} local	-4.6411	1031.73	-4.7580	1232.89	-1.3013	1150.57
Hypernormal local	0.0281	1034.48	0.0122	1234.54	0.0599	1156.83

These results, moreover, would suggest the usefulness of this kind of analysis even if only from the point of view of evaluating the degree to which heterogeneity is present in an application problem and of obtaining a better understanding of the data generating process than would be obtained from a single distribution model that is supposedly valid for the entire input space.

Although the proposed techniques have a simple linear structure, they represent a point of departure for the design of more sophisticated and flexible algorithms that better adapt to different situations. Research in the immediate future will be concerned with new non parametric algorithms (based on local likelihood and neural networks) that will better reflect the evolution of patterns over time in the models. We will also carry out research into other possible optimisation algorithms (e.g. genetic algorithms) which could better tackle the complexity of these new models.

Acknowledgments

The authors wish to thank the anonymous referees for their careful reading of the manuscript and their fruitful comments and suggestions.

References

[1] G.E.P. Box and D.R. Cox, *An analyisis of transformations* (with Discussion). Journal of the Royal Statistical Society B **26** 211-252(1964).

[2] R. Giacomini, A. Gottschling, C. Haefke and H. White, *Hypernormal densities*. University of California at San Diego, Economics Working Paper Series 2002-14, Department of Economics, UC San Diego, 2002.

[3] T.-H. Kim and H. White, *On more robust estimation of skewness and kurtosis: Simulation and application to the S&P500 Index*. Economics Working Paper Series 2003-12, Department of Economics, UC San Diego, 2003.

[4] I.-K. Yeo and R.A. Johnson, *A new familiy of power transformations to improve normality or symmetry*. Biometrika, **87** 954-959 (2000).

VSP International
Science Publishers
P.O. Box 346, 3700 AH Zeist
The Netherlands

*Lecture Series on Computer
and Computational Sciences*
Volume 1, 2004, pp. 370-374

Web-Browsing Collaborative 3-D Virtual Worlds: A Framework for Individuals, Artifacts, and Decorations

Rolando Menchaca-Méndez[1], Leandro Balladares Ocaña[1],
Bruces Campbell[2], Rubén Peredo Valderrama[1]

[1]Computer Science Research Center (CIC) – IPN
Unidad Profesional Adolfo López Mateos, C.P. 07738
Mexico City, Mexico

[2]University of Washington, Human Interface Technology Laboratory (HIT Lab)
215 Fluke Hall, Seattle, WA 98195, USA

Received 6 August, 2004; accepted in revised form 28 August, 2004

Abstract: In this paper, we present an interaction model and architecture for the shared experiences of Collaborative Virtual Worlds (CVWs) over the Internet. By means of a directed graph, the model describes interaction between the entities that populate the virtual world. The nodes and edges of the graph can be easily mapped to entities (classes and interfaces) of the proposed architecture, reducing the time and effort needed to develop Distributed Virtual Worlds (DVWs). The architecture allows for easy distribution management of processes between clients and a server, or otherwise its centralization within a central server. These worlds then enable, through the Web browser, a shared understanding regarding many scientific and engineering subjects concerned with three-dimensional data.

Keywords: Web-based Collaborative Virtual Worlds, Software Engineering, Visualization, Web-based Simulation and Computing, Computational Science and Education.

1. Introduction

Very few people use the Web in the multi-user modes suggested by collaborative systems researchers [1-5]. Perhaps such use would be greatly facilitated if users could enable them via their Web browser by pointing to URLs where a collaborative system was activated upon connection. Results from GIS coordination, for example, seem ripe content for such collaborative worlds. But, both the creation of such worlds for distribution and access to participating in such worlds must continue to become easier tasks. In [3][4], a model for the creation of Distributed Virtual Worlds (DVWs) through the use of distributed objects is proposed. This model aims to reduce the complexity of developing DVWs by separating the communication issues form the virtual reality issues. This work, based upon the proposed concepts, extends the use of DVWs for the creation of Collaborative Virtual Worlds (CVWs). A CVW is a virtual world populated by a set of entitics (individuals, artifacts and decorations) that interact with each other (see Figure 1). Every entity has a role that is defined by its abilities, as well as by the way it interacts with other entities in the world. The collaboration between the entities is then defined by an interaction graph that describes the actions taken when it occurs. The nodes and edges of the graph can be easily mapped to entities of the proposed software architecture (classes or interfaces) reducing the time and effort needed to develop a DVW. One of the most important design goals of the architecture is to enable the use of different actualization schemes. Identifying actualization schemes is very valuable because there is not a best actualization protocol for every situation [1-2][5]. In Section 2, our proposed collaboration model is presented. Also, the elements and the mechanisms that we use to describe the roles of the entities are described. In Section 3, the elements are modeled to work well with a distributed objects technology such as Java RMI. In Section 4, the alternatives for the distribution of the necessary processes to implement the CVW, as well as an analysis of the advantages

---

[1] Corresponding author: Computer Science Research Center (CIC) – IPN, Unidad Profesional Adolfo López Mateos, C.P. 07738, Mexico City, Mexico. E-mail: ballad@cic.ipn.mx; leandro12021975@hotmail.com.

and disadvantages for each alternative, are presented. Finally, in Section 5 a set of conclusions and future work are discussed.

2. Collaboration Model

The set of all existing social groups within a CVW is denoted as S, and the set of all types of artifacts as A. Each group partially defines the way its elements interact with other group elements. In order to totally define the interactions that can be carried out within the CVW, a directed graph $G = (V, E)$ where $V = S \cup A$ and $E \subseteq S \cup A \times S$ is proposed, where the edges represent a service relationship.

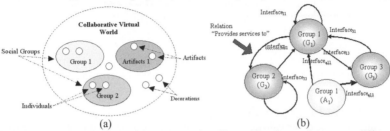

(a)   (b)

Figure 1: a) CVW components: social groups and artifacts. The figure shows two different social groups and one kind of artifact; b) Collaboration graph, a directed graph defining the collaboration relationships in a CVW.

In the interaction graph, we can define the names of the graphic interfaces that the individuals will display to collaborate with other individuals. For example, if a user *I1* pertaining to the group *G1* approaches to *I2* pertaining to *G2*, this will cause both individuals to interchange their collaboration interfaces. Thus, *I2* will send an interface of the type *Interface2* to *I1* and *I1* will send an interface of the type *Interface12* to *I2*. When the individuals move away, it is assumed that the collaboration has been finished, and the graphic interfaces related to *Interface21* and *Interface12* will disappear from the browsers of *I1* and *I2* respectively. Under this scheme, it is possible to divide the CVW in different regions where the individuals acquire new abilities when entering (they can collaborate with the region). The kind of abilities obtained by the individual will depend on its social group and the relation or between its social group and the artifact, defined in the collaboration graph. In order to obtain the previous behavior, it is only necessary to model the regions like artifacts.

3. CVW Structure

The CVW are mainly made up by three classes: *RemoteSoul*, *CollaborativeWorld* and *CollaborativeApplet*. The *RemoteSoul* class represents the state and behavior of the individuals and implements its social groups defined in the collaboration graph. A *CollaborativeWorld* class serves as a meeting point for the participants and, as it is explained further, as a container of souls and references or just as a container of references. The *CollaborativeApplet* class represents the environment of the users. It is the place where the rendering is carried out and, depending on the architecture (centralized or distributed), the processes associated with the soul of the individual that represents the user (see Figure 2a).

Figure 2b shows the summarized class diagram of the graphic interfaces for two hypothetical social groups. *GISubjectType1* and *GISubjectType2*. It defines the set of graphic components used to access the inherent abilities of the subjects (its respective HS methods). There are other two graphic interfaces for group *G1*: *GISubjectType11* for the interaction with other individuals of the group *G1* (because there is and edge from *G1* to *G1* in the collaboration graph) and *GISubjectType12* for the interaction with individuals of the group *G2* (because there is and edge from *G1* to *G2* in the collaboration graph).

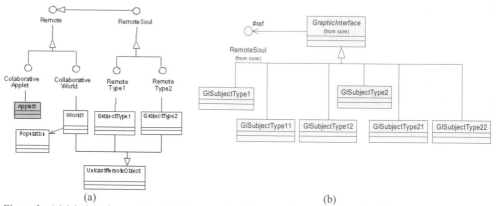

(a)   (b)

Figure 2: a) Main interfaces of the CVW; summarized diagram class of the graphical interfaces for two groups; G1 and G2.

There is a similar case for group *G2*. The collaborative applets do not need to have previous knowledge of the graphic interfaces of the individuals. In the same way, individuals do not need to have previous knowledge of the interfaces that other subjects define. When a collaboration event takes place, souls construct the suitable graphic interface (that corresponding to the type of subject with whom it is collaborating) and send it to the target *CollaborativeApplet*. When a collaboration event takes place, souls construct the suitable graphic interface (those corresponding to the type of subject with whom it is collaborating) and send it to the target *CollaborativeApplet*. In order to dynamically identify the social group of a subject or the type of an artifact, we use the reflection support of the Java language.

The main task of this object is to keep a record of the participants of the CVW. These objects generally reside in the same machine that the Web server that hosts the world. When a new user accesses the Web site, he downloads a *CollaborativeApplet*. The first operation of the applet is to obtain a remote reference to the object that implements the *CollaborativeWorld* interface. Users access the CVW by means of a web browser (see Figure 3a). The browser accesses a HTTP server where it obtains the *CollaborativeApplet* (CA) that is in charge to execute all the necessary code to participate in the CVW. As it is shown in Figure 3b, the CA acts as a bridge between the technologies in charge to render the CVW and the remote objects that implement the CVW behavior. Within the Applet, threads are constructed to obtain the state of the objects that implement the souls of subjects and artifacts. We can use different update schemes depending on the nature of the application [4], such as:

- *Pooling*: a thread is associated with the reference of each active element of the CVW. The thread performs a cycle where it invokes a remote object method to obtain his state. With this information it updates the appearance of the element in the VW.
- *Update*: when the state of a soul is modified, it makes calls to its references (callbacks) so that clients update the appearance of this element in the VW.
- *Both sides processing*: the code that is in the browser implements part of the state and behavior of the souls. Local objects make calculations associated with the behavior of the subjects but they maintain communication between them to be synchronized. This scheme can be implemented so that the standard dis-java-vrml is fulfilled.

Because the update schemes are implemented between the souls and their references, it is possible to find in the same CVW, subjects and artifacts that use different update schemes. In other words, each subject has the flexibility to implement the protocol that better covers its particular necessities. For example, an artifact that implements a collaborative editor could need protocols that preserve the consistency of the state, as well as the intention of the operations. This requires complicated calculations [5], which would be inefficient to apply to the rest of the artifacts of the MVC.

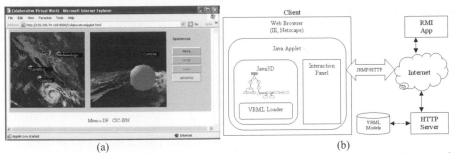

(a)   (b)

Figure 3: a) A collaborative applet; b) The Collaborative Applet serves as bridge between the technologies in charge to render the CVW and the objects that keep their state and implement their behavior.

The proposed architecture allows the implementation of three different schemes to distribute the data and behavior in a CVW. These three schemes are centralized, distributed and hybrid. The distributed scheme has the following advantages over the centralized: The amount of communication resources grows linearly in respect to the number of users; in contrast with the quadratic grow of the centralized schema. Actualizations are made directly from client to client, whereas in the centralized schema the invocations are directed to the server and then to the clients. The feedback is improved because the remote object that implements the user's avatar is in the same address space (in the Web Browser). In the prototypes, we included a check box where the user can locate the soul of his avatar. In the future, this decision can be taken automatically based on objective parameters such as hardware and network characteristics, number of users, world behavior, etc.

4. Conclusions and Future Work

The interactive-entities model defines a CVW based on the abilities of the entities that populate the world as well as in the interactions between them. Hence, these local interactions define the global CVW behavior. One of the most important characteristics of the architecture is that it encapsulates, inside the souls and its references, the algorithms and protocols related to the consistency of the world. In this manner, algorithms and protocols can be easily extended, optimized and replaced without affecting other entities in the world. From the developing process point of view, the goal of the collaboration model is not only to develop CVW faster, but also the resulting CVW have similar structures. They are easier to maintain and eventually to integrate. The proposed framework can be applied to scientific subject matter in the fields of Geography, Astronomy, Ocean Science, and Anatomy for visualization of scientific and engineering data. Currently, the distributed virtual reality group at CIC is developing a CASE tool based on the presented interaction model. We expect that this tool will significantly reduce the time and effort necessary to develop CVWs.

Acknowledgments

Also, the author wishes to thank the anonymous referees for their careful reading of the manuscript and their fruitful comments and suggestions. This work was supported by: The National Polytechnic

Institute of Mexico, the Computer Science Research Center, the Sciences Mexican Academic (Academia Mexicana de Ciencias - AMC) and the United States – Mexico Foundation for Science (Fundación México – USA para la Ciencia – FUMEC) through the program "Researching Stays for Young Researchers 2004". Leandro Balladares O. thanks to Bruce D. Campbell, Dra. Suzanne Weghorst, Dr. Tom Furness, the HIT Lab and the Center for Environmental Visualization (CEV) people for the time he spent at University of Washington during the summer 2004.

References

[1] Hans-Peter Dommel and J. J. Garcia-Luna-Aceves, "Group Coordination Support for Synchronous Internet Collaboration", IEEE Internet Computing, March-April 1999.

[2] Atul Prakash, Hyong Sop Shim, and Jang Ho Lee, "Data Management Issues and Trade-Offs in CSCW Systems", IEEE Transactions on Knowledge and Data Engineering, Vol. 11 No. 1, January/February 1999.

[3] Menchaca R. & Quintero R., "Distributed Virtual Worlds for Collaborative Work based on Java RMI and VRML", Proceedings of the IEEE 6th International Workshop on Groupware CRIWG 2000.

[4] Quintero-Tellez, R., et al, "Implementation Aspects of Synchronous Collaborative Distributed Virtual Worlds", 5th IASTED International Conference, Computer Graphics and Imaging, 2002.

[5] N. Vidot, et-al. "Copies Convergence in a Distributed Real-Time Collaborative Environment", Proceeding of the ACM 2000 Conference on Computer supported cooperative work. Philadelphia, Pennsylvania, United States.

VSP International
Science Publishers
P.O. Box 346, 3700 AH Zeist
The Netherlands

*Lecture Series on Computer
and Computational Sciences*
Volume 1, 2004, pp. 375-378

Multivalued Neural Network for Graph MaxCut Problem

E. Mérida-Casermeiro[1] and D. López-Rodríguez[2]

Department of Applied Mathematics,
E. T. S. I. Informática,
University of Málaga
29071 Málaga, Spain

Received 30 July, 2004; accepted in revised form 28 August, 2004

Abstract: In this paper we have used a multivalued neural model (MREM) in order to solve the maximum cut problem. A new technique, based in the problem, that allows to escape of certain bad local minima has been incorporated in order to improve the goodness of the obtained solutions. Finally, we have made some comparisons with other neural formulations for that problem obtaining better solutions in a reasonable time.

Keywords: Maximum cut problem. Multivalued neural network. Combinatorial Optimization.

Mathematics Subject Classification: 82C32, 68W15

PACS: 120304

1 Introduction

One of the most important combinatorial optimization problems (COP) in graph theory is the maximum cut problem (MAXCUT). The goal of this problem is to find a partition of vertices of a weighted undirected graph $G = (V, E)$ into two disjoint sets, such that the sum of edges with endpoints in different set is as large as possible.

This problem has many applications to pattern recognition, clustering, statistical physics and the design of communication networks, VLSI circuits and circuit layout among them [2].

Although for planar graphs there exists a solution for this problem in polynomial time [5], the problem is known to be NP-complete [4] for a generic graph, even for unweighted graph. So, to obtain the optimum is computationally intractable and it is important to develop an algorithm that achieves near optimum solutions in reasonable time [9, 6, 3].

Neural networks have demonstrated to be a good alternative method to classical algorithms [10]. So, almost every type of COP has been tackled by neural networks, and many of the approaches are very competitive and efficient in terms of quality, time-consuming and resources.

In 1997, Alberti et al. presented a type-Hopfield neural model for MAXCUT [1], but its performance is worse than the presented by Bertoni et al [3]. Takefuyi and his colleagues [11] have developed a powerful neural model labelled 'maximum' and they have shown that their neural model performs better than the rest of algorithms in solving a wide range of COP.

---

[1] Corresponding author. E-mail: merida@ctima.uma.es
[2] E-mail: dlopez@ctima.uma.es

Recently, Galán-Marín et al. proposed a new neural model named OCHOM which obtains much more efficient solutions than 'maximum'. Moreover, it can be used for many COP and it also has the advantage of fast convergence to a valid solution without tuning any parameter.

In order to make OCHOM escape from local minima, Wang et al.[12] have recently proposed a stochastic dynamics for OCHOM, permitting temporary decreases of the objective function.

The model proposed in this paper obtains better solutions than those of Wang and OCHOM for MAXCUT in very short time-consumption. Moreover, the model can be applied to a generalization of MAXCUT and a new technique to scape from local minima is provided for this problem. This technique improves previous solutions and can be easily used with other COP.

2   The Maximum Cut Problem

Let $G = (V, E)$ be an undirected graph without self-connections. $V = \{v_i\}$ is the set of vertices and $E = (e_{i,j})$ is a symmetric binary matrix where $e_{i,j} = 1$, if and only if, an arc with endpoints v_i and v_j exists. For each edge in E there is a weight $c_{i,j} \in \mathbb{R}$. All weights can be expressed by a symmetric real matrix C, with $c_{i,j} = 0$ when it does not exist an arc with endpoints v_i and v_j.

The Maximum Cut Problem (MAXCUT) consists in finding a partition of V into two subsets A_1 and A_2, such that $\sum_{v_i \in A_1, v_j \in A_2, i > j} c_{i,j}$ is maximum.

Generalization of the MAXCUT Problem (K-MAXCUT): It looks for a partition of V into k disjoint sets A_i such that the sum of the weights of the edges from E that have their endpoints in different elements of the partition is maximum. So, the function to be maximized is

$$\sum_{v_i \in A_m, v_j \in A_n, i > j} c_{i,j} \tag{1}$$

3   Neural implementation

In order to solve the K-MAXCUT problem, we have used the MREM neural model since this model has been successfully used for other COP [7, 8].

The MREM neural model: It consists in a series of multivalued neurons, where the state of i-th neuron is characterized by its output (s_i) that can take any value in any finite set M. This set can be a non numerical one, but, in this paper, the neuron outputs only take value in $M \subset \mathbb{Z}^+$.

The state vector $\vec{S} = (s_1, s_2, \ldots, s_N) \in M^N$ describes the network state at any time, where N is the number of neurons in the net. Associated with any state vector, there is an energy function $E : M^N \to \mathbb{R}$, defined by the expression:

$$E(\vec{S}) = \frac{1}{2} \sum_{i=1}^{N} \sum_{j=1}^{N} w_{i,j} f(s_i, s_j) \tag{2}$$

where $W = (w_{i,j})$ is a matrix and $f : M \times M \to \mathbb{R}$ is usually a similarity function since it measures the similarity between the outputs of neurons i and j. At each step, the state vector will be evolving to decrease the energy function.

To solve the K-MAXCUT problem with this neural net, we need as many neurons as number of nodes N in the graph. Each neuron taking value $s_i \in M = \{1, 2, \ldots, k\}$ points to the subset of the partition where the i-th node is assigned to.

The cost function of the K-MAXCUT, given by equation-1, must be identified with the energy function of equation-2. So, $w_{i,j} = -c_{i,j}$ and $f(x, y) = \delta_{x,y}$ (Krönecker delta function).

Initially the state vector is randomly selected. At any time, the net is looking for a better solution than the current one, in terms of the energy function.

Some dynamics can be used for MREM, but we have used one that we have named best-2. It consists in selecting the greatest decrease of the energy function by changing the state of only two neurons at the same time.

Denoting by $s_i = s_i(t)$ the state of the i-th neuron at time t and $s_i' = s_i(t+1)$ the next state of the neuron, an expression for the energy increment when two neurons p and q change their states ($s_p(t) = a$, $s_p(t+1) = a'$, $s_q(t) = b$, $s_q(t+1) = b'$), is given by:

$$\Delta(E) = E(\vec{S}(t+1)) - E(\vec{S}(t)) = \sum_{j=1}^{N} w_{p,j} \left[\delta_{a',s_j'} - \delta_{a,s_j} \right] + \sum_{i=1,i\neq p}^{N} w_{i,q} \left[\delta_{s_i',b'} - \delta_{s_i,b} \right] \quad (3)$$

So, the dynamics of the neural net can be summarized as follows:
- A state vector is randomly assigned.
- Repeat until there is not any change in the state vector
    - At each step, only two neuron outputs can be updated.
    - The scheduling sequentially selects one value $d \in \{1, 2, \ldots, N-1\}$.
    - The next process can be made in parallel. The p-th neuron builds a $k \times k$ matrix B^p whose element $b_{i,j}^p$ equals the decrease of energy resulting from making $s_p(t+1) = i$, $s_q(t+1) = j$, where $q = (p+d) \mod (N)$, $0 < q \leq N$, by applying equation-3.
    - Neuron p then calculates the minimum $\beta_p = b_{i,j}^p$ of the elements in matrix B^p.
    - After this parallel process, the scheduling calculates the minimum of the vector $\vec{\beta}$: $\beta_0 = b_{i_0,j_0}^{p_0}$ and the state vector is updated by changing outputs of neurons p_0 and $q_0 = p_0 + d$ to $s_{p_0}(t+1) = i_0$, $s_{q_0}(t+1) = j_0$.

4 Improving solutions

A new technique to escape from local minima of the energy function has been developed. It improves greatly the quality of the solutions with a little increment of time-consuming.

Given an estimated solution (e.g. the stable state of the previous neural algorithm), a good solution usually has the following property:

"High-weighted arcs must have their endpoints in different subsets".

So, we can study the high-weighted arcs. Let A be the set of arcs with weights greater than a threshold and endpoints in the same subset. Let $V^* \subset V$ be the set of endpoints of arcs in A. Then the current solution is saved and the shake phase begins. It consists in:
- Selecting the nodes that are endpoints of arcs in A and their neighbors: $H = \{v_i / \exists v_j \in V^*, e_{i,j} = 1\}$.
- Nodes in $V - H$ are clamped to their current values, while nodes in H are randomly assigned.
- With this new initial state vector, the network evolves with the usual dynamics, but only nodes in H will be selected in order to be modified, until a new stable state is reached.
- This new solution is compared with the previous saved one and the best one is selected.

5 Simulation Results

We have compared our proposed algorithms to OCHOM and Wang's. All of them have been implemented and tested in MATLAB, on a Pentium IV (3.06 Ghz). More specifically, Wang's network has been tested with its default parameter $\lambda = 30$. In the proposed model, the set A was built by including every edge $e_{i,j}$ whose cost $c_{i,j} > \bar{c} + 3\sigma$, where $\bar{c}$, σ are respectively the mean and the standard deviation of $c_{i,j}$. So, A is forced to include exclusively high-weighted edges.

The test set was built by using the same parameters that Wang et al in [12]. It is formed by random graphs, dependent on two parameters, N and $\rho \in (0,1)$ (density) meaning that the number of edges in E is the closest integer to $\rho \frac{N(N-1)}{2}$. Ten items for each $N \in \{20, 50, 80, 100\}$ and each $\rho \in \{0.05, 0.15, 0.25\}$ have been used. Weights were integers randomly chosen in $[-1, 5]$.

On each graph in the test set, 10 simulations of each algorithm were performed. Both best and average solutions obtained are shown in Table 1. So, we can verify that the proposed algorithm outperforms others, not only giving the best results, but even on average.

Table 1: Best and average performance on test set.

N	ρ	N_a	Wang Best	Wang Av.	Wang t	OCHOM Best	OCHOM Av.	OCHOM t	MREM Best	MREM Av.	MREM t	MREM-shake Best	MREM-shake Av.	MREM-shake t
20	0.05	10	26	21.8	0.003	26	24.4	0.001	26	25.4	0.024	26	25.4	0.023
20	0.15	27	67	29.0	0.002	69	66.5	0.001	69	68.0	0.027	69	68.0	0.027
20	0.25	43	80	63.6	0.002	86	78.5	0.001	86	84.4	0.024	86	84.4	0.025
50	0.05	58	144	113.4	0.016	142	137.4	0.005	149	143.5	0.243	149	143.5	0.265
50	0.15	161	278	248.8	0.015	273	264.6	0.005	284	276.7	0.234	284	277.0	0.369
50	0.25	256	460	397.9	0.012	476	448.8	0.006	469	460.6	0.244	472	463.9	0.482
80	0.05	143	270	238.0	0.031	266	258.6	0.011	279	271.5	0.713	279	271.5	1.025
80	0.15	403	715	702.5	0.034	739	712.4	0.014	754	735.3	0.943	754	742.2	1.954
80	0.25	680	1100	878.2	0.034	1106	1080.5	0.016	1117	1091.0	0.857	1117	1095.1	1.717
100	0.05	204	400	323.7	0.048	407	390.2	0.017	418	406.0	1.539	418	406.6	2.374
100	0.15	631	1071	843.6	0.068	1060	1029.1	0.023	1081	1062.3	1.629	1084	1068.5	3.257
100	0.25	1058	1697	834.4	0.043	1728	1682.3	0.025	1741	1702.7	1.323	1741	1714.8	2.407

References

[1] A. Alberti, A. Bertoni, P Campadelli, G. Grossi and R. Posenato. *A neural algorithm for MAX-2SAT: performance analysis and circuit implementation*, Neural Networks **10-3**, 555-560(1997).

[2] F. Barahona, M. Grotschel, M. Junger and G. Reinelt, *An Application of combinatorial optimization to statistical physics and circuit layout design.* Operat. Research **36** 493-513(1988).

[3] A. Bertoni, P. Campadelli and G. Grossi, *An approximation algorithm for the maximum cut problem and its experimental analysis.* Proceedings: Algorithms and experiments. Trento, **9-11**, 137-143(1998).

[4] M.R. Garey and D.S. Johnson, *Computers and Intractability. A guide to the theory of NP-Completeness.* W. H. Freeman and Company, New York, 1979.

[5] F. Hadlock, *Finding a maximum cut of a planar graph in polinomial time.* SIAM J. Computation **4-(3)**, 221-225(1975).

[6] C.P. Hsu, *Minimum-via topological routing.* IEEE Transaction Computer-Aided Design. Integrated Circuits Systems, **2** 235-246(1983).

[7] E. Mérida-Casermeiro, G. Galán-Marín and J. Muñoz-Pérez. *An Efficient Multivalued Hopfield Network for the Traveling Salesman Problem.* Neural Processing Letters **14** 203-216(2001).

[8] E. Mérida-Casermeiro, J. Muñoz-Pérez and R. Benítez-Rochel *Neural Implementation of Dijkstra's Algorithm.* Lecture Notes in Computer Science **2686** 342-349(2003).

[9] S. Sahni and T. Gonzalez, *NP-complete approximation problems.* J.ACM **23** 555-565(1976).

[10] K. A. Smith, *Neural Networks for Combinatorial Optimization: A review of more than a decade of research.* INFORMS Journal on Computing **11-(1)**, 15-34(1999).

[11] Y. Takefuyi and J. Wang, *Neural computing for optimization and combinatorics.* Singapore, World Scientific, Chapter **3**, (1996).

[12] Jiahai Wang and Zheng Tang. *An improved optimal competitive Hopfield network for bipartite subgraph problems.* Neurocomputing (In press).

VSP International
Science Publishers
P.O. Box 346, 3700 AH Zeist
The Netherlands

Lecture Series on Computer
and Computational Sciences
Volume 1, 2004, pp. 379-382

An Algorithmic Solution of Parameter-Varying Linear Matrix Inequalities in Toeplitz Form

G.B. Mertzios[1], P.K. Sotiropoulos

Department of Mathematics and Informatics,
Technical University of Munich, Germany

Cultural and Educational Technology Institute
Integrated Research for the Information Society (IRIS)
58 Tsimiski str., GR-67100, Xanthi, Greece

Received 29 August, 2004; accepted in revised form 7 September, 2004

Abstract: In this paper the necessary and sufficient conditions are given for the solution of a system of parameter varying linear inequalities of the form $\mathbf{A}(t)\mathbf{x} \geq \mathbf{b}(t)$ for all $t \in T$, where T is an arbitrary set, $\mathbf{x}$ is the unknown vector, $\mathbf{A}(t)$ is a known triangular Toeplitz matrix and $\mathbf{b}(t)$ is a known vector. For every $t \in T$ the corresponding inequality defines a polyhedron, in which the solution should exist. The solution of the linear system is the intersection of the corresponding polyhedrons for every $t \in T$. A decomposition method has been developed, which is based on the successive reduction of the initial system of inequalities by reducing iteratively the number of variables and by considering an equivalent system of inequalities.

Keywords: Linear matrix inequalities; Toeplitz matrices; parameter varying systems; decomposition of inequalities; robust control theory, parameter-varying synthesis.

Mathematics SubjectClassification: Linear inequalities, Linear systems, Robust parameter designs
AMS-MOS: 15A39, 93C05, 62K25

1. Introduction

A wide variety of problems arising in system and control theory can be reduced to constrained optimization problems, having as design constraints a simple reformulation in terms of linear matrix inequalities [1],[2]. Parameter varying Linear Matrix Inequalities (LMIs) have been proved to be a powerful tool, having important applications in a vast variety of systems and control theory problems including robustness analysis, robust control synthesis, stochastic control and identification [3],[4], synthesis of dynamic output feedback controllers [5], analysis and synthesis of control systems [6], error and sensitivity analysis, problems encountered in filtering, estimation, etc. LMI techniques offer the advantage of operational simplicity in contrast with the classical approaches, which necessitate the cumbersome material of Riccati equations [7]. Using LMIs, a small number of concepts and principles are sufficient to develop tools, which can then be used in practice. Also, the LMI techniques are effective numerical tools exploiting a branch of convex programming. In this paper we provide necessary and sufficient existence conditions for the solution of the system of inequalities $\mathbf{A}(t)\mathbf{x} \geq \mathbf{b}(t), \forall t \in T$ and restrictions of this solution, if such exists, in the general case, where T may be an infinite, or even a super countable set. Specifically, t is a variable within an arbitrary set T, which may represent the domain of external disturbances or parameter variations of a system in the most general form, $\mathbf{x} \in \mathbb{R}^N$ is the unknown vector, $\mathbf{A}(t) \in \mathbb{R}^{N \times N}$ is a given triangular Toeplitz matrix dependent on t and $\mathbf{b}(t) \in \mathbb{R}^N$ is a given vector of parameters dependent on t. Every row of the

---

[1] Corresponding authors. E-mail: mertzios@in.tum.de, psotirop@ceti.gr.

vector $\mathbf{A}(t)\mathbf{x}$, where $\mathbf{A}(t)$ is a triangular Toeplitz matrix, is a discrete-time convolution between the sequence of the functions in $\mathbf{A}(t)$ and the sequence in $\mathbf{x}$ and so the inequality $\mathbf{A}(t)\mathbf{x} \geq \mathbf{b}(t)$ represents a convolution that is greater than or equal to a given function, at every moment. So, this choice of this form of the system of inequalities finds many applications in control theory and signal processing.

The case, where $T = \{t_0\}$ is an one-element set, can be solved with various methods, like the ellipsoid-algorithm [8]. Then, the case of a finite set T is a generalization of the latter case, in the sense that one can consider $|T|$ times the special problem on an one-element set. On the other hand, the most general cases, where the set T is infinite and in particular where T is super countable (for example when $T = \mathbb{R}^k, k \in \mathbb{N}$), are of major importance and are considered here. Although the system of equations $\mathbf{A}(t)\mathbf{x} = \mathbf{b}(t), \forall t \in T$ has numerous methods of solutions, there is no available algorithm allowing computing the solutions of a system of inequalities $\mathbf{A}(t)\mathbf{x} \geq \mathbf{b}(t), \forall t \in T$ in the general case of infinite T [9],[10]. The underlying idea in the present paper for the solution of the LMIs $\mathbf{A}(t)\mathbf{x} \geq \mathbf{b}(t), \forall t \in T$ is the decomposition of the involved inequalities into simpler inequalities, considering the cases where each element $a_i(t)$ of $\mathbf{A}(t) \in \mathbb{R}^{N \times N}$ takes zero, positive or negative values. This is possible, since a given inequality is reduced to different simpler inequalities for different ranges of $t \in T$.

The main results of the present contribution are (a) the necessary and sufficient conditions of the existence of a solution $\mathbf{x}$ of the system and (b) the restrictions of the solution, which are expressed in the form of a hypercube, i.e. the upper and lower bound for each unknown variable x_r, $1 \leq r \leq N$, in the case where such a solution exists, which are derived in analytic form.

2. The Decomposition methods

Based on the above approach, in the rest of this work the following results are presented (see [11]) :

1. the Special Decomposition of an arbitrary inequality in $k = 1$ variable, into three equivalent inequalities, the first of them having only known quantities with no variables and the other two expressing explicitly the upper and lower bound for this one variable, in order to satisfy the initial inequality.
2. the necessary and sufficient conditions for the existence of a solution of an arbitrary inequality in $k \geq 2$ variables, in the form of two inequalities each one of them including $k - 1$ variables and
3. the General Decomposition of an arbitrary inequality in $k \geq 2$ variables into four equivalent inequalities, each one of them including $k - 1$ variables.

At first, we partition the set T into three sets S_i^1, S_i^2, S_i^3 for each $i \in \{1, 2, ..., k\}$, on which $a_i(t)$ is zero-, positive- and negative-valued:

$$S_i^1 = \{t \in T : a_i(t) = 0\}, \quad S_i^2 = \{t \in T : a_i(t) > 0\}, \quad S_i^3 = \{t \in T : a_i(t) < 0\}.$$

1. gives us the three necessary and sufficient conditions for the existence of the solution of

$$a_1(t)x_1 \geq b(t), \quad \forall t \in T, \tag{1}$$

which are:

$$0 \geq b(t), \quad \forall t \in S_1^1, \qquad \frac{b(t)}{a_1(t)} \leq x_1, \quad \forall t \in S_1^2, \qquad x_1 \leq \frac{b(t)}{a_1(t)}, \quad \forall t \in S_1^3 \tag{2}$$

2. gives the two necessary and sufficient conditions for the existence of the solution of

$$\sum_{i=1}^{k} a_{k-i+1}(t)x_i = a_k(t)x_1 + a_{k-1}(t)x_2 + \cdots + a_1(t)x_k \geq b(t), \forall t \in T, \quad k \geq 2, \tag{3}$$

which are:

$$\sum_{i=1}^{k-1} a_{k-i+1}(t)x_i \geq b(t), \forall t \in S_1^1, \tag{4}$$

$$\sum_{i=1}^{k-1} \left[\frac{a_{k-i+1}(t_2)}{|a_1(t_2)|} + \frac{a_{k-i+1}(t_3)}{|a_1(t_3)|} \right] x_i \geq \left[\frac{b(t_2)}{|a_1(t_2)|} + \frac{b(t_3)}{|a_1(t_3)|} \right], \forall (t_2, t_3) \in S_1^2 \times S_1^3. \tag{5}$$

Using the equivalence of 2., where only one variable is eliminated, we lose information about the conditions that this variable should satisfy. Indeed, in this elimination the variable x_k has been removed and thus the information about the range of the values that x_k may take in an eventual solution of (3) is lost.

The idea which is used in order to reinstate the information about x_k is the additional elimination of another variable, let x_{k-1}, so that a second couple of equations similar to (4) and (5) are derived, which have a solution if and only if (3) has a solution. Thus, considering the elimination of two variables x_k and x_{k-1}, we arrive at 3., which describes the Decomposition of the initial inequality (3) into a set of four equivalent inequalities, each one of them including $k-1$ variables, without losing information about the range of the variables in the solution.

3. Geometrical representation of the main results

For the illustration of the above, consider the following linear system:

$$\begin{bmatrix} a_1(t) & 0 \\ a_2(t) & a_1(t) \end{bmatrix} \begin{bmatrix} x_1 \\ x_2 \end{bmatrix} = \begin{bmatrix} a_1(t)x_1 \\ a_2(t)x_1 + a_1(t)x_2 \end{bmatrix} \geq \begin{bmatrix} b_1(t) \\ b_2(t) \end{bmatrix}, \forall t \in T = [0,9), \tag{6}$$

where:

$$a_1(t) = \begin{cases} 1, & t \in [0,3) \\ 0, & t \in [3,6) \\ -1, & t \in [6,9) \end{cases}, \; a_2(t) = \begin{cases} 1, & t \in [0,1) \cup [3,4) \cup [6,7) \\ 0, & t \in [1,2) \cup [4,5) \cup [7,8) \\ -1, & t \in [2,3) \cup [5,6) \cup [8,9) \end{cases}, \; b_1(t) = t^2 - 82, \; b_2(t) = t - 10.$$

The functions $a_1(t)$ and $a_2(t)$, for all $t \in T$, are graphically shown in Figure 1.

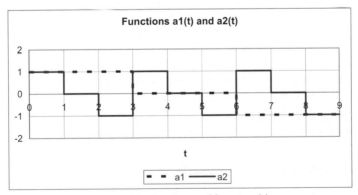

Figure 1: The functions $a_1(t)$ and $a_2(t)$.

The exact set of solutions of (6) and the bounds of these solutions, as given above, are graphically shown in Figure 2. The square produced from these bounds is the smallest possible, since its erosion leads to loss of solutions.

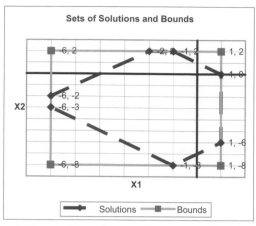

Figure 2: The set of the solutions of the system and their bounds.

References

[1] P. Apkarian, H.D. Tuan and J.Bernussu, Continuous-time analysis, eigenstructure assignment and H_2 synthesis with enhanced linear matrix inequalities (LMI) characterizations. *IEEE Transactions on Automatic Control*, 46:12, (December 2001).

[2] P. Gahinet, Explicit controller formulas for LMI-based H_∞ synthesis. *Proceedings of the American Control Conference*, Baltimore, Maryland, (June 1994).

[3] St. Boyd, L.El Glaoui, E. Feron and V. Balakrishnan, *Linear Matrix Inequalities in System and Control Theory*. SIAM Books, Philadelphia, 1994.

[4] R. Bellman and K. Fan, On systems of linear inequalities in Hermitian matrix variables. *Proceedings of Symposia in Pure Mathematics*, American Mathematical Society, V.L. Klee, ed., 7: 1-11, (1963).

[5] P. Gahinet, A. Nemirovskii, A.J. Laub and M. Chilali, The LMI control toolbox. *Proceedings of the 33rd Conference of Decision and Control*, Lake Buena Vista, Florida, USA, (December 1994).

[6] M. Chilali, P. Gahinet and P. Apkarian, Robust pole placement in LMI regions. *Proceedings of the 36th Conference of Design and Control*, San Diego, California USA, wp06: 1291-1296, (December 1997).

[7] P. Apkarian, H.D. Tuan, and J. Bernussou. Continuous-time analysis, eigenstructure assignment and synthesis with enhanced linear matrix inequalities (LMI) characterizations. *IEEE Transactions on Automatic Control*, 46:1941-1946, (December 2001).

[8] C. Papadimitriou and K. Steiglitz, *Combinatorial Optimization: Algorithms and Complexity*. Dover Publications Inc., Mineola, New York, 1998, p. 170.

[9] G. Strang, A proposal for Toeplitz matrix calculations. *Studies in Applied Mathematics*. 74: 171-176, (1986).

[10] R.M. Gray. Toeplitz and circulant matrices: A review. Technical Report, Information Systems Laboratory. Department of Electrical Engineering. Stanford University, http://ee.stanford.edu/~gray/toeplitz.pdf, August 2002.

[11] G.B. Mertzios, "Solution of parameter-varying linear matrix inequalities in Toeplitz form," *Applied Mathematics and Computation*. Submitted for publication.

VSP International
Science Publishers
P.O. Box 346, 3700 AH Zeist
The Netherlands

*Lecture Series on Computer
and Computational Sciences*
Volume 1, 2004, pp. 383-386

Simulation-based Choice of Strategies
in the Development of Regions[1]

Gergely Mészáros-Komáromy[2]

Department of Information and Knowledge Management,
Faculty of Economy,
Budapest University of Technology and Economics,
H-1111 Budapest, Sztoczek u. 4., Hungary

Received 6 August, 2004; accepted in revised form 21 August, 2004

Abstract: The determination of the optimal strategy for the development of regions both in Europe and in the individual countries as e.g. Hungary is of great importance and has very significant economic and social consequences. This paper outlines a reconstructed economic model intended to promote decision makers in the choice of strategies in the development of regions, based on selected indicators describing social, environmental, infrastructural and economic factors.

Keywords: Agent controlled simulation, soft systems, development of regions

Mathematics Subject Classification: 37F05, 00A72, 65C20, 68U20

PACS: 02.60.Gf

1. Introduction

In the fields of micro and macro economy – like in a number of other fields – highly complex systems are often described by soft models that are only approximated by empirical theories. The situation is even more complicated when other aspects of social sciences, education, environmental, infrastructural problems etc. are in interaction within the model. The optimal solution of such problems may however be very important for the decision makers. The specific question we have dealt with is the development of regions taking into consideration the wide range of various factors interacting and influencing the model behavior. Regions are considered as basic entities, since they are comparable to each other in the whole European Union.

As a tool for our investigations we have applied the CASSANDRA (Cognizant Adaptive Simulation System for Applications in Numerous Different Relevant Areas) [1] simulation system that was developed in the McLeod Institute of Simulation Sciences Hungarian Center and has already been applied with success in a number of European Union and other projects. This simulation system applies intelligent agents (called demons [2]) monitoring the behavior of the initial model during dynamic simulation and modifying it until its behavior matches the historical data obtained from statistics describing the past behavior of the region [3, 4]. The intelligent demons evaluate the model behavior and by comparing it with the historical data change the model structure, parameters and functionalities describing the interaction among the objects of the model. After a final model is determined it can be used to forecast the effects achieved by influencing the system investigated.

2. The problem investigated

---
[1] This work was partly sponsored by the National Scientific Research Fund; Contract number: T38081
[2] Corresponding author. E-mail: meszaros@itm.bme.hu

In our model, we simulated a region of Hungary (consisting of 3 counties: Fejér, Komárom-Esztergom, Veszprém). We collected 29 indicators, covering the field of economy, infrastructure and education, in the last 7 years (from 1996 to 2002), based on the indicators developed for the evaluation of different strategies [5], among them:

Table 1. Sample of historical indicators

Historical indicators
Nr of undergraduate students
Nr of flats built in the preceding year
Nr of phone lines
Investments in education, economy, public service
Nr of employees in education, economy, public service
Unemployment rate
Tax collected per capita
Inland immigration per 1000 capita
Natural reproduction rate
Population rate
Research and development investments

2.1 Model structure

Using the above variables, we defined the list of possible interrelations between these variables. For example, we assume that the number of phone lines depends on the number of flats, the number of flats built in the preceding year, investments in economy and the population rate. For each variable, we defined 3 – 6 possible interrelations, which make about 120 possible interrelations, a very complex graph.

From the 29 variables above, we selected 5 input variables, which means that we wanted to use these 5 variables to change the situation externally.

A simplified version of this model can be seen in Figure 1 (the input variables are on the left side of the picture, each box represents one variable, identified by the number on it):

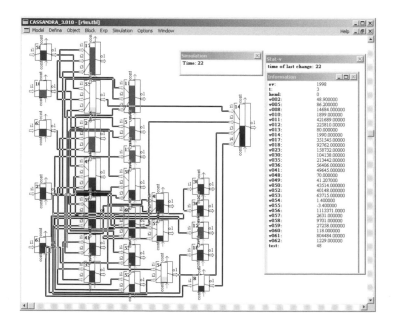

Figure 1. Representation of the simplified model

2.2 Model reconstruction

We built the initial model using the anticipated relationships. The interrelations were defined using simple functions, with some not-exactly-defined parameters, for example:

$$V_{001} = w001\_002 \cdot V_{002} - w001\_003 \cdot V_{003}^2,$$ (1)

where V_{001}, V_{002}, V_{003} are model variables, w001_002 and w001_003 are not-exactly-defined parameters. We can utilize simple multiplying, square, square root, logarithmic, etc., not only linear equations like when we use linear regression. An important advantage of the approach applied is that various delays among the interacting variables can be introduced by the model component structure used (as for example, the effect of capital investments in road construction may have delayed results).

The not-exactly-defined parameters have been defined using the interval of the possible values (these intervals have been set by experts, because other methods, like regression could not be used because of the shortness of the time series).

In the model reconstruction, we use a complex intelligent agent to discover the proper values of these parameters in the given interval of the possible values. The proper value was calculated using the hill-climbing method, where the error function was the sum of the square of the errors for these 7 years for one variable (the agent tries to minimize the value of this function):

$$h = \sum_{t=1996}^{2002} \{v_i(t) - m_i(t)\}^2 ,$$ (2)

where $v_i(t)$ is the historical value of the ith variable in year t, and $m_i(t)$ is the calculated value using our model.

3 Results

After the interrelation between the variables was fully specified, the model was ready to simulate the behavior of the region under various circumstances. For example, the results of the choice between investing into research and development versus investing into national health care. These decisions may have multiple effects on the economic situation of the region and the growth of population, etc.

Results were exported to an Excel sheet, where charts and tables represented the results, and the possibility of additional calculations have been enabled. In Figure 4. – as an example - the results obtained concerning the trajectories of two variables is depicted. The values until 2002 are the historical data used in the process of model identification and the values from 2003 are predicted by simulation.

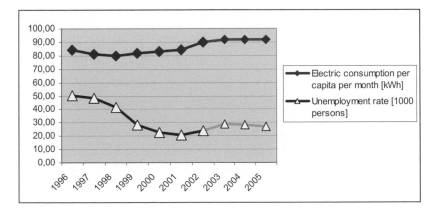

Figure 2. Trajectories of two variables obtained by simulation

The simulation model was tested against historical data in one of the counties, and good results were achieved. Our current model is being examined and tested by the West Hungarian Research Institute, Center of Regional Studies, to enable a wider use of this model, helping the decision-makers to make expectably optimal decisions in the allocation of money resources for regions.

References

[1]   Jávor, A. (1997). Problem Solving by the CASSANDRA Simulation System Controlled by Combined Mobile and Static AI. *Summer Computer Simulation Conference*, Arlington, Virginia, July 13-17, 723-728.

[2]   Jávor, A. (1992). Demon Controlled Simulation. *Mathematics and Computers in Simulation*, 34, 283-296.

[3]   Jávor, A., Mészáros-Komáromy, G. (2002). Investigation of the Influence of the Road Traffic Network Conditions on the Development of Regions by Means of AI Controlled Simulation. *AI, Simulation and Planning in High Autonomy Systems,* Lisbon, Portugal, April 7-10, 86-90.

[4]   Szűcs, G. (2001). Model Identification by a Demon using Stepwise Regression Procedure. *Simulation News Europe*, 31, 18-19.

[5]   Rechnitzer, J. (1999). *Regionális fejlesztési stratégiák (Strategies of Regional Development*, published in Hungarian). Dialog Campus, Budapest Pécs.

VSP International
Science Publishers
P.O. Box 346, 3700 AH Zeist
The Netherlands

*Lecture Series on Computer
and Computational Sciences*
Volume 1, 2004, pp. 387-390

Increasing the Accuracy of Anticipation with Lead-Lag Timing Analysis of Digital Freehand Writings for the Perceptual Environment

Khaireel A. Mohamed

Institut für Informatik, Albert-Ludwigs-Universität Freiburg, Germany 79110
khaireel@informatik.uni-freiburg.de

Received 2 August, 2004; accepted in revised form 18 August, 2004

Abstract: We portray the methods to increase the accuracy of "anticipation" of our current perceptual writing environment based on the perceptual user interface (PUI) model, by means of statistical temporal analysis. We will show that the lead and lag times in freehand writings contribute significantly to the process of singling out command Gestures from normal handwritings, while simultaneously freeing up more processing time for the perceptual writing environment to perform other tasks within its active schedule. We also show our benchmarking results of the analysis compared to our previous works.

Keywords: Digital ink, freehand writing, bivariate lead-lag temporal analysis, perceptual environment.

PACS: 65C60

1. Introduction

A perceptual environment described here for digital freehand writings is one that is able to differentiate between normal handwritings apart from written command-gestures on a same (non-segregated) writing platform, while minimizing the need for command buttons or pull-down menus. Through a transducer device and that of the movements effected from a digital pen, we have a pen-based interface that captures digital ink. Based solely on these input ink traces, we want to be able to decipher users' intentions to correctly classify which of the two domains the ink trace is likely to be in – either as primitive symbolic traces, or some sort of system command.

Our work concentrates on improving modern teaching environments and rallies around Turk and Robertson's [6] paradigm of the Perceptual User Interface (PUI). We attempt to do away with conventional GUIs designed for the desktop, replacing them instead with a clean and uncluttered slate for a rudimentary writing environment. In other words, we want to reproduce the simple, customary blackboard, and still be able to include all other functionalities that an electronic board can offer. But by minimizing the appearance of static menus and buttons on screen, the resultant 'clean' slate becomes the only perceptual input available for users to relate to the background systems.

2. Ink Components in Freehand Writings

We define an *ink Trace* as the rendered graphics of a particular handwritten trace on screen, and an *ink component* as the actual data structure describing a Trace in memory. Traces on the digital board group together to form, what we perceive as freehand sentences. Each Trace, resulting from a pair of successive pen-down and pen-up events, can be categorised in terms of the timings noted for its *duration*, *lead*- and *lag*-time as illustrated in Figure 1.

For a set of contiguous ink components $S_j = \{c_0, c_1, c_2, \ldots, c_n\}$ in a freehand sentence made up of n traces, we note that the lag-

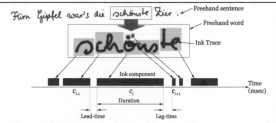

Figure 1: Splitting up freehand writing into ink components on the time line.

time for the ith component is exactly the same as the lead-time of the $(i+1)^{th}$ component; i.e. $\text{lag}(c_i) = \text{lead}(c_{i+1})$. Consequently, the timings that separate one set of ink components apart from another are the first lead-time $\text{lead}(c_0)$ and the last lag-time $\text{lag}(c_n)$ in S_j. These times are significantly longer than their in-between neighbours c_1 to c_{n-1}. Furthermore, if we observe a *complete* freehand sentence made up of a group of freehand words, we can categorise each ink component within those words into one of the following four groups:

- *Beginnings* – Ink components found at the *start* of a freehand word;
- *Endings* – Ink components found at the *end* of a freehand word;
- *In-betweens* – Ink components found in the *middle* of a freehand word; and

- *Stand-alones* – Disjointed ink components.

The groups differ in the demarcations of their lead and lag times, and as such, provide for a way a perceptual system can identify them. Other forms of freehand writings include mathematical equations, characters of various languages, signatures, and graphical illustrations.

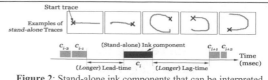

Figure 2: Stand-alone ink components that can be interpreted as a possible gesture.

The temporal relation that singles out stand-alone components (and freehand gestures) is the "longer-than-average" lead and lag times of a single-stroke ink Trace, as shown in Figure 2. There is a pause period between samplings of ink components that results in significantly longer lead *and* lag times. It is not very often that we see people gesturing to a control system in the middle of writing a sentence or drawing a diagram. So we can anticipate, rather convincingly based on the lead and lag times obtained, that the latest ink component might be more of an instance of a Trace, rather than a Gesture. In pursuit of standardising a suitable time-out period, we decided to set the upper limit of both the lead and lag times to 1500 msec, in retrospect after some initial testing. That is to say, if an observed (lead or lag) time exceeds this period, we immediately consider it as "significantly long" and note it down as 1500 msec.

3. Sampling and Categorising Ink Components

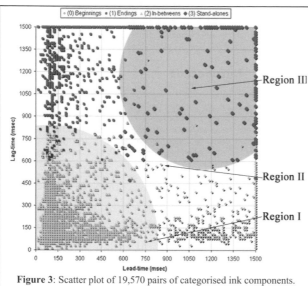

Figure 3: Scatter plot of 19,570 pairs of categorised ink components.

Based on a sample set of 57 InkML files, every single ink component in each file is retrieved and observed for their lead and lag times to be classified into one of the four predefined categories above. Here, InkML is utilised as a standard prescribed by W3C [5] for storing, retrieving and porting ink data between applications – this includes all types of freehand writings.

The scatter-plot diagram in Figure 3 shows the classified ink components in coordinate relation to their individual lead and lag times. We observe three main regions of clustering – the 'near-neighbours' in Region I (made up mainly of the *in-betweens*), the 'extremes' in Region II (made up mainly of the *beginnings* and *endings*), and the 'solitaries' in Region III (made up mainly of the *stand-alones*). The tightness of the clustering decreases as we traverse the plot from Region I towards Region III; where Region I is very tightly clustered, the obvious sparseness in Region III suggests a huge variety of writing styles sampled.

The outline of each regional curve is the estimated resulting decision surface separating the samples between the stand-alones (Region III) and the non-stand-alones (Regions I and II). Their attached correlation coefficients *r* to Regions I, II and III are 0.7699, 0.4819, and 0.1230 respectively. Here, *r* is the measure of strength of the linear relation between the lead and lag time variables where the magnitude of *r* indicates the strength of a linear relation (i.e. a value *r* close to zero means that the linear association is weak), and is calculated with Equation 1, where X refers to the lead-time and Y refers to the lag-time variables.

$$r = \frac{S_{XY}}{\sqrt{S_{XX}}\sqrt{S_{YY}}} \qquad \text{where} \qquad S_{XY} = \sum(x-\bar{x})(y-\bar{y}), \quad S_{XX} = \sum(x-\bar{x})^2, \quad S_{YY} = \sum(y-\bar{y})^2 . \qquad (1)$$

4. Studying the Bivariate Data

When two traits are observed for the individual sampling units and each trait is recorded in some qualitative categories, then the joint behaviour of two random variables, X and Y from each of the traits, is determined by the cumulative distributive function $F(x, y) = P(X \leq x, Y \leq y)$ derived from the fundamental theorem of multivariate calculus [3].

We find it advantageous to know the joint relation between the lead and lag times of each ink component for making generalisations in the future. A lookup probability table constructed from the current samples of categorised ink components, following a bivariate normal density, is expected to expedite our mission of finding the quick probability that an ink component is indeed a symbolic Trace. Equations 2 and 3 describe this density function, and Figure 4 illustrates the probability lookup table using the values computed from the scatter-plot diagram in Figure 3: $\mu_x = 462.0059$, $\mu_y = 462.2614$, $\sigma_x = 682.0767$, and $\sigma_y = 682.6080$, where x and y refer to the lead- and lag-time variables respectively.

$$f(x,y) = \frac{1}{2\pi\sigma_x\sigma_y\sqrt{1-\rho^2}} \exp\left[-\frac{z}{2(1-\rho^2)}\right] \tag{2}$$

$$\text{where } z = \frac{(x-\mu_x)^2}{\sigma_x^2} + \frac{(y-\mu_y)^2}{\sigma_y^2} - \frac{2\rho(x-\mu_x)(y-\mu_y)}{\sigma_x\sigma_y}, \quad \text{and } \rho \equiv cor(x,y) = \frac{\sigma_{xy}}{\sigma_x\sigma_y} \tag{3}$$

For every section on the surface in Figure 4 that gives a high probability that an ink component should be a symbolic Trace, there are other sections on the surface that depicts otherwise. These sections confirm our scatter-plot diagram in Figure 3. The next part of this paper details the use of the graph in Figure 4 through the formal establishment of our statistical hypotheses in making proper inferences of the population samples.

5. Statistical Acceptance

The goal of testing statistical hypotheses is to conjecture about some features of the population, which is strongly supported by the information obtained from the sampled data. Based on key statistical concepts in the context of determining the best solution for our problem definition, we establish the alternative hypothesis H_1 and its nullifying opposite H_0 as stated below, where X and Y refer to the lead- and lag-time variables respectively.

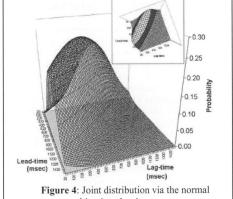

Figure 4: Joint distribution via the normal bivariate density.

H_0: "That an ink component is NOT a symbolic Trace";   i.e. $\overline{X} \geq \mu_x$, $\overline{Y} \geq \mu_y$

H_1: "That an ink component IS a symbolic Trace";   i.e. $\overline{X} < \mu_x$, $\overline{Y} < \mu_y$

Our test statistic refers directly to the joint-distribution described in Equation 2 (and its Figure 4 equivalent), so that Equation 3, the Z-test, forms the rejection criteria needed for a test level of significance of $\alpha/2 = 0.025$. The inset in Figure 5 demonstrates this notion from two cross-sectional views of the normal bivariate density curve. We accept H_1, if and only if $R:|Z| \geq Z_{\alpha/2}$, where R refers to

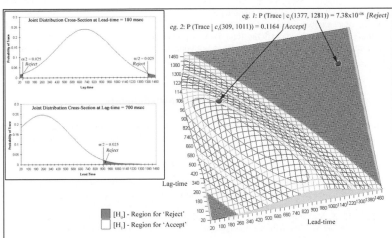

eg. 1: P (Trace | c_i(1377, 1281)) = 7.38x10⁻⁶⁸ *[Reject]*

eg. 2: P (Trace | c_i(309, 1011)) = 0.1164 *[Accept]*

[H₀] - Region for 'Reject'

[H₁] - Region for 'Accept'

Figure 5: (***Inset***: *Cross-sectional view of the joint distribution of the normal bivariate density at lead-time = 180 msec and lag-time = 700 msec respectively*) Incorporating the normal bivariate density to assist with the null and alternative hypotheses.

the rejection regions, and $Z_{\alpha/2} = Z_{0.025}$ is the boundary of acceptance on the joint distribution of the normal bivariate density graph shown in the main Figure 5. Otherwise, we reject H_1 and claim H_0 that an ink component $c_g(X = t_{lead}, Y = t_{lag})$ shall *not* be considered as a symbolic Trace.

Plugging in the values of μ_x, μ_y, σ_x, and σ_y for every time-steps of X and Y from 0 to 1500, we obtain the entire region of acceptance (and rejection) on our bivariate surface. The darkened regions in Figure 5 are the areas for 'rejection' while the lighter region is the area for 'acceptance'. Figure 5 corresponds as our lookup table, with the input parameter c_g (t_{lead}, t_{lag}), and returning an output probability of P (Trace | c_g (t_{lead}, t_{lag})), with an attached H_1 (strong acceptance) or H_0 (strong rejection).

6. Performance in Anticipation and Conclusion

Our investigation of the lead and lag times resulting from freehand writings on digital boards is part of the on-going process of managing, analysing, and reacting to all primitive ink data perceived from the perceptual writing environment. We currently have in place, a background process-model [1, 2], designed to actively assist foreground applications tailored for the PUI paradigm. Part of this model is shown in Figure 6.

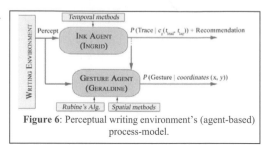

Figure 6: Perceptual writing environment's (agent-based) process-model.

Previously, there were no active mechanisms to temporally track all ink components, and both agents in Figure 6 act upon 'percepts' only on pen-up events issued by the writing environment. Both agents run their computationally expensive diagnoses by building KD-trees and running Rubine's algorithm [1, 4] to determine the user's input strokes should there be the case that a command Gesture was intended.

Now with the addition of temporal methods, the categorisation process is expedited. The Ink Agent only calls on the Gesture Agent when it receives H_1 from our lookup table. Inevitably, we made the overall anticipation process more focused, which results in 6.25% better performance as shown in Figure 7. We believe that the temporal methods highlighted so far are statistically strong for influencing future decisions down the communication chains within the PUI model. Such that, by simply incorporating serious temporal analysis of all individual ink components, we can reduce the number of active parallel processes

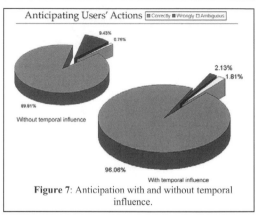

Figure 7: Anticipation with and without temporal influence.

running in the background. Our final lookup probability table gives very reliable outcomes of accepting or rejecting the chances of ink components as symbolic Traces, in our perceptual environment.

Acknowledgments

This research is funded by the Deutschen Forschungsgemeinschaft (DFG) as part of the research project „Algorithmen und Datenstrukturen für ausgewählte diskrete Probleme (DFG-Projekt Ot64/8-3)".

References

[1]   K. A. Mohamed. Fast interpretation of pen gestures with competent agents. In P. Vadakkepat, W.W. Tan, K.C. Tan, and A.P. Loh (editors), _Proceedings of the IEEE Second International Conference on Computational Intelligence, Robotics and Autonomous Systems_, PS09-4, December 2003.

[2]   K. A. Mohamed. Ingrid, Geraldine and Amidala - Competent agents for pen gestures interactivity. _Technical Report No. 209_, Albert-Ludwigs-Universität Freiburg, Institut für Informatik, December 2003.

[3]   J. A. Rice. _Mathematical Statistics and Data Analysis_. 2nd edition. Duxbury Press, 1995.

[4]   D. Rubine. Specifying gestures by example. In _Proceedings of the 18th Annual Conference on Computer Graphics and Interactive Techniques_, pages 329–337. ACM Press, 1991.

[5]   G. Russell, Y.-M. Chee, G. Seni, L. Yaeger, C. Tremblay, K. Franke, S. Madhvanath, and M. Froumentin. Ink markup language [online], Available: http://www.w3.org/tr/inkml/. _W3C Working Draft_, February 2004.

[6]   M. Turk and G. Robertson. Perceptual user interfaces (introduction). In _Communications of the ACM_, 43(3):32 – 34. ACM Press 2000.

VSP International
Science Publishers
P.O. Box 346, 3700 AH Zeist
The Netherlands

Lecture Series on Computer
and Computational Sciences
Volume 1, 2004, pp. 391-395

Fourth order Trigonometrically-fitted and Exponentially-fitted Symplectic Methods for the Numerical Integration of the Schrödinger Equation [1]

Th. Monovasilis

Department of Computer Science and Technology,
Faculty of Sciences and Technology,
University of Peloponnese,
GR-221 00 Tripolis, Greece
Department of International Trade,
Technological Educational Institute of Western Macedonia at Kastoria,
P.O. Box 30, GR-521 00, Kastoria, Greece

Z. Kalogiratou

Department of International Trade,
Technological Educational Institute of Western Macedonia at Kastoria,
P.O. Box 30, GR-521 00, Kastoria, Greece

T.E. Simos[2]

Department of Computer Science and Technology,
Faculty of Sciences and Technology,
University of Peloponnese,
GR-221 00 Tripolis, Greece

Received 5 September, 2004; accepted in revised form 20 September, 2004

Abstract: The solution of the one-dimensional time-independent Schrödinger equation is considered by trigonometrically-fitted and exponentially-fitted symplectic integrators. A modified Yoshida-type fourth order method is proposed. The Schrödinger equation is first transformed into a Hamiltonian canonical equation. Numerical results are obtained for the one-dimensional harmonic oscillator and Morse potential.

Keywords: trigonometrically-fitted, exponentially-fitted, symplectic methods, Schrödinger equation

Mathematics Subject Classification: Here must be added the AMS-MOS or PACS Numbers

PACS: Here must be added the AMS-MOS or PACS Numbers

---

[1] Funding by research project 71239 of Prefecture of Western Macedonia and the E.U. is gratefully acknowledged.
[2] Corresponding author. Active Member of the European Academy of Sciences and Arts. E-mail: simos-editor@uop.gr, tsimos@mail.ariadne-t.gr

1   Introduction

The time-independent Schrödinger equation is one of the basic equations of quantum mechanics. Its solutions are required in the studies of atomic and molecular structure and spectra, molecular dynamics and quantum chemistry. In the literature many numerical methods have been developed to solve the time-independent Schrödinger equation. Symplectic integrators are suitable methods for the numerical solution of the Schrödinger equation, among their properties is the energy preservation, which is an important property in quantum mechanics. Also, exponential-fitting methods have been very widely used for the numerical integration of the Schrödinger equation. In this work we develope modified fourth order symplectic integrators with the exponentially-fitted property based on Yoshida-type symplectic integrators. Our new methods are tested on the computation of the eigenvalues of the one-dimensional harmonic oscillator and Morse potential.

2   The time-independent Schrödinger equation

The one-dimensional time-independent Schrödinger equation may be written in the form

$$-\frac{1}{2}\frac{d^2\psi}{dx^2} + V(x)\psi = E\psi \tag{1}$$

where E is the energy eigenvalue, $V(x)$ the potential, and $\psi(x)$ the wave function. Equation (1) can be rewritten in the form

$$\frac{d^2\psi}{dx^2} = -B(x)\psi$$

where $B(x) = 2(E - V(x))$, or

$$\begin{aligned} \phi' &= -B(x)\psi \\ \psi' &= \phi \end{aligned} \tag{2}$$

3   Numerical methods

3.1   Symplectic numerical schemes

Given an interval $[a, b]$ and a partition with N points

$$x_0 = a, \ x_n = x_0 + nh, \quad n = 1, 2, \ldots, N.$$

a Yoshida-type [4] s-stages method is of the form

$$\begin{aligned} \phi_{n+1} &= a_s p_s - c_{s+1} h B q_s, \\ \psi_{n+1} &= b_s q_s + d_{s+1} h \phi_{n+1} \end{aligned}$$

where

$$\begin{aligned} p_i &= a_i \phi_n - c_i h B \psi_n, \\ q_i &= b_i \psi_n + d_i h p_i, \qquad \text{for} \quad i = 1, 2, \ldots, s \end{aligned}$$

We can write the above method using matrices, each step corresponds to:

$$\begin{pmatrix} 1 & 0 \\ -d_i h & 1 \end{pmatrix} \begin{pmatrix} p_i \\ q_i \end{pmatrix} = \begin{pmatrix} 1 & -c_i h B \\ 0 & 1 \end{pmatrix} \begin{pmatrix} \phi_n \\ \psi_n \end{pmatrix}$$

and the method proceeds as follows

$$\begin{pmatrix} \phi_{n+1} \\ \psi_{n+1} \end{pmatrix} = T \begin{pmatrix} \phi_n \\ \psi_n \end{pmatrix}, \quad \text{where} \quad T = \begin{pmatrix} \alpha(B) & \beta(B) \\ \gamma(B) & \delta(B) \end{pmatrix}$$

The above method is symplectic, since the transformation matrix T satisfies

$$T^T J T = J \quad \text{where} \quad J = \begin{pmatrix} 0 & 1 \\ -1 & 0 \end{pmatrix}$$

The fourth order symplectic Yoshida-type method is a four stage method ($s = 4$) with coefficients [2,4]:

$$\begin{array}{llll}
c_1 &=& 0, & d_1 &=& x + \frac{1}{2}, \\
c_2 &=& 2x + 1, & d_2 &=& -x, \\
c_3 &=& -4x - 1, & d_3 &=& -x, \\
c_4 &=& 2x + 1, & d_4 &=& x + \frac{1}{2}
\end{array}$$

$$\text{where} \quad x = \frac{2^{1/3} + 2^{-1/3} - 1}{6}$$

Now we preserve the conditions up to third order and we want the method to integrate exactly the function $\psi(x) = e^{\pm w\, x}$

$$\begin{aligned}
c_1 + c_2 + c_3 + c_4 &= 1, \\
d_1 + d_2 + d_3 + d_4 &= 1, \\
c_2 d_1 + c_3(d_1 + d_2) + c_4(d_1 + d_2 + d_3) &= \frac{1}{2}, \\
c_2 d_1^2 + c_3(d_1 + d_2)^2 + c_4(d_1 + d_2 + d_3)^2 &= \frac{1}{3}, \\
d_1 c_1^2 + d_2(c_1 + c_2)^2 + d_3(c_1 + c_2 + c_3)^2 + d_4 &= \frac{1}{3},
\end{aligned}$$

$$
\begin{aligned}
e^v - b_1 b_2 b_3 b_4 - (b_2 b_3 b_4 c_1 + a_1 b_3 b_4 c_2 + a_1 a_2 b_4 c_3 + a_1 a_2 a_3 c_4) & \quad v \\
-(b_1 b_3 b_4 c_2 d_1 + a_2 b_1 b_4 c_3 d_1 + a_2 a_3 b_1 c_4 d_1 + b_1 b_2 b_4 c_3 d_2 + a_3 b_1 b_2 c_4 d_2 + b_1 b_2 b_3 c_4 d_3) & \quad v^2 \\
-(b_3 b_4 c_1 c_2 d_1 + a_2 b_4 c_1 c_3 d_1 + a_2 a_3 c_1 c_4 d_1 + b_2 b_4 c_1 c_3 d_2 + & \\
a_1 b_4 c_2 c_3 d_2 + a_3 b_2 c_1 c_4 d_2 + a_1 a_3 c_2 c_4 d_2 + b_2 b_3 c_1 c_4 d_3 + a_1 b_3 c_2 c_4 d_3 + a_1 a_2 c_3 c_4 d_3) & \quad v^3 \\
-(b_1 b_4 c_2 c_3 d_1 d_2 + a_3 b_1 c_2 c_4 d_1 d_2 + b_1 b_3 c_2 c_4 d_1 d_3 + a_2 b_1 c_3 c_4 d_1 d_3 + b_1 b_2 c_3 c_4 d_2 d_3) & \quad v^4 \\
-(b_4 c_1 c_2 c_3 d_1 d_2 + a_3 c_1 c_2 c_4 d_1 d_2 + b_3 c_1 c_2 c_4 d_1 d_3 + a_2 c_1 c_3 c_4 d_1 d_3 + b_2 c_1 c_3 c_4 d_2 d_3 + a_1 c_2 c_3 c_4 d_2 d_3) & \quad v^5 \\
-b_1 c_2 c_3 c_4 d_1 d_2 d_3 v^6 - c_1 c_2 c_3 c_4 d_1 d_2 d_3 & \quad v^7 \quad = 0,
\end{aligned}
$$

$$
\begin{aligned}
e^v - a_1 a_2 a_3 a_4 - (a_2 a_3 a_4 b_1 d_1 + a_3 a_4 b_1 b_2 d_2 + a_4 b_1 b_2 b_3 d_3 + d_4 e^v) & \quad v \\
(a_2 a_3 a_4 c_1 d_1 + a_3 a_4 b_2 c_1 d_2 + a_1 a_3 a_4 c_2 d_2 + a_4 b_2 b_3 c_1 d_3 + a_1 a_4 b_3 c_2 d_3 + a_1 a_2 a_4 c_3 d_3) & \quad v^2 \\
-(a_3 a_4 b_1 c_2 d_1 d_2 + a_4 b_1 b_3 c_2 d_1 d_3 + a_2 a_4 b_1 c_3 d_1 d_3 + a_4 b_1 b_2 c_3 d_2 d_3) & \quad v^3 \\
-(a_3 a_4 c_1 c_2 d_1 d_2 + a_4 b_3 c_1 c_2 d_1 d_3 + a_2 a_4 c_1 c_3 d_1 d_3 + a_4 b_2 c_1 c_3 d_2 d_3 + a_1 a_4 c_2 c_3 d_2 d_3) & \quad v^4 \\
-a_4 b_1 c_2 c_3 d_1 d_2 d_3 v^5 - a_4 c_1 c_2 c_3 d_1 d_2 d_3 & \quad v^6 \quad = 0,
\end{aligned}
$$

$$
\begin{aligned}
e^{-v} - b_1 b_2 b_3 b_4 + (b_2 b_3 b_4 c_1 + a_1 b_3 b_4 c_2 + a_1 a_2 b_4 c_3 + a_1 a_2 a_3 c_4) & \quad v \\
-(b_1 b_3 b_4 c_2 d_1 + a_2 b_1 b_4 c_3 d_1 + a_2 a_3 b_1 c_4 d_1 + b_1 b_2 b_4 c_3 d_2 + a_3 b_1 b_2 c_4 d - 2 + b_1 b_2 b_3 c_4 d_3) & \quad v^2 \\
+(b_3 b_4 c_1 c_2 d_1 + a_2 b_4 c_1 c_3 d_1 + a_2 a_3 c_1 c_4 d_1 + b_2 b_4 c_1 c_3 d_2 + a_1 b_4 c_2 c_3 d_2 + a_3 b_2 c_1 c_4 d_2 & \\
+ a_1 a_3 c_2 c_4 d_2 + b_2 b_3 c_1 c_4 d_3 + a_1 b_3 c_2 c_4 d_3 + a_1 a_2 c_3 c_4 d_3) & \quad v^3 \\
-(b_1 b_4 c_2 c_3 d_1 d_2 + a_3 b_1 c_2 c_4 d_1 d_2 + b_1 b_3 c_2 c_4 d_1 d_3 + a_2 b_1 c_3 c_4 d_1 d_3 + b_1 b_2 c_3 c_4 d_2 d_3) & \quad v^4 \\
+(b_4 c_1 c_2 c_3 d_1 d_2 + a_3 c_1 c_2 c_4 d_1 d_2 + b_3 c_1 c_2 c_4 d_1 d_3 + a_2 c_1 c_3 c_4 d_1 d_3 + b_2 c_1 c_3 c_4 d_2 d_3 + a_1 c_2 c_3 c_4 d_2 d_3) & \quad v^5 \\
-b_1 c_2 c_3 c_4 d_1 d_2 d_3 \, v^6 + c_1 c_2 c_3 c_4 d_1 d_2 d_3 & \quad v^7 \quad = 0,
\end{aligned}
$$

$$
\begin{aligned}
e^{-v} - a_1 a_2 a_3 a_4 + (a_2 a_3 a_4 b_1 d_1 + a_3 a_4 b_1 b_2 d_2 + a_4 b_1 b_2 b_3 d_3 + d_4 e^{-v}) & \quad v \\
-(a_2 a_3 a_4 c_1 d_1 + a_3 a_4 b_2 c_1 d_2 + a_1 a_3 a_4 c_2 d_2 + a_4 b_2 b_3 c_1 d_3 + a_1 a_4 b_3 c_2 d_3 + a_1 a_2 a_4 c_3 d_3) & \quad v^2 \\
+(a_3 a_4 b_1 c_2 d_1 d_2 + a_4 b_1 b_3 c_2 d_1 d_3 + a_2 a_4 b_1 c_3 d_1 d_3 + a_4 b_1 b_2 c_3 d_2 d_3) & \quad v^3 \\
-(a_3 a_4 c_1 c_2 d_1 d_2 + a_4 b_3 c_1 c_2 d_1 d_3 + a_2 a_4 c_1 c_3 d_1 d_3 + a_4 b_2 c_1 c_3 d_2 d_3 + a_1 a_4 c_2 c_3 d_2 d_3) & \quad v^4 \\
+a_4 b_1 c_2 c_3 d_1 d_2 d_3 \, v^5 - a_4 c_1 c_2 c_3 d_1 d_2 d_3 & \quad v^6 \quad = 0
\end{aligned}
$$

where $v = wh$.

Requiring the modified method to integrate exactly $\cos(wx)$ and $\sin(wx)$

$$
\begin{aligned}
\cos(v) - b_1 b_2 b_3 b_4 + (b_1 b_3 b_4 c_2 d_1 + a_2 b_1 b_4 c_3 d_1 + a_2 a_3 b_1 c_4 d_1 + b_1 b_2 b_4 c_3 d_2 + a_3 b_1 b_2 c_4 d_2 + b_1 b_2 b_3 c_4 d_3) & \quad v^2 \\
-(b_1 b_4 c_2 c_3 d_1 d_2 + a_3 b_1 c_2 c_4 d_1 d_2 + b_1 b_3 c_2 c_4 d_1 d_3 + a_2 b_1 c_3 c_4 d_1 d_3 + b_1 b_2 c_3 c_4 d_2 d_3) & \quad v^4 \\
+ b_1 c_2 c_3 c_4 d_1 d_2 d_3 & \quad v^6 \quad = 0,
\end{aligned}
$$

$$
\begin{aligned}
\sin(v) - (b_2 b_3 b_4 c_1 + a_1 b_3 b_4 c_2 + a_1 a_2 b_4 c_3 + a_1 a_2 a_3 c_4) & \quad v \\
+(b_3 b_4 c_1 c_2 d_1 + a_2 b_4 c_1 c_3 d_1 + a_2 a_3 c_1 c_4 d_1 + b_2 b_4 c_1 c_3 d_2 + a_1 b_4 c_2 c_3 d_2 + a_3 b_2 c_1 c_4 d_2 + & \\
a_1 a_3 c_2 c_4 d_2 + b_2 b_3 c_1 c_4 d_3 + a_1 b_3 c_2 c_4 d_3 + a_1 a_2 c_3 c_4 d_3) & \quad v^3 \\
-(b_4 c_1 c_2 c_3 d_1 d_2 + a_3 c_1 c_2 c_4 d_1 d_2 + b_3 c_1 c_2 c_4 d_1 d_3 + a_2 c_1 c_3 c_4 d_1 d_3 + b_2 c_1 c_3 c_4 d_2 d_3 + a_1 c_2 c_3 c_4 d_2 d_3) & \quad v^5 \\
+ c_1 c_2 c_3 c_4 d_1 d_2 d_3 & \quad v^7 \quad = 0,
\end{aligned}
$$

$$
\begin{aligned}
\cos(v) - a_1 a_2 a_3 a_4 + d_4 v \sin(v) & \\
+(a_2 a_3 a_4 c_1 d_1 + a_3 a_4 b_2 c_1 d_2 + a_1 a_3 a_4 c_2 d_2 + a_4 b_2 b_3 c_1 d_3 + a_1 a_4 b_3 c_2 d_3 + a_1 a_2 a_4 c_3 d_3) & \quad v^2 \\
-(a_3 a_4 c_1 c_2 d_1 d_2 + a_4 b_3 c_1 c_2 d_1 d_3 + a_2 a_4 c_1 c_3 d_1 d_3 + a_4 b_2 c_1 c_3 d_2 d_3 + a_1 a_4 c_2 c_3 d_2 d_3) & \quad v^4 \\
+ a_4 c_1 c_2 c_3 d_1 d_2 d_3 & \quad v^6 \quad = 0,
\end{aligned}
$$

$$
\begin{aligned}
\sin(v) - (a_2 a_3 a_4 b_1 d_1 + a_3 a_4 b_1 b_2 d_2 + a_4 b_1 b_2 b_3 d_3 + d_4 \cos(v)) & \quad v \\
+(a_3 a_4 b_1 c_2 d_1 d_2 + a_4 b_1 b_3 c_2 d_1 d_3 + a_2 a_4 b_1 c_3 d_1 d_3 + a_4 b_1 b_2 c_3 d_2 d_3) & \quad v^3 \\
- a_4 b_1 c_2 c_3 d_1 d_2 d_3 & \quad v^5 \quad = 0
\end{aligned}
$$

4   Numerical results

We consider the one-dimesional eigenvalue problem with boundary conditions

$$\psi(a) = 0, \quad \psi(b) = 0$$

We use the shooting scheme in the implementation of the above methods. The shooting method converts the boundary value problem into an initial value problem where the boundary value at the end point b is tranformed into an initial value $y'(a)$, the results are independent of $y'(a)$ if $y'(a) \neq 0$. The eigenvalue E is a parameter in the computation, the value of E that makes $y(b) = 0$ is the eigenvalue computed.

4.1   The Harmonic Oscillator

The potential of the one dimensional harmonic oscillator is

$$V(x) = \frac{1}{2}kx^2$$

we consider $k = 1$. The exact eigenvalues are given by

$$E_n = n + \frac{1}{2}, \quad n = 0, 1, 2, \ldots$$

4.2   The Morse Potential

The Morse potential is

$$V(x) = D\left(\exp\left(-2\alpha x\right) - 2\exp\left(-\alpha x\right)\right)$$

with $D = 12$ and $\alpha = 0.204124$.

The exact eigenvalues are given by

$$E_n = -12 + \left(n + \frac{1}{2}\right) - \frac{1}{48}\left(n + \frac{1}{2}\right)^2, \quad n = 1, 2, \ldots$$

References

[1] Arnold V., *Mathematical methods of Classical Mechanics*, Springer-Verlag, 1978.

[2] Forest E., Ruth R.D., Fourth order symplectic integration, *Physica D*, 43(1990),105-117.

[3] Simos T.E., Numerical methods for 1D, 2D and 3D differential equations arising in chemical problems, *Chemical Modelling: Application and Theory*, The Royal Society of Chemistry, 2(2002),170-270.

[4] Yoshida H., Construction of higher order symplectic integrators, *Physics Letters A* 150(1990),262-268

VSP International
Science Publishers
P.O. Box 346, 3700 AH Zeist
The Netherlands

Lecture Series on Computer
and Computational Sciences
Volume 1, 2004, pp. 396-399

Minimal generator of a non-deterministic operator

A. Mora, P. Cordero, M. Enciso, I. P. de Guzmán[1]

E.T.S.I. Informática.
Universidad de Málaga.
29071 - Málaga, Spain.

Received 31 July, 2004; accepted in revised form 18 August, 2004

Abstract: In [1], we present, in the framework of lattice theory, a new concept of operator, named **non-deterministic ideal operator** (briefly **nd.ideal-o**) that allows us the characterization of concepts frequently used in database. In particular, the **nd.ideal-os** allow us to formalize database redundancy in a more significant way than it was thought of in the literature. In this paper, we present the formalization of the generator of a **nd.ideal-o** that characterize the concept of minimal closure of a set of functional dependencies (FD) in database.

Keywords: Lattice Theory, Closure, Minimal Generator, Database Theory

Mathematics Subject Classification: 03G10, 06A15, 06B10, 68P15

1 Non-Deterministic Operators: Background

We assume that the reader knows the basic concepts of lattice theory (see [4]) and the database concepts of FDs, Amstrong's Axioms, etc. (see [7]).

In [2, 3] the authors show the advantages of use the **non-deterministic operators** as a formal tool in computation. We use this concept in [1] as the tool for formalizing several well-know concepts in database theory: Armstrong's Axioms, f-family, schemes, keys, etc.

In the following, we summarize the basic concepts of the **non-deterministic ideal operators**.

Definition 1.1 *Let A be a non-empty set and $n \in \mathbb{N}$ with $n \geq 1$. If $F : A^n \to 2^A$ is a total application, we say that F is a* **non-deterministic operator with arity** n *in A (henceforth, ndo) We denote the set* ndos *with arity n in A by $\mathcal{N}do_n(A)$ and, if F is a ndo, we denote its arity by $\mathrm{ar}(F)$. As usual, $F(a_1, \ldots, a_{i-1}, X, a_{i+1}, \ldots, a_n) = \bigcup_{x \in X} F(a_1, \ldots, a_{i-1}, x, a_{i+1}, \ldots, a_n)$.*

Definition 1.2 *Let A be a poset, $X \subseteq A$ and $\mathcal{F}$ a family of* ndos *in A. Let us consider the sets $X_0 = X$ and $X_{i+1} = X_i \cup \bigcup_{F \in \mathcal{F}} F(X_i^{\mathrm{ar}(F)})$ We define the* **nd-inductive closure** *of X under $\mathcal{F}$ as $\mathcal{C}\ell_{\mathcal{F}}(X) = \bigcup_{i \in \mathbb{N}} X_i$. We say that X is* **closed** *for $\mathcal{F}$ if $\mathcal{C}\ell_{\mathcal{F}}(X) = X$.*

Theorem 1.3 *Let $\mathcal{F}$ be a family of* ndos *in A. $\mathcal{C}\ell_{\mathcal{F}}$ is a closure operator in $(2^A, \subseteq)$.*

Example 1 *Let $(A, \vee, \wedge)$ be a lattice. The ideal generated by X is $(X] = \mathcal{C}\ell_{\{\vee, \downarrow\}}(X)$ for all $X \subseteq A$.*

---
[1]E-mail: amora@ctima.uma.es. Partially supported by Spanish DGI projects BFM2000-1054-C02-02, TIC2003-08687-C02-01 and Junta Andalucía project TIC115.

In [5, 6] the authors study in lattice theory the FDs. In particular, we characterize the concept of f-family (widely known in bibliography on database) by means of a new concept which we call non-deterministic ideal operator:

Definition 1.4 *Let F be an unary ndo in a poset $(A, \leq)$. We say that F is a **non-deterministic ideal operator** (briefly **nd.ideal-o**) if it is reflexive, transitive and $F(a)$ is an ideal of $(A, \leq)$, for all $a \in A$. Moreover, if $F(a)$ is a principal ideal, for all $a \in A$, then we say that F is **principal**.*

In [1] we shows with an illustrative example the independence of these properties.

Proposition 1.5 *Let F be an nd.ideal-o in a poset $(A, \leq)$ and $a, b \in A$. F is a monotone operator of $(A, \leq)$ to $(2^A, \subseteq)$.*

Proposition 1.6 *Let $(A, \leq)$ be a lattice. The following properties hold: i) Any intersection of nd.ideal-o in A is a nd.ideal-o in A and ii) for all unary ndo in A, F, there exists an unique nd.ideal-o in A, denoted $\widehat{F}$, that is minimal and contains F. This nd.ideal-o is named **nd.ideal-o generated by** F and defined as $\widehat{F} = \bigcap\{F' \mid F'$ is a nd.ideal-o in A and $F \subseteq F'\}$.[2]*

Theorem 1.7 *Let $(A, \leq)$ be a lattice. $\widehat{\ } : \mathcal{N}do_1(A) \to \mathcal{N}do_1(A)$ is the closure operator given by $\widehat{F}(x) = \mathcal{C}\ell_{\{F, \vee, \downarrow\}}(\{x\})$.*

2 Minimal Generator of an nd.ideal-o

Definition 2.1 *Let be $(A, \leq)$ a lattice and F an nd.ideal-o in A. We say that $G \in \mathcal{N}do_n(A)$ is a **generator** of F if $\widehat{G} = F$.*

Particularly, $\forall F \in \mathcal{N}do_n(A)$, F is an **generator** of $\widehat{F}$. Now, we are interested in a generator without redundancy, therefore we search *minimal* generators.

Theorem 2.2 *Let be $(A, \leq)$ a finite Boole algebra and $I \subseteq A$ an ideal in $(A, \leq)$. Then, the ideal generated by $I \cap \text{Átom}(A)$ is $\left(I \cap \text{Átom}(A) \right] = I$.*

Corollary 2.3 *Let be $(A, \leq)$ a finite Boole algebra and F an nd.ideal-o in A. Then, there exists $G : A \to 2^{\text{Átom}(A)}$ generator of F.*

The corollary 2.3 can help us in the search of the minimal generator.

Example 2 *Let be the boole algebra of the positive divisors of 30, D_{30}, with the divisibility relation and the nd.ideal-o $F : D_{30} \to 2^{D_{30}}$ defined by:*

$$F(x) = \begin{cases} (x] & \text{if } x \in \{1, 2, 3, 5\} \\ D_{30} & \text{in other case} \end{cases}$$

the ndo $G(x) = F(x) \cap \text{Átom}(D_{30})$ is described by

$$G(x) = \begin{cases} \varnothing & \text{if } x = 1 \\ \{x\} & \text{si } x \in \{2, 3, 5\} \\ \{2, 3, 5\} & \text{in other case} \end{cases}$$

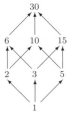

---

[2]If $F, G \in \mathcal{N}do_n(A)$ then $(F \cap G)(a) = F(a) \cap G(a)$.

But G have unnecessary information because the ndo*:*
$$H(6) = \{5\}; \ H(10) = \{3\}; \ H(15) = \{2\}; H(x) = \varnothing \ \text{in other case}$$
satisfies that $H \subset G$ *and* $\widehat{H} = \widehat{G} = F$.

Proposition 2.4 *Let be* $(A, \leq)$ *a boole algebra finite and* F *an nd.ideal-o in* A. *If* $H : A \to 2^{\mathbf{Átom}(A)}$ *is the* ndo *defined by* $H(x) = (F(x) \smallsetminus (x]) \cap \mathbf{Átom}(A)$, *then* H *is a generator of* F.

Definition 2.5 *Let be* $(A, \leq)$ *a lattice and* $F, G \in \mathcal{N}do_n(A)$. *We say that* F *and* G *are* **equivalents** *if* $\widehat{F} = \widehat{G}$. *In this case, we say that* F *is a cover of* G *and vice versa.*

Now, we formalize in the following definition the idea of "to have less information than".

Definition 2.6 *Let be* $(A, \leq)$ *a poset,* $F, G : A^{\mathtt{ar}(F)} \longrightarrow 2^A$, *we define:*

1. $F \preccurlyeq G$ *if* $\mathtt{Grafo}(F) \leq \mathtt{Grafo}(G)$, *that is, if for all* (a, b) *with* $b \in F(a)$ *there exists* (a', b') *with* $b' \in G(a')$ *that satisfies* $a \leq a'$ *and* $b \leq b'$. *Particularly, if* $F \subseteq G$, *we have that* $F \preccurlyeq G$.

2. $F \prec G$ *iff* $F \preccurlyeq G$ *and* $F \neq G$.

Definition 2.7 *Let be* $(A, \leq)$ *a lattice and* $F, G \in \mathcal{N}do_n(A)$. *We say that* F *is* **redundant** *if there exists* G *equivalent to* F *such that* $G \prec F$.

Therefore, a ndo is redundant if there exists another equivalent ndo with less information. The following example shows ndo's with redundancy.

Example 3 *Let be* $U = \{a, b, c\}$, *and* $(2^U, \subseteq)$:

1. *The* ndo $F : 2^U \to 2^{2^U}$, $F(\{a\}) = \{\{a\}, \{c\}\}$, $F(\{a, b\}) = \{U\}$, $F(X) = \varnothing$ *in other case is redundant because the* ndo $G : 2^U \to 2^{2^U}$

$$G(\{a\}) = \{\{a\}, \{c\}\}, \quad G(X) = \varnothing \ \text{in other case}$$

satisfies that $G \prec F$ *and, as* $F(\{a, b\}) \subseteq \widehat{G}(\{a, b\}) = 2^U$, *we have that* $\widehat{G} = \widehat{F}$.

2. *The* ndo $F : 2^U \to 2^{2^U}$, $F(\{a\}) = \{\{a, c\}\}$, $F(X) = \varnothing$ *in other case is redundant because the* ndo $G : 2^U \to 2^{2^U}$

$$G(\{a\}) = \{\{c\}\}, \quad G(X) = \varnothing \ \text{in other case}$$

satisfies that $G \prec F$ *and, as* $F(\{a\}) \subseteq \widehat{G}(\{a\}) = \{\varnothing, \{a\}, \{c\}, \{a, c\}\}$, *we have that* $\widehat{G} = \widehat{F}$.

3. *The* ndo $F : 2^U \to 2^{2^U}$, $F(\{a\}) = \{\{c\}\}$, $F(\{a, c\}) = \{\{b\}\}$, $F(X) = \varnothing$ *in other case is redundant because the* ndo $G : 2^U \to 2^{2^U}$

$$G(\{a\}) = \{\{c\}, \{b\}\}, \quad G(X) = \varnothing \ \text{in other case}$$

satisfies that $G \prec F$ *and* $\widehat{G} = \widehat{F}$.

The following proposition summarize the cases presented in the below example.

Proposition 2.8

1. *there exists* $a \in A$ *and* $b \in F(a)$ *such that* $b \in \widehat{F_{ab}}(a)$, *where* F_{ab} *is given by* $F_{ab}(a) = F(a) \smallsetminus \{b\}$ *and* $F_{ab}(x) = F(x)$ *otherwise.*

2. *there exists $a, b' \in A$ and $b \in F(a)$ such that $b' < b$ and $b \in \widehat{F_{abb'}}(a)$ where $F_{abb'}$ is given by $F_{abb'}(a) = (F(a) \smallsetminus \{b\}) \cup \{b'\}$ and $F_{abb'}(x) = F(x)$ otherwise.*

3. *there exists $a, a' \in A$ and $b \in F(a)$ such that $a' < a$, $b \in \widehat{F}(a')$ and $b \in \widehat{F_{aba'}}(a)$ where $F_{aba'}$ is given by $F_{aba'}(a) = F(a) \smallsetminus \{b\}$, $F_{aba'}(a') = F(a') \cup \{b\}$ and $F_{aba'}(x) = F(x)$ otherwise.*

Finally we remark the following characterization of **minimal generator** of a ndo.

Definition 2.9 *Let be $(A, \leq)$ a lattice and $F \in \mathcal{O}nd_1(A)$. We say that $G \in \mathcal{O}nd_1(A)$ is a* **minimal generator** *of F if:*

- *G is equivalent to F,*

- *$G \prec F$ and*

- *G is not redundant.*

3   Functional Dependencies in lattice theory

In an awful amount of research on databases, the study of FDs is based on the notion of f-family (*Amstrong's Relation*) that we characterize in [1] in the framework of the lattice theory. In [1] we proof the following theorem and turns the proof of well-known properties of FDs in a trivial matter.

Theorem 3.1 *Let A be a non-empty set and F a relation in 2^A. F is a f-family over A if and only if F is a ndo in $(2^A, \subseteq)$.*

In this paper we formalize the database concept de *family non-redundant of FDs* (minimal closure) with the concept of minimal generator of the corresponding ndo.

References

[1] P. CORDERO, M. ENCISO, I. P. DE GUZMÁN, AND A. MORA, *SLFD Logic: Elimination of data redundancy in Knowledge Representation*, LECTURE NOTES - LNAI 2527, Advances in AI, Iberamia 2002, pp. 141-150

[2] J. MARTÍNEZ, P. CORDERO, G. GUTIÉRREZ AND I. P. DE GUZMÁN, *A new algebraic tool for Authomatic Theorem Provers*, Annals of Mathematics and Artificial Intelligence. Kluwer Academic Publisher. To appear.

[3] J. MARTÍNEZ, P. CORDERO, G. GUTIÉRREZ AND I. P. DE GUZMÁN *Generalizations of Lattices Looking at Computation*, Discrete Mathematic., Elsevier. To appear.

[4] G. GRATZER, *General lattice theory*, Birkhauser Verlag. Second Edition. 1998

[5] J. DEMETROVICS AND D. T. VU, *Some results about normal forms for functional dependency in the relational datamodel*, Discrete Applied Mathematics 69, 1996, pp. 61-74

[6] J. DEMETROVICS, L. LIBKIN, AND I. B. MUCHNIK, *Functional dependencies in relational databases: A lattice point of view*, Discrete Applied Mathematics 40, 1992, pp. 155-185

[7] J. PAREDAENS, P. DE BRA, M. GYSSENS, AND D. V. VAN GUCHT, *The Structure of the Relational Database Model*, EATCS Monographs on Theoretical Computer Science, 1989

VSP International
Science Publishers
P.O. Box 346, 3700 AH Zeist
The Netherlands

*Lecture Series on Computer
and Computational Sciences*
Volume 1, 2004, pp. 400-403

Computations of Gas Turbine Blades Heat Transfer Using Two-Equation Turbulence Models

Abdul Hafid M. Elfaghi[†] and Ali Suleiman M. Bahr Enel

Aeronautical Engineering Department,
Engineering Academy Tajoura
Tajoura, Libya

Received 8 June, 2004; accepted in revised form 25 September, 2004

Abstract: Turbine blade heat transfer is a very important engineering problem characterized by high turbulence levels and very complex flow fields. The prediction of turbine blade heat transfer is very important, especially when turbine inlet temperature increases.

This paper is about numerical predictions of flow fields and heat transfer to gas turbine blades using different two-equation turbulence models. Four two-equation turbulence modeles were used, the standard $k - \varepsilon$ model, the modefied Chen-Kim $k - \varepsilon$ model, RNG model and Wilcox standard $k - \omega$ turbulence model. These models are based on the eddy viscosity concept, which determines the turbulent viscosity through time-averaged Navier-Stocks differential equations.

The simulation was performed at Aerospace Engineering Department, University Putrra Malaysia (UPM) using the general-purpose computational fluid dynamics code, PHOENICS, which solves the governing fluid flow and heat transfer equations. An H-type, body-fitted-coordinate (BFC) grid is used and upstream with downstream periodic conditions specified.

The results are compared with the available experimental measurements obtained from a research carried out at the Von Karman Institute of Fluid Dynamics (VKI). A comparison between the turbulence models and their prediction of heat flux on the blade is carried out.

Keywords: CFD, turbine, heat transfer, turbulence models, PHOENICS

1. Introduction

The development of high performance gas turbine requires high turbine inlet temperatures that can lead to sever thermal stresses to the turbine blade particularly in the first stages of turbine. Also, the efficiency of gas turbine depends strongly on the momentum and heat transfer characteristics of the turbulent boundary layers developed on the blade surfaces. Therefore, a major objective of gas turbine design is to determine the thermal and aerodynamical characteristics of the turbulent flow in the turbine cascade.

The two-equation turbulence models have been and will likely continue to be among the most widely used turbulence model for flow predictions in engineering applications. In particular, the k-ε model is the most popular in commercial softwares.

2. Governing Equations

The averaged continuity equation (1), momentum equation (2) and energy equation (3) for compressible fluid, which obtained by averaging the instantaneous equations can be written as folloing:

$$\frac{\partial}{\partial t}\bar{\rho} + \frac{\partial}{\partial x}(\overline{\rho u}) = 0 \qquad (1)$$

[†] Email: hafied@yahoo.com

$$\frac{\partial}{\partial t}(\overline{\rho u}) + \frac{\partial}{\partial x}(\overline{\rho u_i}\overline{u_j} + P\delta_{ij} - \overline{\tau}_{ij}) = 0 \qquad (2)$$

$$\frac{\partial}{\partial t}(\overline{\rho e_0}) + \frac{\partial}{\partial x}(\overline{\rho u_i}\overline{e_0} + \overline{u_j}\overline{P} + \overline{q}_j - \overline{u_i}\overline{\tau}_{ij}) = 0 \qquad (3)$$

The heat flux, q, is given by Fourier's low:

$$q = -k\frac{\partial}{\partial t} \equiv -C_p\frac{\mu}{P_r}\frac{\partial T}{\partial x_i} \qquad (4)$$

The heat flux is treated according to Reynolds analogy. First the heat flux is divided into a laminar and turbulent parts

$$q = q^{lam} + q^{turb} \qquad (5)$$

3. Turbulence Models

A turbulence model can be described as a set of relations and equations needed to determine the unknown turbulent correlations that have arisen from the averaging process.
In the two-equation model, the Reynold's stress are given by Boussinesq's assumption: -

$$-u_i u_j = \upsilon(\frac{\partial U_i}{\partial x_j} + \frac{\partial U_j}{\partial x_i}) - \frac{2}{3}\delta_{ij}(\upsilon_t\frac{\partial U_i}{\partial x_i} + k) \qquad (6)$$

4. Simulation Procedure

The turbine blade chosen for this work is a transonic high loaded cascade from von Karman Institute of fluid dynamics (VKI). Flow and heat transfer experimental measurements at many different operating conditions and for different Mach and Reynolds numbers are available in Arts et al. The case chosen here is referred to as Mur218. it has an inlet free stream turbulence level of 4percent, an outlet Mach number of 0.76 and a chord Reynolds number based on the outlet conditions of 10^6.

A two-dimensional non-orthogonal grid to the x- and z-co-ordinates was laid over the flow domain. A fine grid of 125 cells in stream wise direction were incorporated with 50 cells in the pitch wise direction using PHOENICS input language (PIL). The grid is made to be fine near the wall surface and the inlet and outlet boundaries. H-type, body-fitted co-ordinate (BFC) grid was used and upstream and downstream periodic conditions were specified. Cyclic boundary conditions were used at upstream and downstream regions. The turbulence models are applied using their corresponding commands and boundary conditions. The computational domain and grid system are shown in figure 1.

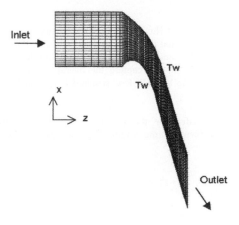

Figure 1: Turbine blade H-grid

5. Grid Independence and Convergence

In order to assess the grid independence of the results, a grid resolution study was undertaken. A grid of 125 in chord wise direction and 50 in pitch wise direction were found to be sufficient for this type of flow. A grid system of 210×70 was tested and found that the results obtained by both grid systems are closed to each other. The time taken by the 125×50 is reduced by 40% compared with 210×70 grid.

6. Results and Discussion

Figure 2 shows that the predicted surface pressure distributions along the blade surface are in good agreement with the experimental data. A strong pressure gradient is observed along the suction side, the pressure failed rapidly up to $S/S_{max}=0.5$ and then increase up to trailing edge.

Figure 3 shows the velocity field along the blade surface. On suction side, the flow is accelerated up to $S^*/S_{max}=0.5$ and then decelerated up to trailing edge. On the pressure side, the flow is a slightly accelerated at the leading edge up to $S^*/S_{max}=0.6$ and then strongly accelerated.

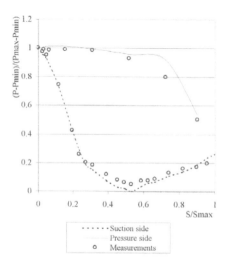

Figure 2: Pressure distribution

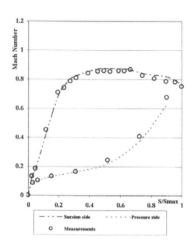

Figure 3: Mach number distribution

Figure 4 shows the predicted wall heat flux distribution on suction side compared with Arts et al. (1990) measurement results. On the initial part of blade, the flow is laminar and the acceleration leads to decrease of heat transfer rate. The standard model starts well, but then gives an over prediction of heat transfer. Chen-Kim and RNG models capture the overall behavior quite well, although the predicted heat transfer level is bet too high. Wilcox standard k-ω model works better in the leading edge region.

Figure 5 shows wall heat transfer for the pressure side. The results agree well with the measurements.there are still problems at the leading edge and the results give high-predicted heat transfer on this region.

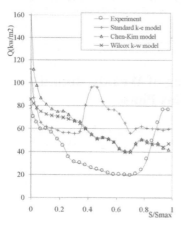

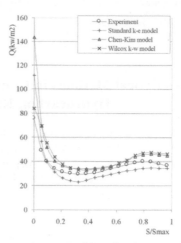

Figure 4: Wall heat flux (Suction side)   Figure 5: Wall heat flux (pressure side)

7. Conclusion

The following points could be pointed out as a concluded remarks of this work: -
1- All models give good heat transfer prediction for pressure side, except close to the leading edge.
2- Standard model slightly under-predicts the heat flux on the pressure side, whereas Chen-Kim and RNG models give high values
3- All k-ε models generate too high turbulence levels in stagnation point region, which give rise to the heat transfer rates.
4- Wilcox standard k-ω model give better heat transfer predictions in the leading edge region.

Acknweldegments

This work was carried out at the department of Aerospace, University Putra Malaysia (UPM), which is gratefully acknowlededge.

References

[1] T. Arts., Lmbert M. Rouvriot and A. W. Rutherford, *"Aero-thermal investigation of high loaded transonic linear turbine guide vane cascade"*, Technical report 174, Von Karman Institute for Fluid Dynamics. (1990)

[2] Y.S. Chen and S.W.Kim, *"Computation of turbulent flows using an extended k-ε turbulence closure model"*, NASA CR-179204 (1987)

[3] J. Larson, L. Eriksson and U. Hall, *"External Heat Transfer Prediction in Supersonic Turbine Using the Reynolds Averaged Navier-Stokes Equations"*, 12th ISABE conference Melbourne, vol. 2, pp. 1102-1112. (1995)

[4] B.E.Launder. and D.B. Sharma, *"The Numerical Computation of Turbulent Flow"*, Methods in Appl. Mech. & Eng., Vol.3, pp.269 (1974)

[5] V. Yakhot and S.A. Orszag, *"Renormalization group Analysis of Turbulence"*, Journal of Sci. Comput., Vol.1, pp.3. (1996)

VSP International
Science Publishers
P.O. Box 346, 3700 AH Zeist
The Netherlands

*Lecture Series on Computer
and Computational Sciences*
Volume 1, 2004, pp. 404-405

Analytical Methods in Computer Modelling of Weakly Bound Hydrocarbon Radicals in the Gas Phase

Marek Narożnik[1]

Faculty of Sciences
Department of Computing
University of Podlasie, 08-110 Siedlce, Poland

Jan Niedzielski

Chemistry Department
Warsaw University, 02-089 Warsaw, Poland

Received 5 July, 2004; accepted in revised form 8 August, 2004

Abstract: The analytical forms for interaction potentials were used to estimate the matrix elements that are necessary to use the perturbation calculus in investigations of the weakly bound hydrocarbon radicals.

Keywords: Perturbation calculus, matrix elements, Schrödinger equation

Mathematics Subject Classification: AMS-MOS: 68-04, 65D05, 65D07, 65D10, 81Q15

1. Theory

An analytical form of the potential interaction energy of hydrocarbon molecules and radicals at long distances is developed. For instance, the interaction energy of methyl radicals can be given in the form [1]

$$V = \varepsilon_0 - \varepsilon_1 \cos^2 \theta_1 - \varepsilon_2 \cos^2 \theta_2 - \\ - A \exp\left(-\frac{Zr}{a_0}\right) \left[f_1(r)\cos\theta_{12} - f_2(r)\cos\theta_1\cos\theta_2\right]^2 \tag{1}$$

where the parameters ε_0, ε_1, ε_2, f_1, f_2 are the spherically symmetrical functions of the distance r between the mass centers of reactants while $\cos\theta_1$, $\cos\theta_2$, $\cos\theta_{12}$ are the directional cosines for mutual orientations of reactants. Using this analytical form we can readily estimate the matrix elements [2]

$$V_{nm} = \int \ldots \int \psi_n \, V \, \psi_m \, d\tau \tag{2}$$

where $d\tau$ is the element of volume while Ψ_n and Ψ_m are the wave functions of the Hamiltonian of the system associated with the rotational states. Using formulae (1) and (2) we can effectively estimate energy levels of the interacting radicals, E_n. To this end the Brillouin-Wigner series is needed. [2,3] The adiabatic curves that are obtained, $E_n(r)$, reveal the presence of loosely bound physicochemical states such as $CH_3\cdots CH_3$, [4] and $CH_3\ldots H$. [5] The prerequisite for the existence of such states is associated with the number l that quantizes the orbital momentum of reactants. l must be smaller than the limiting value $<l_{lim}>$. For instance, in the case of methyl radical interaction $<l_{lim}>=39$. [4] The

---

[1] Corresponding autor: Faculty of Sciences, University of Podlasie, 08-110 Siedlce, Poland
E-mail address: marekn_ap@poczta.onet.pl

energy levels $E_n(r)$ can also be used to estimate the mean radical interaction potential in the hard sphere approximation. [6]

2. Modelling of radical recombinations

The latter method was used to study the following recombinations: $CH_3 + H$, $CH_3 + CH_3$, allyl + allyl, *tert*-C_4H_9 + *tert*-C_4H_9. [6,7] The presence of loosely bonded states can explain the reasons for the negative temperature dependence of the recombination rate constants, k. For the system $CH_3 + CH_3$ we obtain the dependence $k \sim T^{-1/2}$ while for the system *tert*-C_4H_9 + *tert*-C_4H_9 the dependence $k \sim T^{-3/2}$ is obeyed. Both are in perfect agreement with experiment. The statistical sums for the bonded states of radicals that are indispensable in such calculations are estimated with the use of spline functions interpolations.

The existence of loosely bounded states in the radical recombination can throw some light on the little known process of recombination rate constant decrease under very high pressures. [8]

The above mentioned methods were used to study many systems, most recently the interactions and the bonded states of recombining allyl radicals. [9]

References

[1] M.Narożnik and J.Niedzielski, Analytical form of the interaction energy of radicals at short and long distances, *Theoretica Chimica Acta* **94** 257-269 (1996).

[2] M.Narożnik and J.Niedzielski, Energy levels of the weakly interacting radicals, *Theoretica Chimica Acta* **94** 271-285 (1996).

[3] A.G.Makhaneck and W.S.Korolkov, *Analytical Methods in Quantum Mechanical Theory of Interactions.* Nauka y Technika, Minsk, 1982 (in Russian).

[4] M.Narożnik, Recombination of radicals in the high-pressure and high temperature limit. Part 1 Reaction $CH_3 + CH_3$. *Journal of Chemical Society, Faraday Transactions* **94** 2531-2539 (1998).

[5] M.Narożnik and J.Niedzielski, Recombination of radicals in the high-pressure and high temperature limit. Part 2 Reaction $CH_3 + H$. *Journal of Chemical Society, Faraday Transactions* **94** 2541-2547 (1998).

[6] M.Narożnik, Interactions between isobutane molecules and between *tert*-butyl radicals estimated onb the basis of second virial coefficients. *Journal of Molecular Structure (Theochem)* **624** 267-278 (2003).

[7] M.Narożnik and J.Niedzielski, Recombination of *tert*-butyl radicals: the role of weak van der Waals interactions. *Theoretical Chemistry Accounts* **108** 103-112 (2002).

[8] A.G.Zawadzki and J.T.Hynes, Radical recombination rate constants from gas to liquid phase. *Journal of Physical Chemistry* **93** 7031-7036 (1989).

[9] M.Narożnik and J.Niedzielski, to be published

VSP International
Science Publishers
P.O. Box 346, 3700 AH Zeist
The Netherlands

*Lecture Series on Computer
and Computational Sciences*
Volume 1, 2004, pp. 406-409

A Numerical Scheme for a Shallow Water Equation

D. G. Natsis [1], A. G. Bratsos and D. P. Papadopoulos

Department of Mathematics,
Technological Educational Institution (T.E.I.) of Athens,
GR-122 10 Egaleo, Athens, Greece.

Received 30 July, 2004; accepted in revised form 25 August, 2004

Abstract: A finite-difference method is presented for the solution of the two-dimensional Boussinesq-type set of equations as those were introduced by Madsen *et al* [6] in the case of a constant-depth environment. The application of the finite-difference scheme results in an initial value problem and it is proposed the unknown quantities to be calculated implicitly by solving a linear system of equations. The numerical treatment of the system is briefly discussed, while numerical results are the subject of a following work.

Keywords: Boussinesq Model; Wave braking; Finite-difference method.

Mathematics Subject Classification: $65M06, 35L05, 35L45$

PACS: 02.60.Lj

1   Introduction

A finite-difference scheme is proposed for the solution of the two-dimensional Boussinesq-type set of equations in the case of constant depth environment. Following physical laws (see for example Madsen *et al* [6] etc.) the problem in its preliminary form is constructed by the following system of nonlinear equations. The application of the finite-difference scheme results in an initial value problem and it is proposed a method of implicit calculation of the unknown functions by solving a linear system of equations.

$$\frac{\partial w}{\partial t} + \frac{\partial p}{\partial x} + \frac{\partial q}{\partial y} = 0, \tag{1}$$

$$\frac{\partial p}{\partial t} - \tilde{B}h^2 \frac{\partial}{\partial t} \left(\frac{\partial^2 p}{\partial x^2} + \frac{\partial^2 q}{\partial x \partial y} \right) - \frac{h}{3} \frac{\partial}{\partial t} \left(\frac{\partial p}{\partial x} + \frac{1}{2} \frac{\partial q}{\partial y} \right) \frac{\partial h}{\partial x} - \frac{h}{6} \frac{\partial}{\partial t} \left(\frac{\partial q}{\partial x} \right) \frac{\partial h}{\partial y}$$

$$= -\frac{\partial}{\partial x} \left(\frac{p^2}{h+w} \right) - \frac{\partial}{\partial y} \left(\frac{pq}{h+w} \right) - \frac{\partial}{\partial x} \left[\tilde{\delta}(w) \left(c_x - \frac{p}{h+w} \right)^2 \right]$$

$$-\frac{\partial}{\partial y} \left[\tilde{\delta}(w) \left(c_x - \frac{p}{h+w} \right) \left(c_y - \frac{q}{h+w} \right) \right] - gd \frac{\partial w}{\partial x}$$

$$+Bgh^2 \left(2\frac{\partial^2 w}{\partial x^2} + \frac{\partial^2 w}{\partial y^2} \right) \frac{\partial h}{\partial x} + Bgh^2 \frac{\partial^2 w}{\partial x \partial y} \frac{\partial h}{\partial y} + Bgh^3 \left(\frac{\partial^3 w}{\partial x^3} + \frac{\partial^3 w}{\partial x \partial y^2} \right), \tag{2}$$

---

[1] Corresponding author. E-mail: dnatsis@otenet.gr

$$\frac{\partial q}{\partial t} - \tilde{B}h^2 \frac{\partial}{\partial t}\left(\frac{\partial^2 q}{\partial y^2} + \frac{\partial^2 p}{\partial x \partial y}\right) - \frac{h}{3}\frac{\partial}{\partial t}\left(\frac{\partial q}{\partial y} + \frac{1}{2}\frac{\partial p}{\partial x}\right)\frac{\partial h}{\partial y} - \frac{h}{6}\frac{\partial}{\partial t}\left(\frac{\partial p}{\partial y}\right)\frac{\partial h}{\partial x}$$

$$= -\frac{\partial}{\partial y}\left(\frac{p^2}{h+w}\right) - \frac{\partial}{\partial x}\left(\frac{pq}{h+w}\right) - \frac{\partial}{\partial y}\left[\tilde{\delta}\left(w\right)\left(c_y - \frac{q}{h+w}\right)^2\right]$$

$$-\frac{\partial}{\partial x}\left[\tilde{\delta}\left(w\right)\left(c_x - \frac{p}{h+w}\right)\left(c_y - \frac{q}{h+w}\right)\right] - gd\frac{\partial w}{\partial y}$$

$$+Bgh^2\left(\frac{\partial^2 w}{\partial x^2} + 2\frac{\partial^2 w}{\partial y^2}\right)\frac{\partial h}{\partial y} + Bgh^2\frac{\partial^2 w}{\partial x \partial y}\frac{\partial h}{\partial x} + Bgh^3\left(\frac{\partial^3 w}{\partial y^3} + \frac{\partial^3 w}{\partial x^2 \partial y}\right). \qquad (3)$$

We are working in the region $\Omega = \left\{(x,y)\,;\, L_x^0 < x < L_x^1,\, L_y^0 < y < L_y^1\right\}$ for $t > t_0$, where $w = w(x,y,t)$ is the surface elevation, $p = p(x,y,t)$ and $q = q(x,y,t)$ are the density of the flow to the directions x and y respectively, $h = h(x,y)$ the depth of the bottom of the sea, $d = d(x,y)$ is the instantaneous depth with $d = h + w$, $\delta = \delta(x,y,t)$ is the thickness of the surface roller, (c_x, c_y) the components of the roller celerity, $\tilde{\delta}(w) = \delta/\left(1 - \frac{\delta}{h+w}\right)$, and B is the dispersion coefficient in which $\tilde{B} = B + \frac{1}{3}$.

2 The numerical solution

To obtain a numerical solution the region $R = \Omega \times [t > t_0]$ with its boundary ∂R consisting of the lines $x = L_x^0,\ L_x^1,\ y = L_y^0,\ L_y^1$ and $t = t_0$, is covered with a rectangular mesh, G, of points with coordinates $(x,y,t) = (x_k, y_m, t_n) = \left(L_x^0 + kh_x, L_y^0 + mh_y, t_0 + n\ell\right)$ with $k,m = 0,1,...,N+1$ and $n = 0,1,...$, in which $h_x = \left(L_x^1 - L_x^0\right)/(N+1)$ and $h_y = \left(L_y^1 - L_y^0\right)/(N+1)$ represent the discretization into $N+1$ subintervals of the space variables, while ℓ represents the discretization of the time variable. The solution for the unknown functions w, p and q of an approximating finite-difference scheme at the same point will be denoted by $w_{k,m}^n$, $p_{k,m}^n$ and $q_{k,m}^n$ respectively, while for the purpose of analyzing stability, the numerical value of actually obtained (subject, for instance, to computer round-off errors) will be denoted by $\tilde{w}_{k,m}^n$, $\tilde{p}_{k,m}^n$ and $\tilde{q}_{k,m}^n$.

The solution vectors will be

$$\mathbf{w}^n = \mathbf{w}(t_n) = \left[w_{0,0}^n, w_{0,1}^n, ..., w_{0,N+1}^n;\ w_{1,0}^n, w_{1,1}^n, ..., w_{1,N+1}^n;\right.$$
$$\left. ..;\ w_{N+1,0}^n, w_{N+1,1}^n, ..., w_{N+1,N+1}^n\right]^T, \qquad (4)$$

$$\mathbf{p}^n = \mathbf{p}(t_n) = \left[p_{0,0}^n, p_{0,1}^n, ..., p_{0,N+1}^n;\ p_{1,0}^n, p_{1,1}^n, ..., p_{1,N+1}^n;\right.$$
$$\left. ..;\ p_{N+1,0}^n, p_{N+1,1}^n, ..., p_{N+1,N+1}^n\right]^T, \qquad (5)$$

$$\mathbf{q}^n = \mathbf{q}(t_n) = \left[q_{0,0}^n, q_{0,1}^n, ..., q_{0,N+1}^n;\ q_{1,0}^n, q_{1,1}^n, ..., q_{1,N+1}^n;\right.$$
$$\left. ..;\ q_{N+1,0}^n, q_{N+1,1}^n, ..., q_{N+1,N+1}^n\right]^T. \qquad (6)$$

T denoting transpose. Then there are $(N+2)^2$ values to be determined at each time step.

2.1 The boundary conditions

In Eqs. (1)-(3) the boundary conditions will be assumed to be of the form

$$u\left(L_x^0 - h_x, y, t\right) \approx \frac{1}{2}\left[-3u\left(L_x^0, y, t\right) + 6u\left(L_x^0 + h_x, y, t\right) - u\left(L_x^0 + 2h_x, y, t\right)\right], \qquad (7)$$

and

$$u\left(L_x^1 + h_x, y, t\right) \approx \frac{1}{2}\left[-3u\left(L_x^1, y, t\right) + 6u\left(L_x^1 - h_x, t\right) - u\left(L_x^1 - 2h, y, t\right)\right]. \qquad (8)$$

2.2 The finite-difference scheme

Replacing the space derivatives in Eqs. (1)-(3) with appropriate finite-difference approximations we obtain the following system in matrix-vector form

$$\mathbf{w}'(t) = \mathbf{F}_1(\mathbf{p}(t), \mathbf{q}(t)), \tag{9}$$

$$A_1 \mathbf{p}'(t) + A_2 \mathbf{q}'(t) = \mathbf{F}_2(\mathbf{w}(t), \mathbf{p}(t), \mathbf{q}(t)) \tag{10}$$

$$A_3 \mathbf{p}'(t) + A_4 \mathbf{q}'(t) = \mathbf{F}_3(\mathbf{w}(t), \mathbf{p}(t), \mathbf{q}(t)) \tag{11}$$

where

$$\mathbf{w}'(t) = \frac{d\,\mathbf{w}(t)}{dt}, \ \mathbf{p}'(t) = \frac{d\,\mathbf{p}(t)}{dt} \text{ and } \mathbf{q}'(t) = \frac{d\,\mathbf{q}(t)}{dt}. \tag{12}$$

which finally leads to the following initial value problem

$$\mathbf{w}'(t) = \mathbf{F}_1^n, \tag{13}$$

$$A_1 \mathbf{p}'(t) + A_2 \mathbf{q}'(t) = \mathbf{F}_2^n \tag{14}$$

$$A_3 \mathbf{p}'(t) + A_4 \mathbf{q}'(t) = \mathbf{F}_3^n \tag{15}$$

with $\mathbf{w}(0) = \mathbf{w}_0$, $\mathbf{p}(0) = \mathbf{p}_0$, $\mathbf{q}(0) = \mathbf{q}_0$ and $t > 0$, in which A_i; $i = 1, 2, 3, 4$ are block matrices of order $(N + 2)^2$.

This system will give rise to a numerical method for the evaluation of the unknown functions w, p and q. The method is under investigation and the results will subject of a following work.

Acknowledgment

This research was co-funded by 75% from E.E. and 25% from the Greek Government under the framework of the Education and Initial Vocational Training Program - Archimedes, Technological Educational Institution (T.E.I.) of Athens project 'Computational Methods for Applied technological Problems'.

References

[1] A. Bayram & M. Larson, Wave transformation in the nearshore zone: comparison between a Boussinesq model and field data, Coastal Eng. Vol. **39** 149-171(2000).

[2] S. Beji & K. Nadaoka, A formal derivation and numerical modelling of the improved Boussinesq equations for varying depth, Ocean Eng., Vol. **23** 691-704(1996).

[3] J. Larsen & M. Dancy, Open Boundaries in Short Wave Simulations - A New Approach, Coastal Engineering, Vol **7** 285-297(1983).

[4] P. A. Madsen, R. Murray and O. R. Sørensen, A New Form of the Boussinesq Equations with Improved Linear Dispersion Characteristics (Part 1), Coastal Engineering, Vol **15** no 4, 371-388(1991).

[5] P. A. Madsen, O. R. Sørensen, A New Form of the Boussinesq Equations with Improved Linear Dispersion Characteristics, Part 2: A Slowly-Varying Bathymetry, Coastal Engineering, Vol **18** no 1, 183-204(1992).

[6] P. A. Madsen, O. R. Sørensen and H. A. Schäffer, Surf zone dynamics simulated by a Boussinesq type model. Part I. Model description and cross-shore motion of regular waves, Coastal Engineering **32** 255-287(1997).

[7] K. Nadaoka & K. Raveenthiran, A phase-averaged Boussinesq model with effective description of carrier wave group and associated long wave evolution, Ocean Eng., Vol. **29** 21-37(2002).

[8] F.S.B.F. Oliveira, Improvement on open boundaries on a time dependent numerical model of wave propagation in the nearshore region, Ocean Eng. Vol. **28** 95-115(2000).

[9] D. M. Peregrine, Long waves on a beach, J. Fluid Mech., Vol. **27** Part 4(1967).

[10] H. A. Schäffer, P. A. Madsen and R. Deigaard, A Boussinesq model for waves breaking in shallow water, Coastal Eng. Conf. 547-566(1980).

[11] Z. L. Zou, High order Boussinesq equations, Ocean Eng., Vol. **26** 767-792(1999).

VSP International
Science Publishers
P.O. Box 346, 3700 AH Zeist
The Netherlands

Lecture Series on Computer
and Computational Sciences
Volume 1, 2004, pp. 410-413

A Time-Dependent Product Formula
and its Application to an HIV Infection Model

S. Oharu[1], Y. Oharu[2], D. Tebbs[3]

[1,3] Department of Mathematics, Chuo University, Tokyo, Japan
[2] Department of Physics and Astronomy, Arizona State University, Tempe, Arizona, U.S.A.

Received 13 August, 2004; accepted in revised form 29 August, 2004

Abstract: In order to obtain a comprehensive form of mathematical models describing nonlinear phenomena such as HIV infection processes, a general class of time-dependent evolution equations is introduced in such a way that the nonlinear operator is decomposed into the sum of a relatively good operator and a perturbation which is nonlinear in general and satisfies no global continuity condition. An attempt is then made to combine the implicit approach (usually adapted for convective diffusion operators) and explicit approach (more suited to treat continuous-type operators representing various physiological and reaction processes), resulting in a *semi-implicit product formula*. Decomposing the operators in this way and considering their individual properties, it is seen that approximation-solvability of the original equation is verified under suitable conditions. Once appropriate terms are formulated to describe treatment by antiretroviral therapy, the time dependence of the reaction terms appears, and such product formula is useful for generating approximate numerical solutions to the original equation. With this knowledge, a continuous model for HIV infection processes is formulated and physiological interpretations are provided. The abstract theory is then applied to show existence of unique solutions to the continuous model describing the behavior of the HIV virus in the human body and its reaction to treatment by antiretroviral therapy. The product formula suggests appropriate discrete models describing the HIV infection mechanism and to perform numerical simulations based on the model of the HIV infection processes. Finally, the results of our numerical simulations are visualized, and it is observed that our results agree with medical aspects in a qualitative way and on a physiologically fundamental level.

Keywords: Mathematical model, time-dependent evolution equation, semi-implicit product formula, continuous model, discrete model, HIV infection processes, antiretroviral therapy.

Mathematics Subject Classification: 35K57, 47H14, 47J35, 58D25, 62P10, 92C55

1   An HIV infection model

The model given here is based upon the original models treated in [2] and [3]. We here make an attempt to present a mathematical model such that the effects of treatment are taken into account and the suggested mathematical algorithms are used to address practical clinical issues. Once this form of treatment by antiretroviral therapy is intoduced into the model, time-dependence of the reaction terms appear and the equations take the form stated below:

---

[1]Corresponding author.@E-mail:oharu@math.chuo-u.ac.jp

$$
\begin{cases}
\partial_t u &= d_1 \Delta u + \mathbf{b}_1(t) \cdot \nabla u + S - \alpha u + p_1 u(w_s + w_r) - (\eta_1(t)\gamma_s w_s + \gamma_r w_r)u \\
\partial_t v_s &= d_2 \Delta v_s + \mathbf{b}_2(t) \cdot \nabla v_s + \eta_1(t)\gamma_s w_s u - \beta v_s - p_2 v_s(w_s + w_r) \\
\partial_t v_r &= d_3 \Delta v_r + \mathbf{b}_3(t) \cdot \nabla v_r + \gamma_r w_r u - \beta v_r - p_2 v_r(w_s + w_r) \\
\partial_t w_s &= d_4 \Delta w_s + \mathbf{b}_4(t) \cdot \nabla w_s + (1 - \mu)p_3 v_s(w_s + w_r) - \delta u w_s + \eta_2(t)p_4 w_s \\
\partial_t w_r &= d_5 \Delta w_r + \mathbf{b}_5(t) \cdot \nabla w_r + p_3 v_r(w_s + w_r) + \mu p_3 v_s(w_s + w_r) - \delta u w_r + p_5 w_r
\end{cases} \qquad \text{(HIV)}
$$

The function u represents the density of healthy cells, as yet uninfected by the virus. v_s and v_r are the densities of infected cells that are sensitive and resistant, respectively, to the treatment. Finally, w_s and w_r represent the concentration of the virus itself, again divided into sensitive and resistant strains. All cells have a natural mortality, represented by the constants α and β, and we assume that there exists a supply of healthy cells described by S. The supply term S depends on the amount of virus present and takes the form $S = S_0 - S^*(w_s + w_r)/(B_S + w_s + w_r) \leq S_0$. The virus also decays in the presence of the immune system and the terms $-\delta u w_s$ and $-\delta u w_r$ account for this. The infection mechanism is described by means of the constants γ_s and γ_r, which specify the relative proportions of sensitive and resistant cells that are produced when infection occurs. Treatment is represented using the functions η_1 and η_2. These restrict both the reproduction of treatment-sensitive infected cells and the regeneration of the HIV virus. Saturation terms p_k, $k = 1, \cdots, 4$ are given by

$$
p_k = p_k^*/(c_k + w_s + w_r) , \quad k = 1, \cdots, 4, \quad p_5 = g_r(w_s + w_r)/(B + w_s + w_r).
$$

The constants p_k^* and c_k stand for the first and second saturation constants, respectively. In this model it is assumed that the external input of resistant virus from the lymphoid compartment is specified by the threshold function $g_r(w)$: $g_r(w) = 0$ if w is less than the threshold w_0 and $g_r(w) = p_4^*$ if $w \geq w_0$. This means that the capacity of the resistant virus to establish requires that the total population remains above the threshold w_0. It should be noted that $p_k \cdot (w_s + w_r) \leq p_k^*$ for $k = 1, 2, 3, 4$, when $w_s, w_r \geq 0$. The *treatment functions* $\eta_1(\cdot)$ and $\eta_2(\cdot)$ are given

$$
\eta_1(t) = e^{-\widehat{c}_1 t}, \quad \eta_2(t) = \max\{e^{-\widehat{c}_2 t}, \widehat{c}_3\}, \quad \text{for some constants} \quad \widehat{c}_1, \widehat{c}_2 \quad \text{and} \quad \widehat{c}_3.
$$

Here it is assumed that resistant virus is produced through mutation, and that there is no significant level of background resistant virus present to substantially effect on the dynamics before treatment starts. After treatment begins, a proportion $q < 1$ of sensitive virus produces resistant virus because of their mutation, and $1 - q$ is a proportion of sensitive virus which remains sensitive to the treatment. It is expected that numerical simulations based on our model could identify critical points which provide an indication of ineffective antiretroviral therapy and the need to change the treatment regimen [1]. In order to provide useful information to the current guidelines in terms of those critical points, we necessitate adding more parameters in the model to specify the criticl points.

2 Product formula approach

An abstract theory for time-dependent evolution equations can be applied to show existence of unique solutions to the model under suitable boundary and initial conditions. By means of a convective diffusion operator A and nonlinear operators $B(t)$ representing physiological reaction terms on the right-hand sides of (HIV), our HIV infection model above may be formulated as an evolution equation of the form

$$u'(t) = (A + B(t))u(t), \quad 0 < t < \tau; \quad u(0) = v \tag{IM}$$

We then define two functionals φ_0 and φ_1 on X by

$$\varphi_0(u) = \|u\|_\infty, u = (u, v_s, v_r, w_s, w_r) \in X, \varphi_1(u) = \max\{\|u\|_\infty, \|v_s\|_\infty, \|v_r\|_\infty, \|w_s\|_\infty, \|w_r\|_\infty\},$$

for $u = (u, v_s, v_r, w_s, w_r) \in X$, to introduce a secondary bornological structure $\mathfrak{B}$ and specify the class D of admissible initial data. Then it is shown that given any $\tau > 0$ and any $v \in D$, there exists $K \in \mathfrak{B}$ such that the discrete appoximate solution u_i^h, generated by the product formula

$$u_0^h = v ; \quad u_i^h = (I - hA_{ih})^{-1}(I + hB_{(i-1)h})u_{i-1}^h, \quad i = 1, \cdots, [\tau/h], \tag{PF}$$

satisfy the staility condition $u_i^h \in K$ for $0 < h < h_0$ and $ih \in [0, \tau]$. Moreover, it is shown that

$$v \geq 0 \text{ a.e. in } \Omega \text{ implies that } u_i^h \geq 0 \text{ a.e. } \Omega, \ 0 < h < h_0, \ ih \in [0, \tau].$$

This nonnegativity condition is important for validation of the model, since the parameters represent quantities such as density, which loose physical meaning for negative values. It is then shown that under suitable conditions u_i^h converges to the solution to (IM) as $h \to 0$ and $ih \to t$.

Theorem *For any $v \in D$ there exists a strong generalized solution $u(\cdot)$ to (HIV). Moreover, for v having component functions taking nonnegative values almost everywhere, the strong solution $u(\cdot)$ is also nonnegative in the same sense, over all time t.*

3   Numerical simulation

The primary purpose of the numerical simulation is to analyze changes in time of the levels of virus and rates of disease progression, the rates of viral turnover, the relationship between immune system activation and viral replication, and the time to development of drug resistance. The $CD4^+T$ cell count is understood to be the best indicator of the state of immunologic competence of the patient with HIV infection. Our HIV infection model encompass a spectrum ranging from an acute syndrome associated with primary infection to a prolonged asymptomatic state to advanced disease. An attempt has been made to simulate the following four stages of the HIV infection processes: The early and later stages of HIV infection, the median and most crucial stage after antiretroviral therapy has started, and states in the case that treatment stops. More detailed and other important physiological aspects will be taken into account in the subsequent studies.

Figure 1 illustrates the primary stage after infection has been established. With no treatment, infection is taking place. It is observed that treatment-sensitive cells are more prolific than resistant cells. Figure 2 shows the later stage of infection. Treatment-resistant cells are still not as prolific as sensitive cells. This eventually evolves into a stable state. As mentioned in Section 2, it is inferred that mutation of sensitive virus occurs in this stage after the establishment of chronic infection. Figure 3 depicts a state after antiretroviral therapy has started. A stable state is quickly reached, in which the resistant infected cells appears to be extremely dominant. The treatment seems to be effective in the sense that it results in the almost complete destruction of sensitive cells, but the resistant cells remain without competition with the immune system. Figure 4 suggests a state after treatment has been stopped. This result is obtained by means of the same model under the assumption that treatment stops at a specific time. Once treatment stops, resistant cells are still present and motivate rapid increase of sensitive cells. Thus it is likely to reach a state in which both types of cells coexist, and this may cause further treatment to be less effective. These results of our numerical simulations agree with medical aspects. A critical question is when to

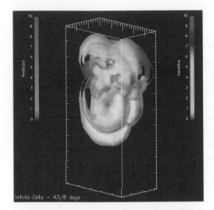

Figure 1: Infection without treatment

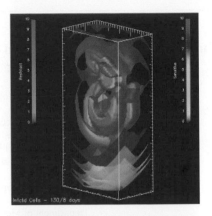

Figure 2: Later stage of infection

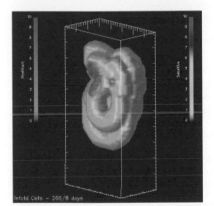

Figure 3: After antiretroviral therapy starts

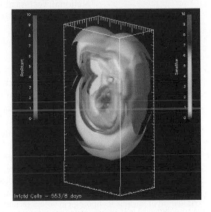

Figure 4: Treatment stops

start antiretroviral therapy in asymptomatic patients who are infected with the HIV virus [1] and our mathematical approach is expected to help in decision-making about this question.

It is best to regard HIV disease as beginning at the time of primary infection and progressing through various stages. Active virus replication and progressive immunologic impairment occur throughout the course of HIV infection in most patients. Our next plan is to develop a new extended model such that all of these aspects are taken into account.

References

[1] Fauci et al. in: E Braunwald, et al., Harrison's Principles of Internal Medicine, 15th Edition, McGraw-Hill, New York, pp. 1852-1913.

[2] D. E. Kirschner and G. F. Webb, Resistance, Remission, and Qualitative Differences in HIV Chemotherapy, Emerging Infectious Diseases, Vol.3, No.3, July-September, 1997, pp.273–283.

[3] D. Tebbs, On the product formula approach to a class of quasi-linear evolution equations, Tokyo J. Math., to appear.

VSP International
Science Publishers
P.O. Box 346, 3700 AH Zeist
The Netherlands

Lecture Series on Computer
and Computational Sciences
Volume 1, 2004, pp. 414-418

Multi Level Decision Support in a Soccer Ticket Club Call Centre with the Use of Simulation

Panayiotou N.A. [1]

Department of Mechanical Engineering,
Sector of Industrial Management and Operational Research,
National Technical University of Athens,
15780 Zografos, Athens, Greece

Evangelopoulos N.P. [2]

Department of Mechanical Engineering,
Sector of Industrial Management and Operational Research,
National Technical University of Athens,
15780 Zografos, Athens, Greece

Ponis S.T.[3]

Department of Mechanical Engineering,
Sector of Industrial Management and Operational Research,
National Technical University of Athens,
15780 Zografos, Athens, Greece

Received 6 August, 2004; accepted in revised form 21 August, 2004

Abstract: This paper presents the results of a study carried out in the call centre of a ticket-selling company. The general purpose of this project was the improvement of the service quality in an environment of highly variable demand. Simulation was initially applied to analyze the present operation of the call centre and extract conclusions concerning its performance. Thereinafter, it was used as a support tool for the evaluation of various technological solutions and for the redesign of its organizational structure. In addition, the staffing needs, the trunk lines' number and the agents' schedules, concerning the following year's anticipations, were estimated by simulation.

Keywords: Call Centre, Simulation, Decision Support, Case Study

Mathematics Subject Classification: 00A72, 37M05, 68U20

1. Introduction

Customer service using the phone has become a very important issue during the last few years. According to Gartner Group [1], more than 70% of business transactions take place over the telephone. As a result, the presence of call centers in the economic life of our society is indispensable for making business. A call centre is defined as any group of employees whose principal business is talking on the telephone to customers or prospects [2] The decisions involved in the successful operation of a call centre are both strategic and tactical. The most important and difficult decision that has to be made by the Top Management is the level of customer service that will be offered compared with the necessary resources, both human and technological, to achieve this level of service. The high cost of the trained

---

[1] Corresponding author. Lecturer at the National Technical University of Athens. E-mail: panayiot@central.ntua.gr

[2] Corresponding author. Mechanical Engineer, NTUA, Greece. E-mail: nevang@softeng.org

[3] Corresponding author. Research Engineer at the National Technical University of Athens. E-mail: staponis@central.ntua.gr

call centre workforce, which can amount up to 65% of the overall budget [1] makes the decision even more difficult.

On the other hand, the Middle Level Management has to utilize the available resources in the best possible way, no matter how complicated and dynamic the operation of the call centre may be, in order to satisfy the strategic constraints imposed by the Top Management. Therefore, the success or failure of a company's customer relationships depends on whether the call center management team coordinates all the resources and the technologies available in service to the company's strategic goals [3].

Call centers have relied historically, on Erlang-C based estimation formulas to help determine the number of agent positions and queue parameters [4]. These estimators have worked fairly well in traditional call centers, however recent trends such as skill-based routing, electronic channels and interactive call handling demand more sophisticated techniques [5]. Discrete event simulation provides the necessary techniques to gain insight into these new trends, and helping to shape their current and future designs [6]. Discrete Event Simulation is a proven methodology that allows to incorporate all of the activities, resources, business rules, workload, assumptions, and other characteristics of a process into one model, and to test the impact of changes in assumptions or other elements on the behavior and performance of the process. Virtually any performance criterion can be examined with simulation [7].

This paper presents a case study demonstrating the use of discrete event simulation in a call center of a company operating in the market of athletic tickets, aiming at the improvement of Management's decision making in both the strategic and tactical levels. The peculiarities of the sector are discussed and specific propositions are made for the improvement of decision making with the use of a simulation-based software tool.

2. The Case Study

The company under study is a small-sized unit operating in the market of athletic tickets and specifically in the soccer market. Although it is a common application in the USA and in many European Countries, in Greece this is the first venture. The field of selling soccer tickets via call centers and the internet is a growing market and especially after the recent success of the Greek national soccer team, the perspectives are extremely promising.

The company collaborates with major Greek soccer clubs and administrates part of and occasionally the whole number of their available tickets. Its main target is to offer high quality services in order to attract social groups who typically abstain from such events. For that purpose, the first step of the company was the development of a web-based application which enables its customers to buy tickets via the internet, and to its agents to sell tickets via a call centre and a sales branch. Consequently the organizational framework of the company was built around the information system, forcing the service processes in the call centre to become of secondary priority.

The ticket market is characterized by the extravagant variation of the demand which is depended on several factors, sometimes visible but others not. Every soccer match acts as a trigger event causing customers to contact the three service points (the call centre, internet & sales branch). The volume of the contacts, the distribution throughout a period of time or among the three service points varies, depending on factors such as the popularity of the event, the general performance of the soccer club, the weather etc. Having to deal with a plethora of different events, some of them scheduled and others not, the demand figures over a year show various periods of average and high demand and even periods of almost no demand. Trying to provide a high level of service under these circumstances, without launching the operational cost, is a tricky and challenging venture.

3. Developing a Flexible Model

The methodological approach which was applied during the project consisted of three major steps:
- Processes analysis and data gathering
- Modeling and verification
- Simulations and output analysis

The processes were recorded along with observations, after conducting interviews with the managers, the call centre supervisor and the agents. The data collection for the call centre was relatively easy as there was a call recorder system and also a database including the customers' requests. On the other hand there was inadequate data concerning the branch and the internet; therefore, they were excluded from the simulation study. During this phase it was revealed that the design of the call centre was far from this of the usual typical call centre. No type of automation was installed apart from the computers and the phones. Furthermore, the number of trunk lines was equal to the number of vacancies, leaving occasionally nil queue space. Some other facts which had a serious impact on the modeling process are listed below:

- The same employees had to work in the call centre and the sales branch which are placed in distant locations. No changes could be made during the day
- The types of customers' requests were: Information, Member Registration, Ticket Purchase and Complaints which had different process times
- In case there was a ticket sale the agents had to prepare it for dispatch with a courier. That process had a low priority and it was interrupted in cases of incoming calls

Passing on to the next phase the goal was to develop a model representing the initial operation of the company, validating it and exporting tangible conclusions concerning the service quality. Owing to the particularities of our case and in order to persuade the management about the efficiency of the simulation study easily, we decided to use detailed modeling techniques in combination with eye catching animation.

From the testing runs of the model the suspicions about the large number of balked customers, due to the lack of queue space, were confirmed. The phenomenon was intense causing the number of customers who were serviced by the clerks to be notably lower than the number of those who attempted to call. In reality the balked customers will probably recall depending on how much they desire to buy a ticket. But from the scope of quality that fact is impermissible, especially in the case of phone sales. Attempting to model the recall process difficulties were encountered in the estimation of factors such as the average waiting time for recalling and the customers' "tolerance", which is the average maximum number of attempts customers would make. A series of simulations was executed for a range of the above factors and a typical sample of our conclusions is shown in Figure 1.

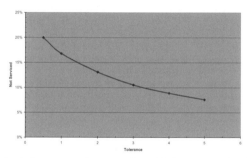

Figure 1: Percentage of aborted customers in connection with the customer's tolerance

As the typical performance measurements (i.e. the percentage of customers reaching an agent within a specific time) couldn't be applied due to the lack of queue space, the recall study was proven a useful tool to persuade the management for organizational changes.

4. Strategic Decisions

After the recognition of the quality problems in the operation of the call centre, simulation was used to plan the next strategic movements of the company, and specifically to compare technological solutions. The key point for the success of such studies is "forecast". The management had strong reasons to believe that the demand would be increased, nevertheless uncertainness was still intense. Therefore, all the solutions which would be examined aimed to increased flexibility and a limited pay back period.

The major operational cost in call centers, estimated up to 65% of the total cost, is related to the personnel. After the essential changes, the initial simulation model was used to compare the staffing needs for three scenarios. The first one was the installation of a single call waiting system. All types of calls would still be handled by the agents. The second one suggested the installation of a VoIP (Voice over IP) IVR (Interactive Voice Response) which would handle part of the calls requiring information. The third scenario consisted of the connection of the above IVR system to the information system of the company, giving the customers the possibility to service themselves. For each case quality measurements, such as the ASA (Average Speed of Answer) and the percentage of the aborted calls, were used as criteria. The scenarios were tested for several volumes of calls and the conclusions are listed below:

- For similar levels of demand compared to those of the previous year, the single IVR solution requires fewer agents than the call waiting system and is a viable investment. The pay back period is estimated to be less than two years
- The second IVR solution is strongly dependent on the customers' acceptance of the automated services. Generally, for it to be viable it requires a significantly increased volume of calls

In addition, sensitivity analysis was conducted to verify that changes on variables such as the acceptance of recorded information didn't have an impact on the above results. Therefore, the second solution was considered more suitable for the company enabling it to gradually set up a connection with the information system, if it was justified by the number of calls.

5. Tactical Decisions

Tactical decisions are closely related to strategic decisions. Having chosen which technological investment would be utilized, simulation was applied to estimate staff size in different occasions through an horizon of a year, the number of trunk lines as well as the schedules of agents.

For the estimation of the staff size, the fact that the management expected an increase in demand of about 50% was taken into account. Under the given circumstances it was concluded that the available at present staff, with the aid of the IVR, would be adequate to offer high level service to customers. On peak days, the above staff could service the increased volume of calls but the percentage of the reneged customers would increase. Assuming that dissatisfied customers recalled, 80% would be serviced only after one attempt. Concerning the trunk lines and given the fact that they are an operational cost, their number was estimated in order to eliminate the possibilities of balked customers.

During the previous year all agents had the same schedule. An alternative schedule was tested in which it was assumed that half the staff would start two hours later every morning. Having no data concerning demand during the two hours after the working schedule, a hypothesis based on the tendency of in-day seasonality was used. The conclusion was that using a rolling schedule wouldn't have any impact on the service level while the operational hours of the call centre would increase along with the volume of the serviced customers.

6. Conclusions and Further Research

Discrete event simulation proved to be an important support tool for different level decision making in the first call centre of soccer ticket selling. Having only the essential statistical data available, simulation is a reliable means to extract valuable conclusions about the call centre operation and its service level. Ordinary call centre measurements, such as agents utilization, the average speed of answer and the abandonment rates are evaluated In addition, several technological investments, along with alternative operational scenarios, are easily tested by simulation, detecting simultaneously the potential bottlenecks and the critical processes of the operation.

Furthermore, the existent model, in combination with forecast application, can evolve into an operational decision support tool. Decisions such as the agents' days off and their extra working hours or the necessity of hiring seasonal staff can be evaluated by simulation experiments.

The modern business processes, including the operations in call centers, tend to be more and more complicated. A flexible and easily applied tool is required to analyze them thoroughly and simulation proved to be an exceptionally efficient solution.

References

[1] Gartner Group, *The Corporate Call Center; Much More Than Call Handling,*1996.

[2] Mehrotra, V., *Ringing Up Big Business*. OR/MS Today, Vol. 24, No. 4, August 1997, pp. 18 – 24, 1997.

[3] Dawson, K., *The Call Center Handbook. The Complete Guide to Starting, Running and Improving Your Call Center*. CMP Books, 3rd Edition, Gilroy, CA, 2001.

[4] Bodin M., Dawson K., *Call Center Handbook*. New York: Flatiron Publishing, Inc., 1996.

[5] Cleveland B., Mayben J., *Call Center Management – On Fast Forward*, Maryland, Call Center Press, 1997.

[6] Tanir O., *Modelling Complex Computer and Communication Systems* – A Domain-Oriented Design Framework, New York, McGraw-Hill, 1997.

[7] Sadowski, R.P., Shannon, R.E., Pedgen,D.P., *Introduction to Simulation: Using SIMAN*, Mc Graw Hill, 1990.

VSP International
Science Publishers
P.O. Box 346, 3700 AH Zeist
The Netherlands

*Lecture Series on Computer
and Computational Sciences*
Volume 1, 2004, pp. 419-423

Flow Analysis of Flush Type Intake Duct of Waterjet

W. G. Park[1]

School of Mechanical Engineering
Pusan National University
Pusan 609-735, Korea

H.H. Chun[2], M.C. Kim[2]

Department of Naval Architecture and Ocean Engineering,
Pusan National University

Received 5 August, 2004; accepted in revised form 20 August, 2004

Abstract: Waterjet propulsion system is widely used to thrust high speed marine vessels in excess of 30-35 knots by virtue of high propulsive efficiency, good maneuverability, and less vibration. Since 7-9% of the total power is approximately lost in intake duct due to flow separation, nonuniformity, etc., detail understanding of flow phenomena occurring within intake duct is essential to reduce the power loss. The present work solved 3-D incompressible RANS equations on multiblocked grid system of the flush type intake duct of waterjet. The numerical results of pressure distributions, velocity vectors, and streamlines were compared with experiments and good agreements were obtained for three jet velocity ratios.
Keywords: RANS equations; Waterjet; Intake duct; Flush type; Boundary layer ingestion

1. Introduction

Waterjet propulsion system is widely used to thrust high speed marine vessels in excess of 30-35 knots by virtue of high propulsive efficiency, good maneuverability, and less vibration. Also, the cavitation can be delayed, or reduced by increasing the static pressure of impeller face through diffusing the cross-sectional area of intake duct of waterjet. Besides application to the high speed ferries, waterjet is installed in military amphibian vehicle to cross the river or land the shore. From the aspect of power loss, approximately 7-9% of the total power is lost in intake duct due to local flow separation, nonuniformity, etc. Thus, detail understanding of flow phenomena occurring within the intake duct is essential to reduce the power loss, as well as noise and vibration. The type of waterjet intake is typically classified into two types: ram (also called as pod or strut) and flush intake. Ram intake is used on hydrofoil crafts. Flush intake is widely used on monohulls, planning crafts, and catamarans. Although a lot of information about the flush intake exists, most of them are confined to manufacturers and are not opened to public domain. The objective of present work is to gain detail flow information of the flush type of waterjet intake duct by solving incompressible RANS equations. The present calculation is also compared with experiments.

2. Governing equations and numerical method

The 3D incompressible RANS equations in a curvilinear coordinate system may be written as:

$$\frac{\partial \hat{q}}{\partial t} + \frac{\partial}{\partial \xi}\left(\hat{E} - \hat{E}_v\right) + \frac{\partial}{\partial \eta}\left(\hat{F} - \hat{F}_v\right) + \frac{\partial}{\partial \zeta}\left(\hat{G} - \hat{G}_v\right) = 0 \tag{1}$$

where $\hat{q} = [0, u, v, w]/J$. $\hat{E}$, $\hat{F}$, and $\hat{G}$ are the convective flux terms. $\hat{E}_v$, $\hat{F}_v$, and $\hat{G}_v$ are viscous flux terms. Eq.(1) is solved by so called "Iterative Time Marching Method"[1-3]. In this method, the

---

[1] Corresponding author. Pusan National University. E-mail: wgpark@pusan.ac.kr
[2] Coauthors. Pusan National University. E-mail: chunahh@pusan.ac.kr , kmcprop@pusan.ac.kr

continuity equation is solved by MAC method and the momentum equation is solved by time marching scheme. The spatial derivatives of convective flux terms are differenced with QUICK scheme. To capture the turbulent flows, low Reynolds number k-ε model[4] was implemented. The initial condition is set by solving equation of fully developed duct flow at each cross-sectional plane. At boundary of nozzle exit, the pressure is obtained from $p_{i\,max} = p_{i\,max-1} + \Delta x \left(\partial p/\partial x\right)_{i\,max-1}$. The velocity is extrapolated from interior nodes and, then, weighted by a factor to satisfy the mass conservation. On the body surface, the no slip condition is applied for velocity components. The surface pressure is determined by setting the zero normal pressure gradient of pressure. The turbulent velocity profile in wind tunnel was given by the 1/7th-power law.

4. Results and discussion

The present method has been applied to the flow within the intake duct of waterjet, as shown in Figure 1 by carrying out three computations of JVR = 6, 7, and 8. Here, Jet Velocity Ratio(JVR) is defined as JVR=V_j/V_∞, where Vj means jet velocity at nozzle exit and V_∞ means vessel velocity. Figure 2 and 3 show surface pressure distribution on ramp and lip side, compared with experiment[5]. The digits of Figure 2(a) and 3(a) mean the locations of pressure tab. The pressure coefficient is defined as: $c_p = p - p_{IN}/\left(\rho V_{IN}^2/2\right)$. The subscript "IN" denotes the value at inlet of the duct. These figures show well agreements with the present computation and experiment. In common with these figures, pressure rapidly increases from tab number 35 and 16, respectively. This is due to the bottleneck effect of the converged nozzle area. This pressure increment becomes severe as JVR increases. This tendency is deserved because high JVR means large mass flow rate through the duct, i.e., high bottleneck effect. Figure 4 shows velocity vectors and streamlines at cross-section of inlet plane at JVR=8, compared with PIV measurement[16]. In these figures, the flow is strongly swallowed through the inlet of the duct, even along the both sides of the inlet. The vortex induced by the separation along the corner of the side wall of inlet is also clearly shown. This flow feature has the same tendency of the flow sketch by Allison[6] in the case of JVR>1.0. From Figure 4(c), the present work gives the ingestion streamtube width around 2 times the physical width of duct inlet. Figure 5(a) shows two locations, "frame 1 and 2", of PIV measurement in vertical plane. Figure 5(b) shows the streamlines in the symmetry plane at JVR=8. Especially, Figure 5(b) shows the stagnation point on the lip side. The position of the stagnation point (or, stagnation line in Figure 4(c)) is an important factor to compute the mass flow rate and the performance of intake duct, because the only upper flow region of the streamline A-A connecting with the stagnation point enters into the intake duct. Therefore, the location of the stagnation point directly affects the mass flow rate of the waterjet. Figure 6 and 7 show streamlines in frame 1 and 2, compared with experiments at JVR=6, 7, and 8. All figures of Figure 6 and 7 give well agreements with the present calculation and experiment. Figure 8 shows the pressure contours and streamlines at JVR=7. Since the flow fields of other JVR are very similar to those of Figure 8, figures of JVR=6 and 8 are omitted here.

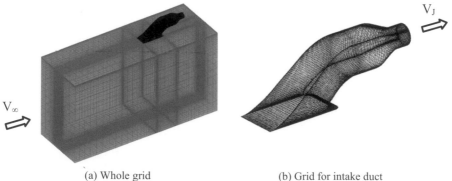

(a) Whole grid   (b) Grid for intake duct

Figure1: Grid system

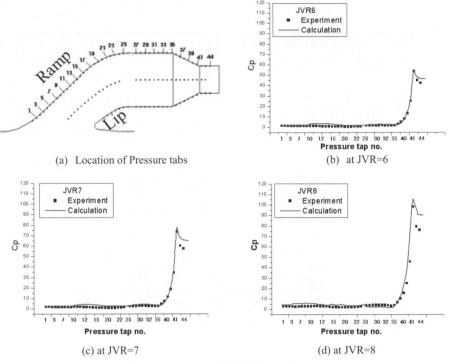

(a) Location of Pressure tabs

(b) at JVR=6

(c) at JVR=7

(d) at JVR=8

Figure 2: Pressure distribution on the ramp

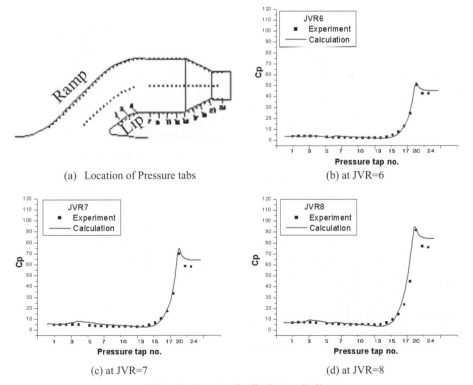

(a) Location of Pressure tabs

(b) at JVR=6

(c) at JVR=7

(d) at JVR=8

Figure 3: Pressure distribution on the lip

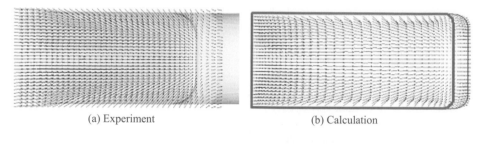

(a) Experiment (b) Calculation

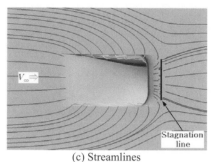

(c) Streamlines
Figure 4: Velocity vectors and streamlines at inlet (JVR=8)

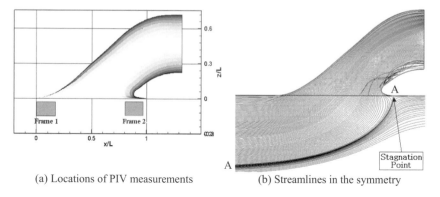

(a) Locations of PIV measurements (b) Streamlines in the symmetry

Figure 5: Location of PIV experimentation

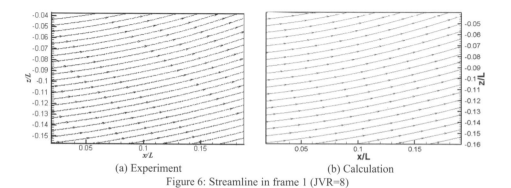

(a) Experiment (b) Calculation
Figure 6: Streamline in frame 1 (JVR=8)

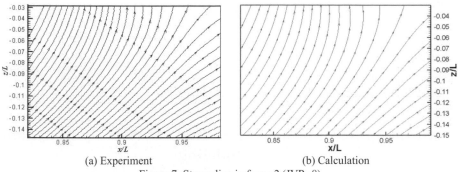

(a) Experiment   (b) Calculation
Figure 7: Streamline in frame 2 (JVR=8)

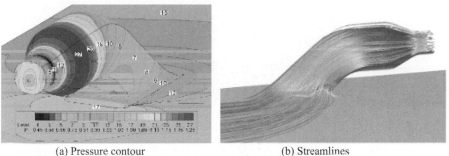

(a) Pressure contour   (b) Streamlines
Figure 8: Pressure contours and streamlines at JVR=7

5. Conclusions

The numerical flow analysis was performed to provide a detail understanding of complicated flow phenomena of an intake duct of waterjet. The 3-D incompressible RANS equations were successfully solved on multiblocked grid system for JVR=6, 7, and 8. The present calculations were compared with experimental data of surface pressure distribution and PIV measurements and attained good agreements with experiments for all values of JVR. From this calculation, strong suction flow through the inlet of the intake duct was shown and the vortex induced by the separation along the corner of the side wall of inlet was also clearly shown. The location of stagnation point on the lip side, an important factor of the performance of intake duct, was well predicted by the present calculation.

References

[1] Park, W. G. and Sankar, L. N., *An iterative time marching procedure for unsteady viscous flows*, ASME-FED **20** (1993)

[2] Park, W. G., Jung, Y. R., and Ha, S. D., *Numerical viscous flow analysis around a high speed train with crosswind effects*, AIAA Journal. **36** 477-479 (1998)

[3] Park, W. G., Jung, Y. R., and Kim, C. K., *Numerical flow analysis of single-stage ducted marine propulsor*, To Be Published, Ocean Engineering (2004)

[4] Chien, K.Y., *Prediction of channel and boundary-layer flows with a low-reynolds number turbulence model*, AIAA Journal, **20** (1982)

[5] Kim, K.C. *Experimentation of waterjet intake duct flow*, PNU Report (in Korean) (2003)

[6] Allison J., *Marine Waterjet Propulsion*, SNAME Transactions, **101** 275-335 (1993)

VSP International
Science Publishers
P.O. Box 346, 3700 AH Zeist
The Netherlands

*Lecture Series on Computer
and Computational Sciences*
Volume 1, 2004, pp. 424-427

An Integrated Mathematical Tool. Part-I: Hydrodynamic Modeling

G. Petihakis[a,b], G. Triantafyllou[a1], G. Korres[a] and A. Theodorou[b]

[a] Hellenic Center for Marine Research, Institute of Oceanography,
PO BOX 712, 19013, Anavyssos, Greece
[b] University of Thessaly, Dept of Agriculture and Water Environment,
Fytoko, Nea Ionia Magnisias (38446), Greece

Received 8 September, 2004; accepted in revised form 20 September, 2004

Abstract: Numerical ecosystem models, as operational tools, may directly apply to problems of environmental management, predicting the response of natural systems to perturbations and modifications of various kinds. Since dynamics are described in some detail, the model may provide insight into the governing mechanisms influencing the functioning of the ecosystem described in a companion paper [1]. Pagasitikos is a semi-enclosed gulf situated on the western part of Aegean Sea north of the island of Evia connected at the south with the Aegean Sea through the 5.5 km wide, narrow channel of Trikeri. The predominant weak winds of the area result in small to moderate water currents while renewal occurs mainly through the deep-water layer of the communication channel with Aegean Sea. During winter months the water mass of Pagasitikos is fairly mixed, forming a two-layer thermocline for the rest of the year with the exception of August when three layers are observed. Inflows of fresh waters in the areas of Volos and Almiros are also observed during winter and spring adding to the complexity of the system. In this work the capability of a high-resolution hydrodynamic model based on Princeton Ocean Model (POM) to describe the phenomenology of Pagasitikos is investigated. Comparison with direct current measurements show good agreement indicating that the model can successfully reproduce the general circulation characteristics of the area aiding in the understanding of the complicated hydrological structure and circulation patterns appearing within the area. In addition the effect of high frequency wind forcing is explored through the comparison with surface forcing specified from monthly mean climatological data.

Keywords: Ocean Modeling, Hydrodynamics, Water masses, Circulation

1 Introduction

Understanding the dynamics of marine ecosystem is a fundamental issue for the exploration, management and protection of coastal areas. Although the study of the underlying processes is rather challenging for the scientific community this often proves to be a difficult task due to the significant complexity of the system. Especially coastal areas are characterized by a high degree of complexity requiring intensive long-term studies capable to reveal relations between the numerous variables and parameters.

Pagasitikos is a semi-enclosed system in the western Aegean Sea north of Evia Island surrounded by high mountains draining their rainwater into the gulf. It is a shallow gulf with a mean depth

---

[1]Corresponding author. E-mail: *gt@ath.hcmr.gr*

of 69 m with the deepest area at the eastern - central part (108 m). The total area is $520 km^2$ and the total volume $36 km^3$. An important feature is the connection with the Aegean Sea and north Evoikos through the narrow (3 - 10 km) and relatively deep (80 m) Trikeri channel modulating the hydrology patterns of the gulf. The microclimate is typical of the Mediterranean basin with two major wind groups, although are particularly weak resulting in the formation of prolonged thermocline. The etesian blow from July to September with a North - West direction exhibiting maximum values during afternoon and minimum during night. The second group is the Southerly warm and dry winds. The mean annual air temperature is $16.5°C$ with maximum $31.0°C$ in July and minimum $11.0°C$ in January.

In this work the phenomenology of Pagasitikos is described using the data from the latest research project Development of an Integrated Policy or the sustainable Management of Pagasitikos Gulf and the capability of a high-resolution hydrodynamic model based on Princeton Ocean Model (POM) to describe the phenomenology of Pagasitikos gulf is investigated. In addition the hydrodynamic model results are used to provide the physical background for an ecosystem model implemented in the same area developing thus an integrated ecosystem operational tool for Pagasitikos gulf. The latter work is presented in a separate scientific article.

2 Materials and Methods

2.1 Model description

The hydrodynamic model (POM) is a primitive equations finite difference model which makes use of the hydrostatic and Boussinesq approximations. The model solves the 3-D Navier-Stokes equations on an Arakawa-C grid with a numerical scheme that conserves mass and energy. The spatial differences are central and explicit in the horizontal and central and implicit in the vertical. Centered differences are used for the time integration (leapfrog scheme) of the primitive equations. In addition, since the leapfrog scheme has a tendency for the solution to split at odd and even time steps, an Asselin filter is used at every time step. The numerical computation is split into an external barotropic mode with a short time step (dictated by the CFL condition) solving for the time evolution of the free surface elevation and the depth averaged velocities, and an internal baroclinic mode which solves for the vertical velocity shear. Horizontal mixing in the model is parameterized according to [2] while vertical mixing is calculated through the Mellor and Yamada 2.5 turbulence closure scheme. The reader is referred to [3] for a detailed description of the POM model equations and the numerical algorithm.

The model state vector contains all prognostic (state) variables of the model at each sea grid point. The state variables consist of the sea surface elevation, the zonal and meridional components of velocity, potential temperature, salinity, the turbulent kinetic energy and the turbulent kinetic energy times the turbulent length scale.

The computational domain for Pagasitikos gulf model covers the area between 22.8125E to 23.3025E and 39.0N to 39.43N. It uses a Cartesian coordinate system consisting of 49×45 elements with a horizontal grid resolution of 0.01 degrees both in latitude and longitude. In the vertical there are 25 elements of variable thickness with logarithmic distribution near the surface and the bottom for greater accuracy where velocity gradients are larger.

For the model's bathymetry direct measurements and naval charts were used. A [4] filter of third order was also applied to the interpolated bathymetry in order to perform the necessary smoothing (elimination of small-scale noise).

The model's climatological run was initialized with spring period objectively analysed field hydrological data while, initial velocities were set to zero. The model was integrated using a 'perpetual year' forcing atmospheric data set. This data set was derived from the 1979 - 1993 6-hour European Centre for Medium-Range Weather Forecasts (ECMWF) reanalysis data (horizontal resolution $1° \times 1°$) by proper averaging in time to get the monthly mean values [5]. It consists of the longitudinal and meridional components of the wind stress, air temperature, air humidity, cloud cover, net upward heat flux, evaporative heat flux and solar insolation at the sea surface. Additionally the precipitation data needed for the freshwater budget were taken from Jaeger (1976) monthly data set (horizontal resolution $5° \times 2.5°$.

3   Results and Discussion

Model results indicate that the general circulation of the basin is a combination of thermohaline circulation driven by surface buoyancy fluxes, river discharges and exchanges through the open boundary and wind-driven circulation.

The general circulation is of the basin is an interplay between a pure cyclonic structure due to the buoyancy forcing and an anticyclonic one triggered by the action of the wind forcing. During winter and spring the barotropic circulation involves a dipole of a cyclonic eddy in the western part of the basin and an anticyclone in the eastern part. During spring the anticyclone weakens while the cyclone intensifies. During summer the anticyclonic circulation prevails into the central part of the basin accompanied by a squeezed cyclonic eddy in the western part of the basin. As already discussed in our previous analysis, the energetics of the basin during summer are dictated by the action of the wind stresses at the surface. The anticyclone is completely absent and the rather intense cyclone developed during spring has taken its place. During autumn the barotropic circulation is very similar with the summer one although the anticyclone progressively moves eastwards and disamplifies.

The temperature distribution along the vertical involves a sharp contrast between the surface and the bottom of the basin during spring and summer and a vigorous homogenization during winter where a temperature inversion is evident at the bottom of the basin. During spring and summer a triple layer system develops, consisting of a warm and shallow surface mixed layer, an intermediate layer of intense stratification (10-40m) and a deep cold layer. The salinity vertical distribution shows a contrast of a surface fresh layer on top of a more saline intermediate and bottom layer. This is actually a prominent characteristic of the basin successfully reproduced by the model run. The salinity contrast is more intense during autumn when the bottom layer salinity is almost 38.0 psu. During autumn and winter the halocline is located at approximately 50 m depth while during spring and summer the halocline up rises to 30 m. Surface salinity minima during autumn and winter are located at the western flag of the basin and can be attributed to fresh water discharges at the Almiros area.

Acknowledgment

This work has been supported by the MFSTEP EU research project. The authors would like to thank A. Pollani for her help during the preparation of this work.

References

[1] G. Petihakis, G. Triantafyllou, G. Korres, and A. Theodorou. An integrated mathematical tool. part-ii: Ecosystem modeling. *Lecture Series on Computer and Computational Sciences*, 1:(submitted), 2004.

[2] J. Smagorinsky. General circulation experiments with the primitive equations, i, the basic experiment. *Mon. Weather Rev.*, 91:99–164, 1963.

[3] A. F. Blumberg and G. L. Mellor. A description of a three-dimensional coastal ocean circulation model. In N.S. Heaps, editor, *Three-Dimensional Coastal Ocean Circulation Models*, Coastal Estuarine Science, pages 1–16. American Geophysical Union, Washington, D.C., 4 edition, 1987.

[4] R. Shapiro. Smoothing, filtering, and boundary effects. *Reviews of Geophysics and Space Physics*, 8(2):359–387, 1970.

[5] G. Korres and A. Lascaratos. An eddy resolving model of the aegean and levantine basins for the mediterranean forecasting system pilot project (mfspp): Implementation and climatological runs. *Annales Geophysicae*, 21:205–220, 2003.

VSP International
Science Publishers
P.O. Box 346, 3700 AH Zeist
The Netherlands

Lecture Series on Computer
and Computational Sciences
Volume 1, 2004, pp. 428-431

An Integrated Mathematical Tool. Part-II: Ecological Modeling

G. Petihakis[a,b]**, G. Triantafyllou**[a1]**, G. Korres**[a]** and A. Theodorou**[b]

[a] Hellenic Center for Marine Research, Institute of Oceanography,
PO BOX 712, 19013, Anavyssos, Greece
[b] University of Thessaly, Dept of Agriculture and Water Environment,
Fytoko, Nea Ionia Magnisias (38446), Greece

Received 8 September, 2004; accepted in revised form 20 September, 2004

Abstract: Pagasitikos gulf is a semi-enclosed basin highly influenced both by anthropogenic activities (inflow of nutrients at the north and west parts) as well as by water exchange between the gulf and the Aegean Sea at its south part (Trikeri channel) resulting in the development of functional sub-areas within the gulf. Thus the inner part is characterized by eutrophic conditions with sporadic formation of harmful algal blooms while the central part acts as a buffer with mesotrophic characteristics influenced by the oligotrophic outer area. In a companion paper [1] the circulation fields and the development of water masses in the Pagasitikos gulf were explored. The aim of this study is to investigate the interactions between the physical and biogeochemical systems in the Pagasitikos gulf by coupling advanced hydrodynamic and ecological models. The simulation system comprises of two on-line coupled sub-models: a three-dimensional hydrodynamic model based on Princeton Ocean Model (POM) and an ecological model adapted from the European Regional Seas Ecosystem Model (ERSEM) for the particular ecosystem. A cost function is used for the validation of model results with field data. Simulation results are in good agreement with in-situ data illustrating the role of the physical processes in determining the evolution and variability of the ecosystem.

Keywords: Numerical modeling, Ecosystem modeling

1 Introduction

Pagasitikos is a rather sensitive ecosystem due to its semi-enclosed nature and the shallow depths. The human activity in the coastal areas is not significant with agricultural farming being the major occupation. However during the last years there has been a shift towards intensive production of cereal and cotton with the use of large quantities of fertilizers rich in nitrogen, phosphate and sulphur. A significant proportion of these chemicals find its way into the marine ecosystem carried by rain waters through a network of periodic small torrents. The only major city is Volos at the north part of the gulf with a population of 120,000 inhabitants and a well-developed industrial sector. It was during 60's when the first heavy industries were built attracting workers from the surrounding areas and leading into a population explosion. The fast growth of the area and the absence of the necessary infrastructure caused significant problems to the gulf, as it became the recipient for major quantities of rural, and industrial effluents. A domestic sewage treatment plant for the domestic effluents was planned as early as 1964 but it took 23 years to become operational.

---

[1]Corresponding author. E-mail: *gt@ath.hcmr.gr*

An extreme perturbation occurred in early 60's during the draining program of lake Karla when large quantities of enriched waters with nutrients were channeled into Pagasitikos via an aqua duct at the north of the gulf. Although this lasted few years only the larger part of the lake was given for cultivation and as a result during winter rainwater washes the soil becoming enriched with fertilizers, pesticides and particulate material a proportion of which is finally poured into Pagasitikos. During 1982 dense mucilage composed by phytoplankton cells, bacteria, zooplankton excretions and detritus covered large areas in the north part of the gulf causing significant problems to the fishing community and to tourism. This phenomenon was greatly reduced both in space and time in the following years just to return with grater severity in 1987 the worst year ever recorded.

2 Materials and Methods

2.1 Model description

An essential feature of aquatic ecosystem models is the combination of biology and physics, which cannot completely be separated (migration with currents, sedimentation due to biological actions, sinking of senescent phytoplankton) however, a rough separation is possible. Thus the 3D ecosystem model initially developed for the Cretan Sea [2] consists of two, highly portable, on-line coupled, sub-models: the three-dimensional Princeton Ocean Model (POM) [3] described in part I, [1], and the ecological model. The physical model describes the hydrodynamics of the area and provides the background physical information to the ecological model, which describes the biogeochemical cycles.

The ecosystem model is based on the 1D European Regional Seas Ecosystem Model [4] and is an extension of the 1D model applied to Pagasitikos [5]. It is a generic model that can be applied in a wide range of ecosystems [6], [7], [5] with a significant degree of complexity allowing thus a good representation of all those processes that are significant in the functioning of the system. A coherent system behavior is achieved by considering the system as a series of interacting physical, chemical and biological complex processes [4]. A 'functional' group approach has been adapted for the description of the ecosystem while organisms are separated into groups according to their trophic level and subdivided to similar size classes and feeding methods. The dynamics between the functional groups include physiological (ingestion, respiration, excretion, egestion, etc.) and population processes (growth, migration and mortality) which are described by fluxes of carbon and nutrients. The physical processes affecting the biological constituents are advection and dispersion in the horizontal, and sedimentation and dispersion in the vertical, with the horizontal processes operating on scales of tens of kilometers and the vertical processes on tens of meters [4]. Biologically driven carbon dynamics are coupled to the chemical dynamics of nitrogen, phosphate, silicate and oxygen. The food web is divided into three main groups, the producers, the decomposers and the consumers, each of which may be defined as having a standard set of processes. The first group includes the phytoplankton, which is further divided to functional groups based on size and ecological properties. These are diatoms P1 (silicate consumers, $20 - 200\mu$), nanophytoplankton P2 ($2 - 20\mu$), picophytoplankton P3 ($< 2\mu$) and dinoflagellates P4 ($> 20\mu$).

In the 3D code the following equation is solved for the concentration of C for each functional group of the pelagic system:

$$\frac{\partial C}{\partial t} = -U\frac{\partial C}{\partial x} - V\frac{\partial C}{\partial y} - W\frac{\partial C}{\partial z} + \frac{\partial}{\partial x}\left(A_H\frac{\partial C}{\partial x}\right)$$

$$+\frac{\partial}{\partial y}\left(A_H\frac{\partial C}{\partial y}\right) + \frac{\partial}{\partial z}\left(K_H\frac{\partial C}{\partial z}\right) + \sum BF$$

where U, V, W represent the velocity field, A_H the horizontal viscosity coefficient, and K_H the vertical eddy mixing coefficient, provided by the POM. $\sum BF$ stands for the total biochemical flux, calculated by the model, for each pelagic group. This equation is approximated by a finite-difference scheme analogous to the equations of temperature and salinity [2] and is solved in two time steps [8]: an explicit conservative scheme [9] for the advection, and an implicit one for the vertical diffusion [10]. Due to the heavy computational cost the benthic-pelagic coupling is described by a simple first order benthic returns module, which includes the settling of organic detritus into the benthos and diffusional nutrient fluxes into and out of the sediment.

2.2 Model validation

The validation of a 3D ecosystem model is usually a difficult task due to the scarce in-situ data. A possible solution is the use of a cost function as described by [11]. It is a mathematical function enabling the comparison of model results with field measurements estimating a non-dimensional value which is indicative of how close or how distant two particular values are.

$$C_{x,t} = \frac{M_{x,t} - D_{x,t}}{sd_{x,t}}$$

where, $C_{x,t}$ is the normalised deviation between model and data for box x and season t, $M_{x,t}$ the mean value of the model results within box x and season t, $D_{x,t}$ the mean value of the in situ data within box x and season t and $sd_{x,t}$ the standard deviation of the in situ data within box x and season t. The cost function results give an indication of the goodness of fit of the model, by providing a quantitative measure of deviation, normalised in units of standard deviation of data. The lower the value of the cost function the better the agreement between model and data. In this work the same categories of cost function results described by [11] ($< 1 = Verygood, 1 - 2 = Good, 2 - 5 = Reasonable, > 5 = Poor$) were used.

3 Results and Discussion

Initially all results from the cost function were grouped (all stations - all seasons - all depths), with 43% falling into the first category (very good), and 23% in the good. Additionally only 11% of the model results are poor, indicating that the model overall simulates very satisfactorily the system.

Considering the expected variability of the system in time and space, model results were also validated for the different seasons and stations. Spring is the season with the most information since there are measurements in all stations and at all depths. The model simulates very well the ecosystem of Pagasitikos with 55% of the model results being very good and only 2% falling in the poor category.

Validation results indicate that the mathematical model adapted into the marine ecosystem of Pagasitikos gulf is a close representation of the real system with only exception the very variable channel area where detailed information on the important processes is necessary.

The particular ecosystem model should be used as a useful tool producing significant information and knowledge on the dynamics and functioning of Pagasitikos gulf under both natural conditions as well as perturbations. It should be used for the exploitation of the response into reasonable changes in parameters or processes which are included in the structure. Such simulations are anyway performed during the sensitivity testing and can provide under appropriate management very useful information. Of course in order to forecast the evolution of the ecosystem after disturbance (e.g. inputs of enriched waters), a monitoring system should be established providing frequent detailed information on the major ecosystem variables. The Pagasitikos nowcasting - forecasting ecosystem model should be the goal of a future operational forecasting system. To achieve the goal a numerical tool was developed. This work can be viewed as a first step towards establishing a system for understanding the dynamics of the ecosystem which, in a second step, through the development of data assimilation techniques, will lead to operational applications. Work in this area is in progress and it will be reported later during the project *Data Integration System for Eutrophication Assessment in Coastal Waters* (InSEA), recently funded by the EU.

Acknowledgment

This work has been supported by the MFSTEP EU research project. The authors wish to thank A. Pollani for her help during the preparation of this work.

References

[1] G. Petihakis, G. Triantafyllou, G. Korres, and A. Theodorou. An integrated mathematical tool. part-i: Hydrodynamic modeling. *Lecture Series on Computer and Computational Sciences*, 1:(submitted), 2004.

[2] G. Petihakis, G. Triantafyllou, J.I. Allen, I. Hoteit, and C. Dounas. Modelling the spatial and temporal variability of the cretan sea ecosystem. *Journal of Marine Systems*, 36(3-4):173–196, 2002.

[3] A. F. Blumberg and G. L. Mellor. A description of a three-dimensional coastal ocean circulation model. In N.S. Heaps, editor, *Three-Dimensional Coastal Ocean Circulation Models*, Coastal Estuarine Science, pages 1–16. American Geophysical Union, Washington, D.C., 4 edition, 1987.

[4] J.W. Baretta, W. Ebenhoh, and P. Ruardij. The european regional seas ecosystem model, a complex marine ecosystem model. *Netherlands Journal of Sea Research*, 33:233–246, 1995.

[5] G. Petihakis, G. Triantafyllou, and A. Theodorou. A numerical approach to simulate nutrient dynamics and primary production of a semi-enclosed coastal ecosystem (pagasitikos gulf, western aegean, greece). *Periodicum Biologorum*, 102(1):339–348, 2000.

[6] J.J. Allen. A modelling study of ecosystem dynamics and nutrient cycling in the humber plume, uk. *Journal of Sea Research*, 38:333–359, 1997.

[7] G. Triantafyllou, G. Petihakis, and J.I. Allen. Assessing the performance of the cretan sea ecosystem model with the use of high frequency m3a buoy data set. *Annales Geophysicae*, 21:365–375, 2003.

[8] G. L. Mellor. An equation of state for numerical models of oceans and estuaries. *Journal Atmospheric Oceanic Technology*, 8:609–611, 1991.

[9] H.J. Lin, S.W. Nixon, D.I. Taylor, S.L. Granger, and B.A. Buckley. Responses of epiphytes on eelgrass, zostera marina l., to separate and combined nitrogen and phosphorus enrichment. *Aquatic Botany*, 52:243–258, 1996.

[10] R.D. Richtmyer and K.W. Morton. *Difference methods for initial-value problems.* Krieger Publishng Company, Malabar Florida, 1994.

[11] A. Moll. Assessment of three-dimensional physical-biological ecoham1 simulations by quantified validation for the north sea with ices and ersem data. *ICES Journal of Marine Science*, 57:1060–1068, 2000.

VSP International
Science Publishers
P.O. Box 346, 3700 AH Zeist
The Netherlands

Lecture Series on Computer and Computational Sciences
Volume 1, 2004, pp. 432-435

Numerical Solution of the Hamilton-Jacobi-Bellman Equation in Stochastic Optimal Control with Application of Portfolio Optimization

Helfried Peyrl[1], Florian Herzog[2], and Hans P. Geering[3]

Measurement and Control Laboratory,
Swiss Federal Institute of Technology,
CH-8092 Zürich, Switzerland

Received 4 August, 2004; accepted in revised form 15 August, 2004

Abstract: This paper provides a numerical solution of the Hamilton-Jacobi-Bellman (HJB) equation for stochastic optimal control problems. The computation's difficulty is due to the nature of the HJB equation being a second-order partial differential equation which is coupled with an optimization. By using a successive approximation algorithm, the optimization gets separated from the boundary value problem. This makes the problem solvable by standard numerical methods. The method's usefulness is shown in an example of portfolio optimization with no known analytical solution.

Keywords: Stochastic Optimal Control; Computational Methods; Computational Finance

1 Introduction

A necessary condition for a solution of stochastic optimal control problems is the *Hamilton-Jacobi-Bellman* (HJB) equation, a second-order partial differential equation that is coupled with an optimization. Unfortunately, the HJB equation is difficult to solve analytically and only for some special cases analytical solutions are known, e.g. the LQ regulator problem.

2 Problem Formulation

Consider the stochastic process $x \in \mathbb{R}^n$ which is governed by the stochastic differential equation

$$dx = f(t, x, u)dt + g(t, x, u)dZ, \tag{1}$$

where dZ denotes k-dimensional uncorrelated standard Brownian motion. The vector u denotes the control variables contained in some compact, convex set $U \subset \mathbb{R}^m$.

Our problem starts at arbitrary time $t \in (0, T)$ and state $x \in G$, where G is an open and bounded subset of $\mathbb{R}^n$.

The final time of our problem denoted by τ is the time when the solution $(t, x(t))$ leaves the open set $Q = (0, T) \times G$:

$$\tau = \inf \{ s \geq t \mid (s, x(s)) \notin Q \}.$$

---

[1] now at Automatic Control Laboratory, Swiss Federal Institute of Technology, E-mail: peyrl@control.ee.ethz.ch
[2] E-mail: herzog@imrt.mavt.ethz.ch
[3] E-mail: geering@imrt.mavt.ethz.ch

Our aim is now to find the admissible feedback control law u which solves the following stochastic optimal control problem leading to the *cost-to-go* (or *value*) function $J(t, x)$:

$$J(t, x) = \max_{u(t,x) \in U} \mathbb{E} \left\{ \int_t^\tau L(s, x, u) ds + K(\tau, x(\tau)) \right\}$$

s.t.

$$dx = f(t, x, u) dt + g(t, x, u) dZ,$$

(2)

where $\mathbb{E}$ denotes the expectation operator and L, K are scalar functions.

The HJB equation turns out to be a necessary condition for the cost-to-go function $J(t, x)$. For a detailed derivation, the reader is referred to [1, 2], or [3]. Using the differential operator $\mathcal{A}(t, x, u) = \frac{1}{2} \sum_{i,j=1}^n \sigma_{ij} \frac{\partial^2}{\partial x_i \partial x_j} + \sum_{i=1}^n f_i \frac{\partial}{\partial x_i}$, where the symmetric matrix $\sigma = (\sigma_{ij})$ is defined by $\sigma(t, x, u) = g(t, x, u) g^T(t, x, u)$, the HJB equation can be written as follows:

$$\begin{aligned} J_t + \max_{u \in U} \{ L(t, x, u) + \mathcal{A}(t, x, u) J \} &= 0, & (t, x) \in Q \\ J(t, x) &= K(t, x), & (t, x) \in \partial^* Q, \end{aligned}$$

(3)

where $\partial^* Q$ denotes a closed subset of the boundary ∂Q such that $(\tau, x(\tau)) \in \partial^* Q$ with probability one:

$$\partial^* Q = ([0, T] \times \partial G) \cup (\{T\} \times G).$$

The HJB equation (3) is a scalar linear second-order PDE which is coupled with an optimization over u. This makes solving the problem so difficult (apart from computational issues arising from problem sizes in higher dimensions).

3 Successive Approximation of the HJB Equation

In this section we will reveal a numerical approach for solving the HJB equation. Solving the PDE and optimization problem at once would lead to unaffordable computational costs. Chang and Krishna propose a successive approximation algorithm which will be used in the following [4]:

1. $k = 0$; choose an arbitrary initial control law $u^0 \in U$.

2. Solve following boundary value problem to compute the problem's value $J^k(t, x)$ for the fixed control law u^k:

$$\begin{aligned} J_t^k + L(t, x, u^k) + \mathcal{A}(t, x, u^k) J^k &= 0, & (t, x) \in Q, \\ J^k(t, x) &= K(t, x), & (t, x) \in \partial^* Q. \end{aligned}$$

(4)

3. Compute the succeeding control law u^{k+1} by solving the optimization problem

$$u^{k+1} = \arg \max_{u \in U} \left\{ L(t, x, u) + \mathcal{A}(t, x, u) J^k \right\}.$$

(5)

4. $k = k + 1$; back to step 2.

To proof convergence of above algorithm following lemmas and theorems are used:

Lemma 1 *Let J^k be the solution of the boundary value problem corresponding to the arbitrary but fixed control law $u^k \in U$:*

$$\begin{aligned} J_t^k + L(t, x, u^k) + \mathcal{A}(t, x, u^k) J^k &= 0, & (t, x) \in Q \\ J^k(t, x) &= K(t, x), & (t, x) \in \partial^* Q. \end{aligned}$$

(6)

Then

$$J^k(t,x) = \mathcal{J}(t,x,u^k),\tag{7}$$

where $\mathcal{J}(t,x,u^k)$ denotes the value of our problem for a fixed control law u^k.

Lemma 2 *Let the sequences of control laws u^k and their affiliated value functionals J^k be generated by the successive approximation algorithm. Then the sequence J^k satisfies*

$$J^{k+1} \geq J^k.\tag{8}$$

Theorem 1 *Let the sequences of control laws u^k and their corresponding value functionals J^k be defined as above. Then they converge to the optimal feedback control law $u(t,x)$ and the value function $J(t,x)$ of our optimal control problem (2), i.e.:*

$$\lim_{k\to\infty} u^k(t,x) = u(t,x) \text{ and } \lim_{k\to\infty} J^k(t,x) = J(t,x).$$

4 Computational Implementation and Issues

4.1 Numerical Solution of the HJB-PDE

Boundary value problem (4) is a scalar second-order PDE with non-constant coefficients and hence can be tackled by standard methods for linear parabolic PDEs. Since the HJB-PDE has a terminal condition $J^k(T,x) = K(T,x)$ rather than an initial condition, it must be integrated backwards in time. We use finite difference schemes as they are both well-suited for simple (rectangular) shaped domains Q and rather easy to implement. Our solver employs an implicit scheme and uses upwind differences for the first order derivatives for stability reasons. Second order and mixed derivatives are approximated by central space differences.

4.2 Optimization

For computing the succeeding control law, the nonlinear optimization problem (5) must be solved. Since we approximate $J^k(t,x)$ on a finite grid, the optimization must be solved for every grid point. This can be accomplished by standard optimization tools. For problems with simple functions, it may be possible to obtain an analytical solution for the optimization step. This is to be preferred in return of less computational time.

4.3 Numerical Issues

Since the number of unknown grid points at which we approximate $J^k(t,x)$ grows by an order of magnitude with dimension (*Bellman's curse of dimensionality*) and grid resolution we have to face issues of memory limitations and computation time and accuracy.

The PDE solvers outlined in Section 4.1 require the solution of large systems of linear equations. The coefficient matrix of these linear systems is banded and therefore strongly encourages the use of sparse matrix techniques to save memory. Furthermore, applying indirect solution methods for linear systems such as *successive overrelaxation* provides higher accuracy and memory efficiency than direct methods.

5 CASE STUDY

The case study presents a portfolio optimization problem in continuous-time with no known analytical solution.

5.1 Portfolio Optimization Problem

We consider a portfolio optimization problem where an investor has the choice of investing in the stock market or to put his money in a bank account. The objective of the investor is to maximize the power utility of his wealth at a finite fixed time horizon T: $\max \frac{1}{\gamma} W^\gamma(T)$. Thus the portfolio optimization problem is

$$\max_{u \in [-1,1]} \frac{1}{\gamma} W^\gamma(T)$$

$$s.t.$$
$$\begin{aligned}
dW(t) &= W(t)(r + u(t)(Fx(t) + f - r))dt \\
&\quad + W(t)u(t)\sqrt{v(t)}dZ_1 \\
dx(t) &= (a_1 + A_1 x(t))dt + \nu dZ_2 \\
dv(t) &= (a_2 + A_2 v(t))dt + \sigma\sqrt{v(t)}dZ_3 ,
\end{aligned} \tag{9}$$

where $\gamma < 1$ is coefficient of risk aversion and $u(t) \in [-1, 1]$. We make the assumption that both of the processes $x(t)$ and $v(t)$ are measurable and we have both of the time series to estimate the model parameters. In Fig. 1 the optimal investment policy as function of $x(t)$ and $v(t)$ is shown.

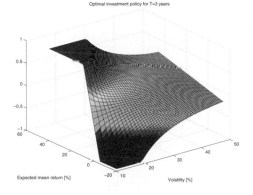

Figure 1: Optimal investment policy

References

[1] Jiongmin Yong and Xun Yu Zhou, *Stochastic Controls, Hamilton Systems and HJB Equations*. Springer Verlag New-York, 1999.

[2] B. Hanzon and H. Schumacher and M. Vellekoop, Finance for Control Engineers. *Tutorial Workshop at the European Control Conference*, Porto, Portugal, 2001.

[3] Wendell H. Fleming and Raymond W. Rishel, *Deterministic and Stochastic Optimal Control*. Springer-Verlag, 1975.

[4] M. H. Chang and K. Krishna, A Successive Approximation Algorithm for Stochastic Control Problems. *Applied Mathematics and Computation*, vol. 18, no. 2, 155–165(1986).

VSP International
Science Publishers
P.O. Box 346, 3700 AH Zeist
The Netherlands

*Lecture Series on Computer
and Computational Sciences*
Volume 1, 2004, pp. 436-439

An Enhanced MEBDF Approach for the
Numerical Solution of Parabolic Partial Differential Equations

G. Psihoyios[1]

Department of Mathematics and Technology,
Anglia Polytechnic University, East Road,
Cambridge CB1 1PT, United Kingdom

Received 21 June 2004; accepted in revised form 12 August 2004

Abstract: In [1] the known MEBDF scheme had been used to solve ordinary differential equations (ODEs) that arise in the method of lines solution of time dependent partial differential equations (PDEs). There, it was shown that MEBDF is superior to the widely used BDF approach in certain important classes of problems. This short paper serves as an introduction to the different logic we have followed in order to improve the stability characteristics of MEBDF. Here wee are also briefly referring to the fact that our new and enhanced approach produced a marked positive effect on both the accuracy and efficiency compared to [1]. Due to space limitations we are unable to present the numerical results that support our claims.

Keywords: Stability, MEBDF, Parabolic partial differential equations, Method of lines.

Mathematics Subject Classification: 65M12, 65L05, 65M20, 65L06, 65l20.

PACS: 02.60.Lj, 44.05.+e

1. Introduction

In [1] the author presented an account of the MEBDF applied to the solution of IVPs that result from time dependent partial differential equations (PDEs). A robust technique for solving time dependent PDEs is the numerical method of lines or simply the method of lines. Let us consider the one dimensional time dependent PDE:

$$\frac{\partial}{\partial t} u(x,t) = g(x,t,u,u_x,u_{xx}), \quad w_1 < x < w_2, \quad t \geq 0$$

with initial condition $u(x,0) = \varphi(x)$ and where the boundary conditions at $x = w_1, x = w_2$ are appropriately selected $\forall\, t \geq 0$. If we semi-discretize the above PDE by making a finite difference approximation to the space derivatives u_x and u_{xx}, we can convert it into a system of ODEs with a specific initial condition. A commonly used second order discretization scheme is the following three-point central difference approximation:

$$\left.\begin{aligned}
\frac{\partial u(x_i,t)}{\partial x} &= \frac{u(x_i + h, t) - u(x_i - h, t)}{2h} + O(h^2) \\[2mm]
\frac{\partial^2 u(x_i,t)}{\partial x^2} &= \frac{u(x_i + h, t) - 2u(x_i,t) + u(x_i - h, t)}{h^2} + O(h^2)
\end{aligned}\right\} \qquad (*)$$

where

$$i = 1, 2, \ldots, N, \qquad h = \frac{w_2 - w_1}{N + 1}, \qquad x_i = w_1 + ih$$

---

[1] Vice-President of the European Society of Computational Methods in Sciences and Engineering.
Correspondence address: 192 Campkin Road, Cambridge CB4 2LH, UK – fax: +44-1223-515349.
E-mail: g.psihoyios@ntlworld.com

and N is the number of interior x-meshpoints. If we use the symbolism $u_i = u(x_i, t)$ and we drop the error term of (*) we obtain a system of ODEs of the form:

$$\frac{dU_i(t)}{dt} = g\left(x_i, t, U_i(t), \frac{U_{i+1}(t) - U_{i-1}(t)}{2h}, \frac{U_{i+1}(t) - 2U_i(t) + U_{i-1}(t)}{h^2}\right)$$

$$\text{i.e. } \frac{dU(t)}{dt} = g(x, t, U(t))$$

where $i=1,2,...,N$ and $U_i(t)$ approximates the solution $u(x_i, t)$. The above approach is widely known as the method of lines.

2. The Modified Extended BDF approach

The MEBDF approach [2, 4] is as follows:

Step 1 : Use a standard BDF to compute the first predictor $\bar{y}_{n+k}$, assuming that approximate solutions y_{n+j} have been computed at x_{n+j}, for $0 \le j \le k-1$

$$\bar{y}_{n+k} + \sum_{j=0}^{k-1} \bar{a}_j y_{n+j} = h\bar{b}_k f\left(x_{n+k}, \bar{y}_{n+k}\right) \tag{1}$$

where $\bar{a}_j$ and $\bar{b}_k$ are the known BDF coefficients.

Step 2 : Use a standard BDF to compute the second predictor $\bar{y}_{n+k+1}$

$$\bar{y}_{n+k+1} + \bar{a}_{k-1}\bar{y}_{n+k} + \sum_{j=0}^{k-2} \bar{a}_j y_{n+j+1} = h\bar{b}_k f\left(x_{n+k+1}, \bar{y}_{n+k+1}\right) \tag{2}$$

Step 3 : Evaluate $\bar{f}_{n+k} = f(x_{n+k}, \bar{y}_{n+k})$, $\bar{f}_{n+k+1} = f(x_{n+k+1}, \bar{y}_{n+k+1})$

Step 4 : Compute a corrected solution of order $(k+1)$ at x_{n+k} using

$$y_{n+k} + \sum_{j=0}^{k-1} a_j y_{n+j} = h\left[b_{k+1}\bar{f}_{n+k+1} + \hat{b}_k \bar{f}_{n+k} + \left(b_k - \hat{b}_k\right)f_{n+k}\right] \tag{3}$$

where $\hat{b}_k = b_k - \bar{b}_k$ and a_j, b_{k+1}, b_k are given in [3].

If the free coefficient $\hat{b}_k$ is chosen appropriately, MEBDF have much better stability than conventional BDF or EBDF [3]. If, for example, $\hat{b}_k = 0$ then (3) reduces to an EBDF. It was observed in [4] that if

$$b_k - \hat{b}_k = \bar{b}_k \text{ or } \hat{b}_k = b_k - \bar{b}_k \tag{4}$$

then computational efficiency was achieved, since the coefficient matrix to be factorised in the modified Newton iteration scheme, $\left(I - h\bar{b}_k J\right)$, is the same for all three sets of non-linear equations to be solved. The choice (4) for $\hat{b}_k$ may not be the optimal one as far as stability is concerned, but it does widen the regions of A(a)-stability compared to EBDF.

3. Stability Considerations

The semi-discretization of a parabolic time dependent PDE usually results in a very large dimension initial value problem (IVP) with a banded Jacobian matrix. It is reasonable to expect that the total CPU time that is allocated to the matrix iterations (or LU factorisations) in the case of a very large dimension stiff IVP with a banded Jacobian matrix, is relatively small. Diamantakis [5] measured that LSODE [6] devotes less than 5% of its total CPU time to the linear algebra for the problems 1 - 4 used in our numerical experimentations. This happens because the modified Newton method involves the use of a sparse Jacobian (in our case tridiagonal) and this requires little computational effort.

This is an excellent indication that the logic followed in [4] and in [1] and which is represented by equation (4), i.e. $\hat{b}_k = b_k - \overline{b}_k$, may need to be revised, since we do not mind solving systems with different coefficient matrices if these matrices are sparse (tridiagonal). Thus, we are now free to select $\hat{b}_k$ solely for stability. For many stiff systems that result from parabolic PDEs excellent stability is often of prime importance and it is now essential to choose $\hat{b}_k$ in order to get the best possible stability. Due to space limitations, it suffices to say that we have achieved noticeable improvements in the absolute stability regions for higher orders in our improved MEBDF approach.

4. Test Problems

We use five different test problems, which model interesting physical phenomena (e.g. a Diffusion-Convection problem, Heat equation, Burgers equation etc), in order to assess the behaviour of our new approach. Due to space restrictions we only present three of them below.

Problem 1
Let us consider the following <u>Diffusion-Convection</u> problem (also used in [1]):

$$\frac{\vartheta}{\vartheta\, t} u(x,t) = D\frac{\vartheta^{\,2} u}{\vartheta x^2} - V\frac{\vartheta u}{\vartheta x}, \quad -2 \le x \le 2, \quad 0 \le t < \infty \qquad (5)$$

$$\text{Initial Conditions}: \quad u(x,0) = e^{-x^2}$$

$$\text{Boundary Conditions}: \begin{cases} u(-2,t) = \dfrac{1}{\sqrt{1+4Dt}}\, e^{-(-2-Vt)^2/(1+4Dt)} \\[3mm] u(2,t) = \dfrac{1}{\sqrt{1+4Dt}}\, e^{-(2-Vt)^2/(1+4Dt)} \end{cases}$$

where D and V are constants. Equations of the form (5) model the way that the concentration (u) of a substance changes as it is carried along in a stream moving with velocity V. As an example we may consider smoke rising in a hot air stream. In such a case the particles of smoke are convected upward and at the same time they diffuse within the air current. The term $D\dfrac{\vartheta^{\,2} u}{\vartheta x^2}$ measures the diffusion contribution in the change of concentration u and the term $-V\dfrac{\vartheta u}{\vartheta x}$ measures the convection contribution. An exact solution of (5) is :

$$u(x,t) = \frac{1}{\sqrt{1+4Dt}}\, e^{-(x-Vt)^2/(1+4Dt)}$$

Problem 2
Cash in [7] presented this artificial test problem which is a form of the <u>heat equation</u> and describes the heat flow in an insulated thin rod with zero temperature at the edges :

$$\frac{\partial}{\partial t} u(x,t) = V\frac{\partial^2}{\partial x^2} u(x,t), \quad 0 \le x \le 1, \quad 0 \le t < \infty \qquad (6)$$

$$\text{Initial Conditions}: \quad u(x,0) = \sin(\pi x) + \sin(k\pi x), \quad k \gg 1$$

$$\text{Boundary Conditions}: \begin{cases} u(0,t) = 0 \\ u(1,t) = 0. \end{cases}$$

Given the above initial conditions, the exact solution to (6) has the following form :

$$u(x,t) = e^{-\pi^2 Vt} \sin(\pi x) + e^{-k^2 \pi^2 Vt} \sin(k\pi x). \tag{7}$$

We see that (7) has a rapidly decreasing highly oscillating component and a relatively slowly decreasing and slowly oscillating component.

Problem 3
The following PDE is a one dimensional linear <u>heat conduction</u> problem and describes the heat flow in a one-dimensional very thin rod:

$$\frac{\partial}{\partial t} u(x,t) = \frac{\partial^2}{\partial x^2} u(x,t) - u(x,t), \quad 0 \le x \le 1, \quad 0 \le t < \infty \tag{8}$$

$$\textit{Initial Conditions}: \qquad u(x,0) = \sin(\pi x)$$

$$\textit{Boundary Conditions}: \begin{cases} u(0,t) = 0 \\ u(1,t) = 0. \end{cases}$$

The heat, u, flows along the rod and along the lateral sides of the rod (non-insulated rod). The temperature at the edges is kept constant at 0 degrees and the temperature at $t=0$ is $u(x,t) = sin(\pi x)$. Thus, initially the temperature peaks at the centre of the rod. The exact solution to (8) is given by :

$$u(x,t) = e^{-(\pi^2 + 1)t} \sin(\pi x).$$

5. Brief Conclusions

After applying appropriate modifications to the MEBDF code and comparing our enhanced approach to the one from [1], epigrammatically, we can say that we have fair evidence indicating that the newly presented enhanced MEBDF displays higher accuracy and efficiency than previously [1]. Furthermore, we have also examined our fresh approach against the widely used LSODE [6] solver, which is based on BDF. Unsurprisingly, in this instance also the results proved significant and we are very encouraged that our enhanced approach may justifiably be considered instead of BDF based solvers, in certain important classes of problems.

References

[1] J.R. Cash and [G.-]Y. Psihoyios: The MOL Solution of Time Dependent Partial Differential Equations, *Computers and Math. Applic.* 31 (11), 69-78, 1996.

[2] J.R. Cash and S. Considine: An MEBDF code for Stiff Initial Value Problems, *ACM Trans. Math. Software*, 142, 1992.

[3] J.R. Cash: On the integration of stiff systems of ODEs using Extended BDF, *Num. Math.*, vol. 34, 235-246, 1980.

[4] J.R. Cash: The integration on stiff IVPs in ODEs using Modified Extended BDF, *Comp. & Maths with Applic.*, vol. 9, 645-657, 1983.

[5] M. Diamantakis, Diagonally extended singly implicit Runge-Kutta methods for stiff IVPs, Ph.D. Thesis, Imperial College, London, 1995.

[6] A.C. Hindmarsh: LSODE and LSODI, two new initial value ordinary differential equation solvers, *ACM-SIGNUM Newsletter*, vol. 15, p.10-11, 1980.

[7] J.R. Cash, Two new finite difference schemes for parabolic equations, *SIAM J. of Num. Analysis*, vol. 21, 433-445, 1984.

VSP International
Science Publishers
P.O. Box 346, 3700 AH Zeist
The Netherlands

Lecture Series on Computer
and Computational Sciences
Volume 1, 2004, pp. 440-443

Towards a General Formula for the Stability Functions of a Family of Implicit Multistep Methods

G. Psihoyios[1]

Department of Mathematics and Technology,
Anglia Polytechnic University, East Road,
Cambridge CB1 1PT, United Kingdom

Received 21 June 2004; accepted in revised form 12 August 2004

Abstract: This short paper serves as an introduction towards obtaining an original general formula that generates the stability functions of a family of methods, named by the author Implicit Advanced Step-point (IAS) methods [1]. The family of IAS methods encompasses the Two IAS methods (TIAS), the Parallel IAS methods (PIAS) and the Extended and Modified Extended BDF schemes (EBDF and MEBDF respectively). A general formula of this sort can substantially facilitate stability analysis and further computational manipulation of the IAS family of methods and analogous schemes. Due to space restrictions we are unable to present a full account of our new general formula.

Keywords: Stability, Implicit advanced step-point methods, Multistep methods, TIAS methods, PIAS methods.

Mathematics Subject Classification: 65L20, 65L06, 65L05, 65M12

1. Introduction

Since Gear's book [2] in 1971, computer codes based on Backward Differentiation Formulae (BDF) have been the most widely used regarding the numerical solution of stiff systems. Actually, BDF had been first introduced by Curtiss and Hirschfelder [3] in 1952. The k-step BDF is given by :

$$\sum_{j=0}^{k} a_j y_{n+j} = h b_k f_{n+k}$$

$$f_{n+k} = f(x_{n+k}, y_{n+k})$$

(1)

In formula (1) the coefficient $a_k=1$ and the order is equal to k. Methods like (1) are routinely employed to solve stiff initial value problems (IVPs).

In [5] the author presented a detailed account of the Parallel Implicit Advanced Step-point (PIAS) methods and in [4] a short introduction was given for the Two Implicit Advanced Step-point (TIAS) methods. Both the TIAS and the PIAS schemes depart rather considerably from methods like the MEBDF or EBDF. The TIAS scheme uses a second advanced step-point and thus an additional step forward, which means that TIAS methods have three predictors and at least one extra function evaluation per step. On the other hand, the PIAS scheme uses two independent of each other predictors (in order to have the chance to compute the predictors in parallel), the second of which is a specially constructed multistep formula.

In [6, 7] Cash investigated a class of "super-future" point schemes for the numerical solution of stiff IVPs. These methods, known as EBDF and MEBDF, use two interdependent BDF predictors and one

---

[1] Vice-President of the European Society of Computational Methods in Sciences and Engineering.
Correspondence address: 192 Campkin Road, Cambridge CB4 2LH, UK – fax: +44-1223-515349.
E-mail: g.psihoyios@ntlworld.com

implicit multistep corrector. Hence, we could say that they predict "two steps forward" and correct "one step back" for each step-point or step-number k.

2. The TIAS and the PIAS schemes

Due to space limitations, we can not enter into a detailed discussion of the TIAS scheme, but we can present the general form of the methods, which is given as follows:

Step 1 : Use a standard BDF of order k to compute the first predictor $\bar{y}_{n+k}$, assuming that approximate solutions y_{n+j} have been computed at x_{n+j}, $0 \le j \le k-1$:

$$\bar{y}_{n+k} + \sum_{j=0}^{k-1} \bar{a}_j y_{n+j} = h\bar{b}_k f\left(x_{n+k}, \bar{y}_{n+k}\right) \tag{1a}$$

where $\bar{a}_j$ and $\bar{b}_k$ are the known BDF coefficients.

Step 2 : Use a standard BDF of order k to compute the second predictor $\bar{y}_{n+k+1}$:

$$\bar{y}_{n+k+1} + \bar{a}_{k-1}\bar{y}_{n+k} + \sum_{j=0}^{k-2} \bar{a}_j y_{n+j+1} = h\bar{b}_k f\left(x_{n+k+1}, \bar{y}_{n+k+1}\right) \tag{1b}$$

Step 3 : Use a standard BDF of order k to compute the third predictor $\bar{y}_{n+k+2}$:

$$\bar{y}_{n+k+2} + \bar{a}_{k-1}\bar{y}_{n+k+1} + \bar{a}_{k-2}\bar{y}_{n+k} + \sum_{j=0}^{k-3} \bar{a}_j y_{n+j+2} = h\bar{b}_k f\left(x_{n+k+2}, \bar{y}_{n+k+2}\right) \tag{1c}$$

Step 4 : Having successfully computed the above 3 steps compute :

$$
\left.
\begin{aligned}
\bar{f}_{n+k} &= \frac{1}{h\bar{b}_k}\left[\bar{y}_{n+k} + \sum_{j=0}^{k-1} \bar{a}_j y_{n+j}\right] \\[2mm]
\bar{f}_{n+k+1} &= \frac{1}{h\bar{b}_k}\left[\bar{y}_{n+k+1} + \bar{a}_{k-1}\bar{y}_{n+k} + \sum_{j=0}^{k-2} \bar{a}_j y_{n+j+1}\right] \\[2mm]
\bar{f}_{n+k+2} &= \frac{1}{h\bar{b}_k}\left[\bar{y}_{n+k+2} + \bar{a}_{k-1}\bar{y}_{n+k+1} + \bar{a}_{k-2}\bar{y}_{n+k} + \sum_{j=0}^{k-3} \bar{a}_j y_{n+j+2}\right]
\end{aligned}
\right\}
$$

Step 5 : Compute a corrected solution of order $(k+1)$ at x_{n+k} using :

$$y_{n+k} - \sum_{j=0}^{k-1} \hat{a}_j y_{n+j} = h\left[\hat{c}_{k+2}\bar{f}_{n+k+2} + \hat{c}_{k+1}\bar{f}_{n+k+1} + c_k\bar{f}_{n+k} + \left(\hat{c}_k - c_k\right)f_{n+k}\right] \tag{2}$$

where $\hat{c}_{k+2}$ and c_k are free coefficients and all the other coefficients, namely $\hat{c}_{k+1}, \hat{c}_k, \hat{a}_j$, are expressed in terms of $\hat{c}_{k+2}$.

The PIAS methods were presented in detail in [5] and it suffices to give their general form:

Step 1 : Use a standard BDF to compute the first predictor $\tilde{y}_{n+k}$, assuming that approximate solutions y_{n+j} have been computed at x_{n+j}, for $0 \le j \le k-1$

$$\tilde{y}_{n+k} + \sum_{j=0}^{k-1} \bar{a}_j y_{n+j} = h\bar{b}_k f\left(x_{n+k}, \tilde{y}_{n+k}\right) \tag{3}$$

where $\bar{a}_j$ and $\bar{b}_k$ are the known BDF coefficients.

<u>Step 2</u> : Use an implicit multistep formula to compute the second predictor $\tilde{y}_{n+k+1}$

$$\tilde{y}_{n+k+1} - \sum_{j=0}^{k-1} \tilde{a}_j y_{n+j} = h[\bar{b}_{k+1} f(x_{n+k+1}, \tilde{y}_{n+k+1}) + \bar{b}_{-1} f_{n+k-1}] \qquad (4)$$

where $\tilde{a}_j, \bar{b}_{k+1}$ are coefficients that can be found in [5] and $\bar{b}_{-1}$ is the free coefficient in terms of which the other coefficients are expressed.

<u>Step 3</u> : Assuming that the Newton iteration converges, evaluate

$$\left.\begin{array}{l} \tilde{f}_{n+k+1} = \dfrac{1}{h\tilde{b}_{k+1}} \left[\tilde{y}_{n+k+1} - \displaystyle\sum_{j=0}^{k-1} \tilde{a}_j y_{n+j} - h\bar{b}_{-1} f_{n+k-1} \right] \\[4ex] \tilde{f}_{n+k} = \dfrac{1}{h\bar{b}_k} \left[\tilde{y}_{n+k} + \displaystyle\sum_{j=0}^{k-1} \bar{a}_j y_{n+j} \right] \end{array}\right\}$$

<u>Step 4</u> : Compute a corrected solution of order $(k+1)$ at x_{n+k} using

$$y_{n+k} + \sum_{j=0}^{k-1} a_j y_{n+j} = h\left[b_{k+1} \tilde{f}_{n+k+1} + \hat{b}_k \tilde{f}_{n+k} + \left(b_k - \hat{b}_k \right) f_{n+k} \right] \qquad (5)$$

where $\hat{b}_k = b_k - \bar{b}_k$ and a_j, b_{k+1}, b_k are discussed in [5] and given in [6].

2. A Stability Analysis Illustration

In order to demonstrate the amount of work needed to perform an analysis of the absolute stability regions of the methods referred to in this short paper, we will provide an illustration via the TIAS methods and for the step-point $k = 2$. Thus, making use of the customary test equation

$$y' = \lambda y \ \text{ or } \ f = \lambda y, \ \ \lambda \in C$$

and substituting it into (1a), (1b) and (1c), with $q = h\lambda$, we get:

The first predictor (1a), for $k = 2$, becomes:

$$\bar{y}_{n+2}(1 - q\bar{b}_2) = -\sum_{j=0}^{1} \bar{a}_j y_{n+j} \ \text{ or } \ \bar{y}_{n+2} = \frac{-\sum_{j=0}^{1} \bar{a}_j y_{n+j}}{(1 - q\bar{b}_2)} = \frac{-A}{(1 - q\bar{b}_2)} \qquad (6a)$$

The second predictor (1b) also using (6a), for $k = 2$, becomes:

$$\bar{y}_{n+3}(1 - q\bar{b}_2) = \bar{a}_1 \left(\frac{A}{1 - q\bar{b}_2} \right) - \sum_{j=0}^{0} \bar{a}_j y_{n+j+1} \ \text{ or } \ \bar{y}_{n+3} = \frac{\bar{a}_1 A}{(1 - q\bar{b}_2)^2} + \frac{-B}{(1 - q\bar{b}_2)} \qquad (6b)$$

where, in this instance $B = \bar{a}_0 y_{n+1}$

The third predictor (1c) also using (6a) and (6b), for $k = 2$, becomes:

$$\bar{y}_{n+4}(1 - q\bar{b}_2) = -\bar{a}_1 \left(\frac{\bar{a}_1 A}{(1 - q\bar{b}_2)^2} + \frac{-B}{1 - q\bar{b}_2} \right) + \bar{a}_0 \left(\frac{A}{1 - q\bar{b}_2} \right) - \sum_{j=0}^{2-3} \bar{a}_j y_{n+j+2}$$

where, in this instance the third quantity on the right hand side becomes zero because the coefficient $\bar{a}_j$ receives a negative subscript (for step-points $k > 2$ this is not the case). Finally, (1c) becomes:

$$\bar{y}_{n+4} = \frac{-\bar{a}_1^2 A}{(1-q\bar{b}_2)^3} + \frac{-B}{(1-q\bar{b}_2)^2} + \frac{\bar{a}_0 A}{(1-q\bar{b}_2)^2} \tag{6c}$$

Naturally we have not completed our work. Actually, we are just starting, since we now need to insert formulae (6a), (6b) and (6c) into the corrector (2). Thus

$$y_{n+2} - \sum_{j=0}^{1} \hat{a}_j y_{n+j} = h\left[\hat{c}_4 \bar{f}_{n+4} + \hat{c}_3 \bar{f}_{n+3} + c_2 \bar{f}_{n+2} + (\hat{c}_2 - c_2)f_{n+2}\right]$$

$$= q\hat{c}_4 (formula \ (6c)) + q\hat{c}_3 (formula \ (6b)) + qc_2 (formula \ (6a))$$

$$+ q(\hat{c}_2 - c_2)y_{n+2} = ...$$

and this last expression of course needs to be markedly further manipulated and ultimately solved for y_{n+2}. Then we need to find the solutions of the resulting polynomial, in order to be able to finally obtain the stability region (just) for the step-point $k = 2$.

3. Brief conclusions

In the previous section we gave an indication of the amount of work required in order to obtain the region of absolute stability for just one step-point of one of the four methods. In order to obtain the stability regions for all four methods and all their respective step-points, we would need to replicate a procedure similar to the one described in section 2 around thirty times, a huge amount of work indeed. It would thus be very helpful if we could have a general formula that could generate the stability polynomials for all our four methods. Such a general formula has indeed been identified and it can produce all the relevant stability functions for the PIAS, TIAS, EBDF and MEBDF methods. The advantages of a general formula of this sort are evident and we could use it to significantly facilitate stability analysis and further computational manipulation of these schemes.

References

[1] G. Psihoyios: "Advanced Step-point Methods for the Solution of Initial Value Problems", *PhD Thesis, Imperial College* - University of London, 1995.

[2] C.W.Gear: Numerical Initial Value Problems in ordinary differential equations, *Prentice Hall*, 253 pp., 1971.

[3] C.F. Curtiss & J.O. Hirschfelder: Integration of Stiff Equations, Proc. of National Academy of Science, vol. 38, pp. 235-243, 1952.

[4] G.-Y. Psihoyios, J.R. Cash: "A Stability Result for General Linear Methods with Characteristic Function having Real Poles only", *BIT Numerical Mathematics*, vol. 38 (no. 3), pp. 612-617, 1998.

[5] G. Psihoyios: "A Class of Implicit Advanced Step-point Methods with a Parallel Feature for the Solution of Stiff Initial Value Problems". To appear in the *J. of Mathematical and Computer Modelling*

[6] J.R. Cash: On the integration of stiff systems of ODEs using Extended BDF, *Num. Math.*, vol. 34, 235-246, 1980.

[7] J.R. Cash: The integration on stiff IVPs in ODEs using Modified Extended BDF, *Comp. & Maths with Applic.*, vol. 9, 645-657, 1983.

VSP International
Science Publishers
P.O. Box 346, 3700 AH Zeist
The Netherlands

*Lecture Series on Computer
and Computational Sciences*
Volume 1, 2004, pp. 444-447

Centre of Excellence and Wild life:
A New Feature of Genetic Algorithm

Jalil Rasekhi[1], Jalil Rashed-Mohassel[2]

Center of Excellence on Applied Electromagnetic Systems,
ECE Department,
Faculty of Engineering,
Univ. of Tehran,
Tehran, Iran

Received 9 June, 2004; accepted in revised form 25 June, 2004

Abstract: A Novel phenomenon in genetic algorithm is presented. In this method, the chromosome population is sub-divided in three categories analogous to real life called CE, NR and WL. Restrictions and rules are assigned for each category and superior chromosomes are selected. The method results in a faster convergence and improved result. Simulation results are verified for optimizing the matching network of a microstrip like antenna. Improvements regarding convergence time are achieved with this new technique.

Keywords: Genetic Algorithm, Microstrip-Like Antenna, MLA

1. Introduction

Nature has always had great hints for man on his way in solving difficult problems. Of these have been examples of plants and animals adapting to their environments, which has led to genetic algorithm. In fact, traditional optimization techniques search for the best solutions, using gradient and/or random guesses[1]. Although gradient methods quickly converge to a minimum, it will not necessarily be a global one. Gradient methods are applicable in problems where optimization parameters are continuous and limited to a few parameters. Random methods however, are capable of handling discrete parameters but tend to be very slow and may stuck in local minima [2]. For these and many other reasons, GA techniques have been preferable in many electromagnetic optimization problems.

Works on GA have generally utilized a very limited number of mutations. Otherwise some potential solutions would be lost. Some papers have tried to overcome this problem by cheating [2] and not letting best chromosomes to mutate. However no one has tried higher rates of mutation in fear of probable divergence in solution set.

In this paper, we have tried to introduce a new feature in genetic optimization to improve convergence speed. As an example of the feature is capabilities, an optimization problem was practiced applying both forms of GA. results indicate a faster convergence when centre of excellence concept is employed.

2. Overview of the technique

GA techniques usually use a limited number of mutations at a time. This is especially important when the cost function has a fast-changing nature and large drifts in genes may result in losing a pre-found global optimum. However if the initial solution chromosomes are far beyond the optimum solution point, it might take long to reach that point.

---

[1] Corresponding author. MSc. student at university of Tehran.. E-mail: j.rasekhi@ece.ut.ac.ir, jrasekhi@yahoo.co.uk

[2] Advisor. Professor at university of Tehran. E-mail: jrashed@ut.ac.ir

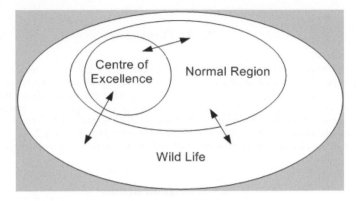

Figure 1: Interrelation between three solution zones

The proposed technique which we call it three-zone GA or simply 3Z, makes use of an important human phenomenon. Every chromosome population is divided into three categories called "centre of excellence", "Normal region" and "wild life" (Fig.1). Centre of Excellence may include just a small number of chromosomes and is in charge of keeping super-characteristic chromosomes until a more appropriate one is recognized in two other regions. Good chromosomes are kept in normal region and include newly born chromosomes and those survived from the previous generation. At the mean time, wild life is out of control and changes in a wild random, that is, wild chromosomes outside the restricted gene zone face substantial changes in their embedded characteristics, however they are not allowed to enter CE or NR until they achieve supreme features which make them a high-ranked chromosome. On the other hand no changes should occur in the centre of excellence.

This technique substantially improves solution searching process and meanwhile keeps optimized chromosomes intact. Chromosomes in the NR are proposed to provide small drifts towards optimal solution in case larger precisions are required. As a result, optimization process will faster converge into the global solution with a preserved accuracy.

Perhaps, the most featured property of this technique is the definition of WL region. In fact, the duty of WL is to random search wherever possible throughout the solution space and is like sending some people to search out there for new undiscovered areas. This method of searching is very common in human communities and has most often led to great findings.

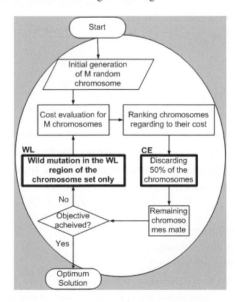

Figure 2: Interrelation between three solution zones

3. 3Z Flowchart

The flowchart of the proposed optimization technique is depicted if Fig.2. As is seen from this figure, CE and WL regions are realized in two ways. First, after ranking chromosomes regarding to their evaluated cost function, we discard 50% of chromosomes, but the remaining 50% of these will not change anyway (and make our CE). This way, we have enough population to mate and produce new offsprings to replace discarded chromosomes. Second, we wild-mutate a pre-defined fraction (e.g. 50%) of the new chromosomes (WL). These are emphasized as bold blocks in Fig.2.
The last region of the chromosome population is NR and includes those chromosomes not categorized as CE or WL.

4. Simulation Results

To verify the ability of 3Z-GA to improve optimization speed, we managed to optimize the matching network of a microstrip-like antenna (Fig.3) using both traditional and proposed 3Z GAs.

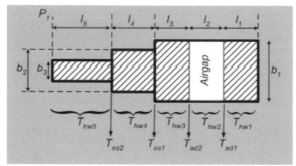

Figure 3: Microstrip-like antenna and its matching network (side view)

The cost function to be optimized was the overall sum-squared Γ_{ss} input reflection of the antenna and its matching network evaluated at five distinct frequencies throughout the design bandwidth (8.5-11 GHz), and is given as follows:

$$\Gamma_{SS} = \sum_{m=1}^{5} \left| \Gamma_{in} \left(8.5 \times 10^{9} + m.\Delta f \right) \right|^{2}, \quad \Delta f = \left(\frac{11-8.5}{4} \right) GHz \tag{1}$$

$$\Gamma_{in} = \frac{T_{11}\Gamma_{ap} + T_{12}}{T_{21}\Gamma_{ap} + T_{22}} \tag{2}$$

$$\Gamma_{ap} = \frac{1-y_{ap}}{1+y_{ap}} \tag{3}$$

where Γ_{ap} and y_{ap} are respectively the aperture reflection and admittance of the MLA antenna available in [3] and are both frequency dependent. The procedure for evaluation of y_{ap} is complex and not given here. However a plot of aperture reflection to be matched is shown in Fig.4. Also, T_{ij} 's are the elements of the matching network transmission matrix found as:

$$T = T_{hw1} T_{ad1} T_{hw2} T_{ad2} T_{hw3} T_{es1} T_{hw4} T_{es2} T_{hw5} \tag{4}$$

where:
T_{es} : E-plane step transmission matrix
T_{hw} : Homogeneous waveguide transmission matrix
T_{ad} : Airgap-to-dielectric transmission matrix
Optimization parameters were homogeneous waveguide lengths $l_1, l_2, ..., l_5$ and E-plane step sizes b_1, b_2 and b_3. These parameters were embedded in 8 genes in each chromosome. The large number of

optimized parameters (5+3) creates a vast solution space, which may cause a very sluggish convergence. To find these optimum parameters we employed both GA and 3Z-GA methods.

In normal simulation we obeyed mutation method proposed in [1] and let just 1% of every chromosome population to mutate. In 3Z simulation though, chromosomes have more freedom to mutate. However mutation can take place in a pre-defined wild zone (50% of the offsprings or 25% of the all chromosomes, in this work). In this region, we have a 50% (wild) mutation, that is, every one out of two bits in every WL chromosome is mutated.

Cost functions evaluated for supreme chromosomes in the CE versus iteration, are depicted in Fig.4 for both methods. As is observed from this figure, 3Z-GA has converged in less iteration compared to that of traditional GA method. Also from the same figure, one can realize that the traditional GA has failed to escape the local minima, however 3Z-GA has easily managed to do so and the result is astonishing: 1dB improvement missed by traditional GA.

Simulations were carried out employing these optimized parameters using HPHFSS. The resulting input reflection is illustrated in Fig.4 and compared to that of mismatched case.

5. Conclusion

A Novel aspect was introduced for GA, which divides chromosome population into three different regions. Simulations have been made regarding convergence speed for both traditional GA and proposed 3Z-GA techniques. The results indicate great improvements achieved via this new method regarding convergence speed and ability to find global minima.

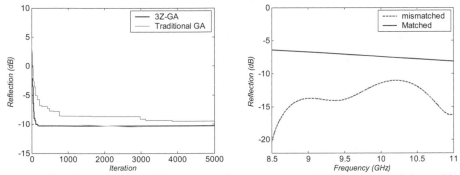

Figure 4: Cost functions evaluated for supreme chromosomes versus iteration (left), and the resulting input reflection compared to that of mismatched case (right)

Acknowledgments

Thanks goes to Dr. N. Hojjat for her helps throughout this work. Also, we appreciate Mr. M. Hajian for his valuable help and the references provided.

References

[1] Goldberg, David E, *Genetic Algorithms*, Addison Wesley, NY, 1989.

[2] R.L. Haupt, *An Introduction to Genetic Algorithms for Electromagnetics*, IEEE Antenna and Propagation Magazine, Vol. 37, No. 2, April 1995.

[3] M. Hajian, D.P. Tran and L.P. Ligthart, *Design of a wideband miniature dielectric-filled waveguide antenna for collision-avoidance radar*, IEEE Trans. AP, Vol. 42, No. 1, February 2000.

VSP International
Science Publishers
P.O. Box 346, 3700 AH Zeist
The Netherlands

*Lecture Series on Computer
and Computational Sciences*
Volume 1, 2004, pp. 448-452

An Efficient Disconnected Operation Protocol in SAN Distributed File System

Md. Abdur Razzaque[1]
Department of Computer
Science and Engineering
University of Dhaka, Bangladesh
Md. Arman Hossain[3]
Department of Information Technology
University of Technology Sydney, NSW,
Australia

Md. Ahsan habib[2]
Department of Computer Science & IT
Ialamic University of Technolgy (IUT),
Bangladesh
Md. Mahfuzur Rahman[4]
Department of Computer Science & IT
Ialamic University of Technolgy (IUT),
Bangladesh

Received 9 June, 2004; accepted in revised form 25 June, 2004

Abstract: Disconnected operation (DO) has become one of the important issues for increasing the performance of distributed computing environment. A good distributed file system should support fundamental system services in case of network or communication failure. In this paper, it has been proposed an efficient disconnected operation protocol in Storage Area Network (SAN) distributed file system. The goal is to provide high performance, high availability, data consistency, increased scalability, automated storage and data management capability to the client, though it is disconnected from the server

Keywords: Storage Area Network, Storage Tank, Disconnected Operation

1. Introduction

Disconnected operation in Storage Area Network (SAN) refers to the ability of a client to continue computation on disk data, being consistent, during the client's isolation from the server. The conventional system designs seriously limit data availability, and that this limitation will become much more severe as computing environments expand in size and evolve in form to incorporate new technologies [1]. This paper advocates the extension of *process management* to support *disconnected operation* as a solution to this problem, and describes the working principle of the protocol built around this idea.

2. Storage area network (SAN) file system

Storage Area Network (SAN) technology allows an enterprise to connect large numbers of devices, including clients, servers, and mass storage subsystems, e.g. disks, tapes etc. to a high-performance network. On a SAN, clients can access large volumes of data directly from storage devices, using high-speed and low-latency connections [2, 4, 5]. Unlike most file systems, in SAN metadata, which includes the location of the blocks of each file on shared storage, are stored in the server. So, the client has to get *lock (permission)* from the server for all the files it uses in a process. In the SAN file system the message passing between client and server is done through Local Area Network (LAN), which is normally known as **control network**. And the client fetches data from disk using Fiber Channel Network as well as new emerging storage network

2.1. Failures in SAN

In SAN, the server takes responsibility for maintaining data integrity and consistent views of data across all clients. This task is significantly complicated when components of the distributed system fail. One such problem arises when a client, that holds *locks* to access data, loses contact with its server because of network partition, as shown in figure 2.

[1] Email: m_a_razzaque@hotmail.com
[3] Email : armandia25@yahoo.com

[2] Email: tareqiut@yahoo.com
[4] Email : m_f_rahman@hotmail.com

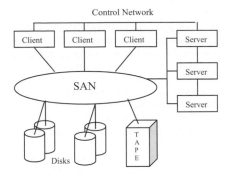

Figure 1: Schematic of the Storage Tank client/server distributed file system.

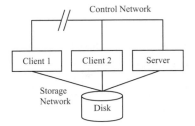

Figure 2: A two Network Storage system with partition in the control.

2.2. Current Solution

A currently accepted solution to this problem is to *fence* the isolated client before stealing its *locks*. To fence a computer, the server instructs the SAN-attached storage devices to no longer accept I/O requests from the isolated computer. Instructing a switch not to route traffic for certain clients may also perform fencing. The SAN devices must enforce this denial of access indefinitely. So, fencing alone is not an adequate solution to this problem. As the only recovery mechanism, fencing violates the sequential consistency and cache coherency guarantees expected from a file system.

2.2.1. Inadequacy of Current Solution

Adding the ability to fence clients and instructing a disk drive to deny service to particular initiators, does not solve all data integrity and cache coherency problems. It also does not allow all file operations to be completed. Returning to our previous example (Figure 2), if the disk drive supported fencing, the server could instruct it not to allow C1 to perform any I/O. Then, the server could grant C2 read/write access to the file. While fencing does prohibit C1 from performing concurrent conflicting writes, it leaves several opportunities for the violation of file system semantics [4]. Problems arise because the isolated client can have unwritten data at the time it is fenced. In this example, after fencing C1, dirty data on C1 are stranded and never reach disk. If C2 reads this data, it reads the old version from persistent storage. This is in violation of consistency guarantee – C2 should properly read the most recently written version of data. Again C1 needs to be allowed to complete all its operation on the file.

2.3. A Proposed Protocol Based on Leases and Processes

It is observed from the above discussion that a safety protocol that supports disconnected operation and protects the inconsistency of data and guarantees cache coherency must perform the following tasks-

- The client should be allowed to renew its lease period.
- The client should know that it is disconnected from the server.
- The server should also know that a particular client is disconnected from the server.

- After disconnection the isolated client would be allowed to do at most file operations as possible.

SAN uses fencing, where the server steals the *lock* only in case of disconnection or when the client informs about release. There is no provision for lease renewal in existing protocol. Again the protocol does not allow client operation and data consistency violating the fourth requirement. The proposed protocol based on leases allows clients to determine that they have become isolated from the server. Upon determining their isolation, clients have the opportunity to complete operations on data before seizing their *locks*. In this lease based protocol the client and server both will be informed about the disconnection and a process named **DO-process** will be created so that after disconnection the isolated client could be able to complete all of its file operation.

2.3.1. Obtaining a lease and the lease interval

A client implicitly obtains or renews a lease with a server on every message it initiates (Figure 3). At time U, the client sends a message to the server. The server receives this message at time V and acknowledges receipt at time W. With synchronized clocks, the client and server have an absolute ordering on events U≤V. Upon receiving the ACK at time X after waiting for time period i which is the lease idle time, the client obtains a lease valid period m. Then the client calculates the lease active time n as follows **$n=m-i$**

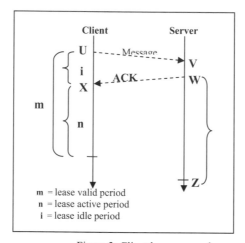

Figure 3: Client lease renewal

This lease period starts from when the client initiates the message, not when it receives the ACK message because initiating the message is known to occur before the server's reply. While this lease is valid for the time interval UY, from the client's perspective it does not activate until time X, when the client receives an ACK message, indicating it is in contact with a server. The client operates under this lease for the period n, i.e. time interval XY. Here, the Server does not need to send any acknowledgement (ACK) messages to the client, if it has already started a counter at the time, W to expire client *locks*. But when the server receives any lease renewal request, it sends an acknowledgement (ACK) message to the client restarting the counter clock. This ensures that the server cannot steal *locks* from a client until after the client lease expires.

Referring to Figure 3, we get, WZ=UY and WZ=UX+XY. Since UX is a positive time interval, so WZ>XY. Therefore, the server cannot steal locks before the client lease expires [4].

The client subdivides the lease valid period, m into three proportional phases: leased idle phase **i**, lease renewal phase **r** and decision phase **d** (where **d <= i**). In the lease idle phase of the lease period, the Client waits for the permission (ACK message) for a particular file from the Server. When the client gets the response from the server along with the lease period for the requested file then the Client enters into the 2nd phase i.e. lease renewal phase. A recently obtained lease allows the Client to access to data objects on disk. Here if the Client does not require the extension of lease period then it sends a lease release message to the Server at the end of lease renewal period. But when the Client requires extending its lease period then it sends a lease renewal request to the server and waits for the ACK message. After obtaining the new lease period the Client rearrange all its time intervals. But if the Client does not get the ACK message within the lease renewal phase then it enters into decision phase without sending any lease release message to the server.

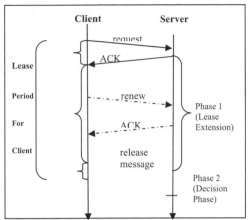

Figure 4: Lease intervals of Client and Server

In this 3rd phase the client decides about the disconnection from the server depending on whether the lease release message is sent or not.

Server subdivides the lease period into two proportional phases: lease extension phase and decision phase. In first phase it is ready to send all renewal ACK message with new lease period to the client. But when it enters into 2nd phase it doesn't accept any renewal request from the client. Rather the server decides about the Client's isolation depending on whether it has received any lease release message from the Client in the previous phase. If the Server finds that the Client is disconnected from it, then an automated process, we name it as Disconnected Operation Process (*DO-process*), is created in the server in this phase, which repeatedly checks the file attributes from disk that are currently accessed by the disconnected client. If *DO-process* finds that the files are still accessed by the isolated client then it prevents server to authorize other requested clients to take the *lock*. When the client completes the file operations, then the process informs the server to steal the *lock* and the server kills the process.

3. Conclusions

Our proposed protocol allows the client to complete functions on data in disconnected state of the SAN distributed file system. In this case data consistency, cache coherency and high availability is maintained in the distributed system even after disconnection. This protocol also supports disconnected operations both in client or server failure. But the clients need to cache all files required for the processes running in it. Another assumption in this protocol is that clocks are rate synchronized in the system. So, the limitations of these considerations may be the fruit of thinking for the fellow researchers to find their working area.

References

[1]. James Jay Kistler. "Disconnected Operation (DO) in Distributed File System (DFS)" CMU-CS-93-156, May 1993

[2]. L. Lamport. How to make a multiprocessor computer that correctly executes multiprocess programs. *IEEE Transactions on Computers*, C-28(9), 1979.

[3]. G. A. Gibson, D. F. Nagle, K. Amiri, F. W. Chang, E. M. Feinberg, H. Gobioff, C. Lee, B. Ozceri, E. Riedel, D. Rochberg, and J. Zelenka. File server scaling with network-attached secure disks. In *Performance Evaluation Review*, volume 25, 1997.

[4]. Randal Chilton Burns, *Data Management in a Distributed File System for Storage Area Network*, University of California, March 2000.

[5]. Randal C. Burns, Robert M. Ress, Darrell D.E.Long; " Safe Caching in a Distributed File System for Network Attatched Storage", March 2000

[6]. Pradeep K Sinha, "Distributed Operating system, concepts and Design", IEEE Press, 1998.

[7]. Andrew S Tanenbaun, "Distributed Operating Systems", Prentice Hall, 1999

[8]. M. Devarakonda, D. Kish, and A. ohindra. Recovery in the Calypso file system. *ACM Transactions on Computer System*s, 14(3), August 1996.

[9]. M. L. Kazar, B. W. Leverett, O. T. Anderson, V. Apostolides, B. A. Bottos, S. Chutani, C. F. Everhart, W. A. Mason, S. Tu, and R. Zayas. DEcorum file system architectural overview. In *Proceedings of the Summer USENIX Conferenc*e, June 1990

VSP International
Science Publishers
P.O. Box 346, 3700 AH Zeist
The Netherlands

*Lecture Series on Computer
and Computational Sciences*
Volume 1, 2004, pp. 453-456

Numerical Solution of Two-Dimensional Contaminant Transport Problem with Dispersion and Adsorption in a Groundwater Layer

M. Remešíková [1]

Department of Mathematical Analysis and Numerical Mathematics
Faculty of Mathematics, Physics and Informatics
Comenius University
Mlynská Dolina, 84248 Bratislava Slovakia

Abstract: In this contribution, an efficient numerical method for solving convection-diffusion problems with adsorption is introduced. The problem considered here comes out from dual-well tests used for soil parameters identification. We use a general two-dimensional mathematical model which includes contaminant transport, mechanical dispersion and molecular diffusion and adsorption in both equilibrium and non-equilibrium modes. First, the original half-plane domain is transformed to a rectangle using an orthogonal transformation. The method of solution is based on time stepping and operator splitting approach. The transport part is 1D due to the transformation and its solution can be found in an analytical form. The dispersion part is solved using finite volume method. For the system of ODE's representing adsorption we derive an implicit scheme.

Keywords: convection-diffusion problem, non-equilibrium adsorption, operator splitting

Mathematics Subject Classification: AMS 76S05, 65M99, 65-06

1 Introduction

Groundwater contamination is one of the most typical hydrogeological and environmental problems. The general model of a groundwater layer includes various hydrogeological processes as contaminant transport, mechanical dispersion, molecular diffusion, sorption, chemical reactions etc. In many practical situations we need to predict the time behaviour of a contaminated groundwater layer. In order to obtain realistic results, it's necessary to have realistic data in the mathematical model. Various soil parameters (porosity, dispersivities, sorption coefficients etc.) can be precisely determined using systems of monitor wells, where by monitoring the contaminant concentration in the wells it's possible to reconstruct the groundwater layer properties. Solving of such inverse problems requires repeated solving of direct problems and therefore it's crucial to have an efficient numerical method for the direct problem.

In this paper we consider a system of two monitor wells and a mathematical model including contaminant transport, dispersion and adsorption. If we use Dupuit-Forchheimer approximation, we have a 2D convection-diffusion-adsorption problem. In e.g. [2], the authors describe an efficient method for solving problems with adsorption in equilibrium mode. The main goal of this paper is to introduce a method also for non-equilibrium sorption problems, based on a similar idea.

---

[1] E-mail: remesikova@fmph.uniba.sk

2   Mathematical model

Let us consider two wells (injection and extraction well) situated at points $(-d, 0)$ and $(d, 0)$ in Cartesian coordinates, with given radii r_1, r_2. Moreover, let us assume that the pumping rate of the injection well is equal to the discharge in the extraction well. The contaminant transport problem with dispersion and adsorption (reaction) in represented by the following system of differential equations

$$h_{eff} \partial_t (C + \Psi_e(C)) - \operatorname{div}(D h_{eff} \nabla C) - \operatorname{div}(h_{eff} \vec{v} C) + h_{eff} \partial_t S = 0$$
$$\partial_t S = K(\Psi_n(C) - S)$$

where $C(x, y, t)$ represents the contaminant concentration and $S(x, y, t)$ is the concentration of adsorbed pollution. We consider the adsorption in both equilibrium mode (represented by sorption isotherm $\Psi_e(C)$) and non-equilibrium mode (represented by $\Psi_n(C)$). Both isotherms are considered to be of Freundlich type, $\Psi(C) = AC^p$, $A > 0$, $0 < p < 1$. Here D is the dispersivity tensor

$$D_{ij} = \left\{ (D_0 + \alpha_T |v|) \delta_{ij} + \frac{v_i v_j}{|v|}(\alpha_L - \alpha_T) \right\}$$

where D_0 is the molecular diffusion coefficient and δ_{ij} the Kronecker symbol. α_L and α_T are longitudinal and transversal dispersivities. $\vec{v}$ is defined by

$$\vec{v} = -\frac{1}{h_{eff} \theta_0} \nabla \Phi$$

where θ_0 is porosity, Φ is the flow potential and h_{eff} is the groundwater acquifer height (if we consider saturated layer) or piezometric head (for unconfined zone).

As the original two-dimensional domain is symmetric along x-axis, we can restrict ourselves only to one of its half-planes. Using a bipolar transformation, this domain can be transformed to a rectangle $(0, \pi) \times (v^{(1)}, v^{(2)})$ (sides $u = 0$, $u = \pi$ corresponding to well borders), where the equipotential curves and streamlines of the flow are parallel to coordinate axes and orthogonal to each other (see [1]). The bipolar coordinates (u, v) are defined by

$$x = \frac{\delta}{2} \frac{\sinh v}{\cosh v - \cos u}, \quad y = \frac{\delta}{2} \frac{\sin u}{\cosh v - \cos u}, \quad \sqrt{r_1^2 + \frac{1}{4}\delta^2} + \sqrt{r_2^2 + \frac{1}{4}\delta^2} = 2d \quad (1)$$

The values $v^{(1)}$, $v^{(2)}$ are obtained from

$$\sinh v^{(1)} = -\frac{\delta}{2r_1}, \qquad \sinh v^{(2)} = \frac{\delta}{2r_2} \quad (2)$$

Applying the transformation described above to the problem in (x, y) coordinates, we obtain the following convection-diffusion-adsorption problem in (u, v) coordinates

$$\partial_t (C + \Psi_e(C)) - F \partial_v C - g(\partial_u(a \partial_u C) + \partial_v(b \partial_v C)) + \partial_t S = 0 \quad (3)$$
$$\partial_t S = K(\Psi_n(C) - S) \quad (4)$$

Terms g, a, b and F are known functions depending on u and v and the soil parameters (α_L, α_T, D_0, θ_0 (see [1])). We consider the boundary conditions

$$C = C_0(t) \text{ on } \Gamma_1, \quad \partial_u C = 0 \text{ on } \Gamma_2 \cup \Gamma_4, \quad \partial_v C = 0 \text{ on } \Gamma_3 \quad (5)$$

where $\Gamma_1 = (0, \pi) \times \{v = v^{(2)}\}$, $\Gamma_2 = \{0\} \times (v^{(1)}, v^{(2)})$, $\Gamma_3 = (0, \pi) \times \{v = v^{(1)}\}$, $\Gamma_4 = \{\pi\} \times (v^{(1)}, v^{(2)})$.

The initial conditions are

$$C((u, v), 0) = 0, \quad S((u, v), 0) = 0 \quad (6)$$

3   Method of solution

The method of solution is based on time stepping and operator splitting approach. This means that first we discretize the considered time interval $< 0, T >$, where the discretization points are denoted by $0 = t_0, t_1, \ldots, t_{m-1}, t_m = T$. Then in each time interval we solve separately the transport part, the diffusion part and the reaction part (the adsorption in non-equilibrium mode).

The transport part presents a hyperbolic problem of the form

$$\partial_t(\phi + \Psi_e(\phi)) - F\partial_v\phi = 0, \qquad t \in (t_{j-1}, t_j), \, j = 1 \ldots m \tag{7}$$

with boundary conditions of the form (5) and initial condition $\phi(u, v, t_{j-1}) = C_{j-1} = C(u, v, t_{j-1})$.

For a piecewise constant initial profile, we have to treat a multiple Riemann problem and the solution can be found in an analytical form (see [2]). By $C_j^{1/3}$ we denote the obtained solution, i.e. $C_j^{1/3} := \phi(u, v, t_j)$.

Now we solve the problem (the diffusion part)

$$\partial_t(\phi + \Psi_e(\phi)) - g(\partial_u(a\partial_u\phi) + \partial_v(b\partial_v\phi)) = 0, \qquad t \in (t_{j-1}, t_j) \tag{8}$$

with the same boundary conditions and initial condition $\phi(u, v, t_{j-1}) = C_j^{1/3}$.

In this case we use a standard finite volume scheme. We obtain a system of non-linear equations that we solve by Newton method. The obtained solution is denoted by $C_j^{2/3}$.

The last part of the procedure is solving the reaction part represented by the system

$$\partial_t(\phi + \Psi_e(\phi)) + \partial_t S = 0 \tag{9}$$
$$\partial_t S = K(\Psi_n(\phi) - S) \tag{10}$$

with the initial conditions $\phi(u, v, t_{j-1}) = C_j^{2/3}$, $S(u, v, t_{j-1}) = S_{j-1}$.

Integrating (9), using initial conditions, solving (10) in S and substituting for S we obtain an integral equation

$$f(\phi(t)) + S_{j-1}e^{-Kt} + K\int_0^t e^{-K(t-z)}\Psi_n(\phi(z))\,dz = f(C_j^{2/3}) + S_{j-1} \tag{11}$$

where $f(\phi) = \Psi_e(\phi) + \phi$.

Using time discretization and approximating the integral in (11) it is possible to derive an implicit numerical scheme for solving (11) (see [3]).

Finally, we put

$$C_j := \phi(u, v, t_j), \qquad S_j := S(u, v, t_j)$$

where $\phi(u, v, t_j)$, $S(u, v, t_j)$ represent the solution of (9)–(10).

4   Numerical experiments

The following experiment in 1D illustrates the efficiency of the described method. For a simple 1D problem

$$\partial_t C + v(x)\partial_x C - \partial_x(D(x, t)\partial_x C) + \partial_t S = 0$$
$$\partial_t S = K(C - S)$$

it is possible to construct an analytical solution according to [4] and compare it with the solution obtained by our numerical scheme. In order to verify the efficiency of the method, we add also a comparison with the solution obtained by a simple finite difference scheme, where we put $\partial_t C \approx (C_i^k - C_i^{k-1})/\tau$, $\partial_x C \approx (C_i^k - C_{i-1}^k)/h$, $\partial_x^2 C \approx (C_{i+1}^k - 2C_i^k + C_{i-1}^k)/h^2$, $\partial_t S \approx (S_i^k - S_i^{k-1})/\tau$, $C_i^k = C(x_i, t_k)$, $S_i^k = S(x_i, t_k)$. In this experiment we set v and D constant, $v(x) \equiv 1$, $D(x,t) \equiv 10^{-4}$ and we use boundary condition $C_0(t) \equiv 1$. Sorption parameter $K = 6.95$. We use time step $\tau = 0.1$ and space grid step $h = 0.1$ for both finite difference and operator splitting schemes. As we can see in fig. 1, even with these relatively large time and space steps, the operator splitting method was able to obtain very precise results, while the finite difference scheme already suffered from a significant numerical dispersion.

Now we take a 2D problem with two wells situated at points $(5,0)$ (injection well) and $(-5,0)$ (extraction well). We consider both adsorption isotherms of the form $\Psi(C) = C^{0.75}$, dispersivities $\alpha_L = \alpha_T = 0.1$, molecular diffusion with $D_0 = 0.02$, sorption coefficient $K = 0.1$. Boundary condition is of pulse shape form, $C_0(t) = 1$ for $t < 2.0$ and $C_0(t) = 0$ otherwise. We use a space grid with 80×400 grid points and time step $\tau = 0.04$. The lines in the picture represent concentration levels ($C = 0.05, 0.1 \ldots, 1.0$) at time $t = 6.0$.

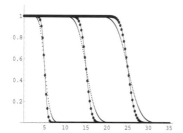
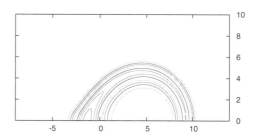

Figure 1: Picture 1: Comparison of analytical solution (solid line) with solution obtained by operator splitting scheme (large dots) and finite difference scheme (dotted line) at time levels 10, 30, 50; Picture 2: Contaminant concentration levels in a 2D domain after 6 days

References

[1] D. Constales, J. Kačur, B. Malengier: Parameter identification by means of dual-well tests. Accepted for publication in *Water Resources Research*

[2] J. Kačur, P. Frolkovič: Semi-analytical solutions for contaminant transport with nonlinear sorption in 1D. *University of Heidelberg, SFB 359*, **24** Preprint 2002, 1-20.

[3] M. Remešíková: Solution of convection-diffusion problems with non-equilibrium adsorption, *Journal of Computational and Applied Mathematics*, Vol 169/1, pp 101-116, 2004

[4] N. Toride, F. J. Leij, M. T. van Genuchten: A comprehensive set of analytical solutions for nonequilibrium solute transport with first-order decay and zero-order production, *Water Resources Research*, 29 (1993), pp. 2167-2182.

VSP International
Science Publishers
P.O. Box 346, 3700 AH Zeist
The Netherlands

Lecture Series on Computer and Computational Sciences
Volume 1, 2004, pp. 457-460

Monte Carlo Study of Polymer Translocation Through a Hole

Piotr Romiszowski[1] and Andrzej Sikorski

Department of Chemistry,
University of Warsaw,
Pasteura 1, 02-093 Warsaw, Poland

Received 2 August, 2004; accepted in revised form 20 August, 2004

Abstract: In this work we studied a simplified model of a polypeptide chain. The model chain was embedded to a [310] lattice. It was represented as a sequence of united atoms located at the positions of alpha carbons. The force field used consisted of the long-range contact potential between united atoms with the excluded volume and with a local preference of forming α-helical states. The chain was built of hydrophilic and hydrophobic segments forming helical septets –*HPPHPP*-. The properties of model chains were determined using the Monte Carlo simulation method based on a Metropolis-like algorithm. During the simulations the translocation of the chain through a hole in an impenetrable wall was observed. The influence of the chain length, the temperature differences on both side of the wall and the size of the hole on the chain properties were investigated.

Keywords: lattice models; Monte Carlo simulations; polymer translocation; polypeptide chains

PACS: 05.10.Ln, 61.25.Hq, 87.15.He

The process of transport of a polymer chains (proteins, DNA, RNA, drug delivery) across a membrane is very important because of its biological implications [1]. These processes are very complicated and therefore, the determining of some general parameters concerning the translocation process can be done studying simplified models of protein chains [2]. Some simplified models of polypeptide were recently studied by the means of computer simulations [3-4]. These simplified models were appeared to be useful for studying the influence of the sequence of amino acid residues in the chain (so-called primary structure) on its secondary structures (α-helices and β-strands). It was shown that the amount of secondary structures formed in chains at low temperatures was rather limited and much lower than in real globular proteins. The extension this model by introducing local preferences for the formation of secondary structures (helical states) showed the importance of this potential. The further extension of this model applying the combining of a classical Metropolis-type simulation method with the Histogram Method led to the full thermodynamic description of the folding transition. The properties of dense polypeptide systems forming a brush based on very similar models we also studied [5]. Theoretical predictions and simulations of homopolymer chains were also recently published [6-7].

In this work we rebuilt the above mentioned models of a polypeptide chains in order to study the translocation of chains through a hole in the impenetrable surface. In our model all atomic details of polypeptide chain were omitted as we were interested in finding some general properties of protein chains independent of its chemical structure. The model chains were built of statistical segments only, which were the approximation of the entire amino acid residues [5]. Each amino acid residue was reduced to a single united atom located at the alpha carbon position. The positions of these united atoms were restricted to vertices of a lattice [3 1 0] consisted of the vectors of the type: [±3, ±1, ±1], [±3, ±1, 0], [±3, 0, 0], [±2, ±2, ±1], [±2, ±2, 0]. In Figure 1 we present a fragment of a structure polypeptide chain with its atomic details as well as its united atom representation. Assuming the lattice unit equal to 1.22 Å the lattice representation of protein chains appeared to be given with the accuracy of 0.6-0.7 Å when compared with real chains [4].

---

[1] Corresponding author. E-mail: prom@chem.uw.edu.pl

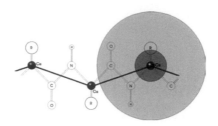

Fig.1 United atom representation of a polypeptide chain.

The model chain was put into a large Monte Carlo box with edge $L = 200$. One surface in the box was made impenetrable for chain's segments (we chosen the surface $x = 0$) The periodic boundary conditions were not imposed because the size of chains were lower that the edge L and simulations were carried out when chains were located near the surface $x = 0$. A hole in the impenetrable surface was made and its size was changed from 5 to 50 lattice units (from the size comparable with a single residue to the size comparable with the size of the entire chain). This surface can be treated as a very crude model of a membrane. The force field used consisted of two parts, long-range and short-range. A pair residues interacted with a contact potential V_{ij} which had the following form:

$$V_{ij} = \begin{cases} \varepsilon_{rep} & for \quad r_{ij} < r_1 \\ \varepsilon_a & for \quad r_1 \le r_{ij} \le r_2 \\ 0 & for \quad r_{ij} > r_2 \end{cases} \tag{1}$$

where r_{ij} is a distance between a pair residues. Basing on previous studies of this model the range of the repulsive part of the potential was assumed $r_1 = 3^{1/2}$ (in lattice units) while the cut-off radius was $r_2 = 5$ (in lattice units). The attractive part of the potential ε_a is the only parameter that distinguish the kind of a residue. In this model we assumed that polypeptides were constructed of two kinds of amino acid residues only: hydrophilic (H) and hydrophilic (P). The attractive part of the potential took the values $\varepsilon_{HH} = -2$, $\varepsilon_{PP} = -1$, $\varepsilon_{HP} = 0$ for a pair HH, a pair PP and a pair of HP respectively. This selection was justified previous theoretical considerations and simulations. In the present study we simulated chains consisted of $N = 10$ up to 100 residues. Chains were built of a typical helical sequence (septets ...÷ $HPP\underline{H}PP\underline{-}...$) [5, 8].

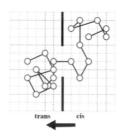

trans   cis

Fig.2 Polypeptide model chain restricted to (310) lattice on *cis* and *trans* sides of a membrane. An arrow shows the direction of the translocation.

The presence of helical states can be determined form the analysis of the following expression:

$$r^{*^2}_{i-1,i+2} = (v_{i-1} + v_i + v_{i-1})^2 \cdot sign((v_{i-1} \times v_i) \cdot v_{i+1}) \tag{2}$$

where v_{i-1}, v_i and v_{i+1} are statistical segments connecting residues number $i-1$, i and $i+1$. Left-handed α-helical states correspond to the values of the above parameter between 9 and 25. In was assumed that the formation of one helical state involved the energy gain ε_{loc}. The simulations were performed for values of $\varepsilon_{loc} = 0$ and -4.

The model chains were simulated by means of the Monte Carlo method. The initial conformation of the chain was chosen at random and the chain was placed near the hole in the impenetrable surface. Its conformation was locally changed using the following set of micromodifications: one-residue motion, two-residue motion and two-residue end reorientations. A new probe conformation was accepted with such transition probability that it was visited with the frequency proportional to its Boltzmann's factor. The temperature was introduced into the model as a parameter that scales the potential. A large number of the above mentioned local changes of chain's conformation were carried out (usually it was of the order of 10^8). The local changes of conformations caused the diffusive motion of the entire chain. We monitored this motion when the chain was passing the hole. In the case the chain wandered out of plane (towards higher values of x) the simulation was suspended, the chain was translated and placed near the hole, and the entire simulation procedure was resumed. In the case the entire chain passed the simulation was also suspended. Simulation runs were performed several times (usually 100) starting from quite different initial conformations of chains.

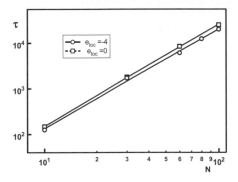

Fig.3 The mean time of chain's translocation τ as a function of the chain length N. The case of the hole size $d = 15$. The values of local potentials are given in the inset.

Basing on the performed simulations we studied the influence of the chain length and the size of the hole on the time of translocation. In Figure 3 we present the translocation times as a function of the number of residues for different helical potential. One can see that the translocation time for both cases scales similarly as $N^{2.16\pm0.11}$ and $N^{2.20\pm0.05}$ for $\varepsilon_{loc}=0$ and -4 respectively. This behavior is in a good agreement with other theoretical findings [6-7]. It was also found that the translocation time depends linearly on the size of the hole when this size is smaller that the mean diameter of the chain. For larger holes the translocation time is rather constant. In Figure 6 the changes of the helical contents (hereafter helicity) along with the simulation time are presented. The helicity was defined as the ratio of number of residues which were involved in forming helical turns to the total number of residues in the chain. In the time range in which the translocation through the pore takes place the helicity of the molecule rapidly growths from the level of a statistical helicity to 0.5 for helical potential –4. In the case of the chain without the helical potential no significant changes of helicity were found during and after the translocation.

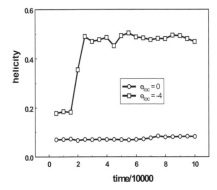

Fig.4 The changes of the helicity of model chain with time. The case of chain consisted of $N = 80$ residues with. The values of local potentials are given in the inset.

The calculations performed for a simplified model of a polypeptide chain crossing the membrane enables one to make the model more complicated and extended.

References

[1] D. K. Lubensky, D. R.Nelson, *Biophysical Journal* 77, 1824 (1999).

[2] Y. Lansac, P. K. Maiti, M. A. Glaser, *Polymer* 45, 3099 (2004).

[3] A. Kolinski, P. Madziar, *Biopolymers* 43, 537 (1997).

[4] P. Romiszowski, A. Sikorski, *Biopolymers* 54, 262 (2000).

[5] A. Sikorski, P. Romiszowski, *Biopolymers,* 69, 391 (2003).

[6] M. Muthukumar, *Journal of Chemical Physics* 111, 10371 (1999).

[7] M. Muthukumar, *Physical Review Letters* 86, 3188 (2001).

[8] A. Sikorski, P. Romiszowski, *Journal of Chemical Information and Computer Sciences* 44, 387 (2004).

[9] E. A. Di Marzio, J. J. Kasianowicz, *Journal of Chemical Physics* 119, 6378 (2003).

VSP International
Science Publishers
P.O. Box 346, 3700 AH Zeist
The Netherlands

*Lecture Series on Computer
and Computational Sciences*
Volume 1, 2004, pp. 461-463

Properties of Star-Branched and Linear Chains in Confined Space. A Computer Simulation Study

P.Romiszowski[1] and A.Sikorski

Department of Chemistry, University of Warsaw
02-093 Warszawa, Poland

Received 6 August, 2004; accepted in revised form 20 August, 2004

Abstract: We studied the properties of simple models of linear and star-branched polymer chains confined in a slit. The polymer chains were built of united atoms and were restricted to a simple cubic lattice. Two macromolecular architectures of the chain: linear and star-branched with 3 branches (of equal length) were studied. The excluded volume was the only potential introduced into the model and thus, the system was athermal. The chains were put between two parallel and impenetrable surfaces. Monte Carlo simulations with a sampling algorithm based on chain's local changes of conformation were carried out. The differences and similarities in the structure and of linear and star-branched chains were shown. ??

Keywords: branched polymers, lattice models, linear polymer, Monte Carlo method

PACS: 02.50.Ng, 03.10.Ln, 61.25.Hq

The presence of the confinement between impenetrable surfaces changes dramatically most of the properties of polymers when compared with the free (unconfined) chains in solution. This problem is also interesting due to its practical applications like lubrication, polymer films, and colloidal stabilization [1, 2].

The purpose of this study was to investigate the influence of the confinement and the chain topology on some static properties of polymers. We studied single polymer chains what corresponded to infinite diluted solution. Two different types of polymer chains having different macromolecular architecture were studied: linear chains and star-branched chains on a simple cubic lattice. Star-branched polymers consisted of three linear chains starting from a common origin (the branching point). The ban of the double occupancy of lattice sites by polymer segments introduced the excluded volume. The chains were put between two impenetrable parallel surfaces [3].

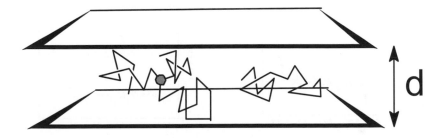

Fig.1 Scheme of the chains confined between the two parallel plates located at distance *d* apart: star-branched (left) and linear (right).

---

[1] Corresponding author. E-mail: prom@chem.uw.edu.pl

The model polymer systems were studied by means of Monte Carlo method. In the simulation algorithm the conformation of a polymer chain was randomly modified using the usual set of micromodifications: 2-bond move, 2-bond move, 3-bond crankshaft move and chain's ends moves. The star-branched chains were also the subjects of the branching point micromodifications what enabled the whole molecule to move freely within the space confined by the two walls and with no limits in xy plane.

The Monte Carlo calculations were performed for star-branched and linear chains models that contained the same number of beads $n = 50, 100, 200$ and 800 beads in order to make the results comparable for both types of chains. We studied properties of our model chains for distances between the two parallel impenetrable surfaces between $d = 3$ and $d = 80$. The resulting trajectories were than analyzed and the properties of the molecules were determined. We have studied the dimensions of molecules such as mean-square end-to-end vector (for star-branched molecules it was the center-to-end vector) $<R^2>$, mean-square radius of gyration $<S^2>$ and their ratio as the function of the distance d between the walls. Our aim was to find a general dependence describing the properties of macromolecules between the walls despite of their length as well as the degree of confinement.

The static properties of confined polymer chains depend on the degree of the confinement what is equivalent to the compression ratio defined as the relation of the chain dimension to the size of the slit. The open question is whether or not the architecture of the polymer have significant influence on the properties of confined chains.

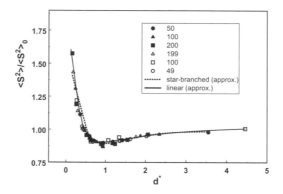

Fig.2 Plot of $<S^2>/<S^2>_0$ as a function of d^* for linear (solid symbols) and star-branched (empty symbols) chains – see inset for the chain length. The line approximate the data..

We calculated the mean-squared radius of gyration of chains of different length and architecture and then we normalized these values dividing them by the appropriate values of $<S^2>_0$ which were found for unconfined chains. The ratio $<S^2>/<S^2>_0$ was plotted against the normalized size of the slit, calculated as $d^*=d/2<S^2>_0^{1/2}$. One can observe that the calculated values for both architectures for one common curve, despite of the chain length. The curve has one minimum in the vicinity of $d^* = 1$.

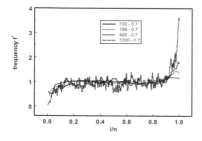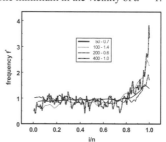

Fig.3 The frequencies of chain-walls interactions for star-branched (left) and linear (right) chains. The length of the chain and the values of d^* are given in insets.

The interactions between the confined chain and the walls can be described by the number of contacts between the chain beads and the wall. The frequency of the contacts were determined for each bead of the chain separately (bead numbering started from the branching point of the chain – the linear chain was treated as the star with two branches of length $n/2$). In order to compare the results for different chain length we normalized the frequencies for all cases as follows:

$$f_i^* = \frac{f_i \cdot n}{\sum\limits_i f_i} \tag{1}$$

where f_i is the number of contacts per one time unit. One can see that the normalized frequency data for different chain length and for different confinement form one common curve. However in the curve for star-branched molecules one can observe three different regimes: one on the vicinity of the branching point ($i/n < 0.1$), then the plateau region ($i/n = 0.1 \div 0.9$) and finally the end of chain region ($i/n > 0.9$). The curve for the linear chains consists only of a plateau and the end chain region ($i/n > 0.8$). This difference between the curves is caused by presence of the relatively dense core in the vicinity of the branching point of the star-branched molecules.

References

[1]   R. Eisenriegler, *Polymers near Surfaces*, World Scientific, Singapore 1993.

[2]   A. Milchev, K. Binder, *European Physical Journal B* 3, 477 (1998).

[3]   P. Romiszowski, A.Sikorski, *Journal of Chemical Physics* 116, 1731 (2002).

[4]   A. Sikorski, P. Romiszowski, *Journal of Chemical Physics* 120, 7206 (2004).

VSP International
Science Publishers
P.O. Box 346, 3700 AH Zeist
The Netherlands

Lecture Series on Computer
and Computational Sciences
Volume 1, 2004, pp. 464-466

A Trigonometrically Fitted Method of Seventh Algebraic Order for the Numerical Solution of the One-Dimensional Schrödinger Equation

D.P. Sakas and T.E. Simos[1]

Department of Computer Science and Technology,
Faculty of Sciences and Technology,
University of Peloponnese,
GR-221 00 Tripolis, Greece

Received 5 August, 2004; accepted in revised form 25 August, 2004

Abstract: In this paper we will develop a trigonometrically fitted Adams-Bashforth-Moulton predictor-corrector method which will be used for the approximate solution of the resonance problem of the Schrödinger equation. Our new trigonometrically fitted P-C scheme is based on the well known Adams-Bashforth-Moulton methods. In particular, they are based on the sixth order Adams-Bashforth scheme (as predictor) and on the seventh order Adams-Moulton scheme (as corrector). The application of the new algorithm to the numerical solution of the Schrödinger equation shows the efficiency of the new developed algorithm.

Mathematics Subject Classification: 65L05

1 Introduction

Equations or systems of equations of the form

$$\mathbf{y}'(\mathbf{x}) = \mathbf{f}(\mathbf{x}, \mathbf{y}), \ \mathbf{y}(\mathbf{x_0}) = \mathbf{y_0}, \tag{1}$$

are used in the mathematical models for the problems arising from several scientific areas such as: celestial mechanics, physical chemistry and chemical physics, quantum mechanics, electronics, nanotechnology, materials sciences and elsewhere. Special attention deserve the class of the above equations with oscillatory/periodic solution (see [1] and [2]).

The numerical solution of the above equation is the subject of extensive research activity the last three decades (see [3]-[4] and references therein). One of the most efficient procedures for the construction of numerical methods for the integration of first order initial value problems with oscillating or periodic solution is the exponential and the trigonometric fitting technique first introduced by Lyche [5]. Psihoyios and Simos [6] had applied trigonometric fitting to a lower algebraic order Adams-Bashforth-Moulton P-C scheme for the solution of (1), with excellent results.

---

[1]Corresponding author. President of the European Society of Computational Methods in Sciences and Engineering (ESCMSE). Active Member of the European Academy of Sciences and Arts. E-mail: simos-editor@uop.gr, tsimos@mail.ariadne-t.gr

The radial Schrödinger equation has the form:

$$y''(r) = [l(l+1)/r^2 + V(r) - k^2]y(r). \tag{2}$$

Models of this type, which represent a boundary value problem, occur frequently in theoretical physics and chemistry, (see for example [7] - [8]). It is known from the literature that during the last decades many numerical methods have been constructed for the approximate solution of the Schrödinger equation (see indicatively [9] - [4]). The aim and the scope of the above activity was the development of fast and reliable methods. The developed methods could be divided into two main categories:

- Methods with constant coefficients.

- Methods with coefficients dependent on the frequency of the problem [2].

2 A new family of trigonometrically fitted seventh algebraic order P-C schemes

Consider the following seventh algebraic order six-step scheme:

$$\bar{y}_{n+1} = y_n + h\left(a_0\,f_n + a_1\,f_{n-1} + a_2\,f_{n-2} + a_3\,f_{n-3} + a_4\,f_{n-4} + a_5\,f_{n-5}\right)$$

$$y_{n+1} = y_n + h\left(c_0\,\bar{f}_{n+1} + c_1\,f_n + c_2\,f_{n-1} + c_3\,f_{n-2} + c_4\,f_{n-3} + c_5\,f_{n-4} + c_6\,f_{n-5}\right) \tag{3}$$

where, in terms of f_{n-i}, a_i, $i = 0(1)5$ are the known Adams-Bashforth coefficients and the c_i, $i = 0(1)6$ coefficients correspond to the Adams-Moulton coefficients.

Our new family of methods has been constructed in order the above method (3) to be exact for any linear combination of the functions:

$$\{1,\ x,\ x^2,\ x^3,\ \cos(\pm vx),\ \sin(\pm vx),\ x\cos(\pm vx),\ x\sin(\pm vx)\} \tag{4}$$

References

[1] L.D. Landau and F.M. Lifshitz: *Quantum Mechanics*. Pergamon, New York, 1965.

[2] I. Prigogine, Stuart Rice (Eds): Advances in Chemical Physics Vol. 93: New Methods in Computational Quantum Mechanics, John Wiley & Sons, 1997.

[3] T.E. Simos: *Numerical Solution of Ordinary Differential Equations with Periodical Solution.* Doctoral Dissertation, National Technical University of Athens, Greece, 1990 (in Greek).

[4] A. Konguetsof and T.E. Simos, On the construction of Exponentially-Fitted Methods for the Numerical Solution of the Schrödinger Equation, *J. Comput. Meth. Sci. Eng.* **1** 143-165(2001).

[5] T. Lyche: Chebyshevian multistep methods for ordinary differential equations, Numerische Mathematik, **10** 65-75(1972).

[6] G. Psihoyios and T.E. Simos: Trigonometrically fitted predictor-corrector methods for IVPs with oscillating solutions, *J. of Computational and Applied Mathematics*, vol. 158 (no. 1), pp. 135-144, 2003

---

[2]In the case of the Schrödinger equation the frequency of the problem is equal to: $\sqrt{|l(l+1)/r^2 + V(r) - k^2|}$

[7] L.Gr. Ixaru and M. Micu, *Topics in Theoretical Physics.* Central Institute of Physics, Bucharest, 1978.

[8] G. Herzberg, *Spectra of Diatomic Molecules*, Van Nostrand, Toronto, 1950.

[9] T.E. Simos, Atomic Structure Computations in Chemical Modelling: Applications and Theory (Editor: A. Hinchliffe, UMIST), *The Royal Society of Chemistry* 38-142(2000).

VSP International
Science Publishers
P.O. Box 346, 3700 AH Zeist
The Netherlands

*Lecture Series on Computer
and Computational Sciences*
Volume 1, 2004, pp. 467-470

Computational Methods in Solving the Additivity Problem in Chemistry: Conformational Energies of Chloroalkanes

D.Šatkovskienė, P. Pipiraitė, R. Jankauskas and J. Šulskus

Department of Theoretical Physics,
Faculty of Physics,
University of Vilnius
LT-10222 Vilnius, Lithuania

Received 23 July, 2004; accepted in revised form 19 August, 2004

Abstract: The additivity of conformational energies for chloro-substituted saturated hydrocarbons has been studied using various quantum mechanical methods. The starting point of investigation was the analytically predicted [1] additivity rule stating that the conformational energy of any saturated compound can be represented as a sum of transferable increments corresponding to the energies of separate conformational segments and their sequences. The quantum mechanical methods (HF anf MP2) with various bases sets available have been used for evaluation of additivity increments of conformational energy E_G^{XY}, E_{GT}^{XY}, E_{GG}^{XY}, $E_{GG'}^{XY}$ ($X=Y= CH_3$; $X= CH_3$, $Y =Cl$ and $X=Y=Cl$). It was shown that the evaluated magnitudes of additive increments do not depend on the size of the molecule, the location of the conformational segment in the molecule and the number of substitutes involved. However they are sensitive to the method and the basis set used. The arrangement of relative conformational energies for various conformers in series of polysubstituted chloroalkanes obtained using the additivity rule coincides both with experimental findings and quantum mechanical calculations.

Keywords: additivity, conformational energies, chloroalkanes.

Physics Subject Classification: PACS: 33.15.Bh

1. Introduction

The concepts of additivity and transferability are closely related to the postulates of classical chemistry about the structure of molecules which state the possibility expressing the physical characteristics of compounds as sums of transferable partial values corresponding to atoms, bonds etc. The "additivity rules" are promising for the simulation of macromolecular systems. However, for this purpose it is necessary to show that the value of given physical characteristic obtained from small molecule is transferable with enough good accuracy to any larger compound of the similar structure. The purpose of present contribution is to study the ability of various quantum mechanical methods and the bases sets available to estimate the conformational energy increments for chloroalkanes for the quantum mechanically based additivity rule [1]. Therein using the perturbation theory for one electron density matrix it was shown that the conformational energy of saturated compounds can be presented as a sum of transferable increments corresponding to the energies of separate conformational segments and their sequences

$$E^{conf} = E - E^T = n_G^{XY} E_G^{XY} + n_{TG}^{XY} E_{TG}^{XY} + n_{GG}^{XY} E_{GG}^{XY} + n_{GG'}^{XY} E_{GG'}^{XY} + \ldots \quad (1)$$

where E is the total energy of molecule in the given conformation; E^T - total energy of the molecule in the lowest energy conformation; E_G^{XY}, E_{GT}^{XY}, E_{GG}^{XY}, $E_{GG'}^{XY}$ (X, $Y = CH_3$, Cl, F, Br, etc.) - energies of corresponding additivity increments; n - denotes the number of corresponding segments in the given conformation.

The following requirements to the calculations were formulated: a) the ability of the method to describe correctly the main characteristics of the molecule (the geometry, dipole moments as well as the relative conformational energies); b) the achievement of the optimal concert at the same level of calculations for obtained values of increments corresponding to the conformational segments as well

468 _____ *D.Šatkovskienė et.al*

as their sequences; c) the possibility to perform calculations for relatively large compounds by means of chosen method and basis set. The mentioned criteria have been applied to the main conformational segments of chloroalkanes: E_G^{XY} and their sequences: E_{GT}^{XY} , E_{GG}^{XY} , $E_{GG'}^{XY}$ ($X=Y=CH_3$; $X=CH_3$, $Y=Cl$ and $X=Y=Cl$) (X, $Y=CH_3$, Cl). The values of additivity increments corresponding to the mentioned conformational segments were calculated using the first members of homologous series of n- and haloalkanes: butane, 1-chloropropane and 1, 2-dichloroethane as difference between total energies of molecules in gauche (G) and anti (A) conformations: $E_G - E_A = E_G^{XY}$ ($X=Y=CH_3$; $X=CH_3$, $Y=Cl$ and $X=Y=Cl$). The additivity increments corresponding to energy of the sequences of conformational segments E_{GG}^{XX}, E_{GT}^{XX} and $E_{GG'}^{XX}$ ($X=Cl$) were calculated from GG, GG', GA and AA forms[1] of 1, 3-dichloropropane using Eq.1:

$$E_{GT}^{XX} = E_{GA} - E_{AA} - E_G^{XY},$$
$$E_{GG}^{XX} = E_{GG} - E_{AA} - 2E_G^{XY},$$
$$E_{GG'}^{XX} = E_{GG'} - E_{AA} - 2E_G^{XY},$$

where $X=Cl$, $Y=CH_3$; E_{GA}, E_{AA}, E_{GG}, $E_{GG'}$ are total energies of corresponding conformers. Whereas the additivity increments E_{GG}^{XY}, $E_{GG'}^{XY}$ and E_{GT}^{XY} ($X=Cl$, $Y=CH_3$) were estimated using total energy calculations of 1-chlorobutane in GG, GG', GA, AG and AA forms using Eq.1:

$$E_{GT}^{XY} = E_{GA} - E_{AA} - E_G^{XY},$$
$$E_{TG}^{XY} = E_{AG} - E_{AA} - E_G^{YY},$$
$$E_{GG}^{XY} = E_{GG} - E_{AA} - E_G^{XY} - E_G^{YY},$$
$$E_{GG'}^{XY} = E_{GG'} - E_{AA} - E_G^{XY} - E_G^{YY},$$

The notation E_{GT}^{XY} means that the substitute X is in *gauche* position with respect to the fixed plane of hydrocarbon chain.

2.Results and discussion

The presented study shows that for satisfying the above formulated requirements it is necessary to perform sufficiently accurate calculations. For example, the calculations performed for 1-chloropropane show that the relative conformational energy (RCE) value obtained in HF calculations has the incorrect sign and almost does not depend on the number of polarization functions included. Thus it can be concluded that influence of electron correlation for conformational energy is very important. On the other hand the MP2 calculations of structural parameters as well as RCE are in sufficiently good agreement with experimental data [3,4] only when polarization functions are taken into account, whereas the results obtained by MP2/6-311G without polarization functions give too high dipole moment (μ) and RCE values. Consequently it might be expected that the only MP2 method with polarization functions for all basis sets used enables to reach the results in the scope of experimental estimates. Therefore for further calculations we confine ourselves by two basis sets: the largest one – 6-311G (3d1f,3p), which gives the dipole moment values close to experimental and demonstrates saturating behavior of RCE when 3p basis functions on hydrogen atoms are added to (3d1f) basis set and the least one- 6-311G (2d,2p) in which the RCE values concert with experimental results [5,6] and which is sufficiently compact for calculations of large molecules. The obtained results for additivity increments and RCE presented in Table 1 demonstrate good agreement with experimental data.

To check the transferability of obtained increments we compared the RCE values calculated using Eq.1 with obtained by direct MP2 calculations and available experimental data for all conformations of 1,2-dichloro – and 1,2,3-trichloropropanes, 2-chlorobutane , 2,3-, 1,2-, and 1,3-dichlorobutanes. In most cases the results show the identical arrangement of RCE. Only for some conformers with close energy values the vice versa arrangement of conformer energy follows from the additivity rule. The observed situation may be related to obtained distortions of geometry of rotamers and attributed to the limiting factors of additivity rule under discussion, especially if considering that the theoretical approach [1] used for obtaining the Eq.1 was based on the assumption that the bond lengths and torsion angles are only little affected by conformational changes. In spite of that the results show that the calculated additivity increments in common are transferable and correctly reflect the arrangement of

_____
[1]All designations of conformations is adequate to those used in [2]

Table 1. RCE and additivity increments (kcal/mol) obtained from ab *initio* calculations

Chloroalkanes (X=Cl, Y=CH₃)	RCE and additivity increments				
	HF/ 6-31G(d)	MP2/6-311G		Exp. data	
		(2d2p)	(3d1f3p)		
1,2-dichloroethane $E_G - E_A = E_G^{XX}$	1.91	1.51	1.18	1.1±0.1 [6]	
1-chloropropane $E_G - E_A = E_G^{XY}$	0.36	-0.15	-0.24	-0.15±0.01 [5]	
1,3-dichloropropane $E_{AA} - E_{GG}$	0.84	1.98	2.18	1.5±0.5	
$E_{GA} - E_{GG}$	0.44	1.09	1.22	-1.1±0.2	
$E_{GG'} - E_{GG}$	4.71	-	4.63	⟩3 [7]	
E_{GT}^{XX}	-0.76	-0.74	-0.72	-	
E_{GG}^{XX}	-1.56	-1.68	-1.70	-	
$E_{GG'}^{XX}$	3.15	-	2.93	-	
1-chlorobutane				Gas[8]	Liquid[9]
$E_{AG} - E_{GA}$	-	0.804	0.805	-	0.65±0.03
$E_{AA} - E_{GA}$	-	0.146	0.295	0±0.4	0.15±0.05
$E_{GG} - E_{GA}$	-	0.085	0	0.7±0.4	0.21±0.10
$E_{GG'} - E_{GA}$	-	2.25	2.13	$E_{GG'} \approx E_{GA}$	-
E_{TG}^{XY}	-	0.06	0.01	-	
E_{GT}^{XY}	-	0.0	-0.05	-	
E_{GG}^{XY}	-	-0.51	-0.55	-	
$E_{GG'}^{XY}$	-	1.65	1.58	-	

different conformational configurations of molecule on the energy scale.

3. Conclusions

The obtained results show that RCE values calculated by MP2 are in agreement with available experimental data when bases sets *6-311G (3d1f,3p)* and *6-311G (2d,2p)* are used. The arrangement of RCE values for large saturated molecules obtained applying Eq.1, with additivity increments calculated using mentioned basis sets, is in agreement both with obtained ones by direct calculations and, if available, with experimental estimates. It follows that the values of additivity increments do not depend on the size of the molecule, number and location of the substitutes and are transferable with enough good accuracy. This allows us to conclude that quantum mechanical methods can be used for determination of the additivity increments necessary for prediction of the relative conformational energies of large chloroalkanes.

References

[1] D.Šatkovskienė, *Int.J.Quant.Chem* **91,** 5 (2003).

[2] R. Stolevik, *J. Mol. Struct.* **352/353,** 23 (1995)

[3] K. Yamanouchi, M.Sugie, K.Takeo, C.Matsumura, K. Kuchitsu *J. Phys. Chem* **88,** 2315 (1984)

[4] W. A. Herrebout and B. J. Van der Vaken *J. Phys. Chem.* **100,** 9671(1996)

[5] J. R. Durig, X. Zhu, S. Shen *J. Mol. Struct.* **570,**1 (2001)

[6] K. B. Wiberg, T. A. Keith, M. J.Frisch, and M.A. Murcko *J. Phys. Chem.* **99,** 9072 (1995)

[7] S. Grindheim and R. Stolevik *Acta Chem. Scand.A* **30,** 625(1976)

[8] S. Fagerland, T.Rydland and R.Stolevik, *J. Mol. Struct.* **96,** 339 (1983)

[9] Y. Ogawa, S.Imazeki, H. Yamanouchi, H. Matsuura, I. Harada and T. Shimanouchi, Bull. Chem. Soc. Jpn., **51,**748(1978)

VSP International
Science Publishers
P.O. Box 346, 3700 AH Zeist
The Netherlands

Lecture Series on Computer
and Computational Sciences
Volume 1, 2004, pp. 471-474

All-Atom Protein Structure Prediction with Stochastic Optimization Methods

A. Schug, T. Herges, A. Verma, W. Wenzel[1]

Research Center Karlsruhe,
Institute for Nanotechnology,
PO Box 3640
76021 Karlsruhe
Germany

Received 30 July, 2004; accepted in revised form 21 August, 2004

Abstract: We recently developed an all-atom free energy forcefield (PFF01) for protein structure prediction with stochastic optimization methods. Here we review recent folding studies, which permitted the reproducible all-atom folding of the (i) 20 amino-acid trp-cage protein, (ii) the 36 amino acid villin headpiece, (iii) the 40-amino acid three-helix HIV accessory protein and (iv) the sixty amino acid bacterial ribosomal protein L20 with a variety of stochastic optimization methods. We could also demonstrate that PFF01 stabilized the native folds of various other proteins ranging from 40-60 amino acids at the all atom level. We compare the efficiency of different stochastic optimization methods for protein folding and give a brief outlook on the perspective of our thermodynamic approach.

Keywords: protein folding, free energy forcefield, stochastic optimization

PACS: 87.15.Cc,02.70.Ns,02.60.Pn

1   Introduction

Ab-initio protein tertiary structure prediction (PSP) and the elucidation of the mechanism of the folding process are among the most important outstanding problems of biophysical chemistry [1, 2]. The many complementary proposals for PSP span a wide range of representations of the protein conformation, ranging from coarse grained models to atomic resolution. The choice of representation often correlates with the methodology employed in structure prediction, ranging from empirical potentials for coarse grained models [3, 4] to complex atom-based potentials that directly approximate the physical interactions in the system. The latter offer insights into the mechanism of protein structure formation and promise better transferability, but their use incurs large computational costs that has confined all-atom protein structure prediction to all but the smallest peptides [5, 6].

It has been one of the central paradigms of protein folding that proteins in their native conformation are in thermodynamic equilibrium with their environment [7]. Exploiting this characteristic the structure of the protein can be predicted by locating the global minimum of its free energy surface without recourse to the folding dynamics, a process which is potentially much more efficient than the direct simulation of the folding process. PSP based on global optimization of the free

---

[1]Corresponding author. E-mail: wenzel@int.fzk.de

Figure 1: Overlay of the native and folded structures of trp-cage protein, the HIV accessory protein and the bacterial ribosomal protein L20.

energy may offer a viable alternative approach, provided that suitable parameterization of the free energy of the protein in its environment exists and that global optimum of this free energy surface can be found with sufficient accuracy [8].

We have recently demonstrated a feasible strategy for all-atom protein structure prediction [9, 10, 11] in a minimal thermodynamic approach. We developed an all-atom free-energy forcefield for proteins (PFF01), which is primarily based on physical interactions with important empirical, though sequence independent, corrections [11]. We already demonstrated the reproducible and predictive folding of four proteins, the 20 amino acid trp-cage protein (1L2Y) [9, 12], the structurally conserved headpiece of the 40 amino acid HIV accessory protein (1F4I) [10, 13], the villin headpiece [14] and the sixty amino acid bacterial ribosomal protein L20 [15]. In addition we could show that PFF01 stabilizes the native conformations of other proteins, e.g. the 52 amino-acid protein A [5, 16], and the engrailed homeodomain (1ENH) from *Drosophilia melangster* [17].

2 Results

Using the PFF01 forcefield we simulated 20 independent replicas of the 20 amino acid trp-cage protein [18, 6] (pdb code 1L2Y) with a modified versions of the stochastic tunneling method [19, 9]. Six of 25 simulations reached an energy within 1 kcal/mol of the best energy, all of which correctly predicted the native experimental structure of the protein(see Fig 1 (left)), in total eight simulations converged to the native structure. We find a strong correlation between energy and RMSD deviation to the native structure for all simulations.

Encouraged by this result, we applied a modified basin hopping or Monte-Carlo with minimization (MCM) strategy [8, 20] to fold the structurally conserved 40-amino acid headpiece of the HIV accessory protein [10] (see Fig 1 (center)). The minimization step in MCM simplifies the potential energy surface (PES), by mapping each conformation to a nearby local minium. For very rugged potential energy surfaces, such as those encountered in protein folding, local minimization yields comparatively little improvement. We have therefore replaced the local minimization by a simulated annealing (SA) [21] run [10] and found the lowest six of twenty independent simulations to converge to the native structure.

We note that both proteins could also be reproducibly folded with a modified parallel tempering method [12, 13]. The parallel (or simulated) tempering technique was introduced to overcome difficulties in the evaluation of thermodynamic observables for models with very rugged potential energy surfaces and applied previously in several protein folding studies [22]. The idea of the PT method is to perform several concurrent simulations of different replicas of the same system at

different temperatures and to exchange replicas (or temperatures) between the simulations i and j
with probability:

$$p = min(1, \exp{(-(\beta_j - \beta_i)(E_i - E_j))}), \tag{1}$$

where $\beta_i = 1/k_B T_i$ and E_i are the inverse temperatures and energies of the conformations respectively. The exchange mechanism, which ensures that all simulations remain in thermodynamic equilibrium at their respective temperatures, improves the conformational averaging of the low-temperature simulations.

While BHT and STUN are essentially sequential algorithms, PT provides some degree of inherent parallelism. Extending this idea we recently succeeded to predictively fold the sixty amino acid bacterial ribosomal protein L20 (Fig. 1 (right)) [15] with a evolutionary strategy, in which the computational work is performed by many of independent client-computers that request tasks from a master-computer. The master maintains a list of open tasks comprising the active conformations of the population. Each client performs an increasingly extensive energy minimization on the conformation it is given. When the client returns a new conformation after completing its task, it may replace an existing conformation following a scheme that maintains a balance between the diversity of the population and the continued energetic improvement of its members. Again, six of the lowest ten conformations approached the native structure of the protein.

3   Discussion

Since the native structure dominates the low-energy conformations arising in all of these simulation, these results demonstrate the feasibility of all-atom protein tertiary structure prediction for a variety of proteins ranging from 20-60 amino acids in length. The methodology used here is predictive and scalable to world-wide distributed computational architectures (GRIDS). The free energy approach emerges as viable trade-off between predictivity and computational feasibility. The computational efficiency of the optimization approach stems from the possibility to visit unphysical intermediate high energy conformations during the search. While sacrificing the folding dynamics, a reliable prediction of its terminus, the native conformation — which is central to most biological questions — can be achieved.

Acknowledgments

We thank the Fond der Chemischen Industrie, the BMBF, the Deutsche Forschungsgemeinschaft (grants WE 1863/10-2,WE 1863/14-1) and the Kurt Eberhard Bode Stiftung for financial support. Part of the simulations were performed at the KIST teraflop cluster.

References

[1] D. Baker and A. Sali. Protein structure prediction and structural genomics. *Science*, 294:93, 2001.

[2] J. Schonbrunn, W. J. Wedemeyer, and D. Baker. Protein structure prediction in 2002. *Curr. Op. Struc. Biol.*, 12:348–352, 2002.

[3] N. Go and H. A. Scheraga. On the use of classical statistical mechanics in the treatment of polymer chain conformation. *Macromolecules*, 9:535–542, 1976.

[4] P. Ulrich, W. Scott, W.F. W. F. van Gunsteren, and A. E. Torda. Protein structure prediction forcefields: Paramterization with quasi newtonian dynamics. *Proteins, SF&G*, 27:367–384, 1997.

[5] C. D. Snow, H. Nguyen, V. S. Panda, and M. Gruebele. Absolute comparison of simulated and experimental protein folding dynamics. *Nature*, 420:102–106, 2002.

[6] C. Simmerling, B. Strockbine, and A. Roitberg. All-atom strucutre prediction and folding simulations of a stable protein. *J. Am. Chem. Soc.*, 124:11258, 2002.

[7] C. B. Anfinsen. Principles that govern the folding of protein chains. *Science*, 181:223–230, 1973.

[8] Z. Li and H.A. Scheraga. Monte carlo minimization approach to the multiple minima problem in protein folding. *Proc. Nat. Acad. Sci. U.S.A.*, 84:6611, 1987.

[9] A. Schug, T. Herges, and W. Wenzel. Reproducible protein folding with the stochastisc tunneling method. *Phys. Rev. Letters*, 91:158102, 2003.

[10] T. Herges and W. Wenzel. Reproducible in-silico folding of a three-helix protein in a transferable all-atom forcefield. http://www.arXiv.org: physics/0310146, 2004.

[11] T. Herges and W. Wenzel. Development of an all-atom forcefield for tertiary structure prediction of helical proteins. Biophysical Journal (in press)), 2004.

[12] A. Schug, T. Herges, and W. Wenzel. All-atom folding of the trp-cage protein in an all-atom forcefield. *Europhyics Lett.*, 67:307–313, 2004.

[13] A. Schug, T. Herges, and W. Wenzel. All-atom folding of the three-helix hiv accessory protein with an adaptive parallel tempering method. Proteins (in press), 2004.

[14] T. Herges, A. Schug, and W. Wenzel. Protein structure prediction with stochastic optimization methods: Folding and misfolding the villin headpiece. (accepted for publication in Lecture Notes in Compuational Science,(Springer, New York)), 2004.

[15] A. Schug, T. Herges, and W. Wenzel. Reproducble folding of a four helix protein in an all-atom forcefield. submitted to J. Am. Chem. Soc., 2004.

[16] H. Gouda, H. Torigoe, A. Saito, M. Sato, Y. Arata, and I. Shimanda. Three-dimensional solution structure of the b domain of staphylococcal protein a: comparisons of the solution and crystal structures. *Biochemistry*, 40:9665–9672, 1992.

[17] U. Mayor, N. R. Guydosh, C. M. Johnson, J. G. Grossmann, S. Sato, G. S. Jas, S. M. V. Freund, D. O. V. Alonso, V. Daggett, and A. R. Fersht. The complete folding pathway of a protein from nanoseconds to micorseconds. *Nature*, 421:863–867, 2003.

[18] J. W. Neidigh, R. M. Fesinmeyer, and N. H. Anderson. Designing a 20-residue protein. *Nature Struct. Biol.*, 9:425–430, 2002.

[19] W. Wenzel and K. Hamacher. Stochastic tunneling approach for global optimization of complex potential energy landscapes. *Phys. Rev. Lett.*, 82:3003, 1999.

[20] J. P.K. Doyle and D. Wales. On potential energy surfaces and relaxation to the global minimum. *J. Chem. Phys.*, 105:8428, 1996.

[21] S. Kirkpatrick, C.D. Gelatt, and M.P. Vecchi. Optimization by simulated annealing. *Science*, 220:671–680, 1983.

[22] C.Y. Lin, C.K. Hu, and U.H. Hansmann. Parallel tempering simulations of hp-36. *Proteins*, 53:436–445, 2003.

VSP International
Science Publishers
P.O. Box 346, 3700 AH Zeist
The Netherlands

Lecture Series on Computer
and Computational Sciences
Volume 1, 2004, pp. 475-478

Presheaf and Sheaf Computation

Elena Sendroiu[1]

Université Paris 7 - Denis Diderot
F-75251, Paris, France

Received 5 August, 2004; accepted in revised form 28 August, 2004

Abstract: The paper presents first how we can compute sheaf tools for accomplishing our goals, i.e. a method of authentication by sheaves. By handling the concepts of Goguen, the behaviour of a system is computed in the internal language of a (pre)sheaf topos. Non-interference by Goguen is also computed in the internal language of a presheaf topos by using predicates.

Keywords: category theory, topos theory, presheaves, sieves, site, sheaves, authentication, non-interference

AMS-MOS *Mathematics Subject Classification:* 18, 18B25, 18F10, 18F20

1. Introduction

We are using topos theory [4] to perform computation for the following considerations. A topos is a kind of generalized set theory in which the logic is intuitionistic instead of classical. The interpretation of the internal language of a topos allows the consideration of a topos as a type system. Thus, the concept of the type has a mathematical definition i.e. a type is an object of a topos. Also, the concept of the subtype is well defined by a subobject [4]. In a topos, a subobject corresponds to a predicate. This is formulated in set theory as follows: a predicate corresponds to a subset. Then to construct objects in topos, we can use not only familiar operations like finite limits and exponentials, but also, the set-formation process $\{x \in X | \phi[x]\}$, where $\phi[x]$ is a formula in its internal language [4]. A topos is a particular bicartesian closed category [3]. Since the subobject classifier provides isomorphisms then a topos is richer in type isomorphisms than a bicartesian closed category. In addition, these isomorphisms are handled in computation.

2. Sheaf Tools for Computation

In [6], we have given a method of authentication by sheaves. This section presents first how we can compute sheaf tools for accomplishing this goals. As necessary background, this section gives, first of all, some ideas about sheaf theory. For more explanations see [4].

So, we consider a base category C. A presheaf P is a contravariant functor $P\colon C^{op} \longrightarrow Sets$. The category of presheaves $Sets^{C^{\circ}}$ is a topos. A Grothendieck topology J, on C, is a collection of sieves that satisfy three axioms, (maximality, stability, and transitivity) [4]. A sieve on a point (node) C is a family of arrows with the same codomain C by definition, and is closed under left composition, i.e. sieve rule. In particular, a sieve of a Grothendieck topology is called covering sieve or cover. A site (C,J) is a pair consisting of a small category C and a Grothendieck topology J on C. A sheaf F for a Grothendieck topology J is a presheaf which satisfies the sheaf condition [4] on the site (C,J). The category of the sheaves $Sh(C,J)$ on a site (C,J) is a subcategory of the category of the presheaves $Sets^{C^{\circ}}$ In addition, $Sh(C,J)$ is also a topos.

---

[1] Corresponding author. E-mail: sendroiu@ccr.jussieu.fr

Now, we explain the construction procedure of sheaves. So, any presheaf is a colimit of a diagram of representable functors [4]. But any diagram of representable functors corresponds to a diagram of the base category. A colimit of a diagram of presheaves can be computed in the same way as in section 3.3 that computes a limit. By using the associated sheaf functor [4] we can associate a sheaf to any presheaf P for a certain Grothendieck topology J. In [6] we have described, starting from the base category, an algorithm for generating Grothendieck topologies and algorithms for construction and validation of sheaves. In order to write an optimal validation algorithm of sheaves, we have introduced the concept of family for matching [6] that allows us to reformulate the sheaf definition presented in [4]. By using this new notion we have given another associated sheaf functor [6]. Hence we have constructed an algorithm that generates efficiently sheaves.

Note that, in our modelling, all computation tools are building from the base category that is provided by a network. So, the base category C, used in our modelling, is generated by a network. The objects are network nodes and the arrows are connections between nodes. More clearly, this is the associated category to the network's graph. Interesting results are provided by a complex architecture of distributed computing systems.

It follows by sheaf properties the following relation: a value is in correspondence with a set of values distributed on certain points. This relation is that we propose to be used in distributed authentication [6,7]. We have seen that we may generate sheaves for this process.

More precisely, a sheaf has the property that there exists a unique amalgamation (an element x in a set) for any matching family [4] of any cover of any object in the category. The element (amalgamation) x can be thought as a seed which gives rise to a unique set of elements obtained by means of a sieve and distributed in the network nodes (the matching family corresponds in fact to this set).

Now, we present our method of authentication by sheaves. So, a password is an amalgamation (i.e. the unique value), hence a matching family (i.e. a set of values) for a covering sieve corresponds to some encryptions of the password on distributed network nodes. Therefore, if x is a password of a user and the matching family is kept by the authentication system, then the authentication process just checks whether or not x gives rise to the same family of matching.

3. Computing of the Non-interference

Non-interference is a definition of security introduced for the analysis of confidential information flow in computer systems. In particular, by using a notion of morphism images of presheaves, Goguen [2] extends non-interference definition to real time distributed concurrent object oriented databases. But, these images are exactly the morphism images in a (pre)sheaf topos. So, this section computes the non-interference by Goguen in the internal language by using predicates.

Note that, our approach is based on Goguen's theory that is the following: Objects are mo-delled by sheaves; inheritance by sheaf morphisms; systems by diagrams. In addition, behaviour of a system is given by limit of its diagram [2].

3.1 Image Definition of Goguen

First, we give the image definition of Goguen [2] to express it in the internal language of a presheaf topos. Given two presheaves O and O' on a category $\Re$, and a morphism $\phi : O \longrightarrow O'$ then the image presheaf of ϕ, denoted $\phi(O) \subseteq O'$, is defined by

$$\phi(O)(U) = \phi_U(O(U)),$$

hence $\phi(O)(U) \subseteq O'(U)$ for each $U \in \Re$.

3.2 Image in the Internal Language

It results from [4] that the images of Goguen [2] are the images of morphisms in a presheaf topos. So $\phi : O \longrightarrow O'$ is of the form $m \circ e$, where $m: \phi(O) \longrightarrow O'$ is a monomorphism and $e: O \longrightarrow \phi(O)$ is an epimorphism. Thus the images of Goguen are the subobjects which play the role of morphism images. Consequently, we can express them in the internal language in the following way

$$\phi(O) \cong \{ y \in O' \mid \exists x \in O \text{ such that } y = \phi(x) \}.$$

3.3 Behaviour in the Internal Language

In this section, we generally use presheaves which correspond to preobjects. So, a system is modelled by a diagram of (pre)sheaves and the behaviour of a system is its limit. But the limit of a diagram is a subobject of the product of the diagram objects. This subobject results from the commutation of the limit cone [4].

Consequently, we express the behaviour in the internal language of a (pre)sheaf topos since any subobject can be expressed in the internal language. The idea is the following. Let X be an object in a topos. Any subobject S of X is in bijection with an arrow $\phi: X \longrightarrow \Omega$, called characteristic arrow. Here Ω is the subobject classifier [4]. But, an arrow $\phi: X \longrightarrow \Omega$, is the same thing as a predicate. It follows by using [4] that we can then build an equivalent subobject expressed in the internal language of the form

$$\{ x \in X \mid \phi[x] \}.$$

So, given a system S with objects S_n for $n \in N$, i.e, $S_n: C^{op} \longrightarrow Set$ where C is the base category. A network of consistent observations on $I \in C$ is a set of observations $f_n \in S_n(I)$ in the sense that for each morphism $\tau: S_n \longrightarrow S_{n'}$, we have $f_{n'} = \tau_I(f_n)$ [2]. But these equalities characterize the cone commutation of the limit definition, and correspond to some formulas of the internal language. From this it results the type's construction L of the object which describes all the possible behaviours of a system. Tony Clark [1] says it clearly: ``The behaviour of the complete system is the object which makes all of the constraints hold.''

Consequently, from a lemma proof in [6] it results that

$$L(I) - \{ 1 \subset \textstyle\prod_n S_n \mid \psi[1] \}(I) \cong \{ (f_1, f_2, ..., f_k) \in \textstyle\prod_n S_n(I) \mid \psi_I[f_1, f_2, ..., f_k] \} \cong$$
$$\cong \{ (f_1, f_2, ..., f_k) \in \textstyle\prod_n S_n(I) \mid \forall_{1 \le i,j \le k} \forall \phi: S_{ni} \longrightarrow S_{nj}, \phi_I(f_i) = f_j \} \; [6].$$

3.4 Non-interference of Goguen

Now, we present the non-interference definition of Goguen [2]. Given a system S with objects S_n for $n \in N$, with morphisms $\phi_e: S_n \longrightarrow S_{n'}$ for $e: n \longrightarrow n'$, and with behaviour (*i.e. limit*) L having projections $\tau_n: L \longrightarrow S_n$, then S_m is non-interfering with S_k if and only if

$$\tau_k(L) = \tau'_k(L'), \tag{1}$$

where L' is the limit of the subsystem of S which does not contain S_m and all morphisms to and from S_m, and $\tau'_n: L' \longrightarrow S_n$, for $n \in N \setminus \{m\}$, are its projections.

3.5 Non-interference of Goguen in the Internal Language

Next, we express the non-interference of Goguen [2] in the internal language. So, $\tau_k(L)$ and $\tau'_k(L')$ are subobjects of S_k and thus we can suppose them isomorph (not necessary equal). By considering the property of subobject classifier Ω[4] its characteristic morphisms (*i.e. predicates*) are thus equal

$$\exists_{x \in L}(eq \circ \langle y, \tau_k \circ x \rangle) = \exists_{x' \in L'}(eq \circ \langle y, \tau'_k \circ x' \rangle),$$

where eq is the internal equality of a topos i.e. the characteristic arrow of the diagonal [4].

In a semantics without variables [5], it results that:

$$\exists_L(eq \circ (1 \times \tau_k)) = \exists_{L'}(eq \circ (1 \times \tau'_k)) \qquad (\text{rel. } S_k) \; [6].$$

For any arrow $h: X \times A \longrightarrow \Omega$, it holds that

$$\exists_A(h) = \forall_\Omega((\forall_A(h \circ (\pi_1 \times 1)) \Rightarrow \pi_2 \circ \pi_1)) \Rightarrow \pi_2),$$

where π_1 and π_2 are canonical projections.

In this case h is both

$$eq \circ (1 \times \tau_k): S_k \times L \longrightarrow \Omega \text{ and } eq \circ (1 \times \tau'_k): S_k \times L' \longrightarrow \Omega.$$

In addition $f \Rightarrow g = eq \circ <f, f \wedge g>$, where $f \wedge g = eq \circ <<f, g>, < \text{T} \circ \star,$

$\text{T} \circ \star>>$ and $\forall_A(h) = eq \circ <\Lambda_A(h), \Lambda_A(\text{T} \circ \star)>$, where $\Lambda_A(h): \Gamma \longrightarrow \Omega^A$ is the abstraction of h [4].

4. Conclusions and Future Work

Consequently, this paper provides computation tools for authentication and for non interference control. We plan to implement this in the future.

Acknowledgments

The author would like to thank Christine Choppy, André Desnoyers and René Guitart for many helpful discussions.

References

[1] Tony Clark: *A Semantics for Object-Oriented Systems*. http://www.dcs.kcl.ac.uk/ staff/tony /docs/OOSemantics.ps (1999)

[2] Joseph A. Goguen: Sheaf Semantics for Concurrent Interacting Objects. *Mathematical Structures in Computer Science*, { 2} (1991) 159--191

[3] J. Lambek and P. J. Scott: *Introduction to higher order categorial logic*. Cambridge Studies in Advanced Mathematics 7, (1989)

[4] Saunders Mac Lane and Ieke Moerdijk: *Sheaves in Geometry and Logic, A First Introduction to Topos Theory*. Springer Verlag (1991)

[5] Alain Prouté: On the Role of Description. *J. Pure Appl. Algebra* 158, 2001

[6] Elena Sendroiu: *Topos, un modèle pour l'informatique*. PhD thesis in preparation.

[7] Elena Sendroiu: From anti-virus to sheaf tools for security. In U.E. Gattiker (Ed), *EICAR 2004 Conference CD-rom*: Other Contributions (ISBN: 67-987271-6-8) 18 pages. Copenhagen: EICAR e.V.

VSP International
Science Publishers
P.O. Box 346, 3700 AH Zeist
The Netherlands

*Lecture Series on Computer
and Computational Sciences*
Volume 1, 2004, pp. 479-482

Molecular Dynamics Simulation of Cis-Trans N-Methylformamide (NMF) Liquid Mixture. Structure and Dynamics

Ioannis Skarmoutsos and Jannis Samios[1]

Laboratory of Physical Chemistry,
Department of Chemistry,
University of Athens,
Panepistimiopolis 15771, Athens,
Greece

Received 15 September, 2004; accepted in revised form 23 September, 2004

Abstract: The liquid mixture cis-trans N-Methylformamide (NMF) (with mole fraction 0.94 for the trans species) has been simulated via canonical Molecular dynamics (MD) simulation techniques at ambient conditions, using a newly developed potential model by us [1]. The properties of the system have been investigated at 298 K and normal pressure and the results obtained are compared with available experimental data. The difference in the local structure and dynamics of these conformers has been observed and discussed.

Keywords: Molecular dynamics, simulation, intermolecular potential, amides, conformers, diffusion

Mathematics SubjectClassification: 82B05, 82D15, 82C70

1. Introduction

Among several classes of organic solvents, liquid amides appear to be very interesting molecular systems. These polar liquid systems are widely used as solvents in several chemical and biological processes and have many applications in medical technology. We have to mention here that specifically NMF has found to exhibit an antitumor activity [2]. Furthermore, these compounds could be used as model systems of the peptide linkage [3]. Among these liquid amide systems liquid N-Methylformamide (NMF) exhibits particularly interesting properties. In liquid NMF the dipolar interactions are expected to dominate the intermolecular forces. Consequently, these interactions should form a short range ordering among the molecules. Furthermore, NMF is an associated hydrogen bonded solvent at liquid conditions [3]. As far as we know, the fundamental question to what extent the hydrogen bonds in NMF could affect the structural and dynamic properties in the liquid, has not yet been definitively answered. From a theoretical point of view, only a limited number of simulation studies on liquid NMF have been reported so far [4-8].
WHY DO WE HAVE TO REINVESTIGATE (THEORETICALLY) THE PROPERTIES OF LIQUID NMF? According to several experimental observations liquid NMF consists of a mixture of cis- and trans-conformers. At normal conditions the mole fraction X_{trans} has been found to be about 0.94 [9]. In all previous studies the system has been modeled as consisted only by trans- species [4-7]. Only in one previous preliminary Monte Carlo (MC) investigation by Krienke et al. [8] the system has been modeled as a mixture consisted of two conformers. Therefore, a more comprehensive MD treatment, in which the system should be modeled as a mixture consisted of two conformers, becomes indispensable.
SCOPE OF OUR WORK: The scope of this work was to investigate the intermolecular structure of the pure liquid as well as the transport properties of the two conformers and to demonstrate that these conformers reveal different local intermolecular ordering and translational dynamic behavior at

[1] Corresponding author. E-mail: isamios@cc.uoa.gr

ambient conditions. To do so, we developed a new effective 9-site interaction potential model for each conformer consisted of LJ 12-6 and Coulombic terms [1]. Using this potential model we performed a NVT-MD simulation study of the system at 298.15 K and normal pressure. We used 256 molecules in a cubic box (N_{trans}=241,N_{cis}=15) with periodic boundary conditions, starting from an initial fcc configuration. The simulation time was 1 ns (t $_{equilibration}$= 200 ps and t $_{after equilibration}$ = 800 ps). We used a time step Δt=1 fs to integrate the translational and rotational equations of motion. The integration algorithm used was a modified Beeman algorithm [10].

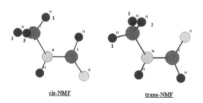

cis-NMF   trans-NMF

2. Results and Discussion

2.1 Intermolecular Structure

We investigated the local structure of liquid NMF in terms of the appropriate site-site radial distribution functions (RDFs) G(r), calculated on the basis of our potential model. Some of the results obtained are depicted in the following Figures 1,2.

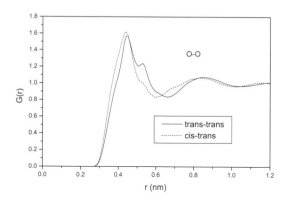

Figure 1: Site-site RDF O-O trans-trans and cis-trans NMF

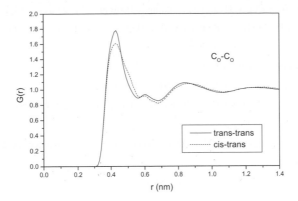

Figure 2: Site-site RDF C_O-C_O (carbonyl carbon) trans-trans and cis-trans NMF

2.2 Translational Dynamics

The translational dynamics of both the NMF conformers were also studied in this treatment by calculating the corresponding linear velocity autocorrelation functions. The normalized velocity autocorrelation functions for the trans and cis conformers at 298 K and normal pressure are presented in Figure 3 (in the insert the difference between the correlation functions in the time scale 0-0.5 ps is clearly depicted).

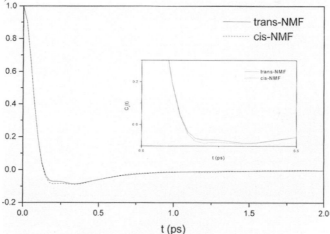

Figure 3: Normalized velocity autocorrelation functions for trans-
and cis-NMF at 298 K and normal pressure.

By integrating the non-normalized velocity autocorrelation functions we have predicted the self-diffusion coeffients for both conformers in a previous publication [1]. The calculated values were $D_{trans}= 7.15\ 10^{-10}\ m^2\ s^{-1}$ and $D_{cis}= 6.76\ 10^{-10}\ m^2\ s^{-1}$. These results are in good agreement with the experimental ones [8].

We also see that the two conformers indeed reveal different local intermolecular ordering and translational dynamic behavior at ambient conditions. Further work to explore the hydrogen-bonding network of the system (static and dynamic) in more details is in progress [11].

References

[1] I. Skarmoutsos and J. Samios, *Chem. Phys. Lett.*, **384**, 108 (2004)

[2] S.P. Langdon and J.A. Hickman, *Toxicology*, **43**, 239 (1987)

[3] G.A. Jeffrey and W. Saenger, Hydrogen Bonding in Biological Structures, Springer, Berlin, 1991

[4] P.C. Schoester, M.D. Zeidler, T. Radnai and P.A. Bopp, Z. Naturforsch. **50**, 38 (1995)

[5] J. Gao, J.J. Pavelites and D. Habibollazadeh, J. Phys. Chem. **100**, 2689 (1996)

[6] J.M.M. Cordeiro, Int. J. Quant. Chem. **65**, 709 (1997)

[7] H. Torii and M. Tasumi, J. Phys. Chem. A **104**, 4174 (2000)

[8] L. Chen, T. Gross, H.D. Lüdemann, H. Krienke and R. Fischer, Naturwissenschaften, **87**, 225 (2000)

[9] L.M. Jackman and E.A. Cotton (eds), Dynamic Nuclear magnetic resonance spectroscopy, Academic, New York, 1975

[10] K. Refson, *Comp. Phys. Comm.*, **126**, 310 (2000)

[11] I. Skarmoutsos, J. Samios and R.L. Mancera (submitted for publication)

VSP International
Science Publishers
P.O. Box 346, 3700 AH Zeist
The Netherlands

*Lecture Series on Computer
and Computational Sciences*
Volume 1, 2004, pp. 483-486

Numerical Simulation of Two-Dimensional Dam-Break Flows

J.V. Soulis and A.J. Klonidis[1]

Department of Civil Engineering,
Fluid Mechanics/Hydraulics Division,
Democritus University of Thrace,
GR-671 00 Xanthi, Greece

Received 3 September, 2004; accepted in revised form 20 September, 2004

Abstract: The present study presents a second order accurate implicit numerical scheme for the calculation of unsteady, two-dimensional depth averaged, free-surface flow problems. The introduction of a non-orthogonal, boundary-fitted coordinate system allows the accurate simulation of irregular geometries. The model is used to analyze dam-break flow in a converging-diverging flume. The computed results are compared with measurements and satisfactory agreement is achieved.

Keywords: Two-dimensional, unsteady, free-surface, implicit, second order, finite volume

PACS: 35Q30

1. Introduction

From the most impressive man-made structures, dams become extremely harmful if sudden collapse occurs. Tons of water surge downstream causing very serious environmental catastrophe in an extended area. From the numerical analysis point of view, the simulation of such rapidly varying water flow has always been attracted to the researchers. Katopodes and Strelkoff[5] presented a numerical model for computing two-dimensional dam-break flood waves. They used the characteristics method to approximate the shallow water flow equations. Fennema and Chaudhry[4] utilized the Beam and Warming implicit scheme to solve the equations describing two-dimensional, free-surface flows. Bellos[2] solved the two-dimensional depth-averaged flow equations using the Mac Cormack, two-step, predictor-corrector explicit scheme. An explicit, finite-volume technique adapted to unsteady, depth-averaged, free-surface flows was presented by Soulis[7]. Extensive comparisons between measured data and numerical results substantiated the validity of the technique. Aureli, Mignosa, and Tomirotti[1] presented a research work where experimental results of 1D dam-break flows with shocks and results found in the literature are compared with those obtained by means of a numerical model based on the well-known McCormack shock-capturing scheme. This paper presents an extension of a previous research work on steady, two-dimensional, free-surface flows developed by Klonidis and Soulis[6]. The model was enhanced in order to be able to solve unsteady flows. The conservative equations of continuity, x-momentum and y-momentum transformed into a non-orthogonal, boundary-fitted coordinate system are numerically solved using a second-order accurate implicit numerical scheme. Numerical results of a two-dimensional dam-break flow simulation are tested successfully against available experimental data.

2. The Unsteady Flow Equations

Under the assumptions of homogenous, 2D, incompressible flow with hydrostatic pressure distribution and with absence of Coriolis and wind forces, the equations used to describe the unsteady flow resulting from the rupture of dam are written in matrix form as

$$Q_t + F_x + G_y = W \tag{1}$$

---

[1] Athanasios J. Klonidis. Research Engineer. E-mail: klonidis@otenet.gr

where

$$Q = \left\{\begin{matrix} h \\ hv_x \\ hv_y \end{matrix}\right\}, \quad F = \left\{\begin{matrix} hv_x \\ hv_x^2 + gh^2/2 \\ hv_x v_y \end{matrix}\right\}, \quad G = \left\{\begin{matrix} hv_y \\ hv_y v_x \\ hv_y^2 + gh^2/2 \end{matrix}\right\}, \quad W = \left\{\begin{matrix} 0 \\ gh(S_{ox} - S_{fx}) \\ gh(S_{oy} - S_{fy}) \end{matrix}\right\} \quad (2)$$

Here h represents the water depth, v_x, v_y are the depth-averaged velocity components along the longitudinal (x) and transverse (y) directions respectively, g is the acceleration due to gravity, S_{ox}, S_{oy} are the bottom slopes along the x and y direction respectively and S_{fx} and S_{fy} stand for the friction slopes in x and y directions.

To overcome difficulties and inaccuracies associated with the determination of flow characteristics near the solid boundaries of channels with complex geometry, a transformation process is introduced through which quadrilaterals that are generated during the discretization procedure of the flow field are mapped onto squares in the computational domain (finite-volume method). This is accomplished through independent transformations from the Cartesian (x, y) to the local coordinate system (ξ, η). The center of each finite-volume is the point where the local coordinate system is initiated (ξ=0, η=0) with ξ and η ranging between $-1 \leq \xi \leq 1$ and $-1 \leq \eta \leq 1$.

Let [J] be the transformation matrix from the physical to local coordinate system; then

$$[J] = \left\{\begin{matrix} x_\xi & x_\eta \\ y_\xi & y_\eta \end{matrix}\right\} \quad \text{and} \quad J^{-1} = [J] \quad (3)$$

The following relations hold:

$$x_\xi = J^{-1}\eta_y, \quad x_\eta = -J^{-1}\xi_y, \quad y_\xi = -J^{-1}\eta_x, \quad y_\xi = J^{-1}\xi_x, \quad (4)$$

The velocity components v_ξ, v_η in the computational domain are related to v_x, v_y in the physical domain with the following relation:

$$\left\{\begin{matrix} v_\xi \\ v_\eta \end{matrix}\right\} = [J]^{-1}\left\{\begin{matrix} v_x \\ v_y \end{matrix}\right\} \quad (5)$$

Under the aforementioned transformation into the local coordinate system, equations (2) take the form:

$$\hat{Q} = \left\{\begin{matrix} J^{-1}h \\ J^{-1}hv_x \\ J^{-1}hv_y \end{matrix}\right\} \qquad\qquad \hat{F} = \left\{\begin{matrix} J^{-1}hv_\xi \\ J^{-1}(hv_\xi v_x + \dfrac{\partial\xi}{\partial x}\dfrac{gh^2}{2}) \\ J^{-1}(hv_\xi v_y + \dfrac{\partial\xi}{\partial y}\dfrac{gh^2}{2}) \end{matrix}\right\}$$

$$\hat{G} = \left\{\begin{matrix} J^{-1}hv_\eta \\ J^{-1}(hv_\eta v_x + \dfrac{\partial\eta}{\partial x}\dfrac{gh^2}{2}) \\ J^{-1}(hv_\eta v_y + \dfrac{\partial\eta}{\partial y}\dfrac{gh^2}{2}) \end{matrix}\right\} \qquad\qquad \hat{W} = \left\{\begin{matrix} 0 \\ J^{-1}gh(S_{ox} - S_{fx}) \\ J^{-1}gh(S_{oy} - S_{fy}) \end{matrix}\right\} \qquad (6)$$

3. Numerical Solution

In the present research work the hydrodynamic equations transformed into the local coordinate system are approximated using a second-order accurate, implicit, finite-difference scheme. The development of this scheme was achieved following a procedure through which equations (7) were linearized by expanding, the time derivatives initially and space derivatives thereafter, to Taylor series. Forward differences were used for time derivatives and central differences for space derivatives. The above procedure leads to the following numerical form:

$$\left(I + \frac{\Delta t}{2}\frac{\delta \hat{A}^n}{\Delta \xi} + \frac{\Delta t}{2}\frac{\delta \hat{B}^n}{\Delta \eta} \right)\Delta \hat{Q}^{n+1} = -\frac{\Delta t}{2}\left(\frac{\delta \hat{F}^n}{\Delta \xi} + \frac{\delta \hat{G}^n}{\Delta \eta} - \hat{W} \right) \tag{7}$$

where, δ denotes central differences, I is the identity matrix 3x3 and $\hat{A} = \frac{\partial \hat{F}}{\partial Q}$, $\hat{B} = \frac{\partial \hat{G}}{\partial Q}$ are the Jacobian matrices resulting from the linearization procedure. Approximate factorisation of equation (7) yields the implicit form of equations (6) which can be implemented in the following sequence, Klonidis et al[6]:

$$\left(I + \frac{\Delta t}{2}\frac{\delta \hat{A}^n}{\Delta \xi} \right)\Delta \hat{Q}^* = -\frac{\Delta t}{2}\left(\frac{\delta \hat{F}^n}{\Delta \xi} + \frac{\delta \hat{G}^n}{\Delta \eta} - \hat{W} \right) \qquad \text{1st step} \tag{8}$$

$$\left(I + \frac{\Delta t}{2}\frac{\delta \hat{B}^n}{\Delta \eta} \right)\Delta \hat{Q}^{n+1} = \Delta \hat{Q}^* \qquad \text{2nd step} \tag{9}$$

$$\hat{Q}^{n+1} = \hat{Q}^n + \Delta \hat{Q}^{n+1} \qquad \text{3rd step} \tag{10}$$

The values of the unknown variables at every point of the field are obtained by solving a block tri-diagonal system.

4. Validation of the Model

In this section numerical results are compared with measured data obtained from a series of experiments that had been performed in a rectangular section converging-diverging flume by Bellos[2]. Figure 1 shows the plan view of the flume configuration. A hypothetical dam consisting of a sluice gate was placed at the throat of the flume. With the gate initially closed, the upstream basin was filled with water at a depth of 30.0 cm. A sudden rising of the gate caused a flood wave to surge, overwhelming the downstream dry bed. Measured stage hydrographs at x=8.5 m along the center-line of the flume are compared with equivalent model results for the test case of h_1=0.30 m, S_{0x}=S_{0y}=0.0, see figure 2. The Manning's roughness coefficient was set equal to 0.012, which is close to the glass-steel material of the tested flume. In all cases computed results and measured data seem to agree well.

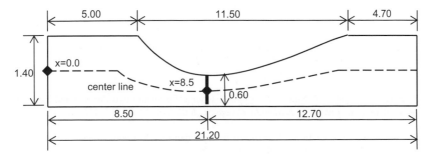

Figure 1: Plan view of the tested flume geometry.

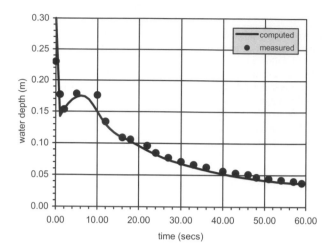

Figure 2: Comparison between computed and measured stage hydrographs at x=8.5 m.

5. Conclusions

An implicit, two-dimensional, finite volume scheme was developed to simulate unsteady, dam-break flows. The depth-averaged flow equations were transformed into a non-orthogonal, boundary-fitted coordinate system. The numerical technique itself turned out to be fast, reliable and flexible concerning its response to handle complex geometry problems with abrupt changes in water level. Reasonable agreement between measured and computed results has been demonstrated.

References

[1] F. Aureli, P. Mignosa and M. Tomirotti, Numerical simulation and experimental verification of dam-break flows with shocks, *Journal of Hydraulic Research, JHR,* Vol. 38, pp.197-206. (2000).

[2] C. V. Bellos, Computation of 2-D water movement in open channels, *Ph.D. Thesis,* Democritus University of Thrace, 1990 (in Greek).

[3] C. V. Bellos, J. V. Soulis, and J. G. Sakkas, Computing 2D unsteady open channel flow by finite-volume method, *Proc. VII Int. Conf. on Computational Methods in Water Resources,* Boston, MIT, pp. 263-276 (1990).

[4] R. J. Fennema and M. H. Chaudhry, Explicit methods for 2-D transient free-surface flows, *J. Hydr. Engrg., ASCE,* 116 (8), pp.1013-1034, (1990).

[5] N. Katopodes and T. Strelkoff, Computing two-dimensional dam-break flood waves, *J. Hydr. Engrg., ASCE,* pp.1269-1288, (1978).

[6] A. J. Klonidis and J. V. Soulis, An implicit scheme for steady two-dimensional free surface flow calculation, *Journal of Hydraulic Research, JHR,* Vol. 39, pp.393-402. (2001).

[7] J. V. Soulis, Computation of two-dimensional dam-break flood flows, *Int. Journal for Numerical Methods in Fluids,* Vol. 14, pp. 631-664 (1992).

VSP International
Science Publishers
P.O. Box 346, 3700 AH Zeist
The Netherlands

*Lecture Series on Computer
and Computational Sciences*
Volume 1, 2004, pp. 487-490

Using Fuzzy Sets, Fuzzy Relations, Alpha Cuts and Scalar Cardinality to estimate the Fuzzy Entropy of a Risk evaluation System: (The case of Greek Thrace Torrential Risk)

S. Spartalis[1], L. Iliadis[2], F. Maris[3]

[1]Department of Agricultural Development,
[2,3]Department of Forestry & Management of the Environment & Natural Resources,
Democritus University of Thrace
GR-682 00, Nea Orestiada, Greece

Received 4 September, 2004; accepted in revised form 19 September, 2004

Abstract: One of the most critical issues of our times is natural disaster Risk management. This study concerns the development of a Decision Support System that has been based on two main *Fuzzy Algebra* frameworks. The first framework applies *Fuzzy Set* theory and various *Fuzzy Algebra Conjunction* operations to perform Risk estimation. Actually the application of the first framework on the problem of Torrential Risk evaluation has been presented thoroughly in another paper (Iliadis et al 2004) and it is described very briefly here in order to give a general hint of the methodology.

On the other hand, the second framework that requires the application of the first, executes two independent main sub-tasks. First it applies *Scalar Cardinality* functions in order to perform a comparative study between the areas under examination. The second sub-task (which is the most crucial) is the calculation of the System's *Entropy* using *Fuzzy Entropy functions*.

The original contribution of this paper is the mathematical model constituting the second framework, its application for the Torrential Risk case (using actual data) and its results. The fact that the model and the Computer System can be applied in any type of natural disaster problem by adjusting the considered Risk factors and also the fact that the System is able to judge itself and to estimate its *Fuzzy Entropy* makes it very flexible, useful and original.

Keywords: Fuzzy Algebra, Fuzzy Sets, Fuzzy Relations, Scalar Cardinality, Fuzzy Entropy, Torrential Risk, Decision Support System.

Mathematics Subject Classification: 03E72, 03B52

1. Introduction

Fuzzy Logic was introduced by Zadeh in 1965 to relax the harsh constraint that everything that can be said about anything is either absolutely true or absolutely false (Kandel 1992). In real life situations this is rarely the case. Zadeh suggested that it is possible to understand a statement as being 0.70 true. The use of Fuzzy numbers and Fuzzy Sets offered the scientists powerful tools for performing classification and ranking tasks.

This study deals with the problem of natural disasters Risk management by using Fuzzy Algebra and it works towards two complementary parallel directions.

The first direction involves the determination of the main n Risk factors affecting a specific Risk problem. Consequently, a number of n *Fuzzy Sets* are formed, each Fuzzy Set corresponding to a Risk factor. *Fuzzy membership functions* (*Trapezoidal, Triangular and Sigmoidal*) are used to estimate the degree of membership of the m watersheds under examination, to each one of the n corresponding *Fuzzy Sets*. In this way each watershed is assigned n real numbers (ranging from 0 to 1) that show its degree of Risk for each of the n Risk factors respectively. Consequently the following two-dimensional Risk matrix 1 (n × m) is formed containing the produced Risk indices.

---

[1] Corresponding author. E-mail: sspart@agro.duth.gr
[2] E-mail: liliadis@fmenr.duth.gr
[3] E-mail: fmaris@fmenr.duth.gr

$$\mu(X_{ij}) = \begin{bmatrix} \mu_{11} & \mu_{12} & \cdots\cdots & \mu_{1m} \\ \mu_{21} & \mu_{22} & \cdots\cdots & \mu_{2m} \\ \cdots\cdots & \cdots\cdots & \cdots\cdots & \cdots\cdots \\ \mu_{n1} & \mu_{n2} & \cdots\cdots & \mu_{nm} \end{bmatrix} \qquad \text{where i = 1 to n and j = 1 to m} \tag{1}$$

The scope is the estimation of a unique Risk Index for each area under examination. This is performed by applying five different types of *Fuzzy Relations* (T-Norms) that operate conjunction between the Partial Risk Indices under different perspectives. This approach and its application for the Torrential Risk estimation of the main watersheds of North-Eastern Greece have already been presented thoroughly in another paper of our research team (Iliadis et. al. 2004).

The second main direction of this research is the development of a comparative study for the Torrential Risk evaluation between the main watersheds of North-Eastern Greece (Rodopi, Xanthi, Northern Evros, Central Evros and Southern Evros areas) by using *Alpha Cuts* and *Scalar Cardinality* functions. However the most important attribute of this direction is the *Fuzzy Entropy* estimation of the Decision Support System that has been developed and applied by our research team. The methodology for the *Fuzzy Entropy* estimation and its application for the Torrential Risk of North-Eastern Greece is described here.

2. Materials and methods

2.1. The multi factor Risk estimation model

Two different algebraic approaches can be applied for the Risk estimation. The one is already established and it uses Crisp Sets (Leondes, 1998) and the other is proposed in this paper and it uses Fuzzy Sets. The following function 2 defines a Crisp Set.

$$\mu_s(X) = \begin{cases} 1, \text{if } X \in S \\ 0, \text{if } X \notin S \end{cases} \tag{2}$$

In Crisp sets a function of this type is also called *characteristic function*. Fuzzy sets can be used to produce the rational and sensible clustering (Kandel, 1992). For Fuzzy sets there exists a degree of membership (DOM) $\mu_s(X)$ that is mapped on [0,1] and every area belongs to the "*Torrential Risky Area*" Fuzzy Set with a different degree of membership (Kandel, 1992). Functions 3 and 4 represent the Triangular and the Trapezoidal Membership functions that determine the Risk indices for each examined area (Kecman 2001).

$$\mu_s(X) = \begin{cases} 0 \text{ if } X < a \\ (X-a)/(c-a) \text{ if } X \in [a,c] \\ (b-X)/(b-c) \text{ if } X \in [c,b] \\ 0 \text{ if } X > b \end{cases} \tag{3}$$

$$\mu_s(X) = \begin{cases} 0, & \text{if } X \le a \\ (X-a)/(m-a), & \text{if } X \in (a,m) \\ 1, & \text{if } X \in [m,n] \\ (b-X)/(b-n), & \text{if } X \in (n,b) \\ 0, & \text{if } X \ge b \end{cases} \tag{4}$$

Various T-Norms were used to unify the Partial Risk Indices and to produce the characteristic Unified Risk Index. The following Table 1 presents the T-Norms that were applied for this purpose.

Table 1: T-Norms used in the project

1. *Minimum Approach* URI $= \text{MIN}(\mu_A(X), \mu_B(X))$
2. *Algebraic Product* URI $= \mu_A(X) * \mu_B(X)$
3. *Drastic Product*   URI $= \text{MIN}(\mu_A(X), \mu_B(X)).if..\text{MAX}(\mu_A(X), \mu_B(X)) = 1$ otherwise URI $= 0$
4. *Einstein Product*   URI $= \mu_A(X) * \mu_B(X) / (2-(\mu_A(X) + \mu_B(X) - \mu_A(X) * \mu_B(X)))$
5. *Hamacher Product*   URI $= \mu_A(X) * \mu_B(X)/(\mu_A(X) + \mu_B(X) - \mu_A(X) * \mu_B(X))$

Their nature proves that each one of them produces a Unified Risk Index under a different perspective. This can be very useful because each Unified Risk Index expresses a valid Risk degree under different circumstances.

2.2. The comparative study methodology

After the estimation of the Partial and the Unified Degrees of Risk for the _m_ areas, calculation and application of the *Scalar Cardinality* for each one of the areas, performs the comparative study. For any fuzzy set A defined on a finite universal set X, we define its *Scalar Cardinality*, $\sum Count(A)$ by the formula (Kandel, 1992):

$$\sum Count(A) = \sum_{i=1}^{m} \mu_A(x_i), \quad \text{for all x in X} \tag{5}$$

where m is the number of the areas under examination, and $\mu_A(\chi_i)$ is the degree of membership of area x to the Fuzzy Set A. Some authors refer to the $\sum Count(A)$ as the *Sigma Count* of A.

The *Scalar Cardinality* is applied first to the partial Degrees of Risk to determine the Degree of Risk for each area and for each one of the Risk factors. In this way the DSS determines the most risky area for each factor affecting the problem.

The *Scalar Cardinality* is applied to the final Unified Degrees of Risk of each area for all of the T-Norms to determine the total overall degree of Risk for each area based on the T-Norm used. The above task offers an overall comparative Risk study of the areas. Finally, *Alpha-Cuts* are used to estimate how severe the Risk is in each area.

An *Alpha-Cut* of the membership function A (denoted *aA*) is the set of all x such that A(x) is greater than or equal to alpha (a). Similarly, a strong Alpha-Cut (denoted *a+A*) is the set of all x such that A(x) is strictly greater than alpha (a). Mathematically,

$$aA = \{x \mid A(x) >= a\} \tag{6}$$
$$a+A = \{x \mid A(x) > a\} \tag{7}$$

"That is, the *Alpha-Cut* (or the strong *Alpha-Cut*) of a fuzzy set A is the crisp set *aA* (or the crisp set *a+A*) that contains all the elements of the universal set X whose membership grades in A are greater than or equal to (or only greater than) the specified value of alpha." *aA and a+A* are crisp sets because a particular value x either is or isn't in the set; there is no partial membership.

"The set of all levels alpha in [0,1] that represent distinct alpha-cuts of a given fuzzy set A os called a level set of A. Formally"

$$L(A) = \{a \mid A(x) = a \text{ for some x in X}\} \tag{8}$$

or

$$L(A) = [min(A), max(A)] \text{ for all x in X} \tag{9}$$

A comparison between the results offers very interesting Rankings and Classifications of the areas under examination.

2.3. The Fuzzy Entropy estimation method

The *Fuzzy Entropy* of the System is measured by the following function 6 (Kandel, 1992). For any fuzzy set A defined on a finite universal set X, we define its *Fuzzy Entropy:*

$$E(A) = \frac{\sum Count(A \cap A^C)}{\sum Coun(A \cup A^C)} \tag{10}$$

where A^C is the complement of Fuzzy Set A. A(x) is defines and the degree to which x belongs to A. Let A^C denote a fuzzy complement of A of type c. $A^C(x)$ is the degree that x belongs to A^C, and the degree to which x does not belong to A. (A(x) is therefore the degree to which x does not belong to A^C.)

$$(A(x))^C = A^C(x) \tag{11}$$

From the left side of the equation above, we have:
Axiom c1. c(0)=1 and c(1)=0 (boundry conditions) $\qquad$ (12)
Axiom c2. for all a,b in [0,1], if a<=b then c(a)>=c(b) $\qquad$ (13)

3. Development of a Decision Support System

A Decision Support System (DSS) was developed using MS Access. All data were stored in properly designed Tables that follow the principles of the Normal forms. The Mathematical model (its logic and its functions) were stored in Structured Query Language (SQL) statements. The system uses a very friendly Graphical user interface and its results are obtained in a straightforward manner.

4. Application of the System for the Torrential Risk case

The research area where the DSS was applied for the estimation of the torrential Risk is located in the Northeastern part of Greece near the border of Greece Bulgaria and Turkey. This area was chosen due to its serious Torrential phenomena that have serious consequences in the life of the local people. The research area was divided in five main sub-areas.

1. The sub-area of Northern Evros includes the torrential streams of the northern part of the Evros prefecture and the torrential streams of river Adras.

2. The sub-area of Central Evros includes the torrential streams of Erithropotamos river and it extends to the south till Likertzotiko stream in the area of Lyras-Laginon.

3. The sub-area of Southern Evros includes the rest part of Evros prefecture.

4. The Prefecture of Rodopi that has common borders to the prefecture of Evros

5. The Prefecture of Xanthi that has common borders to the prefecture of Rodopi

Data was gathered from Greek public services who are responsible for meteorological and map data. Also, our research team gathered important data. Initially the limits of the research areas were estimated. Maps of the Geographical Army Service (GAS) with a scale of 1:250.000 were used for this purpose. The upper and lower limits of the watershed areas are 300 και 2 km² respectively (Kotoulas 1997). For every research area and for each torrential stream the morphometric characteristics were specified. The morphometric characteristics were produced after the process of maps (scale 1:100.000) of the GAS and the accuracy of the data was confirmed by visits of our research teams in the research areas. The results were stored in descending order in matrices. For each research area a matrix was used. According to the bibliography as it is included in the book of "Mountainous Hydronomy" of professor Kotoulas (1969, 1973, 1979, 1987, 1997) and it contains papers of Horton 1932 and 1945, Viessman et al 1989, Stefanidis 1995, the most important morphometric characteristics of the watersheds that influence the torrential risk of an area are the following: The area, the perimeter, the shape of the watershed, the degree of the round shape of the watershed, the maximum altitude, the minimum altitude, the average altitude, the average slope of the watershed, and its maximum altitude. The results have shown that the DSS can be applied successfully and reliably whereas the Fuzzy Entropy has proven to have reasonably low values. The DSS will be tested furtherly in the near future.

Acknowledgments
The author wishes to thank the anonymous referees for their careful reading of the manuscript and their fruitful comments and suggestions.

References

[1] Iliadis L., Spartalis S., Maris F., Marinos D. A Decision Support System Unifying Trapezoidal Function Membership Values using T-Norms: The case of river Evros Torrential Risk Estimation. Proceedings of ICNAAM (International Conference in Numerical Analysis and Applied Mathematics) J. Wiley-VCH Verlag GmbH Publishing co.Weinheim Germany. 2004.

[2] Kandel A. 1992, Fuzzy Expert Systems. CRC Press. USA.

[3] Kecman 2001, Learning and Soft Computing. MIT Press. London England.

[4] Kotoulas D. 1969, The streams of N. Greece, their classification in characteristic types and the main principles of their management. Scientific annals of Agriculture and Forestry, Vol. I3 (Appendix).

[5] Kotoulas D. 1973, The torrential problem in Greece. Report Nr. 47. Laboratory of Silviculture and Mountainous Water Science. School of Agriculture and Forestry. Thessaloniki.

[6] Kotoulas D. 1979, Contribution to the study of the general operational mechanism of a torrential dynamic. Scientific annals for the 50 years of Forestry Department. Thessaloniki.

[7] Kotoulas D. 1987, Research on the characteristics of torrential streams in Greece, as a causal factor for the decline of mountainous watersheds and flooding. Thessaloniki.

[8] Kotoulas D. 1997, Management of Torrents I. Publications of the University of Thessaloniki.

[9] Leondes C.T. 1998, Fuzzy Logic and Expert Systems Applications. Academic Press. California USA.

VSP International
Science Publishers
P.O. Box 346, 3700 AH Zeist
The Netherlands

Lecture Series on Computer
and Computational Sciences
Volume 1, 2004, pp. 491-495

Calculation of Hydrocarbon Bond Dissociation Energies in a Computationally Efficient Way

V. Van Speybroeck [†1], **G.B. Marin** [‡], **M. Waroquier** [†]

[†] Center for Molecular Modeling, Laboratory of Theoretical Physics, Ghent University
Proeftuinstraat 86, B-9000 Gent, Belgium
[‡]Laboratorium voor Petrochemische Techniek, Ghent University
Krijgslaan 286 S-5, B-9000 Gent, Belgium

Received 2 August, 2004; accepted in revised form 22 August, 2004

Abstract: Bond Dissociation Energies of C-H bonds that are located in a variety of hydrocarbons are calculated. A comparative study with experimental data is conducted to determine accurate methods for the determination of BDEs. Various types of radicals at substuted aromatics are considered that typically comprise primary, secondary and tertiary alkylic, allylic, phenylic and vinylic bonds. The influence of the molecular size and local environment on the bond strength is studied in a set of polyaromatic hydrocarbon aryl radicals.

Keywords: Density Functional Theory, hydrocarbons, polyaromatic hydrocarbons, radicals

Mathematics Subject Classification:

PACS:

1 Introduction

Bond Dissociation energies (BDE) are fundamental for understanding many chemical processes because they can be directly related to the reaction enthalpy of homolytic cleavage of considered bond. In this paper we concentrate on BDEs of carbon-hydrogen bonds, which are defined as the enthalpy of the reaction required to break the $C - H$ bond to form two radicals. Such radicals are of primary importance for coal processes. The most important processes of coal are combustion, gasification, pyrolysis and liquefaction [1, 2]. All of these coal processes are directly associated with homolytic bond dissociations of the organic structures of coal into smaller molecules. Thus an understanding of the thermochemistry and reactivity of specific bonds within the coal structure may lead to advances in coal processing.

The accurate calculation of BDEs has been the subject of many debates in recent literature. For some time, it has been recognized that in order to obtain accurate BDEs, i.e. within 1-2 kJ/mol of experimental values, extensive correlations treatments and large basis sets are necessary. For example the G2, CCSD(T) and QCISD(T)/6-311G(d,p) methods were proven to be accurate but unfortunately these methods are only practicable for small molecules. Therefore various attempts have been performed to use density functional theory techniques for the calculations

---

[1]Corresponding author. E-mail: veronique.vanspeybroeck@ugent.be

of BDEs, because of their favorable scaling and lower computational cost. Dilabio and co-workers performed extensive studies to determine the accuracy by which BDEs of X-H, X-Y and X-Y bonds (X,Y=C,N,O,S,halogen) can be calculated using DFT methods [3]. They found that the use of lower level methods such as AM1 and B3LYP/6-31G(d) methods are sufficiently accurate for the determination of optimized structures and vibrational frequencies and yields results which are in excellent agreement with the full (RO)B3LYP/6-311+G(2d,2p) treatment and experiment. The B3LYP methods are however not sufficient for the electronic structure calculations and yield results that systematically underestimate experimental BDEs. Instead the B3P86 functional was found to yield accurate BDEs for X-H,X-X and X-Y bonds. Similar conclusions were found by Yao and co-workers, who explored the accuracy obtained by the B3LYP, B3PW91, MPW1PW91, B3P86, MPW1P86, (RO)B3P86 and CCSD(T) levels using the 6-311+G(d,p) basis set[4]. They found that especially the functional MPW1P86, which is a new combination of the MPW1 for the exchange functional and the P86 nonlocal correlation functional provided by Perdew 86, reduces the systematic underestimation of the other tested functionals. The work of Golden and co-workers further confirms that the commonly B3LYP method is insufficient for an accurate evaluation of BDEs and suggest other hybrid methods such as the KMLYP method, that yield RMS deviations that are close to the experimental error bars [5].

The final goal of this work is the accurate calculation of BDEs of larger polyaromatic structures and substituted aromatics in order to classify various radicals according to their reactivity. Since we are dealing here with relatively large molecules, the highly correlated post-Hartree Fock methods are not an viable alternative and DFT based methods must be applied. In order to determine the best suitable procedure for our systems, we have first performed an extensive comparative study between available experimental data for BDEs of $C - H$ bonds in hydrocarbons and computed values with various functionals (B3LYP, B3P86, MPW1PW91, B3PW91, MPW1B86, KMLYP). This study preveals insight into the level of theory that is suitable for our study but also in the factors that determine the bond strengths of various types of radicals. In the second part of the paper the methods are extrapolated to a set of substituted benzene rings (cfr. fig. 1), in order to obtain further insight into the factors that govern radical stability. Typically, primary, secondary and tertiary alkylic, allylic, phenylic and benzylic radicals are considered.

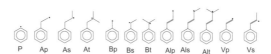

Figure 1: Substitued aromatics for which BDEs are calculated.

Aryl radicals are of particular importance during combustion of coals, since the latter process is initiated by the homolytic cleavage of a carbon-hydrogen bond to form aromatic radicals. Aryl radicals which are formed at large polycyclic aromatic species (PAHs) deserve special attention since they are formed in substantial quantities during incomplete combustion of fossil fuels. Moreover they are also important in a lot of other application fields. The BDEs of such species are not known from experimental point of view and the theoretical calculation of BDEs of PAHs can learn a lot about their reactivity. In this paper BDEs of some larger PAHs are calculated and general trends in reactivity are derived.

2 Results and Discussion

2.1 Comparitive study with experimental results

The C-H bonds of 22 hydrocarbon molecules were calculated with six DFT methods, i.e. B3LYP, MPW1PW91, B3P86, B3PW91, MPW1PW91, KMLYP. B3LYP, MPW1PW91, B3PW91 functionals all underestimate the experimental BDEs and MPW1PW91 and B3PW91 give nearly the same results (largest deviations of about 5 kcal/mol are noticed). The latter functionals have both the same correlation functional and one might conclude that both B3 and MPW1 exchange functionals have the same effect on the BDEs. These conclusions were also found by Yao et al. in an extensive paper about the accurate calculation of Bond Dissociation Enthalpies with Density Functional Methods[4]. The combination of B3 and MPW1 exchange functionals with the P86 correlation functional gives satisfactory agreement with the experiment for C-H bonds. The largest deviations are of the order of 2 kcal/mol and these methods all underestimate the experimental values. The BDEs calculated with the KMLYP functional, that was first introduced by Musgrave and coworkers give about the same accuracy but all overestimate the experimental values [5]. All previous results are in accord with other literature studies.

2.2 Bond Dissociation Energies of substituted aromatics

In this section we present BDEs of radicals located at substuted aromatics. Primary,secondary and tertiary allyl, alkyl and benzyl radicals are considered. The results are schematically shown in fig. 2. The BDE of primary bonds is larger than secondary and tertiary bonds. This can be understood in terms of different polarisability of adjacent substituents. Variations of about 8 kcal/mol are noticed which are somewhat larger for substitued aromatics than for alkyl radicals generated from n-alkanes [5]. Methane has a larger BDE due to the lack of a polarisable C-C bond to stabilize the radical.

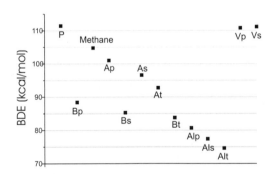

Figure 2: BDEs of substitued aromatics calculated at the B3P86 level of theory

The BDE of primary benzyl radicals is about 10 kcal/mol smaller than for a primary alkylic bond and the BDE of a primary allyl radical is about 20 kcal/mol smaller than its corresponding primary alkylic bond. These effects can be understood in terms of radical stabilization either over the neighboring aromatic ring or neighboring unsaturated bond.

Phenylic and vinylic bonds have the largest BDEs since it concerns here a sigma type radical which has no possiblity to stabilize due to adjacent bonds.

2.3   Bond Dissociation Energies of aryl radicals at PAHs

In this section we present BDEs of a series of aryl radical that are generated at a set of polyaromatics in order to determine the stability of the radicals as a function of the molecular size and radical position. The results are schematically shown in fig. 3.

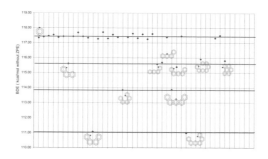

Figure 3: BDEs of of aryl radicals located at PAHs

Examination of the BDE values indicates that the strength of the aryl-bond is essentially independent of the molecular size, but dependent on the neighboring geometry of the C-H bonds. The BDE values are shown in fig. 3 and segregate into four groups. The more steric repulsion induced by the neighboring C-H bonds the weaker the corresponding aryl bond. The variations are of the order of 10 kcal/mol. Similar conclusions were made by various other authors [6].

3   Conclusions

In this work we calculated BDEs of a set of hydrocarbons in order to obtain insight into the factors that govern radical stability. Typically hydrocarbons that include an aromatic ring are considered since these structures are important for a variety of coal processes. The molecules are relatively large and therefore only computationally efficient methods are considered such as Density Functional Theory based methods. Although to assure chemically accurate data we also conducted a comparative study with available experimental data.

Acknowledgment

This work is supported by the Fund for Scientific Research - Flanders (FWO) and the Research Board of Ghent University.

References

[1] V. Van Speybroeck, D. Van Neck, M. Waroquier, S. Wauters, M.Saeys and G.B. Marin, *J.Phys.Chem.A* **104** 10939-10950, 2000.

[2] V. Van Speybroeck, M.F. Reyniers, G.B. Marin, and M.Waroquier, *ChemPhysChem*, **3** , 863-870, 2002

[3] G.A. Dilabio and D.A. Pratt, *J.Phys.Chem A*, **104**, 1938-1943, 2000. E.R. Johnson, O.J. Clarkin and G.A. Dilabio, *J.Phys.Chem.A*, **107**, 9953-9963, 2003.

[4] X. Yao, X. Hou, H. Jiao, H. Xiang and Y. Li, *J.Phys.Chem. A*, **107**, 9991-9996, 2003.

[5] J. P. Senosiain, J.H. Han, C.B. Musgrave, D.M. Golden, *Faraday Discuss.* **119**, 173-189,2001

[6] H.Wang and M. Frenklach, *J.Phys.Chem.*, **97**, 3867-3874, 1993.

VSP International
Science Publishers
P.O. Box 346, 3700 AH Zeist
The Netherlands

*Lecture Series on Computer
and Computational Sciences*
Volume 1, 2004, pp. 496-499

Ion Pulsation in Oxyhydrate Gel System

Yu. I. Sukharev[1], B.A. Markov, I.Yu. Sukhareva

Department of Common and Engineering Ecology,
Department of Mathematical Analysis,
Faculty of Mechanics and Mathematics,
Southern Ural State University, 76 Lenin Av., 454080, Chelyabinsk, Russia

Received 22 July, 2004; accepted in revised form 15 August, 2004

Abstract: Self-organization of gel system is described in time by the Liesegang operator. The periodicity of the spherical radii changing in the field of Van der Waals forces together with the geometrical identity of these curves to the Liesegang operator that changes with time indicate that the forces determining the Liesegang operator (the operator of system self-organization) and the forces determining the Lennard-Jones potential are similar. After the process of sorbtion of ions by gel swirl is finished, the intermolecular forces initiate the instant splash of ions into the external environment. It must correspond to the periodic spontaneous splashes of electroconductivity of oxyhydrate gel systems.

Keywords: Liesegang operator, potential, Lennard-Jones, oxyhydrate, gel systems, sorbtion, Van der Waals forces.

PACS: 02.30.-f, 33.15.Dj + 35.15.Hp

Introduction

Gel systems are metastable systems developing with time. In our previous articles [1,2] we showed that the development of these systems occurs in helical fashion. It is natural that the spiral fragments form complex systems of double electrical layers on their surface.

Theoretical basis for the splash of ions by gel fragments

Let us consider in theory the way Van der Waals forces act in polymer gel fragments against the background of self-organizing pulsating-autosoliton processes of structuring. We will describe these processes with the help of the Liesegang operator [3]. Thus, we will try to give the colloid-chemical interpretation of the Liesegang operator. We suppose that the Liesegang operator determines the processes of gel self-organization.

Let us consider the colloid particle movement in the field of Lennard-Jones potential (intermolecular forces of London – Van der Waals). Let's perform the procedure of Lennard-Jones potential demeasurment. It is known that the Lennard-Jones potential has the form:

$$\Delta U = -\frac{C}{r^6} + \frac{B}{r^{12}} \qquad (1)$$

Let's demeasure the potential ΔU by means of the following operations. Let U_0 be the dimensional and scale constant (suppose that the potential is measured in joules); U' is certain dimensionless quantity. In that case we can write the following product for the potential

$$\Delta U = U_0 U',$$

---

[1] Corresponding author. E-mail: such@susu.ac.ru

and the product for the radius will take the form

$$r = r_0 r'.$$

Let us drop the primes. As a result we obtain the Lennard-Jones potential in the form of the proportion

$$UU_0 = -\frac{C}{r_0^6 r^6} + \frac{B}{r_0^{12} r^{12}}.$$

The equation of the colloid particle under the influence of the Lennard-Jones potential has the form:

$$\frac{r_0^2}{t_0^2}\frac{d^2r}{dt^2} = -U_0\frac{dU}{dr}, \text{ or } \frac{mr_0}{t_0^2}\frac{d^2r}{dt^2} = \left(-\frac{C'}{r_0^7 r^7} + \frac{B'}{r_0^{13} r^{13}}\right),$$

where primed and unprimed quantities are connected by the relationship

$$C' = 6C, \ B' = 12B.$$

As a result we get the relationship of the following form:

$$\frac{mr_0^8}{Ct_0^2}\frac{d^2r}{dt^2} = \left(-\frac{1}{r^7} + \frac{1}{r^{13}}\right). \tag{2}$$

The dimensionless equation has the form:

$$\frac{d^2r}{dt^2} = \left(-\frac{1}{r^7} + \frac{1}{r^{13}}\right). \tag{3}$$

Unfortunately, it is impossible to define the exact solution of the equation. That is why we have to solve this equation numerically.

The remarkable fact is that the particles forming the gel in the field of Van der Waals forces are pulsating with time, that is, they perform complex vibrating movements, thus, periodically changing some dimensionless spherical radius, figure1. These particle movements have two consequences, namely: 1) the particles acquire the physical possibility to rearrange themselves at some state or at some point of time, that is they change their spatial orientation and surrounding (in that case the fragment will have the minimum energy for barrier turns); 2) the period of gel particles vibration is a certain constant of the system of oxyhydrate particles (pacemakers), organized in the oxyhydrate gel which is a certain system of attractors.

If we compare the diagram showing the changing of polymer gel fragments concentration described by the Liesegang operator in [5], (figure 2) with the diagram showing the movement of the colloid particle (Fig.1), we can observe the visual identity of the diagrams.

Self-organization of gel and the Liesegang operator

The self-organization of gel system with time is described by the Liesegang operator. The geometrical identity of the curves on figures 1 and 2 (i.e. figure 1 shows the periodicity of spherical radii changing in the field of Van der Waals forces and figure 2 shows the Liesegang operator that changes with time) very likely indicate that the forces determining the Liesegang operator (the operator of system self-organization) and the forces determining the Lennard-Jones potential are similar. After the process of sorbtion of ions by gel swirl is finished, the intermolecular forces initiate the instant splash of ions into the external environment. When the particle acquires large potential, its movement in the Lennard-Jones potential may sharply change with time.

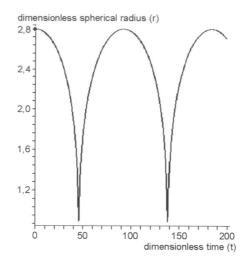

Figure 1: The particle "sticks" in the attraction field, making cyclic movements.

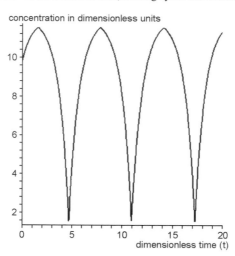

Figure 2: Diagram showing the dependence of concentration on time (the Liesegang operator)

The particle may "run away" into the infinity after having lost contact with the gravitating center (in this connection the frequency iteratively increases within the limits of infinity), figure3.

It is the case that corresponds to the phenomenon of "splash" of ions out of the spiral gel particles (pacemakers), which initiates the process of conformational gel restructuring.

Conclusions

Self-organization of gel system is described in time by the Liesegang operator. The periodicity of the spherical radii changing in the field of Van der Waals forces together with the geometrical identity of these curves to the Liesegang operator that changes with time indicate that the forces determining the Liesegang operator (the operator of system self-organization) and the forces determining the Lennard-Jones potential are similar. After the process of sorbtion of ions by gel swirl is finished, the intermolecular forces initiate the instant splash of ions into the external environment. It must correspond to the periodic spontaneous splashes of electroconductivity of oxyhydrate gel systems.

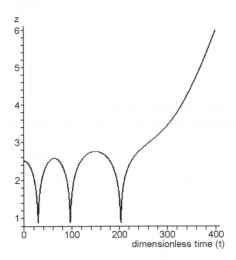

Figure 3: Movement of the electrostatically saturated particle in the Lennard-Jones potential. To simplify the demonstration the charge of the particle is a large one (about 0.1 coulomb).

References

[1] Yu. I. Sukharev, V.A. Potemkin, E.Z. Kurmaev, B.A. Markov, I. Yu. Apalikova, I. V. Antonenko, *Autowave peculiarities of polymerization of the ohyhydarte gels of heavy metals*, J. Inorg. Chem. (Russia) v.61, №6 (1998), pp. 855-863.

[2] Yu. I. Sukharev, V.A. Potemkin, B.A. Markov, *Autowave processes of forming gels as a cause of the coloring of oxyhydrate gels (the chromatic effect) of some rare earth metals (yttrium, gadolinium)*,- Colloid and Surfaces, A: Physiochem. and Eng. Asp., v. 194/1-3, (2001), pp. 75-84.

[3] B.A. Markov, Yu. I. Sukharev. Liesegang operator. *Liesegang rings as the common gross-property of oxyhydrate gel polymer systems*, Chelyabinsk: Bulletin of the scientific center of the Ural branch of RAS, 2002, v. 2(15), pp.54-67 (www.csc.ac.ru/news).

[4] B.A. Markov, Yu.I. Sukharev, *Physical-chemical interpretation of the Liesegang operator*, Chelyabinsk: Bulletin of the scientific center of the Ural branch of RAS, 2002, 3(12), pp. 74-78 (www.csc.ac.ru/news).

[5] Yu.I. Sukharev, B.A.Markov, *Physical-Chemical Polarization Nature of Living Gels of Heavy Metals Oxyhydrates*, Chelyabinsk: Bulletin of the scientific center of the Ural branch of RAS, 2002, 3(16), pp. 79-93 (http://www.csc.ac.ru/news).

VSP International
Science Publishers
P.O. Box 346, 3700 AH Zeist
The Netherlands

*Lecture Series on Computer
and Computational Sciences*
Volume 1, 2004, pp. 500-505

Toward a Pre-Operational Data Assimilation System for the E. Mediterranean using Kalman Filtering Techniques

G.N. Triantafyllou[1a], I. Hoteit[b], A.I. Pollani[a]

[a] National Centre for Marine Research,
Institute of Oceanography,
GR-190 03 Anavyssos, Attica, Greece

[b] Scripps Institution of Oceanography,
Physical Oceanography Research Division #233
8605 La Jolla Shores Drive
La Jolla, 92037, U.S.A

Received 20 August, 2004; accepted in revised form 17 September, 2004

Abstract: This paper describes a sophisticated data assimilation system which has been developed for the estimation of the state of the Eastern Mediterranean ecosystem. The forecast model is based on the complex biochemical ERSEM model coupled with the physical general ocean circulation POM model. The assimilation scheme is a square-root based Kalman filter which makes also use of low rank error covariance matrices to reduce computational cost. The assimilation of only surface sea color data shows a clear improvement in the model's behavior and a continuous decrease in the estimation error.

Keywords: Ocean Modeling, Data assimilation, Kalman filtering.

Mathematics Subject Classification:

PACS: 62L12, 95.75.-z

1. Introduction

Data assimilation aims at incorporating in the most efficient way numerical models and observations to provide the best estimate of the state of a dynamical system. A major goal of the Mediterranean Forecasting System Toward Environmental Predictions (MFSTEP) project, (http://www.bo.ingv.it/mfstep/) is to develop robust ecosystem models and data assimilation techniques for the efficient integration of biochemical observations into biophysical models. Within this project, a complex 3D ecosystem model was developed for the eastern Mediterranean which handles two, highly portable, on-line coupled, sub-models: the three-dimensional Princeton Ocean Model (POM) (Blumberg and Mellor, 1987), which describes the hydrodynamics of the area providing the background physical information to the ecological model, and the Eastern Mediterranean ecosystem model based on the European Regional Seas Ecosystem Model (ERSEM) (Baretta et al., 1995; Ebenhoh et al., 1997) describing the biogeochemical cycles. Different modeling approximations may however drift the model outputs away from reality. To bring the model closer to the available observations, simplified Kalman filter techniques were implemented. These filters actually correct the model forecast each time a new observation is available while taking into account prior information about uncertainties in the prior state forecast and data. The use of simplified filters was necessary to reduce the prohibitive computational burdens mostly due to the huge dimension of the state of the current realistic ecosystem model.

2. Model Description

ERSEM uses a 'functional' group approach to describe the ecosystem where the biota is grouped together according to their trophic level (subdivided according to size classes or feeding methods). The

---

[1] Corresponding author. e-mail: gt@ncmr.gr

ecosystem is considered to be a series of interacting complex physical, chemical and biological processes, which together exhibit coherent system behaviour. Biological functional growth dynamics are described by both physiological (ingestion, respiration, excretion, egestion, etc.) and population processes (growth, migration and mortality). Biologically driven carbon dynamics are coupled to the chemical dynamics of nitrogen, phosphate, silicate and oxygen (Baretta-Bekker et al., 1995; Varela et al., 1995). The food web represents the real system and is described in Figure 1.

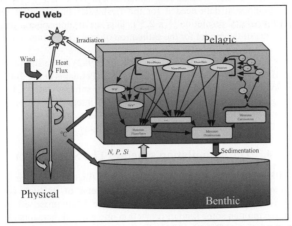

Figure 1: A schematic description of the ecosystem functional groups and their trophic relations.

Pelagic bacteria are assumed to be free-living heterotrophs utilizing particulate and dissolved organic material, produced by the excretion, lysis and mortality of primary and secondary producers, as food. The phytoplankton pool is described by four functional groups based on size and ecological properties. These are diatoms P1 (silicate consumers, 20-200μ), nanophytoplankton P2 (2-20μ), picophytoplankton P3 (< 2μ) and dinoflagellates P4 (> 20μ). All phytoplankton groups contain internal nutrient pools and have dynamically varying C:N:P ratios. The nutrient uptake is controlled by the difference between the internal nutrient pool and external nutrient concentration. The microbial loop contains bacteria B1, heterotrophic flagellates Z6, and microzooplankton Z5, each with dynamically varying C:N:P ratios. Bacteria act to decompose detritus and can compete for nutrients with phytoplankton. Heterotrophic flagellates feed on bacteria and picophytoplankton, and are grazed by microzooplankton. Microzooplankton also consumes diatoms and nanophytoplankton and is grazed by mesozooplankton. The parameter set used in this simulation is the same as in the 3D Cretan Ecosystem Model (Petihakis et al., 2002). The biogeochemical model coupled to the physical model consists of 4-D advection-diffusion-reaction equations, and is solved for the concentration of C for each functional group of the pelagic system:

$$\frac{\partial C}{\partial t} = -U\frac{\partial C}{\partial x} - V\frac{\partial C}{\partial y} - W\frac{\partial C}{\partial z} + \frac{\partial}{\partial x}\left(A_H\frac{\partial C}{\partial x}\right) + \frac{\partial}{\partial y}\left(A_H\frac{\partial C}{\partial y}\right) + \frac{\partial}{\partial z}\left(K_H\frac{\partial C}{\partial z}\right) + \sum BF$$

where U, V, W represent the velocity field, A_H the horizontal viscosity coefficient, and K_H the vertical eddy mixing coefficient, provided by the POM. $\sum BF$ stands for the total biochemical flux, calculated by ERSEM, for each pelagic group.

The benthic-pelagic coupling is described by a simple first order benthic returns module, which includes the settling of organic detritus into the benthos and diffusional nutrient fluxes into and out of the sediment.

3. Model Set-up

The hydrodynamic model used is the ALERMO (Korres and Lascaratos, 2003) which has one open boundary located at 20° E. The computational grid has a horizontal resolution of 1/10° x1/10° and 30 sigma layers in vertical with a logarithmic distribution near the sea surface. The model's bathymetry was obtained from the US Navy Digital Bathymetric Data Base - DBDB5 – (with a nominal resolution of 1/12° x 1/12°) by bilinear interpolation. A Shapiro filter of third order was also applied to the interpolated bathymetry in order to perform the necessary smoothing. The model's climatological run was initialized with the Mediterranean Ocean Data-Base which contains seasonal profiles of temperature and salinity mapped on a $1/4° \times 1/4°$ horizontal grid. Additionally, initial velocities were set to zero. The temperature and salinity profiles at the open boundaries were also derived from the same database.

Atmospheric data sets consist of monthly values for the longitudinal and meridional components of the wind stress, air temperature, air humidity and cloud cover. These monthly values were derived from the 1979 – 1993 6-hour ECMWF re-analysis data (horizontal resolution $1° \times 1°$) by proper averaging in time. Additionally the precipitation data needed for the freshwater budget were taken from Jaeger (1976) monthly data set (horizontal resolution $5° \times 2.5°$). This set of atmospheric data is then used by the air-sea interaction scheme of the model for the estimation of heat, freshwater and momentum fluxes at the sea surface.

The initial conditions for the nutrients are taken from Levitus (1982) while the other biogeochemical state variables from the 3D ecosystem model for the Cretan Sea (Petihakis et al., 2002). The ecosystem pelagic state variables along the open boundary are described by solving water-column 1D ecosystem models at each surface grid point on the open boundary. The integration starts from spring initial conditions (15th of May). The model was run perpetually for four years to reach a quasi steady state and to obtain inner fields fully coherent with the boundary conditions.

4. The assimilation scheme: The SEEK filter

The extended Kalman (EK) filter is an extension of the Kalman filter to nonlinear systems. However, its implementation in realistic ecosystems is not feasible because of its high computational cost. Different degraded forms of the extended Kalman filter have been proposed, which reduce the dimension of the system (n) through some kind of projection into a low dimensional sub-space (Cane et al., 1995; Fukumori and Malanotte-Rizzoli, 1995; Hoang et al., 1997). In this study, we used the Singular Evolutive Extended Kalman (SEEK) which has been developed by Pham et al. (1997). This filter uses low rank estimation error covariance matrices, which make possible the implementation of the EK filter with realistic ocean models. With this assumption, the correction of the filter is made only along the directions for which the error was not sufficiently attenuated by the dynamics of the system. These directions can be further let to evolve in time to follow changes in the model dynamics. The SEEK filter has been successfully implemented in different realistic ocean applications, e.g. Pham et al. (1997) and Hoteit et al. (2001) with physical ocean circulation models, and Triantafyllou et al. (2003) and Carmillet et al. (2001) with ecosystem models. A schematic diagram of the filter's algorithm is illustrated in Figure 2.

The computational cost of the SEEK filter is mainly due to the evolution of its 'correction directions'. Recently, a further simplification of the SEEK has been broadly used, examining the asymptotic approximation of the error covariance matrix. Results from the implementation of the SEEK filter show an immediate reduction of the error level after the first correction. Thus the use of the SFEK filter, a variant of SEEK, which maintains the initial correction directions invariant, is expected to produce reasonable results as well. When the initial error covariance is decomposed into empirical orthogonal functions, it is not easy to determine the truncation level suitable to conduct the assimilation experiments for the given ecosystem. The SFEK filter can be therefore also appropriate to perform sensitivity analysis regarding the relevance of the empirical sub-space for propagating surface observations at the deeper layers avoiding the heavy computation load of the SEEK filter.

5. Determination of the (initial) correction sub-space

The (initial) correction subspace was determined using an empirical orthogonal functions (EOF) analysis (also known as principal components analysis). This analysis is widely used in meteorology and oceanography to determine the principal modes of the models variability. It provides the best approximation of a set of state vectors (and of their sample covariance matrix) in a low dimensional sub-space (with a singular low rank matrix). Since the ecosystem variables are of different nature (not homogeneous), a metric was used to make these variables independent of their units. We actually applied the so-called multivariate EOF analysis. The computation of the EOFs was made through a simulation of the model itself. After three years of model spin-up run to reach a statistically quasi-steady state, an integration of about 2 years was carried out to generate a historical sequence H_s of model realizations. A sequence of 130 state vectors was retained by storing one every six days to reduce the calculations, because successive states are quite similar. The filter's 'directions of correction' were then obtained via a multivariate EOF analysis on the sample H_s. It was found that the first 15 EOFs explain more than 90% of the variance of the system. This is a strong indication that the state of our system evolves in a subspace with a dimension much smaller than the dimension of the original system, which supports the filter's corrections in the directions of the leading modes of the system.

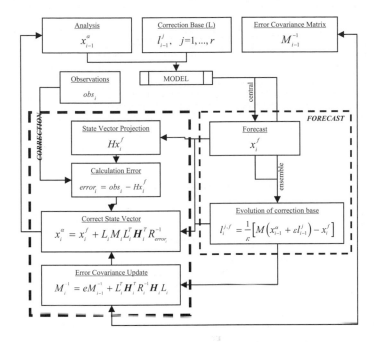

Figure 2. Schematic diagram of the assimilation cycle. In the forecast module the state and the evolution of the correction directions evolve with the model dynamics. In the correction module the forecast is corrected using the new observations and the updated error covariance matrix.

6. Twin experiment results - Conclusions

Prior to the assimilation of real data, the twin experiment approach was used to validate our assimilation system. Within this approach, the true state X^t is assumed to be provided by the model itself. A model run is then performed to generate a set of X^t. These X^t are then used twice: to generate the (pseudo-) observations, and as a reference state to assess the quality of the filter's analysis on non-observed variables. A random error was added to the pseudo-observations to be as close as possible to the real situation. Then the model is initialized with the mean state vector and the performance of the filter can be assessed by its ability to bring the model back into its 'true' trajectory while using ocean colour data only. Therefore, following the two years simulation for the generation of the H_s sequence, the model was run for a half year to produce the reference states set. The assimilation experiments were performed on a three day basis using surface chlorophyll pseudo-measurements, which were extracted from the reference states. The performance of the filter was measured through the comparison of the relative root mean square RRMS error for each state variable, over the whole simulation domain. The definition of the RRMS can be found in Triantafyllou et al. (2003).

The RRMS of the filter using 15 and 25 EOFs are illustrated in Figure 3 and compared with the RRMS of the model free-run for phosphate, nitrate, diatoms, picoplankton, mesozooplankton and bacteria. It can be seen that the error efficiently is reduced and remains relatively low for all the variables. After 17 analysis steps, the estimation error reaches a saturation value for all variables. Regarding the behaviour of the filter in relation to the number of adopted EOFs, the filter is shown to provide very satisfactory results while using 15 EOFs only. The filter's degradation that can be seen in the RRMS of picoplankton, mesozooplankton and bacteria between the 10th and the 17th analysis steps, when using 15 EOFs, suggests that more EOFs must be used during some periods, particularly when the model shows instabilities.

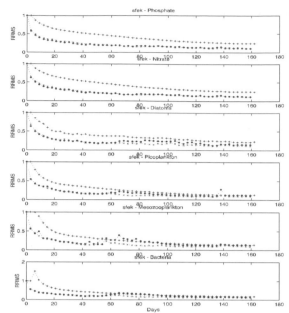

Figure 3. Evolution in time of the RRMS for phosphate, nitrate, diatoms, picoplankton, mesozooplankton, and bacteria from the model free-run (+) and the SFEK filter with rank 15 (*) and 25 (x).

The filter was shown to be very efficient in estimating the state of the ecosystem model. The experiments have been carried out under a "perfect model" assumption. This is very optimistic since in practice any ecosystem model will have deficiencies. When real data will be assimilated, the performance of the filter will highly depends on the incorporation of a realistic model error so that the filter does not completely follow the forecast model. The estimation of this error is however very difficult because of the huge dimension of the system and of the cruel lack of observations. Another point that might improve the filter's behavior when real data are used is to find a way to directly assimilate ocean color data instead of converting them to phytoplankton biomass through chlorophyll. This can be done by modifying the model in order to predict chlorophyll from phytoplankton values and bio-optical algorithm. The mismatch between observed and predicted surface chlorophyll values would then be used to drive the filters. These problems will be investigated in the near future within the framework of the MFSTEP project.

Acknowledgments

This work was carried out within the framework of the project Mediterranean Forecasting System Toward Environmental Predictions (MFSTEP).

References

Baretta, J. W., W. Ebenhoh and P. Ruardij (1995). The European Regional Seas Ecosystem Model, a complex marine ecosystem model, Netherlands Journal of Sea Research, 33: 233-246.

Baretta-Bekker, J. G., J. W. Baretta and E. Rasmussen (1995). The microbial foodweb in the European regional Seas Ecosystem Model, Netherlands Journal of Sea Research, 33: 363-379.

Blumberg, A. F. and G. L. Mellor (1987). A description of a three-dimensional coastal ocean circulation model, In Three-Dimensional Coastal Ocean Circulation Models. N. S. Heaps. Washington, D.C., AGU. 4: 1-16.

Cane, M. A., A. Kaplan, R. N. Miller, B. Tang, E. C. Hackert and A. J. Busalacchi (1995). Mapping tropical Pacific sea level: data assimilation via a reduced state Kalman filter, Journal of Geophysical Research, 101(10): 599-617.

Carmillet, V., J. M. Brankart, P. Brasseur, H. Drange, G. Evensen and J. Verron (2001). A singular evolutive extended Kalman filter to assimilate ocean color data in a coupled physical-biochemical model of the North Atlantic ocean, Ocean Modelling, 3(3-4): 167-192.

Ebenhoh, W., J. G. Baretta-Bekker and J. W. Baretta (1997). The primary production module in the marine ecosystem model ERSEM II with emphasis on the light forcing, Journal of Sea Research, 38: 173-193.

Fukumori, I. and P. Malanotte-Rizzoli (1995). An approximate Kalman filter for ocean data assimilation: an example with an idealized Gulf Stream model, Journal of Geophysical Research, 100(C4): 6777-6793.

Hoang, H. S., P. De Mey, O. Tallagrand and R. Baraille (1997). Adaptive filtering: Application to satellite data assimilation in oceanography, Journal Dynamicas of Atmospheres and Oceans, 27: 257-281.

Hoteit, I., D. T. Pham and J. Blum (2001). A semi-evolutive partially local filter for data assimilation, Marine Pollution Bulletin, 43: 164-174.

Jaeger, L. (1976). Monatskarten des Niederschlags fur die ganze Erde, Ber. Dtsch. Wetterdienste, 18(1839): 1-38.

Korres, G. and A. Lascaratos (2003). An eddy resolving model for the Aegean and Levantine basins for the Mediterranean Forecasting System Pilot Project (MFSPP): Implementation and climatological runs, Annales Geophysicae, 21: 205-220.

Levitus, S. (1982). Climatological Atlas of the World Ocean, No. 13, Washington.
Petihakis, G., G. Triantafyllou, J. I. Allen, I. Hoteit and C. Dounas (2002). Modelling the Spatial and Temporal Variability of the Cretan Sea Ecosystem, Journal of Marine Systems, 36: 173-196.

Pham, D. T., J. Verron and M. C. Roubaud (1997). Singular evolutive Kalman filter with EOF initialization for data assimilation in oceanography, Journal of Marine Systems, 16: 323-340.

Triantafyllou, G., I. Hoteit and G. Petihakis (2003). A singular evolutive interpolated Kalman filter for efficient data assimilation in a 3D complex physical-biogeochemical model of the Cretan Sea, Journal of Marine Systems, 40-41: 213-231.

Varela, R. A., A. Cruzardo and J. E. Gabaldon (1995). Modelling the primary production in the North Sea using ERSEM, Netherlands Journal of Sea Research, 33: 337-361.

VSP International
Science Publishers
P.O. Box 346, 3700 AH Zeist
The Netherlands

Lecture Series on Computer
and Computational Sciences
Volume 1, 2004, pp. 506-509

Computational Science and Engineering Online

T. N. Truong[1]

Henry Eyring Center for Theoretical Chemistry,
Department of Chemistry,
University of Utah,
315 South 1400 East, rm 2020
Salt Lake City, UT 84112, USA

Received 6 August, 2004; accepted in revised form 18 August, 2004

Abstract: We present the development of an integrated extendable Web-based simulation environment for computational science and engineering called Computational Science and Engineering Online (CSEO). CSEO is a collaboratory that allows computational scientists to perform research using state-of-the-art application tools, to query data from personal or public databases, to document results in an electronic notebook, to discuss results with colleagues using different communication tools, and to access grid-computing resources from a web browser. Currently, CSEO provides an integrated environment for multi-scale modeling of complex reacting systems and biological systems. A unique feature of CSEO is in its framework that allows data to flow from one application to another in a transparent manner.

Keywords: Web-based Simulation, Grid computing, Problem Solving Environment.

The discovery of the World Wide Web (often referred to as the web) has been revolutionizing the way we communicate since the last decade. Scientists are also looking into web technology to help revolutionize the way science is being conducted and taught. The last several years have seen tremendous efforts in the creation of 'collaboratories'. These are laboratories without walls, in which researchers can take advantage of web technology to expand their research capabilities and to collaborate in solving complex scientific problems.[1-4] These collaboratories can be classified into two types: data sharing oriented or remote access scientific instrument oriented. The Research Collaboratory for Structural Bioinformatics is an excellent example of the data-oriented collaboratory that provides access to databases of biological structures and tools for determining and analyzing these structures. Most existing collaboratories are instrument driven and provide the capability for real-time data acquisition from remote research instruments through web-accessible servers in a seemingly transparent way and are often focused specifically on a particular complex scientific problem. An example is the Space Physics and Aeronomy Research Collaboratory (SPARC) that provides an internet-based collaborative environment for studies of space and upper atmospheric science facilitating real-time data acquisitions from a remote site in Greenland. The establishment of these collaboratories has undisputed potential for making significant impacts on science and technology in the 21st century.

The possibility of using web technology to provide a new framework for simulation attracted a lot of interest in the mid-1990s [5-9]. However, the progress so far has not made any significant impact on the scientific computing community since it has not been user driven as assessed by Kuljus and Paul [8].

In this study, we describe our current efforts in developing an integrated web-based simulation environment for multi-scale computational modeling of chemical and biological systems. The new environment is called Computational Science and Engineering Online (CSEO). The goal is to create a web-based simulation environment that would benefit the greater simulation community of computational science and engineering while gradually introducing the psycho-social changes to

---

[1] Corresponding author. E-mail: Truong@chemistry.utah.edu.

facilitate the paradigm shift in scientific research.   To do so, we would focus our attention on how research is currently being done and then design the environment that would enhance current research capabilities while not significantly altering the research culture.

The main goal of CSEO is to provide a Web-based grid-computing environment in which a researcher can perform the following functions:
1. Research using a variety of state-of-the-art scientific application tools.
2. Access and analyze information from databases and electronic notebooks.
3. Share and discuss results with colleagues.
4. Access computational resources from the computing grids that are far beyond those available locally.
5. Master subjects in other areas of computational science and engineering asynchronously, without regard to geographical location and schedule.

In designing CSEO, our vision is for it not to be a central web portal but rather a world-wide extensive network of many mirror sites, hosted by different universities, computer centers, national laboratories, even industries, that share their public databases via a secure network.   This will maximize resource utilization, data generation and sharing.

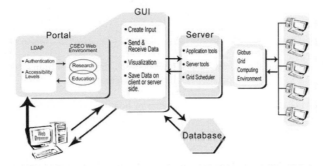

Figure 1: Flowchart showing different components of the CSEO environment and their functionalities.

Figure 1 illustrates how a user from a web browser can login to the CSEO web portal and use its functionalities. The login initiates the authentication process, which also determines the accessibility level of the user since CSEO supports commercial application tools that have different license restrictions.   Each application tool has a graphic-user-interface (GUI) to create input, submit a job to the server or the computing grid, receive and visualize data, and save data either to the database or user directories on the server.   Application tools consist of both commercial and open-source legacy codes residing on the server or on the computing grid.

Below is the roadmap that illustrates the flow of data between different research application tools.   This roadmap also allows us to design a knowledge management system that has filters to extract information from the results of one application to prepare the input for another. This way it eliminates much of the human error in preparing input files by the traditional manual cut-and-paste. From a more philosophical point of view, the knowledge management system of CSEO eliminates the traditional boundaries which define disciplines within the larger scientific computing community such as quantum chemistry, chemical kinetics, reaction engineering, computational biology, material science, etc.

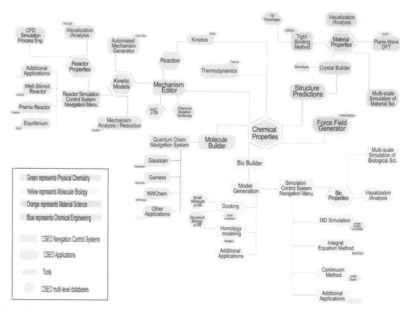

Figure 2: Roadmap that shows how data is flows between applications from different areas of computational science and engineering.

This roadmap does not show the software architecture of how CSEO is constructed nor how data are stored and transferred as well as how a user can access the computing grid. Such issues will be discussed in more details in the presentation.

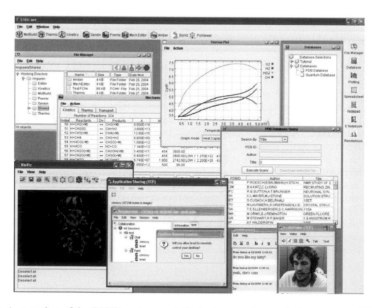

Figure 2: A snap shot of the CSEO environment displaying a variety tools that are accessible to users from a web browser.

The unique feature of CSEO is in its extendable integrative environment, which allows for seamless interface and dataflow between different areas of applications and thus facilitates multi-disciplinary research while it has the look-and-feel of a single desktop application. This is achieved by the use of Java Webstart technology. It allows us to create the CSEO environment that has the look-and-feel and function independent of computer platforms as well as web browsers. We will demonstrate this feature by showing how results from quantum chemistry simulations are used in simulations of a combustion reactor. It is interesting to point out that CSEO is the first application environment that allows multi-scale simulations of complex reaction systems which require bridging length scale from 10^{-8} to 10^{0} meter and time scale from 10^{-15} to 10^{2} second.

Acknowledgments

The author wishes to thank the CSEO development team for their contributions to the project and the USA National Science Foundation for the financial supports.

References

[1] North Carolina Board of Science Technology and National Research Council, *Collaboratories: Improving Research Capabilities in Chemical and Biomedical Sciences: Proceedings of a Multi-site Electronic Workshop*. 1999, Washington, D.C.: National Academy Press.

[2] Agarwal, D.A., Sachs, S.R., and Johnston, W.E., The Realities of Collaboratories. *Comp. Phys. Comm.*, 110 134 (1998).

[3] Kiernan, V., Internet-Based 'Collaboratories' Help Scientists Work Together, in *The Chronicle of Higher Education*. 1999.

[4] Kouzes, R.T., Myers, J.D., and Wulf, W.A., *Collaboratories: Doing Science on the Internet*. IEEE Computer, 29 (1996).

[5] Charnes, J.M., Morrice, D.J., Brunner, D.T., and Swain, J.J., eds. Proceedings of the 1996 Winter Simulation Conference. 1996: *Coronado, CA*.

[6] Fishwick, P.A., Hill, D.R.C., and Smith, R., eds. Proceedings of the 1988 International Conference on Web-Based Modeling and Simulation. 1998, *The Society for Computer Simulation International: San Diego*.

[7] Bruzzone, A.G., Uhrmacher, A., and Page, E.H., eds. Proceedings of the 1999 International Conference on Web-Based Modeling and Simulation. 1999, *The Society for Computer Simulation International: San Diego*.

[8] Kuljis, J. and Paul, R.J., An appraisal of Web-Based Simulation: Whither We Wander. *Simulation Practice and Theory*, 9 37-54 (2001).

[9] Page, E.H., Buss, A., Fishwick, P.A., Healy, K.J., Nance, R.E., and Paul, R.J., Web-Based Simulation: Revolution or Evolution? *ACM Transactions on Modeling and Computer Simulation*, 10 3-17 (2000).

VSP International
Science Publishers
P.O. Box 346, 3700 AH Zeist
The Netherlands

Lecture Series on Computer
and Computational Sciences
Volume 1, 2004, pp. 510-513

Stage Reduction on P-Stable Numerov Type Methods of Eighth Order

Ch. Tsitouras[1]

TEI of Chalkis, Department of Applied Sciences, GR34400, Psahna, Greece

Received 7 September, 2004; accepted in revised form 15 September, 2004

Abstract: We present an implicit hybrid two step method for the solution of second order initial value problem. It costs only six function evaluations per step and attains eighth algebraic order. The method satisfy the P-stability property requiring one stage less. We conclude dealing with implementation issues for the methods of this type and give some first pleasant results from a numerical test.

Keywords: Initial Value Problem, Second Order, Oscillatory solutions.

Mathematics Subject Classification: 65L05, 65L06

1 Introduction.

We are interested in solving the initial value problem of second order

$$y'' = f(x, y), \ y(x_0) = y_0, \ y'(x_0) = y_0'. \tag{1}$$

In this paper we investigate the class of the above problems with oscillatory solutions. Our result are methods which can be applied to many problems in celestial mechanics, quantum mechanical scattering theory, in theoretical physics and chemistry and in electronics.

Implicit hybrid two step methods satisfying P-stability property are used for about twenty years for solving (1), [1, 2, 7]. Their construction is usually based on interpolatory nodes. These nodes carry a lot of information which is useless even for conventional methods. So, an alternative implementation of such methods was introduced in [6, 8, 9], and studied theoretically by Coleman [3] through B-series.

In [7] we used seven stages for achieving P-stability and eighth order of accuracy. Here we propose a six stage method of the form:

$$y_{n+1} = 2y_n - y_{n-1} + h^2 \sum_{j=1}^{s} b_j f(x_n + c_j h, g_j). \tag{2}$$

with

$$g_i = (1 + c_i)y_n - c_i y_{n-1} + h^2 \sum_{j=1}^{s} a_{ij} f(x_n + c_i h, g_j), \ i = 1, 2, \cdots, s = 6,$$

---

[1]E-Mail: tsitoura@teihal.gr

2 Algebraic order of the new method.

When solving (1) numerically we have to pay attention in the algebraic order of the method used, since this is the main factor of achieving higher accuracy with lower computational cost. Thus this is the main factor of increasing the efficiency of our effort. Using the notation of Nyström methods we consider the matrix of the coefficients $A = \{a_{ij}\}$ and the vectors $b = [b_1 \ b_2 \ b_3 \ b_4 \ b_5 \ b_6]$ and $c = [c_1 \ c_2 \ c_3 \ c_4 \ c_5 \ c_6]^T$. Now the method can be formulated in a table like the Butcher tableau,

$$\begin{array}{c|c} c & A \\ \hline & b \end{array}.$$

Under the simplifying assumptions

$$Ae = \frac{1}{2}\left(c^2 + c\right), \ Ac = \frac{1}{6}\left(c^3 - c\right), \ Ac^2 = \frac{1}{12}\left(c^4 + c\right), \ Ac^3 = \frac{1}{20}\left(c^5 - c\right) \tag{3}$$

with $e = [\ 1 \ \ 1 \ \ 1 \ \ 1 \ \ 1 \ \ 1 \ ]^T$ and $c^i = [\ c_1^i \ \ c_2^i \ \ c_3^i \ \ c_4^i \ \ c_5^i \ \ c_6^i \ ]$, we get the following eighth order conditions [3]:

$$b \cdot e = 1, \ b \cdot c = 0, \ b \cdot c^2 = \frac{1}{6}, \ b \cdot c^3 = 0, \ b \cdot c^4 = \frac{1}{15}, \ b \cdot c^5 = 0, \ b \cdot c^6 = \frac{1}{28}$$

$$b \cdot A \cdot c^4 = \frac{1}{840}, \ b \cdot c^7 = 0, \ b \cdot \left(c \ A \cdot c^4\right) = \frac{1}{180} \ , \ b \cdot A \cdot c^4 = 0.$$

Our methods include 48 parameters. Thirty five equations are required assuming order conditions and satisfaction of (3). This leaves thirteen coefficients as free parameters.

3 Periodic problems.

Following Lambert and Watson [5] and in order to study the periodic properties of methods posed for solving (1), it is constructive to consider the scalar test problem

$$y' = -\omega^2 y, \ \ \omega \in \Re. \tag{4}$$

When applying an implicit two step hybrid method of the form (2) to the problem (4) we obtain a difference equation of the form

$$y_{n+1} + S\left(v^2\right) y_n + P\left(v^2\right) y_{n-1} = 0, \tag{5}$$

where $y_n \approx y\left(nh\right)$ the computed approximations at $n = 1, 2, \ldots$, $v = \omega h$, h the step size used, and $S\left(v^2\right), P\left(v^2\right)$ are ratios of polynomials in v^2.

Zero dissipation property is fulfilled by requiring $P\left(v^2\right) \equiv 1$, and helps a numerical method that solves (4) to stay in its cyclic orbit. We observe that $P\left(v^2\right) = 1 - v^2 b \cdot \left(I + v^2 A\right)^{-1} \cdot c$, can be written as an infinite series

$$P\left(v^2\right) = 1 + v^9 b \cdot A^4 \cdot c + v^{11} b \cdot A^5 \cdot c + \cdots.$$

Actually we have to solve $b \cdot A^4 \cdot c = 0$, $b \cdot A^5 \cdot c = 0$, and $b \cdot A^6 \cdot c = 0$ only, demanding another three coefficients and leaving ten free parameters.

The solution of (4) is $y(x) = e^{i\omega x}$, and we may write equation (5) as

$$e^{2iv} + S\left(v^2\right) \cdot e^{iv} + 1 = O.$$

P-stability means that the numerical solution stays in orbit for ever. Thus we want $|S\left(v^2\right)| < 2$ for $v \in (0, +\infty)$. Observing that $S\left(v^2\right) = 2 - v^2 b \cdot \left(I + v^2 A\right)^{-1} \cdot (e - c)$ and after extended search we concluded to a method with the following coefficients.

$a_{11} = -0.33083649953596372, a_{12} = -0.28554560691201376, a_{13} = 0.096020140660069509$
$a_{14} = 0.065976488202945502, a_{15} = 0.93159949396176978, a_{16} = -0.25151621006087468$
$a_{21} = -0.22800572156136017, a_{22} = 0.75775376332106239, a_{23} = 0.044036478175189789$
$a_{24} = 0.044036478175189789, a_{25} = -0.38981527654872163, a_{26} = -0.22800572156136017$
$a_{31} = -0.14560363007308039, a_{32} = -1.9592986015796962, a'_{33} = -0.024082528198053865$
$a_{34} = 0.028081493431889852, a_{35} = 2.1942941895272906, a_{36} = -0.18249268029751431$
$a_{41} = 1.0027874805872521, a_{42} = -1.2518397436149883, a_{43} = -0.092326475684278097$
$a_{44} = -0.14449049731422181, a_{45} = 0.12503961031290593, a_{46} = 1.0396765308116860$
$a_{51} = -0.10278432173227921, a_{52} = 0.18924763112443292, a_{53} = 0.019851517364212751$
$a_{54} = 0.019851517364212751, a_{55} = -0.023382022388299996, a_{56} = -0.10278432173227921$
$a_{61} = 0.28697954935636091, a_{62} = 0.16149513085428000, a_{63} = -0.085716485163353987$
$a_{64} = -0.055672832706229981, a_{65} = -0.62654046521477468, a_{66} = 0.20765925988127187$
$b_1 = 0.29173891914469542, b_2 = 0.12330286145746479, b_3 = 0.084958219397839784$
$b_4 = 0.084958219397839784, b_5 = 0.12330286145746479, b_6 = 0.29173891914469542$
$c_1 = c_6 = 0.33749364930837850, c_2 = c_5 = 0, c_3 = c_4 = -0.76794866228752001$

4 Implementation issues.

First we introduce

$$z_i = g_i - (1 + c_i)y_n + c_i y_{n-1} = h^2 \cdot \sum_{j=1}^{s} a_{ij} f(x_n + c_i h, g_j). \qquad (6)$$

Similar to implicit Runge-Kutta methods [4, p. 118], we observe that $y_{n+1} = 2y_n - y_{n-1} + (d_1 z_1 + d_2 z_2 + d_3 z_3 + d_4 z_4 + d_5 z_5 + d_6 z_6)$, with $d = [d_1\ d_2\ d_3\ d_4\ d_5\ d_6] = b \cdot A^{-1}$. For solving non linear equations (6) we use modified Newton iteration according to the scheme (brackets in the exponent mean number of iteration):

$$(I - h^2 A \otimes J)\Delta Z^{[k]} = -Z^k + h^2(A \otimes I) \cdot F(Z^{[k]})$$
$$Z^{[k+1]} = Z^{[k]} + \Delta Z^{[k]}.$$

Here $Z^{[k]} = [z_1^{[k]}\ z_2^{[k]} \cdots z_s^{[k]}]^T$, is the k-th iteration and $\Delta Z^{[k]}$ are the corresponding increments. $F(Z^{[k]})$ is an abbreviation for

$$F(Z^{[k]}) = [f(x_n + c_1 h, (1 + c_1)y_n - c_1 y_{n-1} + z_1^k),\ f(x_n + c_2 h, (1 + c_2)y_n - c_2 y_{n-1} + z_2^k), \cdots]^T,$$

see [4, pp. 119-120] for details.

A simple choice for the starting value of $Z^{[0]}$ would be $z_i^{[0]} = 0$ for $i = 1, 2, \cdots, s$. A more satisfactory approach uses an $O(h^4)$ interpolation based on known values y_{n-1}, y_n, y''_{n-1} and y''_n. So we may evaluate

$$z_i^{[0]} = -\frac{1}{6}h^2(c_i - 1)(c_i + 1)c_i y''_{n-1} + \frac{1}{6}h^2(c_i + 2)(c_i + 1)c_i y''_n. \qquad (7)$$

In view of (3) we may use high order stage values from previous steps forming more accurate interpolants for $z_i^{[0]}$'s, but (7) is efficient enough to get convergence at one iteration only, for many non-linear problems.

5   Numerical Tests.

In order to perform our tests we choose the well known Duffing problem, $y'' = -y - y^3 + 0.002 \cdot \cos(1.01x)$, $y(0) = 0.200426728067$, $y'(0) = 0$. We solved the above equation in the region $[0, \frac{120.5}{1.01} \cdot \pi]$, and we recorded the running times and the end point global error achieved by our previous method given in [7] and the new method in Table 1. We observe an apparent improvement

Table 1: Accurate digits for Duffing equation

NEW		[7]	
time	accuracy	time	accuracy
2.6s	2.5	2.2s	2.4
5.7s	5.2	4.6s	4.3
9.0s	6.8	7.2s	5.2
12.4s	7.8	9.8s	5.9
15.6s	8.6	12.7s	6.4
19.1s	9.3	15.1s	6.8
22.2s	9.8	17.6s	7.2

of the results.

Acknowledgment

The present work was financed by the program Archimedes for supporting research for TEI.

References

[1] J. Cash, High order P-stable formulae for the numerical integration of periodic initial value problems, *Numer. Math.*, **37** (1981) 355-370.

[2] M. M. Chawla, Two-step fourth order P-stable methods for second order differential equations, *BIT,* **21** (1981) 190-193.

[3] J. P. Coleman, Order conditions for a class of two-step methods for $y'' = f(t, y)$, *IMA J Numer. Anal.*, **23** (2004) 197-220.

[4] E. Hairer and G. Wanner, Solving Ordinary Differential Equations II, *Springer-Verlag,* 2nd ed., Berlin, 1996.

[5] J. D. Lambert and I. A. Watson, Symmetric multistep methods for periodic initial value problems, *J. Inst. Math. Appl.*, **18** (1976) 189-202.

[6] G. Papageorgiou, Ch. Tsitouras and I. Th. Famelis, Explicit Numerov type methods for second order IVPs with oscillating solutions, *Int. J. Mod. Phys. C,* **12**(2001) 657-666.

[7] T. E. Simos and Ch. Tsitouras, A P-stable eighth order method for the numerical integration of periodic initial value problems, *J. Comput. Phys.*, **130** (1997) 123-128.

[8] Ch. Tsitouras, Explicit two-step methods for second-order linear IVPs, *Comput. & Maths with Appl.*, **43** (2002) 943949.

[9] Ch. Tsitouras, Explicit Numerov type methods with reduced number of stages, *Comput. & Maths with Appl.*, **45** (2003) 37-42.

VSP International
Science Publishers
P.O. Box 346, 3700 AH Zeist
The Netherlands

Lecture Series on Computer
and Computational Sciences
Volume 1, 2004, pp. 514-518

Designing Recurrence Sequences: Properties and Algorithms

Evangelos Tzanis [1]

University of Amsterdam
Institute of Logic, Language and Computation

Abstract: Reccurence sequences appear almost everywhere in mathematics and computer science. Their study, is as well, plainly of intrinsic interest and has been a central part of number theory for many years. It is well known that for many recurrence sequences, such as the famous generalized $3 \cdot x + 1$ mapping [6], typical cases are easy to solve; so that computionally hard cases must be rare. This paper shows that the generalized $3x + 1$ mapping can be summarized mainly by at least one parameter: how fast an input number reach a smaller one. Hard problems occur at critical values of such a parameter. The main steps of our method are: classification of numbers, construction of $d-$trees, searching and measuring the density of terminal nodes (numbers which reach a smaller number or a divergent class). We prove that a family of these recurrence sequences has terminal nodes' density 1.

Keywords: $3x + 1$ problem, number theory, recurrence sequences

1 Introduction

Let $d \geq 2$ be a positive integer and $m_0, ..., m_{d-1}$ be non zero integers. Also for $i = 0, ..., d - 1$ let $r_i = im_i \bmod d$. Then the formula ([6]) :

$$T(x) = \frac{m_i \cdot x - r_i}{d}, \text{ if } x \equiv i \bmod d$$

defines a mapping T: $Z \to Z$ called the generalized Collatz function. In this paper we consider the case where $d = 3$ and $m_i = 1(\bmod 3)$ extending also our work in the general case.

The rest of this paper is organized as follows: in section 2 we present an example of our method, an intangible problem as the $3 \cdot x + 1$. Classification of numbers, construction of $d-$trees, searching for terminal nodes are the main steps of our method. While in section 3 we present two algorithms: The first generates the corresponding $d-tree$ for each such class, showing the relation of these trees to Pascal's triangle. The second algorithm provides an efficient way to estimate the total number of steps needed to reach a divergent or a convergent class. This paper combines mathematical methods and programming techniques.

---

[1]. E-mail: etzanis@illc.uva.nl

2 Struggling Recurrence sequences

It is well known [6] that the following recursive function is as struggling as the $3 \cdot x + 1$ (where $d = 3$, $m_0 = 2$, $m_1 = 4$, $m_2 = 4$, $r_0 = 0$, $r_1 = -1$, $r_3 = 1$).

$$Y(x) = \begin{cases} 2 \cdot x_{\frac{1}{3}}, & \text{if x} \equiv 0 \ (mod 3) \\ 4 \cdot x + 1_{\frac{1}{3}}, & \text{if x} \equiv 1 \ (mod 3) \\ 4 \cdot x - 1_{\frac{1}{3}}, & \text{if x} \equiv 2 \ (mod 3) \end{cases}$$

The trajectory $T^k(8)$ appears to be divergent. Following the main steps of our algorithm we classify natural numbers in 3 classes investigating how fast these numbers reach a smaller one. Also, numbers in class 2 (mod 3) (class C) reach a smaller number, since $\frac{2 \cdot x}{3} < x$. The difficult classes are the classes A and B.

A-class	B-class	C-class
1	2	3
4	5	6

Is trivial to see that: Numbers of class A $(1 + 3 \cdot k)$ reach, after one step, the number $(1 + 4 \cdot k)$. Numbers of class B $(2 + 3 \cdot k)$ reach after one step the number $(3 + 4 \cdot k)$.

The Y(x) sequence produced by $x_0 = 2 + k \cdot k$ (B class) is:

Step	0	1
Value	$2 + 3 \cdot k$	$3 + 4 \cdot k$

In case of number $3 + 4 \cdot k$ we do not know immediately which Y(x)'s case to apply, we need to study three cases : $k = 3 \cdot \rho + 1$, $k = 3 \cdot \rho$ and $k = 3 \cdot \rho + 2$. Let us assume that each node of the tree has a complex (Co) and a parametric part (Pa). For instance: $3 + 4 \cdot k$ has Co=3 and Pa=4=4^1. So the pair $(3,4^1)$ is an abbreviation of: $3 + 4^1 \cdot k$.

D's class tree

In the following tree we have three kind of nodes, **Left nodes**: $k = 3 \cdot \rho + 1$, $\rho_i = 3 \cdot \rho_{i-1} + 1$, **Middle nodes**: $k = 3 \cdot \rho$, $\rho_i = 3 \cdot \rho_{i-1}$ and **Right nodes**: $k = 3 \cdot \rho + 2$, $\rho_i = 3 \cdot \rho_{i-1} + 2$

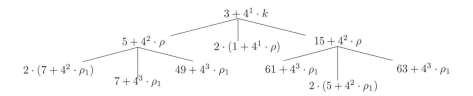

The terminal nodes are: $2 \cdot (1 + 4^1 \cdot \rho)$, (reach a smaller number, since $\frac{2 \cdot 4^1}{3^1} < 3$). So at least of 33.3% numbers of B class are terminal nodes.

3   Structure of parametric part trees

It is interesting to note that if $x = i \bmod d = 1 + k \cdot d$:

$$T(x) = \frac{m_i \cdot i + m_i \cdot k \cdot d - r_i}{d} = \frac{m_i \cdot i - r_i}{d} + \frac{m_i \cdot k \cdot d}{d} = \lambda + m_i \cdot k$$

where λ is a fixed number $\lambda = y \bmod d$. In Appendix A, we prove that independent which are the values of λ and k, the structure of the tree is the following structure (in case of d=3): We extend each non terminal node in three branches. If the non terminal node has `weight` m_0 (m_1, m_2) then the childrens of this node have the following weights: m_0^2, $m_0 \cdot m_1$ and $m_1 \cdot m_2$ (by the same way the other cases).

An example of this parametric $d - tree$ is (d=3):

Denoting by x_l a leaf node of the 3− parametric tree at the depth l and by X_l the whole set of leaf nodes at the depth l then the set Y of the terminal nodes is a subset of the set X_l, satisfying the condition:

$$Y = \{x_l | \frac{x_l}{3^l} < m_0\}$$

A nice conclusion of the previous analysis is that if for each level of the $d-$tree there are $k < d$ terminal nodes then the density of these nodes is 1:

$$\lim_{l \to +\infty} \frac{k \cdot (d-k)^0}{d} + \frac{k \cdot (d-k)^1}{d^2} + \ldots + \frac{k \cdot (d-k)^l}{d^{l+1}} = \frac{k}{d} \cdot \lim_{l \to +\infty} (1 + \frac{(d-k)^1}{k^1} + \ldots + \frac{(d-k)^l}{d^l}) = \frac{k}{d} \cdot \frac{1}{1 - \frac{d-k}{d}} =$$

Kindly note that in **Appendix B** we present a well known open problem as an example of this kind of recurrences.

4   Algorithmic Approach

The main steps of our algorithm are:

- We classify input numbers based on the mod d^a class where $a \in N$. Usually $a = 1$.

- We define which are the terminal nodes. There are two types of terminal nodes: nodes which reach a smaller number or nodes which reach a divergent class.

- If a class of numbers does not reach, quite fast, a terminal nodes class then we construct the $d-$ tree of the parametric part.

- We measure terminal nodes' density.

4.1   Construction of 3-trees

One basic observation about the construction of the $d - trees$ $(d = 3)$ follows from the table below:

Level	Leaf nodes	Sequence	Pascal
0	m_0	1	1
1	$m_0^2, m_0 m_1, m_0 m_2$	1 1 1	1 1
2	$m_0^3, 2m_0^2 m_1, 2m_0^2 m_2, 2m_0 m_1 m_2, m_0 m_1^2, m_0 m_2^2$	1 2 2 2 1 1	1 2 1

In the named collumn sequence the notation 111 corresponds to coefficients of $(m_0^2, m_0 \cdot m_1, m_0 \cdot m_2)$. Can we predict the sequence of the next level? The answer is positive about the general case, the sequence is a cleaved version of 3−binomial sequence. Mainly in this paper we consider the case which $m_1 = m_2$ (as the example of section 2). In this case the sequence of parametric's part tree follows an interesting vesrion of Pascal's Triangle as the table shows below:

Level	Leaf nodes	Sequence	Pascal
0	m_1	1	1
1	$m_1^2, m_1^2, m_1 m_0$	2 1	1 1
2	$4m_1^3, 4m_1^2 m_0, m_1 m_0$	4 4 1	1 2 1

This sequence can be computed as follows:

$$
\text{sequence} = \begin{cases} a[i][j] = 2 \cdot a[i-1][j] + a[i-1][j-1] & \text{if } j \neq 1 \\ \\ a[i][j] = 2 \cdot a[i-1][j-1] & \text{otherwise} \end{cases}
$$

Mainly in this section we suggested an encoding of d-trees using Pascal's Triangle, in order to be more precise a cleaved Pascal's triangle since we use the notion of terminal nodes. This is very efficient in case that $m_1 = m_2$.

4.2   Computing Recursive steps

The $3 \cdot x + 1$ mapping is a special case of a more general class of mapping about which accurate predictions can be made concerning the behaviour of trajectories. Our apprach is similar to mathematical induction. The main steps of our approach is to find an efficient algorithmic approach in order to compute these recurrent sequences:

- Step 1: We verify that a set of input numbers satisfies the conjectures about the specific recursive function that we study. For instance in case of Collatz function we easily can verify that the first 16 numbers reach the number 1.

- Step 2: For each next number we use the information whether this number reach a terminal node's class. The construction of $d−$trees provide us this information.

- Step 3: In case that we do not know whether an input number reach a terminal nodes' class we run all the steps of the recursive function.

4.3   Experiments

The reccurent function is attributed of section 2 is attributed to L. Collatz and using the previous analysis, we get the following results using the source code of Appendix C. (some partial results)

Height	Class	Percent	root
8	A	44.03	$1+4 \cdot k$
8	B	44.03	$3+4 \cdot k$

In case of Collatz function and other similar problems this approach was very efficient [11], [12].

References

[1] Jeffrey C. Lagarias, The 3x+1 problem and its generalizations, The American Mathematical Monthly, vol. 92, no. 1, pp. 3-23, 1985.

[2] David Applegate and Jeffrey C. Lagarias, Lower bounds for the total stopping time of the $3 \cdot x + 1$ iterates, Mathematics of Computation, vol. 72, no. 242, pp. 1035-1049, 2002.

[3] Tomas Oliveira e Silva, http://www.ieeta.pt/~tos/3x+1.html

[4] Tomas Oliveira e Silva, Maximum excursion and stopping time record-holders for the 3x+1 problem: computational results, Mathematics of Computation, vol. 68, no. 225, pp. 371-384, 1999.

[5] Eric Roosendaal, http://personal.computrain.nl/eric/wondrous/

[6] K.R. Matthews, The generalized 3x + 1 mapping, http://www.maths.uq.edu.au/~krm/.

[7] Kenneth G. Monks. $3x + 1$ Minus the +, Department of Mathematics, University of Scranton. Discrete Mathematics and Theoretical Computer Science 5, pp. 47-54.

[8] Peter Cheeseman, Bob Kanefsky, William M. Taylor, Where the Really Hard Problems Are. In Proc. IJCAI'91, pages 163-169, 1991.

[9] Conway, J. H. Unpredictable Iterations. Proc. 1972 Number Th. Conf., University of Colorado, Boulder, Colorado, pp. 49-52, 1972.

[10] John Simons, Bene de Weger, Theoretical and computational bounds for m-cycles of the 3n + 1 problem, www.win.tue.nl/~bdeweger/3n+1v1.0.pdf, November 2003.

[11] Tzanis, E. Collatz Conjecture: Properties and Algorithms, In Proc. of the join program of the 9th Panchellenic Conference in Informatics and the 1st Balcanic Conference in Informatics, pages 700-713 http://skyblue.csd.auth.gr/~bci1/Panhellenic/700tzanis.pdf.

[12] Tzanis, E. Collatz Conjecture and other Similar problems: Properties and Algorithms, In Forthcoming : The Eleventh International Conference on Fibonacci Numbers and Their Applications, July 2004, Braunschweig, Germany.

[13] Terras, R. A Stopping Time Problem on the Positive Integers. Acta Arith. 30, 241-252, 1976

[14] Terras, R. On the Existence of a Density. Acta Arith. 35, 101-102, 1979.

VSP International
Science Publishers
P.O. Box 346, 3700 AH Zeist
The Netherlands

*Lecture Series on Computer
and Computational Sciences*
Volume 1, 2004, pp. 519-522

Nonergodic Behavior of Dissipative Structures – The Cascade of Fracture Centers in Dynamic Fracture of Metals

A.Ya. Uchaev[1], V.T. Punin, S.A. Novikov, Ye.V. Kosheleva, L.A. Platonova, N.I. Selchenkova,

Russian Federal Nuclear Center – VNIIEF,
Mira 37, Sarov, Nizhni Novgorod region,
Russia, 607 188

Received 7 July, 2004: accepted in revised form 1 August, 2004

Abstract: This work considers the mechanism of dynamic fracture of metals caused by high-current beams of relativistic electrons and ultrashort pulses (USP) of laser radiation over $t \sim 10^{-6} \div 10^{-11}$ s longevity range with $dT/dt \sim 10^6 \div 10^{12}$ K/s energy introduction rate.

Keywords: high-current beams of relativistic electrons, ultrashort pulses of laser radiation, dissipative structures, hierarchy of structural levels, ergodic and nonergodic system

During fracture metals are known to experience plastic deformation both over quasi-static ($t > 10^{-3}$ s longevity) and dynamic ($t \sim 10^{-6}$ s) longevity ranges. At low plastic deformation levels the process of plastic deformation was presented as the result of ergodic behavior of the system of defects whose trajectories filled all the phase space with time [1]. In this formulation the relations of the thermodynamic potential of the defect system in the configuration space have the form of "regular" distribution of minima, the least of which corresponds to the stable state of the defect system and the rest are consistent with metastable states (fig. 1).

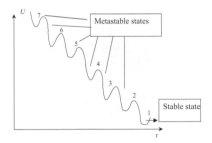

Figure 1: Schematic sketch of the thermodynamic potential at low plastic deformation levels.

At higher plastic deformation levels the density of various lattice defects achieves critical values. Establishing the specific relationship in the ensemble of one structural level creates the prerequisite for autolocalized formation of another structural level that plays the role of the initial structural level for a higher one. Nonequilibrium, hierarchic subordinate systems are nonergodic when the upper state is only possible after the lower states have been achieved.

First, the faster processes responsible for getting over potential barriers having a minimum height proceed in the hierarchic system, where there is a spectrum of relaxation times. Fig. 2 schematically shows the potential energy of the three-level hierarchic nonergodic system.

---

[1] Corresponding author. Head of Department RFNC-VNIIEF. E-mail: uchaev@expd.vniief.ru

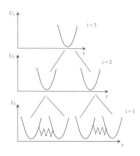

Figure 2: Potential energy of the three-level hierarchic nonergodic system.

For every hierarchic level of dissipative structures – slip bands, fracture centers, the ensemble of fracture centers there is its own value of potential energy U_i that is characterized by its own order parameter n_i, distribution function f_i and relaxation times t_i. In terms of mathematics, description of each level of dissipative structures is a rather labor-intensive procedure. The main objective is to find the common macrofracture process potential that would include the potentials of all hierarchic levels of dissipative structures.

Four self-organization levels of various dissipative structures initiating in the destructing body volume over the dynamic longevity range have been considered.

To reveal the scaling properties of quantitative characteristics of dissipative structures produced during dynamic fracture at the nanoscale level, the data obtained with a scanning tunnel microscope and using mathematical methods for processing the loaded sample surface images were used.

The damaged surfaces resulting from effects produced by high-current beams of relativistic electrons and USP of laser radiation were studied [2-4]. The dimensions of the tested damaged surfaces of metal samples (copper, steel) varied from 701.9 nm to 6.144 µm and the thickness varied from 16.63 nm to 120.8 nm. Fig. 3 shows the damaged surface (STM-image (a) and restored three-dimensional image (b)) for the copper sample exposed to pulsed laser radiation. The two-dimensional image roughness is shown by different blackening levels.

Figure 3: Damaged surface of the copper sample exposed to pulsed effects of USP of laser radiation. D – roughness dimensions; <D> - average roughness dimensions; N(D)/(N(<D>)) – number of roughnesses of D (<D>) dimension.

Using the mathematical package of IIAS programs, the quantitative characteristics of the damaged surfaces have been obtained. Size roughness distribution on STM-image is shown in fig. 4.

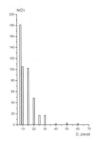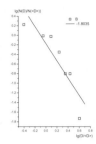

Figure 4: Size roughness distribution in STM-image (fig. 3) for the copper sample exposed to pulsed effects of USP of laser radiation.

The results have shown that it is possible to generate quantitative characteristics of dissipative structures produced during dynamic fracture at the nanoscale level using the mathematical methods for processing the loaded sample surface images (the interactive image analysis system).

In the previous works [2] it was shown that the lattice near the forming fracture centers loses its long-range order.

The structurization of slip bands in the titanium lattice was studied after shock-wave loading [4].

The results of the research show that the cascade of the lattice slip bands, which is formed around fracture centers, is a fractal cluster.

Previously it was shown that the cascade of fracture centers controlling the process of dynamic fracture of metals is a fractal cluster [2]. Distribution of fracture centers for various materials presented in universal coordinates is obtained by the similarity conversion. This testifies to the fact that the process of dynamic fracture of metals proceeds within one primary process of accumulation and growth of fracture centers, which is accounted for by the basic part of longevity.

For the particular body the change in the level of the fracture centers system connectivity from local to global is the principal moment for any fracture process models.

At the final dynamic fracture stage the dimensions of fracture centers and their densities in the destructed body are controlled by the concentration criterion [2]. The results of the carried out studies point to the collective behavior of the ensemble of fracture centers explained by the solid nonequilibrium state (the absorbed energy density is comparable with the lattice energy parameters) [2]. The autolocalized character of behavior of the fracture centers cascade, when the concentration criterion is valid, results from the lost behavior ergodicity of the fracture centers ensemble, where the fracture centers in their turn are connected with the hierarchic subordination (fig. 5) [2].

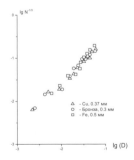

Figure 5: Dependence of the average distance $\langle r \rangle = N^{-1/3}$ between the fracture centers on the fracture center dimension D for some metals.

The results of the studies have made it possible to reveal universal evolution symptoms of the initiating dissipative structures – the roughness of inner surfaces of the fracture centers, the cascade of the lattice slip bands, the cascade of fracture centers. The universal character of behavior of metals in the dynamic fracture phenomenon is explained by the space and time self-organization in ensembles of dissipative structures over the ranges exceeding four dimension magnitudes. This points to no any specific length dimension in the ranges under consideration, which is connected with the dissipative structure formation physics.

The hierarchic properties of dissipative structures determining the dynamic fracture process have been considered at various scale levels such as a nanolevel, I, II, III mesolevels. In terms of mathematics, description of each level of dissipative structures is a rather labor-intensive procedure. The analogy of the dynamic fracture process at four scale levels has been shown. This allows the quantitative description of the dynamic fracture process at the specified scale levels to be greatly simplified.

References

[1] Khorstkhemke V., Lefevr R. *Noise-induced transitions: Theory and application in physics, chemistry and biology*: Transl. from English. – M.: Mir, 1987. – 400 p. ill .

[2] R. I. Ilkayev, V.T. Punin, A. Ya. Uchaev, S. A. Novikov, Ye.V. Kosheleva, L. A. Platonova, N. I. Selchenkova, N.A. Yukina. *Time regularities of dynamic fracture of metals caused by hierarchic properties of dissipative structures – the cascade of fracture centers* // Report of Academy for Science (RAS), 2003, v. 393, № 3.

[3] V.V. Zhmailo, M.G. Vasin, Yu.V. Ignatiev, A.P. Morovov, V.A. Tokarev. *Methods of recording spectral and time characteristics of pulsed x-ray radiation.* // Abstracts of the papers for the 9-th Meeting on diagnostics of high-temperature plasma. –Moscow.: TsNIIatominform, 1997, p. 79.

[4] S.A. Novikov, A.I. Ruzanov, I.R. Trunin, A.Ya. Uchaev. *Study into propagation of stress waves and copper destruction processes under electron beam effects. Applied problems of strength and plasticity.* Collected papers. Publishing House of Nizhni Novgorod University. 1992, pp. 10-14.

VSP International
Science Publishers
P.O. Box 346, 3700 AH Zeist
The Netherlands

Lecture Series on Computer
and Computational Sciences
Volume 1, 2004, pp. 523-526

Scaling Properties of Dissipative Structures Produced at the Metal Nanolevel Under Effects of Ultrashort Laser Radiation Pulses Over a Nanosecond Longevity Range

A.Ya. Uchaev[1], R.I. Ilkayev, V.T. Punin, A.P. Morovov, S.A. Novikov, N.I. Selchenkova,

N. A. Yukina, L.A. Platonova, Ye.V. Kosheleva, V.V. Zhmailo

Russian Federal Nuclear Center – VNIIEF,
Mira 37, Sarov, Nizhni Novgorod region,
Russia, 607 188

Received 7 July, 2004: accepted in revised form 1 August, 2004

Abstract: This work considers up-to-date methods for studies into dynamic fracture processes using a scanning tunnel microscope and a specialized package of mathematical programs, as applied to investigation of scaling, scale-invariant properties of dissipative structures produced in exposing metals to ultrashort laser radiation pulses. As a result of the carried out studies, the hierarchy of structural levels for dissipative structures determining the process of dynamic fracture of metals over $t < 10^{-6}$ s longevity range has been found. It has been shown that formation of dissipative structures is only possible if several structural levels are realized at the same time. Their fractal dimension is a quantitative characteristic of dissipative structures.

Keywords: ultrashort pulses (USP) of laser radiation, high-current beams of relativistic electrons, dissipative structures, roughness, scanning tunnel microscope (STM), interactive image analysis system (IIAS).

A comprehensive approach has been applied to the study into dynamic fracture of metals under effects of ultrashort pulses (USP) of laser radiation and high-power pulses of penetrating radiations over a subnanosecond longevity range ($t \sim 10^{-6} \div 10^{-11} c$). It includes the following research areas:

1. Deriving of time regularities for the metal dynamic fracture under effects of high-power penetrating radiation pulses and high-power ultrashort laser radiation pulses.

2. Study into the pattern of the inner fracture center surfaces using the scanning tunnel microscope (STM).

3. Establishing of quantitative characteristics of dissipative structures (on nanoscales) using the package of mathematical programs of the interactive image analysis system (IIAS) and the fractal geometry methods.

Application of the STM and the specialized package of mathematical programs to the study into scaling properties of dissipative structures produced in exposing metals to USP of laser radiation is a new trend in physics of dynamic fracture, which was implemented by VNIIEF for the first time in the world [1].

The process of dynamic fracture caused, for example, by high-current beams of relativistic electrons over unique temperature and time ranges [1-4] ($t \sim 10^{-6} \div 10^{-10}$ s, $T_0 \sim 4K \div 0.8\ T_{melb}$, $dE/dt \sim 10^{8} \div 10^{12} K/s$) has a hierarchic character and proceeds within one mechanism of initiation, growth and accumulation of fracture centers. In this case the body volume near the produced and growing fracture centers plays the role of a thermostat.

We show the possibility for the dynamic fracture process proceeding over a picosecond range of negative pressure effects.

The metal reactions to laser radiation may be conventionally classified into 3 modes:

1. The duration of the single laser radiation pulse $\tau_i > \tau_h$, ($\tau_h \sim L/c$ – hydrodynamic times, where c is the sound velocity and L is the laser radiation area dimension). If the radiation intensity J is sufficient, a crater is produced. Shock-wave phenomena are not observed.

[1] Corresponding author. Head of Department RFNC-VNIIEF. E-mail: uchaev@expd.vniief.ru

2. $\tau_i \sim \tau_h$ ($\tau_i \sim 10^{-10} \div 10^{-11}s$). In case of the sufficient intensity J a crater is produced and destructive processes result from shock-wave and thermal shock phenomena.

3. $\tau_i \sim 10^{-13} \div 10^{-14}s$. In case of the definite laser radiation intensity J electrons leave the area of laser radiation effects due to light Stoletov pressure, which may lead to Coulomb explosion of the ion core.

After the laser radiation pulse stops affecting (condition 2), a negative pressure is generated due to unloading in the nonequilibrium area equal to $P \sim -\Gamma \rho c_p T_v + \rho c v_M$, which may result in formation of fracture centers cascade having a fractal dimension over a picosecond longevity range. The loading parameters are as follows: the radiation duration is $\tau_0 \sim 10^{-10} \div 10^{-11}$ s, the quantum frequency is $v \sim 10^{14} \div 10^{15}$ Hz, the number of free electrons is $n_e \sim 10^{22}$ cm^{-3}, the velocity of radiation-thermalized electrons is $v_M \sim 10^7$ cm/s.

Using the methods of quantitative fractography and computation facilities, dissipative structures – the cascade of fracture centers were revealed, which shows the possibility for the dynamic fracture process proceeding over a nanosecond longevity range.

Figs. 1 a, b show the sections of copper samples after laser radiation effects.

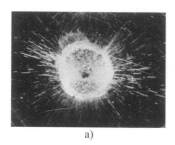 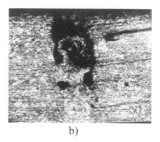

a)   b)

Figure 1: Copper sample sections after laser radiation effects.

Fig. 2 shows the section of a copper sample after effects of high-current beams of relativistic electrons (thermal shock) [1-4].

Figure 2: Copper sample section after effects of high-current relativistic electron beams (thermal shock).

The data in fig. 1, b and fig. 2 point to the analogy of the process of dynamic fracture of metals.

Thus, application of lasers with a picosecond, especially femtosecond pulse duration opens up new possibilities to study the dynamic fracture process over a picosecond range of negative pressure effects, which greatly expands the area of nonequilibrium states under study.

The carried out research has made it possible to establish scaling (fractal) scale-invariant properties of dissipative structures at a nanolevel.

By now a great number of phenomena and tasks, where a fractal structure (dimension) serves as the basic system characteristic, have been determined. This approach is successfully applied [2] for specifying quantitative characteristics of dissipative structures produced in dynamic fracture of metals, explosives and during structural modification of metals and alloys affected by pulsed high-current relativistic electron beams over a dynamic longevity range ($t \sim 10^{-6} \div 10^{-10}$ s). According to the literary data, the systems formed under highly nonequilibrium conditions are fractal systems characterized by a fractal dimension [1-4].

Using a specialized mathematical package of programs of the interactive image analysis system (IIAS), the damaged surface images were processed. The roughness size distribution and the fractal dimension of roughnesses were determined.

A typical view of the copper sample damaged surface after exposure to USP of laser radiation is shown in fig. 3. Its view after effects of high-current relativistic electron beams is shown in fig. 4.

Figure 3: Damaged surface of the copper sample exposed to pulsed laser radiation.

Figure 4: Damaged surface of the copper sample exposed to pulsed effects of high-current relativistic electron beams.

Fig. 5 shows the copper sample damaged surface exposed to pulsed effects of high-current relativistic electron beams. The results of the damaged surface processing after effects of USP of laser radiation and high-current beams of relativistic electrons are shown in figs. 5 and 6.

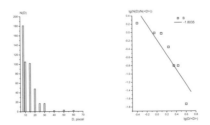

Figure 5: Size roughness distribution of the copper sample exposed to USP of laser radiation. D – rougness dimension; <D> - average roughness dimension; N(D)/(N(<D>)) – number of roughnesses of D (<D>) dimension

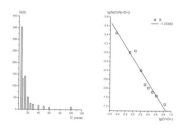

Figure 6: Size roughness distribution of the copper sample exposed to high-current relativistic electron beams.

Thus, ultrashort pulses of laser radiation open up new possibilities to study the behavior of metals under extreme conditions. Application of electron microscopy techniques allows quantitative characteristics of dynamic fracture to be established over a picosecond range of the loading pulse with the amplitude of up to several tens of GPa.

References

[1] R. I. Ilkayev, V.T. Punin, A. Ya. Uchaev, S. A. Novikov, Ye.V. Kosheleva, L. A. Platonova, N. I. Selchenkova, N.A. Yukina. *Time regularities of the dynamic fracture of metals caused by hierarchic properties of dissipative structures – the cascade of fracture centers* // Report of Academy for Science (RAS), 2003, v. 393, No 3.

[2] Ye.K. Bonyushkin, N.I. Zavada, S.A. Novikov, L.A. Platonova, N.I. Selchenkova, A.Ya. Uchaev. *Fractals in applied physics.* Sarov: RFNC-VNIIEF, 1995, pp.123-174.

[3] Ye.K. Bonyushkin, N.I. Zavada, S.A. Novikov, A.Ya. Uchaev. *Kinetics of dynamic fracture of metals under pulsed volumetric heating conditions.* Scientific edition. Ed. by R. I. Ilkayev, Sarov, 1998, 275 p.

[4] Ye.K. Bonyushkin, B.L Glushak, N.I. Zavada, S.A. Novikov, L.A. Platonova, N.I. Selchenkova, A.Ya. Uchaev. // *PMTF*, 1996, No 6, pp.105-115.

VSP International
Science Publishers
P.O. Box 346, 3700 AH Zeist
The Netherlands

*Lecture Series on Computer
and Computational Sciences*
Volume 1, 2004, pp. 527-529

Universal Properties of Metal Behavior in Dynamic Fracture Over a Wide Longevity Range Shown Under Effects of High-Power Penetrating Radiation Pulses

A.Ya. Uchaev[1], R.I. Ilkayev, V.T. Punin, S.A. Novikov, Ye.V. Kosheleva,
N.I. Zavada, L.A. Platonova, N.I. Selchenkova

Russian Federal Nuclear Center – VNIIEF,
Mira 37, Sarov, Nizhni Novgorod region,
Russia, 607 188

Received 7 July, 2004: accepted in revised form 1 August, 2004

Abstract: The invariants of metal behavior in dynamic fracture ($t\sim 10^{-6}$-$10^{-10}s$ longevity range) have been found. The dynamic invariant allows the behavior of untested metals and alloys to be predicted under extreme conditions and new alloys resistant to certain types of pulse effects to be constructed using numerical methods.
Keywords: Thermal shock (TS), quasi-static and dynamic longevity ranges, dissipative structures, fractographic studies, percolation cluster, universal coordinates.

During thermal shock (TS) caused by high-current relativistic electron beams, which is detailed in [1-3], the fracture process proceeds within $t < 10^{-8}$ s time in contrast to conventional loading methods. The energy introduction time is $dT/dt \sim 10^{12}$ K/s and the temperature changes up to T_{melt}. The basic objective is to find general regularities in the behavior of metals, which are invariants with respect to changes in the external environments.

The size distribution of fracture centers for various materials in the sections of loaded samples, which is presented in lg(D/<D>), lg(N(D)/N(<D>)) coordinates, is produced by the similarity conversion [1-3]. This testifies to the fact that the process of dynamic fracture in metals proceeds within one primary process – accumulation and growth of fracture centers, which is accounted for by the basic part of longevity. Spectral size distribution of fracture centers has the form $N(D) \sim D^{-\alpha}$, where D is the fracture center size, $\alpha > 1$ (fig.1, a). The fracture centers accumulation rate may be described by the evolution equation (fig.1, b) of $dN/dt \sim N^{\beta}$, $\beta > 1$ type. The data given in fig.1 show correlated behavior and initiation of self-organization of the fracture centers cascade within the destructed sample. The results of fractographic studies pointed to initiation of plastic flow areas (similar to turbulence – even at initial temperatures ($T \sim 4K$)) causing loss of the long-range lattice order near the forming and increasing fracture centers. These are deformation micro- and mesolevels (fig.2).

Fig.3 shows the result of mathematical simulation of the volumetric percolation cluster of fracture centers and the computer section produced by sectioning the volumetric percolation cluster with Q plane [2, 4]. Taking into account that there is an influence sphere around fracture centers having R size, which is connected with the plastic flow fields, and the fractographic analysis data, the following expression may be obtained

$$N^{-1/3} \cong 1.2R, \tag{1}$$

(N is the density of fracture centers). It allows transition from micro- to macrofracture to be estimated not only qualitatively, but also quantitatively (concentration criterion) [1].

---

[1] Corresponding author. Head of Department RFNC-VNIIEF. E-mail: uchaev@expd.vniief.ru

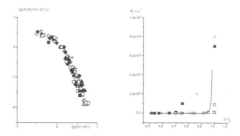

Figure 1: Size distribution of fracture centers in *Fe* (*Δ*=4·10-4m thickness, dark markers) and *Cu* (*Δ*=10-3m) samples in the sections parallel to the damaged surface (δ●(δ○)> δ■(δ□)> δ▲(δΔ)> δ♦ (δ◊)> δ+, where *δ* is the depth from the damaged surface (a) and the fracture centers accumulation rate (calculations and experiments) for various metals on the fracture time scale *tf* (b). ▨ *Pbσ₁* *Δ*=3·10⁻⁴m;   □ *Pbσ₂* *Δ*=4·10⁻⁴m;   ● *Cuσ₁* *Δ*=2·10⁻⁴m;   ○ *Cuσ₂* *Δ*=4·10⁻⁴m; ☆ *Cu* *Δ*=5·10⁻⁵m;   ■ *Pb* *Δ*=2.3·10⁻⁴m, σ₁>σ₂;   ------- curve - solution of equation *dN/dt=N*

Based on the structure-energy analogy of the metal behavior after introduction of thermal and mechanical energy, which in both cases results in the lattice long-range order violation, the longevity data on some metals are presented in universal coordinates (fig.4). The universal coordinates make it possible to establish the time dependence of the relation between the critical density of the absorbed energy causing fracture and the lattice energy parameters (enthalpy - H and melting heat - L_m). The data in fig.4 give the absolute values of dissipative unloading wave losses over the dynamic fracture range, they are close to a single curve and determine the boundary above which there is a fracture area. The insets in fig. 4 show the change in the fracture mechanism from one-site to many-site. The systematized data over the quasi-static range are given for pure metals [5].

Taking into consideration the self-similar character of vulnerability to damage accumulation over the dynamic longevity range, the relation between the critical energy density and the longevity may be obtained: $E^{\gamma} \cdot t = const$ [1-3], where γ ~ 3.8. The derived expression determines the amplitude-and time coordinate irrespective of the loading method and geometry.

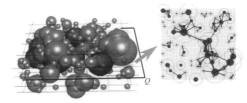

Figure 2: Structurization of the lattice slip bands near the growing fracture centers and tangents to the slip bands

Figure 3: Percolation cluster of the fracture centers and percolation computer section produced by Q plane sectioning.

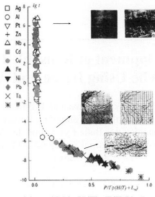

Table: $\rho(T,P)$, $H(T)$, $\Gamma(T,P)$, L_m
Measured: $E=P/\Gamma\rho$,
L_m – melting heat, H- enthalpy, Γ- Grüneisen parameter, ρ- material density

Figure 4: Dependences of *t* longevity for some metals in universal coordinates (*t* is in seconds)

References

[1] Ye.K. Bonyushkin, N.I. Zavada, S.A. Novikov, A.Ya. Uchaev *Kinetics of dynamic fracture of metals under pulse volumetric heating conditions. Scientific edition.* Ed. by R. I Ilkayev, Sarov, 1998, 275 p.

[2] R. I. Ilkayev, A. Ya. Uchaev, S. A. Novikov, N.I. Zavada, L. A. Platonova, N. I. Selchenkova // Report of Academy for Science (RAS), 2002, v. 384, No 3, P. 328-333.

[3] R. I. Ilkayev, V.T. Punin, A. Ya. Uchaev, S. A. Novikov, Ye.V. Kosheleva, L. A. Platonova, N. I. Selchenkova, N.A. Yukina. // DAN, 2003, v. 393, № 3I. Prigogine, Stuart Rice (Eds): Advances in Chemical Physics Vol. 93: New Methods in Computational Quantum Mechanics, John Wiley & Sons, 1997.

[4] Ye.K. Bonyushkin, N.I. Zavada, S.A. Novikov, L.A. Platonova, N.I Selchenkova, A.Ya. Uchaev. Fractals in applied physics. Sarov: RFNC-VNIIEF, 1995, pp.123-174

[5] Regel V.R., Slutsker A.I., Tomashevsky E.I. Kinetic nature of solid strengths. M.: Nauka, 1974. 560 p.

VSP International
Science Publishers
P.O. Box 346, 3700 AH Zeist
The Netherlands

Lecture Series on Computer
and Computational Sciences
Volume 1, 2004, pp. 530-534

Proposal for Development of Reusable Learning Materials for Wbe Using Irlcoo and Agents

R. Peredo Valderrama and L. Balladares Ocaña

Agents Laboratory, Center for Computing Research , IPN, México, D.F.
Av. Juan de Dios Bátiz Esq. Othon de Mendizabal s/n
Col. Nueva Industrial Vallejo, 07738, México, D.F.
Mexico

Received 6 August, 2004: accepted in revised form 26 August, 2004

Abstract: The research in WBE systems is centered in reusability, accessibility, durability and interoperability of didactic materials and environments of virtual education. In this article we make a development proposal based on a special type of labeled materials called Intelligent Reusable Learning Components Object Oriented (IRLCOO), the components conform a fundamental part of the tool CASE of courses authoring tool denominated Authoring IRLCOO, producing learning materials with interface and functionality standardized with SCORM 1.2, rich in multimedia, interactivity and feedback is described. The structuring model for dynamic composition of these components is based on the concept graph knowledge representation model. The multiagent architecture as a middleware for open WBE system is developed for sequencing and delivery of learning materials composed of IRLCOOs.

Keywords: WBE, intelligent reusable learning components object oriented, intelligent components, agents

1.   From RLO to Intelligent Components

In software engineering, a component is a reusable programme building block that can be combined with other components in the same or other computers in a distributed network to form an application. More specifically, a component is a piece of software reusable in binary form you can connect to other components with relatively little effort [13]. Examples of a component include: a single button in a graphical user interface, a small interest calculator, an interface to a database manager. A component within a context called a container and can be deployed different servers in a network and communicate with other for needed services.

1.1 RLCOO

Applying these ideas to the development of RLO, Macromedia's Flash components have been used as the basis of content elements in the format of the Reusable Learning Components (RLC) [14], because Flash is an integrator of media and have a powerful programming language denominated ActionScript 2.0 completely object oriented, is a client that allows multimedia, besides loading of certain media in Run-Time, allowing to have a programmable and adaptive environment to the student's necessities in Run-Time. Flash already has *Smart Clips* for the learning elements denominated LI, so, the original idea was to generate a library multimedia of RLC for WBE. Different templates for RLC as well as for the programmable buttons with *Smart Clips* and *Clip Parameters* interface were generated, for example, generic buttons for audio, video, exams, charts, etc. The interfaces stayed as files *swf* in a static way in the first versions. New learning templates also were created to add the functionality or to improve the existent one. To load components, a new mechanism was generated based on the integration of *loadVars* and *loadMovie* for the personalization of the materials in Run-Time, looking for to separate the contents of the control of the component using different levels inside the Flash player, allowing to generate more specialized components, smaller , reusable, and to be able to integrate them in run-time inside a bigger component. The liberation of *ActionScript* version 2.0 inside Flash MX 2004 allowed to redefine the RLC to Reusable Learning Components Object Oriented (RLCOO), allowing to implement the paradigm object oriented, besides allowing to contain certain common functionalities inside of the Application Programming Interface (API) like: Communication with the LMS, Communication with the Agents and dynamic load of Assets in Run-Time.

As we see later, this mechanism is also used to provide communication of the RLCOO with the LMS. A general scheme of the RLCOO with their classes and the API for load dynamic de Assets is show en Fig. 1 and 2 respectively.

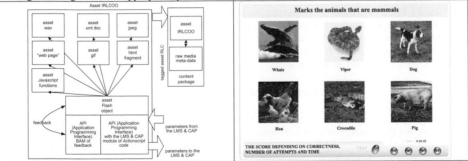

Figure 1. Basic diagram of classes of RLCOO written in *ActionScript* 2.0.	Figure 2. The general schema of RLCOO with load dynamic of Assets.

1.2 Intelligent RLCOO

A new type of components, called IRLCOO, was developed for the advanced templates, adding nonlinear feedback from the users and the functionality of dynamic component generation for the auto-configurable labeled materials. A general scheme of the IRLCOO with nonlinear feedback is shown in Fig. 3.

For the nonlinear feedback, a small Neural Network (NN) of the Bidirectional Associative Memory (BAM) type was used. The NN is composed of two layers, identified by two groups of neurons $\{a_1, a_2, \ldots, a_n\}$ and $\{b_1, b_2, \ldots, b_n\}$. The $n \times p$ generated connections conform a correlation matrix, represented by W [15]:

$$W = \sum_{k}^{m} A_k^T B_k$$

This matrix is generated by means of the vectors A = { $a_1, a_2, \ldots, a_n$ } and B = { $b_1, b_2, \ldots, b_n$ }, being A the input vector that can be composed of different factors that a teacher wants to measure, as for example: time, score and number of intents; while the vector B represents the output or feedback that the student will receive in function of the input vector A (Fig. 4). This NN was programmed as a part of IRLCOO. Since in Flash two-dimensional matrices do not exist, one dimensional matrix was used adding some logic to work appropriately.

Figure 3. The general schema of IRLCOO with dynamic feedback.	Figure 4. An assessment IRLCOO of the Hot Spot type before evaluation.

1.3 Separation of content and control with IRLCOO

The components have three fundamental phases, the first of development, where it is implemented the different components for a course, bound to media, where the goal is to have separate the media of the component, functionality, metadata, and control, using the Authoring IRLCOO by means of files XML like static parameters. The second phase is when the component is mounted in the server and making use of the API for dynamic load of Assets in Run-Time maybe take the control of the component as much to level media as to level control, depending on the metric of the learners in the LMS. The third phase refers to that these components have functionalities to measure metric of the learners in Run-Time without one has to program this. Achieving component the reusables and flexibility in Run-Time.

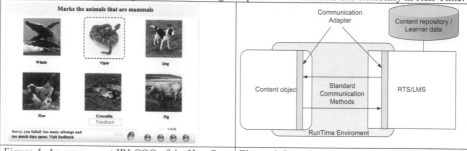

Figure 4. An assessment IRLCOO of the Hot Spot type after evaluation.	Figure 6. Learning content ECMAScript API model (adopted from [16]).

1.4 Communication between IRLCOO and LMS

The common implementation of this model is in an asynchronous RTE (browser-based delivery) and one where a content object initiates all communication. As depicted in Fig. 6, a content object and the LMS implement communication adapters in the content object's RTE. The SCORM communications standard prescribes a rich language whereby the management system and module can communicate. Four of the most important SCORM commands: LMSInitialize, LMSFinish, LMSGetValue, and LMSSetValue. LMSInitialize marks the begin of a module, says to the LMS, "Starting up. Start your clock and begin tracking me". LMSFinish marks the end of a module, says to the LMS, "Stopping up. Stop the clock and cease tracking me". LMSGetValue command enables the module to request information from the LMS. LMSSetValue command can be used to send a wide variety of data to the LMS.

2. Authoring IRLCOO

Once we have components and models for their structuring, development of easy-to-use authoring tools put in the hands of subject matter experts, is imperative to the success of the WBE. In this way, knowledge can be packaged and distributed very quickly. During the development of the didactic materials for EVA, it was decided to develop templates of the components and an authoring tool, called IRLCOO Factory, to facilitate their deployment, since many teachers are not experts in handling of multimedia tools. We have developed the structurer and packer of content multimedia using the IRLCOO as the lowest level of content granularity. The integration of the RLCOO and content aggregation is carried out on the base of the concept graph model that allows to generate materials at different granularity levels. It is important to mention that the IRLCOO are basically components ActiveX. As it can be seen in the Fig. 8.

3. Communication between Agents

In this architecture, agents are embedded in the applications running both at the client and server (modeling the functionality of the LMS) sides and implemented as programmes in Visual Basic .NET (Fig. 9). Each one of these applications contain an agent based on an ActiveX control, by means of which communication among the applications is supported. In the case of a client, this agent is called Personal Assistant (PA) and in the case of a server, Sequencing Agent (SA). These agents inherit the functionality of the BasicAgent of the Component Agent Platform (CAP). The SA agent has an API to manage the RLOs repository (DB), implementing "sequence", "retrieve" and "delivery" services.
Hosting embedded ActiveX agent, client's application has the following additional components. Its essential part is a browser invoked as a COM server that is used to visualize the IRLCOOs. Another component of the client application is an XMLDOM object to manipulate the IRLCOOs required by the user. The interaction between the user and the RLC is provided by the client's GUI. When the user requests an IRLCOO, the client's agent is activated and it generates a FIPA-ACL message (generally using the "request" performative) to be sent to the corresponding LMS's agent. LMS is composed of a

set of (distributed) SAs, each of them controlling a particular domain of the knowledge space defined by the concept graph.

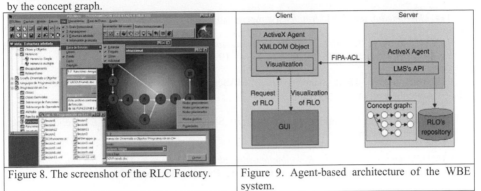

Figure 8. The screenshot of the RLC Factory.	Figure 9. Agent-based architecture of the WBE system.

Acknowledgments

The authors of this article would like to thank the IPN and CIC for partial support for this work within the project: 20040858. The authors would like to acknowledge all their colleagues and students participating in the design and development of the software and learning materials described in this article.

References

[1] W. Veen, B. Collis, S. Santema, & R. Genang, *Learner led learning in an online community*, Proc. of the World Conf. on ED-MEDIA 2001, Tampere, Finland, June 24–30, 2001.

[2] Colin Steed, *Web-based training* (Gower Publishing Limited, 1999).

[3] *Reusable Learning Object Strategy Definition, Creation Process and Guidelines for Building*, v3.1CiscoSystems,Inc.,URL:http:// www.cisco.com/warp/public/10/wwtraining/elearning/implement/rlo_strategy_v3-1.pdf.

[4] *Open Knowledge Initiative*, MIT, URL: http://web.mit.edu/oki/.

[5] *Advanced Distributed Learning Initiative*, http://www.adlnet.org.

[6] Global IMS Learning Consortium, URL: http://www. imsproject.org/.

[7] L. Sheremetov & R. Peredo Valderrama, *Development of labeled multimedia for WBT using SCORM*, Proc. of the IASTED International Conf. CATE '02, Cancún, México, 2002.

[8] L. Sheremetov & V. Uskov, *Hacia la nueva generación de sistemas de aprendizaje basado en la Web* (*Towards the New Generation of Web-based Learning Environments*), Computación y Sistemas, 5 (4), 2002, 356–367 (in Spanish).

[9] L. Sheremetov & R. Peredo Valderrama, *Development of Reusable Learning Materials for WBE using Intelligent Components and Agents*, International Journal of Computers & Applications, Volumen 25 / Number 3 ACTA Press, 2003, 170-178.

[10] D. Cowan, *An object-oriented framework for liveBOOKs*, Technical Report, CS-98, University of Waterloo, Ontario, Canada, 1998.

[11] L. Sheremetov & R. Quintero, *EVA: Collaborative distributed learning environment based in agents*, Proc. of the World Conf. ED-MEDIA 2001, Tampere, Finland, June 24–30, 2001.

[12] L. Sheremetov & A. Guzman Arenas, *EVA: An interactive Web-based collaborative learning environment*, Computers and Education, 39 (2), 2002, 161–182.

[13] G. Eddon & H. Eddon, *Inside distributed COM* (Microsoft Press, 1998).

[14] *Macromedia, Inc.*, URL: http://www.macromedia.com/.

[15] M. Partida, R. Peredo, & F. Córdova, *Simulación de redes neuronales, Memoria Bidireccional Bivalente Adaptativa: BAM*, Polibits, 1 (14), 1995, A˜no VII (in Spanish).

[16] IEEE 1484.2, Draft Standard for ECMAScript API for Content to Runtime Services Communication.

[17] IEEE 1484.11.1, Working Draft 14, Draft Standard for Data Model for Content Object Communication.

[18] U. Vladimir & U. Maria, *Reusable learning objects approach to Web-based education*, Proc. of the IASTED International Conf. CATE '02, Cancún, México, 2002

[19] S. Franklin & A. Graesser, *Is it an agent, or just a program?*, in J.P. Mueller, M.J. Wooldridge, & N.R. Jennings (Eds.), Intelligent agents III (Berlin: Springer, 1997), 21–36.

[20] M. Wooldridge, *Agent-based software engineering*, IEE Proceedings of the Software Engineering, 144 (1), 1997, 26–

[21] N.R. Jennings, *On agent-based software engineering*, Artificial Intelligence, 117, 2000, 277–296.

[22] FIPA XC00061, FIPA ACL *message structure specification*, 2001.

[23] L. Sheremetov & M. Contreras, *Component agent platform*. Proc. of the 2nd International Workshop of Central and Eastern Europe on Multi-Agent Systems, CEEMAS '01, Cracow, Poland, September 26–29, 2001, 395–402.

[24] IMS Simple Sequencing Specification, Version 0.7.5 Public Draft Specification, IMS Global Learning Consortium, Inc., April 2002.

[25] A. Gordon & L. Hall, *Collaboration with agents in a virtual world*, Proc. of the Workshop on Current Trends and Artificial Intelligence in Education, 4, World Congress on Expert Systems, Mexico City, Mexico, 1998.

[26] C.I. Peña de Carrillo, R. Fabregat Gesa, & J.L. Marzo Lázaro, *WWW-based tools to manage teaching units in the PLAN-G distance learning platform,* Proc. of the EDMEDIA 2000, World Conf. on Educational Multimedia, Hypermedia and Telecommunications, AACE, Montreal, July 2000.

VSP International
Science Publishers
P.O. Box 346, 3700 AH Zeist
The Netherlands

Lecture Series on Computer
and Computational Sciences
Volume 1, 2004, pp. 535-538

A Trigonometrically Fitted Runge-Kutta Pair of Orders Four and Five for the Numerical Solution of the Schrödinger Equation

Z.A. Anastassi and T.E. Simos [1] [2] [3] [4]

Department of Computer Science and Technology,
Faculty of Sciences and Technology,
University of Peloponnese,
GR-221 00 Tripolis, Greece

Received 21 September, 2004

Abstract: We are constructing a trigonometrically fitted pair of explicit Runge-Kutta methods of orders four and five. The pair has a variable step size which is determined by a specific algorithm. The algorithm is based on the error as that is expressed by the absolute difference of the values computed by each method separately. The step size control in combination with the trigonometrical fitting result in a very efficient method when compared to well known methods.

Keywords: Runge-Kutta pairs, trigonometrical fitting, radial Schrödinger equation, resonance problem.

PACS: 0.260, 95.10.E

1 Introduction

The radial Schrödinger equation has the following form:

$$y''(x) = \left(\frac{l(l+1)}{x^2} + V(x) - E \right) y(x) \tag{1}$$

where $\frac{l(l+1)}{x^2}$ is the *centrifugal potential*, $V(x)$ is the *potential*, E is the *energy* and $W(x) = \frac{l(l+1)}{x^2} + V(x)$ is the *effective potential*. It is valid that $\lim_{x\to\infty} V(x) = 0$ and therefore $\lim_{x\to\infty} W(x) = 0$.

If we divide $[0,\infty]$ into subintervals $[a_i, b_i]$ so that $W(x)$ is a constant with value $\bar{W}_i$, then problem (1) bedomes

$$y_i'' = (\bar{W} - E)\, y_i, \qquad \text{whose solution is}$$
$$y_i(x) = A_i \exp\left(\sqrt{\bar{W} - E}\, x\right) + B_i \exp\left(-\sqrt{\bar{W} - E}\, x\right), \quad A_i, B_i \in \Re. \tag{2}$$

---

[1]President of the European Society of Computational Methods in Sciences and Engineering (ESCMSE)
[2]Active Member of the European Academy of Sciences and Arts
[3]Corresponding author. Please use the following address for all correspondence: Dr. T.E. Simos, 26 Menelaou Street, Amfithea - Paleon Faliron, GR-175 64 Athens, GREECE, Tel: 0030 210 94 20 091
[4]E-mail: tsimos@mail.ariadne-t.gr

2 Basic theory

2.1 Explicit Runge-Kutta methods

An s-stage explicit Runge-Kutta method used for the computation of the approximation of $y_{n+1}(x)$, when $y_n(x)$ is known, can be expressed by the following relations:

$$y_{n+1} = y_n + \sum_{i=1}^{s} b_i\, k_i$$

$$k_i = h\, f\left(x_n + c_i h,\ y_n + h\sum_{j=1}^{i-1} a_{ij}\, k_j\right),\ i = 1,\ldots,s \tag{3}$$

where in this case $f\left(x, y(x)\right) = (W(x) - E)\, y(x)$.

When solving a second order ODE such as (1) using first order numerical method (3), then problem (1) becomes:

$$z'(x) = (W(x) - E)\, y(x)$$
$$y'(x) = z(x)$$

while we use two pairs of equations (3): one for y_{n+1} and one for z_{n+1}.

The method shown above can also be presented using the Butcher table below:

$$
\begin{array}{c|ccccc}
0 & & & & & \\
c_2 & a_{21} & & & & \\
c_3 & a_{31} & a_{32} & & & \\
\vdots & \vdots & \vdots & & & \\
c_s & a_{s1} & a_{s2} & \cdots & a_{s,s-1} & \\
\hline
& b_1 & b_2 & \cdots & b_{s-1} & b_s
\end{array}
\tag{4}
$$

Coefficients $c_2,\ \ldots,\ c_s$ must satisfy the equations:

$$c_i = \sum_{j=1}^{i-1} a_{ij},\ i = 2,\ldots,s \tag{5}$$

Definition 1 [1] A Runge-Kutta method has algebraic order p when the method's series expansion agrees with the Taylor series expansion in the p first terms: $y^{(n)}(x) = y_{app.}^{(n)}(x),\quad n = 1, 2, \ldots, p$.

2.2 Step size control

The control of the step size is managed by the following algorithm:

We choose two one-step methods of orders q_g and $\widetilde{q}_g$, with $\widetilde{q}_g \geqslant q_g + 1$. Let Y_i and $\widetilde{Y}_i$ be the approximate values of the two one-step methods for the exact solution y at the grid point x_i.

- We compute the approximate solutions Y and $\widetilde{Y}_i$ at $x_i + h$.

- We compute $S = 0.9\, h\left(\dfrac{\varepsilon}{\|Y - \widetilde{Y}\|}\right)^{1/(q_g+1)}$, where $\varepsilon = \|\widetilde{Y}\|\, RELERR + ABSERR$

- If $\|Y - \tilde{Y}\| < \varepsilon$, then $Y_{i+1} = Y$ is accepted as the approximated value at the grid point $x_{i+1} = x_i + h$. The next step size will be $h = \min\{S, 4h\}$.

- If $\|Y - \tilde{Y}\| > \varepsilon$, then the step must be repeated with the new step size $h = \max\{S, 1/4h\}$.

See also [5].

2.3 Exponentially fitted Runge-Kutta methods

Method (3) is associated with the operator

$$L(x) = u(x + h) - u(x) - h\sum_{i=1}^{s} b_i\, u'\,(x + c_i h, U_i)$$

$$U_i = u(x) + h\sum_{j=1}^{i-1} a_{ij} u'\,(x + c_j h, U_j), \quad i = 1, \ldots, s \tag{6}$$

where u is a continuously differentiable function.

Definition 2 [2] Method (6) is called exponential of order p if the associated linear operator L vanishes for any linear combination of the linearly independent functions $\exp(v_0 x)$, $\exp(v_1 x)$, $\ldots$, $\exp(v_p x)$, where $v_i|i = 0(1)p$ are real or complex numbers.

Remark 1 [3] If $v_i = v$ for $i = 0, 1, \ldots, n$, $n \leq p$, then the operator L vanishes for any linear combination of $\exp(vx)$, $x\exp(vx)$, $x^2\exp(vx)$, $\ldots$, $x^n\exp(vx)$, $\exp(v_{n+1}x)$, $\ldots$, $\exp(v_p x)$.

Remark 2 [3] Every exponentially fitted method corresponds in a unique way to an algebraic method (by setting $v_i = 0$ for all i)

Definition 3 [2] The corresponding algebraic method is called the classical method.

3 Numerical results - The resonance problem

In this paper we will study the case of $E > 0$. We will integrate the problem (1) with $l = 0$ at the interval $[0, 15]$ using the well known Woods-Saxon potential

$$V(x) = \frac{u_0}{1 + q} + \frac{u_1 q}{(1 + q)^2}, \qquad q = \exp\left(\frac{x - x_0}{a}\right), \qquad \text{where} \tag{7}$$

$$u_0 = -50, \quad a = 0.6, \quad x_0 = 7 \quad \text{and} \quad u_1 = -\frac{u_0}{a}$$

and with boundary condition $y(0) = 0$.

The potential $V(x)$ decays more quickly than $\frac{l(l+1)}{x^2}$, so for large x (asymptotic region) the Schrödinger equation (1) becomes

$$y''(x) = \left(\frac{l(l+1)}{x^2} - E\right) y(x) \tag{8}$$

The last equation has two linearly independent solutions $k\,x\,j_l(k\,x)$ and $k\,x\,n_l(k\,x)$, where j_l and n_l are the *spherical Bessel* and *Neumann* functions. When $x \to \infty$ the solution takes the asymptotic form

$$y(x) \approx A\,k\,x\,j_l(k\,x) - B\,k\,x\,n_l(k\,x)$$
$$\approx D[sin(k\,x - \pi\,l/2) + \tan(\delta_l)\,\cos(k\,x - \pi\,l/2)], \tag{9}$$

where δ_l is called *scattering phase shift* and it is given by the following expression:

$$\tan(\delta_l) = \frac{y(x_i)\,S(x_{i+1}) - y(x_{i+1})\,S(x_i)}{y(x_{i+1})\,C(x_i) - y(x_i)\,C(x_{i+1})}, \tag{10}$$

where $S(x) = k\,x\,j_l(k\,x)$, $C(x) = k\,x\,n_l(k\,x)$ and $x_i < x_{i+1}$ and both belong to the asymptotic region. Given the energy we approximate the phase shift, the accurate value of which is $\pi/2$ for the above problem. As regards to the frequency w we will use the suggestion of Ixaru and Rizea [4]:

$$w = \sqrt{E - 50}, \quad x \in [0,\ 6.5]$$
$$\sqrt{E}, \quad x \in [6.5,\ 15]$$

References

[1] Hairer E., Nørsett S.P., Wanner G., *Solving Ordinary Differential Equations I, Nonstiff Problems*, Springer-Verlag, Berlin Heidelberg, 1993

[2] Simos T.E., An exponentially-fiited Runge-Kutta method for the numerical integration of initial-value problems with periodic or oscillating solutions, **115** 1-8(1998).

[3] Lyche T., Chebyshevian multistep methods for Ordinary Differential Eqations, *Num. Math.* **19** 65-75(1972).

[4] Ixaru L.Gr., Rizea M., A Numerov-like scheme for the numerical solution of the Schrödinger equation in the deep continuum spectrum of energies, Comp. Phys. Comm. **19** 23-27 1980

[5] Avdelas G., Simos T.E. and Vigo-Aguiar J., An embedded exponentially-fitted Runge-Kutta method for the numerical solution of the Schrödinger equation and related problems, Comp. Phys. Comm. **131** 52-67, 2000.

VSP International
Science Publishers
P.O. Box 346, 3700 AH Zeist
The Netherlands

*Lecture Series on Computer
and Computational Sciences*
Volume 1, 2004, pp. 539-542

A Web-Based Simulator of Spiking Neurons for Correlated Activity Analysis

F.J. Veredas[1]

Departamento de Lenguajes y Ciencias de la Computación,
ETSI Telecomunicación,
Universidad de Málaga,
29071 Málaga, Spain

Received 19 July, 2004; accepted in revised form 14 August, 2004

Abstract: A computational tool to study correlated neural activity is presented. SEN-NECA (Simulating Elementary Neural NEtworks for Correlation Analysis) is a web-based simulator specifically designed to simulate small networks of integrate-and-fire neurons. It implements a model neuron that can reproduce a wide scope of integrate-and-fire models by adjusting the parameter set. Launched simulations run remotely on a cluster of computers. The main features of the simulator are explained, and an example of neural activity analysis is given to illustrate the potential of this tool.

Keywords: Spiking neuron; correlated firing; web-based simulator; computers cluster

1 Introduction

Some general-purpose simulators offer the neuroscientist the possibility of implementing highly realistic models of the membrane electrical behavior, with multicompartmental neuron models like in GENESIS [1, 8] and NEURON [2, 5], among the most popular tools. However, these models are computationally very costly. In particular, tasks that involve analysis of correlated activity with cross-correlograms imply long simulations, since the neurons under study must generate enough spikes for the correlogram shape to emerge from background activity. In these cases, dedicated tools clearly outperform popular simulators.

In this paper we describe SENNECA [9], a specific-purpose web-based neural simulator dedicated to the study of pairwise correlated activity. SENNECA is based on a highly parameterized neuron model to allow the independent study of a wide number of physiological factors, and also to account for most variants of realistic neuron models —leaky *vs.* non-leaky, instantaneous integration *vs.* EPSP (excitatory postsynaptic potential), etc.

Methods

The structure of a simulation has been divided in four sets of parameters that allow the experimenter to flexibly define (i) the network topology, (ii) the cellular electrical behavior, (iii) the synaptic transmission, and (iv) the simulator environmental setting.

---

[1]Corresponding author. Tel.: +34-952-137155; fax: +34-952-131397. E-mail: fvn@lcc.uma.es

Network parameters

A network of neurons is defined by specifying the number of neurons, and the synaptic strength of the connections among them (given by ω_{ij} in equation 2), that can take positive or negative values for excitatory and inhibitory synapses, respectively.

Neuronal parameters

The model neuron embodied in SENNECA is a general-purpose integrate-and-fire model [7]. It has been designed according to two main objectives: (i) to allow the simulation of a wide range of models by setting parameters to proper values, and (ii) to provide the user with enough degrees of freedom to independently study the influence of physiological characteristics in pairwise analysis of neural activity.

The electrical properties of the neuron membrane are expressed in equation 1, parameterized by the following constants: resting potential (v), threshold potential (θ), afterhyperpolarization (ν), membrane time constant (τ) and refractory period (ρ).

The behavior of the dendritic tree depends on five parameters: (i) background noise (r in equation 1) obtained from a distribution selected from a list of different noise sources: uniform noise, Gaussian noise, 1/f noise, Poissonian noise [10], and custom noise; (ii) post-synaptic current injection rise time (α), (iii) decay time (β) and (iv) amplitude (ω); and (v) axonal transmission delay and synaptic jitter.

The model that interrelates all these magnitudes is expressed by three equations that govern the injected synaptic current (1, 2) and the membrane potential (3).

$$I_i(t) \; = \; r_i(t) + \sum_j E_{ji}(t - \hat{t}_j - D_{ji}) \tag{1}$$

$$E_{ij}(u) \; = \; \begin{cases} \omega_{ij} \, \frac{u}{\alpha_j} \, e^{-\frac{u}{\alpha_j}} & if \; u \leq \alpha_j \\ \omega_{ij} \, \frac{\beta_j - \alpha_j + u}{\beta_j} \, e^{-\frac{\beta_j - \alpha_j + u}{\beta_j}} & if \; u > \alpha_j \end{cases} \tag{2}$$

$$V_i(t + \Delta t) \; = \; \begin{cases} \nu V_i(\hat{t}_i) & if \; t \leq \rho + \hat{t}_i \\ \left(1 - \frac{\Delta t}{\tau_i}\right) V_i(t) + \frac{\Delta t}{C} I_i(t) + \frac{\Delta t}{\tau_i} v & if \; t > \rho + \hat{t}_i \end{cases} \tag{3}$$

where $\hat{t}_j$ is the step where neuron j fired last time, and D_{ji} is the overall delay of the last spike in neuron j to reach neuron i, where both, constant transmission delay and variable synaptic jitter, contribute. C is de capacitance of the membrane. Some parameters are not expressed in this formulation for the sake of simplicity.

Simulation parameters

The environment is also parameterized to define the conditions under which the simulation will take place. Two halting conditions are provided: the number of spikes in the target neuron for a correlogram to be significant, and the maximum simulation time, given in simulation steps. The output shape is also affected by a different parameter, the correlogram window size.

User interface

The simulator home-page can be found at http://senneca.geb.uma.es and is structured in several pages for the interaction with the user. The different parameters are grouped into pages according with their functionality into the simulator: (i) topological (number of neurons, and connection weights), (ii) neuronal (membrane and other physiological parameters), (iii) synaptic (background noise and transmission delay and jitters) and (iv) simulation parameters (see above). It is also likely to introduce comments (free-style text) about the simulation.

When a simulation is launched from SENNECA's graphical user interface, it will run remotely on a heterogeneous cluster of computer based on *Mosix 7.0*. The state of each simulation can be monitored and controlled from the web and the results are reported by email and also accessible at SENNECA web pages.

SENNECA is also provided with a simulation data base to file the simulations of each user, which includes the possibility of sharing simulation results and information among different users. Furthermore, the simulator integrates a web-forum to allow scientific discussions among remote users.

SENNECA is compatible with most web browsers (Explorer 5.0, Mozilla 5.0, Netscape 7.0 and Opera 5.0).

Simulation results

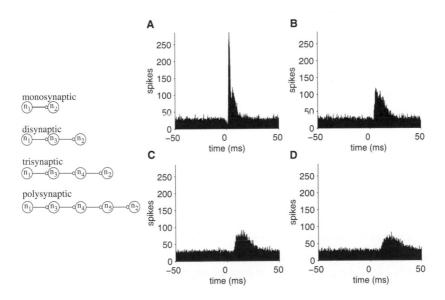

Figure 1: Resulting correlograms (from n_1 to n_2, see circuit schemes on the left) of four simulated circuits with different number of interconnecting neurons: (A) monosynaptic, (B) disynaptic, (C) trisynaptic, and (D) polysynaptic. The amplitude correlates inversely with the number of interneurons, while peak's width and shift are directly correlated.

Since its introduction in the 1960s [4, 6], cross-correlation analysis has been used extensively to study functional connectivity in different parts of the brain (for a review see [3]). Although the technique is not free of limitations, its use has become increasingly popular. It is a general belief that cross-correlation analysis can be reliably used to identify certain type of monosynaptic connections that are strong and excitatory. The computational tool presented here should help to better distinguish between structural and effective connectivity and make more precise inferences about the physiological properties of strong structural connections.

To illustrate how pairwise activity can be studied with this computational tool we have performed an example, where changes in network connectivity reproduce noticeable differences in the correlograms obtained. This experiment is concerned with the number of excitatory interneurons that separate n_1 (presynaptic) from n_2 (postsynaptic). Four circuits were simulated, starting with the monosynaptic connection and adding new relay neurons, up to three interneurons (polysynaptic connection). Figure 1 shows the correlograms obtained from n_1 to n_2, after 10,000 spikes were collected in n_2. As the number of serial interconnected neurons increases, the correlogram peak decreases in amplitude, becomes wider and shifts toward the right.

Conclusions

SENNECA is a user-friendly web-based computational tool that has been optimized to easily specify the neuron model (no scripts, nor new languages) by parameterizing a highly flexible model, that incorporate those physiological features most relevant to correlated activity. SENNECA subscribes to the trend of current programs dedicated to intense computation, by providing a front-end that allows the operation on remote machines.

Acknowledgment

SENNECA has benefited much from the help of our web-designer Juan Miguel Taboada. This research has been supported by NIH-EY 05253, and by the Spanish *Ministerio de Ciencia y Tecnología* (project PTR95.0734.OP).

References

[1] J.M. Bower and D. Beeman. *The Book of GENESIS. Exploring Realistic Neural Models.* Springer-Verlag: New York, 1998.

[2] M.L. Hines and N.T. Carnevale. The NEURON simulation environment. *Neural Computation* **9** 1179-1209 (1997).

[3] P.A. Kirkwood. On the use and interpretation of cross-correlations measurements in the mammalian central nervous system. *J. Neurosci. Methods* **1** 107-132 (1979).

[4] G.P. Moore, D.H. Perkel and J.P. Segundo. Statistical analysis and functional interpretation of neuronal spike data. *Annu. Rev. Physiol.* **28** 493-522 (1966).

[5] Neuron web site at Duke. Retrieved July 7, 2004, from http://kacy.neuro.duke.edu/, n.d.

[6] D.H. Perkel, G.L. Gerstein and G.P. Moore. Neuronal spike trains and stochastic point processes. II. Simultaneous spike trains. *Biophys. J.* **7** 419-440 (1967).

[7] R.B. Stein. Some models of neuronal variability. *Biophys. J.* **7** 37-68 (1967).

[8] The GENESIS Simulator. Retrieved July 7, 2004, from http://www.genesis-sim.org/GENESIS/, 2004.

[9] The SENNECA Simulator. Retrieved July 7, 2004, from http://senneca.geb.uma.es/, 2004.

[10] A. Zador. Impact of synaptic unreliability on the information transmitted by spiking neurons. *J. Neurophysiol.* **79** 1219-1229 (1998).

VSP International
Science Publishers
P.O. Box 346, 3700 AH Zeist
The Netherlands

Lecture Series on Computer
and Computational Sciences
Volume 1, 2004, pp. 543-546

Optimization of a Space Truss Dome - Genetic Algorithms

L. Vermaere[1], E. Lamkanfi, R. Van Impe, W. Vanlaere, P. Buffel

Laboratory for Research on Structural Models,
Department of Structural Engineering,
Faculty of Applied Sciences,
Ghent University,
B-9000 Ghent, Belgium

Received 2 August, 2004; accepted in revised form 22 August, 2004

Abstract: Genetic algorithms are optimization methods based on biological evolution and are applied in this paper to optimize a space truss. The goal is to develop a design, able to withstand the acting loads with a minimum of material. During this process the construction may fail due to different mechanisms. The Euler buckling of a truss member has to be avoided, the stresses in the bars and the displacements of the nodes must be limited. Besides these usual constraints the overall buckling is also taken into account. In this paper a program is developed which will first be validated by means of an example described in the literature. Secondly, this program is extended to optimize a space truss dome.

Keywords: Genetic algorithms, Structural optimization, Space truss, Dome, Buckling

1   Introduction

The main task of strucural engineers is to design and build constructions with a set of desired properties in the design space. They have to consider the load bearing-capacity of the structure as well as the cost to construct them. Material cost is one of the major costs in construction. Therefore designs with a minimum amount of material used are preferable, given that the construction method (e.g. complex node connnections) is not too expensive. To achieve this goal, several optimization techniques can be applied. In this contribution, genetic algorithms are used in the optimization process because of their robustness and ability to work with discrete and continue variables at the same time.

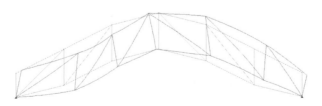

Figure 1: A two-dimensional truss.

---

[1]Corresponding author. E-mail: Lucrese.Vermaere@lid.kviv.be

Trusses are composed out of structural stable modules such as triangles in two dimensions (Figure 1) and tetrahedrons in three dimensions. The bars are only subm to tensile or compressive forces and not to bending moments because of the hinges between the bars. Therefore the loads have to act on the nodes and not in between them.

Genetic algorithms are based on the principle of *survival of the fittest*. This searching method starts from a population of trusses and every truss has his own fitness. Light structures satisfying the given constraints receive a higher fitness. A constraint is always related to the bearing-capacity of the truss structure; the bars mustn't not break, the nodes mustn't not move too much, the construction mustn't not become a mechanism[2] and mustn't not buckle (Figure 1). Euler buckling is of the type normally encountered in the failure of bars. Local buckling involves the snap through of isolated nodes and may initiate general buckling which occurs over a larger area and could result in complete collapse of the dome structure. For information on genetic algorithms we refer to Goldberg [1] and for truss structures we refer to Vandepitte [2]. Optimization of a space truss can be achieved by varying the cross-sectional areas[3] and/or the shape[4] and/or the topology[5]. The program *Space Truss Genetic Algorithm* (*STGA*) was written in FORTRAN. Later on this program was extended to *Space Truss Buckling Genetic Algorithm* (*STBGA*) in order to implement the overall buckling constraint.

2   Comparison of the results with the literature

STGA will be validated by comparing the results of an example derived from the literature[3]-[7]. A plain example is the optimization of the cross-sectional areas of the twenty-five-bar space truss (Figure 2). The truss bar sections are devided into eight groups and given in in^2 in Table1. In comparison with the weight described in other articles, STGA scores relatively well. This result was obtained with the program *Space Truss Buckling Genetic Algorithm*, written in FORTRAN.

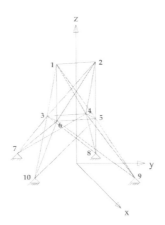

Figure 2: Twenty-five-bar space truss.

---

[2]this is a structure that collapses as soon as a small load acts on it
[3]discrete variables
[4]continue variables
[5]binary variables

Group number	STGA	[3]	[4]	[5]	[6]	[7]
1	0.21	0.2	0.1	0.13	0.1	0.1
2	1.16	1.8	0.5	0.12	0.6	1.0
3	2.97	2.3	3.4	3.48	3.2	3.4
4	0.31	0.2	0.1	0.11	0.2	0.2
5	0.74	0.1	1.5	1.66	1.5	0.6
6	1.06	0.8	0.9	0.84	1.0	1.1
7	0.74	1.8	0.6	0.12	0.6	0.9
8	3.29	3.0	3.4	4.09	3.4	3.0
Mass (*lb*)	510	546	486	470	492	515

Table 1: Comparison of the Space Truss Genetic Algorithm (STGA) results with other articles.

3   Space truss dome optimization

The overall buckling is taken into account for finding the optimum shape of a space truss dome. Without this criterium the genetic algorithm finds an optimal structure with a node in the centre that is lower than the surrounding ones (Figure 3). This is clearly not an optimal design. If the global buckling is considered as a constraint, the result is a realistic dome shape (Figure 4).

Figure 3: Dome without global buckling verification.

Figure 4: Dome with global buckling verification.

4   Conclusion

To optimize a one-layered truss dome, the overall buckling is an important constraint that has to be verified. If this constraint is well implemented in the genetic algorithm, the dome has a realistic optimal shape.

References

[1] D. E. Goldberg, *Genetic algorithms in search, optimization and machine learning.* Addison-Wesley, 1989.

[2] D. Vandepitte, *Berekening van constructies, Boekdeel III, Hoofdstuk 44*, E. Story-Scientia Gent, 1980 (in Dutch).

[3] S.-J. Wu and P.-T. Chow : Steady-State Genetic Algorithms for Discrete Optimization of Trusses, *Computers & Structures* **56** 979-991(1995).

[4] S. Rajeev and C. S. Krishnamoorthy: Discrete optimization of structures using genetic algorithms, *Journal of Structural Engineering* **118** 1233-1250(1992).

[5] C. A. Coello: Constraint-handling using an evolutionary multiobjective optimization technique, *Civil Engineering and Environmental Systems* **17** 319-346(2000).

[6] S.-J. Wu and P.-T. Chow: Integrated discrete and configuration optimization of trusses using genetic algorithms, *Computers and Structures* **55** 695-702(1995).

[7] O. Hasanebi, F. Erbatur, I. Tütüncü and H. Kili: Optimal design of planar and space structures with genetic algortihms, *Computers and Structures* **75** 209-224(2000).

VSP International
Science Publishers
P.O. Box 346, 3700 AH Zeist
The Netherlands

*Lecture Series on Computer
and Computational Sciences*
Volume 1, 2004, pp. 547-550

Development of a Unified Mathematical Framework for Calculating Molecular Weight Distribution in Diffusion Controlled Free Radical Homo-polymerization

G.D. Verros[1,*] T. Latsos[1] and D.S. Achilias[2]

[1]Department of Electrical Engineering,
Technological Educational Institute (TEI) of Lamia,
GR-35100 Lamia, Greece

[2]Department of Chemistry,
Laboratory of Organic Chemical Technology
Aristotle University of Thessaloniki,
GR-54006 Thessaloniki, Greece

Received 14 July, 2004: accepted in revised form 8 August, 2004

Abstract   In the present work a unified mathematical framework for calculating molecular weight distribution in diffusion controlled free radical homo-polymerization is developed. This framework is based on the numerical integration of a large system of ordinary differential equations describing the conservation of mass of macromolecular species in the reactor. Model predictions are in excellent agreement with available experimental data for metlyl methacrylate polymerization.

Keywords: Molecular Weight Distribution, Diffusion Controlled Reactions, Mathematical Modeling

1. Introduction

Bulk free radical homo-polymerization is a process of major financial and scientific interest. In the past thirty years many models with varying degrees of complexity have been published dealing with the mathematical modeling of polymerization kinetics. Most deal with methyl methacrylate polymerization which includes the following reactions:

Chemical Initiation

$$I \xrightarrow{K_d} 2R_1$$

Propagation

$$R_n + M \xrightarrow{K_p} R_{n+1}$$

Termination by Combination

$$R_n + R_m \xrightarrow{K_{tc}} P_{n+m}$$

Termination by Disproportionation

$$R_n + R_m \xrightarrow{K_{td}} P_n + P_m$$

Chain Transfer to Monomer

$$R_n + M \xrightarrow{K_{td}} P_n + R_1$$

where I represents the initiator, M is the monomer, R_n and P_n stand for the "live" (free radicals) and the "dead" polymer having n monomer units, respectively.

Besides the conventional chemical kinetics, physical phenomena related to the diffusion of various reactants play an important role in bulk free radical homo-polymerization. Reactions that are influenced by diffusion phenomena include the termination of radicals, the growth of "live" polymer chains (propagation reaction), and chemical initiation reaction. These reactions are related to the well known phenomena of the gel effect, the cage effect and the glass effect.

The gel effect or Trommsdorff-Norrish effect has been attributed to the decrease in termination rate constants caused by a decrease in the mobility of polymer chains. At the early stages of polymerization the termination kinetic rate constant is equal to the intrinsic constant K_{t0}. As time

---

* Corresponding author. E-mail : gdverros@otenet.gr, verros@vergina.eng.auth.gr

proceeds, the polymer concentration increases and the termination reactions become diffusion controlled. This leads to a decrease in termination rate constants followed by a sharp increase in polymer concentration. This auto-acceleration phenomenon strongly affects the end-use properties of the produced polymer as it leads to broader molecular weight distribution. The glass effect is related to the decrease in the propagation rate constant caused by a decrease in the mobility of monomer units. The cage effect is strongly related to the physical and transport properties of the reaction mixture. As initiator decompose only a fraction of the fragments escape from their "cages" and react with the monomer to form primary radicals. This phenomenon affects not only the initiator efficiency but also monomer conversion and final product molecular weight distribution.

Mathematical models for cage and glass effects have been developed by Gilbert and coworkers [1], Achilias and Kiparissides [2]   However, there are inherent differences between the models describing thegel effect. Two approaches have been proposed to describe the onset of the gel effect, monomer conversion as well as molecular weight developments in the polymerization reactor: the entanglement theory and the free volume approach.

The entanglement models, based on the reptation theory of a polymer chain [3-6], show that the termination rate constant depends on the chain length of individual polymer chains. The free volume approach is based on the Vrentas-Duda equation [7-8] for the self-diffusion of a chain in a polymer solution. These models differ not only in the physical description of the problem but also in the methodology in calculating the entire molecular weight distribution.

Free volume models are based on the method of moments [9-11] that allows the calculation of some characteristic averages of molecular weight distribution such as the number and the weight average of molecular weights. Several mathematical techniques have been proved effective in calculating the entire molecular weight distribution in the presence of the gel effect assuming that the kinetic rate constants are independent of the chain length.   These techniques include z-transforms, weighted residuals, the Galerkin on finite element method, the statistical approach, .the instantaneous molecular weight method and the direct numerical integration method. To calculate the entire molecular weight distribution in the case of chain length dependent termination one has to resort to Galerkin on finite element method or to weighted residuals.

The aim of the present work is to develop a unified framework with respect not only to the physics of the process but also to the mathematical method of calculating MWD. In what follows, the free volume approach as well as the entanglement theory is reviewed and a methodology for calculating MWD in the presence of chain length dependent termination is presented. Finally, results are presented and conclusions are drawn.

2. Model Development

DeGennes [3-6] considered the effect of topological constraints imposed upon the motion of a polymer molecule by its neighbors. In his view the motion of a given macromolecule is confined within a virtual "tube" defined by the locus of its intersections (or points of "entanglement") with adjacent molecules. The molecule is constrained and wriggles, snakelike along its own length by curvilinear propagation of length defects such as kinks or twists along the tube. This mode of motion was termed reptation. In the absence of significant polymer-polymer friction (the semi-dilute regime), application of scaling analysis in the context of the reptation model leads to the following law for the self-diffusion coefficient $D_{s,n}$:

$$D_{s,n} \sim n^{-2}c^{-7/4} \qquad\qquad c > c^{*} \tag{1}$$

Where c is the polymer concentration, n is the chain length of the macromolecules and c^{*} is a critical concentration.

Vrentas and Duda [7-8], based on the free volume theory, derived the following equation to describe the concentration and temperature dependence of the macromolecular mean self-diffusion coefficient in a polymer solution:

$$\overline{D}_{s} = \left(D_{p0}/M_{w}^{2}\right)\exp\left[-\gamma\left(\omega_{m}\overline{V}_{m}^{*} + \omega_{p}\overline{V}_{p}^{*}\xi_{13}\right)/V_{f}\xi_{13}\right] \tag{2}$$

where M_{w} represents the mean molecular weight of the diffused species, ω stands for weight fraction, subscripts p and m represent the "dead" polymer and the monomer, respectively. Other equation parameter values for methyl methacrylate polymerization, as a function of temperature and concentration, have been tabulated by Achilias and Kiparissides [12]

To express the termination kinetic rate constant K_{t} in terms of the free radical self-diffusion coefficient, most free volume models utilized the Smoluchowski equation [13]:

$$\frac{1}{\overline{K}_t} = \frac{1}{K_{t0}} + \frac{1}{4\pi N_A r_t \, \overline{D}_s} = \frac{1}{K_{t0}} + \frac{r_t^2 \lambda_0}{3\overline{D}_s} \tag{3}$$

where N_A is Avogadro number, K_{t0} stands for the intrinsic termination rate constant, $\lambda_0 \left(= \dfrac{4\pi r_t^3}{3N_A} \right)$ is

the total concentration of "live" radicals r_t represents the effective reaction radius for the termination reaction calculated by the excess chain end mobility theory [14] . This equation can be further recast to include the critical concentration c^*, defining the semi-dilute region in the entanglement theory[14]:

$$K_t = K_{t0} \qquad\qquad\qquad c < c^*$$
$$K_{t,n,m} = 4\pi N_A r_t (D_{s,n} + D_{s,m}) \qquad c > c^* \tag{4}$$

Finally, the cage and the glass effect were taken into account following Achilias and Kiparissides model developments [2]

3. Results and Conclusions

To calculate the molecular weight distribution for methyl methacrylate polymerization, a set of differential equations is derived in order to describe the mass conservation of the various reactants G in a batch polymerization reactor:

$$\frac{d(VG)}{dt} = Vr_G \tag{5}$$

Where V is the volume of the reactor, t is the polymerization time, r_G represents the rate functions of the reactant present in the reactor and are given in full detail elsewhere [2,15,16].

The quasi steady state approximation (QSSA) was applied to transform the differential equations system describing the conservation of "live" radicals to an algebraic equation system ($r_{R_n} = 0$). The above non-linear set of coupled algebraic and ordinary differential equations can be solved directly by utilizing standard numerical methods. [17] Additionally, in the entanglement models a trial and error process was employed to calculate total "live" radical concentration.

The model of Achilias and Kiparissides [2], and the model of Tuling and Tirrell [15,16] were utilized as representative models of the free volume approach and the entanglement theory, respectively. Achilias and Kiparissisdes model is based on the method of moments and utilizes the Vrentas-Duda equation for the self diffusion coefficient (Eq. 2) along with the Smoluchowski equation (Eq. 3), to calculate termination kinetic rate constants as a function of polymer concentration, temperature and "dead" polymer weight average molecular weight (M_w). It should be noted that in the original model M_w (see equation 2) represents the weight average molecular weight of the "dead" polymer and in present model M_w stands for the instantaneous number average molecular weight of the "live" polymer. This leads to a significant improvement in agreement between model predictions and experimental data.

The Tulig and Tirrell model is based on the reptation theory (Eq. 1) and the Smoluchowski equation (Eq. 4) is utilized to calculate diffusion controlled termination rate constants in terms of polymer concentration and individual macromolecule chain length.

The adjustable parameters of both models were estimated by fitting predictions to experimental data of conversion history, number and weight average molecular weights vs. time. The reacting conditions and the experimental data of Balke and Hamielec [18] were employed. Values for physical constants of the reactant mixture used in our numerical experiments (kinetic rate constants, physical properties, etc) are given in full detail elsewhere [2,12].

In all cases an excellent agreement with experimental data for conversion history, number and weight average molecular weights vs. time, total radical concentration (λ_0) as a function of time, initiator feed concentration and polymerization temperature was obtained. Moreover, an excellent agreement between MWD experimental data [18] and models predictions is depicted in Figure 1.

These results justify completely the ability of free volume models to describe the complex kinetics of MMA polymerization regardless of chain length dependence. It is believed that the present work may be applied to other diffusion control polymerization thus leading to a more rational design of industrial reactors.

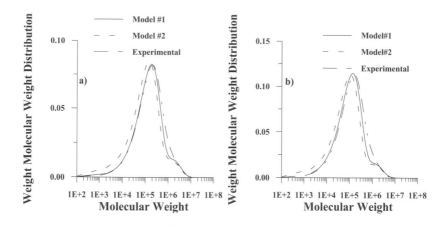

Figure 1: MWD of final product at polymerization temperature 70^0C a) Feed initiator (AIBN) concentration 0.3 % wt b) a) Feed initiator (AIBN) concentration 0.5 % wt. Model#1: free volume. Model#2: chain entanglements. Experimental data obtained by GPC [18].

References

[1] G.T. Russell, D.H. Napper and R.G. Gilbert *Macromolecules* **21** 2141 (1988).

[2] D.S. Achilias and C. Kiparissides, *Macromolecules* **25** 3739 (1992).

[3] P.G. deGennes, *J. Chem. Phys.*, **55** 572 (1971).

[4] P.G. deGennes, *Macromolecules* **9** 587 (1976).

[5] P.G. deGennes, *Macromolecules* **9** 594 (1976).

[6] P.G. deGennes, *Nature (London)* **282** 267 (1979).

[7] J.S. Vrentas and J.L. Duda, *J. Polym. Sci., Part B : Polym. Phys.* **15** 403 (1977).

[8] J.S. Vrentas and J.L. Duda, *J. Polym. Sci., Part B : Polym. Phys.* **15** 417 (1977).

[9] W.H. Ray, *J. Macromol. Sci. Rev. Macromol. Chem.* **C8** 1 (1972).

[10] D.S. Achilias and C. Kiparissides, *J. Macromol. Sci. Rev. Macromol. Chem.* **C32** 183 (1992).

[11] C. Kiparissides, G. Verros and J.F. MacGregor, *J. Macromol. Sci. Rev. Macromol. Chem.* **C33** 437 (1993).

[12] D.S. Achilias and C. Kiparissides, *J. Appl. Polym. Sci.* **35** 1303 (1988).

[13] M. Smoluchowski, *Z. Phys. Chem.* **92** 129 (1918).

[14] S.K. Soh and D.C. Sundberg, *J. Polym. Sci., Polym. Chem. Ed.* **20** 1299 (1982).

[15] T.J. Tulig and M. Tirrell, *Macromolecules* **14** 1501 (1981).

[16] D.J. Coyle, T.J. Tulig and M. Tirrell, *Ind. Engng Chem. Fundam.* **24** 343 (1985).

[17] G.D. Verros, *Polymer*, **44** 7021 (2003).

[18] S. T. Balke and A.E. Hamielec, *J. Appl. Polym. Sci.* **19** 903 (1972).

VSP International
Science Publishers
P.O. Box 346, 3700 AH Zeist
The Netherlands

Lecture Series on Computer
and Computational Sciences
Volume 1, 2004, pp. 551-552

Numerical and Monte-Carlo Calculation of Photon Energy Distribution in Sea Water

D.S. Vlachos[1]

Hellenic Center for Marine Research,
PO BOX 712,
GR-190 13 Anavissos, Greece

Received 5 September, 2004; accepted in revised form 20 September, 2004

Abstract: Photons, when emitted in sea water, are subject to multiple scattering mechanisms, resulting in a shift in their initial energy. Consequently, a problem arises when measured spectra of radioactivity in sea water are processed, because the measured values do not reflect the initial photon energy but a distribution of energies. In this work a detailed analytical formulation of this distribution is performed and numerical results are compared to results from Monte-Carlo calculation.

Keywords: Marine radioactivity, photon energy distribution

PACS: 29.30.Kv, 29.40.Mc

1 Formulation

Assume that radioactive sources are distributed uniformly in sea water. This is a reasonable simplification if we consider the averaging effect of radioactivity over time. Let $n(\alpha)$ be the concentration of photons with energy ϵ and $\alpha = \epsilon/m_0 c^2$ (m_0 is the rest mass of electron and c the speed of light). The change of $n(\alpha)$ over time is given:

$$\frac{dn(\alpha_t)}{dt} = \sum_i G^i \cdot \delta(\alpha_t - \alpha_0^i) + \int_0^\infty n(\alpha) \cdot C(\alpha, \alpha_t) d\alpha + 2 \cdot \delta(\alpha_t - 1) \cdot \int_2^\infty n(\alpha) \cdot R(\alpha) d\alpha - n(\alpha_t) \cdot S(\alpha_t) \quad (1)$$

where

G^i	the rate of photon generation at energy α_0^i
$C(\alpha, \alpha_t)$	the probability per second for a photon with energy α to scatter to the energy α_t due to Compton scattering
$R(\alpha)$	is the probability per second for a photon with energy α to be absorbed leaving in its place an electron-positron pair. Finally, the positron will interact with an electron producing two photons with energy $\alpha = 1$
$S(\alpha_t)$	the probability per second for a photon with energy α_t to scatter to a different energy due to all scattering mechanisms.

[1]Corresponding author. E-mail: dvlachos@ncmr.gr, dvlachos@uop.gr

For clarity, we assume that photons are generated only in one energy, namely α_0. By setting:

$$n(\alpha) = \lambda(\alpha) + k \cdot \delta(\alpha - \alpha_0) + m \cdot \delta(\alpha - 1) \tag{2}$$

we find for k and m

$$k = \frac{G}{S(\alpha_0)} \tag{3}$$

$$m = 2 \cdot \frac{k \cdot R(\alpha_0) + \int_2^{\alpha_0} \lambda(\alpha) \cdot R(\alpha) \cdot d\alpha}{S(1)} \tag{4}$$

and substituting in (1), we find for λ:

$$\frac{G \cdot C(\alpha_0, \alpha_t)}{S(\alpha_0)} + 2 \cdot \frac{\frac{G}{S(\alpha_0)} + \int_2^{\alpha_0} \lambda(\alpha) \cdot R(\alpha) \cdot d\alpha}{S(1)} \cdot C(1, \alpha_t) + \int_0^{\infty} \lambda(\alpha) \cdot C(\alpha, \alpha_t) da - \lambda(\alpha_t) \cdot S(\alpha_t) = 0 \tag{5}$$

where in steady-state $dn(\alpha)/dt = 0$. The above equation can be solved numerically. The functions $R(\alpha)$ and $S(\alpha)$) are calculated from the XCOM software [1]. The function $C(\alpha, \alpha_t)$ is calculated by the Klein-Nishima formula [2]which gives for the angular distribution of scattered photons:

$$p(\theta) = \pi \cdot r_0^2 \left(\frac{\alpha}{\alpha_0} \right)^2 \left(\frac{\alpha}{\alpha_0} + \frac{\alpha_0}{\alpha} - sin^2\theta \right) \tag{6}$$

where

$$\alpha = \frac{\alpha_0}{1 + \alpha_0 \cdot (1 - cos\theta)} \tag{7}$$

Changing the variable θ to α we have:

$$C(\alpha_0, \alpha) = \pi \cdot r_0^2 \frac{1}{\alpha_0^2} \left[\frac{\alpha}{\alpha_0} + \frac{\alpha_0}{\alpha} - 1 + \left(1 - \frac{1}{\alpha} + \frac{1}{\alpha_0} \right) \right] \cdot \left[u(\alpha - \frac{\alpha_0}{1 + 2\alpha_0}) - u(\alpha - \alpha_0) \right] \tag{8}$$

where u is the step function.

Equation (8) is solved numerically. The results are compared with those produced from Monte-Carlo simulation of the distribution of photons in sea water.

References

[1] M.J. Berger and J.H. Hubell: *XCOM: photon cross sections with a personal computer*, NBSIR 87-3597 (1987).

[2] N. Tsoulfanidis: *Measurement and Detection of Radiation*, Hemisphere Publishing Corporation, ISBNJ 0-8916-523-1 (1983).

VSP International
Science Publishers
P.O. Box 346, 3700 AH Zeist
The Netherlands

Lecture Series on Computer
and Computational Sciences
Volume 1, 2004, pp. 553-555

Monte-Carlo Simulation of Clustering in a Plasma-Discharge Source

D.S. Vlachos[1]

Hellenic Center for Marine Research,
PO BOX 712,
GR-190 13 Anavissos, Greece

and

A.C. Xenoulis

N.C.S.R. Demokritos,
GR-153 10 Agia Paraskevi, Greece

Received 5 September, 2004; accepted in revised form 20 September, 2004

Abstract: The assumption that the ionization and coagulation of metal plasma-generating cluster source can be described by the orbital limited motion theory, when it was examined by Monte Carlo calculations, was found inadequate. That assumption cannot firstly predict the modes of cluster ionization observed experimentally. Secondly, the effectiveness of coagulation was found to depend on the initial size of the condensing particles. Specifically, the coagulation of dust particles proceeds in a satisfactory degree. The coagulation of single atoms in a cluster source, however, saturates at about 140 atoms per cluster, in serious underestimation of relevant experimental data.

Keywords: Monte-Carlo, Clustering, Plasma-Discharge

Mathematics Subject Classification: 65Z05

PACS: 02.70.Uu, 81.07.Bc

1 Introduction

Certain of the most important cluster sources (such as the laser-ablation, the magnetron Hnd the hollow-cathode source) generate plasma during particle coagulation. Nevertheless, clustering under plasma conditions is very poorly studied and understood. On the experimental front it has been observed that negatively as well as positively ionized clusters coexist during coagulation [1], suggesting that electromagnetic interactions may be responsible for clustering. On the theoretical front, the coagulation of $10\text{\AA}$-radius dust particles in a steady-state plasma was successfully described in terms of electrostatic and dipole interactions [2]. The untested use of this model to describe clustering in a gas-aggregation source, however, is not safe because in the source the coagulation conditions are significantly less favorable than those previously assumed [2]. The purpose of the present study is to examine whether or under what conditions the model of Ref. [2] is applicable to a plasma-discharge cluster source. The relevant realistic conditions tested in the

---

[1]Corresponding author. E-mail: dvlachos@ncmr.gr, dvlachos@uop.gr

present Monte Carlo calculation is that clustering starts from atoms and that the interaction time is about $1 msec$ [1].

2 The Simulation Algorithm

It may be instructive first to describe qualitatively the evolution of events taking place when Cu atoms (or dust particles) and plasma are mixed. As soon as the interaction is turned on, all the present (initially neutral) Cu atoms are negatively ionized. This happens because the number of collisions with electrons is overwhelming compared to those with Ar^+ ions. As a consequence, a short-lived outburst of clustering observed at the very outset, caused by $Cu^0 - Cu^-$ dipole interaction, stops because, as already mentioned, all the particles become negatively charged and repulse each other. Nevertheless, a few dimmers manage to be formed in the mean time. As the particle-plasma interaction continues, some of the negatively charged metal particles will turn to neutral when they collide with Ar^+. These fresh neutrals are very short lived. They almost instantly interact with either other negatively charged Cu particles leading to larger negative clusters, or with electrons reverting to negatively ionized clusters of the same mass. This clustering process, however, soon fades out because the density of scattering centers is reduced due to clustering. Nevertheless, because in the mean time the dimensions of the clusters have increased, a second co-agulation stage, supported by multiple ionization, becomes possible. During that stage, the mean negative charge per cluster increases significantly causing a commensurate increase in the strength of the dipole interaction giving a fresh outburst of clustering events.

A sample volume, $1O^{-15}m^3$, of the experimental apparatus has been considered with periodic boundary conditions. Monte Carlo calculations take into account the interaction of metal particle with each other and with the plasma components, i.e. electrons and ions. A constant electron density is assumed and the overall neutrality is obtained by adjusting the ion density. All plasma components and metal particles are considered in thermodynamic equilibrium. The scattering rate of a particle with a specific type of scattering center is calculated using the formulae:

$$f_i = \int_0^\infty N(\epsilon) \cdot \nu(\epsilon) \cdot \sigma_i(\epsilon) \cdot d\epsilon \qquad (1)$$

where N is the density of scattering centers, ν the relative velocity of the particle and the scattering center and σ the cross section which is calculated in the center of mass reference system. Both coagulation and ionization cross sections are calculated taking into account electrostatic forces. In particular, a neutral particle interacts with a charged particle via a dipole moment induced by the latter to the former. Finally, the temperature of clusters is considered to be size independent, while the density of all plasma components follows a Maxwell-Boltzmann distribution over energy. More details may be found in Ref.[1]. The calculations were performed for two kinds of particles, differentiated by their initial size. A radius of $1.5\mathring{A}$ represents Cu atoms, while a radius of $10\mathring{A}$ represents the metal dust particles used in the previous calculation [2].

3 Simulation Results

Figure *la* shows the evolution in time of the mean number of atoms per cluster and Figure *lb* of the mean charge per cluster, for two kinds of particles with initial radius $1.5\mathring{A}$ and $10\mathring{A}$. The results of Figure *la* clearly demonstrate that the efficiency of coagulation depends significantly on the initial dimensions of the particles involved. In fact, while the $10\mathring{A}$ particles, in agreement with Ref. [2], coagulate efficiently, the coagulation initiated by single atoms stops relatively soon reaching a plateau at about 140 atoms per clusters. This is a serious size underestimation since the experimental sizes range between $1,000$ and $300,0000$ atoms per cluster [1].

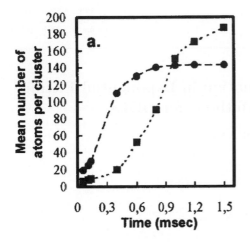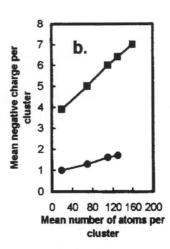

Figure 1: The dependence of mean number of particles per cluster on (a) time and (b) on mean negative charge per cluster for two different atomic radius: circles correspond to $1.5\mathring{A}$ and squares to $10\mathring{A}$.

It should be pointed out that several contradictory effects are competing with each other, and depending on the particular circumstances the one or the other may assume ascendancy. Such a case is, for instance, in the very early stages of clustering, when as Figure 1a reveals single atoms coagulate faster than dust particles. To understand this feature which seems inconsistent with the final outcome, it must be taken into account that, as Figure 1b shows, the small clusters built by atoms sustain smaller negative charge than the clusters composed of dust particles. Consequently, the probability for the former clusters to get neutralized via collisions with Ar^+, and as a consequence to coagulate, is larger than that for the later. This coagulation, however, reduces the number of the atoms and of the atom-composed clusters, leading to the plateau observed in fig 1a. On the other hand, the charge on the dust- composed clusters, as Figure 1b shows, increases faster with size than on the atom-composed clusters. Apparently, this faster increase is sufficient to compensate for the decreasing number of the dust-composed clusters (see Equation 1), ensuring the unhindered coagulation of these particles.

References

[1] A.C. Xenoulis, G. Doukelis, P. Tsouris, A. Karydas, C. Potiriadis, A.A. Katsanos and Th. Tsakalakos: *Vacuum* 49-113 (1998).

[2] V.A. Schweigert and I.V. Schweigert: *J. Phys. D: Appl. Phys.* 29-655 (1996).

VSP International
Science Publishers
P.O. Box 346, 3700 AH Zeist
The Netherlands

Lecture Series on Computer
and Computational Sciences
Volume 1, 2004, pp. 556-557

On Frequency Determination in Exponential Fitting Multi-step Methods for ODEs

D.S. Vlachos

Hellenic Center for Marine Research,
PO BOX 712,
GR-190 13 Anavissos, Greece

and

T.E. Simos[1]

Department of Computer Science and Technology,
Faculty of Sciences and Technology,
University of Peloponnese,
GR-221 00 Tripolis, Greece

Received 5 September, 2004; accepted in revised form 20 September, 2004

Abstract: The frequency determination in an exponential fitting multi-step method is a question without a definite answer. In most of the cases, the estimation of the frequency arises from the nature of the problem, as in the Schrödinger equation. Another approach is to select a frequency which increases the order of the method by zeroing the first non vanishing term of the linear truncation error. In this work, two general methods are exploited: the first is applicable to equations of the form $y''(x) = f(x,y)$ and relates the frequency with the coefficient of y in $f(x,y)$ while the other connects the frequency with the curvature of the solution.

Keywords: Exponential fitting, multi-step methods

Mathematics Subject Classification: 65L06

1 Introduction

Consider a differential equation of the form:

$$y'' = f(x,y) \tag{1}$$

where the first derivative of y is not explicitly appear in f. Now let an exponential fitting method is employed for the numerical integration of the above equation. Recently, Ixaru et al have proposed in [1], that an optimal way to select the frequency for the exponential fitting method is such that the first non vanishing term of the linear truncation error is zeroing. Although this is reasonable, it comes out that this operation needs the evaluation of high order derivatives of the unknown function $y(x)$ making it impractical especially in high order methods.

---

[1]Corresponding author. Active Member of the European Academy of Sciences and Arts. E-mail: simos-editor@uop.gr, tsimos@mail.ariadne-t.gr

On the other hand, Simos in [2] proposed that if

$$f(x, y) = c \cdot y + \dots \tag{2}$$

then we can select for the frequency in the exponential fitting method the parameter:

$$\omega = \sqrt{|c|} \tag{3}$$

which gave very accurate results without further computational effort.

In this work, a different approach is proposed for the calculation of the frequency. Consider for clarity the following differential equation:

$$\frac{d}{dt} \left(\begin{array}{c} x(t) \\ y(t) \end{array} \right) = \left(\begin{array}{c} f_x(t) \\ f_y(t) \end{array} \right) \tag{4}$$

and an exponential fitting method to solve it. It is clear then that the solution $(x(t), y(t))^T$ can be represented in the xy-plane as a curve. It is well known, that the circle that best approximates the curve$(x(t), y(t))^T$ at the point $t = t_0$ has radius equal with the inverse of the curvature at the same point, defined as:

$$\kappa = \frac{|\dot{\mathbf{r}} \times \ddot{\mathbf{r}}|}{|\dot{\mathbf{r}}|^3} \tag{5}$$

where

$$\mathbf{r} = \left(\begin{array}{c} x(t) \\ y(t) \end{array} \right) \tag{6}$$

It follows now, that the frequency for the exponential fitting method can be estimated by:

$$\omega = \kappa \cdot |\dot{\mathbf{r}}| = \frac{|\dot{\mathbf{r}} \times \ddot{\mathbf{r}}|}{|\dot{\mathbf{r}}|^2} \tag{7}$$

The basic advantage of the proposed method is twofold: first that only one derivative has to be calculated and second that the motion of the center of the osculating circle can be similarly considered in order to produce a second frequency and so on.

The proposed method is tested and compared to a set of problems usually encountered in the bibliography.

References

[1] L.Gr. Ixaru, G. Vanden Berghe and H. De Meyer, *Journal of Computational and Applied Mathematics*,140-423 (2002).

[2] T.E. Simos and Jesus Vigo-Aguiar, *Computer Physics Communications*, 140-358 (2001). York, 1965.

VSP International
Science Publishers
P.O. Box 346, 3700 AH Zeist
The Netherlands

Lecture Series on Computer
and Computational Sciences
Volume 1, 2004, pp. 558-561

A Hybrid Adaptive Neural Network for Sea Waves Forecasting

D.S. Vlachos[1]

Hellenic Center for Marine Research,
PO BOX 712,
GR-190 13 Anavissos, Greece

Received 5 September, 2004; accepted in revised form 20 September, 2004

Abstract: The physical process of generation of wind waves is extremely complex, uncertain and not yet fully understood. Despite a variety of deterministic models presented to predict the heights and periods of waves from the characteristics of the generating wind, a large scope still exists to improve on the existing models or to provide alternatives to them. In this work, a hybrid adaptive neural network has been designed and used in order to predict the wave height. The system has been proved to produce a 95% successful 24 hours prediction of wave height after 2 months of operation.

Keywords: Neural networks, fuzzy controller, wave forecast

Mathematics Subject Classification: 68T05

PACS: 84.35.+i, 92.10.Hm

1 Introduction

The knowledge of heights and periods of oscillatory short waves is essential for almost any engineering activity in the ocean. These waves are generated by the action of wind through pressure as well as shear mechanism. Wind-wave relationships have been explored over a period of five decades in the past by establishing empirical equations and also by numerically solving the equations of wave growth [1], [2]. However, the complexity and uncertainty of the wave generation phenomenon is such that despite considerable advances in computational techniques, the solutions obtained are neither exact nor uniformly applicable at all sites and at all times. Moreover, since the numerical models that have been developed for wave prediction are strongly depend on the wind forecasting, any errors or divergences in the wind model have been founded that dramatically amplified in the results of the wave model.

On the other hand, many of the most important properties of biological intelligence arise through a process of self-organization whereby a biological system actively interacts with a complex environment in real time. The environment is often noisy and non-stationary, and intelligence capabilities are learned autonomously and without benefit of an external teacher. For example, children learn to visually recognize and manipulate complex objects without being provided with explicit rules for how to do so. The main problem encountered in the design of an intelligent system

---

[1]Corresponding author. E-mail: dvlachos@ncmr.gr, dvlachos@uop.gr

capable of autonomously adapting in real time to changes in his world, is called the plasticity-stability dilemma, and a theory called adaptive resonance theory is being developed that suggests a solution to this problem [3]. The plasticity-stability dilemma asks how a learning system can be designed to remain plastic, or adaptive, in response to certain events, yet also remain stable in response to irrelevant events. In particular, how can it preserve its previously learned knowledge while continuing to learn new things. The proposed neural network based wave height prediction system has been designed taking into account the above considerations.

The second important consideration of the proposed system is that, since wind forecast is necessary to obtain accurate predictions for the wave height, there must be a mechanism to handle for the reliability of the wind model. The solution that we adopted is to design a hybrid network, in which the wind forecast is used as an input to a fuzzy controller and the overall prediction is produced by a three-layer perceptron.

2   Network Architecture

Figure 1 shows a schematic diagram of the neural system used for the wave height prediction and its subsystems. The input of the system is the significant wave height, the wind speed and the wind direction time series. The data are coming every three hours and already predicted values of wind speed and wind direction are included. These values are produced by the POSEIDON system [4]. The different subsystems are explained in detail in the following.

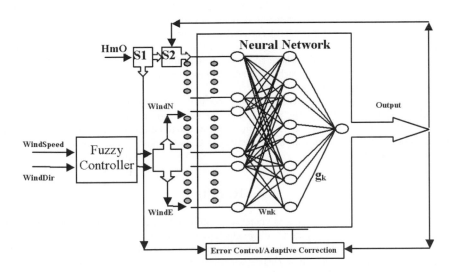

Figure 1: Schematic diagram of the hybrid adaptive predicting system.

The role of the S1 Switch is to feed the actual value of the wave significant height either to the long-term memory or the error control subsystem. When a new value of the significant wave height ($Hm0$) is coming, the S1-Switch feeds this value to the Error Control Subsystem in order to correct the previous predictions of the system. After the correction is performed, the S1-Switch feeds the incoming value to the S2-Switch. The role of the S2-Switch is to feed the long-term memory either with the actual value of the wave height or the predicted one in order to obtain a deeper prediction. More precisely, the value of wave height at the time tn can be used to predict the wave height at time t_{n+1}. Then this value can be used in order to predict the wave height at time t_{n+2} and so on. In our case the prediction is performed up to time t_{n+8} which corresponds to twenty-four hours prediction. The wind speed and wind direction forecasts are the inputs of the fuzzy controller. The output of the controller, which is a fuzzifiedversion of the wind speed and direction, is analyzed in $u - v$ components (WindN, WindE). The neural network used to predict the wave height is a three layer back propagation network [5]. The first layer consists of neurons that simply hold the input values. These values are passed through the weighted connections to the second layer, which consists of neurons that sum their input and pass it to their activation system. The output of the second layer is passed through the weighted connection to the third layer, which have only one neuron. Let I_n be the network input, w_{nk} the weights of the connections between the first and second layer, f_k the activation functions of the neurons of the second layer, g_k the weights of the connections between the second and third layer, N the number of input neurons and K the number of the neurons of the second layer. The activation function of the neurons of the second layer is given by the formulae:

$$f_k(x) = \langle \begin{array}{l} tanh^2(c_k^+ x), x \geq 0 \\ tanh^2(c_k^- x), x < 0 \end{array} \tag{1}$$

The slope of the activation function is very important because the sensitivity of the system depends on it. Moreover, there is a different slope between positive and negative values in order to account for the effect of wind direction in coastal zones. In these cases, the direction of the wind is critical for the wave growth. The weights w_{nk} and g_k and the parameters c_k^+ and c_k^- are the long-term memory of the system. These $(N + 3) * K$ parameters are modulated during adaptive learning. In every correction step, each parameter is modified in such a way that the overall error in prediction will be decreased.

As far as the fuzzy controller is concerned, its role is to distribute the wind in fuzzy bands and register its tension. In this way, the exact values for the wind speed and direction, as they are indicated by the numerical model, is not so important in the dynamics of the system. On the other hand, the tension of the wind in conjunction with the already register values of the wave height is the key to the prediction.

3   Results

Figure 2 shows the prediction results of the system after three months of operation. The results are compared with the measured values of weight height, the values calculated by the arithmetic model of the POSEIDON system and the values produced by an adaptive system without the fuzzy treatment of the wind data [6]. It is clear from this figure, that errors in wind speed and wind direction prediction reflect directly to the wave height prediction by the numeric model. On the other hand, the adaptive system is less sensitive to these errors while the fuzzy treatment of the wind data improves further the predictions.

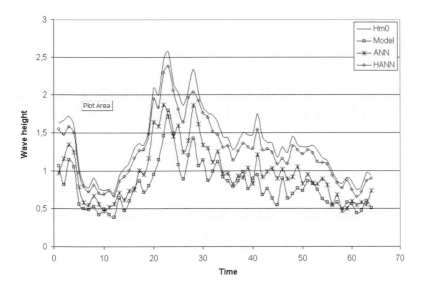

Figure 2: Schematic diagram of the hybrid adaptive predicting system. The solid line is the actual values, the empty squares are the model predictions, the asterisks are the values predicted with the adaptive system and the empty circles are the values predicted by the hybrid adaptive neural network.

References

[1] B. Kissman: *Wind Waves*, Prentice Hall, Englewood Cliffs, NJ, 1965.

[2] World Meteorological Organization, 1988, Guide to Wave Analysis and Forecasting, WMO no. 72. Secretariat of the World Meteorological Organazation, Geneva, Switzerald.

[3] D.S. Vlachos, D.K. Fragoulis and J.N. Avaritsiotis, *Sensors and Actuators B*, 43-1 (1998).

[4] T. Soukissian, G. Chronis and K. Nittis, *Sea Technology*, 40(7)-31 (1999).

[5] B. Kosko: *Neural Networks and Fuzzy Systems*, Prentice Hall, Englewood Cliffs, NJ, 1992.

[6] D.S. Vlachos and A. Papadopoulos, *Elsevier Oceanographic Series*, 69-403, (2003).

VSP International
Science Publishers
P.O. Box 346, 3700 AH Zeist
The Netherlands

Lecture Series on Computer
and Computational Sciences
Volume 1, 2004, pp. 562-565

A Rectangular Trust-Region Approach for Unconstrained and Box-Constrained Optimization Problems

C. Voglis[1] and I.E. Lagaris[2]

Department of Computer Science,
University of Ioannina,
P.O. BOX 1186 - GR 45110 Ioannina, Greece

Received 4 August, 2004; accepted in revised form 21 August, 2004

Abstract: A convex quadratic programming problem with bound constraints is solved at first, using a Lagrange multiplier approach. Then a trust region method for non-linear optimization with box constraints is developed, where the trust region is a hyperbox, in contrast to the usual hyperellipsoid choice. The resulting subproblem is solved using our above mentioned QP technique.

Keywords: Trust-Region, Quadratic programming, Lagrange multipliers, box constraints

Mathematics Subject Classification: 90C20, 90C30, 90C53

1 Introduction

Non-linear optimization plays an important role in many fields of science and engineering, in the industry, as well as in a plethora of practical problems. Frequently the optimization parameters are constrained inside a range imposed by the nature of the problem at hand. Developing methods for bound constrained optimization is hence quite useful. The most efficient optimization methods are based on Newton's method where a quadratic model is adopted as a local approximation to the objective function. Two general approaches have been followed. One uses a line–search along a properly selected descent direction, while the other permits steps of restricted size in an effort to maintain the reliability of the quadratic approximation. The approaches in this second class, bear the generic name Trust-Region techniques a brief description of which is given in Section 2. In this article we will deal with a method of that type. We develop a method that adopts a hyperbox geometry for the trust region. This offers the obvious advantage of dealing with linear instead of quadratic constraints that are imposed by spherical trust regions. In addition it allows effortless adaptation to bound constrained optimization problems. We analyze the approach for the bound-constrained quadratic programming in Section 3, and we present an algorithmic solution in Section 3.1. In section 4, we embed this QP technique in the general setting of the trust region approach, for both unconstrained and bound–constrained problems.

2 Trust Region Methods

Trust region methods[1] fall in the category of sequential quadratic programming. The algorithms in this class are iterative procedures in which the objective function $f(x)$ is represented by a

---

[1]Corresponding author. E-mail: voglis@cs.uoi.gr
[2]E-mail: lagaris@cs.uoi.gr.gr

quadratic model inside a suitable neighborhood (the trust region) of the current iterate, as implied by the Taylor series expansion. This local model of $f(x)$ at the kth iteration can be written as:

$$f(x^k + s) \approx m^k(s) = f(x^k) + s^T g^{(k)} + \frac{1}{2} s^T B^{(k)} s \qquad (1)$$

where $g^{(k)} = \nabla f(x^{(k)})$ and $B^{(k)}$ is a symmetric positive definite approximation to $\nabla^2 f(x^{(k)})$.

The trust region may be defined by:

$$\mathbf{T}^{(k)} = \{x \in \Re^n \mid ||x - x^{(k)}|| \leq \Delta^{(k)}\} \qquad (2)$$

It is obvious that different choices for the norm lead to different trust region shapes. The Euclidean norm $|| \cdot ||_2$, corresponds to a hypershpere, while the $|| \cdot ||_\infty$ norm defines a hyperbox.

Given the model and the trust region, we seek a step $s^{(k)}$ with $||s^{(k)}|| \leq \Delta^{(k)}$, such that the model is sufficiently reduced. Using this step we compare the reduction in the model to that in the objective function. If they agree to a certain extend, the step is accepted and the trust region is either expanded or remains the same. Otherwise the step is rejected and the trust region is contracted. The basic trust region algorithm is sketched in Alg. 1

Algorithm 1 Basic trust region

S0: Pick the initial point and trust region parameter x_0 and Δ_0, and set $k = 0$.

S1: Construct a quadratic model:
$m^{(k)}(s) \approx f(x^{(k)} + s)$

S2: Calculate $s^{(k)}$ with $||s^{(k)}|| \leq \Delta^{(k)}$, so as to sufficiently reduce $m^{(k)}$.

S3: Compute the ratio of actual to expected reduction, $r^{(k)} = \frac{f(x^{(k)}) - f(x^{(k)} + s^{(k)})}{m^{(k)}(0) - m^{(k)}(s^{(k)})}$. This value will determine if the step will be accepted or not and the update for $\Delta^{(k)}$.

S4: Increment $k \leftarrow k + 1$ and repeat from S1.

---

3   Bound-constrained QP

Let $x, d \in R^N$ and B a symmetric, positive definite $N \times N$ matrix and $I = \{1, 2, \cdots, N\}$. Consider then the QP problem:

$$\min_x \frac{1}{2} x^T B x + x^T d, \text{ subject to: } a_i \leq x_i \leq b_i, \forall i \in I \qquad (3)$$

We follow the Lagrange multipliers line and we construct the Lagrangian:

$$L(x, \lambda, \mu) = \frac{1}{2} x^T B x + x^T d - \lambda^T (x - a) - \mu^T (b - x) \qquad (4)$$

The KKT necessary conditions at the minimum x^*, λ^*, μ^* require that:

$$\begin{aligned}
Bx^* + d - \lambda^* + \mu^* &= 0 \\
\lambda_i^* \geq 0, \quad \mu_i^* &\geq 0, \quad \forall i \in I \\
\lambda_i^* (x_i^* - a_i) &= 0, \quad \forall i \in I \\
\mu_i^* (b_i - x_i^*) &= 0, \quad \forall i \in I \\
x_i^* &\in [a_i, b_i], \quad \forall i \in I
\end{aligned} \qquad (5)$$

A solution to the above system of equations (5), can be obtained through an active set strategy described in detail in the following section 3.1.

3.1   The BOXCQP algorithm

Our QP algorithm is sketched in Alg. 2:

Algorithm 2 BOXCQP

S0: Initially set: $k = 0$, $\lambda^{(0)} = \mu^{(0)} = 0$ and $x^{(0)} = -B^{-1}d$.
   If $x^{(0)}$ is feasible, **Stop**, the solution is: $x^* = x^{(0)}$.
   At iteration k, the quantities $x^{(k)}, \lambda^{(k)}, \mu^{(k)}$ are available.

S1: Define the sets:

$$
\begin{aligned}
L^{(k)} &= \{i : x_i^{(k)} < a_i, \text{ or } x_i^{(k)} = a_i \text{ and } \lambda_i^{(k)} \geq 0\} \\
U^{(k)} &= \{i : x_i^{(k)} > b_i, \text{ or } x_i^{(k)} = b_i \text{ and } \mu_i^{(k)} \geq 0\} \\
S^{(k)} &= \{i : a_i < x_i^{(k)} < b_i, \text{ or } x_i^{(k)} = a_i \text{ and } \lambda_i^{(k)} < 0, \\
&\quad \text{or} \quad x_i^{(k)} = b_i \text{ and } \mu_i^{(k)} < 0\}
\end{aligned}
$$

   Note that $L^{(k)} \cup U^{(k)} \cup S^{(k)} = I$

S2: Set:

$$
\begin{aligned}
x_i^{(k+1)} &= a_i, \; \mu_i^{(k+1)} = 0, \quad \forall i \in L^{(k)} \\
x_i^{(k+1)} &= b_i, \; \lambda_i^{(k+1)} = 0, \quad \forall i \in U^{(k)} \\
\lambda_i^{(k+1)} &= 0, \; \mu_i^{(k+1)} = 0, \quad \forall i \in S^{(k)}
\end{aligned}
$$

S3: Solve:

$$
Bx^{(k+1)} + d = \lambda^{(k+1)} - \mu^{(k+1)}
$$

   for the N unknowns:

$$
\begin{aligned}
&x_i^{(k+1)}, \; \forall i \in S^{(k)} \\
&\mu_i^{(k+1)}, \; \forall i \in U^{(k)} \\
&\lambda_i^{(k+1)}, \; \forall i \in L^{(k)}
\end{aligned}
$$

S4: Check if the new point is a solution and decide to either stop or iterate.

   If $(x_i^{(k+1)} \in [a_i, b_i] \; \forall i \in S^{(k)}$ **and** $\mu_i^{(k+1)} \geq 0, \; \forall i \in U^{(k)}$
   and $\lambda_i^{(k+1)} \geq 0, \; \forall i \in L^{(k)})$ **Then**
      Stop, the solution is: $x^* = x^{(k+1)}$.
   Else
      set $k \leftarrow k + 1$ and iterate from **S1**
   Endif

---

4   Rectangular trust region approach

The basic motivation for the development of a robust solver for positive definite quadratic problems with simple bounds, was to solve exactly (and not approximately as in the Dogleg [3] technique) the quadratic subproblem that arises in a trust region framework using the infinite norm $||\cdot||_\infty$. In this case the trust region in which the quadratic model is valid, is a hyperbox.

4.1   Model and norm definition

The problem under consideration is:

$$\min_x f(x)$$
$$\text{subject to: } l_i \le x_i \le u_i \tag{6}$$

Let $x^{(k)}$ be the estimation of the solution at the k-th step of the algorithm. In every step we have to solve the following problem:

$$\min_s m(x^{(k)} + s) = f^{(k)} + g^{(k)^T} h + \frac{1}{2} s^T B^{(k)} s$$
$$\text{subject to: } ||s||_\infty \le \Delta^{(k)} \text{ and } l_i - x_i^{(k)} \le s_i \le u_i + x_i^{(k)} \tag{7}$$

which is equivalent to:

$$\min_s m(s) = g^{(k)^T} h + \frac{1}{2} s^T B^{(k)} s$$
$$\text{subject to: } \max(l_i - x_i^{(k)}, -\Delta^{(k)}) \le s_i \le \min(u_i - x_i^{(k)}, \Delta^{(k)}) \tag{8}$$

where $f^{(k)} = f(x^{(k)}), g^{(k)} = \nabla f(x^{(k)}), B^{(k)}$ is a positive definite approximation to the Hessian matrix. The BFGS formula is used to update $B^{(k)}$. This problem is solved in **S2** step of Alg. 1 laid out in Section 2.

Computational experiments and comparisons with other approaches have hinted evidence of superior performance.

References

[1] A. Conn, N. Gould and P. Toint, *Trust-Region methods*, MPS-SIAM Series on Optimization (2000)

[2] J. E. Dennis and R. B. Schnabel, *Numerical Methods for Unconstrained Optimization and Nonlinear Equations*, SIAM (1996)

[3] M. J. D. Powell, A new algorithm for unconstrained optimization, *Nonlinear Programming*, pp. 31–65, Academic Press, London (1970)

[4] A.R. Conn, Nick Gould and P.L. Toint, Testing a Class of Methods for Solving Minimization Problems with Simple Bounds on the Variables, *Mathematics of Computation*, 50, pp. 399–430 (1988)

VSP International
Science Publishers
P.O. Box 346, 3700 AH Zeist
The Netherlands

*Lecture Series on Computer
and Computational Sciences*
Volume 1, 2004, pp. 566-568

A New Approach to Machine Contour Perception

Vassilios Vonikakis[1], Ioannis Andreadis[1] and Antonios Gasteratos[2]

[1]Laboratory of Electronics,
Section of Electronics and Information Systems Technology,
Department of Electrical and Computer Engineering,
[2]Laboratory of Robotics and Automation,
Section of Production Systems,
Department of Production and Management Engineering,

Democritus University of Thrace,
GR-67100 Xanthi,
Greece.

Received 13 May, 2004; accepted in revised form 24 Jube, 2004

This paper presents a new shape descriptor suitable for machine contour perception, that mimics the orientation selective cells of the human visual cortex. It extracts a description for any 2-dimensional shape, regardless of its size, rotation and position, with affordable computational cost. Furthermore, it can identify different shapes of the same object, distorted shapes and heavily contaminated shapes with up to 20% spike noise. Special queries are also possible, for the detection of certain shape patterns in their contours. The extracted shape description can be used in many applications, such as image understanding, pattern recognition, autonomous systems and shape-based image retrieval systems.

The earliest stage of visual processing in the Human Visual System (HVS) is located in the primary visual cortex in an area known as V1. V1 cells are biological orientation filters, meaning that they respond only to edges with certain orientation. There are 12-24 different groups of such cells in every point of the visual field, each one specialized on the detection of a particular orientation. A stimulated V1 cell carries, among others, two basic pieces of information; its location in the visual field and the orientation that detects. Every point of the visual field contains all 24 groups of V1 cells, in order to detect every possible edge direction. The role of these cells is of great importance for the perception of shape, since they are directional filters, identifying changes in intensity of any direction and at any point of the visual field. Cells from higher levels of the HVS, receive whole visual maps of stimulated V1 cells, and identify certain combinations among them, resulting to the gradual perception of more complex forms.

The algorithm

Similarly to the HVS, the proposed shape descriptor uses 12 groups of artificial V1 cells each one specialized in detecting any edge with a particular orientation within its receptive field. Every cell is a 10×10 kernel containing an oriented straight line segment. Every group comprises 5 cells with the same orientation and different position of the oriented segment, resulting to a set of 60 artificial cells (Table 1). The orientations of the 12 groups are 0°, 15°, 30°, 45°, 60°, 75°, 90°, 105°, 120°, 135°, 150° and 165°. Every cell has a slight tolerance, meaning that can be also stimulated by edges with orientations differing by some degrees from its main orientation. The main stages of the algorithm are:

1. The image is divided to 10×10 non-overlapping regions. In each region all 60 cells are tested to find the one with the higher stimulation. The stimulation process is similar to the biological V1 cells and counts the number of edge pixels of the particular region that fall only inside the straight segment of each cell. If not appropriate stimulation is found in any cell, no edges are contained in the particular region. The whole process aims in finding which cell (orientation

filter) describes better the edges of every image region, and thus, knowing the inclination of the particular edge at that particular section of the image. Previous techniques convolved series of orientation filters with the image, which was computationally intensive. The proposed approach does not use convolution, since not all pixels of the kernel are used in the calculation of its stimulation (only the pixels of the straight line segment) and the kernels are not shifted to every possible position in the image (only to non-overlapping regions). This process demands less computational effort and higher parallelism due to the fact that there are no dependences in the analysis of every region.

2. The main objective of the shape descriptor is to record the relative angles that are formed from the orientations of the stimulated kernels moving in a clockwise fashion on the contour of the shape. In order to do so, the algorithm searches for a kernel, which will be the starting point. It then moves clockwise and calculates the angle that is formed between the orientation of the current kernel and the orientation of the following one. The process continuous until a loop has been completed and the starting kernel is reencountered.

3. The proposed shape description is a chain of the angles formed by the orientations of the stimulated kernels of a certain shape, when moving clockwise. Its novelty is that no proportions in straight line segments are recorded, but only relative angles. This is an important advantage for many image understanding algorithms, because it provides scale and rotation invariance. Additionally, an appropriate normalization of the chain provides interpretation invariance (invariance of the starting kernel). A special metric, examining the dissimilarity of two descriptions, provides an appropriate measure for the comparison of two shapes. For the purposes of shape-based image retrieval, a comparison rule is employed, based on the aforementioned metric, which provides a crisp decision on whether two shapes are similar. Generally, if the result of the metric is less than the length of the chain, then the two shapes match.

Advantages

The method was extensively tested with binary edge images up to 1,300×1,300 pixels, containing various shapes, at different scales and rotations. Experimental results confirm that the method achieves scale, rotation and interpretation invariance. The fact that no proportions are recorded, and only angles, makes the shape descriptor quite flexible for the retrieval of different shapes which are extracted from the same object. Particularly, the method is capable of identifying as similar, different shapes of the same object. This is depicted in Fig. 1, where all shapes have dissimilarity less than 7 (which is the length of their description) and for that reason they are classified by the comparison rule as being similar. Furthermore, for the same reasons, stretch distortions on both axes of the shape, do not almost affect the relative angles and result to small changes in dissimilarity metric, thus not affecting the final result. Another advantage that derives from the form of the shape description is the fact that the proposed shape descriptor can search for certain patterns in the contour of a shape or partial similarities, rather than searching for a match with the whole contour. An important capability of the new shape descriptor is its ability not to be affected by noise, and additionally to reconstruct the original noisy image to a new filtered one. Noisy images with up to 20% spike noise can be handled successfully. The reconstruction can be achieved by tiling the stimulated kernels of every image region, making them part of the actual image and not useful only for the computation of edge orientations. The computational demands of the proposed method are quite affordable considering its advantages. Implemented in C code and executed by an Intel Celeron Processor, running at 1 GHz, the typical execution time for an image of 700×700 pixels, is approximately 0.5 seconds.

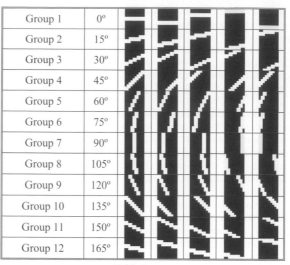

Group 1	0°	
Group 2	15°	
Group 3	30°	
Group 4	45°	
Group 5	60°	
Group 6	75°	
Group 7	90°	
Group 8	105°	
Group 9	120°	
Group 10	135°	
Group 11	150°	
Group 12	165°	

Table 1

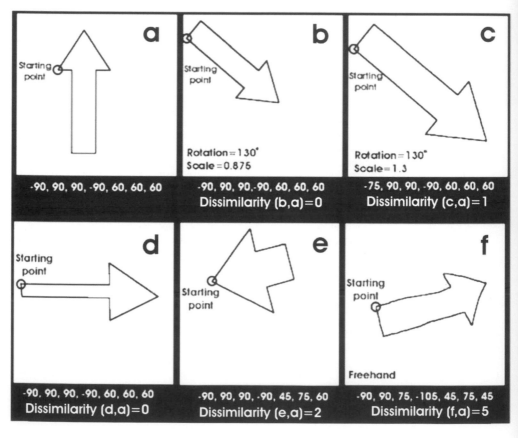

Fig. 1

VSP International
Science Publishers
P.O. Box 346, 3700 AH Zeist
The Netherlands

*Lecture Series on Computer
and Computational Sciences*
Volume 1, 2004, pp. 569-572

Isotropic Periodical Sum: An Efficient and Accurate Approach to the Calculation of Long-Range Interactions

Xiongwu Wu[1] and Bernard R. Brooks

Laboratory of Biophysical Chemistry,
National Heart, Lung, and Blood Institute,
National Institute of Health,
Bethesda, Maryland, USA

Received 6 August, 2004; accepted in revised form 14 August, 2004

Abstract: This work proposes an approach to calculate long-range interactions for molecular modeling and simulation. Based on the isotropic character of homogenous systems, for each particle this method describes a molecular system as a local region around the particle and many images of the local region distributed on shells around the particle. The isotropic and periodic distribution of these images makes the summation of long-range interactions over all images an easy task, which we call the isotropic periodic sum (IPS). Analytic solutions of IPS for electrostatic and Lennard-Jones potentials have been worked out. The same approach can be applied to potentials of any functional form and to any type of periodic systems. Using an argon fluid and a $CaCl_2$ ionic system, we demonstrate that the IPS method gives very close results to lattice sum. For macromolecular systems, this method is better than lattice sum by avoiding the symmetry imposed from periodic boundary conditions. For non-periodic systems, IPS is better than cutoff methods if the system size is large.

Keywords: Isotropic periodic sum, Long-range interaction, periodic boundary condition, electrostatic energy, Lennard-Jones potential, Lattice sum, Ewald summation.

1. Introductions

Molecular simulation has been widely used in the study of many-particle systems. Long-rang interactions, such as electrostatic and van der waals interactions, are the most costly part. Electrostatic interaction is especially troublesome because its range reaches far beyond the size of a typical simulation system. There are many approaches for long range interaction calculation[1], such as cutoff, minimum image, reaction field, and lattice sum. Cutoff or minimum image methods cannot provide accurate description of long-range interactions. The reaction field approach needs an *a priori* knowledge of dielectric constants of the simulation system. Lattice sum uses periodic images to calculate long-range contribution, which not only is time consuming, but also imposes symmetries of periodic boundary conditions into a homogenous system.

In this work, we propose an approach called isotropic periodic sum (IPS) to calculate long-range interactions. The underlying concept is based on the homogenous behavior of molecular systems. On the contrast to lattice sum where the long-range contribution is from the images generated from a periodic boundary condition, the isotropic periodic sum calculates the long-range contributions from images isotropically distributed around each particle. This isotropic distribution of images makes the summation an easy task to be solved mathematically. In the following sections, we will first explain the concept of this method, then we present comparison of this method with cutoff and Ewald summation methods, and finally we show the simulation results for example systems.

---

[1] Corresponding author. E-mail: wuxw@nhlbi.nih.gov

2. Theories and Methods

For a homologous system consists of many particles, there is no structural preference at any direction. Two regions far away have the possibility to be similar to each other. For any particle, assuming any region far away from it be an image of it local region is a reasonable approximation. The widely used periodic boundary conditions are a special case of this approximation, which assumes the regions around each lattice points are images of the local region. Lattice sum calculates long-range contributions by summing interactions over all these lattice images. The anisotropically distributed images make lattice sum a difficult computational task, and even worse, overemphasize lattice symmetry to homogenous systems. Instead of using the images created by periodic boundary conditions, we use isotropically distributed images of a local region to calculate long-range interactions as shown in Fig.1. For each particle, the near region around it is defined as its local region, and the rest of the system is represented by a distribution of images of this local region. Based on periodicity, the images are distributed on many spherical shells around the particle, and images of different shells are bounded by each other. It should be noted that these isotropic periodic images are used only for long-range interaction calculation, not actually present in the system.

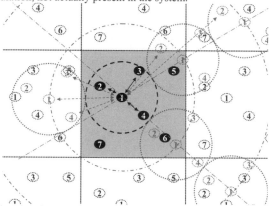

Figure 1: The local region (black circle) of particle 1 and its isotropic periodic image regions (red circles). The system is surrounded by images created by a square periodic boundary condition. Particle 1 interacts only with the particles (black) in its local region, 2, 3, and 4, and all particles (red) in its isotropic periodic image regions.

For any two particles A and B within the local region cutoff distance. the long range interaction is the sum of the interaction between A and B and between them and all their images:

$$\varepsilon'(r_{AB}) = \varepsilon(r_{AB}) + \frac{1}{2}\sum_m \left(\varepsilon(r_{AB_m}) + \varepsilon(r_{A_mB})\right) = \varepsilon(r_{AB}) + \phi(r_{AB}) \tag{1}$$

Here, A_m and B_m are the *mth* image of particle A and B respectively and $\phi(r_{AB})$ is the image contribution to the A-B interaction, which we calculated as the isotropic periodic sum (IPS) of $\varepsilon(r_{AB})$. The isotropism of the image distribution makes $\phi(r_{AB})$ easy to be analytically solved. For electrostatic interaction of the form:

$$\varepsilon(r) = \frac{e_1 e_2}{r} \tag{2}$$

We have a very neat analytic expression of the IPS:

$$\phi(r) = -\frac{e_1 e_2}{2r_c}\left(2\gamma + \psi(1 - \frac{r}{2r_c}) + \psi(1 + \frac{r}{2r_c})\right) \tag{3}$$

where, $\gamma = \lim_{m\to\infty}\left(\sum_{k=1}^m \frac{1}{k} - \log m\right) \approx 0.577216$, $\psi(z) = \frac{\Gamma'(z)}{\Gamma(z)}$, $\Gamma(z) = \int_0^\infty t^{z-1}e^{-t}dt$, and r_c is the radius of the local region, or the cutoff distance. The analytic expression of IPS is also derived for Lennard-Jones potential. The IPS method has been extended to 2-D homogenous systems like membranes and 1-D homogenous systems like DNAs.

3. Results and Discussions

The isotropic periodic sum (IPS) method introduces a distance-dependent long-range term to particle interactions. Like other cutoff methods, IPS only considers particle pairs within certain distances. Therefore, it can be calculated the as efficient as the cutoff methods while takes into account long range contributions. There are many ways to perform cutoff energy calculation, typically, energy switch, energy shift, force switch, and force shift[2]. Using electrostatic potential as an example, we calculated the energies and forces of a pair of atoms as a function of their distance using different methods (Fig.2). Obviously, these cutoff based methods significantly changed the energy and force results. The IPS results are very close to the Ewald summations at cubic and truncated octahedral periodic boundary conditions, which are calculated from the isotropic approximation functions fitted by Adams and Dubey[3].

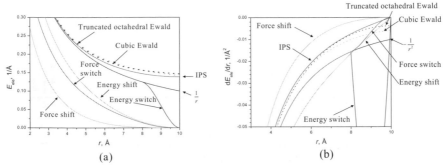

Figure 2: Electrostatic energies (a) and forces (b) between a pair of particles calculated from different methods. The cutoff distances for Energy switch, energy shift, force switch, and force shift methods or the radius of local region for the IPS method are $r_c = 10$ Å. The Ewald results are taken from Adams and Dubey[3] with box size set to have the same particle density as the other methods: cubic:

$$L = \sqrt[3]{\frac{4}{3}\pi r_c^3} , \text{ truncated octahedral: } L = \sqrt[3]{\frac{8}{3}\pi r_c^3}$$

Using a $CaCl_2$ ionic system, we examined the electrostatic and Lennard-Jones energies at various densities under 3-D, 2-D, and 1-D periodic sceneries, and found that the IPS produces very close results to the Ewald summation.

For non-periodic systems, we demonstrated that the IPS method produces more accurate long-range interaction than the cutoff methods if the cutoff distance (radius of local region) is small or the system is large. This result indicates that it is better to assume the region beyond cutoff is filled with image regions than to assume there is nothing beyond the cutoff distance.

For macromolecular system in periodic systems, our calculation shows Ewald summation exaggerated the orientational dependence of macromolecules, while IPS can well suppress the symmetry effect imposed by a periodic boundary condition.

We applied IPS to *NPT* simulations of an argon fluid to evaluate the description of the long range VDW interaction. Fig.3 (a) shows the average densities and potential energies at different cutoff distances using different methods. As can be seen, using the cutoff method the density decreases as the cutoff distance becomes shorter. With the long range correction (LRC)[1], the dependence of the density on cutoff distances is much smaller but is still significant. Using the IPS method, the density is almost the same for all the cutoff distances. Similar behaviors are also found for the potential energies. These results clear demonstrate that IPS can better describe long-range VDW interactions with short ranged cutoffs.

To evaluate the description of charged systems, we performed *NVT* simulations of a highly charged ionic $CaCl_2$ fluid. Fig.3 (b) shows the L-J energy and the electrostatic energy averaged over the simulations using cutoff method, Ewald method, and IPS method. As can be seen, the average L-J energy from IPS calculation is almost constant when cutoff distance is 7 Å or up, and the average electrostatic energies from IPS at different cutoff distances are very close to the Ewald summation. Clearly, IPS can reproduce Ewald summation results while is as efficient as the cutoff methods.

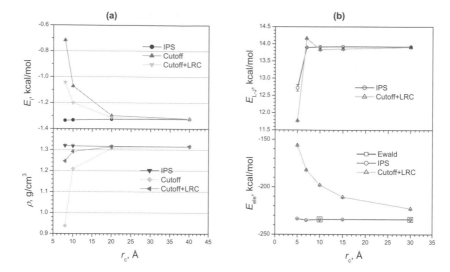

Figure 3: Comparison of simulation results with different energy calculation methods. (a). *NPT* simulations of an argon fluid. The Lennard-Jones parameters of argon atoms are ε=-0.238 kcal/mol and σ=3.405Å. 9450 argon atoms are simulated in a cubic periodic boundary condition at 1 atom and 100 K. (b) *NVT* simulation of a $CaCl_2$ ionic system. Ions interact through Lennard-Jones and electrostatic potentials. 2048 Ca^{2+} and 4096 Cl^- ions are simulated at 10,000 K in a 60×60×60 $Å^3$ cubic periodic box.

4. Conclusions

Isotropic periodic sum (IPS) is proposed for calculating long-range interaction of any potential forms and under any periodic boundary condition. For homogenous systems, it has been demonstrated that IPS produces very close result to Ewald summation. While for macromolecular systems, IPS is better than Ewald summation by avoiding the symmetry effect imposed from periodic boundary conditions. For non-periodic systems, IPS is more accurate than the cutoff methods if the system size is large or the cutoff distance is small. IPS is suitable for many fast computing techniques, like parallelization, multi-cutoff methods, etc. In summary, IPS is a better choice than lattice sum for calculating long-range interactions.

References

1. M.P.Allen and D.J.Tildesley, "Computer Simulations of Liquids", Clarendon Press, Oxford (1987).

2. P.J.Steinbach and B.R.Brooks, New Spherical-Cutoff Methods for Long-Range Forces in Macromolecular Simulation, *Journal of Computational Chemistry* **15** 667-683 (1994).

3. D.J.Adams and G.S.Dubey, Taming the Ewald Sum in the Computer Simulation of Charged Systems, *Journal of Computational Physics* **72** 156-176 (1987).

VSP International
Science Publishers
P.O. Box 346, 3700 AH Zeist
The Netherlands

Lecture Series on Computer
and Computational Sciences
Volume 1, 2004, pp. 573-577

Nonlinear Excitation and Dynamic Features
of Deoxyribonucleic ACID (DNA) Molecules

Pang Xiao-feng[(1)(2)] and Feng Yuan Ping[(3)]
(1)Institute of Life Sceince and Technplogy, University of Electronic Science and Technology of
China, Chengdu 610065, P. R. China and [(2)] International Center for Materials Physics, Chinese
Academy of Sciences, Shenyang 110015 and [(3)]Department of Physics, National University of
Singapore, Singapore 119260, Singapore

Received 7 June, 2004; accepted in revised form 10 August, 2004

Abstract: According to the features of molecular structure conformation changes of DNA we
rebuild the dynamic model of nonlinear excitations in it. We stress the important role of the
hydrogen atom in the hydrogen bonds in DNA and take into account its motion. This model has
three degrees of freedom per base: two vibrational variables related to the vibration of the
hydrogen atom in the hydrogen bonds and base (nucleotide) , respectively , and an angular one
related to the rotation of each base. We give the Hamiltonian of the systems and found out some
solutions of the equations of motion. This model claims an energy-localization process which
is helpful to explain the small-amplitude expansion and " melting" of the hydrogen bonds and
DNA base-pairs open and local unwinding of the helixs.

Keywords: DNA , nonlinear excitation , dynamic feature , soliton , transcription , denaturation

PACS numbers: 87.10.+e, 63.20.Pw, 87.15.He, o5.90.+m, 66.90.+r

The structure of DNA and its biological function have been one of the most fascinating problem of
modern biophysics, much attentions to it have been paid up to now because it is at the basis of life.
The idea that nonlinear excitation could play a role in the dynamics of DNA has become increasingly
popular. Yomosa[1] proposed a soliton model of plane base-rotator that was further refined by Takino
and Homma[2]. Although the models can exhibit solitary wave solutions that could show open states,
they were not related with the thermal denaturation of DNA since temperature effect was not considered. In this model the base binds together with the hydrogen bond or hydrogen atoms to rotate only
likely a rigid rods, ignoring completely the individual peculiarity of the hydrogen bond or hydrogen
atom and changes of states of the base. Prohofsky et al[3]pointed out the essential role of strong nonlinearities and treated the dynamics of DNA in a vibrational soliton model, devoting one's attention to
hydrogen bond stretching modes. Peyrard and Bishop[4] (PB model) employ a transfer integral way to
analyze the statistical mechanics of the model and determine the interstrand separation in the double
helix as a function of temperature. Their model allows a local "melting" of the hydrogen bonds and
formation of denaturation bubbles. The local "melting" can be analytically described as breather-like
objects of small amplitude, they can move along the chain, collect energy and grow. However the bases
was thought to be a point masses, or decanting only the total dynamic features, e.g., the rotation of the
bases containing hydrogen bonds. Recent development [5] has introduced other freedoms in the more
sophisticated case, such as the helicoidal structure. Complete idea is to take into account both vibration
of hydrogen atom or hydrogen bond and the rotation of base-pairs (A-T and C-G) because the former
was verified in the experiments of exchange between the hydrogen and deuterium or other ions in the
solutions and infrared absorption of DNA, the latter was demonstrated in duplication and transcription
of DNA in which there is always open state. Thus we proposed a new dynamic model of the nonlinear
excitations in DNA in which we introduce three variable (three degrees of freedom) per base, displacement the hydrogen atom, u_n , describing its vibrations, the displacement of the base (nucleotide), R_n,
denoting its harmonic state and the rotation angle of the base, φ_n, exhibiting its rotation motion. The
Hamiltonian of DNA systems can be represent by[6]

$$H = H_h(u_n) + H_t(\varphi_n) + H_b(R_n) + H_{int} = \sum_n [\frac{1}{2}m\dot{u}_n^2 + \frac{1}{2}m\omega_0^2 u_n^2 + V(u_n) - J u_n u_{n-1}] +$$

$$+ \sum_n \{\frac{1}{2}I\dot{\varphi}_n^2 + \beta_1[3(1 - \cos\varphi_n \cos\varphi_{n-1}) - (1 - \cos(\varphi_n - \varphi_{n-1}))] +$$

$$B[1 - \cos(\varphi_n - \varphi_{n-1})] + \lambda[(1 - \cos\varphi_n) + (1 - \cos\varphi_{n-1})]\}$$

$$\sum_n [\frac{1}{2}M\dot{R}_n^2 + \frac{1}{2}W(R_n - R_{n-1})^2] + \sum_n [m\chi_1 u_n^2 (R_n - R_{n-1})] \tag{1}$$

here u_n is displacement of the hydrogen atom in the hydrogen bond, $p_n = m\dot{u}_n$ is its conjugate momentum, m is mass of the hydrogen atom, ω_0 is its frequency of harmonic vibration, $V(u_n)$ is nonlinear potential the hydrogen atom lie in , provided by complementary bases in nth base -pair, J is a constant of interaction between neighboring hydrogen atoms, including dipole –dipole interactions[12-15]. J can be evaluated and denoted by $J = q'^2/4\varepsilon_o\pi r'^3$, q' is the charge transfer due to the displacement of the hydrogen atom, r' is the distance between neighboring hydrogen atoms, ε_o is the dielectric constant. R_n

and $P_n = M\dot{R}_n$ are displacement and conjugate momentum of the base at nth site, respectively, M is its mass, W is spring constant of the sugar- phosphate backbone. I is the moments of inertia of the bases for the rotation around the axe, which pass through the point where the base B_n attach to the strand and is parallel to the z axis. The direction of base B_n in horizontal plane is specified by rotation angle φ_n around the axe, $\beta_1[3(1 - \cos\varphi_n \cos\varphi_{n-1}) - (1 - \cos(\varphi_n - \varphi_{n-1}))]$ denotes the permanent dipole-dipole interaction between two neighboring bases, it is obtained from $[\mathbf{u_1}.\mathbf{u_2} - 3(\mathbf{u_1}. \mathbf{r})(\mathbf{u_2}.\mathbf{r})/r^2]/r^3$, where $\mathbf{u_1}$ and $\mathbf{u_2}$ are the dipole moments of the transversally neighboring bases G and A , or C and T, which are about 5.755-6.44 Bebyes for $\mathbf{u_A}$ and $\mathbf{u_T}$ for A-T base pair and 6.004-6.483 Bebyes for $\mathbf{u_G}$ and $\mathbf{u_C}$ for G-C base pair, respectively, r is the distance between them, the interaction constant β_1 can be represented by $\beta_1 = u_1 u_2/r^3$, $B[1-\cos(\varphi_n - \varphi_{n-1}))$ is the sum of stacking energy between intrastrand adjacent bases and torsional energy of the nucleotide strand[2] which are both functions of the relative torsional angles between adjacent bases and are given in the same functional form, B is interaction constant, $\lambda[(1 - \cos\varphi_n) - (1 - \cos\varphi_{n-1})]$ is inducting dipole-dipole interaction between transversally neighboring bases, λ is a coupling constant. χ_1 is the coupling constant. From Eq.(1) we can find equations of motion in continuum approximation for $u_n(t), R_n(t)$ and $\varphi_n(t)$ as follows

$$m\ddot{u}(z,t) = (2J - m\omega_0^2)u + Jr_0^2\frac{\partial^2 u}{\partial z^2} - \frac{\partial V(u)}{\partial u} - 2mr_0\chi_1 u \frac{\partial R}{\partial z} \tag{2}$$

$$M\ddot{R}(z,t) = r_0^2 W \frac{\partial^2 R}{\partial z^2} - r_0 m\chi_1 \frac{\partial R^2}{\partial z} \tag{3}$$

$$I\ddot{\varphi}(z,t) = r_0^2 (B - \beta_1)\frac{\partial^2\varphi}{\partial z^2} - 3\beta_1 \sin 2\varphi - 2(\lambda_0 + r_0\chi_2 \frac{\partial R}{\partial z})\sin\varphi \tag{4}$$

....

where $\lambda = \lambda_0 + \chi_2(R_n - R_{n-1})$. Let $\xi = z - vt$, from Eq.(3) we can get

$$\frac{dR}{d\xi} = \frac{R}{z} = -\frac{mr_0\chi_1}{Mv_0^2(1 - s^2)}u^2 + g \tag{5}$$

where $s = v/v_0$, $v_0 = r_0(W/M)^{1/2}$, r_0 is distance between neighboring base-pairs at z direction, g is integral constant. We choose $V(u)$ in Eq.(2) as double Morse–potential with two different well-depth, $V(u) = U_1[1- \exp(-\alpha u_n)]^2 + U_2\{1- \exp[-\alpha(u_n -b)]\}^2$ to simulate the potential acted on the Hydrogen atom in the hydrogen bonds, where U_1 and $U_2 < U_1$ are the depths of the two wells at u=0 and u=b (b>0), respectively. Inserting Eq.(5) into Eqs.(2)-(5) and expanding V(u) to quartic power we get

$$\ddot{u}(z,t) = A_1 + C_1 u + B_1 u^2 + v_1^2 \frac{\partial^2 u}{\partial z^2} + D_1 u^3 \tag{6}$$

$$I \ddot{\varphi}(z,t) = r_0^2 (B - \beta_1) \frac{\partial^2 \varphi}{\partial z^2} - 3\beta_1 \sin 2\varphi - 2(gr_0 \chi_1 + \lambda_0) \sin \varphi + N_0 u^2 \sin \varphi \tag{7}$$

where $A_1 = bU_2 \alpha^2 (6\alpha b - 2 - \alpha^2 b^2)/m$, $D_1 = [(\alpha^4 (U_1 + U_2) - D_0)/m$, $v_1 = r_0 (J/m)^{1/2}$,

$B_1 = \alpha^3 (U_2 - U_1 - U_2 b\alpha)/m$, $C_1 = \{[U_1 + U_2 (1 - 3b\alpha + 3b^2\alpha^2/2)] - C_0\}/m$,

$D_0 = 2m^2 r_0^2 \chi_1^2 / Mv_0^2 (1-s^2)$, $C_0 = (m\omega_0^2 - 2J + 2gm\chi_1)$, $N_0 = m\chi_1 \chi_2 r_0^2 / Mv_0^2 (1-s^2)$,

The soliton-like solution of Eq.(6) can be represented by

$$u(z,t) = \frac{(u_j - u_i)}{\{1 + \exp[(k\gamma \sqrt{D_1}/v_0)(z - ut - z_0)]\} + u_i} \tag{8}$$

here $k = (\varepsilon'/\sqrt{2})(u_i - u_j)$, $v = \varepsilon'/2(u_i + u_j - 2u_k)$, $\varepsilon' = \pm 1$,, $u_3 \neq 0$,, I.j.k = 1,2,3, but

$u_i \neq u_j$, u_k are three solution of u in $F(u) = (u - u_1)(u - u_2)(u - u_3)$, z_0 is constant. Thus

$$R(z,t) = [\frac{mr_0 \chi_1 u_0^2}{Mv_0^2 (1-s^2)k}]\{(\gamma \sqrt{D_1}/v_0)(z - vt - z_0') - \tanh[(\gamma \sqrt{D_1} k/v_0)(z - vt - z_0)] - \tag{9}$$

$$.........................Ln \cosh^2[(\gamma \sqrt{D_1} k/v_0)(z - vt - z_0)]\}$$

Then the soliton-like solution of Eq.(7) can be approximately expressed in the following form

$$\varphi(Z,t) = 2 \tan^{-1}\{\pm(1-4\rho)^{-1/2} \cos ech[(1+4\rho)^{1/2} Z]\} \tag{10}$$

where + and – corresponding to the soliton and antisoliton, respectively.

$$Z = (\gamma_1/q)(z - ct), ..q^2 = r_0^2 (B - \beta_1)/(gr_0 \chi_1 + \lambda_0 - N_0 u_0^2/2), ..\gamma_1 = (1 - c^2/c_0^2),.$$

$$c = [(B - \beta_1)/I]^{1/2} r_0, ...\rho = 3\beta_1 /[4gr_0 \chi_1 + 4\lambda_0 - 2N_0 u_0^2].$$

These soliton-like solutions are helpful to explain denaturation, duplication and transcription of DNA, for example , " melting" of the hydrogen bonds and DNA base-pairs open and local unwinding of the helix , which are related to temperature of the system, T. Thus we write effective Hamiltonian of the hydrogen atoms corresponding to Eq.(6) as

$$H_{eff} = \frac{1}{r_0} \int dz m[\frac{1}{2} u_t^2 + \frac{1}{2} v_1^2 u_z^2 + \frac{D_1}{4} u^4 - \frac{C_1}{2} u^2] \tag{11}$$

The discrete form corresponding it can also be written. Using statistical formulae , $Z(\beta, L) = \int dp \int due^{-\beta H_{eff}} = Z_p Z_u$, .here.. $\beta = 1/K_B T$, L is length of the system, K_B is Boltzmann constant, p(z,t)= $mu_t(z,t)$, and transfer integral way and statistical physics [7-9], we find out free energy per particle and specific heat of DNA as follows

$$f = -\frac{1}{2} K_B T \{Ln[2^2 \pi^2 (K_B T)^2 / v_1^2] + Ln(2q_1 \cosh g_0)\} \tag{11}$$

$$C_v/K_B = 1 + q_3 q_4 [2 + \frac{1}{2} \beta q_3] e^{-\beta q_3} \tanh g_0 + \frac{1}{2} \beta^2 g_0^2 q_3^2 \sec h^2 g_0 \tag{12}$$

$$q_1 = \exp[-4q_4], g_0 = q_4 \exp[-\beta q_3], q_4 = r_0 \sqrt{2C_2}/4v_1, q_3 = \frac{w_0}{r_0}(m^2 v_1^2 C_2^2 / 2D_2)^{1/2},$$

the specific heat can be approximately represented by

$$c_v = -\frac{\partial e}{\partial T} = K_B + \frac{8MC_2^2}{15}(K_B T) + O[(K_B T)^2], \text{ which concords to experimental data}[10].$$

From the definition[3] $<u^2> = \dfrac{1}{m}\dfrac{\partial f_u}{\partial C_2}$ of average square-displacement of the hydro-gen atom, which

designates the character of melting of DNA. we can find

$$<u^2> = (K_B T / 2mC_2)\ln(mv_1 r_0 \sqrt{C_2} / 2\pi K_B T) +$$

$$\frac{C_2}{4D_2} + \frac{r_0 K_B T}{\sqrt{2C_2 mv_1}} - \frac{3}{16}\frac{(K_B T)^2}{r_0^2 m^2 D_2 v_1^2} - \frac{17 r_0^3 D_2^2 (K_B T)^3}{128\sqrt{2} m^3 v_1^3 C_2^{7/2}} - \frac{4C_2^{1/2}(K_B T)^{1/2}}{(\sqrt{2}\pi v_1^2 D_2 m)^{1/2}}\exp[-\frac{2mv_1 C_2^{3/2}}{3D_2 r_0 K_B T}]$$

The values of physical parameters for DNA use usually[1-6] $m = m_p$, $M = 300m_p$, $J = (0.11\ -0.25)$ ev/ $\overset{0}{A}{}^2$,

$u_0 = (3.4 - 6)\ \overset{0}{A}$, $\omega_0 = (10 - 300)cm^{-1}$, $U_1 \sim U_2 = (0.15-0.33)$ ev, $W = (0.1\ 1- 0.31)ev/\overset{0}{A}{}^2$, $r_0 = 1\overset{0}{A}$ and

$\chi_1 = (1-3) \times 10^{37}\ (S^2\ m)^{-1}$, thus we can find out that $<u^2>$ increases gradually with increasing of $K_B T$.

Therefore the melting of DNA is loca- lized and extended gradually, and $<u^2> = 0.0442\ \overset{0}{A}{}^2$ at biolo-

gical temperature T =300 K, and $<u^2> = 0.0447\ \overset{0}{A}{}^2$ at T=310K. This show that in normal biological condition DNA could melt partially and go on normal transcription. The critical value of $<u^2>$ of a hydrogen bond[3], which designates beginning of melting or transcription of DNA, is chosen as $<u^2> =$

$0.040\ \overset{0}{A}{}^2$. Thus we can find down-critical temperature is $T_{down} = 285K$ in our model. If assuming that

at $<u^2>_{cr} = 0.06\ \overset{0}{A}{}^2$ all base- pairs are melted, then we can find up-critical temperature is $T_{up} \cong 400K$.

Acknowledgments

*One of authors, Pang Xiao-feng, would like to acknowledge National Natural Science Foundation of China (grant No:19974034) .

References

[1], S.Yomosa,Phys.Rev.A27(1983)2120;30(1984)474; J.Phys.Soc.JPN 64(1995)1917;62(1993) 2186;62(1993)1075;59(1990)3765:

[2],S.Takeno and S.Homma, Prog.Theor.Phys.70(1983)30877(1987)548; 59(1990)1890; S. Homma,physica D114(1998)202; J.Biol. Phys. 24(1999)115;Phys. Lett.A133(1988)275

[3]E.W.Prohofsky, Statistical mechanics and stability of macromolecules, Cambridge Univ. Press,Cambridge, 1995; Phys.Rev.A38(1988)1538;Comments Mol.Cell.Biophys.2(1983)65; E.W.Prohofsky,K.C.Lu,L.Van Zandt and B.E.Putnam, Phys.Lett.70A(1979)492

[4]M.Peyrard and A.R.Bishop, Phys.Rev.Lett.62(1989)2755

[5], M.Barbi, S.Cocco and M.Peyrard, Phys. Lett.A253(1999)358;S.Cocco, M.Barbi and M.Peyrard, Phys. Lett.A253(1999)161; M.Barbi, S.Cocco, M.Peyrard and S.Ruffo, J.Biol. Phys.24(1999)97;S.Cocco and R.Monasson, Phys.Rev.Lett.83(1999)5178

[6]Pang Xiao-feng, Chin. J. Atom. Mol. Phys. 19(2002)417; Physics, Prog. Phys. (Chinese) 22(2002)218

[7],S.W.Englander and J.J.Englander, Proc.Natl.Acad.Sci.USA,53(1965)370

[8],J.J.Englander and P.H.Von.Hippel, J.Mod.Bio.63(1972)171

[9],S.W.Englander and J.J.Englander, methods Enzymol.49(1978)24

[10],G.M.Mrevlislivil, Sov. Phys. USP.22(1979)433;

VSP International
Science Publishers
P.O. Box 346, 3700 AH Zeist
The Netherlands

Lecture Series on Computer
and Computational Sciences
Volume 1, 2004, pp. 578-581

Influences of Structure Disorders in Protein Molecules on the Behaviors of Soliton Transferring Bio-Energy

Pang Xiao-feng[1][2], and Yu Jia-feng[1] and Lao Yu-hui[1]

[1]Institute of High-Energy Electronics, University of Electronic Science and Technology of China, Chengdu 610065, P. R. China and [2]International Center for Materials Physics, Chinese Academy of Sciences, Shenyang 110015, P.R.China

Received 11 July, 2004; accepted in revised form 8 August, 2004

Abstract: Collective effects of the structure disorder of the protein molecules, containing inhomogeneous distribution of masses for the amino acids and fluctuations of the spring constant, of dipole-dipole interaction constant and of exciton-phonon coupling constant and diagonal disorder, resulting from nonuniformities and aperriodicities, on the soliton in the improved model have been numerically simulated. The results obtained shows that the structure disorders can change the states of the solitons, but the soliton is quite robust against these disorder effects, it is only dispersed or disrupted at larger structure disorders. According to properties of structure of normal proteins we can conclude from these results that the soliton in the improved model is stable, it is possibly a carrier of bio-energy transport in the protein molecules.

Keywords: Protein molecule , structure disorder , collective effect , siliton , bioenergy transport

PACS numbers: 87. 22As, disorder 03. 65-w, 05. 40+j, 71.38+j

We know that Bio-energy transport is a fundamental process in the biology, a lot of biological phenomena, for example, muscle contraction, DNA replication, nervous information transfer along cell membrane and work of sodium and calcium pumps, are associated with it, where the energy is released by hydrolysis of adenosin triphosphate. However understanding the mechanism of the transport is a long-standing problem that remains of great interest up to now. Following Davydov's idea[1], proposed first by Davydov in the 1970s[1],one can take into account the coupling between the amide-I vibrational quantum (exciton) and the deformation of amino acid lattice. Through the coupling, nonlinear interaction appears in the motion of the vibrational quanta, which could lead to a self-trapped state of the vibrational quantum, thus a soliton occurs in the systems[1]. It can move together with deformational lattice over macroscopic distances along the molecular chains retaining the wave shape and energy and momentum and other properties of quasiparticle. The Davydov model have been extensively studied by many scientists from the 1970s[2]. However, characteristic of Davydov's solitonlike quasiparticles occur only at low temperatures about T<10K, the soliton's lifetime is too small (about 10^{-12}-10^{-13}Sec.) to be useful in the biological processes. This shows clearly that the Davydov soliton is not a true wave function of the systems. We[3] improve and extend the Davydov model, in which new coupling interaction between the acoustic phonon and amide-I vibrational modes, was added and the one-quantum (exciton) state in the Davydov's wave function by a quasi-coherent two-quantum state. They was represented by[3]

$$| \Phi(t) >=| a(t) >| \beta(t) >= \frac{1}{\lambda} \left[1 + \sum_n a_n(t)B_n^+ + \frac{1}{2!}(\sum_n a_n(t)B_n^+)^2 \right] | 0 >_{ex} .$$

$$\exp\left\{-\frac{i}{\hbar}\sum_n [q_n(t)P_n - \pi_n(t)u_n]\right\} | 0 >_{ph} \qquad (1)$$

$$H = H_{ex} + H_{ph} + H_{int} = \sum_n \left[\varepsilon_0 B_n^+ B_n - J(B_n^+ B_{n+1} + B_n B_{n+1}^+) \right] + \sum_n \left[\frac{P_n^2}{2M} + \frac{1}{2}W \cdot \right.$$

$$\left. (u_n - u_{n-1})^2 \right] + \sum_n [\chi_1 (u_{n+1} - u_{n-1}) B_n^+ B_n + \chi_2 (u_{n+1} - u_n)(B_{n+1}^+ B_n + B_n^+ B_{n+1})] \tag{2}$$

where B_n^+ and B_n are Boson creation and annihilation operators for the exciton, u_n and P_n are the displacement and momentum operators of lattice oscillator in site n, respectively. λ is a normalization constant, we here choose $\lambda = 1$. Where $\varepsilon_0 = \hbar \omega_0 = 1665 \text{cm}^{-1} = 0.2035 \text{ev}$ is excitation energy of an isolated amide-I oscillator or energy of the exciton (the C=0 stretching mode). Present non-linear coupling constants are χ_1 and χ_2, they represent the modulations of the one-site energy and resonant (or dipole-dipole) interaction energy for the excitons caused by the molecular displacements, respectively. M is the mass of an amino acid molecule, and W is the elasticity constant of the protein. J is the dipole-dipole interaction energy between neighboring sites. Average values of the parameters are

$\bar{J} = 0.967$ meV, $\bar{W} = 13$ N/m, $\bar{M} = 114 m_p$, $\bar{\varepsilon}_0 = \varepsilon_0$, $\bar{\chi}_1 = 62$ pN and $\bar{\chi}_2 = 10\text{-}15$ pN, respectively. From our Hamiltonian and wave function in Eqs. (1)-(2) we obtained that the new soliton is thermal stable, has so enough long lifetime (about $10^{-9}\text{-}10^{-10}$ ps) at biological temperature 300K, its critical temperature is 320K. This shows that the new soliton is possibly a carrier of bio-energy transport in the protein. However, the proteins, which consist of 20 different amino acid residues with molecular weights between 75 m_p (glycine) and 204 m_p (tryptophane), are not periodic, but nonuniform, in which there is structure disorder. The inhomogeneous distribution of masses for amino acid residues result necessarily in fluctuations of the spring constant, dipole-dipole interaction constant and exciton-phonon coupling constant and ground state energy or diagonal disorder of the protein molecule. Thus, it is very necessary to study influences of different structure disorders on the solitons. We now investigate this problem by fourth order Runge-Kutta way[4].

From Eqs. (1)-(2)and Schrodinger and Heisenbeger equations we can get the following simulation equations

$$\dot{ar}_n = -J(ai_{n+1} + ai_{n-1}) + \chi_1(q_{n+1} - q_{n-1})ai_n + \chi_2(q_{n+1} - q_n)(ai_{n+1} + ai_{n-1}) \tag{3}$$

$$-\hbar \, \dot{ai}_n = -J(ar_{n+1} + ar_{n-1}) + \chi_1(q_{n+1} - q_{n-1})ar_n + \chi_2(q_{n+1} - q_n)(ar_{n+1} + ar_{n-1}) \tag{4}$$

$$\dot{q}_n = \pi_n / M \tag{5}$$

$$\pi_n = W(q_{n+1} - 2q_n + q_{n-1}) + 2\chi_1(ar_{n+1}^2 + ai_{n+1}^2 - ar_{n-1}^2 - ai_{n-1}^2)$$
$$+ 4\chi_2[ar_n(ar_{n+1} - ar_{n-1}) + ai_n(ai_{n+1} - ai_{n-1})] \tag{6}$$

where $a_n(t) = a(t)r_n + ia(t)i_n$, $|a_n|^2 = |ar_n|^2 + |ai_n|^2$, $a(t)r_n$ and $a(t)i_n$ are real and imaginary parts of $a_n(t)$.
In the simulation by the fourth order Runge-Kutta way we should first make discrete treatment for the variables in above equations, the time be represented by j, the step size of the spacing variable is denoted by h. The system of units ev for energy, $\overset{0}{A}$ for length, and ps for time proved to be suitable for the numerical solutions with a time step size of 0.0195. Note that we here used fixed chains and at the size N, where N is the number of units, we choose to be N=200 or 50 .The initial excitations are $a_n(o) = A \text{sech}[(n-n_0)$ ($\chi_1 + \chi_2$)$^2/4JW$] (where A is normalization constant), $q_n(0) = \pi_n(0) = 0$. The simulation was performed by data parallel algorithm and MALAB langrage.

In the case of periodicity proteins. Using above average values for the physical parameters, numerically simulation shows that the solution can always retain basically constant to move without dispersion along the molecular chains in the range of spacings of 200 amino acid residues and the time of 50ps, i.e. Eqs.(3)-(6) have exactly soliton solution. In order to study influence of a random series of masses on the soliton, we introduce a parameter a_k, which is a random -number generator with equal probability within a prescribed interval and can denote the mass at each point in the chain, i. e., $M_k = a_k \bar{M}$. We show numerically that up to a larger intervals for a_k, for example, $0.67 \leq a_k \leq 300$, the stability of the soliton does not change, but in the case of the large intervals as $0.67 \leq \alpha_k \leq 700$, the vibrational energy is dispersed.

In general, the disorder of masses of amino acids will result in the changes of the spring constant W, dipole-dipole interaction constant J, and coupling constant($\chi_1 + \chi_2$) and ground state energy ε_0 which are represented by $\Delta W = W - \overline{W}$, $\Delta J = J - \overline{J}$, $\Delta(\chi_1 + \chi_2) = (\chi_1 + \chi_2) - (\overline{\chi}_1 + \overline{\chi}_2)$ and $\Delta\varepsilon_0 = \varepsilon_0 - \overline{\varepsilon}_0 = \varepsilon|\beta_n|$, respectively, here β_n is a random number generators to designate the random features of the ground state energy. We see that up to a random variation of $\pm 40\% \overline{W}$ or $\Delta J \leq 9\% \overline{J}$, or $\Delta(\chi_1 + \chi_2) < 25\%(\overline{\chi}_1 + \overline{\chi}_2)$, or $\Delta\varepsilon_0 = \varepsilon|\beta_n|, |\beta_n| \leq 0.5$, $\varepsilon < 1$meV, change in the dynamics of the new soliton occurs not, but for $\pm 50\% \overline{W}$, or $\Delta J = \pm 15\% \overline{J}$, or $\Delta(\chi_1 + \chi_2) = 35\%(\overline{\chi}_1 + \overline{\chi}_2)$, the soltion disperse somewhat. Therefore the soliton will disperse with increasing fluctuations of the physical parameters due to the structure disorder, but it is more sensitive to the variation in J, when compared with the other parameters, i. e., for variation in J alone the soliton is stable up to $\pm 9\% \overline{J}$, it disperse at change $\Delta J = \pm 15\% \overline{J}$. If we consider simultaneously effects of disorder of mass sequence and fluctuation of dipole-dipole interaction constant J, the states of the soliton change. For mass interval of $0.67 \leq \alpha_k \leq 2$ and changes $\Delta J = \pm 5\% \overline{J}$, $= \pm 10\% \overline{J}$, the soliton is stable, but it disperses at change $\Delta J = \pm 15\% \overline{J}$, it disperses significantly at $\Delta J = \pm 20\% \overline{J}$. For $0.67 \leq \alpha_k \leq 2$ and $\Delta J = \pm 10\% \overline{J}$, we find that up to $\pm 30\% \overline{W}$ the dynamics of the new soliton change not, but for $\pm 40\% \overline{W}$, the soltion disperses somewhat, its velocity is somewhat diminished. Finally, for $\pm 50\% \overline{W}$, the soliton's propagation is irregular. In $0.67 \leq \alpha_k \leq 2$, $\Delta W = \pm 30\% \overline{W}$ and $\Delta(\chi_1 + \chi_2) = 20\%(\overline{\chi}_1 + \overline{\chi}_2)$ the soliton is stable, at $\Delta W = \pm 70\% \overline{W}$ the soliton disperses gradually, but at $\Delta(\chi_1 + \chi_2) = 25\% (\overline{\chi}_1 + \overline{\chi}_2)$ and $\Delta W = \pm 50\% \overline{W}$, propagation of the soliton is irregular. When $0.67 \leq \alpha_k \leq 2$, $\Delta(\chi_1 + \chi_2) = 10\%(\overline{\chi}_1 + \overline{\chi}_2)$, $\Delta W = \pm 20\% \overline{W}$ and $\Delta J = \pm 2.5\% \overline{J}$, the soliton is stable, but it disperses at $\Delta J = \pm 10\% \overline{J}$, disperses obviously at $\Delta J = \pm 20\% \overline{J}$, destroys at $\Delta J = \pm 25\% \overline{J}$. In general, the soliton is very sensitive to the diagonal disorder which is just change of ground state energy, $\Delta\varepsilon_0$, caused by different amino acid side groups and corresponding local geometric distortions due to the impurities imported. In the case of a random sequence of the impurity, the soliton can pass the chain at $\Delta\varepsilon_0 = \varepsilon|\beta_n|, |\beta_n| \leq 0.5$, $\varepsilon < 1$meV. For higher values of ε the soliton is dispersed. When the disorder occurs together with the fluctuations of other four parameters, the states of the new soliton change. When $0.67 \leq \alpha_k \leq 2$ and $\Delta(\chi_1 + \chi_2) = 5\%(\overline{\chi}_1 + \overline{\chi}_2)$ and $\Delta J = \pm 5\% \overline{J}$ and $\Delta\varepsilon_0 = \varepsilon|\beta_n|, |\beta_n| \leq 0.5$, $\varepsilon < 1$meV, the soliton is stable. Five parameters vary simultaneously the maximal possible disorder that would still occur the soliton motion are $\Delta W = \pm 10\% \overline{W}$, $\Delta J = \pm 5\% \overline{J}$, $\Delta(\chi_1 + \chi_2) = \pm 5\%(\overline{\chi}_1 + \overline{\chi}_2)$, $0.67 \overline{M} \leq M \leq 2 \overline{M}$ and $\Delta\varepsilon_0 = \varepsilon|\beta_n|, |\beta_n| \leq 0.5$, $\varepsilon = 1$meV. In other cases the soliton disperses or is reflected. The soliton is reflected by the impurities at $\Delta W = \pm 20\% \overline{W}$, $\Delta J = \pm 5\% \overline{J}$, $\Delta(\chi_1 + \chi_2) = \pm 7\%(\overline{\chi}_1 + \overline{\chi}_2)$, $0.67 \overline{M} \leq M \leq 2 \overline{M}$ and $\Delta\varepsilon_0 = \varepsilon|\beta_n|, |\beta_n| \leq 0.5$, $\varepsilon = 1$meV, it is disrupted at $\Delta W = \pm 20\% \overline{W}$, $\Delta J = \pm 10\% \overline{J}$, $\Delta(\chi_1 + \chi_2) = \pm 9\%(\overline{\chi}_1 + \overline{\chi}_2)$, $0.67 \overline{M} < M < 2 \overline{M}$ and $\Delta\varepsilon_0 = \varepsilon|\beta_n|, |\beta_n| \leq 0.5$, $\varepsilon = 1$meV.

As known, protein molecules are a bio-self-organization with high-order. The order feature is a necessary condition for the protein to perform its biological functions. Any large disorder in the protein means the degeneration of structure and the disappearance of the functions of the protein molecules and the disease of the living systems to occur which belong not to the problems we here should discuss. Therefore, in normal conditions the structure disorders of the protein molecules are small. Meanwhile, we know that the natural amino acids are not free particles, but are covalently bound in the main polypeptide chains, hence the naturally occurring disorders for other parameters which are mainly due to small influences of the side groups on the geometry of the main chains, should be smaller although the natural disorder may interfere with the soliton only when the J and the ε_0 are varied. The above results show clearly that the new soliton is very stable and robust in the cases of larger structure disorders, at least, in the random disorder cases of $\Delta W = \pm 10\% \overline{W}$, $\Delta J = \pm 5 \overline{J}$, $\Delta(\chi_1 + \chi_2) = \pm 5\%(\overline{\chi}_1 + \overline{\chi}_2)$, $0.67 \overline{M} \leq M \leq 2 \overline{M}$ and $\Delta \varepsilon_0 = \varepsilon |\beta_n|$, $|\beta_n| \leq 0.5$, $\varepsilon = 1\text{meV}$, which are enough great than the natural disorders of the protein molecules. Therefore, the soliton is a good carrier, the improved model proposed by us[24] is possibly a relevant and available mechanism for bio-energy transport in the protein molecules.

Acknowledgments

*One of authors, Pang Xiao-feng, would like to acknowledge National Natural Science Foundation of China (grant No:19974034) .

References

[1], A. S. Davydov, J. Theor. Biol. 38, 559 (1973); Phys. Scr. 2, 387 (1979); Physica 3D, 1(1981); Sov. Phys. USP. 25 .898 (1982); Biology and quantum mechanics (Pergamon, New York, 1982); The solitons in molecular systems (Reidel, Dordrecht, 1985, 2nd, ed, 1991);

[2], P. L. Christiansen and A.C. Scott, Self-trapping of vibrational energy (Plenum Press, New York,1990)

[3], Pang Xiao-feng, Phys. Rev. E 62 (2000) 9846; 49, 4747 (1994);European Phys. J.19,297(2001); 10, 415(1999);Commun.Theor.Phys.35,323(2001);37(2002)187:J.Int.J.Infr.Mill.Waves,22,291(2001);J.Ph ys.Chem.Solids,62,793 (2001); Chin.J.BioMed.Engineering 8,39(1999);10,26(2001); Phys. Stat.Sol.(b)1397(2002) J. Phys. condensed matter 2, 9541 (1990); 12 (2000) 885; Physica D 154(2001)138; Chin. Phys. Lett. 10, 381 and 437 and 573 (1993); Chin. Sci.Bulletin38,1572 and 1665 (1993); Chin. J. Biophys. 9, 637 (1993), 10, 133 (1994); Acta Math. Sci. 13, 437 (1993) and 16(supp), 1(1996); Acta phys. Sinica 42, 1856 (1993); 46, 625 (1997); The theory for non linear quantum mechanics (Chinese Chongqing Press, Chongqing. 1994) p415—686, Acta Phys. Slovaca 47, 89 (1998); soliton physics; Chinese Sichuan Science and Technology Press, (Chengdu, 2002, P648-710).

VSP International
Science Publishers
P.O. Box 346, 3700 AH Zeist
The Netherlands

Lecture Series on Computer
and Computational Sciences
Volume 1, 2004, pp. 582-585

A Two-Dimensional Thin-Film Transistor Simulation Using Adaptive Computing Algorithm

Yiming Li[a,b,1], Cheng-Kai Chen[c], and Shao-Ming Yu[c]

[a]Department of Computational Nanoelectronics, National Nano Device Laboratories
[b]Microelectronics and Information Systems Research Center, National Chiao Tung University
[c]Department of Computer and Information Science, National Chiao Tung University

Received 31 July, 2004; accepted in revised form 31 August, 2004

Abstract: In this paper we simulate physical properties of thin-film transistor (TFT) using adaptive computing algorithm. Based on the finite volume approximation, the monotone iterative method, a posteriori error estimation, and the one-irregular meshing technique, a two-dimensional drift diffusion (DD) model of TFT is numerically solved. The grain boundary effect is considered and modelled when solving the TFT's DD model. For each decoupled partial differential equation, our numerical scheme converges monotonically and adaptively captures variations of the computed solutions. Various biasing conditions have been verified on the simulated device to demonstrate the accuracy of method. The proposed adaptive computing algorithm also shows very good performance in terms of the mesh refinements.

Received 31 July, 2004; accepted in revised form 28 August, 2004

Keywords: Thin-Film Transistor, Drift Diffusion Model, Grain Boundary Effect, Semiconductor Device Simulation, Adaptive Computing Algorithm, and Numerical Method

Mathematics Subject Classification: 65M60, 65M50, 65M12

PACS: 71.20.Mq, 73.61.Cw, 02.60.-x, 02.60.Cb, 02.60.Lj, 85.30.De, 85.30.Tv

1 Introduction

Development of the thin-film transistor (TFT) has recently been of great interest in modern micro- and opto-electronics industries [1]. It is known that the drift-diffusion (DD) model consisting of the Poisson equation and the electron-hole current continuity equation has successfully been applied to describe the carriers' transport phenomena. Compared with conventional metal-oxide-semiconductor field effect transistors (MOSFETs), TFT possesses significant grain structures in the substrate which lead to different impurities and affect the charge distribution as well as transport current. In this paper, using an adaptive computing technique we successfully extend our recent work to solve the DD model with grain boundary traps. Based on the error estimation, the gain boundary traps induced potential variation has been accurately calculated and tracked. Testing on various problems, the proposed computing technique shows the simulation accuracy and numerical robustness. This paper is organized as follows. In Sec. 2, we introduce the studied TFT model. In Sec. 3, we state the adaptive computing algorithm. In Sec. 4, we report and discuss the simulation results. In Sec. 5, we draw the conclusions.

---

[1]Corresponding author. Postal Address: P.O. BOX 25-178, Hsinchu 300, Taiwan. E-mail: ymli@mail.nctu.edu.tw

2   A Mathematical Model of Thin-Film Transistor

It is still widely employed in modern semiconductor device simulation [5]. By considering the grain boundary effects, the DD model for TFT simulation is

$$\Delta\phi = \frac{q}{\varepsilon_s}(n - p + D(x,y) + BT(n,p)) \tag{1}$$

$$D(x,y) = -(N_D^+(x,y) - N_A^-(x,y)) \tag{2}$$

$$BT(n,p) = \sum_{E_t}(N_{Dt} - nN_{Dt}) - \sum_{E_t}(N_{At} - pN_{At}) \tag{3}$$

$$\frac{1}{q}\nabla\cdot(-q\mu_n n\nabla\phi + qD_n\nabla_n) = R(n,p) \tag{4}$$

$$\frac{1}{q}\nabla\cdot(-q\mu_p p\nabla\phi + qD_p\nabla_p) = -R(n,p) \tag{5}$$

Eq. (1) is so-called the Poisson equation. The unknown $\phi = \phi(x,y)$, in Eq. (1), to be solved is the electrostatic potential, n and p are electrons and holes concentrations. Eq. (2) is the specified ionized net doping profile. Eq. (3) is the grain boundary concentration distribution function which occurs on the grain boundary. It sums all trap energy levels E_t, and N_{Dt} is the donor trap concentration, N_{At} is the acceptor trap concentration. R of n,p is carrier's recombination rate. Eqs. (4) and (5) derived from the charge conservation law are the electron and hole continuity equations, respectively. Fig. 1 shows the cross-section view of TFT. The boundaries $\overline{ca}$, $\overline{hj}$, $\overline{de}$, and $\overline{fg}$ are the Neumann boundary conditions. The boundaries $\overline{cd}$, $\overline{gh}$, and $\overline{ja}$ are the Dirichlet boundary conditions. The boundaries $\overline{ef}$ and $\overline{bi}$ are Robin boundary conditions. Moreover, the segments $\overline{X_0X_0'}$, $\overline{X_1X_1'}$..., $\overline{X_nX_n'}$ are the grain boundaries of the device.

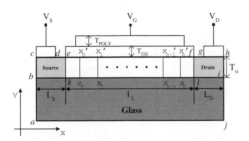

Figure 1: A cross-section view of the simulated TFT.

3   Adaptive Computing Algorithms

The adaptive computational method includes three procedures; the Gummel's decoupling algorithm [2], the adaptive finite volume method [3], and the monotone iterative method [4]. To solve the corresponding system of nonlinear algebraic equations for each decoupled and finite volume discretized differential equations, we use the following monotone iteration scheme

$$(D + \lambda I)Z^{(m+1)} = (L + U)^{(m)} - F(Z^{(m)}) + \lambda I Z^{(m)}, \tag{6}$$

where Z is the unknown vector, F is the nonlinear vector form, and D, L, U, I, and m are diagonal, lower triangular, upper triangular, identity matrices, and index of monotone iteration, respectively.

The monotone iterative parameter is determined node-by-node depending on the device structure, doping concentration, grain boundary concentration, bias condition, and nonlinear property of each decoupled equation. The monotone iterative method applies here for semiconductor device simulation is a global method in the sense that it does not involve any Jacobian matrix [5].

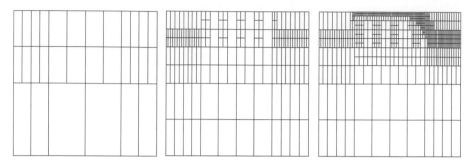

Figure 2: The left figure is the initial mesh which contains 52 nodes, the middle one is the third refined mesh containing 1086 nodes, and the right one contains 3555 nodes is the 5th mesh.

4   Results and Discussion

We now present several simulation results to demonstrate accuracy and performance of the proposed numerical algorithm. The device is a typical poly-TFT with 100 nm gate oxide thickness, and the junction depth is 50 nm. It has a LDD elliptical Gaussian profile with $2*10^{20}cm^{-3}$. Fig. 2 shows the process of mesh refinements. The mechanism of one-irregular mesh refinement is based on the estimation of solution error element by element. Fig. 3-(a) reports the relationship of the number of nodes (and elements) versus the refinement levels. The increasing rate of the number of nodes (and elements) gradually becomes slow when the refinements increase. Fig. 3-(b) shows the monotone iterative convergence behavior when solving the electrostatic potential with (w/) and without (w/o) grain boundary traps in the simulated device, and the biasing conditions are $V_D = 1.0\ V$ and $V_G = 1.0\ V$. Fig. 4 shows the computed potential and electron density for the device under bias conditions $V_D = 1.0\ V$ and $V_G = 1.0\ V$, respectively. In the channel region of the device, obviously, the simulated electron density reveals the grain boundary effects.

5   Conclusions

In this paper, we have successfully applied our adaptive computing methodology to the poly-TFT simulation. This solution scheme mainly relies on adaptive one-irregular mesh, finite volume, and monotone iterative methods. Numerical results and benchmarks for a TFT are also presented to show the robustness and efficiency of the method.

Acknowledgment

This work is supported in part by the National Science Council of Taiwan under contracts NSC-93-2215-E-429-008 and NSC 93-2752-E-009-002-PAE, the grant of the Ministry of Economic Affairs, Taiwan under contract No. 92-EC-17-A-07-S1-0011, and the grant from Toppoly Optoelectronics Corp, Miao-Li County, Taiwan.

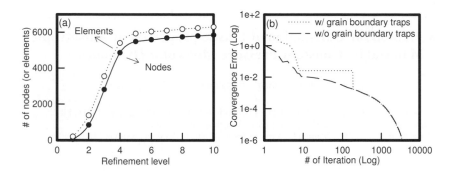

Figure 3: (a) The number of nodes and elements versus the refinement levels. (b) A monotone iterative convergence behavior w/ and w/o applying grain boundary traps.

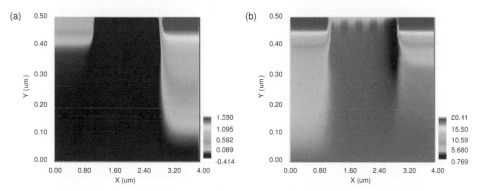

Figure 4: Contours of (a) the potential and (b) electron density for the TFT with $V_D = V_G = 1.0\,V$.

References

[1] R. W. Dutton and A. J. Strojwas, Perspectives on Technology and Technology- Driven CAD, *IEEE Trans. CAD* 19, 2, 1544-1560(2000).

[2] D. L. Scharfetter, H. K. Gummel, Large-Signal Analysis of a Silicon Read Diode Oscillator, *IEEE Trans. Elec. Dev.* ED-16, 66-77(1969).

[3] Y. Li, S-M Yu, and P. Chen, A Parallel Adaptive Finite Volume Method for Nanoscale Double Gates MOSFETS Simulation, in *Proceedings of the International Conference of Computational Methods in Sciences and Engineering*, 12-16 September, 382-386(2003).

[4] Y. Li, S. M. Sze, and T-S Chao, A Practical Implementation of Parallel Dynamic Load Balancing for Adaptive Computing in VLSI Device Simulation, *Engineering with Computers* 18, 2, 8(2002), 124-137.

[5] Y. Li, A Parallel Monotone Iterative Method for the Numerical Solution of Multidimensional Semiconductor Poisson Equation, *Computer Physics Communications* 153, 3, 7(2003), 359-372.

VSP International
Science Publishers
P.O. Box 346, 3700 AH Zeist
The Netherlands

*Lecture Series on Computer
and Computational Sciences*
Volume 1, 2004, pp. 586-588

A New Method for Calculation of Elastic properties of Anisotropic Material by Constant Pressure Molecular Dynamics

Kailiang Yin[*1,2], Heming Xiao[3], Jing Zhong[1] and Duanjun Xu[2]

1.   Department of chemical engineering, Jiangsu Polytechnic University, Changzhou 213016, PRC;
2.   Department of chemistry, Zhejiang University, Hangzhou 310027, PRC;
3.   Department of chemistry, Nanjing University of Science and Technology, Nanjing 210094, PRC.

Received 30 July, 2004; accepted in revised form 27 August , 2004

Abstract: 1,3,5-tri-amino-2,4,6-tri-nitrobenzene (TATB) is a typical and wide- studied explosive molecular and its single crystal is a typical triclinic lattice with a, b, c respectively 9.010, 9.028, 6.812 angstrom and α, β, γ 108.59, 91.82, 119.97. This explosive crystal is a stiff and anisotropic material. Packing of polymer on its crystal surface can obviously improve its mechanics properties. As a powerful tool, molecular dynamics (MD) simulation can be used to calculate the mechanics properties mainly elastic properties. But when applied the software package of Materials Studio (MS) to carry out MD, only the isotropic elastic properties which were averaged in x, y and z directions of materials can be obtained. To calculate the elastic properties in one direction of anisotropic materials, we developed a method which is similar to the experimental determination of elastic properties in one direction. Firstly, a P1 periodic super cell of TATB with 34.06×36.04 ×28.838 angstroms which ab plane was designated as (0 1 0) planar of the crystal was constructed. After several fixing and relaxing steps, the cell was pre-equilibrated 500 ps and performed 100 ps MD at 298 K within *NVT* ensemble. The averaged isotropic tensile modulus and Poisson' ratio were calculated by MS analysis module of elastic properties (static) and they were 1796±90 GPa and 0.231 respectively. Secondly, to obtain the elastic properties in each direction, *NPT* ensemble was chosen and different stresses in six different directions were added through many tries to keep the cell parameters fluctuating around those values in *NVT* ensemble while performing constant pressure MD. The MS averaged isotropic tensile modulus and Poisson' ratio of well equilibrated *NPT*'s system were then 2361±30 GPa and 0.273 respectively. Finally, while carrying out subsequent constant pressure MD, different magnitude of compressive or tensile stress was applied to the cell in one direction. The elastic properties were then obtained via the strain-stress profile. The calculated anisotropic tensile modulus was separately 1423, 2558 and 1955 GPa in x, y and z direction and the averaged value was 1979 GPa while averaged Poisson's Ratio was 0.217. These two averaged values were well agreed with the MS calculated ones within about 20% deviation. The result revealed that our method can be applied to calculate elastic properties of anisotropic materials.

Keywords: TATB, constant pressure molecular dynamics, anisotropic, elastic properties
PACS: 46. 15. -x

Theoretic Section

1.   Calculation of the elastic properties of isotropic material by MS

As we known, the static elastic properties of isotropic material are same in each direction and can be calculated by the software package of Materials Studio (MS). The method is as follow: for each configuration submitted for static elastic constants analysis, a total of 13 minimizations are performed. The first consists of a conjugate gradients minimization of the undeformed amorphous system. The target minimum derivative for this initial step is 0.1 kcal/Å. Following this initial stage, three tensile and three pure shear deformations of magnitude ±0.0005 are applied to the minimized undeformed system and the system is reminimized following each deformation. The internal stress tensor is then obtained from the analytically-calculated virial and used to obtain estimates of the six columns of the elastic stiffness coefficients matrix.   If more than a single configuration has been analyzed, the

---

[*] Corresponding author.. E-mail: mat_studio@jpu.edu.cn, klyin@263.net

averages and standard deviations for all the elastic constants and averages for all elastic properties such as tensile or Young modulus, Poisson's Ratio, shear modulus, bulk modulus, λ and μ. For an anisotropic material, the MS considers it as an isotropic material and the averaged elastic properties in each direction but not the elastic properties in one direction will be obtained.

2. Calculation of the elastic properties of anisotropic material by constant pressure MD

To calculate the elastic properties of anisotropic materials by MS, we developed a method which is similar to the experimental determination of elastic properties in one direction. The key of this method is that the simulated cell must be firstly adjusted to equilibrium in *NPT* ensemble by applying six different external stress σ_x, σ_y, σ_z, σ_{xy}, σ_{xz}, σ_{yz} to the cell. After the equilibration, different external stress in one direction was applied to the cell and a constant pressure MD within certain time period was performed. By averaging the equilibrated box length in each direction, averaged l_a, l_b, l_c were obtained, then corresponding change of each of three box length Δl and the strain $\Delta l/l_0$ could be calculated. By plotting the strain-stress profile and fitting line by least squares technique, we can calculate the tensile modulus and Poisson's ratio in one direction though the slopes of three lines according to the following formula (x as the direction which external stress σ_x applied to):

Longitudinal strain $\quad \varepsilon_n = \dfrac{\Delta l_a}{l_{0a}}$ $\hfill (1)$

Lateral strain $\quad \varepsilon_b = \dfrac{\Delta l_b}{l_{0b}}, \quad \varepsilon_c = \dfrac{\Delta l_c}{l_{0c}}$ $\hfill (2)$

Tensile or Young modulus $\quad E_a = \dfrac{\sigma_x}{\varepsilon_n} = \dfrac{\sigma_x}{\Delta l_a / l_{0a}}$ $\hfill (3)$

Poisson' ratio $\quad \gamma_b = -\dfrac{\varepsilon_b}{\sigma_x} E_a = -\dfrac{\Delta l_b / l_{0b}}{\sigma_x} E_a, \quad \gamma_c = -\dfrac{\varepsilon_c}{\sigma_x} E_a$ $\hfill (4)$

Table 1: Calculated anisotropic elastic tensile modules and Poisson' ratios of TATB in P1 periodic cell with (0 1 0) planar and calculated isotropic ones of equilibrated TATB in *NPT* ensemble by MS

Direction a		Direction b		Direction c		Averaged	MS value
E_a (GPa)		E_b (GPa)		E_c (GPa)		E (GPa)	E (GPa)
1423		2558		1955		1979	2395
γ_b	γ_c	γ_a	γ_c	γ_a	γ_b	γ	γ
0.253	0.233	0.309	0.189	0.134	0.185	0.217	0.273

Acknowledgments

The author wishes to thank the anonymous referees for their careful reading of the manuscript and their fruitful comments and suggestions. This work was supported by NSFC 29836150 and JSNFS BK2003402.

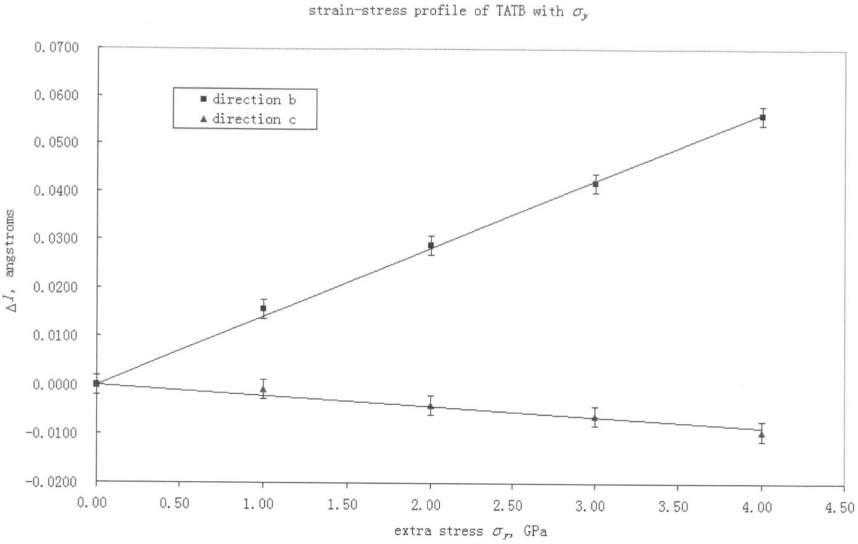

strain-stress profile of TATB with σ_y

Figure 1: The strain-stress profile of the TATB cell in direction *b* and *c* when applied different extra external stress σ_y to the cell

References

[1] K. L. Yin, Q. Xia, H. T. Xi, D. J. Xu, X. Q. Sun and C. L. Chen, *J. Mol. Struct. (Theochem)* v674, 19-59-165 (2004).

[2] K. Yin, Q. Xia, D. Xu, H. Xi, X. Sun and C. Chen, *Macromol. Theory Simul.* V12, 593-598 (2003).

[3] G. Zhang, Y. Pei, J Ma, K. Yin and C.-L. Chen, *J. Phys. Chem. B* v108, 6988-6995 (2004).

[4] *Materials Studio* 2.0; Discover/Accelry: San Diego, CA, 2001.

[5] C. Chen, Y. Wang, W. Ma and J. Li, *Elastic Mechanics*, High Educational Press, Beijing, 1999, p113-131 (in Chinese).

VSP International
Science Publishers
P.O. Box 346, 3700 AH Zeist
The Netherlands

*Lecture Series on Computer
and Computational Sciences*
Volume 1, 2004, pp. 589-591

A Brand New Reactive Potential RPMD Based on Transition State Theory Developed for Molecular Dynamics on Chemical Reaction

Kailiang Yin [1,2*], Duanjun Xu [2], Chenglung Chen [3]

[1] Department of Chemical Engineering, Jiangsu Polytechnic University, Changzhou Jiangsu 213016, China

[2] Department of Chemistry, Zhejiang University, Hangzhou Zhejiang 310027, China

[3] Department of Chemistry, National Sun Yat–Sen University, Kaohsiung, Taiwan, 80424

Abstract: A force field was developed to integrate a new reactive potential form within molecular dynamics (MD) that is defined as RPMD. This force field is based on transition state theory and considers bond breaking and forming that can be applied to simulate chemical reactions using MD techniques. We report the force field, CRACK which splines together a harmonic well and a shallow Lennard-Jones well: the harmonic well models C-C covalent bond while the Lennard-Jones well models long-range van der Waals forces and these two wells are separated by an energy barrier which controls dissociation and recombination, for studying n-alkane pyrolysis processes and compare to experimental trends. To validate the reactive potential form RPMD as well as the force field CRACK, the dissociation probabilities and product distributions of two n-alkane systems were analyzed and their time history of pyrolysis was illustrated. The research on dissociation probabilities shows that the defined pyrolytic temperature could be altered by changing the displacement constant *de* which controls the height of dissociation barrier of C-C bond and it was also found that the pyrolytic temperature of *n*-octane was higher than of *n*-decane at the same *de*, which is consistent with the pyrolysis fundamental. The time history of decane pyrolysis revealed an inspirer result that the CRACK force field certainly reflected the recombination between free radicals. Another exciting result is that, by analysis of the distribution of pyrolytic products, the main products distribution was observed to transfer from mid-short to mid-long chains and chain distribution tended to shorter or longer with the elevated temperature which is well consistent with the regularity of thermal cracking. We announce that molecular dynamics can be certainly used to simulate the process of chemical reaction as long as the reactive potential is proper.

Keywords: RPMD, CRACK force field, n-alkane, pyrolysis, molecular dynamics
PACS: 47.70.Fw

RPMD and CRACK force field Section

1.   Two-body potential term *V(r)* in RPMD

V(r) which contains long-distance van der Waals interaction and bond stretching potential is adapted to express the interaction between two special carbon particles: two adjacent carbon particles in the same chain in the dissociation reactant region or in the recombination product region and two carbon particles which are able to participate in bond breaking or forming in different chains in the dissociation product region or in the recombination reactant region. The potential value is set to zero when two carbon particles are separated infinite far. The expression of *V(r)* is:

$$V(r) = V_1(r) + V_0 \tag{1}$$

where V_0 is a constant introduced to ensure that *V(r)* trends to zero when two carbon particles are away from infinite distance.
$V_1(r)$, a subsection function, has different function forms in six different regions:

$$\left[\frac{1}{2} k_s (r - r_0)^2 - de \right] \qquad\qquad r \leq r_0$$

---

* Corresponding author, Tel: +86-519-3290253, E-mail: mat_studio@jpu.edu.cn or klyin@263.net

$$
V_1(r) = \begin{cases} \left| \left[\dfrac{1}{2}k_s(r-r_0)^2 - de\right]SW(r,r_0,r_1) & r_0 \le r \le r_1 \\[2mm] \{ k_a(r-r_1)(\sigma-r)\left[1-SW(r,r_1,r_2)\right], & r_1 \le r \le r_2 \\[2mm] |\, k_a(r-r_1)(\sigma-r) & r_2 \le r \le \sigma \\[2mm] |\, 4\varepsilon_A\left[\dfrac{\sigma^{12}}{r^{12}} - \dfrac{\sigma^6}{r^6}\right] & \sigma \le r \le r_{min} \\[2mm] \{ 4\varepsilon\left[\dfrac{\sigma^{12}}{r^{12}} - \dfrac{\sigma^6}{r^6}\right] - (\varepsilon_A - \varepsilon) & r_{min} \le r \le r_s \end{cases} \tag{2}
$$

where SW is a switch function, by multiplying it any function can be controlled on or off. The expression of SW is chosen as:

$$
SW(r,r_a,r_b) = \begin{cases} 1 & r \le r_a \\[2mm] (r_b - r)^2(r_b + 2r - 3r_a)/(r_b - r_a)^3 & r_a \le r \le r_b \\[2mm] 0 & r > r_b \end{cases} \tag{3}
$$

In equation (2), the former two parts are bond stretching terms, k_s is C-C bond force constant, r_0 is 'equilibrium' or referenced bond length, r_1 is the limited bond length beyond which the bond will be considered as near breaking state, de is an alterable displacement constant which can be used to adjust the depth of potential energy $V_1(r)$ trap at $r = r_0$ so can be used to control the dissociation activate energy. The $r \le r_1$ region is considered as the bonded region. The third and forth regions are transition region, in which two-body potential transfers from bonded to non-bonded. In this region, $V_1(r)$ reaches its maximum when $r = r_2$, which corresponds to the energy of activated dissociation or recombination transition state. Where, k_a is also a parameter used to control the height of energy barrier. The last two parts are $V_1(r)$ in non-bonded *i.e.* van der Waals interaction region, in which $V_1(r)$ includes the normal Lennard-Jones potential (LJ 12-6). Here, $V_0 = \varepsilon_A - \varepsilon$, ε_A is another adjustable parameter. In CRACK force field, non-bond cutoff is used and worked by multiplying a switch function:

$$
V(r) = (V_1(r) + V_0)SW(r,r_s,r_c) \qquad\qquad r_s \le r \le r_c \tag{4}
$$

where r_s and r_c are respective cut-on and cut-off distance. The values of boundary points in $V(r)$ are primarily set as follow according to the properties of carbon particles: $r_1 = 0.6\,\sigma$, $r_2 = 0.8\sigma$, $r_{min} = \sqrt[6]{2}\sigma$, $r_s = 2.3\sigma$, $r_c = 2.5\sigma$. To ensure the continuity of potential function $V_1(r)$ at every boundary point, here should be $k_a = \dfrac{24\varepsilon_A}{\sigma(\sigma - r_1)}$.

Table 2. The number of pyrolyzed molecules of *n*-decane at different simulation time under varied temperatures （*de*=125kJ/mol）

Force Field	Temperature / K	Time / ps					
		5	10	15	20	25	30
Toxvaerd	500 – 900	0	0	0	0	0	0
CRACK	700	0	0	0	0	0	0
CRACK	710	1	0	1	0	1	1
CRACK	715	3	4	6	5	6	7
CRACK	800	6	8	11	10	12	14

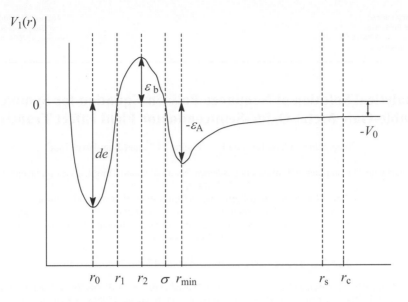

Figure 1. Illustration of $V_1(r)$ potential curve

Acknowledgments

The author wishes to thank the anonymous referees for their careful reading of the manuscript and their fruitful comments and suggestions.

References

[1] Leach, A. R. *Molecular Modelling-Principles and Applications*. Pearson Education Limited: Harlow, England, (2001)

[2] Lipkowitz, K. B. *Chem. Rev.* **5,** 1829 (1998).

[3] Hummer, G., Rasaiah, J. C. & Noworyta J. P. *Nature* **414**, 188 (2001).

[4] Yin, K. L., Xu, D. J. & Chen, C. L. *Chinese J. Inorg. Chem.* **19**, 480 (2003).

[5] Yin, K. L., Xia, Q., Xu, D. J., Xi, H. T., Sun, X. Q. & Chen, C. L. *Macromol. Theor. Simul.* **12,** 593 (2003).

[6] Yin, K. L., Xu, D. J., Xia, Q., Ye, Y. J., Wu, G. Y. & Chen, C. L. *Acta Phys.-Chim. Sin.* **20**, 302 (2004).

[7] Yin, K. L., Xia Q., Xi, H. T., Xu, D. J., Sun, X. Q. & Chen, C. L. *J. Mol. Struct.* (Theochem) **674**, 157 (2004).

[8] Bounaceur, R., Warth, V., Marquaire, P., Scacchi, G., Domine, F., Dessort, D., Pradier, B. & Brevart, O. *J. Anal. Appl. Pyrolysis* **64**, 103 (2002).

[9] Vandenbroucke, M., Behar, F. & Rudkiewicz, J.L. *Org. Geochem.* **30**, 1105 (1999).

[10] Savage, P. E. *J. Anal. Appl. Pyrolysis* **54**, 109 (2000).

[11] Franz, J. A., Camaioni, D. M., Autrey, T., Linehan, J. C. & Alnajjar M.S. *J. Anal. Appl. Pyrolysis* **54**, 37 (2000).

[12] Fierro, V., Schuurman, Y., Mirodatos, C., Duplan, J.L. & Verstraete, J. *Chem. Eng. J.* **90,** 139 (2002).

[13] Brenner, D. W., Shenderova, O. A., Harrison, J. A., Stuart, S. J., Ni, B. & Sinnott, S. B. *J. Phys.:Condens. Matter* **14**, 783 (2002).

[14] Toxvaerd, S. *J. Chem. Phys.* **87,** 6140 (1984).

[15] Van Gunsteren, W. E. & Berendsen, H. .J C. *Angew. Chem., Int. Ed.* **29,** 992 (1990).

[16] Brown, D. & Clarke, H. R. *Mol. Phys.* **51,** 1243 (1984).

[17] Zhang, J. F. & Shan, H. H. *Fundamentals of Oil Refining Technic*s (China Petrochemical Press, Beijing, Chinese) (1994).

VSP International
Science Publishers
P.O. Box 346, 3700 AH Zeist
The Netherlands

*Lecture Series on Computer
and Computational Sciences*
Volume 1, 2004, pp. 592-595

Analytical Solution of Nonlinear Poisson Equation for Symmetric Double-Gate Metal-Oxide-Semiconductor Field Effect Transistors

Shao-Ming Yu[a], Shih-Ching Lo[b], Yiming Li[cd1], and Jyun-Hwei Tsai[b]

[a] Department of Computer and Information Science, National Chaio Tung University, Hsinchu 300, Taiwan
[b] National Center for High-Performance Computing, Hsinchu 300, Taiwan
[c] Department of Computational Nanoelectronics, Nano Device Laboratories, Hsinchu 300, Taiwan
[d] Microelectronics and Information Systems Research Center, National Chaio Tung University, Hsinchu 300, Taiwan

Received 31 July, 2004; accepted in revised form 28 August, 2004

Abstract: In this paper, an analytical solution of Poisson equation for double-gate MOSFET is presented. An explicit surface potential function is also derived to make the whole solution be a fully analytical one. The resulting solution provides an accurate description for partially and fully depleted devices in regions of operation. Doping concentration in silicon region is also considered. A comparison with numerical data shows that the solution gives an accurate approximation of potential distribution for a nano-scale double-gate MOSFET in all regions of operation.

Keywords: analytical solution, Poisson equation, surface potential, double-gate MOSFET.

Mathematics Subject Classification: 34B15, 35F30

PACS: 02.60.-x, 02.60.Cb, 02.60.Lj, 85.30.De, 85.30.Tv

1. Introduction

Double gate metal oxide semiconductor field effect transistor (DG-MOSFET) are of interest today mainly because of their inherent suppression of short-channel effects (SCEs), high transconductance and ideal subthreshold swing (S-swing) [1-4]. Thus, the scalability of semiconductor devices is intimated to nanoscale. A key question is how to calculate the characteristics of device efficiently and accurately, especially for circuit simulation. Among the characteristics of devices, potential is a basic and important one. As potential is obtained, electron density is derived. The purpose of this work is to introduce a new 1-D closed-form analytical approximation of potential distribution in the silicon film of a DG-MOSFET in regions of operation (depletion, weak inversion and strong inversion). Previous studies can only considered analytical solution of DG-MOSFET with given surface potential [2-3] or undoped silicon film [4]. To make the solution be a fully analytical one, the explicit form of surface potential is also derived. A comparison with numerical data shows that the solution gives an accurate approximation of potential distribution for a nano-scale double-gate MOSFET in all regions of operation.

2. Explicit Solution of Surface Potential

In this study, 1-D symmetric double-gate MOSFET, which is illustrated in Fig. 1, is considered. The Poisson equation and the boundary conditions can be written as

$$\Delta\phi = qN_a\left[1 - \exp(-\phi/V_t) + (n_i/N_a)^2(\exp(\phi/V_t)-1)\right]/\varepsilon_{si}, \quad -t_{si}/2 < x < t_{si}/2, \qquad (1)$$

---
[1] Corresponding author. Postal address: P.O. BOX 25-178, Hsinchu 300, Taiwan, E-mail: ymli@faculty.nctu.edu.tw

$\phi = \phi_s$ and $\partial\phi/\partial x = \pm C_{ox}(V_G - V_{FB} - \phi_s)/\varepsilon_{si}$, for $x = \pm t_{si}/2$. For $x = 0$, $\phi = \phi_b$ and $\partial\phi/\partial x = 0$. ϕ is potential, ϕ_s is surface potential, ϕ_b is potential at the center of silicon region. V_G is gate voltage, V_{FB} is flat-band voltage and V_t is thermal voltage, which is considered as a constant. n_i is intrinsic concentration and N_a is doping concentration. t_{ox} is thickness of oxide, t_{si} is thickness of silicon film, ε_{ox} is the dielectric constant of oxide, ε_{si} is dielectric constant of silicon and C_{ox} is the capacitance of oxide. q is electron charge. $\partial\phi/\partial x = 0$ is obtained from the assumption of symmetric applied bias [4].

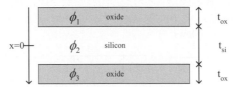

Figure 1: The profile of double-gate MOSFET

Multiplying $(\partial\phi/\partial x)dx$ on both side of the Poisson equation of Eq. (1) and integrating from the surface toward the center of silicon film, i.e., $x = -t_{si}/2$ to 0, and ignoring ϕ_s-ϕ_b, which is much smaller than $V_t \exp(\phi_b/V_t) - V_t \exp(\phi_s/V_t)$, one obtains

$$(V_G - V_{FB} - \phi_s)^2/\gamma^2 = (\phi_s - \phi_b) + [V_t \exp(-\phi_s/V_t) - V_t \exp(-\phi_b/V_t)] + [V_t \exp((\phi_s - 2\phi_f)/V_t) - V_t \exp((\phi_b - 2\phi_f)/V_t)], \quad (2)$$

where $\gamma = \sqrt{2q\varepsilon_{si}N_a}/C_{ox}$ and $\phi_f = V_t \ln(N_a/n_i)$. To solve for ϕ_s, another expression relating ϕ_s and ϕ_b is required. The relation is obtained from the condition of full depletion for body charge and the simplification of discretization of Eq. (1) [5-7]. It is given as $\phi_b = \phi_s - a t_{si} C_{ox}(V_G - V_{FB} - \phi_s)/2\varepsilon_{si} - V_{xd}$, where a is a fitting parameter. The critical voltage V_G–V_C at which is changed from partially depleted (PD) to the fully depleted (FD) device needs to be calculated before deriving the analytical solution. It will be derived from the condition that $\phi_b = 0$ and $\phi_s = V_{xd}$.

$$V_C = V_{FB} + V_{xd} + \gamma\sqrt{V_{xd} + [V_t \exp(-V_{xd}/V_t) - V_t] + [V_t \exp((V_{xd} - 2\phi_f)/V_t) - V_t \exp((-2\phi_f)/V_t)]} \quad (3)$$

Before deriving the analytical solution of surface potential, two cases are discussed. Case (I): depletion and weak inversion, in this case, $0 < \phi_s < 2\phi_f$, $\phi_s \gg V_t$, therefore, $\exp(-\phi_s/V_t) \approx 0$ and $\exp(\phi_s - 2\phi_f/V_t) \approx 0$. Case (II): strong inversion, in this case $0 < \phi_s < 2\phi_f$, $\phi_s \gg V_t$, $\exp(-\phi_s/V_t) \approx 0$ and $\exp(\phi_s - 2\phi_f/V_t) \approx 0$. Firstly, surface potential of PD devices is discussed, i.e., $\phi_b = 0$. In case (I), Eq. (2) can be approximated and solved as $\phi_{si}^P = V_{GB} - V_{FB} + \gamma^2/2 - \gamma\sqrt{\gamma^2/4 + V_G - V_{FB} + A}$, where $A = V_t(1 + \exp(-2\phi_f/V_t))$. According to the studies [6-7], surface potential of case (II) is given by $\phi_{si}^P = f_\phi + V_t \ln\{[(V_G - V_{FB} - f_\phi)^2/\gamma^2 - f_\phi + 1]/V_t\}$, where $f_\phi = \left(2\phi_f + \phi_{ss}^P - \sqrt{(\phi_{ss}^P - 2\phi_f)^2 + 4\delta^2}\right)/2$. As mentioned above, surface potential equation can be linked by the following smooth function [3, 4], which is given as $\phi_{st}^P = \phi_{si}^P - V_t \ln\{1 + \exp((\phi_{si}^P - \phi_{sd}^P)/V_t)\}$.

Next, surface potential of fully depleted device is derived. In case (I), Eq. (2) can be approximated and solved as $\phi_{sd}^f = V_{GB} - V_{FB} - \gamma\sqrt{V_{xd}}$. In case (II), the surface potential is given as $\phi_{si}^f = 2\phi_f + V_t \ln\{[(V_G - V_{FB} - f_\phi)^2/\gamma^2 - V_{xd}]/V_t[1 - \exp(-V_{xd}/V_t)]\}$. Also, surface potential equation can be linked by the following smooth function, which is given as $\phi_{st}^f = \phi_{si}^f - V_t \ln\{1 + \exp((\phi_{si}^f - \phi_{sd}^f)/V_t)\}$.

If a device can change the operation mode from PD device to FD device as V_G increases. Then the smooth function is employed to link the operation mode [4]:

$$\phi_s^m = \frac{\phi_{st}^P}{1 + l_1 \exp((V_G - V_C)/n_1 V_t)} + \frac{\phi_{st}^f}{1 + l_1 \exp(-(V_G - V_C)/n_1 V_t)}, \quad (4)$$

where l_1 and n_1 are fitting parameters.

3. Analytical Solution of Poisson equation

After deriving the explicit function of surface potential, Poisson equation of Eq. (1) could be solved. Integrating it from surface to center of silicon film (considering $x = -t_{si}/2$ to 0), we have

$$\partial\phi/\partial x = \pm\sqrt{\frac{2qN_A(x)}{\varepsilon_{si}}\left\{\phi - \phi_b + V_t\exp(-\phi/V_t) - V_t\exp(-\phi_b/V_t)\right] + (n_i/N_A(x))^2\left[\phi_b - \phi + V_t\exp(\phi/V_t) - V_t\exp(\phi_b/V_t)\right]\right\}} \quad (5)$$

The positive sign is for $0 \leq x \leq t_{si}/2$ and the negative sign is for $-t_{si}/2 \leq x \leq 0$. In case (I), Eq. (5) can be simplified and rewritten as follows

$$\partial\phi/\sqrt{(\phi - \phi_b)} = \pm\sqrt{2qN_a\left[1 - (n_i/N_a)^2\right]/\varepsilon_{si}}\,\partial x. \quad (6)$$

Integrating Eq. (11) from top surface to center of the silicon film, we have

$$\phi = \phi_b + (\phi_s - \phi_b)\left[1 - \sqrt{qN_a\left[1 - (n_i/N_a)^2\right]/(2(\phi_s - \phi_b)\varepsilon_{si})}\left(x + t_{si}/2\right)\right]^2 - \widetilde{E}_1(x) \text{ for } -t_{si}/2 \leq x \leq 0, \quad (7)$$

$$\phi = \phi_b + (\phi_s - \phi_b)\left[1 + \sqrt{qN_a\left[1 - (n_i/N_a)^2\right]/(2(\phi_s - \phi_b)\varepsilon_{si})}\left(x - t_{si}/2\right)\right]^2 - \widetilde{E}_2(x) \text{ for } 0 \leq x \leq t_{si}/2, \quad (8)$$

where $\widetilde{E}_1(x) = (\widetilde{a}_1 x^6 + \widetilde{b}_1 x^2)\ln[\widetilde{c}_1/(x + \widetilde{d}_1)]$ and $\widetilde{E}_2(x) = (\widetilde{a}_2 x^6 + \widetilde{b}_2 x^2)\ln[\widetilde{c}_2/(x + \widetilde{d}_2)]$. $\widetilde{E}_1(x)$ and $\widetilde{E}_2(x)$ are the adjust term, which are introduced into the solution of Eq. (7). The function form is obtained by comparing Eqs (5) and (6), we know the slope of Eq. (5) is larger (smaller) than or equal to the slope of Eq. (6) for $x > 0$ ($x < 0$). $\widetilde{a}_1$, $\widetilde{a}_2$, $\widetilde{b}_1$, $\widetilde{b}_2$, $\widetilde{c}_1$, $\widetilde{c}_2$, $\widetilde{d}_1$ and $\widetilde{d}_2$ are fitting parameters. In case (II), Eq. (5) can be approximated and solved as follows

$$\phi = \phi_b - V_t\ln\left[\cos\left(\sqrt{qn_i^2/2V_t\varepsilon_{si}N_a}\,\exp(\phi_b/2V_t)x\right)^2\right] - \hat{E}_1(x), \text{ for } -t_{si}/2 \leq x \leq 0, \quad (9)$$

$$\phi = \phi_b - V_t\ln\left[\cos\left(\sqrt{qn_i^2/2V_t\varepsilon_{si}N_a}\,\exp(\phi_b/2V_t)x\right)^2\right] - \hat{E}_2(x), \text{ for } 0 \leq x \leq t_{si}/2, \quad (10)$$

where $\hat{E}_1(x) = (\hat{a}_1 x^6 + \hat{b}_1 x^2)\ln[\hat{c}_1/(x + \hat{d}_1)]$ and $\hat{E}_1(x) = (\hat{a}_1 x^6 + \hat{b}_1 x^2)\ln[\hat{c}_1/(x + \hat{d}_1)]$. $\hat{E}_1(x)$ and $\hat{E}_2(x)$ are the adjust term, $\hat{a}_1$, $\hat{a}_2$, $\hat{b}_1$, $\hat{b}_2$, $\hat{c}_1$, $\hat{c}_2$, $\hat{d}_1$ and $\hat{d}_2$ are fitting parameters. Then, we have analytical solution and surface potential of 1-D DG-MOSFET.

4. Calibration and Verification

In this study, a 20nm double-gate NMOSFET with $t_{ox} = 2$nm, $t_{si} = 20$nm and $Na = 10^{17}$ cm^{-3} is simulated for $V_G = 0.05, 0.1, 0.7$ and 1.0 V. Eq. (4) is considered to estimate the surface potential. From the results, the analytical solution of surface potential has an approximation with error, which is small than 0.005% in depletion and weak inversion region. In the strong inversion region, a larger error, which is smaller than 0.01% is obtained. The results, which are given in Table 1, show that the explicit function gives good approximations of surface potential. With the explicit surface potential, potential distributions are also simulated under four given gate applied biases, i.e., $V_G = 0.05, 0.1, 0.7$ and 1.0 V. The results are illustrated in Figs 2.

Table 1: Numerical and analytical surface potential of the simulated device

	Numerical (A)	Analytical (B)	Difference (C)=(B)-(A)	Error (C)/(A)
$V_G = 0.05$ V	0.940459	0.9405	0.000041	0.00436 %
$V_G = 0.1$ V	0.953597	0.953588	-0.000009	-0.00094 %
$V_G = 0.7$ V	1.03872	1.03866	-0.00006	-0.00578 %
$V_G = 1.0$ V	1.06509	1.06498	-0.00011	-0.010 %

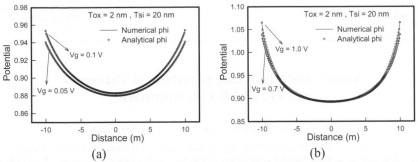

(a)   (b)

Figure 2: Comparison of numerical and analytical results, (a) V_G=0.05 and 0.1 V; (b) V_G =0.7 and 1.0 V.

5. Conclusion

In this study, a 1-D analytical solution of Poisson equation for DG-MOSFET is derived successfully. The solution can be used for PD and FD devices in all regions of operation. In addition, the solution is workable for devices with doping concentration in silicon region. According to the numerical comparison, the results are very accurate. Therefore, the solution can be employed to couple with quantum effects so as to apply to simulation of transport characteristics for devices and circuit simulation.

Acknowledgments

The work is partially supported by the National Science Council (NSC), Taiwan, under Contracts NSC-92-2215-E-429-010, NSC-93-2215-E-429-008, and NSC 93-2752-E-009-002-PAE. It is also partially supported by Ministry of Economic Affairs, Taiwan, under contract No. 92-EC -17-A-07-S1-0011.

References

[1] D. Hisamoto, et al., FinFET-A Self-Aligned Double-Gate MOSFET Scalable to 20nm, *IEEE Transaction on Electron Device* **47** 2320-2325 (2000).

[2] G. Baccarani and S. Reggiani, A Compact Double-Gate MOSFET Model Comprising Quantum-Mechanical and Nonstatic Effects, *IEEE Transaction on Electron Device* **46** 1656-1666 (1999).

[3] A. Rahman and M. S. Lundstrom, A Compact Scattering Model for the Nanoscale Double-Gate MOSFET, *IEEE Transaction on Electron Device* **49** 481-489 (2002).

[4] Y. Taur, Analytical solutions of charge and capacitance in symmetric and asymmetric double-gate MOSFETs, *IEEE Transactions on Electron Device* **48** 2861-2869 (2001).

[5] S. -L. Jang; B. –R. Huang; J. –J. Ju, A unified analytical fully depleted and partially depleted SOI MOSFET model, *IEEE Transactions on Electron Devices* **46** 1872-1876 (1999).

[6] R. van Langevelde and F. M. Klaassen, An explicit surface-potential-based MOSFET model for circuit simulation, *Solid-State Electronics* **44** 409-418 (2000).

[7] Y. S. Yu, S. W. Hwang, and D. Ahn, A unified analytical SOI MOSFET model for fully- and partially-depleted SOI devices, 2001 *Asia-Pacific Workshop on Fundamental and Application of Advanced Semiconductor Devices* (2001), 329-334.

VSP International
Science Publishers
P.O. Box 346, 3700 AH Zeist
The Netherlands

*Lecture Series on Computer
and Computational Sciences*
Volume 1, 2004, pp. 596-599

Atomic and Electronic Structure of Vacancies
in UO$_2$: LSDA+U Approach

Younsuk Yun[1†], Hanchul Kim[2], Heemoon Kim[3], Kwangheon Park[1]

1 Department of Nuclear Engineering, Kyung-Hee University, Suwon, , 449-701, Korea
2 Korea Research Institute of Standard and Science, P.O.Box 102, Yuseong, Daejeon, 305-600 Korea
3 Korea Atomic Energy Research Institute , P.O.Box 105, Yuseong, Daejeon, 305-353 Korea

Received 5 August, 2004; accepted in revised form 24 August, 2004

Abstract: We present the density functional theory calculations of UO$_2$ within the LSDA+U approach. For the bulk UO$_2$, the electronic structures of UO$_2$ obtained both the LDA and the LSDA+U approaches in order to elucidate the strong correlation effect. Then we performed supercell calculations to investigate the structure of different defects in UO$_2$, the defect formation energy, and the defect-induced changes in the electronic structure. Finally, we deduce the activation energy of Xe diffusion, which is one of fission products, through the defects in UO$_2$.

Keywords: 5f electrons, LSDA+U approach, defect formation energy, fission products

Uranium dioxide is an important fuel material for nuclear industry and the electronic structure of 5f electrons of uranium has not been reliably described so far. The 5f electrons of uranium is one of the representative strongly-correlated systems, and thus a proper treatment of electron-electron correlations in essential for the description of the electronic structure on the antiferromagnetic (AFM) state of UO$_2$. During irradiation UO$_2$ undergo nuclear fission and as a result, the noble gas Xe is formed. Xe diffuses into the gap between the cladding and pellets of the fuel, and cause swelling of the fuel. Point defects are thought to be the major channel of Xe diffusion in UO$_2$ [3,4]. Therefore, it is important to understand defects in order to investigate the diffusion mechanism in nuclear reactor.
The density functional theory (DFT) calculations within the local spin density approximation (LSDA) often fail to describe systems with strongly correlated electrons, predicting to metallic behaviour contrary to the observed insulating behaviour [5]. Orbital-dependent functionals like the LSDA+U approach are known to correct this kind of problems. The LSDA+U method was proposed by Anisimov *et al.* in order to bridge the gap between the DFT-LDA and the many-body approaches for the strongly correlated electronic systems [6,7]. Dudarev et al. also showed that the LSDA+U method of Anisimov *et al.* improves the calculated ground state [8-10].
In this paper, we investigate the localized nature and the effect of the Coulomb repulsion in the U-5f shells on the electronic structure of UO$_2$ using LSDA+U method. We used the simplified LSDA+U energy functional, due to Dudarev *et al.*, which is expressed the following form

$$E_{LSDA+U} = E_{LSDA} + \frac{1}{2}(\overline{U} - \overline{J})\sum_{\sigma}(n_{m,\sigma} - n_{m,\sigma}^2) \qquad (1)$$

where $n_{m,\sigma}$ is the occupation number of the mth f states, σ is the projection of spin. $\overline{U}$ and $\overline{J}$ are the spherically averaged matrix elements of the screened Coulomb interaction between f electrons. This approach is to add a penalty functional to the LSDA total energy, which forces the on-site Coulomb repulsion. It is important to be aware of the fact that when using the LSDA+U, the total energy depends on the parameters $\overline{U}$ and $\overline{J}$. In Eq. (1), the parameters $\overline{U}$ and $\overline{J}$ do not enter separately, only the difference $(\overline{U} - \overline{J})$ is meaningful [8-10]. The 5f Coulomb correlation energy $\overline{U}$ was determined to be 4.6±0.8eV from the energy difference in the two spectroscopies of X-ray Photoemission Spectroscopy and Bremsstrahlung Isochromat Spectroscopy [1]. We chose the parameters of $\overline{U}$ = 4.5eV and $\overline{J}$ = 0.5 from Dudarev *et al.* and compare them with other experimentally observed data [1,10]. All the

---

[†] Corresponding author. E-mail : ysyun4301@empal.com

calculations have been performed using the VASP package, that is based on projected-augmented-wave and the plane wave basis set [11-13].

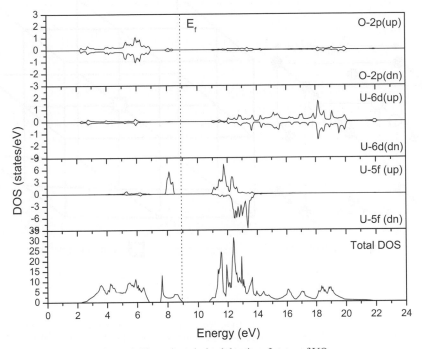

Figure 1. The spin polarized density of states of UO$_2$

The calculated equilibrium lattice constant of UO$_2$ is found to be 5.44 Å. It is underestimated by about 0.55 % of the experimental value of 5.47 Å [14]. The calculated cohesive energy per UO$_2$ molecule is 26.9 eV, that is larger than experimental value of 22.3 eV. It seems to be overestimated about 20 %, because it does not contains spin-polarization energy of atoms. The underestimation of lattice constant and the overestimation of cohesive energy are typical of DFT-LDA calculations [15]. The spin polarized local density of states (LDOS) and total density of states (DOS) of AFM bulk UO$_2$ obtained from LSDA+U calculations shown Fig. 1. The LSDA+U calculations predict the correct insulating ground states with the band gap of 1.9 eV. In contrast, calculations employing original LDA result in metallic electronic structure where the uranium $5f$ bands are partially occupied. The insulating nature of the ground state originates from the presence of Hubbard-type correlations in the bands of uranium $5f$ states. The band gap of 1.9 eV is opened up due to the split $5f$ bands, and agrees with the experimental result of 2.0 eV [16]. The LDOS in Fig. 1 shows noticeable difference between up- and down-spins only 5f states. This implies that the AFM ground state of UO$_2$ is governed by the $5f$ electrons.

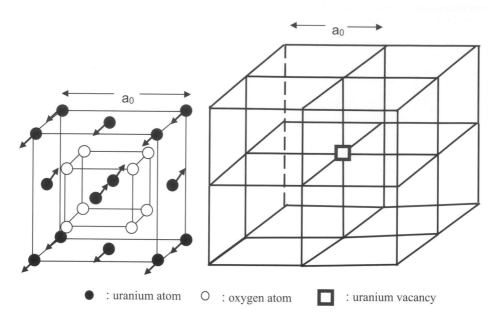

● : uranium atom   O : oxygen atom   □ : uranium vacancy

Figure 2. (left) Anitiferromagnetic ordering in the (100) direction of bulk UO_2 and (right) 2x2x2 supercell containing uranium vacancy

    In order to investigate the defects in UO_2, we used the 2x2x2 supercells containing 98 atoms as shown in Fig, 2. We calculated the atomic and electronic structure of UO_2 containing the four kinds of defects which are an uranium vacancy, an oxygen vacancy, di-vacancy of uranium-oxygen and tri-vacancy of uranium-oxygen-oxygen vacancies.

References

[1] Y. Baer and J. Schoenes, Phys. Rev. B **47**, 888 (1980)

[2] A.P. Cracknell, M.R. Daniel, Proc. Phys. Soc. **92**, 705 (1967)

[3] J. P. Crocombette, F. Jollet, L.Thien Nga, and T.Petit, Phys. Rev. B **64**, 104107

[4] Richard G. J. Ball, Robin W. Grimes, J. Chem. Soc. Faraday Trans., 1990, **86**(8), 1257-1261

[5]. G. A. Sawatzky and J. W. Allen, Phys. Rev. Lett. **53**, 2339 (1984)

[6] V. I. Anishimov, J. Zaanen, and O. K. Andersen, Phys. Rev. B **44**, 943(1991)

[7] V. I. Anishimov, I. V. Solovyev, M. A. Korotin, M. T. Czyzyk, and G. A. Sawatzky , Phys. Rev. B **48**, 16929(1993)

[8] S.L. Dudarev, D. Nguyen-Manh and A.P. Sutton, Philos. Mag. B. vol. 75, 613(1997)

[9] S. L. Dudarev, A. I. Liechtenstein, M. R. Castell , G. A. D. Briggs , and A. P. Sutton , Phys. Rev. B **56**(8), 4900(1997).

[10]. S. L. Dudarev, G. A. Botton, S. Y. Savrasov, C. J. Humphreys, A. P. Sutton, Phys. Rev. B, **57**(3), 1505(1998).

[11] G. Kresse and J. Hafner, Phys. Rev. B **47**, RC558 (1993)

[12] G. Kresse and J. Hafner, Phys. Rev. B **48**, 13115 (1993)

[13] G. Kresse and J. Hafner, Phys. Rev. B **47**, 14251 (19931994)

[14] L. Lynds, J.Inorg. Nucl. Chem. **24**, 1007 (1962)

[15] Paxton, A. T., Methfessel, M., and Polatoglou, H. M., 1990. Phys. Rev. B **41**, 8127

[16] Schoenes, J., 1990, J. Solid st. Chem., **88**, 2.

VSP International
Science Publishers
P.O. Box 346, 3700 AH Zeist
The Netherlands

*Lecture Series on Computer
and Computational Sciences*
Volume 1, 2004, pp. 600-603

First-Principles Investigations of the Electronic Structure
of Β-Lani4alh$_x$ (X=4, 4.5, 5, 7)

R.J. Zhang[1], C.H. Hu, M.Q. Lu, Y.M. Wang, K. Yang, D.S. Xu

Institute of Metal Research, Chinese Academy of Sciences, Shenyang 110016, China

Received 6 September, 2004; accepted in revised form 20 September, 2004

Abstract: The electronic structures of β-LaNi$_4$AlH$_x$ (x=4, 4.5, 5, 7) have been investigated using a plane-wave pseudo-potential method. The results of calculated formation energy and equilibrium unit cell volume show that LaNi$_4$AlH$_{4.5}$ is the most favorable structure to the real composition of LaNi$_4$AlH$_x$. LaNi$_4$AlH$_7$ should not be formed as a stable hydride due to the reduction in the number of H-Ni bonding states associated with the reduction in the number of d states near the Fermi surface in LaNi$_4$AlH$_7$. LaNi$_4$Al can not absorb hydrogen as much as LaNi$_5$.

Keywords: electronic structure; plane-wave pseudo-potential method; formation energy

1. Introduction

A large number of investigations on LaNi$_5$ and related compounds as representative rare alloys have attracted much attention all over the world since Philips Company in Holland firstly found them in 1959. And the study about them has acquired important developments since the early 70's in relation to their exceptional hydriding properties[1, 2]. In recent years, hydrogen storage alloys based on La are commercially used as negative electrode materials for nickel–metal hydride (Ni–MH$_x$) battery[3, 4]. For practical application, substitutions at the La and Ni sites have been extensively used to improve the hydrogen absorption and desorption characteristics, since they affect the stability and the hydrogen content of the hydride. As observed in experiments, the substitutions of aluminum for nickel induce a decrease of the hydrogen content for the hydrides. In this paper, in order to investigate the role played by Al atoms in the decrease of the hydrogen content, we performed a first-principles calculations for β-LaNi$_4$AlH$_x$ (x=4, 4.5, 5, 7), in which imaginary LaNi$_4$AlH$_7$ is at the structure of LaNi$_5$H$_7$.

2. Computational details

The calculations presented in this work were performed using a plane-wave pseudo-potential (PW-PP) method[5] within the generalized-gradient approximation (GGA) of Perdew–Wang form[6] to density-functional theory. A finite basis set correction was applied to the total energy and stress tensor when a geometric optimization of variable cell parameters was made. The pseudopotentials were constructed for neutral atoms as described in the report[7]. The La-4f orbital was therefore not included. In fact, the contributions of La-4f to the interactions between atoms were proved to be so small that they were ignored in the PW-PP calculations and the covalent density analysis in their study[8]. The much more transferable ultrasoft pseudopotentials[9] were constructed for all ions, i.e., H, La, Ni, and Al. The ultrasoft potential for these elements was recently reported to have a great advantage for an accurate description of them in complex solids[8]. With these pseudopotentials, the plane-wave cutoff, E_{cut} was chosen to be 400 eV in the present study. This was confirmed to achieve a good convergence with respect to the total energy E_t and the heat of formation[7, 8].

The formation energy ΔH of LaNi$_4$AlH$_x$ can be calculated by the following expression:

$$\Delta H = E_t[LaNi_4AlH_x] - \left(E_t[LaNi_4Al] + \frac{x}{2}E_t[H_2] \right) \tag{1}$$

where E_t is the total energy calculated for the equilibrium unit formula shown in the parenthesis.

[1] Corresponding author. Ph.D. candidate. Tel.: +86-242-397-1641; fax: +86-242-389-1320. E-mail: rjzhang@imr.ac.cn

We chose four structures of β-LaNi₄AlHₓ (x=4, 4.5, 5, 7) for study the hydrogen content of LaNi₄AlHₓ. Calculations of these hydrides were made by optimizing all degrees of freedom including cell parameters and internal coordinates within a given space group. For carrying out efficient calculation, the supercells were made; of its sizes were multiple of the unit cell. Zero-point energies were not taken into account because their effects on the total energy were smaller than the computational accuracy, in agreement with those reported by the theoretical calculation in Ref.[10].

3. Results and discussion

3.1. Favorable structure of β-LaNi₄AlHₓ

In the structures of LaNi₄AlHₓ, Al atoms only occupy the 3g sites[11, 12]. The hydrogen in LaNi₄Al has been reported to occupy only two types of sites, which are 6m and 12n[13]. Obviously, Al atoms additive induce the decrease in the number of the Ni atoms and Ni-H bonds. And since the deviation of H from the 3f to the off-center 12n position is driven by the increase in number of Ni-H bonds[7], the occupied sites of hydrogen in LaNi₄Al can be considered as 6m and 3f for the structures of LaNi₄AlHₓ (x=4, 4.5, 5) for simplicity. LaNi₄AlH₇ was supposed to have the same structure as LaNi₅H₇ with the space group $P63mc$[7] for comparing them each other. The occupied sites of hydrogen in LaNi₄AlHₓ (x=4, 4.5, 5) and LaNi₄AlH₇ are shown in Fig.1. The calculated formation energy, equilibrium unit cell volume of β-LaNi₄AlHₓ (x=4, 4.5, 5, 7) and its error to the experimental value are listed in Table 1. It can be seen that both the formation energy and the volume of LaNi₄AlH₄.₅ are most in agreement with the value of experiments. It indicates that LaNi₄AlH₄.₅ is closer and more favorable to the real composition of LaNi₄AlHₓ than the others, though the hydrogen content of LaNi₄AlH₅ looks more similar than LaNi₄AlH₄.₅ to that of LaNi₄AlH₄.₈ observed in experiment.

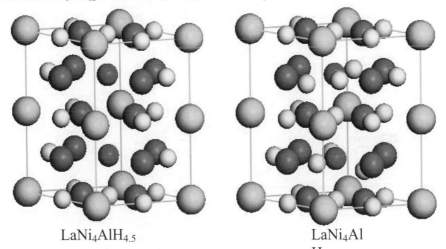

LaNi₄AlH₄.₅   LaNi₄Al

Fig. 1. Schematic supercell of the LaNi₄AlH₄.₅ and LaNi₄AlH₇. Left: The structure with two different H sites (6m and 3f) and a double unit cell (Z=2). Right: The structure reported by Kazuyoshi Tatsumi *et al.* (Ref. 22) with three different H sites (6m, 3f and 4h) and a double unit cell (Z=2). From the largest to smallest atoms denote La, Ni, Al and H atoms, respectively.

Table 1 The computed formation energy (ΔH), equilibrium unit cell volume and its error to the experimental value of β-LaNi₄AlHₓ (x=4, 4.5, 5, 7)

	LaNi₄AlH₄	LaNi₄AlH₄.₅	LaNi₄AlH₅	LaNi₄AlH₇
ΔH (kJ/molH₂) Calc.	- 63.33	- 46.22	- 33.78	- 33.88
ΔH (kJ/molH₂) Expt.[14]		- 47.88		
Volume (Å³)	100.52	102.86	104.81	110.41
Volume (Å³) Expt.[15]		103.3		

3.2. Electronic structures of β-LaNi₄AlHₓ

The total electronic densities of states (TDOSs) for β-LaNi$_4$AlH$_x$ are plotted in Fig.2. The partial density of states (PDOS) plots for a H 1s part, a Ni 3d part, a La 5d part, and an Al 3sp part, which are the main contributors to the local density of states (LDOS), are given in Fig.3. There are two strong PDOS peaks originating from Ni 3d and about five PDOS peaks originating from H 1s. Moreover, the positions of the two strong PDOS peaks from Ni 3d and the strong H 1sσ peaks (approximately −5.4 eV) for LaNi$_4$AlH$_x$ (x=4, 4.5, 5) are not quite different, while those for LaNi$_4$AlH$_7$ has an obvious shift to the low energy region from the others. The anti-bonding state is formed near but below Fermi level E_F (that is set to zero as a reference) by the interaction of H 1sσ* band with a hybridization orbital of metal d, s and p orbitals. The scaled bond order, BO^s $_{H\text{-}Ni\,(or\,La,\,Al)}$ =BO$_{H\text{-}Ni\,(or\,La,\,Al)}$ / BL$_{H\text{-}Ni\,(or\,La,\,Al)}$ between H-Ni, H-La, and H-Al for LaNi$_4$AlH$_x$, which is used to evaluate the covalent bonding strength, are listed in Table 2. BL is the average bond length between H and Ni (or La, Al) atoms. We can consider that the smallest BO^s$_{H\text{-}Ni}$ in LaNi$_4$AlH$_7$ just confirms the decreasing levels of covalency. According to the hydrogen content increasing with the number of H-Ni bonding states, LaNi$_4$AlH$_x$ (x=4, 4.5, 5, 7) should have more H-Ni bonding states with increasing x. However, in the two strong PDOS of Ni 3d, the DOS of the one near to the Fermi surface (approximately −1.4 eV) increases from LaNi$_4$AlH$_4$ to LaNi$_4$AlH$_5$ and then decreases to LaNi$_4$AlH$_7$. Therefore, we can conclude that LaNi$_4$AlH$_7$ should not be formed as a stable hydride due to the reduction in the number of H-Ni bonding states associated with the reduction in the number of d states near the Fermi surface in LaNi$_4$AlH$_7$. So, LaNi$_4$Al can not absorb hydrogen as much as LaNi$_5$.

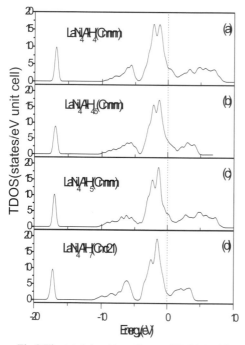

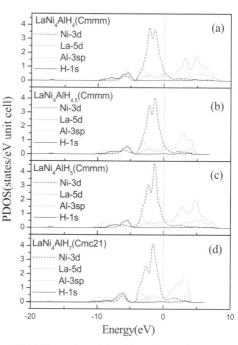

Fig.2 The total densities of states (TDOSs) of β -LaNi$_4$AlH$_x$ (x=4, 4.5, 5, 7). E_F is referenced as zero.

Fig.3 The projected densities of states (PDOSs) of β -LaNi$_4$AlH$_x$ (x=4, 4.5, 5, 7). E_F is referenced as zero.

Table 2 The average bond order (BO), bond length (BL) and scaled bond order (BO^s) between H-Ni and H-La for β -LaNi$_4$AlH$_x$ (x=4, 4.5, 5, 7)

	H-Ni			H-La			H-Al		
	BO	BL	BO^s	BO	BL	BO^s	BO	BL	BO^s
LaNi$_4$AlH$_4$	0.196	2.017	0.097	-0.167	2.535	-0.066	~0.000	3.206	0.000
LaNi$_4$AlH$_{4.5}$	0.184	1.971	0.093	-0.188	2.618	-0.072	0.200	2.029	0.099
LaNi$_4$AlH$_5$	0.204	1.877	0.109	-0.182	2.673	-0.068	0.200	2.006	0.100
LaNi$_4$AlH$_7$	0.183	2.064	0.089	-0.124	2.423	-0.051	0.227	2.010	0.113

4. Conclusions

The electronic structures of β-LaNi$_4$AlH$_x$ (x=4, 4.5, 5, 7) have been investigated by first-principles calculations. The results show as following:
(1) LaNi$_4$AlH$_{4.5}$ is the most favorable structure to the real composition of LaNi$_4$AlH$_x$.
(2) LaNi$_4$AlH$_7$ should not be formed as a stable hydride due to the reduction in the number of H-Ni bonding states associated with the reduction in the number of d states near the Fermi surface in LaNi$_4$AlH$_7$. LaNi$_4$Al can not absorb hydrogen as much as LaNi$_5$.

Acknowledgments

The authors are grateful to the anonymous referees for their careful reading of the manuscript and their fruitful comments and suggestions.

References

[1] H.H. Van Mal, K.H.J. Bushow, F.A. Kuijpers, Hydrogen absorption and magnetic properties of LaCo$_{5x}$Ni$_{5-5x}$ compounds, *J. Less-Common Met* 32 (1973) 289 296

[2] H.H. Van Mal, K.H.J. Bushow, A.R. Miedema, Hydrogen absorption in LaNi$_5$ and related compounds: Experimental observations and their explanation, *J. Less-Common Met* 35 (1974) 65-76

[3] K.H.J. Buschow, P.C.P. Bouten, A.R. Miedema, Hydrides formed from intermetallic compounds of two transition metals: a special class of ternary alloys, *Rep. Prog. Phys.* 45 (1982) 937-1039

[4] A. Anani, A. Visintin, K. Petrov, *et al.*, Alloys for hydrogen storage in nickel/hydrogen and nickel/metal hydride batteries, *J. Power Sources* 47 (1994) 261-275

[5] M D Segall, Philip J D Lindan, M J Probert, *et al*, First-principles simulation-ideas, illustrations and the CASTEP code, *J. Phys.: Condens. Matter* 14 (2002) 2717-2744

[6] J.P. Perdew, J.A. Chevary, S.H. Vosko, *et al*, Atoms, molecules, solids, and surfaces: Applications of the generalized gradient approximation for exchange and correlation, *Phys. Rev. B* 46 (1992) 6671-6687

[7] Kazuyoshi Tatsumi, Isao Tanaka, Haruyuki Inui, *et al*, Atomic structures and energetics of LaNi$_5$-H solid solution and hydrides, *Phys. Rev. B* 64 (2001) 184105

[8] Kazuyoshi Tatsumi, Isao Tanaka, Katsushi Tanaka, *et al*, Elastic constants and chemical bonding of LaNi$_5$ and LaNi$_5$H$_7$ by first principles calculations, *J. Phys.: Condens. Matter* 15 (2003) 6549-6561

[9] D. Vanderbilt, Soft self-consistent pseudopotentials in a generalized eigenvalue formalism, *Phys. Rev. B* 41 (1990) 7892-7895

[10] L.G. Hector Jr., J.F. Herbst, T.W. Capehart, Electronic structure calculations for LaNi$_5$ and LaNi$_5$H$_7$-energetics and elastic properties, *J Alloy Comp* 353 (2003) 74-85

[11] A. Percheron-Guegan, C. Lartigue, J.C. Achard, Neuron and x-ray diffraction profile analyses and structure of LaNi$_5$,LaNi$_{5-x}$Al$_x$, and LaNi$_{5-x}$Mn$_x$ intermetallics and their hydrides (deuterides), *J Less-Common Met* 74 (1980) 1-12

[12] C. Lartigue, A. Percheron-Guegan, J.C. Achard and F. Tasset, Thermodynamic and structural properties of LaNi$_{5-x}$Mn$_x$ compounds and their related hydrides, *J Less-Common Met* 75 (1980) 23-29

[13] A. Percheron-Guegan, C. Lartigue, J.C. Achard, Correlations between the structural properties, the stability and the hydrogen content of substituted LaNi$_5$ compound, *J Less-Common Met* 109 (1985) 287-309

[14] H. Diaz, A. Percheron-Guegan, J.C. Achard, *et al*, Thermodynamic and structural properties of LaNi$_{5-y}$Al$_y$ compounds and their related hydrides, *Int. J. Hydrogen Energy* 4 (1979) 445-454

[15] R.T.Walters, Helium dynamics in metal tritides I. The effect of helium from tritium decay on the desorption plateau pressure for La-Ni-Al tritides, *J Less-Common Met* 157 (1990) 97-108

VSP International
Science Publishers
P.O. Box 346, 3700 AH Zeist
The Netherlands

*Lecture Series on Computer
and Computational Sciences*
Volume 1, 2004, pp. 604-607

A Nonlinear Elastic Unloading Damage Model for Soft Soil and its Application to Deep Excavation Engineering

Xi Hong Zhao

Department of Geotechnical Engineering,
Tongji University, Shanghai 200092, China

Bei Li

Taihu Basin Authority of the Ministry of Water Resources of China
Shanghai 200434, China

Received 14 July, 2004; accepted in revised form 14 August, 2004

This paper is divided into two parts:

1. First part, based on damage soil mechanics [1], combined with test results for unloading stress path in Shanghai soft soil, presents a non-linear elastic unloading damage model for soft soil.

This part deals with a nonlinear elastic unloading model for soft soil and a nonlinear elastic unloading damage model for soft soil in detail.

The constitutive equation for nonlinear elastic unloading model for soft soil is expressed as follows:

$$\{\sigma\} = [C]\{\varepsilon\}$$

in which $\{\varepsilon\} = \{\varepsilon_x \ \varepsilon_y \ \varepsilon_z \ \gamma_{xy} \ \gamma_{yz} \ \gamma_{zx}\}^{\mathrm{T}}$, $\{\sigma\} = \{\sigma_x \ \sigma_y \ \sigma_z \ \tau_{xy} \ \tau_{yz} \ \tau_{zx}\}^{\mathrm{T}}$ and elastic matrix $[C]$

The constitutive equation for nonlinear elastic unloading damage model for soft soil is expressed in incremental form as follows:

$$d\sigma_{ij} = \widetilde{C}_{ijkl} d\varepsilon_{kl} + d\widetilde{C}_{ijkl}\varepsilon_{kl}$$

in which $\widetilde{C}$ is effective elastic matrix for damage soil, and $d\widetilde{C}$ is its increment. According to the principle of virtual work, this incremental FEM equation could be expressed as:

$$\{\Delta F\}^e = \left(\int_{\Omega} [B]^T [\widetilde{C}][B] d\Omega\right)\{\Delta\delta\}^e + \int_{\Omega} [B]^T [d\widetilde{C}]\{\varepsilon\} d\Omega$$

where $\{\Delta F\}^e$ = increment of node force vector for an element, $[B]$ = strain matrix of an element, and $\{\Delta\delta\}^e$ = increment of node displacement vector for an element.

$$[d\widetilde{C}] = \frac{1}{2\widetilde{\tau}} \frac{\partial G}{\partial \widetilde{\tau}} [\widetilde{C}][L_1][B]\{\Delta\delta\}^e$$

where $[L_1] = -\dfrac{\dfrac{\partial G}{\partial\{\varepsilon\}}}{\dfrac{\partial G}{\partial L_d}\dfrac{\partial L_d}{\partial\{D\}}\dfrac{\partial G}{\partial\{\varepsilon\}}}$ and $\{\Delta D\} = -\dfrac{\dfrac{\partial G}{\partial\{\varepsilon\}}\dfrac{\partial G}{\partial\{\varepsilon\}}}{\dfrac{\partial G}{\partial L_d}\dfrac{\partial L_d}{\partial\{D\}}\dfrac{\partial G}{\partial\{\varepsilon\}}}[B]\{\Delta\delta\}^e$

Thus, the incremental FEM equation can be also expressed as:

$$\{\Delta F\}^e = \left(\int_\Omega [B]^T[\tilde{C}][B]d\Omega\right)\{\Delta\delta\}^e + \left(\int_\Omega [B]^T\frac{1}{2\tilde{\tau}}\frac{\partial G}{\partial\tilde{\tau}}[\tilde{C}][L_1][B]\{\varepsilon\}d\Omega\right)\{\Delta\delta\}^e$$

Now, let $[\tilde{K}]^e = \int_\Omega [B]^T[\tilde{C}][B]d\Omega$, the element stiffness matrix coupling with damage,

$[\tilde{K}_d]^e\{\Delta\delta\}^e$, the damage dissipation matrix due to nonlinear elastic damage, in which

$$[\tilde{K}_d]^e = \int_\Omega [B]^T\frac{1}{2\tilde{\tau}}\frac{\partial G}{\partial\tilde{\tau}}[\tilde{C}][L_1][B]\{\varepsilon\}d\Omega$$

then, the incremental FEM equation can be simplified as:

$$\{\Delta F\}^e = [\tilde{K}]^e\{\Delta\delta\}^e + [\tilde{K}_d]^e\{\Delta\delta\}^e$$

It is the incremental FEM equation for non-linear elastic unloading damage model. According to the incremental FEM equation, an FEM program can be complied. Thus, this program can be applied to unloading problem, such as excavation engineering.

2. Second part mainly describes the application of non-linear elastic unloading damage model to an engineering case with the depth of 30.4m. This case is one of most important parts of the project "The Outer Ring Tunnel Project of Shanghai", whose size is the first one in Asia and the second one in the world. In this part FEM is introduced to analyze this specially big and deep excavation engineering in deformation of wall and earth pressure on wall etc.

Next, the outline and assumption of FEM model are introduced as follows:

(1) The constitutive relation of soil is adopted as the nonlinear elastic unloading damage model. Soil behind wall is assumed as unloading horizontally and load invariable vertically, while soil beneath excavation bottom in front of wall is assumed as unloading vertically and load invariable horizontally. The damage of soil is assumed as across-anisotropic with damage variables in horizontal direction as $D_1 = D_2$ and damage variable in vertical direction as D_3.

(2) The diaphragm wall is considered as a linear elastic material while the braces are simulated as spring elements.

(3) The excavation is analyzed as a plane strain problem.

(4) The depth of calculation field is taken as 2 times the depth of the excavation under its bottom. The width of calculation field is taken as half the width of the excavation inside and 2 times the depth of the excavation outside.

(5) The displacement boundary condition is that the bottom boundary is a fixed one and the lateral boundary, fixed horizontally and free vertically.

(6) The water level is assumed consistent with ground level outside the excavation and consistent with the excavation surface inside the excavation. This assumption keeps a constant in one working case.

(7) Initial displacement of the soil is assumed to be zero, i.e. the stress-strain state of the soil is assumed not changed during the construction of diaphragm wall before excavating.

(8) An 'air element' is adopted to simulate any removed element, which means the original element is still remained and the soil is replaced by air with very small modulus.

(9) Incremental method is used to simulate excavation in steps. In every step the element stiffness matrix is calculated using current stress level of soil.

(10) The load of excavation is calculated as follows: the incremental displacement caused by the former excavation step is used to calculate the corresponding incremental stress, accordingly the corresponding incremental node load is obtained as the incremental excavation load of this step.

From the FEM results of macro-analysis and micro-analysis some conclusions could be drawn as follows:

a) Based on damage soil mechanics, combined with test results for unloading stress path in Shanghai soft soil, this paper presents a nonlinear elastic unloading damage model for soft soil. It should be pointed out that this model is the companion model of the non-linear elastic damage model for soft soil.

b) The FEM results reasonably consistent with the measured ones for deformation of wall in every step of excavation, which confirms the feasibility and applicability of the unloading damage model of soil to deep excavation engineering.

c) FEM calculation and analysis show that the distribution of earth pressure along depth and the relationship between earth pressure and displacement of wall could be well described qualitatively, and the reinforced soil beneath excavation bottom could be simulated with increase of initial unloading modulus of the soil. All these facts illuminate the broad application future for the unloading damage model of soil to deep excavation engineering.

d) The FEM results for damage variables show that the damage develops with step of excavation displayed in the expanding of damaged area and increasing of damage variable value. A kind of concentrated damage occurs near the corner of excavation and extends to a certain depth, where the maximum value of damage variable and the measured maximum horizontal displacement occur in these regions.

e) The different evolution rules of damage appear in different directions and different regions of excavation. That is, The different damage for soil in different directions and in different regions of excavation reflects the notable influence for stress path on deformation of soil。

f) In general, the degree of damage in passive region in front of wall is more serious than that in active region behind wall.

Therefore, the method of macro analysis integrated with micro analysis may be a good way for studying the deep excavation engineering

References

[1] X.H. Zhao, H. Sun and K.W. Lo, *Damage Soil Mechanics* (in Chinese and English). Tongji University Press, Shanghai, China(2000).

[2] G.B. Liu and X.Y. Hou, The unloading stress strain characteristic for soft soil. Chinese J. *China Underground Engineering and Tunnel*(in Chinese), 2 16-23(1997).

[3] X.H. Hu, Study on unloading time effect of soft soil in Shanghai, *Dissertation for M.S., Tongji University*(in Chinese), Shanghai, China, 16-23(1999).

[4] S.D. Li, S.Q. Zhang, B.T. Wang et al, Modulus formula for foundation soil under lateral unloading during excavation, Chinese J. *China Civil Engineerin*(in Chinese), 35(5) 70-74(2002).

[5] G.B. Liu, A study on unloading deformation characteristic for soft soil, "*Tongji Geotechnical Engineering Facing to 21st Century–Proceeding of 90th Anniversary of the Founding of Tongji University*" (in Chinese), 320-329(1997).

[6] D.E. Daniel and R.E. Olson, Stress-strain properties of compacted clay. *Proceeding of ASCE*, JGTD, 100(GT10), 1123(1974).

[7] C.L. Chow and J.Wang, An anisotropic theory of elasticity for continuum damage mechanics. *International Journal of Fracture*, 33 3-16(1987).

[8] J. Lemaitre, Evolution of dissipation and damage in metals, submitted to dynamic loading. *Proc. I.C.M.1*, Kyoto Japan(1971).

[9] X.H. Zhao, B. Li, K. Li, et al, A study on theory and practice for specially big and deep excavation engineering—deep excavation engineering in Puxi, Outer Ring Tunnel Project of Shanghai. Chinese J. *Geotechnical engineering(*in Chinese), 25(3) 258-263(2003).

[10] G.X.Yang,, K. Li, X.H. Zhao, et al, A study on IT construction for specially big & deep excavation engineering—Deep excavation engineering of Outer Ring Tunnel Project in Puxi, Shanghai. Chinese J. of *Geotechnical Engineering*(in Chinese) , 25(4) 483-487(2003).

[11] B. Li, Theory and practice on specially big & deep excavation engineering in soft soil areas. *Doctoral degree dissertation* (in Chinese), Tongji University, Shanghai, China (2003).

VSP International
Science Publishers
P.O. Box 346, 3700 AH Zeist
The Netherlands

*Lecture Series on Computer
and Computational Sciences*
Volume 1, 2004, pp. 608-611

Computational Aided Analysis and Design of Chassis

X. Zhao[1]*   J.S Liu*   M. Brown**   C. Adams**

*Department of Engineering,
University of Hull,
HU6 7RX Hull, UK

**Bankside-Patterson Ltd
Barmston Road, Swinemoor Road
HU17 OLA Beverley, UK

Received 6 August, 2004; accepted in revised form 26 August, 2004

Abstract: In this work, a computer-aided design and finite element analysis methodology has been developed for the caravan chassis. 3-D models are created by software ANSYS to predict the stress and deflection of the chassis under various working conditions. The optimal design techniques developed by the group helps to find the best shapes and sizes of the chassis elements. Computer-aided manufacturing has been employed for the components and assembly procedures. The computational techniques have been applied for both the traditional and the new generation chassis. For the traditional chassis, the weak parts of the structure have been discovered with improvements proposed to enhance the four ground beams and reducing the material in other beams. Pre-stress has been reduced for the manufacturing process. For the new generation chassis, a versatile, high quality and cost effective chassis has been successfully designed and launched into UK market. It is envisaged that benefits will gain from the improved products for better performance and less cost and the computational techniques will play a key role in the product design.
Keywords: computer aid design, finite element analysis, optimal design, caravan chassis

Mathematics SubjectClassification: 65F30 73K25

1. Introduction

The old design method for caravan chassis is based on experience and tests, which could be inefficient, conservative, unreliable and costly. With the development of high performance computing techniques, numerical simulation of the caravan chassis is envisaged as an efficient and accurate method to assist design. Market analysis of the chassis shows a trend for a versatile chassis with less production time and cost, better performance and longer warrantee, easier assembly and transportation. Therefore, the development of suitable computational methods for the chassis is proposed to meet the market request [1].

The traditional chassis is the main products in UK caravan market. It employs rolled-steel section beams and weld techniques to assemble the beams into structure. Those beams are composed of four groups: longitudinal channel, cross member, brace (under, side and knee brace) and subsidiary beam. The subsidiary beam consists of tow bar, axle and corner steady. It generally provides 2 years warrantee by painting or galvanizing the chassis. As the result of computer aid engineering, the new generation chassis aims at the standardization of the design and manufacture for the product catalogue. By employing the galvanized beams and riveting techniques, the NG chassis is composed of middle longitudinal channel, tapered wings and cross beams. It would be much stronger than the traditional product and a longer warranty could be offered [2].

2. Computer aided analysis and optimal design

In the caravan industry, the computer aided design techniques are widely used, such as Auto CAD drawings. However, the information of the geometry and configuration provided by Auto CAD is not

---

[1] Corresponding author. Dept of Engineering, University of Hull, E-mail: x.zhao@hull.ac.uk, zhaoalice@hotmail.com

adequate to predict the structure performance under certain load conditions. Theoretical calculation is not applicable for the complex geometry of the chassis. A computational model has more advantages in accuracy and efficiency. Finite element analysis has been used in this paper for the simulation. The numerical method is to replace the analytical equations describing the solid material (stress-strain equations) by discretised approximations and the structural response is obtained numerically at a finite number of chosen distributed points. The possibility of achieving a design that efficiently optimizes multiple performances, coupled with the difficulty in selecting the values of a large set of design variables, makes structural optimization an important tool for the design of chassis. 3-D models have been created by the commercial software ANSYS and applied on the caravan chassis for the first time. Carbon steel BS EN 10025 has been selected for the chassis. Its Young's modulus is 2.05×10^{11} N/(m^2) and Poisson's ratio is 0.285 Ref [3]. Beam element 188 [4] is selected for the simulation. Generally, a computer model is composed of 14 types of beam sections, around 100 nodes and 130 elements. A multi-objective optimisation computer program, MOST (multifactor optimisation of structures technique), has been used to accommodate and implement the optimization [5]. The optimization methodology can be stated as:

$$\text{Minimize compliance } f_1(T_i) \text{ and/or mass } f_2(T_i) \tag{1}$$

$$\text{Subject to geometric and response constraints} \tag{2}$$

For the design of chassis elements, although the constraint and load differ, the objective of the design is the same: find the best possible shapes and sizes of the structural members, i.e. minimising stress, deformation and mass [5].

3. Results and discussions

The performance of the traditional chassis in Bankside-Patterson Ltd Ref [1] has been studied. A uniformly distributed load (UDL) is applied on the cross beam of the chassis to represent the 5.5 ton weight of caravan. In order to meet the Code of Caravan [6] and less pre-stress request from the customers, 3-D models have been tested on 5 working conditions [1]. Calculation shows that the working condition at 3-point support (caravan on wheels and tow bar), is the worst case with the biggest deformation and highest stress. Weak parts are found on the ground brace near the axle (see figure 1). Although the maximum stress 239 N/mm² is bellow the material safety stress (275 N/mm²), it reflects that it may enhance the chassis performance by improving the four beams and other beams may work far away from their fully capacity. Material can be saved from the longitudinal channel and cross beam. In order to meet the customer request for less pre-stress during the manufacturing process, existing traditional chassis has been tested under no pre-stress, current pre-stress and other pre-stress situation with results shown in table 1. It is realized that amount of pre-stress is necessary but can be reduced from 20 mm to 10mm.

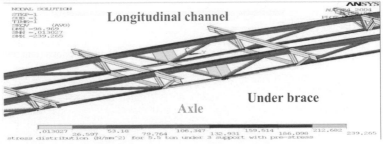

Figure 1: The maximum stress distribution at under brace near the axle

Table 1: The deformation and highest stress in the 3-point support traditional chassis

	Max stress (N/mm²)	Front/back deformation (mm)	
		Final	Pre-stress
1. No pre-stress	285	-60/-117	0
2. Less pre-stress	250	-41/-104	13/13
3. Uneven less pre-stress	261	-75/-95	2/22
4. Uneven less pre-stress	256	-53/-97	5/20
5. Existing pre-stress	236	-38/-97	19/20

The effects of design parameters on traditional chassis performance have been studied (i.e., the beam cross section types and sizes) and applied in finding the optimal design for the new generation chassis. The numerical analysis show that the G shape would be the best cross-sectional shape for the longitudinal channels and the wings, considering load bearing and anti-buckling capabilities [7]. Increasing the height and width of the cross sections and reducing the plate thickness is found efficient to improve the stiffness with less material. Punched holes are designed on the longitudinal channels and wings to assemble the cross members. It reduces the weight of chassis and the galvanising cost. The angle bracket is designed to joint the cross members with longitudinal channels. Four choices in vertical position and various choices in horizontal position have been provided for the cross beams to give flexibility of heights and positions of cross members. A series of holes are punched near the bottom flange of the longitudinal channel to offer a flexible bolt joints from one to three axles. To join the middle longitudinal channel, the wing is flapped and folded. Therefore, two G sections are jointed using the riveting technique in cooperation with specialist in fastening systems. It is a cold formed joint which creates a very tight and strong mechanical interlock with no clearance around fastener. It is a sealed joint without heating and holes [8]. Figure 2 shows the deflection of the NG chassis under 7 ton load. The maximum deflection happened at both ends of the chassis. The final catalogue of the NG chassis has been formed with a standard length of middle longitudinal channel, a set of length wings, and the cross members. It covers all the existing chassis, sized from 7 to 14 meters in length, weighted from 4.2 to 7 ton, and axle from 1 to 3 in number. Any customer request can be fit into 4 catalogues of the chassis and therefore no more design work needed. Current prototype production samples indicate that the production speed has been greatly increased with cost reduction [9]. Figure 3 shows the final product.

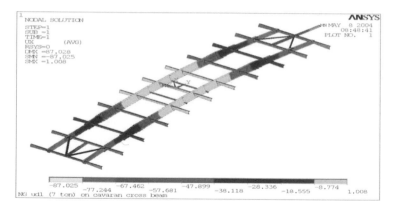

Figure 2: The deflection of chassis under 7 ton load on 5 supports (unit mm)

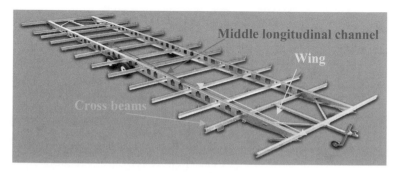

Figure 3: The picture of NG chassis

4. Conclusions

The computational technology (i.e. computer aid design, finite analysis and optimal design) has been applied for the caravan chassis design. For the traditional product, the weak parts of the chassis have been found in the ground brace by computer model. Improvements have been proposed to enhance the under brace beams and reduce material in other elements. Detailed pre-stress has been proposed to meet the customer request. The design helps to cut down manufacturing time and provides competitive market advantages. Furthermore, the design criteria are derived and applied in finding an optimal design. A new generation chassis is proposed with better performance at competitive cost. It is composed of G section longitudinal beams, tapered wings and C section cross beams jointed by the rivet technology. The improved product makes it possible to form a chassis catalogue in terms of suitable range of caravan sizes and loads, it also reduces the design and production process. The design of a flexible axle and cross beam position will provide customers with more choices. Two transportation methods can be selected for NG chassis: (a) shipping the assembled chassis, or (b) shipping the components and assembling the shipped components at the customer's site. An additional benefit is that pre-stress has been significantly reduced from an industry norm of 20 to 25 mm, down to 5 mm currently with the prospect to be eliminated entirely. Customer's responses to date have been very positive and the product has been viewed as a significant advance in the caravan industry.

Acknowledgments

The author wishes to thank the KTP and Bankside-Paterson Ltd for funding of this project.

References

[1] X. Zhao, J. Liu, K Swift, M. Brown and C. Adam, Finite element analysis and optimal for caravan chassis, *the 4th International Conference on Advanced Engineering Design*, 5th-8th Sept 2004, 1-6, Glasgow, UK.

[2] D. A. Brewer, A Chassis for use with Static caravan homes, *UK patent Application,* Bankside-Patterson Ltd., GB 2371026A, 1-9 (2001).

[3] S. P. Timoshenko, *Mechanics of Material*, Van Nostrand Reinhold Company, 1972.

[4] ANSYS User Manual, version 6.1, *ANSYS Company*, 2004.

[5] J.S. Liu and L. Hollawy, Design optimization of composite panel structures with stiffening ribs under multiple loading cases, *Computers and Structures,* **78**, 637-647(2000).

[6] Code of Practice 501 Draft, Specification for undergear of caravan holiday homes and residential park homes, *National Caravan Council Limited,* **2,** 1-12(2003).

[7] X. Zhao, NG II structural analysis, *Technical report,* Bankside-Paterson Ltd, 1-48, 2004.

[8] X. Zhao, J. Liu, K Swift, M. Brown and C. Adam, The design of new generation chassis by advanced techniques, *the 4th International Conference on Advanced Engineering Design*, 5th-8th Sept 2004, 1-6, Glasgow, UK.

[9] C. Adam, New Generation (NG) Chassis, *Launch presentation*, Beverley, UK, March 13, 2004.

VSP International
Science Publishers
P.O. Box 346, 3700 AH Zeist
The Netherlands

*Lecture Series on Computer
and Computational Sciences*
Volume 1, 2004, pp. 612-615

A Transient 3-D Numerical Model for Fluid and Solid Interactions

X. Zhao[1]*   T. G. Karayiannis**   M. Collins**

*Department of Engineering,
University of Hull,
HU6 7RX Hull, UK

**Department of Engineering Systems
London South Bank University
SE1 0AA London, UK

Received 6 August, 2004; accepted in revised form 26 August, 2004

Abstract: In this paper, a transient 3-D numerical model has been developed for the general utilization for the fluid solid interaction problem. Iterative coupling method has been involved to retain the advantages of finite element method (FEM) and finite volume method (FVM) by using commercial software ANSYS6.0 and CFX4.4. The data exchange and mesh control of the fluid solid interface have been developed by user subroutines and FOTRAN programs. It has been applied on the study of water flow in flexible tubes, blood flow in arteries and single phase flow in Coriolis mass flowmeters, satisfying results have been achieved agree to theoretical, numerical and experimental data.

Keywords: Fluid solid interaction, iterative coupling method, finite element method, finite volume method

Mathematics SubjectClassification: 65F30 73K25

1. Literature review

The numerical method for fluid solid interaction is to replace the analytical equations describing the flow (e.g. Navier Stokes equations) or the solid material (stress-strain equations) by discretised approximations and the solutions of the flow properties are obtained by numerical means at a finite number of chosen distributed points. Typically, the numerical schemes in fluid-solid coupling can be classified as one of three types [1]: simultaneous, hybrid and iterative methods. Both simultaneous and hybrid methods limit the involvement of commercial packages and restricts its extension to complex wall behavior. The iterative method ensures that fluid and solid equations are solved separately and then coupled externally in an iterative manner, using the boundary solution from one as a boundary constraint for the other. In its latest developments, Penrose et al's [2] developed a piece of software BLOODSIM to control fluid simulation by CFX5.5 and structure analysis by ANSYS. Further application of BLOODSIM on artery can be found in [3] by the author. However, its limitation on elastic material and inconvenience for user subroutine makes the group to develop a novel coupled fluid-flow and stress-analysis numerical model for the general utilization. Based on the work [1-7], a novel coupling method using CFX4.4 and ANSYS is created. Data transference and the iterative simulation were realized by manual operation between ANSYS and CFX4.4

2. Formulation of the computing

The key in the coupling of the fluid flow discretized equations to the structural discretized equations lies to how the interface conditions are imposed. According to ref [8], the equations (1) of displacement compatibility, traction (velocity) equilibrium and stress equilibrium are applied on the interface

$$d_f^S = d_s^S, \quad f_f^S = f_s^S, \quad s_f^S = s_s^S \tag{1}$$

Where d_f^S, f_f^S and s_f^S (d_s^S, f_s^S and s_s^S) are the displacement, tractions and stress of the fluid (solid) on the interface respectively.

---

[1] Corresponding author. Dept of Engineering, University of Hull, E-mail: x.zhao@hull.ac.uk, zhaoalice@hotmail.com

The numerical model developed in this paper is based on Arbitrary Langarian-Eulerian method and designed for the general utilization of FSI problems, with no limitation of deformation requirement and material property. Interface of fluid and solid domain was selected as data transformation base. Hexahedral element was employed both for fluid and solid domain with identical network on interface. Fluid mesh comply the solid deformation on the moving boundary. Mesh movement of the interface and inside fluid domain is controlled by CFX4.4 subroutine USRGRD, data transformed by USRTRN, boundary condition by USRBCS. Particular FOTRAN programs (Fromctoa.f and Fromatoc.f) have been created to transfer the data format between CFX and ANSYS, i.e. stress on interface element faces and interface nodes relocation. Figure 1 scheme the details of method 1.

3. Results and discussions

First application of the numerical model is to study the wave propagation in an elastic tube. A water flow in elastic tube is selected where the fluid inlet pressure is proportional to time. Figure 2 shows the comparison of numerical model present in this paper compared with wave speed calculated by analytical [9] and published numerical result [2]. The calculation shows that the numerical result of this model under predict the speed while the published numerical model over predict the speed compared with analytical result.

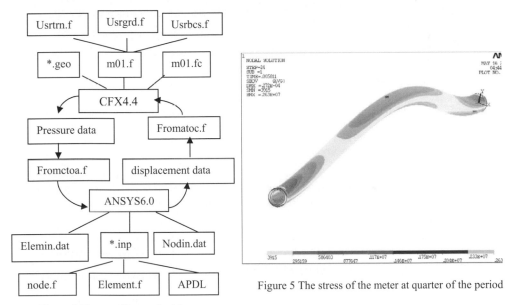

Figure 1 The coupling method 1

Figure 5 The stress of the meter at quarter of the period

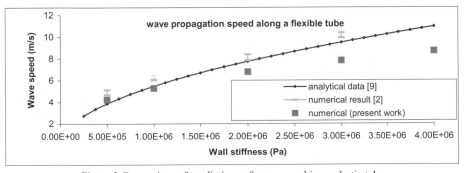

Figure 2 Comparison of predictions of wave speed in an elastic tube

Second application of the numerical model is to study the pulsatile blood flow in simple artery. The artery is assumed as simple axis symmetric geometry with inner radius 5mm, outer radius 6mm and length 50mm. Blood is treated as a incompressible laminar, Newtonian fluid, with 10^3 kg/m^3 in density and 4×10^{-3} kg/(s·m) in viscosity. The decoupled solid and fluid models were first validated before the running the coupled calculation. The coupling model has been tested for the following benchmarks: (1) steady flow in rigid artery, (2) steady flow in linear elastic artery, (3) pulsatile flow in rigid artery, (4) pulsatile flow in linear elastic artery and (5) pulsatile flow in nonlinear elastic artery [2, 7]. Figure 3 shows the comparison of the numerical result with published data. A satisfying agreement has been achieved.

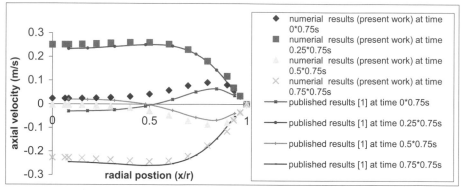

Figure 3 Velocity comparison for transient flow in linear elastic artery

Finally the single-phase flow in straight Coriolis meter has been simulated [6]. Figure 4 shows the time delay of the flow speed in comparison with the numerical result, where D is the sensor distance of the meter. A linear relationship of the flow rate and time delay between the sensors can be seen. Numerical results agree to the experimental [10] and analytical data [11]. Figure 5 shows the stress maximum distribution of the meter at quarter of the vibration period, which agrees to published numerical result [12] .Furthermore, the coupling model has been applied to study the design parameter and the application parameter influence on the meter performance. It helps to investigate the transient fluid pressure and solid stress distribution. Results show the meter's accuracy is restricted in a certain range of flow conditions.

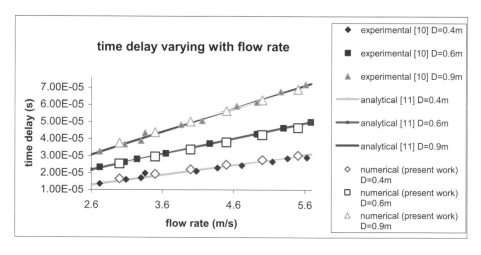

Figure 4 The time delay varying with flow speed and sensor locations of the meter

4. Conclusions

The comprehensive numerical method combining two commercial codes (CFX4.4 and ANSYS) for solving coupled solid/fluid problems has been developed and applied to the simulation of water flow in tube, blood flow in arteries and single-phase flow in Coriolis meters. Those benchmarks have shown that the transient 3-D model capacity for accuracy modeling fluid solid interaction with wide application. As a novel research result, the fluid pressure and solid stress distribution of the Coriolis meter has been investigated and the design and application factors of the meter have been tested by the numerical model.

References

[1] S.Z. Zhao, The numerical study of fluid-solid interactions for modelling blood flow in Arteries, *Ph.D. thesis,* City University, UK (1999).

[2] J. M. T. Penrose, D. R. Hose, C. J. Staples, I. S. Hamill, I. P. Jones, and D. Sweeney, Fluid structure interactions: coupling of CFD and FE, 18. *CAD-FEM Users' Meeting, Internationale FEM-Technologietage,* Graf-Zeppelin-Haus, Friedrichshafen, 20-22 September (2000).

[3] X. Zhao, The numerical study of moving boundary problems: application in arteries and Coriolis meters, *Ph.D. thesis in preparation,* London South Bank University, UK (2004).

[4] S. Z. Zhao, X. Y. Xu and M. W. Collins, The numerical analysis of fluid-solid interactions for blood flow in arterial structures, part 2: development of coupled fluid-solid algorithms. *Proc Instn Mech Engrs,* **212,** 241-252 (1998).

[5] S. Z. Zhao, X. Y. Xu and M. W. Collins, A novel numerical method for analysis of fluid and solid coupling, *Numerical Methods in Laminar and Turbulent Flow,* **10,** Editors Taylor, C., 525-534 (1997), also in *the Proceedings of the Tenth International Conference* Held in Swansea, 21-25 July (1997).

[6] X. Zhao, M. Collins and T. G. Karayiannis, Numerical coupling method for moving boundary problems: application in Coriolis meters, *Proceeding of WSEAS and IASME International Conferences on Fluid Mechanics,* Corfu, Greece, August 17-19, 1-6 (2004).

[7] X. Zhao, J. M. T. Penrose, S. Z. Zhao, T. G. Karayiannis and M. Collins, Numerical study of fluid solid interaction (FSI) in arteries, *Proceeding of WSEAS and IASME International Conferences on Fluid Mechanics,* Corfu, Greece, August 17-19, 1-6 (2004).

[8] K. J. Bathe, H. Zhang, S. H. Ji, Finite element analysis of fluid flows fully coupled with structural interactions, *Computers and structures,* **72,** 1-16(1999).

[9] McDonald, *Blood Flow in Arteries,* Edward Arnold (1990).

[10] Sultan, G., Theoretical and Experimental Studies of the Coriolis Mass Flowmeter, *Ph.D. Thesis,* Cranfield University, UK (1990).

[11] H. Raszillier, and F. Durst, Coriolis-effect in mass flow metering, *Archive of Applied Mechanics,* **61,** 192-214 (1991).

[12] R. M. Watt, Modelling of Coriolis Mass Flowmeters Using ANSYS, *ANSYS users Conf.* Pittsburgh, U.S. No. 10, 67-78(1991).

VSP International
Science Publishers
P.O. Box 346, 3700 AH Zeist
The Netherlands

*Lecture Series on Computer
and Computational Sciences*
Volume 1, 2004, pp. 616-619

Reliability Analysis of Concrete Structure
by Monte Carlo Simulation

Zhigen Zhao [a,b 1]   Yingping Zheng [a]   Wei-ping Pan [b]

a. Department of Resources and Environmental Engineering, Anhui University of Science and
Technology, Huainan, P. R. China 232001;   b. Material Characterization Center,
Western Kentucky University, Bowling Green, KY, USA 42101

Received 6 August, 2004; accepted in revised form 15 August, 2004

Abstract: Traditionally, factor of safety has always been used to evaluate the stability of a given structure.
However, factor of safety is a definite value and is established with no consideration of any uncertainties
and variability of the concerned objective. In this paper, the difference between reliability and factor of
safety is briefly introduced. In addition, taking a given concrete structure as an example, the Monte Carlo
simulation method is introduced for the reliability analysis. This research may be helpful to provide a
method for evaluating the safety degree more effectively and completely.

Keywords: safety degree, factor of safety, reliability, Monte Carlo simulation, concrete structure

Mathematics Subject Classification: 65C05

1 Introduction

In the past, factor of safety has always been used to evaluate the safety degree of a given structure.
However, the factor of safety is limited to a definite value and is established with no consideration of
the uncertainties and variability of the concerned objective. Therefore, factor of safety cannot
effectively and completely reflect the safety degree of a given structure. For example, the safety degree
is not 120% when the factor of safety is 1.2. Moreover, the same factors of safety may not mean the
same degrees of safety to different given structures.

In fact, the safety degree is a random variable with uncertainty and variability. Therefore, it is
important to take this kind of randomness into consideration when the safety degree of a given structure
is studied. In the recent years, the reliability analysis is rapidly developed and widely used to evaluate
the safety degree in many fields, and is generally accepted as an effective way to make up for the
shortcomings of factor of safety in concerning the randomness of safety degree.

The theory of Reliability Analysis is a frontier science based on Probability and Mathematical Statistics.
Because the reliability of a given structure can be measured by probability, so the mathematic method
can be used to evaluate the safety degree of a concerned project.

In this paper, the difference between the reliability and factor of safety is briefly introduced. In addition,
taking a given concrete structure as an example, the Monte Carlo simulation method of reliability
analysis is introduced.

2 Reliability and Factor of Safety

2.1 Reliability

Reliability (R) can be defined as the probability of a device performing its purpose adequately for the
period of time intended under the specified conditions encountered. This probability is actually called
probability of success (P_s) while the opposite is the probability of failure (P_f). Their relationship

---

[1] Corresponding author. A PhD and a Professor of Anhui University of Science and Technology. And at present, also a visiting
scholar at Western Kentucky University. E-mail: zhgzhao@aust.edu.cn or zhigen.zhao@wku.edu.

is $R = P_s = 1 - P_f$. A device as mentioned above can be anything, and in this paper, it refers to a given concrete structure.

The reliability of a given concrete structure can be defined as the probability when the random variable of strength (S) is greater than the random variable of load (L) of this given concrete structure. That is $R = P_s = P(S > L) = P(S - L > 0)$.

2.2 Factor of Safety

Factor of safety can be defined as the ratio of the available to the required, or the Capability of a device to the Demand on this device. In this paper, the available or the Capability means the Strength of a given concrete structure while the required or the Demand means the Load on this given concrete structure. Therefore, the factor of safety of a given concrete structure (F_s) can be defined as the ratio of average Strength ($\overline{S}$) to average Load ($\overline{L}$). That is $F_s = \overline{S} / \overline{L}$.

3 A Given Concrete Structure

3.1 Distribution of the Strength and the Load

Listed in Table 1 and Figure 1 are the distributions of the Strength and the Load of a given concrete structure.

Table 1: Distribution of the Strength and the Load of a given concrete structure

Class		Load(L)		Strength(S)	
Interval (Mpa) (I_i)	Midpoint (Mpa) (M_i)	Relative Frequency (f_i)	Relative Cumulative Frequency (F_i)	Relative Frequency (f_i)	Relative Cumulative Frequency (F_i)
14.5-15.5	15.0	0.000	0.000		
15.5-16.5	16.0	0.078	0.078		
16.5-17.5	17.0	0.095	0.173		
17.5-18.5	18.0	0.118	0.291		
18.5-19.5	19.0	0.131	0.422		
19.5-20.5	20.0	0.126	0.548	0.000	0.000
20.5-21.5	21.0	0.112	0.660	0.075	0.075
21.5-22.5	22.0	0.101	0.761	0.092	0.167
22.5-23.5	23.0	0.091	0.852	0.107	0.274
23.5-24.5	24.0	0.080	0.932	0.123	0.397
24.5-25.5	25.0	0.068	1.000	0.143	0.540
25.5-26.5	26.0			0.150	0.690
26.5-27.5	27.0			0.131	0.821
27.5-28.5	28.0			0.101	0.922
28.5-29.5	29.0			0.078	1.000

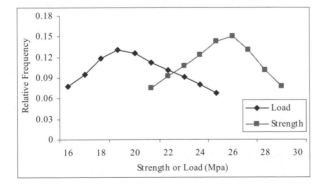

Figure 1: Relative Frequency distribution of Strength and Load

3.2 Factor of Safety

Based on table 1, when the midpoint stands for the corresponding interval, the mean values of Load and Strength can be calculated as follows:

$$\bar{L} = \sum L_i \times f_i = 16.0 \times 0.078 + 17.0 \times 0.095 + \ldots + 25.0 \times 0.068 = 20.283$$

$$\bar{S} = \sum S_i \times f_i = 21.0 \times 0.075 + 22.0 \times 0.092 + \ldots + 29.0 \times 0.078 = 25.114$$

Therefore, the factor of safety is $F_s = \bar{S} / \bar{L} = 25.114 / 20.283 = 1.238$

3.3 About Probability of Failure

Although factor of safety is 1.238 based on the above calculation, the probability of failure is obviously existed when the Strength is less than the Load. In the following, the Monte Carlo Method is used to calculate this probability of failure so as to obtain the probability of success.

4 Reliability Analysis by Monte Carlo Simulation

The term "Monte Carlo" was introduced by von Neumann and Ulam during World War II, as a code word for the secret work at Los Alamos; it was suggested by the gambling casinos in the city of Monte Carlo in Monaco; however, the Monte Carlo simulation goes way beyond gambling applications. Currently, the Monte Carlo simulation is widely used in many fields. The principle behind this method is to develop a computer-based analytical model that predicts the behavior of a system. That is this method provides approximate solutions to a variety of mathematical problems by performing statistical sampling experiments on a computer.

4.1 Random Number Generation

Many techniques for generating random numbers have been suggested, tested, and used in past years. The most commonly used present-day method for generating pseudorandom numbers is one that produces a nonrandom sequence of numbers according to a recursive formula that is based on calculating the residues modulo of an integer m of a linear transformation. Although these processes are completely deterministic, it has been shown that the numbers generated by the sequence appear to be uniformly distributed and statistically independent.

Congruential methods are based on a fundamental congruence relationship, which may be expressed as $X_{i+1} = (aX_i + c)(\mathrm{mod}\, m), i = 1, \cdots, n$, where the *multiplier* a, the *increment* c, and the mod*ulus* m are nonnegative integers. Given an initial starting value X_0 (also called the *seed*), the sequence $\{X_i\}$ with any value i can be yielded. The random numbers on the unit interval $(0,1)$ can be obtained by $U_i = X_i / m$.

4.2 Random Variables ($\{S_i\}$, $\{L_i\}$) Generation

Take Strength as an example to illustrate how to get the random variable $\{S_i\}$.

Based on Table 1, the relative cumulative frequency $\{F_i\}$ is known. For a certain U_i, it falls into $[F_i, F_{i+1})$, and corresponding No. (i+1) class, and M_{lower}, M_{upper} represent lower class boundary value and upper class boundary value. So, the $S_i = M_{lower} + (M_{upper} - M_{lower}) \times (U_i - F_i) / (F_{i+1} - F_i)$

For instance, suppose $U_i = 0.600$, it falls into $[0.540, 0.690)$, corresponding $M_{lower} = 25.5$, $M_{upper} = 26.5$. So, $S_i = 25.5 + (26.5 - 25.5) \times (0.600 - 0.540) / (0.690 - 0.540) = 25.9000$.

Likewise, for Load, it is the same way with the Strength. Suppose $U_i = 0.600$, it falls into $[0.548, 0.660)$, corresponding $M_{lower} = 20.5$, $M_{upper} = 21.5$. Therefore,

$S_i = 20.5 + (21.5 - 20.5) \times (0.600 - 0.548) / (0.660 - 0.548) = 20.9643$.

4.3 Flow Chart of Monte Carlo Method

The flow chart of Monte Carlo Simulation is given in Figure 2.

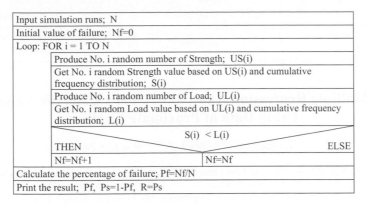

Figure 2: Flow chart of Monte Carlo Method

4.4 Simulation Results

From Table 2, the results are relatively stable with the increase of simulation runs. Therefore, the percentage of success can be deemed as probability of success, and that is reliability. The probability of success is about 0.94, which means that there is about 6% probability of failure for this concrete structure although the factor of safety is 1.238.

Table 2 Results of Monte Carlo Simulation

Simulation runs	200	300	400	500	600	700	800	900	1000
Reliability	0.935	0.933	0.940	0.940	0.940	0.943	0.941	0.941	0.944
Simulation runs	2000	3000	4000	5000	6000	7000	8000	9000	10000
Reliability	0.939	0.942	0.940	0.942	0.939	0.940	0.942	0.942	0.942

5 Summary

Concerning the research about the safety degree of a device, factor of safety is always useful but with no consideration of the randomness. Reliability analysis is generally accepted as an effective and more complete way to make up for the shortcomings of factor of safety. The Monte Carlo Method is a practical approach in analyzing the reliability.

Acknowledgments

The author wishes to thank Stanley Herren for his careful reading of this manuscript and Dr. Boshu He for his fruitful comments and suggestions.

References

[1] Bird, J. O. *Basic engineering mathematics*. Oxford, Newnes, 2000

[2] B. M. Ayyub, R. H. McCuen. *Probability, statistics, and reliability for engineers and scientists*. Chapman & Hall/CRC, Boca Raton, Florida, 2003

[3] D. P. Landau, K. Binder. *A guide to Monte Carlo simulations in statistical physics*. Cambridge University Press, Cambridge, New York, 2000.

[4] G. Muscolino, G. Ricciardi, P. Cacciola. Monte Carlo simulation in the stochastic analysis of non-linear systems under external stationary Poisson white noise input. *International Journal of Non-Linear Mechanics*, **38** (8) 1269-1283(2003)

[5] M. Christian: *Design of concrete structures*. Prentice Hall, Upper Saddle River, New Jersey, 1996.

VSP International
Science Publishers
P.O. Box 346, 3700 AH Zeist
The Netherlands

*Lecture Series on Computer
and Computational Sciences*
Volume 1, 2004, pp. 620-623

Multiple Regression Analysis of Coal's Calorific Value Using Data of Proximate Analysis

Zhigen Zhao [a,c 1]   Boshu He [b,c]   Wei-ping Pan [c]

a. Department of Resources and Environmental Engineering, Anhui University of Science and Technology, Huainan, P. R. China 232001;   b. Beijing Jiaotong University, Beijing, P. R. China 100044;   c. Combustion Lab, Department of Chemistry, Western Kentucky University, Bowling Green, KY, USA 42101

Received 6 August, 2004; accepted in revised form 15 August, 2004

Abstract: 210 pairs of coal's calorific value and proximate analysis are collected in this paper. And, the multiple linear regression equation is established between coal's calorific value and proximate analysis groups. The regression coefficients are rational in the theory of coal science. And, this equation is significant by the F-test and R-test. The work reveals that multiple regression analysis provides an effective way to research their relationship quantitatively and with multi-variables.

Keywords: coal's calorific value, proximate analysis, multiple regression analysis

Mathematics SubjectClassification: 62J12

1 Introduction

The calorific value of a coal is the combined heats of combustion of the carbon, hydrogen, etc. The energy released upon combustion is of primary interest to coal producers and users. In addition, proximate analysis is the most often used type of analysis for characterizing coals in connection with their utilization. The proximate analysis of coal separates the products into four groups: moisture; ash; volatile matter; and fixed carbon.

The relationship between coal's calorific value and proximate analysis groups is not well researched. In the past, it is common to discuss their relationship qualitatively instead of quantitatively or to discuss the relationship between coal's calorific value and only one proximate analysis group, instead of all of the proximate analysis groups. It is known that coal's calorific value is complex and influenced by many factors, therefore single predictor studies are limited in their predictive power. Multiple regression analysis, focused on the establishment of the equation and predicting a dependent variable from a set of predictors, provides an effective solution to this problem.

The purpose of this paper is based on the data collected, to use multiple regression analysis to establish the equation between coal's calorific value and proximate analysis groups to investigate their relationship quantitatively and with multi-variables.

2 About Data Set

2.1 About Data Source

The data in this paper are collected from Combustion Lab in Department of Chemistry, Western Kentucky University. All coal samples were properly collected from several power plants and prepared according to ASTM D2013. The groups of Proximate Analysis of the coal were tested by the method established with ASTM D5142. In addition, the Calorific Value of Coal was tested according to ASTM D5865. The data concerning Proximate Analysis and Coal's Calorific Value given in this paper are based on the same basis, i.e, as-determined basis.

---

[1] Corresponding author. A PhD and a Professor of Anhui University of Science and Technology. And at present, also a visiting scholar at Combustion Lab of Western Kentucky University. E-mail: zhgzhao@aust.edu.cn or zhigen.zhao@wku.edu.

2.2 About Independent Variables and Dependent Variable

The proximate analysis of coal includes moisture, ash, volatile matter and fixed carbon. Among these, fixed carbon is the resultant of the summation of the percentage of moisture, ash, and volatile matter subtracted from 100. Specifically, fixed carbon is a calculated value, not an independent factor.

Therefore, the independent variables are moisture (X_1), ash (X_2) and volatile matter (X_3). The dependent variable is Calorific Value of Coal (Y).

2.3 Data Set

The sample size is 210. The data set consists of a set of 210 observations on 3 independent variables and 1 dependent variable. The data set can be a matrix having dimensions of 210 by 4. Table 1 gives the illustrated format of the data set, while Table 2 gives the brief information about the characters of the 4 variables.

Table 1: Illustrated format of data set

Sample No.	Independent variables			Dependent variable
	Moisture (%)	Ash (%)	Volatile matter (%)	Calorific Value (btu/lb)
	X_1	X_2	X_3	Y
1	5.61	10.76	39.07	13803.87
...	...	...	...	...
210	7.81	12.02	56.32	13952.84

Table 2: Characters of concerned variables

Characters	Independent variables			Dependent variable
	Moisture (%)	Ash (%)	Volatile matter (%)	Calorific Value (btu/lb)
Average	5.52	10.94	39.38	13535.42
Max	10.41	15.57	59.61	14440.27
Min	2.65	5.35	29.76	12091.00

3 Multiple Regression Analysis

In general, linear associations are evident, although the scatter about the apparent linear relationships is significant. In addition, the linear models have been used in the past research, linear models are easily applied, and the statistical reliability is easily assessed.

3.1 Three Components of Regression Analysis

The three components of regression analysis are the model, the objective function, and the data set. In this paper, there are 3 independent variables and 1 dependent variable, and the linear association is proposed. Therefore, the multiple regression model is:

$$\hat{Y} = b_0 + b_1 X_1 + b_2 X_2 + b_3 X_3 \tag{1}$$

In which $X_j (j = 1,2,3)$ are the independent variables, $b_j (j = 1,2,3)$ are the partial regression coefficients, b_0 is the intercept coefficient, and Y is the dependent variable.

Using the least-squares principle and the above model, the objective function becomes:

$$F_{obj} = \min \sum_{i=1}^{n} e_i^2 = \min \sum_{i=1}^{n} (\hat{y}_i - y_i)^2 = \min \sum_{i=1}^{n} \left(b_0 + \sum_{j=1}^{3} b_j x_{ij} - y_i \right)^2 \tag{2}$$

In which, F_{obj} is the value of the objective function. In addition, the independent variables include two subscripts, with i indicating the observation and j the specific independent variable, and n is the number of observations (i.e., the sample size).

3.2 Regression Coefficients

The solution method to minimize the objective function is to take the 4 derivatives of the objective function, Equation 2, with respect to the unknowns, $b_j (j = 0,1,2,3)$; setting the derivatives equal to zero; and solving for the unknowns. A set of 4 normal equations is an intermediate result of this process. The solution of the 4 simultaneous equations yields values of b_0, b_1, b_2, and b_3.

In this paper, $b_0 = 12954.68, b_1 = -162.7956, b_2 = -81.7136, b_3 = 60.2579$

Given the estimates b_0, b_1, b_2, and b_3, the estimated regression equation is:

$$\hat{Y} = 12954.68 - 162.7956X_1 - 81.7136X_2 + 60.2579X_3 \qquad (3)$$

4 Reliability of the Regression Equation

After a multiple linear regression equation has been established, we may ask how well the model represents the observed data. It is natural to seek measures of correlation that reflects the "goodness of the fit."

4.1 The Rationality of the Coefficients

The moisture absorbs the heat until it vaporizes in combustion. Coal ash is the remaining residue derived from the mineral matter after the combustion of coal under the specified condition. The volatile matter obtained during the pyrolysis of coal consists mainly of combustible gases such as hydrogen, carbon monoxide, and methane plus other hydrocarbons. According to the theory of coal science, usually, the higher percentage the moisture or the coal ash, the lower the calorific value of the coal, and the higher percentage the volatile matter, the higher the calorific value.

In this paper, the regression coefficients are rational: $b_1 = -162.7956$, corresponding to moisture (X_1), $b_2 = -81.7136$, corresponding to coal ash (X_2). Both b_1 and b_2 are negative numbers, which means moisture and coal ash are negative in relation to the Calorific Value of Coal. In addition, $b_3 = 60.2579$, corresponding to volatile matter (X_3) and is a positive number, which means volatile matter is positive in relation to the Calorific Value of Coal.

4.2 Test of Significance

Table 3: Analysis of variance

Source of Variation	Sum of Squares	Degrees of Freedom	Mean Square	F Value
Regression	40148390	3	13382796.67	182.90
Residual	15073170	206	73170.73	
Total	55221560	209		

4.2.1 F-test

Based on the hypothesis test $H_0 : b_1 = b_2 = b_3 = 0$, H_1 : not all $b_k = 0$, $k = 1,2,3$, We use the test statistic $F = MS_{reg} / MS_{res}$ to determine if there is a regression relationship between the dependent variable Y and the independent variables (X_1, X_2, X_3), in which, MS_{reg} is mean square due to regression, MS_{res} is mean square due to residual.

$$F = MS_{reg} / MS_{res} = 13382796.67 / 73170.73 = 182.90 \qquad (4)$$

The critical value F_c is taken from an F-table. At the $\alpha = 0.01$ level of significance, $F_c(3,206) = 3.88$. Because F is significantly larger than $F_c(3,206)$, the "null" hypothesis is rejected. That is to say there is a regression relationship between Y and (X_1, X_2, X_3).

4.2.2 R-test

The correlation coefficient (R) measures the degree to which the measured and predicted values agree and is used as a measure of the accuracy of future predictions.

$$R^2 = \sum_{i=1}^{n}(\hat{y_i} - \overline{Y})^2 / \sum_{i=1}^{n}(y_i - \overline{Y})^2 = (4.014839 \times 10^7)/(5.522156 \times 10^7) = 0.7270 \qquad (5)$$

$$R = \sqrt{R^2} = \sqrt{0.7270} = 0.8527 \qquad (6)$$

The critical value R_c is taken from an R-table. At the $\alpha = 0.01$ level of significance and with 4 variables, $R_c(208) = 0.234$.

Because R is significantly larger than R_c, the regression equation is significant and thus is practical to use for the study of relationship between coal's calorific value and proximate analysis groups and also for future predictions of coal's calorific value by proximate analysis groups.

5 Summary

Coal's calorific value is complex and influenced by many factors, and single predictor studies are limited to interpret the multi-variables relationship and also limited in their predictive power. Multiple regression analysis provides an effective solution to this problem.

In this paper, 210 pairs of coal's calorific value and proximate analysis are collected. Next, the multiple linear regression equation is established between coal's calorific value and groups of proximate analysis. The regression coefficients reveal moisture and coal ash are negative to Calorific Value of Coal while volatile matter is positive, which is rational in the theory of coal science. Moreover, this equation is proven significant by the F-test and R-test.

Acknowledgments

The author wishes to thank Stanley Herren for his careful reading of this manuscript and Dr. Zhongxian Cheng for his fruitful comments and suggestions.

References

[1] L. Grainger and J. Gibson. Coal utilization: technology, economics and policy. Halsted Press: New York, 1981. 3-30

[2] J. G. Speight. The chemistry and technology of coal. M. Dekker: New York, 1994. 3-20

[3] B. M. Ayyub, R. H. McCuen. Probability, statistics, and reliability for engineers and scientists. Chapman & Hall/CRC, Boca Raton, Florida, 2003.

[4] P. C. Meier, R. E. Zund. Statistical methods in analytical chemistry. John Wiley & Sons, Inc: New York, 2000

[5] J. P. Stevens. Intermediate statistics: a modern approach. Lawrence Erlbaum Associates, Publishers: Mahwah, New Jersey, 1999

[6] C. J. Huberty. Multiple Correlation Versus Multiple Regression. *Educational and Psychological Measurement*, **63 (2)** 271-278 (2003)

VSP International
Science Publishers
P.O. Box 346, 3700 AH Zeist
The Netherlands

*Lecture Series on Computer
and Computational Sciences*
Volume 1, 2004, pp. 624-627

A Possible Organic Half-Metallic Magnet: 2-(5-pyrimidinyl)-4,4,5,5-tetramethyl-4,5-dihydro-1H-3-oxoimidazol-1-oxyl

Wei-Dong Zou[1†], Minghu Wu[2]

[1]Department of Physics, Xianning College, Xianning, Hubei 437005, China
[2] Department of Chemistry, Xianning College, Xianning, Hubei 437005, China

Received 6 August, 2004; accepted in revised form 21 August, 2004

Abstract: Ab initio calculations have been performed to study the magnetic and half metallic properties in the pure organic materials: 2-(5-pyrimidinyl)-4,4,5,5-tetramethyl-4,5-dihydro- 1H-3-oxoimidazol-1-oxyl. The total and partial density of states and atomic spin magnetic moments are calculated and discussed. The calculations revealed that the unpaired electrons in this compound are localized in a molecular orbital constituted primarily of $\pi^*(NO)$ orbital, and the main contribution of the spin magnetic moment comes from the NO free radicals. It is predicted that this compound is half-metallic magnetism. It is also found that there exists ferromagnetic intermolecular interaction in the compound.

Keywords: An initio method; The organic half-matellic magnet; Electronic structure; Ferromagnetic properties

PACS: 75.25.+z, 71.20.-b, 72.25.-b, 75.10.Lp

1. Introduction

The synthesis of ferromagnets using only organic constitutes is a challenge for molecular chemists. In general, covalent bonding between atoms in organic compounds pairs off all available electrons resulting in a net spin S = 0, and a diamagnetic materials. However, there are a number of examples of pure organic compounds made from molecules containing free radicals with one unpaired electron and thus S = 1/2. The first pure organic compound which exhibits ferromagnetic behavior (βphase of (*p*-nitrophenyl) nitronyl nitroxide) has been reported by Kinoshita[1]. With this context, Takahashi, Chiarelli and Rey et al have also succeeded in synthesizing this kind of pure organic ferromagnets [2-8]. At the same time, the mechanism of ferromagnetic interactions attracts interest of research and a good number of radical compounds are still studied now.

In 1993, Rey reported the structural characterization and magnetic properties of 2-(5-pyrimidinyl)-4,4,5,5-tetramethyl-4,5-dihydro-1H-3-oxoimidazol-1-oxyl[5]. To understand deeply the mechanism of the ferromagnetic interactions in this compound, a more detailed knowledge of the electronic band structure and the magnetic properties is still required. In this paper, we adopt the density-functional theory (DFT) with generalized gradient approximation (GGA) [9] to calculate the electronic band structure and the ferromagnetic properties of this compound by the full potential linearized augmented plane wave (FPLAPW). We obtain the electronic structure, the partial density of states and atomic spin magnetic moments of this compound. From the partial density of states and atomic spin magnetic moments, the mechanism of the ferromagnetic interactions and the half-metallic property in this compound is discussed.

2. Results and Discussion

The first-principles electronic structure calculations employ the well known full-potential linearized augmented-plane-wave method. In this method, no shape approximation either on the potential or on the electronic charge density is made. The calculations were performed using the WIEN [10]. The code was developed by P. Blaha and allows inclusion of local orbits in basis. Thus it improves upon linearization and makes possible a consistent treatment of semicore and valence in one

---

† Corresponding author. Department of Physics, Xianning College, Xianning, Hubei 437005, China E-mail: zouwd@yahoo.com.cn

energy window. The exchange and correlation effects are treated with the generalized gradient approximation (GGA) according to Perdew-Burke-Ernzerhof [9]. In order to achieve a satisfactory degree of convergence, the plane-wave cutoff energy is 154eV. Thirty k-points in the irreducible Brillouin zone (IBZ) are used in the spin-polarized approximations. According to experiment, the compound have tetragonal lattice. The lattice parameters we used in the present calculation are

$a=18.245\ \overset{0}{A}$, $b=18.245\ \overset{0}{A}$, $c=7.427\ \overset{0}{A}$.

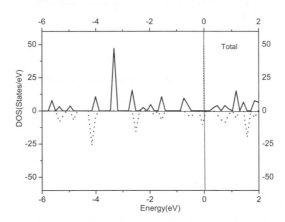

Fig.1 The calculated total density of states (DOS) for molecule. The solid and dot line denote majority and minority spin, respectively. The Fermi levels are located at 0 eV.

To study the electronic structure and the magnetic properties of this compound, the total density of states (DOS) of the molecule are shown in Fig.1. Because the DOS distribution near the Fermi level determines the magnetic properties, we concentrate our attention upon the DOS in the vicinity of the Fermi level, which range from –3.0eV to 1.0eV. From Fig.1, we found that the total DOS below the Fermi level are sharp peaks, this means that all energy bands below the Fermi level are narrow and flat and the electrons are localized. In the vicinity of the Fermi level, the total DOS distribution of the up- and down-spin electrons are obviously split. One valence band is split into two subbands; one is the up-spin valence band, the other one is the down-spin band. So the ordered spin arrangement of electrons are formed by the exchange-correlation of the electrons, which provide the static magnetic moment of this compound. To find out the contributions of atoms to the magnetism, we calculate the spin moments on atoms, which defined as the difference of the average occupied ions between spin up and spin down in the muffin tin sphere. We found that the value of the two free radicals are -0.16994 μ_B/O1N1 group and -0.10419μ_B/O2N2 group respectively, but the value of carbon atoms is below 0.031164μ_B. These results suggest that most of the contribution to the magnetism comes from the NO groups, and the carbon atoms give the weak contributions.

From the total DOS in Fig.1, we also found that the energy gap in the spin-up is opened and the value of the energy gap is o.633eV, while the spin-down total DOS is continuous in the vicinity of the Fermi level. So the spin-down subbands exhibit metallic properties and the spin-up subbands show insulator properties. The metallic and insulator behaviors coexist in the 2-(5-pyrimidinyl)-4,4,5,5-tetramethyl-4,5-dihydro-1H-3-oxoimidazol-1-oxyl, so this compound may be a half-metallic magnet. To our best knowledge, such a organic half-metallic magnet consisting only of the light elements H, C, N and O has not been reported yet. If this compound is really a half-metallic magnet, it is important to spin electronics.

In order to understand deeply the mechanism of the magnetic interactions in this compound, we give the Fig. 2 (a) and Fig. 2 (b) shown the partial DOS of 2-p orbital of O1, N1 and O2, N2 atoms. It is found that the partial DOS of 2-p orbital of O1 and N1 atoms (O2 and N2 atoms) have similar peaks and character, which means that there is hybridization between the 2-p orbital of O1 and N1 atoms (O2 and N2 atoms). The unpaired electron in the compound is localized in the molecular orbital constituted primarily of the π^*(NO) orbital, which is formed from 2p (π) atomic orbital of N1 and O1 (O2 and N2 atoms). In this way, the net spin magnetic moment is formed in the two free radicals.

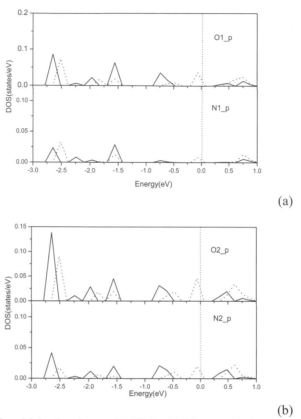

(a)

(b)

Fig.2 The calculated partial density of states (PDOS) for O1, N1 atoms (a), O2, N2 atoms (b). The solid
and dot line denote majority and minority spin, respectively. The Fermi levels are located at 0 eV.

In the present calculation, we calculate the spin magnetic moments on atoms, The value of the spin magnetic moments of the two free radicals are: -0.16994 μ_B/O1N1 group and -0.10419μ_B/O2N2 group. Further consideration, we calculate the spin magnetic moments of C1 atom which give the spin magnetic moment of 0.022884μ_B /C1 atom. Then opposite sign spin magnetic moment is formed between the two nitroxide free radicals. This result suggests that there exist ferromagnetic intermolecular interaction in the compound. To understand the mechanisms for exchange coupling between the individual nitroxide molecules, we rationalized the rationalization of ferromagnetic behavior in conjugated πradicals on the basis of an intermolecular coupling mechanism suggested by McConnell [11], which involves spin polarization effect leading to alternating positive and negative spin densities on the carbon backbone of these molecules. In its original form the first McConnell mechanism refers to the intermolecular ferromagnetic coupling between organic radicals with alternating positive and negative spin densities in the conjugated π system of their carbon backbones. According to McConnell two neighboring molecules have to be positioned in such a way that a site with positive spin density on one molecule is in registry with a site with negative spin density on the neighboring molecule and vice versa in order to obtain ferromagnetic coupling. It can be seen that the opposite sign spin magnetic moment is arranged in the molecule because of the spin polarization effect and formed the intermolecular ferromagnetic interactions.

In conclusion, we have studied the electronic band structure and the ferromagnetic properties of the pure organic solid: 2-(5-pyrimidinyl)-4,4,5,5-tetramethyl-4,5-dihydro-1H-3- oxoimidazol-1-oxyl by employing DFT with GGA. It is shown that the most of the contribution to the magnetization of this compound comes from the NO groups. The spin magnetic moments of C1 atom bridges the two ON free radicals, and in this way there formed ferromagnetic intermolecular interaction in the compound. It is predicted that this compound is half-metallic magnetism.

Acknowledgments

This work was supported by the National Natural Science Foundation of China under the grant No: 10174023 and 90103034. And supported by the group of Young Scientist innovation Foundation of Hubei Province under the grant No: *2004T006*.

References

[1] M. Tamura, Y. Nakazawa, D. Shiomi, K. Nozawa, Y. Hosokoshi, M. Ishikawa, M. Takahashi, M. Kinoshita, Chem. Phys. Lett. 186 (1991) 401

[2] M. Takahashi, P. Turek, Y. Nakazawa, M. Tamura, Phys. Rev. Lett., 67 (1991) 746

[3] R. Chiarelli, M. A. Novak, A. Rassat, Nature. 363 (1993) 147

[4] K. Mukai, K. Nedachi, J. B. Jamali, N. Achiwa, Chem. Phys. Lett., 214 (1993) 559

[5] F. Lanfranc de Panthou, DF. Luneau, J. Laugier. P. Rey J. Am. Chem. Soc. 115 (1993) 9095

[6] F. Palacio et al., Phys. Rev. Lett., (1997) 2336

[7] F. Romero et al., Adv. Mater. 8 (1998) 826

[8] D. Luneau, Current Opinion in Solid State and Materials Sciences, 5 (2001) 123

[9] J. P. Perdew, S. Burke, M. Ernzerhof, Phys. Rev. Lett., 77 (1996) 3865

[10] P. Blaha, K. Schwarz, J. Luitz, WIEN97, A Full Potential Linearized Augmented Plane Wave Package for Calculation Crystal Properties (Karlheinz Schwarz, Techn. Universit Wien, Austria), 1999.isbn 3-9501031-0-4

[11] H. M. McConnell, J. Chem. Phys., 39 (1963) 1910

VSP International
Science Publishers
P.O. Box 346, 3700 AH Zeist
The Netherlands

Lecture Series on Computer
and Computational Sciences
Volume 1, 2004, pp. 628-629

Symposium on
Industrial and Environmental Case Studies

Preface

The Symposium on 'Industrial and Environmental Case Studies' includes research works pertaining to at least one of the following domains: (I) Industrial Operation/Process, (II) Environmental Impact, (III) Modelling/Simulation, (IV) Data/Information Processing – AI System. The kind of colligation is quoted in the simplified ontological diagram shown below. As the first two domains refer to content, the semantic links are mainly of the *is-a* or *has-a* type; as the last two domains refer to methodology, the semantic link *is-examined-by* (*means of a*) dominates.

Regarding content, the contributions [2,5,7,8] pertain to domain (I), the contributions [9,10,11] pertain to domain II, while the contributions [1,4,6,12] refer or allude (even with perspective) to both. In terms of methodology, all contributions colligate to domains (III) and (IV), either in combination with at least one content domain or solely (paper [3]). Consequently, consistency and coherence have been achieved to a satisfactory degree. However, there is still ground for effecting improvements to be achieved by discussions during the Symposium. To further next year's Symposium, a well in advance interactive communication with the participants is planned, so that there will be time for constructive impact and fruitful cooperation.

I would like to thank Prof. Theodore Simos, who very perceptively and masterly embedded the Symposium in the Conference, and Dr. Christina Siontorou for her contribution in the interactive communication with the participating research team leaders, as well as with the teams that, although strongly interested in, were unable to attend this year but they are going to send a research work well in advance for the next year's Symposium.

The current Symposium is hopefully anticipated to establish the conditions and requisites for both (a) a general and wide multi-lateral forum for exchanging ideas and experience in the field of Computer Aided Industrial/Environmental Applications, and (b) certain partial cooperative schemes in specialised topics where the participating teams have similar or complementary interests.

Fragiskos Batzias
Symposium Organiser

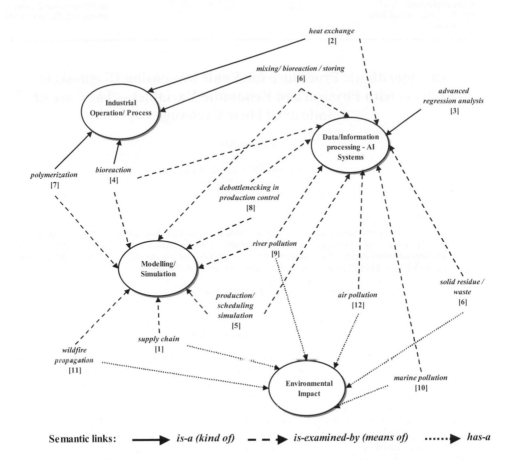

Semantic links: ⟶ *is-a (kind of)* - - ▶ *is-examined-by (means of)* ⋯⋯▶ *has-a*

Simplified ontology designed ad hoc by induction at a surface phenomenological level on the basis of the contributed works; only certain semantic links are depicted

[1] Empowering Financial Tradeoff in Joint Financial & Supply Chain Scheduling & Planning Modeling
[2] An Algorithmic Procedure for Fault Dimension Diagnosis in Systems with Physical and Economic Variables – The Case of Industrial Heat Exchange
[3] A General Procedure for Linear and Quadratic Regression Model Identification
[4] Optimal Design of Batch Plants under Economic and Ecological Considerations : Application to a Biochemical Batch Plant
[5] Automatic Generation of Production Scheduling Problems in Single Stage Multi-Product Batch Plants: Some Numerical Examples
[6] Integration of a Reactor/Mixer/Internal-Inventories Subsystem with Upstream External Inventories distributed over a time/space domain
[7] Receding Horizon Control for Polymerization Processes
[8] Evaluation of Debottlenecking Strategies for a Liquid Medicine Production Utilising Batch Process Simulation
[9] Simulation of River Streams: Comparison of a new Technique to QUAL2E
[10] GIS-based Discrimination of Oil Pollution Source in an Archipelago –The Case of the Aegean
[11] Modeling and Simulation of Wildfires based on Artificial Intelligence Techniques
[12] GIS-based Computer Aided Air Pollution Biomonitoring for Impact Assessment – Application in the Case of Materials Deterioration

VSP International
Science Publishers
P.O. Box 346, 3700 AH Zeist
The Netherlands

*Lecture Series on Computer
and Computational Sciences*
Volume 1, 2004, pp. 630-634

An Algorithmic Procedure for Fault Dimension Diagnosis in Systems with Physical and Economic Variables – The Case of Industrial Heat Exchange

F.A. Batzias[1]

Department of Industrial Management and Technology,
University of Piraeus,
Karaoli & Dimitriou 80, GR-18534, Piraeus, Greece

Abstract: An algorithmic procedure has been designed/developed for fault dimension diagnosis in systems with physical and economic variables/parameters. The implementation presented refers to the economotechnical design of the industrial heat exchangers and proved the suitability of the proposed methodology to identify error sources and contribute to system rectification.

Keywords: dimensional analysis, traceability, diagnosis, heat exchangers

1. Introduction

Dimensional Homogeneity (DH) of the terms of an algebraic expression relating physical/economic variables and parameters is a necessary yet not sufficient condition for connecting these terms rationally. Consequently, automatic checking of algebraic terms for DH can be used as a complementary filter before starting calculus with relations which may be either ill-defined or deduced by following an irrational procedure, not to mention simple errors in writing/transferring/copying mathematical expressions, especially in cases of computer programming. In these situations, enhancing traceability without human intervention is quite difficult because back searching to find out the source of error requires semantic analysis of terms and their constituents.

The aim of the present study is to provide an algorithmic procedure for fault dimension diagnosis in systems with physical and economic variables/parameters. This procedure has been implemented through a computer program and a paradigm is presented concerning the economotechnical optimization of industrial heat exchangers.

2. Methodology

The algorithmic procedure designed for the purpose described in Introduction consists of the following stages, including activity and decision nodes. Their interconnections, via executive and information lines are shown in Fig. 1.

1. Input of variables, parameters, constants, expressions, constraints, assumptions, simplifications, approximations, relevant to the equation (or inequality) under examination.
2. Testing of each term of the equation/inequality based on the dimensional matrix constructed according to the nomenclature given within the original relation source used in stage 1.
3. Testing of each term of the relation under examination, according to the nomenclature given by the Knowledge Base (KB).
4. Comparison of the new variables/parameters that have reestablished dimensional homogeneity with the old ones, found responsible for heterogeneity.
5. Remedial proposal for replacing the old with the new ones.

---

[1] Corresponding author. E-mail: fbatzi@unipi.gr

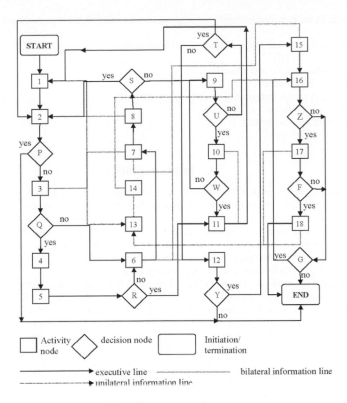

Figure 1. Flow chart of the algorithmic procedure designed/developed for fault dimension diagnosis in systems with physical and economic variables.

6. Ranking of terms in order of increasing recognizability/reliability: T_1, T_2, ..., T_h,..., T_q expected to form s non empty discrete sets with member-terms of the same dimension, called hereafter 'Representative Dimension' or RD_g of the set R_g, $g = 1$, ..., s.
7. Ranking of variables/parameters V_{gj} within each set R_g in order of increasing recognizability.
8. Discrimination of the successive deductive steps/layers formed for the derivation of the relation under examination; if m is the number of these steps (not necessarily known *a priori*), then, by moving backwards, we symbolize E_m the step when starting the algorithmic procedure, E_1 the initial step (usually a definition or a widely accepted generalized relation), and E_k the set of relations/constraints/assumptions forming the k step/layer ($1 \leq k \leq m$).
9. Inductive search to find out identical terms with different dimensions in similar relations.
10. Comparison of the dimensions of new terms with the dimension of other R in the relation under examination.
11. Corrective action over the relation.
12. Processing of collected unsuccessful results to find out whether they might be considered either part of a conventional ontology or members within relevant categorical sets (partonomy/taxonomy approach) constituting a new ontology.
13. Creation/enrichment of the KB.
14. Intra- and inter- net searching via an intelligent agent based on an ontological adaptable interface [1].
15. Reconstruction of the successive deductive steps formed in stage 8 so that each term to maintain its original semantic content.
16. Inductive search to find out terms with the same semantic content in similar ontologies.
17. Comparison of the dimensions of new terms with the dimensions of other R in the relation under examination within step k.
18. Corrective action over all relations/constraints/assumptions.
P. Are these terms dimensionally homogeneous?

Q. Are the same terms now dimensionally homogeneous according to the nomenclature given by the KB?

R. Does the remedial proposal establish dimensional homogeneity?

S. Is the inequality $k < m$ valid ?

T. Are there any other R left unexamined?

U. Have such terms been found?

Y. Are the processed results in favour of an ontological approach?

Z. Have such terms been found?

F. Has identification been achieved?

G. Is the inequality $k < m$ valid?

3. Implementation

The application of the herein described methodology is based on a custom-developed software that has been developed using Microsoft's Visual Studio .Net 2003. Visual Studio .Net is the first development environment fully supporting the development of application using the .Net suite of technologies. The most significant advantages of the .NET architecture are the following: (a) Interoperability with many languages (C++, VB, Pascal, COBOL, REXX etc) applications and systems through the use of widely accepted standard technologies (HTTP, XML, Web Services, SOAP, WSDL), and (b) easy of deployment and robustness of operation.

More specifically, the application exploits the following technologies of the .Net architecture [2,3]: (a) the .NET Framework that includes the Common Language Runtime (CLR) and the .NET Framework class library. The CLR provides a common set of data types and other services (e.g. Garbage Collector) that can be used by all .NET languages, while the .NET Framework class library includes a large set of standard classes and other types that can be used by any .NET Framework application written in any language, and (b) the ADO.NET that includes classes for accessing data stored in relational database management systems (RDBMS) and in other formats (e.g. XML). In our application all the results of the computation are stored in a Dataset object, a relational in-memory cache for data.

The Software package incorporates two classes: the Non linear regression class and the MyMatrix class. The former class is responsible for the non-linear parameter estimation procedure where as the latter incorporates the matrix computations. Furthermore, a Dimensional Analysis Library incorporates the dimensional analysis computational core, which includes the calculation and sorting of dimensional terms.

The algorithmic procedure and the software described above has been implemented in the case of optimum design of heat exchangers. The relation to be examined for DH has been extracted from the classical work of Peters and Timmerhaus [4], which is widely used in chemical engineering practice. Due to lack of space, only one of the expressions used for optimization at layer E_5 and the objective function (which coincides with the relation of the initial step E_1, as quoted in stage 8 of Fig. 1) are given herein as eqs (1) and (2), depicting the initiation and determination of a vertical search path:

$$T(1,1,5) + T(2,1,5) + T(3,1,5) - T(4,1,5) = 0 \tag{1}$$

$$T(1,1,1) - T(2,1,1) - T(3,1,1) - T(4,1,1) - T(5,1,1) = 0 \tag{2}$$

where

$$T(1,1,5) = 2.5h_{io}^{3.5}\Psi_i H_y C_i \, , \; T(2,1,5) = 3.5h_{io}^{4.5}\Psi_i H_y C_i D_i R_{dw}/D_o \, ,$$

$$T(3,1,5) = 2.9h_{io}^{3.72}\left(\Psi_i C_i D_i/D_o\right)^{0.83}\left(\Psi_o C_o\right)^{0.17} H_y \, , \; T(4,1,5) = K_F C_{Ao}$$

$$T(1,1,1) = C_T \, , \; T(2,1,1) = A_o K_F C_{Ao} \, , \; T(3,1,1) = w_u H_y C_u \, , \; T(4,1,1) = A_o E_i H_y C_i \, , \; T(5,1,1) = A_o E_o H_y C_o$$

Ψ_o dimensional factor for evaluation of power loss outside tubes, $\in \left[M^{3.75}T^{11.25}\Theta^{4.75}\right]$

Ψ_i dimensional factor for evaluation of power loss inside tubes, $\in \left[M^{-2.5}T^{7.5}\Theta^{3.5}\right]$

K_F annual fixed charges including maintenance, expressed as a fraction of initial cost for completely installed unit dimensionless $\in [1]$

C_{Ao} installed cost of heat exchanger per unit of outside-tube heat-transfer area, $\in \left[NL^{-2} \right]$

C_o cost for supplying one energy-unit to pump fluid flowing through shell side of unit, $\in \left[NM^{-1}L^{-2}T^{2} \right]$

C_i cost for supplying one energy-unit to pump fluid flowing through inside of tubes, $\in \left[NM^{-1}L^{-2}T^{2} \right]$

C_u cost of utility fluid, $\in \left[NM^{-1} \right]$

R_{dw} combined resistance of tube wall and scaling or dirt factors, $\in \left[M^{-1}T^{3}\Theta \right]$

E_i power loss inside tubes per unit of outside tube area, $\in \left[MT^{-2} \right]$

E_o power loss outside tubes per unit of outside tube area, $\in \left[MT^{-2} \right]$

H_y coefficient for time units transformation, dimensionless $\in [1]$

h_{io} tube-inside coefficient of heat transfer optimised, $\in \left[MT^{-3}\Theta^{-1} \right]$

C_T total annual variable cost for heat exchanger and its operation, $\in \left[NT^{-1} \right]$

w_u flow rate of utility fluid, $\in \left[MT^{-1} \right]$ A_o area of heat transfer, $\in \left[L^{2} \right]$

D_i inside tube diameter, $\in [L]$ D_o outside tube diameter, $\in [L]$

$L = length$ $M = mass$ $T = time$ $\Theta = temperature$ $N = money$

The order that the variables/parameters have been written is in accordance with stage 7, while the five successive deductive layers, in order of backwards searching, are:

E_5: deduction of an equation with h_{io} as the only unknown, by eliminating the rest independent variables, already fixed at their optimal values.

E_4: deduction of a set of equations by setting the partial derivatives of the objective function under its Lagrangian form with respect to each independent variable equal to zero

E_3: incorporation of the constraint into the objective function to apply the Lagrange multiplier method

E_2: objective function expressed in terms of primary variables

E_1: objective function by definition (terms with the original semantic content)

For the technical implementation of stages 6-8, we constructed the $R(g,e,k)$-matrix, where g is the order of set of dimensionally homogeneous terms within the same algebraic expression and e is the order of algebraic expression within the same deduction layer; the order of g is determined according to decreasing number of included terms within each set while the order of e is defined arbitrarily as all algebraic expressions within the same deduction layer are finally examined by executing the corresponding loop in the computer software. To illustrate the technique we give the following examples:

For $k=5$ $\{T(1,1,5), T(2,1,5), T(3,1,5)\} \in R(1,1,5)$, $\{T(4,1,5)\} \in R(2,1,5)$

For $k=1$ $\{T(1,1,1), T(3,1,1), T(4,1,1), T(5,1,1)\} \in R(1,1,1)$, $\{T(2,1,1)\} \in R(2,1,1)$

$$RD(1,1,5) = [NL^{-2}T^{1}], \ RD(2,1,5) = [NL^{-2}], \ RD(1,1,1) = [NT^{-1}], \ RD(2,1,1) = [N]$$

In these examples, $m=5$, $s=2$, while q depends on the expression each time (i.e., line in the $R(g,e,k)$-matrix) examined. In the screenshot of Fig. 2, the coordinates of each cell are k, e, h while g is represented by the value on the cell, which is active: by clicking on a cell its representative dimension $RD(g,e,k)$ appears in the small window at the top. The dimensional deviation appears as the ratio $RD(1,e,k)/RD(2,e,k)$ if $s=2$ and in combinatorial form if $s>2$. The exponents of each term dimensions enter the program as fractions to avoid malfunctioning.

Traceability is proved to be quite satisfactory as the dimensional deviation T^{-1} is finally attributed to mischaracterization of K_F as 'dimensionless'. Theoretically, this error could be avoided by visual inspection, as the adjective 'annual' at the beginning of the definition of K_F could be taken into account to lead, through comparison with the dependent variable C_T, to the right dimensional input. In practice, this cannot be done as the designer/engineer begins calculating with layers $g>1$, i.e., one does not build from the grass or from first principles. Moreover, the target of the present study is to develop a system that enters the required variables/parameters automatically or with minimal human supervision. On the other hand, when a semantic network is applied to minimize human intervention, it always recognizes (at least in our experiments) the adjective 'dimensionless' (i.e. $K_r \in [1]$) against the adjective 'annual' (i.e. $K_r \in [T^{-1}]$) because of its placement at the end of the corresponding definition as it holds for all definitions of the same nomenclature section.

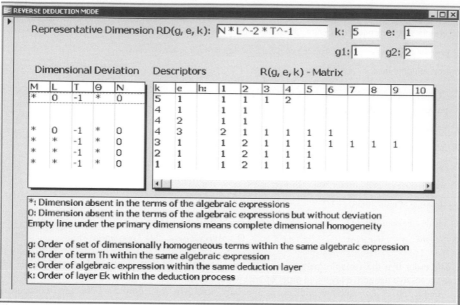

Figure 2. Screenshot of the computer program developed for fault dimension diagnosis in the economotechnical optimization of industrial heat exchangers. The initial cause recognized is the lack of the T^{-1} dimension (quoted as exponential value -1 in column T within the Dimensional Deviation Table) in the original term $T(2,1,1) = A_o K_F C_{Ao}$ at $k=1$ layer, which is the fixed cost by definition.

It is worthwhile noting that the result was achieved by reverse deduction as the steps 1-5 exhibit algebraic consistency in the derivation process through the successive layers. However, backward searching for fault detection becomes difficult in case empirical relations are used in the intermediate derivation layers, e.g. when introducing external flow relations in heat exchangers design [5,6].

4. Conclusions

The algorithmic procedure we have designed and the corresponding software we have developed for fault dimension diagnosis in systems with physical and economic variables give satisfactory results, as it is shown in the case of heat exchangers optimal design. In this case, starting from a dimensionally fault term used for calculating the optimal value of tube-inside film coefficient of heat transfer at the fifth step of the deduction process the program localized the initial cause in the fixed cost term at the first layer. Therefore, the application of this procedure enhances traceability in complicated computing systems, where a dimensional fault may be found in a quantitative expression derived after many transformations. Consequently, the user, who needs not to be aware of the derivation process, will obtain the variable responsible for the fault and the corresponding deductive after-path rectified.

References

[1] F. A. Batzias and E. C. Markoulaki, Restructuring the keywords interface to enhance CAPE knowledge via an intelligent agent, *Computer-Aided Chemical Engineering*, **10** 829-834 (2002).
[2] D. Vitter: *Designing Visual Basic .NET Applications.* Coriolis Group Books, 2001.
[3] P. Thorsteinson and R. J. Oberg: *.NET Architecture and Programming Using Visual C++.* Pearson Education, N.Y., 2002.
[4] M.S Peters and K.D. Timmerhaus: *Plant Design and Economics for Chemical Engineers,*.McGraw-Hill, New York, 574-582 (1968).
[5] A. Zhukauskas: *Advances in Heat Tranfser.* Academic Press, 1987.
[6] O.S. Paitoonsurikarn, N. Kasagi, Y. Suzuki, Design Study of High-Performance Heat Exchanger with Possible Use of Micro bare Tubes, www.thtlab.t.u.-tokyo.ac.jp/Docs/JSME99MHEx.pdf

VSP International
Science Publishers
P.O. Box 346, 3700 AH Zeist
The Netherlands

Lecture Series on Computer
and Computational Sciences
Volume 1, 2004, pp. 635-641

Integration of a Reactor/Mixer/Internal-Inventories Subsystem with Upstream External Inventories Distributed Over a Time/Space Domain

F. A. Batzias[1], A. S. Kakos and N. P. Nikolaou

Department of Industrial Management and Technology,
University of Piraeus,
GR-185 34 Piraeus, Greece

Abstract: Mixing is a very common industrial process; the resulting mixture may be either a final or an intermediate product of predetermined content for each constituent, according to certain quality specifications. During production within an agro-industrial complex, external inventories, distributed in the time/space domain, hosting raw materials (usually highly inhomogeneous and sensitive by nature) should cater for internal ones that feed the mixer. The present work aims to the computer-aided integration of such a combined industrial process by (a) continuously allocating transhipment points and determining the optimal routing of raw materials, (b) maintaining steady-state conditions in the reactor, (c) constantly calibrating the mixture proportions through a feedback control mechanism checking the quality standards at the reactor output and (d) obtaining the minimal operating cost provided through a custom-developed software module based on a hybrid Genetic Algorithm. A case study is also presented referring to the design of an ethanol production unit located at Crete in Greece where the number of inventories varies.

Keywords: Mixing, Inventories, Optimization.

Mathematics Subject Classification: 81T80, 90B10, 90C11, 90C29

1. Introduction

Mixing is a process, which is frequently met in agro-industrial plants; the resulting mixture may be either (a) a final product which should be characterized by predetermined content for each constituent, according to specifications set by the client or the market, or (b) an intermediate product, used as a feed for a downstream process e.g. a chemical reactor, in which case the composition should be estimated according to the conditions required for the optimal operation of this downstream process. From the view point of process integration, the two units, i.e. the mixer and the reactor, should be optimised as a whole; in practice, however the importance of maintaining steady state conditions in the reactor dominates over the total operation. Consequently, the optimisation procedure is limited to the combination of the internal inventories with the mixer, especially when the raw materials are highly inhomogeneous, e.g. quarry products for making a mixture to feed a cement clinker rotary kiln or agricultural residues for making a mixture to feed an ethanol bio-reactor. The design of such an optimisation/integration procedure on a quasi-online real-time basis is the main theme of the present work and the ethanol production from agricultural residues is the case study (see Figure 1). A custom-made software module, based on a hybrid Genetic Algorithm (GA), is (a) continuously fed with data concerning the location of transhipment points and availability of raw materials in the time/space domain, (b) calculating the optimal routing of raw materials from the external inventories to the industrial area, (c) reshuffling the external inventory list through a 'death/birth' process (where inventories are abandoned, merged, relocated or created because of various situations), and (d) obtains

---

[1] Corresponding author. E-mail: fbatzi@unipi.gr

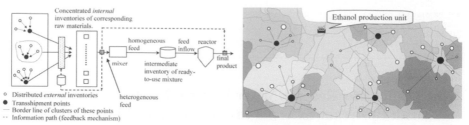

Figure 1. Schematic representation of the GIS-assisted network, providing information for raw materials collection/transportation/storing under the constraints set by the industrial process to be integrated; the feedback control mechanism assures the optimization of the system as a whole.

Figure 2: The allocation suggested by the External Inventory Cost Optimizer at $t=5$ (day) for the case study of an ethanol production unit located in northern Crete at Greece.

the minimal operating cost provided through a custom-developed module, namely the External Inventory Cost Optimiser. This module has been encapsulated in a DSS (Decision Support System) that interacts with the expert to determine the necessary system conditions to produce a final (and/or intermediate) product of acceptable cost and quality. The main concern of the design is to maintain steady state conditions for the reactor subsystem through a feedback control mechanism which checks and proposes the 'closest working conditions' (to the current steady-state conditions) in order to adjust the quality of the final or intermediate product by calibrating the proportions at the mixture.

2. Methodology

Let $C_{j,0,t}$ be the *optimal* concentrations for each constituent j ($j = 1, ..., m$) in the feed at time unit t, considered as fractions determined by the cost and quality optimization taking place in the mixer/reactor/final product downstream process; let C_{jt} be the variable representing the actual concentration of the j constituent at time unit t in the feeder, allowing them to vary only within certain corresponding tolerance intervals $[C_{j,1,t}, C_{j,2,t}]$. Let m_{it} be the mass quantity drawn out from the inventory i ($i = 1, ..., n$) at time unit t; since the feed is formed by mixing different raw materials from n inventories maintained in the industrial unit under consideration, the following equations hold for C_{jt}:

$$C_{jt} = \sum_{i=1}^{n} b_{ji} m_{it} / M , C_{j,1,t} \leq C_{jt} \leq C_{j,2,t} , \sum_{j=1}^{m} b_{ji} = 1 , M = \sum_{i=1}^{n} m_{it} \qquad (1)$$

where M is the total feed mass and b_{ji} is the content of constituent j in the raw material i, kept in inventory i. In the present formulation, M is a time-invariant constant; in case M varies over time, the model can be easily expanded. $C_{j,1,t}$ and $C_{j,2,t}$ are each time adjusted through a feedback control mechanism checking the quality of the final product and the stability of the downstream process.

2.1 Model Parameters

Due to the nature of lignocellulosic raw materials and the periodicity in their production, optimization should be performed on a time period covering at least a full cycle of the supply side. The model herein proposed, is capable of performing optimization on a *full cycle* as well as on a time *snapshot* in case the short-term cost is to be minimized. The time period, in question, is subdivided in equal time-units. In our model formulation, T denotes the total number of time-units, D denotes the useful life of raw material (in time units), p denotes the number of external inventories, and r denotes the number of (potential) transhipment points. Each external inventory and transhipment point hosts only one type of raw material; however, many external inventories or transhipment points can be co-located.

As limited by space, only the objective functions of our model will be presented; thus, the nomenclature is restricted only to the indispensable constants and variables:

$T_{ex}(l, k, t, Q)$ transportation cost of carrying Q units of raw material from the l external inventory to the k transhipment point at time unit t.

$T_{tr}(k, i, t, Q)$ transportation cost of carrying Q units of raw material from the k transhipment point to the i internal inventory at time unit t.

$Q_{ex}(l, k, t, d)$ quantity of raw material that has been left to deteriorate for d time units transported from the l ext. inventory to k transhipment point at time unit t.

$Q_{tr}(k, i, t, d)$ quantity of raw material that has been left to deteriorate for d time units transported from the k trans. point to the i internal inventory at time unit t.

$v(l, t)$ the unit cost of the raw material at external inventory l at time t based either on past experience or on real-time data.

$Res_{ex}(l, t, d)$ quantity of raw material that has been left to deteriorate for d time units at the l external inventory at time unit t.

$Res_{tr}(k, t, d)$ quantity of raw material that has been left to deteriorate for d time units at the k transhipment point at time unit t.

$A(l, t)$ quantity of newly arrived/fresh raw material at ext. inventory l at time t.

2.2 Objective functions

One of the objective functions to be minimized is the total cost *TOTC* of buying and transporting the raw materials:

$$TOTC = \sum_{l=1}^{p}\sum_{t=1}^{T} A(l,t)\cdot v(l,t) + \sum_{t=1}^{T}\sum_{k=1}^{r}\sum_{l=1}^{p} T_{ex}(l,k,t,\sum_{d=1}^{D}Q_{ex}(l,k,t,d)) + \sum_{t=1}^{T}\sum_{k=1}^{r}\sum_{i=1}^{n} T_{tr}(k,i,t,\sum_{d=1}^{D}Q_{tr}(k,i,t,d)) \tag{2}$$

The transportation cost varies according to the quantities transported as follows:

$$T_{tr}(k,i,t,Q) = f_{tr}\left(Q - \left[\frac{Q}{F_{tr}}\right]\cdot F_{tr}\right) + \left[\frac{Q}{F_{tr}}\right]\cdot F_{tr}\cdot TC_{min} \tag{3}$$

where F_{tr} is the maximum capacity of the vehicle used to transport the quantity Q, TC_{min} is the unit transportation cost when the vehicle is fully loaded, and f_{tr} is a function of Q that represents the transportation cost of a not fully-loaded truck. A simple linear formulation for f_{tr} is given by $f_{tr}(Q) = TC_{max} + (TC_{min}\cdot F_{tr} - TC_{max})\cdot Q/F_{tr}, Q \le F_{tr}$, where TC_{max} is the cost of employing a truck. A similar formulation holds for the transportation cost T_{ex} from external inventories.

The second objective function *TDL*, to be minimized, quantifies, in cost terms, *the loss of quality* at the transhipment points and external inventories due to deterioration:

$$TDL = \sum_{t=1}^{T}\sum_{l=1}^{p}\sum_{d=1}^{D} Res_{ex}(l,t,d)\cdot DL_{ex}(l,t,d) + \sum_{t=1}^{T}\sum_{k=1}^{r}\sum_{d=1}^{D} Res_{tr}(k,t,d)\cdot DL_{tr}(k,t,d) \tag{4}$$

where $DL_{ex}(l, t, d)$ and $DL_{tr}(k, t, d)$ are the cost/penalty of deterioration per unit quantity at the external inventories and transhipment points, accordingly.

The third objective function is a *quality measure* of the mixture feeding the mixer and takes into account the feedback of the downstream quality control mechanism in order to minimize the dispersion from the optimal concentrations (also refer to Figure 1):

$$TQM = \sum_{j=1}^{m} w_j(t)\cdot\left((C_{j,0,t} - C_{jt})/C_{j,0,t}\right)^2 \tag{5}$$

where $w_j(t)$ is a *significance index* of the corresponding constituent and is, each time, adjusted/calibrated from the feedback control mechanism which takes into account the quality of the intermediate and final product at time unit t and proposes the 'closest conditions', i.e. the sub-optimal conditions that improve the quality and at the same time approximate the current working conditions of the overall agro-industrial process so as to maintain the steady state conditions. In this way, the feedback controlled mechanism triggers, at each time unit, a global optimization of the system as a whole.

An economic incentive for maximizing the use of transhipment points is expressed by the following objective function:

$$TTU = \sum_{k=1}^{r}\sum_{t=1}^{T}\left(1 - \sum_{d=1}^{D}\sum_{i=1}^{n}Q_{tr}(k,i,t,d)\Big/L_{tr}(k,t)\right)^{a}\cdot IC_{kt}, a > 0 \tag{6}$$

that should be minimized; $L_{tr}(k, t)$ and IC_{kt} represent the capacity and the significance of fully-utilizing the transhipment point k at time unit t, respectively.

The concept behind this objective function is to give a vehicle leaving an external inventory an incentive to prefer routing to a transhipment point that is already semi-loaded over an empty one, even if such a route is not optimal in terms of distance. *TTU* is a measure of the achievement of scale economies in the location/operation of transhipment points over the time/space domain.

3. Implementation and Discussion

The methodology presented herein was implemented in the design of a medium-scale ethanol production unit, based on the Simultaneous Saccharification and Extractive Fermentation (SSEF) method, located at northern Crete in Greece, where $m=3$, $n=3$, $p=34$, $r=5$, $D=7$ days, $M=150$ tonnes/day, and $T=365$ days. The allocation of the 34 external inventories to the 5 transhipment points at day 5 is exhibited in Figure 2.

The objective was to carry the agricultural residues to bioreactors so as to achieve optimal processing of biomass waste in the vicinity and a high degree of integration with the mixer/reactor downstream process. The latter is achieved through a feedback control mechanism checking the quality of the intermediate and/or final product (as it is schematically represented in Figure 1).

The results obtained herein exhibit optimal or near-optimal features and are significantly better than the ones suggested by the empirical strategies used so far in Crete and in most major agricultural prefectures of Greece. Managing to achieve a 25% reduction in transportation cost (which sums to a total transportation cost up to of € 125,000/year) and taking into account that, in such agro-industrial units, the transportation cost is the most significant overhead, the significance of our methodology as a cost cutting factor is thereby proven.

The application of our model revealed that some raw materials are not as cost efficient as they were thought to be and their constituents can be replenished by other sources. However, in case there are serious environmental or social reasons for choosing a particular raw material or external inventory, our approach can be easily adopted to cater for such a need.

Although Biethahn and Volker [1] suggest certain different solutions for this category of problems, Srinivas et al [2] point out that the GA approach seems to have some unique features in real industrial applications such as the locally based (quasi-) real-time optimization of truck allocation to transhipment points. In such cases, GA gives a set of near-optimal solutions instead of one so as the decision maker evaluates the set of solution in order to implement the most suitable one [3]. Moreover, multi-criteria optimization can be employed through GA instead of a typical single-objective traditional algorithm [4, 5]. In an analogous manner, our model managed to cope with the annual variability and periodicity and revealed its merits, in comparison with the simple operational research procedures usually followed, especially when the deterioration of residues quality, mainly due to enzymatic hydrolysis (and/or other unpredictable events on raw materials concentration) is taken into account.

4. Conclusions

An abundance of agricultural byproducts is produced each year; these quantities remain unexploited, usually forming an environmental risk due to the high biodegradability when stored in open air. The need for exploitation of such residues by an industrial complex forms a multi-criteria optimization problem. One of the targets is to maximize recycling/reusing of such by-products by facilitating demand to meet the supply.

In complex industrial systems, such as the dynamic one presented in Figure 1, our methodology suggests the coupled use of a Geographical Information System (GIS) with a Global Positioning System (GPS) to: (a) locate the distributed -in time/space domain- external inventories, (b) feed the External Inventory Cost Optimizer subsystem with real-time data, (c) propose the optimal allocation of transhipment points, (d) target to the simultaneous minimization of more than one criterion, and (e) recommend a set of alternative sub-optimal solutions to the decision maker. Then, a tailored algorithm is used to (a) reshuffle the inventory list through a 'death/birth' process, (b) obtain the minimal operating cost and (c) resolve the optimal mass quantity m_{it} drawn out from the i internal inventory to prepare optimal mixing for feeding the downstream processes.

Acknowledgments

Unclassified data/information by the Greek Ministries of Development and Agriculture as well as financial support provided by the Research Center of the University of Piraeus are kindly acknowledged.

References

[1] J. Biethahn, N. Volker (Eds): *Evolutionary Algorithms in Management Science*, Springer, Berlin, 1995.

[2] M. Srinivas et al, Genetic algorithms: A Survey, *IEEE Computer* 27 (6) 17-26 (1994).

[3] A. J. Chin, H. W Kit., and A. Lim., A New GA Approach for the Vehicle Routing Problem, *11th IEEE Inter. Conf. on Tools with AI* 307 (1999).

[4] A. Santos and A. Dourado, Constrained GA applied to production and energy management of a pulp and paper mill, *Proc. of the 1999 ACM symposium on Applied computing* (1999).

[5] T. Back, *Evolutionary Algorithms in Theory and Practice*, Oxford Univ. Press, Oxford, 1996.

Appendix

The complete model formulation is the following:

Let $C_{j,0,t}$ be the *optimal* concentrations for each constituent j ($j = 1, ..., m$) in the feed at time unit t, considered as fractions, as determined by the cost and quality optimization taking place in the mixer/reactor/final product downstream process; let C_{jt} be the variable representing the actual concentration of the j constituent at time unit t in the feeder allowing them to vary only within certain corresponding tolerance intervals $[C_{j,1,t}, C_{j,2,t}]$. Let m_{it} be the mass quantity drawn out from the inventory i ($i = 1, ..., n$) at time unit t; since the feed is formed by mixing different raw materials from n inventories maintained in the industrial unit under consideration, the following equations hold for C_{jt}:

$$C_{jt} = \sum_{i=1}^{n} b_{ji} m_{it} / M \,, \quad C_{j,1,t} \leq C_{jt}, C_{j,0,t} \leq C_{j,2,t} \,, \quad \sum_{j=1}^{m} b_{ji} = 1 \,, \quad M = \sum_{i=1}^{n} m_{it} \tag{1}$$

where M is the total feed mass and b_{ji} is the content of constituent j in the raw material i, kept in the corresponding store (inventory i). In the present formulation, M is a time-invariant constant; in case M varies over time, the model can be easily expanded. $C_{j,1,t}$ and $C_{j,2,t}$ are each time adjusted through a, neural network driven, feedback control mechanism checking the quality of the final product and the stability of the downstream process.

A.1 Model Parameters

Due to the sensitive nature of the lingo-cellulosic raw materials and the periodicity in their production cycle, optimization should be performed on a time period covering at least a full cycle of the supply side. The model herein proposed, is capable of performing optimization on a *full cycle* as well as on a time *snapshot* in case the short-term cost is to be minimized. The time period, we discuss, is subdivided in equal time-units. In our model formulation T denotes the total number of time-units, D denotes the useful life or raw material (in time units), p denotes the number of external inventories, and r denotes the number of (potential) transshipment points. Each external inventory and transhipment point hosts only one type of raw material; however, many external inventories or transhipment points can be co-located.

The system is considered working under steady-state conditions; thus, the optimal concentration of the constituents in the feed will be almost constant at each time unit. However, there are other quantities of the model that vary over time/space:

$T_{ex}(l, k, t, Q)$ transportation cost of carrying Q units of raw material from the l external inventory to the k transhipment point at time unit t.

$T_{tr}(k, i, t, Q)$ transportation cost of carrying Q units of raw material from the k transhipment point to the i internal inventory at time unit t.

$Q_{ex}(l, k, t, d)$ quantity of raw material that has been left to deteriorate for d time units transported from the l external inventory to the k transhipment point at time unit t.

$Q_{tr}(k, i, t, d)$ quantity of raw material that has been left to deteriorate for d time units transported from the k transhipment point to the i internal inventory at time unit t.

$L_{ex}(l, t)$ maximum capacity of external inventory l at time unit t.

$L_{tr}(k, t)$ maximum capacity of transhipment point k at time unit t.

$e_{ex}(l, t)$ a binary variable representing the utilization of the external inventory l at time unit t (i.e. $eex(l, t)=1$ means that l will be utilized at t).

$e_{tr}(k, t)$ a binary variable representing the utilization of the transhipment point k at time unit t.

$v(l,t)$ the unit cost of the raw material at external inventory l at time t. The unit cost $v(l,t)$ is a known quantity based either on past experience or on real-time data coming from the network of ext. inventories.

$Res_{ex}(l, t, d)$ Quantity of raw material that has been left to deteriorate for d time units at the l external inventory at time unit t.

$Res_{tr}(k, t, d)$ Quantity of raw material that has been left to deteriorate for d time units at the k transhipment point at time unit t.

$A(l,t)$ quantity of newly arrived/fresh raw material at external inventory l at time unit t; this quantity is known at every time-unit because of the local information network of GPS appliances established in the area.

$W(l, t, d)$ Quantity of raw material, that has been left to deteriorate for d time units, leaving external inventory l at time unit t; it holds that

$$W(l,t,d) = \sum_{k=1}^{r} Q_{ex}(l,k,t,d) \tag{2}$$

A.2 Model Constraints
The model equations are as follows:

$$\sum_{d=1}^{D}\sum_{l=1}^{p}Q_{ex}(l,k,t,d)\leq e_{tr}(k,t)\cdot L_{tr}(k,t),\forall(k,t) \tag{3}$$

i.e. all outgoing raw material from the external inventories to a transhipment point k, when employed, should not overcome its current capacity.

$$\sum_{d=1}^{D}W(l,t,d)\leq e_{ex}(l,t)\cdot L_{ex}(l,t),\forall(l,t) \tag{4}$$

i.e. all outgoing raw material from an external inventory l should not overcome its storing capacity.

$$\sum_{d=1}^{D}\sum_{k=1}^{r}Q_{tr}(k,i,t,d)=m_{it},\forall(i,t) \tag{5}$$

i.e. all incoming raw material to the i internal inventory should, exactly, cover the demand. In case $\sum_{d=1}^{D}\sum_{k=1}^{r}Q_{tr}(k,i,t,d)<m_{it}$ then a shortage of raw materials occurs; otherwise a surplus of raw material will accumulate in internal inventories and eventually will exceed the node capacity.

$$\operatorname{Re}s_{tr}(k,t,d)=\sum_{l=1}^{p}Q_{ex}(l,k,t,d)-\sum_{i=1}^{n}Q_{tr}(k,i,t,d)+\operatorname{Re}s_{tr}(k,t-1,d-1),\forall(k,t,d) \tag{6}$$

i.e. the residuals at a transhipment point k at time unit t hat has been left to deteriorate for d time units equals the quantity entered minus the quantity leaving plus the quantity already left at the transhipment point from the previous time unit. It holds that

$$\operatorname{Re}s_{tr}(k,t,d)\geq 0,\forall(k,t,d) \tag{7}$$

$$\sum_{d=1}^{D}\operatorname{Re}s_{tr}(k,t,d)\leq e_{tr}(k,t)\cdot L_{tr}(k,t),\forall(k,t) \tag{8}$$

i.e. the total residuals left at the transhipment point k at time unit t should not exceed the corresponding capacity.

$$Q_{tr}(k,i,t,d)=0,\text{ if the trans. point } k \text{ transports a raw material different from } i \tag{9}$$

$Q_{ex}(l,k,t,d)=0,$ if the ext. inventory l stores a raw material different from thattransported to the transhipment point k at time unit t (10)

$$W(l,t,d)\geq e_{ex}(l,t),\forall(l,t,d) \tag{11}$$

$$\sum_{i=1}^{n}Q_{tr}(k,i,t,d)\geq e_{tr}(k,t),\forall(k,t,d) \tag{12}$$

i.e. in case an ext. inventory or a trans. point is not used there is no need for employed it in the model

$$\operatorname{Re}s_{ex}(l,t,d)\geq 0,\forall(l,t,d) \tag{13}$$

i.e. the quantity left at the external inventory l is a non-negative number.

$$\sum_{d=1}^{D}\operatorname{Re}s_{ex}(l,t,d)+A(l,t)<e_{ex}(l,t)\cdot L_{ex}(l,t),\forall(l,t) \tag{14}$$

i.e. if external inventory l is used, the quantity accumulated at l should be less than its maximum capacity. Furthermore, the quantity of raw material $Res_{ex}(l, t, d)$ that has been left to deteriorate for d time units at the l external inventory at time unit t is given by the following formula:

$$\operatorname{Re}s_{ex}(l,t,d)=\begin{cases}\operatorname{Re}s_{ex}(l,t-1,d-1)-W(l,t,d),\text{if }d\geq 1\\ A(l,t)-W(l,t,0),\text{if }d=0\end{cases},\forall(l,t) \tag{15}$$

A.3 Objective functions
One of the objective functions to be minimized is the total cost *TOTC* of buying and transporting the raw materials:

$$TOTC=\sum_{l=1}^{p}\sum_{t=1}^{T}A(l,t)\cdot v(l,t)+\sum_{t=1}^{T}\sum_{k=1}^{r}\sum_{l=1}^{p}T_{ex}(l,k,t,\sum_{d=1}^{D}Q_{ex}(l,k,t,d))+\sum_{t=1}^{T}\sum_{k=1}^{r}\sum_{i=1}^{n}T_{tr}(k,i,t,\sum_{d=1}^{D}Q_{tr}(k,i,t,d)) \tag{16}$$

It should be noted that the transportation cost varies according to the quantities transported as follows:

$$T_{tr}(k,i,t,Q)=f_{tr}(Q-\left[\frac{Q}{F_{tr}}\right]\cdot F_{tr})+\left[\frac{Q}{F_{tr}}\right]\cdot F_{tr}\cdot TC_{min} \tag{17}$$

where F_{tr} is the maximum capacity of the truck/vehicle used to transport the quantity Q, TC_{min} is the unit transportation cost when the vehicle is fully loaded, and f_{tr} is a function of Q that represents the transportation of a not fully-loaded truck. A simple linear formulation for f_{tr} is given by

$$f_{tr}(Q) = TC_{\max} + (TC_{\min} \cdot F_{tr} - TC_{\max}) \cdot \frac{Q}{F_{tr}}, Q \le F_{tr},$$ $TC_{\max}$ is the cost of employing a truck. A similar

formulation holds for the transportation cost T_{ex} from transhipment points.

The second objective function *TDL*, to be minimized, quantifies, in cost terms, *the loss of quality* at the transhipment points and external inventories due to the deterioration of the raw materials:

$$TDL = \sum_{t=1}^{T}\sum_{l=1}^{P}\sum_{d=1}^{D} \mathrm{Re}\, s_{ex}(l,t,d) \cdot DL_{ltd} + \sum_{t=1}^{T}\sum_{k=1}^{r}\sum_{d=1}^{D} \mathrm{Re}\, s_{tr}(k,t,d) \cdot DL_{ktd} \qquad (18)$$

where DL_{ltd} and DL_{ktd} are the cost/penalty of deterioration per unit quantity at the external inventories and transhipment points accordingly.

The third objective function to be minimized is a *quality measure* of the mixture feeding the mixer and takes into account the feedback of the quality control held at the downstream process (intermediate product/reactor/final product) in order to minimize the dispersion from the optimal concentrations:

$$TQM = \sum_{j=1}^{m} w_j(t) \cdot \left(\frac{C_{j,0,t} - C_{jt}}{C_{j,0,t}} \right)^2 \qquad (19)$$

where $w_j(t)$ is a *significance index* of the corresponding constituent and is each time adjusted/calibrated from the feedback control mechanism which takes into account the quality of the intermediate and final product at time unit t and proposes the 'closest conditions', i.e. the sub-optimal conditions that improve the quality and at the same time is near to the current working conditions of the overall agro-industrial process so as to maintain the steady state conditions. In this way, the feedback controlled mechanism triggers, at each time unit, a global optimization of the integrated system.

An economic incentive for maximizing the use of transhipment points is expressed by the following objective function:

$$TTU = \sum_{k=1}^{r}\sum_{t=1}^{T} e_{tr}(k,t) \cdot \left(1 - \frac{\sum_{d=1}^{D}\sum_{i=1}^{n} Q_{tr}(k,i,t,d)}{L_{tr}(k,t)} \right)^a \cdot IC_{kt}, a > 0 \qquad (20)$$

that should be minimized. IC_{kt} represents the significance of fully-utilizing the transhipment point k at time unit t.

The concept behind this objective function is to give an incentive to a vehicle leaving an external inventory to prefer routing to a transhipment point that is already semi-loaded to an empty one, even if such a route is not optimal in terms of distance. *TTU* is a measure of the achievement of scale economies in the location/operation of transhipment points over the space/time domain.

A.4 Alternative formulation

The cost of deterioration in the preceding model formulation is handled by equation (16) that associates an economic cost to the product deterioration. An alternative formulation, however, would take into account the fact that an implicit cost in the process is that of using/ordering more quantity out of the deteriorated agricultural raw materials in order to produce the same quantity (and quality) of final product. This can be employed in the model by reformulating the content b_{ji} of constituent j in the raw material i with a coefficient that takes into account the effect of the deterioration in the content of the constituents, i.e. b_{ijd} is the content of constituent j in the raw material i after leaving it to deteriorate for d time units and it holds that $b_{ij,d-1} < b_{ijd}$. In such a formulation, equation (16) is no more needed and the deterioration cost is encapsulated in the transportation cost.

VSP International
Science Publishers
P.O. Box 346, 3700 AH Zeist
The Netherlands

*Lecture Series on Computer
and Computational Sciences*
Volume 1, 2004, pp. 642-646

GIS-based Discrimination of Oil Pollution Source in an Archipelago –The Case of the Aegean

A. F. Batzias, N. P. Nikolaou, D. K. Sidiras[*]

Department of Industrial Management and Technology,
University of Piraeus,
Karaoli & Dimitriou 80, GR-18534, Piraeus, Greece

Abstract: This work is a knowledge-based approach of an integrated system including oil spill dispersion modeling in real marine environment with a Geographical Information System (GIS). This approach permits the application of Case Based Reasoning techniques for optimal selection of dispersion models leading to high resolution of (moving) potential oil pollution sources. Such an approach is particularly valuable in the case of an archipelago where the great number of islands, the local sea currents and the wind speed/direction changes contribute to the creation of a heterogeneous distribution of oil spills; under these conditions, knowledge extracted from facing similar pollution cases in the past may offer considerable help for solving present problems.

The implementation of the software developed in a case set in the Aegean Archipelago in East Mediterranean has proven the applicability of this approach in the successful localization of the potential oil-spilling sources.

Keywords: Geographical Information System, marine environment, sea pollution, oil spill dispersion

1. Introduction

Oil releases are a worldwide problem, which is more important in closed seas like the Mediterranean. Coastal authorities have the right and the responsibility to ensure compliance with national laws and international treaties, and therefore the detainment of vessels suspected of causing marine pollution is justified. Moreover, the International Convention on Arrest of Ships recognizes a broad variety of claims by salvers, by people who respond to spills, and by victims of pollution as grounds for detaining a ship or restricting its movement. The prerequisite for taking such measures is to localize the suspected vessel in time because many oil spills occur in areas of high ship-traffic and often are detected only once the suspected vessel has left the area that is under the jurisdiction of the country that suffers the damage.

In recent years certain Geographical Information Systems (GIS) have been introduced in conjunction with oil spill dispersion models for the determination of possible sources of oil discharge when this kind of marine pollution is observed at a particular site. Such a conjunction leads either to loose coupling, in which the only interaction is to export/import data via an exchange format, or to tight coupling, in which there is a high degree of interoperability (such as Windows compliancy) and sharing of data formats, or to full integration, where dispersion models are incorporated as GIS functionality [1]. The present work is a knowledge-based approach of the integrated type, putting simple functionality at a lower/executive level. This approach permits the application of Case Based Reasoning (CBR) techniques [2,3] for optimal selection of dispersion models leading to high resolution of (moving) potential oil pollution sources. It is worthwhile noting that such an approach is particularly valuable in the case of an archipelago where the great number of islands, the local sea currents and the wind speed/direction changes contribute to the creation of a heterogeneous distribution of oil spills; under these conditions, knowledge extracted from facing similar pollution cases in the past may offer considerable help for solving present problems. An application set in the Aegean Archipelago in East Mediterranean serves as a paradigm.

---

[*] Corresponding author. E-mail: sidiras@unipi.gr

2. Methodology

The following GIS-based algorithmic procedure has been designed/developed to solve the problem stated in the introduction. Due to lack of space, only the static part is presented herein (the dynamic part incorporates measurement/prevention-of-dispersion activities in the time course), including the stages described below. Fig. 1 depicts the interconnection of these stages, under the form of activity and decision nodes, as well as the communication lines with the Knowledge Base (KB), which is the heart of the system.

1. Collection of information that may influence dispersion of oil or chemical spill on a permanent real-time basis.
2. Selection of dispersion models and conditions under which they apply.
3. Design of GIS layers with heterogeneous grid suitable for spatially distributed data reception/storing/retrieval on a real time basis.
4. Extension of the network for more satisfactory coverage of the grid of each GIS layer.
5. Testing of network operation.
6. Reporting from an observation point on evidence of possible incident an emerging threshold.
7. Activation of the network for collecting relevant data and storing in an emergency file within the KB.
8. Extraction of most likely incidents from the KB through CBR.
9. Ranking of models in order of expected decreasing fitness to the situation under investigation.
10. Selection of the first available model for which data are sufficient.

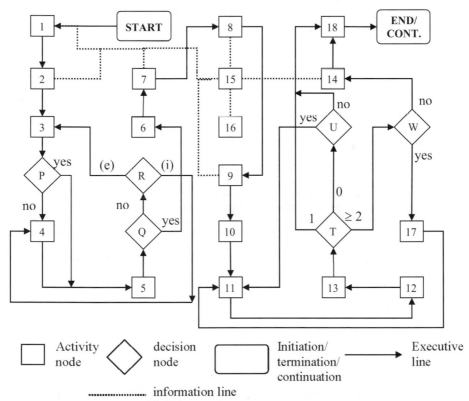

Figure 1. Flow chart of the knowledge based algorithmic procedure (static part) designed/developed for locating the source of pollution; the intelligent agent in stage (16) keeps the system open.

11. Computer aided running of the model in the integrated (with the GIS) environment.
12. Illustration of results on the GIS layer where the most likely suspected area including where the source of pollution is quoted under the form of contours at different confidence levels.

13. Searching for ships that were in the suspected area during the critical time period of pollutant discharge.
14. Statistical processing of information available to locate the vessel suspected most of all.
17. Formation of the new data input.
18. Preparation and dispatch of the final conclusions report.
15. Development/enrichment of a Knowledge Base (KB).
16. Communication with external information sources via an intelligent agent.

P. Is the extent of the network established for observing/measuring/reporting adequate?
Q. Are the numerical data sufficient to ensure validation of at least one model over the whole marine area under consideration?
R. Is the required additional information extensive (e) or intensive (i)?
T. What is the number of suspected ships?
U. Is there any other model (left unexamined) for which data are sufficient?
W. Is there any report from another observation point indicating pollution parameter-values (even if they are not higher than the emergency threshold)?

3. Implementation

The implementation of the herein presented methodology was based on an extension of the tailor-made software package designed/developed by our research group within the framework of a LIFE/EU programme by using the latest edition of Microsoft's Visual Studio .NET™, and more specifically Visual Basic.NET [2]. The application exploits the following technologies of the .Net architecture [3]: (a) The .NET Framework that includes the Common Language Runtime (CLR), (b) XML that is used as a storage format, and (c) the GDI+ which includes classes for working with pens, brushes, vector graphics, controlling printers, manipulating fonts and other drawing tool.

In more detail, the application consists of the main application part and the component part. The former includes the computational skeleton of the simulation of the oil spill behavior; the latter includes three components: (a) the Mathlibrary which incorporates a custom-developed non linear regression class, as well as a unit conversion class, (b) the OilspillLibrary (Oil spill class) which hosts computational components of the simulation procedure (e.g. calculation of the oil-spill spreading area, emulsification percentage, and computation of the horizontal displacement and vertical dispersion over time), and (c) the ReverseEstimation class library that makes an estimation of the possible position of the original oil leakage based on a reverse non-deterministic version of the model employed in the OilspillLibrary component. Last but not least, both components have been created in a Dynamic Link Library (.dll) form and are referenced by the main application part which interacts in real-time performance with an on-line data providing system [4].

The afore-described software has been implemented in the case of the leakage of 12,000 m^3 of ADCO (Abu Dhabi Company for Onshore Oil Operations) oil type at (23.08N, 36.98E) which is located upon an oil sea transportation route of heavy traffic at the Aegean Sea north of the *Evia* and *Andros* islands. More specifically, Figure 2 (a)-(d) illustrate four snapshots of the oil dispersion at 0, 21, 57 and 106 hours as simulated by our application. Figure (e)-(f) exhibit the implementation of the 'reverse simulation' algorithmic procedure used to predict possible leakage areas based on two spot measurements. The superimpose of trajectory estimations through the described algorithmic procedure suggests the area (of 90% and 95% confidence level) that the suspect ships should have been located at the time of the leakage.

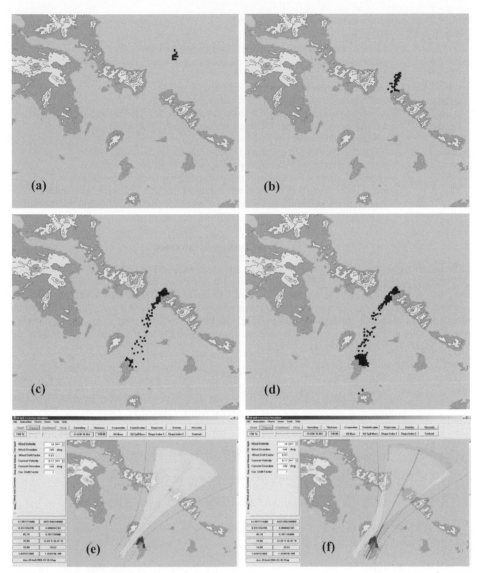

Figure 2. GIS-based diagrams showing (a) an oil spill appearance of approximately 12,000 m^3 ADCO oil type (ρ_0= 0.832 t/m^3, D = 0.2, K_{en} = 0.0079, T_{ref} = 15°C, V_{ref} =62 cP, Y_{max} = 0.7, Kem = 2 10^{-6} sec/m^2) north of Andros island where strong NE wind prevails (10 m/sec), (b) its dispersion after 21 hours, (c) its dispersion 57 hours, (d) its dispersion after 106 hours, when the observation points at (21.91N, 36.63E) and at (21.94N, 36.60E) (see diagrams e and f) give a signal that activates the reverse procedure described in Fig.1; the area in gray is the most likely to include the source of pollution (excluding land areas).

It is evident that the determination of the exact oil type leaked is of utmost interest since its natural characteristics influence the behavior of the oil spill. This determination can take place with spot measurements of oil residues on sea surface or upon the shore by measuring oil type characteristics such as the oil spill thickness, the density ρ and the viscosity V of the oil residues. More specifically, the nomenclature of the physical quantities used for oil identification is summarized in Table 1:

Table 1: Physical quantities that characterize oil type.

ρ_0 (t/m^3)	oil density
D (%)	fraction as percentage (by weight) distilled at 180 degrees C
K_{en}	empirical constant dependent on the oil type and weathered state (for vertical dispersion)
con_1 (1/K)	empirical constant for density calculation (K = °C + 273)
con_2	empirical constant for density calculation
v_{ref} (cP)	reference viscosity at some reference temperature T_{ref} (for viscosity calculation)
P_{ref} (t/m^3)	reference density at some reference temperature T_{ref} (for density calculation)
T_{ref} (°C)	reference temperature (for viscosity and density calculation)
con_{Temp}	empirical constant (Temperature Coefficient) for viscosity calculation
con_{Evap}	empirical constant (Evaporation Coefficient) for viscosity calculation
con_{Emul}	empirical constant (Emulsification Coefficient) for viscosity calculation
Y_0	initial water fraction of the oil-water emulsion (for emulsification calculation)
Y_{max}	maximum water fraction of the oil-water emulsion (for emulsification calculation)
K_{em} (sec/m^2)	emulsification coefficient (for emulsification calculation)

4. Concluding Remarks

The knowledge-based approach we have adopted and the corresponding software we have developed for localizing sources of marine pollution due to oil spill dispersion are successfully applied in a real case set in the Aegean Archipelago in East Mediterranean. The characteristics of this approach/software are: (a) flexibility, as it can change the basic environmental model it uses within the integrated GIS platform, (b) high resolution, as it can discriminate a vessel suspected as potential source of pollution among other nearby vessels, and (c) extensibility, as it can easily obtain dynamic features to operate in the time domain, as well.

References

[1] Y. Li, A.J. Brimicombe, and M.P. Ralphs, Spatial data quality and sensitivity analysis in GIS and environmental modeling: the case of coastal spills, *Computers, Environment and Urban Systems* **24** 95-108 (2000).

[2] J. Kolodner: *Case-Based Reasoning.* San Mateo: Morgan Kaufmann, 1993.

[3] C, Riesbeck and R. Schank: *Inside Case-Based Reasoning.* Northvale, NJ, 1989.

[4] D. Vitter: *Designing Visual Basic .NET Applications,* Coriolis Group Books, 2001.

[5] P. Thorsteinson and R. J. Oberg: *.NET Architecture and Programming Using Visual C++,* Pearson Education, N.Y., 2002.

[6] B. P. Douglass: *Doing Hard Time: Developing Real-Time Systems with UML, Objects, Frameworks and Patterns*, Addison-Wesley Professional, 1999.

VSP International
Science Publishers
P.O. Box 346, 3700 AH Zeist
The Netherlands

*Lecture Series on Computer
and Computational Sciences*
Volume 1, 2004, pp. 647-652

GIS-based Computer Aided Air Pollution Biomonitoring for Impact Assessment – Application in the Case of Materials Deterioration

C.G. Siontorou[*1], A. S. Kakos[1], G. Batis[2]

[1]Department of Industrial Management and Technology,
University of Piraeus,
Karaoli & Dimitriou 80, GR-18534, Piraeus, Greece

[2]Department of materials Science and Engineering,
School of Chemical Engineering,
National Technical University of Athens,
9 Iroon Politechniou St. GR-15780, Greece

Abstract: This work presents a framework for GIS-based air pollution biomonitoring in the vicinity of buildings and monuments with a view of (a) facilitating pollution distribution patterns around constructions, (b) providing information that may lead to confirmation of construction material decay mechanisms, and (c) contributing to the prevention of materials deterioration. The framework is realized through an algorithmic procedure and the monitoring system utilizes a permanent, nature-derived, cheap system (like lichens serving as bioindicators) coupled to a more accurate, precise and expensive measuring system (like conventional biosensors) for controlling and assessing bioindicator pollution estimates.

Keywords: GIS, biosensors, bioindicators, construction decay

1. Introduction

Atmospheric pollution is a significant factor that accelerates construction materials deterioration. The pollutants attack differs according to substrate, that may cover a wide range of materials, like steel reinforcement in modern buildings [1], stones in mediaeval monuments [2], and marbles in ancient ruins [3]. Biological monitoring can be very effective as an early warning system to detect environmental changes due to atmospheric pollution. This approach is based on the assumption that any changes taking place in the environment have a significant effect on the biota [4]. Sentinel organisms are species capable of accumulating persistent pollutants and are used to measure the biologically available amount of a given pollutant in a given ecosystem [5]. A calibration process measuring the relationship between the sentinel and its source is generally required [5]. Biological monitoring is usually qualitative or semi-quantitative, employing single indicator species or using community changes, and it is complementary and in conjunction with field monitoring devices [4].

Biosensors, combining a biological recognition element (enzyme, antibody, receptor protein, nucleic acid, etc.) and a suitable transducer (electrochemical, optical, piezoelectric, etc.) are useful analytical instruments, able to measure quantitatively target analytes [6]. The significant technological advances over the past decades have facilitated the environmental applications of these highly sensitive and selective devices.

The aim of the present work is to provide a suitable framework for pollution monitoring in the vicinity of buildings and monuments, based on a permanent, nature-derived, cheap system (like lichens serving as bioindicators) periodically controlled, assured and assessed by a more accurate and precise measuring system (like conventional biosensors). This framework should be incorporated within a Geographical Information System (GIS) in order to (a) depict each pollutant species spatial distribution

---

[*] Corresponding author. E-mail: csiontor@unipi.gr

and (b) correlate their impact on materials under examination with pollutant concentrations taking separately and in combination in order to reveal any synergistic effects. The coupling of cheap bioindicators to expensive biosensors can be proven quite efficient in providing a cost-effective, reliable and simple GIS-based monitoring system for (a) facilitating pollution distribution patterns around the construction under consideration, (b) providing useful information that may lead to confirmation of suggested physicochemical mechanisms of degradation through multivariate statistical analysis, and (c) contributing to the prevention of construction materials deterioration.

2. Methodology

The algorithmic procedure especially designed and implemented for the monitoring program, includes the stages shown below. Fig. 1 illustrates the connection of stages, represented by the corresponding number or letter, in case of activity or decision node, respectively.

1. Determination of the limits of the system under consideration.
2. Estimation of extent and intensity of material deterioration within the system's limits.
3. Collection of deteriorated materials samples and in depth physicochemical examination.
4. Selection of relevant phenomenological models of the identified materials deterioration.
5. Ranking of models in order of decreasing likelihood (to be valid in the certain situation), according to technical literature and experts' opinion.
6. Determination of the species to be measured and the corresponding significance level (for each species) required for model validation.
7. Choice (from the Knowledge Base) of the vegetation species (indigenous or foreign) to be utilized as bioindicators.
8. Choice (from the Knowledge Base), preparation and laboratory testing of the dedicated biosensor considered to be most suitable for each species.
9. Biosensor development for operating in real environment.
10. Testing by operating in simulated environment under laboratory controlled conditions; bioindicator calibration and assignment of sensitivity indices; biosensor/bioindicator correlation.
11. Testing by operating in real environment under most likely natural conditions; determination of accumulation levels of monitoring pollutant species in bioindicators.
12. Parallel measurement of all species at the same time by using the biosensors passed the tests so far.
13. Selection of the most likely tested model by relaxing the statistical significance criterion.
14. Packaging the set of approved biosensors within a kit for protection and convenience (i.e. easiness of handling during inspection/maintenance/replacement).
15. Cost estimation per kit, including placement/checking/maintenance/replacement.
16. Construction of grid within each GIS layer, under the constraint of the budget available, corresponding to each pollutant species considered as input variable of the model.
17. Optimal placement of biosensors according to experimental design techniques.
18. Small scale preliminary operation for testing and design of a maintenance/replacement program.
19. Full scale operation, long enough (at least one year) to permit adequate output data collection of both, biosensors and bioindicators, in the same course.

Performance of stages 20-29, as well as the intermediate decision nodes V, X, X and W; each iteration shall be executed serially or in parallel, for each pollutant layer Gi, $(i = 1, ..., k)$:

20. Extension of the full operation period to obtain additional biosensor measurements to be compared with corresponding estimated values based on bioindicator accumulation levels and the learning set obtained in the previous stage.
21. Comparison between biosensor measurements and corresponding estimated values for the Gi pollutant GIS layer (based on the pollution prediction model outcome and the interpolation method currently employed).
22. Application of a program for saving economic resources by removing most biosensors from this system (to be used within a similar system, taking advantage of the knowledge/experience acquired) but keeping a number of them required to maintain predictability within an acceptable region.
23. Estimation of time periods Ti (duration di and frequency fi) that biosensors are established to execute a full program of measurements together with the bioindicators which act on a permanent basis.
24. Realization of Ti, di ,and fi.

25. Comparison between biosensor measurements and corresponding estimated values within each *Gi* pollutant GIS layer. At time period *Ti*.
26. Creation/enrichment of a Knowledge Base (KB).
27. Intra- and inter- net searching via an intelligent agent based on an ontological adaptable interface [7].
28. Periodic comparison of measured deterioration with the corresponding values obtained by the model estimations, as depicted on a separate GIS layer (based on the pollution prediction model outcome and the interpolation method currently employed).
29. Sampling of bioindicators and lab testing for resistance/diversification

P. Is the operation in simulated environment satisfactory?
Q. Is the operation in real environment satisfactory?
R. Is there another species left unexamined?
S. Is the model used valid at a predetermined significance level?
T. Is there another model left unused?
U. Are the results of preliminary testing satisfactory?
V.Is the comparison satisfactory, for each Gi pollutant GIS layer, at a predetermined significance level?
W.Is the comparison satisfactory, for this Gi pollutant GIS layer, at a predetermined significance level?
X. Is the comparison satisfactory, or should we extent the biosensor measurement period?
Y. Is the packaged kit performance satisfactory?
Z. Are species resistant/diversified?

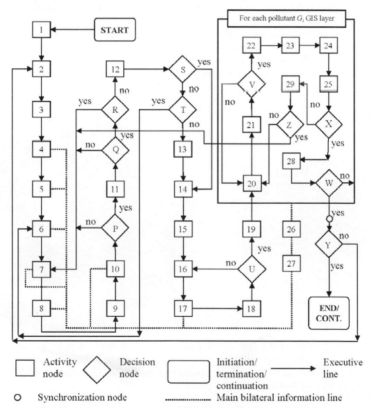

Figure 1: The algorithmic procedure designed/developed for the air pollution monitoring in the vicinity of buildings and monuments.

3. Implementation

The methodology described above has been implemented in the case of SO$_2$ monitoring around buildings and monuments (stage 6). Sulfur compounds have been indicted as the most critical environmental factor regarding masonry decade, mainly because they are often acidic and can have high concentrations in city and suburban air. Fluxes of sulfur dioxide can be high, especially when promoted by biological activity. Dissolution by chemical reaction with contaminants contained in precipitation is one of the most familiar eroding processes, particularly in the case of carbonaceous stones. Among the reactions that favour the degradation of calcium carbonate, sulfation is the most significant, producing a gypsum layer on the material surface:

$$SO_2 \xrightarrow{\frac{1}{2}O_2} SO_3 \xrightarrow{H_2O} H_2SO_4 \xrightarrow{CaCO_3, H_2O} CaSO_4.2H_2O + CO_2$$

As regards biosensor selection (stage 8), the devices considered are electrochemical and are based on biocatalysis or bioaffinity [8,9]; they require no sample preparation, offer fairly low detection limits, increased sensitivity and selectivity towards sulfur dioxide, and short response time, whereas interpretation of results can be performed by computer software.

The candidate sentinel organisms (stage 7) are indigenous lichen species. Lichens are perennial, slow-growing, organisms that maintain a fairly uniform morphology in time and are highly dependent on the atmosphere for nutrients [10]; the lack of a waxy cuticle and stomata allows many contaminants to be absorbed over the whole thallus surface. It has been demonstrated that the concentrations of elements in lichen thalli are directly correlated with their environmental levels [11], and as a consequence lichens are very useful for monitoring spatial and/or temporal deposition patterns of pollutant species. Lichens also meet other characteristics of the ideal sentinel organism [5]: they are long-lived, provide abundant material to be harvested, and a large body of scientific knowledge is available. As a result, lichens are very effective in trapping substances from the surrounding environment, especially sulfur dioxide [4,5,10,11], whereas the toxicity effect produces marked external changes (change in thallus colour, reduction in thallus size and changes in the thickness of the thallus) [11].

Concerning the implementation of methodological stages 21 and 28 the authors have employed: (a) EPA's Industrial Source Complex Dispersion Model - Short Term Version 3 (ISCST3) model for the estimation of air pollutants within the area of interest [12] and (b) an interpolation method such as the Inverse Distance Weighted (IDW) or the Kriging method [13] for the creation of a pollution map such as those depicted in Fig. 1 and Fig. 2 (see Appendix for details).

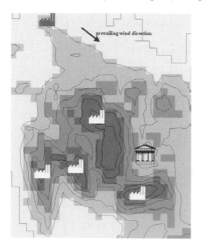 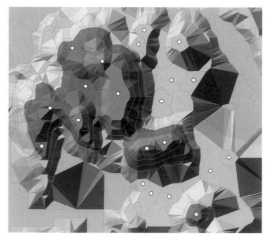

Figure 1. Digital elevation Model (DEM) derived out of the calculation of the downwash plume and other air pollutants in the immediate vicinity of buildings based on EPA's ISCST3 model.

Figure 2. The points selected by the algoritmic procedure as candidate locations for the (reduced number) of biosensors (depicted as white dots in the picture).

4. Concluding Remarks

The framework designed/developed for GIS-based computer aided air pollution biomonitoring can be used for impact assessment, especially in the case of construction materials decay. Its characteristic functions are (a) optimization of economic resources by maximizing the monitoring potential at a predetermined significance level in more than one locations of interest, (b) utilization of a flexible/dynamic computational platform, where the atmospheric pollution-material deterioration models can be modified in the time course, (c) maintenance and update of a knowledge base for biosensor and bioindicator ontologies, and (d) technology acquisition that can be transferred/adapted in similar locations of the application network.

References

[1] G. Batis and E. Rakanta, Corrosion of steel reinforcement due to atmospheric pollution, *Cement Concrete Composites,* forthcoming.

[2] A. Moropoulou, P. Theoulakis, and T. Chrysophakis, Correlation between stone weathering and environmental factors in marine atmosphere, *Atmospheric Environment* **29** 895-903 (1995).

[3] Th. Skoulikidis, and P. Vassiliou, Corrosion and Conservation of the Building Materials of the Monuments; The Acropolis Case, *Corrosion Reviews* **17** 295-332 (1999).

[4] P.A Keddy, Biological monitoring and ecological prediction: from nature reserve management to national state of the environment indicators, *Monitoring for Conservation and Ecology* (Editor: F.B. Goldsmith), Chapman and Hall, 1991.

[5] A. Beeby, What do sentinels stand for?, *Environmental Pollution* **112** 285-298 (2001).

[6] A.P.F. Turner, I. Karube and G. Wilson, *Biosensors, Fundamentals and Applications.* NY, Oxford University Press, 1987.

[7] F. A. Batzias and E. C. Markoulaki, Restructuring the keywords interface to enhance CAPE knowledge via an intelligent agent, *Computer-Aided Chemical Engineering,* **10** 829-834 (2002).

[8] E. Kress-Rogers, *Handbook of Biosensors and Electronic Noses, Medicine, Food, and the Environment.* CRC Press, 1997.

[9] U. Bilitewski and A.P.F. Turner (Eds): *Biosensors for Environmental Monitoring.* Harwood Academic Publishers, 2000.

[10] M.E. Hale, *The Biology of Lichens.* Edward Arnold, London, 1983.

[11] M. Kovács (Ed): *Biological indicators in environmental protection.* Ellis Horwood, New York, 1992.

[12] U.S. Environmental Protection Agency, Office of Solid Waste, *User's Guide For The Industrial Source Complex (ISC3) Dispersion Models,* Washington DC, 1999.

[13] E.H. Isaaks and R.M. Srivastava, *An Introduction to Applied Geostatistics.* Oxford University Press, 1989.

Appendix

Stage 1 of the methodology is accompanied by the definition of the terrain domain and the consequent store of its morphology in Digital Elevation Model (DEM) format. The same format can be used to store data, estimated or collected *in situ,* regarding air pollutants within the defined area canvas. Air pollutant measurements/estimations for each canvas cell is stored in a separate GIS layer.

On the other hand, the determination of biosensor initial placement and bioindicator positions within the area of interest, should take into consideration the nearest-neighbor index R, which measures the degree of spatial dispersion of features based on the minimum distance between individual features (point & line). It holds that

$$R = d_{obs}/d_{ran} \tag{1}$$

where d_{obs} is the observed mean nearest-neighbour distance and d_{ran} is the expected nearest-neighbor distance for a random arrangement of points: i.e.

$$d_{ran} = 0.5\sqrt{\frac{A}{n}} \tag{2}$$

where n is the total number of biosensors (or bioindicators) and A is the terrain area. The nearest-neighbor index can have values between 0.0 (indicating a completely clustered pattern) and 2.15 (indicating a completely dispersed pattern). A random arrangement is indicated by a value of 1.0.

An associated Z-test $Z = \dfrac{d_{obs} - d_{ran}}{SE(d_{ran})}$ indicate the significance of randomness where

$SE(d_{ran}) = \sqrt{\dfrac{(4-\pi)A}{4\pi n^2}}$.

For the estimation of air pollutants in terrain locations where no biosensors have been placed, an interpolation method can be employed; such are the Inverse Distance Weighted (IDW) and the Kriging method. IDW estimates cell values by averaging the values of sample data points in the vicinity of each cell. The closer a point is to the center of the cell is being estimated, the more influence, or weight, it has in the averaging process. Kriging, on the other hand, is a method of interpolation which predicts unknown values from data observed at known locations by using variograms to express the spatial variation, and it minimizes the error of predicted values which are estimated by spatial distribution of the predicted values. In ordinary kriging, which estimates the unknown value using a weighted linear combinations of the available sample:

$$\hat{u} = \sum_{j=1}^{n} w_j * u, \sum_{j=1}^{n} w_j = 1 \tag{3}$$

The error of ith estimate, r_i, is the difference of estimated value and true value at that same location:

$$r_i = \hat{u} - u_i \tag{4}$$

The average error of a set of k estimates is:

$$m_r = \frac{1}{k}\sum_{i=1}^{k} r_i = \frac{1}{k}\sum_{i=1}^{k}(\hat{u} - u_i) \tag{5}$$

The error variance is:

$$\delta_r^2 = \frac{1}{k}\sum_{i=1}^{k}(r_i - m_r)^2 = \frac{1}{k}\sum_{i=1}^{k}[\hat{u} - u_i - \frac{1}{k}\sum_{i=1}^{k}(\hat{u}-u_i)]^2 \tag{6}$$

Unfortunately, the equation cannot be used because true value $V_1,...,V_k$ is not known. In order to solve this problem, a stationary random function is applied that consists of several random variables, $V(x_i)$. x_i is the location of observed data for $i > 0$ and $i \leq n$. (n is the total number of observed data). The unknown value at the location x_0 that must be estimated is $V(x_0)$. The estimated value represented by random function is:

$$\hat{V}(x_0) = \sum_{i=1}^{n} w_i * V(x_i), R(x_0) = \hat{V}(x_0) - V(x_0) \tag{7}$$

The error variance is:

$$\hat{\delta}_r^2 = \hat{\delta}^2 + \sum_{i=1}^{n}\sum_{j=1}^{n} w_i w_j \widetilde{C}_{ij} - 2\sum_{i=1}^{n} w_i \widetilde{C}_{i0} + 2\mu(\sum_{i=1}^{n} w_i - 1) \tag{8}$$

where $\hat{\delta}$ is the covariance of the random variable $V(x_0)$ with itself and all of our random variables are assumed to have the same variance, μ is the Lagrange parameter.

In order to get the minimum variance of error, the partial first derivatives of the former equation are calculated for each w and setting the result to 0.

A variogram can be used instead of covariance to calculate each weight of the equation. The variogram is:

$$\gamma_{ij} = \hat{\delta}^2 - \widetilde{C}_{ij} \tag{9}$$

The minimized estimation variance is:

$$\hat{\delta}_r^2 = \sum_{i=1}^{n} w_i \gamma_{i0} + \mu \tag{10}$$

VSP International
Science Publishers
P.O. Box 346, 3700 AH Zeist
The Netherlands

*Lecture Series on Computer
and Computational Sciences*
Volume 1, 2004, pp. 653–656

Empowering Financial Tradeoff in Joint Financial & Supply Chain Scheduling & Planning Modeling

M.Badell, E.Fernández, J.Bautista, L.Puigjaner [1]

Dept. of Chemical Engineering, Dept. of Operations Research, Dept. of Enterprise Management
Universitat Politècnica de Catalunya,
Diagonal 647, Barcelona 08028, Spain

Received 17 July, 2004; accepted in first revised form 22 July, 2004

Abstract: This paper addresses the integration of scheduling and planning with financial management in the chemical process industries. A prototype is described coupling cash flow management and a budgeting model with an advanced scheduling and planning procedure. This paper suggests that a new conceptual approach in enterprise management systems – consisting on the integration of enterprise finance models with the company operations models – is a must to improve the firm's reliability and its overall value.

Keywords: Programming model, integration, scheduling, planning, budgeting, supply chain,

Mathematics SubjectClassification: AMS-MOS or PACS

PACS: AMS-MOS or PACS Numbers

1. Integration in Supply Chain Programming Models

All supply chain (SC) has in parallel a financial chain that creates it, lets it grow and multiplies it, or on the contrary, destroys it. Modeling the SC by estimating its performance in space and time dimensions is no longer sufficient. The level of integration achieved by ERP systems is not enough. It is also required a simultaneous functional integration in order to optimize decision making when assets or financings are assigned. During the decision making the plant operation equipment and resources compete against manpower, supplies, marketing, distribution, post services, investments, maintenance, technology, new product research and the remaining enterprise activities. It is necessary a dynamic integrated financial and operative planning in order to change and achieve the supremacy of the financial logic over the already exhausted material logic. The material logic and hierarchical planning (HPP) will be defeated owing to its inefficiency when trying to support the globalization race. Further, at present it is not enough to integrate SC measuring information of its performance by financial key performance indicators (KPI) that in this position assume a passive informative role. It is anymore feasible to continue using intermediate cost related process KPI as makespan, whose assumed minimization by engineers is supposed "equivalent" to the maximum profit being this not always true due to the fact that the ranking of products is schedule dependant [1]. It is no longer sufficient to use time to compare expenses or earnings when deciding what production will be done in which multiplant.

2. The Sequencing & Performance Measurement Approach

The theory of scheduling has used different types of methodologies like combinatorial, simulation, graph techniques, network methods and heuristics for addressing efficiently the sequencing problem. Nevertheless, the timing assignment of initial and final times to units/operations still requires many calculations within schedulers. For engineers a detailed timing is certainly necessary. However, during the enumerative search of sequences these calculations can be simplified taking care that at the end the solution must recover all details. For business actions these details are not necessary, the timing at unit level is meaningless and what is interesting is the sequence of whole batches to check bottlenecks and optimize the interferences to other batches and resources. For this reason chief financial officers (CFO) only require to consider the batch as a black box, without internal process details.
Reviewing literature we observed that TSP formulations were mainly used when sequencing operative schedules for managing programming models by indirect parameters as cleaning cost matrixes or

---

[1] Corresponding author. E-mail: Luis.Puigjaner@upc.es

quality weight coefficients. Here we use a methodology that considers directly a temporal distance matrix, the matrix of not overlapping times. Initially we calculate all possible binary production sequences in multiproduct plants and construct a matrix of not overlapping times with the intention of minimizing a production path visiting all nodes; each *city* represents a binary sequence of production processes of two products. With this approach it is required only one match between batches, as in a puzzle. A TSP/ATSP algorithm was developed to encourage more flexible solutions of larger problems in less time using as data the not overlapping times matrix. The solver developed fulfils this expectative.

Schedules are generally evaluated by aggregate quantities that involve information about jobs resulting in one dimensional performance measures that are usually function of the set of completion times of schedules and hence are regular performance measures. Here we measure the performance by means of a two dimensional aggregate quantity that in unison takes into account the timing and the economic/financial value added of production sequences in the SC. This permits to decide the dominance of the most valuable assignments as a function of *money & time*.

3. APS Systems Technology to CFO Users

The modeling framework elaborated was also adapted to work in advanced planning and scheduling systems (APS) because APS systems can support an integrated interactive decision support tool. The APS approach relies on the creation of computer models that permits to schedule resources assuming finite capacity. With this aim in mind the financial activities were represented similarly as production unit operations in APS systems. Suppose a 5 hours chemical process in a unit operation, i.e. a reactor. This process is established by the production recipe. Similarly a purchase of raw materials that costs 700 € with a given delivery time can correspond to a financial recipe using the fictitious unit operations that we created for this aim.

Cash management is part of CFO's work. Long range decisions, such as capital budgeting, funds, dividend policy, merges or fusions are usually of higher priority than cash management. Also the capital budget outcome usually is an input of short term cash management tasks. However, the low prioritized work of short term budgeting is the most controversial and time consuming management task in decision making and could be the first step on the way to firms' insolvency. *But how can CFO control solvency if in this day and age we still have a cash open loop regulation cycle?* See section 5.

For the reason of the long time horizon of capital investments, they are usually divided into annual periods. Hence, the first period of the capital budget is the horizon for a cash management model. This joint time period means that the two types of financial decisions are interrelated and that an independent solution of each model may result in sub-optimal decisions. The vision of both underlying functions is therefore very convenient so these financial decisions should be derived jointly as will be shown in section 6. The capital budget problem gives the first year input data to short term plan/ budget.

Unlike capital budgeting, the management of working capital concentrates on short term financial decisions and therefore it is closely related to cash management. The management of working capital is defined as the management of liquid assets, which can be divided into cash, marketable securities, accounts receivable and inventories. Cash and marketable securities offer the way to make useful decisions with current receivables and payables and the APS framework offers the technique to support the decision making with a holistic vision of the firm. In this scenario Gantt charts could show financial operations jointly with production operations to perform plans or budgets that could be handled interactively as in electronic charts. Moreover, new profiles of resources can be displayed: of money, earnings, value added or even the balance sheet corresponding with the plan or budget on trial.

4. The Joint Modeling Framework

In addition to the *money-time* models and simulations that represent batches instead of the detailed timing of equipment, the liquidity control is undertaken retaining cash at a minimum level and searching an option to invest the surplus if an opportunity exists. Due to the fact that in this joint environment are freely permitted the movements of the events related with the financial activities of the supply chain that create the monetary flows, it is feasible to flat the cash peaks achieving an optimal synchronization of cash flows. While engineers in operative scheduling used to work with incremental analysis now this is not possible in the integrated approach because financial managers need to control liquidity. It is evident that in order to control liquidity it is necessary to know the total liquid assets in the working capital of the firm. That is why we formalize the financial model using the typical financial variables of the balance sheet, that is, the current assets, fixed assets, current liabilities, long term debt and equity. This mathematical framework supports an updated formulation of the integrated functional model. The balance sheet model allows the calculation of the historic inventory of money taking into account the whole set of economic categories that are always updated in databases of companies and always ready to be used at a mouse click. Further, the work develops a new approach to solve hierarchical production planning pitfalls (Figure 1). The extension of the process to consider provisions and sales facilitates the inclusion of the MRP system for inventory management in the same APS. With this approach can be suppressed the delay when the feasibility of the master plan in hierarchical planning is needed. This innovation has been applied in practice.

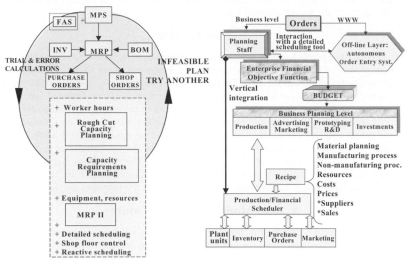

Figure 1: Hierarchical production planning and the joint functional integration approach

5. A Web-based Order System Considering Liquidity

Taking into account that production computer aided tools and its guidelines today in force are not transparent to CFO, added to the fact that usually they are not consulted when at scheduling level is applied the due dates fulfillment policy in make to order systems or a marketing policy stretching receivables receipt, CFO repeatedly becomes slave of supply chain decisions. Here is the *open loop* of section 3. Frequently CFO can be surprised with unexpected stress in liquidity caused by the plant floor or sales negotiations. An appropriate web based order system can solve this controversy. In this work a web- based system with intelligent agents for orders' management integrates finances and supply chain during the routine production practice. A central agent attempts to preserve liquidity by a financial tradeoff on due dates and closes negotiations considering the human staff guidelines organized in a customer system of rules with knowledge base format.

6. Previous Work and Present Guidelines

Several authors appraise in the literature the integration coupling finances and operations in planning models as Klein, 1998 and Shapiro, 2002. Shapiro places the focal point of linkage between finances and supply chain models at the strategic level. In his opinion it is recommendable a top-down approach, e.g. from the strategic model to the tactical, not considering the short term due to the ambiguity in timing at this level. In our opinion it is necessary to make cross walks in both up and down directions and vice versus, including in first place the short term planning and budgeting models that have the privilege of being reality. This doesn't mean that a single integrated model can play the role in the three terms mentioned. The granularity of each level must be different adjusting the different sizes of periods of time in the planning and budgeting model. The timing or inter-temporal aspects of the problem are captured by the unequal multiple period model where the consequences of the decisions of one period are related to all relevant decisions in subsequent periods. During the strategic planning analysis, when defining the capital budget, it is necessary to consider the interactions of the technological processes and resources in plant floor to refine which previsions can be realistically produced and hence be in plan and budget. Certainly it is known that in the chemical companies there is a high percent of failure in strategic plans and capital budgets when not being able to produce a considerable part of the production foreseen in the prevision. This is due to the ignorance of the interactions and bottlenecks that arise at plant level between the teams and resources when joining the processes defined at strategic level. The eradication of this defect requires fostering the development of a modeling tool capable of adjusting the high level production plans to the real possibilities of the plants without loosening any time as in the HPP offline trial and error (Figure 1). In the short term initial periods could be of days or weeks with a total certainty while the final periods with uncertainty must be of months. The number of the inter-period equations depends on the planning horizon and its division into smaller intervals. Further, if it is a budget with twelve periods in a year horizon and you apply the first period of one week, when the week elapsed, the model must be recomputed to a year horizon again, as a rolling horizon, only applying the periods planned with certainty.

7. Case Studies

The approach described herein has been implemented and proven by several case studies [1]-[6].
One of them presents a novel approach to request and implement water reuse in industry within a realistic forward looking financial scenario. Frequently environmental investments are delayed/ refused when competing with other investments that produce more money. A global financial synchronization can aid the arguments. An integrated model in the framework of the enterprise balance sheet is offset and synchronized including the proposed potential investment for water reuse. To simulate cash flow management when coupling exploitation (current assets) and investments (fixed assets) taking into account taxes, a budget model connected with net present value (NPV) formulation is integrated. The model safeguards liquidity by the cash flow synchronization. The case study suggests that a joint decision-making procedure could improve the acceptance of investments to face sustainability demonstrating that the effect on the firm liquidity smoothes when it is managed satisfactorily.

8. Joint Modeling and the Future

Even though the advantages of joint functional integration have not been valued, there is a positive consensus in literature. The approaches reviewed about integrated models are based in the inclusion of fixed offline links between operation and financial models through the input of fixed exogenous data pre-calculated. So the operation of the supply chain is not included in the same model and the synergy between corporate financial and supply chain management has not been exploited up to now. As a result production managers cannot appreciate the potential impact that can provoke in corporate finances and conversely CFOs cannot become conscious of the force of optimizing the combined effects of financial decision making. In fact, it is not adequate to continue leaving open the regulation loop in the financial short term policy of a firm. According to our knowledge, a similar approach as the one presented here was not found in literature. This contribution to solve this old challenge has been recognized by diverse authorities that reviewed this work. But, are we talking about a beloved breakthrough to CFOs/users and software developers or a challenge impossible to achieve, as happened in the last century?

Acknowledgements

Financial EC support by project VIPNET (G1RD-CT-2000-003181), from MCYT by OCCASION (DPI2002-00856) and from CIRIT by GICASA (I0353) are gratefully acknowledged.

References

[1] M. Badell, L. Puigjaner. *"A New Conceptual Approach for ERM Systems"*, AIChE Series No.320. 94, 217-223, Eds. Pekny & Blau, NY, 1998.

[2] Badell, M., Puigjaner, L., Discover a powerful tool for scheduling in ERM systems. *Hydrocarbon Processing* 80 (3), 160 (2001).

[3] Badell, M. *Integration of financial and production aspects within the supply chain*. Internal Report of the EC project VIPNET (G1RD-CT-2000-003181). June 2002.

[4] Badell, M., J. Romero, L. Puigjaner. Optimal budget and cash flows during retrofitting periods in batch chem. process ind. *Intern. Journal of Production Economic*, Received 28 May 2003; accepted 16 June 2003, available online at http://www.sciencedirect.com (2003).

[5] Badell, M., J. Romero, L. Puigjaner. Planning, scheduling and budgeting value-added chains *Computers & Chemical Engineering,* 28 45-61(2004).

[6] Romero, J., Badell, M. Bagajewicz, M.,L. Puigjaner. Integrating Budgeting Models into Scheduling & Planning Models for Chem. Batch Ind.. *Ind. Eng. Chem. Research 42,* 6125-6134 (2003).

[7] Klein, M. *"Coupling Financial Planning and Production Planning Models"*, Xth Intern. Working Seminar on Production Economics, Innsbruck, Vol.3, 28-39 (1998).

[8] Shapiro, J.F. *"Modeling Supply Chains"*, Ed. Duxbury, 2001.

VSP International
Science Publishers
P.O. Box 346, 3700 AH Zeist
The Netherlands

*Lecture Series on Computer
and Computational Sciences*
Volume 1, 2004, pp. 657-660

Automatic Generation of Production Scheduling Problems in Single Stage Multi-Product Batch Plants: Some Numerical Examples

M. Yuceer, R. Berber[1] and Z. Ozdemir

Department of Chemical Engineering,
Faculty of Engineering,
Ankara University,
06100 Ankara,Turkey

Abstract: This work considers optimal scheduling of a set of orders in a multi-product batch plant with non-identical parallel processing units where the process is single stage. The allocation of orders to the production units was formulated as an MILP problem in continuous time. Starting from the basic model proposed earlier [5], the new formulation solves the problem with a different objective function which considers the total production time or total production cost of the set of orders, without resorting to the application of any heuristic rules. A special MATLAB program has been developed for automatic creation of the optimization model, which may be a very time consuming task prone to errors. The results indicate the importance of the proposed modifications and effectiveness of the automated generation of the model, and present slightly better solution.

Keywords: MILP, production scheduling

1.   Introduction

Production scheduling involves the distribution of process operations to the available equipment in order to meet a specified performance criteria. In some fields, like the production of specialty chemicals and polymers, range of products and product specifications is very large and also rapidly changing. For this reason, these types of processes can not afford the high cost of a large inventory, and thus the production needs to be based on customer orders only. Production scheduling under such circumstances is known to be a difficult problem. Non pre-emptive production scheduling problem in one stage parallel processing units is NP-complete when maximum flow time or makespan is to be minimized. In other words, the number of constraints grows exponentially as the dimensions of the problem increases. Formulation of zero wait flowshop problem as an asymmetric travelling salesman problem has drawn attention in chemical engineering, with probably the first attempt by Pekny and Miller [1] who developed a branch and bound algorithm for parallel processing to speed up the computation. Another approach is the use of heuristics, which has been proposed by Musier and Evans [2] and an MILP formulation in discrete time for such problems was given by Kondili *et al.* [3] and Shah *et al.* [4]. A continuous time MILP mathematical formulation for the batch scheduling problem involving a single processing stage for every product has been given by Cerda *et al.* [5]. Their model accounted for a number of practical issues and constraints, and the use of heuristics to allow elimination of a subset of feasible predecessors for each customer order, reducing the model size was proposed. However, heuristic algorithms result in sub-optimal solutions whereas the rigorous mathematical formulations create MILP models that may become unmanageable for large size problems.

2. Problem Formulation and Solution

---

[1] Corresponding author. E-mail: berber@eng.ankara.edu.tr

In this work, the short-term scheduling problem in single-stage multi-product plants with nonidentical parallel production units is revisited. The problem has the following features:

- Each customer order is composed of only one product.
- Each order can be processed in a subset of the available processing units.
- The plant is operated in 'campaign' mode, i.e. batches of the same order are processed successively in the same unit.
- A changeover period for equipment cleaning and set-up is considered between campaigns.
- Some jobs may not be processed in every processing unit because of incompatibilities.
- Some units may not be available at the beginning of the planning period.
- A new campaign cannot be started in a unit unless the previous campaign has been completed.

The model was mathematically defined by a group of problem sets and parameters. The allocation of orders to units was represented by binary variables comprising three subscripts, like X_{iml} being the binary variable denoting that the processing of order i takes place in unit l just before the campaign for m. The formulation incorporated a number of constraints that might be encountered in practice. For most of the constraints the same definition and formulation proposed by Cerda *et al.*[5] was followed. However, we added a new constraint such that "*each order can be first processed in only one processing unit*" in this work, because the above constraints, which had also been employed in previously reported models, could lead to the results that one order might be assigned to more than one unit as a first order to be processed. The objective of the problem we propose here is the minimization of a function involving the total processing time of orders being considered.

A MATLAB program has been developed to generate the model in MILP form. The solution to the problem has been found by a branch and bound algorithm via automatically created input files to a Fortran subroutine. Different constraints have been accommodated into the formulation and the performance of the algorithm has been tested numerically under different conditions. As a measure of the economic performance of the plant, the total cost of production of the given set of orders is chosen as the objective function. A predefined makespan is considered as a constraint among others and a new constraint was added.

A special MATLAB program, called MPS (Modelling for Production Scheduling), was developed to formulate the MILP problem automatically without any user intervention. A standard branch-and-bound algorithm embedded in MINOS [6] was used to solve the optimization problem in Fortran environment. The output of the branch and bound algorithm was again automatically interpreted in MATLAB environment to gather the information for the Gantt chart. The program takes the data related to the properties of the orders and production units, creates the model, i.e. objective function and all the constraints, produces the input files for MINOS optimization package to be run in Fortran environment and necessary other files for easy interpretation of the MINOS output. The user, then, easily extracts the results of the optimum schedule, i.e. allocation of orders into the production units and the schedule in each unit, to be drawn as a Gantt chart. With some additional developments, the program can be converted into a MATLAB toolbox.

Working examples are provided and compared to previously reported literature data to show the identified problems and the effectiveness of the new formulation, which resulted in optimal solutions without applying any heuristic rules.

3. Results and Conclusion

Out of a number of runs that have been made to test the algorithm, one will be reported here as an exemplary case. The example has 3 units and a customer demand of 10 products. There are no limitations for the predecessors, i.e. any product can be manufactured before each other. The orders should be completed in a time period of 100 hours. The quantities of each product, the batch times and capacities of units are given in Table 1. Cleaning times between the batches are given in Table 2. The objective is to manufacture the orders at minimum possible total production time. The problem has 211 binary variables and 831 constraint equations. The optimum results are shown in Figure 1. As the results depict, the orders are completed within 99.4 hours, which is considerably less than the scheduling horizon, 100 hours. Total production time obtained is 223.6 hours, the number of iterations is 172 and Cpu times are 3.6 s and 175.4 s for DOS environment and MATLAB respectively.

Table 1. Requested quantities of orders (kg), production times (h/batch) and capacity (kg/batch) of each order in processing units

Products	Quantity	BATCH TIMES			CAPACITIES		
		Unit 1	Unit 2	Unit 3	Unit 1	Unit 2	Unit 3
O1	1000	3.7	2.5	-----	80	70	-----
O2	620	5.0	4.0	3.5	150	130	135
O3	550	3.5	2.3	4.0	180	-----	175
O4	650	4.5	5.0	-----	120	105	100
O5	970	4.0	3.0	5.0	130	110	-----
O6	870	1.0	1.4	1.8	120	115	125
O7	620	6.5	3.0	1.6	250	300	275
O8	730	6.0	3.0	6.3	75	50	65
O9	940	3.0	5.0	-----	300	190	-----
O10	540	-----	5.0	4.0	-----	130	160

Table 2. Cleaning times (h) between orders

Predecessors   Successors ⟶

	O1	O2	O3	O4	O5	O6	O7	O8	O9	O10
O1	-----	1.5	0.6	0.8	1.9	0.2	1.0	1.1	0.7	1.0
O2	0.4	-----	0.6	0.2	1.9	0.9	1.0	1.2	1.0	0.1
O3	0.5	1.8	-----	0.1	1.8	1.2	1.3	1.4	0.8	0.1
O4	0.4	1.1	0.7	-----	1.3	0.8	1.4	1.0	0.2	0.5
O5	0.2	1.4	1.0	1.0	-----	0.8	1.4	0.1	0.4	0.6
O6	0.3	1.5	0.4	0.8	1.0	-----	1.3	0.4	0.3	0.4
O7	0.7	1.7	0.2	0.7	1.1	0.4	-----	1.2	1.6	0.4
O8	0.4	1.7	0.5	0.4	1.2	0.8	1.5	-----	0.6	0.7
O9	0.6	1.9	2.5	3.2	0.7	0.8	1.3	1.4	-----	1.8
O10	0.1	1.1	0.4	0.5	1.5	0.7	1.7	1.5	0.8	-----

The results, obtained without applying any heuristic rules, indicate that starting from moderately large problems, the 'cycle constraint' becomes the main constraint and increases the complexity of the problem more than any other constraint. Therefore, contrary to the previous literature reports, this work showed that the cycle constraint was 'compulsory' and emphasized furthermore that the size of the problem was primarily, and the most, affected by this constraint. The suggested formulation has been tested with numerical examples to prove its effectiveness.

In conclusion, an effective and improved solution, with reasonable computer time and branch-and-bound iterations has been provided for production scheduling in single stage multi-product batch plants. A MATLAB tool has been generated to automatically create the MILP formulation without resorting to heuristic rules. Otherwise, manual creation of such formulation may become a difficult task for large size problems, and prone to errors.

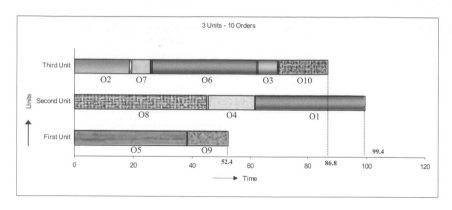

Figure 1. Gantt chart for the solution of the example problem

References

[1] J.F. Pekny and D.L. Miller, Exact Solution of the No-Wait Flowshop Scheduling Problem with a Comparison to Heuristic Methods, *Comp. & Chem. Engng.* **15** 741-748(1991).

[2] R. Musier and L. Evans, An Approximate Method for the Production Scheduling of Industrial Batch Processes with Parallel Units. *Comp. & Chem. Engng.* **13** 229-238(1989).

[3] E. Kondili, C.C. Pantelides and R.W.H. Sargent, A general algorithm for short-term scheduling of batch operations - I. MILP formulation, *Comp. & Chem. Engng.* **17** 211-227 (1993).

[4] N. Shah, C.C. Pantelides and R.W.H. Sargent, A general algorithm for short-term scheduling of batch operations - II. Computational issues, *Comp. & Chem. Engng.* **17** 229-244 (1993).

[5] J. Cerda, G.P. Henning and I.E. Grossmann, A Mixed – Integer Linear Programming Model for Short –Term Scheduling of Single Stage Multiproduct Batch Plants with Parallel Lines, *Ind. Eng. Chem. Res.*, **36** 1695-1707(1997).

[6] MINOS Ver. 5.4 User's Guide (by B.A. Murtagh and M.A. Saunders), Systems Optimization Laboratory, Stanford University, Stanford, CA, 1995.

VSP International
Science Publishers
P.O. Box 346, 3700 AII Zeist
The Netherlands

*Lecture Series on Computer
and Computational Sciences*
Volume 1, 2004, pp. 661-665

Simulation of River Streams: Comparison of a New Technique to QUAL2E

M. Yuceer[1], E. Karadurmus[2] , R. Berber[1*]

[1]Department of Chemical Engineering,
Faculty of Engineering,
Ankara University,
06100 Ankara,Turkey
[2]Corum Engineering Faculty,
Gazi University,
19200 Corum, Turkey

Abstract: Predictions and quality management issues for environmental protection in river basins rely on water-quality models. These models can be used to simulate conditions in or near the range of the calibrated or verified conditions. In this respect, estimation of parameters, which is still practiced by heuristic approaches (i.e. manually), seems to be the point where the attention needs to be focused. The authors' research group developed a systematic approach for dynamic simulation and parameter estimation in river water quality models [1,2], which eliminated the cumbersome trial-end-error method. The approach is based on simulating the river stream like a series of CSTRs with embedded parameter estimation capability by using dynamic data. For the implementation of the suggested technique, they have later reported a user-interactive software named as RSDS (River Stream Dynamics and Simulation) [3]. This study provides a comparative investigation of the suggested modeling approach to a well established and worldwide known water quality software, QUAL2E. Experimental data collected in field observations along the Yesilirmak river basin in Turkey were checked against the predictions from the both software. The results indicated that much better agreement with the experimental data could be obtained from the RSDS with respect to QUAL2E. Thus, the systematic procedure suggested in the present work provides an effective means for reliable estimation of model parameters and dynamic simulation for river basins, and therefore, contributes to the efforts for predicting the extent of the effect of possible pollutant discharges in river basins.

Keywords: River water quality, parameter estimation, dynamic simulation

1.   Introduction

Water quality models generally require a relatively large number of parameters to define their functional relationship, and since prior information on parameter values is limited, they are commonly defined by fitting the model to observed data.

The model can be used to simulate conditions in or near the range of the calibrated or verified conditions. In this respect, estimation of parameters, which is still practiced by heuristic approaches (i.e. manually), seems to be the point where the attention needs to be focused.

The state of the art in river water quality modeling was summarized by Rauch *et al.* [4] who addressed some issues related to the practical use of the river quality models upon comparison of 10 important software products they indicated that only two of them offered limited parameter estimation capability. Mullighan *et al* .[5] also noted that practitioners often resorted to manual trial-and-error curve fitting for calibration. Generally accepted software for river water quality modeling worldwide is U.S. EPA's

---

[1*] Corresponding author. E-mail: berber@eng.ankara.edu.tr

QUAL2E [6]. However, this software does not address a number of practical problems such as the issue of parameter estimation. In a previously study [1], we have suggested a dynamic simulation and parameter estimation strategy so that the heavy burden of finding reaction rate coefficients was overcome. Modeling was based on the fact that the segment of river between sampling stations was assumed as a completely stirred tank reactor (CSTR). This work was later extended to the series of CSTR approach [2], and furthermore a MATLAB-based user-interactive software was developed for easy implementation of the technique [3]. The program, named as RSDS (River Stream Dynamics and Simulation), was coded in MATLAB^{TM} 6.5 environment.

This work extends the previous work in the sense that a comparative study is provided to assess the capabilities and effectiveness of the suggested technique with respect to the QUAL2E software. The model was constituted from the dynamic mass balances for different forms of nitrogen and phosphorus, biological oxygen demand, dissolved oxygen, coliforms, non-conservative constituent and algae for each computational element. Model parameters conforming to those in QUAL2E water quality model, were estimated by an SQP algorithm by minimizing an objective function. As QUAL2E model is almost the standard for river water quality modeling, we have chosen it for comparing the predictions from the suggested new methodology.

2. Computing Methodology

In dynamic modeling, serially connected CSTRs are assumed to represent the behavior of river stream. Each reactor forms a computational element and is connected sequentially to the similar elements upstream and downstream. The following assumptions were employed for model development:
- Well-mixed dendritic stream
- Well mixing in cross sections of the river
- Constant stream flow and channel cross section
- Constant chemical and biological reaction rates within the computational element.

Physical, chemical and biological reactions and interactions that might occur in the stream have all been considered. The modeling strategy stems from that of QUAL2E water quality model [6]. The dynamic mass balances for 11 constituents (i.e. state variables) were written, and additionally several algebraic equations describing various phenomena such as conversion of different forms of nitrogen were involved.

A nonlinear constrained parameter estimation strategy has been incorporated into the simulation so that a number of techniques (involving Gauss-Newton, Levenberg-Marquardt and Sequential Quadratic Programming SQP algorithms) can be employed. SQP, being probably the most effective of all, updates the Hessian matrix of the Lagrangian function, solves quadratic programming subproblem and uses a line search and merit function calculation at each iteration. The estimation strategy was based on minimizing an objective function defining the difference between the predictions and observed data during the transient period of observations.

In many practical applications of water quality models, parameter values are chosen manually rather than by automated numerical techniques, or heuristic approaches are resorted to [4]. It is known that automated methods are associated with some difficulties depending on the model structure, optimization method, number of variables and parameters, type of measurements and so on. So far no automated estimation of parameters for river water quality models has been reported.

In a most recent work, Sincock *et al.* [7] reported a detailed study involving the identification of model parameters. However, relatively too short of data was considered, and although temperature was predicted accurately; nitrate, BOD and DO predictions were less so.

Therefore, this work provides extension over the previous approaches particularly in terms of automatically generating reliable estimates of water quality model parameters without resorting to the trial-and-error simulations.

3. Experimental measurements

For both of field observations and data collection, the concentrations of many water-quality constituents indicative of the level of pollution in the river were determined either on-site by portable analysis systems or in laboratory after careful conservation of the samples. The experimental data for parameter estimation was obtained from two sampling stations along the Yesilirmak River around the city of Amasya in Turkey. The concentrations of many water-quality constituents, corresponding to the state variables of the model and indicative of the level of pollution in the river, were determined in 30 minutes intervals either on-site by portable analysis systems or in laboratory after careful conservation of the samples.

Water quality constituents of the river were determined at various locations along a 7 km long section of the river. The sampling location matched with the river stream such that the volume element of water whose quality was sampled at the location zero (i.e. starting point, Durucasu gauging station) was followed with the stream. This was just like dynamically keeping track of an element in the river flowing the same velocity as the main stream. The location of sampling points were determined such that available measurement and would suffice to make such a study possible. After the starting point, sampling was done at locations 0.225 km., 4 km, 5 km and 7 km downstream. The industrial wastewater of a baker's yeast production plant was being discharged right after the starting point. Therefore, the results of the study would indicate the extent of the pollution caused by the discharge from this industrial plant. In the simulations the addition of this discharge was considered as a continuous disturbance to the system and its effect on the water quality, thus, were determined.

4. Results

Predictions from the RSDS and QUAL2E were then compared to field data for the river reach of 7 kms. Figure 1 and 2 show the profiles of the two most important pollution variables, BOD and Dissolved Oxygen (DO), in a 7 kms section of the river after point source input. For quantitative evaluation, Absolute Average Deviation (AAD) values were calculated for both softwares, which are given for some pollution variables as follows:

	Dissolved Phosphor		BO D$_5$		DO	
	RSDS	QUAL2E	RSDS	QUAL2E	RSDS	QUAL2E
AAD (%)	5.26	6.92	1.61	3.14	1.0	0.85

The predictions from the RSDS indicated a much better agreement with the experimental data, compared to QUAL2E. Thus, the systematic procedure suggested in the present work provides an effective means for reliable estimation of model parameters and dynamic simulation for river basins, and therefore, contributes to the efforts for predicting the extent of the effect of possible pollutant discharges in river basins.

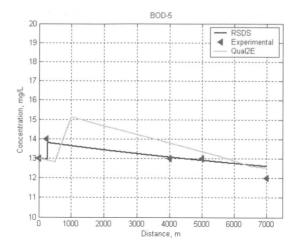

Figure 1. BOD profile in a 7 kms section

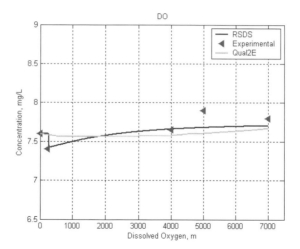

Figure 2. DO profile in a 7 kms section

References

[1] E. Karadurmus and R. Berber, Dynamic simulation and parameter estimation in river streams, *Environmental Technology* **25** 471-479(2004).

[2] M. Yuceer, E. Karadurmus and R. Berber, Dynamic modelling of river streams by a series of CSTR approach, AIChE Annual Meeting, 16-21. November 2003, San Francisco (Session: CAST 10- Environmental Performance Monitoring and Metrics), Paper 455e.

[3] M. Yuceer and R. Berber, Effective verification of river water quality models through optimum parameter estimation: A new software. Joint Conf. of American & Indian Inst. of Chem. Engineers, Session on Environment, Mumbai, India, December 28-30, 2004 (Submitted for presentation).

[4] W. Rauch, M. Henze, L. Koncsos, P. Reichert, P. Shanahan, L. Somlyody, and P. Vanrolleghem, River water quality modelling: I. State of the art, IAWQ Biennial Int. Conf. Vancouver-Canada, 21-26 June 1998.

[5] A.E. Mulligan and L.C. Brown, Genetic algorithms for calibrating water quality models, *J. Environmental Engineering* **124**(3) 202-211(1998).

[6] L. C. Brown and Jr. T. O Barnwell, The enhanced stream water quality model QUAL2E and QUAL2E-UNCAS, Documentation No. EPA/600/3-87/007, Environmental Research Laboratory, Office of Research and Development, U.S. Environmental Protection Agency, Athens, Georgia, 1987.

[7] A.M. Sincock, H.S. Wheather, and P.G. Whitehead, Calibration and sensitivity analysis of a river water quality model under unsteady state conditions, *J. of Hydrology* **277** 214-229(2003).

VSP International
Science Publishers
P.O. Box 346, 3700 AH Zeist
The Netherlands

*Lecture Series on Computer
and Computational Sciences*
Volume 1, 2004, pp. 666-670

Evaluation of Debottlenecking Strategies for a Liquid Medicine Production Utilising Batch Process Simulation

Jully Tan, Dominic Chwan Yee Foo[1], Sivakumar Kumaresan, Ramlan Abdul Aziz,
Chemical Engineering Pilot Plant, Universiti Teknologi Malaysia
81310 Skudai, Johor, Malaysia.
Tel: +607-5531662, Fax: +607-5569706
E-mail addresses: jully2181@hotmail.com, cyfoo@cepp.utm.my, shiva@cepp.utm.my,
ramlan@utm.my

Alexandros Koulouris
Intelligen, Inc
Thessaloniki Technology Park
PO Box 328, 57001 Thermi-Thessaloniki, Greece
Tel: +30-2310-498292, Fax: +30-2310-498280
Email Address: akoulouris@intelligen.com

Abstract: Computer Aided Process Design (CAPD) and simulation tools are used to debottleneck an integrated pharmaceutical production producing a liquid medicine, LIQMED. The process is comprised of several sections: 3 pre-blending sections A, B and C, a syrup making section, a main blending section, an intermediate storage and a packaging section. The bottleneck was found to be the cartoning process in the packaging section. Eight equipment alternatives were proposed to remove this bottleneck and were evaluated based on the process throughput and economic criteria such as Cost Benefit Ratio (CBR) and cost of investment. The best debottlenecking alternative was found to have the highest cost benefit ratio of 1.4 and the lowest cost of investment of $1.2 M.

Keywords: Batch process, pharmaceutical production, modelling and optimisation, throughput analysis, debottlenecking.

1. Introduction

Computer Aided Process Design (CAPD) and simulation tools have been successfully used in the bulk chemical industry since the early 1960s (Westerberg *et al.*, 1979). However, the use of these tools has only begun in the biochemical-related production industry throughout the past decade (Ernst *et al.*, 1997). Process simulation can be applied in several stages of process development. Once process ideas are conceived, process simulators can be used for project screening and selection for strategic planning based on economic analysis or any other critical process requirement. When process development nears completion on the pilot level, simulation tools are used to systematically design and optimise the large-scale process for the commercialisation of products. Good process simulators can facilitate the transfer of process technology as well as to facilitate design. They can be used to estimate the required capital investment of the process. In large-scale batch manufacturing, process simulation is primarily used for process scheduling, debottlenecking and on-going process optimisation. It is capable of tracking equipment use for overlapping batches and identifying process bottlenecks (Petrides, *et al.*, 2002).

2. Case study

---

[1] Corresponding author. Postgraduate Researcher. Chemical Engineering Pilot Plant. Universiti Teknologi Malaysia. E-mail: cyfoo@cepp.utm.my

A case study is modelled based on the operation condition of an existing pharmaceutical process for the production of an oral liquid medicine, *LIQMED*. Figure 1 shows the simulation flowsheet for the case study modelled in SuperPro Designer 5.0. Fifty two weeks of operation (with five working days a week and sixteen hours a day) are taken as the basis of this work. The dispensing and blending process is divided into 5 main parts, i.e. pre-blending sections A, B and C, the syrup making section and the main blending section.

The batch size of the process is 2000 L, with the final product being packed into 90 mL bottles. Twelve ingredients are used to produce *LIQMED*. Each of these ingredients plays its own role either as an active reagent or excipient. Active ingredients 1, 2 and 3 are ingredients that are responsible for *LIQMED*'s therapeutic action. Excipients such as food additives and colouring are inactive or inert substances which are added to provide stability of the drug formation in *LIQMED*.

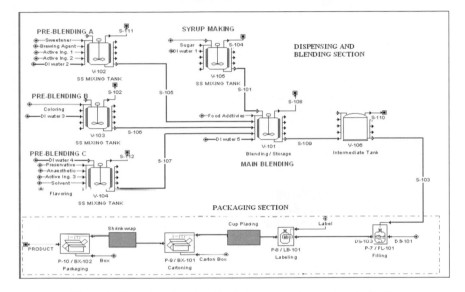

Figure 1: Simulation flowsheet for the base case of production *LIQMED*

The syrup making procedure is the leading process in the production of *LIQMED*. The syrup ingredients (sugar and water) are charged into a 2000 L blending tank (V-105) and agitated for 3 hours to ensure a well-mixed solution. The syrup is then transferred into the main blending tank (V-101), followed by the food additives. While the mixture is stirred in tank V-101, the pre-blending procedure takes place. Three different pre-blending procedures will take place. In pre-blending A, the blending of the sweetener, brewing agent, active ingredient 1 and active ingredient 2 take place in tank V-102. The colouring agents is heated and melted with the presence of deionised water at 100°C and is blended in V-103 in Section B. This colouring agent requires a separate blending procedure as the raw material is in creamy form. Pre-blending of the preservative agents, anaesthetic agents, solvent and flavouring agents takes place in tank V-104 in Section C.

When all pre-blending procedures are completed, the ingredients are transferred into tank V-101 starting with the mixture in V-102. This is followed by the mixture in V-103. After the completion of V-104 product transfer, agitation is repeated for 10 minutes to ensure all ingredients are well mixed. Finally, deionised (DI) water is added to adjust the final mixture volume to the required amount and the mixture is stirred again for 15 minutes.

Upon completion of the main blending process, the mixture in the V-101 is transferred to the intermediate tank, V-106 before packaging take place. In the base case study, the filling machine has a capacity of filling 28 bottles per minute. The Labeller, on the other hand, has a labelling speed of 30 bottles per minute. The cup placing procedure is carried out manually in the base case. The cartoning machine is able to carton up to 24 bottles per minute. Finally, the products are packed into boxes with 72 bottles each for shrink wrapping.

Figure 2 shows the operation Gantt chart for the case study. Note that the operating time of the labeller is shorter than the filler, due to the different speed of the two machines. Thus, the cartoning machine with the longest operating time of 15.69 h, is the overall process bottleneck.

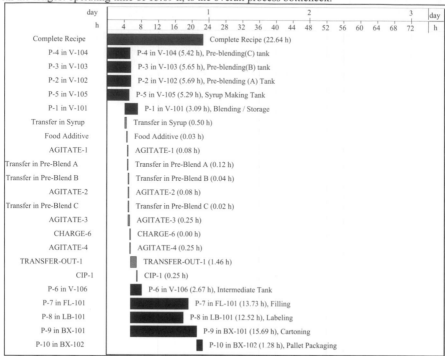

Figure 2: Operational Gantt chart for the base case study

3. Throughput analysis and debottlenecking alternatives

From the base case simulation the batch cycle time is estimated to be 22.64 hours. The annual throughput is calculated based on an estimate of 264 batches per year. Due to the increasing market demand for this product, options for increasing the plant throughput are needed. The ability to identify and remove equipment and resource bottlenecks will increase plant throughput and fulfill customer orders in time. This can be done by performing throughput analysis on existing production facility (Koulouris *et al.*, 2000). Figure 3 displays the capacity, time and combined utilisation of all procedure/equipment pairs in the base case flowsheet. Figure 3 reveals that the cartoning section is the overall process bottleneck with a capacity, time and combined utilisation of 100%.

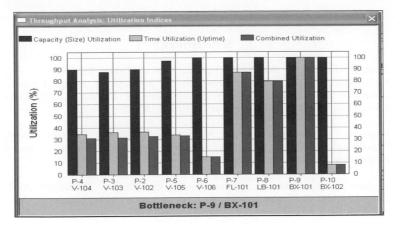

Figure 3: Capacity, time and combined utilisation chart for the base case

Various debottlenecking strategies are proposed for the evaluation of eight alternatives for the cartoning process proposed by different equipment suppliers. A comparison of these different alternatives is carried out based on the achievable annual throughput, cost benefit ratio (CBR) as well as the cost of investment of the project. CBR is calculated as the ratio of the benefits to the cost associated with the particular project. The CBR for this case study is given in Equation 1,

$$CBR = \frac{\text{Revenue of the altenatives} - \text{revenue of current process}}{\text{Operating cost} + \text{investment cost} - \text{operating cost of current process}} \qquad (1)$$

4. Results and discussion

Figure 4 shows the comparison of the eight alternatives for eliminating the process bottleneck by replacing existing equipment either with automatic or semi-automatic functions. The achievable annual throughput (number of batches per year) and the investment cost are shown for every alternative.

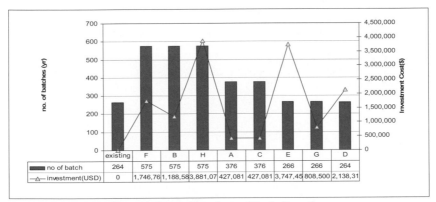

Figure 4: Comparison charts of the 8 alternatives

As shown in Figure 4, alternatives B, F and H have the same annual throughput of 575 batches. This is approximately a 100% improvement compared to the annual throughput of the base case study. The selection of the best debottlenecking candidate hence lies with the two other criteria, i.e. the CBR and the cost of investment for the project. Alternative B has the highest CBR of 1.4 while the CBR of alternative F and H are 1.3 and 1.1 respectively. In addition, alternative B has a much lower cost of investment of $1.2 M as compared to Alternative F at $1.7 M and Alternatives H at $3.9M. As a result, Alternative B is finally selected as the debottlenecking strategy for the study.

5. References

[1] Ernst, S., Garro, O. A., Winkler, S., Venkataraman, G., Langer R., Cooney, C. L. and Ram Sasisekharan, (1997). Process Simulation for Recombinant Protein Production: Cost Estimation and Sensitivity Analysis for Heparinase I Expressed in *Escherichia coli. Biotechnology and Bioengineering*, Vol. 53, No.6: 575-582.

[2] Koulouris, A., Calandranis, J. and Petrides, D. P. (2000). Throughput Analysis and Debottlenecking of Integrated Batch Chemical Processes. *Computers and Chemical Engineering*. Vol. 24, S1387-S1394.

[3] Petrides, D. P., Koulouris, A. and Siletti, C. (2002). Throughput Analysis and Debottlenecking of Biomanufacturing Facilities: A Job for Process Simulators. *BioPharm,* 2-7, August 2002.

[4] Westerberg, A. W., Hutchison, H. P., Motard, R. L. and Winter, P. (1979). *Process Flowsheeting*. Cambridge University Press, Cambridge, United Kingdom.

VSP International
Science Publishers
P.O. Box 346, 3700 AH Zeist
The Netherlands

*Lecture Series on Computer
and Computational Sciences*
Volume 1, 2004, pp. 671-673

Receding Horizon Control for Polymerization Processes

Nida SHEIBAT-OTHMAN[1], Rachid AMARI, Sami OTHMAN

LAGEP-Université Lyon1 -CNRS/ESCPE
Bât 308, 43 Blvd. du 11 Nov. 1918, 69622 Villeurbanne CEDEX, France

Received 13 July, 2004; accepted in revised form 19 July, 2004

Abstract: Online control of chemical process is an important issue in order to ensure the process security and good product quality. Polymerization processes are part of those processes that can be exothermic, rapid and sensitive to impurities. For this reason, modelling monitoring, and control of these processes is of high importance in order to optimize the process productivity. Modelling gives a good comprehension of the process and allows us to relate the process inputs (or manipulating variables) to the process outputs (or the process measurements). Online monitoring is important to obtain real information on the evolution of the process and to detect any deviation in its course. Online control allows then the correction of this deviation by manipulating the process parameters. In this work, nonlinear geometric control is used to control the heat produced by the reaction, the concentration of monomer in the reactor and the reaction rate coefficient in order to minimize the process time and ensure its security.

Keywords: Moving/receding horizon control, optimizing process productivity, emulsion polymerization.

1. Process mathematical model

One interesting problem in emulsion polymerization processes, is optimizing the process productivity by increasing the reaction rate under specific constraints, that are usually due to physical limitations such as the heat that can be exchanged by the jacket and the saturation in monomer. Several works were proposed for maximizing the process productivity using the monomer concentration as given by the following equation:

$$\dot{N} = F - R_P \tag{1}$$

wher F is the monomer flow rate (mol/s), N is the number of moles of monomer in the reactor (mol), and R_P is the reaction rate (mol/s) given by the following equation:

$$R_P = k_P(T)\underbrace{\frac{\overline{n}N_P}{N_A}}_{\mu s}[M^P]V \tag{2}$$

where μ and $[M_P]$ are the concentration of radicals and monomer in the polymer particles respectively (mol/cm^3), V is the reaction medium volume (cm^3) and k_P is the reaction rate coefficient (cm^3/mol/s) that depends on the reaction temperature, T(K). The concentration of monomer in the particles is given by the following equation:

$$[M^P] = \begin{cases} \dfrac{\left(1-\phi_P^P\right)\rho_m}{MW_m} & if\ \dfrac{MW_m N}{\rho_m} - \dfrac{1-\phi_P^P}{\phi_P^P}\left(\dfrac{MW_m}{\rho_m}\left(N^T - N\right)\right) \geq 0 \\[2em] \dfrac{N}{MW_m\left(\dfrac{\left(N^T - N\right)}{\rho_P} + \dfrac{N}{\rho_m}\right)} & else \end{cases} \tag{3}$$

[1] Corresponding author. E-mail: nida.othman@lagep.cpe.fr

where ϕ_p^P is the volume fraction of polymer in the particles under saturation, MW_m the monomer molecular weight (g/mol), ρ_m and ρ_p the monomer and polymer density respectively (g/cm³). These parameters are constant with time. N^T is the total number of moles of monomer introduced to the reactor:

$$N^T = \int_t F \, dt \qquad (4)$$

Equation 2 clearly shows that maximizing the reaction rate can be done by manipulating one of the thee parameters: the concentration of monomer or radicals in the polymer particles or the reaction temperature. In fact, controlling the concentration of monomer in the polymer particles can easily be done by manipulating the flow rate of monomer. However, m cannot be manipulated in many cases. Therefore, some papers treated maximizing productivity by manipulating the concentration of monomer in the polymer particles. By doing this, one can maintain $[M_P]$ at the maximum allowed value in order to maximize R_P. During this time (saturation of the polymer particles), this technique is limited and R_P cannot be maximized more. In order to do better, another parameter must be used in order to maximize R_P.

In this work, we propose to manipulate the reaction temperature with the concentration of monomer in order to maximize the reaction rate. The system constraints are the saturation of polymer particles (first equation in 2) and the maximum allowable heat produced by the reaction given by the following equation:

$$Q_R^{max} = (-\Delta H) R_P \qquad (5)$$

where ΔH is the reaction enthalpy (J/mol) and Q_R(W) can be obtained from the process heat balance:

$$MC_p \dot{T} = Q_R + UA(T_j - T) + \underbrace{F_{feed} C_{Pfeed}(T_{feed} - T)}_{Q_{feed}} \qquad (6)$$

where M, is the reaction medium mass(Kg), CP the heat capacity of the reaction medium (J/Kg/K), U the heat transfer coefficient (J/K/s/cm²), A the heat transfer area (cm²), Tj is the jacket temperature, F_{feed} the input flow rate, T_{feed} is the input temperature and CP_{feed} is the heat capacity of the feed.

2. Receding horizon control

In order to control the process a multi input/ multi output nonlinear controller must be applied. To do that, the process is first linearised in an input-output manner. The input/output linearization gives the following system:

$$\begin{bmatrix} \dot{N} \\ \dot{T} \end{bmatrix} = \begin{bmatrix} w_1 \\ -\dfrac{UA}{MC_P} T + \dfrac{UA}{MC_P} w_2 \end{bmatrix} \qquad (7)$$

The new inputs (w_1 and w_2) can be calculated by any linear controller. Linearising the process allows us to decouple the coupling variables u_1 and u_2. However, since the process model might be inaccurate, it is preferable to construct a robust controller that takes into account model uncertainties. For this purpose, we used a sliding horizon controller that is based on the minimization of an objective function as summarized in the following block diagram:

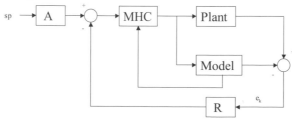

where A and R are two reference models related to the set point and the regulation respectively; MHC represents the moving horizon control algorithm.

The following figure shows that controlling the reaction temperature and the concentration of monomer in the polymer particles allows us to maintain the heat produced by the reaction close to the desired one. The advantage of adding a new control variable allow a good control of Q_R even under saturation of polymer particles. Both objectives are therefore realized: $[M_P]$ and Q_R are maximized with the given constrains

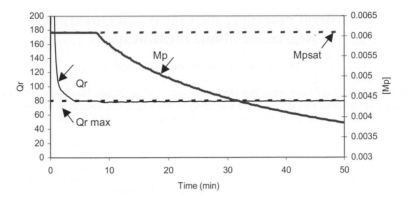

Figure 1: Simultaneous control of the heat produced by the reaction (Q_R) and the concentration of monomer in the polymer particles [Mp].

References

[1] Astorga Zaragoza, C.M., S. Othman, H. Hammouri, Moving horizon control of a dentrification reactor. WSES/ INTERNATIONAL CONFERENCES AMTA 2000- MCBC 2000-MCBE 2000, JAMAICA, pp 931-936, 2000.

[2] Alamir, M., and G. Bornard, Stability of a truncated infinite constrained receding horizon scheme: the general discrete nonlinear case. Automatica, 31 (9) 1353-1356, 1995.

[3] Sh'eibat Othman, N., G. Févotte and T.F. McKenna, Biobjective control of emulsion polymerizations: control of the polymer composition and the concentration of monomer in the polymer particles, Chemical Engineering Journal, 98 (1-2), 69-79, 2004.

[4] Buruaga, I. S., Ph. D. Armitage, J. R. Leiza, J. M. Asua. Nonlinear control for maximum production rate latexes of well-defined polymer composition. Ind. Eng. Chem. Res., **36** 4243-4254, 1997.

VSP International
Science Publishers
P.O. Box 346, 3700 AH Zeist
The Netherlands

*Lecture Series on Computer
and Computational Sciences*
Volume 1, 2004, pp. 674-676

A General Procedure for Linear and Quadratic Regression Model Identification

M. Shacham[1*], H. Shore[2] and N. Brauner[3]

[1]Department of Chemical Engineering, Ben-Gurion University of the Negev, Beer-Sheva 84105, Israel
[2]Dept. of Industrial Eng., Ben-Gurion University of the Negev, Beer-Sheva 84105, Israel
[3]School of Engineering , Tel-Aviv University, Tel-Aviv 69978, Israel

Abstract: A new regression procedure is developed for identification of linear and quadratic models. The new procedure uses indicators based on signal-to-noise ratio, as well as more traditional indicators, to validate the models. Various traditional stages in the modeling process, like stepwise regression, outlier detection and removal and variable transformations, are pursued, however the interdependence between these stages is accounted for to ensure detection of the best model (or subset of models).

Three examples are presented, where the proposed procedure is implemented. Some of the models identified have better goodness-of-fit than those reported in the literature. Furthermore, for two of the examples, complex quadratic models were identified that in fact model also the stochastic experimental error. While traditional indicators failed to signal the invalidity of these models, signal-to-noise ratio indicators, based on realistic noise estimates detected such over-fitting.

Keywords: stepwise regression, signal-to-noise ratio, variable selection, quadratic model

Identification of the most appropriate regression model to represent given data is often a complex task. In particular, difficulties are encountered when there is no prior information from theoretical considerations and/or empirical experience regarding which of the explanatory variables (regressors) to include in the regression model. The process may involve stepwise regression, using a bank of regressors, transformation of variables and outlier detection and removal. These stages are not necessarily independent. For example, an outlier associated with a particular explanatory variable may cause this variable to be excluded from the model by a stepwise regression procedure. However, its exclusion from the model may prevent the outlier detection. Similarly, variables selected to be included in the model may influence the required transformation of the dependent variable (such as the Box-Cox transformation). Yet setting the parameter of this transformation affects the variables that would be included in the model.

Because of the complexity of the task of selecting the most appropriate models, this task must be carried out in an orderly, procedural fashion, or else some solutions may be overlooked. Furthermore, the models must be arranged in increasing order of accuracy and complexity, so that the user can select a model that best fits his/her needs.

Modeling a non-linear relationship via linear regression may often be problematic. Introducing into the model nonlinear terms (such as quadratic or higher-degree polynomial terms) may render traditional indicators for the validity and stability of the model (such as the *t*–test ratio and confidence intervals for estimates of the parameter values) inaccurate and unfounded. In some cases more variables may be included in the model than is justifiable on the basis of the apparent data accuracy and the sample size. Such 'over-fitting' implies that the noise in the data is also being modeled.

In this paper a new stepwise regression procedure is introduced. This procedure uses Shacham and Brauner's [1] signal-to-noise based indicators instead of the traditional statistical indicators. QR decomposition with Gram-Schmidt orthogonalization is used for regression as this method can give significant results even in cases of evident over-fitting.

---

* Corresponding author. E-mail: shacham@bgu.ac.il

A flow diagram of the proposed procedure is displayed in Appendix A. The data required to implement this procedure comprise the sample observations, and error estimates for the different variables.
The procedure starts with the selection of the first-type model (model 1.1), according to the model definition-hierarchy. The model definition hierarchy contains six types of models: linear model (of a non-zero or zero intercept) and a quadratic model (of a non-zero or zero intercept) with standardization or without. The options regarding the intercept were explicitly included, since stepwise regression programs often cannot determine whether a free parameter is needed. Standardization option is not included with a linear model, as it complicates the model, but rarely makes any difference. The first model to be selected is a linear model with a free parameter (non-zero intercept). The initial value of the Box-Cox parameter is set to $\lambda = 1$.

Next, a stepwise regression procedure is implemented to identify the independent variables that will be included in the model. The SROV program of Shacham and Brauner [2] is used. Results were also verified using the JMP^{TM} linear regression procedure. After identifying a possible subset of independent variables, an outlier detection procedure is carried out. If outliers are detected, they are removed, and the selection by stepwise regression of explanatory variables from the full set is reiterated. Once a model without outliers has been derived, it is checked for goodness-of-fit. Several criteria are used for this purpose. If the model is unsatisfactory, a search is carried out to find a Box-Cox transformation for the response. If the resulting model (after transformation) is still unacceptable, the next model type is selected from the model-definition hierarchy, and the whole procedure is repeated.

The proposed procedure was tested with the "Modeling of Biological Activity" example of Wold *et al.*[3], the "Cloud-seeding experiment" of Miller [4] and the "Operation of a Petroleum Refining Unit" example of Gorman and Toman [4]. Although the stepwise regression algorithm does not rely on traditional statistical indicators, the algorithm generated sequences of stable linear and quadratic models with steadily decreasing variances, in all three examples. One or more satisfactory models (in terms of confidence intervals, *t* tests, linear correlation coefficients and residual and normal probability plots) were identified. Furthermore, stable models with error variances lower than those reported in the literature were obtained.

In two of the examples, the traditional statistical indicators failed to diagnose cases of over-fitting, resulting in models that represent the noise in the data. By contrast, the signal-to-noise-ratio based indicators (Shacham and Brauner [2]), with realistic noise level estimates, have proven to be helpful in detecting over-fitting.

References

[1] M. Shacham and N. Brauner, Considering Precision of Experimental Data in Construction of Optimal Regression Models, *Chemical Engineering and Processing*, **38**, 477-486(1999).

[2] M. Shacham and N. Brauner, The SROV Program for Data Analysis and Regression Model Identification, *Computers and Chemical. Engineering.* **27**(5), 701-714(2003).

[3] S. Wold, A. Ruhe, H. Wold and W. J. Dunn III, The Collinearity Problem in Linear Regression. The Partial Least Squares (PLS) Approach to Generalized Inverses, *Siam J. Stat. Comput.*, **5**(3), 735-743 (1984).

[4] A. J. Miller, *Subset Selection in Regression*, Chapman and Hall, London, 1990.

[5] J. W. Gorman and R. J. Toman, Selection of Variables for Fitting Equations to Data, *Technometrics*, **8**, 27-51(1966).

Appendix A

Flow Diagram of the Proposed Procedure

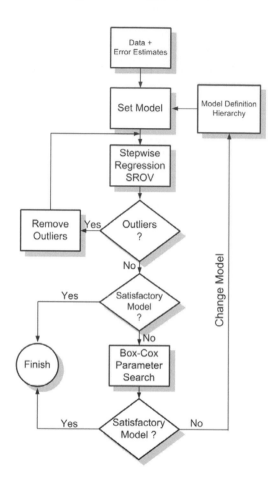

<u>Model Definition Hierarchy</u>

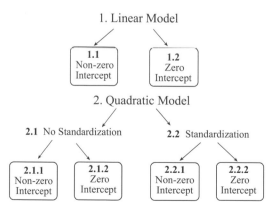

VSP International
Science Publishers
P.O. Box 346, 3700 AH Zeist
The Netherlands

Lecture Series on Computer
and Computational Sciences
Volume 1, 2004, pp. 677-682

Optimal Design of Batch Plants
under Economic and Ecological Considerations :
Application to a Biochemical Batch Plant

A. Dietz, C. Azzaro-Pantel[1], L. Pibouleau, S. Domenech

Laboratoire de Génie Chimique- UMR 5503 CNRS/INP/UPS
5, Rue Paulin Talabot BP1301
31106 TOULOUSE Cedex 1, France

Abstract: This work deals with the multicriteria cost-environment design of multiproduct batch plants, where the design variables are the equipment item sizes as well as the operating conditions. The case study is a multiproduct batch plant for the production of four recombinant proteins. Given the important combinatorial aspect of the problem, the approach used consists in coupling a stochastic algorithm, indeed a Genetic Algorithm (GA) with a Discrete Event Simulator (DES). To take into account the conflicting situations that may be encountered at the earliest stage of batch plant design, i.e. compromise situations between cost and environmental consideration, a Multicriteria Genetic Algorithm (MUGA) was developed with a Pareto optimal ranking method. The results show how the methodology can be used to find a range of trade-off solutions for optimizing batch plant design.

Keywords: Multicriteria optimization, Genetic Algorithm, Batch plant design, Environmental impact

1. Introduction

The design of multiproduct and multipurpose batch plants is a key problem in chemical engineering. The problem formulation generally involves mathematical programming methods such as MINLP (Mixed-Integer Non-Linear Programming). The main limitation of such methodologies is the difficulty, even impossibility, to describe with a high degree of sophistication, the real constraints that may be encountered (various storage policies or operator shift, for instance ...). Moreover, the number of equations to take as constraints often renders the problem impossible to solve. An alternative proposed in [1] consists in coupling a Discrete Event Simulator (DES) in order to evaluate the feasibility of the production at medium term scheduling, with a master optimization procedure based on a Genetic Algorithm (GA). The optimization variables take only discrete values and the problem exhibits a marked combinatorial feature (the equipment sizes are considered as discrete values). This approach was then generalized in [2] to consider multicriteria design and retrofitting. The choice of a hybrid method GA/DES was then all the more justified as several criteria were simultaneously taken into account: a trade-off between investment cost, equipment number and a flexibility index based on the number of campaigns necessary to reach a steady state regime was thus investigated. The Multicriteria Genetic Algorithm (MUGA) developed was based on the combination of a Monocriterion Genetic Algorithm (MOGA) and a Pareto Sort (PS) procedure.

This work is thus motivated by the need to take into account the capital cost as well as the environmental impact from the earliest design stage. A simplest version of the previously developed DES model has been implemented to model multiproduct batch plant features.

The originality of the global model is that it takes into account computed values for operating times deduced from embedded local models for unit operations. Let us recall that the constant time and size factor model [3] is the most widespread to design multiproduct batch processes. These models are used

---

[1] Corresponding author. E-mail: Catherine.AzzaroPantel@ensiacet.fr

to optimize the plant design by proper selection of batch sizes of each product, the operating times of semi-continuous units and the structure of the plant. Only the works presented in [4, 5, 6, 7] include process performance models to compute time and size factors and select process variables as optimization variables.

In this perspective, the approach proposed in this work is to offer a general methodology for ecological and economic assessment for batch plant design problems.

This paper is organized as follows: section 2 presents the basic principles of the general framework. Section 3 discusses some key points of the implementation and displays some significant results obtained for a case study, i.e., the design of a batch plant dedicated to the production of proteins. In the final section, the conclusion is presented and the guidelines established.

2. Methodology

The framework for batch plant design proposed in this study (see Figure 1) integrates simple unit operation models into the batch plant wide model, which is then embedded in an outer optimization loop. The approach adopted in this work [8] consists in coupling a stochastic algorithm, indeed a Multicriteria Genetic Algorithm (MUGA) with a Discrete Event Simulator (DES). The objective of the master GA involved is to propose several good and even optimal solutions, whereas the DES checks the feasibility of the proposed configuration and evaluates different criteria with both economic and ecological targets.

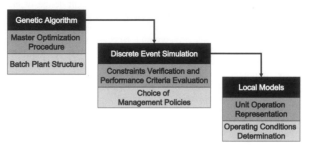

Figure 1: General methodology for optimal batch plant design

Indeed, engineering design problems are usually characterized by the presence of many conflicting objectives that the design has to fulfill. Therefore, it is natural to look at the engineering design problem as a multiobjective optimization problem (MOOP). References to multiobjective optimization could be found in [9, 10, 11]. As most optimization problems are multiobjective by nature, there are many methods available to tackle these kinds of problems. Lately there has been a large development of different types of multiobjective genetic algorithms, which is also reflected in the literature. The big advantage of genetic algorithms over other methods is that a GA manipulates a population of individuals. It is therefore tempting to develop a strategy in which the population captures the whole Pareto front in one single optimization run. This approach was adopted in this study.

The MUGA developed in this study involves different procedures :

1) A method for encoding solutions in strings of digits (or chromosomes); here, a chromosome represents a workshop configuration and the corresponding operating conditions. The encoding procedure will be presented in detail in what follows.

2) An initial population has to be randomly generated.

3) An evaluation function which takes a string as input and returns different fitness values which measure the quality of the solution that the chromosome represents relative to each criterion. Since this work is related to minimization cases (investment cost, limitation of pollution), the individual fitness F_i is calculated by:

$$F_i = C_{max} - C_i$$

where C_i is one of the objective function value for individual i, C_{max} is the maximum objective function value computed on the current population for the corresponding criterion.

4) An adaptive plan involving evolution and mutation, based on string crossover and mutation operators.

The cycle {Evaluation, Selection, Crossover and Mutation} is repeated until a stop criterion is reached. After this cycle, the Pareto sort is applied. Concerning selection and multicriteria aspects involved, it must be emphasized that for a given survival rate, the selection process is achieved via a classical

Goldberg's biased roulette wheel [12], relative to each criterion. For this purpose, the initial population is randomly partitioned into sub-populations (the number of sub-populations corresponds to the number of criteria considered simultaneously). A same number of individuals is chosen for each sub-population.

A binary system was chosen for encoding, as it simplifies the genetic operators, i.e., crossover and mutation. This encoding method presented was developed for the cases where the equipment items are identical at a given stage. The continuous variables were discretized and encoded in a binary way by a variable change (Figure 2), using the same bit number (i.e., eight bits). Figure 3 shows a code part used for operating stage encoding. For each stage, the equipment item number was encoded by a binary way (part A in Figure 3). The number of bits reserved to this variable set the maximum equipment item at the stage. For equipment size (L for large, M for Medium, S for Small, a number of bits equal to the available sizes for the equipment items was reserved (part B in Figure 3): the chosen size takes a positive value whereas zero is allocated to the other places. When equipment items are composed of several parts, the same approach is repeated for each component (part B and B' in Figure 3).

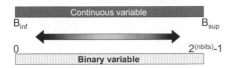

Figure 2 Continuous variables encoding

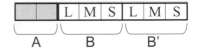

Figure 3 Operating stage encoding method

4. Implementation and results

The previous methodology was applied to a batch plant for the production of proteins taken from the literature [5]. More detail can be found in Appendix A.

The cost criterion considered in this study is classically based on investment minimisation. Considering environmental impact (EI), let us recall that several methodologies are available in the literature (see Appendix B). The most important concept refers perhaps to the Life Cycle Assessment (LCA) [13]considering all the wastes generated in order to produce the different products in the upstream stages (i.e., raw material production, energy generation, etc.), in the study stage (i.e. solvents, non-valuable by-products, etc) and in the downstream steps (i.e. recycling, incineration, etc). The aim of LCA is to consider the wide chain in order to prevent pollution generation and to compare the different alternatives to produce a product. Another concept used the Pollution Balance (PB) principle to carry out a pollution balance [14] equivalent to the balance made for mass or energy. It means that a process can not only pollute but also consume a polluting product and will be a benign process.

Finally, the Pollution Vector (PV) methodology [15] consists in evaluating the environmental impact by means of an impact vector over different environments (i.e. water, air, etc) defined as the mass emitted on an environment divided by the standard limit value in this environment.

Since the conflicting behaviour between the three criteria (investment cost, solvent used and biomass released) was previously demonstrated [8], the multicriteria cost-environment batch plant design was carried out while keeping the two environmental criteria independent. Figures 4 to 6 show the results obtained. Instead of showing a three dimensional graph, it was preferred to project over a plane the criteria by pairs to facilitate result analysis. In all cases, the points are more concentrated near the compromise zone, as it preferred for the final decision.

5. Conclusions

A methodology was proposed for batch plant design, considering both investment cost and environmental impact minimization. An optimization scheme has been implemented using a multiobjective genetic algorithm with a Pareto optimal ranking method. This technique is ideally suited to this type of problem, where a number of conflicting considerations must be taken into account. The use of MUGA makes possible a robust optimization technique, across a non-linear search space (the objective functions are computed by the use of a discrete event simulator (DES) integrating shortcut unit operations models) linking multiple variables and objectives. The paper clearly shows that opportunities for process optimization and environmental impact minimization must be considered at the early stages of process development before the process is frozen due to regulatory reasons.

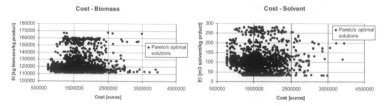

Figures 4 and 5 Pareto's optimal solutions Cost - IE

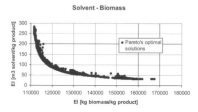

Figure 6 Pareto's optimal solutions Biomass - Solvent

References

[1]   Bérard F., C. Azzaro-Pantel, L. Pibouleau, S. Domenech, D. Navarre, M. Pantel, Towards an incremental development of discrete-event simulators for batch plants : use of object-oriented concepts, *Comp. and Chem. Eng. Supplements*, S565-S568 (1999).

[2]   Dedieu S., Pibouleau L., Azzaro-Pantel C., Domenech S., Design and Retrofit of Multiobjective Batch Plants via a Multicriteria Genetic Algorithm, *Computers. and Chemical Engineering,* 27 1723-1740 (2003).

[3]   Biegler, L.T., Grossmann, I.E. & Westerberg, A.W., Systematic Methods of Chemical Process Design, Prentice-Hall (1997).

[4]   Salomone H.E., Montagna J.M., Irribarren O.A., Dynamic simulations in the design of batch processes, *Computers and Chemical Engineering,* 18, 191-240 (1992).

[5]   Montagna J.M., Iribarren O.A., Galiano F.C., The design of multiproduct batch plans with process performance models, Trans IChemE, *72, Part A, 783-791 (1994).*

[6]   Chiotti O.J., Salomone H.E., Iribarren O.A., Batch plants with adaptive operating policies, *Computers & Chemical Engineering,* 20, 1241-1256 (1996)

[7]   Asenjo, Montagna, Vecchietti, Iribarren, Pinto, Strategies for the simultaneous optimisation of the structure and process variables of a protein production plant. *Computers and Chemical Engineering,* 24, 2277-2290 (2000).

[8]   Dietz A., Azzaro-Pantel C., Pibouleau L., Domenech S., Integrating Environmental Impact Minimization Into Batch Plant Design: Application To Protein Production ESCAPE-14: European Symposium on Computer Aided Process Engineering Lisbon, Portugal, May 16-19, 1033-1038 (2004).

[9]   Bhaskar V, Gupta S.K., Ray A.K., Applications of Multiobjective Optimisation in Chemical *Engineering, Reviews in Chemical Engineering*, Vol. 16 (2000).

[10]   Coello A.C., An Updated Survey of GA-based Multiobjective Optimisation Techniques, ACM Computing Survey, Vol. 32, 109-143 (2000).

[11]   Ehrgott M., Lecture Notes in Economics and Mathematical Systems – Multicriteria Optimization*, Springer-Verlag Berlin Heidelberg* (2000).

[12]   Goldberg D.A., Algorithmes Génétiques, *Addison-Wesley, MA* (1994).

13]   Burgess A. A., Brennan D. J., The application of life cycle assessment to process optimisation. *Computers and Chemical Engineering* 23, 1509-1526 (1999).

[14]   Cabezas H., J. C. Bare and S. K. Mallick, Pollution prevention with chemical process simulators: the generalized waste reduction (WAR) algorithm—full version, *Computers and Chemical Engineering*, Volume 23, Issues 4-5, Pages 623-634 (1999)

[15]   Stefanis S. K., A. G. Livingston and E. N. Pistikopoulos, Minimizing the environmental impact of process Plants: A process systems methodology. *Computers and Chemical Engineering*, Volume 19, Supplement 1, 39-44 (1995).

Appendix A

The biochemical multiproduct batch plant involves four products to be manufactured by fermentation and eight treatment stages. This example is used as a test bench since short-cut models describing the unit operations involved in the process are available. Eight stages are considered for the production of four recombinant proteins, on the one hand two therapeutic proteins, Human insulin (I) and Vaccine for Hepatitis B (V), and on the other hand a food grade protein, Chymosine (C) and a detergent enzyme, cryophilic protease (P). Figure A1 shows the flowsheet of the multiproduct batch plant. All these proteins are produced as cells grow in the fermentor (Fer).

Vaccine and protease are considered as being intracellular, hence, for these two products, the first microfilter (Mf1) is used to concentrate the cell suspension, which is then sent to the homogeniser (Hom) for cell disruption to liberate the intracellular proteins. The second microfilter (Mf2) is used to remove the cell debris from the solution proteins.

The ultrafiltration (Uf1) prior to extraction is designed to concentrate the solution in order to minimize the extractor volume. In the liquid-liquid extractor (Ext), salt concentration (NaCl) is used to first drive the product to a poly-ethylene-glycol (PEG) phase and again into an aqueous saline solution in the back extraction.

Ultrafiltration (Uf2) is used again to concentrate the solution. The last stage is finally chromatography (Chr), during which selective binding is used to separate better the product of interest from the other proteins.

Insulin and chymosin are extracellular products. Proteins are separated from the cells in the first microfilter (Mf1), where cells and some of the supernatant liquid stay behind. To reduce the amount of valuable product lost in the retentate, extra water is added to the cell suspension.

The homogenizer (Hom) and microfilter (Mf2) for cell debris removal are not used when the product is extracellular. Nevertheless, the ultrafilter (Uf1) is necessary to concentrate the dilute solution prior to extraction. The final step of extraction (Ext), ultrafiltration (Uf2), and chromatography (Chr) are common to both the extracellular and intracellular products.

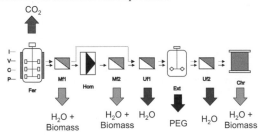

Figure A1 Multiproduct batch plant for proteins production and environmental impact evaluation

Appendix B Environmental Impact Evaluation

Given the production recipes for the different products and the general flowsheet, the first step consists in applying the LCA methodology to determine all the products contributing to the environmental impact. For information availability reasons the study was reduced to the process being studied, which is limited application of LCA. Products (i.e. vaccine) and raw materials (glucose, NH3) were considered not having an environmental impact. After that, a PB is applied, using the PV to quantify the environmental impact. In this case, a different vector of pollution was defined, because the standard limit values for the polluting product were not found in the literature. The vector has two components; the first one is the total biomass quantity released, and the second one is the PEG volume used. Even if the solvent can be recycled, it cannot be done at 100%, so the environmental impact will be proportional to this quantity. The pollution indexes were defined as the quantities divided by the mass of products elaborated. Let us remark that the environmental impact minimisation is a multicriteria problem in itself (see also Figure A1).

VSP International
Science Publishers
P.O. Box 346, 3700 AH Zeist
The Netherlands

Lecture Series on Computer
and Computational Sciences
Volume 1, 2004, pp. 683-686

Modeling and Simulation of Wildfires Based on Artificial Intelligence Techniques

D. Vakalis, H. Sarimveis[1], C.T. Kiranoudis, A. Alexandridis, G. Bafas

School of Chemical Engineering,
National Technical University of Athens,
GR-157 80 Athens, Greece

Abstract: A forest fire propagation simulation tool is presented based on discrete contour propagation modeling for front evolution and fuzzy systems technology for the estimation of fire front. To bust calculations the fuzzy system's output is interpolated by means of a radial basis function neural network model. Simulations of the system under different scenarios concerning the mountain area of Penteli in the prefecture of Attiki are finally presented.

Keywords: Fire propagation modeling, fuzzy logic, fuzzy systems, neural networks, fire simulation

1. Introduction

In recent years, advances have been made in the development of mathematical models and simulation techniques for tracing the progress of wildland fire perimeters through curve expansion models. A number of scientists have undertaken research on fire propagation, whose understanding and adequate estimation is the most important factor for successful forest fire simulations[1]. Weber[2] presents a non-exhaustive list of the research groups on this field. He defines three sorts of models: statistical[3], empirical[4] and physical[5,6,7]. This classification is based on the physical mechanisms involved in these different models. Albini[5] proposed a fire spread model in which the radiant heat from the flame is the dominant heat transfer mechanism. Weber[8] proposed a reaction-diffusion equation for the conservation of energy in order to model fire spread. Di Blasi[9], Dik and Selikhovkin[10], Grishin[11] and Larini et al.[12] proposed more complex models including pyrolysis mechanisms, air motion and combustion modeling. Richards[13] through an analysis of parameter growth for homogeneous conditions over a surface of constant slope developed a mathematical model for fire growth for heterogeneous conditions and variable surface and solved it using a numerical scheme.

The cited literature which due to space limitations is only representative, shows that for flat land problems the art of simulation has reached the stage of being a practical tool for fire managers and local authorities. However, for fires over variable topography the problem is considerably more complex due to the fact that the slope of the surface makes a marked effect on the fire's behaviour. It is well known that fires tend to burn faster uphill and slower downhill than on the flat terrain and, depending on its direction, the wind can either reinforce or counteract the effect of the slope. In order to develop an efficient and complete mathematical model for fire spread behaviour and fire perimeter growth, that takes into account the above mentioned situations, we need a curve growth formulation as well as a reliable expression for correlating fire spread over the factors influencing it such as terrain slope, vegetation growth and density and current meteorological conditions. The simulation tool we are presenting in this work is able to cover the above mentioned complex situations and is based on two fold reasoning; a discrete contour propagation model for estimating fire consequences and a fuzzy / neural system for the estimation of fire spread as a function of the influencing factors.

---

[1] Corresponding author. E-mail: hsarimv@chemeng.ntua.gr

2. Modeling the fire spread rate as a fuzzy system

Fuzzy logic theory has emerged over the last years as a useful tool for modeling processes which are too complex for conventional quantitative techniques or when the available information from the process is qualitative, inexact or uncertain. The Mamdani type of fuzzy modelling was employed where the fuzzy rules are formulated as follows:

$$\forall r \in R: \; \textit{if} \; \bigwedge_{1 \le i \le n} (x_i \in A_i^r) \quad \textit{then} \; \bigwedge_{1 \le j \le m} (y_j \in B_j^r) \tag{1}$$

In order to develop such a system we employed human experts who built the system based on their intuition, knowledge and experience, but also utilizing a database that contains the following information:

- Measurements carried out in actual conditions as well as in the laboratory, cited in the literature.
- Information concerning approximately 15000 forest fires that occurred in Greece from 1983 until 1996. The data include the location, the duration of the fire (time of ignition and extinction), the burnt area, the type of vegetation, the slope steepness of the ground and the climate conditions (meteorological data) during the fire such as the air temperature, wind force and relative humidity.

The selection of the input parameters was made so that all the important influencing factors are considered, while maintaining the system at a reasonable size. In the process of choosing the proper input variables, the issue of data availability was also taken into account, meaning that only variables for which data can be easily available were selected as inputs to the fuzzy system. Geographical information databases, meteorological information and remote sensors can possibly become sources of data. Eventually, the list of input parameters was selected as follows: Vegetation flammability, wind speed, air temperature, slope steepness, vegetation density and relative humidity of air. Triagonal fuzzy sets were defined on the domains of all input variables. Examples are shown in Figure 1 below.

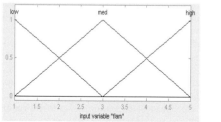 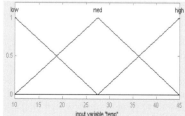

Figure 1: Fuzzy sets defined on the domains of the variables flammability and temperature

The experts then developed a rule base of 162 fuzzy rules, using logical reasoning as far as the combined effect of the different input parameters on the spread rate is concerned. The fuzzy rules were fine tuned, so that they can relate successfully the input conditions to fire characteristics contained in the database of fires. The *min-max* inferencing technique was utilized where the output membership function of each rule is clipped off at a height corresponding to the rule premise's computed degree of truth and the combined fuzzy output membership function is constructed by combining the results of all the fuzzy rules. As far as the defuzzification technique is concerned, the centroid method was utilized, where the crisp value of the output variable is computed by finding the center of area below the combined membership function.

3. Modeling the fire spread rate as a neural network system

The fuzzy model described previously is computationally intensive and results in a rather slow system, since all fuzzy rules must be evaluated at each time point and location. Given the need for fast computations, which can help the decision makers to manage a fire crisis more effectively, it is very important that the system can perform the same simulation in a much shorter time period. This was achieved by training a neural network model based on an input-output data set, which were generated using the fuzzy system. For the development of the neural network model, we employed the Radial Basis Function (RBF) network architecture, which has certain advantages over other types of artificial neural networks, including better approximation capabilities, simpler network structures and faster learning algorithms.

The RBF network was developed using a method that is based on the subtractive clustering technique[14], which is able to determine both the structure of the network and the hidden node centres using only one pass of the training data. The method measures the potential of each training example to become the center of a cluster. After a training example is selected, the potentials of all the non-selected examples are revised, so that the generated centers are not close to each other. Based on a data set consisting of 5000 examples, training of the neural network was completed in less that 1 min CPU time in a Pentium IV 2800 Mhz machine running Matlab. The produced network was very compact in size since it consisted of only 8 hidden nodes. Obviously the neural network was orders of magnitude faster compared to the fuzzy system.

4. Discrete contour propagation model

The RBF network served as the basis for developing the fire propagation model. More specifically, the fire was taken to burn over a three-dimensional surface, which is represented in a rectangular coordinate system $\vec{r} = (x, y, z) \in R^3$, such that any point on the surface satisfies:

$$z = S(x, y) \tag{2}$$

The complete terrain surface can be expressed through an appropriate three-dimensional point set:

$$\Omega = \{\vec{r}(x, y, z) \in R^3 : z = S(x, y)\} \tag{3}$$

If $\vec{C}(t) : [0, \infty) \to \Omega$ is a marching three-dimensional closed curve superimposed on the surface, the evolution of its state can be formulated as follows, given its normal speed:

$$\frac{\partial \vec{C}}{\partial t} = \vec{S}\hat{n} \tag{4}$$

The curve propagation problem was solved using a variation of the Minimal Path Critical Tree computational procedure after creating and appropriately addressing a suitable mathematical grid.

5. Case study

The operational system was tested in a pilot application concerning the mountain area of Penteli in the prefecture of Attiki. The selection of Penteli is of great interest as it is one of the most hit by fires mountain area in the country of Greece. It is on the northeast of the city of Athens and on its slopes there are a lot of built up areas. In the Penteli area there have been registered a lot of conflagrations with detrimental effects such as loss of human lives, forests, properties and homes. Two scenarios are presented: In the first scenario the start of the fire is placed on the Kifisia region. The temperature (45 C^0) and the wind (speed 30m/s, origin west) are very high. The west wind directs the fire towards the top of the mountain. We can observe the high speed of the fire spread with the help of the slope of the ground. Fortunately, inhabited areas are not endangered. In the second scenario the fire follows the wind direction (because of its high speed) and is going to fire the region of Dionysos. We can see the reduction of rate of spread when the fire approached the top of the mountain and the front begins to advance downhill (it corresponds to 50% of the spreading rate in horizontal ground without wind).

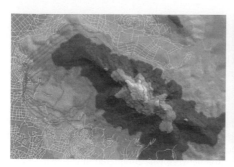
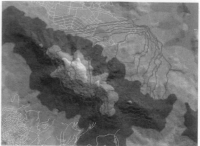

Figure 2: Scenario 1   Figure 3: Scenario 2

References

[1] W.L. Fons, Analysis of fire spread in light forest fuels, *Journal of Agricultural Research* **72** 93-121 (1946).

[2] R.O. Weber, Modeling fire spread through fuel beds, *Progress in Energy and Combustion Science* **17** 67-82 (1991).

[3] A.G. McArthur, *Weather and grassland fire behaviour*, Australian Forest and Timber Bureau Leaflet N° 100, Canberra.

[4] R.C. Rothermel, *A mathematical model for predicting fire spread in wildland fuels*, United States Department of Agriculture, Forest Service Research, paper INT-115, 1972.

[5] F.A Albini, A model for fire spread in wildland fuels by radiation, *Combustion Science and Technology* **42** 229-258 1985.

[6] A.M. Grishin, A.D. Gruzin, V.G. Zverev, Mathematical modeling of the spreading of high-level fires, *Soviet Physics Doklady* **28** 328-330 1983.

[7] R.O. Weber, Analytical models for fire spread due to radiation, *Combustion and Flame*, **78** 398-408 1989.

[8] R.O. Weber, Toward a comprehensive wildfire spread model, *International Journal of Wildland Fire*, **1** 245-248 1991.

[9] C. Di Blasi, Modeling and simulation of combustion processes of charring and non charring solid fuels, *Progress in Energy and Combustion Science*, **19** 71-104 1993.

[10] I.G. Dik, A.M. Selikhovkin, A model of ignition and transition to combustion in a condensed matter gasified in the course of combustion, *Mathematical Modeling and Computational Experiment*, **1** 25-33 1993.

[11] A.M. Grishin, *Mathematical modeling of forest fires and a new method of fighting them*, Publishing House of the Tomsk State University, 1997.

[12] M. Larini, F. Giroux, B. Porterie, J.C. Loraud, A multiphase formulation for fire propagation in heterogeneous combustible media, *International Journal of Heat and Mass Transfer*, **41** 881-897 1998.

[13] G.D. Richards, The mathematical modeling and computer simulation of wildland fire perimeter growth over a 3-dimensional surface, International *Journal of Wildland Fire*, **9** 213-221 1999.

[14] H. Sarimveis, A. Alexandridis, G. Bafas, A fast training algorithm for RBF networks based on subtractive clustering, *Neurocomputing*, **51** 501-505 2003.

VSP International
Science Publishers
P.O. Box 346, 3700 AH Zeist
The Netherlands

Lecture Series on Computer
and Computational Sciences
Volume 1, 2004, pp. 687-687

Computational Methods in Molecular Biology and Medicine

Christos Makris and Athanasios Tsakalidis

Computer Engineering and Informatics Department,
University of Patras, 26500 Patras, Greece
AND
Research Academic Computer Technology Institute,
61 Riga Feraiou Str., 26221 Patras, Greece

Preface

Computational Methods in Molecular Biology and Medicine provide important tools for solving many of the key problems in Bioinformatics including determining the function of a newly discovered genetic sequence; determining the evolutionary relationships among genes, proteins, and entire species; and predicting the structure and function of proteins.

The purpose of the proposed session is to bring together scientists from different fields of expertise in order to discuss biological problems and propose new ideas and techniques in today's applications, in the areas of Molecular Biology and Medicine. The growing number of Molecular Sequence Data requires the development of faster and more complex mathematical models and computer algorithms. Moreover our goal is to study the effectiveness of different approaches that utilize different data sources, including molecular sequence data and abstracts of research papers.

Two contributions in this session focus on Biological Weighted Sequences and provide efficient algorithms to cope with the problems of identification of known or unknown motifs or regions involved in various biological processes such as initiation of transcription, gene expression and translation, or the discovery of various types of repeats. An accurate identification and localization of such elements will allow biologists to perform deeper studies of the structure, function and evolution of genomes.

Another contribution presents a new approach for gene finding based on a Variable-Order Markov Model, while another one discuss the problem of Synonymous Codons in the Expression of Proteins. An application is also presented in the area of Concept Discovey and Classification of scientific Abstracts in Molecular Biology.

We wish to thank all the authors, who submitted their papers to this session and thus contributed to the creation of a high-quality program and fascinating meeting. We are extremely grateful to Prof. T. Simos for hosting the session under the International Conference of Computational Methods in Science and Engineering 2004.

VSP International
Science Publishers
P.O. Box 346, 3700 AH Zeist
The Netherlands

*Lecture Series on Computer
and Computational Sciences*
Volume 1, 2004, pp. 688-691

Concept Discovery and Classification of Cancer Specific Terms from Scientific Abstracts of Molecular Biology

E. Giannoulatou, K. Perdikuri[1] and A. Tsakalidis

Computer Engineering and Informatics Department,
University of Patras, 26500 Patras, Greece
AND
Research Academic Computer Technology Institute,
61 Riga Feraiou Str., 26221 Patras, Greece
A. Papavassiliou

Department of Biochemistry, School of Medicine,
University of Patras, 26500 Patras, Greece

Abstract: The vast accumulation of electronically available literature in the fields of Biology and Medicine has raised new challenges in Knowledge Discovery technology and provides increasingly attractive opportunities for Text Mining. In this work we present a methodology for concept discovery from the Molecular Biology Literature. Our approach combines Natural Language Processing Techniques and Clustering Methods in order to produce clusters of biological abstracts based on term co-occurence. Experiments show that the resulting document clusters are meaningful as assesed by cluster-specific terms. The application of this method to a collection of abstracts relevant to *transcription factors* provided a shallow description of the document corpus and supported classification of cancer specific terms.

Keywords: biological literature mining, text mining, document clustering, concept discovery

1   Introduction

The last years Text Mining has gain the interest of Biologists and Computer Scientists focused on knowledge discovery from scientific papers such as: the detection and extraction of relations between genes and proteins [1], the discovery of new functional relationships in protein families [2], the discovery of biological pathways [3], the functional annotation of proteins [4], etc.

In the fields of Biology and Medicine, abstracts are collected and maintained in MEDLINE, a project supported by U.S. National Library of Medicine. MEDLINE constitutes a valuable resource that allows scientists to retrieve articles of interest, based on keyword searches. The real body of biological knowledge comes in the form of abstracts, published in knowledge repositories such as *PubMed Central*. *PubMed Central*, is a service of the National Library of Medicine, providing access to over 12 million MEDLINE citations back to the mid-1960's.

In this work we present the methodology followed for concept discovery from the Molecular Biology Literature. It is based on statistical treatment of terms, using suitable linguistic tools and two different kinds of weighting schemes. This method is applied to a collection of MEDLINE

---

[1]Corresponding author. E-mail: perdikur@ceid.upatras.gr

abstracts relevant to transcription factors. The pivotal role of specific transcription factors in certain cancers, either as mutants or in overexpressed level, highlights them as rational targets for chemotherapeutic intervention [5]. The large volume of data available on the molecular anatomy of transcription factors and the biochemical pathways that modulate their function offer opportunities for the analysis of textual data and the classification of cancer.

2 Methodology

In this section we describe every step of our methodology. The overall architecture of the system employing the proposed methodology is summarized in figure 1.

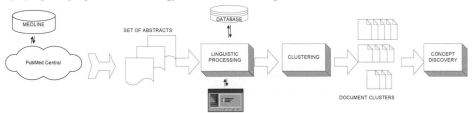

Figure 1: System Architecture.

Step 1: Abstract Retrieval. A set of abstracts is retrieved through *PubMed Central* using a keyword-based query.

Step 2: Tokenization. Every abstract is divided into a set of lexical units, called *tokens*. The purpose of this stage is to determine the lexical units of each retrieved abstract, and to filter out punctuation and symbols such as brackets, quotation marks, as well as numbers. Special consideration was given in the identification of special tokens such as chemical formulas and equations.

Step 3: Part-of-Speech Tagging. In this step we identify the morpho-syntactic category of each word in the abstract (noun, verb, adjective, etc). The overall objective of this step is to filter out non-significant words on the basis of their morpho-syntactic category (i.e. prepositions, pronouns, articles, etc).

Step 4: Stemming. In this step we restrict the morphologic variation of the textual data by reducing each of the different inflections of a given word form to a unique canonical representation (or lemma). For example the words *treatment* and *treat* identified as NOUN and VERB respectively (from the previous step) could be assigned under the same morpho-syntactic tag VERB. This process is user-driven.

Step 5: Filtering out words. To eliminate common English words we have employed two well-known term weighting schemes, the *TF-IDF* metric and the *Shannon* metric.

The first scheme is based on the assumption that terms that appear frequently in a document (TF = Term Frequency), but rarely in the collection of abstracts (IDF = Inverse Document Frequency) are more likely to be specific to the document. For the IDF we have used the following variant: $w_i = log_2(N_i/n_i)$, where w_i is the weight of term i in the abstract, N_i is the frequency of term i in the collection of abstracts L, and n_i is the number of documents in L that term i occurs in. Terms with a high *TF-IDF* value (*the product TF*IDF*) are retained for further processing.

For the *Shannon* metric, we have used the following variant $w_i = N_{ij} * s_i$, where w_i is the weight of term i, N_{ij} if the frequency of term i in abstract j and s_i is the signal of term i in the set of abstracts. The signal and the noise of a term are calculated on the basis of its frequency and its probability of occuring. Terms that appear with the same probability in the set of abstracts have high noise, while terms with different probabilities of appearance in the set of abstracts are meaningful.

Step 6: Creation of a go-list. In order to extract the *go-list* (final vocabulary of terms) we have used the **union** of the results produced by the application of the above two metrics. Thus we have tried to retain the most representative terms for each abstract, namely, the frequently occuring terms with zero noise. To further reduce the vocabulary size, the user has the ability to select the word categories as identified by the part-of-speech-tagging and restrict the analysis to specific word categories (i.e. nouns, verbs, adjectives).

We subsequently encode the textual information (obtained from the previous steps) into a lexical table, where each row corresponds to a scientific abstract and each column to a term of the *go-list*. The lexical table is represented as a frequency table, where the $cell_{i,j}$ contains the number of occurrences of token j in the abstract i.

Step 7: Clustering. In the clustering process we have used the *K-means Clustering* algorithm. In more detail having as an input a set of K abstracts to be clustered, the clustering process starts by assigning randomly each point to a specific cluster. In each step the distances between the points in each cluster and the centroid of the cluster are computed. Every point (abstract) is assigned in the nearest cluster. The process is repeated until no regroupings of the abstracts takes place.

Step 8: Concept Discovery. In order to obtain the final set of terms W that are specific and highly descriptive for a given cluster of documents, we employ the well-known log-odds formula: $q_{ij} = log_2(f_{ij}/f_i)$, where q_{ij} represents the preference of term i in a document cluster j, f_{ij} represents the frequency of term i in cluster j and f_i represents the frequency of term i in the total set of abstracts. A term is considered significant for a cluster if its presence in the cluster is considered more significant than its presence in the total population.

3 Results

To evaluate the reliability of our approach, we have performed various case studies, one of which is presented here. We have retrieved 158 abstracts from the *PubMed Central* using the keyword-query: "corepressor AND co-activator". After the abstract retrieval process we applied the linguistic processing methodology. The frequency threshold for the *TF-IDF* technique was set to 15 and for the *Shannon* metric we retain the terms wiht positive signal and zero noise. In the particular experiment we kept the word categories of nouns, verbs and adjectives.

The iterative *K-means* clustering produced two very distict clusters, with descriptive terms for two important types of cancer: breast cancer and prostate cancer. Characteristc terms with high log-odds values are shown in Table 1. It is interesting to note that many terms describe the clusters with high fidelity and immediately imply the transciption factors and the biochemical pathways that modulate different types of cancer. In more detail *Cluster 1* contains descriptive terms of endocrine -hormonal- therapies in breast cancer while *Cluster 2* contains descriptive terms of biology and therapeutic -hormonal- interventions in prostate cancer.

These results are actually promising for the use of our methodology since the researches concerning the biological mechanisms of carcinogenesis are closely related to altered functions of co-repressors and co-activators gene-proteins, which are revealed in this case study.

Table 1: Representative terms describing two clusters in a case study

*Cluster*1	*Cluster*2
antagonist	ablation
antiestrogenic	agonist-dependent
breast	androgen
receptor-beta	ARNIP
tamoxifen	prostate
tumor	PCa
ERalpha-positive	AR
ERalpha	androgen-independent

4 Discussion and Further Work

It appears that term co-occurence and the processing steps that we have implemented generate reliable document clusters that not only associate *PubMed Central* abstracts into meaningful groups but also provide the *labels* for a shallow content analysis in a rapid and reliable way.

The above methodology is based on the statistical treatment of words. All the threshold values that have been used in our case studies were otpimized empirically, by extensive experimentation, and can be set as parameters by the user. Thus the method is sufficiently flexible and parameter-based to allow extensive exploration of various document collections.

Our approach was to keep the methodology as general as possible, without encoding facts pertinent to a specific biological process. We are particularly interested in ontology induction experiments for transcription profiling, as textual analysis in scientific abstracts of Transcription Factors has not been analyzed yet.

References

[1] T. Ono, H. Hishigaki, A. Tanigami, T. Takagi: Automated extraction of information on protein-protein interactions from the biological literature. *Bioinformatics* 2001, 17:155-161.

[2] T. Sekimizu, H. Park, J. Tsujii: Identifying the interaction between genes and gene products based on frequently seen verbs in MEDLINE abstracts. *Genome Informatics* 1998.

[3] C. Fredman, P. Kra, H. Yu, M. Krauthammer, A. Rzhetsky: "GENIES": a natural-language processing system for the Extraction of Molecular Pathways from Journal Articles. *In Proceedings of the 10th International Conference on Intelligent Systems for Molecular Biology*, (ISMP 2001), S74-S82

[4] A. Renner, A. Aszodi: High-throughput functional annotation of novel gene products using document clustering. *In Proceedings of the Pacific Symposium on Biocomputing*, (PSB2000), 54-68, (2000).

[5] A.G. Papavassiliou: Transcription Factor-Based Drug Design in Anticancer drug Development, *Molecular Medicine*, Vol. 3 (12), 799-810, (1997).

VSP International
Science Publishers
P.O. Box 346, 3700 AH Zeist
The Netherlands

*Lecture Series on Computer
and Computational Sciences*
Volume 1, 2004, pp. 692-695

Do Synonymous Codons Point Towards a Thermodynamic Theory of Tissue Differentiation?

G. Anogianakis[1], A. Anogeianaki, V. Papaliagkas

Department of Physiology,
Faculty of Medicine,
University of Thessaloniki,
GR-54124 Thessaloniki, Greece

Abstract: Deciphering the human genome has held a number of surprises including an unexpectedly low number of genes to account for our preconceptions about human structure and function; especially when one has to take the existence of control mechanisms into consideration. Although a DNA molecule can be thought of as a string over an alphabet of four characters (the nucleotides) and proteins as strings over an alphabet of twenty characters (amino acids), recent discoveries have brought forth the possibility that control mechanisms of tissue differentiation which are based on purely thermodynamic principles may exist. It is proposed that the definition and quantification of such potential mechanisms requires algorithms that combine string computation with thermodynamic evaluation of the DNA translation and transcription processes.

Keywords: Codon, differentiation, strings, synonyms, thermodynamics

1. Introduction

A DNA strand can be thought of as a string over an alphabet of four characters, the nucleotides, i.e., molecular entities consisting of a phosphate group and a pentose sugar (deoxyribose) linked to a nitrogenous base. As only four nitrogenous bases (Adenine or "A," Guanine or "G," Cytosine or "C" and Thymine or "T") are used in the DNA structure, it is the bases that give the nucleotides their individuality (and name). Codons, i.e. triplets of nucleotides that code for the same amino acid are the "words" in this "four letter" language. One of the corollaries of the DNA coding mechanism is that DNA sequence alterations in a gene can change the structure of the protein that it codes for. Indeed, given the fact that the genetic alphabet uses four letters, that the words it uses have, all, three letters each but the language of genetics contains twenty one "semantically different words" (i.e., the twenty amino acids and a "terminating codon" which serves as the punctuation mark) it is evident that there must be many different ways of "spelling," for at least some, of the words of the language of genetics. The actual correspondences are tabulated in Table 1. Thus the codon UCC codes for the amino acid Serine, CCU for Proline etc, but also CCU, CCC, CCA and CCG all code for Proline. The corresponding frequencies are presented in Table 2.

2. Implications of synonymous codons

The genomes of species from bacteria [1] to Drosophila [2] show unique biases for particular synonymous codons and, recently, it was shown that such codon preferences exist in mammals [3]. Systematic differences in synonymous codon usage between genes selectively expressed in six adult human tissues were reported, while the codon usage of brain-specific genes is, apparently, selectively

---

[1] Corresponding author. E-mail: anogian@auth.gr

preserved throughout the evolution of human and mouse from their common ancestor [3]. In particular, when genes that are preferentially expressed in human brain, liver, uterus, testis, ovary, and vulva were analyzed, synonymous codon biases between gene sets were found. The pairs that were compared were brain-specific genes to liver-specific genes; uterus-specific genes to testis-specific genes; and ovary-specific genes to vulva-specific genes. All three pairs differed significantly from each other in their synonymous codon usage raising the possibility that codon biases may be partly responsible for determining which genes are expressed in which tissues. Such a determination may, of course, take place at the level of transcriptional control. However, given the relatively low number of genes identified in the human genome, "vis-à-vis" our preconceptions about human structure and function, one is tempted to explore whether other control mechanisms operate (alone, in tandem or in parallel with transcriptional level mechanisms) in tissue differentiation.

Table 1: The 64 possible combinations of the four bases and the codons they represent. X stands for the terminating codon.

1ST BASE	2ND BASE				3RD BASE
	U	C	A	G	
U	PHE	SER	TYR	CYS	U
	PHE	SER	TYR	CYS	C
	LEU	SER	X	X	A
	LEU	SER	X	TRP	G
C	LEU	PRO	HIS	ARG	U
	LEU	PRO	HIS	ARG	C
	LEU	PRO	GLN	ARG	A
	LEU	PRO	GLN	ARG	G
A	ILE	THR	ASN	SER	U
	ILE	THR	ASN	SER	C
	ILE	THR	LYS	ARG	A
	MET	THR	LYS	ARG	G
G	VAL	ALA	ASP	GLY	U
	VAL	ALA	ASP	GLY	C
	VAL	ALA	GLU	GLY	A
	VAL	ALA	GLU	GLY	G

Table 2: The number of codons used to code for each amino acid and the relative frequency of codons used by each amino acid.

AMINO ACID	CODONS USED	FREQUENCY (%)
ALA	4	6.2500
ARG	6	9.3750
ASN	2	3.1250
ASP	2	3.1250
CYS	2	3.1250
GLN	2	3.1250
GLU	2	3.1250
GLY	4	6.2500
HIS	2	3.1250
ILE	3	4.6875
LEU	6	9.3750
LYS	2	3.1250
MET	1	1.5625
PHE	2	3.1250
PRO	4	6.2500
SER	6	9.3750
THR	4	6.2500
TRP	1	1.5625
TYR	2	3.1250
VAL	4	6.2500
X	3	4.6875
TOTAL	64	100.0000

Assuming that some kind of parsimony principle governs the development of control mechanisms at the DNA transcription level (a very strong but necessary assumption that is required in order to limit and focus the subsequent discussion), there are two obvious mechanisms that can be used to link synonymous codon choice and tissue-specific gene expression:

- The first mechanism depends on local transfer RNA abundance. The tRNA pools in the brain, e.g., may differ from the pools in liver, and so if the codon usage of a gene is calibrated to the tRNA pools that exist in the brain, that gene will be translated more efficiently in brain.

- The second mechanism would make use of the different chemical affinities between different tRNAs coding for the same amino acid and the underlying DNA structure. In other words, if certain codons have larger affinities with their corresponding tRNAs than their synonymous codons, it stands to reason that they will be expressed more readily. A corollary of this argument is that differentiation will occur when the appropriate genes find themselves in an energetically appropriate environment to be expressed.

3. Ways to approach the problem

It is interesting that the scientists that announced the discovery of tissue-specific codon usage and the expression of human genes offer the first mechanism as an explanation of their observations [3, 4]. However, this is still speculative and although there is evidence for it [5], it may be thought of as leading to a circular argument: a gene is expressed (to abundantly produce specific tRNA) so that another gene may be 'induced to induce" yet another gene, and so on ad infinitum. Although such a process cannot be ruled out without in vitro identification of gene induction sequences and experimentation on their functioning under different tRNA pool composition, it appears to be an extremely inefficient, albeit accurate, procedure that sacrifices an unordinary number of genes for control purposes.

In order to resolve the question of whether the second mechanism that was proposed as the link between synonymous codon choice and tissue-specific gene expression has any theoretical (or practical) merit, it is necessary to, first, associate each codon with a value reflecting its potential for expression. To illustrate this point let us assume, e.g., that a gene is coding a peptide with the sequence:

ALA-ALA-ALA-ALA-ALA-ALA-ALA-ALA-ALA-ALA.

Let us, further, assume that codon GCU has twice the affinity for its corresponding tRNA than codon GCC has for its own corresponding tRNA. Similarly, codon GCC has twice the affinity for its corresponding tRNA than codon GCA has for its own corresponding tRNA and, finally that codon GCA has twice the affinity for its corresponding tRNA than codon GCG has for its own corresponding tRNA. It is evident that, if chemical affinities alone were the determining factor and if gene expression was a linear function of the product of its codons' affinities for their corresponding tRNAs, then a gene represented by the sequence:

GCUGCUGCUGCUGCUGCUGCUGCUGCUGCU,

would be expressed 2^{30} times more readily than the gene represented by the sequence:

GCGGCGGCGGCGGCGGCGGCGGCGGCG,

despite the fact that these two hypothetical genes are synonymous. Real life, of course, is not, usually, that generous to scientists or simple enough to be easily simulated but the example above serves as an illustration of the approach that is required to resolve the problem of amino acid coding redundancies, codon synonyms and their possible involvement in tissue differentiation. However, once the problem is formulated, it will be possible to utilize computer models of tRNAs to estimate the thermodynamics of the codon – tRNA interaction and to proceed to derive reasonable estimates of the propensity of a gene to express itself.

The second step, of the proposed approach to the problem, is to translate gene sequences into sequences of "gene potentials for expression" and to attempt to characterize the different tissues by those genes (and their potential for expression) that are actually expressed in them. In order to do this, databases of all possible permutations of a gene sequence (in terms of all the synonymous codons to the codons that are actually used), should be calculated and compared to the actually occurring codon sequences. We could speculate that, given the immense number of permutations in terms of its composing codons that a gene represents (i.e., the potentially synonymous genes), sequences that actually occur will have specific, thermodynamic (?) reasons for being selected.

The final step of the proposed approach is to attempt to resolve the inverse problem, i.e., based on protein structure, to attempt identify which genes give rise to particular proteins. Although, conceptually, resolving the inverse problem does not differ from the translation of gene sequences into sequences of "gene potentials for expression," this step is much more specific in that it requires protein sequencing before one proceeds to take it. On the other hand, given that many proteins are produced by enzymatic sectioning of precursor molecules, it also represents a unique opportunity to link genes coding for such protein molecule precursors with the tissues where they are expressed.

4.   Significance of the problem

We should, at this point, underscore that linking tissue differentiation with thermodynamic constraints, in synonymous codon expression, has profound philosophical and even aesthetic implications: It reduces intellectual dependence on models of molecular evolution where extremely detailed designs in outcome have to be "designed" by blind evolutionary forces and introduces a view whereby some of the most basic and best understood natural principles (those of thermodynamics) undertake to assist evolution in shaping the biological order at a more fundamental level. At the same time, its aesthetic implications sprout from the fact that it introduces a vision of differentiation very similar to that observed in a garden at spring.

Apart from its philosophical significance, however, the answer to the problem of whether the existence (and expression characteristics) of synonymous codons, also implies a thermodynamic theory of tissue differentiation, will have a significant impact on cancer research. Indeed a central issue of cancer research has to do with the mechanisms which cancer cells dedifferentiate, lose the ability to perform the normal functions of the normal cell type from which they mutated and often resemble embryonic cells. It is plausible, therefore, that in case cancer development can be associated with usage of the wrong codon synonyms, that gene therapies can be devised in a much more logical fashion.

Acknowledgments

The corresponding author, a non-mathematician, very early in his adult life, left the green pastures of molecular biology for brain electrophysiology. From this place, he wishes to thank Joshua B. Plotkin of the Bauer Center for Genomic Research at Harvard and his colleagues, for rekindling his, by now ancient, interest in synonymous codons with their recent paper which appeared in PNAS [3]. He also wants to thank Prof. Athanasios Tsakalidis Head of the Computer Engineering and Informatics Dept., in the University of Patras, (Greece) for constantly stimulating his interests in strings, and databases.

References

[1] T. Ikemura, Codon usage and tRNA content in unicellular and multicellular organisms. *Mol Biol Evol* **2(1)** 13-34 (1985)

[2] J.R. Powell and E.N. Moriyama, Evolution of codon usage bias in *Drosophila*. *PNAS* **94** 7784-7790 (1997)

[3] J.B. Plotkin, H. Robins and A. J. Levine, Tissue-specific codon usage and the expression of human genes. *PNAS* **101** 12588-12591 (2004)

[4] M. Phillips, Different codons, same amino acid.
http://www.biomedcentral.com/news/20040817/01 (last visited Sept 02, 2004)

[5] D. B. Carlini and W. Stephan, In vivo introduction of unpreferred synonymous codons into the Drosophila Adh gene results in reduced levels of ADH protein. *Genetics* **163(1)** 239-43 (2003)

VSP International
Science Publishers
P.O. Box 346, 3700 AH Zeist
The Netherlands

*Lecture Series on Computer
and Computational Sciences*
Volume 1, 2004, pp. 696-700

Gene-Finding with the VOM Model

K.O. Shohat-Zaidenraise^, A. Shmilovici*[1], I. Ben-Gal^

*Dept. of Information Systems Eng.
Ben-Gurion University
P.O.Box 653, Beer-Sheva, Israel

^Department of Industrial Engineering Tel-
Aviv University
Ramat-Aviv, Tel-Aviv 69978, Israel

Accepted 31 August, 2004

Abstract: We present the architecture of an elementary gene-finding algorithm that is based on the Variable Order Markov model (VOM). Experiments with the gene-finder on three Prokaryotic genomes indicate that it has advantage on the detection of short genes.

Keywords: Gene Finding, Variable Order Markov Models, Sequence Analysis

Mathematics Subject Classification: 92D20, 60J20, 91B82, 60-08

1. Introduction

Though many gene-finder programs exist, there are numerous unresolved issues, such as uncovering genes in recently sequenced organisms when the gene-finding program is matched to the characteristics of different known sequences. Here, we present a new approach for gene finding based on a Variable-Order Markov (VOM) Model. The VOM model is a generalization of the traditional Markov model, which is more efficient in terms of its parameterization, thus, can be trained on relatively short sequences. As a result, the proposed VOM gene-finder outperforms traditional gene-finders that are based on fifth-order Markov models for newly sequenced bacterial genomes.

This paper presents two contributions: First, it demonstrates the use of VOM models to annotate DNA sequences into coding and non-coding regions. Second, we showcase the ability to predict short genes that may be undetectable by other gene-finder programs.

2. Introduction to the VOM Algorithm

The VOM model is a close relative of the context tree algorithm [1] that was recently demonstrated [2] as a universal compression algorithm that operates in the non-asymptotic domain. A VOM construction algorithm, is an algorithm that computes from a given training sequence the probability estimates for every context. The VOM model is used to evaluate the probability of any testing sequence (generated by the source that generated the training sequence). A context in the context tree is represented by a node where lengths (depths) of various contexts (branches in the tree) do not need to be equal.

The context-tree algorithm of [3] contains two distinct phases: In the tree growing phase, the counts of all the sub-sequences that are shorter than a predefined depth K_{max} are used to update the symbol counters in the nodes. In the tree pruning phase, probability estimates are computed for every context and pruning rules keep a descendent node only if the entropy of its symbols is sufficiently different from the entropy of the symbols of its parent node. The distribution of symbols in the nodes of the pruned tree defines the VOM model that is used to estimate $P(X_1 \mid X_{-K}^0)$ - the probability of any symbol X_1 given the context string $X_{-K}^0 \equiv X_{-K}, X_{-K+1}, \ldots, X_0$. Note that for a Markov chain model, the order is fixed to K_{max} so there is no pruning phase. The Markov chain model suffers from

---

[1] Corresponding author. E-mail: armin@bgumail.bgu.ac.il

exponential growth of the number of parameters to be estimated. For small data-sets this results in over-fitting to the training set and poor variance-bias tradeoff. The VOM algorithm was implemented in the MATLAB scripting language setting $K_{max} = 9$.

3. Introduction to Gene-finding

A gene recognition algorithm takes as input a DNA sequence and produces as output a feature table describing the location of all genes present in the sequence. Nevertheless, the reliability of the gene prediction must be questioned since only a relatively small number of genes have been verified in laboratories. A gene-finder usually contains two steps: a) coding-region recognition - recognizing Open Reading Frames (ORF) - sections of DNA that contains a series of codons (base triplets) coding for amino acids located between the start codon (initiation codon) and a stop codon. An ORF represents a candidate gene that encodes for a protein. b) Gene parsing due to recognition of motifs and start/stop codons.

3.1. The VOM Based Gene-Finder

The key idea behind the VOM based gene-finder is to use alternative VOM models to compress sliding windows of DNA sequences. DNA compressibility is used as the measures of interest. It is well known that DNA sequences are neither chaotic nor random. Thus, DNA sequences should be reasonably compressible. However, it is well known that the compression of DNA sequences is a very difficult task [4].

The VOM-based gene-finding algorithm includes the following steps:

1. The annotation algorithm was adapted from [5]. Given a sequence and four VOM models: three phased non-homogeneous models (since the first codon can starts from the first, second or the third nucleotide) and one non-coding homogeneous model, the sequence is annotated such that each nucleotide receives a symbol 1, 2, 3, or N, corresponding to the coding phases 1, 2, or 3 or to a noncoding region N. The four VOM models are constructed from the training genome. A running window of W=54 nucleotides, (starting 26 nucleotides upstream and ending 27 nucleotides downstream of the current nucleotide) was compressed with the four possible VOM models. Neighboring transitions in the reading frames are eliminated with a Viterbi-like algorithm: each nucleotide was annotated according to the best compression (the compression, which achieved the minimum number of bits per sequence window of size 54). The annotation of a sequence is achieved by the path that minimizes the total compression for all the nucleotides in the sequence and the transition penalties. A penalty cost P=100 is used to eliminate frequent frame shifts.
2. Identify the boundaries of the predicted coding segments in the transitions in the annotation.
3. Find the location of potential start and stop codons in the three different phases.
4. Use search algorithms to match each putative coding region found in step 1 with proximal start and stop codons.
5. Create the complementary string in the translating direction (as if it were the main string) by converting every 'A' to 'T' and vice versa, and every 'C' to 'G' and vice versa. Match each putative coding region to the proximal start and stop codons. No information containing potential promoter sites or other motifs is used.
6. Repeat steps 1-4 for the complementary sequence.

3.2. Optimization Experiments for the VOM Based Gene-Finder

The VOM based Gene-finder contains several algorithms that need to be optimized. Here, we briefly describe the results of several optimization experiments. The full details are presented in [6].

The purpose of the first set of experiments was to improve the accuracy of distinguishing coding and non-coding DNA segments. It was conducted on the dataset of [7], which contains representative segments of the human genome. First, the effect of the pruning constant C in the VOM algorithm was investigated. It was concluded that a pruning constant of C=2 provides a superior accuracy to the Hexamer model (Markov 5) while using only about half the number of parameters.

The purpose of the second set of experiments was to optimize the gene annotation algorithm. The experiments were conducted on the genome of Synechocystis PCC6803. Each version of the VOM based gene-finder annotated the genome, and the Sensitivity (Sn) and Specificity (Sp) of each annotation was computed by comparing it to the annotation published in *GenBank* that is considered accurate. Experiments were conducted with three running window sizes (54,78,102), four penalty costs

(50,100,200,300), two methods for considering neighboring coding segments, and four methods for matching the start/stop codons with the coding segments. It was concluded that the most accurate annotation is produced with window size of 54, penalty of 100, and the start/stop codons are the closest to the coding segment boundary.

4. Comparative Genome Experiments

In the following experiments the VOM based gene-finder - with the best parameters found above - was used to annotate the five bacteria genomes presented in Table 1, whose annotated sequence.

In the first set of experiments, the VOM based gene-finder with a VOM trained on the GENIE [8] human genome was compared with two other gene-finders. The dicodon gene-finder of [9] is based on an algorithm similar to ours, except that there the VOM is replaced by a dicodon (Hexamer, Markov 5) distribution for the coding segments. The *GeneMark.fbf* [10] is a state-of-the-art gene-finder that combines various algorithms, including motif identification. The results of the first group of experiments are presented in Table 2. The inferiority of the VOM based gene-finder can be attributed to the lack of a learning mechanism that the others have, and to the use of a inappropriate VOM trained on the human genome.

Table 1: Genomes used for comparison experiments

Species	Accession Number	Length
Synechocystis PCC6803 (SP)	NC_000911	3,573,470
Pyrococcus horikoshii (PH)	NC_000961	1,738,505
Mycobacterium tuberculosis(MT)	AL123456	4,411,529
Helicobacter pylori(HP)	AE000511	1,667,867
Bacillus subtilis(BS)	AL009126	4,214,814

Table 2: Comparing to [9] and to *GeneMark.fbf* [10]

Gen ome	Gene-finder	Sn	Sp	Sn+Sp
SP	VOM G-F	.9627	.8774	1.8401
	Kim [9]	.9640	.9850	1.9490
	GeneMark	.9696	.9970	1.9666
PH	VOM G-F	.7356	.8769	1.6125
	Kim [9]	.9730	.9390	1.9120
	GeneMark	.9777	.9027	1.8808
BS	VOM G-F	.7971	.9630	1.7601
	Kim [9]	.9750	.9770	1.9520
	GeneMark	.9554	.9917	1.9471

Table 3: Short Genes Summary

Genome	Identification	The VOM gene-finder	Gene-Mark
SP	Full identification	14	3
	Partial identification	1	0
	Un-identified genes	0	12
PH	Full identification	24	0
	Partial identification	17	0
	Un-identified genes	7	48
BS	Full identification	65	5
	Partial identification	28	1
	Un-identified genes	12	99

Table 4: Comparing the FP Rates

	SP		PH		BS	
	FP	FN	FP	FN	FP	FN
Gene Mark	0.26	2.65	8.87	1.87	0.72	3.99
VOM G-F	11.7	3.24	8.68	22.25	2.74	18.13

Comparing the genes detected to the true genes in GeneBank, (Table 3) revealed an advantage for short genes. Many more short genes were discovered by our gene-finder that were undetected by *GeneMark.fbf.* Table 4 compares the false positive (FP) and the false negative (FN) rates of the VOM-based gene-finder and *GeneMark.fbf.* The rate is computed by dividing the FP count by the total examined length (#nucleotides for each gemone). The FP rate of the VOM gene-finder is higher or equal to that of the *GeneMark.fbf.* The FN rate is much higher. Therefore, it is recommended to combine the current gene-finding algorithm with other identification features (such as motif detection) to improve the total accuracy.

In the analysis of a new genome, the best statistical models of the coding and non-coding regions are unknown. Using an inappropriate model reduces the accuracy of the gene-finder. The purpose of the second set of experiments is to analyze the performance of Prokaryotic VOMs implemented on a new prokaryotic genome. The following cross-validation experiment was performed for each genome in Table 1: a VOM is trained from the combination of *four* genomes and the output of the VOM based gene-finder on the *fifth* genome is compared to the true genes in *GeneBank*. Surprisingly, Comparing the results to that of table 2, it turns out that the VOM trained on the human dataset produced higher accuracy then the VOMs trained on the prokaryotic datasets. We postulate that one possible reason for the poor performance in the experiments is that the diversity between two prokaryotic genomes might be greater than the diversity between prokaryotic and eukaryotic genomes. To verify this assumption, further experiments were conducted in which we cross-tested the VOM models for two different genomes – *Mycobacterium tuberculosis* and, *Bacillus subtilis* – belonging to the same family of bacterium (gram-positive bacterium). A significant improvement was detected for the *Bacillus subtilis*, and no improvement for the *Mycobacterium tuberculosis*. Note that the training sets for training the different VOMs do not have the same sizes - which may affect the accuracy of the VOM models.

5. Conclusions

We presented a new approach for gene finding based on the VOM Model. Although the VOM-based gene-finder is simple - it does not consider motifs other than the start and the stop codons - we demonstrated that the VOM model can be applied for gene finding instead of the commonly used Markov models. The VOM based gene-finder excels in detecting short genes when only a small training set is available, therefore, it is suggested that it is used as a complementary tool to other gene-finders that excel in detecting longer genes. The current version of the gene-finder is only a prototype - integrating it with other learning algorithms is expected to further improve its accuracy.

References

[1] J. Rissanen, "A Universal Data Compression System", *IEEE Transactions on Information Theory*, **29** (5), 656- 664(1983).

[2] J. Ziv, "A Universal Prediction Lemma and Applications to Universal Data Compression and Prediction", *IEEE Transactions on Information Theory*, **47**(4), 1528- 1532(2000).

[3] I. Ben-Gal, Morag G., Shmilovici A., "CSPC: A Monitoring Procedure for State Dependent Processes", *Technometrics*, **45**(4), 293-311(2003).

[4] S. Grumbach, F. Tahi, "A new challenge for compression algorithms: genetic sequences", *J. of Information Processing and Management*, **30**(6), 866-875(1994).

[5] A. Shmilovici, I. Ben-Gal, "Using a Compressibility Measure to Distinguish Coding and Noncoding DNA", *Far East Journal of Theoretical Statistics*, **13**(2), 215-234(2004).

[6] K.O. Shohat-Zaidenraise, "Gene finding via Context Learning Models", Thesis submitted to Tel-Aviv University, Israel, February 2004.

[7] J.W. Fickett, C.S. Tung, "Assessment of Protein Coding Measures", *Nucleic Acids Research*, **20**(24), 6441-6450,(1992).

[8] GENIE data-sets, from *Genbank* version 105, 1998. Available: www.fruitfly.org/seq_tools/datasets/Human/intron_v105/

[9] J. Kim, "A Study on Dicodon-oriented Gene Finding using Self-Identification Learning", A thesis submitted to School of Knowledge Science, Japan Advanced Institute of Science and Technology, February 2000

[10] A.M. Shmatkov, A. A. Melikyan, F.L. Chernousko, M. Borodovsky, "Finding Prokaryotic Genes by the 'Frame-by-Frame' Algorithm: Targeting Gene Starts and Overlapping Genes", *Bioinformatics*, **15**(11), 874-886(1999).

VSP International
Science Publishers
P.O. Box 346, 3700 AH Zeist
The Netherlands

*Lecture Series on Computer
and Computational Sciences*
Volume 1, 2004, pp. 701-704

Searching for Regularities in Weighted Sequences

M. Christodoulakis, C. Iliopoulos, K. Tsichlas
Department of Computer Science, King's College, Strand, London WC2R 2LS, England
{manolis,csi,kostas}@dcs.kcl.ac.uk

K. Perdikuri
Department of Computer Engineering and Informatics, University of Patras, 26500 Patras, Greece
perdikur@ceid.upatras.gr

Abstract: In this paper we describe algorithms for finding regularities in weighted sequences. A weighted sequence is a sequence of symbols drawn from an alphabet Σ that have a prespecified probability of occurrence. We show that known algorithms for finding repeats in solid sequences may fail to do so for weighted sequences. In particular, we show that Crochemore's algorithm for finding repetitions cannot be applied in the case of weighted sequences. However, one can use Karp's algorithm to identify repeats of specific length. We also extend this algorithm to identify the covers of a weighted sequence. Finally, the implementation of Karp's algorithm brings up some very interesting issues.

1 Introduction

Weighted sequences are used for representing relatively short sequences such as binding sites as well as long sequences such as profiles of protein families (see [2], 14.3). In addition, they are also used to represent complete chromosome sequences ([2], 16.15.3) that have been obtained using a whole-genome shotgun strategy with an adequate cover. The cover is the average number of fragments that appear at a given location. Usually, the cover is large enough so that errors as well as SNPs are clearly spotted and removed by the consensus step.

By keeping all the information the whole-genome shotgun produces, we would like to dig out information that has been previously undetected after being faded during the consensus step (for example the consensus step wrongly chooses a symbol for a specific position than another). As a result, errors in the genome are not removed by the consensus step but remain and a probability is assigned to them based on the frequency of symbols in each position.

In this paper we present efficient algorithms for finding repetitions and covers in a weighted sequence. In solid sequences the algorithms of Crochemore [1] and Karp [5] are well known and have a $O(n \log n)$ time complexity. Their difference is that the first algorithm computes repetitions of all possible lengths while the second can compute repetitions of prespecified length.

There was already an attempt [4] to apply Crochemore's algorithm to weighted sequences. However, as we show in this paper the algorithm fails to find repetitions in $O(n \log n)$ time. In fact it needs $O(n^2)$ time to be able to compute all repetitions. However, Karp's algorithm has already been applied for this problem [3] successfully. In this paper we extend this algorithm to compute covers on weighted strings while we experimentally investigate its efficiency.

The structure of the paper is as follows. In Section 2 we give the basic definitions to be used in the rest of the paper. In Section 3 we argue why Crochemore's algorithm is not suitable for weighted sequences while in Section 4 we sketch the algorithm for finding repetitions and covers based on Karp's algorithm. Finally, in Section 5 we provide experimental results.

2   Preliminaries

In this work we concentrate on the identification of repetitions and covers of fixed length in a weighted biological sequence with probability of appearance $\geq 1/k$, where k is a small fixed constant determined by biologists (for example $k \leq 10$). The size of k is chosen small in order to represent the restricted ambiguity in the appearance of several characters in a biological sequence.

Assume an alphabet $\Sigma = \{1, 2, \ldots, \sigma\}$. A word s of length n is represented by $s[1..n] = s[1]s[2] \cdots s[n]$, where $s[i] \in \Sigma$ for $1 \leq i \leq n$, and $n = |s|$ is the length of s. A factor f of length p is said to occur at position i in the word s if $f = s[i, \cdots i + p - 1]$. A word has a repetition when it has at least two equal factors. A repetition is a cover when each position of the word s belongs in this repetition. A weighted sequence is defined as follows:

Definition 1 *A weighted sequence $s = s_1 s_2 \cdots s_n$ is a set of couples $(q, \pi_i(q))$, where $\pi_i(q)$ is the occurrence probability of character $q \in \Sigma$ at position i. For all positions $1 \leq i \leq n$, $\sum_{q=1}^{\sigma} \pi_i(q) = 1$.*

A factor is *valid* when its probability of occurrence is $\geq \frac{1}{k}$, where k is a small fixed constant. The probability of occurrence of a factor $f = f[1] \ldots f[m]$ occurring at position i in weighted sequence s is the product of probabilities of occurrence of the respective symbols of f in s, i.e. $\prod_{j=1}^{m} \pi_{i+j-1}(f[j])$. A weighted sequence has a repetition when it has at least two identical occurrences of a factor (weighted or not). In biological problems scientists are interested in discovering all the repetitions as well as covers of all possible words having a probability of appearance larger than a predefined constant. In the algorithms we provide we can always find the largest repetition or cover in a weighted sequence by a simple exponential and binary search on possible lengths. As a result we focus only for a prespecified length d.

3   Why Crochemore's Algorithm Fails for Weighted Sequences

This section assumes that the reader is familiar with the algorithm of Crochemore for finding repetitions [1]. The algorithm uses integers to represent factors. The E_i vector holds the factors of length i that start at each position of the text. The algorithm works in stages, each of which corresponds to the computation of repetitions of length larger by one with respect to the previous stage, while at the first stage repetitions of length 1 are computed. At each stage the algorithm chooses small classes to work on, without processing large classes. All classes that have not been processed in stage i, *implicitly* specify longer factors at stage $i + 1$. This is the crucial property of this algorithm that results in an $O(n \log n)$ time complexity.

The problem on weighted sequences is clear: we *cannot* increment factors implicitly; we have to update their probabilities of occurrence at each step so that we know whether these repetitions have a probability $\geq \frac{1}{k}$. As a result, we are obliged to process all classes which leads to an $O(n^2)$. The authors of [4] did not notice this problem so they claimed that the complexity is $O(n \log n)$, which is wrong. Alternatively, one could try find all the repetitions without computing probabilities, and then compute the probabilities of the actual repetitions. This is no better because the length of each repetition can be $O(n)$ (thus, $O(n)$ multiplications) for each repetition. Moreover, if we don't compute probabilities throughout the algorithm we might end up with $O(|\Sigma|^n)$ factors. As a result, it seems that adopting this approach for finding repetitions in weighted sequences will probably not lead to and $o(n^2)$ algorithm.

4   Karp's algorithm

Karp's algorithm computes equivalence classes, similar to Crochemore's, but it computes them using $\log n$ steps of $O(n)$ time each. It has been successfully applied to weighted sequences [3]. The following lemma is the basic mechanism of Karp's algorithm.

Lemma 1 *For integers i, j, a, b with $b \leq a$ we have $iE_{a+b}j$ precisely when $iE_a j$ and $i + bE_a j + b$ (1) or, equivalently, when $iE_a j$ and $i + aE_b j + a$ (2).*

Based on Karp's algorithm, we will briefly sketch how it is applied to weighted sequences.

Definition 2 *Given a weighted sequence s, positions i and j of s are k-equivalent ($k \in \{1, 2, \ldots, n\}$ and $i, j \in \{1, 2, \ldots, n - m + 1\}$) —written $iE_k j$— if and only if there exists at least one substring f of length m, that appears at (starts at) both positions i and j.*

The equivalence class E_k is represented as a vector $v_1^{(k)} v_2^{(k)} \ldots v_{n-k+1}^{(k)}$ of sets of integers, where each set $v_i^{(k)}$ contains the labels of the equivalence classes of E_k to which each factor starting at position i belongs. The implementation of the algorithm is based on Equation (2) from Lemma 1 while its description in [3] was based on Equation (1). A detailed description of the algorithm follows. The new algorithm will be using $e_a + e_b$ pushdown stores, $P(1), \ldots, P(e_a)$ and $Q(1), \ldots, Q(e_b)$.

1. Sort the vector $v^{(a)}$ using the P-pushdown stores; that is, run through $v^{(a)}$, and for each factor x at each position i, push i into $P(x)$. Note that the same position i may be pushed into more than one stacks. So far, having the same position i in more than one P-stack causes no problem, since these stacks are distinct. But, for the sake of the explanation, let's distinguish them in the following way: we will use i_x to denote the position i when it refers to the factor x starting at i; thus, $i_{x'}$ will denote the same position i but referring to it's second factor x' ($\neq x$).

2. In success, pop each $P(x)$ until it is empty. As the number i_x is popped from $P(x)$, push it into the Q-pushdown stores $Q(y)$ $y \in v_{d+a}^{(b)}$ provided that $d + a \leq n - (a + b - 1)$. Note that there may be more than one factors y of length b starting at position $d + a$. Therefore, position i_x will be pushed into all appropriate Q stacks. However, when another P stack, say $P(x')$, is popped, it is possible that the same position $i_{x'}$ (referring to different factor) will be pushed into the same Q stacks as i_x. Had we not distinguished i_x from $i_{x'}$, we would end up with the same position i appearing more than once in the same stack $Q(y)$, and of course the ambiguity as to which factor starting at position i this refers to, would make it impossible to go on to the third step.

3. Finally, construct $v^{(a+b)}$: Successively pop each Q-stack until empty. Start with a variable class counter c initially set to 1. As each i_x is popped from a given stack $Q(y)$ test whether or not x is equal to x', where x' is the factor represented by position j just previously popped from the same stack; that is, element $j_{x'}$ was previously popped. If this is so, then $iE_a j$ and $i + bE_a j + b$, so $iE_{a+b}j$; therefore, insert c in the set $v_d^{(a+b)}$. Otherwise we have $i + bE_a j + b$ but *not* $iE_a j$; therefore, insert $c + 1$ into $v_d^{(a+b)}$ and increment c to $c + 1$. When stack $Q(y)$ is exhausted, increment c to $c + 1$ before beginning to pop the next Q-stack. Whenever i_x is the first element from a Q-stack, insert c into $v_d^{(a+b)}$ automatically.

4.1 Covers

Covers in weighted sequences fall into two categories: **(a)** allow overlaps to pick different symbols from one single and **(b)** factors that overlap choose the same symbols for overlapping regions. Notice that the first kind of covers allows border-less covers to overlap while the second does not.

Assume that we are interested only in length-d covers. This problem is solved in $O(n)$ extra time (thus $O(n \log n)$ total time) for either case:

1. Scan E_d for every factor occurring at position 1 (there is a constant number of them); if the distance of consecutive occurrences of the same factor is always $\leq d$ then a cover has been found.

2. Compute the border array of the candidate cover ($O(d)$ time). Scan E_d, like in (1), only now reject occurrences that start at positions other than some border of the previous occurrence.

Obviously we can do this for every class E_i that is computed on the way to create E_d. The computation of type (a) covers is straightforward. On the other hand, type (b) covers face a difficulty:

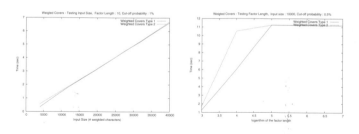

Figure 1: Weighted Covers. Left: The running time with respect to n. Right: The running time with respect to $\log_2 d$.

how can the border array of a factor be computed, since factors are only represented by integers, and the actual factors (strings) are never stored? Obviously, there is no other way than storing the actual strings that correspond to the numbers that represent factors. The space complexity for this is $O(nd)$, since there are at most $O(n)$ factors of length d. The time complexity remains unaffected by the fact that the actual factors are identified and stored. For example, consider that we will combine the equivalence relations E_a and E_b to obtain E_{a+b}. The identification of a factor in E_{a+b} takes only constant time since it can be constructed by the concatenation of one factor from E_a with one factor from E_b.

5   Experimental Results

The algorithms were implemented in C++ using the Standard Template Library (STL), and run on a Pentium-4M 1.7GHz system, with 256MB of RAM, under the Red Hat Linux operating system (v9.0). The datasets used for testing the performance of our algorithms consisted of many copies of a small random weighted sequence. We chose this repeated structure, rather than totally random files, in order to get a fair comparison of the running times.

The running time for locating weighted covers is shown in Figure 1. As expected, weighted covers of type (b) need more time to be computed since, in contrast with type (a) covers, a border array has to be constructed, and overlapping between consecutive occurrences of the same factor needs to be tested. Nevertheless, the asymptotic growth is still $O(n \log d)$. An interesting aspect of the algorithm is being revealed in the right graph: the running time tends to become constant for larger values of d. The reason is simple: as the length of the factor increases, there is a point at which the number of factors (with adequate probability) gets to zero.

References

[1] M. CROCHEMORE. An Optimal Algorithm for Computing the Repetitions in a Word. *Information Processing Letters*, 12:244–250, 1981.

[2] D. Gusfield. Algorithms on strings, trees, and sequences. Cambridge University Press, 1997.

[3] C.ILIOPOULOS, K. PERDIKURI, A. TSAKALIDIS AND K. TSICHLAS. The Pattern Matching Problem in Biological Weighted Sequences. *In the Proc. of FUN with Algorithms*, 2004.

[4] C. ILIOPOULOS, L. MOUCHARD, K. PERDIKURI, A. TSAKALIDIS. Computing the repetitions in a weighted sequence. *In the Proc.of the Prague Stringology Conference (PSC)*, pp. 91-98, 2003.

[5] R. KARP, R. MILLER, A. ROSENBERG. Rapid Identification of Repeated Patterns in Strings, Trees and Arrays. *In the Proc. of the Symposium on Theory of Computing (STOC)*, 1972.

VSP International
Science Publishers
P.O. Box 346, 3700 AH Zeist
The Netherlands

*Lecture Series on Computer
and Computational Sciences*
Volume 1, 2004, pp. 705-708

Efficient Algorithms for Handling Molecular Weighted Sequences and Applications

C. Makris[1], Y.Panagis[2] and E. Theodoridis[2]

[1] Department of Applied Informatics in Management & Finance,
Technological Educational Institute of Mesolonghi, Greece
makri@ceid.upatras.gr
[2] Computer Engineering & Informatics Dept. of Univesity of Patras
and Research Academic Computer Tecnology Institute (RACTI).
Rio, Greece, P.O. BOX 26500.
{panagis, theodori}@ceid.upatras.gr

Abstract: In this paper we present the Weighted Suffix Tree, an efficient data structure for computing string regularities in weighted sequences of molecular data. Molecular Weighted Sequences can model important biological processes such as the DNA Assembly Process or the DNA-Protein Binding Process. Thus pattern matching or identification of repeated patterns, in biological weighted sequences is a very important procedure in the translation of gene expression and regulation. We present time and space efficient algorithms for constructing the weighted suffix tree and some applications of the proposed data structure to problems such as pattern matching, repeats discovery, discovery of the longest common subsequence of two weighted sequences.

Keywords: Molecular Weighted Sequences, Suffix Tree, Pattern Matching, Identifications of repetitions, Covers.

1 Introduction

Molecular Weighted Sequences appear in various applications of Computational Molecular Biology. A molecular weighted sequence is a molecular sequence (either a sequence of nucleotides or aminoacids), where each character in every position is assigned a certain weight. This weight could model either the probability of appearance of a character or the stability that the character contributes in a molecular complex.

Definition *A weighted word $w = w[1]w[2] \cdots w[n]$ is a sequence of positions, where each position $w[i]$ consists of a set of ordered pairs. Each pair has the form $(s, \pi_i(s))$, where $\pi_i(s)$ is the probability of having the character s at position i. For every position w_i, $1 \leq i \leq n$, $\sum_{\forall s} \pi_i(s) = 1$.*

Position	1	2	3	4	5	6	7	8	9	10	11
	A	C	T	T	(A,0.5)	T	C	(A,0.5)	T	T	T
					(C,0.5)			(C,0.3)			
					(G, 0)			(G,0)			
					(T, 0)			(T,0.2)			

Figure 1: A weighted word w.

Thus in the first case a molecular weighted sequence can be the result of a DNA Assembly process. The key problem today in sequencing a large string of DNA is that only a small amount of DNA can be sequenced in a single read. That is, regardless of whether the sequencing is done by a fully automated machine or by a more manual method, the longest unbroken DNA substring that can be reliably determined in a single laboratory procedure is about 300 to 1000 (approximately 500) bases long [1],[2]. A longer string can be used in the procedure but only the initial 500 bases will be determined. Hence to sequence long strings or an entire genome, the DNA must be divided into many short strings that are individually sequenced and then used to assemble the sequence of the full string. The critical distinction between different large-scale sequencing methods is how the task of sequencing the full DNA is divided into manageable subtasks, so that the original sequence can be reassembled from sequences of length 500.

Reassembling DNA substrings introduces a degree of uncertainty for various positions in a biosequence. This notion of uncertainness was initially expressed with the use of "don't care" characters denoted as "*". A "don't care" character has the property of matching against any symbol in the given alphabet. For example the string $p = AC*C*$ matches the pattern $q = A*GCT$ under the alphabet $\Sigma = \{A, C, G, T, *\}$. In some cases though, scientists are able to go one step further and determine the probability of a certain character to appear at the position previously characterised as wildcard. In other words, a "don't care" character is replaced by a probability of appearance for each of the characters of the alphabet. Such a sequence is modelled as a *weighted sequence*.

In the second case a molecular weighted sequence can model the binding site of a regulatory protein. Each base in a candidate motif instance makes some positive, negative or neutral contribution to the binding stability of the DNA-protein complex [4], [10]. The weights assigned to each character can be thought of as modeling those effects. If the sum of the individual contributions is greater than a treshold, the DNA-protein complex can be considered stable enough to be functional.

Thus we need new and efficient algorithms in order to analyze molecular weighted sequences. A fundamental problem in the analysis of Molecular Weighted Sequences is the computation of significant repeats which represent functional and structural similarities among molecular sequences. In [7] authors presented a simple algorithm for the computation of repeats in molecular weighted sequences. Although their algorithm is simple and easy to be implemented, it is not efficient in space needed. In this paper we present an efficient algorithm, both in time and space limitations, to construct the Weighted Suffix Tree, an efficient data structure for computing string regularities in biological weighted sequences. The Weighted Suffix Tree, was firstly intoduced in [6]. In this work, which is primarily motivated by the need to efficiently compute repeats in a weighted sequence, we further extend the use of the Weighted Suffix Tree to other applications on weighted sequences.

2   The Weighted Suffix Tree

In this section we present a data structure for storing the set of suffixes of a weighted sequence with probability of appearance greater than $1/k$, where k is a given constant. We use as fundamental data structure the suffix tree, incorporating the notion of probability of appearance for every suffix stored in a leaf. Thus, the introduced data structure is called the *Weighted Suffix Tree* (abbrev. WST).

The weighted suffix tree can be considered as a generalisation of the ordinary suffix tree to handle weighted sequences. We give a construction of this structure in the next section. The constructed structure inherits all the interesting string manipulation properties of the ordinary suffix tree. However, it is not straightforward to give a formal definition as with its ordinary

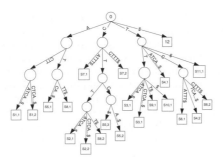

Figure 2: A Weighted Suffix Tree example.

counterpart. A quite informal definition appears below.

Definition *Let S be a weighted sequence. For every suffix starting at position i we define a list of possible weighted subwords so that the probability of appearance for each one of them is greater than $1/k$. Denote each of them as $S_{i,j}$, where j is the subword rank in arbitrary numbering. We define $WST(S)$ the weighted suffix tree of a weighted sequence S, as the compressed trie of a portion of all the weighted subwords starting within each suffix S_i of $S\$$, $\$ \notin \Sigma$, having a probability of appearance greater than $1/k$. Let $L(v)$ denote the path-label of node v in $WST(S)$, which results by concatenating the edge labels along the path from the root to v. Leaf v of $WST(S)$ is labeled with index i if $\exists j > 0$ such that $L(v) = S_{i,j}[i..n]$ and $\pi(S_{i,j}[i \cdots n]) > 1/k$, where $j > 0$ denotes the j-th weighted subword starting at position i. We define the leaf-list $LL(v)$ of v as a list of the leaf-labels in the subtree below v.*

We will use an example to illustrate the above definition. Consider again the weighted sequence shown in Fig. 1 and suppose that we are interested in storing all suffixes with probability of appearance greater than a predefined parameter. We will construct the suffix tree for the sequence incorporating the notion of probability of appearance for each suffix.

For the above sequence and $k \geq 1/4$ we have the following possible prefixes for every suffix: prefixes for suffix $x[1 \cdots 11]$: $S_{1,1} = ACTT\underline{A}TC\underline{A}TTT$, $\pi(S_{1,1}) = 0.25$, and $S_{1,2} = ACTT\underline{C}TC\underline{A}TTT$, $\pi(S_{1,2}) = 0.25$, prefixes for suffix $x[2 \cdots 11]$: $S_{2,1} = CTT\underline{A}TC\underline{A}TTT$, $\pi(S_{2,1}) = 0.25$, and $S_{2,2} = CTT\underline{C}TC\underline{A}TTT$, $\pi(S_{2,2}) = 0.25$, etc. The weighted suffix tree for the above subwords appears in Fig. 2.

In [11] we present $O(N)$ time algorithm for constructing the Weighted Suffix Tree of a weighted sequence of length N.

3   Applications of the WST

In this section we present compendiously some applications of the Weighted Suffix Tree.

Pattern matching in weighted sequences: in this problem given a pattern p and a weighted sequence x, we want to find the starting positions of p in x, each with probability of appearance greater than $1/k$. Firstly, we build the WST for x with parameter k. If p consists entirely of non-weighted positions we spell p from the root of the tree until at an internal node v, either we have spelled the entire p, in which case we report all items in $LL(v)$, or we cannot proceed further and thus we report failure. If p contains weighted positions we decompose it into solid patterns each with $\Pr\{occurence\} > 1/k$ and match each one of them using the above procedure. This solution takes $O(m + \alpha)$ time, $m = |p|$, where α is the output size, with $O(n)$ reprocessing.

Computing the Repeats: in this problem given a weighted sequence x and an integer k we are searching for all the repeats of all possible words having a probability of appearance greater than $1/k$ and do not overlap. We build again a WST with parameter k and traverse it bottom-up. At each internal node v, which has more than 1 leafs at his subtree we report the leafs beneath it, in pairs. This process requires $O(n \log n + \alpha)$, where α is the output size, with $O(n)$ reprocessing. When we allow overlaps, the repetitions have to present the same characters at the positions that overlap. For this case we need nearest common ancestor queries to check this restriction in constant time. This process requires $O(n^2 + \alpha)$, where α is the output size, with $O(n)$ reprocessing

Longest Common Substring in Weighted Sequences: in this problem given two weighted strings S_1 and S_2, find the longest common substring with probability of appearance greater than $1/k$ in both strings. To find the longest common substring in two given weighted strings S_1 and S_2 a generalised weighted suffix tree for S_1 and S_2 is built. The path label of any internal node is a substring common to both S_1 and S_2 with probability of appearance greater than $1/k$. The algorithm merely finds the node with greatest string-depth with a preorder traversal. The above procedure runs in $O(n)$ time.

References

[1] Celera Genomics: The Genome Sequence of Drosophila melanogaster. Science, V(287), (2000)2185–2195.

[2] Celera Genomics: The Sequence of the Human Genome. Science, Vol. 291, (2001) 1304–1351.

[3] Crochemore, M.: An Optimal Algorithm for Computing the Repetitions in a Word. Inf. Proc. Lett., Vol. 12. (1981) 244–250.

[4] G. Grillo, F. Licciuli, S. Liuni, E. Sbisa, G. Pesole. PatSearch: a program for the detection of patterns and structural motifs in nucleotide sequences. *Nucleic Acids Res.*,3608-3612,**31**(2003).

[5] Gusfield, D.: Algorithms on Strings, Trees, and Sequences: Computer Science and Computational Biology. Cambridge University Press, New York (1997)

[6] Iliopoulos, C., Makris, Ch, Panagis, I., Perdikuri, K., Theodoridis, E., Tsakalidis, A.: Computing the Repetitions in a Weighted Sequence using Weighted Suffix Trees, In European Conference on Computational Biology (ECCB 2003), Posters' Track.

[7] Iliopoulos, C., Mouchard, L., Perdikuri, K., Tsakalidis, A.,: Computing the repetitions in a weighted sequence, Proceedings of the Prague Stringology Conference (PSC 2003), 91-98.

[8] Kolpakov, R., Kucherov, G.. Finding maximal repetitions in a word in linear time. In Proc. FOCS99, pp. 596–604, (1999).

[9] Kurtz, S., Schleiermacher, C.,: REPuter: fast computation of maximal repeats in complete genomes. Bioinformatics, Vol. 15, (1999) 426–427.

[10] H. Li, V. Rhodius, C. Gross, E. Siggia Identification of the binding sites of regulatory proteins in bacterial genomes *Genetics*, **99** (2002), 11772–11777.

[11] Iliopoulos C., Makris C., Panagis Y., Perdikuri K., Theodoridis E. and Tsakalidis A.. IFIP International Conference on Theoretical Computer Science. Toulouse August 2004.(To appear).

VSP International
Science Publishers
P.O. Box 346, 3700 AH Zeist
The Netherlands

*Lecture Series on Computer
and Computational Sciences*
Volume 1, 2004, pp. 709-710

Preface to Symposium--
Performance Measurement Based
on Complex Computing Techniques

D. Wu[1]   K. Womer [2]

[1]School of Business, University of Science and Technology of China,
230026 Jinzhai Road Hefei, PR China

[2]College of Business Administration,
University of Missouri -- St Louis, One University Blvd
St Louis, MO 63121-4400

Abstract: this preface presents the purpose, content and results of one of the ICCMSE 2004 symposiums organized by Dr. D. Wu and Professor K. Womer, who is also Dean of College of Business Administration, University of Missouri -- St Louis.

Keywords: Performance evaluation, Complex computing

Performance evaluation is a very complex task since economic organizations are complex multidimensional systems [1]. They may compose different sub-systems characterized by different measures of effectiveness and cost. Moreover, information is inherently incomplete and uncertain in these organizations. For example, the tradeoffs and relationships among different measures are often completely unknown. Any successful method has to handle the complexity of the system and the inherent uncertain and subjective information.

Analyzing performance factors in such systems has proved to be a challenging task that requires innovative performance analysis tools and methods to keep up with the rapid evolution and ever increasing complexity of such systems. Three major evaluation techniques are important here: measurement, analytic modeling and simulation. Complex computing methodologies such as neural networks have been widely adopted for this purpose. They are systems of adaptable nodes that learn by example, store the learned information, and make it available for later use[2,3,4]. Programming techniques such as Data Envelopment Analysis (DEA) are more transparent techniques that are also promising ways to to deal with this task [5].

We feel that the time is ripe to offer, in an international symposium, a selection of good papers having, as their principal theme, performance evaluation theory, practice, or impact in a large-scale application.

This symposium will cover various topics on quantitative modeling, simulation and measurement/testing of complex systems. The topics of interest include, but are not limited to:

· Integrated performance tool environments
· Analytical modeling
· Performance measurement and monitoring using neural networks
· Performance metrics
· Performance predictions
· Performance of memory systems
· Performance-directed system design
· Performance implications of parallel and distributed systems
· Case studies of in-depth performance analysis on existing systems
· System case studies showing the role of performance tools in the design of
  systems

---
[1] Corresponding author. E-mail: dash@mail.ustc.edu.cn, dash_wu@hotmail.com [2] keithwomer@umsl.edu

The manuscripts submitted were reviewed by highly qualified expert referees in a thorough two-stage review procedure. The articles that we have been happy to accept all enjoy a combination of originality, high technical merit, and relevance to the topic.

We would like to thank all the authors who submitted manuscripts to this special issue. We regret that only a fraction of the submissions, all of them interesting, could be included. We owe a special debt of gratitude to the many able reviewers who generously commented, often is extraordinary detail, on the submissions. Finally, we would like to acknowledge the good-hearted patience of the authors and of the organizers in spite of this international symposium's long gestation.

References

[1] O. Marta, Lorenzo D. Ambrosio, Ra.aele Pesenti, Walter Ukovich . Multiple-attribute decision support system based on fuzzy logic for performance assessment. European Journal of Operational Research .in press, 2003.

[2] Dimla, S. (1999). Application of perceptron neural networks to tool-state classification in a metal-turning operation. Engineering Applications of Artificial Intelligence, 2(4), 471–477.

[3] Kusiak, A., & Lee, H. Neural computing-based design of components for cellular manufacturing. International Journal of Production Research Society, 34(7), 1777–1790 (1996).

[4] L. Liang and D. Wu . An application of pattern recognition on scoring Chinese corporations financial conditions based on backpropagation neural network Computers & Operations Research, In Press, Available online 14 November(2003)

[5]A. Charnes, W.W. Cooper, and E. Rhodes, Measuring the efficiency of decision making units, European Journal of Operational Research 2, 429-444(1978)

VSP International
Science Publishers
P.O. Box 346, 3700 AH Zeist
The Netherlands

Lecture Series on Computer
and Computational Sciences
Volume 1, 2004, pp. 711-713

Workforce Schedule and Roster Evaluation Using Data Envelopment Analysis

Y. Li[1]   D. Wu[2]

[1] School of Business University of Alberta, Edmonton, AB, Canada T6G 2R6

[2] School of Business, University of Science and Technology of China,
HeFei, AnHui, P.R.China, 230052

Abstract: A framework composed by simulation modeling and data envelope analysis (DEA) is proposed to evaluate the overall service quality of various schedules and rosters that satisfy a time-varying demand. Workforce scheduling is a process of generating employee shifts to match customer demands for service while keeping costs under control and satisfying all applicable regulations (e.g., shift lengths and spacing of breaks). Workforce rostering is to assign shifts to employees preferably with the consideration of employees' satisfaction to their working time. The proposed approach provides an effective and innovative tool to estimate the performance of schedules and rosters through multiple objectives that management has concerns with, other than merely service level. A cone-ratio DEA model is used to differentiate various schedules or rosters according to a decision maker's preferences to their performance indicators produced by a simulation model.

Keywords: Workforce scheduling; Multiple Objectives; DEA; cone ratio

1. Introduction

Workforce scheduling is a process of generating employee shifts to match customer demands for service while keeping costs under control and satisfying all applicable scheduling regulations (e.g., shift lengths and spacing of breaks). One of the reasons that make this problem a difficult one to solve is the complexity of evaluating service quality. Traditional workforce scheduling approaches use service level – the percentage of customers served within a predetermined threshold time, as the only target to achieve while minimizing the staffing cost.

Glover and McMillan (1986) employed techniques that integrate management science methods and artificial intelligence to solve general shift scheduling problems. Balakrishnan and Wong (1990) solved a rotating workforce scheduling problem as modeling a network flow problem. Smith and Bennett (1992) combined constraint satisfaction and local improvement algorithms to develop schedules for anesthetists. Schaerf and Meisels (1999) proposed a general local search method for employee scheduling problems. Recently, Laporte (1999) considered developing rotating workforce schedules by hand and showed how the constraints can be relaxed to get acceptable schedules. All of those among many other proposed workforce scheduling algorithms used only one criterion service level to indicate service quality.

However, unless service level is set to be a hundred percent, which is not common in practice, it is never clear what happens to the customers who wait more than the proposed threshold time. There are certainly more indicators than service level exist to reveal service quality from different angles (Castillo, Joro and Li 2003). For example maximum wait time indicates the worst scenario; balking and reneging rate show the percentage of the customers who attempted to but did yet receive services; average and max queue length provide information for the physical capacity of service facilities etc. Using service level only to represent service quality, at least the above mentioned information has to be ignored. The service quality indicators considered in this paper include but are not limited to service level, average wait time, maximum wait time, average queue length, maximum queue length, balking

---

[1] yongyue@ualberta.ca
[2] Corresponding author: dash@ustc.edu

rate, reneging rate, and personnel utilization.

A simulation model is used to generate these indicators given various proposed schedules because of two main reasons. First, simulation models are able to accommodate complicated realistic situations without being imposed on strong assumptions such as M/M/s and single independent period by period (SIPP). In addition, most analytical models have difficulty to generate certain indicators such as maximum queue length etc. Considering current available techniques, we decide to adopt simulation instead of analytical models to achieve high flexibility in various situations and performance indicators with the expense of speed.

Workforce rostering assigns shifts in a schedule to employees while satisfying various constrains such as one employee working at most one shift in a day and employee availability. In many operations, hospitals in particular, satisfying employees' preferences for working time has been a chronic problem in workforce rostering. Poor employee satisfaction often leads to high personnel turnover, absenteeism, resentment, poor job performance and unfit mental and physical conditions—situations that translate to loss of productivity, quality and even safety. Therefore, indicators denote employees' satisfaction for working time should be considered to evaluate workforce rosters compare to schedules. An aggregated employee preference score is used in the data sample to represent overall employee preference to a roster; however, more indicators can be included if necessary.

A framework composed by simulation modeling and data envelope analysis (DEA) is proposed to evaluate the overall service quality of various schedules and additionally overall employee satisfaction for rosters. This approach provides an effective and innovative tool to estimate the performance of schedules and rosters through multiple objectives that management has concerns with, other than merely service level. The cone-ratio DEA model is used here to aggregate the decision maker's preference information to the simulated performance indicators and employee satisfaction indicators.

2. Cone-ratio DEA

The methodology we use here is cone-ratio Data Envelopment Analysis (DEA), a weight-restricted DEA developed based on the CCR model (Charnes et al. 1978). DEA is considered a robust tool for the evaluation of relative efficiencies as well as for the establishment of goals (or benchmarks) for the entities out of the efficiency border (or envelope). The analyzed entities or DMU's (for Decision Making Units) are compared under Farrel's concept of efficiency (Farrel *et al.*, 1962), that consists of a ratio of the weighted sum of the outputs over the weighted sum of the inputs of each DMU.

Suppose that there are n DMUs, that is DMU_j ($j = 1,2,\cdots,n$) to be evaluated. Each DMU_j has m different inputs x_{ij} and s different outputs y_{rj}. Let the observed input and output vectors of DMU_j be $X_j = (x_{1j}, x_{2j}, \cdots, x_{mj})^T > 0$, $j = 1,2,\cdots,n$ and $Y_j = (y_{1j}, y_{2j}, \cdots, y_{sj})^T > 0$, respectively.

The Cone-ratio model can be written as

$$Max \quad \frac{U^T Y_0}{V^T X_0} \tag{1}$$

$$s.t. \quad \frac{U^T Y_j}{V^T X_j} \le 1 \quad j = 1,2,\cdots,n$$

$$U \in P, V \in Q$$

where $P \subset E_+^m$ is a closed convex cone, and $Int\ P \neq \emptyset$

$Q \subset E_+^s$ is a closed convex cone, and $Int\ Q \neq \emptyset$

When P and Q are polyhedral cones given by the "intersection-form", that is $A_{p_1 \times m} \omega \le 0_{p_1}$ defines an input cone P and $B_{p_2 \times s} \mu \le 0_{p_2}$ defines an output cone Q, where $p_1 =$2(m-1) and $p_2 =$2(s-1).

In the cone-ratio DEA model, assurances are specified by defining bounds on weights that reflect the relative importance of different inputs and outputs. These weight constraints are represented in the form of linear inequalities and appended to the ratio DEA model. This model is solved for each unit j_0 to determine its DEA efficiency. Theoretically, it is known that units, which receive full efficiency score 1.0, are on the Pareto-Koopman efficient frontier defined by the data, and that the DEA weights define hyperplanes tangent to the frontier. These weights are therefore not necessarily useful as measures of the relative importance of the respective inputs and outputs. By changing the DMU under evaluation for n times, we may gain the DEA efficiency given to DMU_j ($j = 1,2,\cdots,n$.

3. Results

In order to have a comparison of model power, four DEA models are used to compute four groups of scores. The four DEA models are CCR model, common cone-ratio DEA, super-DEA and cone-ratio super-DEA. For inefficient DMUs the super-efficiency score coincides with the standard score defined above. For efficient DMUs a score is computed which indicates the maximal radial change which is feasible such that the DMU remains efficient. Formally, it is defined like the standard score but the DMU under evaluation is excluded from the constraints (i. e. the definition of the technology set). See Andersen and Petersen (1993).

In this research, we have 110 DMUs (working schedules) which are carefully proposed by the managers in practice and each DMU has 5 inputs and 3 outputs which are from the simulation step. Results show that the number of the efficient DMUs will decrease if the decision-makers' preference information is reflected in the DEA model. Ranking order given by the common cone-ratio DEA is the same as that by the cone-ratio super-DEA. We also demonstrate the benchmarks by the common cone-ratio DEA.

Acknowledgments

The authors wish to thank the anonymous referees for their careful reading of the manuscript and their constructive comments and suggestions.

References

[1]   P. Andersen and N. C. Petersen (1993), A Procedure for Ranking Efficient Units in Data Envelopment Analysis, Management Science, 39, 1261–1264.

[2]   N. Balakrishnan, R.T. Wong, A network model for the rotating workforce scheduling problem, Networks 20 (1990) 25–42.

[3]   A. Charnes, W.W. Cooper, and E. Rhodes, 1978, Measuring the efficiency of decision making units, European Journal of Operational Research 2, 429-444.

[4]   A. Charnes, W.W. Cooper, Z.M. Huang, and D.B. Sun, 1990, Polyhedral cone-ratio DEA models with an illustrative application to large commercial banks, Journal of Econometrics 46, 73-91.

[5] W.W. Cooper, L.M. Seiford and K. Tone, 2000, Data Envelopment Analysis: A Comprehensive Text with Models, Applications, References, Kluwer Academic Publishers, Boston.

[6]   F. Glover, C. McMillan, The general employee scheduling problem: an integration of MS and AI, Comput. Oper. Res. 13 (5) (1986) 563–573.

[7] G. Laporte, The art and science of designing rotating schedules, J. Oper. Res. Soc. 50 (1999) 1011–1017.

[8] A. Schaerf, A. Meisels, Solving employee timetabling problems by generalized local search, in: Paper presented at AI*IA'99, Bologna, Italy, 1999.

[9] B.M. Smith, S. Bennett, Combining constraint satisfaction and local improvement algorithms to construct anaesthetist's rotas, in: Proc. CAIA-92, 8th IEEE Conf. on Arti>cial Intelligence Applications, 1992, Monterey, CA, pp. 106–112.

VSP International
Science Publishers
P.O. Box 346, 3700 AH Zeist
The Netherlands

*Lecture Series on Computer
and Computational Sciences*
Volume 1, 2004, pp. 714-716

Isometry Algorithm for the Plane Loop in Shape Machining

Wei Jiang [1], Changan Zhu[2], Yi Zhang[3], Xiaoqiang Zhong[4], Desheng Wu[5]

[1,2,3,4]Department of Precision Machinery and Precision Instrumentation,
University of Science and Technology of China, Anhui Hefei, China
[5]School of Business, University of Science and Technology of China, Anhui Hefei, China

Received 17 March, 2004; accepted in revised form 10 Aug. 2004

Extended Abstract

To plan the tool path in the CAD /CAM system, we generally adopt the cutter center as the input point, but the real cutter contact locates the cutter end, that is the edge of the cutter. As a result, it is necessary to take the isometry operation on the contour of the parts projected to some surface while machining a curved surface. So the isometry algorithm to a complicated contour loop is one of the important and basic algorithms in the CAD /CAM system, whether the algorithm is perfect or not determines the basic performance of CAD /CAM system. The paper presents a new isometry algorithm to the complicated contour loop. We have adopted the algorithm in our IE-CAM software and received a good effect. Fig.1 shows the method of prevention chopping leakage in our IE-CAM software.

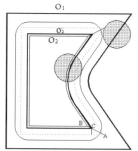

Fig.1. The method of prevention chopping leakage in our IE-CAM software

The paper illustrates the algorithm in detail by six sections.

(1) Several basic conceptions on shape CL Engine: what are the cutter location and the cutter contact, and how to describe the cutter location.

(2) Managing to deal with the isometry loop: In the CAD /CAM system, the contour loop is usually marked as a direction loop, which is helpful to distinguish the inner contour loop from the outer one. Once marking the contour loop as a vector loop, we can conveniently control the inner or outer isometry operation by setting the symbol of isometry parameters. There are splits when taking the isometry operation on the original contour, and we should deal with the different forms of splitting accordingly. That is to construct two parameter lines $L^i = P_e^i + \lambda^i \vec{T}_e^i$ and $L^{i+1} = P_s^{i+1} + \lambda^{i+1} \vec{T}_s^{i+1}$, and calculate the parameter λ^i and λ^{i+1} at the intersection point. According to the different symbols of λ^i and λ^{i+1}, we can sum up three

---
[1] Corresponding author. Wei Jiang   Changan Zhu   Yi Zhang   Xiaoqiang Zhong   Desheng Wu   E-mail: jiangw@ustc.edu
changan@ustc.edu.cn , zyzy@mail.ustc.edu.cn

situations (that are $\lambda^i < 0, \lambda^{i+1} > 0$, $\lambda^i > 0, \lambda^{i+1} < 0$ and $\lambda^i \lambda^{i+1} > 0$) and deal with them respectively.

(3) Dealing with self-intersection isometry loops: To take isometry operation on a closed contour, it possibly results of self-intersection, that is to say, there occurs several least closed loops. When self-intersection happens with the isometry loop, firstly we make subsections with the self-intersection contour in the isometry loop according to the self-intersections, and pick up information of the new closed loop, then eliminate the loops whose orientation is opposite to that of the original loop by judging the orientation of them. Also it's necessary to point out that it is easy to judge the orientation of a protruding closed loop, the only thing needed is to calculate the intersection product of the vectors of two arbitrary bordered contour segments, and judge the symbol of the result. While it is more complex with a concave one, and we adopt the corresponding algorithm.

(4) Dealing with mutual-intersection of the isometry loops: A common case is that a real part has mould pocket, inside which there are more than one islands. while taking the isometry operation on the mould pocket and the contour of the island at one time, chances happen on the intersection of the isometry, which are of the mould pocket and the island, or two islands. In this case, it needs to have an aggregation operation on the intersection closed loops. In order to do that efficiently, we let the outer loop be counter clock wise (CCW), in reverse clock wise (CW). Also, we attach the each intersection point of the two direction lines a denotation eigenvalue except the geometry coordinate, which corresponds to the geometry of the intersection element.

(5) The ending condition of the isometry operation: All the inside contour will dwindle after the isometry operation, even if the island contour increases, it will join with the outer contour and decrease in the end. Hence, the operation must be stopped when the result of the isometry operation on the inner contour reduces to a point. However, the common case is not reducing to a point after the inner contour loop isometry operation but an unexpected contour loop. So it's unmeaningful to have the isometry operation based on that. We can find the condition of ending the isometry operation, that is, the orientation of the isometry loop is opposite to that of the original loop.

(6) The examination and prevention of chopping leakage: The cutter removes the remainder along the isometry loops of the contour-parallel isometry machining on the 2D mould pocket. In this process, there will be chop leakage in milling. If there occurs interspace between the wrapped line of the cutter feeding along two arbitrary bordered isometry line, finally it results of some unexpected islands. In IE-CAM pocket machining, we adopt a new method to examine and prevent the chopping leakage.

Acknowledgments

The research is supported by the Professor Changan Zhu and Chuanqi Li, through the National Research Lab Program.

References

[1] K. Morishige, Y. Takeuchi, K. Kase. *Tool Path Generation Using C-Space for 5-axis Control Machining*, Transaction of the ASME, 1999, 121(2): 144-149

[2] Chih-Ching Lo. *Two-stage cutter-path scheduling for ball-end milling of concave and wall bounded surfaces*. Computer-Aided Design, 2000, 32: 597-603

[3] C-C. Lo. *Efficient cutter-path planning for five-axis surface machining with a flat-end cutter.* Computer-Aided Design, 1999, 31: 557-566

[4] Yuan-Shin Lee, Bahattin Koc. *Ellipse-offset approach and inclined zig-zag method for multi-axis roughing of ruled surface pockets*. Computer-Aided Design, 1998, 30(12): 957-971

[5] Yong Seok, Kunwoo Lee. *NC milling tool path generation for arbitrary pockets defined by sculptured surfaces*. Computer-aided Design, 1990, 22(5): 273-284

[6] S.C. Park, B.K. Choi. *Uncut free pocketing tool-paths generation using pair-wise offset algorithm.* Computer-aided Design, 2001, 33: 739-746

[7] Daoyuan Yu, Jianling Zhong, Zhuang Xiong. *The Computational method of the cutter path planning in space surface NC and its problem.* Mechanic Automation, 1997，19(1): 21-27

[8] Solid Edge 2000 API Online Help

[9] SolidWorks 2001 API Online Help

VSP International
Science Publishers
P.O. Box 346, 3700 AH Zeist
The Netherlands

*Lecture Series on Computer
and Computational Sciences*
Volume 1, 2004, pp. 717-721

A Strategy of Optimizing Neural Networks by Genetic Algorithms and its Application on Corporate Credit Scoring

Desheng Wu[1*], Liang Liang[1], Hongman Gao[2] and Y. Li[3]

[1]School of Business,University of Science and Technology of China,
230026 Jinzhai Road Hefei, PR China
[2] School of Business, University of Mississippi,USA
[3] School of Business, University of Alberta, Edmonton, AB, Canada T6G 2R6

Abstract: When the ordinary backpropagation neural network is trained, the problem of seriously local minimum often occurs. This makes net-training unfinished and weights distributed in the network immature. The current paper proposes a strategy of optimizing neural networks by genetic algorithm to deal with this problem. If the initially decided conditions (including training time and error precision) cannot be satisfied, i.e., the neural networks settle in local minimum, the unaccomplished network turns to be optimized by the genetic algorithm through adjusting the immature weights. Then the evolutionary weights which are believed to grasp the property of the sample construct a new network model which shows strong discriminant power. By use of the proposed evolutionary strategy, a credit scoring model applied to Chinese business corporations is developed and the scoring power of the model is tested in this study. Finally conclusions are presented.

Keywords: Corporate credit scoring; backpropagation neural networks; genetic algorithms; weight

1. Introduction

The separation of ownership and management poses the problem of potential conflict between the shareholders, i.e. the owners of the firm, and the managers who run it. By exposure of corporate information quickly and accurately, credit analysis is viewed as a good way to assist to solve the problem of moral hazard and adverse selection. It is widely believed that the neural network is an accurate technique to score credit conditions[1−6]. Since Odom firstly used neural networks to analyze credit risk in 1990, the application of the neural network in this field has been attached much importance by the applicants and researchers [1]. Coats and Fant contrast a neural network with linear discriminant analysis for a set of cases labeled viable or distressed obtained from COMPUSTAT for the period 1970-1989 [2]. They find the neural network to be more accurate than linear discriminant analysis. Tam and Kiang studied the application of a BP neural network model to Texas bank failure prediction for the period 1985-1987[3]. The neural network prediction accuracy is compared to linear discriminant analysis, logistic regression, *k* nearest neighbor, and a decision tree model. Their results suggest that the multilayer perceptron is most accurate. Altman et al. employs both linear discriminant analysis and a neural network to score corporate credit for 1000 Italian firms. The authors conclude that neural networks are not a clearly dominant mathematical technique compared to traditional statistical techniques such as discriminant analysis [4]. Piramuthu studies credit scoring of Belgium corporation, comparing a BP neural network and a neural-fuzzy model, preprocessing the data sample by primary component analysis(PCA) [5]. Mu-Chen and Shih-Hsien Huang construct a NN-based credit scoring model, which classifies applicants as accepted (good) or rejected (bad) credits. The GA-based inverse classification technique is used to reassign rejected credits to the preferable accepted class[6].

All of the models of these studies are based on special variables and data set of their relative countries. Thus, the adaptation to Chinese conditions of the results for assessing distress potential of

---

[1*] Corresponding author.. E-mail: dash@ustc.edu

firms is questionable because the representative credit ratios and data set are different from country to country and even from company to company. Hence, it is critical to do experiential research on credit scoring with variables and data set of Chinese firms. A few scholars in China such as Baoan Yang, Hai Ji and Chende Lin[8] [17], dipped into the fields. However, firm-specific factors, macroeconomic trends, and industrial factors are omitted from the models, as well as the decision-maker's individual influence from experience, wisdom and information priority. Thus, the results of these studies are not generalizable due to these reasons as well as the data accuracy of their samples. Baoan Yang, Hai Ji et. al.[8] suggested a BP model to diagnose credit risk conditions with a set of data from the credit bank of Suzhou city in China ,which were not selected carefully, namely, their study is not real experiential research. Similar to those of [8], cursory demonstration results by use of neural networks and comparison with some other methods are given in [17]. The tactic to deal with problems such as local minimum, which are commonly met, is seldom detailed in these previous researches.

In order to overcome the suffering from data redundancy, it is always necessary to preprocess the data by primary component analysis[2-5]. But this may causes seriously local minimum, which will be found in this study. In this research, after a preprocessing technique by primary component analysis, an evolutionary layer is suggested by the authors to optimize the BP neural network. In the evolutionary layer, genetic algorithm is used to assist training the network by adjusting the weights and biases. Thus the deficiency of serious local minimum is well dealt with. Whilst the research work was undertaken in the field of credit scoring, many, if not all, of the procedures and findings described may be applicable to other areas of classification and trend analysis.

2.   Variable Selection and Sample Design

We refer to the index system that is currently used in domestic relative industry. Then the representative variables employed in this study are zeroed in on four areas of different businesses that may need attention: variables estimating profitability(V1,V2,V3), variables estimating operational efficiency(V4,V5), variables estimating solvency power (V6,V7,V8)and variables estimating development ability(V9,V10) which are paid attention by three different principal parts(shown in Table 1).So these variables are comprehensive to account for corporate credit conditions.

Table 1: Corporate credit scoring variables

Total-object	Sub-0bject	variables	Principal caring part
Corporate credit scoring	profitability	Earnings per stock(V1)	Examined carefully by the investors
		Return on assets(V2)	
		Net operating profit rate of return(V3)	
	operational efficiency	Receivables turnover(V4)	Examined carefully by the loaner
		inventory turnover(V5)	
	solvency efficiency	total-debts-to-total-assets(V6)	
		Current ratio (V7)	
		Acid-test ratio(V8)	
	Development ability	market domain(V9)	Examined carefully by the nation
		Final-equity-to-initial-equity (V10)	

In the process of collecting data, the key problems are the representation and reliability of sample as well as sample errors. So the case companies are all public companies from the Chinese business industry in 1996. 30 case companies are carefully selected, balance sheet and income statement data as well as specialist scoring values are collected. The modeling data set are composed of 24 case companies, testing data set composed of the other 6 case companies.

3.   Model

Generally speaking, the whole discriminating process mainly follows four stages: pre-processing stage, network training stage, model testing stage, i.e., credit scoring stage and post-processing stage. Specially, in the network training stage, we propose an evolutionary network to improve the learning process. The so-called evolutionary network is a kind of neural network combing the GA to deal with local minimum. When the neural network is being trained, the problem of local minimum often occurs and makes the training of the neural network unfinished. Then the initially decided precision cannot be achieved and the unaccomplished network constructed by the immature weights cannot effectively grasp the genuine property of cases being scored. Evidently, weights need to be further optimized as well as biases. This is the focus of the current study. Figure 1 plots the scoring flow by the evolutionary network. In figure 1, it is shonwn that if the initially decided conditions (including training time and error precision) cannot be satisfied, as is said in the rhombus, the unaccomplished network turns to be optimized by GA through adjusting the immature weights. On the contrary, if the initially decided conditions are satisfied (either after GA optimization or before it), the scoring can be executed and results can ten be yielded.

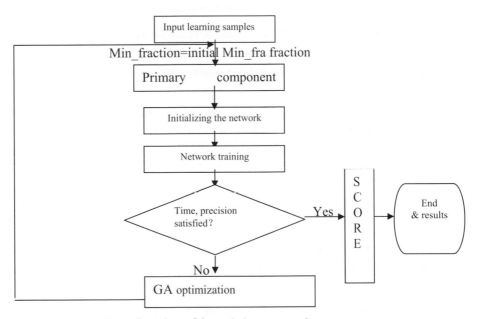

Fig.1: flow chart of the evolutionary network

Step 1-6 give the outline of the GA used in the evolutionary network as follows.

Step 1: when the network set in a local minimum, it enters the evolutionary layer following our algorithm and randomly generates an initial population of chromosomes limited within the interval of maximal and minimal weight values (these represent our trial weights, and are normally encoded as a parameter binary string).
Step 2: transfer the binary population to weights population and evaluate each chromosome to establish its fitness (success at solving the problem).
Step 3: select pairs of individuals for recombination based on their fitness.
Step 4: combine the individuals, using genetic operators such as crossover and mutation.
Step 5: create a new population from the newly-produced individuals.
Step 6: repeat steps 2 through 5 until the satisfactory solution is found,namely, optimized weights and biases are found.

4. Summery and conclusion

The current paper mainly focuses on the validation of a strategy of combining the BP neural networks and genetic algorithms on corporate credit scoring. This strategy seeks to deal with local minimum occurred during net-training. By use of the strategy, the unaccomplished network will turn to be optimized by the genetic algorithm through adjusting the immature weights if the initially decided conditions (including training time and error precision) cannot be satisfied, i.e., the neural networks settle in local minimum. Then the evolutionary weights which are believed to grasp the property of the sample construct a new neural network model named as the evolutionary neural network. It is demonstrated to be better than ordinary BP neural networks in scoring credit conditions. Also, the high accurate rate of the supposed model validates the strong scoring power of the 10 variables in this research. Whilst the research work was undertaken in the field of credit scoring, many, if not all, of the procedures and findings described may be applicable to other areas of classification and trend analysis.

References

[1] Odom, MD, and Sharda, R. A Neural Network Model for Bankruptcy. Prediction. Proceedings of the IEEE International Joint Conference on Neural Networks. 1990 ,2: 163-168

[2] Coats P, Pant L . Recognizing financial distress patterns using a neural network tool.Financial Management.1993:142-155

[3] K. Tam and M. Kiang, Managerial applications of the neural networks: The case of bank failure predictions. Management Science, 1992, 38:416–430

[4] E. Altman, G. Marco, and F. Varetto. Corporate distress diagnosis: Comparisons using linear discriminant analysis and neural networks. Banking and Finance. 1994, 18:505–529.

[5] Piramuthu, S. Financial credit-risk evaluation with neural and neurofuzzy systems. European Journal of Operational Research.1999,112(2): 310-321

[6] Mu-Chen Chen, Shih-Hsien Huang. Credit scoring and rejected instances reassigning through evolutionary computation techniques. Expert Systems with Applications. 2003(24) :433–441

[7] Jinmei Wu. Credit Analysis. Audit publishing company of China.2001

[8] Baoan yang，Hai Ji. A Study of Commercial Bank Loans Risk Early Warning Based on BP Neural Network. Systems Engineering——Theory and Practice 2001, 5

[9] Brill, J. The importance of credit scoring models in improving cash flow and conditions. Business Credit. 1998 (1): 16–17.

[10] J. E. Baker J. Adaptive Selection Methods for Genetic Algorithms.Proc. ICGA 1. 1985:101-111

[11] Hecht Nielsen R. Neural computing. Addison Wesley,1990;124-133

[12] Matlab. Matlab Users Guide. The Mathworks Inc,South Natick,MA,USE,1994

[13] K. A. De Jong. Analysis of the Behaviour of a Class of Genetic Adaptive Systems, Dept. of Computer and Communication Sciences,University of Michigan, Ann Arbor, 1975.

[14] J. E. Baker. Reducing bias and inefficiency in the selection algorithm. Proc. ICGA 2, 1987:14-21

[15] L. Booker.Improving search in genetic algorithms. In Genetic Algorithms and Simulated Annealing, L. Davis (Ed.). Morgan Kaufmann Publishers .1987:61-73

[16] H. Mühlenbein and D. Schlierkamp-Voosen. Predictive Models for the Breeder Genetic Algorithm Evolutionary Computation.1993,1(1):25-49

[17] Chen Xionghua , Lin Chengde,YE Wu.Credit risk assessment of enterprise basing on neural network. Journal of Systems Engineering. 2002 ,17(6).570-576.

[18] Liang Liang, Desheng Wu. An Application of Pattern Recognition on Scoring Financial Conditions Based on Backpropagation Neural Network. Computers & Operations Research. under publication.

VSP International
Science Publishers
P.O. Box 346, 3700 AH Zeist
The Netherlands

*Lecture Series on Computer
and Computational Sciences*
Volume 1, 2004, pp. 722-725

The Valuation of Quanto Options Under Stochastic Interest Rates

Li ShuJin[1], Li ShengHong[2], Desheng Wu[3]

[1,2] Department of Mathematics,
University of ZheJiang,
310027 ZheJiang Province, HangZhou

[3] School of Business, University of Science and Technology of China,
230026 Hefei, PR China
dash@mail.ustc.edu.cn

Extended Abstract

The quanto option is a contingent claim whose investor has to consider to hedge the risk from both the foreign stock price and exchange rate simultaneously. This paper discussed two cases of quanto options: (1) an option of floating exchange rate, written on foreign stock in terms of foreign currency, whose value is transformed to domestic currency value by exchange rate at maturity time; (2) an option of fixed exchange rate, by which the value of this option is transformed to domestic currency value at maturity time.

Quanto options were introduced by Reiner in 1992. When he or she invests foreign stock, investor is not only concerned about the risk of foreign stock price but also that of the exchange rate. Therefore, the investor demands for controlling the risk of foreign stock price and exchange rate simultaneously due to the unfavorable changes in the shapes of yield curves, or avoid one risk.

There are a number of author who have discussed the pricing of options on interest rates, for instance, Flesaker (1991) and Viswanathan (1991) have derived a closed-form solution for a European option on an interest rate using various of the one-factor Heath et al. (1992) model. Dravid et al. (1993) considers to price foreign index contingent claims and Reiner (1993) introduces the price of the quanto option under deterministic interest rate. To our knowledge, there are few literatures about the quanto options pricing under stochastic domestic and foreign spot rates.

We consider a financial market operating in continuous time and described by a probability space $(\Omega, \Im, P)$, a time horizon T and a filtration $\{\Im_t\}_{0 \le t \le T}$, representing the information available at time t. One central problem in financial mathematics is the pricing and hedging of contingent claims by means of dynamic trading strategies based on S, which is the asset price for trading. We assume the market is complete. That is, in this market, there are neither transaction costs nor taxes, borrowing and short-selling are allowed without restrictions, and borrowing rate is the same as the lending rate. A European contingent claim is a $\Im_T$-measurable random variable describing the payoff at time T of some derivative security.

To address the importance of correlation, this paper employs a multi-factor valuation model. To begin with, we assume that given the initial curves, $f(0, \tau)$ and $g(0, \tau)$, $\forall \tau$, the domestic and foreign forward rate, $f(t, T)$ and $g(t, T)$, and the foreign stock price, $S(t)$, and the exchange rate, $C(t)$, follow the following diffusion process of the form under the measure P:

$$d_t f(t, T) = \alpha(t, T) dt + \sigma_d(t, T) dW_1(t)$$

$$d_t g(t,T) = \beta(t,T)dt + \sigma_f(t,T)dW_2(t)$$

$$dS_t = S_t\left[\mu_S(t)dt + \sigma_S(t)dW_3(t)\right]$$

$$dC_t = C_t\left[\mu_C(t)dt + \sum_{i=1}^{4}\rho_i\Gamma(t)dW_i(t)\right]$$

where $W(t) = \left(W_1(t), W_2(t), W_3(t), W_4(t)\right)$ is a 4-dimensional standard P-Brownian motion, and $\alpha(t,T)$, $\beta(t,T)$, $\mu_S(t)$ and $\mu_C(t)$ are the drift coefficients of the domestic and foreign forward rate, $f(t,T)$ and $g(t,T)$, and the foreign stock price, $S(t)$, and the exchange rate, $C(t)$, respectively, and $\rho_i, i = 1,2,3,4$ are constants satisfying

$$\rho_1^2 + \rho_2^2 + \rho_3^2 + \rho_4^2 = 1$$

Furthermore, $\sigma_d(t,T)$, $\sigma_f(t,T)$, $\sigma_S(t)$, and $\Gamma(t)$ are their volatilities, respectively.

It is easy to testify that under the measure P,

$$\rho_1 W_1(t) + \rho_2 W_2(t) + \rho_3 W_3(t) + \rho_4 W_4(t)$$

is also a standard P-Brownian motion.

In section 2 of this paper, we find an equivalent martingale measure, Q with P, i.e., under which all processes

$$X_t = B_d^{-1}(t)P(t,T)$$

$$Y_t = B_d^{-1}(t)C_t Q(t,T)$$

$$V_t = B_d^{-1}(t)C_t B_f(t)$$

$$Z_t = B_d^{-1}(t)C_t S_t$$

are Q-martingales, where

$$B_d(t) = \exp^{\int_0^t r_d(u)du}$$

$$B_f(t) = \exp^{\int_0^t r_f(u)du}$$

$$P(t,T) = \exp^{-\int_t^T f(t,u)du}$$

$$Q(t,T) = \exp^{-\int_t^T g(t,u)du}$$

In effect, in order to find an equivalent measure Q with P, we find a previsible vector

$$\gamma(t) = \left(\gamma_1(t), \gamma_2(t), \gamma_3(t), \gamma_4(t)\right),$$

called market price of risk, and a new measure Q under which

$$\tilde{W}(t) = \left(\tilde{W}_1(t), \tilde{W}_2(t), \tilde{W}_3(t), \tilde{W}_4(t)\right)$$

is a new 4-dimensional standard Brownian motion, where

$$\tilde{W}_i(t) = W_i(t) + \int_0^t \gamma_i(v)dv, \ i = 1,2,3,4$$

At last, we find out the martingale measure, Q, under which domestic and foreign forward rate, $f(t,T)$ and $g(t,T)$, foreign stock price, $S(t)$, and the exchange rate, $C(t)$, satisfy the following diffusion processes, respectively:

$$d_t f(t,T) = \sigma_d(t,T)d\tilde{W}_1(t) - \sigma_d(t,T)\Sigma_d(t,T)dt$$

$$d_t g(t,T) = \sigma_f(t,T)d\tilde{W}_2(t) - \sigma_f(t,T)\left(\rho_2\Gamma(t) + \Sigma_f(t,T)\right)dt$$

$$dS_t = S_t\left[\sigma_S(t)d\tilde{W}_3(t) + \left(r_f(t) - \rho_3\sigma_S(t)\Gamma(t)\right)dt\right]$$

$$dC_t = C_t\left[\left(r_d(t) - r_f(t)\right)dt + \sum_{i=1}^{4}\rho_i\Gamma(t)d\widetilde{W}_i(t)\right]$$

In section 3 of this paper, we discuss the two cases of quanto options: (1) an option of floating exchange rate;

The payoffs of the call options at maturity time, T, is

$$V_{QC}^1(T) = C_T \max\{S_T - K, 0\}$$

In fact, its payoffs at maturity time, T, is the one of the call options computed by means of foreign currency, then transformed to domestic currency by exchange rate. Therefore, the investor who invests the quanto option attaches importance not only to the risk of foreign stock price but also to that of exchange rate; (2) an option of fixed exchange rate; The payoffs of the call option at maturity time, T, is

$$V_{QC}^2(T) = \overline{C}\max\{S_T - K, 0\}$$

where K is strike price in term of foreign currency and $\overline{C}$ is pre-agreed exchange rate by which the value of the call options is transformed to domestic currency value at maturity time in the second case.

Making use of martingale theory, we derive the close-form pricing formulas of call quanto options and put-call parity formulas under stochastic and foreign interest rates in a complete market, respectively. Furthermore, the pricing formulas of put quanto options are obtained.

Our main result in this paper is

Theorem 1 The price of the first quanto call option at time 0, denoted by V_{QC}^1, whose exercise price is K (foreign currency value) and maturity time is T is

$$V_{QC}^1 = C_0 S_0 N\left(d_1 + \frac{1}{2}\sigma\right) - C_0 KQ(0,T)N\left(d_1 - \frac{1}{2}\sigma\right)$$

where $N(.)$ is the cumulative probability function for a standard normal stochastic variable, and

$$d_1 = \frac{-\ln Q(0,T) + \ln S_0 - \ln K}{\sigma}, \quad \sigma^2 = \int_0^T \Sigma_f^2(v,T)dv + \int_0^T \sigma_S^2(v)dv$$

Theorem 2 (parity formula) At time 0 the price of the first quanto put option, denoted by V_{QP}^1, and the price of call one, denoted by V_{QC}, which have same strike price K and maturity time T, have the following relation:

$$V_{QC}^1 - V_{QP}^1 = C_0 S_0 - KC_0 Q(0,T)$$

Furthermore, we can get the price of the first quanto put option at time 0 is

$$V_{QP}^1 = C_0 KQ(0,T)N\left(-d_1 + \frac{1}{2}\sigma\right) - C_0 S_0 N\left(-d_1 - \frac{1}{2}\sigma\right)$$

where $N(.)$, d_1 and σ refer to theorem 1.

Theorem 3 The price of the second quanto call option at time 0, denoted by V_{QC}^2, whose exercise price is K (foreign currency value) and maturity time is T and pre-agreed exchange rate is $\overline{C}$ is

$$V_{QC}^2 = \overline{C}S_0 \frac{P(0,T)}{Q(0,T)}\exp^b N\left(d_2 + \frac{1}{2}\sigma\right) - \overline{C}KP(0,T)N\left(d_2 - \frac{1}{2}\sigma\right)$$

where $N(.)$ and σ refer to theorem 1 and

$$b = \int_0^T \Sigma_f^2(v,T)dv + \int_0^T \left(\rho_2\Sigma_f(v,T) - \rho_3\sigma_S(v)\right)\Gamma(v)dv$$

$$d_2 = \frac{-\ln P(0,T) + \ln C_0 S_0 - \ln K + b}{\sigma}$$

Theorem 4 (parity formula) At time 0 the price of the second quanto put option, denoted by V_{QP}^2, and the price of call one, denoted by V_{QC}^2, which have same strike price K and maturity time T, have the following relation:

$$V_{QC}^2 - V_{QP}^2 = \overline{C} S_0 \frac{P(0,T)}{Q(0,T)} \exp^b - \overline{C} K P(0,T)$$

Furthermore, we can get the price of the second quanto put option at time 0 is

$$V_{QP}^2 = \overline{C} K P(0,T) N\left(-d_2 + \frac{1}{2}\sigma\right) - \overline{C} S_0 \frac{P(0,T)}{Q(0,T)} N\left(-d_2 - \frac{1}{2}\sigma\right)$$

where $N(.)$, d_2 and σ refer to theorem 3.

Only if there are formulas of the quanto options, we may know how to hedge risk from both foreign stock price and exchange rate and foreign and domestic spot interest rate. In this paper, we won't discuss hedging problem duo to the length of the article.

VSP International
Science Publishers
P.O. Box 346, 3700 AH Zeist
The Netherlands

*Lecture Series on Computer
and Computational Sciences*
Volume 1, 2004, pp. 726-729

Evaluation of the Collected Pages and Dynamic Monitoring of Scientific Information

Yanping ZHAO[†] and Donghua ZHU[‡]

School of Management & Economics, Beijing Inst. Of Tech.（Beijng 100081）

Abstract: This paper deals with the problems faced in gathering and analyzing of large amount of topic-specific scientific information for enterprises, governments or militaries. We propose a framework for efficient use of collected papers and dynamically monitoring the changes occurred in the specific-topic. A composite index is built for ranking the collected scientific papers with objective and authority criteria. The index can be automatically updated according to computer collected parameters for their changes or for the user demand. An assisted module for detecting the changes are discussed. Initial test show that our prototype is promising, and has potential value in many information gathering and retrieval application prospects.

Keywords: technology forecasting, scientific information gathering, PageRank, dynamic algorithm.

1. Introduction

The fast development makes scientific and engineering literature grow explosively, all traditional analysis of information entirely depend on human experts to complete will become impossible and may cause delay in decision making or technology information assessment. So some intelligent methods to assist human experts to do analysis work, access and process varieties of topics of scientific information are becoming more welcome and face challenges. The challenges are:

(1) New technology intercrosses with many other research areas.

(2) The fast the speed of change and the large amount of scientific information.

(3) How to make the collected information possess authoritative, relevance, and freshness, how to make the collection process efficient and guarantee the precision and comprehensive.

(4) The published ones are often in modifications, and new hot spots emerge at any time and somewhere. How to track them.

Therefore, we make some contributions and introducing some intelligent methods to partly solve these problems, and propose a methods for arranging the collected information.

This paper is organized as follows, section 2, the related works; section 3, methods and techniques for establishing dynamic ranking and updating models; section 4 construction of a dynamic monitoring module; section 5 initial results and future works.

2. Related Works

Recently, the research search engines pages rank and for topic-specific page gathering has grown rapidly. Many new methods and systems have been developed.

Chakrabarti, (1999) proposed focused crawling concept and build a system with Bayesian classifier as the target documents selector.

Aggarwal et al.(2001) have proposed a statistical learning model for crawlers to discover topic-specific resources.

Kleinberg (1998) proposed HITS algorithm, which can provide hub pages and authorities. But the method is for a general search engine and need initial large amount of document collections and intensive iteration to calculate.

Page L. et al. (1998) presented a popular algorithm called PageRank makes it very successful.

[†] E-mail: zhaoyp@sun.ihep.ac.cn
[‡] E-mail: zhudh111@sohu.com

And Haveliwala, (1999) also provides an efficient algorithm.

Recently, Taher H. Haveliwala (2002) proposed a topic-sensitive PageRank approach, by computing a set of PageRank vectors, each biased with a different topic, which can generate more accurate rankings than with a single, generic PageRank vector. And gave some suggestions for improving the results further.

Nowadays, many authors or groups supply the on-line service for the tracking changes.

AT&T 's Internet Difference Engine(AIDE), a system, can automatically compare the two pages before and after changes, then construct a "incorporate" page, to tell the specific change.

There are also systems like Netmind and so on, can detect the new pages containing the keywords user given.

Khoo Khyou Bun et al.(2003) proposed a method which can summarizes the changes in the page, the advantage of it is not only to tell user the newly hot spot, but also to show the newly emerged topic in the specific domain.

We use and modify those models to give the topic change or shifting, and arrange the collected papers for efficient use.

3. Dynamic Ranking And Updating Models

Users are not interested in all collected papers, but the pages relevant or deemed important. Moreover, papers from different tunnels will have different ranking policies, they couldn't be comparative to each other. So we have to find ourselves own ranking measures as an overall equity scores for all collected pages and also provide different profiles of them. We considered four factors:

(1) The context or semantic measure;

(2) The Link structure(which means citations) measure;

(3) The official statistical data for authorities(such as survey paper, top rated author(s), organization(s), journal(s));

(4) Freshness and others (such as page age or publication date).

Those factors are for the prioritized pages ranked in the top list if there are large amount of available material of relevance.

3.1 The Context Or Semantic Measure

In feature selection stage, one of the major problems is the treatment of terminology, which represents documents domain specific knowledge or concepts. Therefore, the selection of target features is very important for collecting relevant documents. Here we take the advantage of the keywords from authoritative scientific citation thesaurus, such as EI or INSPECT etc. to make them most representative and efficient for specific scientific topic. If user couldn't supply accurate search terms, our system has a learning ability for expending the co-occurrence terms for the suitable target feature generation. And if for Chinese documents, Chinese Journal Networks (CNKI) will take the place of EI, Chinese word group(s) will be extracted from the CNKI database, and for Chinese document analysis, a NLP module uses a Chinese segmentation program, ICTCLAS system from the Institute of Computing Technology of Chinese Academy of Science, to parse the co-occurrence terms from Chinese character documents. Then each collected page have its IS(P) greater than a threshold will be saved.

3.2 The Link Structure Measure

Since collected papers only have similarities to the target, but can not tell which paper is more authoritative than the others. From scientometrics, the more citations to one paper, the more authoritativeness of the paper. So we introduce IP (P) based on famous Google's PageRank algorithm with personalized (Topic-specific) feature. The efficient PageRank algorithm with our personalized initialization uses the following formula.

$$Rank = d\,M\,Rank + (1\text{-}d)\,Per \tag{1}$$

Where Rank is the $(PR(P_1),\ PR(P_2),...,PR(P_n))^{\tau}$ vector, M is the link graph matrix with m_{ij} in M as $1/n(P_j)$, if Page P_j pointing to P_i, otherwise m_{ij} is 0. *Per* is the Personalized initial vector, with *Per* = $[1/N]_{N \times 1}$ for high quality computation.

Here we introduce a dynamic approach: to modify the Per after one round of the collection, and Let the initial Per be the real scores got from cited pages, which our system can pick automatically from the web, such as the "cited by" number of the page from NEC's virtual Library or the Web of Science because they are authority scientific databases. And the initial vector is standardized to the PageRank initial condition as the sum of the components of the vector is equal to 1.

Since the PageRank scores is computed once and saved with the titles, so this kind of work is not overburden for the system by periodically updated.

3.3 The Official Statistical Data For Authorities

For this case, we use statistical data as real scientific citations, Journal Impact Factors from JCR or others relevance ranks from the searched scientific databases, and the survey papers and influencing authors or affiliations on the top list feed back from the our monitoring module, All these are collected and associated with the paper title. And can be combined together with weighted indicators. For comparability of the contributions among them, consideration such as making them normalized and use weights like:

$$IA(P) = \beta_1 I_{Citation}(P) + \beta_2 I_{Survey}(P) + \beta_3 I_{Author}(P)$$
$$+ \beta_4 I_{Journal}(P) + \beta_5 I_{Affil}(P) \tag{2}$$

3.4 The Indicator of the Freshness

For this factor, we use the date of the paper publication, the Index as

$$IF(P) = 1 - \frac{published\ \ year - 1990}{Current\ ye\ ar - 1990} \tag{3}$$

3.5 The Composite Indicator for Importance

We make the 4 indicators normalized for comparability and with the consideration of simplicity and fast in computation. Also make the indicator give some profiles of the collected papers.

$$IC(P) = \alpha_1 IS\ (P) + \alpha_2 IP\ (P) + \alpha_3 IA\ (P) + \alpha_4 IF(P) \tag{4}$$

Perhaps some documents may not have scores of some of the Indexes such as IP(P) or IA(P), we assume the corresponding indicators as zero, at least they have IS(P) for their contents since they are similar to the topic.

4. Dynamic Monitoring Module

We modified an approach from Adam Jatowt and Khoo Khyou Bun et al. proposed and construct a module to accomplish the dynamic monitoring task. Our module contains three components: the new domain viewer, crawler, and change extractor. The viewer combines our dynamic feature generating algorithm, selects feature terms from collected papers and sends them to EI, CNKI, or GOOGLE, and collect those results meeting our target, Secondly, the crawlers is set by timer to crawl the pages in those domains containing the results to collect the updated part of the original pages or newly emerged pages. Thirdly, the change extractor makes the comparison of the new and old one, and extract the changed part in the pages.

4.1 the Domain Fitness

The aim of the new domain viewer is to get the domains which contain newly emerged user interested topic or new domains. First, it uses the dynamic feature selecting method to send the user target to the search engines, then recognizes the returned sites significance by calculating the domain Fitness scores of the returned results. The fitness scores is as follows:

$$Fitness = \frac{\#\ of\ \{\ pages.contain.feature.terms\ \}}{Total\{\ pages.in.the.domain\ \}} + \frac{\#\ of\ \{\ links.to.other.domains\ \}}{Total\{\ links.of.the.domain\ \}} \tag{5}$$

where # stands for the number, and if the Fitness score bigger enough, then choose the site as a fitness domain.

4.2 the Crawlers

We use depth first algorithm go to 3 depths level of the fitness domains, by sending the sites http header request, and get the "last modified" value, to judge if the collected page is changed. If it is changed, then pick it up, and find out changed part, if it has new links, then pick the new pages for similarity check.

4.3 Change Detection and Extraction Methods

Change detection is made by Sentence extractor and sentence analyzer. The sentence extractor find two sets of the sentences in new and old pages, and the sentence analyzer compares them and find out those not inside the old page. And incorporate the new ones to construct a changed part. Then uses TF-PDF algorithm to find the important sentences. The formula is as follows:

$$W_j = \sum_{d=1}^{d=D} |F_{jd}|\ exp(\frac{n_{jd}}{N_d}); and\ |F_{jd}| = \frac{F_{jd}}{\sum_{k=1}^{k=K} F_{kd}} \tag{6}$$

Where Wj is the weight of jth term, Fjd is the frequency of the jth term occurred in the domain d, n_{jd} is the number of documents in the domain d and containing jth term, and N_d is the total number of the documents belong to the domain d, and K is the total number of the terms. Finally uses sentence extractor to compute the average sentence weights for each sentence in the changed part by summarizing the terms weights inside the sentence, and to pick out the significant sentences to construct the final changed part of the topics.

5. Initial Results And Future Works

The system in VC++ 6.0. (Called BIT). We track some topics we are interested, for instance "PageRank" , From Fig.1 we see that the top one papers is from L.Page, who gave the famouse GOOGLE search algorithm. And other indicators show some profiles of the collected papers.

Fig 2 shows the terms with the heavy weights picked by sentence extractor. It tells the newly emerged top terms. We can see topic such as "PageRank" is related to "Information retrieval" etc and also shows the newly related with topics such as "availability", "reliability", "data mining" "personalization" etc. It is meaningful for the technology monitoring! These initial tests are promising. And we can suggest that it can be stronger and more efficient if it gets more thorough collections of the topic related areas. Which means the change detection make dynamic monitoring of scientific topic possible.

VSP International
Science Publishers
P.O. Box 346, 3700 AH Zeist
The Netherlands

*Lecture Series on Computer
and Computational Sciences*
Volume 1, 2004, pp. 730-733

Research on Expert's Colony Decision Method of Gradual Refinement Based on Theory of Evidence

Weidong Zhu[1] and Yong Wu[2]

School of Management
Hefei University of Technology, Hefei, 230009,China

Abstract: This paper discusses decision method based on theory of evidence, it analyses the problem of the existing decision method, and puts forward analyzing the focal element of basic probability number by gradual refinement method, in order to solve basic probability number which forms state factor of focal element. This method establishes contacts between the solution of basic probability number which forms state factor of focal element and the process of expert's colony decision, It makes decision method based on theory of evidence more scientific and more rational.

Keywords: Theory of evidence, dempster combination rule, neural network

1.Introduction

In decision analyzing, it's very difficult to provide future state of the system accurately in general cases. The general expert analyses the system state according to its knowledge and experience mastered, provides the subjective estimation of the system future state. When utilizing expert's colony to estimate the future state of the system, it's necessary to combine the prediction suggestion of expert's colony. The combination method based on theory of evidence [1 - 9] and neural network offers the effective method for combining expert's colony prediction suggestion. It is expressed as basic probability number M of Θ that this method combines [10] the prediction suggestion of expert's colony (Θ is system state set). Suppose decision set as $D = \{d_1, d_2, \cdots, d_m\}$, System state i decision scheme j and corresponding remuneration function is $r(d_i, x_j)$ $(i=1,\cdots,m, \quad j=1,\cdots,n)$. Under known above-mentioned conditions, how to carry on decision, it is the problem that decision should be solved on the basis of theory of evidence.

2. Decision method based on theory of evidence and existing problem

XinSheng Duan provides two kinds of decision methods in document [11,12], one is utilize plausibility function to solve subjective probabilities of different system state, then ask out the expectation remuneration of each decision scheme.

Utilize formula of plausibility function $pl(\{x\}) = \sum_{x \in A} m(A)$

Ask out point plausibility $pl(\{x_1\}), \cdots, pl(\{x_n\})$

$$p(x_i) = \frac{pl(\{x_i\})}{pl(\{x_1\}) + \cdots + pl(\{x_n\})}$$
$$= \frac{pl(\{x_i\})}{\sum_{i=1}^{n} pl(\{x_i\})} \qquad i = 1, \cdots, n \tag{2.1}$$

---

[1] Weidong Zhu, Ph.D, Professor, E-mail: gdzwd@mail.hf.ah.cn
[2] Yong Wu, Postgraduate, E-mail: wuyong127@tom.com

$p(x_1)$, $\cdots$, $p(x_n)$ is subjective probability that various kinds of states appear. We choose the decision scheme that is corresponding to the maximum expectation remuneration as the best scheme. Namely:

$$V = \max_{d_i} \left\{ \sum_{j=1}^{n} r(d_i, x_j) * p(x_j) \right\} \qquad i = 1, \cdots, m \qquad (2.2)$$

Another kind of decision method is known as M decision method. The thought of this method is that we ask out firstly remuneration function of each focal element, then we regard basic probability number as subjective probability in order to ask out the expectation remuneration of every decision scheme as the decision basis.

$$\overline{r}(d_i, A_j) = \frac{1}{|A_j|} * \sum_{x_k \in A_j} r(d_i, x_k) \qquad (2.3)$$

$$V = \max_{d_i} \left\{ \sum_{j=1}^{h} \overline{r}(d_i, A_j) * m(A_j) \right\} \qquad i = 1, \cdots, m \qquad (2.4)$$

We choose the decision scheme that is corresponding to the maximum expectation remuneration as the best scheme.

We should pay attention to the assumption of each kind of method while using in above-mentioned two kinds of decision methods. Suppose it as follows:

The first kind of method uses (2. 1) formula to confirm various kinds of subjective probability that each state appears, it means that the more plausibility, the more subjective probability. And generally speaking, point function $pl(\{x_i\})$ can't show all information that is concluded by basic probability number M and belief function Bel. The bigger $pl(\{x_i\})$ is, might not be big $m(\{x_i\})$ and $Bel(\{x_i\})$.

The second kind of method uses (2.3) formula to calculate remuneration function of corresponding focal element, this formula ask out the simple average of remuneration function value, which is included by focal element state factor. It means that the appearance probability that every state factor included by each focal element is equal, and generally speaking, the appearance probability in every state is not equal.

Because of the existence of above-mentioned assumption, it makes above-mentioned two kinds of decision methods have limitation. In order to solve this problem scientifically, we must know how to ask and solve the appearance probability and basic probability number of system state on terms that we have already known to the basic probability assignment in each focal element.

3. Decision method that ask and solve basic probability number of each state based on focal element analyzing

From above-mentioned analysis we can know, if we want to carry on decision analyzing with theory of evidence, we should direct concrete problem and explore new decision method in order to solve problem that above-mentioned assumptions bring.

The method is that asks and solves basic probability number and belief function of each state based on focal element analyzing. We can use this method to analyze the focal element of basic probability number which combine colony's expert opinion, it is divided into two situations, first situation is that we know basic probability number of all state factors of system, namely we know $m(\{x_i\})$ $i = 1, \cdots, n$. The second kind of situation is that we are unknown basic probability number of all state factors of system.

The first kind of situation: Basic probability number of each system state can be regarded as the basis that we analyze and calculate appearance probability of each state. $m(\Theta)$ may not equal zero at this moment, $m(\Theta)$ express the suspicion to the evidence.

When we believe the basic probability number of each system state in fully, we can define the appearance probability of each state. As follows:

$$p(x_i) = \frac{m(\{x_i\})}{m(\{x_1\}) + m(\{x_2\}) + \cdots + m(\{x_n\})} \qquad i = 1, \cdots, n \qquad (3.1)$$

When we are suspicious of the basic probability number of each system state, we can give $m(\Theta)$ to each state. We can define the appearance probability of each state. As follows:

$$p(x_i) = \frac{m(\{x_i\}) + \frac{1}{n}*m(\Theta)}{m(\{x_1\}) + m(\{x_2\}) + \cdots + m(\{x_n\}) + m(\Theta)} \qquad i = 1, \cdots, n \qquad (3.2)$$

The second kind of situation: We analyze the focal element of basic probability number by gradual refinement method, in order to solve basic probability number which forms state factor of focal element.

First of all, if focal element that is formed by individual state factor, we can confirm that the basic probability number of individual state factor is the basic probability number of corresponding focal element. Secondly, if focal element contained the focal element that is formed by individual state factor which we have already known basic probability number, we can eliminate individual state factor which we have already known basic probability number and corresponding basic probability number from focal element and basic probability number. Finally, we need to introduce new information to judge and solve the individual state factor which we haven't confirm basic probability number yet. We can use expert consult method.

State factors which is included by focal element A_i are much fewer than state factors of Θ, This method is to resolve the system, and the gradual refinement method, so this method is feasible. This method establishes contacts between the solution of basic probability number which forms state factor of focal element and the process of expert's decision, It makes decision method based on theory of evidence more scientific and more rational.

After the ones that ask out the basic probability number $m(\{x_i\})$ of each system state, we can ask out the probability that each state takes place by means of the first situation method. After that, we can calculate expectation remuneration of each decision scheme and choose the optimum decision scheme. Its calculation formula as follows:

$$V = \max_{d_i}\left\{\sum_{j=1}^{n} r(d_i, x_j) * p(x_j)\right\} \qquad i = 1, \cdots, m \qquad (3.3)$$

4. Empirical Research

This paper chooses the expert prediction on the security market as the instance, carries on the evidence combination by means of combination method based on theory of evidence and neural network, we get the basic probability number M which reflected future state situation of security market, then use expert colony decision method of gradual refinement and M decision method to ask out the best decision scheme separately, we can conclude that the former scheme is obviously superior to the latter through the calculation result

Acknowledgments

I gratefully acknowledge the support of NSFC (70171033) and natural science fund of Anhui Province (00043607); I also wish to thank the anonymous referees for their careful reading of the manuscript and their fruitful comments and suggestions.

References

[1] Shafer G, A Mathematical Theory of Evidence[M], Princeton University Press, Princeton, New Jersey, 1976,p1-16

[2] Voorbraak F. On the Justification of Dempster's Rule of Combination [J], Artificial Intelligence, 1991,48: 171-197

[3] Renbing Xiao Xue Wang, Qi Fei, The research of relevant evidence combination method[J], Pattern Recognition and Artificial Intelligence, 1993,9:227-234

[4] Chun Yang, Huaizu Li, A n Evidence Reasoning Model with Its Application to Expert Opinions Combination[J] ,System Engineering-Theory and Practice ,2001.4:43-48

[5] Wenji Du, Yanhui Chen, Weixin Xie, A Weighted Dempster's Rule of Combination [J], Journal of

XiDian University, 1999,10:549-551

[6] Ronald R.Yager, On the dempster-shafer framework and new combination rules[J], Information Sciences, 1987,41:93-137

[7] Quan Sun, Xiuqing Ye, Weikang Gu, A New Combination Rules of Evidence Theory[J], Acta Electronica sinica, 2000,8 :117-235

[8] Shanlin Yang,Weidong Zhu,Minglun Ren,Research on Combination Method for Unequal Conflicted Evidence Based Optimal Adjustment Coefficient[J],Chinese MIS,2003,3:55-60

[9] Weidong Zhu,Shanlin Yang,Minglun Ren Research of Evidence Combination Based on Neural Network ,Proceedings of the Eightth PACIS,2004,8

[10] Weidong Zhu, Shanlin Yang, Minglun Ren ,Expert's Group Forecasting System Based on Learning and Theory of Evidence[J], Forecasting ,2003,1

[11] Weidong Zhu, Research on Theory of Evidence Based and Internet Oriented Intelligent Decision Support Systems, Dissertation of Heifei University of Technology 2003.4:42-45

[12] Xinsheng Duan, Theory of Evidence and Decision, Artificial intelligence [M], China Renming University Press, 1993.3:13-19

VSP International
Science Publishers
P.O. Box 346, 3700 AH Zeist
The Netherlands

Lecture Series on Computer
and Computational Sciences
Volume 1, 2004, pp. 734-736

Preface of the Symposium :
Stochastic Methods and Applications

D. T. Hristopulos[1]

Department of Mineral Resources Engineering,
Technical University of Crete,
GR-73100 Chania, Greece

When we express in mathematical language a physical process that involves random quantities, we use essentially a stochastic model. Many scientific fields employ such models, even if their name does not involve the word stochastic (e.g., the physics of disordered media, geostatistics). The mathematical background of stochastic analysis is well developed and described in excellent textbooks (e.g., [1, 7, 13]). There is today a strong interest in applications of stochastic methods, because all natural and industrial processes in non-idealized environments involve random components due to intrinsic variabilities, which defy complete characterization, as well as the inescapable presence of experimental noise. Stochastic analysis has applications in fields such as structural engineering [11], physical processes in environmental media [3, 9], natural resources exploration [5], environmental health [4], finance [10], materials fracture [2], filtering theory [6], signal analysis [8], image analysis [12], etc. This session aims to facilitate an exchange of ideas between such disciplines. In this spirit, I will briefly outline some issues of terminology and of current research interest.

The term stochastic typically refers to processes that are distributed in time (e.g., financial option prices), space (e.g., mineral concentrations), or both (e.g., groundwater pollutant concentrations). For spatially distributed variables the term *spatial random fields* is used, while for time evolution the term *stochastic process* is preferred. For space-time dependence the term *space-time random fields* is common. I will use the general term *random field (RF)* for both space and time dependence. RFs are used in the study of *random heterogeneity*, i.e., spatial variability that can not be captured by deterministic models. In the physics literature, the word *disorder* is often used for the same purpose, and it indicates departure from the perfectly orderly arrangement of crystal structures. Various classifications of RFs exist, depending on the property of interest. Regarding spatial (temporal) *dependence*, three main categories exist: (1) *Uncorrelated RFs*, i.e., white noise or fluctuations that can not be resolved by the measurement procedures. (2) *Weakly correlated RFs*, in which the correlations decay 'fast' with the distance between points (lag), so that the integral of the correlation function remains finite. Then, the RF is said to have *short-range order*. Note that disorder and short-range order are not mutually exclusive, since it is possible to have the latter in a system that exhibits long-range disorder. (3) *Long-range correlated RFs* in which the correlations decay slowly (e.g., as a power-law function of the lag).

Correlations are important for many engineering applications, because they allow statistical prediction at points where measurements are not available, giving engineers the ability to design optimal exploitation strategies for mines and to estimate the environmental risk from the dispersion of toxic pollutants in the groundwater. Hence, randomness does not exclude predictability but

---

[1] Corresponding author. E-mail: dionisi@mred.tuc.gr

implies a statistical distribution of the predictions, which must be accompanied by a measure of the associated uncertainty. If the probability density function (pdf) of the estimated variable is not available, one is limited to estimates of the mean (expected) value and the standard deviation. Uncertainty quantification is crucial in the analysis of industrial and environmental, processes. The related term *reliability*, gives a measure of belief in the nominal value of the process. In connection with product quality the term *uniformity* is used. Uniformity is inversely proportional to the width of the pdf for the specific product property. Presence of long-range correlations is important, because it invalidates mean-field approaches and has a significant impact on the statistics of extreme events.

A random process is stationary (in the weak sense) if its correlation function depends only on the time lag but not the specific times. Similarly, spatial RFs are called stationary or *statistically homogeneous*. A space-time RF is then characterized as, e.g., stationary and statistically homogeneous. Notably, stationarity in time does not imply statistical homogeneity and vice versa. There is currently strong interest in the development of space-time correlation functions for environmental and other geostatistical applications. *Quenched* randomness occurs if the fluctuations are frozen in the system, if they are invariant during the time scales of interest and can be probed only empirically (e.g., fluid permeability of geological formations). Solutions of stochastic partial differential equations represent RFs that can be static (e.g., fluid pressure in porous media) or dynamic (e.g., tracer concentrations in groundwater transport). A significant distinction is between Gaussian and non-Gaussian RFs. The former are standard for many engineering applications (e.g., geostatistics), while the latter are necessary if the details of the tail are important (e.g., in the weak-link scaling model of fracture).

It is hoped that this symposium will provide a forum for the fruitful exchange of ideas on new developments and applications of stochastic methods.

References

[1] R.J. Adler, *The Geometry of Random Fields*, Wiley, New York (1981).

[2] B.K. Chakrabarti and L.G. Benguigui, *Statistical Physics of Fracture and Breakdown in Disordered Systems*, Clarendon Press, Oxford (1997)

[3] G. Christakos, *Random Field Models in Earth Sciences*, Academic Press, San Diego, (1992).

[4] G. Christakos and D. T. Hristopulos, *Spatiotemporal Environmental Health Modelling*, Kluwer Adademic Publ., Boston (1998).

[5] P. Goovaerts, *Geostatistics for Natural Resources Evaluation*, Oxford, NY (1997).

[6] A. H. Jaswinski, *Stochastic Processes and Filtering Theory*, Academic Press (1970).

[7] B. Øksendal, *Stochastic Differential Equations*, Springer, Berlin (1998).

[8] A. Papoulis, *Probability Random Variables, and Stochastic Processes*, McGraw-Hill, NY (2001).

[9] Y. Rubin, *Applied Stochastic Hydrogeology*, Oxford University Press, New York (2003).

[10] A. N. Shiryaev, *Essentials of Stochastic Finance*, World Scientific, Singapore (1999).

[11] J. Sólnes, *Stochastic Processes and Random Vibrations*, J. Wiley, New York (1997).

[12] G. Winkler, *Image Analysis, Random Fields and Dynamic Monte Carlo Methods: A Mathematical Introduction*, Springer-Verlag, New York (1995).

[13] M. Yaglom, *Correlation Theory of Stationary and Related Random Functions I*, Springer, New York (1987).

VSP International
Science Publishers
P.O. Box 346, 3700 AH Zeist
The Netherlands

Lecture Series on Computer
and Computational Sciences
Volume 1, 2004, pp. 737-740

Effects of Uncorrelated Noise on the Identification of Spatial Spartan Random Field Parameters

D. T. Hristopulos[1]

Department of Mineral Resources Engineering,
Technical University of Crete,
GR-73100 Chania, Greece

Received 4 September, 2004; accepted in revised form September 10, 2004

Abstract: Spartan Spatial Random Fields (SSRF's) provide a new class of models for spatial dependence. Their main advantages compared to classical geostatistical techniques are computational efficiency and parametric parsimony. This paper focuses on the identification of the SSRF parameters from data contaminated with uncorrelated random noise, known by the name 'nugget effect' in geostatistical applications.

Keywords: geostatistics, inference, inverse problem, spatial dependence, optimization

Mathematics Subject Classification: 63M30, 62P12, 62P30, 62M40

1   Introduction

Spatial Random Fields (SSRF's) are widely used as models of spatial dependence for environmental and geophysical processes [2, 3, 4, 11]. A spatial random field (SRF) [1, 13] includes a number of possible states (realizations) so that the observation frequency of a particular state is determined from a multivariate probability density function (pdf), which depends on the point values of the field but also on the spatial configuration of each state. In general, an SRF state can be decomposed into a trend $m_X(\mathbf{s})$ and a fluctuation $X'(\mathbf{s})$, so that

$$X(\mathbf{s}) = m_X(\mathbf{s}) + X'(\mathbf{s}). \tag{1}$$

The trend represents the large-scale variations of the field, obtained by averaging the realizations over the ensemble, i.e. $m_X(\mathbf{s}) = E[X(\mathbf{s})]$, while the fluctuation corresponds to fast variations, which may appear as random changes. For many applications in Geostatistics one can assume that the fluctuation is a second-order stationary SRF, or a random field with second-order stationary increments [13]. In the following, we will assume that the initial SRF has been detrended, and we will use the symbol $X(\mathbf{s})$ for the fluctuation. In addition, for short we will refer to second-order stationary fields as stationary.

Spartan spatial random fields (SSRFs) [6], belong to the class of Gauss-Markov random fields [12]. They aim to provide computationally efficient tools for geostatistical spatial distributions, including environmental pollutant or mineral concentrations, transport coefficients of heterogeneous media, environmental health factors, etc. The term *Spartan* indicates a parsimonious set of model parameters. In [6] the fluctuation-gradient-curvature (FGC) SSRF model was studied. Its energy

---

[1]Corresponding author. E-mail: dionisi@mred.tuc.gr

functional embodies Gaussian fluctuations and involves three terms that measure the magnitude of the fluctuations, as well as the fluctuation gradient and the curvature. The main mathematical properties, including expressions for the covariance function, and permissibility conditions [13] for the model parameters were discussed for the continuum and square-lattice models. The covariance spectral density of the SSRFs is cut off by construction at a frequency that corresponds to physical resolution constraints. Since the SSRFs are band-limited, their realizations are differentiable in the mean square sense [9].

The probability density function (pdf) of SSRFs is determined from an energy functional $H[X_\lambda(\mathbf{s})]$, according to the familiar in statistical physics expression for the Gibbs pdf

$$f_{\mathrm{x}}[X_\lambda(\mathbf{s})] = Z^{-1} \exp\left\{-H[X_\lambda(\mathbf{s})]\right\}. \tag{2}$$

The constant Z(partition function) is the pdf normalization factor obtained by integrating $exp(-H)$ over all degrees of freedom. The subscript λ denotes the fluctuation resolution scale. The energy functional determines spatial variability in terms of interactions between neighboring sites. The pdf of *stationary Gaussian SRFs* used in classical geostatistics can be expressed in a similar form as follows

$$H[X_\lambda(\mathbf{s})] = \frac{1}{2}\int d\mathbf{s}\int d\mathbf{s}'\, X_\lambda(\mathbf{s})\, G_{\mathrm{x};\lambda}^{-1}(\mathbf{s},\mathbf{s}')\, X_\lambda(\mathbf{s}'), \tag{3}$$

where $G_{\mathrm{x};\lambda}(\mathbf{s},\mathbf{s}')$ is the covariance function; the latter needs to be determined from the data for all distance vectors $\mathbf{s}-\mathbf{s}'$. In the following, we suppress λ for notational convenience.

In SSRF models, spatial dependence is determined from interactions, which are physically motivated (e.g., in the cases of space-time models that evolve dynamic evolutions of the pdf) or they represent plausible geometric constraints (e.g., in the case of quenched geological disorder). The pdf of the FGC model involves three main parameters: the scale factor η_0, the covariance-shape parameter η_1, and the correlation length ξ. The introduction of these physically meaningful parameters simplifies the identification problem and allows intuitive initial guesses. A factor that adds flexibility is the coarse-graining kernel that determines the fluctuation resolution scale λ [6]. The resolution is directly related to the smoothness properties of the SSRF. In [6, 7], a kernel with a boxcar spectral density with a sharp cutoff in frequency/wavevector space at $k_c \propto \lambda^{-1}$ was assumed. While the cutoff frequency is treated as a constant, it is also possible to consider it as an additional model parameter. An implication of the interaction-based energy functional is that the model parameters follow from simple sample constraints. This feature permits fast, practically linear in N, solution of the parameter inference problem. For SSRFs sampled on inhomogeneous lattices, the interactions between 'near neighbors' are not uniquely defined. One alternative [6] is to superimpose a regular square lattice on the sample area. Each lattice cell then contains a variable number of sample points. The neighbor interactions in this scheme are between points in neighbor cells. Determining the neighbor structure increases the computational effort [6], but the inference process is still fast. Methods for non-constrained simulation of FGC SSRFs were presented in [7]. In both cases the simulations are *exact*, in the sense that the simulated states respect Gaussian statistics and the spatial structure by construction. In the lattice case, spatial correlations are imposed by filtering independent Gaussian random numbers with the square root of the covariance spectral density function. In the case of irregular distributions, correlations are imposed by sampling an adequate number (i.e., the order of 10^4) of wavevectors from a probability distribution obtained by integrating the covariance spectral density. Geometric anisotropy has been addressed with a systematic method [5], which allows determining the main parameters of anisotropy and then transforming into an isotropic coordinate system. The method was applied to real data from a mining site in Greece in [8].

2 SSRF with Random Noise

Eq.1 assumes that the SSRF represents a *measurable property*, the spatial variation of which consists of a deterministic trend and a correlated fluctuation. In practice, the observable $X^*(\mathbf{s})$ is likely to be contaminated by noise, which may represent unresolved fluctuations or measurement errors. Then, the observable SRF is given by the sum of two terms,

$$X^*(\mathbf{s}) = X(\mathbf{s}) + e(\mathbf{s}) \tag{4}$$

where $e(\mathbf{s})$ is a *Gaussian white noise* SRF that is *statistically independent* of the SSRF $X(\mathbf{s})$. The covariance function of the observable field $X^*(\mathbf{s})$ is also additive due to the independence of the noise and property fields, i.e., $G_x^*(\mathbf{r}) = \sigma_e^2 \delta(\mathbf{r}) + G_x(\mathbf{r})$). The additive property holds for the covariance spectral densities as well. The observable spectral density includes an additional parameter (the variance of the noise field). Based on [6], if d is the spatial dimension and $\tilde{K}_\lambda(\mathbf{k})$ the kernel spectral density, it follows that

$$\tilde{G}_x^*(\mathbf{k}) = \sigma_e^2 + \frac{\tilde{K}_\lambda(\mathbf{k})\,\eta_0\,\xi^d}{1 + \eta_1\,(k\xi)^2 + (k\xi)^4} \tag{5}$$

3 Parameter Identification

To determine the model parameters, one needs to define experimental constraints that capture the main features of the SSRF and the noise. The constraints used in [6] are based on two-point products of the field in d orthogonal directions, i.e., $S_0(\mathbf{s}) = X_\lambda^2(\mathbf{s})$, $S_1(\mathbf{s}) = \sum_{i=1}^d [\nabla_i X_\lambda(\mathbf{s})]^2$, and $S_2(\mathbf{s}) = \sum_{i,j=1}^d \Delta_2^{(i)} [X_\lambda(\mathbf{s})]\, \Delta_2^{(j)} [X_\lambda(\mathbf{s})]$, where $\Delta_2^{(i)}$ denotes the centered second-order difference operator. The experimental constraints are given by $\overline{S_0(\mathbf{s})}$ (sample variance), $\overline{S_1(\mathbf{s})}$ (average square gradient) and $\overline{S_2(\mathbf{s})}$, where the bar denotes the sample average. The procedure defined in [6] for the calculation of the experimental constraints does not change in the presence of noise.

The respective stochastic constraints are $E[S_m(\mathbf{s})]$, $m = 0, 1, 2$, where $E[\cdot]$ denotes the ensemble average. The goal is to obtain relations for the stochastic constraints in terms of the parameters of the *observable* covariance function. Let $E^*[S_m]$, $m = 0, 1, 2$ denote the moments of the noisy field. Since the noise term has an impact on the covariance only at zero lag, we obtain:

$$E^*[S_0] = E[S_0] + \sigma_e^2 \tag{6}$$

$$E^*[S_1] = E[S_1] + d\frac{\sigma_e^2}{2\alpha^2} \tag{7}$$

$$E^*[S_2] = E[S_2] + 4d^2\frac{\sigma_e^2}{\alpha^4} \tag{8}$$

where the SSRF moments $E[S_m]$ are given in [6], and α is the lattice step. Matching the stochastic and experimental constraints is formulated as an optimization problem in terms of a *distance functional* that measures the deviation between the two sets of constraints [6]. Suitable grouping of the constraints allows eliminating η_0. This corresponds to using the normalized spectral density $\tilde{G}_x^*(\mathbf{k}) = \tilde{G}_x^*(\mathbf{k})/\eta_0$. Minimization of the distance functional leads to a set of optimal values $\hat{\eta}_1, \hat{\xi}, \hat{\sigma}_e^2/\hat{\eta}_0$. The value of $\hat{\eta}_0$ follows then from the ratio

$$\hat{\eta}_0 = (2\pi)^{2d}\,\overline{S_0(\mathbf{s})}/\int d\mathbf{k}\,\tilde{G}_x^*(\mathbf{k}). \tag{9}$$

Using k_c as an additional parameter in the optimization needs further investigation. The efficiency of the parameter identification problem is due to two factors: Firstly, formulation of the

FGC pdf in terms of a clearly defined and physically meaningful correlation length and a dimensionless shape coefficient implies straightforward interpretation of the model parameters. Rough estimates for the correlation length can be obtained, either by visual inspection of preliminary maps or from the plot of the dispersion variance of block-averaged SRF values versus the block size [10]. Secondly, the distance functional involves scaled moments that are independent of the global scaling parameter η_0. This reduces the number of free parameters from four to three, i.e., $\eta_1, \xi, \sigma_e^2/\eta_0$. The parameters η_0 and σ_e^2 are then evaluated from the total variance of the sample.

We will investigate the framework outlined above for the SSRF parameter identification problem in the presence of noise using synthetic SSRF realizations mixed with uncorrelated noise.

References

[1] R.J. Adler, *The Geometry of Random Fields*, Wiley, New York (1981).

[2] G. Christakos, *Random Field Models in Earth Sciences*, Academic Press, San Diego, CA (1992).

[3] G. Christakos and D. T. Hristopulos, *Spatiotemporal Environmental Health Modelling*, Kluwer Adademic Publ., Boston (1998).

[4] P. Goovaerts, *Geostatistics for Natural Resources Evaluation*, Oxford University Press, NY (1997).

[5] D. T. Hristopulos, New anisotropic covariance models and estimation of anisotropic parameters based on the covariance tensor identity, *Stochastic Environmental Research and Risk Assessment*, **16** (1) 43-62 (2002).

[6] D. T. Hristopulos, Spartan Gibbs random field models for geostatistical applications, *SIAM Journal in Scientific Computation* **24** 2125-2162 (2003).

[7] D. T. Hristopulos, Simulations of Spartan random fields, in Proceedings of the International Conference of Computational Methods in Sciences and Engineering (Editor: T. E. Simos) 242-247 *World Scientific, London, UK* (2003).

[8] D. T. Hristopulos, Anisotropic Spartan Random Field Models for Geostatistical Analysis, in Proceedings of 1st International Conference on Advances in Mineral Resources Management and Environmental Geotechnology (Editors: Z. Agioutantis and K. Komnitsas) 127-132 *Heliotopos Conferences* (2004).

[9] D. T. Hristopulos, Spartan Random Fields: Smoothness Properties of Gaussian Densities and Definition of Certain Non-Gaussian Models, in Interfacing Geostatistics, GIS and Spatial Data Bases, Proceedings of International Workshop STATGIS 2003 (Editor: J. Pilz), *Springer, Berlin, Heidelberg*, forthcoming.

[10] C. Lantuejoul, *Geostatistical Simulation: Models and Algorithms*, Springer, Berlin, Germany (2002).

[11] Y. Rubin, *Applied Stochastic Hydrogeology*, Oxford University Press, New York (2003).

[12] G. Winkler, *Image Analysis, Random Fields and Dynamic Monte Carlo Methods: A Mathematical Introduction*, Springer-Verlag, New York (1995).

[13] M. Yaglom, *Correlation Theory of Stationary and Related Random Functions I*, Springer, New York (1987).

VSP International
Science Publishers
P.O. Box 346, 3700 AH Zeist
The Netherlands

*Lecture Series on Computer
and Computational Sciences*
Volume 1, 2004, pp. 741-744

An Application of Spartan Spatial Random Fields in Geostatistical Mapping of Environmental Pollutants

Manolis Varouchakis and Dionissios T. Hristopulos[1]

Department of Mineral Resources Engineering,
Technical University of Crete,
GR-731 00 Chania, Greece

Received 8 August, 2004; accepted in revised form 16 September, 2004

Abstract: This paper presents a preliminary application of Spartan Spatial Random Fields (SSRFs) to a real geostatistical data set. The SSRFs bypass calculation and fitting of the experimental variogram and thus provide a computationally fast alternative to the classical geostatistical approach. The study focuses on the concentration of the heavy metal chromium (Cr) in the Jura region (Switzerland). A map of Cr concentration on a square prediction grid is generated based on a set of irregularly spaced measurements. A new estimation method that minimizes the SSRF interaction functional is applied instead of the commonly used kriging estimators.

Keywords: geostatistics, estimation, chromium concentration, prediction, validation

Mathematics Subject Classification: 63M30, 62P12, 62P30, 62M40

1. Introduction

The distribution of heavy metals in the soil over an area of 14.5km^2 in the Swiss Jura region [1] is studied. The available data consist of 359 samples that contain concentrations in parts per million (ppm) for seven heavy metals. The aim of this study is to generate from the data a map describing the dispersion of pollutant concentrations in the entire area. This can be accomplished with the aid of a geostatistical model. Such maps are useful tools for planning soil remediation actions in compliance with EU regulations. Typically, a variant of the kriging algorithm [1] is used to generate geostatistical maps. In contrast, we apply the recently introduced Spartan Spatial Random Field (SSRF) model. The initial data are segregated in two sets: The *training set* used for the parameter identification (inference) process contains 259 samples at the locations $s_t, t = 1, ..., N_t$. The *validation set* contains 100 points $s_v, v = 1, ..., N_v$, where the estimated concentrations are compared with the actual values, to obtain a quantitative measure of the method's accuracy. At this stage, the geostatistical analysis is applied to chromium (Cr), which follows approximately the normal distribution.

2. Geostatistical Model

Spartan Spatial Random Fields (SSRFs) have been introduced and their mathematical properties have been investigated [2]-[5], mostly for Gaussian probability density functions. However, the SSRF models have not been applied to real data sets. The SSRF models are determined by a small number of parameters, the inference of which from the sample does not require the variogram calculation. The SSRF parameter inference is based on relatively simple statistical restrictions that are calculated effectively. Let us assume that the variable of interest (e.g., concentration) is represented by the spatial random

---

[1] Corresponding author. E-mail: dionisi@mred.tuc.gr

field, $X_\lambda(\mathbf{s})$, where λ is the resolution. The SSRF probability density function (pdf) includes all the information regarding the spatial dependence. General SSRF models are subsets of Gibbs random fields and thus can be expressed with the following mathematical relation:

$$f_x[X_\lambda(\mathbf{s})] = Z^{-1}\exp\{-H[X_\lambda(\mathbf{s})]\} \tag{1}$$

In the above equation, $f_x[X_\lambda(\mathbf{s})]$ is the pdf, Z is a normalization constant, and $H[X_\lambda(\mathbf{s})]$ is a *spatial interaction* functional that expresses the dependence of the random field values at different locations in space. The innovation that is imported by the SSRFs is the dependence of $H[X_\lambda(\mathbf{s})]$ on physical quantities related to the spatial distribution of values, i.e., the local gradient and curvature of the "topography", as well as the complete determination of the interaction from a small number of parameters. An example of local interdependence is given by the following functional:

$$H[X_\lambda] = \frac{1}{2\eta_0\xi^d}\int d\mathbf{s}\left[\left\{X_\lambda(\mathbf{s}) - E[X_\lambda(\mathbf{s})]\right\}^2 + \eta_1\xi^2\left\{\nabla X_\lambda(\mathbf{s})\right\}^2 + \xi^4\left\{\nabla X_\lambda(\mathbf{s})\right\}^2\right] \tag{2}$$

The integral implies that all the locations are included (in practice, discretized versions are used). The scale parameter η_0 determines the total variance of the fluctuations, the coefficient η_1 the shape of the covariance function, and ξ the correlation radius (i.e., the range of the spatial dependence).

In geostatistical applications, the first objective is solving the inverse problem that leads to the identification of the model parameters from the available data. For the SSRF model, *statistical constraints* that can be calculated from the sample are defined. It is possible (but not constraining) for the restrictions to correspond to the terms of the interaction functional. For example, for the interactions given by the equation (1), the constraints include the sample variance, as well as the average gradient square and curvature square. The respective *ensemble (mean) values* of these quantities are calculated from the pdf (1), and they involve the unknown parameters η_0, η_1, ξ. The optimal values of the parameters are determined by minimizing the difference between the sample and the ensemble values of the constraints [2]. The calculation of the statistical constraints and the solution of the optimisation problem are typically faster than the calculation of the variogram [2].

3. Estimation and Validation

It is assumed that the Cr distribution corresponds to a Gaussian, second-order stationary and isotropic random field, the spatial dependence of which can be modelled by means of the SSRF given by equations (1) and (2). The parameters η_0, η_1 and ξ, are determined from the Cr concentration at the points of the training set, using the procedure described in [2]. In The optimal values are $\eta_1 = 394$ and $\xi = 1.3$ (km). A *correlation neighbourhood* $B(\mathbf{s}_v)$ is defined around each validation point $\mathbf{s}_v$, based on the correlation radius. In general, two options are available for the estimation: (i) The SSRF covariance function is used in a kriging algorithm. (ii) The estimates $X_\lambda^*(\mathbf{s}_v)$ are determined from local solutions of the equation $H[X_\lambda(\mathbf{s})] = 0$, $\mathbf{s} \in B(\mathbf{s}_v)$ in the neighbourhood of $\mathbf{s}_v$. The solutions of this equation can be expressed in closed form for the interaction functional of equation (2). Here we use the approach (ii), which will be described in detail elsewhere. Explicit expressions for the uncertainty of the estimates can also be obtained. For values of the shape coefficient $\eta_1 \geq 2$, as is the case here, the estimate is given by the equation

$$X_\lambda^*(\mathbf{s}_v) = 4\pi(A_v + B_v), \ v=1,\ldots,N_v. \tag{3}$$

The local parameters A_v, B_v are determined by minimizing the average square error of the explicit solutions $X_\lambda^*(\mathbf{s}_i)$ with respect to the actual values $X_\lambda(\mathbf{s}_i)$. The average is calculated over all the training points $\mathbf{s}_i \in B(\mathbf{s}_v)$ that are located inside $B(\mathbf{s}_v)$. The parameter η_0 does not influence the estimated concentration values, but it does affect the uncertainty of the estimates.

4. Analysis of SSRF Model Performance

A comparison between the estimated values (i.e., predictions) and the actual values of Cr concentration at the validation points is shown in Fig.1. The plot reveals deviations between the SSRF estimates and the actual values, and the overall variability of the estimates is less than the actual. The deviations is measured by means of the average absolute value of the relative error, ε_v, which is defined by means of

$$\varepsilon_v = \frac{1}{N_v} \sum_{v=1}^{N_v} \left| \frac{X_\lambda^*(s_v) - X_\lambda(s_v)}{X_\lambda(s_v)} \right|. \tag{4}$$

The calculated value of ε_v is approximately 28%. Further examination of the training and validation sets shows that in the former the concentration values range between 18 ppm and 65ppm, while in the latter between 3.3 ppm and 70 ppm. This observation explains why the estimated variability (based on the training set) underestimates the actual variability. The observed deviation between the actual values and the SSRF estimates is partly due to this factor. In particular, the SSRF predictions are significantly off at validation points where the concentration values are outside the range of the training set. For example, if the two points with the lowest and highest concentrations are removed from the validation set, the error decreases to 19%. The performance is reasonable, given the fact that a simple SSRF model is used, which corresponds to a normal (Gaussian), stationary and isotropic distribution. A map of Cr concentration on a regular grid that covers the study area is shown in Fig 2.

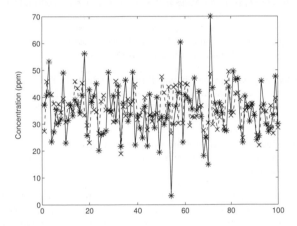

Figure 1: Plot of the actual Cr concentration (*, continuous line) and the SSRF estimates (×, dashed line) at the locations of the validation set.

5. Conclusions

This paper presents a preliminary application of a specific SSRF model in environmental pollutant mapping, using a data set of Cr soil concentration from the Jura region of Switzerland. Reasonable agreement between the actual and estimated values at the points of the validation data set is observed, in spite of several conceptual simplifications implied by the specific model. This supports the feasibility of using SSRFs for the generation geostatistical maps. Future research will focus on (i) testing the conceptual assumptions and (ii) methodological improvements of the SSRF inference and estimation algorithms, aiming to obtain better discretization for the training and prediction grids, optimal values of the correlation radius ξ and the shape coefficient, which are expected to improve the prediction accuracy. In addition, the estimation uncertainty will be investigated, and comparisons with kriging methods will be conducted.

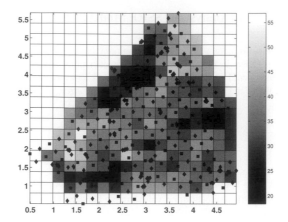

Figure 2: Gray scale map of Cr concentrations generated by the SSRF method on a prediction grid. The training set locations are marked by diamonds, and the validation set locations are marked by squares.

References

[1] P. Goovaerts, *Geostatistics for Natural Resources Evaluation*. Oxford Univ. Press, New York (1997).

[2] D. T. Hristopulos, Spartan Gibbs random field models for Geostatistical Applications, *SIAM Journal in Scientific Computation*, **24**, 2125-2162 (2003).

[3] D. T. Hristopulos, Anisotropic Spartan Random Field Models for Geostatistical Analysis, *Proceedings of 1ˢᵗ International Conference on Advances in Mineral Resources Management and Environmental Geotechnology* (editors: Z. Agioutantis and K. Komnitsas, TUC), Heliotopos Conferences (2004), 127-132.

[4] D. T. Hristopulos, Numerical Simulations of Spartan Gaussian Random Fields for Geostatistical Applications on Lattices and Irregular Supports, *Journal of Computational Methods in Sciences and Engineering*, in print.

[5] D. T. Hristopulos, Spartan Random Fields: Smoothness Properties of Gaussian Densities and Definition of Certain Non-Gaussian Models, *Interfacing Geostatistics, GIS and Spatial Data Bases, Proceedings of International Workshop STATGIS 2003* (Editor: J. Pilz, Klagenfurt), Springer, Berlin, in print.

VSP International
Science Publishers
P.O. Box 346, 3700 AH Zeist
The Netherlands

*Lecture Series on Computer
and Computational Sciences*
Volume 1, 2004, pp. 745-748

On the Application of Statistical Process Control to the Stochastic Complexity of Discrete Processes

A. Shmilovici*[1], I. Ben-Gal^

*Dept. of Information Systems Eng.
Ben-Gurion University
P.O.Box 653, Beer-Sheva, Israel

^Department of Industrial Engineering Tel-
Aviv University
Ramat-Aviv, Tel-Aviv 69978, Israel

Accepted 29 August, 2004

Abstract: A change in a process would change its description length - as measured by its stochastic complexity. The key idea here is to monitor the statistics of the stochastic complexity (the equivalent code length) of a data sequence A context tree model is used as the universal device for measuring the stochastic complexity of a state dependent discrete process. The advantage of this method is in the expected reduction in the number of samples needed for reliable monitoring.

Keywords: Process Control, Control charts, Stochastic Complexity, Context Tree Algorithm

Mathematics Subject Classification: 62M02, 93E35, 62M10, 62P30

1. Introduction

Most statistical process control (SPC) methods construct a Shewhart type control chart for the distribution of some attributes of the process (e.g., mean) that is being monitored. Often the observations (or attributes) are assumed to be independent and identically distributed (i.i.d.) and traditionally they are assumed to be normally distributed. As evidenced by the wide spread implementation of the Shewhart control charts, this practice has proved to be very useful.

However, there are important situations where the i.i.d., and the normality assumptions are grossly inaccurate, and can lead to false alarms and late detection of faults. For example, due to the wide-spread of digital controllers, many industrial processes are being controlled by a feedback controller that intervenes whenever the process deviates significantly from a pre-specified set-point. This introduces dependency between consecutive samples and deviation from the normality assumption.

The need to find substitutes for the traditional Shewhart control charts has been recognized in the literature. Model specific methods (such as Exponential Weighted Moving Average) unfortunately need relatively large amounts of data to produce accurate results, and can not capture more complex process dynamics, such as Hidden Markov Models (HMM). Model generic methods (such as the ITPC[1] and the CSPC[2]) use asymptotic properties from information theory as a replacement for explicit distribution assumption about the process characteristics. In practice, large amounts of data are needed for the application of ITPC, either for deriving an empirically based control limit, or for using an analytically derived control limit that is based on asymptotic considerations.

The stochastic complexity of a sequence is a measure of the number of bits needed to represent and reproduce the information in a sequence. It is a statistic commonly used as a yardstick for choosing between alternative models for an unknown time series.

Universal coding methods have been developed to compress a data sequence without prior assumptions on the properties of the generating process. Universal coding algorithms - typically used for data compression - model the data for coding in a less redundant representation. The length of the data after compression is a practical measure of the stochastic complexity of the data. Some universal algorithms are known to have asymptotic performance as good as the optimal non-universal algorithms.

---

[1] Corresponding author. E-mail: armin@bgumail.bgu.ac.il

This means that for long sequences, the model provided by the universal source behaves like the "true" system for all tasks we wish to use it for, such as coding, prediction, and decision making in general.

Here we use the context tree algorithm of Rissanen [3], which is a universal coding method, for measuring the stochastic complexity of a time series. The advantage of this specific algorithm, in contrast to other universal algorithms known only to have asymptotic convergence, is that it has been also to have the best non-asymptotic convergence rate. Thus, it can be used to compress even relatively short sequences - like the ones available from industrial processes.

The key idea in this paper is to monitor the statistics of the stochastic complexity (the equivalent code length) of a data sequence A context tree model is used as the universal device for measuring the stochastic complexity of a state dependent discrete process. The advantage of this method is in the expected reduction in the number of samples needed for reliable monitoring.

2. The Context Tree Method

Following the notation in [4], let us consider a discrete sequence with $N+1$ symbols, $X_{-N}^0 \equiv X_{-N}, X_{-N+1}, \ldots, X_0$, where each symbol X_i belongs to an alphabet A of cardinality, $|A|$ and where the sequence is emitted by a stationary source. In the estimation problem, given X_{-N}^0, you need to estimate $P(X_1 \mid X_{-N}^0)$, the unknown conditional probability distribution of X_1 given X_{-N}^0.

Consider the class of universal conditional probability measures that count the recurrence of the longest suffix of X_1 in X_{-N}^0. The suffix $- X_{-K_o(X_{-N}^0)}^0 -$ is a subsequence of the past sequence X_{-N}^0 termed also as the context.

In the context tree algorithm, probability estimates are computed for every context and a weighted probability is evaluated. A context is represented by the path of branches starting at the root until it reaches a specific node. The context order is reversed with respect to the order of observance, such that deeper nodes correspond to previously observed symbols. The lengths (depth) of various contexts (branches in the tree) do not need to be equal. Given a context tree, compression can be obtained as a result of recurring patterns in the data. Each node in the tree is related to a specific recurring context (sub-sequence). Hence, the original sequence can be coded by the sub-sequences in the context tree. Using an arithmetic coder, it is guaranteed that the redundancy does not exceed two bits per sequence. A sequence that does not belong to the same class of sequences from which the context tree was generated (trained) is expected to obtain a lower compression rate when using the context tree probabilities from the training set.

3. SPC for the Stochastic Complexity Derived from the Context Tree Model

Within the context of developing a SPC procedure, we have to set up three parameters for the algorithm: N - the series length from which the reference tree will be computed; K_{max} - the maximal context tree depth; and $\hat{N}$ - the sequence length for which the compression statistic is computed with the context tree. According to Ziv [5], if we set the depth of the context tree such $P(X_1 \mid X_{-K_{max}}^0) \geq \frac{1}{AT_1}$, then $\hat{N} \geq K_{max}^3$ and $N > \hat{N} A T_1^3$.

The stochastic complexity of any sequence $X_1^{\hat{N}}$ that was prefixed by the sequence X_{-D+1}^0 (the context), and was generated by the same information source that generated the sequence Y_{-N+1}^0 can be measured with the universal algorithm that uses the context tree $T \equiv T(S, \Theta_S)$ that has a structure S and parameters Θ_S. The context tree will found from the sequence Y_{-N+1}^0 (training sequence), There is a recursive method to calculate the stochastic complexity measure

$$-\log_2\left(\Pr\left(X_1^N \mid X_{-D+1}^0, T\right)\right) = -\log_2\left(\prod_{j=1}^N \Pr\left(X_j \mid X_{j-D}^{j-1}, T\right)\right) = -\sum_{j=1}^N \log_2\left(\Pr\left(X_j \mid X_{j-K_1(j)}^{j-1}\right)\right)$$

where $K_1(j) \equiv K_1\left(X_j, X_{j-D}^{j-1}, T\right)$.

The context tree T with $|S|$ leafs describes a multinomial distribution with values $\Pr\left(X_j \mid X_{j-K_1(j)}^{j-1}\right)$. When the context tree model is estimated from a sufficiently long sequence Y_{-N+1}^0, $N \gg |S|$, than $\Pr\left(X_j \mid X_{j-K_1(j)}^{j-1}\right)$ are also independent and identically distributed. In that case, the expression is approximately the sum of i.i.d. random variables. The value of the stochastic complexity of the sequence X_1^N prefixed by the sequence X_{-D+1}^N is a random variable with $|S|^N$ possible discrete values. For a sufficiently large N, the distribution of the stochastic complexity can be approximated with the Central Limit Theorem.

Now, we are ready to devise a recipe for the SPC of a process, using its stochastic complexity as the measure for the process stability:

- Use a sequence of observation Y_{-N+1}^0 from the in-control state of the system to develop the context tree model $T \equiv T(S, \Theta_S)$.
- Calculate the first two moments from T
- Denote by q_1 and q_2 the required False Alarm Rates (FARs) with respect to the Upper Probability limit (UPL) and Lower Probability Limit (LPL). Construct the Shewart control charts with $q_1 = q_2 = 0.00135$.

- For every sequence of length $\hat{N}$ use the context tree to compute its stochastic complexity. Insert a point in the Shewart control charts and observe the UCL and LCL.

.

4. Example: SPC for a single stage production system

Consider a system of two machines: M1, M2 separated by a finite-capacity buffer B. Machine M2 attempts to process a part whenever it is not starved (i.e., whenever there are parts in its input buffer) and machine M1 attempts to process a part whenever it is not blocked. Figure 1 presents the state transition diagram for the system with in-control production probabilities of the machines to be $p_1 = 0.9$ and $p_2 = 0.8$, respectively, and a buffer capacity to be limited to C=3.

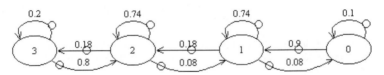

Figure 1: The State transition Diagram of the in-control production system

Figure 2 presents the 'in control' distribution of buffer levels in the referenced process, $\Pr_1\left(X_j \mid X_{j-1}\right)$, in the form of an analytical context tree Ta. Note that it is a single-level tree with $|S| = 4$ contexts and a symbol set of size $|A| = 4$. The root node presents the system steady-state probabilities and the leaves presents the transition probabilities given the current state.

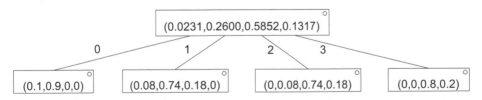

Figure 2: In-control context tree – Ta - based on the "in-control" state

Figure 3 presents a second context tree $\Pr_2\left(X_j \mid X_{j-1}\right)$ - Tb - with different values of the production probabilities, namely $p_1 = 0.7, p_2 = 0.9$. It is used for simulating the "out of control" distribution. As we can see, it had captured the probability difference.

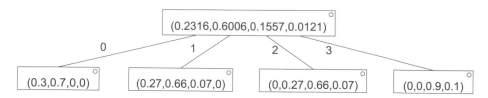

Figure 3: Context tree *Tb* for the modified system parameters to $p_1 = 0.7, p_2 = 0.9$.

5. Conclusions

We proposed a method for using the stochastic complexity of a sequence as the statistical measure for SPC. The stochastic complexity measure uses a context tree model as a universal model that can capture complex dependencies in the data.

The advantage of the proposed method is two-fold: a) It is generic and suitable for many types of discrete processes with complex unknown dependencies. b) It is suitable for relatively short data sequences. The viability of the proposed SPC method and its advantages were demonstrated with numerical experiments.

References

[1]   L.C. Alwan, Ebrahimi N., Soofi E.S, "Information Theoretic Framework for Process Control", *European Journal of Operations Research*, **111**, 526-542(1998).

[2]   I. Ben-Gal, Morag G., Shmilovici A., "CSPC: A Monitoring Procedure for State Dependent Processes", *Technometrics*, **45**(4), 293-311(2003).

[3]   J. Rissanen, "A Universal Data Compression System", *IEEE Transactions on Information Theory*, **29** (5), 656- 664(1983).

[4]   J. Ziv, "A Universal Prediction Lemma and Applications to Universal Data Compression and Prediction", *IEEE Transactions on Information Theory*, **47**(4), 1528- 1532(2000).

[5]   J. Ziv, "An Efficient Universal Prediction Algorithm for Unknown Sources with Limited Training Data", 2002, Available: www.msri.org/publications/ln/ msri/2002/infotheory/ziv/1/

VSP International
Science Publishers
P.O. Box 346, 3700 AH Zeist
The Netherlands

*Lecture Series on Computer
and Computational Sciences*
Volume 1, 2004, pp. 749-749

Preface of the Symposium : Mathematical Chemistry

T.E. Simos
Chairman of ICCMSE 2004

Department of Computer Science and Technology,
Faculty of Sciences and Technology,
University of Peloponnese,
GR-221 00 Tripolis, Greece

This symposium has been created after a proposal of Sonja Nikolić of the Rugjer Bošković Institute (Croatia). After a successful review, we have accepted her proposal.

The organizer of the symposium, Sonja Nikolić, has selected, after international peer review three papers:

- *"Graphical Matrices as Sources of Double Invariants for Use in QSPR"* by Sonja Nikolić, Ante Milićević and Nenad Trinajstić
- *"Chemical Elements: a topological approach"* by Guillermo Restrepo, Eugenio J. Llanos and Héber Mesa and
- *"Molecular Simulation Studies of Fast Vibrational Modes"* by D. Janežič and M. Penca

I want to thank the symposium organizer for her activities and excellent editorial work.

VSP International
Science Publishers
P.O. Box 346, 3700 AH Zeist
The Netherlands

*Lecture Series on Computer
and Computational Sciences*
Volume 1, 2004, pp. 750-752

Graphical Matrices as Sources of Double Invariants for Use in QSPR

Sonja Nikolić[1] Ante Miličević and Nenad Trinajstić

The Rugjer Bošković Institute,
P.O.B. 180, HR-10002 Zagreb, Croatia

Received 7 July, 2004; accepted in revised form 4 September, 2004

Abstract: Graphical matrices are presented. Their construction via selected sets of subgraphs and the replacement of subgraphs by numbers representing graph invariants are discussed. The last step of the procedure is to apply the method of choice for obtaining the desired double invariant from the graphical matrix in the numerical form. It is also pointed out that many so-called special graph-theoretical matrices from the literature are rooted in the corresponding graphical matrices.

Keywords: graphical matrices, chemical graph theory, double invariants, QSPR

Mathematics Subject Classification: 05C10, 68R10

Introduction

Graphical matrices are matrices whose elements are subgraphs of the graph rather than numbers. Since the elements of these matrices are (sub)graphs, they are called the *graphical matrices* [1]. Thus far a very little work has been done on these matrices [1,2]. However, many so-called special matrices [3], such as the Wiener matrices [4,5] and the Hosoya matrices [6], may be regarded to be the numerical realizations of the corresponding graphical matrices. Therefore, the graphical matrices appear to be a promising area of research in chemical graph theory.

The advantage of a graphical matrix lies in the fact that it allows great many possibilities of numerical realizations. In order to obtain a numerical form of a graphical matrix, one needs to select a graph invariant and replace all the graphical elements (subgraphs of some form) by the corresponding numerical values of the selected invariant. In this way, the numerical form of the graphical matrix is established and one can select another or the same type of invariant – this time an invariant of the numerical matrix. Graph invariants generated in this way are *double invariants* [2] in view of the fact that two invariants are used in constructing the targeted molecular descriptor.

Construction of the graphical matrix

We present several graphical matrices that lead to Wiener-Wiener indices, Hosoya-Wiener indices, Hosoya-Balaban indices, Randić-Wiener indices, etc. These indices were considered because the values of the Randić indices, Wiener indices, Balaban indices and Hosoya indices for smaller acyclic fragments (trees) are readily available [e.g., 7]. As an example, we give the construction of the graphical matrix, based on the concept of the edge-Wiener matrix [8], that leads to the edge-Wiener-Wiener index ^eWW. It should be noted that the edge-Wiener graphical matrix is constructed by the consecutive removal of the edges from the graph. This is shown below for the hydrogen-depleted graph G representing 2,2,3-trimethylpentane. Since a graphical matrix is a square, $V \times V$, symmetric matrices, it is enough to give for the demonstrative purposes only the upper triangle of the matrix. For graphs without loops, the corresponding graphical matrices have zeros diagonal elements.

---

[1] Corresponding author. E-mail: sonja@irb.hr

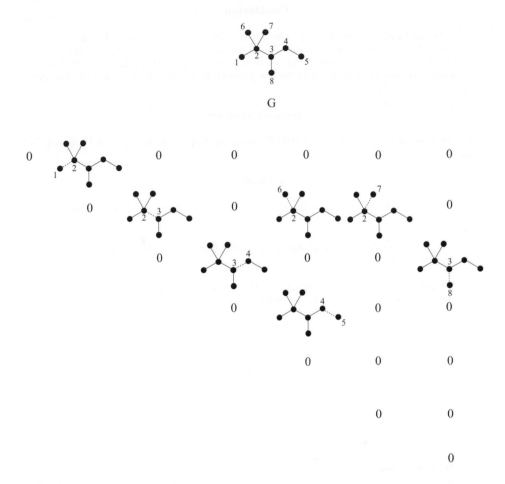

G

Numerical realization of the graphical matrix that leads to a double invariant

The next step is to replace (sub)graphs with their Wiener numbers obtained by summing up the Wiener numbers of acyclic fragments. Below is given the numerical realization of the above edge-Wiener graphical matrix.

$$
\begin{array}{cccccccc}
0 & 46 & 0 & 0 & 0 & 0 & 0 & 0 \\
 & 0 & 19 & 0 & 0 & 46 & 46 & 0 \\
 & & 0 & 29 & 0 & 0 & 0 & 46 \\
 & & & 0 & 46 & 0 & 0 & 0 \\
 & & & & 0 & 0 & 0 & 0 \\
 & & & & & 0 & 0 & 0 \\
 & & & & & & 0 & 0 \\
 & & & & & & & 0 \\
\end{array}
$$

The summation of the matrix-elements in the above matrix-triangle gives the edge-Wiener-Wiener number ^{e}WW of 2,2,3-trimethylpentane (^{e}WW =278).

Conclusion

The application of this kind of molecular descriptors to QSPR modeling is described for octanes since the modeling properties of this class alkanes is well-studied in the literature [e.g., 1,9,10] and thus we have a standard against which we tested our models. Our modeling is based on the CROMRsel procedure so devised to give the best possible model for a given number of descriptors [11-13].

Acknowledgments

This work was supported by Grant No. 0098034 from the Ministry of Science and Technology of Croatia.

References

[1]   M. Randić, N. Basak and D. Plavšić, Novel Graphical Matrix and Distance-Based Molecular Descriptors, *Croat. Chem. Acta* **77** (2004).

[2]   M. Randić, D. Plavšić and M Razinger, Double Invariants *MATCH – Comm. Math. Comput. Chem.* **35** (1997) 243-259.

[3]   A. Miličević, S. Nikolić, N. Trinajstić and D. Janežić, Graph-Theoretical Matrices in Chemistry, a review in preparation. Draft free of charge is available on the e-mail address: sonja@irb.hr

[4]   M. Randić, Novel Molecular Descriptor for Structure-Property Studies, *Chem. Phys. Lett.* **211** (1993) 478-483.

[5]   M. Randić, X. Guo, T. Oxley and H. Krishnapryan, Wiener Matrix: Source of Novel Graph Invariants, *J. Chem. Inf. Comput. Sci.* **33** (1993) 709-716.

[6]   M. Randić, Hosoya Matrix – A Source of New Molecular Descriptors, *Croat. Chem. Acta* **67** (1994) 415-429.

[7]   N. Trinajstić, S. Nikolić, J. von Knop, W. R. Müller and K. Szymanski, *Computational Chemical Graph Theory – Characterization, Enumeration and Generation of Chemical Structures by Computer Methods,* Horwood, New York, 1991, pp. 263-266.

[8]   M. Randić, Novel Molecular Descriptor for Structure-Property Studies, *Chem. Phys. Lett.* **211** (1993) 478-483.

[9]   M. Randić and N. Trinajstić, In Search for Graph Invariants of Chemical Interest, *J. Mol. Struct. (Theochem)* **300** (1993) 551-572.

[10]   G. Rücker and C. Rücker, On Topological Indices, Boiling Points and Cycloalkanes, *J. Chem. Inf. Comput. Sci.* **39** (1999) 788-802.

[11]   B. Lučić and N. Trinajstić, Multivariate regression outperforms several robust architectures of neural networks, *J. Chem. Inf. Comput. Sci.* **39** (1999) 121-132.

[12]   B. Lučić, N. Trinajstić, S. Sild, M. Karelson and A.R. Katritzky, A.R. A new efficient approach for variable selection based on multiregression: Prediction of gas chromatographic retention times and response factors, *J. Chem. Inf. Comput. Sci.* **39** (1999) 610-621.

[13]   B. Lučić, D. Amić and N. Trinajstić, Nonlinear multivariate regression outperforms several concisely designed neural networks on three QSPR data sets, *J. Chem. Inf. Comput. Sci.* **40** (2000) 403-413.

VSP International
Science Publishers
P.O. Box 346, 3700 AH Zeist
The Netherlands

*Lecture Series on Computer
and Computational Sciences*
Volume 1, 2004, pp. 753-755

Chemical Elements: A Topological Approach

Guillermo Restrepo[1], Eugenio J. Llanos and Héber Mesa

Laboratorio de Química Teórica, Universidad de Pamplona, Pamplona, Colombia
Observatorio Colombiano de Ciencia y Tecnología, Bogotá, Colombia

Received 28 August, 2004; accepted in revised form 4 September, 2004

Abstract: We carried out a topological study of 72 chemical elements taking advantage of chemometric tools and general topology. We defined every element as an ordered set of 128 properties (physico-chemical and chemical), then applied PCA and cluster analysis CA (4 similarity measures and 5 grouping methodologies). Afterwards we took dendrograms (complete binary trees) of the CA and developed a mathematical methodology to extract neighbourhood relationships of those trees. By means of this approach we built up a basis for a topology on the set of chemical elements and then provided with a topology the same set. Finally we calculated some topological properties (closures, derived sets, boundaries, interiors and exteriors) of several subsets of chemical elements such as alkaline metals, alkaline earths, noble gases, metals and non-metals. We found that alkaline metals, alkaline earths and noble gases appear not related to the rest of elements and found that the boundary of non-metals are some semi-metals and its shape is like a "stair".

Keywords: Mathematical chemistry, Chemical topology, Chemical elements, Periodic Table

Mathematics Subject Classification: 92E99, 54A10

1. Extended Abstract

Mendeleev's work on the classification of chemical elements was the result of taking the whole set of elements, not one by one [1]. In this way he found, looking on the set, several relationships among properties of chemical elements. A picture that represents these similarities is the periodic table and exactly the groups of elements within it. Our aim in this work is to take the chemical elements defined not by means of their electrons but their properties and study their similarity relationships. In this way we collect a set of 128 phisico-chemical and chemical properties and define every element as a 128-tuple. It means that every element is a point in a space of properties of 128 dimensions.

Once we have a mathematical representation of chemical elements we start to calculate similarities among them by means of cluster analysis CA [2,3]. This procedure starts calculating a similarity matrix, which commonly is a distance matrix and then builds up groups or clusters by means of a grouping methodology [4,5]. CA procedure finishes with a 2-dimensional representation of clusters called dendrogram [6], which in mathematical terms is a complete binary tree. We use 4 similarity functions, 3 of them metrics [7], and 5 grouping methodologies. In this way we have 20 dendrograms that in a mathematical point of view show similarity relationships among elements (neighborhoods). Afterwards, with the aim of having a way to represent the information of 20 dendrograms in only one mathematical object, we calculate the ultrametric matrix to every dendrogram and then we do the sum of all of them. Thus, we have an ultrametric matrix that represents all dendrograms. After, we do a density plot of that matrix and in it was evident the similarity among elements of several groups of the conventional periodic table [8].

On the other hand, no wanting to lose the tree character of the dendrograms, we calculated a consensus tree [9] (Adams consensus), that collect the information of the 20 dendrograms in only one tree. This

---

[1] Corresponding author. Professor at Universidad de Pamplona, Colombia. E-mail: grestrepo@unipamplona.edu.co

consensus, as every dendrogram, shows similarity relationships. With this tree, and according to Villaveces' conjecture [1,6] and some of our results [10,11], we start a mathematical description of the information shown by the consensus tree. This description takes advantage of cuts of "branches" of the tree. In this part, we describe a mathematical procedure to put the intuitive idea of cut on the tree in mathematical terms. This procedure starts considering a tree as a graph and every branch as a subtree of restricted cardinality [10].

Once we have the collection of subtrees or branches we take them to build up a basis for a topology and provide with a topology the set of chemical elements. With this topology on the set of elements we study the following topological properties of several subsets of chemical interest [12] within chemical elements: closure, derived set, boundary, interior and exterior [13]. Subsets of chemical interest that we study are: all groups of the conventional periodic table, main elements, transition elements, metals, non-metals, semimetals, hydrogen, boron, carbon, silicon, the group of three: germanium, tin and lead, nitrogen, phosphorus, the group of three: arsenic, antimony and bismuth, oxygen, sulfur and the group of three: selenium, tellurium and polonium.

Among topological results it is important to remark that the boundary of alkaline metals, alkaline earths, noble gases, hydrogen, scandium group, titanium group, boron, carbon, silicon and oxygen is empty which means that this subsets are not related to other elements. In other words, that these elements appear far from the others in the space of properties. On the other hand we found that the boundary of metals is made of some elements that have been considered as semimetals many years ago. We found too, that this boundary has a "stair" shape in the conventional periodic table.

According to these results we can talk about a mathematical structure on the set of chemical elements, a topological structure which reproduce several aspects of the chemical understanding about chemical elements. Besides, the mathematical procedure to provide with a topology this set is not only to chemical elements, it is possible to apply this methodology to whatever set, in particular a chemical set that can be defined by means of its properties as we showed recently [14].

Acknowledgments

Authors wish to thank the great help of Dr. Villaveces at Observatorio Colombiano de Ciencia y Tecnología for his excellent ideas. In the same way the mathematical support of Dr. Isaacs at Universidad Industrial de Santander, on the other hand Dr. Sneath at Leicester University for having discuss his work, Dr Carbó-Dorca at Universitat de Girona and Dr. Bultinck at Gent Universitet for their comments and suggestions. We thank to Universidad de Pamplona, especially its rector Dr. González for all his support during this research.

References

[1] J. L. Villaveces, Química y Epistemología, una relación esquiva, *Rev. Colomb. Fil. Cienc.* **1** 9-26(2000).

[2] P. H. A. Sneath, Numerical Classification of the Chemical Elements and its relation to the Periodic System, *Found. Chem.* **2** 237-263(2000).

[3] P. Willett, Chemical Similarity Searching, *J. Chem. Inf. Comp. Sci.* **38** 983-996(1998).

[4] R. G. Brereton, *Chemometrics: Applications of Mathematics and Statistics to Laboratory Systems,* Ellis Horwood, Chichester, 1993.

[5] M. Otto, *Chemometrics: Statistics and Computer Application in Analytical Chemistry,* Wiley-VCH, Weinheim, 1999.

[6] G. Restrepo, H. Mesa, E. J. Llanos and J. L. Villaveces, Topological Study of the Periodic System, In: B. King and D. Rouvray, *The Mathematics of the Periodic Table*, Nova, New York, 2004 (in press).

[7] B. Mendelson, *Introduction to Topology,* Dover, New York, 1990.

[8] W. C. Fernelius and W. H. Powell, Confusion in the Periodic Table of the Elements, *J. Chem. Educ.* **59** 504-508(1982).

[9] R. D. M. Page, *COMPONENT User's manual (Release 1.5) University of Auckland,* Auckland, 1989 (http://taxonomy.zoology.gla.ac.uk/rod/cpw.html).

[10] G. Restrepo, H. Mesa, E. J. Llanos and J. L. Villaveces, Topological Study of the Periodic System, *J. Chem. Inf. Comp. Sci.* **44** 68-75(2004).

[11] G. Restrepo, Los elementos químicos, su matemática y relación con el sistema periódico, *Bistua.* **2** 91-98(2004).

[12] N. N. Greenwood and A. Earnshaw, *Chemistry of the Elements*, Butterworth-Heinemann, Oxford, 1998.

[13] S. Lipschutz, *General Topology*, McGraw-Hill, New York, 1965.

[14] H. A. Contreras, M. C. Daza and G. Restrepo: *Estudio Topológico de los Alcanos.* Proceedings of the Primer Encuentro Nacional de Químicos Teóricos, Universidad de Pamplona, Colombia, 2004.

VSP International
Science Publishers
P.O. Box 346, 3700 AH Zeist
The Netherlands

Lecture Series on Computer
and Computational Sciences
Volume 1, 2004, pp. 756-759

Molecular Simulation Studies of Fast Vibrational Modes

D. Janežič[1]and M. Penca

National Institute of Chemistry,
Hajdrihova 19,
SI-1000 Ljubljana, Slovenia

Received 1 September, 2004; accepted in revised form 15 September, 2004

Abstract: The survey of our past and present work on molecular simulation studies of fast vibrational modes will be presented. In particular, new symplectic integration algorithms for numerical solution of molecular dynamics equation and methods for the determination of vibrational frequencies and normal modes of large systems will be described.

Keywords: molecular dynamics simulation, normal mode analysis, symplectic integration methods, Hamiltonian systems, Lie algebra, vibrational modes, large systems

Mathematics Subject Classification: 65C20, 65L20, 70H05, 70H15

PACS: 31.15.Qg

A Introduction

Many physical problems, particularly in chemical and biological systems, involve processes that occur on widely varying time scales. Such problems have motivated the development of new methods for treating fast vibrational modes.

B Molecular Dynamics

Among the main theoretical methods of investigation of the dynamic properties of biological macromolecules, such as proteins, are molecular dynamics (MD) simulation and harmonic analysis. MD simulation is a technique in which the classical equation of motion for all atoms of a molecule is integrated over a finite period of time. The resulting trajectory is used to compute time-dependent properties of the system. Harmonic analysis is a direct way of analyzing vibrational motions. Harmonicity of the potential function is a basic assumption in the normal mode approximation used in harmonic analysis. This is known to be inadequate in the case of proteins because anharmonic effects, which MD has shown to be important in protein motion, are neglected. When anharmonic effects are incorporated quasiharmonic analysis may be applied. In this method, the MD simulation is utilized to obtain effective modes of vibration from the atomic fluctuations about an average structure. These modes include the anharmonic effects neglected in a normal mode calculation [1].

---

[1]Corresponding author. E-mail: dusa@cmm.ki.si

C Harmonic Dynamics

Harmonic analysis - normal mode calculation has been used for many years in the interpretation of vibrational spectra of small molecules. It provided the motivation for the application of harmonic approximation to large molecules, particularly proteins. For macromolecules, normal mode analysis focuses on the low frequency modes which are frequently associated with biological function.

The role of low frequency normal modes involving global conformation changes and which have been theoretically determined for several proteins is emphasized. Low frequency modes of proteins are particularly interesting because they are related to functional properties. The analysis of these motions in the limit of harmonic dynamics lends insight into the behavior and flexibility of these molecules. The modes presented here include the lowest modes of Bovine Pancreatic Trypsin Inhibitor (BPTI) [2, 3].

For a typical macromolecular system, the problem may be so intractable that it becomes desirable to reduce the computation cost of a harmonic analysis by making some approximations concerning the nature of the motions. One such approximation involves reducing of the size of the secular equation by partitioning the Hessian matrix into relevant and irrelevant parts. This can be done through the use of an appropriate unitary transformation which approximately block diagonalizes the full Hessian matrix. The irrelevant part is subsequently ignored, and the relevant part is represented by a smaller matrix. Another approximation is the reduced basis harmonic analysis techniques which allow the study of motions of interest in the harmonic limit. The reduction of the problem can be viewed as either the removal of unwanted motion through the use of constraints, or as the inclusion of desired motion.

Once the normal modes have been obtained, a great variety of analyses can be performed [1, 4].

HARMONIC ANALYSIS OF LARGE SYSTEMS

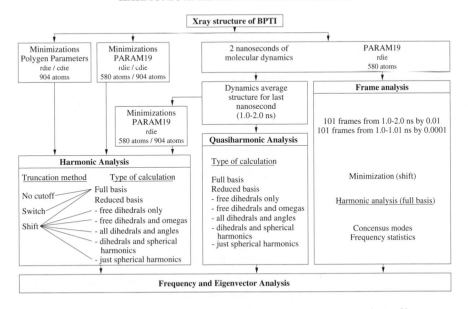

Figure 1: Shematic representation of procedures for performing harmonic analysis of large systems.

D Symplectic Dynamics

Harmonic analysis also proved useful in developing an efficient symplectic MD integration methods. Symplectic integration methods are often the right way of integrating the Hamilton equations of motion. Recent advances in development of SISM (Split Integration Symplectic Method) and HANA (Hydrogens ANAlytical) for combined analytical and numerical solution of the Hamiltonian system based on a factorization of the Liouville propagator are presented [5, 6].

These techniques, derived in terms of the Lie algebraic language, are based on the splitting of the total Hamiltonian of the system into two pieces, each of which can either be solved exactly or more conveniently than by using standard methods. The individual solutions are then combined in such a way as to approximate the evolution of the original equation for a time step, and to minimize errors [7].

The SISM and HANA use an analytical treatment of high frequency motions within a second order generalized leap-frog scheme. The computation cost per integration step for both methods is approximately the same as that of commonly used algorithms, and they allow an integration time step up to an order of magnitude larger than can be used by other methods of the same order and complexity [8].

The SISM and HANA have been tested on a variety of examples. In all cases they posses long term stability and the ability to take larger time steps while being economical in computer time.

The approach developed here is general, but illustrated at present by application to the MD integration of the model system of linear chain molecules and a box of water molecules.

Further improvements in efficiency were achieved by implementing the SISM to computers with highly parallel architecture [9]. The SISM performs in parallel as standard leap-frog Verlet method and the speedup is gained due to the larger time step used [10].

E Conclusions

The present work provides an overview of a variety of methods for the molecular, harmonic, and symplectic dynamics of large systems, such as biological macromolecules, in the case when all degrees of freedom were taken into account.

If the same macromolecular system employing the same potential function is used in all of the calculations, then a direct comparison of all the methods is possible. Through the combination of methods presented, insight can be gained into the dynamics and flexibility of large molecular systems.

Acknowledgment

This work was supported by the Ministry of Education, Science and Sports of Slovenia under grants No. P1-0002 and J1-6331.

References

[1] BROOKS Bernard R., JANEŽIČ Dušanka, KARPLUS Martin, Harmonic analysis of large systems. I. Methodology. *J. Comput. Chem.* **16** 1522-1542 (1995).

[2] JANEŽIČ Dušanka, BROOKS Bernard R. Harmonic analysis of large systems. II. Comparison of different protein models. *J. Comput. Chem.* **16** 1543-1553 (1995).

[3] JANŽIČ Dušanka, VENABLE Richard M., BROOKS Bernard R. Harmonic analysis of large systyems. III. Comparison with molecular dynamics. *J. comput. Chem.* **16** 1554-1566 (1995).

[4] JANEŽIČ Dušanka, Some multiple-time-scale problems in molecular dynamics. *Cell. Mol. Biol.* **7** 78-81 (2002).

[5] JANEŽIČ Dušanka, MERZEL Franci. Split integration symplectic method for molecular dynamics integration. *J. Chem. Inf. Comput. Sci.* **37** 1048-1054 (1997).

[6] JANEŽIČ Dušanka, PRAPROTNIK Matej. Symplectic Molecular Dynamics Integration Using Normal Mode Analysis. *Int. J. Quant. Chem.* **84** 2-12 (2001).

[7] PRAPROTNIK Matej, JANEŽIČ Dušanka. Symplectic Molecular Dynamics Integration Using Normal Mode Analysis. *Cell. Mol. Biol.* **7** 147-148 (2002).

[8] JANEŽIČ Dušanka, PRAPROTNIK Matej. Molecular Dynamics Integration Time Step Dependence of the Split integration Symplectic method on System Density. *J. Chem. Inf. Comput. Sci.* **43** 1922-1927 (2003).

[9] BORŠTNIK, Urban, HODOŠČEK, Milan, JANEŽIČ, Dušanka. Improving the performance of molecular dynamics simulations on parallel clusters. *J. Chem. Inf. Comput. Sci.* **44** 359-364 (2004).

[10] TROBEC, Roman, ŠTERK, Marjan, PRAPROTNIK, Matej, JANEŽIČ, Dušanka. Parallel programming library for molecular dynamics simulations. *Int. J. Quant. Chem.* **96** 530-536 (2004).

VSP International
Science Publishers
P.O. Box 346, 3700 AH Zeist
The Netherlands

*Lecture Series on Computer
and Computational Sciences*
Volume 1, 2004, pp. 760-761

Preface for the Mini-Symposium on:
"Simulations Of Nonlinear Optical Properties For The Solid State"

Benoît Champagne [1]

Laboratoire de Chimie Théorique Appliquée
Facultés Universitaires Notre-Dame de la Paix
61, rue de Bruxelles
B-5000 Namur (BELGIUM)

As reviewed recently [1], the calculation of the properties of individual molecules as found in an infinitely dilute gas has for long been of great interest to quantum chemists. This curiosity has been spurred in recent decades by the increasing importance of the communications industry in the world and the parallel need for materials having specific properties for electronic, optical and other devices. In particular, the nonlinear-optical quantities, defined at the microscopic level as hyperpolarizabilities and at the macroscopic level as nonlinear susceptibilities, have played a key role in determining the suitability of substances for practical use, for example, in electro-optical switching and frequency mixing.

With few exceptions, a useful nonlinear optical material will be in the solid phase, for example, a single crystal or a poled polymer embedded in a film. Ironically, the quantum chemical calculations of nonlinear optical properties have for the most part been concerned with a single microscopic species. Much has been learnt in this way about appropriate molecular construction but the ultimate goal must be to investigate the nonlinear optical (NLO) properties in the solid phase.

In the physics arena the theoretical determination of NLO properties of solids has been more advanced though not to the degree that has been achieved for simple gas-phase molecules using modern quantum-chemical practices. For example, density functional theory in its crudest form has been frequently adopted to find some NLO properties for semi-conductors. A glaring example of lack of progress is the third-order susceptibility of quartz. There is, as yet, no rigorous calculation of this quantity, even though it is the reference point for nearly all NLO measurements. It is our opinion that in the next few decades this situation is going to change, that the field of single molecule calculations will be saturated and attention will turn to the more practically relevant solid phase. This makes it an opportune time to review and discuss in this minisymposium what computational strategies have been already developed.

As the situation currently stands, there are two extreme approaches: (i) the oriented-gas model, and (ii) the supermolecule model - and everything in between. For (i), in its simplest disguise, single molecules lie side-by-side and the properties of the solid are just an appropriate combination of the molecular ones. For (ii), the whole solid is considered as a giant molecule and the computational approach is that of standard quantum-chemical calculations. An important variation of (ii) is to make use of the translational symmetry that exists in a crystal or a periodic polymer and this leads to so-called crystal orbital methods. Here is an important break from the conventional molecular calculations and one that is likely to see a great deal more use in the NLO field in the near future. Refinement of the oriented-gas model, including more and more precise accounting for inter-species interactions, is another likely avenue for progress.

The macroscopic optical responses of a medium are given by its linear and nonlinear susceptibilities, which are the expansion coefficients of the material polarization, P, in terms of the

---

[1] Corresponding author. benoit.champagne@fundp.ac.be

Maxwell fields, E. For a dielectric or ferroelectric medium under the influence of an applied electric field, the defining equation reads:

$$P = P_0 + \chi^{(1)}E + \chi^{(2)}E^2 + \chi^{(3)}E^3 + ... \tag{1}$$

where P_0 is the electric dipole moment per unit volume in zero electric field, $\chi^{(1)}$ is the (electric dipole) linear susceptibility, $\chi^{(2)}$ and $\chi^{(3)}$ are the second- and third-order nonlinear susceptibilities. For the sake of clarity, the tensor nature and frequency-dependence of the susceptibilities have been omitted. In Eq. 1, the polarization and Maxwell fields are macroscopic quantities in the sense they are derived from their microscopic counterpart by a suitable averaging procedure over a region which is large with respect to the atomic dimensions but small with respect to the wavelengths of the optical waves. For an isolated molecule, the electric dipole moment, μ, is expanded in a Taylor series of the external field:

$$\mu = \mu_0 + \alpha E + \frac{1}{2}\beta E^2 + \frac{1}{6}\gamma E^3 + ... \tag{2}$$

where μ_0, α, β, and γ are the permanent dipole moment, the polarizability, and the first and second hyperpolarizabilities, respectively. These microscopic responses are, in many schemes, used as a starting point to evaluate the macroscopic responses and the effects of the surroundings are introduced subsequently. This is the case for ionic and molecular solids where disjoint volumes can easily be associated with the constituting entities and it also applies to bond/atom charge models. In such cases, the effects of the surroundings can be decomposed into two contributions. First, the surroundings modify the molecule/ion wavefunction and its properties through permanent electrostatic interactions as well as by polarization, dispersion and exchange-overlap contributions. The corresponding (microscopic) NLO properties are generally referred to as "dressed" or effective hyperpolarizabilities. The second effect is the difference between the (local) electric field which (hyper)polarizes the entities and the macroscopic field which defines the bulk susceptibility. The inhomogeneity and the non-locality of the responses to the applied fields are consequently crucial aspects to be included in these determinations.

The experts who have kindly accepted the invitation to contribute to this minisymposium address different aspects of the methodology, of the calculations, and of the characterization of the NLO properties of solid phases, showing pros and cons of the different methods and pointing out directions of investigations for the coming years.

[1] B. Champagne and D.M. Bishop, Calculations of Nonlinear Optical Properties for the Solid State, Adv. Chem. Phys. 126 (2003), 41-92.

International
Science Publishers
P.O. Box 346, 3700 AH Zeist
The Netherlands

*Lecture Series on Computer
and Computational Sciences*
Volume 1, 2004, pp. 762-764

Modeling of the (Non)Linear Optical Properties of an Infinite Aggregate of *All-Trans* Hexatriene Chains by the Supermolecule Approach and the "Electrostatic Interactions" Model

Maxime Guillaume

Laboratory of Applied Theoretical Chemistry
Faculty of Science
Facultés Universitaires Notre-Dame de la Paix
5000 Namur BELGIUM

*Abstract :*A classical electrostatic polarization scheme using the additive distribution procedure, and a supermolecule approach, have been applied to determine the longitudinal polarizability of an all-trans hexatriene molecule in an infinite stretched fiber. The parameters have been derived from ab initio CPHF/6-31G calculations and the electrostatic scheme has been validated via comparison with supermolecule *ab initio* results on small clusters. Upon packing the polarizability of all-trans hexatriene increases by 7%. This small increase results from the balance between the enhancement of the polarizability due to collinear packing and the reduction associated with lateral packing.

Keywords: (hyper)polarizability, crystal packing effects, interaction scheme, clusters, all-trans hexatriene

In this article, the approach of the electrostatic interactions scheme proposed in Refs. 1 and 2 and the supermolecule, or cluster, approach used in Refs 3 and 4 are applied to the evaluation of the polarizability of aggregates of *all-trans* hexatrienes that grow in the three spatial directions. The PA fibers have been experimentally-characterized by Zhu *et al.* [5]. A projection of the structure in the plane perpendicular to the chain direction is sketched in Fig. 1. Each chain has 6 nearest-neighbors, two are parallel, four are almost perpendicular and are displaced in the longitudinal direction by halve of a unit cell. The second shell contains 12 chains, the third shell, 18, … the nth shell, $6 \times n$. Although the lateral distances between the hexatriene chains are determined by the unit cell parameters (Fig. 1), the intermolecular distance between collinear chains has been fixed at the van der Waals contact distance (d_{CC} = 4.289 Å). Up to 5 shells of 29 collinear chains have been considered in this study, *i.e.* more than 10 000 carbon atoms. This corresponds to numbers of laterally-packed chains, $n\perp$, of 1, 7, 19, 37, and 61. A perspective view of the collinear packing of the chains along the longitudinal axis is also presented in Fig. 1 for the first shell. The laterally packed molecules form what is called bundles.

The validity of this scheme to reproduce the longitudinal polarizability (as well as second hyperpolarizability) of small bundles of hexatrienes has been checked using a supermolecule approach. The data reported in Table 1 display the comparison between *ab initio* and electrostatic calculations of polarizability tensors for collinear arrangements of hexatrienes. For the longitudinal component, which is the main focus of this investigation, the agreement between the two schemes is very good with an error smaller than 0.3%.

The evolution of the longitudinal polarizability component of the central hexatriene chain, as a function of the size of the system for the different bundles sizes is given at Fig. 2 Using extrapolation schemes [4] for each bundles, the effective polarizability longitudinal component has been estimated to 151.8 atomic units, i.e. 7% larger than the polarizability of the isolated molecule.

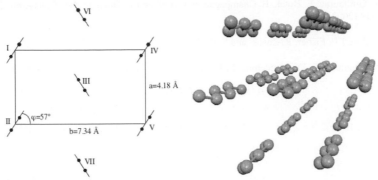

Figure 1. Lateral cut representation and perspective projection of the *all-trans* hexatriene chain fibers, as given by Ref [5]. The points on the left figure represent the projection of the six carbon atoms on the (y,z) plane of the system, while the figure on the right shows the collinear arrangement of the fiber for the first shell

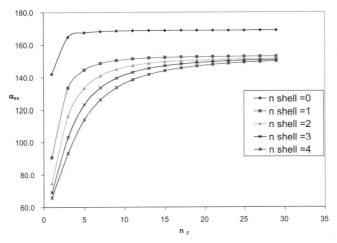

Figure 2. Evolution of the longitudinal polarizability of the central all-trans hexatriene molecule as a function of the number of collinear chains, $n_{//}$, for different number of shells (n shell). The polarizability is expressed in atomic units. (1 a.u. of α= 1.648778 10^{-41} C^2 m^2 J^{-1} = 0.14818 10^{-24} cm^3).

Acknowledgments

M. Guillaume thanks the "Fonds pour la Formation à la Recherche dans l'Industrie et dans l'Agriculture" (FRIA) for financial support. The calculations have been performed on the Interuniversity Scientific Computing Facility (ISCF), installed at the Facultés Universitaires Notre-Dame de la Paix (Namur, Belgium), for with the authors gratefully acknowledge the financial support of the FNRS-FRFC and the "Loterie Nationale" for the convention n° 2.4578.02 and of the FUNDP.

References

[1] C.E. Dykstra, *J. Comp. Chem.* **9**, 476 (1988).

[2] B. Kirtman, C.E. Dykstra, and B. Champagne, *Chem. Phys. Lett.* **305**, 132 (1999).

[3] E. Botek and B. Champagne, *Chem. Phys. Lett.* **370**, 197 (2003).

[4] M. Guillaume, E. Botek, B. Champagne, F. Castet, and L. Ducasse, *Int. J. Quantum Chem.* **90**, 1378 (2002)

[5] Q. Zhu, J.E. Fisher, R. Zusok, and S. Roth, *Solid State Commun.* **83**, 179 (1992).

[6] B. Champagne, D. Jacquemin, J.M. André, and B. Kirtman, *J. Phys. Chem. A* **101**, 3158 (1997).

VSP International
Science Publishers
P.O. Box 346, 3700 AH Zeist
The Netherlands

*Lecture Series on Computer
and Computational Sciences*
Volume 1, 2004, pp. 765-768

First-Principles Study of Non-Linear Optical Properties of Ferroelectric Oxides

M. Veithen[1] and Ph. Ghosez

Département de Physique (B5), Université de Liège, Allée du 6 août 17,B-4000 Sart Tilman,
Belgium

X. Gonze

Unité PCPM, Université Catholique de Louvain, Place Croix du Sud 1, B-1348 Louvain-la-Neuve,
Belgium

Abstract: We describe recent advances in condensed matter physics that make it possible to study non-linear optical properties of periodic solids from first-principles density functional theory. These methods are either based on perturbation theory in the framework of the $2n+1$ theorem, finite electric fields or effective hamiltonians. We apply them to the study of non-linear optical properties of various ferroelectric ABO_3 compounds such as $BaTiO_3$, $PbTiO_3$ or $LiNbO_3$.

Keywords: Electric fields, non-linear optical properties, density functional theory

1   Introduction

In order to study the non-linear optical (NLO) properties of a system from first principles, one has to determine its response to an electric field $\mathcal{E}$. In periodic solids, usually described within Born-von Karman boundary conditions, the electric field perturbation requires a careful treatment. In fact, the scalar potential $\mathcal{E} \cdot r$, where r is the position operator, breaks the periodicity of the crystal lattice. Moreover, it is unbound from below: it is always possible to lower the energy of the system by transferring electrons from the valence states to the conduction states in a distant region (Zener breakdown). Both problems can be overcome by using a modified energy functional

$$E\left[\psi_{n,\mathbf{k}}, \mathcal{E}\right] = E_0\left[\psi_{n,\mathbf{k}}\right] - \Omega \mathcal{E} \cdot \mathcal{P} \tag{1}$$

where E_0 is the zero field energy and $\mathcal{P}$ the polarization that can be computed as a Berry phase of field-polarized Bloch functions $\psi_{n,\mathbf{k}}$. On the one hand, using a perturbation expansion of Eq. (1) [1], we obtain analytic expressions of electric field derivatives of the energy and related physical quantities such as NLO susceptibilities, Raman tensors or electrooptic (EO) tensors. On the other hand, for small fields, Eq. (1) can be minimized with respect to the $\psi_{n,\mathbf{k}}$ to study the response of a solid to a finite electric field [2]. Both approaches will be discussed below in the framework of the Kohn-Sham density functional theory (DFT). Then, we use them in the study of ferroelectric ABO_3 compounds, a class of materials with high EO and NLO coefficients that are of direct interest for various technological applications [3]. In addition, we present a method to compute the temperature dependence of the EO coefficients and the refractive indices of these materials within a first-principles effective hamiltonian approach [4].

---

[1]Corresponding author. E-mail: Marek.Veithen@ulg.ac.be

2 First-principles study of non-linear optical properties from density functional perturbation theory

Many important properties of solids can be expressed as derivatives of an appropriate thermodynamic potential with respect to perturbations such as atomic displacements, electric fields or macroscopic strains. For example, the NLO susceptibilities $\chi_{ijl}^{(2)}$ can be computed as third-order derivatives of the energy functional in Eq. (1) with respect to three electric fields [5] while the Raman susceptibilities are related to third-order derivatives of E with respect to two electric fields and one atomic displacement [6]. These derivatives can be computed from density functional perturbation theory by applying the so-called $2n + 1$ theorem [7] to Eq. (1) [1]. This theorem says that it is possible to compute energy derivatives up to the order $2n + 1$ from the derivatives of the wavefunctions up to the order n. As a consequence, third-order energy derivatives can be computed from the knowledge of the ground-state wavefunctions and of their first-order derivatives with respect to the corresponding perturbations.

The NLO susceptibilities are determined by the response of the valence electrons to an electric field. At the opposite, the electrooptic coefficients involve the response of the electrons and the crystal lattice. In the Born and Oppenheimer approximation, these coefficients can be decomposed into (i) a bare *electronic* part, (ii) an *ionic* contribution and (iii) a *piezoelectric* contribution [8]. The *electronic* part describes the response of the valence electrons to a (quasi-)static macroscopic electric field when the ions are considered as artificially clamped at their equilibrum positions, and can be deduced from the NLO susceptibilities. The *ionic* contribution is produced by the relaxation of the atoms within the quasi-static electric field. It can be computed from the knowledge of the frequency, polarity and Raman susceptibility of the zone-center transverse optical phonon modes. Finally, the *piezoelectric* contribution is related to the modifications of the unit cell shape induced by the electric field. It can be computed from the piezoelectric strain coefficients and the elasto-optic coefficients.

We apply this methodology to the study of the EO response of various ferroelectric oxides such as $LiNbO_3$, $BaTiO_3$ and $PbTiO_3$, with the aim of identifying the origin of the large electrooptic coefficients of these compounds. At high temperature, these materials are in a highly-symmetric paraelectric phase. As the temperature is lowered, they undergo one or several ferroelectric phase transitions driven by the condensation of an unstable phonon mode. At the phase transition, this unstable mode transforms into a *low energy* and *highly polar* mode in the ferroelectric phase that can strongly interact with an electric field. It is therefore apt to generate a large EO response if it exhibits, in addition, a *large Raman susceptibility*. This is the case in the ferroelectric phase of $LiNbO_3$ and $BaTiO_3$, where we observe a strong contribution of the successor of the soft mode to coefficients r_{13} and r_{33}. At the opposite, in the tetragonal phase of $PbTiO_3$, the successor of the soft mode plays a minor role due to its lower Raman susceptibility.

3 Finite electric field techniques

As mentioned in the introduction, because of the interband tunneling, an insulator in an electric field does not have a true ground state. In many practical situations however, tunneling is negligible on the relevant time scale and, for relatively small fields, the system remains in a polarized long lived metastable state. It is therefore possible to minimize the energy functional in Eq. (1) and to study the response of a solid to a finite electric field. This method offers an alternative approach to the one described in the previous section. In particular, it allows an accurate computation of the energy, the forces, the stress tensor and the polarization in the presence of an electric field [2]. It can therefore be applied to study the structural and electronic response of an insulator to an electric field and to compute electric field derivatives of the energy up to any order from finite differences.

4   Temperature dependence of the EO coefficients

Based on the dominant contribution of the soft mode to the EO tensor at 0 K, we develop an effective Hamiltonian [4] approach to simulate its temperature dependence in BaTiO$_3$. This approach is based on a Taylor expansion of the total energy and the electronic dielectric susceptibility around the paraelectric reference structure. The only degrees of freedom included in this expansion are the macroscopic strain and atomic displacements along the lattice Wannier function associated to the soft mode. The finite-temperature dependence of the model is studied from Monte Carlo simulations.

In the tetragonal phase, we observe a divergence of the EO coefficients r$_{13}$ and r$_{33}$ at the transition to the cubic phase and a divergence of r$_{42}$ at the transion to the orthorhombic phase. Moreover, the formalism used in the effective Hamiltonian provides a microscopic interpretation of the model of DiDomenico and Wemple [9] that identifies the linear EO effect in ferroelectrics to a quadratic effect biased by the spontaneous polarization.

Acknowledgments

MV and XG acknowledge financial support from the FNRS Belgium. This work was supported by the Volkswagen Stiftung project "Nano-sized ferroelectric Hybrids" (I/77 737), FNRS (grants 9.4539.00 and 2.4562.03), the Région Wallonne (Nomade, project 115012), the PAI 5.01 and EU Exciting network (European Union contract HPRN-CT-2002-00317).

References

[1] R. W. Nunes and X. Gonze, Berry-phase treatment of the homogeneous electric field perturbation in insulators, *Phys. Rev. B* **63** 155107 (2001).

[2] I. Souza, J. Íñiguez, and D. Vanderbilt, First-Principles Approach to Insulators in Finite Electric Fields, *Phys. Rev. Lett.* **89** 117602 (2002).

[3] M. E. Lines and A. M. Glass, *Principles and applications of ferroelectrics and related materials* Clarendon Press, Oxford, 1977.

[4] W. Zhong, David Vanderbilt and K. M. Rabe, Phase Transitions in BaTiO$_3$ from First Principles, *Phys. Rev. Lett.* **73** 1861-1864 (1994).

[5] A. Dal Corso, F. Mauri and A. Rubio, Density-functional theory of the nonlinear optical susceptibility: Application to cubic semiconductors, *Phys. Rev. B* **53** 15638-15642 (1996).

[6] G. Deinzer and D. Strauch, Raman tensor calculated from the 2n + 1 theorem in density-functional theory, *Phys. Rev. B* **66** 100301 (2002).

[7] X. Gonze and J.-P. Vigneron, Density-functional approach to nonlinear-response coefficients of solids, *Phys. Rev. B* **39** 13120-13128 (1989).

[8] M. Veithen, X. Gonze and Ph. Ghosez, unpublished.

[9] M. DrDomenico Jr. and S. H. Wemple, Oxygen-Octahedra Ferroelectrics. I. Theory of Electro-optical and Nonlinear optical Effects, *J. Appl. Phys.* **40** 720-734 (1969).

VSP International
Science Publishers
P.O. Box 346, 3700 AH Zeist
The Netherlands

*Lecture Series on Computer
and Computational Sciences*
Volume 1, 2004, pp. 769-770

Mixed Electric-Magnetic Second Order Response of Helicenes

Edith Botek[*,1], Jean-Marie André[1], Benoît Champagne[1], Thierry Verbiest[2], and André Persoons[2]

[1] Laboratoire de Chimie Théorique Appliquée, Facultés Universitaires Notre-Dame de la Paix, rue de Bruxelles, 61, B-5000 Namur (Belgium).
[2] KU Leuven, Laboratory of Chemical and Biological Dynamics, University of Leuven, Celestijnenlaan 200D, B-3001 Leuven (Belgium)

Received June 26 2004; accepted July 15 2004

Abstract : The mixed electric-magnetic second-order NLO responses of helicenes are evaluated. The microscopic responses are determined at the RPA level whereas the macroscopic responses are obtained by averaging over the microscopic response for the various orientations of the molecule.

Keywords: second-order NLO response, mixed electric-magnetic response, helicenes

PACS: 31.15.-p, 31.25.-v, 33.15.Kr, 33.55.Fi

1. Introduction

Chiral systems are particularly interesting to study second-order nonlinear optical (NLO) effects due to their intrinsic non-centrosymmetry that allows NLO processes to be observed even in highly symmetric media [1]. Hicks and co-workers and Persoons and co-workers, both independently, demonstrated the presence of anomalously large chiral effects in NLO responses measurements of surface films of chiral molecules and chiral polymers [2-3]. Schanne-Klein et al. have reported the presence of magnetic-dipole contributions of the same order of magnitude as electric-dipole contributions performing SHG-ORD/CD/LD experiments in a nonresonant configuration for an isotropic spin-coated layer of a chiral salt [4]. Hence, the possibility of enhancing NLO properties by properly optimizing such magnetic-dipole contribution appears to become an exploitable tool to obtain new NLO materials. Persoons and co-workers have developed experimental techniques, based on SHG, to study magnetic-dipole contributions to the nonlinearity in thin films of chiral materials. They are able to detect magnetic-dipole contributions on the order of 5% of the highest electric-dipole counterparts [5-7].

In the present work the relation between the magnetic contributions (up to first order) to the first susceptibility with respect to the pure electric contribution is investigated from a theoretical frame in a series of chiral systems.

2. Mixed electric-magnetic nonlinear responses

At the microscopic (molecular) level, the second-order electric-dipole moment is:

$$\mu_i(2\omega) = \sum_{jk} \left[\beta_{ijk}^{eee} E_j(\omega) E_k(\omega) + \beta_{ijk}^{eem} E_j(\omega) B_k(\omega) \right] \tag{1}$$

---

[*] Corresponding author. e-mail: edith.botek@fundp.ac.be

and the second-order magnetic-dipole moment is:

$$m_i(2\omega) = \sum_{jk} \beta_{ijk}^{mee} E_j(\omega) E_k(\omega) \tag{2}$$

acting both as sources of radiation. The subscripts i, j and k are the Cartesian coordinates, meanwhile the superscripts refer the electric-dipole (e) and the magnetic-dipole (m) interactions. β_{ijk} is a component of the first hyperpolarizability and ω is the frequency of the incident light.

The microscopic hyperpolarizabilities are calculated employing the Green functions formalism at the Random Phase Approximation (RPA) level [8] using the 6-31G* basis set by means of the quadratic response functions [9]. The macroscopic response $\chi^{(2)}$ is obtained by averaging over the microscopic response for the various orientations of the molecule in space with respect to the laboratory coordinate system.

3. Second-order NLO responses of helicenes

Families of substituted helicenes (as well as the corresponding non-substituted species) presenting the potentiality of optimizing the magnetic contribution for maximizing the NLO response were considered as examples to investigate the relation between pure electric and mixed electric-magnetic macroscopic responses. The electric response of some helicenes investigated in this work has been studied in a previous work [10] by tuning the first hyperpolarizability with a proper addition of donor/acceptor groups. For the present work the calculations predict mixed electric-magnetic responses which, in some cases, are not negligible with respect to the pure electric contributions. Such predictions could be useful for complementing investigations related to the NLO processes in several field of sciences.

References

[1] M. Kauranen, T. Verbiest, and A. Persoons, J. Nonlinear Optical Phys. & Materials **8**, 171 (1999).

[2] T. Petralli-Mallow, T.M. Wong, J.D. Byers, H.I. Yee, and J.M. Hicks, J. Phys. Chem. **97**, 1383 (1993).

[3] M. Kauranen, T. Verbiest, E.W. Meijer, E.E. Havinga, M.N. Teerenstra, A.J. Schouten, R.J.M. Nolte, and A. Persoons, Adv. Mater. **7**, 641 (1995).

[4] M.C. Schanne-Klein, F. Hache, A. Roy, C. Flytzanis, and C. Payrastre, J. Chem. Phys. **108**, 9436 (1998).

[5] M. Kauranen, J.J. Maki, T. Verbiest, S. Van Elshocht, and A. Persoons, Phys. Rev. B 55, 1985 (1997).

[6] S. Sioncke, T. Verbiest, and A. Persoons, Optical Materials **21**, 7 (2002).

[7] T. Verbiest, S. Sioncke, and A. Persoons, J. Photochem. And Photobiology A: Chemistry **145**, 113 (2001).

[8] T. H. Dunning and V. McKay, J. Chem. Phys. **47**, 1735 (1967); **48**, 5263 (1968).

[9] See for example: W.A. Parkinson, and J. Oddershede, J. Chem. Phys. **94**, 7251 (1991).

[10] B. Champagne, J.M. André, E. Botek, E. Licandro, S. Maiorana, A. Bossi, K. Clays, and A. Persoons, CHEMPHYSCHEM, in press.

VSP International
Science Publishers
P.O. Box 346, 3700 AH Zeist
The Netherlands

*Lecture Series on Computer
and Computational Sciences*
Volume 1, 2004, pp. 771-774

Determination of the Macroscopic Electric Susceptibilities $\chi^{(n)}$ from the Microscopic (hyper)polarizabilities α, β and γ

M. Rérat[1]

Laboratoire de Chimie Théorique et de Physico-Chimie Moléculaire, UMR5624,
Université de Pau, FR-64000 Pau, France

R. Dovesi

Dipartimento di Chimica, IFM,
Università di Torino, I-10125 Torino, Italy

Received 31 July, 2004

Abstract: In this work, we will show that it is possible to evaluate the bulk dielectric constant $\epsilon^{(0)}$ and the electric susceptibilities $\chi^{(n)}$ of a crystalline system in two different ways, namely from the variation of the cell energy as a function of the applied electric field, or from the macroscopic response field in the solid. The two ways are obviously equivalent from a formal point of view; the numerical equivalence was till now undocumented. The use a saw-tooth like finite potential (so that a positive and a negative field is applied at the same time), required for maintaining the periodicity of the lattice, introduces some complication in the formulation of the working equations.

Keywords: Dielectric constant, non linear optical properties, ab initio calculation, periodic system

PACS: 71.15.Ap and 71.15.Mb, 77.22.Ch, 42.65.An

1   Formulas

If we expand the total energy of the system in terms of the external applied field F_0, that for simplicity we suppose to be oriented along the z-direction, we get:

$$E(F_0) = E_0 - \mu_0 F_0 - \frac{1}{2!}\alpha_0 F_0^2 - \frac{1}{3!}\beta_0 F_0^3 - \frac{1}{4!}\gamma_0 F_0^4 - \cdots \tag{1}$$

The total dipole moment is minus the first derivative of the total energy with respect to F_0, so we have:

$$\mu = \alpha_0 F_0 + \frac{1}{2}\beta_0 F_0^2 + \frac{1}{6}\gamma_0 F_0^3 + \cdots \tag{2}$$

as $\mu_0 = 0$ when periodic boundary conditions are applied.

From Eq. (1), it results that: $\alpha_0 = -(\frac{\partial^2 E}{\partial F_0^2})_{F_0 \to 0}$, where E is the energy of the unit cell.

---

[1]Corresponding author. E-mail: michel.rerat@univ-pau.fr

If we define the macroscopic polarization vector P in terms of the microscopic dipole induced by the applied field F_0, we have:

$$P = \mu/V = \frac{\alpha_0}{V} F_0 + \frac{\beta_0}{2V} F_0^2 + \frac{\gamma_0}{6V} F_0^3 + \cdots = \epsilon_0 (\chi^{(1)} F + \chi^{(2)} F^2 + \chi^{(3)} F^3 + \cdots) \tag{3}$$

where the previous equation defines the linear susceptibilities $\chi^{(i)}$; V is the volume of the unit cell, ϵ_0 is the vacuum permittivity, and F is the local macroscopic field (to be distinguished from the external applied field F_0) such as:

$$F = F_0 - P/\epsilon_0 \tag{4}$$

The dielectric constant being defined as $\epsilon = F_0/F$, then we get from the two previous equations:

$$
\begin{aligned}
\epsilon &= \frac{F_0}{F} = \frac{F + P/\epsilon_0}{F} = (1 + \chi^{(1)}) + \chi^{(2)} F + \chi^{(3)} F^2 + \cdots \\
&= (1 + 4\pi \frac{\alpha_0}{V} \frac{F_0}{F}) + 4\pi \frac{\beta_0}{2V} (\frac{F_0}{F})^2 F + 4\pi \frac{\gamma_0}{6V} (\frac{F_0}{F})^3 F^2 + \cdots
\end{aligned} \tag{5}
$$

remembering that the vacuum permittivity ϵ_0 is equal to $1/4\pi$ when atomic units are used.

Comparing the two expansions, using: $\epsilon^{(0)} = 1 + \chi^{(1)}$, and the definition: $\epsilon^{(i)} = \chi^{(i+1)}$ for $i > 0$, we obtain:

$$
\begin{aligned}
\epsilon^{(0)} + \epsilon^{(1)} F + \epsilon^{(2)} F^2 + \cdots &= 1 + 4\pi \frac{\alpha_0}{V} (\epsilon^{(0)} + \epsilon^{(1)} F + \epsilon^{(2)} F^2) \\
&\quad + 4\pi \frac{\beta_0}{2V} (\epsilon^{(0)} + \epsilon^{(1)} F + \epsilon^{(2)} F^2)^2 F \\
&\quad + 4\pi \frac{\gamma_0}{6V} (\epsilon^{(0)} + \epsilon^{(1)} F + \epsilon^{(2)} F^2)^3 F^2 + \cdots
\end{aligned} \tag{6}
$$

Equating the coefficients of the same power of F on both sides, we obtain:

$$\epsilon^{(0)} = 1 + \chi^{(1)} = \frac{1}{1 - 4\pi\alpha_0/V} \tag{7}$$

$$\epsilon^{(1)} = \chi^{(2)} = \frac{2\pi}{V} \frac{\beta_0 \epsilon^{(0)2}}{1 - 4\pi\alpha_0/V} = \frac{2\pi}{V} \beta_0 \epsilon^{(0)3} \tag{8}$$

$$\epsilon^{(2)} = \chi^{(3)} = \frac{4\pi}{V} \frac{\beta_0 \epsilon^{(0)} \epsilon^{(1)} + \frac{\gamma_0}{6} \epsilon^{(0)3}}{1 - 4\pi\alpha_0/V} = \frac{4\pi}{V} \epsilon^{(0)2} (\beta_0 \epsilon^{(1)} + \frac{\gamma_0}{6} \epsilon^{(0)2}) \tag{9}$$

If symmetry is such that β_0 in Eq. (1) is null, $\epsilon^{(1)} = 0$ and the first addendum to $\epsilon^{(2)}$ desappears. This is the case, for example, of a cubic system such as LiF.

2 Application to LiF

In order to explore the effect of the applied external field on the dielectric properties of the cubic LiF crystal, we considered:

a) the total energy of the system as a function of F_0, reported in Table 1, and,

b) the macroscopic response of the electron density and field using Poisson's equation as discussed in Ref. [1].

The SCF-LCAO wave functions and cell energies have been obtained at the LDA level of calculation with the CRYSTAL program [2], using the middle-size b)-basis set given in Ref.[1], a saw-tooth electric potential describing the finite field (FF) perturbation has been added to the Fock operator in the SCF process; periodic boundary conditions have been maintained.

Let us see how to proceed in case a).

The total energy of the supercell with $N = 4$, $N = 6$ and $N = 8$ unit cells, given in Table 1, are fitted according to Eq. (1):

$$E_4 = -854.181907 - 14.6256F_0^2 - 24.3298F_0^4$$

$$E_6 = -1281.27286 - 23.1749F_0^2 - 40.2897F_0^4$$

$$E_8 = -1708.36381 - 31.7250F_0^2 - 56.3536F_0^4$$

Odd terms are not included due to symmetry reasons (this means that β_0 and $\epsilon^{(1)}$ are null by symmetry).

As we are trying to simulate the effect of a constant electric field (for example $F_0 = +0.005$), the results obtained with the sawtooth potential cannot be used as such, because such potential corresponds to the application of a field $F_0 = +0.005$ to one half of the infinite lattice, and of a field $F_0 = -0.005$ to the other half, with a large perturbation at the border zone among the two values. On the other hand, if we consider the supercells with $N = 4$ and $N = 6$, we can look at the latter as the former to which the zones where the field is constant have been increased by one cell. The difference between the E_4 and E_6 fits represents then the effect of F_0 on parts of the crystal where the field is constant, because discontinuities have been eliminated through the difference.

Remembering Eq. (1), the difference between the second and fourth order terms provides α_0 and γ_0 once the appropriate factorials and volume effects are taken into account:

$$\alpha_0 = 2! \times (23.1749 - 14.6256)/M$$

$$\gamma_0 = 4! \times (40.2897 - 24.3298)/M$$

where M is the difference in the number of LiF units in the $N = 6$ and $N = 4$ calculations ($M = 4$ in the present case). The resulting value for α_0 is: 4.27465 a.u.. In order to obtain $\epsilon^{(0)}$, we use Eq. (7) (the volume of the unit cell is: 16.20487 $\mathring{A}^3$) to get the value given in Table 2. For γ_0, we obtain: 95.7594 a.u.. In order to obtain $\epsilon^{(2)}$ or $\chi^{(3)}_{zzzz}$, we apply Eq. (9) where the β_0 contribution is null by symmetry.

$$\chi^{(3)}_{zzzz} = 2\pi \times (1.9655)^4 \times 95.7594/(3 \times 16.20487) = 27.368 \text{ a.u.}$$

This value is given in Table 2. In cgs units, $\chi^{(3)}_{zzzz} = 27.368/(1.7153 \times 10^7)^2 = 0.93 \times 10^{-13}$ esu^{-2} cm^4.

Similar calculations have been performed for E_8 and E_6; the results are also reported in Table 2.

b) The dielectric constant calculated from the averaged charge density for the three supercells and values F_0 are given in Table 1. The fit of these values provides the following coefficients:

$$\epsilon_4 = 1.9731 + 6.8784F_0^2 = 1.9731 + 26.7789F^2$$

$$\epsilon_6 = 1.9704 + 7.0249F_0^2 = 1.9704 + 27.2746F^2$$

$$\epsilon_8 = 1.9704 + 6.9467F_0^2 = 1.9704 + 26.9706F^2$$

In Table 2, the ϵ_0 and $\chi_{zzzz}^{(3)}$ values as obtained directly from the fitting are reported. Good agreement between the values obtained with the two strategies is observed, with differences smaller than 1% for $\chi_{zzzz}^{(3)}$ and 0.2% for ϵ_0.

Table 1: Total energy and dielectric constant as a function of the applied field F_0 for two different supercells. The field is applied along the z-direction.

$N = 4$	F_0 (a.u.)	E (a.u.)	ϵ
	0.000	-854.18190657	-
	0.005	-854.18227222	1.973301
	0.010	-854.18336937	1.973806
	0.015	-854.18519856	1.974675
$N = 6$	F_0 (a.u.)	E (a.u.)	ϵ
	0.000	-1281.27285994	-
	0.005	-1281.27343933	1.970589
	0.010	-1281.27517782	1.971129
	0.015	-1281.27807632	1.971996
$N = 8$	F_0 (a.u.)	E (a.u.)	ϵ
	0.000	-1708.36381314	-
	0.005	-1708.36460626	1.970584
	0.010	-1708.36698616	1.971105
	0.015	-1281.37095407	1.971851

Table 2: Static dielectric constant and second order non linear susceptibility as obtained with the two different schemes

method	$\epsilon^{(0)}$	$\chi_{zzzz}^{(3)}$ (a.u.)
$E_6 - E_4$	1.9655	27.368
$E_8 - E_6$	1.9656	27.556
ϵ_4	1.9731	26.779
ϵ_6	1.9704	27.275
ϵ_8	1.9704	26.971

References

[1] C. Darrigan, M. Rérat, G. Mallia and R. Dovesi, *J. Comp. Chem.*, **24** (2003) 1305

[2] http://www.crystal.unito.it

VSP International
Science Publishers
P.O. Box 346, 3700 AH Zeist
The Netherlands

Lecture Series on Computer
and Computational Sciences
Volume 1, 2004, pp. 775-778

Elongation Method at Semi-empirical and *ab initio* Levels for Large Systems

Y. Aoki[1,2], F.L. Gu[2] and J. Korchowiec[2,3]

[1]Department of Molecular and Material Sciences,
Interdisciplinary Graduate School of Engineering Sciences,
Kyushu University, 6-1 Kasuga-Park,
Fukuoka, 816-8580, Japan

[2]Group, *PRESTO*,
Japan Science and Technology Corporation (JST),
Kawaguchi Center Building,
Honcho 4-1-8, Kawaguchi,
Saitama 332-0012, Japan

[3]K. Gumiński Department of Theoretical Chemistry,
Jagiellonian University,
R. Ingardena 3,
30–060 Cracow, Poland

Abstract: We are developing the elongation method that makes it possible to obtain the electronic states of large aperiodic polymers, treating only a few units at a time, of the total system using molecular orbitals(MOs) localized in a specific region of the system. This treatment has been developed at the semi-empirical, *ab initio*, and density functional theory (DFT) levels of theory. The accuracy and efficiency of this method are demonstrated by polyglycine model systems.

Keywords: Elongation method, Quantum chemical approach, Polyglycine, Ab initio method, DFT

1. Basic Concept of the Elongation Method

The conventional quantum chemical approach based on the molecular orbital method is difficult to apply to large systems like biopolymers though it is useful for small molecules. From the 1990s, we have been developing an elongation method for efficient calculations of large polymers [1]. This treatment is so called as O(*N*) quantum chemical approach because constant number of localized molecular orbitals (LMOs) as well as atomic orbitals (AOs) is involved in diagonalization through the elongation process. This makes possible to investigate huge systems like proteins or DNA even by highly accurate MO treatment with large basis sets.

The basic idea of this method is that the electronic structure of a polymer is synthesized step by step only for the reaction site between the oligomer and an attacking fragment analogous to polymerization reaction as shown schematically in Figure 1. For this purpose, some of MOs in the starting cluster are localized into region B (active LMOs) near to the attacking fragment and the others are into region A (frozen LMOs) apart from the interaction site. To produce the specific LMOs that are

---

[1] Corresponding author. Email: aoki@cube.kyushu-u.ac.jp.

$$\begin{pmatrix} \psi'_j \\ \psi'_i \end{pmatrix} = \begin{pmatrix} \sin\theta & \cos\theta \\ -\cos\theta & \sin\theta \end{pmatrix} \begin{pmatrix} \psi_j \\ \psi_i \end{pmatrix}$$

localized in the different regions, the unitary transformation is performed by using

specific θ in the following rotation:

(1)

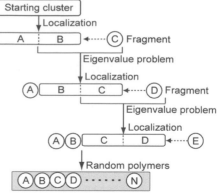

Figure 1: Schematic illustration for Elongation method

where θ is defined so that the frozen LMOs and the active LMOs are separated as much as possible as shown in Figure 2. The eigenvalue problem is solved on the basis of the active LMOs and the attacking fragment C. This means that the diagonalization is only required for small space of active LMOs and attacking fragment CMOs during the whole elongation process. The LMO based Fock matrix for the regions A, B, and C is shown in the right-hand side of Figure 3. Since this matrix is almost block diagonal, it is possible to consider only the sub-matrix (surrounded by bold solid lines) of ψ'_j and fragment CMOs instead of AO based Fock matrix (F) that includes all AOs on the regions A, B and C.

Figure 2: Schematic illustration for two LMOs after unitary transformation

In the off-diagonal blocks, the elements between occupied orbitals and between vacant orbitals have some non zero values though those elements between occupied and vacant orbitals are zero by definition. Nevertheless, we can omit the off-diagonal blocks completely to get correct total energy as well as density matrix. This is due to the fact that the density matrix is unchanged provided the orbitals between occupied and vacant are not mixed. This LMO basis expression of the total system provides us a big advantage to keep the interactive matrix to be constant even if the system is increased.

This method is implemented into *ab initio* program package GAMESS and becomes available for semi-empirical, *ab initio* and DFT [2] methods for the restricted closed-shell Hartree-Fock (RHF), restricted open-shell Hartree-Fock (ROHF) and unrestricted Hartree-Fock (UHF) methods. Besides the two-by-two unitary transformation given by Eq. (1), a novel strategy [3] has been introduced into the elongation method to obtain well and fast localized orbitals in specific regions. Furthermore, for constructing AO basis Fock matrix, an efficient technique [4] in getting two-electron integrals is also incorporated in our method. Applications to polyglycines will be presented in the Section 2 and further development toward more efficient calculations will be outlined in Section 3.

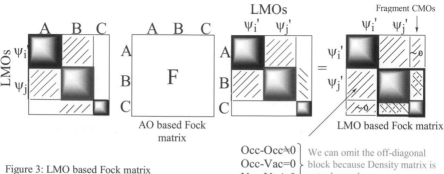

Figure 3: LMO based Fock matrix

2. Applications to Polyglycines

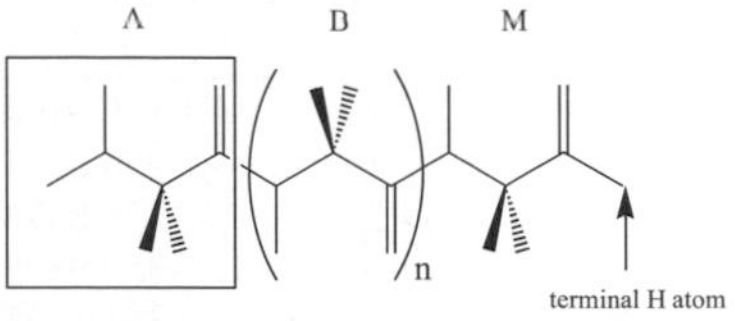

Let us now show some illustrative results. As a model system we have taken an idealized structure of polyglycine (Fig. 4), namely its C5 conformer (5 atoms in the ring enclosed the hydrogen bond) corresponding to β-sheet. Table 1 collects energies obtained at RHF and UHF level of theory with PM3 semiempirical

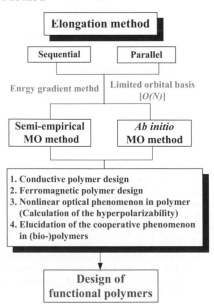

Figure 4: C5 conformer of polyglycine

hamiltonian. The *ab initio* (STO/3G) results are displayed in Table 2 for RHF and ROHF-type calculations. In the case of ROHF and UHF, we enlarge our system as a radical; the terminal hydrogen atom is excluded (see Fig. 4). However, the final termination step includes the terminal H atom.

In the elongation procedure, one monomer unit (A) is frozen and another one (M) is added at the same time. In semiempirical calculations the starting cluster is built from 10 residues. One can observe that both elongation approaches (RHF and UHF) give the total energies very close to the reference RHF energies. The initial errors are 1×10^{-9} and 4×10^{-9}, respectively. These values increase to -1.4×10^{-6} and 1.6×10^{-7} when N approaches 20. The UHF energies are closer to the reference values. For N=13 the quality of the localization for RHF calculations is not good enough and in the next elongation step the obtained energy is lower than HF one. This is a consequence of a few effects (1) initial orbital selection, (2) the requirement that the frozen and active parts should be closed-shell. The second condition by definition isn't applied for the UHF elongation.

At *ab initio* level of theory, we have performed ROHF calculations instead of UHF. The energies of the latter are strongly spin contaminated. This problem does not appear for semiempirical methods due to strong orthogonality requirement. The energies of polyglicine, with starting cluster built from 5 units, are depicted in Table 2. One can observe that except for N=6 (ROHF) the elongation energies approach to the RHF reference from upper side. This indicates that the quality of localization is very good, so the orthogonality between the frozen and active MOs is not violated. At the final N=20 step the elongation errors are 6.4×10^{-6} and 2.2×10^{-6} for RHF and ROHF treatments, respectively. Slightly lower error for ROHF is due to the terminal hydrogen atom. This atom for RHF calculations should be removed since it will be replaced by new monomer. Therefore, when one removes AOs and MOs belonging to the terminal hydrogen atom, the intra-fragment orthogonality is slightly violated. At ROHF level this problem is avoided.

3. Further Development of Elongation Method

The crucial point of the elongation method is the MO localization procedure. We have recently developed the new localization scheme. The scheme does not depend on the initial MO selection and is very efficient. Therefore, the elongation calculations can be performed at *ab initio* level for any extended basis sets. The next simplification that we have included to the elongation method is so called cut-off technique. This allows us to disregard the frozen parts, which are weekly coupled to the active one in direct computations. In this way we avoid calculating huge amount of two-electron integrals, thus substantially saving the computation time. The application of the elongation method to very huge biopolymers is planned.

As the foundations of the elongation method are established, this treatment can be a powerful tool to design functional polymers with conducting, ferromagnetic, non-linear optical properties and so on, as shown in the Figure 5. Furthermore, incorporation of post-Hartree-Fock method is in progress in our laboratory to get more accurate ground states as well as excited states in large systems.

Figure 5: Development of efficient molecular design integrated system of functional polymers by elongation method

Table 1   PM3 total energies of polyglicine, and the energy difference between conventional and elongation
methods at RHF and UHF level of theory.   (all in a. u.)

N	RHF(Conventional)	ΔE (RHF)	ΔE (UHF)
11	-299.69314916	1.00E-09	4.00E-09
12	-326.83761046	5.00E-09	1.20E-08
13	-353.98207488	1.00E-08	2.30E-08
14	-381.12654287	-1.69E-06	3.80E-08
15	-408.27101267	-1.57E-06	5.50E-08
16	-435.41548472	-1.52E-06	7.40E-08
17	-462.55995792	-1.49E-06	9.40E-08
18	-489.70443263	-1.46E-06	1.16E-07
19	-516.84890811	-1.43E-06	1.40E-07
20	-543.99338465	-1.41E-06	1.65E-07

Table 2 *Ab initio* (STO/3G) total energies of polyglicine, and the energy difference between conventional and
elongation methods at RHF and ROHF level of theory.   (all in a. u.)

N	RHF (Conventional)	ΔE (RHF)	ΔE (ROHF)
6	-1225.75192047	-5.50E-07	3.20E-07
7	-1429.86140248	6.00E-08	4.20E-07
8	-1633.97089955	1.40E-07	7.40E-07
9	-1838.08040603	2.50E-07	1.12E-06
10	-2042.18991907	3.90E-07	1.51E-06
11	-2246.29943637	5.30E-07	1.95E-06
12	-2450.40895707	7.00E-07	2.38E-06
13	-2654.51848003	8.70E-07	2.85E-06
14	-2858.62800499	1.02E-06	3.30E-06
15	-3062.73753128	1.20E-06	3.89E-06
16	-3266.84705884	1.45E-06	4.33E-06
17	-3470.95658724	1.67E-06	4.84E-06
18	-3675.06611649	1.90E-06	5.37E-06
19	-3879.17564632	2.06E-06	5.89E-06
20	-4083.28517676	2.22E-06	6.44E-06

Acknowledgments

This work was supported by the Research and Development Applying Advanced Computational Science and Technology of the Japan Science and Technology Agency (ACT-JST) and the Ministry of Education, Culture, Sports, Science and Technology (MEXT). The calculations were performed on the Linux PC cluster in our laboratory.

References

[1] (a) A. Imamura, Y. Aoki, and K. Maekawa, J. Chem. Phys., 95, 5419-5431 (1991); (b) Y. Aoki and A. Imamura, J. Chem. Phys., 97, 8432-8440 (1992); (c) A. Imamura, Y. Aoki, K. Nishimoto, Y. Kurihara, and A. Nagao, Int. J. Quantum Chem., 52, 309-320 (1994); (d) A. Imamura and Y. Aoki, Advances in Colloid and Interface Science, 71-72, 147-164 (1997); (e) Y. Kurihara, Y. Aoki, and A. Imamura, J. Chem. Phys., 107, 3569-3575 (1997);(f) Y. kurihara, Y. Aoki and A. Imamura, J. Chem. Phys., 108, 10303-10308(1998); (g) G. Raether, Y. Aoki and A. Imamura, Int. J. Quantum Chem., 74, 35-47(1999).
[2] Y. Aoki, S. Suhai, and A. Imamura, J. Chem. Phys., 101, 10808-10823 (1994).
[3] F. L. Gu, Y. Aoki, J. Korchowiec, A. Imamura, and B. Kirtman, J. Chem. Phys., sumbitted.
[4] J. Korchowiec, F. L. Gu, A. Imamura, B. Kirtman, and Y. Aoki, Int. J. Quantum Chem. In preparation for publication.

VSP International
Science Publishers
P.O. Box 346, 3700 AH Zeist
The Netherlands

Lecture Series on Computer
and Computational Sciences
Volume 1, 2004, pp. 779-782

Evaluation of Nonlinear Susceptibilities of 3-Methyl-4-nitropyridine 1-oxide Crystal: An Application of the Elongation Method to Nonlinear Optical Properties

Feng Long Gu [a,1], Benoît Champagne [b], and Yuriko Aoki [a,c]

a) *Group, PRESTO, Japan Science and Technology Agency (JST), Kawaguchi Center Building, Honcho 4-1-8, Kawaguchi, Saitama, 332-0012, Japan,*

b) *Laboratoire de Chimie Théorique Appliquée, Facultés Universitaires Notre-Dame de la Paix, 61 rue de Bruxelles, B-5000 Namur, Belgium*

c) *Department of Molecular and Material Sciences, Interdisciplinary Graduate School of Engineering Sciences, Kyushu University, 6-1 Kasuga-Park, Fukuoka, 816-8580, Japan*

Abstract: The elongation finite-field method is applied to the evaluation of the nonlinear susceptibilities of 3-Methyl-4-nitropyridine 1-oxide (POM) crystal. Static (hyper)polarizabilities of one-dimensional POM clusters, built along each of three crystal axes, are obtained by the elongation-FF method for the clusters containing up to 15 POM unit cells. It is found that the crystal packing effects along the b axis are the most prominent. For the first hyperpolarizability tensor component β_{abc}, packing along the b axis has an increase by about 8% whereas packing along the a and c axis shows a decrease by 2 and 23%, respectively. For the second hyperpolarizabilities the crystal packing effects are much stronger than for the polarizability and first hyperpolarizability. The dominant γ_{bbbb} component displays a large increase (by 120%) when packing along the b axis or a decrease (by 64%) when packing along the c axis.

Keywords: nonlinear optics, POM, elongation finite-field method

1. The elongation finite field method

The elongation method [1] builds up super clusters by adding a monomer unit stepwise to a starting oligomer while keeping the variational degrees of freedom fixed. In contrast to the conventional Hartree-Fock treatment, the elongation method works in a localized molecular orbital (LMO), instead of canonical molecular orbital (CMO), basis. The advantage of an LMO representation is that one part of the system, which is far away from the interactive site, is excluded from the SCF procedure. The quality of the localization has a strong impact on the accuracy that is achievable. Recently, a new localization scheme for efficiently determining the LMOs from CMOs has been developed [2].

Among the different properties, the elongation method can be employed to determine the static linear and nonlinear responses of large molecules or clusters to electric fields, *i.e.* the polarizability (α), first and second hyperpolarizabilities (β and γ). This is achieved by adopting a novel finite field (FF) approach based on the elongation method [3], which has been implemented and linked to the GAMESS program package [4].

One-dimensional POM clusters have been built up along each of three crystal axes. The starting cluster for the elongation contains two unit cells with one in the frozen region and the other one in the active region. For the first elongation step, a POM unit cell is attacking this starting cluster. The Hartree-Fock equation is solved for the interactive region, which consists of the active region of the starting cluster and the attacking monomer. On the other hand, the frozen part of the starting cluster is excluded from the current SCF procedure though its contribution is accounted in the total density matrix. These elongation steps are repeated until the size of the POM cluster is reached. Then, the (hyper)polarizabilities are obtained by numerically differentiating the field-dependent dipole moments or total energies.

---

[1] Corresponding author. E-mail: gu@cube.kyushu-u.ac.jp

2. Results and discussions

In this paper, the static α, β, and γ of POM molecular crystal are evaluated by the above mentioned elongation-FF method at the semi-empirical AM1 level. The POM crystal parameters are taken from Ref. [5] with $a = 20.890$Å, $b = 6.094$Å, and $c = 5.135$Å. The POM unit cell structure (see Figure 1) consists of four POM molecules, distributed in such a way that the only non-zero component of $\chi^{(2)}$, the macroscopic analog of β, is $\chi^{(2)}_{abc}$.

Static α's of one-dimensional POM clusters built along each of three crystal axes are listed in Table I. The crystal packing effect on the (hyper)polarizabilities is characterized by the ratio R defined as

$$R = \frac{P^{cluster}}{NP^{unit}} \quad , (P = \alpha, \beta, \gamma) \tag{1}$$

The crystal packing effects when extending the cluster along the b axis are the most prominent. For the first hyperpolarizability tensor component, β_{abc}, listed in Table II, packing along the b axis leads to an increase by about 8% whereas packing along the a and c axis is associated with a decrease by 2 and 23%, respectively. For the second hyperpolarizabilities, displayed in Table III, the crystal packing effects are stronger than for the polarizability and first hyperpolarizability. They are characterized by larger increase (by 120% when packing along b) or decrease (by 64% when packing along c) of the dominant γ_{bbbb} component. The slow convergence of R – especially for the nonlinear responses – as a function of the number of stacked unit cells shows the performance of the elongation method for addressing the NLO properties of molecular crystals.

TABLE I. Diagonal components of the static polarizability tensors and crystal packing rations (R) for POM clusters elongated along three crystal axes as a function of the number (N) of stacked unit cells.

N	α_{aa}	R	α_{bb}	R	α_{cc}	R
			Along *a* axis			
1	315	1.000	411	1.000	263	1.000
2	644	1.022	807	0.982	516	0.981
3	973	1.030	1202	0.975	768	0.973
4	1303	1.034	1597	0.971	1020	0.970
5	1632	1.036	1992	0.969	1273	0.968
6	1962	1.038	2388	0.968	1525	0.966
7	2292	1.039	2783	0.967	1778	0.966
8	2621	1.040	3178	0.967	2030	0.965
9	2951	1.041	3573	0.966	2282	0.964
10	3280	1.041	3968	0.965	2535	0.964
11	3610	1.042	4363	0.965	2787	0.963
12	3940	1.042	4758	0.965	3039	0.963
13	4269	1.042	5153	0.964	3292	0.963
14	4599	1.043	5548	0.964	3544	0.963
15	4929	1.043	5944	0.964	3796	0.962
∞		1.044		0.963		0.960
			Along *b* axis			
1	315	1.000	411	1.000	263	1.000
2	618	0.981	859	1.045	492	0.935
3	918	0.971	1320	1.071	718	0.910
4	1218	0.967	1789	1.088	943	0.896
5	1517	0.963	2262	1.101	1168	0.888
6	1815	0.960	2738	1.110	1392	0.882
7	2114	0.959	3214	1.117	1617	0.878
8	2412	0.957	3692	1.123	1841	0.875
9	2710	0.956	4170	1.127	2066	0.873
10	3008	0.955	4648	1.131	2290	0.871
11	3306	0.954	5127	1.134	2514	0.869
12	3604	0.953	5607	1.137	2738	0.868
13	3902	0.953	6086	1.139	2963	0.867
14	4200	0.952	6566	1.141	3187	0.866
15	4498	0.952	7046	1.143	3411	0.865
∞		0.950		1.150		0.862

		Along *c* axis				
1	315	1.000	411	1.000	263	1.000
2	623	0.989	749	0.911	546	1.038
3	929	0.983	1079	0.875	835	1.058
4	1233	0.979	1407	0.856	1127	1.071
5	1536	0.975	1734	0.844	1422	1.081
6	1839	0.973	2060	0.835	1717	1.088
7	2141	0.971	2385	0.829	2014	1.094
8	2443	0.969	2711	0.825	2311	1.098
9	2745	0.968	3036	0.821	2609	1.102
10	3046	0.967	3361	0.818	2906	1.105
11	3348	0.966	3686	0.815	3204	1.108
12	3649	0.965	4011	0.813	3502	1.110
13	3951	0.965	4336	0.812	3800	1.111
14	4252	0.964	4660	0.810	4099	1.113
15	4554	0.964	4985	0.809	4397	1.115
∞		0.963		0.805		1.120

TABLE II. Static β_{abc} tensor and crystal packing rations (R) for POM clusters elongated along three crystal axes as a function of the number (N) of stacked unit cells.

	Along *a* axis		Along *b* axis		Along *c* axis	
N	β_{abc}	R	β_{abc}	R	β_{abc}	R
1	-813	1.000	-813	1.000	-813	1.000
2	-1610	0.990	-1643	1.010	-1429	0.879
3	-2407	0.987	-2501	1.025	-2047	0.839
4	-3204	0.985	-3370	1.036	-2664	0.819
5	-4002	0.985	-4244	1.044	-3281	0.807
6	-4799	0.984	-5120	1.050	-3897	0.799
7	-5596	0.983	-5999	1.054	-4513	0.793
8	-6393	0.983	-6878	1.058	-5129	0.789
9	-7190	0.983	-7759	1.060	-5744	0.785
10	-7989	0.983	-8639	1.063	-6360	0.782
11	-8784	0.982	-9520	1.065	-6975	0.780
12	-9581	0.982	-10402	1.066	-7590	0.778
13	-10378	0.982	-11283	1.068	-8206	0.776
14	-11175	0.982	-12165	1.069	-8821	0.775
15	-11972	0.982	-13047	1.070	-9437	0.774
∞		0.981		1.080		0.770

TABLE III. Diagonal components of the static second-order hyperpolarizability tensors and crystal packing rations (R) for POM clusters elongated along three crystal axes as a function of the number (N) of stacked unit cells.

N	γ_{aaaa}	R	γ_{bbbb}	R	γ_{cccc}	R
			Along *a* axis			
1	18389	1.000	133199	1.000	9605	1.000
2	43821	1.192	244726	0.919	18661	0.971
3	69517	1.260	355912	0.891	27697	0.961
4	95271	1.295	466996	0.877	36728	0.956
5	121047	1.317	578038	0.868	45757	0.953
6	146833	1.331	689058	0.862	54784	0.951
7	172626	1.341	800065	0.858	63811	0.949
8	198421	1.349	911065	0.855	72838	0.948
9	224219	1.355	1022059	0.853	81864	0.947
10	250019	1.360	1133050	0.851	90890	0.946
11	275820	1.364	1244037	0.949	99915	0.946
12	301622	1.367	1355023	0.848	108941	0.945
13	327424	1.370	1466007	0.847	117966	0.945
14	353228	1.372	1576990	0.846	126992	0.944
15	379031	1.374	1687972	0.845	136017	0.944
∞		1.380		0.842		0.943
			Along *b* axis			
1	18389	1.000	133199	1.000	9605	1.000
2	47761	1.299	358558	1.346	15034	0.783
3	76876	1.394	618643	1.548	20642	0.716

4	105589	1.435	897992	1.685	26212	0.682
5	134050	1.458	1188388	1.784	31757	0.661
6	162361	1.472	1485480	1.859	37285	0.647
7	190579	1.481	1786857	1.916	42804	0.637
8	218738	1.487	2091097	1.962	48317	0.629
9	246861	1.492	2397319	2.000	53824	0.623
10	274956	1.495	2704958	2.031	59329	0.618
11	303030	1.498	3013637	2.057	64831	0.614
12	331087	1.500	3323099	2.079	70332	0.610
13	359133	1.502	3633160	2.098	75830	0.607
14	387168	1.504	3943690	2.115	81328	0.605
15	415196	1.505	4254592	2.129	86825	0.603
∞		*1.510*		*2.200*		*0.600*
Along *c* axis						
1	18389	1.000	133199	1.000	9605	1.000
2	36349	0.988	173010	0.649	25998	1.353
3	53616	0.972	217252	0.544	44234	1.535
4	70398	0.957	261491	0.491	63529	1.654
5	86936	0.946	305522	0.459	83447	1.738
6	103341	0.937	349382	0.437	103753	1.800
7	119665	0.930	393116	0.422	124314	1.849
8	135939	0.924	436759	0.410	145048	1.888
9	152178	0.919	480334	0.401	165904	1.919
10	168393	0.916	523859	0.393	186849	1.945
11	184589	0.913	567344	0.387	207859	1.967
12	200771	0.910	610798	0.382	228920	1.986
13	216943	0.907	654228	0.378	250019	2.002
14	233107	0.905	697637	0.374	271149	2.016
15	249264	0.904	741030	0.371	292302	2.029
∞		*0.901*		*0.364*		*2.100*

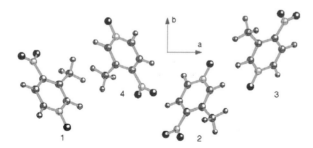

Figure 1. Projection of the POM unit cell in the plane (*a, b*) and numbering of the molecules.

Acknowledgments

Financial supports from the Research and Development Applying Advanced Computational Science and Technology of the Japan Science and Technology Agency (ACT-JST) and the Ministry of Education, Culture, Sports, Science and Technology (MEXT) are acknowledged. The calculations were performed on the Linux PC cluster in our laboratory. BC thanks the Belgium National Fund for Scientific Research for his Senior Research Associate position.

References

[1] A. Imamura, Y. Aoki, and K. Maekawa, J. Chem. Phys., **95**, 5419 (1991).
[2] F. L. Gu, Y. Aoki, J. Korchowiec, A. Imammura, and B. Kirtman, J. Chem. Phys., submitted.
[3] F. L. Gu, Y. Aoki, A. Imamura, D. M. Bishop, and B. Kirtman, Mol. Phys., **101**, 1487 (2003).
[4] GAMESS/Version 14, Jan. 2003 (R2) from Iowa State University, M.W. Schmidt, K.K. Baldridge, J.A. Boatz, S.T. Elbert, M.S. Gordon, J.H. Jensen, S. Koseki, N. Matsunaga, K.A. Nguyen, S.J. Su, T.L. Windus, together with M. Dupuis, J.A. Montgomery: J. Comput. Chem., **14**, 1347 (1993).
[5] F. Baert, P. Schweiss, G. Heger, and M. More, J. Molec. Struct., **178**, 29 (1988).

VSP International
Science Publishers
P.O. Box 346, 3700 AH Zeist
The Netherlands

*Lecture Series on Computer
and Computational Sciences*
Volume 1, 2004, pp. 783-786

Evaluation of Second-Order Susceptibilities of 2-methyl-4-nitroaniline (MNA) and 3-methyl-4-nitropyridine-1-oxyde (POM) Crystals

F. Castet[1#], L. Ducasse[#], M. Guillaume[*], E. Botek[*], B. Champagne[*]

[#]Laboratoire de Physico-Chimie Moléculaire, Université Bordeaux I,
351 cours de la libération, 33405 Talence cedex, France
[*]Facultés Universitaires Notre-Dame de la Paix
61 Rue de Bruxelles, Namur, Belgique

Received 5 August, 2004; accepted in revised form 28 August, 2004

Abstract: The nonlinear optical properties of the 2-methyl-4-nitroaniline (MNA) and of the 3-methyl-4-nitropyridine-1-oxyde (POM) crystals are determined using the supermolecule approach within ab initio and semi-empirical quantum chemistry techniques. The static and dynamic (SHG) second-order susceptibilities are evaluated from the knowledge of the first hyperpolarizability of the constituting one-dimensional molecular clusters by using a simple multiplicative scheme.

Keywords: Hyperpolarizability, Second-order nonlinear susceptibility, Second harmonic generation (SHG), Time-Dependent Hartree-Fock (TDHF), semi-empirical AM1 parametrization

1. Objectives

Whereas a mass of calculations has been carried out to evaluate the nonlinear optical (NLO) properties of isolated molecules, only few studies were devoted to materials [1]. Beyond the simplest oriented gas approximation [2], in which the molecular hyperpolarizabilities are assumed to be additive and the surrounding effects are accounted for by using local field factor corrections, the supermolecule approach remains the straightforward strategy to evaluate NLO properties of molecular assemblies. However, the required computational times only allow one to deal with reduced molecular aggregates, and new strategies are still to be developed to go beyond such limitations, and to simply relate the properties of finite molecular clusters to the bulk properties. Recently, we proposed an original scheme to evaluate NLO properties of crystals including non additive intermolecular effects from the knowledge of the properties of the constitutive molecular aggregates [3-5]. We present here an application of this new scheme to the evaluation of the second-order susceptibilities of the 2-methyl-4-nitroaniline (MNA) and 3-methyl-4-nitropyridine-1-oxyde (POM) crystals, which present large NLO responses and radically different structures (Fig. 1).

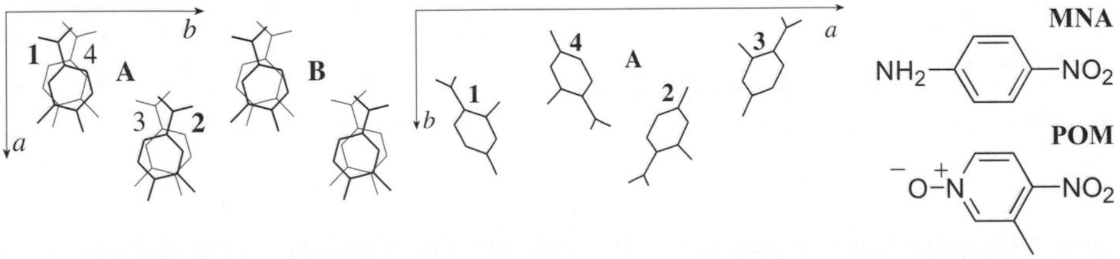

Figure 1: Left: MNA crystal structure [6]. *a*, *b*, and *c* refer to the crystallographic axes for a nonstandard setting (*Ia*) of the space group *Cc* ($a = 7.58 \pm 0.01$, Å, $\beta = 94.08°$). Right: POM crystal structure [7]. The crystal is orthorhombic with space group $P_{2_1 2_1 2_1}$ and unit cell parameters a = 21.359, b = 6.111, and c = 5.132 Å.

[1] Corresponding author. E-mail: f.castet@lpcm.u-bordeaux1.fr

2. General Methodology and Results

The crystal packing effects upon the static and dynamic (SHG, λ=1064 nm) first hyperpolarizabilities of MNA and POM clusters have been calculated within the Time-Dependent Hartree-Fock (TDHF) method [8] associated with the AM1 [9] hamiltonian. This semiempirical parametrization has been shown to reproduce fairly well the quality of ab initio correlated calculations while drastically reducing the computational cost [3,5]. The crystal packing effects are measured by the ratios:

$$R = \beta^{cluster}/N\beta^{ref}$$

where $\beta^{cluster}$ and β^{ref} are the first hyperpolarizabilities of the cluster and of the reference packing unit, respectively. In the MNA crystal, the (1,4) stacked dimer (Fig. 1), of which the dipole moment is parallel to the a crystallographic axis, has been shown to be the natural building block, since its β_{aaa} tensor component is more than one order of magnitude than all the other components [3]. In the case of the POM crystal, the unit cell constitutes the reference packing unit. Due to the symmetry of the crystal packing, $\chi^{(2)}_{abc}$ is the only non-vanishing component of the macroscopic second-order nonlinear susceptibility tensor.

The R ratios were calculated for increasingly large one-dimensional arrays (Fig. 2) and extrapolated to the infinite array size (R_∞) following the procedure described in Ref. [10]. By using a simple multiplicative scheme, which has been validated in our previous works [3-5], the crystal packing ratios of the full three dimensional crystal ($R^{crystal}$) can be estimated, and, in turn, the effective first hyperpolarizabilities of the reference packing unit in its crystal environment:

$$\beta^{eff} = \beta^{ref} \times R^{crystal} = \beta^{ref} \times R^a(N \to \infty) \times R^b(N \to \infty) \times R^c(N \to \infty)$$

The macroscopic second-order susceptibilities are eventually estimated by the relation [11] :

$$\chi^{(2)} = \frac{\beta^{eff}}{\varepsilon_0 \, V^{cell}}$$

All the calculated values are reported in Table 1.

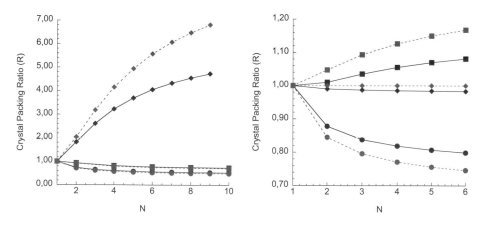

Figure 2: Crystal packing ratios (eq. 1) for MNA (left) and POM (right) clusters extended along the a (rhombs), b (squares) and c (circles) crystal directions as a function of the number N of reference stacking units. The plain and dotted lines are related to the static and SHG cases, respectively.

MNA						
	Static			SHG (λ = 1064 nm)		
	a axis	*b axis*	*c axis*	*a axis*	*b axis*	*c axis*
R_∞	5.1	0.72	0.51	7.9	0.68	0.47
β_{aaa}^{eff}	3241			4093		
$\chi_{aaa}^{(2)}$	41.1			76.1		
$\chi_{aaa}^{(2)}$ corrected	53.4			98.9		
POM						
	Static			SHG (λ = 1064 nm)		
	a axis	*b axis*	*c axis*	*a axis*	*b axis*	*c axis*
R_∞	0.982	1.111	0.787	1.000	1.211	0.789
β_{abc}^{eff}	700			1173		
$\chi_{abc}^{(2)}$	3.8			6.4		
$\chi_{abc}^{(2)}$ corrected	3.6			6.1		

Table 1: Effective unit cell first hyperpolarizabilities (β^{eff}, in a.u.) deduced using the multiplicative scheme with crystal packing ratios extrapolated to infinite 1D arrays (R_∞). The last lines of both the static and dynamic cases report the AM1-based macroscopic nonlinear second-order susceptibility $\chi^{(2)}$ (pm/V) as well as the $\chi^{(2)}$ values that have been corrected to account for electron correlation (see text).

3. Discussion

The theoretical dynamic $\chi^{(2)}$ estimations underestimate the experimental values of 300 pm/V [6] and 12 pm/V [12] related to MNA and POM, respectively (using d_{11} = 0.3 pm/V for the quartz [13]). The relative magnitude of the NLO response of MNA with respect to POM is also underestimated whereas it is qualitatively reproduced. The ratio between the theoretical second-order susceptibilities of MNA and POM is equal to 11.9, while the experimental ranges from 19 to 31, accounting for the reported experimental errors [6].

The discrepancies between theory and experiment can have several origins. To account for the missing electron correlation effects, *ab initio* calculations of the static hyperpolarizability tensor of MNA and POM molecular stacks have also been carried out at the HF and MP2 levels by using different atomic basis sets. The *ab initio* data listed in Table 2 show that the HF and MP2 β values for the monomer and dimer differ from the static AM1 quantities whereas the corresponding crystal packing ratios are almost independent of the calculation method. These results demonstrate the adequacy of using the AM1 semi-empirical scheme to address the crystal packing effects, and enable to improve the AM1-based effective β values by using a correcting multiplicative factor, defined as the ratio between the static MP2 β value and the corresponding AM1 result. By using correction factors of 1.3 (MNA) and 0.95 (POM), the corrected dynamic $\chi^{(2)}$ nonlinear second-order susceptibility amounts to 98.9 and 6.1 for MNA and POM, respectively, leading to a ratio between the SHG responses of the two crystals equal to 16.2.

	MNA		POM	
	β_{aaa} monomer	β_{aaa} dimer (1,4)	β_{abc} monomer	β_{abc} dimer (4,2)
AM1	1351	1741 (0.644)	205	398 (0.97)
HF	1112	1481 (0.666)	158	314 (0.99)
MP2	1702	2260 (0.664)	195	370 (0.95)

Table 2 : Static β values for MNA and POM monomer and dimers calculated at the AM1, HF and MP2 levels of approximation. *Ab initio* calculations were carried out within the 6-31G (MNA) and aug-cc-vdz (POM) basis sets.

Another correction that should be included arises from the nature of the electric field considered in the calculations. Indeed, the field-derivatives are calculated with respect to the external field, whereas the macroscopic nonlinear susceptibilities are taken with respect to the Maxwell field [14]. The ratio between these two field amplitudes is however difficult to evaluate because it is sensitive to the shape of the cluster.

4. Conclusion and Prospects

Using the multiplicative scheme with one-dimensional crystal packing ratios extrapolated to infinite size arrays, the effective first hyperpolarizability dressed by the crystal environment have been determined and used to estimate the macroscopic second-order nonlinear responses. Comparison with experiment shows that this approach is suitable for assessing the amplitude of $\chi^{(2)}$ as well as to analyze the crystal packing effects. Nevertheless, it would be of interest to know how these macroscopic responses are reproduced by using other simulation schemes including the crystal orbital methods [15] and the interaction schemes [16] that have already been applied to determine the linear and nonlinear macroscopic responses of organic and inorganic crystals and polymers.

Acknowledgments

This work has benefited from a Tournesol scientific cooperation established and supported by the Centre National de la Recherche Scientifique (CNRS), the Belgian National Fund for Scientific Research (FNRS), and the Commissariat Général aux Relations Internationales (CGRI) de la Communauté Wallonie-Bruxelles. E.B. thanks the Interuniversity Attraction Pole on "Supramolecular Chemistry and Supramolecular Catalysis" (IUAP N°. P5-03) for her postdoctoral grant. B.C. thanks the Belgian National Fund for Scientific Research for his Senior Research Associate position. The calculations were performed thanks to computing time made available by the SiMoA (Simulation et Modélisation en Aquitaine, France) as well as by the Interuniversity Scientific Computing Facility (ISCF), installed at the Facultés Universitaires Notre-Dame de la Paix (Namur, Belgium), for with the authors gratefully acknowledge the financial support of the FNRS-FRFC and the "Loterie Nationale" for the convention n° 2.4578.02 and of the FUNDP.

References

[1] See for example : B. Champagne and D.M. Bishop, Adv. Chem. Phys. **126**, 41 (2003).

[2] D.S. Chemla, J.L. Oudar, and J. Jerphagnon, Phys. Rev. B **12**, 4534 (1975).

[3] F. Castet and B. Champagne, J. Phys. Chem. A **105**, 1366 (2001).

[4] M. Guillaume, B. Champagne, F. Castet, and L. Ducasse, in *Computational chemistry, Reviews of current trends*, edited by J. Leszczynski, (World Scientific Publishing, Singapore), Vol. VIII, Chap. 2, p. 81 (2003).

[5] M. Guillaume, E. Botek, B. Champagne, F. Castet, L. Ducasse, to be published in J. Chem. Phys. (2004).

[6] G. F. Lipscomb, A. F. Garito, R. S. Narang, J. Chem. Phys. **75**, 1509 (1981).

[7] F. Baert, P. Schweiss, G. Heger, and M. More, J. Mol. Struct. **178**, 29 (1988).

[8] H. Sekino and R.J. Bartlett, J. Chem. Phys. **85**, 976 (1986); S.P. Karna and M. Dupuis, J. Comp. Chem. **12**, 487 (1991).

[9] M.J.S. Dewar, E.G. Zoebisch, E.F. Healy, and J.J.P. Stewart, J. Am. Chem. Soc. **107**, 3902 (1985) ; MOPAC2000, © Fujitsu Limited, 1999; J.J.P. Stewart, Quantum Chemistry Program 7 Exchange, n°. 455.

[10] B. Kirtman, J.L. Toto, K.A. Robins, and M. Hasan, J. Chem. Phys. **102**, 5350 (1995); B. Champagne, D. Jacquemin, J.M. André, and B. Kirtman, J. Chem. Phys. A **101**, 3158 (1997).

[11] R.W. Boyd, *Nonlinear Optics*, (Academic Press, San Diego, 1992), Appendix A.

[12] J. Zyss, D.S. Chemla, J. F. Nicoud, J. Chem. Phys. **74**, 4800 (1981).

[13] D. A. Roberts, IEEE Quant. Elect. QE **28**, 2057 (1992); A. Mito, K. Hagimoto and C. Takanashi, Nonlinear Opt. **13**, 3, (1995).

[14] R. W. Boyd, Non Linear Optics, (Academic Press, San Diego, 1992), Appendix A.

[15] J.L.P. Hugues and J.E. Sipe, Phys. Rev. B **58**, 7761 (1998); B. Kirtman, F.L. Gu, and D.M. Bishop, J. Chem. Phys. **113**, 1294 (2000); M. Rérat, W.D. Cheng, and R. Pandey, J. Phys. Condens. Matter **13**, 343 (2001); E.K. Chang, E.L. Shirley, and Z.H. Levine, Phys. Rev. B **65**, 35205 (2001); M. Veithen, Ph. Ghosez, and X. Gonze, arXiv:cond-mat/0311240.

[16] R.W. Munn, Mol. Phys. **64**, 1 (1988); H. Reis, M.G. Papadopoulos, and R.W. Munn, J. Chem. Phys. **109**, 6828 (1998); B. Kirtman, C.E. Dykstra, and B. Champagne, Chem. Phys. Lett. **305**, 132 (1999); L. Jensen, *Modelling of Optical Response Properties: Application to Nanostructures*, Rijksuniversiteit Groningen, 2004.

VSP International
Science Publishers
P.O. Box 346, 3700 AH Zeist
The Netherlands

*Lecture Series on Computer
and Computational Sciences*
Volume 1, 2004, pp. 787-790

Structural Dependence of Second Hyperpolarizability of Nanostar Dendritic Systems

R. Kishi,[1] M. Nakano[1] T. Nitta[1] and K. Yamaguchi[2]

[1]Department of Materials Engineering Science, Graduate School of Engineering Science,
Osaka University, Toyonaka, Osaka 560-8531, Japan
[2]Department of Chemistry, Graduate School of Science,
Osaka University, Toyonaka, Osaka 560-0043, Japan

Abstract: We examine the structural dependence of second hyperpolarizability (γ) for nanostar dendritic systems using quantum master equation method including exciton-phonon coupling. The $\gamma(-\omega, \omega, \omega, -\omega)$ values in the DFWM are calculated by the nonperturbative approach. We reveal the effects of structural changes of nanostar dendritic systems, i.e., modification in core monomer and the change in the divergent angle, on the exciton migration and the off- and near-resonant γ values.

Keywords: hyperpolarizaibility; nanostar dendirmer; qunatum master equation; exciton; relaxation
PACS: 36.40.-c; 33.80.-b; 33.15.-e; 34.30.+h

1. Introduction

Dendritic systems have attracted much attention in the field of novel functional materials such as catalysis, drug delivery and energy transfer due to their high controllability of structures and feasibility of chemical modification [1-3]. For excitation energy transfer, in particular, the relation between the transfer mechanism and unique architecture of dendritic systems has been clarified using the exciton migration dynamics based on the relaxation theory [4-8]. On the other hand, the off-resonant nonlinear optical (NLO) properties of dendritic systems have been investigated in view of a new control scheme of NLO properties by adjusting the monomer configuration and the intermolecular interaction. For example, the static second hyperpolarizabilities (γ) of phenylacetylene dendrimers have been investigated by the finite-field (FF) approach in the semi-empirical molecular orbital (MO) approximation [9-12]. The spatial contributions to γ values are found to be well localized in the linear leg regions segmented at the *meta*-substituted benzene rings. A a result, the exciton migration and the spatial contributions to γ in the phenylacetylene dendrimers are turned out to be closely related to the segmentation of π-conjugation at the *meta*-substituted benzene rings and the fractal architecture concerning the linear-leg lengths. In this study, the dynamic third-order NLO properties, i.e., $\gamma(-\omega, \omega, \omega, -\omega)$ in the degenerate four-wave mixing (DFWM), are examined for the nanostar dendritic systems (Fig. 1) and their structural dependences are revealed using several modified nanostar dendritic models.

2. Quantum master equation approach

We consider model molecular aggregates composed of ladder-type three-state monomers (Fig. 1). The dipole-dipole intermolecular coupling is assumed between monomers. The one-exciton states $\{|\psi_\alpha\rangle\}$ with energies $\{\omega_\alpha\}$ obtained by a diagonalization of aggregate hamiltonian H_S are expressed as

---

[1] Corresponding author. E-mail: mnaka@cheng.es.osaka-u.ac.jp

$$|\psi_\alpha\rangle = \sum_i^M |i\rangle\langle i|\psi_\alpha\rangle \equiv \sum_i^M C_{\alpha i}|i\rangle, \quad (\alpha = 2, \dots, M) \tag{1}$$

where $\{|i\rangle\} \equiv \{|i_1\dots i_N\rangle\}$ (aggregate basis , $i = 1,\dots,M$) and M is equal to $2N+1$ (N: the number of monomers) for the one-exciton model composed of three-state monomers. An exciton on monomer represented by basis $|i\rangle$ is assumed to interact with a nuclear vibration, i.e., a phonon state. Considering the weak exciton-phonon coupling and the interaction between exciton and external electric field F, we obtain a quantum master equation [7] for exciton density matrix ρ under the Born-Markov approximation:

$$\dot{\rho}_{\alpha\alpha} = -\sum_m^M \Gamma_{\alpha\alpha;mm}\rho_{mm} - F\sum_n^M (\mu_{\alpha n}\rho_{n\alpha} - \rho_{\alpha n}\mu_{n\alpha}), \tag{2}$$

and

$$\dot{\rho}_{\alpha\beta} = -i(\omega_\alpha - \omega_\beta)\rho_{\alpha\beta} - \sum_{m,n}^M \Gamma_{\alpha\beta;mn}\rho_{mn}. \quad (\alpha \neq \beta) \tag{3}$$

The explicit form of the relaxation factors ($\Gamma_{\alpha\beta;mn}$) are expressed in our previous papers [7,8]. We solve Eqs. (2) and (3) numerically using the fourth-order Runge-Kutta method and calculate the time series of poalrization $p(t)$.

3. Nonperturbative calculation scheme of γ in the DFWM

In the DFWM process, the polarization $p(\omega)$ with a frequency ω is induced by three incident beams with amplitudes $\varepsilon(\omega)$ and frequencies $(\omega,\omega,-\omega)$. The response $p(\omega)$ can be expressed by using the intensity-independent (hyper)polarizabilities:

$$p(\omega) = 3\alpha(-\omega;\omega)\varepsilon(\omega) + 36\gamma(-\omega;\omega,\omega,-\omega)\varepsilon^2(\omega)\varepsilon\ (-\omega) + \ \dots \ . \tag{4}$$

We must eliminate the lowest order term involving $\alpha(-\omega;\omega)$ in order to obtain the nonperturbative $\gamma(\omega,\omega,-\omega)$ in the DFWM. Therefore, we first calculate the intensity-independent polarizability $\alpha(-\omega;\omega)$ with good precision. The $\alpha(-\omega;\omega)$ can be approximately given by

$$\alpha(-\omega;\omega) \cong \frac{p(\omega)}{\varepsilon(\omega)} \quad \text{for a weak field } \varepsilon(\omega), \tag{5}$$

where $p(\omega)$ is induced by a weak electric field with an amplitude $\varepsilon(\omega)$ and a frequency ω, and $\varepsilon(\omega)$ and $p(\omega)$ are obtained by the Fourier transformation of incident electric field and induced polarization time series, respectively. From Eqs. (4) and (5), the nonperturbative $\gamma(-\omega;\omega,\omega,-\omega)$ in the DFWM can be defined by [12,13]

$$\gamma^{(3)}(-\omega;\omega,\omega,-\omega) \ (\text{DFWM}) = \frac{p(\omega) - 3\alpha(-\omega;\omega)\varepsilon(\omega)}{36\varepsilon^2(\omega)\varepsilon(-\omega)}, \tag{6}$$

where $p(\omega)$ is induced by three intense electric fields with amplitudes $\varepsilon(\omega)$ and frequencies $(\omega,\omega,-\omega)$. The γ value (complex quantity) is also described by the Fourier transformation of real parts of density matrices $\rho_{ba}^{\text{real}}(t)$ in the aggregate basis:

$$\gamma(-\omega;\omega,\omega,\omega) = \sum_{a>b}^M \gamma_{a-b} = \sum_{a>b}^M \frac{2\mu_{ab}\rho_{ba}^{\text{real}}(\omega) - 6\mu_{ab}(\rho_{ba}'^{\text{real}}(\omega)/\varepsilon'(\omega))}{36\varepsilon^2(\omega)\varepsilon(-\omega)}, \tag{7}$$

where ρ', ε' are density and field amplitude concerned with Eq. (5). Using this equation, we can partition total γ into the contribution of each one-exciton generation represented by a-b. In case of $a = |121\dots1\rangle$ and $b = |111\dots1\rangle$, for example, we can elucidate the spatial contribution of one-exciton generation to γ by showing the one-exciton distribution γ_{a-b} on the second monomer.

4. Calculation model

We briefly explain the calculated models in this study. Figure 1 shows two types of model aggregates which mimics the nanostar dendrimers. We firstly examine the effect of changing the core monomer on γ since the relative energy level of core monomer with respect to that of **G1** is predicted to significantly affect the exciton migration rate. Second, the effects of angle between the divergences of linear legs on γ are investigated. In fact, such difference in the divergent angle is observed in different solutions and is predicted to affect the interaction between neighboring linear legs and exciton migration rate. Details of our results will be explained in our presentation.

(a) Model A with different core

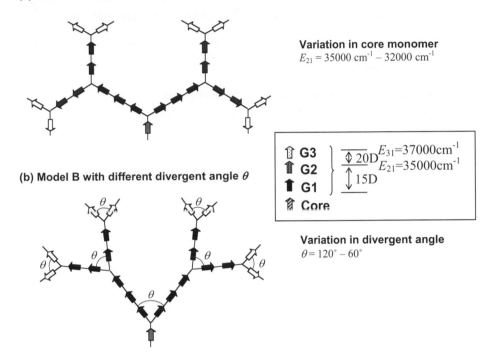

Variation in core monomer
$E_{21} = 35000 \text{ cm}^{-1} - 32000 \text{ cm}^{-1}$

(b) Model B with different divergent angle θ

⇕ **G3**
↥ **G2**
↑ **G1**
↥ **Core**

↕ 20D $E_{31} = 37000 \text{cm}^{-1}$
        $E_{21} = 35000 \text{cm}^{-1}$
↧ 15D

Variation in divergent angle
$\theta = 120° - 60°$

Figure 1: Two nanostar dendritic aggregate models.

Acknowledgments

This work was supported by Grant-in-Aid for Scientific Research (No. 14340184) from Japan Society for the Promotion of Science (JSPS).

References

[1] R. S. Knox, in *Primary Processes of Photosynthesis*, J. Barber (Ed.) Vol.2 (Elsevier, Amsterdam, 1977).

[2] A. Archut, G. C. Azzellini, V. Balzani, L. De Cola, F. Vögtle, J. Am. Chem. Soc. **120**, 12187 (1998).

[3] M. R. Shortreed, S. F. Swallen, Z-Y. Shi, W. Tan, Z. Xu, C. Devadoss, J. S. Moore, R. Kopelman, J. Phys. Chem. B **101**, 6318 (1997).

[4] S. Tretiak, V. Chernyak, S. Mukamel, J. Phys. Chem. B 102 (1998) 3310.

[5] J.C. Kirkwood, C. Scheurer, V. Chernyak, S. Mukamel, J. Chem. Phys. **114**, 2419 (2001).

[6] M. Nakano, M. Takahata, H. Fujita, S. Kiribayashi, K. Yamaguchi, Chem. Phys. Lett. **323**, 249 (2000).

[7] M. Takahata, M. Nakano, H. Fujita and K. Yamaguchi, Chem. Phys. Lett. **363**, 422 (2002).

[8] M. Nakano, M. Takahata, S. Yamada, K. Yamaguchi, R. Kishi, T. Nitta, J. Chem. Phys. **120**, 2359 (2004).

[9] M. Nakano, H. Fujita, M. Takahata, K. Yamaguchi, J. Am. Chem. Soc. **124**, 9648 (2002).

[10] M. Nakano, H. Fujita, M. Takahata, K. Yamaguchi, J. Chem. Phys. **115**, 1052 (2001).

[11] M. Nakano, H. Fujita, M. Takahata, K. Yamaguchi, J. Chem. Phys. **115**, 6780 (2001).
[12] M. Nakano, K. Yamaguchi, *Advances in Multi-photon Processes and Spectroscopy*, vol. 15 (World Scientific, 2003) pp.1.

[13] M.Nakano and K. Yamaguchi, Phys. Rev. A. **50**, 2989 (1994).

VSP International
Science Publishers
P.O. Box 346, 3700 AH Zeist
The Netherlands

Lecture Series on Computer
and Computational Sciences
Volume 1, 2004, pp. 791-794

Quantum Master Equation Approach to the Second Hyperpolarizability of Nanostar Dendritic Systems

M. Nakano,[1] R. Kishi,[1] T. Nitta[1] and K. Yamaguchi[2]

[1]Department of Materials Engineering Science, Graduate School of Engineering Science,
Osaka University, Toyonaka, Osaka 560-8531, Japan
[2]Department of Chemistry, Graduate School of Science,
Osaka University, Toyonaka, Osaka 560-0043, Japan

Abstract: We develop the calculation method of the second hyperpolarizability (γ) of nanostar dendritic systems using the quantum master equation approach. In the nanostar dendritic systems composed of three-state monomers, the multi-step exciton states are obtained by the dipole-dipole interactions, and the directional energy transport, i.e., exciton migration, from the periphery to the core is predicted to occur by the relaxation between exciton states originating in the exciton-phonon coupling. We examine the effects of the intermolcecular interaction and/or the exciton migration on the second hyperpolarizability (γ) for the nanostar dendritic systems. Further, the method for analysis of spatial contributions of excitons to γ is presented by partitioning the total γ into the one-exciton contributions. It turns out that the intermolecular interaction and/or relxation between exciton states are significantly affect the γ values of the nanostar dendritic systems.

Keywords: hyperpolarizaibility; dendirmer; qunatum master equation; exciton, relaxation
PACS: 36.40.-c; 33.80.-b; 33.15.-e; 34.30.+h

1. Introduction

Excitation energy transport is one of the essential processes in photosynthesis in green plants on earth and also finds an important application in photonics and biology [1,2]. The efficient and controllable energy transport is known to be one of the fascinating properties of dendritic systems with ordered fractal-like architecture, which exhibits a directed, multistep energy transport of absorbed light. In particular, there have been lots of studies on the efficient light-harvesting properties of phenylacetylene dendrimers and dendritic aggregates with a fractal-antenna (Cayley-tree) structure [3-9]. It has been found that the efficient energy transport is carried out by the multistep exciton migration from the periphery to the core, and is related to two peculiar structural features: (I) increase in the lengths of linear-legs involved in each generation as going from the periphery to the core and (II) *meta*-branching points (*meta*-substituted benzene rings). These features are predicted to provide multistep exciton states, in which the exciton distribution is spatially well localized in each generation and the exciton energy decreases as going from the periphery to the core region [3,7,8]. In our previous studies using the dipole-coupled dendritic aggregate models [10-12], we have also pointed out the necessity of the relaxation effect between exciton states, originating in exciton-phonon coupling, for such efficient exciton migration in addition to multi-step exciton states. The efficient multistep relaxation turns out to require partial spatial overlaps of exciton distributions between neighboring exciton states, which are respectively distributed in adjacent generations linked with *meta*-branching points. Our treatment has been extended and applied to real dendrimeric molecules based on the *ab initio* molecular orbital (MO) configuration interaction (CI) calculations [13]. As a result, structural features (I) and (II) are found to satisfy these two conditions, (A) well-segmented exciton distribution in each multistep exciton state

---

[1] Corresponding author. E-mail: mnaka@cheng.es.osaka-u.ac.jp

and (B) the existence of partial overlaps of exciton distributions between neighboring exciton states, in order to realize efficient directed multistep exciton migration.

In this study, we examine the second hyperpolarizability (γ) of an associated dendrimer derivative, "nanostar", terminated at the focal point with a core monomer, e.g., a perylene luminophor. This compound exhibits efficient energy transport to the core monomer, which eventually emits the fluorescence enormously enhanced as compared to the case of isolated core monomer. In this study, we develop a calculation scheme of γ for such dendritic nanostar systems including the exciton migration effects and investigate the effects of the multistep exciton states and the relaxation between them on the γ using the one-exciton dipole-coupled aggregate model with the nanostar structure. The relaxation between exciton states are taken into account in the quantum master equation approach including exciton-phonon coupling.

2. Calculation scheme

We consider a molecular aggregate composed of three-state monomers with excitation energies $\{E_{i_k}^k\}$ and magnitudes of transition moments $\{\mu_{i_k,i_k'}^k\}$ ($k = 1, 2, \cdots, N$, N: the number of monomers; $i_k = 1, 2, 3$). The Hamiltonian, H_S, for the molecular aggregate is expressed in atomic units by

$$H_S = \sum_k^N \sum_{i_k}^3 E_{i_k}^k \left| ...i_k... \right\rangle \left\langle ...i_k... \right| + \frac{1}{2} \sum_{k<l}^N \sum_{\substack{i_k,i_k' \\ i_l,i_l'}}^3 J_{i_k,i_k',i_l,i_l'} \left| ...i_k...i_l... \right\rangle \left\langle ...i_k'...i_l'... \right|, \tag{1}$$

where $\left| i_1 i_2 ... i_N \right\rangle$ indicates the aggregate basis, in which i_k means the excited state i_k of monomer k. We assume the dipole-dipole intermolecular coupling between monomers k and l with an intermolecular distance R_{kl}:

$$J_{i_k,i_k',i_l,i_l'} = \frac{1}{4\pi\varepsilon_0 R_{kl}^3} \mu_{i_k,i_k'}^k \mu_{i_l,i_l'}^l \{\cos(\theta_{k_l} - \theta_{i_k}) - 3\cos\theta_{k_l}\cos\theta_{i_l}\}, \tag{2}$$

where $\theta_{k_l}(\theta_{i_l})$ is the angle between the transition moment of monomer $k(l)$ and the vector drawn from monomer k to l. The one-exciton states $\{|\psi_\alpha\rangle\}$ with energies $\{\omega_\alpha\}$ obtained by a diagonalization of H_S are expressed as

$$\left|\psi_\alpha\right\rangle = \sum_i^M |i\rangle\langle i|\psi_\alpha\rangle \equiv \sum_i^M C_{\alpha i}|i\rangle, \quad (\alpha = 2, ... , M) \tag{3}$$

where $\{|i\rangle\} \equiv \{|i_1...i_N\rangle\}$ ($i = 1,...,M$) and M is equal to $2N+1$ for the one-exciton model composed of three-state monomers.

An exciton on monomer represented by basis $|i\rangle$ is assumed to interact with a nuclear vibration, i.e., a phonon state $\{|q_i\rangle\}$ with a frequency $\{\Omega_{q_i}\}$. The Hamiltonian, H_R, for the phonon is given by

$$H_R = \sum_i^M \sum_{q_i} \Omega_{q_i} c_{i,q_i}^+ c_{i,q_i}, \tag{4}$$

where c_{i,q_i}^+ and c_{i,q_i} represent the creation and annihilation operators concerning a phonon state $|q_i\rangle$, respectively. The interaction Hamiltonian, H_{SR}, for weak exciton-phonon coupling is assumed to be

$$H_{SR} = \sum_i^M \sum_{q_i} |i\rangle\langle i| (\kappa_{i,q_i}^* c_{i,q_i}^+ + \kappa_{i,q_i} c_{i,q_i}), \tag{5}$$

where κ_{i,q_i} denotes a coupling constant between an exciton on monomer represented by basis $|i\rangle$ and a phonon state $|q_i\rangle$. We investigate the dynamics of exciton system involved in total system composed of exciton (H_S), phonon (H_R), exciton-phonon interaction (H_{SR}) and the interaction between exciton and external electric field F (H_{int}). Using the standard method of relaxation theory [12], we obtain a quantum master equation [11] for exciton density matrix ρ under the Born-Markov approximation:

and

$$\dot{\rho}_{\alpha\alpha} = -\sum_{m}^{M} \Gamma_{\alpha\alpha;mm}\rho_{mm} - F\sum_{n}^{M}(\mu_{\alpha n}\rho_{n\alpha} - \rho_{\alpha n}\mu_{n\alpha}),$$ (6)

$$\dot{\rho}_{\alpha\beta} = -i(\omega_\alpha - \omega_\beta)\rho_{\alpha\beta} - \sum_{m,n}^{M}\Gamma_{\alpha\beta;mn}\rho_{mn}. \quad (\alpha \neq \beta)$$ (7)

The relaxation factors are expressed by

and

$$\Gamma_{\alpha\alpha;mm} = 2\delta_{\alpha m}\sum_{k}^{M}\sum_{i}^{N}|C_{\alpha i}|^2|C_{ki}|^2\gamma_{(ij)}(\omega_m - \omega_k) - 2\sum_{i}^{N}|C_{\alpha i}|^2|C_{mi}|^2\gamma_{(ij)}(\omega_m - \omega_\alpha),$$ (8)

$$\Gamma_{\alpha\beta;mn} = \sum_{k}^{M}\sum_{i}^{N}\left[\delta_{\beta n}C_{\alpha i}^*|C_{ki}|^2 C_{mi}\gamma_{(ij)}(\omega_m - \omega_k) + \delta_{\alpha m}C_{ni}^*|C_{ki}|^2 C_{\beta i}\gamma_{(ij)}(\omega_n - \omega_k)\right]$$
$$- \sum_{i}^{N}\left[C_{\alpha i}^* C_{mi}C_{ni}^* C_{\beta i}\{\gamma_{(ij)}(\omega_m - \omega_\alpha) + \gamma_{(ij)}(\omega_n - \omega_\beta)\}\right]$$ (9)

where

$$\gamma_{(ij)}(\omega) = \frac{2\gamma_{(ij)}^0}{1 + \exp(-\omega/k_B T)}.$$ (10)

This factor $\gamma_{(ij)}(\omega)$ is taken to satisfy the thermal equilibrium condition: $\gamma_{(ij)}^0$ indicates the high-temperature limit of $\gamma_{(ij)}(\omega)$, and k_B is the Boltzmann constant. We solve Eqs. (6) and (7) numerically using the fourth-order Runge-Kutta method.

Using the definition of nonperturbative (hyper)polarizabilities [9,14], we can obtain, for example, the $\gamma(-3\omega;\omega,\omega,\omega)$ in the THG as

$$\gamma(-3\omega;\omega,\omega,\omega) = \frac{p(3\omega)}{27\varepsilon^3(\omega)},$$ (11)

where $p(3\omega)$ is obtained by the Fourier transformation of induced polarization time series. The γ value (complex quantity) is also described by the Fourier transformation of real parts of density matrices $\rho_{ba}^{real}(t)$ in the aggregate basis:

$$\gamma(-3\omega;\omega,\omega,\omega) = \sum_{a>b}^{M}\gamma_{a-b} = \sum_{a>b}^{M}\frac{2\mu_{ab}\rho_{ba}^{real}(3\omega)}{27\varepsilon^3(\omega)}.$$ (12)

In the one-exciton model, the total γ can be partitioned into the virtual excitation contribution (γ_{a-b}) between bases a and b, in which only non-zero elements are the matrix elements between one exciton state, e.g., $|121...1\rangle$, and $|11...1\rangle$ (1: the ground state of monomer), and those between one exciton states, e.g., $|121...1\rangle$ and $|131...1\rangle$ (2 and 3 indicate the first and the second excited states of monomer). For example, in the case of $a = |121...1\rangle$ and $b = |111...1\rangle$, we can elucidate the spatial contribution of one-exciton generation to γ by showing the one-exciton distribution γ_{a-b} on the second monomer.

3. Calculation model

Figure 1 shows a model aggregate which mimics the nanostar dendrimer. Firstly, we apply our calculation scheme to the investigation of the exciton migration of the nanostar dendritic aggregate. Second, the γ values in the DFWM are calculated for the non-interacting model and the non-relaxation model by using the above nonperturbative scheme in order to elucidate the effects of intermolecular interaction and the relaxation (exciton migration) on the γ values for nanostar dendritic systems. Details of our results will be explained in our presentation.

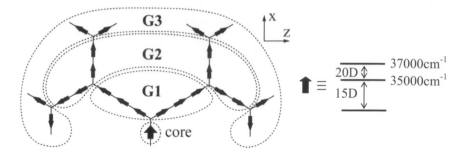

Figure 1: Nanostar dendritic aggregate models.

Acknowledgments

This work was supported by Grant-in-Aid for Scientific Research (No. 14340184) from Japan Society for the Promotion of Science (JSPS).

References

[1] R. S. Knox, in *Primary Processes of Photosynthesis*, J. Barber (Ed.) Vol.2 (Elsevier, Amsterdam, 1977).

[2] A. Archut, G. C. Azzellini, V. Balzani, L. De Cola, F. Vögtle, J. Am. Chem. Soc. **120**, 12187 (1998).

[3] M. R. Shortreed, S. F. Swallen, Z-Y. Shi, W. Tan, Z. Xu, C. Devadoss, J. S. Moore, R. Kopelman, J. Phys. Chem. B **101**, 6318 (1997).

[4] M. Nakano, H. Fujita, M. Takahata, K. Yamaguchi, J. Am. Chem. Soc. **124**, 9648 (2002).

[5] M. Nakano, H. Fujita, M. Takahata, K. Yamaguchi, J. Chem. Phys. **115**, 1052 (2001).

[6] M. Nakano, H. Fujita, M. Takahata, K. Yamaguchi, J. Chem. Phys. **115**, 6780 (2001).

[7] S. Tretiak, V. Chernyak, S. Mukamel, J. Phys. Chem. B 102 (1998) 3310.

[8] J.C. Kirkwood, C. Scheurer, V. Chernyak, S. Mukamel, J. Chem. Phys. **114**, 2419 (2001).

[9] M. Nakano, K. Yamaguchi, *Advances in Multi-photon Processes and Spectroscopy*, vol. 15 (World Scientific, 2003) pp.1.

[10] M. Nakano, M. Takahata, H. Fujita, S. Kiribayashi, K. Yamaguchi, Chem. Phys. Lett. **323**, 249 (2000).

[11] M. Takahata, M. Nakano, H. Fujita and K. Yamaguchi, Chem. Phys. Lett. **363**, 422 (2002).

[12] M. Takahata, M. Nakano, K. Yamaguchi, J. Theoret. Comp. Chem. **2**, 459 (2003).

[13] M. Nakano, M. Takahata, S. Yamada, K. Yamaguchi, R. Kishi, T. Nitta, J. Chem. Phys. **120**, 2359 (2004).

[14] M.Nakano and K. Yamaguchi, Phys. Rev. A. **50**, 2989 (1994).

VSP International
Science Publishers
P.O. Box 346, 3700 AH Zeist
The Netherlands

*Lecture Series on Computer
and Computational Sciences*
Volume 1, 2004, pp. 795-795

Multiscale Modeling of Irradiation Effects in Solids and Associated Out-of-Equilibrium Processes

Marjorie Bertolus
CEA Cadarache, Département d'Etude du Combustible,
SESC/LLCC, Bâtiment 151,
13108 Saint-Paul-lez-Durance, France

Mireille Defranceschi
CEA Saclay, Direction de l'Energie Nucléaire,
DSOE/RB, Bâtiment 121,
91191 Gif-sur-Yvette, France

Keywords: modeling, irradiation, out-of-equilibrium processes

Theoretical materials chemistry is rapidly emerging as a key component of contemporary science. The emphasis of materials science has traditionally been on physical aspects of solid state behavior and concepts have been developed to rationalize physical properties of ideal solids. Many efficient codes relying either on *ab initio* methods or on semi-empirical approaches are currently available and allow the calculation of a wide range of properties from atomistic characteristics.

In the real world materials are far from being ideal and the nature and the degree of structural order which they exhibit vary largely depending on their usage conditions. For instance, solids incorporating radioactive elements are submitted to radioactive decays which induce modifications of their structural and electronic properties. Modeling the long term behavior of such materials is a challenge which requires the development of a multiscale approach, in length as well as in time. Another example is the breakdown of insulators experiencing a high intensity electric field.

Typically, the most interesting physical phenomena for realistic modeling are out-of-equilibrium processes, the main driving force being the relaxation of constraints originating in the storage of energy in the materials. Some particular aspects are known and modeling is possible but understanding the evolution of a given materials under specific constraints is largely out of reach of present calculations. Nevertheless thanks to the improvement of theoretical methods and to the progresses in computational science, a multiscale description of the overall evolution of materials may be considered as rapidly feasible. Many groups in the scientific community are now starting to tackle this problem.

The purpose of the present session is to contribute to this challenge of multiscale modeling of out-of-equilibrium processes. Three main aspects are considered: the improvement of formalism, both from the physical and mathematical viewpoints; the speeding up of numerical methods; and finally the emergence of new models. The present session gives the opportunity to bring together scientists from different communities contributing to the construction of such a modeling in the particular case of out-of-equilibrium processes, namely physicists, chemists, mathematicians experts in numerical analysis, and computer scientists. A selection of relevant topics related to this subject will be presented: point and extended defects, electronic excitations, charged defects, irradiation effects, thermodynamics of out-of-equilibrium processes. Results of a*b initio* methods, empirical potentials, kinetic Monte Carlo or continuous mechanical models will be discussed as well as their mathematical and numerical aspects.

VSP International
Science Publishers
P.O. Box 346, 3700 AH Zeist
The Netherlands

Lecture Series on Computer
and Computational Sciences
Volume 1, 2004, pp. 796-796

Some Mathematical Issues Arising in the Ab Initio Modeling and Simulation of Crystalline Materials

E Cances [1]

CERMICS, Ecole Nationale des Ponts et Chaussées
6 et 8 rue Blaise Pascal, Cité Descartes, Champs sur Marne
77455 Marne la Vallée Cedex 2 (France)

Keywords: Ab initio methods, algorithms, linear scaling

PACS : 31.15.-p, 71.15.-m

I shall address various mathematical and numerical issues related to Quantum Chemistry simulations of crystalline materials. In the first part of the talk, the mathematical properties of some *ab initio* models (Hartree-Fock, Kohn-Sham, orbital-free) will be investigated. In the second part, numerical issues will be dealt with. I shall present in particular recent results on SCF algorithms and linear scaling methods, and some attempts to couple models of different nature (quantum/classical, Kohn-Sham/orbital free, ...).

---

[1] corresponding author : E. Cances - coudray@lps.u-psud.fr

VSP International
Science Publishers
P.O. Box 346, 3700 AH Zeist
The Netherlands

*Lecture Series on Computer
and Computational Sciences*
Volume 1, 2004, pp. 797-799

Some Properties of Isolated Defects in Charged Mgo Clusters

C. Coudray [1]

Laboratoire de Physique des Solides, Université Paris-Sud
Bâtiment 510 - 91405 Orsay (France)

Keywords: Ab-initio calculations, DFT-LDA, charged clusters, defect clusters, ionisation potentials

PACS : 31.15 Ar ; 33.15 Ry ; 36.40 Wa ; 61.72 Ji

In order to try to understand some of the processes of charge in crystalline insulators, MgO clusters containing a central defect are studied. These clusters are obtained by optimising cubic pieces of a crystal with the help of the DFT-LDA method [1,2] DMol [3,4]. Two types of defects are studied : oxygen vacancies [5] and substitutional atoms [6].

1. Oxygen vacancies

The influence of the cluster size is first examined. Two MgO cubes :$(3 \times 3 \times 3)$ and $(5 \times 5 \times 5)$, containing a central oxygen vacancy, are optimized and compared. These comparisons need a common reference. In a neutral crystal, when a neutral vacancy (F center) is present, the valence band is full, the conduction band is empty and the vacancy level, located in the gap, contains two electrons. The density of states of a neutral vacancy cluster must be similar. Then F^+ and F^{++} centers are obtained by removing one or two electrons.

The high value of their surface to volume ratio provides to small clusters a non negligible sphericity. Big clusters are almost cubic, with interatomic distances very close to those of the crystal. The properties of their central atom - or of their central vacancy - are likely not far from those of an atom - or of a vacancy - of a crystal. In the following only big clusters will be considered.

The suppression of the central oxygen atom (F center) leads to a sharing of its negative charge between its six first neighbours and to a slight increase of the sphericity of the inner part of the cluster. The F center so obtained does not accept more than about one electron, the other one being spread in the rest of the cluster. When one electron is suppressed, only a fraction of this electron comes from the vacancy. It is principally taken off from the same magnesium atoms first neighbours of the vacancy. Simultaneously, the deformation of the inner part of the cluster increases. The removal of the second electron gives rise to a similar process and the F^{++} center still contains about two thirds of electron. These operations give rise to polarisation effects along the <110> axes passing through the vacancy.

2. Substitutional atoms

The central atoms of the cluster, either O or Mg, is replaced by another atom, O by atoms of negative valences : C, N, F and Ne, and Mg by atoms of positive valences : Ne, Na, Al and Si. In both cases, the cluster contains an equal number of Mg and O atoms.

As previously, a comparison between these clusters requires a common reference. In the absence of level in the bandgap, a neutral crystal has a full valence band and an empty conduction band. The charge $Q = 0$ will be attributed to a cluster with a similar density of states. A cluster contains 62 pairs of Mg and O atoms, and the impurity, which gives rise to a level, generally included either in the valence, or in the conduction band and containing electrons. According to the sign of the valence v of

---

[1] corresponding author : C. Coudray - coudray@lps.u-psud.fr

this atom, electrons have to be removed (v > 0), or to be added (v < 0) in order to obtain a cluster equivalent to a neutral crystal (Q = 0). This cluster has a Coulomb charge Q_{Cl} = v. Then, the charges Q = ± 1, ± 2 are obtained by further addition or suppression of electrons. The Coulomb charges become Q_{Cl} = Q + v.

For each value of Q, the ionisation potentials of these clusters follow a linear law with respect to v, i.e. with respect to Q_{Cl} (cf. Fig. 1). This law results from the Coulomb interaction between the cluster and the departing or arriving electron. As the curves do not pass through the origin, another energy is present : the energy needed by electrons to jump from the top of the valence band to the ionisation level, or the energy gained when electons fall from this last level to the bottom of the conduction band. For a given value of Q this energy is the same for all clusters, because the presence of the impurity does not modify the gap nor the distances between the limits of the bands and the ionisation level.

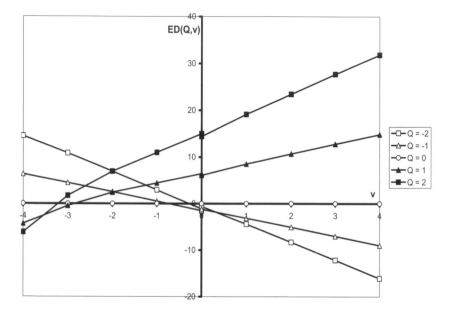

Figure 1: The ionisation potentials ED(Q,v) with respect to the valence v of the central atom.

However, Fig. 1 shows that for v < - 2 and positive values of Q, the ionisation potentials are less than expected. This results from the position of the impurity level, which stands in the bandgap for these clusters. For Q = 0, this level is full ; when an electron is removed, starting from this level, it has only to cross the upper part of the bandgap. Simultaneously, the Mulliken charges of the impurity, which remain approximatively constant in all the other clusters, show a decrease in these clusters, indicating that at least a fraction of the removed electron comes from the central atom. The atoms for which v is less than - 2 are lighter than O. Their nucleus is less positive, so their last atomic level is higher than that of O, giving rise to a level higher than the valence band in the clusters.

The same computations have been performed with other substitutional atoms, giving rise to ionisation potentials which do not differ from more than 4 % from the previous ones.

3. Conclusion

This study shows that the ionisation potentials and the electron affinities of clusters including a substitution atom have no direct relation with those of this atom. However they can be foreseen with a precision of some percents as long as they do not give rise to a level in the bandgap. Besides, it is likely that in MgO crystals, localised positive charging can only occur either in atoms of negative valence giving rise to a level in the bandgap, or in oxygen vacancies. The same result is probably true in other crystalline clusters of insulators, and in crystalline insulators.

References

[1] P. Hohenberg, W. Kohn,, *Phys. Rev.* **B136**, 864 (1964).

[2] W. Kohn, L.J. Sham, *Phys. Rev.* **A140**, 1133 (1965).

[3] B. Delley, *J. Chem. Phys.* **92**, 508 (1990).

[4]   B. Delley, *J. Chem. Phys.* **94**, 7245 (1991).

[5] C. Coudray and G. Blaise, *Eur. Phys. J.* **D27**, 115 (2003).

[6] C. Coudray, accepted in *International Journal of Quantum Chemistry*.

VSP International
Science Publishers
P.O. Box 346, 3700 AH Zeist
The Netherlands

*Lecture Series on Computer
and Computational Sciences*
Volume 1, 2004, pp. 800-801

Electronic Excitations: *ab initio* Calculations of Electronic Spectra and Application to Zirconia, ZrO$_2$

Louise Dash

Laboratoire des Solides Irradiés, Ecole Polytechnique,
91128 Palaiseau, France

In this talk, we address the question of the theoretical prediction and interpretation of experimental spectra obtained by the excitation of valence electrons, such as valence photoemission spectra, bremsstrahlung isochromat spectra (BIS), electron energy-loss spectra (EELS) or absorption spectra. Within density functional theory (DFT), the density of occupied states $\sum_v \delta(E - E_v)$, empty states $\sum_c \delta(E - E_c)$ or the joined density of states $\sum_{v,c} \delta(E - E_c - E_v)$ can be calculated and compared directly to photoemission, BIS or EELS spectra. However, we go one step further and present computational methods to calculate more accurate electronic spectra with *ab initio* techniques beyond DFT [1].

We will first discuss the widely-used random phase approximation (RPA). For many systems, however, it is also necessary to include crystal local-field effects, i.e. to take into account the microscopic variation upon excitation of the Hartree potential (RPA with local fields) and also of the exchange and correlation potential (adiabatic time-dependent DFT (TDDFT)). Quasiparticle effects using the GW approximation will then be discussed. For absorption spectra, excitonic effects can be calculated using the Bethe-Salpeter equation, or within TDDFT using a newly developed exchange and correlation kernel either exactly [2] or approximately [3]. We will show applications of these methods to simple test systems such as silicon, and then concentrate on application to a more complex material, zirconia (ZrO$_2$) [4].

Zirconia is currently of considerable technological interest due to its high strength and stability, even under irradiation, and its excellent dielectric properties. We have thus performed calculations of the structural and electronic properties, and make a direct comparison between the theoretical loss function and new EELS experiments performed at low momentum transfer.

The atomic and electronic structures of zirconia are calculated within density functional theory, and their evolution is analyzed as the crystal-field symmetry changes from tetrahedral (cubic (*c*-ZrO$_2$) and tetragonal (*t*-ZrO$_2$) phases) to octahedral (notional rutile ZrO$_2$), to a mixing of these symmetries (monoclinic phase, *m*-ZrO$_2$).

We calculate the relative phase stabilities and bulk properties and find that the theoretical bulk modulus in pure *c*-ZrO$_2$ is 30% larger than the experimental value for yttria-doped cubic zirconia, showing that the introduction of yttria to stabilize the cubic phase has a significant effect on the elastic properties. We find a fingerprint of tetrahedral symmetry in the electronic structure, being the presence of a gap in the conduction band separating the empty E$_g$ and T$_{2g}$ states in the cubic and tetragonal phases which is not present in the monoclinic phase.

We have carried out electron energy-loss spectroscopy experiments at low momentum transfer and compared these results to the theoretical spectra calculated within the random phase approximation. We show a dependence of the valence and $4p$ ($N_{2,3}$ edge) plasmons on the crystal structure, the dependence of the latter being brought into the spectra by local field effects. Lastly, we attribute low energy excitations observed in EELS of *m*-ZrO$_2$ to defect states 2 eV above the intrinsic top valence band, and the EELS fundamental band gap value is reconciled with the 5.2 or 5.8 eV gaps determined by vacuum ultraviolet spectroscopy.

Finally, we present new theoretical results for the optical absorption spectra for the different phases, both within the random phase approximation and using time-dependent density functional theory.

References

[1] Giovanni Onida, Lucia Reining, and Angel Rubio, Rev. Mod. Phys. **74**, 601 (2002)

[2] Francesco Sottile, Valerio Olevano, and Lucia Reining, Phys. Rev. Lett. **91**, 056402 (2003)

[3] Silvana Botti, Francesco Sottile, Nathalie Vast, Valerio Olevano, Lucia Reining, Hans-Christian Weissker, Angel Rubio, Giovanni Onida, Rodolfo Del Sole, and R. W. Godby, Phys. Rev. B **69**, 155112 (2004)

[4] L. K. Dash, Nathalie Vast, Philippe Baranek, Marie-Claude Cheynet, and Lucia Reining, submitted to PRB (2004)

VSP International
Science Publishers
P.O. Box 346, 3700 AH Zeist
The Netherlands

*Lecture Series on Computer
and Computational Sciences*
Volume 1, 2004, pp. 802-805

Ab Initio Modelling of the Behaviour of Helium in Americium and Plutonium Oxides

M. Freyss, T. Petit

Commissariat à l'Energie Atomique (CEA)
Centre de Cadarache,
DEN/DEC/SESC/LLCC,
Bâtiment 151,
13108 Saint-Paul lez Durance, France
Michel.Freyss@cea.fr

Received 7 March, 2004; accepted in revised form 10 March, 2004

Abstract: By means of an *ab initio* plane wave pseudopotential method, plutonium dioxide and americium dioxide are modelled, and the behaviour of helium in both these materials is studied. The formation energies of point defects (vacancies and interstitials) are calculated, as well as the incorporation and solution energies of helium in PuO_2 and AmO_2. The results are discussed according to the incorporation site of the gas atom in the lattice and to the stoichiometry of $PuO_{2\pm x}$ and $AmO_{2\pm x}$.

Keywords: ab initio, actinide oxides, point defects.

PACS: 61.72.Bd, 61.72.Ji

1. Introduction

Actinide oxide compounds such as $(Pu, Am)O_2$ are candidates as possible fuels to be irradiated in order to transmute long-lived minor actinides and reduce the radiotoxicity of nuclear waste. However few knowledge exists on such compounds and a significant R&D is still required. The scope of this work is to determine basic physical properties of actinide oxide compounds by *ab initio* calculations. *Ab initio* calculations, based on the Density Functional Theory (DFT) [1], enables to solve the Schrödinger equation for electrons in solids and therefore to determine electronic properties such as cohesive and mechanical properties. As it requires no adjustable parameters, *ab initio* calculations provide more accurate results than empirical methods.

As a first step, we focuss here on the actinide dioxides PuO_2 and AmO_2 and study the stability of point defects as well as the behaviour of helium in these oxides. Helium is produced mainly by alpha disintegrations and can form bubbles which could lead to a swelling and/or a cracking of the material. The solubility of helium is therefore a key safety parameter.

In order to compute the ground-state properties of the system, an *ab initio* plane wave pseudopotential method [2] based on the Density Functional Theory (DFT) [1] is used. The electronic exchange-correlation interactions are taken into account in the Generalized Gradient Approximation (GGA) as parametrised by Perdew-Burke-Ernzerhof [3]. Norm-conserving pseudopotentials for plutonium, americium, oxygen and helium were generated according to the Troullier-Martins scheme [4]. The pseudopotentials of americium, oxygen and helium were generated with the FHI98PP code [5]. The plutonium pseudopotential was provided by F. Jollet and S. Bernard (CEA-Bruyères-le-Châtel). All calculations are performed using the ABINIT code [6] in the scalar relativistic approximation. A 160 Ry energy cut-off in the expansion of the plane wave basis is used to calculate the cohesive properties of bulk PuO_2 and AmO_2. Point defects and helium incorporation in PuO_2 and AmO_2 are modelled in a supercell geometry: a box containing a limited number of atoms is considered, this box being repeated in space to form an infinite solid. A 12 atom supercell is used here. A 120 Ry cut-off energy in the expansion of the plane wave basis is found to be enough to determine formation energies of defects and incorporation energies with 0.1 eV accuracy. A 10 *k*-point mesh is used to sample the irreducible Brillouin zone.

2. Stability of point defects in PuO_2 and AmO_2

The formation energy of the following defects are calculated: vacancies, interstitials at the octahedral site, Frenkel pairs and Schottky defects. Frenkel pairs consist of a vacancy and an interstitial of the same chemical element. The vacancy and the interstitial are supposed to be non-interacting. A Schottky defect is a larger defect consisting of an actinide vacancy and two oxygen vacancies, all of which are again supposed to be non-interacting. The values obtained for the formation energies of point defects in PuO_2 and AmO_2 are given in Table 2: oxygen and actinide vacancies (V_O and V_{An}), interstitials at an octahedral site (I_O and I_{An}), Frenkel pairs (FP_O and FP_{An}), and Schottky defects (S).

Table 2: Formation energies E^F (in eV) of point defects.

E^F(eV)	V_O	I_O	V_{An}	I_{An}	FP_O	FP_{An}	S
PuO_2	5.3	0.1	9.2	4.9	5.3	14.1	9.1
AmO_2	3.7	1.4	11.9	4.7	5.1	16.6	8.4

An interesting point shown in Table 2 is that the formation energies of oxygen interstitials are small but positive in both PuO_2 and AmO_2. This is in contrast to oxygen incorporation in UO_2: the formation energy of an oxygen interstitial in UO_2 was found strongly negative (about -2.5 eV) [11], in agreement with the well known first step of oxidation of UO_2 occurring by oxygen incorporation at an interstitial site. A different oxidation mechanism in PuO_2 and AmO_2 compared to UO_2 is expected. Oxidation of PuO_2 experimentally conducted by P. Martin *et al.* [12] revealed that hyper-stoichiometric PuO_{2+x} is indeed difficult to elaborate and that the oxidation of PuO_2 might only occur under very specific conditions.

It is possible to calculate the variation of the defect concentration with the stoichiometry in $PuO_{2\pm x}$ and $AmO_{2\pm x}$. The so-called Point Defect Model [13,14] links the point defect concentrations to both the formation energies and the deviation from stoichiometry *x*:

$$[V_O].[I_O] = \exp{-\frac{E^F_{FP_O}}{k_B T}} \qquad [V_{An}].[I_{An}] = \exp{-\frac{E^F_{FP_{An}}}{k_B T}}$$

$$[V_O]^2.[V_{An}] = \exp{-\frac{E^F_S}{k_B T}} \qquad 2.[V_{An}] + [I_O] = 2.[I_{An}] + 2.[V_O] + x$$

where $E^F_{FP_O}$, $E^F_{FP_{An}}$, and E^F_S are respectively the formation energies of oxygen Frenkel pairs, actinide Frenkel pairs, and Schottky defects. *T* is an arbitrary temperature, which was fixed here at 1700 K. By solving this set of equations using the formation energies previously calculated (Table 2), the evolution of the point defect concentrations as a function of the deviation from stoichiometry *x* can be determined. The concentration variations obtained in $PuO_{2\pm x}$ and in $AmO_{2\pm x}$ are very similar since the Frenkel pairs and Schottky defects formation energies are of the same order of magnitude in both materials (see Table 2). In the hyperstoichiometric dioxides and for small deviation from stoichiometry, the concentration of oxygen interstitials is the largest. For larger deviations ($x > 10^{-4}$), the concentration of actinide vacancies becomes dominant. In the hypostoichiometric dioxides, the concentration of oxygen vacancies is found to rapidly dominate as *x* increases.

3. Incorporation of helium in PuO_2 and AmO_2

The incorporation energy is defined as the energy required to incorporate an helium atom at a pre-existing vacancy or at an interstitial site:

$$E_{inc} = E_{He}^{N,N+1} + E^{N-1,N} - E_{He}$$

where $E_{He}^{N,N+1}$ is the energy of the system with incorporated helium, $E^{N-1,N}$ is the energy of the system with an empty trap site, and E_{He} is the energy of an isolated He atom. The incorporation energies obtained for helium in AmO_2 and PuO_2 are reported in Table 3 for the different incorporation sites: a cation site (An), an oxygen site (O) and an octahedral interstitial site (Int.).

Table 3: Incorporation energies (in eV) of helium in PuO_2 and AmO_2.

E_{inc} (eV)	He (An)	He (O)	He (Int.)
PuO_2	0.7	- 0.5	0.4
AmO_2	0.4	0.5	1.1

In all cases, the incorporation energy is very small: it is hardly larger than 1 eV. The incorporation energy is found negative only in PuO_2, at an oxygen site. In AmO_2, the incorporation energy is always positive, the smallest value being obtained for incorporation at an americium site.

The incorporation energy as defined above does not take into account any thermodynamic equilibrium between the different trap sites. This means that even though the incorporation energy at a given trap site is the lowest, if the concentration of this trap is very small, helium is unlikely to be actually incorporated at such a trap site. It is therefore required to take into account the concentration of the trap sites in the discussion of the solubility of helium. This is done by defining the solution energy.

4. Solubility of helium in PuO_2 and AmO_2

The solution energy is defined by:

$$E_{sol} = E_{inc} + E_{Vac}^{Fapp}$$

where E_{inc} is the incorporation energy as calculated in the previous section, and E_{Vac}^{Fapp} is the apparent formation energy of the trap at which helium is incorporated. The apparent formation energy depends on the vacancy concentration as $E_{Vac}^{Fapp} = -kT \ln \left([V_X] \right)$.

For helium at an interstitial site, the solution energy equals the incorporation energy since no trap site needs to be created. On the other hand, for helium at a substitution site, one needs to determine the concentration of actinide and oxygen vacancies $[V_X]$. This is made possible by means of the Point Defect (see previous section). Helium is found soluble only in hypostoichiometric PuO_2: the solution energy is negative for helium incorporation at an oxygen site. This result can be related to the small but negative incorporation energy found for helium at an oxygen vacancy and to the rapidly increasing concentration of oxygen vacancies with x in hypostoichiometric PuO_2. The solution energy is however very small (less than half an eV). For stoichiometric and hyperstoichiometric PuO_2, helium is found not soluble: the smallest values for the solution energy is found in both cases for an interstitial site. The solution energies are again small. In AmO_2, contrary to PuO_2, helium is found not soluble whatever the stoichiometry and whatever the incorporation site. Small values (less than 1 eV) for the solution energy are however obtained for helium at an oxygen vacancy in hypostoichiometric AmO_2, as well as for helium at an americium vacancy in hyperstoichiometric AmO_2. In stoichiometric AmO_2, helium is more clearly not soluble, with a solution energy above 1 eV.

5. Conclusion

The *ab initio* pseudopotential method in the Generalized Gradient Approximation (GGA) which is used here for plutonium and americium dioxides gives satisfactory results for the lattice parameter, the bulk modulus and the cohesive energy of these actinide oxides. The calculation of the formation energies of point defects in PuO_2 and AmO_2 shows that these materials probably have an oxidation mechanism different to the one of UO_2, since oxygen incorporation at an interstitial site is not found stable like in the case of UO_2. To study the behaviour of helium in PuO_2 and AmO_2, the incorporation energy and the solution energy of helium were calculated. Helium is found soluble only in hypostoichimetric PuO_2, but with a small solution energy. On the other hand in AmO_2 helium is found not soluble whatever the stoichiometry and the incorporation site. The small values obtained for the solution energies (of the order of 1 eV or less) in both PuO_2 and AmO_2 makes it however difficult to conclude on the stability of helium in a diluted form or as bubbles.

Acknowledgments

The author wish to thank F. Jollet, S. Bernard and J.P. Crocombette for fruitful discussions. The European Commission is acknowledged for its financial support through the FUTURE (Fuel for Transmutation of Transuranium Elements) project.

References

[1] P. Hohenberg, W. Kohn, Phys. Rev. **136**, B864 (1964). W. Kohn, L.J. Sham, Phys. Rev. **140**, A1133 (1965).

[2] M.C. Payne, M.P. Teter, D.C. Allan, T.A. Arias, J.D. Joanopoulos, Rev. Mod. Phys. **64**, 1045 (1992).

[3] J.P. Perdew, K. Burke, M. Ernzerhof, Phys. Rev. Lett. **77**, 3865 (1996).

[4] N. Troullier, J.L. Martins, Phys. Rev. B **43**, 1993 (1991).

[5] http://www.fhi-berlin.mpg.de/th/ fhi98md/fhi98PP/index.html

[6] http://www.abinit.org. X. Gonze *et al.*, Comp. Mat. Science **25**, 478 (2002).

[7] U. Benedict, W.B. Holzapfel, *Handbook on the Physics and Chemistry of Rare Earths,* vol 17, ed. K.A. Gschneidner Jr, L. Eyring, G.H. Lander, G.R. Chopin, Elsevier, Amsterdam, pp 245-300 (1993).

[8] G. H. Lander, M. Idiri, private communication.

[9] L.R. Morss, J. Less Common Mct. **93**, 301 (1983). P. Brix, G. Herzberg, J. Chem. Phys. **21**, 2240 (1953).

[10] P.E. Raison, R.G. Haire, *Progress in Nuclear Energy*, vol. 38, No 3-4, pp 251-254 (2001).

[11] M. Freyss, J.P. Crocombette, T. Petit, F. Jollet, *Proceedings of the CIMTEC 2002 – 10th International Ceramics Congress* – July 14-18th 2002, Florence (Italy), volume D, pp 191 (2002).

[12] P. Martin, S. Grandjean,, M. Ripert, M. Freyss, P. Blanc, T. Petit, J. Nucl. Mat. **320**, 138 (2003).

[13] H.J. Matzke, J. Chem. Soc. Faraday Trans. II, **83**, 1171 (1987).

[14] J.P. Crocombette, F. Jollet, L. Thien NGa, T. Petit, Phys. Rev. B **64**, 104107 (2001).

VSP International
Science Publishers
P.O. Box 346, 3700 AH Zeist
The Netherlands

*Lecture Series on Computer
and Computational Sciences*
Volume 1, 2004, pp. 806-807

Specific Problems Raised by the Modelisation of Actinides under Irradiation : Case of Delta Plutonium

**G. Robert[1], G. Jomard[1], B. Amadon[1],
B. Siberchicot[1], F. Jollet[1] and A. Pasturel[2]**

[1] Centre d'Etudes de Bruyères le Châtel, BP12, 91680 Bruyères le Châtel
[2] Laboratoire de Physique Numèrique, CNRS, 25 Avenue des Martyrs, BP 166, 38042 Grenoble

The study of actinides under irradiation raised specific problems related to the more or less localized character of f electrons. From this point of view, the most interesting material is plutonium, for which the localization effects depends upon the crystallographic phase: the f electrons are the most localized in the delta phase, that is stabilized by an alloy element. The plutonium nucleus decays principally by alpha decay which produces an alpha particle and a recoil uranium nucleus. This induces defects that evolve with time in the material. The question we therefore want here to raise is: how does a plutonium alloy evolve with time under self- irradiation? This question is typically a multiscale problem in space and time: point defects are produced within a nanosecond, but the macroscopic time can spread over several years. They can lead to spread defects, like dislocations or to swelling of the material.

Ideally, experimental data could be provided, but they are very scarce in the litterature, due to the difficulty to manipulate plutonium. That is why we need to have a theoretical framework adapted to the specificity of actinides. Indeed, in delta plutonium, electronic correlations are very strong, the specific heat coefficient is high, the Cauchy pressure (C12-C44) is negative, the anisotropy of elastic constants is high (C44/C'=7) and the thermal dilatation coefficient is negative.

We have therefore developed a modelisation chain, starting from ab initio calculations, going on with classical molecular dynamics and kinetic Monte Carlo approaches.

Quantum scale studies:

We worked in the frame of the Density Functional Theory (DFT) that gave very good results on very different systems. Unfortunately, the commonly used Local Density Approximation (LDA) is not appropriate in delta plutonium, due to the strong localization of f electrons. We have therefore followed two directions to overcome this difficulty:

- the first one is that, surprisingly, the Generalized Gradient Approximation (GGA), associated with a magnetic order in plutonium gives quite good results for ground state properties, although no magnetic order is observed in plutonium. We shall present the main results of this study[1]:

  - a coherent treatment of the different phases of plutonium is possible and a (P,T) phase diagram of pure plutonium has been calculated using a simple Debye Grneisen model and a corrected term

  - Energies of formation of five intermediate phases which occur on the Pu-Ga phase diagram have been calculated, in good agreement with experimental data.

- The calculations were performed for various superstructures based on the fcc lattice to get the interaction parameters to describe the energetics of the delta Pu-Ga solid solution. The solid part of the Pu rich side phase diagram involving phase equilibria between Pu phase, delta Pu-Ga solid solution and Pu3Ga compound is calculated. We show the importance of chemical short range order on Ga solubility in delta Pu

- In this frame, energies of formations of point defects are calculated.

- The second one is to go beyond the LDA approximation. This is done thanks to the Dynamical Mean Field Theory (DMFT) that is applied to delta plutonium to better take into account electronic correlations. First results on plutonium will be presented.

Classical molecular dynamic studies:

Quantum scale studies are limited to less than one hundred of atoms, which is not enough to account for collision cascades that are responsible for the production of defects in plutonium alloys. We therefore need to give up the quantum approach to obtain the nature and the quantities of defects produced under irradiation, and to perform atomistic simulations. This supposes to have a reliable interaction potential for pure plutonium and alloys. To account for the negative Cauchy pressure in delta plutonium, a Modified Embedded Atom Model (MEAM) potential is needed. Our MEAM simulations reveal that plutonium seems to behave more like semiconductors tan fcc metals against irradiation. Indeed, at room temperature defects in the core of the cascades form part of an amorphous damage pocket which is stable for few tens of nanoseconds. To decribe alloys (Pu-Ga), a MEAM potential is adjusted on ab initio calculations performed on intermediate phases which occur on the Pu-Ga phase diagram. First results will be presented as well as energetics of defects calculated by atomistic simulations.

These first results are the first step to obtain a set of calculated data that will be used as input data for kinetic Monte Carlo calculations, to have an idea of the evolution of plutonium alloys against irradiation on a longer time scale.

VSP International
Science Publishers
P.O. Box 346, 3700 AH Zeist
The Netherlands

Lecture Series on Computer
and Computational Sciences
Volume 1, 2004, pp. 808-810

Computational Methods for Atomistic Simulation of the Molecular and Electron Dynamics in Shocked Materials

D. Mathieu[1]

Commissariat à l'Energie Atomique,
Centre d'Etudes du Ripault,
BP 16,
F-37260 Monts, France

Abstract: Recent computational methods for the atomistic simulations of shocked materials are presented. The review focusses especially on the aspects of shockwave simulations of possible relevance for the study of irradiation damages in materials. Two challenges are especially discussed : the need for an efficient description of interatomic forces under unusual conditions, and the assessment and modeling of the concentration and role of electronic excited states.

Keywords: Shockwave, simulation, molecular dynamics, electronic structure

PACS: 62.50.+p ; 82.20.Wt

1 Introduction

Shock waves provide an approach to the study of the response of materials to extreme pressures. Their ability to initiate chemical decomposition demonstrates the presence of so-called hot spots in shocked materials. In such hot spots, the energy concentration is sufficient to induce chemical decomposition. Thus, high local temperatures arise even before any release of chemical energy. In fact, in energetic materials, a self-sustained decomposition process may develop and eventually lead to an explosion. The need to avoid accidental initiation of such systems has recently motivated atomistic simulations of shock waves, with the aim to better understand the mechanisms involved at the molecular scale. In a first part, methodological advances specific to the treatment of shocked systems in atomistic simulations are outlined. The second part discusses recent interatomic potentiels suitable for such simulations. Finally, a third part is devoted to the controversial role of electrons in shock initiation of chemical reactions in energetic materials.

2 Shock wave simulations

A shock is a (quasi)-discontinuity in the thermodynamic variables that propagates through the sample: the front width may be comparable to molecular dimensions. The intial pressure P_0, temperature T_0 and energy E_0 correspond the the equilibrium state of the material, which may be simulated using standard techniques. However, simulation of the shocked state (P, V) behind the front is not so straightforwards. Indeed, while it depends on the shock intensity, this state cannot be arbitrary. Indeed, a given choice for (P, V) imposes the energy E through the equation of state.

---

[1]Corresponding author. E-mail: didier.mathieu@cea.fr

However, conservation of energy implies a further relationship between P, V and E. The latter may be written as $H(P, V, E, P_0, V_0, E_0) = 0$ where H is the so-called Hugoniot expression. Therefore, many simulations are necessary to find a satisfactory shocked state [1]. Another straighforwards approach consists in propagating a shock wave in a material using a piston moving with a constant velocity [2]. The shocked state is thus obtained from a single molecular dynamics simulation, but a very large system is required to allow a proper averaging of the properties behind the shock. In fact, the Hugoniot relation $H = 0$ naturally suggests modified molecular dynamics equations constraining the shocked material to satisfy this condition [3]. More recently, using running averages for the thermodynamic variables, a periodic adjustment of the pressure to satisfy the Hugoniot condition has been put forwards [4].

3   Interatomic potentials

The simulation of shock waves bring about constraints regarding the interatomic potential used. Indeed, semi-empirical potentials may be unreliable as they are usually derived from experimental data measured in quite different conditions. In other words, the potential should be derived from first-principles. In addition, for shocks sustained be chemical reactions (detonations) the potential should be able to describe chemical reactions. This usually implies the use of a quantum-mechanical formalism. Finally, the potential should allow the simulation of extended systems over sufficiently long periods of time. A first-principle reactive potential that is reasonably accurate is provided by the Self-Consistent-Charges Tight-Binding approach [5]. Recently, it proved successful to account for the behaviour of nitric acid at very high pressures [6]. However, it remains somewhat costly for extended molecular dynamics runs [7] despite the publication of tight-binding simulations of shocks. Therefore, analytic reactive potentials are of special interest. Such potentials (REBO, ReaxFF) [8] will be presented and directions for improvement suggested.

4   Electronic excitations

Photon emission, presumably by electronic excited states, are observed in shock experiments. The occurence of excited state in a detonating medium is no surprise in view of the extreme conditions that prevail. However, the fraction of excited electrons in non-reactive shock waves and whether they may contribute to initiate the chemical decomposition is controversial [9]. In recent years, classical path simulations have been carried out to study how the ions velocities in shocks may be consistent with significant non-adiabatic effects, that could account for shock-induced excitations [10, 11]. On the basis of a fully coherent propagation of the electron wavefunction, values as high as 10^{-3} may be obtained for the fraction of excited electrons, in defective regions of the material where the bandgap is about one eV. At the other extreme, the incoherent picture with electrons in thermal equilibrium with the ions leads to corresponding values of 10^{-11}. Considering the fact that the valence electrons under shock may interact with a macroscopic electron bath, especially when the gap is as small as 1 eV, the time-dependent orbitals considered are likely to undergo significant decoherence [11]. Introducing an empirical decoherence time adjusted in order to reproduce thermal electron populations at equilibrium bring about a dramatic collapse of non-adiabatic effects. The limitations of such simulations will be outlined.

References

[1] J. J. Erpenbeck, *Phys. Rev. A* **46** 6406 (1992)

[2] D. H. Roberson, D. W. Brenner *et al.*, *Phys. Rev. Lett.* **67** 3132 (1991)

[3] J.-B. Maillet, M. Mareschal *et al.*, *Phys. Rev. E* **63** 016121 (2000)

[4] J. K. Brennan and B. M. Rice, *Mol. Phys.* **101** 3309 (2003)

[5] M. Elstner, D. Porezag *et al.*, *Phys. Rev. B* **58** 7260 (1998)

[6] R. Méreau, D. Mathieu *et al.*, *Phys. Rev. B* **69** 104101 (2004)

[7] M. R. Manaa, L E. Fried *et al.*, *J. Phys. Chem. A* **106** 9024 (2002)

[8] A. C. T. van Duin, S. Dasgupta *et al.*, *J. Phys. Chem. A* **105** 9396 (2001)

[9] L. E. Fried, M. R. Manaa *et al.*, *Annu. Rev. Mater. Res.* **31** 291 (2001)

[10] D. Mathieu and P. Martin, *Comput. Mater. Sci.* **17** 347 (2000)

[11] D. Mathieu, P. Martin *et al.*, *Physica Scripta* in press (2004)

VSP International
Science Publishers
P.O. Box 346, 3700 AH Zeist
The Netherlands

Lecture Series on Computer
and Computational Sciences
Volume 1, 2004, pp. 811-811

Investigation of Amorphization-Induced Swelling in SiC:
A Classical Molecular Dynamics Study

Fabienne Ribeiro, Marjorie Bertolus
CEA Cadarache, Département d'Etude du Combustible,
SESC/LLCC, Bâtiment 151,
13108 Saint-Paul-lez-Durance, France

Mireille Defranceschi
CEA Saclay, Direction de l'Energie Nucléaire,
DSOE/RB, Bâtiment 121,
91191 Gif-sur-Yvette, France

Keywords: Silicon carbide, irradiation-induced swelling, classical molecular dynamics

PACS numbers: 61.43.Dq; 61.82.Fk; 31.15.Qg

SiC is a promising material for nuclear applications, as well as relatively simple to model accurately using *ab initio* methods or classical Molecular Dynamics simulations thanks to the availability of successful interatomic potentials, such as Tersoff-type potential [1-2].

Irradiation-induced swelling of SiC is investigated using a molecular dynamics simulation-based methodology. To mimic the effect of heavy ion irradiation extended amorphous areas of various shapes and sizes are introduced in a crystalline SiC sample either by *in situ* melting and quenching or by introduction of antisite defects. The resulting configurations are relaxed using molecular dynamics at constant pressure. A previous investigation [3] showed that this way of proceeding yields results that compare very well with data from existing ion implantation experiments.

The objective of this study is to get further insight into the exact swelling mechanisms and into the relationship between chemical and structural disorders in the material. In particular, the disorder created is analyzed, and the influence of the interface and of the characteristics of the amorphous zone is investigated. Furthermore, our results are compared with results of elastic calculations.

References

[1]   J. Tersoff, Phys. Rev. B 39, 5566 (1989)

[2]   J. Li, L.J. Porter, S. Yip, J. Nucl. Mater. 246, 53 (1997)

[3]   A. Romano, M. Bertolus, M. Defranceschi, S. Yip, Nucl. Instrum. and Methods B 202, 100 (2003)

VSP International
Science Publishers
P.O. Box 346, 3700 AH Zeist
The Netherlands

*Lecture Series on Computer
and Computational Sciences*
Volume 1, 2004, pp. 812-812

Computational Vibrational Spectroscopy

Currently, there has been intense interest in the pursuit of new computational methods for the calculation of vibrational anharmonic spectra in order to characterize and assign the observed bands of high resolution IR and Raman spectra often obtained with great success through very sophisticated experiments. This theoretical approach requires, in general, two restricting steps: the construction of the potential energy surface and the solution of the vibrational equation in order to obtain the anharmonic energy levels.

For the potential energy surface, the determination of the potential function is based on electronic energy data calculated by sophisticated post-Hartree-Fock methods, as CCSD(T), in a grid in which the required number of points grows drastically with increasing molecule size. Thus it becomes difficult to determine accurately a complete anharmonic force field for molecules containing more than four atoms. In order to calculate the potential function for medium size molecules the efforts of theoretical chemists have investigated two possible alternatives. The first relies on the possibility of density functional theory (DFT) methods to approximate high-level CCSD(T) calculations. The second follows the development of least-squares-based methods in order to fit both the energy and its gradient thus reducing the number of required calculations. Two contributions to this Symposium focus on this subject.

The second step concerns the solution of the vibrational equation. One can attack this problem by perturbational, variational or variation-perturbational methods. The selection of method usually depends on the nature of the problem and the vibrational levels (fundamental bands, harmonics, combinations) of interest. The remaining contributions focus on the solution of the vibrational equation with special emphasis on new methods and the parallelization of the algorithms.

Claude Pouchan
Laboratoire de Chimie Structurale UMR 5624
FR 2606 IPREM
Rue Jules Ferry
64075 Pau
France

VSP International
Science Publishers
P.O. Box 346, 3700 AH Zeist
The Netherlands

*Lecture Series on Computer
and Computational Sciences*
Volume 1, 2004, pp. 813-816

Accurate Vibrational Spectra and Magnetic Properties of Organic Free Radicals in The Gas Phase and in Solution

Vincenzo Barone

Laboratorio di Struttura e Dinamica Moleculare
Dipartimento di Chimica, Universita "Federico II",
Complesso Universitario Monte Sant'Angelo, Via Cintia
I-80126 Napoli, Italy

Received 4 September, 2004; accepted in revised form 10 September, 2004

Abstract: An integrated quantum mechanical approach for the structural, vibrational, and magnetic characterization of free radicals in the gas phase and in solution is introduced and validated for a set of representative systems. Hyperfine couplings computed by hybrid density functionals are in fair agreement with experiment and their reliability is significantly improved by combining CCSD(T) equilibrium values with vibrational corrections obtained at the DFT level. Experimental values in aqueous solution are then recovered using a mixed discrete-continuum solvent model. The g-tensors computed by hybrid density functionals are reasonably accurate and show a small dependence on the specific form of the density functional, the extension of the basis set over a standard 2z+polarization level, vibrational averaging, and bulk solvent effects. However, hydrogen bridges with solvent molecules belonging to the first solvation shell play a significant role.

Keywords: Density Functional, Free Radicals, Vibrations, Magnetic Properties, Solvent
Mathematics SubjectClassification: Here must be added the AMS-MOS or PACS

PACS: Here must be added the AMS-MOS or PACS Numbers

1. Introduction

Methods rooted in the Density Functional Theory (DFT) coupled to purposely tailored basis sets are generally able to reproduce with good accuracy the hyperfine coupling constants (hcc) of organic free radicals in the gas phase [1], although really accurate results can be obtained only by the most sophisticated post-Hartree-Fock methods (e.g. CCSD(T)) with purposely tailored or very large conventional basis sets [2].

The development of the ONIOM method [3] then allows to combine refined computations of small model systems with DFT evaluation of smaller polarization and delocalisation effects by more distant parts of the systems [1]. In the same spirit bulk solvent effects can be taken into account by continuum models like the so called polarizable continuum model, PCM [4,5], whereas inclusion of some explicit solvent molecules (together with the continuum) is mandatory in the case of hydrogen bonding solvents [1]. This latter approach is particularly effective in view of the very low number of solvent molecules to be considered and of the effectiveness of the latest implementation of PCM.

Despite the good results obtained by the above procedures, vibrational averaging effects cannot be neglected for obtaining really quantitative results and in some cases dynamical fluctuations of the system become crucial. In the last years computation of harmonic force fields by quantum mechanical methods has provided an unvaluable aid in this connection thanks to the development of more reliable models with good scaling properties. With the progress in hardware, it is becoming feasible to investigate if effective approaches going beyond the harmonic level can offer further significant improvements. The most effective models are at present based on truncated two- or three-mode potentials followed by an effective second order perturbative treatment (PT2) [6] or diagonalization of configuration interaction matrix built on harmonic reference states (Var) [7]. One of the most significant issues of recent studies is that reliable spectroscopic properties of semi-rigid molecules can be computed by the PT2 approach using last generation density functionals and the results can be further improved coupling structures and harmonic force fields computed at higher levels with DFT anharmonic corrections. The same kind of computations provide also a first order evaluation of vibrational averaging effects on observables, which is largely sufficient for semi-rigid molecules. Flexible systems require a different approach, which can be based either on effective quantum Hamiltonians describing a reduced number of large amplitude motions or on direct dynamics simulations by, e.g., the extended Lagrangian approach. In this contribution we will show that coupling of ONIOM/PCM and effective vibrational approaches defines a very reliable computational approach for the completely ab-initio prediction of structures, vibrations, and magnetic properties of organic free radicals in the gas phase and in solution.

2. Methods and computational details

Most of the computations have been performed with a locally modified version of the Gaussian 03 package [8]. Starting from an optimized geometry, it is possible to build automatically third and semidiagonal fourth derivatives w.r.t normal coordinates Q for any computational model for which analytical second derivatives are available. Next, fundamental transitions (v_i), overtones ($2v_i$), combination bands (v_i+v_j) are evaluated using second order perturbation theory. Vibrational averaging effects on geometries and physico-chemical properties can be obtained at the same level of perturbation computing the properties at the same points used for the evaluation of cubic and quartic force constants and computing first and diagonal second property derivatives by a numerical approach. Computation of isotropic and dipolar hyperfine couplings and of g tensors is implemented in Gaussian03 and described in detail elsewhere [1,9]. Bulk solvent effects are taken into account by the latest implementation of the polarizable continuum model (PCM) coupling remarkable efficiency with proper account of escaped charge effects [5]. Large amplitude intramolecular motions have been taken into account by a proper one-dimensional numerical treatment [1] or by a Car-Parrinello Molecular Dynamics (CPMD) [10] approach. The last kind of computations has been performed using a parallel version of the original CP code [11] and the PBE density functional [12]. The wave-functions (density) were expanded in a plane-wave basis set up to an energy cut-off of 25 (200) Ry; core states are projected out using "ultra-soft" pseudopotentials. Equations of motion were integrated using a time step of 10 au (0.242 fs) with a fictitious electron mass μ = 1000 au. The considered systems have been equilibrated for 1.5 ps, and then the systems have been observed for a total time of 4.0 ps, during which statistical averages were taken.

3. Results and discussion

As a first example, we have considered the H_2CN radical. The results collected in Table I show that fundamental frequencies computed by the PT2 approach are close to their variational counterparts and represent a significant improvement with respect to harmonic frequencies. As a matter of fact, PT2 results can be compared to experiment without any ad hoc scaling. In the present case, the agreement is very good except for the antisymmetric CH and the CN stretchings. In the former case, the experimental assignment was quite indirect and we think that the computed value is much more realistic. Concerning CN stretching, we do not have any simple explication of the discrepancy between theory and experiment. Table II shows that very accurate hyperfine splittings can be obtained by refined CCSD(T) computations, but the effect of vibrational averaging is not negligible especially for hydrogen, and further improves the agreement with experiment.

Very recently, high field EPR spectra of the glycyl radical in several enzymes have been recorded and compared to those of the irradiated crystal of N-acetylglycine; however, a straightforward interpretation of the experimental results is not always possible, since overall spin-dependent properties often result from a delicate balance between different structural, electronic and environmental effects. We have thus undertaken a comprehensive study of the magnetic properties of N-formylglycylamide radical in aqueous solution. We recall that, among all the low-frequency motions, only those involving out-of-plane displacements at the radical center have significant effects on π-radicals. We have thus traced an energy and property profile for the out-of-plane displacement of H^α by the so called distinguished coordinate approach and next evaluated vibrational averaging effects by an one-dimensional numerical integration of the effective Hamiltonian governing motion along this path. The results collected in Table III show that solvent effects and, especially, vibrational averaging from the out of plane vibration cannot be neglected and that, after proper inclusion of these contributions, DFT results come into close agreement with experiment. The same applies also to g-tensors (see Table IV) except that, in this case, vibrational averaging and bulk solvent effects can be neglected, whereas first-shell water molecules must be explicitly included in view of the significant H-bonds they form with the solute.

The results obtained from minimized clusters involving a reduced number of solvent molecules could overestimate solvent shifts irrespective of the proper inclusion of bulk effects. In such circumstances, improved estimates can be obtained repeating the spectroscopic property calculations for different frames spanning the phase space of the configurations energetically accessible to the system. Di-methyl nitroxide radical (DMNO) is the simplest realistic model for the family of nitroxide radicals that have been largely used as magnetic probes for the study of macromolecular systems of both biological and technological interest [1]. We have performed dynamic simulations of DMNO both in the gas phase and in aqueous solution within the Car-Parrinello (CP) scheme [10]. A number of frames from these simulations were employed to set up the best computational procedure to compute solvent effects on the nitrogen hcc. It is well known that hcc's of nitroxide radicals are very sensitive to the out of plane displacement of the nitrogen atom (measured by the CNO...C improper dihedral angle) [1]. In this context, it is remarkable that geometry optimisation leads to values of this angle smaller than its dynamical average (152.8 vs.165° at the PBE/6-31+G(d) level in vacuo), which leads, in turn, to an overestimation of average hcc (15.45 vs. 13.3 G). Replacement of the PBE functional by its more reliable hybrid PBE0 counterpart [13], does not modify this conclusion, but leads to a lower hcc (14.99) due to both an intrinsic difference and to a shorter NO distance. Of course, vibrational averaging by the out of plane nitrogen motion can be taken into account by a proper effective one-dimensional treatment (see above), but, in the present context we prefer to use the average values provided by MD simulations. In particular, it is of interest to compare the results issuing from the present MD simulations in aqueous solution to those obtained from geometry optimisations of suitable clusters. In several previous studies we have shown that a cluster formed by the nitroxide radical and two water molecules bound to the oxygen lone pairs provides, when immersed in a dielectric continuum with the dielectric bulk constant of water, remarkably accurate solvent shifts of nitrogen hcc's [1]. The results delivered by this this model are also in good agreement with MD averages provided that the CNO...C improper dihedral is fixed at its average value (3.4 and 3.2 G from MD and optimisation, respectively at the PBE/6-31+G(d) level). This paves the route for the evaluation of improved absolute values performing single point computations at suitable optimised geometries [1].

4. Conclusion

This paper deals with the capability of an integrated CCSD(T)/DFT computational approach completed by an effective evaluation of vibrational averaging and solvent effects to provide accurate structural, vibrational, and magnetic properties of organic free radicals. Very accurate results of isotropic hyperfine splittings can be obtained coupling CCSD(T) equilibrium values with B3LYP vibrational averaging corrections and solvent shifts issuing from a mixed discrete continuum solvent model. Since also dipolar hyperfine couplings, g-tensors, and vibrational frequencies can be computed with remarkable accuracy, a fully ab-initio description of organic free radicals in the gas phase and in solution appears feasible.

Table I. Fundamental frequencies (cm^{-1}) of H_2CN

	Harmonic	PT2	Var.	Exp.
$\nu 1$	3001	2831	2828	2820
$\nu 2$	1693	1663	1661	1725
$\nu 3$	1386	1350	1343	1337
$\nu 4$	988	970	965	954
$\nu 5$	3070	2888	2902	3103
$\nu 6$	941	922	914	913

Table II. Different contributions to the isotropic hyperfine splittings (in G) of H_2CN.

	equil.value	$<\Delta a>_{298}$[b]	best estimate	Exp.
N	9.6	-0.03	9.6	10.2
C	-28.3	-1.1	-29.4	-29.0
H	79.1	2.9	82.0	84.7

a) CBS3 values from ref.2; b) PT2//B3LYP/EPR-III values, this work.

Table III. Different contributions to the isotropic hyperfine splittings (in G) of N-formylglycylamide radical computed at the PBE0 level, using the EPR-II basis set and the PBE0/6-31+G(d,p) geometry.

	equil.value	$<\Delta a>_{298}$	Δa_{solv}	best estimate[a]	Exp.
C^a	18.8	5.0	0.0	13.8	16-21;24
H^a	-18.7	3.4	0.5	14.9	14;15;17
H	-2.6	0.3	1.3	-1.0	6-5;1-2

a) columns 2+3+4.

Table IV. Different contributions to the principal components of the g tensor (in ppm) of N-formylglycylamide radical computed at the PBE0/EPR-II level at PBE0/6-31+G(d,p) geometries. The experimental values have an estimated incertitude of 400 ppm.

	equil.value	$<\Delta g>_{298}$	Δg_{solv}	best estimate[a]	Exp.
δg_{xx}	-173.3	4.4	-5.4	174.3	-300/400
δg_{yy}	1524.9	-28.9	-489.4	1006.6	800/900
δg_{zz}	2363.9	-26.4	320.9	2016.6	1900/2200
$\Delta(g_{zz}-g_{xx})$	2537.2	-22.0	315.5	2199.7	1500/2500

a) columns 2+3+4.

References

[1] R.Improta, V.Barone, Chem.Rev. 2004, 104, 1231.

[2] A.R.Al Derzi, S.Fau, R.J.Bartlett, J.Phys.Chem. A 2003, 107,6656.

[3] S.Dapprich, I.Komaromi, K.S.Byum, K.Morokuma, M.Frisch, Theochem 1999, 461, 1.

[4] S.Miertus, E.Scrocco, J.Tomasi, Chem.Phys. 1981, 55,117.

[5] M.Cossi, N.Rega, G.Scalmani, V.Barone, J.Chem.Phys. 2002, 117, 43.

[6] V. Barone, J.Phys.Chem.A 2004, 198, 4146.

[7] P.Carbonniere,V.Barone, Chem.Phys.Lett. 2004, 392, 365.

[8] Gaussian 03, Revision C.02 M.J. Frisch et all., Gaussian.Inc., Pittsburgh PA, 2003.

[9] I.Ciofini, C.Adamo, V.Barone, J.Chem.Phys. 2004, in press.

[10] R. Car, M Parrinello, Phys.Rev.Lett. 1985, 55, 2471.

[11] P. Giannozzi, F. De Angelis, R. Car, J.Chem.Phys. 2004, 120, 5903.

[12] J.P.Perdew, K.Burke, M.Ernzerhof, Phys.Rev.Lett. 1996, 77, 3865.

[13] C.Adamo,V.Barone, J.Chem.Phys. 1999, 110, 6158.

VSP International
Science Publishers
P.O. Box 346, 3700 AH Zeist
The Netherlands

*Lecture Series on Computer
and Computational Sciences*
Volume 1, 2004, pp. 817-819

A Parallel Approach for a Variational Algorithm in the Vibrational Spectroscopy Field

D. Begue[†], N. Gohaud and C. Pouchan

Laboratoire de Chimie Structurale.
UMR 5624– Fédération de Recherche IPREM 2606
I.F.R. Rue Jules Ferry,
64075 PAU, Cedex (France)

Received 7 March, 2004; accepted in revised form 10 March, 2004

1. Introduction

Parallel computing is still nowadays one of the biggest challenge for quantum chemistry, as regard to the evolution of computational world: Hardware situation has improved dramatically following Moore's law, efficient mathematical libraries are available (BLAS, LAPACK,...) , and standard parallel protocol exist and are actively improved (MPI). The area of *ab intio* quantum chemistry too is alight with much parallelization activity (Gaussian03, NWChem, GAMESS, ADF...), with successful efforts to parallelize electron correlation methods such as Møller-Plesset perturbation theory, coupled-cluster, Multi Configurational SCF, and Configuration Interaction.
Development of codes based on parallel calculations in the vibrational spectroscopy field seems to be very promising as regards to the dimension of chemical systems that are now available to model [1].
Even if it is nowadays possible to perform a full calculation of an anharmonic force field for molecular systems under six atoms [2], and even if several work in progress seems to be promising for systems of greater dimensions [1,3], it is not obvious to success in doing such calculations for 10 atoms systems without symmetry. Difficulty of such computations is not only the potential energy surface (PES) determination but also resolution of vibrational Schrödinger equation for which the number of vibrational states to take into account in a variational process increases drastically (for example, 170000 states for a simple system like CH_3Li, whereas it is more than 26 millions for its dimer). The aim of this work is to show it is now possible to reach vibrational spectrum of large system in a reasonable computing time thanks to a coupled variational and parallel approach. We present here our new code P_Anhar, which was developed by using Fortran 90 and MPI protocol

2. Computational methods

a - PES

Whatever the dimension of the system, calculations of molecular vibrations require, in a first step, an analytical expression of the potential function. All the methodological details are lastly proposed in ref [4].

b-Vibrational treatment

Regarding the treatment of the Schrödinger vibrational equation, the most current methods found in literature is the perturbational approach [5]. In this case, the vibrational part of the energy of a symmetric top is given by the well known formula:

[†] E-mail: didier.begue@univ-pau.fr

$$E_{v,\ell} = \sum_s \omega_s\left(v_s + \tfrac{1}{2}\right) + \sum_t \omega_t(v_t + 1) + \sum_{s \neq s'} \chi_{ss'}\left(v_s + \tfrac{1}{2}\right)\left(v_{s'} + \tfrac{1}{2}\right) + \sum_{s=s'} \chi_{ss}\left(v_s + \tfrac{1}{2}\right)^2$$
$$+ \sum_{s,t} \chi_{st}\left(v_s + \tfrac{1}{2}\right)(v_t + 1) + \sum_{t \neq t'} \chi_{tt'}(v_t + 1)(v_{t'} + 1) + \sum_{t=t'} \chi_{tt}(v_t + 1)^2 + \sum_{t \neq t'} g_{tt'}\ell_t\ell_{t'} + \sum_{t=t'} g_{tt}\ell_t^2 \tag{3}$$

where the indices s and t refer to the non-degenerated and doubly-degenerated modes respectively, x and g the constants of anharmonicity that depend on cubic and quartic force constants and ℓ the quantum number associated with vibrational angular momentum. If we take into account that each matrix element of the vibrational Hamiltonian can be directly evaluated with respect to the vibrational wavefunction, the equation expression (3) is more useful:

$$E_{v,\ell} = \sum_s \omega_s\langle v_s | q_s^2 | v_s \rangle + \sum_t \omega_t \sum_\ell \langle v_t, \ell | q_t^2 | v_t, \ell \rangle + \sum_{s \neq s'} k_{ss's'}\langle v_s | q_s^2 | v_s \rangle\langle v_{s'} | q_{s'}^2 | v_{s'} \rangle + \sum_{s=s'} k_{ssss}\langle v_s | q_s^4 | v_s \rangle$$
$$+ \sum_{s,t} k_{stt}\langle v_s | q_s^2 | v_s \rangle\left(\sum_\ell \langle v_t, \ell | q_{t_-} q_{t_-} | v_t, \ell \rangle\right)$$
$$+ \sum_{t \neq t'} k_{tt't'}\left(\sum_t \langle v_t, \ell | q_{t_-} q_{t_-} | v_t, \ell \rangle\right)\left(\sum_{t'} \langle v_{t'}, \ell | q_{t'_-} q_{t'_-} | v_{t'}, \ell' \rangle\right) + \sum_{t=t'} k_{tttt}\left(\sum_\ell \langle v_t, \ell | q_{t_-}^2 q_{t_-}^2 | v_t, \ell \rangle\right) - \frac{1}{2}\sum_{t=t'} k_{ttt}\ell_t^2$$

This procedure, easy to implement, have the strong advantage to have independent calculations for each vibrational level (E_v, ℓ). Thus, the development of a parallel algorithm enable us to treat a problem as large as $(CH_3Li)_2$ (24 modes for 26 millions of configurations $\psi_{v_i, v_j, ..., v_{3n-6}}$ and $v_{i(max)} = 7$) in few seconds. Despite of being very efficient, perturbation theory gives sensitive results, in particular with regard to the study of resonances such as, for example, Fermi's and Darling-Denisson's. When these type of accidental degeneracy occurs between two vibrational levels it is no longer legitimate to use the perturbation theory. In order to circumvent this problem, the energies of the two perturbed levels could be obtained by direct diagonalization of the corresponding part of the energy matrix corrected to the first order as mentioned in ref [6]. These conditions of calculation are sufficient to determine (from a highly correlated wave functions by using at least a triple ζ basis set) at less than 2% the position of the fundamental bands of small and medium [3] organic compounds. Generally, it isn't sufficient to study both combination and hot bands.

For a more comprehensive treatment of complete vibrational spectra, it is better to use a variational approach. However, for systems containing more than five atoms, it is generally difficult to use this solution because the treatment of the gigantic matrices of IC is required. During the last years, an intermediate solution consists of the use of mixed methods of variation-perturbation [7], which select iteratively the vibrational states by using the perturbation method in a reduced size under space containing the useful information. Nowadays, it's possible, thanks to the development of average data processing, to work with the whole vibrational information and then develop pure variational methods for the treatment of medium size systems. It is, on the other hand, necessary to optimize in a different way the variational algorithm. One of the ways to reach this challenge is to use parallel calculations. All the details concerning the method are now available in ref [8].

This approach consists of:
- Taking an inventory of the vibrational states potentially needed for the description of the problem. This vibrational anharmonic states are constructed by the linear combinations of products of wavefunctions expressed in the harmonic oscillators basis.
- Taking into account the symmetry of each state
- Cutting out the process into several spectral windows.
- Use a massive parallel algorithm for the treatment of the variational approach

3. Overall structure and method

We decided to do calculations on a small heterogeneous cluster, in order to debug and optimize the code. This cluster runs under Linux O/S.
The ten-processor system used to carry out the calculations reported in this paper has the following configuration:

- 8 1GHz Intel Pentium III processors on dual-processors motherboards with 512 Mo RAM memory per motherboard;
- 2 2GHz Intel Xeon processors on dual-processors motherboard with 4 Go RAM;

- 100 Mbits/s fast ethernet for network communications;
- Linux O/S (Red Hat Linux release 7.1 for the PIII and 7.2 for the Xeon Kernel 2.4.xx);
- LAM-MPI v7.0.4 software for parallelism communications;
- PBS software for job soumission.
- BLAS & LAPACK libraries (latest releases)
- Intel Fortran Compiler version 7.1

The MPI message passing library was chosen for parallelisation, as this is currently the most portable parallelisation model and implementations are available for almost every computer on the market, as well as clusters of workstations or personal computers. The overall structure of the code (P_Anhar) consists of single software, which is made up of several modules and many other minor routines. All the details concerning the code are now available in ref [8].

References

1. P.K. Berzigiyarov, V.A. Zayets, L.Y. Ginzburg, V.F. Razumov, E.F. Sheka, Int. J. Of Quant. Chem 96(2), 73-79 (2004)
   V. Barone, G. Festa, A. Grandi, N. Rega, N. Sanna, Chem. Phys. Lett. 388, 279-283 (2004).

2. P. Carbonniere, D. Bégué, C. Pouchan, Chem. Phys. Lett. 393(1-3), 92-97 (2004) and references therein.

3. V. Barone, Chem. Phys. Lett. 383, 528-532 (2004)

4. P. Carbonniere, D. Bégué, A. Dargelos, C. Pouchan, Chem. Phys. 300, 41 (2004)

5. A. Willetts, N.C. Handy, Chem. Phys. Lett. 235, 286 (1995)
   P. Carbonniere, V. Barone, Chem. Phys. Lett. 392, 365-371 (2004)

6. R. Burel, S. Carter, N.C. Handy, Chem. Phys. Lett. 373, 357 (2003)

7. C. Pouchan, K. Zaki, J. Chem. Phys. 107, 342 (1997)

8. N. Gohaud, D. Bégué, C. Darrigan, C. Pouchan *submitted*

VSP International
Science Publishers
P.O. Box 346, 3700 AH Zeist
The Netherlands

*Lecture Series on Computer
and Computational Sciences*
Volume 1, 2004, pp. 820-824

Vibrational Computation Beyond the Harmonic Approximation: An Effective Tool for the Investigation of Large Semi-Rigid Molecular Systems

P. Carbonnière[1] and V. Barone[2]*

[1]Laboratoire de Chimie Theorique et Physico-Chimie Moleculaire, UMR 5624
Federation de recherché IPREM 2606,
Universite de Pau et des Pays de l'Adour, IFR Rue Jules Ferry,
F-64000, Pau, France.

[2]Laboratprio di Struttura e Dinamica Moleculare
Dipartimento di Chimica, Universita "Federico II",
Complesso Universitario Monte Sant'Angelo, Via Cintia
I-80126 Napoli, Italy

Received 2 August, 2004; accepted in revised form 20 August, 2004

Abstract: This paper deals with the computation of vibrational transitions by quantum mechanical methods. It shows the capability of density functional theory based methods (DFT) to determine anharmonic force fields able to provide reliable interpretation of IR spectra via an effective second order perturbative procedure.
The study based on an investigation of 6 semi-rigid molecules containing from 4 to 12 atoms reviews several pure and hybrid functionals as well as basis sets in order to propose an efficient route for the vibrational calculation of larger semi-rigid molecules.

Keywords: hybrid force fields, DFT, basis set, vibrational frequencies, performances.

Mathematics SubjectClassification: Here must be added the AMS-MOS or PACS

PACS: Here must be added the AMS-MOS or PACS Numbers

1. Introduction

Infrared (IR) and Raman spectroscopies are among the most powerful techniques for characterizing medium size molecules, but proper assignment of fundamental frequencies is often not straightforward due to the large number of (often overlapping) transitions present in high resolution experimental spectra. In the last years computation of harmonic force fields by quantum mechanical methods has provided an unvaluable aid in this connection thanks to the development of more reliable models with good scaling properties. With the recent impressive progresses in hardware and software, it is becoming feasible to investigate if effective approaches going beyond the harmonic level can offer further significant improvements. For small molecules, converged ro-vibrational levels can be obtained by fully variational methods[1-3]. However, for larger molecules some approximation becomes unavoidable concerning both the form of the potential and the ro-vibrational treatment. The most effective models are at present based on truncated two- or three-mode potentials followed by an effective second order perturbative treatment (PT2)[4,6] or by diagonalization of configuration interaction matrix built on self-consistent vibrational states (VSCF-CI)[7] and/or guessed by a preliminary perturbative approach (CI-PT2)[8]. Of course, the cheap PT2 approach cannot give the same results of a converged variational computation. For instance, the PT2 expressions are (at least in one dimension) exact for a Morse oscillator, and therefore certainly not correct for the incomplete quartic development, which is the most practical representation of the potential for large molecules. Thus, the PT2 predictions can be closer to experiment than their variational counterpart[9,10]. Nearly degenerate vibrational states (e.g Fermi and Darling-Dennison resonances) are, of course, a problem for a straightforward PT2 approach, but a well-established improvement is obtained by removing the strongest interactions in the second order treatment and treating them in a proper fashion by diagonalizing the corresponding part of the Hamiltonian matrix. On these grounds, the effective PT2 approach can achieve an accuracy of the order of 10 cm^{-1} for fundamental transitions in the case of small size systems with high level quartic force fields (CCSD(T)/spdf basis set quality)[6]. As a matter of fact, a fully automated implementation of the highly cost effective PT2 vibrational approach has been recently coded[4] in the Gaussian package[11]. At this point, the only obstacle to investigate vibrational spectra of large molecules comes from the determination of reliable quartic force fields, keeping in mind that the CCSD(T) approach is still limited to hexa-atomic systems, due to its very unfavourable scaling with the number of active electrons. One of the most significant issues of our most recent investigations is that reliable spectroscopic properties of semi-rigid molecules can be computed by last generation density functionals and that the results can be further improved coupling structures and harmonic force fields computed at higher levels with DFT anharmonic corrections. Considering the very promising results obtained recently by DFT computations of the vibrational spectra for small[12,13] and medium size molecules[10,14], we have undertaken a series of systematic studies aimed to propose the most effective computational model for semi-quantitative investigations of large molecules[15,16]. The final results of this investigation are summarized in the present paper.

2. Method and computational details

DFT quartic force fields of H_2CO (formaldehyde), H_2CS (thioformaldehyde), H_2CN (methylenimine), C_2H_4 (ethylene), s-tetrazine, and benzene have been computed by the Gaussian 03 package[11]. The performances of the widely used B3LYP functional in the computation of harmonic and anharmonic frequencies has been investigated using 14 standard basis sets summarized in Table 1. Other conventional (BLYP, HCTH,TPSS) and hybrid (B97-1, B1B95) functionals were also tested with 3 different basis sets: the cheap 6-31+G(d,p), the TZ2P which performs fairly well with the B97-1 functional and the cc-pVTZ which is particularly accurate for post Hartree-Fock methods. The quality of the results is assessed by comparison with the most reliable CCSD(T) computations available in the literature.

Details about the implementation of effective PT2 anharmonic computations in the Gaussian 03 package are given elsewhere[4]. Here, we only recall that starting from an optimized geometry, it is possible to build automatically third and semidiagonal fourth derivatives w.r.t normal coordinates Q for any computational model for which analytical second derivatives are available.

Note that for non linear molecules these computations require at most the Hessian matrices at 6N-11 different points, N being the number of atoms in the molecule. Note also that the best compromise between different error sources is obtained using a stepsize of 0.01 Å in the numerical differentiation of second derivatives, tight geometry optimizations and fine grids (at least 99 radial and 590 angular points) in both SCF and CPKS computations. Next, fundamental transitions (v_i), overtones ($2v_i$), combination bands (v_i+v_j) are evaluated using second order perturbation theory. Rotational contributions to anharmonicity, which cannot be neglected for quantitative studies, have been added to the procedure. As shown in a recent study, a perturbative evaluation of these terms leads to results in very close agreement with their variational counterparts[16]. Furthermore, as mentioned above, standard PT2 computations are not sufficient for strongly interacting states. The usual empirical criterium based on their energetic difference (Δ_1) does not take into account that the error in the perturbative treatment comes also from the strength of the coupling. A viable solution to this problem has been suggested by Martin and coworkers[6], who derived simple formulae giving fairly good estimates of the differences between explicitly including a Fermi resonance and absorbing it into the anharmonicity constant (Δ_2). $\Delta_2=1$cm^{-1} is the default value in the latest version of the implemented algorithm and may be manually set up by the keyword DelVPT[4].

Table 1: Basis sets used in this study together with cc-pVDZ and cc-pVTZ. (+ diff X : sp diffuse functions on each heteroatom if any.

Acronym	**Dzp**	**dzpd**	**dzpdT**	**dzpp**	**dzppd**	**dzppdT**
Description	6-31G*	6-31G* + diff X	6-31+G*	6-31G**	6-31G** + diff X	6-31+G**
Acronym	**tzp**	**tzpd**	**tzpdT**	**tzpp**	**tzppd**	**tzppdT**
Description	6-311G*	6-311G* + diff X	6-311+G*	6-311G**	6-31G** + diff X	6-311+G**

3. Results and discussion

Several studies have shown that diffuse basis functions are much more important for optimizing the performances of DFT than larger expansions of the valence orbitals and/or additional polarization functions[10,17,18]. Figure 1 shows the same trend in the case of harmonic computations driven by the B3LYP model where the accuracy of calculations is improved by a factor of 2 when using the cheap 6-31+G(d,p) basis set, compared to the cc-pVTZ results. Thus, the B3LYP/6-31+G(d,p) model leads to an average error approaching 10 cm^{-1} on harmonic frequency calculations for the trial set of molecules. It is noteworthy that the larger discrepancies generally come from the small size systems: considering only medium size molecules (s-tetrazine and benzene) the average absolute error is lowered by about 30 %.

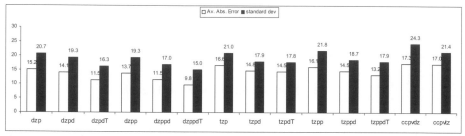

Figure 1: Average absolute error and standard deviation w.r.t. CCSD(T)/cc-pVTZ results for B3LYP harmonic frequencies (cm^{-1}) of H$_2$CO, H$_2$CS, H$_2$CN, C$_2$H$_4$, s-tetrazine and benzene.

Concerning the very costly calculations beyond the harmonic level, the question which arises is how well DFT anharmonic force fields fit their CCSD(T) counterparts. A first answer can be given adding B3LYP anharmonic corrections to CCSD(T) harmonic frequencies and then comparing the resulting anharmonic frequencies to the most reliable full CCSD(T) anharmonic computations available in the literature. The results collected in figure 2 show that the B3LYP functional is able to reproduce CCSD(T) anharmonicities with an error lower than 4 cm^{-1} and that basis set extension above the 6-31G(d) level leads to a modest variation on computed values.

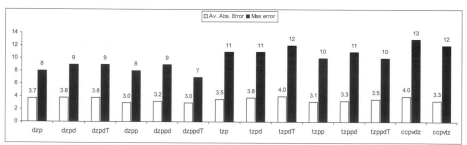

Figure 2: Average absolute error and maximum error w.r.t. CCSD(T)/cc-pVTZ results for B3LYP anharmonicities (cm^{-1}) of the small size systems H$_2$CO, H$_2$CS, H$_2$CN, C$_2$H$_4$.

As a matter of fact, several medium size semi-rigid molecules (pyrrole, furan, uracil, 2-thiouracil, azabenzenes, benzene) were investigated by hybrid force fields in which harmonic and anharmonic parts are computed at different level of theory[10,14]. In the framework of this study, figure 3 shows that the combination B3LYP/6-31+G(d,p)//B3LYP/6-31G(d) yields results in fair agreement with experimental data, paving the route towards the semi-quantitative study of larger semi-rigid molecules. Although the CCSD(T) method is definitely more accurate, the cases of benzene and s-tetrazine confirm that this level of theory is required only at the harmonic level, at least for the precision (say 10 cm^{-1}) we are pursuing.

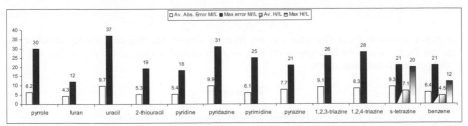

Figure 3: Average absolute error and maximum error w.r.t. experimental results of the fundamental transitions calculated by differents hybrid force fields for larger semi-rigid molecules. The harmonic//anharmonic hybrid force fields are:
H: CCSD(T)/ANO-4321; M: B3LYP/dzppdT; L: B3LYP/dzp.
For clarity, maximum errors are given without inclusion of kekulé and C=O modes when present.

B3LYP is the most implemented hybrid density functional in user friendly packages, and, therefore, the most widely used. Nevertheless, it's worth examining if other DFT based models may outperform the B3LYP results. Systematic studies for small size systems have already shown that among current density functionals (LDA, BLYP, BP86, B3LYP, B97-1) the two last hybrid functionals perform a very good job. In our recent work, the analysis has been extended to medium size molecules with functionals based on the generalized gradient approximation, GGA (HCTH, TPSS, BLYP) and their hybrid counterparts (B97-1, B3LYP, B1B95). Figure 4 shows the relative accuracy of the different methods, and point out that GGA functionals are not sufficiently accurate to drive vibrational computations. Furthermore, in view of the significantly lower errors of TPSS w.r.t. BLYP, one may expect very promising results from its corresponding hybrid. While work is in progress in this connection in our laboratories, B3LYP and B97-1 functionals combined with the 6-31+G(d,p) or TZ2P basis sets already perform a remarkably good job and represent at present the best compromise between computation time and quality of the results.

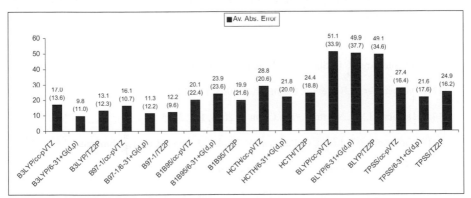

Figure 4: Average absolute error w.r.t. CCSD(T)/cc-pVTZ results for different DFT harmonic frequencies (cm^{-1}) of H_2CO, H_2CS, H_2CN, C_2H_4, s-tetrazine and benzene. In parenthesis : analysis without inclusion of CH stretching modes.

4. Conclusion

This paper deals with the capability of DFT functionals to approach CCSD(T)/spdf computations of harmonic and anharmonic frequencies. The results based on an investigation of 6 semi-rigid molecules containing from 4 to 12 atoms show that B3LYP and B97-1 models perform a very good job using the TZ2P or the 6-31+G(d,p) basis sets. Concerning the B3LYP computations, it has been observed that B3LYP anharmonicities come in close agreement with their CCSD(T) counterparts and that basis set extension above the 6-31G(d) level leads to a modest variation on computed frequencies. This leads to the proposal of less demanding hybrid force fields able to provide reliable results, paving the route toward the investigation of larger molecules.

Acknowledgments

The author wishes to thank Gaussian Inc. for finantial support, the Regional Center Bioteknet for computer facilities, and the anonymous referees for their careful reading of the manuscript and their fruitful comments and suggestions.

References

[1] S. Carter, S.J. Culik, J.M. Bowman, J. Chem. Phys. 1997, 107, 10458.

[2] J. Koput, S. Carter, N.C. Handy, J. Chem. Phys. 2001, 115, 8345.

[3] P. Cassam-Chenai, J. Lievin, J. Quantum Chem. 2003, 93, 245.

[4] V.Barone, J.Chem.Phys. 2004, 120, 3059; V. Barone, J. Chem. Phys. Submitted.

[5] R.D. Amos, N. C. Handy, W.H. Green, D. Jayatilaka, A. Willets, P. Palmieri, J. Chem. Phys. 1991, 95, 8323.

[6]   J.M.L. Martin, T.J. Lee, P.R. Francois, J. Chem. Phys. 1995, 102, 2589.

[7] S. Carter, J.M. Bowman, N.C. Handy, Theor. Chem. Acc. 1998, 100, 191.

[8] P. Carbonniere, D. Begue, C. Pouchan, J. Phys. Chem. A. 2002, 106, 9290 ; I. Baraille, C. Larrieu, A. Dargelos, M. Chaillet, Chem. Phys. 2001, 273, 91.

[9] R. Burel, N.C. Handy, S. Carter, Spectrochimica. Acta A, 2003, 59, 1881; R. Burel, S. Carter, N.C. Handy, Chem. Phys. Lett. 2003, 373, 357.

[10] V.Barone, Chem.Phys.Lett. 2004, 383, 528; V.Barone, J.Phys.Chem. A 198, 4146 (2004); V.Barone, G.Festa, A.Grandi, N.Rega, N.Sanna, Chem.Phys.Lett. 2004, 388, 279.

[11] Gaussian 03, Revision C.01 M.J. Frisch ct all., Gaussian.Inc., Pittsburgh PA, 2003.

[12] M.O. Sinnokrot, C.D. Sherill, J. Chem. Phys. 2001, 115, 2439.

[13] J. Neugebauer, A.B. Hess, J. Chem. Phys. 2003, 118, 7215.

[14] A.D.Boese, J.M.L.Martin, J.Phys.Chem. A 2004, 108, 3085.

[15] P. Carbonniere, T. Lucca, C. Pouchan, V. Barone, J. Comp. Chem., submitted.

[16] P. Carbonniere, V. Barone, Chem.Phys.Lett., submitted.

[17] P. Carbonniere, V. Barone, Chem. Phys. Lett. 2004, 392, 365.

[18] B.J. Lynch, Y. Zhao, D.G. Truhlar, J. Phys. Chem. 2003, 107, 1384.

[19] M.D. Halls, H.B. Schlegel, J. Chem. Phys. 1999, 111, 8819.

VSP International
Science Publishers
P.O. Box 346, 3700 AH Zeist
The Netherlands

*Lecture Series on Computer
and Computational Sciences*
Volume 1, 2004, pp. 825-827

Hyper-Raman Spectroscopy and Quantum Chemistry Simulations

Benoît Champagne[1]

Laboratoire de Chimie Théorique Appliquée,
Facultés Universitaires Notre-Dame de la Paix,
Rue de Bruxelles, 61, B-5000 Namur, Belgium

Received 23 July, 2004

Abstract: This paper surveys successive methods which have been elaborated for simulating the hyper-Raman spectra of molecules and concentrates on a recently-developed Time-Dependent Hartree-Fock procedure. The later is illustrated by simulating spectra of small molecules for which experimental spectra are available.

Keywords: hyper-Raman effect, time-dependent Hartree-Fock, Fermi resonance.

PACS: 31.15.-p, 31.25.-v, 33.20.Fb, 33.20.Tp

1. Introduction

The hyper-Raman effect was first observed in 1965 by Terhune and co-workers [1]. It is a nonlinear inelastic scattering phenomenon where the exchange of energy between light and matter leads, typically, to rotational and vibrational transitions. The Hyper-Raman effect is the second-order nonlinear counterpart of the linear inelastic scattering phenomenon called Raman effect. Hyper-Raman spectroscopy provides complementary information to IR and Raman spectroscopies. Indeed, whereas the IR and Raman activities are associated with the first-order derivatives of the dipole moment and the dynamic polarizability with respect to vibrational normal mode coordinates, $(\partial\mu/\partial Q_a)_0$ and $(\partial\alpha(-\omega,\omega)/\partial Q_a)_0$, respectively, the hyper-Raman signal is driven by the first-order derivatives of the dynamic first hyperpolarizability with respect to normal mode coordinates, $(\partial\beta(-2\omega;\omega,\omega)/\partial Q_a)_0$. Of course, this pertains to the double harmonic oscillator approximation. In particular, hyper-Raman spectroscopy is helpful for studying i) low-frequency vibrational transitions, which are difficult to detect with the IR and Raman spectroscopies, and ii) modes that are silent in both IR and Raman spectroscopies [2]. So far, the use of hyper-Raman spectroscopy to determine the molecular structures has however been hampered by the weak scattering cross sections but the latest improvements in the spectra detection allow now to overcome this intrinsic difficulty as demonstrated by the recent developments of V. RODRIGUEZ in Bordeaux [3].

Due to the large number of active modes, the hyper-Raman spectra are rich in information on the molecular structure. Nevertheless, this large number of modes combined with the nonlinear character of the response makes the hyper-Raman spectra complex to interpret and the recourse to simulation tools turns out to be a promising solution. The subject of the talk is a presentation of the successive theoretical tools that have been elaborated to simulate the hyper-Raman spectra with an emphasis on the latest developments. When possible, comparison between theory and experiment is discussed.

[1] Corresponding author. benoit.champagne@fundp.ac.be

2.   Successive simulation schemes since 1988

In fact, the quantum chemical simulations of hyper-Raman spectra are scarce, probably due to a double computational requirement, i) the high order of the derivatives of the energy or the density matrix with respect to external perturbations (dynamic electric field and nuclear displacements) and ii) the frequency dependence of the response. The first theoretical simulation is nevertheless already 15 years old and has been carried out by Golab *et al.* [4] in order to address the orientation of pyridine molecules adsorbed onto silver electrodes. This simulation invokes the harmonic oscillator approximation and is based on the application of the π-electron Pariser-Parr-Pople (PPP) method to isolated molecules. Later, the theoretical treatment was improved either by adopting the semi-empirical intermediate neglect of differential overlap (INDO) scheme or by performing *ab initio* Hartree-Fock calculations [5]. Although the finite frequency of the incident beam was taken into account in the semi-empirical (PPP and INDO) treatments, the Hartree-Fock simulations were based on normal coordinate derivatives of the static first hyperpolarizability, *i.e.* $(\partial\beta(0;0,0)/\partial Q_a)_0$ where Q_a is the ath vibrational normal mode coordinate and the subscript 0 means that the derivatives are taken at equilibrium. However, it should be pointed out that in the PPP and INDO treatments the Q_a which are used to evaluate the $(\partial\beta/\partial Q_a)$ terms were obtained either from empirical force field [4-5] or from the corresponding *ab initio* Hartree-Fock calculations [5]. Other semi-empirical simulations of hyper-Raman spectra were carried out by Nørby Svendsen and Stroyer-Hansen by using a variational approach within the CNDO/2 approximation [6]. An extended basis set with polarization functions was used and the importance of taking into account the frequency of the incident light on the hyper-Raman intensities for methane, ethylene, and ethane was stressed.

3.   The time-dependent Hartree-Fock method

More recently, analytical procedures based on the time-dependent Hartree-Fock (TDHF) scheme were elaborated to evaluate the first-order derivatives of the dynamic first hyperpolarizability with respect to atomic Cartesian coordinates and subsequently to simulate the nonresonant hyper-Raman spectra within the double harmonic approximation [7]. In order to reduce the computational task, the algorithms have been developed to satisfy the *2n + 1* rule [9] and to take advantage of the interchange relations [10]. The *2n + 1* rule states that, for a variational wave-function, the *2n+1*st energy derivatives can be determined from the nth-order (and lower orders) derivatives of the LCAO coefficients. The first-order derivatives of the first hyperpolarizability with respect to vibrational normal coordinates being fourth-order derivatives of the energy, first- and second-order derivatives of the LCAO coefficients are sufficient. On the other hand, the interchange relations enable to replace derivatives taken with respect to the awkward *3N* atomic Cartesian coordinates by derivatives taken with respect to the more convenient *3* components of the electric fields in such a way that the hyper-Raman intensities can be evaluated from calculating i) the first-order derivatives of the LCAO coefficients with respect to either dynamic electric fields or atomic Cartesian coordinates and ii) the second-order derivatives of the LCAO coefficients with respect to dynamic electric fields. These methods have been implemented in the GAMESS quantum chemistry package [8].

So far, few applications of these approaches have been reported. They were first applied to characterize the frequency-dispersion effects on the hyper-Raman intensities of three prototypical molecules, H_2O, NH_3, and CH_4 [7]. For instance, it was found that when going from $\hbar\omega = 0$ to $\hbar\omega = 1.165$ eV ($\lambda = 1064$ nm) the hyper-Raman intensity can increase by more than 40% and that the frequency dispersion is not monotonic in all cases. Furthermore, it was demonstrated that, for "large intensity" modes, the aug-cc-pvdz and Sadlej POL basis sets provide similar hyper-Raman intensities whereas the intensities calculated with the 6-31++G** and cc-pvdz basis sets are generally larger by a factor that can be as large as 2. Subsequent investigations have addressed the ethylene, ethane, dimethylether, and tetrachloromethane molecules [11]. Comparison with experimental spectra shows that the TDHF method reproduces most of the experimental features, including the variations of intensities associated with changing the polarization of the incident and detected light as well as the position of these modes that are silent or almost silent in IR and Raman spectroscopies. In the case of tetrachloromethane, the effects of a Fermi resonance (between modes 1 and 4 on the one hand and mode 3 on the other hand) on the position and intensity of the peaks have been determined by considering a perturbation theory approach including terms up to second order in electrical and/or mechanical anharmonicity.

4. time-dependent DFT as a means for including electron correlation

In addition to anharmonicity effects, which also play a central role for characterizing overtones and combination bands, including electron correlation effects is a second direction for improving the simulations. Different levels of theory can be adopted ranging from wave-function to density functional theory approaches. As a first account of such studies, electron correlation effects were estimated at the time-dependent density functional theory (TDDFT) level using several exchange-correlation functionals [12]. For some of the systems studied so far, it was found that the electron correlation effects on the intensities and depolarization ratios are not negligible and are related to the relative position of the excitation wavelength to the electronic excitation energies.

Acknowledgments

The author thanks the Belgium National Fund for Scientific Research for his Senior Research Associate position.

References

[1]   R.W. Terhune, P.D. Maker, and C.M. Savage, Phys. Rev. Lett. **14**, 681 (1965).

[2]   S.J. Cyvin, J.E. Rauch, and J.C. Decius, J. Chem. Phys. **43**, 4083 (1965).

[3]   V. Rodriguez, Thèse d'Habilitation à Diriger des Recherches, Bordeaux (France), 2003.

[4]   J.T. Golab, J.R. Sprague, K.T. Carron, G.C. Schatz, and R.P. Van Duyne, J. Chem. Phys. **88**, 7942 (1988).

[5]   W.H. Yang and G.C. Schatz, J. Chem. Phys. **97**, 3831 (1992); W.H. Yang, J. Hulteen, G.C. Schatz, and R.P. Van Duyne, J. Chem. Phys. **104**, 4313 (1996).

[6]   E. Nørby Svendsen and T. Stroyer-Hansen, J. Molec. Struct. **266**, 423 (1992).

[7]   O. Quinet and B. Champagne, J. Chem. Phys. **117**, 2481 (2002).

[8]   M.W. Schmidt, K.K. Baldridge, J.A. Boatz, S.T. Elbert, M.S. Gordon, J.H. Jansen, S. Koseki, M. Matsunaga, K.A. Nguyen, S.J. Su, T.L. Windus, M. Dupuis, and J.A. Montgomery, J. Comput. Chem. **14**, 1347 (1993).

[9]   J.L. Silverman and J.L. van Leuven, Phys. Rev. **162**, 1175 (1967); T.S. Nee, R.J. Parr, and R.J. Bartlett, J. Chem. Phys. **64**, 2216 (1976).

[10]   A. Dalgarno and A.L. Stewart, Proc. R. Soc. London, Ser. A **242**, 245 (1958). J.O. Hirschfelder, W. Byers Brown, and S.T. Epstein, Adv. Quantum Chem. **1** (1964).

[11]   O. Quinet and B. Champagne, Theor. Chim. Acta, **111**, 390 (2004); O. Quinet, B. Champagne, and V. Rodriguez, J. Chem. Phys., in press.

[12]   O. Quinet, B. Champagne, and S.J.A. van Gisbergen, Int. J. Quantum Chem., submitted.

VSP International
Science Publishers
P.O. Box 346, 3700 AH Zeist
The Netherlands

*Lecture Series on Computer
and Computational Sciences*
Volume 1, 2004, pp. 828-830

How to Process a Vibrational Problem
for a Medium Size Molecule:
Application To CH₃N, CH₃Li AND (CH₃Li)₂

N. Gohaud, D. Begue[†] and C. Pouchan[‡]

Laboratoire de Chimie Structurale.
UMR 5624– Fédération de Recherche IPREM 2606
I.F.R. Rue Jules Ferry,
64075 PAU, Cedex (France)

Received 7 March, 2004; accepted in revised form 10 March, 2004

Abstract: A generalized least-squares fitting procedure which jointly uses the energy and gradient data obtained at B3LYP/cc-pVTZ level is applied to construct the anharmonic force field of acetonitrile, methyllithium and its dimer. From a pure variational method we obtain all vibrational energy transitions expected in the medium IR region. For acetonitrile our results are in perfect agreement with the observed data. For methyllithium the disagreements observed suggest a new interpretation of the experimental spectra.

1. Introduction

Currently, there has been intense interest in the pursuit of new computational methods for the calculation of vibrational anharmonic spectra in order to characterize and assign the observed bands on IR spectrum. This theoretical approach requires generally two restricting steps: the first one is the knowledge of the potential energy function; the second is the resolution of the vibrational equation in order to obtain the anharmonic energy levels.

For systems in which anharmonicity is small, one of the possible ways is to write the potential function as a Taylor expansion series terms of the curvilinear displacement coordinates. This expansion series often is limited at the fourth order. Quadratic, cubic, and quartic force constants are generally obtained either by fitting the electronic energy data calculated by the *ab initio* methods for various nuclear configurations close to the optimized geometry, or by a finite difference procedure of first or second derivatives of the electronic energy with respect to the nuclear coordinates. In both cases, the required number of *ab initio* data points grows drastically with the increasing size of the molecule, such that it becomes difficult to determine accurately a complete quartic force field for systems without symmetry containing more than four atoms. In order to reduce the number of calculations, we used in this presentation generalized least-squares procedure which simultaneously utilizes energy and gradients data in a well-suited simplex sum grid [1-3].

Regarding the treatment of the Schrödinger vibrational equation, the most current methods found in literature is the perturbational approach [4]. The procedure is simple but the results are sensitive in particular with regard to the study of resonances such as, for example, Fermi's and Darling-Denisson's. It is possible to use too a variational approach but for systems containing more than five atoms this process is limited by the treatment of gigantic matrices. An intermediate solution consists by using variation-perturbation method [5] which iteratively selects the vibrational states to be diagonalized from a perturbative approach and reduce like this the size of the space containing the useful information.

In this presentation we show that it is possible, thanks to the development of average data processing, to develop a pure variational method for the treatment of medium size molecules.

[†] E-mail : didier.begue@univ-pau.fr
[‡] E-mail : claude.pouchan@univ-pau.fr

We will illustrate the results obtained for the vibrational spectra of acetonitrile, methyllithium monomer and dimer by combinig these two original development for obtention of a quartic force field and vibrational energy levels.

2. Computational details

The quartic force fields for all systems under consideration are obtained by including in the same process of linear regression [1] the values of the energy and analytical gradients obtained in each point of a well-suited grid corresponding to curvilinear displacement coordinates. Points are chosen following a simplex-sum design [6] known for its efficiency and accuracy and taking into account the symmetry for each system.

For solving the Schödinger vibrational equation, the initial basis set is made up as products of monodimensional harmonic oscillators.

$$\left|\Phi_{(V,\ell)_i}^{(0)}\right\rangle = \prod_{i=1}^{n}\left|\phi_{v_i,\ell_i(q_i)}\right\rangle$$

with $\left|\phi_{v_i,\ell_i}(q_i)\right\rangle \propto H_{v_i}(q_i)$ if the normal mode is non-degenerate, and where ℓ_i is not defined,

and $\left|\phi_{v_i,\ell_i}(q_i)\right\rangle \propto L_{v_i,\ell_i}(q_i)$ if the normal mode is doubly degenerate.

The vibrational anharmonic states are constructed by the linear combinations of products of wavefunctions expressed in the harmonic oscillators basis :

$$\left|\Psi_{(V,\ell)_i}\right\rangle = C_i^{(0)}\left|\Phi_{(V,\ell)_i}^{(0)}\right\rangle + \sum_j C_{ij}^{(1)}\left|\Phi_{(V,\ell)_j}^{(1)}\right\rangle + \sum_j C_{ij}^{(2)}\left|\Phi_{(V,\ell)_j}^{(2)}\right\rangle + \sum_j C_{ij}^{(3)}\left|\Phi_{(V,\ell)_j}^{(3)}\right\rangle + \sum_j C_{ij}^{(4)}\left|\Phi_{(V,\ell)_j}^{(4)}\right\rangle$$

where $\left|\Phi_{(V,\ell)_j}^{(1)}\right\rangle$, $\left|\Phi_{(V,\ell)_j}^{(2)}\right\rangle$, $\left|\Phi_{(V,\ell)_j}^{(3)}\right\rangle$, $\left|\Phi_{(V,\ell)_j}^{(4)}\right\rangle$ are the wavefunctions that correspond to, respectively, the mono,

di, tri, and quadri excitations of each state, $\left|\Phi_{(V,\ell)_i}^{(0)}\right\rangle$, and $C_i^{(0)}$, $C_{ij}^{(1)}$, $C_{ij}^{(2)}$, $C_{ij}^{(3)}$, and $C_{ij}^{(4)}$, their

coefficients in the development. In our examples, although the number of multi excitations is fairly large, it is possible to treat the space thus built variationally in the basis of the active configurations because the spectral area studied was cut into several spectral windows, with each containing a few hundreds of states; thus, the size of each matrix to be diagonalized does not exceed 15,000 configurations for obtaining the first 50 converged eigenvalues with an accuracy of 1 cm^{-1}.

For CH$_3$CN, CH$_3$Li and (CH$_3$Li)$_2$ calculations were carried out at B3LYP/cc-pVTZ level, using the GAUSSIAN 98 suit of programs.

3. Results and discussion

The wavenumbers calculated for acetonitrile are in good agreement with the experimental data. For the fundamental bands the mean deviation reaches 1.6%. In the same way the Fermi resonances are fairly described and particularly those between the methyl deformation mode ν_6 and the overtone $\nu_7 + \nu_8$ calculated respectively at 1476 and 1419 cm^{-1} comparatively to the observed bands (1453 and 1409 cm^{-1}). These results give good confidence about our B3LYP/cc-pVTZ force field to calculate vibrational spectra from a variational method.

At the sight of these results we calculate using the same basis sets and the same methods the vibrational spectra for the methyllithium. Indeed, recent calculations published by Breidung *et al* [7] show a large discrepancy between experiment [8] and theory especially for the C-Li stretching (ν_3) and the methyl umbrella (ν_2) modes. Our results, reported in Table 1, confirm for the two modes the disagreement previously observed and explained by a possible aggregation between the methyllithium molecules.

From the quartic force field calculated for the dimer we have obtained the characteristic wavenumbers expected in the range 500-600 cm^{-1} and 1100-1200 cm^{-1}. Two bands with large intensity are calculated in these regions (table 2) respectively at 524 and 1174 cm^{-1} giving a new interpretation of the experimental data.

Table 1. Vibrational spectra (in cm^{-1}) calculated for CH_3Li from a quartic B3LYP/cc-pVTZ force-field and a variational method.

Mode	Calculated	Exp.
ν_6	413	409
ν_3	606	530
ν_2	1069	1158
ν_5	1421	1387
ν_1	2823	2780
ν_4	2861	2820

Table 2. Calculated values obtained for the dimer in the range 500-600 and 1100-1200 cm^{-1}. Intensities in km/mol

Exp[8]	Monomer		Dimer	
	Wavenumber	Intensity	Wavenumber	Intensity
530	606	23	524	202
1158	1069	13	1174	42

References

1. P. Carbonniere, D. Bégué, A. Dargelos, C. Pouchan, Chem. Phys. 300, 41 (2004)

2. P. Carbonniere, D. Bégué, C. Pouchan, Chem. Phys. Lett. 393(1-3), 92-97 (2004)

3. P. Carbonniere, D. Bégué, A. Dargelos, C. Pouchan, REGRESS EGH Code -LCTPCM-UMR5624 (2001)

4. A. Willetts, N.C. Handy, Chem. Phys. Lett. 235, 286 (1995)

5. C. Pouchan, K. Zaki, J. Chem. Phys. 107, 342 (1997)

6. G.E.P. Box, D.W. Behnken, Ann. Mat. Stat. 31, 838 (1960)

7. J. Breidung, W. Thiel, J. Mol. Struct. 599, 239 (2001)

8. L. Andrews, G.C. Pimentel, J. Chem. Phys. 47, 3637 (1967)

VSP International
Science Publishers
P.O. Box 346, 3700 AH Zeist
The Netherlands

Lecture Series on Computer
and Computational Sciences
Volume 1, 2004, pp. 831-833

A Jacobi Wilson description coupled to a Block-Davidson Algorithm : An efficient scheme to calculate highly excited vibrational spectrum

Christophe Iung [1], Fabienne Ribeiro , Claude Leforestier

LSDSMS (UMR 5636), CC 014, Université Montpellier II
34095 Montpellier Cedex 05, France

Received 7 March, 2004; accepted in revised form 10 March, 2004

1 Introduction

High resolution spectroscopists developed with great success very sophisticated experiments in order to obtain fully resolved spectra of highly excited polyatomic systems, such as $H_2CO^{(1)}$, $HFCO^{(2)}$, $CF_3H^{(3)}$, $CH_3OH^{(4)}$ or $C_6H_6^{(5)}$, for instance. However, it should be emphasized that such very accurate data can not be fully understood using basic models. We have recently proposed[6] a new method based on a Jacobi-Wilson description[7] of the system coupled to a modified Davidson algorithm which provides eigenvalues and eigenvectors of selected highly excited ro-vibrational states. We have applied it to the determination of all the vibrational states of H_2CO up to $9500 cm^{-1(6,8)}$. Convergence of the levels can be checked during the iteration process by looking at the residual $||(\mathbf{H} - E_\alpha)\Psi_\alpha||$. The analysis of this residual constitutes a relevant criterion of convergence of the eigenvalues and eigenvectors. We found that the scheme proposed always converges the eigenvector of largest projection onto the initial reference state u_0. Consequently, two different reference states u_0 and u'_0 can lead to the same eigenvector Ψ_α if u_0 and u'_0 are strongly coupled together. In such a case, some eigenvectors are never obtained. It is the reason why we propose an alternative method based on *a Block-Davidson Scheme* to calculate high excited vibrational spectra. The Davidson scheme is applied to a block containing coupled states and provides all the eigenstates and eigenvalues, even when the states density becomes important or when intermode couplings are strong. The efficiency of this numerical strategy has been illustrated by the calculation of high excited vibrational spectra in H_2CO and $HFCO$.

2 Block-Davidson Algorithm

The regular Davidson method we used relies on a one-to-one correspondence between initial guesses $|\mathbf{v}^{(2)}\rangle$ and eigenvectors. When this condition is no longer met, e.g. in the denser part of the spectrum or in case of Fermi resonances, different initial vectors can lead to the same eigenvector, or conversely some eigenvectors will never be unveiled by our modified Davidson scheme.

One way to cope with this problem consists in switching to a block-Davidson scheme. In this block version, which is described in the following Table, one starts from an initial subspace (or block) $\mathcal{H}$ (of dimension b) and propagates b vectors at *each* iteration. Choice of the b eigenvectors

---

[1] E-mail: iung@univ-montp2.fr

$\{\Psi_r^{(M)}, r = 1, \ldots, b\}$ to be propagated can be made from those displaying the largest projections onto the initial subspace (or block) $\mathcal{H}$. Once an eigenvector $\Psi_r^{(M)}$ is converged ($||\mathbf{q}_r|| < \epsilon$), one can skip its propagation, thus reducing by one the number of vectors to be propagated.

0)	Initialization (M = b)	Define the initial subspace $\mathcal{H} = \{u_0 \ldots u_{b-1}\}$
1)	Diagonalization	Diagonalize $\mathbf{H}$ in the $\{u_0, \ldots, u_{M-1}\}$ basis set
		Select the b eigenvectors $\{\Psi_m^{(M)}, m = 1 \ldots b\}$ with largest
		projections onto $\mathcal{H}$
2)	Convergence	Form residual $\mathbf{q}_m = (\mathbf{H} - E_m^{(M)})\Psi_m^{(M)}$
3)	Propagation	Only the eigenvectors $\Psi_m^{(M)}$ for which $||\mathbf{q}_m|| > \epsilon$
	Preconditioning	with zero order Hamiltonian $\mathbf{H}^\circ : \bar{\mathbf{q}}_m = \left(E_m^{(M)} - \mathbf{H}^\circ\right)^{-1} \mathbf{q}_m$
	Orthonormalization	$\xi = \left\{ 1 - \displaystyle\sum_{m=0}^{M-1} |u_m\rangle\langle u_m| \right\} \bar{\mathbf{q}}_m;\, u_M = \xi/||\xi||;\, M := M + 1$
4)	back to 1)	

In principle, the initial subspace $\mathcal{H}$ (of dimension b) should be spanned by the different guesses $\{|\mathbf{v}_p^{(1)}\rangle\}$ strongly coupled together. We will propose two alternative ways to select this subspace $\mathcal{H}$:

- (i) *Perturbative approach to determine* $\mathcal{H}$: The vectors can be identified *a priori* from second order perturbation theory applied to the $|\mathbf{v}_p^{(1)}\rangle$ states and then systematically treated by a block version of the Davidson algorithm. This approach is very efficient when intermode couplings are strong. In such a case, a Block version of the Davidson scheme is imperative.

- (ii) *Regular Davidson Scheme to determine* $\mathcal{H}$: Another possibility consists in first running a regular Davidson scheme on some initial guess u_0. If this guess displays significant projections onto several eigenvectors, the root $E_r^{(M)}$ (i.e. the eigenvalue associated to the eigenvector with the largest projection onto u_0) will display oscillations along the iterations. Those oscillations are the signature of such a situation, and are associated to a very slow convergence, or to no convergence after a given maximum number N of iterations. Examination of the corresponding eigenvectors $\{\Psi_m^{(N)}\}$ reveals that some of them in fact share a significant projection ($\langle u_0|\Psi_m^{(N)}\rangle \geq 0.3$ typically) onto the initial guess. Those eigenvectors are then chosen to span the initial subspace $\mathcal{H}$ of a subsequent block-Davidson scheme.

Because of a possible drift of the reference energy, one has to select the energy E_{ref} used in the preconditioner from the overlap between the $\Psi_m^{(M)}$ eigenvectors and some *fixed* reference u_0, that is to choose $E_{ref} = E_r^{(M)}$ such that $\left|\langle \Psi_r^{(M)}|u_0\rangle\right|$ be maximum. For this scheme to work out, one has to improve the zero order description of the initial guess u_0 such that it can serve as a reference even when considering highly excited levels. This is realized by means of a prediagonalization-perturbation scheme.

3 Discussion and perspectives

We have presented in this work a modified Block-Davidson algorithm aimed at calculating highly excited vibrational states of polyatomic molecules. We show that this method is efficient even when the state density is significant or when the intermode couplings are strong. Its new features concern the definition of an initial guess through a prediagonalition-perturbation scheme, which allows us to locate the root during the course of the Davidson iterations. The preconditioning step, inherent to the Davidson scheme, was also improved by means of a second-order Gauss-Seidel resolution of a approximate Green's function. This scheme is also well suited to provide selected energy levels on their zero-order description. The results obtained in this study show that the Block-Davidson scheme can greatly outperform the basic Lanczos algorithm, which is still the reference method due to its simplicity of use[9]. The scheme relies on a good zero-order description of the molecular basis set, subsequently refined through a prediagonalization-perturbation step. This initial description was achieved by the Jacobi-Wilson method that we developed earlier[7], and which constitutes a key ingredient of the method.

This method has also the great advantage in that it is very easy to use because it depends on only one parameter, ϵ, whose physical meaning is obvious. This parameter plays a central role because it allows the user to fix the accuracy of the result. Consequently, this method presents the advantages of being robust, easy to use, very efficient numerically and adapted to focus on one part of the spectrum. We have proposed some systematic ways to determine the Block of states used in the Block Davidson scheme. This method does not require large memory because less than 200 vectors have to be stored. Also, this method could be easily parallelized because the energics are calculated one block by one block. Our goal is now to optimize this general numerical strategy to provide with a satisfactory accuracy high excited overtones $|n\nu_i>$.

References

[1] R.J. Bouwensi, J.A. Hammerschmidt, M.M. Grzeskowiak, T.A. Stegink, P.M. Yorba and W.F. Polik, *J. Chem. Phys*, **104**,460

[2] Y. Choi and C.B. Moore, *J. Chem. Phys*, **94** , 5414 (1991)

[3] O. Boyarkin and T. Rizzo, *J. Chem. Phys*, **105** , 6285 (1996)

[4] O. Boyarkin, M. Kowalszyk and T. Rizzo, *J. Chem. Phys*, **118** , 93 (2003)

[5] A. Callegari, U. Merker, P. Engels, H. Srivastava and G. Scoles, *J. Chem. Phys*, **113**, 10583 (2000)

[6] F. Ribeiro, C. Iung and C. Leforestier *Chem. Phys. Letters*, **362**, 199 (2002)

[7] C. Leforestier, A. Viel and C. Leforestier *J. Theo Comp. Chem.*, **2**, 609 (2003)

VSP International
Science Publishers
P.O. Box 346, 3700 AH Zeist
The Netherlands

*Lecture Series on Computer
and Computational Sciences*
Volume 1, 2004, pp. 834-837

The Vibrational Mean Field Configuration Interaction (VMFCI) Method: A Flexible Tool for Solving the Molecular Vibration Problem

J. Liévin[1] and P. Cassam-Chenaï[$]

[1]Laboratoire de Chimie Quantique et Photophysique
Université Libre de Bruxelles
50, av. F.D Roosevelt
B-1050 Belgium

[$]Laboratoire d'Etude théorique des milieux extrêmes
Faculté des Sciences, Parc Valrose
F-06108 Nice cedex 2

Abstract: In this talk we present a general variational scheme to find approximate solutions of the spectral problem for the vibration molecular Hamiltonian. It is called the "vibrational mean field configuration interaction" (VMFCI) method and includes as particular cases the vibrational self-consistent field (VSCF) and the vibrational configuration interaction (VCI) methods. The flexibility and efficiency of this new approach are demonstrated on molecules of atmospheric interest: water, ozone and methane.

Keywords: anharmonic vibrations, variational calculations, SCF method, mean field, water, ozone, methane

1. Variational approaches to the vibrational problem

The strict application of the variational method to solve the vibrational problem is restricted to small systems or to a small number of simultaneously coupled vibrational degrees of freedom [1-3]. The so-called vibrational configuration interaction (VCI) method, which consists to diagonalize the vibrational Hamiltonian matrix in a direct product basis set of the $3N$-6 degrees of freedom of the system, is limited by the blow up of this matrix when the number N of nuclei goes up. If converged solutions can be obtained routinely for triatomics [4], it is necessary to use special techniques to render the variational approach feasible for larger systems. Among others efficient approaches, let us cite the discrete variable representation (DVR) [5] and the pseudo-spectral [6] methods. Another family of methods, which will be reviewed in this talk, are those based on the "mean field" or "self consistent field" (SCF) concept. These approaches aim at optimizing variationally the so-called modals[2], which are described as vibrating in the mean-field created by the others. Mean--field optimizations present the advantage to provide compact basis sets to be exploited in further VCI treatments. The standard vibrational SCF (VSCF) method [7-9], based on a trial function writen as a single configuration (i.e. a single product of modals), has been generalized to multiconfigurational versions, like the VMCSCF [10] and VCASSCF [11] methods. Note that the latter techniques are directly migrated from the expertise acquired in the last decades in the field of electronic *ab initio* calculations [12]. The vibrational mean field configuration interaction (VMFCI) method [13, 14], presented here, is a more general and flexible mean-field approach. It encompass both the VSCF and VCI techniques as particular cases, but it especially introduces the coupling between the different degrees of freedom in a hierarchical way, allowing a mean-field treatment and a basis set truncation at each step of the coupling. A summary of the VMFCI method is presented in the next section.

---

[1] Corresponding author: E-mail: jlievin@ulb.ac.be
[2] A modal is a wave function describing a single internal degree of freedom of the system.

2. The VMFCI method

The general vibrational Hamiltonian of an n modes system can be expressed in terms of vibrational coordinates Q_i and their conjugate momenta P_i as:

$$H = h_0 + \sum_{i_1} h_1(Q_{i_1}, P_{i_1}) + \sum_{i_1, i_2} h_2(Q_{i_1}, P_{i_1}, Q_{i_2}, P_{i_2}) + \ldots + h_n(Q_{i_1}, P_{i_1}, \ldots, Q_{i_n}, P_{i_n}) \qquad (1)$$

where the operator h_p gathers all terms involving an explicit coupling between p oscillators, regardless of the actual origin of the coupling (kinetic or potential energy).

Consider a given partition of the n modes $\{1, 2, \ldots, n\}$ into q subsets or contractions of respectively p_1, $p_2, \ldots, p_q$ modes:

$$(I_1, I_2, \ldots, I_q) = (\{i_1^1, i_2^1, \ldots, i_{p_1}^1\}, \{i_1^2, i_2^2, \ldots, i_{p_2}^2\}, \ldots, \{i_1^q, i_2^q, \ldots, i_{p_q}^q\}) \qquad (2)$$

For each contraction I_j, we define a partial Hamiltonian gathering all the terms of H involving the modes included in I_j:

$$H_j = \sum_{i_1 \in I_j} h_1(Q_{i_1}, P_{i_1}) + \sum_{i_1, i_2 \in I_j} h_2(Q_{i_1}, P_{i_1}, Q_{i_2}, P_{i_2}) + \ldots \quad . \qquad (3)$$

The corresponding product basis set is of modals ϕ_{v_i} belonging to I_j is also formed:

$$\phi(Q_{i_1^j}, \ldots, Q_{i_{p_j}^j}) = \prod_{i \in I_j} \phi_{v_i}(Q_i), \qquad (4)$$

where v_i is the quantum number associated to mode i.

All the "spectator" modes, which do not belong to the considered contraction I_j, are described by their wave function in a given reference state, for instance the ground state ϕ_0. The mean-field Hamiltonian results from an averaging of H_j over all the spectator modes:

$$H_j + \left\langle \prod_{I_k \neq I_j} \phi_0(Q_{i_1^k}, \ldots, Q_{i_{p_j}^k}) \middle| H - H_j \middle| \prod_{I_k \neq I_j} \phi_0(Q_{i_1^k}, \ldots, Q_{i_{p_j}^k}) \right\rangle. \qquad (5)$$

This Hamiltonian is diagonalized in the basis set (4), possibly truncated according to energy criteria. Repeating the above steps for all the contractions defining partition (2) provides a new basis set for all the modes. These basis sets are <u>VCI functions</u> , because they are linear combinations of vibrational configurations (4), but also <u>mean-field functions</u>, because they were obtained by diagonalization a mean-field Hamiltonian. Obtaining such basis sets from a given partition (2) is a VMFCI step. Such steps can then be iterated keeping the same definition of the partition. At convergence one obtain self consistent field solutions.

A new partition, contracting subsets of the previous one, can then be defined, and new VMFCI steps can be performed to further optimize the basis set. Successive contraction steps can thus be designed before reaching the final VCI calculation, in which all vibrational degrees of freedom are coupled. The Hamiltonian to be diagonalized in this last step is thus the full Hamiltonian (1).

The multi-step method we just described is a flexible tool. It allows indeed a progressive introduction of the inter-modes coupling in a hierarchical way. A basis set truncation being carried out at each intermediate contraction step, one can expect a drastic reduction of the basis set size. The efficiency of the method will be illustrated in the talk by many examples.

As in any mean-field context, the variational content of the resulting wave functions can be analyzed by means of the Brillouin theorem [15]. Such an analysis, found to be worthwhile in the framework of electronic structure calculations [16], has been extended to the vibrational VSCF and VMCSCF context [10, 11]. Application of this theorem on the VMFCI scheme will be discussed in detail.

3. Implementation of the VMFCI method in the CONVIV program

The VMFCI method has been implemented in the CONVIV computer code for the $J=0$ part of the Watson Hamiltonian [17] in normal coordinates:

$$H = \frac{1}{2}\sum_k P_k^2 + V + \frac{1}{2}\sum_{\alpha\beta}\mu_{\alpha\beta}\pi_\alpha\pi_\beta - \frac{1}{8}\sum_\alpha \mu_{\alpha\alpha} , \qquad (6)$$

where α and β run over (x, y, z), μ is a matrix depending on the inertia tensor at equilibrium, V is the Born-Oppenheimer potential and π is a vibrational angular momentum. This Hamiltonian can be cast in the form of (1) by expanding μ and V in Taylor series truncated at given finite orders. All h_p terms occuring in (1) are then polynomial functions in the Q_i and P_i.

The program allows a flexible definition of the successive contraction steps. It also calculates all the integrals required to another program calculating the rotational energy levels from a perturbational approach based on the generalized Rayleigh-Schrödinger theory [13, 18].

4. Application of the VMFCI method

Systematic test calculations are presented on three molecules of atmospheric interest: water, ozone and methane. This choice is dictated by different reasons. The first one is the availability of accurate *ab initio* potential energy surfaces in the literature. The second one is to provide test calculations on systems of different numbers of internal degrees of freedom, different kinds of inter-mode couplings involving resonances and different symmetries. Full-VCI calculations, only possible for the triatomic systems, will provide a reference level for studying the convergence of the results with the different parameters of the method (order of truncation of the μ-matrix and of the potential V, energy truncation at the different VMFCI steps). The convergence is in all cases obtained at less than 1 cm^{-1}. Methane, with its nine internal degrees of freedom, is a challenge for the method. Isotopologues CH_4, CD_4, CMu_4, CH_3D, CHD_3 and CH_2D_2 are studied. They belong to different point group symmetries, including the non-abelian T_d and C_{3v} groups, with two- and three-fold degeneracies. Such systems gives us the opportunity to show how the full-symmetry can be taken into account by the method. Alternative contraction schemes are systematically tested, by changing the hierachical order of mode the coupling and determining in this way the more efficient coupling scheme. The zero point energy value is shown to be a good indicator for determining the best n-modes contraction scheme. A general conclusion is that all the intermode coupling can be accounted for in a satisfactory manner by contracting at most two modes at a time in the mean-field of the others, despite the existence of non negligible 3- or 4-mode coupling terms in the potential. VMFCI is thus a promising approach for tackling larger systems. The calculation of rotational energy levels will be examplified on the Q-branch of the vibrational ground state of CH_4 at room temperature and at 500K [18].

References

[1] B.T. Sutcliffe, in *Chemical modelling, Applications and Theory*, Volume 3, The Royal Society of Chemistry, 2004.

[2] T. Carrington, Jr., in *Encyclopedia of Computational Chemistry*, P. von R. Schleyer, N.L. Allinger, T. Clark, J. Gasteiger, P. Kollmann, and H.F. Shaefer III, eds, Wiley, New York, 1998.

[3] M. Herman, J. Liévin, J. Vander Auwera and A. Campargue, Adv. Chem. Phys. 108 (1999) 1.

[4] B.T. Sutcliffe and J. Tennyson, Int. J. Quant. Chem. 39 (1991) 183.

[5] J.C. Light and T. Carrington, Adv. Chem. Phys. 114 (2000) 263.

[6] A. Viel and C. Leforestier, J. Chem. Phys. 112 (2000) 1212.

[7] G.D. Carney, L.L. Sprandel and C.W. Kern, Adv. Chem. Phys. 37 (1978) 37.

[8] R.B. Gerber and M.A. Ratner, Chem. Phys. Lett. 68 (1979) 195.

[9] H. Romanovski, J.M. Bowman and L.B. Harding, J. Chem. Phys. 82 (1985) 4155.

[10] F. Culot and J. Liévin, Theoret. Chim. Acta 89 (1994) 227.

[11] F. Culot, F. Laruelle and J. Liévin, Theoret. Chim. Acta 92 (1995) 211.

[12] B. Roos, Adv. Chem. Phys. 67 (1987) 399.

[13] P. Cassam-Chenaï and J. Liévin, Int. J. Quant. Chem.93 (2003) 245.

[14] P. Cassam-Chenaï and J. Liévin, submitted.

[15] B. Levy and G. Berthier, Int. J. Quant. Chem. 2 (1968) 307.

[16] J. Liévin and N. Vaeck, Int. J. Quant. Chem. 62 (1997) 521.

[17] J.K.G. Watson, Mol. Phys. 19 (1970) 465.

[18] P. Cassam-Chenaï, J. Quant. Spectrosc. Radiat. Transfer 82 (2003) 251.

VSP International
Science Publishers
P.O. Box 346, 3700 AH Zeist
The Netherlands

Lecture Series on Computer
and Computational Sciences
Volume 1, 2004, pp. 838-842

Computational Approaches to Artificial Intelligence: Theory, Methods and Applications

Organizer: Michael N. Vrahatis, University of Patras, Greece.

Co-organizers: George D. Magoulas, University of London, U.K.
Gerasimos C. Meletiou, ATEI of Epirus, Greece.
Vassilis P. Plagianakos, University of Patras, Greece.

1 Introduction

Scientists have been trying to implement human intelligence in computers in various ways. Artificial Intelligence (AI) may be defined as the branch of computer science that is concerned with the automation of intelligent behaviour. There are two main paradigms adopted in AI: (i) the symbolic, and (ii) the subsymbolic.

The symbolic paradigm is based on the theory of physical symbolic systems. The symbols have semantic meanings and they represent concepts or objects. Propositional logic, predicate logic and the production systems facilitate dealing with symbolic systems. AI implementations of these systems include rule-based systems, the logic programming and production languages and they have been applied to various fields, such as natural language processing, expert systems, machine learning, cognitive modelling and so on. Unfortunately, symbolic systems encounter difficulties in certain tasks, when inexact, missing or uncertain information is used, when only raw data are available for knowledge acquisition, or parallel solutions need to be elaborated. Nevertheless, these tasks are not considered difficult for the human brain.

The subsymbolic paradigm is based on the idea that intelligent behaviour is performed at a subsymbolic level (this is higher than the neuronal level in the human brain but is different from the symbolic one) by performing appropriate computations.

Neural computing is inspired by information processing and computation in the brain. Nodes, or artificial neurons, in models of neuronal network are usually considered as simplified models of biological neurons, i.e. real nerve cells, and the connection weights between nodes resemble to synapses between neurons. In fact, artificial neurons are much simpler than biological neurons. However, it is far from clear how much of this simplification is justified because of our poor understanding of neuronal functions when embedded in networks. Networks of these simple processing units are called Artificial Neural Networks (ANNs) offer a powerful and distributed computing architecture equipped with significant learning abilities. They help represent highly nonlinear and multivariable relationships. ANNs have the ability to adapt and learn new data, to self organize, to operate on-line (in real-time) and they can be implemented parallel and in VLSI (Very Large Scale Integrated Systems). They have already been successfully used in many real life applications.

In many applications areas, such as medical imaging, control, and decision making, learning from data encounters several difficulties. In certain case, the data sets are characterised by incompleteness (missing parameter values), incorrectness (systematic or random noise in the data), sparseness (few and/or non-representable records available), and inexactness (inappropriate selection of parameters for the given task). ANNs are able to handle these data sets and are mostly used for their pattern matching abilities and their human like characteristics (generalisation, robustness to noise).

Probabilistic neural networks (PNNs) constitute a special class of ANNs, which is known in the statistical literature as kernel discriminant analysis. PNNs are used in scientific fields such as bioinformatics, medicine, and engineering with promising results. However, their heavy dependence on their parameters limits their applicability. For this purpose, evolutionary and swarm intelligence algorithms that are described below can be applied to provide proper parameter configuration, thereby enhancing their performance and applicability.

Additional benefits in handling uncertainty can be gained by adopting fuzzy logic-based approaches. Fuzzy Logic theory forms a key methodology for representing and processing linguistic or, in general, qualitative information. It supports a diversity of mechanisms for knowledge representation focusing on a relevant selection of information granularity. Fuzzy Logic exploits imprecision in an attempt to make intelligent systems' complexity manageable. Along a similar line rough sets were developed they have already been found to be essential in coping with ambiguity. Fuzzy Logic has been applied with great success in control applications and robotics.

Another paradigm that belongs to this class of computational approaches to AI is Evolutionary Computation. This paradigm embraces genetic algorithms (GA), evolutionary computation and evolutionary strategies, Swarm Intelligence, Differential Evolution Algorithms which are biologically and nature inspired methodologies aimed at global optimization. These methodologies are particular effective when the function that has to be minimized is non-differentiable and discontinuous.

Evolutionary computing has been inspired by the learning and adaptation characteristics that are exhibited by the biological systems. Evolutionary Algorithms (EAs) can be incorporated in learning modules as part of information processing systems. Their major characteristics are their heuristic nature and their ability to learn. They are mainly considered as heuristic methods for search and optimisation that may not lead to the perfect solution but to a near perfect one. In nature, the criteria for perfection keep changing and what seems to be close to the perfection according to one "goodness" criterion may be far from it according to another. Thus, EAs operate as search heuristics for the "best" instance in the space of all possible instances. They begin only with a "fitness" criterion for selecting and storing partial solutions that are "good" and dismissing those that are "not good". EAs are also used for learning from data, especially in complex multi-optional optimisation problems. In this case, they do not need in-depth problem knowledge but a "fitness" or "goodness" criterion for the selection of the most promising individuals (they may be partial solutions to the problem). In this context, they exhibit adaptive behaviour through leaning, and accumulate facts and knowledge without having any previous knowledge. Genetic algorithms are one paradigm in the area of evolutionary computation. One application of this area is creating distributed AI systems with emergent collective computational abilities or even a certain level of intelligence, called Artificial Life.

Swarm Intelligence (SI) is strongly related to evolutionary computation. However, instead of imitating nature's evolutionary procedures (such as EAs), SI algorithms are based on the simulation of social behaviour. Many paradigms from nature justify that the exchange of information among members of a population provides them with an evolutionary advantage. There are a few swarm intelligence algorithms developed for solving combinatorial and numerical optimization problems. Ant Colony Optimization (ACO) and Particle Swarm Optimization (PSO) are the most known methods and they have gained increasing attention due to their ability to solve efficiently and effectively a plethora of problems in various scientific and technological fields, including real-life applications, giving also rise to the field of Swarm Engineering.

An emerging area in which computational approaches have been found extremely essential recently is Data Mining and Knowledge Discovery. With the growing interest in sifting through a flood of data collected nowadays, there is a genuine need for Knowledge Discovery. The current way in which the Internet facilities are exploited along with an almost exponentially growing number of links calls for an evident use of data mining. An analysis of time series, going through huge databases, identification of trends therein, building patterns out of alarm records are just a number of other representative examples. A fundamental issue in data mining is clustering. Clustering can be defined as the process of partitioning a set of patterns into disjoint and homogeneous meaningful groups, called clusters. Clustering is fundamental in knowledge acquisition, and has been applied in numerous fields.

From the previous discussion it becomes apparent that all of the technologies discussed above should be used concurrently rather than separately in order to alleviate their complexity requirements. Consider, for instance, the design of neural networks. Here, fuzzy sets deal with interfacing and preprocessing information in neural networks, especially if it comes in a nonnumeric format. Evolutionary techniques are instrumental in determining not only the connections of the network but, more importantly, an entire topology of the network as well as its size. The important observation is that computational approaches retain their generality while being flexible enough to address the needs and specificity of particular applications.

2   The papers

The Session on "Computational Approaches to Artificial Intelligence" is part of the *International Conference of Computational Methods in Sciences and Engineering 2004 (ICCMSE 2004)* that was held between 19-23 November 2004 at Vouliagmeni-Kavouri, Attica, in Greece. The Session aimed at bringing together a variety of research works concerned with computational methods for knowledge engineering, data modelling and complex problem solving, and to show how computational approaches can be applied to challenging real-world problems, such as applications in economic geography, adaptive networks, cryptography and cryptanalysis.

In particular, the Session comprised eleven referred papers, which looked at the theories, methods and applications of computational intelligence. Next, a brief description of these contributions is presented.

2.1   Theory and Methods

Anastasiadis et al., [1], build on a mathematical framework to propose new globally convergent first-order batch training algorithm for neural networks. Their approach ensures that the direction of search is always a descent one and provides accelerated learning, outperforming other recently proposed methods.

Couceiro et al., [2], focus on knowledge representation methods presenting compositions of clones of Boolean functions. These can be interpreted as representation theorems, providing representations of Boolean functions analogous to the disjunctive normal form, the conjunctive normal form, and the Zhegalkin polynomial representations.

Georgiou et al, [3], focus on self–adaptive probabilistic neural networks and investigate the sensitivity of the networks performance to the spread parameter's value. They propose algorithms for the selection of an optimized value that employ state-of-the-art computational intelligence optimization algorithms, like Particle Swarm Optimization, Differential Evolution and Evolution Strategies.

Parsopoulos and Vrahatis, [8], introduce Unified Particle Swarm Optimization, a new scheme that harnesses the local and global variants of the standard Particle Swarm Optimization algorithm, combining their exploration and exploitation abilities. Convergence in probability can be proved for the new approach in unimodal cases and preliminary results justify its superiority against the standard Particle Swarm Optimization.

Pavlidis et al., [10], explore nonlinear systems theory as applied to modeling the emergence of economic agglomerations. They apply topological degree theory to investigate the existence of a fixed–point to a system of equations that describes their model. The fixed–point corresponds to the short–run equilibrium of the model. They further study its uniqueness, and efficient computational methods to determine it.

Tasoulis and Vrahatis, [11], propose a modification of the k-windows clustering algorithm that they have proposed recently. The k-windows algorithm attempts to enclose all the patterns that belong to a single cluster within a d–dimensional window. In this contribution they propose to modify this algorithm by using semi algebraic data structures instead of windows.

2.2 Applications

Ghinea and Magoulas, [4], address the problem of integrating user preferences with network Quality of Service parameters for the streaming of media content, and suggest protocol stack configurations that satisfy user and technical requirements to the best available degree. Their approach is able to handle inconsistencies between user and networking considerations, formulating the problem of construction of tailor-made protocols as a prioritisation problem that can be solved using fuzzy programming.

Laskari et al., [5], study cryptographic systems whose security relies on the hypothesis that the underlying mathematical problems are computationally intractable, in the sense that they cannot be solved in polynomial time. The authors study the performance of artificial neural networks on the approximation of data derived from the use of elliptic curves in cryptographic applications. In another contribution, [6], the same authors propose an optimisation approach to a problem introduced by the cryptanalysis of Feistel cryptosystems. They study the performance of Evolutionary Computation methods in addressing this problem for a representative Feistel cryptosystem, the DES. This approach is applicable to all Feistel cryptosystems that are amenable to the differential cryptanalysis method.

Manetos and Photis, [7], argue that contemporaneous and interdependent spatial phenomena cannot be analyzed and represented efficiently with conventional deterministic techniques. They propose the development of a new integrated spatial toolbox that provides an integration of Data Mining methods like neural networks and fuzzy clustering under a consistent methodological framework to model urban growth dynamics. They apply their approach focusing on the greater region of Athens, Greece. Their interpretation of results reveals the efficiency of the proposed framework in capturing spatial trends and revealing future tendencies.

Parsopoulos et al., [9], consider data fitting schemes that are based on different norms to determine the parameters of curve–models that model landslides in dams. The Particle Swarm Optimization method is employed to minimize the corresponding error norms. The method is applied on real–world data with promising results.

References

[1] A.D. Anastasiadis, G.D. Magoulas, and M.N. Vrahatis, A Globally Convergent Jacobi-Bisection Method for Neural Network Training, *Proc. Int. Conf. Computational Methods in Sciences and Engineering, Lecture Series on Computer Science and Computational Sciences,* this volume, VSP International Publishers, 2004.

[2] M. Couceiro, S. Foldes, and E. Lehtonen, On Compositions of Clones of Boolean Functions, *Proc. Int. Conf. Computational Methods in Sciences and Engineering, Lecture Series on Computer Science and Computational Sciences,* this volume, VSP International Publishers, 2004.

[3] V.L. Georgiou, N.G. Pavlidis, K.E. Parsopoulos, Ph.D. Alevizos, and M.N. Vrahatis, Evolutionary Adaptive Schemes of Probabilistic Neural Networks, *Proc. Int. Conf. Computational Methods in Sciences and Engineering, Lecture Series on Computer Science and Computational Sciences*, this volume, VSP International Publishers, 2004.

[4] G. Ghinea and G.D. Magoulas, Integrating Perceptech Requirements through Intelligent Computation of Priorities in Multimedia Streaming, *Proc. Int. Conf. Computational Methods in Sciences and Engineering, Lecture Series on Computer Science and Computational Sciences*, this volume, VSP International Publishers, 2004.

[5] E.C. Laskari, G.C. Meletiou, Y.C. Stamatiou, and M.N. Vrahatis, Assessing the Effectiveness of Artificial Neural Networks on Problems Related to Elliptic Curve Cryptography, *Proc. Int. Conf. Computational Methods in Sciences and Engineering, Lecture Series on Computer Science and Computational Sciences*, this volume, VSP International Publishers, 2004.

[6] E.C. Laskari, G.C. Meletiou, Y.C. Stamatiou, and M.N. Vrahatis, Applying Evolutionary Computation Methods for the Cryptanalysis of Feistel Ciphers, *Proc. Int. Conf. Computational Methods in Sciences and Engineering, Lecture Series on Computer Science and Computational Sciences*, this volume, VSP International Publishers, 2004.

[7] P. Manetos and Y. N. Photis, Integrating Data Mining Methods for Modeling Urban Growth Dynamics, *Proc. Int. Conf. Computational Methods in Sciences and Engineering, Lecture Series on Computer Science and Computational Sciences*, this volume, VSP International Publishers, 2004.

[8] K.E. Parsopoulos and M.N. Vrahatis, UPSO: A Unified Particle Swarm Optimization Scheme, *Proc. Int. Conf. Computational Methods in Sciences and Engineering, Lecture Series on Computer Science and Computational Sciences*, this volume, VSP International Publishers, 2004.

[9] K.E. Parsopoulos, V.A. Kontogianni, S.I. Pytharouli, P.A. Psimoulis, S.C. Stiros, and M.N. Vrahatis, Nonlinear Data Fitting for Landslides Modeling, *Proc. Int. Conf. Computational Methods in Sciences and Engineering, Lecture Series on Computer Science and Computational Sciences*, this volume, VSP International Publishers, 2004.

[10] N.G. Pavlidis, M.N. Vrahatis, and P. Mossay, Economic Geography: Existence, Uniqueness and Computation of Short-Run Equilibria, *Proc. Int. Conf. Computational Methods in Sciences and Engineering, Lecture Series on Computer Science and Computational Sciences*, this volume, VSP International Publishers, 2004.

[11] D.K. Tasoulis and M.N. Vrahatis, Unsupervised Clustering Using Semi-Algebraic Data Structures, *Proc. Int. Conf. Computational Methods in Sciences and Engineering, Lecture Series on Computer Science and Computational Sciences*, this volume, VSP International Publishers, 2004.

VSP International
Science Publishers
P.O. Box 346, 3700 AH Zeist
The Netherlands

Lecture Series on Computer
and Computational Sciences
Volume 1, 2004, pp. 843-848

A Globally Convergent Jacobi-Bisection Method for Neural Network Training

A.D. Anastasiadis[1], G.D. Magoulas

School of Computer Science and Information Systems,
Birkbeck College, University of London,
Malet Street, London WC1E 7HX, United Kingdom.

M.N. Vrahatis

Department of Mathematics, University of Patras
Artificial Intelligence Research Center (UPAIRC),
University of Patras, GR-26110 Patras, Greece.

Received 31 July, 2004; accepted in revised form 5 September, 2004

Abstract: In this paper a new globally convergent first–order batch training algorithm is proposed, which is equipped with the sign–based updates of the composite nonlinear Jacobi-Rprop method. It is a Jacobi-Rprop modification that builds on a mathematical framework for the convergence analysis ensuring the direction of search is always a descent one. This approach led to accelerated learning and outperformed a recently proposed modification, the Jacobi-bisection method, in all cases tested.

Keywords: Supervised learning, nonlinear iterative methods, nonlinear Jacobi, pattern classification, feedforward neural networks, convergence analysis, global convergence.

Mathematics Subject Classification: 65H10, 68Q32, 68T05, 90C30, 92B20.

1 Introduction

Gradient descent is the most widely used class of algorithms for supervised learning of neural networks. The most popular training algorithm of this category is the batch Back-Propagation with constant step-size [10]. It is a first order method that minimizes the error function by updating the weights using the steepest descent method. Adaptive gradient-based algorithms with local step-sizes try to overcome the inherent difficulty of choosing the right learning rates for each region of the search space depending on the application [4, 5]. This is done by controlling the weight update of each weight in order to minimize oscillations and maximize the length of the overall step-size. One of the best algorithms of this class, in terms of convergence speed, accuracy and robustness with respect to fine tuning its parameters, is the Resilient Backpropagation (Rprop) algorithm [9].

Recently a modification of the Rprop, the so-called Jacobi-Rprop method (JRprop) has been proposed. Empirical evaluations of JRprop gave good results, showing that JRprop outperforms in several cases the Rprop and Conjugate Gradient algorithms [1]. This paper proposes a new globally convergent modification of the JRprop, named GJRprop, and presents a theoretical justification for its development.

---

[1]Corresponding author. E-mail: aris@dcs.bbk.ac.uk

This paper is organized as follows. First, we give a brief outline of the Jacobi-Rprop algorithm and discuss its parameters. Next, the new globally convergent modification of the JRprop is presented. Then we conduct an empirical evaluation of the new algorithm by comparing it with the classic Rprop, and the JRprop [1, 2]. Finally our results are discussed and conclusions are drawn.

2   The Jacobi-Bisection Learning Method

For completeness purposes let us briefly describe the composite Jacobi-Rprop method that belongs to the class of Jacobi-Bisection methods proposed in [2]. Jacobi-Bisection methods combine "individual" information about the network error surface, i.e. the sign of the partial derivative of the error function with respect to a weight, with more "global" information, i.e. the magnitude of the network learning error, in order to decide whether or not to revert/reduce a correction step for each weight individually.

Following the nonlinear Jacobi prescription, one-dimensional subminimization is applied along each weight direction in order to compute a minimizer of an objective function $f : \mathcal{D} \subset \mathbb{R}^n \to \mathbb{R}$ [15]. More specifically, starting from an arbitrary initial vector $x^0 \in \mathcal{D}$, one can subminimize at the kth iteration the function $f(x_1^k, \ldots, x_{i-1}^k, x_i, x_{i+1}^k, \ldots, x_n^k)$, along the ith direction and obtain the corresponding subminimizer $\hat{x}_i$. Obviously for the subminimizer $\hat{x}_i$ holds that $\partial_i f(x_1^k, \ldots, x_{i-1}^k, \hat{x}_i, x_{i+1}^k, \ldots, x_n^k)/\partial x_i = 0$. This is a one–dimensional subminimization because all the components of the vector x^k, except from the ith component, are kept constant. Then the ith component is updated according to:

$$x_i^{k+1} = x_i^k + \tau_k(\hat{x}_i - x_i^k), \tag{1}$$

for some relaxation factor τ_k. The objective function f is subminimized in parallel for all i.

In neural network training we have to minimize the error function E with respect to the weights w_{ij}, E is the batch error measure defined as the sum of squared differences error function over the entire training set. Assuming that along a weight's direction an interval is known which brackets a local minimum $\hat{w}_{ij}$. When the gradient of the error function is available at the endpoints of the interval of uncertainty along this weight direction, it is necessary to evaluate function information at an interior point in order to reduce this interval. This is because it is possible to decide if between two successive epochs $(t-1)$ and (t) the corresponding interval brackets a local minimum simply by looking the function values $E(t-1)$, $E(t)$ and gradient values $\partial E(t-1)/\partial w_{ij}$, $\partial E(t)/\partial w_{ij}$ at the endpoints of the considered interval (see [11] for a general discussion on the problem). The conditions that have to be satisfied are [11]:

$$\frac{\partial E(\mathcal{S}_1)}{\partial w_{ij}} < 0 \quad \text{and} \quad \frac{\partial E(\mathcal{S}_2)}{\partial w_{ij}} > 0,$$

$$\frac{\partial E(\mathcal{S}_1)}{\partial w_{ij}} < 0 \quad \text{and} \quad E(\mathcal{S}_1) < E(\mathcal{S}_2), \tag{2}$$

$$\frac{\partial E(\mathcal{S}_1)}{\partial w_{ij}} > 0 \quad \text{and} \quad \frac{\partial E(\mathcal{S}_2)}{\partial w_{ij}} > 0 \quad \text{and} \quad E(\mathcal{S}_1) > E(\mathcal{S}_2),$$

where $\mathcal{S}_1$ and $\mathcal{S}_2$ determine the sets of weights for which the coordinate that corresponds to the weight w_{ij} is replaced by $a_i = \min\{w_{ij}(t-1), w_{ij}(t)\}$, and $b_i = \max\{w_{ij}(t-1), w_{ij}(t)\}$ correspondingly. Notice that, at this instance, between two successive epochs $(t-1)$ and (t) all the other coordinates remain the same. The above three conditions lead to the conclusion that the interval $[a_i, b_i]$ includes a local subminimizer along the direction of weight w_{ij}. A robust method of interval reduction called bisection can now be used. We will consider here the bisection method which has been modified to the following version described in [14]:

$$w_i^{p+1} = w_i^p + h_i \operatorname{sign} \partial_i E(w^p) / 2^{p+1}, \tag{3}$$

where $p = 0, 1, \ldots$ is the number of subminimization steps and $w_i^0 = a_i$; $h_i = \operatorname{sign} \partial_i E(w^0) (b_i - a_i)$; w^0 determines the set of weights at the $(t-1)$ epoch while w^p is obtained by replacing the coordinate of w^0 that corresponds to the weight w_{ij} by w_i^p, sign defines the well known triple valued sign function and $\partial_i E$ denotes the partial derivative of E with respect to the ith coordinate. Of course, the iterations (3) always converge with certainty to $\hat{w}_i \in (a_i, b_i)$ if for some w_i^p, $p = 1, 2, \ldots$, the first one of the conditions (2) holds.

The reason for choosing the bisection method is that it always converges within the given interval (a_i, b_i) and it is a globally convergent method. Also, the number of steps of the bisection method that are required for the attainment of an approximate minimizer $\hat{w}_i$ of (1) within the interval (a_i, b_i) to a predetermined accuracy ε is known beforehand and is given by $\nu = \left\lceil \log_2[(b_i - a_i)\varepsilon^{-1}] \right\rceil$. Moreover it has a great advantage since it is worst-case optimal, i.e. it possesses asymptotically the best possible rate of convergence in the worst-case [12]. This means that it is guaranteed to converge within the predefined number of iterations and moreover, no other method has this property. Therefore, using the value of ν it is easy to have beforehand the number of iterations that are required for the attainment of an approximate minimizer $\hat{w}_i$ to a predetermined accuracy. Finally, it requires the algebraic signs of the values of the gradient to be computed.

3   The Globally Convergent JRprop Method

The aim of the new algorithm is to improve the learning speed as well as to ensure subminimization of the error function along each weight direction. The term global convergence in our context is used "to denote a method that is designed to converge to a *local* minimizer of a nonlinear function, *from almost any starting point*" [3, p.5]. Dennis and Schnabel also note that "it might be appropriate to call such methods *local* or *locally convergent*, but these descriptions are already reserved by tradition for another usage". Moreover, Nocedal, [7, p.200], defines a globally convergent algorithm as an algorithm with iterates that converge from a remote starting point. Thus, in this context, global convergence is totally different from global optimisation [13].

In our approach JRprop's convergence to a local minimiser is treated using principles from unconstrained minimisation theory. Suppose that x^0 is the starting point of the iterative scheme:

$$x^{k+1} = x^k + \tau^k d^k, \quad k = 0, 1, \ldots \tag{4}$$

where d^k is the search direction and $\tau^k > 0$ is a step–length. Suppose further that

(i) $f : \mathcal{D} \subset \mathbb{R}^n \to \mathbb{R}$ is the function to be minimized and f is bounded below in $\mathbb{R}^n$;

(ii) f is continuously differentiable in a neighborhood $\mathcal{N}$ of the level set $\mathcal{L} = \{x : f(x) \leqslant f(x^0)\}$;

(iii) ∇f is Lipschitz continuous on $\mathbb{R}^n$ that is for any two points x and $y \in \mathbb{R}^n$, ∇f satisfies the Lipschitz condition with constant L, $\|\nabla f(x) - \nabla f(y)\| \leqslant L\|x - y\|$, $\forall x, y \in \mathcal{N}$.

Convergence of the general iterative scheme (4) requires that the adopted search direction d^k satisfies the condition $\nabla E(w^k)^\top d^k < 0$, which guarantees that d^k is a descent direction of $f(x)$ at x^k. The step–length in (4) can be defined by means of a number of rules, such as the Armijo's rule [3], the Goldstein's rule [3], or the Wolfe's rule [16], and guarantees the convergence in certain cases. For example, when the step–length is obtained through Wolfe's rule [16]:

$$f(x^k + \tau^k d^k) - f(x^k) \leqslant \sigma_1 \tau^k \nabla f(x^k)^\top d^k, \tag{5}$$
$$\nabla f(x^k + \tau^k d^k)^\top d^k \geqslant \sigma_2 \nabla f(x^k)^\top d^k, \tag{6}$$

where $0 < \sigma_1 < \sigma_2 < 1$, then a theorem by Wolfe [16] is used to obtain convergence results. Moreover, the Wolfe's Theorem [3, 7] suggests that if the cosine of the angle between the search direction d^k and $-\nabla f(x^k)$ is positive then $\lim_{k\to\infty} \nabla f(x^k) = 0$, which means that the sequence of gradients converges to zero. For an iterative scheme (4), $\lim_{k\to\infty} \nabla f(x^k) = 0$ is the best type of global convergence result that can be obtained (see [7] for a detailed discussion). Evidently, no guarantee is provided that (4) will converge to a global minimiser, x^*, but only that it possesses the global convergence property, [3, 7], to a local minimiser.

Theorem 1: Suppose that Conditions **(i)-(iii)** are fulfilled. Then, for any $x^0 \in \mathbb{R}^n$ and any sequence $\{x^k\}_{k=0}^{\infty}$ generated by the JRprop's scheme:

$$x^{k+1} = x^k - \tau^k \operatorname{diag}\{\eta_1^k, \dots, \eta_i^k, \dots, \eta_n^k\} \operatorname{sign}\left(\nabla f(x^k)\right), \tag{7}$$

where $\operatorname{sign}\left(\nabla f(x^k)\right)$ denotes the column vector of the signs of the components of $\nabla f(x^k)$, $\tau^k > 0$, η_m^k, $m = 1, 2, \dots, i-1, i+1, \dots, n$ are small positive real numbers generated by the JRprop learning rates' schedule:

> **while** $f(x^k) \leqslant f(x^{k-1})$ **then**
>
> **if** $\left(\partial_m f(x^{k-1}) \cdot \partial_m f(x^k) > 0\right)$ **then** $\eta_m^k = \min\left\{\eta_m^{k-1} \cdot \eta^+, \Delta_{\max}\right\}$, $\qquad$ (8)
>
> **if** $\left(\partial_m f(x^{k-1}) \cdot \partial_m f(x^k) < 0\right)$ **then** $\eta_m^k = \max\left\{\eta_m^{k-1} \cdot \eta^-, \Delta_{\min}\right\}$, $\qquad$ (9)
>
> **if** $\left(\partial_m f(x^{k-1}) \cdot \partial_m f(x^k) = 0\right)$ **then** $\eta_m^k = \eta_m^{k-1}$, $\qquad$ (10)

where $\partial_m f(x_1, \dots, x_m, \dots, x_n)$ denotes the partial derivative of f with respect to the mth coordinate, $0 < \eta^- < 1 < \eta^+$, $\Delta_{\max}$ is the learning rate upper bound, $\Delta_{\min}$ is the learning rate lower bound, and

$$\eta_i^k = -\frac{\sum_{\substack{j=1 \\ j \neq i}}^{n} \eta_j^k \, \partial_j f(x^k) + \delta}{\partial_i f(x^k)}, \qquad 0 < \delta \ll \infty, \qquad \partial_i f(x^k) \neq 0, \tag{11}$$

holds that $\lim_{k\to\infty} \nabla f(x^k) = 0$.

Proof: The sequence $\{x^k\}_{k=0}^{\infty}$ generated by the iterative scheme (7) follows the direction

$$d^k = -\operatorname{diag}\{\eta_1^k, \dots, \eta_i^k, \dots, \eta_n^k\} \operatorname{sign}\left(\nabla f(x^k)\right),$$

which is a descent direction if η_m^k, $m = 1, 2, \dots, i-1, i+1, \dots, n$ are positive real numbers derived from Relations (8-10), and η_i^k is given by Relation (11), since $\nabla f(x^k)^\top d^k < 0$. Following the proof of [15, Theorem 6], since d^k is a descent direction and f is continuously differentiable and bounded below along the radius $\{x^k + \tau d^k \mid \tau > 0\}$, then there always exist τ^k satisfying Relations (5)–(6) [3, 7]. Moreover, the Wolfe's Theorem [3, 7] suggests that if the cosine of the angle between the descent direction d^k and the $-\nabla f(w^k)$ is positive then $\lim_{k\to\infty} \nabla f(x^k) = 0$. In our case, indeed $\cos\theta_k = \frac{-\nabla f(x^k)^\top d^k}{\|\nabla f(x^k)\| \, \|d^k\|} > 0$. $\square$

In batch training, E is bounded from below, since $E(w) \geqslant 0$. For a given training set and network architecture, if w^* exists such that $E(w^*) = 0$, then w^* is a global minimiser; otherwise, w with the smallest $E(w)$ value is considered a global minimiser. Also, when using *smooth enough* activations (the derivatives of at least order p are available and continuous), such as the well known hyperbolic tangent, the logistic activation function etc., the error E is also smooth enough.

The globally convergent modification of the JRprop, named Globally-Jacobi-Rprop (GJRprop), is implemented through Relations (7)–(11). The role of δ is to alleviate problems with limited precision that may occur in simulations, and should take a small value proportional to the square root of the relative machine precision. In our tests we set $\delta = 10^{-6}$ in an attempt to test the convergence accuracy of the proposed strategy. Also $\tau^k = 1$ for all k allows the minimisation step

Table 1: Results for the cancer and diabetes problems.

Algorithm	Cancer		Average		Diabetes		Average	
	Epochs	Speed (sec)	Class. Success (%)	Conv. (%)	Epochs	Speed (sec)	Class. Success (%)	Conv. (%)
Rprop	225	1.58	97.60	100	312	1.90	75.60	90
JRprop	194	1.36	97.64	100	290	1.77	75.64	95
GJRprop	137	1.20	97.66	100	260	1.60	75.78	100

along the resultant search direction to be explicitly defined by the values of the local learning rates. The length of the minimisation step can be regulated through τ^k tuning to satisfy (5)–(6). Checking (6) at each iteration requires additional gradient evaluations; thus, in practice (6) can be enforced simply by placing the lower bound on the acceptable values of the learning rates [5, p.1772], i.e. $\Delta_{\min}$.

4   Experimental study and results

In this section, we evaluate the performance of the GJRprop algorithm, and compare it with the Rprop and the JRprop algorithms. We have used well–studied problems from the UCI Repository of Machine Learning Databases of the University of California [6], as well as problems studied extensively by other researchers in an attempt to reduce as much as possible biases introduced by the size of the weights space. In all cases we have used networks with classic logistic activations. Below, we report results from 50 independent trials for four UCI problems. These 50 random weight initializations are the same for the two learning algorithms, and the training and testing sets were created according to *Proben1* [8].

The first benchmark is the *breast cancer diagnosis* problem which classifies a tumor as benign or malignant based on 9 features [6, 8]. We have used an FNN with 9–4–2–2 nodes (a total of 56 weights) as suggested in [8]. The termination criterion is $E \leqslant 0.02$ within 2000 iterations. The second problem is the *diabetes1* benchmark. It is a real-world classification task which concerns deciding when a Pima Indian individual is diabetes positive or not [6, 8]. There are 8 features representing personal data and results from a medical examination. The Proben1 collection suggests a 8–2–2–2 FNN (34 weights overall). The termination criterion is $E \leqslant 0.14$ within 2000 iterations.

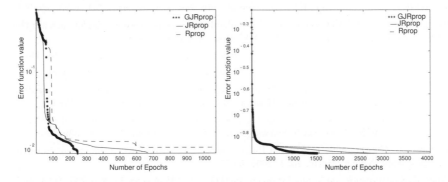

Figure 1: GJRprop, JRProp and Rprop learning curves: Cancer (left) and Diabetes (right)

The results for the cancer and diabetes problems are summarized in Table 1. The new algorithm performs significantly better than the JRprop and Rprop. In Table 1, we represent the average time ("Speed", measured in secs), the classification success with the testing set ("Class. Success", measured by the percentage of testing patterns that were classified correctly), and the convergence success in the training phase ("Conv.", measured by the percentage of simulation runs that converged to the error goal) for an algorithm. The increased convergence speed does not seem to affect the classification success of the new method in testing. Figure 1 illustrates a case where GJRprop converges to a minimiser faster than JRrpop, while Rprop gets stuck to a minimiser with higher error value, for the two tested problems.

References

[1] Anastasiadis A. D., Magoulas G. D., Vrahatis M. N., An efficient improvement of the Rprop algorithm, In: Proceedings of the *1st International Workshop on "Artificial Neural Networks in Pattern Recognition (IAPR 2003)"*, University of Florence, Italy, pp.197–201, 2003.

[2] Anastasiadis A. D., Magoulas G. D., Vrahatis M. N., Sign-based learning schemes for pattern classification, *Pattern Recognition Letters*, submitted.

[3] Dennis J. E. and Schnabel R. B., *Numerical Methods for Unconstrained Optimization and nonlinear equations*, SIAM, Philadelphia, 1996.

[4] Magoulas, G. D., Vrahatis, M. N., and Androulakis, G. S., Effective back-propagation training with variable stepsize, Neural Networks, 10, 1, 1997, 69-82.

[5] Magoulas, G. D., Vrahatis, M. N., and Androulakis, G. S., Improving the convergence of the backpropagation algorithm using learning rate adaptation methods, Neural Computation, 11, 7, 1999, 1769-1796.

[6] Murphy, P. M. and Aha, D. W., UCI Repository of machine learning databases, Irvine, CA: University of California, Department of Information and Computer Science, 1994, http://www.ics.uci.edu/~mlearn/MLRepository.html

[7] Nocedal J., Theory of algorithms for unconstrained optimization, *Acta Numerica*, pp.199–242, 1992.

[8] Prechelt, L., PROBEN1-A set of benchmarks and benchmarking rules for neural network training algorithms, T.R. 21/94, Fakultät für Informatik, Universität, Karlsruhe, 1994.

[9] Riedmiller, M. and Braun, H., A direct adaptive method for faster backpropagation learning: The RPROP algorithm, Proc. Int. Conf. Neural Networks, San Francisco, CA, 1993, 586-591.

[10] Rumelhart, D. E. and McClellend, J. L., Parallel distributed processing: Explorations in the microstructure of cognition, Cambridge, MIT Press, 1986, pp.318-362.

[11] Scales, L. E., Introduction to non-linear optimization, MacMillan Publishers, 1985, pp. 34-35.

[12] Sikorski, K., Optimal Solution of Nonlinear Equations, Oxford Univ. Press, New York, 2001.

[13] Treadgold, N. K. and Gedeon, T. D., Simulated annealing and weight decay in adaptive learning: The SARPROP algorithm, IEEE Trans. Neural Networks, 9, 4, 1998, 662-668.

[14] Vrahatis, M. N., Solving systems of nonlinear equations using the nonzero value of the topological degree, ACM Trans. Math. Software, 14, 1988, 312–329; ibid: 14, 1988, 330–336.

[15] Vrahatis, M. N., Magoulas, G. D., and Plagianakos, V. P., From linear to nonlinear iterative methods, Appl. Numer. Math., 45, 1, 2003, 59-77.

[16] Wolfe P., Convergence conditions for ascent methods, *SIAM Review*, 11, 226–235, 1969; ibid: 13, 185–188, 1971.

VSP International
Science Publishers
P.O. Box 346, 3700 AH Zeist
The Netherlands

Lecture Series on Computer
and Computational Sciences
Volume 1, 2004, pp. 849-851

On Compositions of Clones of Boolean Functions

Miguel Couceiro[1], Stephan Foldes[2], Erkko Lehtonen[3]

Received 31 July, 2004; accepted in revised form 5 September, 2004

Abstract: We present some compositions of clones of Boolean functions that imply factorizations of Ω, the clone of all Boolean functions, into minimal clones. These can be interpreted as representation theorems, providing representations of Boolean functions analogous to the disjunctive normal form, the conjunctive normal form, and the Zhegalkin polynomial representations.

Keywords: Function class composition, Clones, Boolean functions, Post Classes, Class factorization, Normal forms, DNF, CNF, Zhegalkin polynomial, Applications of universal algebra in computer science

Mathematics Subject Classification: 06E30, 08A70, 94C10

1 Introduction

It is a well-known fact that every Boolean function can be represented as a disjunction of conjunctions of literals, as a conjunction of disjunctions of literals, and as a multilinear polynomial over GF(2). These representations are called the *disjunctive normal form* (DNF), the *conjunctive normal form* (CNF), and the *Zhegalkin polynomial*, respectively.

These facts can be reformulated by means of composition of clones of Boolean functions. Moreover, the clones occurring in these compositions are minimal. In other words, these theorems represent different factorizations of the clone Ω of all Boolean functions into minimal clones.

In this work, we present more such factorizations of Ω. Furthermore, these factorizations can be interpreted as representation theorems similar to the DNF, CNF and Zhegalkin polynomial representation theorems.

2 Definitions

Let A, B, E, G be arbitrary non-empty sets. A *function on A to B* is a map $f : A^n \to B$, for some positive integer n called the *arity* of f. A *class* of functions on A to B is a subset $\mathcal{F} \subseteq \cup_{n \geq 1} B^{A^n}$. For a fixed arity n, the n different *projection maps* $\mathbf{a} = (a_t \mid t \in n) \mapsto a_i$, $i \in n$, are also called *variables*. For $A = B = \{0, 1\}$, a function on A to B is called a *Boolean function*.

If f is an n-ary function on B to E and $g_1, \ldots, g_n$ are all m-ary functions on A to B, then the *composition* $f(g_1, \ldots, g_n)$ is an m-ary function on A to E, and its value on $(a_1, \ldots, a_m) \in A^m$

---

[1]Department of Mathematics, Statistics and Philosophy, University of Tampere, FI-33014 Tampereen yliopisto, Finland. Partially supported by the Graduate School in Mathematical Logic MALJA. Supported in part by grant #28139 from the Academy of Finland.

[2]Institute of Mathematics, Tampere University of Technology, P.O. Box 553, FI-33101 Tampere, Finland. Corresponding author. E-mail: stephan.foldes@tut.fi

[3]Institute of Mathematics, Tampere University of Technology, P.O. Box 553, FI-33101 Tampere, Finland.

is $f(g_1(a_1, \ldots, a_m), \ldots, g_n(a_1, \ldots, a_m))$. Let $\mathcal{I} \subseteq \cup_{n \geq 1} E^{B^n}$ and $\mathcal{J} \subseteq \cup_{n \geq 1} B^{A^n}$ be classes of functions. The *composition of* $\mathcal{I}$ *with* $\mathcal{J}$, denoted $\mathcal{I} \circ \mathcal{J}$, is defined as

$$\mathcal{I} \circ \mathcal{J} = \{f(g_1, \ldots, g_n) \mid n, m \geq 1, \ f \text{ } n\text{-ary in } \mathcal{I}, \ g_1, \ldots, g_n \text{ } m\text{-ary in } \mathcal{J}\}.$$

A *clone* on A is a set $\mathcal{C} \subseteq \cup_{n \geq 1} A^{A^n}$ that contains all projections and satisfies $\mathcal{C} \circ \mathcal{C} \subseteq \mathcal{C}$ (or equivalently, $\mathcal{C} \circ \mathcal{C} = \mathcal{C}$).

The following is a corollary to the Associativity Lemma in [1]: if $\mathcal{I}$, $\mathcal{J}$, $\mathcal{K}$ are clones on A, then $(\mathcal{I} \circ \mathcal{J}) \circ \mathcal{K} = \mathcal{I} \circ (\mathcal{J} \circ \mathcal{K})$.

The clones of Boolean functions form an algebraic lattice. These clones and the lattice are often called the Post Classes and the Post Lattice, respectively. For general background on the Post Lattice and generating sets, see [3]. We use the notation of [2]. I_c is the smallest element of the lattice, Ω is the largest, and the lattice operations are defined as follows: meet is the intersection, join is the smallest clone containing the union.

We note the following fact: If $\mathcal{I}$ and $\mathcal{J}$ are clones and $\mathcal{I} \circ \mathcal{J} = \mathcal{J} \circ \mathcal{I}$, then $\mathcal{I} \circ \mathcal{J}$ is a clone and $\mathcal{I} \circ \mathcal{J} = \mathcal{I} \vee \mathcal{J}$.

3 Factorization Lemmas

The DNF, CNF, and Zhegalkin polynomial representation theorems can be stated as $\Omega = V_c \circ \Lambda_c \circ I^*$ (DNF), $\Omega = \Lambda_c \circ V_c \circ I^*$ (CNF), $\Omega = L \circ \Lambda_c$ (Zhegalkin). V_c is the class of disjunctions of variables, Λ_c is the class of conjunctions of variables, L is the class of linear (affine) functions, and I^* is the class of variables and negations of variables.

We mention some other clones of Boolean functions. I_c is the class of variables, I_0 (I_1) is the class of variables and the constant functions having value 0 (1, respectively) everywhere, I is the class of variables and all constant functions, $\Omega(1)$ is the class of variables, negations of variables and all constant functions, L_c is the class of constant-preserving linear functions, S_c is the class of constant-preserving self-dual functions, SM is the class of self-dual monotone functions.

It is easy to verify that $L = L_c \circ I_0 \circ I_1$, and that composition of any two of I_0, I_1, Λ_c commute, so the Zhegalkin polynomial theorem can actually be stated as $\Omega = L_c \circ (I_0 \circ I_1 \circ \Lambda_c)$, where the factors in parenthesis can be written in any order.

It should be noted that in the above-mentioned factorizations of Ω the factors are minimal, i.e. they have I_c as the only proper subclone. We present a few lemmas that imply more such factorizations of Ω.

Lemma 1 $I = I_0 \circ I_1 = I_1 \circ I_0$.

Lemma 2 $\Omega(1) = I \circ I^* = I^* \circ I_0 = I^* \circ I_1$.

Lemma 3 $SM \circ \Omega(1) = \Omega$.

Lemma 4 $SM \circ L_c = S_c$.

Lemma 5 $L_c \circ SM = S_c$.

Lemma 6 $S_c \circ I = \Omega$.

4   Conclusions

Lemmas 1–6 give us the following factorizations of Ω into minimal clones:

$$\Omega = SM \circ (I_0 \circ I_1) \circ I^* = SM \circ I^* \circ I_0 = SM \circ I^* \circ I_1,$$
$$\Omega = (SM \circ L_c) \circ (I_0 \circ I_1),$$

where the factors in parenthesis can be written in any order.

These can be interpreted as representation theorems, providing representations of Boolean functions analogous to the CNF, DNF, and Zhegalkin polynomial representations. Questions of computational complexity associated with such alternative representations of Boolean functions, concerning in particular transformation algorithms between algebraic expressions and function evaluation, can be approached in a universal-algebraic framework.

References

[1] M. Couceiro and S. Foldes, Function Class Composition, Relational Constraints and Stability under Compositions with Clones, Rutcor Research Report 22-2004, Rutgers University, http://rutcor.rutgers.edu/.

[2] S. Foldes and G. R. Pogosyan, Post Classes Characterized by Functional Terms, *Discrete Applied Mathematics* **142** (2004), p. 35–51.

[3] S. W. Jablonski, G. P. Gawrilow, W. B. Kudrjawzew, *Boolesche Funktionen und Postsche Klassen*, Vieweg, Braunschweig, 1970.

VSP International
Science Publishers
P.O. Box 346, 3700 AH Zeist
The Netherlands

Lecture Series on Computer
and Computational Sciences
Volume 1, 2004, pp. 852-855

Evolutionary Adaptive Schemes of Probabilistic Neural Networks

V.L. Georgiou[1], N.G. Pavlidis[2], K.E. Parsopoulos[3], Ph.D. Alevizos[4], M.N. Vrahatis[5]

Department of Mathematics,
University of Patras Artificial Intelligence Research Center (UPAIRC),
University of Patras, GR–26110 Patras, Greece

Received 31 July, 2004; accepted in revised form 5 September, 2004

Abstract: Self–adaptive probabilistic neural networks have already been proposed in the literature. Typically, the kernel of the probabilistic neural network is an n–dimensional identity matrix multiplied by the scalar spread parameter, σ. This prevents the network from properly fitting the data. Another approach is to use a diagonal spread matrix, allowing each dimension to assume its own spread value. This approach increases the degrees of freedom of the network and thus allows it to fit better to the available data. However, since the optimization procedure is now multivariate instead of univariate, it is harder and computationally more demanding. To address this optimization problem we employ state-of-the-art computational intelligence optimization algorithms, like Particle Swarm Optimization, Differential Evolution and Evolution Strategies.

Keywords: Probabilistic Neural Networks, Spread Parameter, Particle Swarm Optimization, Differential Evolution Algorithm, Evolution Strategies.

Mathematics Subject Classification: 65C35, 65C60, 92B20.

1 Introduction

In this contribution an effort is attempted to combine a classical statistical technique with state-of-the-art computational intelligence methods. A common task in numerous scientific fields is the classification and prediction of class membership. Many approaches have been proposed to this end, including discriminant analysis, artificial neural networks, probabilistic neural networks, naive Bayes classifiers, and projection pursuit regression among others.

Probabilistic neural networks (PNNs) are known in the statistical literature as kernel discriminant analysis [9]. PNNs are used in many fields such as bioinformatics, medicine, and engineering with not always very promising results. Recently, PNNs have been successfully applied for the prediction of protein cellular localization sites [2]. For this purpose, the Particle Swarm Optimization (PSO) algorithm has been employed for the optimization of a spread parameter which is crucial for the PNN's operation.

---

[1] Corresponding author. E-mail: vlg@math.upatras.gr
[2] E-mail: npav@math.upatras.gr
[3] E-mail: kostasp@math.upatras.gr
[4] E-mail: philipos@math.upatras.gr
[5] E-mail: vrahatis@math.upatras.gr

2   Probabilistic Neural Networks

The PNN was introduced by Specht [9] as a new neural network type, although it was already widely known in the statistical literature as kernel discriminant analysis [3]. What Specht introduced was the neural network approach of kernel discriminant analysis which incorporates the Bayes decision rule and the non–parametric density function estimation of a population [7].

The training procedure of the PNN requires only a single pass of the patterns of the training data, which results in a very small training time. In fact, the training procedure is just the construction of the PNN from the available data. The structure of the PNN has always four layers; the *input layer*, the *pattern layer*, the *summation layer*, and the *output layer*. An input feature vector, $X \in \mathbb{R}^n$, is applied to the n input neurons and is passed to the pattern layer. The pattern layer is organized into K groups, where K is the number of classes present in the data set. The ith neuron in the jth group of the pattern layer computes its output using a Gaussian kernel of the form,

$$f_{i,k}(X) = \frac{1}{(2\pi\sigma^2)^{n/2}} \exp\left(-\frac{\|X - X_{i,k}\|^2}{2\sigma^2}\right),\tag{1}$$

or in matrix form,

$$f_{i,k}(X) = \frac{1}{(2\pi)^{n/2} |\Sigma|^{1/2}} \exp\left(-\frac{1}{2}(X - X_{i,k})^T \Sigma^{-1}(X - X_{i,k})\right),\tag{2}$$

where $X_{i,k} \in \mathbb{R}^n$ is the center of the kernel and σ is the spread (smoothing) parameter of the kernel. In the matrix form of the kernel, Σ equals $\sigma^2 I_n$ so there is only one parameter to estimate. The summation layer has K neurons and estimates the conditional class probabilities as follows:

$$G_k(X) = \sum_{i=1}^{M_k} w_{ik} f_{i,k}(X), \quad k \in \{1, \dots, K\},\tag{3}$$

where M_k is the number of patterns of class k and w_{ik} is the prior probability of class k. So a vector X is classified to the class that has the maximum output of its summation neuron.

3   Established Computational Intelligence Algorithms

Particle Swarm Optimization: Particle Swarm Optimization is a population–based, stochastic, optimization algorithm [6]. It exploits a population of individuals to synchronously probe promising regions of the search space. The population is called a *swarm* and the individuals (i.e., the search points) are called *particles*. Each particle moves with an adaptable velocity within the search space, and retains a memory of the best position it ever encountered. In the *global* variant of PSO, the best position ever attained by all individuals of the swarm is communicated to all the particles at each iteration [5].

Let X_i be the n–dimensional feature vector of the ith particle, $X_i \in S \subset \mathbb{R}^n$ and V_i its velocity in the search space. Then, the swarm is manipulated by the equations,

$$V_i(t+1) = w\,V_i(t) + c_1\,r_1\big(P_i(t) - X_i(t)\big) + c_2\,r_2\big(P_{best_i}(t) - X_i(t)\big),\tag{4}$$
$$X_i(t+1) = X_i(t) + V_i(t+1),\tag{5}$$

where $i = 1, \dots, NP$; P_i is the best position it ever attained; w is a parameter called *inertia weight*; c_1 and c_2 are two positive constants called *cognitive* and *social* parameter, respectively; and r_1, r_2, are random vectors uniformly distributed within $[0,1]^n$. Alternatively, the velocity update can be performed through the following equation [1],

$$V_i(t+1) = \chi \left[V_i(t) + c_1\, r_1 \left(P_i(t) - X_i(t) \right) + c_2\, r_2 \left(P_{best_i}(t) - X_i(t) \right) \right], \tag{6}$$

where χ is a parameter called *constriction factor*, giving rise to a different version of PSO.

Differential Evolution: Differential Evolution (DE) is also a population–based, stochastic algorithm that exploits a population of potential solutions, *individuals*, to probe the search space. Each new individual is generated by the combination of randomly chosen individuals of the population. This combination will be called mutation. So, for each individual w_g^k, $k = 1, \ldots, NP$, where g denotes the current generation, a new individual v_{g+1}^i is generated according to one of the following equations:

$$
\begin{align}
v_{g+1}^i &= w_g^{best} + F\left(w_g^{r_1} - w_g^{r_2} \right), \tag{7}\\
v_{g+1}^i &= w_g^{r_1} + F\left(w_g^{r_2} - w_g^{r_3} \right), \tag{8}\\
v_{g+1}^i &= w_g^i + F\left(w_g^{best} - w_g^i \right) + F\left(w_g^{r_1} - w_g^{r_2} \right), \tag{9}\\
v_{g+1}^i &= w_g^{best} + F\left(w_g^{r_3} - w_g^{r_2} \right) + F\left(w_g^{r_3} - w_g^{r_4} \right), \tag{10}\\
v_{g+1}^i &= w_g^{r_1} + F\left(w_g^{r_2} - w_g^{r_3} \right) + F\left(w_g^{r_4} - w_g^{r_5} \right), \tag{11}\\
v_{g+1}^i &= \left(w_g^{r_1} + w_g^{r_2} + w_g^{r_3} \right)/3 + (p_2 - p_1)\left(w_g^{r_1} - w_g^{r_2} \right) + (p_3 - p_2)\left(w_g^{r_2} - w_g^{r_3} \right) + \\
&\quad (p_1 - p_3)\left(w_g^{r_3} - w_g^{r_1} \right), \tag{12}
\end{align}
$$

where w_g^{best} is the best individual of the previous generation; $F > 0$ is a real parameter, called mutant constant and r_1, r_2, r_3, r_4, r_5 and $r_6 \in \{1, 2, \ldots, k-1, k+1, \ldots, NP\}$ are random integers mutually different.

Evolution Strategies: Evolution Strategies (ES) are population based search algorithms that have been developed by Rechenberg and Schwefel [11]. ES exploit a population of μ individuals to probe the search space. At each iteration of the algorithm, λ offsprings are produced by stochastic variation, called mutation, of recombinations of a set of individuals (called the *parents*) from the current population. Mutation is typically carried out by adding a realization of a normally distributed random vector. After the creation of the offspring individuals, selection takes place, and either the μ best individuals among the offspring population, or the μ best individuals among both the parent and the offspring populations are selected to form the parents in the next generation. These two selection schemes are denoted as (μ, λ)–ES and $(\mu + \lambda)$–ES, respectively.

ES use a set of parameters, called *strategy parameters*, to parameterize the normal distribution used in the mutation procedure. These parameters can either be fixed, or evolve during the evolution process resulting in self–adaptive ES. A special case of ES is the Covariance Matrix Adaptation Evolution Strategies (CMA–ES), which has been developed by Hansen and Ostermeier [4]. CMA–ES adapt the parameters of the algorithm such that strategy parameter settings that produce individuals that are selected are favored [4].

4   The Proposed Approach

The main disadvantage of PNNs is their sensitivity with respect to the choice of a parameter, called *spread parameter*, which influence the receptive field of each kernel. Usually, this parameter is set to standard default values suggested in the literature. However, this approach is not always appropriate due to the dependence of the spread parameter on the specific data at hand. There-fore, techniques for the automatic determination and adaptation of this parameter can be proved very useful. We investigate the sensitivity of the PNNs' performance with respect to the spread parameter and propose algorithms for the selection of an optimized value. The PSO, DE and ES algorithms are considered for this purpose.

Alternatively, a matrix of spread parameters can be used to enhance PNNs' performance by increasing the network's degrees of freedom. This approach, however, imposes a heavier computa-tional burden since the optimization procedure becomes multivariate instead of univariate. Thus, although the matrix of spread parameters offers more flexibility to the model, the high dimension-ality of the problem renders the optimization over–identified [8]. For example, if the problem is 100–dimensional, then the matrix would be 10000–dimensional, and huge amount of data would be necessary to obtain a reliable estimate of the matrix. To overcome this problem, a diagonal form of the matrix of spread parameters can be used. This matrix can be efficiently computed through the PSO, DE and ES algorithms.

Acknowledgment

We thank European Social Fund (ESF), Operational Program for Educational and Vocational Training II (EPEAEK II) and particularly the Program IRAKLEITOS for funding the above work.

References

[1] Clerc, M. and Kennedy, J., The particle swarm–explosion, stability, and convergence in a multidimensional complex space, *IEEE Trans. Evol. Comput.*, **6**(1), pp. 58–73, 2002.

[2] Georgiou, V. L., Pavlidis, N. G., Parsopoulos, K. E., Alevizos, Ph. D. and Vrahatis, M. N., Optimizing the performance of probabilistic neural networks in a bioinformatics task, In *Proceedings of the EUNITE 2004 Symposium*, pp. 34–40, Aachen, Germany, 2004.

[3] Hand, D. J., *Kernel Discriminant Analysis*, Research Studies Press, Chichester, 1982.

[4] Hansen, N. and Ostermeier, A., Completely derandomized self-adaptation in evolution strate-gies. *Evolutionary Computation*, **9**(2), pp. 159–195, 2001.

[5] Kennedy, J. and Eberhart, R. C., *Swarm Intelligence*, Morgan Kaufmann Publishers, 2001.

[6] Parsopoulos, K. E. and Vrahatis, M. N., On the computation of all global minimizers through particle swarm optimization, *IEEE Trans. Evol. Comput.*, **8**(3), pp. 211–224, 2004.

[7] Parzen, E., On the estimation of a probability density function and mode, *Annals of Mathe-matical Statistics*, **3**, pp. 1065–1076, 1962.

[8] Ripley, B. D., *Pattern Recognition and Neural Networks*, Cambridge University Press, Cam-bridge, 1996.

[9] Specht, D. F., Probabilistic neural networks, *Neural Networks*, **1**(3), pp. 109–118, 1990.

[10] Storn, R. and Price, K., Differential evolution–a simple and efficient heuristic for global optimization over continuous spaces, *J. Global Optimization*, **11**, pp. 341–359, 1997.

[11] Schwefel., H. P., *Evolution and Optimum Seeking*, Wiley, New York, 1995.

VSP International
Science Publishers
P.O. Box 346, 3700 AH Zeist
The Netherlands

*Lecture Series on Computer
and Computational Sciences*
Volume 1, 2004, pp. 856-859

Integrating Perceptech Requirements through Intelligent Computation of Priorities in Multimedia Streaming

Gheorghita Ghinea[1]* and George D. Magoulas[+]

*Department of Information Systems and Computing,
Brunel University, U.K.
[+]School of Computer Science and Information Systems and Computing,
Birkbeck College, University of London, U.K

Received 31 July, 2004; accepted in revised form 5 September, 2004

Abstract: We propose a mechanism for the integration of combined perceptual and technical parameters towards the streaming of multimedia content. Our approach, based on Multi-criteria decision making is able to adapt according to user and networking requirements for the perceptual benefit of the user.

Keywords: Communication protocols, Fuzzy logic, Multimedia communications, Quality of service.

Mathematics SubjectClassification: 68M14, 68T05, 62C12, 68U35, 90C70

1. Introduction

Integrating user-level expectations with parameters characterising underlying network performance is a problem seldom studied in multimedia streaming, for it attempts to bridge the gap existing between user perceptions of multimedia quality and the *Quality of Service* (QoS) with which multimedia is transmitted over the network. Work in this respect has focused on the effects that different video frame rates have on human satisfaction with the multimedia presentation [11], on the perceptual impact of errors [9], delay [1] and jitter [4], or, alternatively, on the development of metrics for assessing subjective multimedia quality based on models of the human visual system [8].

However, only rarely is such research carried forward in the development of adaptive streaming applications. Accordingly, the QUASAR project [5] exploits human perceptual tolerance to media losses and frame dropping, In contrast, [9] uses identified subjective acceptance of occasional media errors to develop a streaming mechanism in which such errors are distributed without negative perceptual impact. However, such approaches generally fall short on two counts: firstly, they ignore multimedia's *infotaiment* (i.e. combined information-entertainment) duality; secondly, many of them assume that users of distributed multimedia technology have considerable technical skills, which, given the proliferation of the Web, represents the exception rather than the norm today.

In our work, we address both issues. Thus, our approach to user-perceived QoS encompasses not only the traditional view of a user's satisfaction with the presentation quality of a multimedia application, but also his/her ability to understand, analyse and synthesise their informational content. Accordingly we evaluated user-perceived QoS through a series of empirical tests, whose results [3] indicated that technical-oriented QoS must also be specified in terms of perception, understanding and absorption of content if multimedia presentations are to be truly effective from both a user as well as technical perspective. Based on these results, we propose a scheme for intelligent multimedia streaming in which users do not necessarily need to specify their preferences. We formulate the problem in terms of Multicriteria Decision Making [2] [11], and propose a solution in terms of fuzzy programming [11] to resolve inconsistencies between user and network considerations and derive streaming priorities of multimedia content geared towards ensuring an optimum user experience.

2. Computing streaming priorities using Fuzzy Programming

In integrating user preferences with network considerations for the streaming of media content, there are practically no universal elementary measurable properties in terms of which user perception can be defined, nor is there a measurable functional relationship relating measurable properties of multimedia

---

[1] Corresponding author. E-mail: George.Ghinea@brunel.ac.uk

presentations to a user's perception. Nevertheless, our evaluation of user-perceived QoS revealed that multimedia perceptual quality varies with the number of media flows, the type of medium and application, as well as the context-dependent relative importance of each medium [3]. Thus, each multimedia application can be characterised by the relative importance of the Audio (*A*), Video (*V*) and Textual (*T*) components as conveyors of information, as well as the Dynamism (*D*) of the presentation. On the other hand, 5 commonly considered network level QoS parameters have been incorporated in our model: Bit Error (*BER*), Segment Loss (*SL*), Segment Order (*SO*), Delay (*DEL*) and Jitter (*JIT*). The approach adopted in our paper is based on a formulation of the problem in terms of Multicriteria Decision Making-MDM [2] [6] [11]. The decision making problem examined here is as follows:

Let a set of K alternative microprotocols {micro1, micro2, micro3, ..., microK} and a set of P decision criteria {BER, SO, SL, DEL, JIT, A, V, T, D } that relate to user and network considerations be given. Then the problem is to rank the microprotocols in terms of their total preferences when all the decision criteria are considered simultaneously, and ultimately find a protocol stack configuration that satisfies all user and technical requirements to the best available degree. There are two steps in solving this problem: 1) Attach numerical measures to the relative importance of the criteria and to the impacts of the alternatives on these criteria. This consists of two sub-procedures: (i) determine the relative importance of the criteria; (ii) determine the relative standing of each alternative with respect to each criterion. 2) Process the numerical values to determine a ranking priority of each alternative.

In our approach, streaming priorities are derived by (i) determining *weighted priorities* w_i (i=1,...9) for the criteria, and (ii) determining *relative weights* $w_{j,i}$, which denote how preferable is microprotocol *j* with respect to criterion *i*. These are then synthesised to yield the *overall ranking priority* of each microprotocol, and thus determine their inclusion/exclusion in the protocol stack.

$$
A = \begin{bmatrix}
1 & [l_{12}, u_{12}] & [l_{13}, u_{13}] & \cdots & [l_{1K}, u_{1K}] \\
[l_{21}, u_{21}] & 1 & [l_{23}, u_{23}] & \cdots & [l_{2K}, u_{2K}] \\
[l_{31}, u_{31}] & [l_{32}, u_{32}] & 1 & \cdots & [l_{3K}, u_{3K}] \\
\vdots & \vdots & \vdots & \vdots & \vdots \\
[l_{K1}, u_{K1}] & [l_{K2}, u_{K2}] & [l_{K3}, u_{K3}] & \cdots & 1
\end{bmatrix}
$$

(a)

$$
R = \begin{bmatrix}
1 & -u_{12} & 0 & 0 & \cdots & 0 \\
-1 & l_{12} & 0 & 0 & \cdots & 0 \\
1 & 0 & -u_{13} & 0 & \cdots & 0 \\
-1 & 0 & l_{13} & 0 & \cdots & 0 \\
\cdots & \cdots & \cdots & \cdots & \cdots & \cdots \\
1 & 0 & 0 & 0 & \cdots & -u_{1p} \\
-1 & 0 & 0 & 0 & \cdots & l_{1p} \\
0 & 1 & -u_{23} & 0 & \cdots & 0 \\
0 & -1 & l_{23} & 0 & \cdots & 0 \\
\cdots & \cdots & \cdots & \cdots & \cdots & \cdots \\
\vdots & \vdots & \vdots & \vdots & \vdots & \vdots \\
0 & 0 & 0 & \cdots & 1 & -u_{p-1,p} \\
0 & 0 & 0 & \cdots & -1 & l_{p-1,p}
\end{bmatrix}
$$

(b)

Figure 1. (a) General form of an interval comparison matrix; (b) General form of a constraints matrix.

In a real-world context, it is not uncommon for there to be inconsistencies between technical and perceptual information represented in the judgement matrices. This leads to matrix entries that are just estimations of ideal judgements. To alleviate this situation, the Fuzzy Programming Method (FPM), capable of solving even highly inconsistent matrices, can be used [6] [7]. This approach formulates the prioritization problem as a maximin fuzzy programming problem, which finds a crisp priority vector, maximizing the overall decision-maker's satisfaction with the final solution. Along this line, the FPM enables judgements to be expressed either as crisp, intervals or fuzzy numbers. In order to deal with the uncertainties in the estimation of the judgements and avoid computational overheads we describe the information with interval judgments. This can be done by introducing tolerance parameters d_{ij} that will consider each judgment as an interval with lower and upper bounds $a_{ij} = (l_{ij}, u_{ij})$ [6]. Thus, a comparison matrix can be formulated as shown in Figure 1(a). The idea behind the FPM is a geometrical representation of the prioritisation process as an intersection of hyperlines, by seeking priority vectors that satisfy:

$$
l_{ij} \leq \frac{w_i}{w_j} \leq u_{ij}, \quad i = 1, 2, \ldots K - 1, \ j = 2, 3, \ldots K, \ j > i
\tag{1}
$$

In case of inconsistent matrices, the hyperlines have no common intersection point, i.e. the intersection set is empty. Thus, the FPM represents the hyperlines as fuzzy lines and finds the solution of the approximate priority assessment problem, as an intersection point of these fuzzy lines with values for the priorities that satisfy all judgements "as well as possible". Previous work gives evidence that FPM is able to produce better results than other methods when the degree of inconsistency is high [7]; this is a valuable property in our case. By following [6], the problem can be formulated as the standard linear programming problem of Equation (2), where the objective is to maximise λ, a measure of intersection region of the fuzzy lines, subject to a set of $m=p(p-1)$ constraints, which are given in matrix form, $R \in \Re^{m \times p}$, in Figure 1(b).

$$\text{maximize } \lambda, \text{ subject to } \lambda d_k + R_k w \le d_k, \ k = 1,\dots, m, \ 1 \ge \lambda \ge 0, \text{ and } \sum_{i=1}^{p} w_i = 1 \ , \ w_i > 0, \ i = 1,\dots, p \quad (2)$$

In Equation (2), p is the number of elements compared; $w = (w_1, w_2, \dots, w_p)^T$ is the vector of priority weights; k indicates the k-th row of matrix R, and the values of the tolerance parameters d_k represent the admissible interval of approximate satisfaction of the crisp inequalities $R_k w < 0$. For the practical implementation of the FPM, it is reasonable for all these parameters, d_k, to be set equal [7]. The optimal solution to the problem is a vector (w^*, λ^*), whose first component maximises the degree of membership of the fuzzy feasible region set, whilst the second gives the value of the maximum degree of satisfaction. The method is explained in detail in [6].

After deriving the underlying weights from the comparison matrices through the FPM technique, the weighted priority, w_i, and the relative weight, $w_{j,i}$, are synthesised using weighted sum aggregation in order to find the preference of microprotocol j with respect to all criteria/requirements simultaneously. Preference is denoted by W_j and determines the overall ranking priority, or weight, of microprotocol j (obviously, the microprotocol with the maximum overall value W_j will be chosen):

$$W_j = \sum_{i=1}^{P} w_i \cdot w_{j,i}. \quad (3)$$

3. Application example

We use a scenario which illustrates the ability of our approach to select appropriate microprotocols and construct a suitably-tailored protocol stack depending on the prevailing operating network environment and user perceptual preferences. The FPM was applied and the relative scores, $w_{j,i}$, thus obtained are presented in Table 1, where for example, one can notice that the first four microprotocols considered have an equal importance with respect to managing segment loss (SL). However, the most important microprotocol for segment loss is microprotocol 6, which has the highest relative score.

Table 1: Relative scores, $w_{j,i}$, of the alternatives comparison matrices with respect to each criterion

	micro1	micro2	micro 3	micro 4	micro 5	micro 6	micro 7	micro 8	micro 9
BER	0.0574	0.0574	0.0574	0.0574	0.0574	0.0574	0.0574	0.1541	0.4441
SO	0.1082	0.2754	0.0787	0.0787	0.1049	0.1180	0.0787	0.0787	0.0787
SL	0.0881	0.0881	0.0881	0.0881	0.1872	0.2753	0.0617	0.0617	0.0617
DEL	0.2280	0.0638	0.2280	0.0638	0.0517	0.0456	0.2280	0.0456	0.0456
JIT	0.1614	0.1958	0.0771	0.3335	0.0474	0.0316	0.0743	0.0474	0.0316
V	0.0842	0.2254	0.0842	0.1883	0.0979	0.0818	0.0462	0.0694	0.1226
A	0.2175	0.0658	0.2175	0.0658	0.0590	0.0568	0.2529	0.0354	0.0293
T	0.1761	0.0833	0.1761	0.0833	0.1250	0.0833	0.1761	0.0625	0.0341
D	0.0843	0.0826	0.0826	0.3710	0.0759	0.0759	0.0759	0.0759	0.0759

In this example, no assumptions are made about the underlying network conditions, or about the multimedia content to be transported over the network. As such, the a priori judgement values which arise from technical considerations, as well as user judgements resulting out of our user-perceptual evaluation experiments, [3], are used. As can be observed from the table, delay and audio are the most important criteria from a technical and user point of view, respectively. This is because distributed multimedia applications have an essential real-time characteristic, which makes delay the primordial factor from the technical point of view. On the other hand, our work on perceptual aspects of multimedia has confirmed previous experiments in highlighting that the most important medium in a multimedia presentation, from a user perspective, is the audio component.

Furthermore, from a technical angle, the Segment Loss (SL) criterion has the same importance as the Segment Order (SO) criterion, and thus in Figure 1(b) these two criteria are shown to be *"equally important"*. Similarly, for the user, the Text (T) criterion is as important as Dynamism (D). Moreover, as can be observed, QoS and perceptual parameters are considered to be equal to unity, which reflects a balance between perceptual and QoS considerations in this initial scenario.

Table 2: Priority weights, w_i, with respect to the criteria (i=1,...9)

BER	SO	SL	DEL	JIT	V	A	T	D
0.0858	0.0951	0.0951	0.1573	0.1105	0.0936	0.1495	0.1066	0.1066

By applying the FPM, the priority weights w_i are derived (see Table 2). Finally, by synthesising the relative scores and the priority vector, using Equation (3), the overall priorities of the alternatives are obtained (see Figure 2). Micro1 is suggested as the best alternative, as it has a high relative score for both delay (the most important parameter from a technical/QoS point of view) as well as audio (the most important parameter from a user point of view). Although micro7 is even better suited to manage these two parameters, if one considers the overall set of parameters, it is micro1 that achieves at least equal or higher relative scores than micro7 for eight out of the nine parameters/criteria.

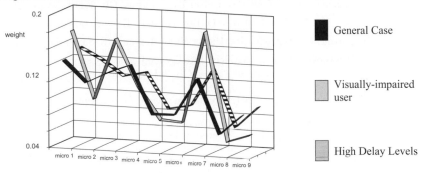

Figure 2: Microprotocol priority weights for three cases

References

[1] Bouch, A., Kuchinsky, A., and Bhatti, N., Quality is in the eye of the beholder, *Proc. CHI 2000*, The Hague, The Netherlands, 2000, pp. 297-304.

[2] Chen, S.J. and C.L. Hwang, Fuzzy Multiple Attribute Decision Making: Methods and Applications, *Lecture Notes in Economics and Mathematical Systems*, 375, 1992.

[3] Ghinea G. and Thomas J.P., Quality of Perception: User Quality of Service in Multimedia Presentations, accepted for publication, *IEEE Transactions on Multimedia*.

[4] Hikichi, K., Morino, H., Matsumoto, S., Yasuda, Y., Arimoto, I., Ijume, M. and Sezaki, K., Architecture of Haptics Communication System for Adaptation to Network Environments, *Proc. IEEE Int. Conference on Multimedia and Expo*, Tokyo, Japan, 2001, pp. 744-747.

[5] Krasic, C., Walpole, J., Feng, W. Quality-Adaptive Media Streaming by Priority Drop, *Proc. NOSSDAV'03*, Monterey, California, USA, 2003, pp. 307 – 310.

[6] Mikhailov, L. A fuzzy programming method for deriving priorities in the analytic hierarchy process, *Journal of the Operational Research Society*, 51, 341-349, 2000.

[7] Mikhailov, L., and Singh, M.G., Comparison Analysis of Methods for Deriving Priorities in the Analytic Hierarchy Process, *Proc. IEEE International Conference on Systems, Man and Cybernetics*, Tokyo, Japan, 1999, pp. 1037-1042.

[8] Sermadevi, Y., Masry, M.A., Hemami, S.S. MINMAX Rate Control with a Perceived Distortion Metric, *Proc SPIE - Visual Communications and Image Processing*, San Jose, CA, January 2004

[9] Varadarajan, S., Ngo, H.Q., Srivastava, J. An Adaptive, Perception-Driven Error Spreading Scheme in Continuous Media Streaming, *Proc ICDCS 2000*, Taiwan, 2000, pp. 475-483.

[10] Varadarajan, S., Ngo, H.Q., Srivastava, J. Error spreading: a perception-driven approach to handling error in continuous media streaming, *IEEE/ACM Transactions on Networking*, 10(1), 139-152, 2002

[11] Yadavalli, G., Masry, M. A., Hemami, S. S. Frame Rate Preferences in Low Bit Rate Video, *Proc IEEE ICIP*, 2003, Barcelona, Spain, 2003, pp. 441 – 444.

[12] Zimmermann, H.-J., *Fuzzy Set Theory and Its Applications*, Kluwer, 2nd ed., Boston, 1991.

VSP International
Science Publishers
P.O. Box 346, 3700 AH Zeist
The Netherlands

*Lecture Series on Computer
and Computational Sciences*
Volume 1, 2004, pp. 860-863

Assessing the Effectiveness of Artificial Neural Networks on Problems Related to Elliptic Curve Cryptography

E.C. Laskari[*,‡1], **G.C. Meletiou**[†,‡2], **Y.C. Stamatiou**[§,‡3] **and M.N. Vrahatis**[*,‡4]

[*]Department of Mathematics, University of Patras, GR–26110 Patras, Greece

[†]A.T.E.I. of Epirus, P.O. Box 110, GR–47100 Arta, Greece

[§]Department of Mathematics, University of Aegean, GR–83200 Samos, Greece

[‡]University of Patras Artificial Intelligence Research Center (UPAIRC),
University of Patras, GR–26110 Patras, Greece

Received 31 July, 2004; accepted in revised form 5 September, 2004

Abstract: Cryptographic systems based on elliptic curves have been introduced as an alternative to conventional public key cryptosystems. The security of both kinds of cryptosystems relies on the hypothesis that the underlying mathematical problems are computationally intractable, in the sense that they cannot be solved in polynomial time. Artificial neural networks are computational tools, motivated by biological systems, which have the inherent ability of storing and making available experiential knowledge. These characteristics give to the artificial neural networks the ability to solve complex problems. In this paper, we study the performance of artificial neural networks on the approximation of data derived from the use of elliptic curves in cryptographic applications.

Keywords: Artificial neural networks, approximation, elliptic curves, discrete logarithm.

Mathematics Subject Classification: 68T05, 68T20, 68T30, 33F05, 14H52, 11G20, 11T71.

1 Introduction

Cryptographic systems based on Elliptic Curves (ECs) have been proposed in [5, 12] as an alternative to conventional public key cryptosystems. Their main advantage is that they use smaller parameters compared to the conventional cryptosystems (e.g. RSA). This is due to the apparently increased difficulty of the underlying mathematical problem, the *Elliptic Curve Discrete Logarithm Problem* (ECDLP). This problem is believed to require more time for its solution than the time required for the solution of its finite field analogue, the *Discrete Logarithm Problem* (DLP) that ensures the security of a number of cryptosystems (e.g. ElGamal). The security of cryptosystems that rely on both types of discrete logarithms, is based on the hypothesis that the underlying mathematical problems are computationally intractable, in the sense that they cannot be solved in polynomial time. Numerous techniques have been proposed to speed up the solution of these two types of the discrete logarithm problem, relying on both algebraic and number theoretic methods, software oriented methods and as well as approximation and interpolation techniques [2, 9, 10, 17].

---

[1]Corresponding author. E-mail: elena@math.upatras.gr
[2]E-mail: gmelet@teiep.gr
[3]E-mail: stamatiu@aegean.gr
[4]E-mail: vrahatis@math.upatras.gr

Artificial neural networks are computational tools that are motivated by biological systems, which have the inherent ability of storing experiential knowledge and make it available to use. These characteristics give to the artificial neural networks the ability of solving complex problems. In this paper, we study the performance of artificial neural networks when applied to the approximation of data produced by cryptographic applications that employ elliptic curves.

2 A Brief Introduction to Elliptic Curves

An *Elliptic Curve* over a prime finite field $\mathbb{F}_p$, $p > 3$ and prime, is denoted by $E(\mathbb{F}_p)$ and it is defined as the set of all pairs $(x, y) \in \mathbb{F}_p$ (points in affine coordinates) that satisfy the equation $y^2 = x^3 + ax + b$ where $a, b \in \mathbb{F}_p$, with the restriction $4a^3 + 27b^2 \neq 0$. These points, together with a special point denoted by $\mathcal{O}$ called *point at infinity* and a appropriately defined point addition operation form an Abelian group. This is the *Elliptic Curve group* and the point $\mathcal{O}$ is its identity element (see [1, 16] for more details on this group).

The *order m of an elliptic curve* is defined as the number of the points in $E(\mathbb{F}_p)$. According to Hasse's theorem (see e.g., [1, 16]) it holds that

$$p + 1 - 2\sqrt{p} \leqslant m \leqslant p + 1 + 2\sqrt{p}.$$

The *order of a point* $P \in E(\mathbb{F}_p)$ is the smallest positive integer n for which $nP = \mathcal{O}$. From Lagrange's theorem, it holds that the order of a point cannot exceed the order of the elliptic curve.

We will, now, describe the discrete logarithm problem. Let G be any group and y one of its elements. The discrete logarithm problem for G to the base $g \in G$ consists in determining an integer x such that $g^x = y$ when the group operation is written as multiplication or $xg = y$ when the group operation is written as addition. In groups formed by elliptic curve points the group operation is the addition. Therefore, the definition of the discrete logarithm problem for elliptic curves is as follows. Let E be an elliptic curve over a finite field $\mathbb{F}_q$, P a point on $E(\mathbb{F}_q)$ of order n and Q a point on a $E(\mathbb{F}_q)$ such that $Q = tP$, with $0 \leqslant t \leqslant (n-1)$. The discrete logarithm problem for elliptic curves consists in determining the value of t. Groups defined on elliptic curves are special since the best algorithms that solve the discrete logarithm problem for them require have an exponential expected number of steps. In contrast, for the discrete logarithm problem defined over the multiplicative group of $\mathbb{F}_q^*$ the best algorithms known today requires sub-exponential time in the size of the used group.

3 Approximation through Artificial Neural Networks

Artificial Neural Networks (ANNs) have been motivated by biological systems and more specifically by the human brain. Formally, "an ANN is a massively parallel distributed processor made of simple processing units, the *neurons*, which has a natural propensity for storing experiential knowledge and making it available to use. It resembles the brain in two ways. First, knowledge is acquired by the network from its environment through a learning process, and second, interneuron connection strengths, called *weights*, are used to store the acquired knowledge" [3].

Each artificial neuron is characterized by an input/output (I/O) relation and implements a local computation. Its output is determined by its I/O characteristics, its interconnection to other neurons and possible external inputs. The overall functionality of a network is determined by its topology, the training algorithm applied, and its neuron characteristics. In this paper from the wide family of neural network types, we focus on Feedforward Neural Networks (FNN). In FNNs all neuron's connections lead to one direction and the neurons can be partitioned in layers. This kind of networks can be described with a series of integers that denote the number of neurons at each layer. The operation of such networks consists of iterative steps. The input layer neurons are generally assigned to real inputs, while the remaining hidden and output layer neurons are passive. In the next step the neurons of the first hidden layer collect and sum their inputs and compute their output, which is applied to the second layer. This procedure is propagated to all layers until the final outputs of the network are computed. The computational power of neural networks derives from their parallel distributed structure and their inherent ability to adapt to specific problems, learn and generalize. These characteristics give to the ANNs the ability to solve complex problems [4, 14].

The training process involves modification of the network weights by presenting to it training samples, called patterns, for which the desired outputs are *a priori* known. The ultimate goal of training is to assign to the weights (free parameters) of the network W, values such that the difference between the desired output (target) and the actual output of the network is minimized. The adaptation process starts by presenting all the patterns to the network and computing a total error function $E = \sum_{k=1}^{P} E_k$, where P is the total amount of patterns used in the training process and E_k is the partial network error with respect to the kth training pattern, computed by summing the squared discrepancies between the actual network outputs and the desired values of the kth training pattern. The training patterns can be applied several times to the network but in a different order. Each full pass of all the patterns that belong to the training set, T, is called a *training epoch*. If the adaptation method succeeds in minimizing the total error function then it is obvious that its aim has been fulfilled. Thus training is a non-trivial minimization problem. The most popular training method is *back propagation* method [3], which is based on the steepest descent optimization method. The back propagation learning process applies small iterative steps which correspond to the training epochs. At each epoch t the method updates the weight values proportionally to the gradient of the error function $E(w)$. The whole process is repeated until the network reaches a steady state, where no significant changes in the weights are observed, or until the overall error value drops below a pre-determined threshold. At this point we conclude that the network has learned the problem "satisfactorily". The total number of epochs required can be considered as the speed of the method. More sophisticated training techniques can be found in [3, 6, 7, 8, 15, 18].

4 Problem Formulation and Preliminary Results

In this contribution, we consider the approximation of the elliptic curve discrete logarithm problem using ANNs. More specifically, we consider a specific elliptic curve along with several instances of the discrete logarithm over it and study the performance of ANNs in approximating the value t of the corresponding discrete logarithm. The training methods considered in this study are the Resilient Back Propagation (RPROP) [15] and the Scaled Conjugate Gradient (SCG) method [13]. Relative to the network architecture, we test a variety of topologies with different number of hidden layers and various numbers of neurons at each layer. To make the adaptation of the network easier, the data are normalized before the training. Assuming that the data presented to the network are in $\mathbb{Z}_p$, where p is prime, the interval $S = [-1, 1]$, is split in p-subintervals. Thus, numbers in data are transformed to analogous ones in S.

To evaluate the network performance we measure the percentage of training data for which the network was able to compute the exact target value t. This measure is denoted as μ_0. Next, we employed the $\mu_{\pm v}$ measure, which represents the percentage of the data for which the difference between desired and actual output does not exceed $\pm v$ of the real target. As input patterns of the ANN, the tetrad of the components of the two points P, Q defined by the discrete logarithm over $E(\mathbb{F}_q)$, is presented to the network and the corresponding value t of the discrete logarithm forms the target value of the network. Preliminary results for small primes p, i.e. primes from 1009 to 5003, indicate that ANNs are able to adapt to the training data, even for the measure μ_0. However, the topology and the number of epochs required for the adaptation of the data of the ECDLP, are quite large in comparison with the corresponding requirements for the DLP [11]. Further examination of the ability of ANNs on adapting to the training data for larger primes, as well as, study on the generalization performance of the networks to unknown data will be presented.

Acknowledgment

We acknowledge the partial support by the "Archimedes" research programme awarded by the Greek Ministry of Education and Religious Affairs and the European Union.

References

[1] Blake, I., Seroussi, G., and Smart, N., *Elliptic Curves in Cryptography*, London Mathematical Society Lecture Notes Series 265, Cambridge University Press, 1999.

[2] Coppersmith, D. and Shparlinski, I., On polynomial approximation of the discrete logarithm and the diffie–hellman mapping, *J. Cryptology*, 13, pp. 339–360, 2000.

[3] Haykin, S., *Neural Networks*, Macmillan College Publishing Company, 1999.

[4] Hornik, K., Multilayer feedforward networks are universal approximators, *Neural Networks*, 2, pp. 359–366, 1989.

[5] Koblitz, N., Elliptic curve cryptosystems, *Math. Comp.*, 48, pp. 203–209, 1987.

[6] Magoulas, G.D., Plagianakos, V.P., and Vrahatis, M.N., Adaptive stepsize algorithms for on-line training of neural networks, *Nonlinear Analysis T.M.A.*, 47(5), pp. 3425–3430, 2001.

[7] Magoulas, G.D., Vrahatis, M.N., and Androulakis, G.S., Effective backpropagation training with variable stepsize, *Neural Networks*, 10(1), pp. 69–82, 1997.

[8] Magoulas, G.D., Vrahatis, M.N., and Androulakis, G.S., Increasing the convergence rate of the error backpropagation algorithm by learning rate adaptation methods, *Neural Computation*, 11(7), pp. 1769–1796, 1999.

[9] Maurer, U. and Wolf, S., The relationship between breaking the diffie-hellman protocol and computing discrete logarithms, *SIAM J. Computing*, 28, pp. 1689–1721, 1999.

[10] Meletiou, G.C. and Mullen, G.L., A note on discrete logarithms in finite fields, *Appl. Algebra Engrg. Comm. Comput.*, 3(1), pp. 75–79, 1992.

[11] Meletiou, G.C., Tasoulis, D.K., and Vrahatis, M.N., Cryptography through interpolation approximation and computational intelligence methods, *Bull. Greek Math. Soc.*, 2004, in press.

[12] Miller, V., Uses of elliptic curves in cryptography, *LNCS*, 218, pp. 417–426, 1986.

[13] Møller, M.F., A scaled conjugate gradient algorithm for fast supervised learning, *Neural Networks*, 6, pp. 525–533, 1993.

[14] Pincus, A., Approximation theory of the MLP model in neural networks, In *Acta Numerica*, Cambridge University Press, pp. 143–195, 1999.

[15] Riedmiller, M. and Braun, H., A direct adaptive method for faster backpropagation learning: The RPROP algorithm, In *Proceedings of the IEEE International Conference on Neural Networks*, pp. 586–591, San Francisco, CA, 1993.

[16] Silverman, J.H., *The Arithmetic of Elliptic Curves*, Springer-Verlag, 1986.

[17] Winterhof, A., Polynomial interpolation of the discrete logarithm, *Des. Codes Cryptogr.*, 25(1), pp. 63–72, 2002.

[18] Vrahatis, M.N., Androulakis, G.S., Lambrinos, J.N., and Magoulas, G.D., A class of gradient unconstrained minimization algorithms with adaptive stepsize, *J. Comput. Appl. Math.*, 114(2), pp. 367–386, 2000.

VSP International
Science Publishers
P.O. Box 346, 3700 AH Zeist
The Netherlands

Lecture Series on Computer
and Computational Sciences
Volume 1, 2004, pp. 864-867

Applying Evolutionary Computation Methods for the Cryptanalysis of Feistel Ciphers

E.C. Laskari[*,‡1], **G.C. Meletiou**[†,‡2], **Y.C. Stamatiou**[§,‡3] **and M.N. Vrahatis**[*,‡4]

[*]Department of Mathematics, University of Patras, GR–26110 Patras, Greece

[†]A.T.E.I. of Epirus, P.O. Box 110, GR–47100 Arta, Greece

[§]Department of Mathematics, University of Aegean, GR–83200 Samos, Greece

[‡]University of Patras Artificial Intelligence Research Center (UPAIRC),
University of Patras, GR–26110 Patras, Greece

Received 31 July, 2004; accepted in revised form 5 September, 2004

Abstract: Evolutionary Computation algorithms are stochastic optimization methods inspired by natural evolution and social behavior. These methods have been proven effective in tackling difficult problems that involve discontinuous objective functions and disjoint search spaces. In this contribution a problem introduced by the cryptanalysis of Feistel cryptosystems by means of differential cryptanalysis is formulated as an optimization problem and the performance of Evolutionary Computation methods in addressing this problem is studied for a representative Feistel cryptosystem, the DES. This approach is applicable to all Feistel cryptosystems that are amenable to the differential cryptanalysis method.

Keywords: Computational Intelligence, Evolutionary Computation, DES, Differential Cryptanalysis, Feistel Ciphers

Mathematics Subject Classification: 90C59, 90C30, 90C26, 94A60.

1 Introduction

Evolutionary Computation algorithms are stochastic optimization methods inspired by natural evolution and social behavior. The most known paradigms of such methods are Genetic Algorithms, Evolution Strategies and the Differential Evolution algorithm, which are based on the principles of natural evolution, and the Particle Swarm Optimization, which is based on the simulation of social behavior [4, 5, 11]. These methods have been applied in several scientific fields that include optimization problems, such as mathematics, economy, medicine, engineering and others, and have proven effective in tackling difficult problems that involve discontinuous objective functions and disjoint search spaces.

In this contribution we consider a problem introduced by the cryptanalysis of a Feistel cryptosystem and formulate it as an optimization one. Then the performance of Evolutionary Computation methods in addressing this problem is studied. More specifically, induced by Differential

---

[1]Corresponding author. E-mail: elena@math.upatras.gr

[2]E-mail: gmelet@teiep.gr

[3]E-mail: stamatiu@aegean.gr

[4]E-mail: vrahatis@math.upatras.gr

Cryptanalysis, we investigate the problem of finding some bits of the key that is used to a simple Feistel cipher, the Data Encryption Standard (DES) with reduced number of rounds. Our preliminary results on DES reduced to four rounds are encouraging since the method managed to address the problem at hand using a smaller amount of function evaluations than the brute force approach.

2 Background and Problem Formulation

A *Feistel cipher* is a cryptosystem based on a sequential r times repetition of a function, the *round function*, that maps an n-bit plaintext P, to a ciphertext C. In a Feistel cipher the current n-bit word is divided into $(n/2)$-bit parts, the left part L_i and the right part R_i [3]. Then round i, $1 \leqslant i \leqslant r$ has the following effect:

$$L_i = R_{i-1}, R_i = L_{i-1} \oplus F_i(R_{i-1}, K_i),$$

where K_i is the subkey used in the ith round (derived from the cipher key K), and F_i is an arbitrary round function for the ith round. The output of a Feistel cipher is ordered as (R_r, L_r), i.e. after the last round function has been applied, the two halves are swapped.

An interesting characteristic of Feistel based cryptosystems is that the decryption function is identical to the encryption function except that the subkeys, K_i, and the round functions, F_i, are applied in reverse. This makes the Feistel structure an attractive choice for both software and hardware implementations.

The best known and most widely used Feistel block–cipher cryptosystem is DES, which is the outcome of the collaboration between the government of the United States and IBM in the '70s. It is a *symmetric algorithm*, meaning that the parties exchanging information possess the same key. DES processes plaintext blocks of $n = 64$ bits, producing 64-bit ciphertext blocks, with effective key size $k = 64$ bits, 8 of which can be used as parity bits. The plaintext block is divided into the left and right parts of 32 bits each. The main part of the round function is the F function, which works on the right half of the data, using a subkey of 48 bits and eight S-boxes. The *S-boxes* are mappings that transform 6 bits to 4 bits in a nonlinear way and constitute the only nonlinear part of DES. The 32 output bits of the F function are XORed with the left half of the data and the two halves are exchanged. A more detailed description of the DES algorithm can be found in [10, 12].

Two of the most powerful cryptanalytic attacks for Feistel based ciphers, that were first applied with success to the cryptanalysis of DES, depend critically on the exploitation of specific weaknesses of the S-boxes of the target cryptoalgorithm. These attacks are the *Linear Cryptanalysis* (see [9, 8]) and the *Differential Cryptanalysis* (see [1, 2]). Differential Cryptanalysis (DC) is a chosen plaintext attack which uses only the resultant ciphertexts. The basic tool of the attack is the *ciphertext pair* which is a pair of ciphertexts whose plaintexts have particular differences. The two plaintexts are chosen at random, as long as they satisfy the difference condition. DC analyzes the effect of particular differences in plaintext pairs on the differences of the resultant ciphertext pairs. These differences can be used to assign probabilities to the possible keys and to locate the most probable key. This method usually works on a number of pairs of plaintexts with the same particular difference using only the resultant ciphertext pairs. For cryptosystems similar to DES, the difference is chosen as a fixed XORed value of the two plaintexts.

The most important component in DC is the use of a *characteristic*, which can be informally defined as follows [1]: "Associated with any pair of encryptions are the XOR value of its two plaintexts, the XOR of its ciphertexts, the XORs of the input of each round in the two executions and the XORs of the outputs of each round in the two executions. These XOR values form an r-*round characteristic*. A characteristic has a probability, which is the probability that a random pair with the chosen plaintext XOR has the round and ciphertext XORs specified in the characteristic".

Each characteristic allows the search for a particular set of bits in the subkey of the last round: the bits that enter some particular S-boxes, depending on the chosen characteristic. The characteristics that are most useful are those that have a maximal probability and a maximal number of subkey bits whose occurrences can be counted.

The DC is a statistical method and can fail in rare instances. A more extended analysis on DC and its results on DES for different numbers of rounds can be found in [1].

For the Differential Cryptanalysis of DES reduced to four rounds, a one–round characteristic with probability 1 can be used. This characteristic at the first step of cryptanalysis provides 42 bits of the subkey of the last round. In the case where the subkeys are calculated with the DES key scheduling algorithm, the 42 bits given by DC are actual key bits of the 56 key bits and there are 14 key bits still missing. A suggestion for finding these key bits was to try all the 2^{14} possibilities in decrypting the given ciphertexts, using the resulting keys. The right key should satisfy the known plaintext XOR value for all the pairs that are used by DC. An alternative way is to use a second characteristic that corresponds to the missing bits and try a more careful counting on the key bits of the last two rounds.

Instead of using the aforementioned approaches to find the missing key bits, we formulate the problem of the missing 14 bits as an optimization problem as follows. We consider each one of the 14 bits as a component of a 14-dimensional vector. Such a vector represents a possible solution of the problem. Assume that the right 42 key bits found by DC were suggested using np pairs. We use these np pairs for the evaluation the possible solutions provided by the optimization method. More specifically, for each possible solution, X_i, suggested by the optimization algorithm, we construct the 56 bits of the key, using the 42 bits which are known by DC and the 14 components of X_i in proper order. With the resulting key, we decrypt the np ciphertext pairs that where used by DC and count the number of decrypted pairs that satisfy the known plaintext XOR value, denoted as cnp_{X_i}. Thus, the evaluation function f, is the difference between the desired output np and the actual output cnp_{X_i}, i.e., $f(X_i) = np - cnp_{X_i}$. The global minimum of the function f is zero and the global minimizer provided, will be the actual key with high probability.

3   Preliminary Results and Discussion

In a recent work [6], we have studied the performance of the PSO method formulated for Integer Programming [7], in addressing the prescribed problem of cryptanalysis. Some indicative results of this study using 20 ciphertext pairs, for both global (PSOCG) and local (PSOCL) neighborhood

Table 1: Results for six different keys using $np = 20$ ciphertext pairs.

key	Method	Suc.Rate	Function Evaluations	
			mean	min
k_1	PSOCG	98%	1146	200
k_1	PSOCL	100%	2020	200
k_2	PSOCG	99%	854	200
k_2	PSOCL	100%	2079	200
k_3	PSOCG	97%	1542	200
k_3	PSOCL	100%	2300	200
k_4	PSOCG	97%	1698	200
k_4	PSOCL	100%	1884	300
k_5	PSOCG	93%	1870	200
k_5	PSOCL	100%	1788	300
k_6	PSOCG	100%	740	200
k_6	PSOCL	100%	1717	200

variants of the PSO method with constriction factor and population size of 100 particles, are reported in Table 1. The notation k_i, for $i = 1, \ldots, 6$, is for the six different keys that were used in the experiments and as a measure of the performance of the proposed approach, the number of function evaluations required by the method to locate the global minimum were counted. Each function evaluation corresponds to the decryption of all 20 ciphertext pairs, using a particular key. The success rates of each algorithm, that is the proportion of the times it achieved the global minimizer within a prespecified threshold is also reported.

Our first results were encouraging since the applied method was able to locate the 14 missing bits in an average of 1500 function evaluations as opposed to the 2^{14} required by brute force. Furthermore, the considered methods are simple and can be readily adapted to handle more complex Feistel based cryptosystems. Thus, in this contribution, we extend our study providing more detailed results and interesting conclusions for the cryptanalysis of Feistel–based cryptosystems using Evolutionary Computation methods.

Acknowledgment

We acknowledge the partial support by the "Archimedes" research programme awarded by the Greek Ministry of Education and Religious Affairs and the European Union.

References

[1] Biham, E. and Shamir, A., Differential cryptanalysis of DES–like cryptosystems, *Journal of Cryptology*, 1991.

[2] Biham, E. and Shamir, A., *Differential Cryptanalysis of the Data Encryption Standard*, Springer–Verlag, 1993.

[3] Feistel, H., Cryptography and computer privacy, *Scientific American*, 1973.

[4] Fogel, D.B., *Evolutionary Computation: Towards a New Philosophy of Machine Intelligence*, IEEE Press, Piscataway, NJ, 1995.

[5] Kennedy, J. and Eberhart, R.C., *Swarm Intelligence*, Morgan Kaufmann Publishers, 2001.

[6] Laskari, E.C., Meletiou, G.C., Stamatiou, Y.C., and Vrahatis, M.N., Evolutionary computation based cryptanalysis: A first study, In *Proceedings of the World Congress of Nonlinear Analysts* WCNA-2004, in press.

[7] Laskari, E.C., Parsopoulos, K.E., and Vrahatis, M.N., Particle swarm optimization for integer programming, In *Proceedings of the IEEE 2002 Congress on Evolutionary Computation*, pp. 1576–1581, Hawaii, HI, 2002, IEEE Press.

[8] Matsui, M., Linear cryptanalysis method for DES cipher, *Lecture Notes in Computer Science*, 765, pp. 386–397, 1994.

[9] Matsui, M. and Yamagishi, A., A new method for known plaintext attack of FEAL cipher, *Lecture Notes in Computer Science*, pp. 81–91, 1992.

[10] Menezes, H., van Oorschot, P., and Vanstone, S., *Handbook of applied cryptography*, CRC Press series on discrete mathematics and its applications, CRC Press, 1996.

[11] Schwefel, H.-P., *Evolution and Optimum Seeking*, Wiley, New York, 1995.

[12] Stinson, D., *Cryptography: Theory and Practice (Discrete Mathematics and Its Applications)*, CRC Press, 1995.

VSP International
Science Publishers
P.O. Box 346, 3700 AH Zeist
The Netherlands

*Lecture Series on Computer
and Computational Sciences*
Volume 1, 2004, pp. 868-873

UPSO: A Unified Particle Swarm Optimization Scheme

K.E. Parsopoulos[1] and M.N. Vrahatis[2]

Department of Mathematics, University of Patras Artificial Intelligence
Research Center (UPAIRC), University of Patras, GR–26110 Patras, Greece

Received 31 July, 2004; accepted in revised form 5 September, 2004

Abstract: We introduce Unified Particle Swarm Optimization, a new scheme that harnesses the local and global variants of the standard Particle Swarm Optimization algorithm, combining their exploration and exploitation abilities. Convergence in probability can be proved for the new approach in unimodal cases and preliminary results justify its superiority against the standard Particle Swarm Optimization.

Keywords: Optimization, Particle Swarm Optimization, Stochastic Algorithms, Swarm Intelligence

Mathematics Subject Classification: 90C26, 90C30, 90C59

1 Introduction

Particle Swarm Optimization (PSO) is a stochastic, population–based optimization method. Up–to–date it has been applied successfully on a plethora of test problems in diverse scientific fields [1, 5, 6, 7]. Its efficiency can be attributed to the information exchange among the search points that constitute the population. There are two main variants of PSO with respect to the information exchange scheme that is used, each with different exploration and exploitation characteristics. Practitioners usually select the most proper variant based on their experience as well as on the special characteristics of the problem at hand. Unified Particle Swarm Optimization is a new scheme that harnesses the two variants of PSO in a unified scheme that combines their exploration and exploitation capabilities. Under assumptions, convergence in probability can be proved for the new approach. Preliminary results on a widely used set of benchmark problems are indicative of the new scheme's efficiency.

2 Unified Particle Swarm Optimization

Emergent behavior in socially organized colonies constituted a great source of inspiration for computer scientists. Ant colonies, bird flocks and fish schools that could tackle efficiently combinatorial and numerical optimization problems were modeled and applied successfully on numerous benchmark and real–life problems, giving rise to the class of *swarm intelligence* algorithms [3].

PSO is a swarm intelligence optimization algorithm developed by Eberhart and Kennedy [3]. It employs a population called a *swarm*, $\mathbb{S} = \{x_1, \ldots, x_N\}$, of search points called *particles*, $x_i = (x_{i1}, x_{i2}, \ldots, x_{in})^\top$, $i = 1, \ldots, N$, which probe the search space, $S \subset \mathbb{R}^n$, simultaneously. The

---

[1] Corresponding author. E-mail: kostasp@math.upatras.gr
[2] E-mail: vrahatis@math.upatras.gr

algorithm works iteratively. Each particle is initialized to a random position in the search space. Then, at each iteration, each particle moves with an adaptable *velocity*, $v_i = (v_{i1}, v_{i2}, \ldots, v_{in})^\top$, while retaining in a memory the best position, $p_i = (p_{i1}, p_{i2}, \ldots, p_{in})^\top \in S$, it has ever visited in the search space. In minimization problems, best positions have lower function values. The particle's movement is also influenced by the experience of the rest particles, i.e., by their best positions. This is performed through the concept of neighborhood. More specifically, each particle is assigned a neighborhood which consists of some prespecified particles. Then, the particles that comprise the neighborhood share their experience by exchanging information. There are two main variants of PSO with respect to the number of particles that comprise the neighborhoods. In the *global* variant, the whole swarm is considered as the neighborhood of each particle, while, in the *local* variant, smaller neighborhoods are used. Neighboring particles are determined based rather on their indices than their actual distance in the search space.

Let g_i be the index of the best particle in the neighborhood of x_i, i.e., the index of the particle that attained the best position among all the particles of the neighborhood. The particles are considered in a ring topology. Thus, their indices are considered in a cyclic order, i.e., 1 is the index that follows after N. At each iteration, the swarm is updated according to the equations [1, 7],

$$v_i^{(k+1)} = \chi \left[v_i^{(k)} + c_1 r_1 \left(p_i^{(k)} - x_i^{(k)} \right) + c_2 r_2 \left(p_{g_i}^{(k)} - x_i^{(k)} \right) \right], \tag{1}$$

$$x_i^{(k+1)} = x_i^{(k)} + v_i^{(k+1)}, \tag{2}$$

where $i = 1, \ldots, N$; k is the iterations' counter; χ is a parameter called *constriction factor* that controls the velocity's magnitude; c_1 and c_2 are positive acceleration parameters, called *cognitive and social parameter*, respectively; and r_1, r_2 are random vectors that consist of random values uniformly distributed in $[0, 1]$. All vector operations in Eqs. (1) and (2) are performed componentwise. A stability analysis of PSO, as well as recommendations regarding the selection of its parameters are provided in [1, 7].

The performance of a population–based algorithm depends on its ability to perform global search of the search space (exploration) as well as more refined local search (exploitation). Proper balance between these two characteristics result in enhanced performance. In the global variant of PSO, all particles are attracted by the same overall best position, converging faster toward specific points. Thus, it has better exploitation abilities. On the other hand, in the local variant, the information of the best position of each neighborhood is communicated slowly to the other particles of the swarm through their neighbors. Therefore, the attraction to specific best positions is weaker, hindering the swarm from getting trapped in locally optimal solutions. Thus, the local variant of PSO has better exploration ability. Proper selection of the neighborhood's size affects the trade–off between exploration and exploitation. The selection of the most proper neighborhood size is an open problem. In practice, it is up to the practitioner and it is based solely on his experience.

Unified Particle Swarm Optimization (UPSO) is a new scheme that harnesses the global and the local variant of PSO, thereby combining their exploration and exploitation capabilities. Let $\mathcal{G}_i^{(k+1)}$ denote the velocity update of the ith particle, x_i, in the global PSO variant, while $\mathcal{L}_i^{(k+1)}$ denotes the corresponding velocity update for the local variant. Then, according to Eq. (1),

$$\mathcal{G}_i^{(k+1)} = \chi \left[v_i^{(k)} + c_1 r_1 \left(p_i^{(k)} - x_i^{(k)} \right) + c_2 r_2 \left(p_g^{(k)} - x_i^{(k)} \right) \right], \tag{3}$$

$$\mathcal{L}_i^{(k+1)} = \chi \left[v_i^{(k)} + c_1 r_1' \left(p_i^{(k)} - x_i^{(k)} \right) + c_2 r_2' \left(p_{g_i}^{(k)} - x_i^{(k)} \right) \right], \tag{4}$$

where k denotes the iteration number; g is the index of the best particle of the whole swarm (global variant); and g_i is the index of the best particle in the neighborhood of x_i (local variant). These two search directions can be combined in a single equation, resulting in the main UPSO scheme,

$$U_i^{(k+1)} = u\,\mathcal{G}_i^{(k+1)} + (1-u)\,\mathcal{L}_i^{(k+1)}, \tag{5}$$

$$x_i^{(k+1)} = x_i^{(k)} + U_i^{(k+1)}, \tag{6}$$

where $u \in [0,1]$ is a parameter called the *unification factor*, which determines the influence of the global and local components in Eq. (5). For $u = 1$, Eq. (5) is equivalent to the global PSO variant, while for $u = 0$ it is equivalent to the local PSO variant. For all intermediate values, $u \in (0,1)$, we obtain composite variants of PSO that combine the exploration and exploitation characteristics of its global and local variant.

UPSO can be further enhanced by incorporating a stochastic parameter in Eq. (5) that imitates the mutation of evolutionary algorithms, however, it is directed toward a direction which is consistent with the PSO dynamics. Thus, Eq. (5) can be written as,

$$U_i^{(k+1)} = r_3\,u\,\mathcal{G}_i^{(k+1)} + (1-u)\,\mathcal{L}_i^{(k+1)}, \tag{7}$$

which is mostly based on the local variant or, alternatively,

$$U_i^{(k+1)} = u\,\mathcal{G}_i^{(k+1)} + r_3\,(1-u)\,\mathcal{L}_i^{(k+1)}, \tag{8}$$

which is mostly based on the global variant, where $r_3 \sim \mathcal{N}(\mu, \sigma^2 I)$ is a normally distributed parameter, and I is the identity matrix.

A proof of convergence in probability can be given for the schemes of Eqs. (7) and (8). The proof follows the analysis of Matyas [4] for stochastic optimization algorithms. Assume that $F : S \to \mathbb{R}$ is a unimodal objective function, x_{opt} is its unique minimizer in S, and $F_{\text{opt}} = F(x_{\text{opt}})$. Also, let $x_i^{(k)}$ be the ith particle of the swarm and $p_i^{(k)}$ be its best position in the kth iteration. The proof does not take into consideration the index i, therefore we will refer to them as $x^{(k)}$ and $p^{(k)}$, respectively. The level set of F at a constant value, K, is defined as $G[K] = \{x : F(x) < K\}$. We assume that $G[K] \neq \emptyset$, for all $K > F_{\text{opt}}$. Let $A^{(k+1)} = (1-u)\mathcal{L}^{(k+1)}$, $B^{(k+1)} = u\,\mathcal{G}^{(k+1)}$, and $f(z)$ be the probability distribution of $r_3\,B^{(k)}$, with $r_3 \sim \mathcal{N}(\mu, \sigma^2 I)$. The choice of the Normal distribution as the probability distribution of r_3, guarantees that $f(z) \neq 0$, for all z, although the proof holds for any choice of probability distribution for r_3, as long as this relation holds. We define as a *successful step* of UPSO at iteration k, the fact that $F\left(x^{(k+1)}\right) < F\left(p^{(k)}\right) - \varepsilon$, for a prescribed $\varepsilon > 0$. The probability of a successful step from $x^{(k)}$ is given by

$$P_F(x) = \int_{G[F(p)-\varepsilon]} f(z-x)\,dz.$$

Then, based on the analysis of Matyas [4], the following theorem is straightforwardly proved:

Theorem 1 *Let $F(x)$ have a unique minimum in S, $G[K] \neq \emptyset$, for all $K > F_{\text{opt}}$, and $f(z) \neq 0$ for all z. Then, at least one sub–sequence of best positions, $\{p^{(k)}\}$, of any particle, x, of the swarm in UPSO tends in probability to x_{opt}.*

Proof. Let $\delta(x) = \{z : \varrho(z,x) < \delta\}$, $\delta > 0$, be the δ–neighborhood of a point x. We will prove that for any $\delta > 0$ it holds that,

$$\lim_{k \to \infty} P\left\{\varrho\left(p^{(k)}, x_{\text{opt}}\right) > \delta\right\} = \lim_{k \to \infty} P\left\{p^{(k)} \notin \delta(x_{\text{opt}})\right\} = 0,$$

i.e., the probability that the distance $\varrho\left(p^{(k)}, x_{\text{opt}}\right) > \delta$, or equivalently that $p^{(k)} \notin \delta(x_{\text{opt}})$, tends to zero. If we denote by F_δ the minimum value of F on the boundary of $\delta(x_{\text{opt}})$, we shall have

$F_\delta > F_{\mathrm{opt}}$. We can now define $\varepsilon = \varepsilon(\delta)$ such that $0 < \varepsilon(\delta) < F_\delta - F_{\mathrm{opt}}$. For all previous best positions, $p \notin \delta(x_{\mathrm{opt}})$, of the particle under consideration, the inequality $F(p) - \varepsilon > F_{\mathrm{opt}}$, is valid. Furthermore, from the assumptions of the theorem, $G[F(p) - \varepsilon]$ is a non–empty region. Since $f(z) > 0$ for all z, there will exist an $\alpha > 0$, such that $P_F(x) \geqslant \alpha$, i.e., the probability of a successful step from x is positive (although in some cases it may become very small).

Let $F\left(x^{(1)}\right) = F\left(p^{(1)}\right)$ be the initial function value of x and p, respectively (recall that the initial position and the initial best position of a particle coincide). We denote $\tau = \left(F\left(p^{(1)}\right) - F_\delta\right)/\varepsilon$, and $m = \lfloor \tau \rfloor$, i.e., m is the largest integer less than τ. From the design of the PSO and UPSO algorithm, if even $m+1$ steps turn out to be successful, then all the subsequent points of the sequence $\{p^{(k)}\}$ lie in $\delta(x_{\mathrm{opt}})$. Consequently, the probability $P\{p^{(k)} \notin \delta(x_{\mathrm{opt}})\}$ is less than or equal to the probability that the number of successful steps does not exceed m, i.e.,

$$P\left\{p^{(k)} \notin \delta(x_{\mathrm{opt}})\right\} \leqslant P\left\{\sum_{i=1}^{k} y^{(i)} \leqslant m\right\},$$

where, $y^{(i)} = 1$, if there was a successful step in iteration i, and $y^{(i)} = 0$, otherwise. The latter probability increases with a decrease in the probability of successful steps, and since $P_F(x) \geqslant \alpha$, it obeys the well–known Newton's theorem (on the binomial probability distribution),

$$P\left\{\sum_{i=1}^{k} y^{(i)} \leqslant m\right\} \leqslant \sum_{i=0}^{m} \binom{k}{i} \alpha^i (1-\alpha)^{k-i},$$

where k is the number of steps (iterations) taken. Further, when $k > 2m$ and $\alpha < 0.5$,

$$\sum_{i=0}^{m} \binom{k}{i} \alpha^i (1-\alpha)^{k-i} < (m+1)\binom{k}{m}(1-\alpha)^k = \frac{m+1}{m!}k(k-1)(k-2)\cdots(k-m+1)(1-\alpha)^k$$

$$< \frac{m+1}{m!}k^m(1-\alpha)^k.$$

Consequently, $P\{\varrho(p^{(k)}, x_{\mathrm{opt}}) > \delta\} < \frac{m+1}{m!}k^m(1-\alpha)^k$. Thus, for $\alpha > 0$, it is clear that,

$$\lim_{k \to \infty} k^m(1-\alpha)^k = 0,$$

and the theorem is proved. ∎

We must note that the best position p of the particle x may be replaced by the overall best position of the whole swarm, with minor modifications in the proof.

3 Experimental Results and Discussion

The performance of UPSO was investigated on the test set used by Trelea in [7], and it consists of the Sphere, Rosenbrock, Rastrigin, and Griewank function in 30 dimensions, as well as the Schaffer's function in 2 dimensions. The test functions are denoted as F_1, F_2, F_3, F_4, and F_5, respectively. For comparison purposes, the PSO configuration reported in [7] was also adopted here. Specifically, two sets of parameters were used, denoted as Set 1 and Set 2, respectively. Set 1 consists of $\chi = 0.6$ and $c_1 = c_2 = 2.833$, which is the equivalent of the set $a = 0.6$ and $b = 1.7$ in [7], and Set 2 consists of $\chi = 0.729$, $c_1 = c_2 = 2.05$, which is the equivalent of the set $a = 0.729$ and $b = 1.494$ in [7]. The maximum number of iterations was 10000. The swarm was initialized in the range $[-100, 100]^{30}$ for the Sphere function, $[-30, 30]^{30}$ for the Rosenbrock function, $[-5.12, 5.12]^{30}$ for the Rastrigin function, $[-600, 600]^{30}$ for the Griewank function, and $[-100, 100]^2$ for Schaffer's

Table 1: Results for UPSO. Success rates (in parenthesis) and expected number of function evaluations are reported.

	Success Rates & Expected Number of Function Evaluations							
	φ = 0.9				φ = 0.1			
	μ = 0		μ = 1		μ = 0		μ = 1	
N	Set 1	Set 2	Set 1	Set 2	Set 1	Set 2	Set 1	Set 2
	F_1 (Sphere). Best value reported in [2, 7] is (1.00)10320							
15	(1.00)**3078**	(1.00)3754	(1.00)**3981**	(1.00)5268	(1.00)**2892**	(1.00)3658	(1.00)**3876**	(1.00)5401
30	(1.00)6588	(1.00)6924	(1.00)6852	(1.00)8807	(1.00)6278	(1.00)7011	(1.00)6579	(1.00)8427
60	(1.00)12174	(1.00)14679	(1.00)11370	(1.00)14238	(1.00)12771	(1.00)14640	(1.00)11433	(1.00)14202
	F_2 (Rosenbrock). Best value reported in [2, 7] is (0.50)15930							
15	(1.00)**2138**	(1.00)2413	(1.00)**4950**	(0.95)10451	(1.00)**2193**	(1.00)2570	(1.00)**3656**	(1.00)10333
30	(1.00)4385	(1.00)4835	(1.00)6765	(1.00)17714	(1.00)4223	(1.00)5330	(1.00)8601	(1.00)11466
60	(1.00)8505	(1.00)9936	(1.00)12186	(1.00)16626	(1.00)8040	(1.00)9765	(1.00)15450	(1.00)14688
	F_3 (Rastrigin). Best value reported in [2, 7] is (0.90)4667							
15	(0.90)16541	(1.00)8412	(1.00)**4373**	(1.00)11315	(1.00)3820	(1.00)**2889**	(1.00)**6846**	(1.00)7506
30	(1.00)8766	(1.00)**7233**	(1.00)8744	(1.00)11673	(1.00)5805	(1.00)9294	(1.00)8196	(1.00)10256
60	(1.00)11475	(1.00)11640	(1.00)15882	(1.00)19437	(1.00)8433	(1.00)10293	(1.00)14013	(1.00)17934
	F_4 (Griewank). Best value reported in [2, 7] is (1.00)9390							
15	(1.00)**3161**	(1.00)3275	(1.00)**4157**	(1.00)5473	(1.00)**3312**	(1.00)3620	(1.00)**4343**	(1.00)5383
30	(1.00)6507	(1.00)7229	(1.00)6914	(1.00)8114	(1.00)6009	(1.00)6870	(1.00)6516	(1.00)8300
60	(1.00)11844	(1.00)14019	(1.00)10725	(1.00)13230	(1.00)12150	(1.00)14640	(1.00)10839	(1.00)13458
	F_5 (Schaffer). Best value reported in [2, 7] is (0.75)6440							
15	(0.90)24153	(0.85)20927	(0.95)21922	(0.95)37092	(1.00)37790	(1.00)**18591**	(1.00)36664	(0.95)**19234**
30	(1.00)**16716**	(1.00)19437	(1.00)22254	(1.00)36020	(1.00)25433	(1.00)26211	(1.00)29874	(1.00)22680
60	(1.00)25425	(1.00)24072	(1.00)21531	(1.00)**19203**	(1.00)33411	(1.00)20115	(1.00)21456	(1.00)22764

function, while the corresponding error goals were 0.01, 100, 100, 0.1, and 10^{-5} [7]. For each function, 20 experiments were performed, using three different swarm sizes, 15, 30, and 60. The particles were allowed to move anywhere in the search space without constraints on their velocity. Regarding the configuration of UPSO, two cases were investigated, namely the case of Eq. (8) with unification factor $u = 0.9$, which is closer to the global PSO variant used in [7], and Eq. (7) with $u = 0.1$, which is closer to the local PSO variant. For both cases, two different configurations of the distribution of r_3 were considered, namely, one with mean value $\mu = 0$, and one with mean value $\mu = 1$. The standard deviation of r_3 was $\sigma = 0.01$ in all cases, to avoid wide deterioration of the PSO dynamics. The neighborhood radius for the determination of the local PSO search direction, $\mathcal{L}_i$, in Eqs. (7) and (8), was 1, i.e., the neighbors of the ith particle, x_i, were the particles x_{i-1} and x_{i+1}. This selection was made in order to take full advantage of the properties of the local version, since the larger the neighborhoods the closer is the local variant to the global one.

For each parameter configuration and test function, the success rate of UPSO, namely the fraction of the experiments in which the error goal was achieved, as well as the corresponding expected number of function evaluations, defined as (Number of Particles)×(Average Number of Iterations)/(Success Rate) [7], were recorded and they are reported in Table 1. The results were very promising. The success rate never fell under 0.90 (i.e., 90%), while (in most cases) the expected number of function evaluations was smaller than the values reported in [7] and [2]. UPSO achieved success rates of 100% even in cases where the plain PSO had very low success rates. For example, in the cases of F_4 and F_5 and a swarm of 15 particles, the plain PSO with the parameters of Set 1 had a success rate of just 0.35 and 0.45, respectively, as reported in [7], while UPSO's success rate was higher than 0.90 for any set of parameters. Furthermore, the success rate and the lowest expected number of function evaluations for the test functions F_1–F_5, as reported in [7], was (1.0)10320, (0.50)15930, (0.90)4667, (1.00)9390, (0.75)6440, respectively. In UPSO, the corresponding numbers were (1.00)2892, (1.00)2138, (1.00)2889, (1.00)3161, and (1.00)16716 (for each case the lowest expected number of function evaluations is bold faced in the table). We must notice that in the case of F_5, the number 6440 reported in [7] corresponds to a success rate equal to 0.75, while the number 16716 of UPSO corresponds to a success rate equal to 1.00.

Regarding the different configurations of UPSO, the version of Eq. (7) with $u = 0.1$ and $\mu = 0$ had the better overall performance, probably due to the better exploration ability of the local PSO variant, which is favored in this scheme. Moreover, the Set 1 proposed by Trelea in [7] outperformed the (most popular) Set 2. Also, UPSO with $\mu = 0$ outperformed that with $\mu = 1$, in most problems.

Further experiments were performed using the plain UPSO scheme of Eq. (5), both in static and dynamic optimization problems, revealing that the values $u = 0.5$ and $u = 0.2$ result in enhanced performance of the algorithm. The mathematical properties behind this effect are still under investigation, along with possible correlations between the UPSO scheme and PSO variants with adaptive neighborhood size.

4   Conclusions

A Unified Particle Swarm Optimization (UPSO) that aggregates the local and the global variant of PSO in a unified scheme has been introduced. The proposed approach seems to exploit the good properties of both variants and preliminary experiments on the test set used in [7] justify its efficiency. Further investigation is required to analyze the dynamics of UPSO. A self–adaptive scheme that will exploit knowledge of the characteristics of the objective function, as well as the performance of the algorithm, to control the unification factor is currently under development, along with an analysis of its convergence rates.

Acknowledgment

The authors wish to thank the anonymous referees, as well as Dr. M. Clerc and Dr. I.C. Trelea for their careful reading of the manuscript and their fruitful comments and suggestions.

References

[1] Clerc, M. and Kennedy, J., The particle swarm–explosion, stability, and convergence in a multidimensional complex space, *IEEE Transactions on Evolutionary Computation*, **6**(1), pp. 58–73, 2002.

[2] Eberhart, R.C. and Shi, Y., Comparing inertia weights and constriction factors in particle swarm optimization, In *Proc. 2000 IEEE CEC*, pp. 84–88, Piscataway, NJ, 2000. IEEE Service Center.

[3] Kennedy, J. and Eberhart, R.C., *Swarm Intelligence*, Morgan Kaufmann Publishers, 2001.

[4] Matyas, J., Random optimization, *Automatization and Remote Control*, **26**, pp. 244–251, 1965.

[5] Parsopoulos, K.E. and Vrahatis, M.N., Recent approaches to global optimization problems through particle swarm optimization, *Natural Computing*, **1**(2–3), pp. 235–306, 2002.

[6] Parsopoulos, K.E. and Vrahatis, M.N., On the computation of all global minimizers through particle swarm optimization, *IEEE Transactions on Evolutionary Computation*, **8**(3), pp. 211–224, 2004.

[7] Trelea, I.C., The particle swarm optimization algorithm: Convergence analysis and parameter selection, *Information Processing Letters*, **85**, pp. 317–325, 2003.

VSP International
Science Publishers
P.O. Box 346, 3700 AH Zeist
The Netherlands

*Lecture Series on Computer
and Computational Sciences*
Volume 1, 2004, pp. 874-879

Nonlinear Data Fitting for Landslides Modeling

K.E. Parsopoulos[*1], **V.A. Kontogianni**[†2], **S.I. Pytharouli**[†3], **P.A. Psimoulis**[†4],
S.C. Stiros[†5] **and M.N. Vrahatis**[*6]

[*]Department of Mathematics, University of Patras Artificial Intelligence
Research Center (UPAIRC), University of Patras, GR–26110 Patras, Greece

[†]Department of Civil Engineering, University of Patras, GR–26500 Patras, Greece

Received 31 July, 2004; accepted in revised form 5 September, 2004

Abstract: We consider data fitting schemes that are based on different norms to determine
the parameters of curve–models that model landslides in dams. The Particle Swarm Optimization method is employed to minimize the corresponding error norms. The method is
applied on real–world data with promising results.

Keywords: Curve Fitting, Optimization, Particle Swarm Optimization, Swarm Intelligence,
Landslides Modeling

Mathematics Subject Classification: 65D10, 90C26

1 Introduction

A common problem in physics, earth sciences and engineering is the optimal fitting of a curve to
a set of observations of certain parameters versus time (or space). The observations are usually
contaminated by various types of errors. The usual procedure that is followed to solve such
problems is to test various empirically selected model–curves and estimate the parameters of the
curve that minimize the difference between the values obtained through the model and the observed
ones.

Traditionally, this task is accomplished using the well–known Least Squares Method (LSQR).
More specifically, linear or linearized equations are used and the sum of squares of differences
among observations and the corresponding model–curve values is minimized. Therefore, the practitioner has to decide only regarding the most appropriate curve–model (e.g. polynomial, periodic,
exponential, mixed, etc.) such that an acceptable fit is obtained.

In some cases, however, the available data are noisy, unevenly distributed versus time, there is
no *a priori* knowledge of the variance–covariance matrix or they do not correspond to rather smooth
curves (for instance they include offsets, a usual case in tectonic and geotechnical studies [4]). In
such cases, the LSQR approach may not be successful, resulting in complex curve–models that
lack physical significance and ability to be incorporated to further modeling and analysis. In such
cases, the use of different data fitting approaches has been proved very useful [1].

[1]Corresponding author. E-mail: kostasp@math.upatras.gr
[2]E-mail: vkont@civil.upatras.gr
[3]E-mail: spitha@upatras.gr
[4]E-mail: papsimouli@upnet.gr
[5]E-mail: stiros@upatras.gr
[6]E-mail: vrahatis@math.upatras.gr

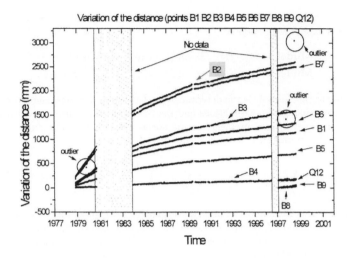

Figure 1: The record of observations for the Polyfyto Dam.

Evolutionary and Swarm Intelligence algorithms have been successfully applied on several data fitting problems [5, 6]. Their ability to work using solely function values even for discontinuous and non–differentiable functions renders them a promising alternative in cases where traditional algorithms, such as LSQR, fail. The aim of this paper is to investigate alternative curve fitting techniques based on the Particle Swarm Optimization (PSO) algorithm and three different norms to cope with a real–life curve fitting problem from the field of Civil Engineering. Results are reported and discussed.

Section 2 is devoted to the description of the problem, while the employed optimization algorithm, PSO, is briefly described in Section 3. Experimental results are reported and discussed in Section 4.

2 Description of the Curve Fitting Problem and Models

The problem investigated here is the monitoring of a landslide of the Polyfyto Dam in the Aliakmonas river in north Greece. A record of observations has been collected in collaboration with the Greek Public Power Corporation s.a.. The record consists of a large number of observations of distance changes obtained by monitoring 7 control points, denoted as $B1 - B7$, on the landslide relative to a stable reference station on stable ground, over a period of 20 years. The record is depicted in Figure 1, along with observations for 3 auxiliary points, $B8$, $B9$ and $Q12$. As we can see, the control point $B2$ exhibited the largest displacement.

The first step in the analysis of the landslide is the determination of a mathematical model, which captures the pattern of the landslide movement and can be used to estimate its future trends [7]. For this purpose, the movement of each control point was individually investigated. In Figure 1 it is clear that almost all points are moving faster in early years, while their movement tends to be stabilized in late years. This effect can be described using different mathematical models, although, just a few models retain the physical meaning of the specific phenomenon. The simplest model that could be used is a polynomial of degree four. However, it exhibits some upward and downward branches that do not fit the observations, and for this purpose, two types of an exponential decay model were adopted,

$$\textbf{Model 1:} \qquad f(t) \;=\; A\,(1 - \exp\,(-t/B)) + C, \qquad\qquad (1)$$

$$\textbf{Model 2:} \qquad f(t) \;=\; A\,(1 - \exp\,(-t/B)) + K\,t + C. \qquad\qquad (2)$$

The next step in the analysis is the determination of the unknown parameters A, B, C and K, such that the error among the observations and the corresponding values provided by the model is minimized. For the error measurement, several norms can be used. The most common choices are the ℓ_1, ℓ_2 and ℓ_∞–norms, which are defined as,

$$\|\varepsilon\|_1 = \sum_{i=1}^{m} |\varepsilon_i|, \qquad \|\varepsilon\|_2 = \left(\sum_{i=1}^{m} |\varepsilon_i|^2 \right)^{1/2}, \qquad \|\varepsilon\|_\infty = \max_{1 \leqslant i \leqslant m} |\varepsilon_i|,$$

respectively, where m is the number of observations and $\varepsilon_i = M_i - O_i$, $i = 1, \dots, m$, with O_i being the ith observed value and M_i be the corresponding value implied by the model.

The ℓ_1–norm is the most "fair" norm since it uses the absolute values of the errors. However, it results in non–differentiable minimization problems, therefore, it cannot be used with traditional gradient–based minimizers. On the other hand, the ℓ_2–norm results in differentiable minimization problems but the assumed error values are not always consistent with the actual ones. For example, an absolute error value equal to 10^{-3} becomes 10^{-6}, while an absolute error equal to 10^2 becomes 10^4. The ℓ_∞–norm constitutes the most proper choice in cases where outliers that must be taken seriously into consideration appear in the set of observations, since it minimizes the maximum among all absolute errors.

The performance of LSQR for the determination of the unknown parameters A, B, C and K, is rather poor with the deviation being larger at the edge of the curve where indeed a good fitting is sought. This happens due to the ℓ_2–norm, on which LSQR is based. Thus, alternative fitting techniques that use different norms are of great interest in order to provide more reliable results.

3 Particle Swarm Optimization

Particle Swarm Optimization (PSO) is a swarm intelligence optimization algorithm developed by Eberhart and Kennedy [3]. It employs a population, called a *swarm*, $\mathbb{S} = \{x_1, \dots, x_N\}$, of search points, called *particles*, $x_i = (x_{i1}, x_{i2}, \dots, x_{in})^\top$, $i = 1, \dots, N$, which probe the search space, $S \subset \mathbb{R}^n$, simultaneously. The algorithm works iteratively. Each particle is initialized to a random position in the search space. Then, at each iteration, each particle moves with an adaptable *velocity*, $v_i = (v_{i1}, v_{i2}, \dots, v_{in})^\top$, while retaining in a memory the best position, $p_i = (p_{i1}, p_{i2}, \dots, p_{in})^\top \in S$, it has ever visited in the search space. In minimization problems, best positions have lower function values. The particle's movement is also influenced by the experience of the rest particles, i.e., by their best positions. This is performed through the concept of neighborhood. More specifically, each particle is assigned a neighborhood which consists of some prespecified particles. Then, the particles that comprise the neighborhood share their experience by exchanging information. There are two main variants of PSO with respect to the number of particles that comprise the neighborhoods. In the *global* variant, the whole swarm is considered as the neighborhood of each particle, while, in the *local* variant, smaller neighborhoods are used. Neighboring particles are determined based rather on their indices than their actual distance in the search space [6].

Let g_i be the index of the best particle in the neighborhood of x_i, i.e., the index of the particle that attained the best position among all the particles of the neighborhood. The indices of the particles are considered in a cyclic order, i.e., 1 is the index that follows after N. At each iteration,

Table 1: Computed solutions for the two models.

Model	Norm	A	B	C	K
Model 1	ℓ_1	2377.10	2447.27	236.68	
	ℓ_2	2391.74	2468.11	246.08	
	ℓ_∞	2397.33	2588.62	283.42	
Model 2	ℓ_1	1626.50	1431.98	0.106	178.76
	ℓ_2	1624.57	1464.39	0.106	187.87
	ℓ_∞	1686.59	1582.49	0.096	202.47

the swarm is updated according to the equations [2, 8],

$$v_i^{(k+1)} = \chi \left[v_i^{(k)} + c_1 r_1 \left(p_i^{(k)} - x_i^{(k)} \right) + c_2 r_2 \left(p_{g_i}^{(k)} - x_i^{(k)} \right) \right], \qquad (3)$$

$$x_i^{(k+1)} = x_i^{(k)} + v_i^{(k+1)}, \qquad (4)$$

where $i = 1, \ldots, N$; k is the iterations' counter; χ is a parameter called *constriction factor* that controls the velocity's magnitude; c_1 and c_2 are positive acceleration parameters, called *cognitive* and *social* parameter, respectively; and r_1, r_2 are random vectors that consist of random values uniformly distributed in $[0, 1]$. All vector operations in Eqs. (3) and (4) are performed componentwise. A stability analysis of PSO, as well as recommendations regarding the selection of its parameters are provided in [2, 8].

PSO has been applied on ℓ_1–norm errors–in–variables data fitting problems with very promising results, exhibiting superior performance even than the well–known Trust Region methods [5]. Therefore it was selected for the error minimization in our problem using the ℓ_1, ℓ_2 and ℓ_∞–norms.

4   Results and Discussion

The PSO algorithm was used for the determination of parameters of the two models defined in Eqs. (1) and (2), minimizing the error defined through the ℓ_1, ℓ_2 and ℓ_∞–norms, which will be denoted as *L1*, *L2* and *L3*, respectively. We concentrated on the case of the control point *B2*, which had the largest displacement in our set of observations. The data set for *B2* consisted of 404 observations. For the PSO, the default parameters, $\chi = 0.729$ and $c_1 = c_2 = 2.05$ were used. The swarm size was equal to 60 for Model 1 and 80 for Model 2. The algorithm was let to run for 5000 iterations. We conducted 100 independent experiments for each model and norm.

In all experiments, the same solutions (model parameters) were computed and they are reported in Table 1. The absolute error for each observation was also recorded for the detected model parameters. The mean value and the standard deviation of these absolute error values as well as the typical error for a single observation,

$$S0 = \sqrt{\frac{\varepsilon^2}{m - n}},$$

where m is the number of observations and n is the dimension of the problem were computed for the three norms. For Model 1, the plot of the actual data along with the corresponding model values for each norm, a boxplot with the distribution of the absolute error for the 404 observations

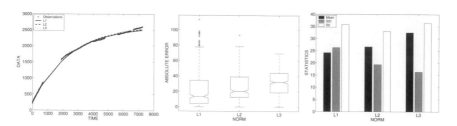

Figure 2: Plot of the actual data along with the corresponding model values for each norm (left), boxplot with the distribution of the absolute error for all observations for the computed model parameters (center), and statistics of absolute error (right) for Model 1. Labels $L1$, $L2$ and $L3$ correspond to the norms ℓ_1, ℓ_2 and ℓ_∞, respectively.

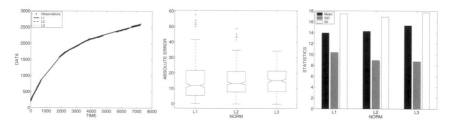

Figure 3: Plot of the actual data along with the corresponding model values for each norm (left), boxplot with the distribution of the absolute error for all observations for the computed model parameters (center), and statistics of absolute error (right) for Model 2. Labels $L1$, $L2$ and $L3$ correspond to the norms ℓ_1, ℓ_2 and ℓ_∞, respectively.

for the computed model parameters, as well as a bar plot with the mean value and standard deviation of absolute error and the quantity $S0$, are depicted in Figure 2. Figure 3 reports the corresponding graphs for Model 2. The boxplot produces a box and whisker plot for the sample of 404 absolute error values. The box has lines at the lower quartile, median, and upper quartile values. The whiskers are lines extending from each end of the box to show the extent of the rest of the data. Outliers are data with values beyond the ends of the whiskers. Notches represent a robust estimate of the uncertainty about the medians for box to box comparison. All displacement units in figures are in millimeters (mm).

In the case of Model 1, the ℓ_1–norm exhibited the smallest mean absolute error, followed by ℓ_2 and ℓ_∞. The latter norm had the smallest standard deviation of absolute error, which implies its robustness. Finally, the best value of $S0$ was obtained using the ℓ_2–norm. Model 2 provided a far better fit, although the corresponding model is more complex and harder to be incorporated in further analysis. The same comments with Model 1 can be made for the mean value, standard deviation and $S0$, although the differences between the different norms are smaller than in the case of Model 1. The ℓ_∞–norm is much better, especially at the edges of the intervals covered by observations.

Concluding, the three different approaches through PSO using ℓ_1, ℓ_2 and ℓ_∞–norms resulted in an efficient scheme that optimizes the exponential decay models considered for the curve fitting problem of the Polyfyto Dam, providing further intuition on tackling similar problems. Further work is needed toward the direction of estimating the future trends of the landslide.

Acknowledgment

This work was partially supported by the PENED 2001 Project awarded by the Greek Secretariat of Research and Technology.

References

[1] Amiri–Simkooei, A., Formulation of ℓ_1 norm minimization in gauss–markov models, *J. Surveying Engineering, ASCE*, **129**(1), pp. 37–43, 2003.

[2] Clerc, M. and Kennedy, J., The particle swarm–explosion, stability, and convergence in a multidimensional complex space, *IEEE Transactions on Evolutionary Computation*, **6**(1), pp. 58–73, 2002.

[3] Kennedy, J. and Eberhart, R.C., *Swarm Intelligence*, Morgan Kaufmann Publishers, 2001.

[4] Liu, Q.W. and Chen, Y.Q., Combining the geodetic models of vertical crustal deformation, *J. Geodesy*, **72**, pp. 673–683, 1998.

[5] Parsopoulos, K.E., Laskari, E.C. and Vrahatis, M.N., Solving ℓ_1 norm errors-in-variables problems using particle swarm optimizer, In M.H. Hamza, editor, *Artificial Intelligence and Applications*, pp. 185–190. IASTED/ACTA Press, 2001.

[6] Parsopoulos, K.E. and Vrahatis, M.N., On the computation of all global minimizers through particle swarm optimization, *IEEE Transactions on Evolutionary Computation*, **8**(3), pp. 211–224, 2004.

[7] Stiros, S., Vichas, C. and Skourtis, C., Landslide monitoring based on geodetically–derived distance changes, *J. Surveying Engineering, ASCE*, 2004. in press.

[8] Trelea, I.C., The particle swarm optimization algorithm: Convergence analysis and parameter selection, *Information Processing Letters*, **85**, pp. 317–325, 2003.

VSP International
Science Publishers
P.O. Box 346, 3700 AH Zeist
The Netherlands

Lecture Series on Computer
and Computational Sciences
Volume 1, 2004, pp. 880-883

Economic Geography: Existence, Uniqueness and Computation of Short-Run Equilibria

N.G. Pavlidis[1], M.N. Vrahatis[2]

Department of Mathematics,
University of Patras Artificial Intelligence Research Centre (UPAIRC)
University of Patras, GR-26110 Patras, Greece.

P. Mossay[3]

Departamento de Fundamentos del Análisis Económico,
Universidad de Alicante, E-03080 Alicante, Spain.

Received 31 July, 2004; accepted in revised form 5 September, 2004

Abstract: The new economic geography literature provides a general equilibrium framework that explains the emergence of economic agglomerations as a trade-off between increasing returns at the firm level and transportation costs related to the shipment of goods. We apply topological degree theory to investigate the existence of a fixed–point to this system of equations, which corresponds to the short–run equilibrium of the model. Further we study its uniqueness, and efficient computational methods to determine it.

Keywords: Economic Geography, Spatial Economy, Short–Run Equilibria, Fixed Points, Topological Degree.

Mathematics Subject Classification: 55M25, 65H10, 74G15, 91B72

1 The topological degree and its computation

Computing fixed points, or equivalently, solving systems of nonlinear equations has been a topic of great interest for researchers in the field of mathematics, engineering, economics, and many other professions. Numerous problems such as finding an equilibrium, a zero point, or a fixed point, can be formulated as the problem of finding a solution of an equation of the form $F(x) = p$ in an appropriate space. Topological degree theory provides means for examining this solution set and obtaining information on the existence of solutions, their number and their nature.

We outline topological degree theory for determining the exact number of zeros of a system of nonlinear transcendental equations by computing the value of the topological degree using Kronecker's integral [9] on Picard's extension [2, 7].

Suppose that a function $F_n = (f_1, f_2, \ldots, f_n)\colon \overline{\mathcal{D}}_n \subset \mathbb{R}^n \to \mathbb{R}^n$ is defined and twice continuously differentiable in an open and bounded domain $\mathcal{D}_n$ of $\mathbb{R}^n$ with boundary $\vartheta\mathcal{D}_n$. Suppose further that the zeros of the equation $F_n(x) = p$, where $p \in \mathbb{R}^n$ is a given vector, are not located on $\vartheta\mathcal{D}_n$, and

---

[1]Corresponding author. E-mail: npav@math.upatras.gr
[2]E-mail: vrahatis@math.upatras.gr
[3]E-mail: mossay@merlin.fae.ua.es

that they are simple, i.e., the determinant, $\det J_{F_n}$, of the Jacobian matrix of F_n at these points is non–zero.

Definition: The topological degree of F_n at p relative to $\mathcal{D}_n$ is denoted by $\deg[F_n, \mathcal{D}_n, p]$ and is defined by the following sum:

$$\deg[F_n, \mathcal{D}_n, p] = \sum_{x \in F_n^{-1}(p) \cap \mathcal{D}_n} \text{sgn}\big(\det J_{F_n}(x)\big), \tag{1}$$

where $\text{sgn}(\psi)$ defines the well known three valued sign function.

The topological degree is invariant under changes of the vector p in the sense that, for any $q \in \mathbb{R}^n$, it holds that: $\deg[F_n, \mathcal{D}_n, p] \equiv \deg[F_n - q, \mathcal{D}_n, p - q]$, where $F_n - q$ denotes the mapping $F_n(x) - q$, $x \in \mathcal{D}_n$ [6, p.157]. Thus, for simplicity, we consider the case where the topological degree is defined at the origin $\Theta_n = (0, \dots, 0)$ in $\mathbb{R}^n$.

The topological degree $\deg[F_n, \mathcal{D}_n, \Theta_n]$ can be represented by the Kronecker integral which is defined as:

$$\deg[F_n, \mathcal{D}_n, \Theta_n] = \frac{\Gamma(n/2)}{2\pi^{n/2}} \int \int \cdots \int_{\vartheta \mathcal{D}_n} \frac{\sum_{i=1}^{n} A_i dx_1 \dots dx_{i-1} dx_{i+1} \dots dx_n}{\left(f_1^{\,2} + f_2^{\,2} + \cdots + f_n^{\,2}\right)^{n/2}}, \tag{2}$$

where A_i define the following determinants:

$$A_i = (-1)^{n(i-1)} \det \left[F_n \quad \frac{\partial F_n}{\partial x_1} \quad \cdots \quad \frac{\partial F_n}{\partial x_{i-1}} \quad \frac{\partial F_n}{\partial x_{i+1}} \quad \cdots \quad \frac{\partial F_n}{\partial x_n} \right], \tag{3}$$

where, $\frac{\partial F_n}{\partial x_k} = \left(\frac{\partial f_1}{\partial x_k}, \frac{\partial f_2}{\partial x_k}, \dots, \frac{\partial f_n}{\partial x_k} \right)$ is the kth column of the determinant $\det J_{F_n}$ of the Jacobian matrix J_{F_n}.

The topological degree can be generalized when the function is only continuous [6]. In this case, Kronecker's theorem [6] states that $F_n(x) = \Theta_n$ has at least one zero in $\mathcal{D}_n$ if $\deg[F_n, \mathcal{D}_n, \Theta_n] \neq 0$. Furthermore, if $\mathcal{D}_n = \mathcal{D}_n^1 \cup \mathcal{D}_n^2$ where $\mathcal{D}_n^1$ and $\mathcal{D}_n^2$ have disjoint interiors and $F_n(x) \neq \Theta_n$ for all $x \in \vartheta \mathcal{D}_n^1 \cup \vartheta \mathcal{D}_n^2$, then the topological degree is additive, i.e.:

$$\deg[F_n, \mathcal{D}_n, \Theta_n] = \deg[F_n, \mathcal{D}_n^1, \Theta_n] + \deg[F_n, \mathcal{D}_n^2, \Theta_n].$$

Since $\deg[F_n, \mathcal{D}_n, \Theta_n]$ is equal to the number of zeros of $F_n(x) = \Theta_n$ that give positive determinant of the Jacobian matrix minus the number of zeros that give negative determinant of the Jacobian matrix, the total number $\mathcal{N}^r$ of zeros of $F_n(x) = \Theta_n$ would be equal to the value of $\deg[F_n, \mathcal{D}_n, \Theta_n]$ had all these zeros yielded the same sign of the determinant of the Jacobian matrix. Note that, by assumption, all the zeros of $F_n(x) = \Theta_n$ are simple. To this end, Picard proposed the following extension of the function F_n and the domain $\mathcal{D}_n$:

$$F_{n+1} = (f_1, \dots, f_n, f_{n+1}) \colon \mathcal{D}_{n+1} \subset \mathbb{R}^{n+1} \to \mathbb{R}^{n+1},$$

where $f_{n+1} = y \det J_{F_n}$, and $\mathcal{D}_{n+1}$ is the direct product of the domain $\mathcal{D}_n$ with an arbitrary interval of the real y–axis containing the point $y = 0$. Then the zeros of the following system of equations:

$$\begin{aligned} f_i(x_1, x_2, \dots, x_n) &= 0, \quad i = 1, \dots, n, \\ y \det J_{F_n}(x_1, x_2, \dots, x_n) &= 0, \end{aligned} \tag{4}$$

are the same as the zeros of $F_n(x) = \Theta_n$ provided that $y = 0$. Moreover, the determinant of the Jacobian matrix of (4) is equal to $[\det J_{F_n}(x)]^2$ which is always nonnegative (positive at the simple zeros). Thus we may conclude:

Theorem [7]: The total number $\mathcal{N}^r$ of zeros of $F_n(x) = \Theta_n$ is given by

$$\mathcal{N}^r = \deg[F_{n+1}, \mathcal{D}_{n+1}, \Theta_{n+1}], \tag{5}$$

under the hypotheses that F_n is twice continuously differentiable and that all the zeros are simple and lie in the strict interior of $\mathcal{D}_{n+1}$.

Several methods for the computation of the topological degree have been proposed [5, 9, 8]. These methods are based on Stenger's method that is an almost optimal complexity algorithm for some classes of functions [8].

2   Economic Geography

Lately the increasing interest in the field of economic geography has attracted many scientists from various disciplines ranging from economics to regional science and geography. There is no doubt that the building of the European Union and the several policy issues which come along have contributed to boost interest in the field.

The New Economic Geography has emerged from the long-existing need to explain concentrations of economic activity and thus of people. The literature in the field provides a general equilibrium framework explaining the emergence of economic agglomerations as a trade-off between increasing returns at the firm level and transportation costs related to the shipment of goods.

We consider a standard new economic geography model involving a finite number of regions (see [1]). This model can be viewed as the extension of Krugman's core-periphery model [3] to the case of a spatial economy consisting of N regions. Like in Krugman's original work, there are two sectors in the economy [3]. The agricultural sector employs farmers and produces a single homogeneous good under constant returns to scale. The manufacturing sector employs workers and produces a differentiated good, giving rise to manufacturing varieties. Consumers (workers and farmers) buy the agricultural good on a perfectly competitive national market and manufacturing varieties on monopolistically competitive regional markets. In addition, transporting manufacturing varieties from their production place to the place where they are consumed, is costly.

Economic equilibria define economic allocations and prices derived from optimal behaviors of firms and consumers that are compatible with market clearing. On the one hand, short-run equilibria are obtained under the assumption of no spatial adjustment. These short-run equilibria are thus viewed as implicitly determined by some given spatial distribution of labor. On the other hand long-run equilibria refer to steady states of a spatial economy where workers are allowed to adjust their location over time. In the case of a spatial economy consisting of 2 regions, a short-run equilibrium has been shown to exist and to be unique (see [4]), and the number and stability of steady states have been studied (see [1]). However, in the case of 3 regions or more, no analytical result concerning short- or long-run equilibria has been derived so far.

3   Proposed Approach

In this paper we investigate short–run equilibria of the general N–regional model. Regions are denoted by $i = 1, ..., N$. Consider some spatial distribution of labor L_i across these regions. The proportion of the labor force in region i is denoted as

$$\lambda_i = L_i / \sum_{j=1}^{N} L_j.$$

The variables of the model are y_i, θ_i, and w_i representing the income, the manufacturing price index, and the manufacturing wage in region i, respectively. The system of equations defining the short–run equilibria of the spatial economy can be written in the following reduced form:

$$y_i = \frac{1-\mu}{N} + \mu\lambda_i w_i,$$

$$\theta_i = \frac{1}{\left\{\displaystyle\sum_{j=1}^{N} \frac{\lambda_j}{w_j^{(\sigma-1)}} \exp\left[-\tau(\sigma-1)d(i,j)\right]\right\}^{\frac{1}{(\sigma-1)}}},$$

$$w_i = \left\{\sum_{j=1}^{N} y_j \theta_j^{\sigma-1} \exp\left[-\tau(\sigma-1)d(i,j)\right]\right\}^{1/\sigma},$$

where:

$d(i,j)$	:	distance between locations i and j,
σ	:	elasticity of substitution among manufacturing varieties,
μ	:	share of manufacturing expenditure,
τ	:	transportation cost per unit of distance for manufacturing goods.

In this contribution we propose to investigate the existence of a fixed point to this system of equations, to verify its uniqueness, and to apply efficient computational methods to determine it using tools from topological degree theory [5, 9, 8].

References

[1] Fujita, M., Krugman, P., and Venables, A., *The spatial economy, cities, regions and international trade*, MIT Press, 1999.

[2] Hoenders, B.J. and Slump, C.H., On the calculation of the exact number of zeros of a set of equations, *Computing*, 30, 1983, pp.137–147.

[3] Krugman, P., Increasing returns and economic geography, *The Journal of Political Economy*, 99, 3, (1991), pp.483–499.

[4] Mossay, P., The core–periphery model: Existence of short–run equilibria, Technical Report, Universidad de Alicante, Spain, 2004.

[5] Mourrain, B., Vrahatis, M.N., and Yakoubsohn, J.C., On the complexity of isolating real roots and computing with certainty the topological degree, *J. Complexity*, 18, 2, 2002, pp.612–640.

[6] Ortega, J.M. and Rheinbolt, W.C., *Iterative Solution of Nonlinear Equations in Several Variables*, Academic Press, New York, 1970.

[7] Picard, E., *Traité d'analyse*, 3rd ed., chap. 4.7., Gauthier–Villars, Paris, 1922.

[8] Sikorski, K., *Optimal Solution of Nonlinear Equations*, Oxford Press, 2000.

[9] Stenger, F., Computing the topological degree of a mapping in $\mathbb{R}^n$, *Numer. Math.*, 25, 1975, pp.23–38.

VSP International
Science Publishers
P.O. Box 346, 3700 AH Zeist
The Netherlands

*Lecture Series on Computer
and Computational Sciences*
Volume 1, 2004, pp. 884-887

Unsupervised Clustering Using Semi-Algebraic Data Structures

D.K. Tasoulis[1] and M.N. Vrahatis[2]

Department of Mathematics,
University of Patras Artificial Intelligence Research Center (UPAIRC),
University of Patras, GR-26110 Patras, Greece

Received 31 July, 2004; accepted in revised form 5 September, 2004

Abstract: A clustering algorithm named k-windows clustering algorithm has been recently proposed [14]. The k-windows algorithm attempts to enclose all the patterns that belong to a single cluster within a d–dimensional window. In this contribution we propose to modify this algorithm by using semi algebraic data structures instead of windows.

Keywords: Unsupervised clustering, cluster analysis, range searching, semi-algebraic data structures.

Mathematics Subject Classification: 68P05, 68T10, 62H30.

1 Introduction

Clustering techniques were originally conceived by Aristotle and Theophrastos in the fourth century B.C. and in the 18th century by Linnaeus [8], but it was not until 1939 that one of the first comprehensive foundations of these methods was published [13]. Clustering can be defined as the process of partitioning a set of patterns into disjoint and homogeneous meaningful groups, called clusters. Clustering is fundamental in knowledge acquisition, and has been applied in numerous fields.

A fundamental issue in cluster analysis, independent of the particular clustering technique applied, is the determination of the number of clusters present in a data set. This issue remains an unsolved problem in cluster analysis. For instance well–known and widely used iterative techniques, such as the k-means algorithm [7], require from the user to specify the number of clusters present in the data prior to the execution of the algorithm.

Even the simplest clustering problems are known to be NP-Hard [1]. The Euclidean k-center problem in the plane is NP-Hard [9]. This problem can be defined as: *Given a set S of n points in a d–dimensional metric space $(\mathbb{R}^d, \rho)$ and an integer k, compute a partition Σ of S into k subsets $S_1, \ldots, S_k$, such that Σ has the smallest possible size.* We define the *size* of a cluster S_i to be the maximum distance (under the ρ-metric) between a fixed point c_i called *center of the cluster* and a point of S_i. The size of a partition is defined to be the maximum size of a cluster in the partition.

Recently the k-windows clustering algorithm [14], has been extended [3, 4, 11, 12] in order to be able to automatically determine the number of clusters present in a dataset. The k-windows algorithm attempts to place a window over all the patterns that belong to a single cluster. In this

---

[1]Corresponding author. Email: dtas@math.upatras.gr
[2]Email: vrahatis@math.upatras.gr

contribution we propose to modify this algorithm by using semi algebraic data structures instead of windows.

2 Unsupervised k-windows clustering algorithm

For completeness purposes we briefly describe the workings of the original k-windows algorithm and its extension to automatically determine the number of clusters. Intuitively, the k-windows algorithm tries to capture all the patterns that belong to one cluster within a d-dimensional window (frame, box) [14]. To meet this goal it uses two fundamental procedures "movement" and "enlargement". During the movement procedure each window is centered at the mean of patterns that are included in it. The movement procedure is iteratively executed as long as the distance between the new and the previous center exceeds the user–defined variability threshold, θ_v. On the other hand, the enlargement process tries to augment the window to include as many patterns from the current cluster as possible. To this end, enlargement takes place at each coordinate separately. Each range of a window is enlarged by a proportion θ_e/l, where θ_e is user–defined and l stands for the number of previous successful enlargements. To consider an enlargement successful firstly the movement procedure is invoked, and after it terminates the proportional increase in the number of patterns included in the window is calculated. If this proportional increase exceeds the user–defined coverage threshold, θ_c, then the enlargement is considered successful. In this case, if the successful enlargement was for coordinate $c' \geqslant 2$, then all coordinates c'', such that $c'' < c'$, undergo enlargement assuming as initial position the current position of the window. Otherwise, the enlargement and movement steps are rejected and the position and size of the d–range are reverted to their prior to enlargement values. In Figure 1 the two processes are illustrated.

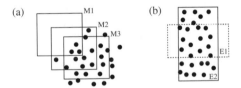

Figure 1: (a) Sequential movements M2, M3 of initial window M1.
(b) The enlargement process. E1 enlargement is rejected while E2 is accepted.

A fundamental issue in cluster analysis, independent of the particular clustering technique applied, is the determination of the number of clusters present in a dataset. This issue remains an unsolved problem in cluster analysis. For instance well–known and widely used iterative techniques, such as the k-means algorithm [7], require from the user to specify the number of clusters present in the data prior to the execution of the algorithm. On the other hand, the unsupervised k-windows algorithm generalizes the original algorithm by endogenously determining the number of clusters. The key idea to achieve this is to apply the k-windows algorithm using a "sufficiently" large number of initial windows. The windowing technique of the k-windows algorithm allows for a large number of initial windows to be examined, without any significant overhead in time complexity. Once movement and enlargement of all windows terminate, all overlapping windows are considered for merging. The merge operation is guided by two thresholds the merge threshold, θ_m, and the similarity threshold, θ_s. Having identified two overlapping windows, the number of patterns that lie in their intersection is computed. Next, the proportion of this number to the total patterns included in each window is calculated. If the mean of these two proportions exceeds θ_s, then the two windows are considered to be identical and the one containing the smaller number of points is deleted. Otherwise, if the mean exceeds θ_m, then the windows are considered to belong to the same cluster and are merged. This operation is illustrated in Figure 2; the extent of overlapping between windows W_1 and W_2 exceeds the threshold criterion and the algorithm considers both to belong to a single cluster, unlike windows W_3 and W_4, which capture two different clusters. On the other hand the extent of overlapping of windows W_5 and W_6 exceeds the θ_s threshold thus W_6 is deleted. For a comprehensive description of the algorithm and an investigation of its capability to automatically identify the number of clusters present in a dataset, refer to [3, 4, 11, 12].

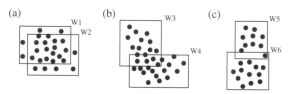

Figure 2: (a) W_1 and W_2 satisfy the similarity condition and W_1 is deleted.
 (b) W_3 and W_4 satisfy the merge operation and are considered to belong to the same cluster.
 (c) W_5 and W_6 have a small overlapment and capture two different clusters.

The computational complexity of the algorithm depends on the computational complexity of the range searches. To make this step time efficient a technique from Computational Geometry [2, 10] is employed. This technique constructs a multi–dimensional binary tree (kd-tree) for the data at a preprocessing step and traverses this tree to solve the Orthogonal Range Search Problem. From the performance viewpoint, the kd-tree requires, optimally, $\theta(dn)$ storage and can be optimally constructed in $\theta(dn\log n)$ time, where d is the dimension of the data and n is the number of patterns. The worst–case behavior of the query time is $\mathcal{O}(|A| + dn^{1-1/d})$ [10] where A is the set containing the points belonging to the specific d–range.

3   Clustering Using Semi-Algebraic Data Structures

The aim of this paper is to replace the windowing technique of k–windows algorithm, by using data structures different from d–ranges. Using semi–algebraic structures like simplices, spheres or ellipsoids, the clustering ability of the algorithm can be enhanced since these kind of structures are able to capture clusters with non–trivial shapes. The difficulty that arises, though is the efficiency of the range searching.

More specifically unlike the orthogonal range searching no simplex range searching data structure is known that can answer a query in polylogarithmic time using near–linear storage. Thus,

Range	Storage	Query Time
Simplex	$n\log(n)$	$\log n$
Ball	m	$n/m^{1/\lceil d/2\rceil}\log^c(n) + |A|$
Γ_f	n	$n^{1-1/(2d-3)+\epsilon}$

Table 1: Lower bounds for range searching.

by considering firstly how fast a simplex range search query can be answered using a linear–size data structure and secondly how large a data structure should be in order to answer a query in polylogarithmic time, a space to query time trade-off can be computed. Chazelle [5] in a series of papers, proved nontrivial lower bounds on simplex range searching using various mathematical techniques (see Table 1). By allowing the ranges to be bounded by nonlinear functions, balls and ellipsoids can be used by the clustering procedure. In detail, ball range searching in $\mathbb{R}^d$ can be formulated as half-space range searching in $\mathbb{R}^{d+1}$ [6] and the corresponding complexity bounds are reported in Table 1. A natural class of more general ranges is the family of Tarski cells [6], defined as $\Gamma_f = \{x \in \mathbb{R}^d | f(x) \geqslant 0\}$, where $f(x)$ is a d-variate polynomial specifying the type of ranges (disks, cylinders, cones, ellipsoids, etc.). For this kind of ranges the lower bounds are also reported in Table 1.

References

[1] Agarwal, P.K. and Procopiuc, C.M., Exact and approximation algorithms for clustering, In: Proceedings of the *9th Symposium on Discrete Algorithms*, pp.658–667, 1998.

[2] Alevizos, P., An algorithm for orthogonal range search in $d \geqslant 3$ dimensions, In: Proceedings of the *14th European Workshop on Computational Geometry*, Barcelona, 1998.

[3] Alevizos, P., Boutsinas, B., Tasoulis, D.K., and Vrahatis, M.N., Improving the orthogonal range search k-windows clustering algorithm, In: Proceedings of the *14th IEEE International Conference on "Tools with Artificial Intelligence"*, pp.239–245, Washington D.C., 2002.

[4] Alevizos, P., Tasoulis, D.K., and Vrahatis, M.N., Parallelizing the unsupervised k-windows clustering algorithm, In: R. Wyrzykowski *et al.* (Eds.), *Lecture Notes in Computer Science*, vol. **3019**, pp.225–232, Springer-Verlag, 2004.

[5] Chazelle, B., A functional approach to data structures and its use in multidimensional searching, *SIAM J. Comput.*, **17**, 427–462, 1988.

[6] Goodman, J.E. and O'Rourke, J., (Eds.), *Handbook of Discrete and Computational Geometry*, CRC Press LLC, Boca Raton, FL, 1997.

[7] Hartigan, J.A. and Wong, M.A., A k-means clustering algorithm, *Applied Statistics*, **28**, 100–108, 1979.

[8] Linnaeus, C., *Clavis Classium in Systemate Phytologorum in Bibliotheca Botanica*, Amsterdam, The Netherlands: Biblioteca Botanica, 1736.

[9] Megiddo, N. and Supowit, K.J., On the complexity of some common geometric problems, *SIAM J. Comput.*, **13**, 182–196, 1984.

[10] Preparata, F. and Shamos, M., *Computational Geometry*, Springer Verlag, NY, Berlin, 1985.

[11] Tasoulis, D.K., Alevizos, P., Boutsinas, B., and Vrahatis, M.N., Parallel unsupervised k-windows: an efficient parallel clustering algorithm, In: V. Malyshkin (Ed.), *Lecture Notes in Computer Science*, vol. **2763**, pp.336–344, Springer-Verlag, 2003.

[12] Tasoulis, D.K. and Vrahatis, M.N., Unsupervised distributed clustering. In: Proceedings of the *IASTED International Conference on "Parallel and Distributed Computing and Networks"*, pp.347–351, Innsbruck, Austria, 2004.

[13] Tryon, C., *Cluster Analysis*, Ann Arbor, MI: Edward Brothers, 1939.

[14] Vrahatis, M.N., Boutsinas, B., Alevizos, P., and Pavlides, G., The new k-windows algorithm for improving the k-means clustering algorithm, *J. Complexity*, **18**, 375–391, 2002.

VSP International
Science Publishers
P.O. Box 346, 3700 AH Zeist
The Netherlands

*Lecture Series on Computer
and Computational Sciences*
Volume 1, 2004, pp. 888-889

Approaches and Methods of Security Engineering

Tai-hoon Kim[1]

Sansung Gongsa,
San 7, Geoyeo-Dong, Songpa-Gu, Seoul, Korea

Abstract: The general systems of today are composed of a number of components such as servers and clients, protocols, services, and so on. Systems connected to network have become more complex and wide, but the researches for the systems are focused on the 'performance' or 'efficiency'. While most of the attention in system security has been focused on encryption technology and protocols for securing the data transaction, it is critical to note that a weakness (or security hole) in any one of the components may comprise whole system. Security engineering is needed for reducing security holes may be included in the software. There are very many approaches or methods in software development or software engineering. Therefore, more security-related researches are needed to reduce security weakness may be included in the software and complement security-related considerations of general software engineering.

Keywords: Security engineering, Security hole, Security weakness.

1. Security Engineering for Information Assurance

There are many standards, methods and approaches which are used for assuring the quality of software. ISO/IEC TR 15504, the Software Process Improvement Capability Determination (SPICE), provides a framework for the assessment of software processes [1-2]. But, in the ISO/IEC TR 15504, considerations for security are relatively poor to others. For example, the considerations for security related to software development and developer are lacked.

When we are making some kinds of software products, ISO/IEC TR 15504 may provide a framework for the assessment of software processes, and this framework can be used by organizations involved in planning, monitoring, controlling, and improving the acquisition, supply, development, operation, evolution and support of software. But, in the ISO/IEC TR 15504, considerations for security are relatively poor to other security-related criteria such as ISO/IEC 21827, the Systems Security Engineering Capability Maturity Model (SSE-CMM), or ISO/IEC 15408, Common Criteria (CC) [3-6]. Security-related software development is concerned with many kinds of measures that may be applied to the development environment or developer to protect the confidentiality and integrity of the IT product or system developed.

It is essential that not only the customer's requirements for software functionality should be satisfied but also the security requirements imposed on the software development should be effectively analyzed and implemented in contributing to the security objectives of customer's requirements. Unless suitable requirements are established at the start of the software development process, the resulting end product, however well engineered, may not meet the objectives of its anticipated consumers. The IT products like as firewall, IDS (Intrusion Detection System) and VPN (Virtual Private Network) which made to perform special functions related to security are used to supply security characteristics. But the method using these products may be not the perfect solution. Therefore, when making some kinds of software products, security-related requirements must be considered.

2. Approaches and Methods of Security Engineering

Assurance methods are classified in Fig.1 according to the three assurance approach categories. Depending on the type of assurance method, the assurance gained is based on the aspect assessed and the lifecycle phase. Assurance approaches yield different assurance due to the deliverable (IT

---

[1] Corresponding author. Sansung Gonsa, San 7, Geoyeo-Dong, Songpa-Gu, Seoul, Korea, E-mail: taihoonn@empal.com

component or service) aspect examined. Some approaches examine different phases of the deliverable lifecycle while others examine the processes that produce the deliverable (indirect examination of the deliverable). Assurance approaches include facility, development, analysis, testing, flaw remediation, operational, warranties, personnel, etc. These assurance approaches can be further broken down; for example, testing assurance approach includes general testing and strict conformance testing assurance methods [3].

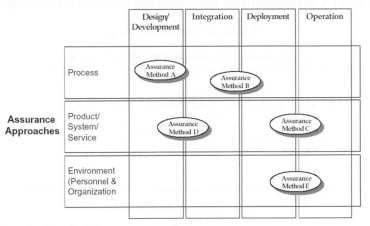

Figure 1: Categorization of existing assurance methods.

References

[1] ISO. ISO/IEC TR 15504-2:1998 Information technology – Software process assessment – Part 2: A reference model for processes and process capability

[2] ISO. ISO/IEC TR 15504-5:1998 Information technology – Software process assessment – Part 5: An assessment model and indicator guidance

[3] ISO. ISO/IEC 21827 Information technology – Systems Security Engineering Capability Maturity Model (SSE-CMM)

[4] ISO. ISO/IEC 15408-1:1999 Information technology - Security techniques - Evaluation criteria for IT security - Part 1: Introduction and general model

[5] ISO. ISO/IEC 15408-2:1999 Information technology - Security techniques - Evaluation criteria for IT security - Part 2: Security functional requirements

[6] ISO. ISO/IEC 15408-3:1999 Information technology - Security techniques - Evaluation criteria for IT security - Part 3: Security assurance requirements

[7] Tai-Hoon Kim, Byung-Gyu No, Dong-chun Lee, Threat Description for the PP by Using the Concept of the Assets Protected by TOE, ICCS 2003, LNCS 2660, Part 4, pp. 605-613

[8] Tai-hoon Kim and Haeng-kon Kim: The Reduction Method of Threat Phrases by Classifying Assets, ICCSA 2004, LNCS 3043, Part 1, 2004.

[9] Tai-hoon Kim and Haeng-kon Kim: A Relationship between Security Engineering and Security Evaluation, ICCSA 2004, LNCS 3046, Part 4, 2004.

[10] Eun-ser Lee, Kyung-whan Lee, Tai-hoon Kim and Il-hong Jung: Introduction and Evaluation of Development System Security Process of ISO/IEC TR 15504, ICCSA 2004, LNCS 3043, Part 1, 2004

VSP International
Science Publishers
P.O. Box 346, 3700 AH Zeist
The Netherlands

*Lecture Series on Computer
and Computational Sciences*
Volume 1, 2004, pp. 890-892

Security Considerations for Securing Software Development

Eunser Lee[1], Sunmyoung Hwang[2]

[1]Chung-Ang University, 221, Huksuk-Dong, Dongjak-Gu, Seoul, Korea
[2]Daejeon University, 96-3 Yongun-dong, Tong-gu, Taejon 300-716, South Korea

Abstract: Security and assurance are achieved when there is confidence that information and information systems are protected against attacks through the appropriate security services or solutions in such areas as availability, integrity, authentication, confidentiality, and non-repudiation. When developing some software, security requirements should be considered at the same time customers' requirements are processed. This paper proposes some security considerations which may be included when software is developed.

Keywords: Security Requirement, Software Development, Common Criteria, Security Function.

1. Introduction

When we are making some kinds of software products, ISO/IEC TR 15504 may provide a framework for the assessment of software processes, and this framework can be used by organizations involved in planning, monitoring, controlling, and improving the acquisition, supply, development, operation, evolution and support of software. But, in the ISO/IEC 15504 [1], considerations for security are relatively poor to other security-related criteria such as ISO/IEC 21827, the Systems Security Engineering Capability Maturity Model (SSE-CMM) [2], or ISO/IEC 15408, Common Criteria (CC) [3]. Security-related software development is concerned with many kinds of measures that may be applied to the development environment or developer to protect the confidentiality and integrity of the IT product or system developed.

It is essential that not only the customer's requirements for software functionality should be satisfied but also the security requirements imposed on the software development should be effectively analyzed and implemented in contributing to the security objectives of customer's requirements. Unless suitable requirements are established at the start of the software development process, the resulting end product, however well engineered, may not meet the objectives of its anticipated consumers. The IT products like as firewall, IDS (Intrusion Detection System) and VPN (Virtual Private Network) which made to perform special functions related to security are used to supply security characteristics. But the method using these products may be not the perfect solution. Therefore, when making some kinds of software products, security-related requirements must be considered.

This paper proposes a security engineering based approach considering security when developing software. And this paper proposes a concept of security requirement appended to customer's requirement.

2. Security requirements for Software

2.1 General Software Development Process

There are many methodologies for software development, and security engineering does not mandate any specific development methodology or life cycle model. Fig.1 depicts underlying assumptions about the relationship between the customer's requirements and the implementation. The figure is used to

[1] Chung-Ang University, 221, Huksuk-Dong, Dongjak-Gu, Seoul, Korea, E-mail : eslee@object.cau.ac.kr
[2] Daejeon University, 96-3 Yongun-dong, Tong-gu, Taejon 300-716, South Korea, E-mail : sunhwang@dju.ac.kr

provide a context for discussion and should not be construed as advocating a preference for one methodology (e.g. waterfall) over another (e.g. prototyping).

It is essential that the requirements imposed on the software development be effective in contributing to the objectives of consumers. Unless suitable requirements are established at the start of the development process, the resulting end product, however well engineered, may not meet the objectives of its anticipated consumers. The process is based on the refinement of the customer's requirements into a soft-ware implementation. Each lower level of refinement represents design decomposition with additional design detail. The least abstract representation is the software implementation itself.

In general, customer does not mandate a specific set of design representations. The requirement is that there should be sufficient design representations presented at a sufficient level of granularity to demonstrate where required:

a) that each refinement level is a complete instantiation of the higher levels (i.e. all functions, properties, and behaviors defined at the higher level of abstraction must be demonstrably present in the lower level);

b) that each refinement level is an accurate instantiation of the higher levels (i.e. there should be no functions, proper-ties, and behaviors defined at the lower level of abstraction that are not required by the higher level).

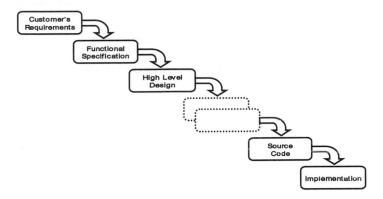

Fig.1 the relationship between the customer's requirements and the implementation

2.2 Append Security Requirements

For the development of software, the first objective is the perfect implementation of customer's requirements. And this work may be done by very simple processes. How-ever, if the software developed has some critical security holes, the whole network or systems that software installed and generated are very vulnerable.

Therefore, developers or analyzers must consider some security-related factors and append a few security-related requirements to the customer's requirements. Fig.2 depicts the idea about this concept.

The processes based on the refinement of the security-related requirements are considered with the processes of soft-ware implementation.

2.3 Implementation of Security Requirements

Developers can reference the ISO/IEC 15408, Common Criteria (CC), to implement security-related requirements appended.

The multipart standard ISO/IEC 15408 defines criteria, which for historical and continuity purposes are referred to herein as the CC, to be used as the basis for evaluation of security properties of IT products and systems. By establishing such a common criteria base, the results of an IT security evaluation will be meaningful to a wider audience.

The CC will permit comparability between the results of independent security evaluations. It does so by providing a common set of requirements for the security functions of IT products and systems and for assurance measures applied to them during a security evaluation. The evaluation process establishes a level of confidence that the security functions of such products and systems and the assurance measures applied to them meet these requirements. The evaluation results may help consumers to determine whether the IT product or system is secure enough for their intended application and whether the security risks implicit in its use are tolerable.

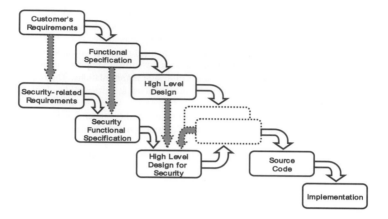

Fig.2 Append security-related requirements

3. Conclusion and Future Work

This paper proposes a method appending some security-related requirements to the customer's requirements. For the development of software, the first objective is the perfect implementation of customer's requirements. However, if the software developed has some critical security holes, the whole network or systems that software installed and generated may be very vulnerable. Therefore, developers or analyzers must consider some security-related factors and append a few security-related requirements to the customer's requirements.

For the future work, the processes based on the refinement of the security-related requirements must be considered with the processes of software implementation.

References

[1] ISO, ISO/IEC TR 15504-5:1998 Information technology – Software process assessment – Part 5: An assessment model and indicator guidance.

[2] ISO, ISO/IEC 21827 Information technology – Systems Security Engineering Capability Maturity Model (SSE-CMM).

[3] ISO, ISO/IEC 15408: Information technology - Security techniques - Evaluation criteria for IT security, 1999.

[4] Tai-hoon Kim and Haeng-kon Kim: A Relationship between Security Engineering and Security Evaluation, ICCSA 2004, LNCS 3046, Part 4, 2004.

[5] Eun-ser Lee, Kyung-whan Lee, Tai-hoon Kim and Il-hong Jung: Introduction and Evaluation of Development System Security Process of ISO/IEC TR 15504, ICCSA 2004, LNCS 3043, Part 1, 2004.

VSP International
Science Publishers
P.O. Box 346, 3700 AH Zeist
The Netherlands

*Lecture Series on Computer
and Computational Sciences*
Volume 1, 2004, pp. 893-895

Common Development Security Requirements to Improve Security of IT Products

Sang ho Kim[1], Choon seong Leem[2]

[1]KISA, 78, Garak-Dong, Songpa-Gu, Seoul, Korea
[2]Yonsei University, 134, Shinchon-Dong, Seodaemun-Gu, Seoul, Korea

Abstract: T Development security has been more important to make secure IT products maintaining integrity and confidentiality during development process. However, standards which specify development security requirements are not detailed to use and differences also exist among requirements of each standard. In this paper, we present common development security requirements to maintain confidentiality and integrity of desired IT products in development environment. These requirements are induced through the analysis of security standards such as Common Criteria, BS 7799, and SSE-CMM. We believe that common development requirements we suggest contribute on improving security of IT products.

Keywords: Security Requirements, Security Improvement, Common Criteria

1. Introduction

Security has been a crucial issue for IT products. As the use of IT products arises, the demands to secure these products also arise. This is the reason that IT products usually contain private data which have to be available only to authorized user. Developers usually have been defined security functional requirements which should be in-luded in IT Products for given particular environment against expected threats identified by risk analysis. However, Non-functional security requirements like development security are still ignored. It causes security breaches to IT products.

In this paper, we focus on development security among non-functional security requirements. There are at least two reasons for the lack of development security in applying to development of IT products. Firstly, development security requirements are generally difficult to define. Standards which specify development security requirements are not detailed to use and differences also exist among requirements of each standard. Common Criteria(CC, ISO/IEC15408) [1], international standard for evaluation of security properties of IT products and systems, specify development security requirements in ALC_DVS components of assurance requirements but these requirements are not specific and even ambiguous. BS7799-1 [2], SSE-CMM [3] also shows requirements for the development security management, focusing on security management of organization. Those can't be directly applied to development security of IT products. Secondly, developer lacks expertise for development security. Many developers tend to consider security functional features but rarely review development security issues.

In this paper, we present common development security requirements to maintain confidentiality and integrity of desired IT products during development process. These requirements are induced through the analysis of security standards such as Common Criteria, BS 7799 and SSE-CMM.

2. Related Works

2.1 Common Criteria

A Common Criteria is an international standard to be used as the basis for evaluation of security properties of IT products. It defines development security in ALC_DVS of assurance requirements.

---

[1] KISA, 78, Garak-Dong, Songpa-Gu, Seoul, Korea, E-mail : shkim@kisa.or.kr
[2] Yonsei University, 134, Shinchon-Dong, Seodaemun-Gu, Seoul, Korea

Items	Requirements
ALC_DVS.2. 1	Describe all the physical, procedural, personnel, and other security measures that are necessary to protect the confidentiality and integrity of the TOE(Target of Evaluation) d in its development environment.
ALC_DVS.2. 2	Provide evidence that these security measures are followed during the development and maintenance of the TOE.
ALC_DVS.2. 3	Security measures provide the necessary level of protection to maintain the confidentiality and integrity of the TOE

2.2 BS 7799-1

BS7799-1 is also an international standard for controls of information security management of organization. Requirements for development security are specified in personnel security and physical & environmental security section repetitively.

Personnel security	Physical & environmental security section
·Security roles and responsibilities of job & employment ·Personnel screening, confidentiality agreement ·Security education and training	·Physical security perimeter ·Physical entry controls ·Securing offices, rooms and facilities ·Working in secure area ·Equipment security

2.3 SSE-CMM

SSE-CMM is a standard related to security engineering practices in software life cycle. Development security requirements are described in BPs(Base Practices) of PA(Process Area)01, PA09, PA10, PA20.

PA	BPs
PA01;Administer Security Control	BP.01.01 , BP01.03, BP01.04
PA09; Provide security Input	BP.09.05, BP.09.06
PA10; Specify Security Needs	BP.10.02, BP.10.05
PA20; Manage System Engineering Support Environment	BP02, BP03, BP06

3. Common requirements for development security

We propose common requirement for development as a blow framework. Development security requirements are composed of three categories and their elements.

Figure1. Common development security Framework

3.1 Personal Security

Requirements of personal security are defined as PES1, PES2, and PES3.

PES.1 Security management	1.1 Recruit & Retire developer management 1.2 Developer role & responsibility definition 1.3 Outsourced developer management

PES2. Security awareness	2.1 Security awareness program
	2.2 Periodical program execution
PES3. Security account management	3.1 Account create/modification/delete
	3.2 Password guideline
	3.3 Access privileges

3.2 Physical Security

Requirements of physical security are defined as PHS1, PHS2, and PHS3.

PHS1 Develop site access control	1.1 Access limited area setting
	1.2 Access control equipments
	1.3 System protection measures
PHS2. Network security	1.1 Network security policy
	1.2 Network protection equipment
	1.3 Periodic vulnerability analysis & Audit
PHS3. Documents protection facility	1.1 Access control equipment for
	1.2 Periodic back-up & audit

3.3 Procedural Security

Requirements of procedural security are defined as PRS1, PRS2, and PRS3.

PRS 1. Visitor access control	1.1 Visitor lists & escorts
	1.2 Security badges
PRS 2. Product distribution	1.1 Production line security
	1.2 Secure distribution rout
	1.3 Stock security management
PRS 3. Incident Handling	1.1 Emergency contact points
	1.2 Response & Recovery procedure

4. Conclusions and Future works

In this paper, common development security requirements have been provide to maintain integrity and confidentiality during development process of IT products. We believe that common development security requirements we suggest contribute on improving security of IT products and reduce a security breach. More research for detailed metrics of security developments is future works.

References

[1] ISO. ISO/IEC 15408-3:1999 Information technology - Security techniques - Evaluation criteria for IT security - Part 3: Security assurance requirements

[2] BSI (UK), BS7799-1: Information security management-Part1: Code of practice for information security management, 1999.

[3] Carnegie Mellon University, Systems Security Engineering Capability Maturity Model, Version 3.0, 2003.

[4] ISO. ISO/IEC TR 13335, Information Technology Guidelines for the management of IT Security, 1996.

[5] Sang ho Kim, SSE-CMM BPs to Meet the Requirements of ALC_DVS.1 Component in CC, Page 1069-1075, Springer LNCS. 2003.

[6] Sang ho Kim, Supplement of Security-Related Parts of ISOIEC TR 15504, Page 1084-1089, Springer LNCS. 2003.

VSP International
Science Publishers
P.O. Box 346, 3700 AH Zeist
The Netherlands

*Lecture Series on Computer
and Computational Sciences*
Volume 1, 2004, pp. 896-899

Decision Supporting Method with the Analytic Hierarchy Process Model for the Systematic Selection of COTS-based Security Controls

Sangkyun Kim[1]
Somansa,
Woolim e-Biz center, 16, Yangpyeongdong 3-ga, Yeongdeungpogu, Seoul 150-103, Korea

Choon Seong Leem[2]
Department of Computer and Industrial Engineering,
Yonsei University,
134, Shinchondong, Seodaemoongu, Seoul 129-749, Korea

Received 31 July, 2004; accepted in revised form 25 August, 2004

Abstract: The successful management of information security within an organization is vital to its survival and success. The effective implemenation of COTS (Commercial, Off-The-Shelf)-based security controls is one of the critical success factors of information security managements. This paper presents the formal method which provides the process of selection and the criteria for evaluation of COTS-based security controls for the effectiveness and the efficiency of decision making of corporate managers. A case study proves practical values of this paper.

Keywords: AHP; Decision Support; COTS; Security Control

Mathematics SubjectClassification: 68U99
PACS: 89.70.+c

1. Previous Researches

The previous methodologies which are famous as a selection methodology of information systems such as METHOD/1, ASAP and VIP2000 do not consider security issues, but only focus on reliability and usability issues [1, 2, 3, 4]. TCSEC, ITSEC and CC deal with issues of functionality or effectiveness of security product itself. Because the objective of these evaluation schemes is to supply official certification of particular security controls, it's difficult to use these evaluation schemes when an organization evaluates and selects security controls which have similar levels of certification. Several methods have been proposed in previous researches to characterize and to guide selection of COTS-based security tools. A summary of these researches is shown in table 1 [5, 6, 7, 8].

Table 1: Previous researches on the selection of security controls

Research	Objective	Limits
Gilbert, 1989	Guide for selecting automated risk analysis tools	Only focus on specific security products
Polk, 1992	Guide for selecting anti-virus tools and techniques	
Barbour, 1996	SCE(Software Capability Evaluation) implementation guide for supplier selection	Absence of an evaluation and selection criteria
Henze, 2000	Implementation procedures of IT security safeguards	

---

[1] Corresponding author, Member of the Research Group of Security Engineering, E-mail: saviour@somansa.com
[2] E-mail: leem@yonsei.ac.kr

2. Process Model

In this paper we propose the process model which includes eight steps. The logic of these steps which provide tactics-level plannings is derived from the researh of METHOD/1, ASAP and VIP2000 [1, 2, 3]. The activities and tasks which provide operational plannings are derived from the research of Gilbert, Polk, Barbour and Henze [5, 6, 7, 8]. Key characteristics of eight steps are: Step 1) Requirement analysis: Technical and administrative requirements are consolidated. Technical requirements include the types of database, communication protocol, manipulation structure, interoperability, and functional list. Administrative requirements include the allocatable resources, legal liability, business needs, and constraints; Step 2) Introduction planning: An introduction plan is based on a requirement analysis. It arranges resources and provides time plans from a RFP development to operation phase. The introduction team is organized with experienced and trained members. Roles and responsibilities of the team are defined considering an introduction plan and characteristics of each member; Step 3) RFP development: A RFP is developed considering internal requirements and market environments. Internal requirements are consolidated during a requirement analysis phase. Market environments include informations on what kinds of vendors and products are available, and the industrial best practices. Criteria might be used in this phase; Step 4) Proposal receipt: The introduction team identifies solution providers who have a public confidence, and judges bidders who will receive a RFP. A NDA must be signed before sending a RFP and other related materials; Step 5) Bidders' briefing: The introduction team conducts a review on presented proposals, vendor presentations, interviews, and benchmarking test; Step 6) Judgement & contract: The introduction team consolidates data, makes judgement on vendor and product, and produces a report. Finally, the introduction team makes a contract considering technical and administrative requirements; Step 7) Introduction: A prototype may be required to assure an operational performance in production environments. An installation and data migration are conducted with detailed testing and revising activities. The administrators should be trained before an operation phase. Finally, an accreditation is required based on contract terms; Step 8) Operation: An awareness training of users, auditing of violation, compliance checking of legal liability and security policy, and change management of security related features should be performed.

3. Evaluation Criteria

We make the first level of evaluation criteria with a concept of S3IE provided in VIP2000. The factors of the second and third level of evaluation criteria are composed of the factors of Kavanaugh, Beall and Firth [9, 10, 11]. The software quality attributes of Mario and ISO9126 are also included in the third level of evaluation criteria [12]. The evaluation criteria are described in table 2.

Table 2: Evaluation criteria

1st level	2nd level	3rd level
Credibility of supplier	track record	market share, certification, relationship
	speciality	security expertise, solution lineup, best practice, turnkey solution
	coverage	geographic coverage
Competitiveness of product	sales condition	price, marketing program, maintenance, support services
	architecture	H/W requirement, OS supported, source language, source code available, NOS supported, protocols supported, component model supported
	function	preventive function, detective function, deterrent, recovery function, corrective function
	performance	functionality, reliability, usability, efficiency, maintainability, portability
Continuity of service	vendor stability	financial stability, vision and experience of the management staff
	contract terms	warranty, product liability

4. A Case Study

In this case study, the AHP method was applied to a particular project in which XYZ Co. Ltd. wanted to implement information security systems. XYZ had a planning to introduce a firewall. In this study, there was no relationship between vendors and XYZ, and vendors and products were treated independently.

The process model provided in section 2 was used to support introduction steps. XYZ used selection criteria for a RFP development, judgement, and making a contract. The first step was to develop a hierarchical structure of the problem. It classifies the goal, all decision criteria and variables into four major levels. The first level of the hierarchy is the overall goal to select the best firewall. Level 2 and Level 3 represent the evaluation criteria in selecting the firewall. Level 4 contains the decision alternatives that affect the ultimate selection of chosen firewalls. The judgements were elicited from the security experts in the security solution providers and government agencies. Expert Choice was used to facilitate comparisons of priorities. For example, the competitiveness of product was the most important criteria in Level 2. After inputting the criteria and their importance into Expert Choice, the priorities of each set of judgements were found and recorded in figure 1.

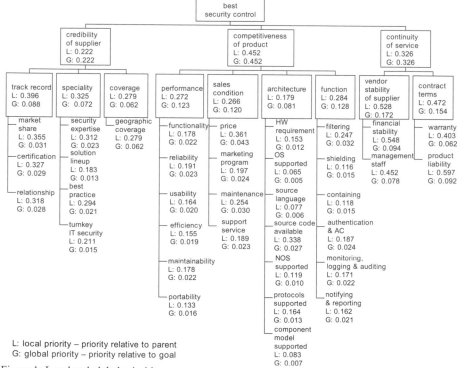

L: local priority – priority relative to parent
G: global priority – priority relative to goal

Figure 1: Local and global priorities

The overall consistency of the input judgements at all levels is within the acceptable ration of 0.1, as recommended by Saaty [13]. The overall priorities of firewall alternatives were calculated by multiplying their global priority with the corresponding weight along the hierarchy. When we synthesised all elements using Expert Choice, we obtained the result shown in figure 2. It shows that Firewall-C scored the highest in the result, followed by Firewall-S and Firewall-N.

Synthesis with respect to:
Best Security Control(firewall)

Firewall- C 0.361
Firewall- N 0.317
Firewall- S 0.321

Overall inconsistency = .01

Figure 2: Synthesis for firewall selection problem

The analysis of sensitivity was conducted to check the impact of change in the input data or parameters of the proposed firewalls. We found that relatively small changes in the hierarchy or judgement might lead to different outcomes. Figure 3 shows a gradient sensitivity analysis of the alternative priorities with respect to changes in the relative importance of the criteria: level 2 elements of decision tree.

The validation of the model was carried out with the help of five consultants and three planners of security controls. The model was checked to ensure that the results reflected what was happening in the real world and that reasonable solutions were produced. The functionality of the model was examined by comparison of the results between the judgements of security consultants with their own business knowhow and the judgements of planners of XYZ with the methodology provided in this paper. The results were found to match their judgements in most instances. The usefulness of the model was examined through the observation of its effect on the decision-making process in selection of security controls. The approach follows a systematic decision-making process. The security consultants and planners found that the developed methodology was very useful to support managers in selection of security controls.

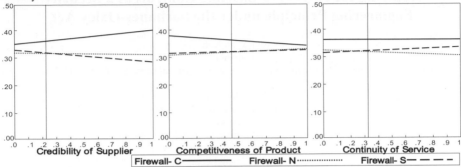

Figure 3: Sensitivity analysis

References

[1] Martin D. et al., A formal specifications maturity model, *Communications of the ACM* **40** (1997).

[2] Monheit M. and A. Tsafrir, Information systems architecture: a consulting methodology, *Proceeding of the 1990 IEEE International Conference on Computer Systems and Software Engineering* (1990).

[3] Leem, C. S. and Kim, S., Introduction to an integrated methodology for development and implementation of enterprise information systems, *Journal of Systems and Software* **60** (2002).

[4] Choi, S: *A Study on the Methodology to Establish the Security Systems for e-business.* Mater thesis, Yonsei University, Korea, 2000.

[5] Gilbert I. E., *Guide for Selecting Automated Risk Analysis Tools (SP 500-174.* NIST, 1989.

[6] Polk W. T. and Bassham L. E., *A Guide to the Selection of Anti-Virus Tools and Techniques.* NIST, 1992.

[7] Barbour, R., *Software Capability Evaluation: Implementation Guide for Supplier Selection.* Software Engineering Institute, Carnegie Mellon University, 1996.

[8] Henze, D., *IT Baseline Protection Manual, BSI* (2000).

[9] Kavanaugh, K., *Security Services: Focusing on User Needs*, Gartner, 2001.

[10] Beall. S. and Hodges, R., *Protection & Security: Software Comparison Columns*, Gartner, 2002.

[11] Firth, R. et al., *An Approach for Selecting and Specifying Tools for Information Survivability.* Software Engineering Institute, Carnegie Mellon University, 1998.

[12] Barbacci, M. et al., *Quality Attributes.* Software Engineering Institute, Carnegie Mellon University, 1995.

[13] Saaty: *The Aalytic Hierarchy Process.* McGraw-Hill, NY, 1980.

VSP International
Science Publishers
P.O. Box 346, 3700 AH Zeist
The Netherlands

*Lecture Series on Computer
and Computational Sciences*
Volume 1, 2004, pp. 900-903

Applying the ISO17799 Baseline Controls as a Security Engineering Principle under the Sarbanes-Oxley Act

Sangkyun Kim[1]
Somansa,
Woolim e-Biz center, 16, Yangpyeongdong 3-ga, Yeongdeungpogu, Seoul 150-103, Korea

Hong Joo Lee[2]
Department of Computer and Industrial Engineering,
Yonsei University,
134, Shinchondong, Seodaemoongu, Seoul 129-749, Korea

Choon Seong Leem[3]
Department of Computer and Industrial Engineering,
Yonsei University,
134, Shinchondong, Seodaemoongu, Seoul 129-749, Korea

Received 31 July, 2004; accepted in revised form 25 August, 2004

Abstract: The Sarbanes-Oxley Act arising from the Enron bankruptcy contains a number of provisions directly affecting the structure and governance of information security. Because the relationship of the act to the information security arena is not quite as intuitive, security managers of the enterprise suffer from hard time to solve the problems regarding this Act. This paper provides the logical analysis and useful comments: (1) the security issues of the Act, (2) the introduction of the ISO/IEC 17799, and (3) the application methods of ISO/IEC 17799 to the Act. With this paper, the security managers could manage the brain-teasing works of information security management to meet the requirements of the SOA.

Keywords: Sarbanes-Oxley Act; ISO/IEC 17799; Security Engineering; Audit

Mathematics SubjectClassification: 68U99
PACS: 89.70.+c

1. The Sarbanes-Oxley Act and Information Security

There are several laws which address information security directly or indirectly as financial governance, privacy, or reporting requirements [1]. A summary of these laws is provided in Table 1.

The Enron and WorldCom scandals made the Bush Administration and the U.S. Congress to enact the law, the Sarbanes-Oxley Act of 2002 ("SOA") [2]. The SOA is one of the most important piece of legislation affecting corporate governance, financial disclosure and the practice of public accounting since the US securities laws of the early 1930s. The primary intent of the SOA is to increase corporate accountability to investors and creditors [3]. Additionally, the SOA imposes sanctions, particularly on accountants and corporate officers who participate in or foster corporate irresponsibility to investors and creditors. The SOA includes both monetary and non-monetary sanctions for those individuals and firms found guilty of violating the SOA's provisions. The monetary sanctions included in the SOA seem not to fit in the magnitude of the misdeeds. It is often the case that the fines imposed for criminal acts seem not to fit the crime. In the case of violation of public trust, there are also civil penalties. The civil penalties are, of course, not addressed in the SOA.

Table 1: Information security legislations

---

[1] Corresponding author, Member of the Research Group of Security Engineering, E-mail: saviour@somansa.com
[2] E-mail: blue1024@yonsei.ac.kr
[3] E-mail: leem@yonsei.ac.kr

Legislation	Target	Security issues	Penalties
Sarbanes-Oxley Act of 2002	All public companies subject to US security laws	Internal controls and financial disclosures	Criminal and civil penalties
Gramm-Leach-Bliley Act of 1999	Financial institutions	Security of customer records	Criminal and civil penalties
Health Insurance Privacy and Accountability Act (HIPAA)	Health plans, health care clearinghouses, and health care providers	Personal health information in electronic form	Civil fines and criminal penalties

According to the 2004 CSI/FBI Computer Crime and Security Survey, the respondents in the financial, utility and telecommunications sectors believe the SOA has an impact on their organizations' information security [4]. The relationship of the SOA to the information security arena is not quite as intuitive. For years, information security professionals have struggled to vault their information security practices into positions of prominence and influence, ones that have strategic value to their organization. As staff within organizations tasked with achieving compliance with SOA studied the provisions of this Act, it became increasingly apparent that the adequacy of controls depends substantially on the mainstream issues for information security professionals [5].

2. ISO/IEC 17799

The ISO/IEC 17799 standard gives recommendations for information security management for use by those who are responsible for initiating, implementing or maintaining security in their organization. ISO 17799 is "a comprehensive set of controls comprising best practices in information security," and is essentially an internationally recognized generic information security standard. It has ten sections of security controls with various perspectives. Each section has two levels of sub-items. The key objectives of these sections are described in below [6].

Sec.4.1 - Security policy: to provide management direction and support for information security.

Sec.4.2 - Security organization: to manage information security within the company; to maintain the security of organizational information processing facilities and information assets accessed by third parties; to maintain the security of information when the responsibility for information processing has been out-sourced to another organization.

Sec.4.3 - Asset classification & control: to maintain appropriate protection of corporate assets and to ensure that information assets receive an appropriate level of protection.

Sec.4.4 - Personnel security: to reduce risks of human error, theft, fraud or misuse of facilities; to ensure that users are aware of information security threats and concerns, and are equipped to support the corporate security policy in the course of their normal work; to minimize the damage from security incidents and malfunctions and learn from such incidents.

Sec.4.5 - Physical & environmental security: to prevent unauthorized access, damage and interference to business premises and information; to prevent loss, damage or compromise of assets and interruption to business activities; to prevent compromise or theft of information and IPFs.

Sec.4.6 -Communication & operation management: to ensure the correct and secure operation of information processing facilities; to minimize the risk of systems failures; to protect the integrity of software and information; to maintain the integrity and availability of information processing and communication; to ensure the safeguarding of information in networks and the protection of the supporting infrastructure; to prevent damage to assets and interruptions to business activities; to prevent loss, modification or misuse of information exchanged between organizations.

Sec.4.7 - Access control: to control access to information; to prevent unauthorized access to information systems; to ensure the protection of networked services; to prevent unauthorized computer access; to detect unauthorized activities; to ensure information security when using mobile computing and tele-networking facilities.

Sec.4.8 - System development & maintenance: to ensure security is built into operational systems; to prevent loss, modification or misuse of user data in application systems; to protect the confidentiality, authenticity and integrity of information; to ensure IT projects and support activities are conducted in a secure manner; to maintain the security of application system software and data.

Sec.4.9 - Business continuity planning: to counteract interruptions to business activities and to critical business processes from the effects of major failures or disasters.

Sec.4.10 - Compliance: to avoid breaches of any criminal or civil law, statutory, regulatory or contractual obligations and of any security requirements; to ensure compliance of systems with organizational security policies and standards; to maximize the effectiveness of and to minimize interference to/from the system audit process.

3. Applying ISO17799 to the SOA

Although SOA focuses on corporate governance and does not specify internal controls, it requires management to certify the accuracy of their financial reports. To do so with confidence, all security controls and processes associated with the financial systems and any related systems need to be tested, validated, documented, and improved [7].
There are three sections of the SOA in particular that raise security-related issues for companies [8, 9, 10].
1) Section 204 (Auditor reports to audit committees) – Public accounting firm that performs for any issuer any audit shall timely report to the audit committee of the issuer including material written communication between the registered public accounting firm and the management of the issuer.
2) Section 302 (Corporate responsibility for financial reports) - Signing Officers (CEOs and CFOs) are required to certify that periodic reports are accurate and are not misleading in any way. They need to ensure that an internal control structure is established and maintained and that they are made aware of any pertinent information relating to the well-being of the company.
3) Section 404 (Management assessment of internal controls) - An Internal Control Report must be produced as part of the Annual Report. This report needs to reflect the Internal Control framework that is in place and show that it is being maintained. In this report Management are required to make an assessment as to the effectiveness of the company's internal controls. Finally, external auditors must report on and attest to the accuracy of the assessment as provided by Management.
We analyzed each section of the SOA and ISO/IEC 17799 to assist the security managers to establish, maintain and update security controls for sections 204, 302 and 404 of the SOA. Table 2 describes the results of our analysis.

Table 2: Applying sections of the ISO/IEC 17799 to the sections 204, 302 and 404 of the SOA

Security Issues of the SOA			ISO/IEC 17799	
Section	Sub-item	Security Issue	Section	Application Methods
SEC.204.	(k)(3)	Record retention and reporting	4.3.2.2 4.6.4.1 4.6.4.2 4.6.6 4.7.7.2	1) The information should be labeled, handled with predefined procedures. 2) Operator logs and monitored data of system usage should be recorded and duplicated as a backup copy. 3) The marking, handling, storage and disposal of media should be managed properly.
SEC.302.	(a)(4)(A)	Establishing & maintaining of internal controls	4.1.1.1 4.2.1	With the cooperation of security organizations, the signing officer should establish and maintain an information security policy.
	(a)(4)(C) (a)(4)(D)	Evaluation of the internal controls	4.10.1 4.10.2	An independent review of information security should be performed periodically.
SEC.404	(a)(2) (b)(a)	Auditing and reporting for the internal controls	4.1.1.2 4.2.1.7	The signing officer should: 1) check the compliance with legal requirements such as an identification of applicable legislation, safeguarding of organizational records, data protection, and collection of evidence; 2) review the security policy and technical compliance.

The certification mechanism of ISO/IEC 17799 could be used as an auditing tool which verifies the compliance between the requirements of information security and the SOA. As an example, the sections of ISO/IEC 17799 and their supplement could be used as auditing lists: section 4.1.1, 4.2.1, 4.3.2.2, 4.6.4.1, 4.6.4.2, 4.6.6, 4.7.7.2, 4.10.1 and 4.10.3.

4. Summary

The primary intent of the SOA is to increase corporate accountability to investors and creditors. The relationship of the SOA to the information security arena is not quite as intuitive, but it's apparent that both are inseparably related to each other. This paper provides the review of the SOA, its implications of information security, and applicability of the ISO 17799 to it. With this paper, the security managers could manage the brain-teasing works of information security management to meet the requirements of the SOA.

Acknowledgments

We would like to appreciate for help of Dr. Kuma for his careful reading and fruitful comments on the earlier draft of this paper.

References

[1] Conner, B., Noonan, T. and Robert W. Holleyman, II, *Information Security Governance: Toward a Framework for Action, BSA* (2004).

[2] Bequai, A., Safeguards for IT managers and staff under the Sarbanes-Oxley Act, *Computers & Security* **22** 124-127(2003).

[3] Ashley W. Burrowes, Kastantin, J. and Milorad M. Novicevic, The Sarbanes-Oxley Act as a hologram of post-Enron disclosure: a critical realist commentary, *Critical Perspectives on Accounting* **15** 797-811(2004).

[4] Lawrence A. Gordon, Martin P. Loeb, Lucyshyn, W. and Richardson, R., *2004 CSI/FBI Computer Crime and Security Survey.* Computer Security Institute, 2004.

[5] E. Eugene Schultz, Sarbanes-Oxley-a huge boon to information security in the US, *Computers & Security* **23** 353-354(2004).

[6] Kim, S., Choi, B. K., Lee, H. J. and Leem, C. S., Analytic pespective of ISO17799 ISMS, *Proceeding of ICOQM-V* (2004).

[7] *Sarbanes-Oxley Act: Data Security and Encrypyion*, PGP Corporation, 2004.

[8] *Sarbanes-Oxley Act of 2002.* Thomas Legislative Information on the Internet: The Library of Congress (http://thomas.loc.gov), 2002.

[9] White, T. and Greig, J., Sarbanes-Oxley Act of 2002: A new regime of corporate governace, *International Business Lawyer* **30** (2002).

[10] Winston, United States: corporate governance, risk management and compliance: best practices after Sarbanes-Oxley, *International Financial Law Review* **10** (2002).

VSP International
Science Publishers
P.O. Box 346, 3700 AH Zeist
The Netherlands

*Lecture Series on Computer
and Computational Sciences*
Volume 1, 2004, pp. 904-907

Cost Efficient Group Communication & Management for Intrusion Tolerant System

Chaetae Im[1], HyoungJong Kim[1], GangShin Lee[1]

Korea Information Security Agency
78 Garak-Dong, Songpa-Gu, Seoul, Korea 138-803

Abstract: Intrusion tolerance technology is aim to guarantee the service despite of errors occurred by attackers or arbitrary faults. Group communication in ITS(Intrusion Tolerant System) should satisfy the reliability and total-ordered properties to maintain consistency among members. But it requires additional communication overhead. To apply ITS in the real, such an overhead should be minimized not to affect the target service much. To solve the problem, we suggest newly-defined ITS group communication scheme that requires little communication overhead with satisfying properties for ITS. Our proposed "cost efficient group communication & management" mechanism is based on NACK control message to notice which message isn't received. The proposed NACK-based mechanism shows that it requires little communication overhead especially when the system environment is stable. To evaluate the suggested mechanisms, modeling and simulation is performed, and the enhancement is achieved

Keywords: Intrusion Tolerant System, Replicated member, Reliable multicast, Total-ordered, SITIS

1. Introduction

The ITS is emerged to guarantee the service by dealing unknown attacks that are not treated by conventional prevention and detection technologies. Since the ITS is accomplished by combination of fault tolerant and security technologies, the architecture and its operation is very complex. Because of the complexity of the ITS, it can degrade performance of the target service. However, to deploy the ITS to the real world, performance degradation due to intrusion tolerant service should be minimized. Since the computer systems have high computing power lately, processing overhead for a intrusion tolerance service isn't much. However, communications among replicated group members need heavy cost since most of messages should be delivered to all members reliably and orderly. Moreover, communications among group members affect the service latency directly. To accomplish these two controversial design aspects such as low overhead and high complexity in communication, we suggest newly-defined ITS group communication scheme which satisfy both reliability and total-ordered properties.

In this paper, the *cost efficient group communication & management* scheme is designed to be used among replicated members. It is NACK-based reliable multicast that satisfies both reliability and total-ordered properties and requires low communication overhead. The proposed mechanisms are evaluated through simulations and compared with other popular communication schemes. Our implementation of the suggested scheme is used to our scalable ITS-middleware, SITIS (Scalable Intrusion Tolerance System for Information Service Survivability). SITIS is a well-defined ITS middleware framework for guaranteeing quality of essential service.

2. Related Works

There are two representative programs about ITS; OASIS(The Organically Assured and Survivable Information Systems) and MAFTIA(Malicious and Accidental-Fault Tolerance for Internet Applications). The goal of MAFTIA is to research the tolerance paradigm and support by proposing and developing the integrated tolerance architecture that many application programs have dependability. The OASIS program has goal to develop, experiment, and validate every elements such as architecture, tools, and techniques for survivability of the information systems. The program has sponsored many projects and the representative projects are HACQIT, ITUA, SITAR, and so on. [1~4]

There are research works about group communication mechanism for ITS. As one of the representative works, the *securing* protocol has the basic concept to deliver message reliably by making the logical ring in broadcast domain. Processor which owns the token can broadcast messages and the token is passed to the next member periodically. If the received message were out of sequence, then the member

---

[1] Korea Information Security Agency, E-mail: {chtim, hjkim, kslee}@kisa.or.kr

may request the retransmission. This protocol satwisfies the reliability and total-ordered properties, but it uses the channel inefficiently. Group communication protocol in ITUA and Rampart make use of *echo-reliable* multicast mechanism. To deliver messages, exchange of *init* and *echo* control messages is required. If the sequence number in next messages is out of sequence, the node requests the missing by sending NACK control message. But multiple steps; *init-echo-message-nack*; for delivery of messages require high cost. One of the well-known reliable multicast protocol, SRM(Scalable Reliable Multicast) is for general multimedia broadcast. It is based on NACK mechanism for the reliable delivery and supports efficient delivery in large scale of network. But it doesn't consider the order of message arrivals and assumes the large distributed network. So it is not proper to adapt to ITS. [4~8]

3. Cost efficient Group Communication & Management Mechanisms for ITS

Figure 1 shows the overall procedure of the intrusion tolerant service that we are considering. Many ITS functions are not represented in this figure. As shown in the figure, service request from client is propagated to all group members, and the service reply from the application server is sent to voter which determines the correct result by comparing all replies. Finally, the client gets the correct service reply. The number of member is based on the theory to solve the Byzantine problem that the number of node should be at least 3f+1, where f is the number of faults which can occur simultaneously.

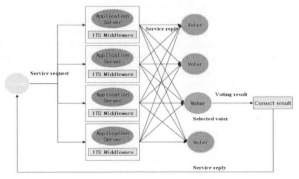

Figure 1: Overall Procedure of the Intrusion Tolerant Service

The role of a leader in a group is to check each member's status by using heartbeat mechanism, to propagate the updated group information, and so on. We proposed *distributed random leader selection* scheme for selecting a leader among group members. The random selection method is for an attacker not to know which one is a leader. Also, periodic change of the leader makes more difficult to predict.

Reliability of multicast protocol is generally achieved by acknowledgement procedure and sequence number. As shown in the figure 2.b), in the ACK-based mechanism, sender can recognize the missing messages by *ACK* control message from receivers. When a sender received all expected *ACK* packets, it sends the *commit* control packet. Members which receive the *commit* message from the sender pass the message to proper module. The *commit* control message makes multicast mechanism preserve the total-ordered property and it is also used to check the sender's status before processing the message. From the figure 2.a), in the NACK-based mechanism, receivers can recognize the missing message by checking the sequence number in succeeding messages. Members which receive the *confirm* control message pass it to the destined module. By the *confirm* control message, receivers can identify the missing immediately and the total-ordered property can be satisfied. If the sequence number in a *confirm* message is out of order, the member sends the *NACK* packet to the sender directly. If other receivers neither receive the message but they hear the *NACK* message from other receiver, it doesn't send the *NACK* message since the sender will retransmit the message right after. In both mechanisms, the maximum number of retransmission trials exists to avoid unlimited service delay and a retransmitted packet is simply dropped on members received it already.

In the figure 2, m is the duration for sending message and is directly proportional to size of the message. c and a are the duration for sending *confirm* and *ACK* control packet respectively. d_{NA} is the duration for waiting *NACK* control packet. For d_{NA}, if there is no *NACK* message, then the sender transmits a *confirm* message. n_r is the number of retransmission trials. p is the channel propagation delay and it is ignored since it is very short time.

$$T_{NACK} = m + c + d_{NA} + n_r (m + n\_a + c) \qquad (1)$$

$$T_{ACK} = m + (n-1) \cdot a + n_r (m + d_A) + cm \qquad (2)$$

Where the transmission time for messages, $t = packet\_size(byte) \cdot 8 / chanel\_rate$

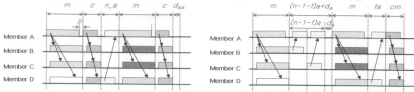

Figure 2: NACK and ACK based reliable multicast

As shown in the equation (1), (2), Cost for the ACK-based mechanism is proportional to number of member since all receivers should send *ACK* control message. In case of the NACK-based, cost is independent on number of members. The NACK-based mechanism is more efficient than the ACK-based since the NACK-based usually require only *confirm* message when a network is stable and errors doesn't occur in intrusion tolerant domain.

When a sender fails right after sending a message, *confirm* and *commit* control message can't be sent in NACK- and ACK-based mechanisms respectively. In such cases, one of the receivers sends the *request_confirm* control message in NACK-based mechanism and the *request_commit* control message in the ACK-based. If the number of trials for requesting *commit* or *confirm* message reached the predefined number, then all receivers drop the message and reconfigure the group by eliminating the erroneous sender. If the number of retransmission increases, the NACK-based mechanisms take more overhead than the ACK-based since *confirm* and *NACK* control messages should be sent on each retransmission. However we should notice that the maximum number of retransmission trials exist, so the NACK-based mechanism is usually more efficient than the ACK-based.

4. Performance Evaluation

In our simulation to evaluate the proposed schemes, we analyzed the cost for one message transmission and one service transaction. In measuring cost for one transaction, we ignored processing time such as comparing value on voter, processing service requests on application server. Table 1 shows the parameters for simulation. We compared 4 multicast protocols; unreliable multicast, proposed NACK-based, proposed ACK-based, and echo reliable multicast schemes. Unreliable multicast has no acknowledgement and sequence numbering mechanism. Echo reliable multicast requires *init-echo* procedure before sending message and is based on NACK but has no *confirm* control message as in our proposed mechanism.

Table 1: Parameters for Simulation

LAN	100Mbps, Ethernet	*init, echo*	60 (byte)
# of member	4, 7, 10 (members)	*confirm, commit*	4 member: 64, 7 member:76, 10 member:88 (byte)
Request	72 (byte)	d_{NA}, d_A	2 (μs)
Service Reply	0~1500 (byte)	*ACK, NACK*	51(byte)

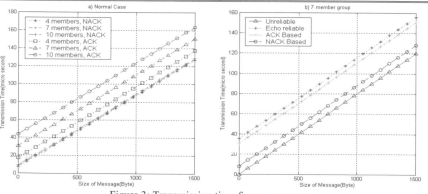

Figure 3: Transmission time for one message

Figure 3 shows the cost for transmitting a packet. In the figure 3.a), cost for the ACK- and NACK-based reliable multicast mechanism are compared. From the figure, cost of the ACK-based is proportional to number of member since it needs ACK messages as the number of members. Cost of the NACK-based is almost independent on number of the member. From the figure 3.b), echo reliable multicast mechanism takes biggest cost since it requires more steps for delivering a packet than others.

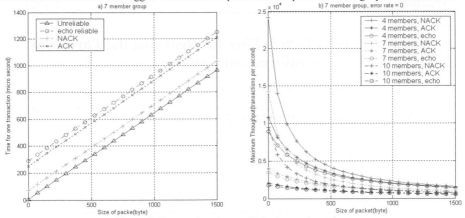

Figure 4: Transaction time and Maximum throughput

Figure 4.a) shows the cost for one service transaction which take the steps as follow; service request from client, processing the request on application server, validating the reply on voter, and returning the correct service reply to client. Since several reliable deliveries are required in a transaction, difference of cost between each mechanism becomes bigger. Figure 4.b) represents the maximum throughput that means processed transactions per second. In this simulation, we assumed that error doesn't occur and there are only messages related to target service and ITS in the service domain. Since the cost for transmitting a big size of packet take long time, the bigger size of service reply packet is transmitted, the less number of transactions is processed in a second. Therefore, the advantage of the NACK-based can be greater where the size of message is short.

5.　Conclusion

To deploy the ITS in the real world, performance degradation due to intrusion tolerant service should be minimized. So, we proposed the group communication and management mechanism that requires relatively low cost and satisfy properties for ITS such as reliable and total-ordered communication. Through the modeling and simulation, we showed that the proposed communication mechanism requires less cost than other reliable multicast protocols. Also, to strengthen the ITS against an attacker, distributed random leader selection scheme is suggested.

References
[1] David Powell and Robert Stroud, et. Al: Conceptual Model and Architecture of MAFTIA, *Project MAFTIA IST-1999-11583 deliverable D2*, Nov. 2001.
[2] Feiyi wang, Fengmin Gong, and Chandramouli Sargor, SITAR, A Scalable Intrusion- Tolerant Architecture for Distributed Services, *Workshop on Information Assurance and Security 2002*, Jun. 2001.
[3] F. Wang and Charles Killian, Design and Implementation of SITAR Architecture: A Status Report, *Intrusion Tolerant System Workshop*, C-3-1, Supplemental Volume on 2002 International Conference on Dependable System & Networks, Jun. 2002
[4] Courtney Tod, Lyons James, et.al, Providing Intrusion Tolerance with ITUA, *Proceedings of the ICDSN 2002*, Jun. 2002.
[5] K. Kihlstrom, L. Moser, P.M. Mellia-Smith, The SecureRing Protocols for Securing Group Communication, *Proceedings of the Hawai'i Int'l Conference on System Sciences*, Jan. 1998.
[6] Michael K. Reiter and Kenneth P. Birman, How to Securely Replicate Services, *ACM Transactions on Programming Languages and Systems, vol.16, no.3*, May. 1994, pp.986-1009.
[7] Michael K. Reiter, The Rampart Toolkit for Building High-Integrity Services, *Theory and Practice in Distributed Systems, LNCS 938, pp.99-110. Springer*, 1995.
[8] Floyd, S., Jacobson, V., Liu, C., McCanne, S., and Zhang, L., A Reliable Multicast Framework for Light-weight Sessions and Application Level Framing, *IEEE/ACM Transactions on Networking, Volume 5, Number 6, pp. 784-803.*, Dec 1997.

VSP International
Science Publishers
P.O. Box 346, 3700 AH Zeist
The Netherlands

*Lecture Series on Computer
and Computational Sciences*
Volume 1, 2004, pp. 908-910

Countermeasure Design Flow for Reducing the Threats of Information Systems

Tai-hoon Kim[1], Chang-wha Hong[1], Sook-hyun Jung[1]

[1]San 7, Geoyeou-dong, Songpa-gu, Seoul, South Korea

Abstract: The developers of the security policy recognize the importance of using both technical and non-technical countermeasures, such as personnel and operational facts, in formulating an effective overall security solution to address threats at all layers of the information infrastructure. This paper uses the security engineering principles for determining appropriate technical security countermeasures. It includes information on threats, security services, robustness strategy, and security mechanism. This paper proposes a countermeasure design flow which may reduce the threats to the information systems.

Keywords: Threat, Security countermeasure, Security Engineering

1. Introduction

In general, threat agents' primary goals may fall into three categories: unauthorized access, unauthorized modification or destruction of important information, and denial of authorized access. Security countermeasures are implemented to prevent threat agents from successfully achieving these goals. This paper proposes a countermeasure design flow considering attacks (including threat), security services, and appropriate security technologies.

Security countermeasures should be considered with consideration of applicable threats and security solutions deployed to support appropriate security services and objectives. Subsequently, proposed security solutions may be evaluated to determine if residual vulnerabilities exist, and a managed approach to mitigating risks may be proposed.

2. Threats Identification

A 'threat' is an undesirable event, which may be characterized in terms of a threat agent (or attacker), a presumed attack method, a motivation of attack, an identification of the information or systems under attack, and so on. In order to identify what the threats are, we need to answer the following questions [1-2]:

- What are the assets that require protection? (e.g., sensitive data or secret),
- Who or what are the threat agents? (e.g., an authorized user who wants to access to the system),
- What are the motivations of attack? (e.g., delete some information),
- What attack methods or undesirable events do the assets need to be protected from? (e.g., impersonation of an authorized user of the TOE).

Threat agents come from various backgrounds and have a wide range of financial resources at their disposal. Typically Threat agents are thought of as having malicious intent. However, in the context of system and information security and protection, it is also important to consider the threat posed by those without malicious intent. Threat agents may be Nation States, Hackers, Terrorists or Cyber terrorists, Organized Crime, Other Criminal Elements, International Press, Industrial Competitors, Disgruntled Employees, and so on.

Most attacks maybe aim at getting inside of information system, and individual motivations of attacks to "get inside" are many and varied. Persons who have malicious intent and wish to achieve

---

[1] San 7, Geoyeou-dong, Songpa-gu, Seoul, South Korea, E-mail : taihoon@empal.com

commercial, military, or personal gain are known as hackers (or cracker). At the opposite end of the spectrum are persons who compromise the network accidentally. Hackers range from the inexperienced Script Kiddy to the highly technical expert.

3. Determination of Robustness Strategy

The robustness strategy is intended for application in the development of a security solution. An integral part of the process is determining the recommended strength and degree of assurance for proposed security services and mechanisms that become part of the solution set. The strength and assurance features provide the basis for the selection of the proposed mechanisms and a means of evaluating the products that implement those mechanisms.

Robustness strategy should be applied to all components of a solution, both products and systems, to determine the robustness of configured systems and their component parts. It applies to commercial off-the-shelf (COTS), government off-the-shelf (GOTS), and hybrid solutions. The process is to be used by security requirements developers, decision makers, information systems security engineers, customers, and others involved in the solution life cycle. Clearly, if a solution component is modified, or threat levels or the value of information changes, risk must be reassessed with respect to the new configuration.

Various risk factors, such as the degree of damage that would be suffered if the security policy were violated, threat environment, and so on, will be used to guide determination of an appropriate strength and an associated level of assurance for each mechanism. Specifically, the value of the information to be protected and the perceived threat environment are used to obtain guidance on the recommended strength of mechanism level (SML) and evaluation assurance level (EAL) [3].

4. Consideration of Strength of Mechanisms

SML (Strength of Mechanism Levels) are focusing on specific security services. There are a number of security mechanisms that may be appropriate for providing some security services. To provide adequate information security countermeasures, selection of the desired (or sufficient) mechanisms by considering particular situation is needed. An effective security solution will result only from the proper application of security engineering skills to specific operational and threat situations. The strategy does offer a methodology for structuring a more detailed analysis. The security services itemized in these tables have several supporting services that may result in recommendations for inclusion of additional security mechanisms and techniques.

5. Selection of Security Services

In general, primary security services are divided five areas: access control, confidentiality, integrity, availability, and non-repudiation. But in practice, none of these security services is isolated from or independent of the other services. Each service interacts with and depends on the others. For example, access control is of limited value unless preceded by some type of authorization process. One cannot protect information and information systems from unauthorized entities if one cannot determine whether that entity one is communicating with is authorized. In actual implementations, lines between the security services also are blurred by the use of mechanisms that support more than one service.

6. Application of Security Technologies

An overview of technical security countermeasures would not be complete without at least a high-level description of the widely used technologies underlying those countermeasures. Next items are some examples of security technologies.

- Application Layer Guard, Application Program Interface (API).
- Common Data Security Architecture (CDSA).
- Circuit Proxy, Packet Filter, Stateful Packet Filter.
- CryptoAPI, File Encryptors, Media Encryptors.
- Cryptographic Service Providers (CSP), Certificate Management Protocol (CMP).
- Internet Protocol Security (IPSec), Internet Key Exchange (IKE) Protocol.

- Hardware Tokens, PKI, SSL, S/MIME, SOCKS.
- Intrusion and Penetration Detection.
- Virus Detectors.

7. Determination of Assurance Level

The discussion of the need to view strength of mechanisms from an overall system security solution perspective is also relevant to level of assurance. While an underlying methodology is offered by a number of ways, a real solution (or security product) can only be deemed effective after a detailed review and analysis that consider the specific operational conditions and threat situations and the system context for the solution.

Assurance is the measure of confidence in the ability of the security features and architecture of an automated information system to appropriately mediate access and enforce the security policy. Evaluation is the traditional method which ensures the confidence. Therefore, there are many evaluation methods and criteria exist. In these days, many evaluation criteria such as ITSEC and TCSEC are replaced by the ISO 15408, Common Criteria [4].

The Common Criteria provide assurance through active investigation. Such investigation is an evaluation of the actual product or system to determine its actual security properties. The Common Criteria philosophy assumes that greater assurance results come from greater evaluation efforts in terms of scope, depth, and rigor.

8. Conclusion and Future Work

This paper proposes a countermeasure design flow considering threats, robustness, strength of mechanism, security service, technology, and appropriate assurance level. But this is not a framework yet. More efforts are needed for refining the flow and making framework.

References

[1] Tai-Hoon Kim, Byung-Gyu No and Dong-chun Lee: Threat Description for the PP by Using the Concept of the Assets Protected by TOE, ICCS 2003, LNCS 2660, Part 4, pp. 605-613, 2003.

[2] Tai-hoon Kim and Haeng-kon Kim: The Reduction Method of Threat Phrases by Classifying Assets, ICCSA 2004, LNCS 3043, Part 1, 2004.

[3] ISO. ISO/IEC 15408: Information technology - Security techniques - Evaluation criteria for IT security, 1999

[4] NSA, The Information Assurance Technical Framework, September, 2002.

[5] Tai-hoon Kim and Haeng-kon Kim: A Relationship between Security Engineering and Security Evaluation, ICCSA 2004, LNCS 3046, Part 4, 2004.

[6] Eun-ser Lee, Kyung-whan Lee, Tai-hoon Kim and Il-hong Jung: Introduction and Evaluation of Development System Security Process of ISO/IEC TR 15504, ICCSA 2004, LNCS 3043, Part 1, 2004

VSP International
Science Publishers
P.O. Box 346, 3700 AH Zeist
The Netherlands

*Lecture Series on Computer
and Computational Sciences*
Volume 1, 2004, pp. 911-914

Authentication for Single Domain in Ubiquitous Computing Using Attribute Certification

Deok-Gyu Lee[1], Hee-Un Park[2], Im-Yeong Lee[3]

Division of Information Technology Engineering,
Soonchunhyang University,
#646 Eup-nae ri, Shin-chang myun, A-san si, Choongchungnam-do, Korea

Abstract: The Ubiquitous computer environment is thing which invisible computer that is not shown linked mutually through network so that user may use computer always is been pervasive. Intend computing environment that can use easily as user wants and it is the smart environment that user provides context awareness that is wanting computing environment. This Ubiquitous computing contains much specially weak side in security. Masquerade attack of that crawl that is quoted to user or server among device that is around user by that discrete various computing devices exist everywhere among them become possible. Hereupon, in this paper, proposed method that have following characteristic. Present authentication model through transfer or device. Suggest method that realize authentication through device in case of moved user's direct path who device differs.

Keywords: Authentication, Ubiquitous computing, Attribute certification, Single Domain

1. Introduction

Ubiquitous computing aims at an environment in which invisible computers interconnected via the network exist. In this way, computers are smart enough to provide a user with context awareness, thus allowing the user to use the computers in the desired way. Ubiquitous computing has the following features: Firstly, a variety of distributed computing devices exist for specific users. Secondly, computing devices that are uninterruptedly connected via the network exist. Thirdly, a user sees only the personalized interface because the environment is invisible to him. Lastly, the environment exists in a real world space and not in a virtual one.[3][4][7][9] However, the ubiquitous environment is weak in security. Since distributed computing devices are spread out in the environment, it is possible to launch disguised attacks against the environment from a device authenticated by a user or a server. Also, although a user approves of the installation of only authenticated applications into the devices, there is a chance that a malicious code will be transmitted to surrounding devices that do not have computing capability. Since many ubiquitous computing devices do not provide efficient memory protection, the memory where user information (authentication information) is stored can easily be attacked. These problems can be resolved by using encryption or electronic signature for computing devices. But non-computing devices cannot be protected using encryption codes or an electronic signature, which brings up potential security issues. Also, when a device is moved out of the domain into a new user space, user authentication must be performed smoothly in the new space. This is so because a different device in the new user's space can be authenticated with the user authentication information in the new user space, and not with the previous user authentication information.[8] It is this paper's purpose to propose an authentication model through the movement of a smart device. The conceptualized model proposes two methods. The other method is to implement the authentication through a device when another device is moved into the single domain.

---
[1] Corresponding author. Division of Information Technology Engineering. Soonchunhyang University. E-mail: hbrhcdbr@sch.ac.kr

[2] Korea Information Security Agency. E-mail: hupark@kisa.or.kr

[3] Division of Information Technology Engineering. Soonchunhyang University. E-mail: imylee@sch.ac.kr

This work was supported by grant No. R05-2003-000-12019-0 from the Basic Research Program of the Korea Science & Engineering Foundation

2.   Proposed Scheme

In the previous chapters, we have gone over the existing ubiquitous environment, JARM method, and PMI. Although many researches have been done regarding ubiquitous computing, the most active area of research is on communication rather than on security. Security is often researched only as part of the project, not as the main research topic. This paper has selected the JARM method as its research topic since the JARM method exclusively studies authentication in ubiquitous computing. In reviewing current existing researches, the researcher believes that several researches regarding security have been accomplished and published. At the time of research, the researcher discovered that ubiquitous computing must have mobility, entity authentication, corresponding entity authentication, outgoing data authentication, and connection/non-connection confidentiality as basic requirements. Thus, this paper proposes the adoption of PMI to meet the requirements listed above and to implement the authentication for the smart device. The ubiquitous computing devices lacked computing, storing, and other capabilities. But since a device must meet the requirements discussed earlier, applying PMI on top of the currently used encryption system will satisfy the device capabilities and requirements. Since all devices can carry out the authentication and access control with the PMI certificate, only activities authorized to the devices will be allowed. I will also propose a method that uses a PMI certificate for a device.

2.1.   Consideration for Proposed Scheme

The goal of the proposed method is to provide a device retaining the user information of the previous domain even when the device is moved into the single domain. Thus, the following must be considered for the proposed method.

- A user device alone can be moved and this user device can be linked to other devices. : This means that when a user's smart device is moved into the single domain, the smart device can be linked to devices in the single domain to receive services. User information must therefore be extracted from the smart device, as other devices exist for services only.

2.2.   System Parameters

Next, system parameters used in this method are explained. Each parameter is distinguished according to its components. The components create and transmit the parameters.

· *: (SM: SMart device, D: Device, SD: Single Domain, MDC: Multi Domain Center, A: Alice, B: Bob, MDCM: MDC Manager, ASC: Active Space Center)   · pw : password   · $ID*$: * of Identity

· $Cert*$: public key of * including Certification   · $R*$: * of authority   · $Hub*$: * of Hub

· $PCert*$: public key of * including PMI Certification   · AP : Available Period

· n : PMI Certification maximum issues number   · $E_*(\ )$: * key with Encryption

· r : user Hub generated random number   · $H(\ )$: Secure Hash Function   · i : user issued device

2.3.   Proposed Scheme

The detailed flow of these proposed methods is described below. In the first method, when a user moves to his domain with his smart device and attempts to use devices in the new domain, the user is authenticated using the smart device in which the user authentication information is stored.

1) User registration and device registration

A user must have authentication information for all devices in the initial stage to use the devices in the single domain. A user receives a certificate from the MDC (Multi Domain Center), and his devices are granted a PMI (Privilege Management Infrastructure) certificate through a Hub. PMI certificates are granted according to the mutually agreed methods with the MDC. Granted PMI certificates are stored in a smart device.

Step 1. The following processes are required to create Device_A authentication information for User A. The MDC grants User A a certificate which allows User A to create n number of PMI certificates. A PMI certificate consists of the User A ID, privilege, and effective period of the certificate.

$$MDC \to Hub_A : Cert_A[ID_A, R_A, n, AP]$$

Step 2. User A grants PMI certificates to Device_A and SM_A using the granted certificate. The certificate contains the path to a higher-level certificate.

$$Hub_A \to SD_A(orDevice_A) : PC_A = PCert_A[ID_A, H(Cert_A\|r), i]\|AP$$

$$SD_A : E_{PK_{DDC}}[PC_A], \quad SD_A install : E_{pw(orPIN)}[E_{PK_{DDC}}[PC_A]]$$

Step 3. User A informs the MDC of the certificate granted to his Device_A. Afterwards, User A's PMI certificate is used, and User A is authenticated using the PMI certificate path within SM_A.

$$Hub_A \to MDC : E_{PK_{DDC}}[H(Cert_A\|r), r, i]$$

2) Authentication in the single domain.
When SM_A on User A's Hub attempts to use Device_A2 in AS_A2, SM_A uses existing information as is.
Step 1. SM_A exists in the active space AS_A1 and sends the movement signal to Device_A1 when the movement occurs.

$$SD_{A_1} \to Device_{A_1} : Signal(Outgoing)$$

Step 2. Device_A1 notifies Hub A of SM_A movement.

$$Device_{A_1} \to Hub_A : E_{PK_{Hub}}[Device_{A_2}, E_{PK_{Hub}}(PC)]$$

Step 3. Device_A1 also transmits SM_A information to Device_A2.

$$Device_{A_1} \to Device_{A_2} : [Device_{A_1}\|E_{pw(orPIN)}E_{PK_{Hub}}(PC)]$$

Step 4. Device_A2 uses the authentication information received from Device_A1 to send its information to Hub A.

$$Device_{A_2} \to Hub_A : [Device_{A_2}\|E_{pw(orPIN)}E_{PK_{Hub}}(PC)]$$

Step 5. Hub A also confirms the authentication information received from Device_A1 by comparing it to the information received from Device_A2, and then approving the SM_A authentication.

$$Hub_A : E_{pw(orPIN)}E_{PK_{Hub}}(PC) = E_{PK_{Hub}}(PC)$$

$$Compare : (Device_{A_1})E_{PK_{Hub}}(PC) \overset{?}{=} (Device_{A_1})E_{PK_{Hub}}(PC)$$

Step 6. Hub A completes the confirmation and accepts the authentication for SM_A.

$$Hub_A \to Device_{A_2} : [Device_{A_2}\|Auth_{SD}]$$

Step 7. After SM_A provides its values and compares the values, it approves the use of Device_A2 in the active space.

3. Comparison with Proposed Scheme and JARM Scheme

This chapter will attempt to analyze the proposed protocol by classifying the user and device registration and the authentication in the single domain, and compare the protocol to the existing method. In the existing method, SM_A is not authenticated by moving to AS1. Therefore, the research in this paper seeks to put emphasis on how to authenticate a user who wishes to use Device_B by using user information A when SM_A is moved to Domain_B. The existing methods attempt to solve the problem by assigning different authentication information to the devices. But the weak point of this approach is to require all devices to be available when the entire authentication information is obtained. This method raises a problem in that authentication information cannot be obtained if a device is lost. The proposed method and the existing method are compared below. The details of the proposed method will also be discussed below.

    a) Corresponding Entity Authentication
When a device is located in Domain_A, corresponding entity authentication is provided to verify that Device_B and B are identical entities. This authentication method implements device authentication through the entity of the previous user when multiple devices are connected to one domain. This authentication can provide different levels of protection. Even when a smart device is moved, the authentication can be done $E_{pw(orPIN)}[E_{PK_{Hub}}(PC)]$ using what is stored in the smart device. Also, a

PMI certificate, which is an internal certificate and identical to the certificate from User A, can be used when performing the corresponding entity authentication.

4. Conclusion

Rapid expansion of the Internet has required a ubiquitous computing environment that can be accessed anytime anywhere. In this ubiquitous environment, a user ought to be given the same service regardless of connection type even though the user may not specify what he needs. Authenticated devices that connect user devices must be used regardless of location. If a device is moved to another user space from a previous user space, the authentication must be performed well in the transferred space. This is so because a device is not restricted to the previous authentication information, but can use new authentication information in the new space. This paper attempts to solve the problems discussed earlier by utilizing such entities as the Hub, ASC, and MDC in order to issue PMI certificates to devices that do not have computing capability. This provides higher-level devices the authentication information of the smart devices in order to authenticate the movement of these smart devices. With this proposed method, if a smart device requests the authentication after moving to the single domain, the authentication is performed against the devices in the domain where the smart device belongs, and the smart device requests the authentication from the MDC. In the user domain, the authentication is performed through the Hub. However, the authentication is performed through MDC when a device is moved to the single domain environment. This proposed method, therefore, attempts to solve the existing authentication problem. With regard to the topics of privacy protection, which is revealed due user movement key simplification (i.e., research on a key that can be used for a wide range of services), and the provision of smooth service for data requiring higher bandwidth, the researcher has reserved them for future researches.

Acknowledgments

The author wishes to thank the anonymous referees for their careful reading of the manuscript and their fruitful comments and suggestions.

References

[1] A. Aresenault, S. Tuner, Internet X.509 Public Key Infrastructure, Internet Draft, 2000. 11

[2] ITU-T, Draft ITU-T RECOMMANDATION X.509 version4, ITU-T Publications, 2001. 5.

[3] Jalal Al-Muhtadi, Anand Ranganathan, Roy Campbell, and M. Dennis Mickunas,"A Flexible, Privacy-Preserving Authentication Framework for Ubiquitous Computing Environments," ICDCSW '02, pp.771-776, 2002

[4] Mark Weiser,"Hot Topics: Ubiquitous Computing," IEEE Computer, October 1993

[5] M. Roman, and R. Campbell, "GAIA: Enabling Active Spaces," 9th ACM SIGOPS European Workshop, September 17th-20th, 2000, Kolding, Denmark

[6] S. Farrell, R. Housley, An Internet Attribute Certificate Profile for Authorization, Internet Draft, 2001.

[7] Geun-Ho Lee, "Information Security for Ubiquitous Computing Environment", Symposium on Information Security 2003. pp.629-651, 2003

[8] Deok-Gyu Lee, Hee-Un Park, Im-Yeong Lee, " A Study on Authentication for Differential

[9] Im-Yeong Lee, Jae-Kwang Lee, Woo-Yong So, Young-Rak Choi, "Cryptography and Network Security Principles and Practice", Press Green , 2001

VSP International
Science Publishers
P.O. Box 346, 3700 AH Zeist
The Netherlands

*Lecture Series on Computer
and Computational Sciences*
Volume 1, 2004, pp. 915-918

Authenticated IP Traceback Mechanism on IPv6 for Enhanced Network Security Against DDoS Attack

Hyung-Woo Lee[1], Sung-Hyun Yun[2], Nam-Ho Oh, Hee-Un Park, Jae-Sung Kim[3]

(1) Dept. of Software, Hanshin University, Yangsan, Osan, Gyunggi, 447-791, Korea
(2) Div. of Info. and Comm. Engineering, Cheonan University, Anseo, Cheonan, 330-704, Korea
(3) Korea Information Security Agency, Garak, Songpa, Seoul, 138-803, Korea

Abstract: IPv6 offers a number of other key design improvements over IPv4. IPv6 provide improved efficiency in routing and packet handling. And IPv6 can provide end-to-end security services such as access control, confidentiality, and data integrity with less impact on network performance. However, we can also prospect that there will be much more dangerous and tremendous type of attack on IPv6 than that on IPv4. So, we must consider on the advanced IP traceback function for tracing the spoofed real source on IPv6 packet using authentication mechanism. Proposed authentication mechanism supports the key disclosure mechanism on packet marking on the process of IP traceback.

Keywords: IPv6, Traceback, DDoS, Network Security, Packet Marking, Authentication.

1 Introduction

IPv4 has proven remarkably robust, easy to implement, and interoperable with a wide range of protocols and applications. However, the ongoing explosive growth of the Internet and Internet services has exposed deficiencies in IPv4. The global need for IP addresses has even added political force to the drive for IPv6 implementation. Aside from the increased address space, IPv6 offers a number of other key design improvements over IPv4. (1) IPv6 provide improved efficiency in routing and packet handling. And (2) IPv6 format allows plug-and-play deployment of Internet-enabled devices such as cell phones, wireless devices, and home appliances. (3) By providing globally unique addresses and embedded security, IPv6 can provide end-to-end security services such as access control, confidentiality, and data integrity with less impact on network performance. Finally, (4) IPv6 Multicast saves network bandwidth and improves network efficiency because IPv6 multicast completely replaces IPv4 broadcast functionality, by handling IPv4 broadcast functions such as router discovery and router solicitation requests[1].

Although IPv6 is powerful as above compared with existing IPv4 mechanism, we can also prospect that there will be much more dangerous and tremendous type of attack on IPv6 than that on IPv4[2]. So, we must consider on the advanced *traceback* function for tracing the spoofed real source on IPv6 packet using authentication mechanism.

In this paper, We overviewed the possible attack mechanism on IPv6 by considering of the vulnerability of proposed IPv6 structure[4,5,6,7]. IPv6 can provide end-to-end security services

---

[1]Corresponding author. Dept. of Software, Hanshin University, Korea. E-mail: hwlee@hs.ac.kr
[2]Div. of Info. and Comm. Engineering, Cheonan University, Korea. E-mail : shyoon@cheonan.ac.kr
[3]Korea Information Security Agency, Seoul, Korea. E-mail : {nhooh, hupark, jskim}@kisa.or.kr

such as access control, confidentiality, and data integrity using IPSec(AH/ESP) functions[1]. However, *spoofing attack on IPv6 is also possible by forging the real source*[2,3]. The reason is that we can commonly use IPv6 over IPv4 tunneling mechanism for encapsulating IPv6 traffic within IPv4 packets, to be sent over an IPv4 backbone[2]. Therefore, this enables similar diverse DDoS attack[4] on IPv6 to be applicable through an existing IPv4 infrastructure.

As a solution, we suggest *authenticated IP traceback mechanism on IPv6* against DDoS attack. IPv6 improves productivity by enabling network connectivity via a wider range of media and delivery mechanisms. But for general acceptance, the new IPv6 networks must demonstrate responsiveness at least equal to that of IPv4. Therefore, we considerate on the authenticated IP Traceback mechanism with hashed *MAC* function on IPv6 header. We suggest ideal form of authentication supported by signed disclosure of the keys on the process of IP traceback with the key disclosure mechanism on packet marking.

2   Secure Mechanism and Possible Attacks on IPv6

The features which IPv6 protocol can be summarized as follows: (1) New header format, (2) Large address space, (3) Efficient and hierarchical addressing and routing infrastructure, (4) Stateless and stateful address configuration, (5) Security, (6) Better Quality of Service(QoS) support, (7) New protocol for neighboring node interaction, (8) Extensibility[1,3].

IPv6 Header : A new format is designed to minimize header overhead. The IPv6 header is only twice the size of the IPv4 header. This is achieved by moving both nonessential and optional fields to extension headers that are placed after the IPv6 header.

Security features which are developed for IPv6 are known under the name of IP Security or shortly IPSec. IPSec is today most commonly used in IPv4, where it is optional, while it is mandatory to use it in the IPv6 protocol. IPSec consists of enhancements to original IP protocol which provide authenticity, integrity, confidentiality and access control to each IP packet through usage of the two new headers: AH (authentication header) and ESP (Encapsulations Security Payload)[1,3].

AH provides mechanisms for applying authentication algorithms to an IP packet, whereas ESP provides mechanisms for applying any kind of cryptographic algorithm to an IP packet including encryption, digital signature, and/or secure hashes. ESP can be used in tunnel or transport mode, similar to the Authentication Header. *However, complicated attacks also possible on IPv6 network.*

IP Duplication Attack : In the IPv4 world there is simple no standardized mechanism for detecting and reacting at the moment of duplicate address appearance on the network segment, but in the IPv6 the problem is more serious, as the interface ID portion of the L3 IPv6 address is directly derived from L2 device address.

Within RFC 2462 it is defined that if during the neighbor discovery process IPv6 device which received response that some other particular device is already using its proposed address - it must not use it. Therefore, it is fairly easy for a malicious user to craft the address which already exists on the local segment and hence achieve Denial of Service(DoS) attack on particular IPv6 device trying to initially obtain the stateless IPv6 address and start the IPv6 communication.

In both versions of the MIP protocol (v4 and v6) there are strong security requirements for tunnel authentication and optional tunnel confidentiality of the re-routed traffic from the mobile device to its home network. Mobile IPv6 can be considered as a mobility extension for the basic IPv6 functionality.

Mobile IP Attack : Mobile IPv6 correspondent node functionality can be used to launch reflection attacks against other parties and the tunnels between the mobile node and the home agent are attacked to make it appear like the mobile node is sending traffic while it is not[5].

3   Authenticated Traceback Mechanism on IPv6

3.1   IP Traceback on IPv6

IPv6 headers consists of 8 fields spread over 40 bytes[1,2,3]. *Differentiated Services(DS) field* contains a 6-bit value. Aggregates or aggregated flows may also be referred to as classes of packets. This study defines the unused 1 bits out of 6 DS field as *TM(traceback marking flag).* Flow *Lable(FL) field* is a 20-bit value used to identify packets that belong to the same flow. We mark router's information on FL field for IPv6 packet traceback. Finally, *Hop Limit(HL) field* in IPv6 packets is an 8-bit field, which is set at 255 in ordinary packets. We use this HL field value for calculating traceback distance.

Specifically because the maximum network hop count is 32 in general, the distance of packet transmission can be calculated only with the lower 6 bits out of the 8 bits of HL field in packet P_x arrived at router R_x. That is, the router extracts information of the lower 6 bits T_x from the HL field of packet P_x. T_x value indicates the distance of the packet from the attack system.

Authentication Mechanism for Traceback Two parties can share a secret key K. When party A sends a message M to party B, A appends the message with the MAC of M using key K. When B receives the message, it can check the validity of the MAC.

HMAC Function Let f denote a MAC function and f_K the MAC function using key K. Router R_i can apply a MAC function to its IP address and some packet-specific information with shared unique secret key K_i.

We could use the entire packet content in the MAC computation, i.e. encode R_i with its IP address A_i on packet P_i as $f_{K_i}(<P_i; A_i>)$.

Time-released Key Each router R_i first generates a sequence of secret keys, $\{K_{j;i}\}$ where each key $K_{j;i}$ is an element of a hash chain. By successively applying a one-way function g to a randomly selected seed, $K_{N;i}$, we can obtain a chain of keys, $K_{j;i} = g(K_{j+1;i})$.

Because g is a one-way function, anybody can compute forward (backward in time), e.g. compute $K_{0;i}$; ⋯ ; $K_{j;i}$ given $K_{j+1;i}$, but nobody can compute backward (forward in time) due to the one-way generator function.

3.2   Authenticated Traceback Path Marking

When informed of the occurrence of abnormal traffic, router R_x performs marking for packet P_x corresponding to congestion notification on IPv6 network. If the router received a congestion packet on which the Explicit Congestion Notification (ECN) field is set, it resets TM field in DS field as 1. And we can define K_x as the router R_x's authentication key. K_x can be used in hashed MAC function for packet authentication. We can calculates T_x for 8-bit HL field of packet P_x and stores it in the 6 bits of DS field, P_x^{TF}.

Router R_x commits to the secret key sequence through a standard commitment protocol, e.g. by signing the first key of the chain $K_{0;x}$ with its private key, and publish the commitment out of band, e.g. by *posting it on a web site.* We assume that each router has a certified public key. The time is divided into intervals. Router R_x then associates its key sequence with the sequence of the time interval, with one key per time interval. It is also sent by reserved 8-bit filed of AH header to the victim V.In time interval t, the router R_x uses the key of the current interval, $K_{t;x}$, to mark packets in that interval. The router uses $K_{t;x}$ as the key to compute the hashed MAC. R_x will then reveal/disclosure the key $K_{t;x}$ after the end of the time interval t. And the interval value t can also sent by reserved field of AH header.

Step 1 : Authenticated Packet Marking $P_x^{MF1} = H(f_{K_{t;x}}(<P_x^{TF}; A_x>))$. the router R_x's authentication chain key information t on $K_{t;x}$ can be sent by AH header.

Now when router R_y checks P_x^{TM} the value of TM(traceback marking flag) field in the packet and finds its flag is set by 1, the router applies the hashed MAC function to the value obtained by subtracting 1 from P_x^{TF}, and marks the resulting value on $f_{K_{t:y}}(< P_x^{TF} + 1; A_y >)$ by using MAC function f with chain key $K_{t:y}$ on its own IP address A_y. Then the router also calculates 8-bit hashed MAC value $H(f_{K_{t:y}}(< P_x^{TF} + 1; A_y >))$, and marks this value on P_x^{MF2}, the second 8 bits of FL field. After marking, the router sends the packet to the destination. On transmission path, intermediate router does not perform marking if finding TM is set as 1, because the packet has been marked by the previous router.

3.3 Authenticated Traceback Path Reconstruction

For a packet transmitted through the network, victim system V restructures the malicious DDoS attack path with authenticated verification process. First of all, let's say P_v is a set of packets arrived at victim system V. P_v is a set of packets corresponding to DDoS attacking, and M_v is a set of packets within P_v, which were marked by routers.

First it obtains $B_x{}'$ and K_x respectively included in the authenticated header. Now it is possible to obtain K_x by generating $B_x{}'$, which is authentication information in packet. And then extract K_x using A_x and timestamp information $t; x$.

Step 2 : Authenticated Reconstruction $M_i^{MF1} == H(f_{t;x}(M_i^{TF}; R_x), (R_x \in D(M_i) == 2)$
and $M_i^{MF1} == H((f_{t;x}((HLofM_i \wedge 00111111 + 2); R_x), (R_x \in D(M_i) == 2)$, where $D(M_i) = M_i^{TF} - (HLofM_i \wedge 00111111)$.

Now the victim system can restructure the actual attack path through which packets in malicious DDoS attack packet set P_v were transmitted by repeating the same process for M_j satisfying $D(M_j) == n, (n \geq 3)$. We can verify the packet's integrity with authentication process on IP traceback packet. And we can also verify hashed MAC value with chain key $K_{t:x}$ and $K_{t:y}$ from AH header.

4 Conclusions

We overviewed the possible attack mechanism on IPv6 by considering of the vulnerability of proposed IPv6 structure. As a solution, we propose a new traceback function on IPv6 packet. If a victim system is under attack, it identifies the spoofed source of the hacking attacks using the generated and collected traceback path information. Thus this study proposed a authenticated technique to trace back the source IP of spoofed DDoS IPv6 packets with a traceback function.

Acknowledgment

This work is supported by KISA and University IT Research Center(ITRC) Project of Korea.

References

[1] S. Deering, R. Hinden, "Internet Protocol, Version 6 (IPv6) Specification", RFC2460, IETF drafe, December 1998.
[2] Franjo Majstor, "Does IPv6 protocol solve all security problems of IPv4?", Information Security Solutions Europe, 7-9 October 2003.
[3] Pete Loshin, "IPv6 : Theory, Protocol, and Practice", Second edition, Morgan Kaufmann, 2003.
[4] John Elliott, "Distributed Denial of Service Attack and the Zombie and Effect", IP professional, March/April 2000.
[5] Timo Koskiahde, "Security in Mobile IPv6", 8306500 Security protocols, Tampere University of Technology, 2002.
[6] Tatsuya Baba, Shigeyuki Matsuda, "Tracing Network Attacks to Their Sources", IEEE Internet Computing, pp. 20-26, March, 2002.
[7] Hassan, Aljifri, "IP Traceback: A New Denial-of-Service Deterrent?", IEEE SECURITY & PRIVACY, pp.24-31, May/June, 2003.

VSP International
Science Publishers
P.O. Box 346, 3700 AH Zeist
The Netherlands

Lecture Series on Computer
and Computational Sciences
Volume 1, 2004, pp. 919-922

The Undeniable Digital Copyright Protection Scheme Providing Multiple Authorship of Multimedia Contents

Sung-Hyun Yun[1], Hyung-Woo Lee[2], Hee-Un Park, Nam-Ho Oh, Jae-Sung Kim[3]

(1) Div. of Info. and Comm. Engineering, Cheonan University, Anseo, Cheonan, 330-704, Korea

(2) Dept. of Software, Hanshin University, Yangsan, Osan, Gyunggi, 447-791, Korea

(3) Korea Information Security Agency, Garak, Songpa, Seoul, 138-803, Korea

Abstract: A digital copyright is a signature added to a digital multimedia contents. The copyright has been used to establish authorship and to check if the data has been tampered with. In existing copyright protection schemes based on ordinary signature scheme, some one can make counterfeited copyright and embeds it to the copyrighted digital contents to assert multiple claims of rightful authorship. In this case, an original author wants to make the copyright which can not be verified without help of the author so that only the intended verifier can be convinced about the validity of it.

In undeniable signature scheme, the signature can be verified and disavowed only with cooperation of the signer. If the copyright has property of undeniable signature, the verifier can not distinguish between valid and invalid copyright without help of the copyright owner. Generally, the digital multimedia contents is made by many authors. In application that requires many authors and a designated verifier, an algorithm for undeniable multi-signature is needed.

In this paper, the digital copyright protection scheme based on undeniable multi-signature scheme is proposed. The copyright can be verified and disavowed only in cooperation with all signers. Our scheme is secure against active attacks such as signature modification and repudiation of signature by signers. It can be applied to fair on-line sales of co-authoring copyrighted digital multimedia contents.

Keywords: Digital Copyright Protection Scheme, Undeniable Signature, Undeniable Multi-Signature, Multimedia Contents Security

1 Introduction

A digital copyright is generally used to establish author's right on digital multimedia contents. The digital copyright is a signature added to the digital multimedia contents so as to check if the contents has been tampered with or to identify its intended recipient. In the conventional digital signature schemes, a signer can make original signature on the digital data in which the signer can not repudiate the signature. Everybody can certify the signature since the ordinary signature has self-verification property. Therefore, in existing copyright protection schemes based on ordinary signature schemes, some one can make counterfeited copyright and embeds it to the copyrighted digital multimedia contents to assert multiple claims of rightful authorship. In this case, an original author wants to make the copyright which can not be verified without help of the author so that only the intended verifier can be convinced about the validity of it.

---

[1]Corresponding author. Div. of I&C Engineering, Cheonan University, Korea. E-mail : shyoon@cheonan.ac.kr
[2]Dept. of Software, Hanshin University, Korea. E-mail: hwlee@hs.ac.kr
[3]Korea Information Security Agency, Seoul, Korea. E-mail : {hupark, nhooh, jskim}@kisa.or.kr

An undeniable digital signature scheme is proposed by D.Chaum at first[2]. In the undeniable signature scheme, the digital signature can be verified and disavowed only with cooperation of the signer. There are many applications which conventional digital signature schemes can not be applied to. In on-line sales of digital contents, an owner of the digital contents wants to know whether a distributor sales the contents to customers fairly. In this case, the owner can satisfy on-line sales model if the model provides the mechanism which the customer can not buy the digital contents without help of the owner. The digital copyright based on undeniable signature scheme differ from ordinary copyright in that the customer can not distinguish between valid and invalid copyright without help of the copyright owner. Only the original owner can confirm this copyright as authentic to the customer.

Existing copyright protection schemes are mainly focused on protection of single owner's authorship[7]. Generally, the digital multimedia contents is made by many authors with cooperative works. The new digital copyright protection scheme is required to provide equal rights to co-workers. In application that requires many authors and a designated verifier, an algorithm for undeniable multi-signature is needed.

In this paper, the digital copyright protection scheme providing multiple authorship of multimedia contents is proposed. The proposed scheme is based on undeniable multi-signature scheme. The multi-signature can be verified and disavowed only in cooperation with all signers. In the proposed scheme, the El-Gamal signature scheme is modified to satisfy properties of undeniable signature and to extends it to make multi-signature[1,5]. Our scheme is secure against modification of multi-signature and repudiation of it by signers. It also can be applied to protection of digital copyright on co-authoring digital multimedia contents. In case of dispute between authors, the proposed scheme can resolve it by launching disavowal protocol to identify whether authors have cheated.

2   The Undeniable Digital Copyright Protection Scheme

To make the proposed scheme, we modify the El-Gamal signature equation[1], let $k(m_h + s) \equiv xr \ (mod \ p - 1)$, and extends it to accept undeniable properties of D.Chaum's scheme[2] and multi-signature properties[3,4]. The proposed scheme consists of multi-signature generation, multi-signature confirmation and disavowal protocols. In multi-signature generation protocol, a copyright maker can make an undeniable digital copyright. In multi-signature confirmation protocol, the digital copyright can be verified only with the help of all authors. If the copyright verification fails, the disavowal protocol is used to identify whether the copyright is invalid or some authors have cheated. The following parameters are used in the proposed scheme.

Authors: $u_1, u_2, \ldots, u_n$, Multimedia Contents: $m \in Z_{p-1}$
Author i's Private Key: $x_i \in Z_{p-1}$,   Author i's Public Key: $y_i \equiv g^{x_i} \ (mod \ p)$

(1) Multi-signature Generation Protocol

A copyright maker sends digital multimedia contents to all authors. Each author computes the undeniable signature and sends it to the copyright maker. The copyright maker uses each author's undeniable signature to compute the undeniable multi-signature.

Step 1: The copyright maker generates hash value $m_h = h(m, hpr)$ and sends (m, hpr) to the first author u_1. The hash parameter hpr is adjusted to make m_h as a primitive root of p.

Step 2: To make the common public key Y, the u_1 let $Y_1 = y_1$. The u_1 chooses a random number k_1. k_1 and $p - 1$ are relatively prime integers. The u_1 computes $r_1 \equiv m_h{}^{k_1} \ (mod \ p)$ and sends it to the second author u_2.

Step 3: The intermediate author $u_i \ (2 \leq i \leq n)$ receives (r_{i-1}, Y_{i-1}) from the u_{i-1}. The u_i chooses a random number k_i and computes $r_i \equiv r_{i-1}{}^{k_i} \equiv m_h{}^{\prod_{j=1}^{i} k_j} \ (mod \ p)$, $Y_i \equiv Y_{i-1}{}^{x_i} \equiv g^{\prod_{j=1}^{i} x_j} \ (mod \ p)$. The u_i sends (r_i, Y_i) to the next author u_{i+1}. If u_i is the last author, the u_i computes $R \equiv r_{n-1}{}^{k_n} \equiv$

$m_h{}^{\prod_{j=1}^{n} k_j} \pmod p$, $Y \equiv Y_{n-1}^{x_n} \equiv g^{\prod_{j=1}^{n} x_j} \pmod p$ and sends it to all authors as well as the copyright maker.

Step 4: Each author u_i $(1 \leq i \leq n)$ computes the undeniable signature s_i and sends it to the copyright maker. Since k_i and $p-1$ are relatively prime integers, there exists s_i satisfying the following equation, $k_i \cdot s_i \equiv x_i \cdot R - k_i \cdot m_h \pmod{p-1}$.

Step 5: The copyright maker computes the undeniable multi-signature $S \equiv \prod_{j=1}^{n}(m_h + s_j) \pmod p$.

(2) Multi-signature Confirmation Protocol

Step 1: The verifier chooses two random numbers (a, b) and computes the challenge $ch \equiv R^{S \cdot a} \cdot Y^{R^n \cdot b} \pmod p$. The challenge ch is delivered to the first author u_1.

Step 2: The u_1 computes the response $rsp_1 \equiv ch^{x_1^{-1}} \pmod p$ and sends it to the second author u_2.

Step 3: The intermediate author u_i $(2 \leq i \leq n)$ receives the response rsp_{i-1} from the u_{i-1}. Then the u_i computes the response $rsp_i \equiv rsp_{i-1}^{x_i^{-1}} \pmod p$ and sends it to the next author u_{i+1}. If the u_i is the last author, the response rsp_n is delivered to the verifier.

Step 4: If the equation $rsp_n \equiv m_h^{R^n \cdot a} \cdot g^{R^n \cdot b} \pmod p$ holds, the verifier ensures that multi-signature is valid. Otherwise, the disavowal protocol is launched to identify whether multi-signature is invalid or some authors have cheated.

(3) Disavowal Protocol

The verifier chooses two random numbers (c, d) and computes the second challenge $ch' \equiv R^{S \cdot c} \cdot Y^{R^n \cdot d} \pmod p$, $a \cdot d \neq b \cdot c \pmod{p-1}$. If the second response rsp'_n is not equal to $m_h^{R^n \cdot c} \cdot g^{R^n \cdot d} \pmod p$, additional step 5 is required.

Step 5: The verifier makes the discrimination equations. If $R_1 \equiv (rsp_n \cdot g^{-R^n \cdot b})^c \pmod p$ equals to $R_2 \equiv (rsp'_n \cdot g^{-R^n \cdot d})^a \pmod p$, the verifier confirms that multi-signature is invalid. Otherwise, at least more than one signer have cheated on valid multi-signature.

3 Undeniable Property Analysis

In this section, undeniable properties of the proposed scheme is analyzed. In theorem 1 and 2, we show correctness of our disavowal protocol.

Definition 1 *The valid multi-signature and the invalid multi-signature on the message m are defined as follows.* $X \equiv \prod_{j=1}^{n} x_j \pmod{p-1}$ *contains private keys of all signers and m_h is the hash result on m.*

- *Valid multi-signature :* $R \equiv m_h^{\prod_{j=1}^{n} k_j} \pmod p$, $\prod_{j=1}^{n} k_j(m_h + s_j) \equiv R^n \cdot X \pmod{p-1}$

- *Invalid multi-signature :* $R' \equiv m_h^{\prod_{j=1}^{n} k_j'} \pmod p$, $\prod_{j=1}^{n} k_j'(m_h + s_j) \neq R'^n \cdot X \pmod{p-1}$

We also define X' satisfying following equation, $\prod_{j=1}^{n} k_j'(m_h + s_j) \equiv R'^n \cdot X' \pmod{p-1}$

Theorem 1 *The proposed disavowal protocol can identify that signers have compute the invalid response on the valid multi-signature.*

Proof: If more than one signer have cheated during the multi-signature confirmation protocol, X^{-1} is modified. We assume that X^{-1} is the valid inverse of X and X'^{-1} is the invalid inverse of X. The verifier computes the challenge $ch \equiv R^{S \cdot a} \cdot Y^{R^n \cdot b} \pmod p$. If more than one signer compute the response improperly on the valid multi-signature, the response made by all signers becomes $rsp_n \equiv ch^{X'^{-1}} \pmod p$. Since the response rsp_n is not equal to $m_h^{R^n \cdot a} \cdot g^{R^n \cdot b} \pmod p$, the verifier launches disavowal protocol with new challenge value (c, d) as follows.

$$ch' \equiv R^{S \cdot c} \cdot Y^{R^n \cdot d} \pmod p, \quad rsp'_n \equiv ch'^{X'^{-1}} \pmod p$$

Since the second response rsp'_n is not equal to $m_h^{R^n \cdot c} \cdot g^{R^n \cdot d} \pmod p$, the verifier makes following discrimination equations.

$$R_1 \equiv (rsp_n \cdot g^{-R^n \cdot b})^c \pmod p, \quad R_2 \equiv (rsp'_n \cdot g^{-R^n \cdot d})^a \pmod p$$

From the above equations, R_1 is not equal to R_2. Therefore, we prove the correctness of the proposed disavowal protocol in case that more than one signer have cheated on the valid multi-signature. Q.E.D.

Theorem 2 *The proposed disavowal protocol can identify that the multi-signature is invalid.*

Proof: The first challenge and response on the invalid multi-signature is as follows.

$$ch \equiv R'^{S \cdot a} \cdot Y^{R'^n \cdot b} \equiv m_h^{a \cdot R'^n \cdot X'} \cdot g^{b \cdot R'^n \cdot X} \ (mod\ p), \quad rsp_n \equiv m_h^{a \cdot R'^n \cdot X' \cdot X^{-1}} \cdot g^{b \cdot R'^n} \ (mod\ p)$$

The second challenge and the response on the invalid multi-signature is as follows.

$$ch' \equiv R'^{S \cdot c} \cdot Y^{R'^n \cdot d} \equiv m_h^{c \cdot R'^n \cdot X'} \cdot g^{d \cdot R'^n \cdot X} \ (mod\ p), \quad rsp'_n \equiv m_h^{c \cdot R'^n \cdot X' \cdot X^{-1}} \cdot g^{d \cdot R'^n} \ (mod\ p)$$

The verifier makes the following discrimination equations. In the following equation, R_1 is equal to R_2.

$$R_1 \equiv (rsp_n \cdot g^{-R'^n \cdot b})^c \ (mod\ p), \quad R_2 \equiv (rsp'_n \cdot g^{-R'^n \cdot d})^a \ (mod\ p)$$

Therefore, we prove the correctness of the proposed disavowal protocol in case that the multi-signature is invalid. Q.E.D.

4   Conclusion

Many authors can participate jointly in authoring of a digital multimedia contents. In this case, the copyright of the digital contents must be shared by all participants. In this paper, we propose the undeniable digital copyright protection scheme based on undeniable multi-signature scheme and undeniable property of the proposed scheme is proved. For sales on digital multimedia contents by on-line, a customer can buy it by launching multi-signature confirmation protocol. Without the consent of all authors, the customer can not buy digital contents. Especially, in case of dispute between authors, the proposed disavowal protocol can discriminate whether authors have cheated or the digital copyright is invalid.

Acknowledgment

This work is supported by KISA and University IT Research Center(ITRC) Project from Korea.

References

[1] T.Elgamal, "A Public Key Cryptosystem and a Signature Scheme Based on Discrete Logarithms," IEEE Transactions on Information Theory, Vol. IT-31, No. 4, pp.469-472, 1985.

[2] D.Chaum, "Undeniable Signatures," Advances in Cryptology, Proceedings of CRYPTO'89, Springer-Verlag, pp.212-216, 1990.

[3] L.Harn, "(t,n) Threshold Signature and Digital Multisignature," Workshop on Cryptography & Data Security, pp.61-73, 1993.

[4] S.H.Yun, T.Y.Kim. "A Digital Multisignature Scheme Suitable for EDI Message," Proceedings of 11th International Conference on Information Networking, pp.9B3.1-9B3.6., 1997.

[5] S.H.Yun, T.Y.Kim, "Convertible Undeniable Signature Scheme," Proceedings of IEEE High Performance Computing ASIA'97, pp. 700-703, 1997.

[6] S.H.Yun, S.J.Lee, "An electronic voting scheme based on undeniable signature scheme," Proceedings of IEEE 37th carnahan conference on Security Technology, pp.163-167, 2003.

[7] Andre Adelsbach, Birgit Pfitzmann, Ahmad-Reza Sadeghi : Proving Ownership of Digital Content, 3rd International Information Hiding Workshop (IHW '99), LNCS 1768, Springer-Verlag, 117-133, 1999.

VSP International
Science Publishers
P.O. Box 346, 3700 AH Zeist
The Netherlands

*Lecture Series on Computer
and Computational Sciences*
Volume 1, 2004, pp. 923-928

Field Test-based Propagation Path-Loss Models for Terrestrial ATSC Digital TV

Seung youn Lee[1], Myong chul Shin[1], Sang yule Choi[2],
Seung won Kim[3], Jae sang Cha[4].

[1]SungKyunKwan Univ., Department of Information & Communication Eng.
Kyonggi, 440-746, Korea
[2]Induk Institute of Technology, San76, Wolgye-dong, Nowon-Gu, Seoul 139-749, Korea.
[3]Radio and Broadcasting Research Lab., ETRI, Daejeon, 305-350, Korea.
[4]SeoKyeong Univ., Dept. of. Information & Communication Eng., Seoul, 136-704, Korea.

Abstract: In this paper, we propose Propagation Path-loss models for terrestrial ATSC (Advanced Television Systems Committee) DTV (Digital Television) using the results of ATSC DTV field-test measurements in Seoul, Korea in 2001. Numerical formulae of proposed Path-loss models are derived from the measured values of received field strength. Newly proposed Path-loss models have closer correctly agreement with the field-measured results than these of conventional Path-loss models (e.g., Free space model, Friss model, Hata model, 2-Ray model, ITU-R 526-3 model and Lee model) in the LOS(Line of Sight) and Non-LOS areas. Path-loss models presented in this paper can be readily utilized usefully for efficient ATSC DTV system implementation requiring accurate link-budget calculation.

Keywords: ATSC DTV, path-loss model, link margin, link budget.

1. Introduction

The studies of development of ATSC DTV technology have been progressed actively in various areas of the world so far. In comparison with conventional NTSC(National Television System Committee)-type TV, ATSC DTV could implement HDTV(high definition television) broadcasting easily with robust signal propagation property. From the results of ATSC DTV field measurements[2][3], we certified the propagation distance such as cell-coverage-radius concept of mobile communication could be reached to father than 100km or more when the transmitted output power is given as 1kW. Since one of the most important characteristics of ATSC DTV signal can be found in the longer propagation distance (i.e., longer cell coverage property) than that of cellular mobile communication or NTSC TV, the path loss model which means large scale channel characteristics can be one of the most important components to be considered in the system design or performance analysis. However, new path-loss model that present terrestrial ATSC DTV have not been proposed till now, because DTV's concept itself appeared just nowadays. Thus, in this paper, we prove that the large deviation between the measured result and conventional path-loss [4]-[8] exists. Moreover, we proposed new Propagation Path-loss models for terrestrial ATSC DTV using the results of ATSC DTV field measurements as shown in Table 1.

2. Proposed Field Test-based Path-loss models for terrestrial ATSC DTV

As shown in Figs.1-4, we certified the large deviation between the Filed measured result and theoretical values of conventional Path-loss models [4]-[8] existed in the LOS and Non-LOS area, where conventional Path-loss models are selected that have suitable property for NTSC TV or mobile communications to be compared. Table 1 summarized new Path-loss models proposed in this paper. Thus, as shown in Figs. 1-4, newly proposed Path-loss models are more suitable for the field-measured results than that of conventional Path-loss models in the LOS and Non-LOS areas.

2.1 Proposed LOS model

---

[4] Corresponding author: Dept. of Information & communication Engineering Seokyeong Univ., E-mail: chajs@skuniv.ac.kr

Measured path loss is compared with existing propagation models such as Free space model, Friss Model, Hata model and 2-Ray model. The Fig. 1 shows the compared result. Free space model is similar to the one of measured path loss' average. Friss model has a low average value of path loss, the changing pattern of Hata model is quite different [4][8].

There is a difference between 2-Ray model and measured path loss in numerical value. But, both of them have a similarity in a changing pattern. Therefore, by using the similarity of changing pattern, we made our suggested LOS model based on the various revision of 2-Ray model.

Table 1. Equations of Path-Loss Models proposed in this paper

		Equations of Proposed Path loss Model
LOS area	Basic Eq.	PL[dB] = 40log(d[km]) − {10log(G_t) + 10log(G_r) + 20log(h_t) + 20log(h_r)} + 120 + α(k)
	Suburban Eq.	In this case, α(k) of Basic Eq = 22.17dB
	Urban Eq.	In this case, α(k) of Basic Eq = 37.32dB
	Massed Urban	In this case, α(k) of Basic Eq = 55.58dB
Non-LOS area	Single Knife-edge Diffraction Model	$$v = h\sqrt{\frac{2}{\lambda}\left(\frac{1}{d_1}+\frac{1}{d_2}\right)}$$
		$v \le -1$: $L = 0$ *(dB)*
		$-1 \le v \le 0$: $L = 20\log(0.5 - 0.62v)$ *(dB)*
		$0 \le v \le 1$: $L = 20\log(0.5e^{0.95})$ *(dB)*
		$1 \le v \le 2.4$: $L = 20\log(0.4 - \sqrt{0.1184-(0.1v-0.38)^2})$ *(dB)*
		$v \ge 2.4$: $L = 20\log(-0.225/v)$ *(dB)*
	Mutiple Knife-edge Diffraction Model	$L_d = \sum_{i=1}^{N} L'_i + L''(wx)_1 + \sum_{i=1}^{N} L''(yz)_i - 10\log C_N$ *(dB)*

P_r : Received power , P_t : Transmitted Power,
h_t: Height of Transmitter(Tx), h_r: Height of Receiver(Rx),
G_t : Tx Antenna Gain, G_r : Rx Antenna Gain, λ : wave-length,
d (m) : Distance between Tx and Rx, f: frequency,
h : (Height of Knife-edge obstacle)-(Height of Line between Tx and Rx Antenna)
d1,: Distance between Tx Antenna and Knife-edge obstacle.
d2,: Distance between Knife-edge obstacle and Rx Antenna.
L'i:diffraction loss over the I-th cylinder(dB),
L''(wx)₁ : sub-path diffraction loss for the section between points w and x for the first cylinder
L''(wx)₁ : sub-path diffraction loss for the section between points y and z for the all cylinders.
C_N: correction factor to account for spreading loss diffraction over successive cylinders.

Conventional 2-Ray model is a path loss model considered direct wave as well as reflected wave. When the receiver is far from the transmitter by distance d, the path loss equation is as follows

$$PL(dB) = 40\log d - (10\log G_t + 10\log G_r + 20\log h_t + 20\log h_r)$$

d : the distance of transmitter to receiver(m)
G_t : transmitter antenna gain
G_r : receiver antenna gain
h_t : the height of transmitter antenna
h_r : the height of receiver antenna

The numerical formula of conventional 2-ray model is considered only distance and height of transmitter and receiver. Moreover, frequency and other environmental factors are left out of consideration in the conventional 2-ray model.

Based on the basic equation, various new LOS models are proposed, which can be applied to Suburban, Urban, Massed urban. According to the regional characteristics, we followed process that basic equation is added to various average constants calculated with consideration of obstacles by regional groups.

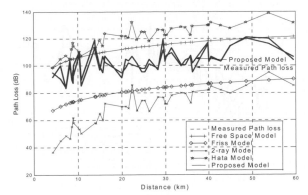

Figure. 1 Comparison of Measured value and Theoretical values of Path-Loss models in the LOS areas.

The criteria of classification of regional groups are defined as follows in this paper:

- Massed-urban : In this paper, it is defined as a downtown area where high-rise apartment buildings and skyscrapers higher than 15 stories are clustered close together around measurement area
- Urban : In this paper, it is defined as a downtown area where lower apartment buildings and residential buildings lower than 15 stories are scattered around measurement area.
- Sub Urban : In this paper, it is defined as a suburbs where buildings higher than 3 stories seldom exist around measurement area, come in sight.

Fig. 1 shows the comparison between measured value and theoretical values of path-Loss models in the LOS areas. In the Fig 1, we can find path-loss pattern of the proposed model is more similar to the actual measured result than that of other conventional path-loss models.

Fig. 2 is a graph that is compared between standard deviation of measurement result and that of calculation result calculated in each propagation model. We can see that standard deviation between measured path loss and the Standard deviation of conventional path-loss models exists between 8.43dB and 13.52dB. However, Standard deviation of the suggested model in this paper is just only 5.45dB. So, we can confirm that the model suggested in this paper corresponds to the measured result more than that of conventional path-loss models.

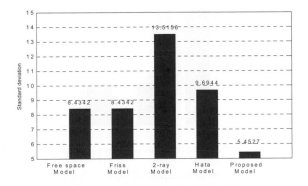

Figure. 2 Standard deviation to the Measured value of Propagation Models in the LOS areas.

2.2 Proposed Non-LOS model

Non-LOS area is the area where wave doesn't come to directly but come to after being diffracted because there are obstacles between receiver and transmitter. Particularly, In the Korea having a lot of mountain regions, we can find easily many Non-LOS regions.

In this paper, measured path loss data was compared to the conventional Non-LOS models such as ITU-R 526-3 model[5] and Lee model[7].

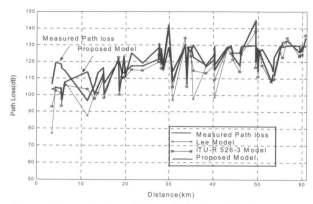

Figure. 3   Comparison of Measured value and Theoretical values of Path-Loss models in the Non-LOS areas.

As shown in the Fig.3, changing pattern of the Lee model and the ITU-R 526-3 model is similar to that of measured path loss. The reason why it doesn't match completely is that the only topographical information is considered without factors like artificial construction. The path loss of Lee model fluctuated a lot. On the other hand, the path loss of ITR-R 526-3 model is similar to that of measured path loss and a little changed comparing to the Lee model.

In this paper, we presented a Non-LOS model by altering distance element in the Lee model in order to reflect the characteristics that the location of terrestrial DTV transmitter antenna in this country is much higher than any other. The numerical equation of a presented Non-LOS model in this paper is corresponded to the Lee model[7]. However, the distance factors of a presented model is obtained by using different method compared to the conventional Lee model[7]. In the presented Non-Los model, distance factors are obtained by the following method. That is, draw a straight line between transmitter and receiver. And then draw a orthogonal line from the top of the obstacle to the line connected line transmitter and receiver. Let an effective distance d_1 and d_2 are the length from orthogonal point to transmitter and receiver respectively, and effective height h is the length from the orthogonal point to the top of obstacle.

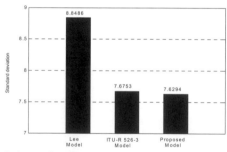

Figure. 4 Standard deviation to the Measured value of Propagation Models in the Non-LOS areas.

Figureg.4 show the Standard deviation to the measured value of Propagation Models in the Non-LOS areas. According to the new distance factors, we confirmed that the standard deviation value of the presented Non-LOS Model is less than that of existing propagation models, as written in the Figure.4

3.   Estimated example of Link budget

Table.2 shows the calculated example of Link budget for the ATSC DTV system using our proposed path loss model.

In the Table 2. data rate is 19.39Mbps and bandwidth is 6MHz. The frequency of 473MHz, which is in the middle of the MBC (470~476MHz), was used, and other transmission parameters were also based on the frequency of MBC transmission antenna.

The real values of KBS test equipment [3] were applied to the gain of the receiver antenna and to the implementation loss.

Recommended values in FCC regulations were applied to the noise figure and critical SNR.

P_t (Average T_x power) used for our field tests is 30dB(1kW) as shown in Table 2, however, Pt of a normal transmitter also can be considered as a 2.5kW or 34 dB.

P_N is equal to N plus N_f, where N(Average noise power per bit) equres to -106.2 dBm and N_f is 10dB. Recommend values in FCC regulations[10] are used for Minimum E_b/N_o(S) and Implementation Loss(I).

The results of estimated link budget using the suggested path loss model and varying the distance from the transmitter, shown that performance of data transmission is good up to a distance of 80km from the transmitter with sufficient margin.

And also, if normal transmission power of 2.5kW is used as the power of the transmitter, data transmission will be more stable since margin of 4dB is increased.

The result of this investigation for margin of link budget at various distances, concludes that the path loss model suggested in this document is very useful for estimating or calculating actual Link budget.

4. Conclusions

In this paper, we proposed new Propagation Path-loss models for terrestrial ATSC DTV using the results of ATSC DTV field measurements in Seoul, Korea. Newly proposed Path-loss models agree to the field-measured results correctly more than that of conventional Path-loss models in the LOS and Non-LOS areas.

Table.2 The calculated example of Link budget for the ATSC DTV system

Parameter	Unit	Value				
Bandwidth	MHz	6				
Information data rate	Mbps	19.39				
Distance (d)	km	10	20	40	60	80
Average Tx power(Pt=1kW)	dBm	30	30	30	30	30
Tx antenna gain (G_t)	dBi	6.6				
geometric center frequency (f_C)	MHz	473				
Proposed Path Loss at d(km) (PL)	dB	72.89	84.93	96.97	104.02	109.01
Rx antenna gain(G_r)	dBi	17.00				
Rx power ($P_r=P_t$-PL)	dBm	-42.89	-54.93	-66.97	-74.02	-79.01
Average noise power per bit(N)	dBm	-106.2				
Rx noise figure (N_f)	dB	10.00				
Average noise power per bit($P_N=N+N_f$)	dBm	-96.20				
Minimum E_b/N_o(S)	dB	15.20				
Implementaion Loss(I)	dB	3.50				
Link Margin M(M=P_r-P_N-S-I)	dB	34.61	22.57	10.53	3.48	-1.51

References

[1] http://www.atsc.org/standards.html
[2] http://tri.kbs.co.kr/publish.html
[3] Korea Broadcasting System, Ministry of Information and Communication, "A Study on the Financial Support For Digital Terrestrial TV Broadcasting Testbed Implementation". 2000.10.31
[4] Theodore S. Rappaport, "Wireless Communications – Principles & Practice", Prentice Hall, pp. 69-138, 1996
[5] International Telecommunication Union, "ITU-R Recommendations", PN.526-3, "Propagation By Diffraction". 1994PN series Volume, ITU propagation in non-ionized media

[6] Electronics and Telecommunications Research Institute, " Development of utilization Technologies for Radio Resources", pp.326-334, , 1996.12.

[7] Lee, W.C.Y., "Mobile Communications Engineering", Mcgraw hill publications, New York, 1985

[8] Hata, Masaharu, "Empirical Formula for Propagation Loss in Land Mobile Radio Services", IEEE Transaction on Vehicular Technology, Vol. VT-29, No. 3, pp. 317-325, August 1980.

[9] Young-Woo Suh, Ha-Kyun Mok, Tae-Hoon Jwon, " Analysis of field that result of digital TV On-Channel repeaters", Journal of Korean Society of Broadcast Engineers, 2002, vol 7 No.1 pp.10~20.

[10] Gerald W. Collins, Fundamentals of digital Television Transmission, A Wiley-Interscience Publication JHONE WILEY @ SONS, Inc.,2001,pp.23

VSP International
Science Publishers
P.O. Box 346, 3700 AH Zeist
The Netherlands

*Lecture Series on Computer
and Computational Sciences*
Volume 1, 2004, pp. 929-936

Application of Binary Zero-correlation-duration Sequences to Interference-cancelled WPAN system

*Jaesang Cha, **Kyungsup Kwak, **Jeongsuk Lee and ***Chonghyun Lee

*Dept. of Information and Communication Eng. Seokyeong Univ. Seoul, Korea
**UWB Wireless Communications Research Center (UWB-ITRC), Inha Univ. Incheon, Korea
***Dept. of Electronic Engineering Seokyeong Univ.
16-1 Jung-nung dong Sungbuk-ku, Seoul, 136-704, Korea

Abstract: In this paper, we present an interference-cancelled wireless personal area network (WPAN) system based on ultra wideband (UWB) wireless communication system using binary zero-correlation-duration (ZCD) spreading sequences. The proposed interference-cancelled WPAN system is based on the direct sequence - ultra wideband (DS-UWB) scheme using special binary spreading sequences with zero correlation duration (ZCD) characteristics. After presenting a mathematical model for interference-cancelled WPAN system, we present theoretical bit error rate (BER) performance. Through computer simulation, we verify high system capacity of the proposed system and superior performance over the conventional Walsh-based DS-UWB systems in multi-path and in multiple access environments. Furthermore, outstanding performance of the proposed system is provided under multiple access interference (MAI) and multi-path interference (MPI). Finally, we show that perfect interference cancellation and high system capacity can be achieved without adopting expensive multi user detection (MUD) scheme or other interference cancellation schemes.

Keywords: ZCD; UWB; spreading code, MAI, MPI

PACS: 84.40.Ua Telecommunications: signal transmission and processing; communication satellites

1. INTRODUCTION

The UWB technique has been utilized for military radar technique so far. Recently, the technique has been paid much attention and been debated by IEEE 801.15.3 [1] for standardization. Even though the technique is limited for indoor use such as Pico cell, the UWB wireless communication technique can accommodate about 1Gbps transmit data rate. Therefore, the technique can be a strong candidate of 4G mobile communication system which is concerned with wireless multimedia transmission.

Conventional UWB wireless communication technique has been mainly focused on PPM(pulse position modulation) and TH(time hopping) method as a multiple access scheme[2],[3]. However, this time-domain schemes could have vital demerits and large performance degradation in multi-path environment since position synchronization is so difficult in muti-path environment. Therefore, recently, another code domain UWB scheme such as DS-UWB[4] has been studied by some researchers. In the DS-UWB, since the short pulse signal is modulated by PSK(phase shift keying) and spreaded using spreading code, BER performance of DS-UWB can be largely affected by the orthogonal characteristic of the spreading codes. Furthermore, system performance of DS-UWB based multiple access system or conventional CDMA cellular systems are determined by MAI(multiple access interference) and MPI(multi-path interference) environment . In order to solve this MPI and MAI problems of CDMA applications, many researchers including authors[5]-[14] have proposed ZCD spreading codes maintaining the orthogonality within a local duration around the origin. However, the conventional ZCD codes used for CDMA are not appropriate for the high data rate transmission due to the low code capacity and their narrowband environment.

Thus in this paper, we apply ZCD code to the new system with very enlarged frequency band; i.e., UWB system. Thus, we propose a ZCD-UWB transmission system based on DS-UWB using ZCD

*** Corresponding author: Dept. of Electronic Engineering Seokyeong Univ., E-mail: chonglee@skuniv.ac.kr

code and present the performance results with computer simulation. Furthermore, we show that with the proposed system, complete interference cancellation is possible without adopting expensive MUD scheme or other interference cancellation schemes. The performance of the ZCD-UWB system is analyzed and is verified with computer simulation in order to show the outstanding performance of proposed system in multi-user and multi-path environment.

2. SYSTEM MODEL

We define ZCD-UWB as the DS-based UWB system using ZCD spreading code. Thus, in this section, we consider ZCD-UWB system by assuming antipodal modulation for transmitted binary symbols. Then UWB transmitted waveform of ZCD-UWB is defined at the following :

$$s^k(t) = \sum_{i=-\infty}^{\infty} \sum_{n=0}^{N-1} \sqrt{P_k} b_i^k c_n^k z(t - iT_b - nT_c) \tag{1}$$

where, N is the Period of spreading code, $b_i^k \in \{\pm 1\}$ are the modulated data symbols for the kth user, $c_n^k \in \{\pm 1\}$ are the spreading code for the kth user, $z(t)$ is the transmitted pulse waveform, T_b is the bit period and, T_c is the chip period.

In this model, the UWB pulse denoted by $z(t)$ is assumed to include the differential effects in the transmitter and receiver antenna systems. A typical pulse employed in the literature [2],[3] is the second derivative of a Gaussian pulse given by

$$p_2(t) = \left[1 - 4\pi\left(\frac{t}{T_m}\right)\right] \exp\left[-2\pi\left(\frac{t}{T_m}\right)^2\right] \tag{2}$$

In the receiver, for simplicity, we assume that the multi-path components arrives at the some integer multiple of a minimum path resolution time. By assuming the minimum path resolution time T_m ($T_m \sim$ 1/Bs), we can write the received waveform as follows:

$$r(t) = \sum_{l=0}^{L-1} c_l^0 s^0(t - lT_m - \tau^o) + \sum_{k=1}^{K} \sqrt{P_k} \sum_{l=0}^{L-1} c_l^k s^k(t - lT_m - \tau^k) + n(t) \tag{3}$$

where L is the number of multi-paths, c_l^k is the amplitude of the lth path, T_m is the pulse period, and $n(t) \sim N(0,1)$ is the AWGN(additive white gaussian noise). The multipath delay is described as $\tau^k = q_k T_m$, q_k being an integer uniformly distributed in the interval [0, $N_r N_c$-1] and N_r is the processing gain of spreading code and $N_c = T_c/T_m$ which results in $0 \le \tau^k < T_r$. Here note that T_r is the maximum time delay to be considered. For the detection of received signal, we could form the vector of sufficient statistics **y** obtained by collecting the outputs of the K individual matched filters over one symbol with the input r(t), written as follows:

$$\mathbf{y} = \mathbf{RWCb} + \mathbf{n} \tag{4}$$

where the cross-correlation matrix of normalized signature waveform vector is

$$\mathbf{R} = \int_0^{T_r} \mathbf{d}(t)\mathbf{d}^H(t)dt \tag{5}$$

Here, n is a Gaussian zero-mean K-vector with covariance matrix equal to R and the A is a multi-channel matrix of Gaussian random variables. Here, the matrix of C , W and the vector d are given by

$$C = \begin{bmatrix} \mathbf{c}^0 & 0 & 0 & \cdots \\ 0 & \mathbf{c}^1 & 0 & \cdots \\ 0 & 0 & \ddots & \cdots \\ \cdots & 0 & 0 & \mathbf{c}^K \end{bmatrix} \tag{6}$$

$$W = \begin{bmatrix} \mathbf{W}^0 & 0 & 0 & \cdots \\ 0 & \mathbf{W}^1 & 0 & \cdots \\ 0 & 0 & \ddots & \cdots \\ \cdots & 0 & 0 & \mathbf{W}^K \end{bmatrix} \tag{7}$$

$$\mathbf{d}(t) = [\mathbf{a}_0^T(t) \quad \mathbf{a}_1^T(t) \quad \mathbf{a}_K^T(t)] \tag{8}$$

where $\mathbf{W}^k = \sqrt{P_k}\mathbf{I}_L$, $\mathbf{a}_k(t) = [a_0{}^k z(t) \quad a_1{}^k z(t - T_c) \quad \cdots \quad a_{Nr}{}^k z(t - (Nr-1)T_c)]^T$ and $\mathbf{c}_k(t) = [c_0{}^k \quad c_1{}^k \quad \cdots \quad c_{L-1}{}^k]^T$.

3. ANALYSIS

In this section, we present the theoretical BER(bit error rate) of the proposed system by using the results in [15]. By referring the BER value of the single-user detection, we rewrite BER of the kth user as

$$P^k(\sigma) = \frac{1}{2} \sum_{e1 \in \{-1,1\}} \cdots \sum_{\substack{ej \in \{-1,1\} \\ j \neq k}} \cdots \sum_{ek \in \{-1,1\}} Q(\frac{c_k}{\sigma} + \sum_{j \neq k} e_j \frac{c_k}{\sigma} \beta \rho_{jk}) \tag{9}$$

where $Q(x)$ is the complementary cumulative distribution function of the unit normal variable defined as

$$Q(x) = \int_x^\infty \frac{1}{\sqrt{2\pi}} e^{-t^2/2} dt \tag{10}$$

and $\rho_{jk} = \mathbf{R}_{j,k}$,

In Rayleigh fading channel, the bit error probability can be written as

$$P^{Fk}(\sigma) = \frac{1}{2}(1 - \frac{c_k}{\sqrt{\sigma^2 + \sum_j c_j (\beta \rho_{jk})^2}}) \tag{11}$$

Since the cross-correlations of our proposed scheme are zeros, the performance of proposed ZCD-UWB system is as good as single user case, or no MAI and thus the ber probabilities both in AWGN and Rayleigh fading can be written as follows:

$$P^k(\sigma) = \frac{1}{2}Q(\frac{c_k}{\sigma}) \tag{12}$$

$$P^{Fk}(\sigma) = \frac{1}{2}(1 - \frac{c_k}{\sqrt{\sigma^2 + c_k}}) \tag{13}$$

Since the $Q(x)$ function is monotonically decreasing function with respect to x, the probabilities of the proposed system are always smaller than the system using code with cross-correlation.

Next, suppose MUD scheme is applied to our proposed ZCD-UWB system. As described in [15] the probabilities of error both in AWGN can be written as

$$P^k{}_d(\sigma) = \frac{1}{2}Q\left(\frac{c_k}{\sigma}\sqrt{1 - \mathbf{r}_k'\mathbf{R}_k^{-1}\mathbf{r}_k}\right) \tag{14}$$

where $\mathbf{a}_k$ is the kth column of $\mathbf{R}$ without the diagonal element, and $\mathbf{R}_k$ is the (KL-1)x(KL-1) matrix that by striking out the kth row and column from $\mathbf{R}$. With the property of ZCD(zero correlation duration) of the proposed system, $\mathbf{a}_k^T\mathbf{R}^{-1}\mathbf{a}_k$ becomes zero and thus the error probability of MUD applied system becomes as follows:

$$P^k{}_d(\sigma) = \frac{1}{2}Q\left(\frac{c_k}{\sigma}\right) \tag{15}$$

which is the same as that of SUD(single user detection) system.

Similarly, the BER performance of MUD applied system under Rayleigh fading can be written as follows:

$$P^k{}_d(\sigma) = \frac{1}{2}(1 - \frac{1}{\sqrt{1 - \dfrac{\sigma^2}{c_k^2}\dfrac{1}{(1 - \mathbf{r}_k'\mathbf{R}_k^{-1}\mathbf{r}_k)}}}) = \frac{1}{2}(1 - \frac{c_k}{\sqrt{\sigma^2 + c_k}}) \tag{16}$$

, which is the same as that of SUD system.

Comparing the results obtained MUD and SUD applied system, we can conclude that the proposed system eliminates the necessity of the MUD while keeping the same performance obtained with MUD.

4. RESULTS

In this section, we present the performance of the proposed system via computer simulation. We assume the bit rate of system to be 100 Mbps around and the channel to be Rayleigh fading channel which is generated by using Jake's model. We use Walsh-Hadamard code and ZCD code for comparison and present the BER performance in Figure 1 through 4. In Figure 1, we present the BER performance under MPI condition by using SUD scheme. Also, we present the BER performance under MAI of two users and MPI of two multi-paths condition by using SUD scheme in Figure 2. In these figure, we can observe that the proposed ZCD-UWB system exhibits better performance that that of Walsh-Hadamard UWB system. The results obtained using SUD and MUD under MAI of two users and MPI of two multi-paths are presented in Figure 3. In Figure 4, the results obtained using SUD and MUD under MAI of four users and MPI of four multi-paths are presented. In these figures, we observed that the proposed ZCD-UWB system exhibits same BER performance both in SUD and MUD conditions, which agrees with the theoretical results discussed in last section.

5. ZCD-UWB DESIGN EXAMPLE FOR WPAN

We consider a ZCD-UWB system for WPAN with a MAI free Pico-cell. For Uplink under MAI environment, the time delay differences of signals arriving at access point (AP) within a Pico-cell are due to the different propagation delay between the mobile stations and AP. The implementation of MAI-cancelled system is corresponds to the construction of intra-cell without MAI. Thus, the MAI-free Pico-cell of WPAN can be designed as

$$R = \frac{c \cdot \delta}{2} = \left(\frac{c \cdot (ZCD - 1)}{(4 \cdot R_c)}\right) = \frac{c \cdot T_p \cdot PF \cdot (ZCD - 1)}{4} \tag{17}$$

where R is the radius of MAI-free Pico-cell, δ is the maximum propagation delay time in the cell, R_c is chip rate, c is the speed of light, T_p is the pulse width, and PF is defined as Pulse factor, i.e., Pulse number/Chip.

Next, we can implement MPI cancelled system by designing the ZCD-length covering the delay-path-length of MPI. In other word, if the delay-path-length of MPI is included in the $0.5 \times (ZCD-1)$

duration, MPI cancelled system can be obtained by using ZCD property. The system capacity proposed system could be represented by the peak bit rate as

$$Rb = \frac{Mr \cdot rT \cdot M \cdot B_N}{N_r \cdot T_p \cdot PF} \tag{18}$$

And the terminology definition in (18) and the specification examples for ZCD-UWB systems are listed in the Table1 in detail. In the system examples of Table 1, TDD-based Multi-band ZCD-UWB systems with ZCD property for MPI cancellation are considered. From the results of estimated system specification, we can verify the usability of the proposed system and thus the proposed system have ZCD capability and Peak bit rate of 117Mbps to 700Mbps using simple BPSK modulation within the WPAN Pico-net area.

6. CONCLUSIONS

In this paper, we propose a ZCD-UWB wireless communication system with which complete interference cancellation property. By incorporating DS-UWB scheme and ZCD spreading code with ZCD characteristics, we built a mathematical model for ZCD-UWD wireless system and we present theoretical BER performance. To verify the system performance, we perform computer simulation and we show that performance of the proposed system is superior to the systems of conventional scheme and is compatible with the MUD system in multi-path and in multiple access environment. Finally, we show that the proposed ZCD-UWD system can remove interference signal perfectly with low computational complexity.

Acknowledgments

This work was supported in part by University IT Research Center Project (INHA UWB-ITRC) in Korea.

References

[1] http://www.ieee802.org/15/pub/TG3.html

[2] M. Win and R. Scholtz, "Ultra-Wide Bandwidth Time-Hopping Spread Spectrum Impulse Radio for Wireless Multiple-Access Communications," IEEE Transactions on Communications, Vol. 48, No. 4, April 2000.

[3] R. A. Scholtz, "Multiple access with time-hopping impulse modulation, "Proc. MILCOM '93, vol. 2, 1993, pp. 447-450.

[4] J. Foerster, "The Performance of a Direct-Sequence Spread ULTRA- Ultra-Wideband System in the Presence of Multipath, Narrowband Interference, and Multiuser Interference," IEEE UWBST Conference Proceedings, May, 2002.

[5] Cha,J.S.,Kameda,S.,Takahashi,K.,Yokoyama,M.,Suehiro,N.,Masu,K.and Tsubouchi, K , "Proposal and Implementation of Approximately synchronized CDMA system using novel biphase sequences", Proc. IEICE ITC-CSCC 99, Vol. 1, pp.56-59, Sado Island, Japan, July13-15, 1999.

[6] Cha, J.S., Kameda, S., Yokoyama, M., Nakase, H., Masu, K., and Tsubouchi, K.: 'New binary sequences with zero-correlation duration for approximately synchronized CDMA '. Electron. Lett., 2000, Vol. 36, no.11, pp. 991–993

[7] Fan, P, Suehiro, N., Kuroyanagi, N and Deng, X.M.: "Class of binary sequences with zero correlation zone," Electron. Lett., 1999, Vol. 35, no.10, pp. 777–779

[8] Deng, X., and Fan, P.: "Spreading sequence sets with zero correlation zone," Electron. Lett., 2000, Vol. 36, no.11, pp. 993–994

[9] Cha,J.S. and Tsubouchi, K, "Novel binary ZCD sequences for approximately synchronized CDMA", Proc. IEEE 3G Wireless01, Sanfransisco, USA, Vol. 1, pp.810-813, May 29, 2001.

[10] Cha,J.S, "Class of ternary spreading sequences with zero correlation duration", IEE Electronics Letters , Vol. 36, no.11, pp. 991-993, 2001.5.10

[11] Cha,J.S. and Tsubouchi, K, "New ternary spreading codes with with zero-correlation duration for approximately synchronized CDMA", Proc. IEEE ISIE 2001, Pusan, Korea, Vol. 1, pp.312-317, June 12, 2001.

[12] Cha,J.S., Song,S.I.,Lee,S.Y.,Kyeong,M.G. and Tsubouchi, K, "A class of Zero-padded spreading sequences for MAI cancellation in the DS-CDMA systems", Proc. IEEE VTC01 Fall, Atlantic City, USA,October 6, 2001.

[13] Jae-sang Cha, Sang-yule Choi, Jong-wan Seo, Seung-youn Lee, and Myung-chul Shin, "Novel Ternary ZCD Codes With Enhanced ZCD Property and Power-efficient MF Implementation", Proc. IEEE ISCE'02, Erfurt, Germany, Vol. 1, pp.F117-122, 2002.

[14] Mun Geon Kyeong, Suwon park, Jae Kyun Kwon, Dan Keun Sung, and Jae-sang Cha, "3G Enhancements with a view towards 4G", Tutorial. CIC 2002, Seoul, Korea, Tutorial-3, pp.91~245,2002.

[15] Sergio Verdu, "Multi User Detection", Cambridge university press, 1998

APPENDIX

Table 1. Specification Example for the ZCD-UWB system

	System A	System B	System B
Access, Duplex	Multi-band DS-UWB, TDD		
Given BW(bandwidth)	Low band(3.15 to 5GHz) + High band(5.825 to 10.6GHz) [2]		
BW/channel	500MHz plus Guard Band/channel		
Data modulation	BPSK		
Error correction coding	No		
Pulse width Tp	2 nsec		
Rake combining	EGC or MRC		
Network, Cell radius R	Pico-net, within 10m		
BW number for Ch. B_N	14 (4Ch./Low band + 10Ch./High band)		
Mono pulse type	Gaussian pulse with 2 nsec		
Pulse Factor	1		
Chip rate R_C	500Mcps		
Spreading sequence	Enhanced ZCD preferred pair		
	$N_r = 12$ ZCD = 11	$N_r = 32$ ZCD = 31	$N_r = 72$ ZCD = 71
Receptible Time to the Delay path / cell radius	10nsec / 1.5m	30nsec/ 4.5m	70nsec/ 10.5m
Peak Bit rate R_b	700Mbps	263 Mbps	117 Mbps

ZCD： Zero correlation duration, δ： Maximum propagation delay

time Gc: Guard chip,

M： Family size of sequence, Tp : Pulse width, PF : Pulse factor, i.e.

Pulse number/Chip, Nr : Spreading factor = Sequence period N R_c :

Chip rate, M_r : Mary-Phase level factor; i.e, M_r of BPSK case =1,

rT : Time share ratio of TDD, e.g., 0.6,

R: Cell radius, R_b: Peak bit rate, B_N : BW number for channelization.

$Gc = (ZCD-1)/2, \ \delta = Gc/Rc, \ R = (c \cdot \delta)/2, \ R_c = 1/(T_p \cdot PF)$
$R_b = (Mr \cdot rT \cdot M \cdot B_N)/(N_r \cdot T_p \cdot PF)$

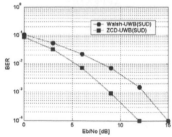

Figure 1: Comparison of BER performance of ZCD-UWB, Walsh/DS-UWB, in the MPI environment with SUD (1user 4path, Uplink, Spreading Factor=64chips, Path delays=[0 2 5 7]chips).

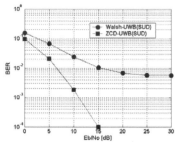

Figure 2: Comparison of BER performance of ZCD-UWB, Walsh/DS-UWB, in the MAI and MPI environment with SUD (2user 2path, Uplink, Spreading Factor=64chips, Path delays for Each users =[0 1], [1 2]chips).

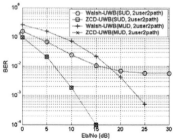

Figure 3: Comparison of BER performance of ZCD-UWB, Walsh/DS-UWB, in the MAI and MPI environment with SUD and MUD (2user 2path, Uplink, Spreading Factor=64chips, Path delays for Each users =[0 1], [1 2]chips).

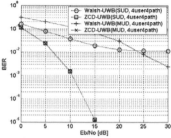

Figure 4: Comparison of BER performance of ZCD-UWB, Walsh/DS-UWB, in the MAI and MPI environment with SUD and MUD (4user 4path, Uplink, Spreading Factor=64chips, Path delays for Each users=[0 1 2 4], [1 2 3 4], [0 1 3 2], [1 3 2 4]).

VSP International
Science Publishers
P.O. Box 346, 3700 AH Zeist
The Netherlands

*Lecture Series on Computer
and Computational Sciences*
Volume 1, 2004, pp. 937-951

Numerically Stable and Efficient Algorithm for Vector Channel Parameter and Frequency Offset Estimation

Chong Hyun Lee†

Jae Sang Cha§[1]   **Kyungsup Kwak**‡ and **Jeongsuk Lee**‡
†Department of Electronics,
§Dept. of Information and Communication Engineering , Seokyeong Univ. Seoul, Korea
‡UWB Wireless Communications Research Center (UWB-ITRC), Inha Univ. Incheon, Korea

Abstract: In this paper, we present a novel and numerically efficient algorithm for vector channel and calibration vector estimation, which works when frequency offset error caused by either unstable oscillator or Doppler effect is present in Spread Spectrum antenna system. We propose an estimation algorithm based on Gauss-Seidal algorithm rather than using eigen-decomposition or SVD in computing eigen-values and eigen-vectors at each iteration. The algorithm is based on the two step procedures, one for estimating both channel and frequency offset and the other for estimating the unknown array gain and phase. Consequently, estimates of the DOAs, the multi-path impulse response of the reference signal source, and the carrier frequency offset as well as the calibration of antenna array are provided. The analytic performance improvement in multiplications number is presented. The performance of the proposed algorithm is investigated by means of computer simulations. Throughout the analytic and computer simulation, we show that the proposed algorithm reduces the number of multiplications by order of one.

Keywords: Gauss-Seidal algorithm, Vector Channel Estimation, Eigen-Decomposition, Singular Value Decomposition

PACS: 84.40.Ua Telecommunications: signal transmission and processing; communication satellites

1   Introduction

Space-time processing techniques employing multiple antennas are used to increase spectrum efficiency and capacity for future cellular communications. The goal of the space-time processing is to combine spatial and temporal information. For the downlink beamforming, which is one of the most fundamental space-time processing techniques, accurate channel estimation such as direction of arrivals(DOAs) and time delays is essential. Most high-resolution DOA estimation algorithms, however, require perfect knowledge of the array manifold, which is not feasible in practice. The gain and phase responses of a channel (or an antenna) vary according to temperature and humidity changes from day to day [1], and therefore *online* calibration is preferable in wireless cellular communications. Various array calibration methods with or without known source directions have been proposed [2]-[5]. In [5], the direction-independent array gain and phase are estimated by using knowledge of the true field covariance at the sensor locations. In [2], one first estimates the DOAs with the unknown sensor parameters being set at their nominal values. These estimated DOAs are

---

[1]Corresponding author. Department of Information and Communication Engineering E-mail: chajs@skuniv.ac.kr

then used to estimate the unknown sensor parameters with an optimization technique. This two-step process continues iteratively until a certain convergence criterion is met. The other methods for the case where the DOAs are known have been studied in [3] and [4] The methods in [2]-[5], however, require that the number of signal sources should be less than the number of antennas, and are not applicable to direct-sequence code-division multiple access (CDMA) communication systems. Furthermore, deploying known signals at a known location may not be tolerable since it would increase multiple access interference and decreases the channel capacity as well.

In general, most digital communication systems undergo frequency offset errors caused by either unstable oscillator or Doppler effect. It is well known that this error causes degradation of the system performance. To solve this error, many algorithms have been reported [6] and [7] by utilizing the training sequences to obtain the frequency offset.

In this paper, we present a new estimation calibration algorithm which estimates channel parameters and the carrier offset as well as calibration vector. The algorithm does not use training sequences and date model is made by incorporating the frequency offset term with the model for an asynchronous CDMA-based antenna array in [8] and [10]. The proposed algorithm efficiently removes the necessity of eigen-decomposition in estimating channel parameters. The algorithm is based on Gauss-Seidal algorithm [9] and is proved that the total amount of computation is reduced by order of one. To verify the performance of the algorithm, computer simulations have been done by changing the parameters.

2 Data Formulation

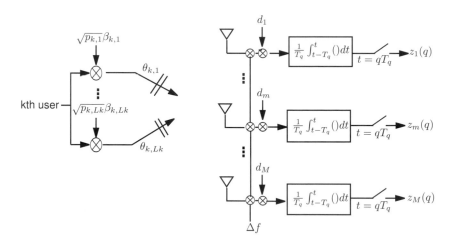

Figure 1: The channel and receiver front end including frequency offset.

Assume that antenna array is composed of M elements and K_a users are in a cell. Suppose that the received signals at the array are sampled at chip rate T_c and have the identical frequency offset, Δf for all K_a users. The multipath channel and the receiver front-end including frequency offset is shown in Figure 1. Then the obtained complex sequence with an unknown complex antenna gain d_m can be expressed as:

$$z_m(q) = d_m \exp(j2\pi\Delta f q T_c) \sum_{k=1}^{K_a} \sum_{l=1}^{L_k} \sqrt{P_{k,l}} \exp(j\phi_m^{k,l})\beta_{k,l}(q)y_{k,l}(q) + n(q)$$

where $P_{k,l}$ and $\beta_{k,l}(q)$ are the received power and the envelope of the path fading, L_k is the number of multipaths from the kth user, and $\phi_m^{k,l}$ is the phase delay due to the signals coming from the angle of $\theta_{k,l}$ (for the lth path from the kth user). The term $n(q)$ represents additive white Gaussian noise with zero-mean and covariance σ_n^2 at the receiver. The term $y_{k,l}(q)$ represents the chip matched filter output of the transmitted signal from the kth user.

Suppose the signal is collected for one bit interval T and formed into a vector. Then, by incorporating Δf with the asynchronous CDMA model in [8] and [10], we can write the signal of the lth path from the kth user as follows:

$$x_m^{k,l}(i) = d_m \exp(j\phi_m^{k,l})\mathbf{F}(\Delta f)\left[\mathbf{u}_{k,l}^R,\ \mathbf{u}_{k,l}^L\right]\left[\begin{array}{c}\gamma_{k,l}(i-1)\\\gamma_{k,l}(i)\end{array}\right]$$

where $\gamma_{k,l}(i-1)$ and $\gamma_{k,l}(i)$ are complex constants which involve the power, the fading and the symbol of transmitted signals. The matrix $\mathbf{F}(\Delta f)$ is a diagonal matrix defined as

$$\mathbf{F}(\Delta f) = diag(1, \exp(j2\pi\Delta f T_c), \cdots, \exp(j2\pi\Delta f(N-1)T_c)$$

where N represents the processing gain defined as $N = T/T_c$.

The vector pair $[\mathbf{u}_{k,l}^R,\ \mathbf{u}_{k,l}^L]$ has the form

$$\mathbf{u}_{k,l}^R = U_k^R \mathbf{h}_{k,l},$$
$$\mathbf{u}_{k,l}^L = U_k^L \mathbf{h}_{k,l},$$

where $\mathbf{h}_{k,l}$ is a vector of non-integer time delay and

$$U_k^R = [\mathbf{p}_k^R(0)\ldots\mathbf{p}_k^R(N-1)], \quad U_k^L = [\mathbf{p}_k^L(0)\ldots\mathbf{p}_k^L(N-1)],$$

The vector $\mathbf{p}_k^R(\tau_{k,l})$ and $\mathbf{p}_k^L(\tau_{k,l})$ are vectors of code sequence, which can be written as

$$\mathbf{p}_k^R(\tau_{k,l}) = [0,\ldots,0,\ c_k(N-\tau_{k,l}),\ldots,c_k(N-1)]^H,$$
$$\mathbf{p}_k^L(\tau_{k,l}) = [c_k(0),\ldots,c_k(N-\tau_{k,l}-1),\ 0,\ldots,0]^H$$

where $c_k(t)$ is the spreading code of the kth user and the superscript H represents the complex conjugate transpose. The integer $\tau_{k,l}$ is time delay such that $\tau_{k,l} \in \{0,\cdots,N-1\}$. Here, note that non-integer time delay can be expressed by choosing appropriate values in vector $\mathbf{h}_{k,l}$. After forming the matrix $Z = [\mathbf{z}_1(i),\cdots,\mathbf{z}_M(i)]$, and stacking the *row vectors* of Z into an $MN \times 1$ single composite snapshot vector $\mathbf{z}(i)$, we obtain

$$\mathbf{z}(i) = \mathbf{A}\mathbf{s}(i) + \mathbf{n}(i),$$

where

$$\mathbf{A} = [\mathbf{a}_{1,1}^R(\Delta f, \theta_{1,1}, \mathbf{h}_{1,1}, \mathbf{d}), \mathbf{a}_{1,1}^L(\Delta f, \theta_{1,1}, \mathbf{h}_{1,1}, \mathbf{d}),\ldots,\mathbf{a}_{K,L_K}^L(\Delta f, \theta_{K_a,L_K}, \mathbf{h}_{K_a,L_K}, \mathbf{d})],$$

$$[\mathbf{a}_{k,l}^R(\cdot), \mathbf{a}_{k,l}^L(\cdot)] = \mathbf{F}(\Delta f)[\mathbf{u}_{k,l}^R,\ \mathbf{u}_{k,l}^L] \otimes (\mathbf{b}(\theta_{k,l}) \odot \mathbf{d}),$$

$$\mathbf{b}(\theta_{k,l}) = [e^{j\phi_1^{k,l}}, e^{j\phi_2^{k,l}},\ldots,e^{j\phi_M^{k,l}}]^H, \quad \mathbf{d} = [d_1,\ldots,d_M]^H,$$

$$\mathbf{s}(i) = [\gamma_{1,1}(i-1), \gamma_{1,1}(i),\cdots,\gamma_{K,L_K}(i-1), \gamma_{K,L_K}(i)]^H.$$

Here, $\odot$ represents element-by-element multiplication, and $\otimes$ the Kronecker product.

3 Vector Channel Estimation

Without loss of generality, let us assume that the user using code vector $\mathbf{p}_1(\tau)$ is reference user and $\mathbf{h}$ is the vector of time delay to be estimated. By using the received signal vector $\mathbf{z}$, we perform the eigen-decomposition of R_{zz}:

$$R_{zz} = E\{\mathbf{z}\mathbf{z}^H\} = \mathbf{V}\mathbf{D}\mathbf{V}^H = [\mathbf{E}_S, \mathbf{E}_N]\mathbf{D}[\mathbf{E}_S, \mathbf{E}_N]^H$$

where $\mathbf{E}_S$ and $\mathbf{E}_N$ span the signal space and the noise space, respectively. Then, we define MUSIC-like cost function as follows:

$$J_1(\Delta f, \mathbf{h}, \theta, \mathbf{d}) = \parallel \mathbf{E}_N^H \mathbf{v}_1^R(\Delta f, \theta, \mathbf{d}) \parallel_2^2 + \parallel \mathbf{E}_N^H \mathbf{v}_1^L(\Delta f, \theta, \mathbf{d}) \parallel_2^2, \tag{1}$$

where

$$\mathbf{v}_1^R(\Delta f, \theta, \mathbf{d}) = \mathbf{F}(\Delta f)\mathbf{u}_1^R \otimes (\mathbf{b}(\theta) \odot \mathbf{d}),$$
$$\mathbf{v}_1^L(\Delta f, \theta, \mathbf{d}) = \mathbf{F}(\Delta f)\mathbf{u}_1^L \otimes (\mathbf{b}(\theta) \odot \mathbf{d}).$$

The vectors $\mathbf{u}_1^R$ and $\mathbf{u}_1^L$ can be written as

$$\mathbf{u}_1^R = U_1^R \mathbf{h}, \quad \mathbf{u}_1^L = U_1^L \mathbf{h},$$

where

$$U_1^R = [\mathbf{p}_1^R(0) \dots \mathbf{p}_1^R(N-1)], \quad U_1^L = [\mathbf{p}_1^L(0) \dots \mathbf{p}_1^L(N-1)],$$

For simplicity, let us assume that estimate of $\mathbf{d}$ is available. Then, in order to find the channel parameters and frequency offset, a multi-dimensional search would be needed in the space of $\mathbf{h}$, Δf and θ. Instead, by using the "Mixed Product Rule" of kronecker product of $(AC \otimes BD) = (A \otimes B)(C \otimes D)$ [11], we rewrite $\mathbf{v}_1^R()$ and $\mathbf{v}_1^L()$ as

$$\mathbf{v}_1^R(\Delta f, \theta, \mathbf{d}) = C_1^R(\Delta f)B(\theta)\mathbf{h}, \quad \mathbf{v}_1^L(\Delta f, \theta, \mathbf{d}) = C_1^L(\Delta f)B(\theta)\mathbf{h}$$

where

$$C_1^R(\Delta f) = [\mathbf{F}(\Delta f)U_1^R] \otimes I_M, \quad C_1^L(\Delta f) = [\mathbf{F}(\Delta f)U_1^L] \otimes I_M. \tag{2}$$

The matrix I_M is the identity matrix of size M. The matrix $B(\theta)$ which contains both angle and calibration information can be written as:

$$B(\theta) = I_N \otimes (\mathbf{b}(\theta) \odot \mathbf{d}). \tag{3}$$

Then the cost function $J_1(\Delta f, \mathbf{h}, \theta, \mathbf{d})$ can be written, with respect to $\mathbf{h}$ and θ as

$$J_1(\Delta f, \mathbf{h}, \theta; \mathbf{d}) = \mathbf{h}^H \mathbf{Q}_1(\theta, \Delta f)\mathbf{h}, \tag{4}$$

where

$$\mathbf{Q}_1(\theta, \Delta f) = B^H(\theta)[C_1^R(\Delta f)^H E_N E_N^H C_1^R(\Delta f) + C_1^L(\Delta f)^H E_N E_N^H C_1^L(\Delta f)]B(\theta).$$

Since transmitted multi-path fading is unknown, we can only determine $\mathbf{h}$ to within a complex constant. Thus, by introducing the constraint of $\parallel \mathbf{h} \parallel_2 = 1$, the solution for $\mathbf{h}$ is given by

$$\mathbf{h} = \nu_{min}(\mathbf{Q}_1(\theta, \Delta f)), \tag{5}$$

where ν_{min} denotes the eigenvector of $\mathbf{Q}_1(\theta, \Delta f)$ associated with the minimum eigenvalue $\lambda_{min}(\mathbf{Q}_1(\theta, \Delta f))$. By substituting (5) to (4), we get the function of angle parameter θ,

$$\theta = \arg\min_{\theta} \lambda_{min}(\mathbf{Q}_1(\theta, \Delta f)). \tag{6}$$

Here, note that $\mathbf{Q}_1(\theta, \Delta f)$ is a function of θ and Δf. Thus in order to find $\mathbf{h}$, we need to find the eigenvector corresponding the minimum eigenvalue by searching the space of θ and Δf. However, by using the fact that θ and Δf are the irrelevant parameters, a suboptimal algorithm is possible, which finds the one parameter (say θ) by fixing the other parameter (say Δf) or vice versa.

In short, the proposed algorithm can be expressed as follows:
First, with initial calibration vector $\mathbf{d}$ and frequency offset Δf, we obtain estimates of $\mathbf{h}_1$ and θ by finding the values minimizing the J_1 in (4). Next by using the obtained channel parameters, we find channel offset (Δf) by finding the root of the polynomial obtained by replacing $\exp(j2\pi\Delta f T_c)$ by z.

Finally, we find the calibration vector $\mathbf{d}$ which minimizes a certain cost function J_3 using the obtained channel parameters and the frequency offset. The above steps are iterated until the cost function converges J_3 to the minimum value. By assuming that the first user is the reference user, we shall summarize the estimation procedure below

A. *Estimation of channel parameters and frequency offset*

Procedure 1
With estimated (Δf) and $\mathbf{d}$,

(A) Formulate $\mathbf{Q}_1(\theta_i, \Delta f)$ using (2) and (3) for each discrete $0 \le \theta_i \le \pi$, $(1 \le i \le I_\theta)$;

(B) Find and record the minimum eigenvalue $\lambda_{min}^{(i)}$ of $\mathbf{Q}_1(\theta_i, \Delta f)$;

(C) Plot $\{\lambda_{min}^{(i)}\}_{i=1}^{I_\theta}$ with respect to $\{\theta_i\}_{i=1}^{I_\theta}$, and select L_1 local minima $\{\theta_{1,i}\}_{i=1}^{L_1}$; Compute the $\{\mathbf{h}_{1,i}\}_{i=1}^{L_1}$ corresponding $\{\theta_i\}_{i=1}^{I_\theta}$ from the plot;

(D) Using the estimated $\{\theta_{1,i}\}_{i=1}^{L_1}$ and $\{\mathbf{h}_{1,i}\}_{i=1}^{L_1}$, formulate a cost function J_2

$$J_2(\Delta f, \mathbf{h}_1, \theta; \mathbf{d}) = \sum_{i=1}^{L_1} (\| \mathbf{E}_N^H \hat{\mathbf{v}}_1^R(\theta_{1,i}, \mathbf{h}_{1,i}, \Delta f) \|_2^2 + \| \mathbf{E}_N^H \hat{\mathbf{v}}_1^L(\theta_{1,i}, \mathbf{h}_{1,i}, \Delta f) \|_2^2) \qquad (7)$$

where

$$\hat{\mathbf{v}}_1^R(\theta_{1,i}, \mathbf{h}_{1,i}, \Delta f) = [\mathbf{F}(\Delta f)U_1^R \mathbf{h}_{1,i}] \otimes (\mathbf{b}(\theta_{1,i}) \odot \mathbf{d}),$$
$$\hat{\mathbf{v}}_1^L(\theta_{1,i}, \mathbf{h}_{1,i}, \Delta f) = [\mathbf{F}(\Delta f)U_1^L \mathbf{h}_{1,i}] \otimes (\mathbf{b}(\theta_{1,i}) \odot \mathbf{d}).$$

(E) By replacing $\exp(j2\pi\Delta f T_s)$ in matrix $\mathbf{F}(\Delta f)$ by z, reformulate the $J_2(\cdot)$ in (7) into a polynomial function. Find the roots of $J_2(\cdot)$ and choose a root, z_o which is closest to unit circle. Then the frequency offset estimate can be obtained as follows:

$$\Delta f = \frac{1}{2\pi T_c} \arg(z_o)$$

End of Procedure 1

B. *Estimation of calibration vector* $\mathbf{d}$

With the estimates $\mathbf{h}_{1,i}$, Δf and $\theta_{1,i}$ obtained in the procedure 1, we can find the frequency compensated code vectors $\hat{\mathbf{u}}_i^R$ and $\hat{\mathbf{u}}_i^L$, $(i = 1, \dots, L_1)$ by using the following equation

$$\hat{\mathbf{u}}_i^R = \mathbf{F}(\Delta f)U_1^R \mathbf{h}_{1,i}, \quad \hat{\mathbf{u}}_i^L = \mathbf{F}(\Delta f)U_1^L \mathbf{h}_{1,i}. \qquad (8)$$

By substituting the vectors of $\hat{\mathbf{u}}_i^R$, $\hat{\mathbf{u}}_i^L$ and $\theta_{1,i}, (i = 1, \ldots, L_1)$ into equation (7), we define a cost function J_3 defined as follows

$$J_3(\mathbf{d}; \mathbf{h}_1, \theta, \Delta f) = \sum_{i=1}^{L_1} (\| E_N^H \hat{\mathbf{v}}_1^R(\mathbf{d}) \|_2^2 + \| E_N^H \hat{\mathbf{v}}_1^L(\mathbf{d}) \|_2^2)$$

where

$$\hat{\mathbf{v}}_1^R(\mathbf{d}) = \hat{\mathbf{u}}_i^R \otimes (\mathbf{b}(\theta_{1,i}) \odot \mathbf{d}),$$
$$\hat{\mathbf{v}}_1^L(\mathbf{d}) = \hat{\mathbf{u}}_i^L \otimes (\mathbf{b}(\theta_{1,i}) \odot \mathbf{d}).$$

This function represents the current cost value obtained with the current estimates of channel parameters and calibration vector $\mathbf{d}$. Note that $\hat{\mathbf{v}}_1^R(\mathbf{d})$ and $\hat{\mathbf{v}}_1^L(\mathbf{d})$ can be rewritten as

$$\hat{\mathbf{v}}_1^R(\mathbf{d}) = \hat{U}_R \hat{B}_i \mathbf{d}, \quad \hat{\mathbf{v}}_1^L(\mathbf{d}) = \hat{U}_L \hat{B}_i \mathbf{d},$$

where

$$\hat{U}_i^R = diag([\underbrace{u_1^R, u_1^R, \ldots, u_1^R}_{M} \cdots \underbrace{u_N^R, u_N^R, \ldots, u_N^R}_{M}])$$

$$\hat{U}_i^L = diag([\underbrace{u_1^L, u_1^L, \ldots, u_1^L}_{M} \cdots \underbrace{u_N^L, u_N^L, \ldots, u_N^L}_{M}]) \tag{9}$$

$$\hat{B}_i = \left[\tilde{\mathbf{b}}^H(\theta_{1,i}), \ldots, \tilde{\mathbf{b}}^H(\theta_{1,i}) \right]^H.$$

Here, $\tilde{\mathbf{b}}(\theta_{1,i}) = diag(\mathbf{b}(\theta_{1,i}))$, and u_j^R and $u_j^L, (j = 1, \ldots, N)$ are the jth elements of $\hat{\mathbf{u}}_i^R$ and $\hat{\mathbf{u}}_i^L$, respectively. Then the cost function J_3 can be simplified as

$$J_3(\mathbf{d}; \Delta f, \mathbf{h}_1, \theta) = \mathbf{d}^H \mathbf{Q}_2 \mathbf{d},$$

where

$$\mathbf{Q}_2 = \sum_{i=1}^{L_1} B_i^H [\hat{U}_i^{RH} E_N E_N^H \hat{U}_i^R + \hat{U}_i^{LH} E_N E_N^H U_i^L] B_i.$$

Now we can find $\mathbf{d}_{new}$ which minimizes $J_3(\cdot)$ with respect to either one of the constraints below

$$(a)\ \mathbf{d}^H \mathbf{w} = 1, \quad (b)\ \| \mathbf{d} \|_2 = 1, \tag{10}$$

where $\mathbf{w} = [1, 0, \cdots, 0]^H$.

We shall summarize the procedure below:

Procedure 2

(A) Formulate $\mathbf{Q}_2$ using (8), (9) and $\{\theta_{1,i}\}_{i=1}^{L_1}$;

(B) Find $\mathbf{d}_{new}$ which satisfies (a) or (b) in (10):

$$(a)\ \mathbf{d}_{new} = \mathbf{Q}_2^{-1} \mathbf{w} / \mathbf{w}^H \mathbf{Q}_2^{-1} \mathbf{w};$$
$$(b)\ \mathbf{d}_{new} = \nu_{min}(\mathbf{Q}_2);$$

(C) Compute one of

$$(a)\ J_{new} = \mathbf{d}_{new}^H \mathbf{Q}_2 \mathbf{d}_{new};$$
$$(b)\ J_{new} = \lambda_{min}(\mathbf{Q}_2);$$

If $J_2 - J_{new} > Threshold$, then go to Procedure 1, and repeat Procedure 2;
If $J_2 - J_{new} \leq Threshold$, then terminate;

End of Procedure 2

C. *Efficient Estimation Algorithm*

Major computational load for the channel estimation comes from the eigen-decompositions of $\mathbf{Q}_1(\theta, \Delta f)$ corresponding to θ. On the other hand, in order to find calibration vector $\mathbf{d}$, eigen-decomposition or matrix inverse of $\mathbf{Q}_2$ is needed as well as a few matrix-vector manipulations. For each θ, iteration, $o(N^3)$ flops are needed for eigen-decomposition of $\mathbf{Q}_1(\theta, \Delta f)$ and $o(M^3)$ flops needed for the eigen-decomposition or the inverse of $\mathbf{Q}_2$ [9]. This amount of complexity is tolerable since calibration process could be done once in a while.

However, if we assume that the curve obtained by plotting minimum eigenvalue of $\lambda_{min}(\mathbf{Q}_1(\theta, \Delta f))$ for each θ is smooth, then we can find fast algorithm for finding $\nu_{min}(\mathbf{Q}_1(\theta, \Delta f))$, the eigenvector associated with the minimum eigenvalue of $\mathbf{Q}_1(\theta, \Delta f)$. The power method of $\mathbf{Q}_1(\theta, \Delta f)^{-1}$ is the one of fast method [9]. However, direct application of the power method is not desirable, since it requires matrix inverse of $\mathbf{Q}_1(\theta, \Delta f)$. Suppose that vectors ν_i and ν_{i+1} are the eigenvectors associated with the minimum eigenvalues of $\mathbf{Q}_1(\theta_i, \Delta f)$ and $\mathbf{Q}_1(\theta_{i+1}, \Delta f)$, respectively. By using the assumption of smoothness of curve, we may write the relationship of two vectors as follows

$$\nu_{i+1} = \mathbf{Q}_1(\theta_{i+1}, \Delta f)^{-1} \nu_i$$

or

$$\mathbf{Q}_1(\theta_{i+1}, \Delta f) \nu_{i+1} = \nu_i. \tag{11}$$

Suppose that ν_i and $\mathbf{Q}_1(\theta_{i+1}, \Delta f)$ are obtained already, then ν_{i+1} can be found by solving the linear equation in (11). To obtain a fast adaptive solution, we use the Gauss-Seidel iteration for solving the linear equation and we summarize the procedure as follows:

Procedure 3

A Initialization

    1. $\theta_0 = 0$;

    2. $\theta = \theta_0$;

    3. $\lambda_0 =$ minimum eigenvalue of $(\mathbf{Q}_1(\theta, \Delta f))$

    4. $\nu^{old} =$ eigenvector corresponding to λ_0

B Loop

    1. for $(k = 1 :$ Maximum Iteration $)$

    2. for $(i = 1 : N)$

$$\nu_i^{k+1} = (\nu_i^{old} - \sum_{j=1}^{i-1} a_{i,j} \nu_j^k - \sum_{j=i+1}^{N} a_{i,j} \nu_j^k)/a_{i,i}$$

    where the ν_i^{old} and $a_{i,j}$ are the ith and (i,j) elements of ν^{old} and $(\mathbf{Q}_1(\theta, \Delta f))$, respectively.

C Termination Check

    If $\theta \leq \theta_{max}$, the maximum angle, then compute the following;

    1. $\theta = \theta + \Delta\theta$, where $\Delta\theta$ is the increment in angle.

    2. Compute $\mathbf{Q}_1(\theta, \Delta f)$

3. $\nu^{old} = \nu^{K+1}$

4. go to Procedure [B];

If $\theta > \theta_{max}$, then terminate;

End of Procedure 3

In order to compare the computational cost between the algorithm based on eigen-decomposition and the one based on Gauss-Seidal, we examine the number of multiplications in computing $(\mathbf{Q}_1(\theta, \Delta f))$ for each θ. The number of multiplications required to obtain the matrix is expressed as follows:

$$T(n) = 4Nn^2 + (2N^2 + 2N - 4NK)n - 2(N^2 + N)K \tag{12}$$

where the $n = NM$ and K is the dimension of noise subspace in $\mathbf{R}_{zz}$. By setting K to be 90% of n, we plot the number of multiplications according to the number of antennas and N, the length of code sequence in Figure (2).

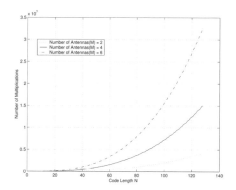

Figure 2: Number of multiplication in computing $(\mathbf{Q}_1(\theta, \Delta f))$

By simple evaluation of the Gauss-Seidal algorithm in [C], we can find that the algorithm requires N^2 multiplications for single iteration k. This amount of computations is fairly smaller since $12N^3$ is required in performing eigen-decomposition of $(\mathbf{Q}_1(\theta, \Delta f))$. By adding these numbers into (12), we plot the number of multiplications required for single angle θ and for total angle $\theta_0 \leq \theta \leq \theta_{max}$ in Figure (3) and (4). We can observe that the fast algorithm works with fairly small amount of computations.

4   EXPERIMENTAL RESULTS

The performance of the proposed algorithm with either one of the constraints in (10) is similar, and therefore, the result for the case of $\mathbf{d}(1) = 1$ is presented in this chapter. We use a uniform circular array with six antennas separated by half a wavelength. We use random BPSK modulated data streams, and the Gold codes with the processing gain of $N = 31$. We assume that 15 users $(K = 15)$ produce two multi-path signals $(L_k = 2)$. For simplicity, we consider azimuthal angle only. The DOAs of the reference user are assumed to be $[40°, 85°]$ and the delays be $[19.3, 25.3]$

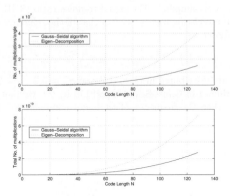

Figure 3: Number of multiplications using antenna number = 4

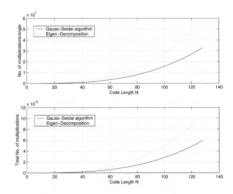

Figure 4: Number of multiplications using antenna number = 6

chips. The DOAs and delays of the rest of users are randomly generated between $[0, 180]$ and $[0, 31]$, respectively. The frequency offset Δf is assumed to be 0.1.

We assume that the number of multi-path signals of the reference user is known. We use 400 observation symbols with no over-sampling. The signal-to-noise ratio (SNR) is assumed to be 20 dB, and the total power of an interfering user is twice of that of the reference user. The gain-phase $\alpha_m e^{j\psi_m}$ of each antenna is chosen as

$$\alpha_m = 1 + \sqrt{12}\sigma_\alpha u_m, \quad \psi_m = \sqrt{12}\sigma_\psi u_p$$

where u_m and u_p are uniformly distributed in $[-0.5, 0.5]$ and $\sigma_\alpha = 0.2$ $\sigma_\psi = 20°$.

By fixing the iteration number in Gauss-Seidal algorithm by 20, we plot the minimum eigenvalues. The result obtained when the frequency offset is not compensated, is shown in Figure 5 and Figure 6.

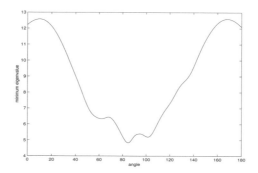

Figure 5: Minimum eigenvalues obtained using un-calibrated array of un-compensated frequency offset

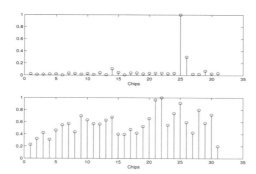

Figure 6: time delays obtained using un-calibrated array of un-compensated frequency offset

The result obtained when the frequency offset is compensated but the array is not calibrated, is shown in Figure7 and Figure 8. The result obtained when the array is calibrated and the frequency offset is compensated, is shown in Figure9 and Figure 10.

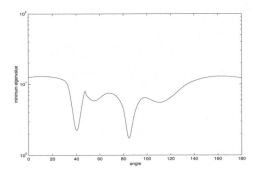

Figure 7: Minimum eigenvalues obtained using un-calibrated array of frequency compensation

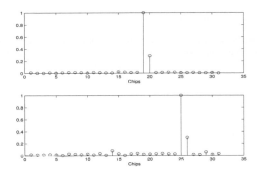

Figure 8: time delays obtained using un-calibrated array of frequency compensation

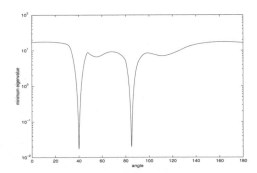

Figure 9: Minimum eigenvalues obtained using calibrated array of frequency compensation

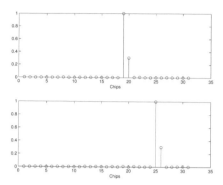

Figure 10: time delays obtained using calibrated array of frequency compensation

In these figures, we observe that when frequency offset is not compensated, we cannot estimate channel parameters in both space and time as well as the calibration vector **d**. The convergence behavior of DOA according to iteration is shown in Figure 11.

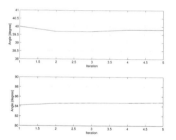

Figure 11: DOA Estimation vs. iteration

Figure 12 shows the normalized calibration error defined as $(\| \mathbf{d}_j - \mathbf{d}_t \|_2 \; / \; \| \mathbf{d}_t \|_2)$, where $\mathbf{d}_j$ and $\mathbf{d}_t$ are the gain-phase vector at the jth iteration and the true gain-phase vector, respectively.

From these figures, we observe that the proposed algorithm performs well even when the there is carrier offset error.

Finally, we present the result obtained by using the iteration number (maximum iteration) by 2 and 20 in Gauss-Seidal algorithm, respectively. The minimum eigenvalues at the first iteration and the fifth iteration is shown in Figure 13 and Figure 14.

The estimated time delays obtained using 2 iterations are shown in Figure 15. From the results shown in the Figures, we verify that the proposed fast algorithm perform well even with small number of iteration in Gauss-Seidal algorithm.

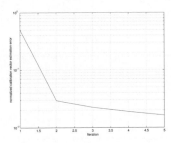

Figure 12: DOA Estimation vs. iteration

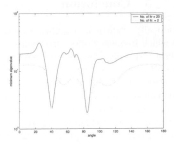

Figure 13: Minimum eigenvalues obtained at first iteration

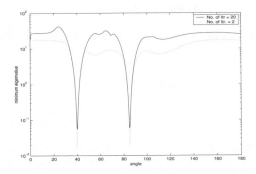

Figure 14: Minimum eigenvalues obtained at fifth iteration

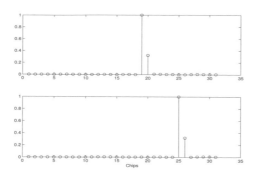

Figure 15: time delays obtained using iteration of 2 at 5th. iteration.

5 Conclusion

In this paper, we presented a numerically efficient and stable estimation algorithm that can be used in Spread spectrum system in which frequency offset error exists. The proposed algorithm is not dependent on the structure of data and array geometry and requires only binary code sequence of an arbitrary reference. By using the binary sequence, the algorithm provides us with vector channel estimates of the frequency offset, the DOA and delay of multi-path signals.

The efficient algorithm is based on Gauss-Seidal algorithm rather than using eigen-decomposition or SVD in computing eigen-values and eigen-vectors at each iteration. The algorithm is based on the two step procedures, one for estimating both channel and frequency offset and the other for estimating the unknown array gain and phase. Consequently, estimates of the DOAs, the multi-path impulse response of the reference signal source, and the carrier frequency offset as well as the calibration of antenna array are provided. The performance of the proposed algorithm is investigated by means of computer simulations. The analytic and simulation results reveals that proposed algorithm is reduces the number of multiplications by order of one.

Acknowledgment

This work was supported in part by University IT Research Center Project (INHA UWB-ITRC) in Korea

References

[1] G. Tsoulos and M. Beach, "Calibration and linearity issues for an adaptive antenna system," In *Proc. 47th IEEE Vehicular Technology Conference*, pp. 1597 -1600 Volume: 3 , May 1997.

[2] B. Friedlander and A. Weiss, "Eigenstructure Methods for Direction Finding with Sensor Gain and Phase Uncertainties," *Proc. IEEE ICASSP*, pp. 2681-2684, January 1988.

[3] A. Paulraj, T. Shan, V. Reddy and T. Kailath, "A subspace approach to determining sensor gain and phase with applications to array processing", In *SPIE, Adv. Algorithms Architectures Signal Process.*, , vol. 696 , pp 102-109, Aug. 1986.

[4] V. Soon, L. Tong, Y. Huang, and R. Liu, "A subspace method for estimating sensor gains and phase", *IEEE Trans. Signal Processing*, , vol. 42 , pp 973-976, Apr. 1994.

[5] D. Fuhrmann, "Estimation of sensor gain and phase", *IEEE Trans. Singal Processing*, vol. 42, pp 77-87, 1994

[6] M. Hebley and P. Taylor, "The effect of diversity on a burst-mode carrier-frequency estimator in the frequency selective mulipath channel", *IEEE Tras. Comm.*, Vol. 46, pp 553 - 560, Apr. 1998.

[7] M. Morelli and U. Mengali, "Carrier frequency estimation for transmissions over selective channels", *IEEE Tras. Comm.*, Vol. 48, pp 1580 - 1589, Sep. 2000.

[8] M. Eric, S. Parkvall, M. Dukic and M. Obradovic, "An Algorithm For Joint Direction Of Arrival, Time-Delay and Frequency-Shift Estimation in Asynchronous DS-CDMA Systems", *1998 IEEE 5th International Symposium on Spread Spectrum Techniques and Applications*, Vol. 2, pp 595 -598, Sep. 1998.

[9] G. Golub and C. Van Loan, *Matrix Computations*, Baltimore, Johns Hopkins Uni. Press, 1996

[10] S. Bensley and Behnaam Aazhang, "Subspace-Based Channel Estimation for Code Division Multiple Access Communication Systems", *IEEE Tras. Comm.*, Vol. 44, pp 1009 - 1020, Aug. 1996.

[11] A. Graham, *Kronecker Products and Marix Calculus: with Applications*, New York, John Wiley & Sons, 1981

[12] D. Astely, A. Lee Swindlehurst and Bjorn Ottersten, "Spatial Signature Estimation for Uniform Linear Arrays with Unknown Receiver Gains and Phases", *IEEE Truns. Signal Proccessing*, Vol. 47, NO. 8, pp 2128 - 2138, Aug. 1999.

[13] C. H. Lee, S. Kim and J. Chun, "An Online Calibration Algorithm for the CDMA based Adaptive Antenna Array," *in Proc. 34th Asilomar Conf.*, Pacific Grove, CA, Oct. 2000.

[14] S. Kobayakawa, M. Tsutsui, and Y. Tanaka, "A Blind Calibration Method for an Adaptive Array Antenna in DS-CDMA Systems Using an MMSE Algorithm," *Proc. IEEE VTC*, May 2000.

[15] Ying-Chang Liang and Francois P.S. Chin, " Coherent LMS algorithms," *IEEE Signal Proc. letter*, pp. 92-94, March 2000.

[16] Simon Haykin, *Adaptive Filter Theory*, 3rd ed., Prentice-Hall, Englewood CLiffs, N.J. 1996.

VSP International
Science Publishers
P.O. Box 346, 3700 AH Zeist
The Netherlands

Lecture Series on Computer
and Computational Sciences
Volume 1, 2004, pp. 952-954

Security Requirements of Development Site

Eunser Lee[1], Sunmyoung Hwang[2]

[1]Chung-Ang University, 221, Huksuk-Dong, Dongjak-Gu, Seoul, Korea
[2]Daejeon University, 96-3 Yongun-dong, Tong-gu, Taejon 300-716, South Korea

Abstract: The IT products like as firewall, IDS (Intrusion Detection System) and VPN (Virtual Private Network) are made to perform special functions related to security, so the developers of these products or systems should consider many kinds of things related to security not only design itself but also development environment to protect integrity of products. When we are making these kinds of software products, ISO/IEC TR 15504 may provide a framework for the assessment of software processes, and this framework can be used by organizations involved in planning, monitoring, controlling, and improving the acquisition, supply, development, operation, evolution and support of software. But, in the ISO/IEC TR 15504, considerations for security are relatively poor to other security-related criteria such as ISO/IEC 21827 or ISO/IEC 15408. In fact, security related to software development is concerned with many kinds of measures that may be applied to the development environment or developer to protect the confidentiality and integrity of the IT product or system developed. This paper proposes some measures related to development process security by analyzing the ISO/IEC 21827, the Systems Security Engineering Capability Maturity Model (SSE-CMM) and ISO/IEC 15408, Common Criteria (CC). And we present a Process of Security for ISO/IEC TR 15504.

Keywords: Site Security, Software Development, Common Criteria, Security Process.

1. Introduction

ISO/IEC TR 15504, the Software Process Improvement Capability Determination (SPICE), provides a framework for the assessment of software processes [1]. This framework can be used by organizations involved in planning, monitoring, controlling, and improving the acquisition, supply, development, operation, evolution and support of software. But, in the ISO/IEC TR 15504, considerations for security are relatively poor to others. For example, the considerations for security related to software development and developer are lacked.

When we are making some kinds of software products, ISO/IEC TR 15504 may provide a framework for the assessment of software processes, and this framework can be used by organizations involved in planning, monitoring, controlling, and improving the acquisition, supply, development, operation, evolution and support of software. But, in the ISO/IEC TR 15504, considerations for security are relatively poor to other security-related criteria such as ISO/IEC 21827 or ISO/IEC 15408 [2-3]. In fact, security related to software development is concerned with many kinds of measures that may be applied to the development environment or developer to protect the confidentiality and integrity of the IT product or system developed.

In this paper, we propose a process related to security by comparing ISO/IEC TR 15504 to ISO/IEC 21827 and ISO/IEC 15408. The proposed scheme may be contributed to the improvement of security for IT product or system. And in this paper, we propose some measures related to development process security by analyzing the ISO/IEC 21827, the Systems Security Engineering Capability Maturity Model (SSE-CMM) and ISO/IEC 15408, Common Criteria (CC). And we present a Process for Security for ISO/IEC TR 15504.

[1] Chung-Ang University, 221, Huksuk-Dong, Dongjak-Gu, Seoul, Korea, E-mail : eslee@object.cau.ac.kr
[2] Daejeon University, 96-3 Yongun-dong, Tong-gu, Taejon 300-716, South Korea, E-mail : sunhwang@dju.ac.kr

2. A New Process for Development Site Security

For example, we want to deal the security for the site where the software is developed. In the ISO/IEC TR 15504-5, there is the Engineering process category (ENG) which consists of processes that directly specify, implement or maintain the software product, its relation to the system and its customer documentation. In circumstances where the system is composed totally of software, the Engineering processes deal only with the construction and maintenance of such software.

The processes belonging to the Engineering process category are ENG.1 (Development process), ENG.1.1 (System requirements analysis and design process), ENG.1.2 (Software requirements analysis process), ENG.1.3 (Software design process), ENG.1.4 (Software construction process), ENG.1.5 (Software integration process), ENG.1.6 (Software testing process), ENG.1.7 (System integration and testing process), and ENG.2 (Development process).

These processes commonly contain the 52nd work product (Requirement specification), and some of them have 51st, 53rd, 54th work products separately. Therefore, each process included in the ENG category may contain the condition, 'Identify any security considerations/constraints'. But the phrase 'Identify any security considerations/constraints' may apply to the 'software or hardware (may contain firmware) development process' and not to the 'development site' itself.

In this paper we will present a new process applicable to the software development site. In fact, the process we propose can be included in the MAN or ORG categories, but this is not the major fact in this paper, and that will be a future work. We can find the requirements for Development security in the ISO/IEC 15408 as like;

Development security covers the physical, procedural, personnel, and other security measures used in the development environment. It includes physical security of the development location(s) and controls on the selection and hiring of development staff.

Development security is concerned with physical, procedural, personnel, and other security measures that may be used in the development environment to protect the integrity of products. It is important that this requirement deals with measures to re-move and reduce threats existing in the developing site (not in the operation site). These contents in the phrase above are not the perfect, but will suggest a guide for development site security at least.

The individual processes of ISO/IEC TR 15504 are described in terms of six components such as Process Identifier, Process Name, Process Type, Process Purpose, Process Outcomes and Process Notes. The style guide in annex C of ISO/IEC TR 15504-2 provides guidelines which may be used when extending process definitions or defining new processes.

Next is the Development Security process we suggest.

(1) Process Identifier: ENG.3

(2) Process Name: Development Security process

(3) Process Type: *New*

(4) Process purpose:
The purpose of the Development Security process is to protect the confidentiality and integrity of the system components (such as hardware, software, firmware, manual, operations and network, etc) design and implementation in its development environment. As a result of successful implementation of the process:

(5) Process Outcomes:
- access control strategy will be developed and released to manage records for entrance and exit to site, logon and logout of system component according to the released strategy
- roles, responsibilities, and accountabilities related to security are defined and released
- training and education programs related to security are defined and followed
- security review strategy will be developed and documented to manage each change steps

(6) Base Practices:
ENG.3.BP.1:   Develop physical measures. Develop and release the physical measures for protecting the access to the development site and product.

ENG.3.BP.2:   Develop personnel measures. Develop and release the personnel measures for selecting and training of staffs.
ENG.3.BP.3:   Develop procedural measures. Develop the strategy for processing the change of requirements considering security.

ENG.3 Development Security process may have more base practices (BP), but we think these BPs will be the base for future work. For the new process, some work products must be defined as soon as quickly. Next items are the base for the definition of work products

WP category number	WP category	WP classification number	WP classification	WP type
1 ORGANIZATION		1.1	Policy	Access control to site and so on
		1.2	Procedure	Entrance and so on
		1.3	Standard	Coding and so on
		1.4	Strategy	Site open and so on
2 PROJECT		Future work	Future work	Future work
3 RECORDS		3.1	Report	Site log and so on
		3.2	Record	Entrance record and so on
		3.3	Measure	Future work

3. Conclusion and Future Work

This paper proposed a new Process applicable to the software development site. In fact, the Process we proposed is not perfect not yet, and the researches for improving going on. Some researches for expression of Base Practice and development of Work Products should be continued. But the work in the paper may be the base of the consideration for security in ISO/IEC TR 15504.
ISO/IEC TR 15504 provides a framework for the assessment of software processes, and this framework can be used by organizations involved in planning, monitoring, controlling, and improving the acquisition, supply, development, operation, evolution and support of software. Therefore, it is important to include considerations for security in the Process dimension.
In this paper we did not contain or explain any component for Capability dimension, so the ENG.3 Process we suggest may conform to capability level 2. Therefore, more research efforts will be needed. Because the assessment cases using the ISO/IEC TR 15504 are increased, some processes concerns to security are needed and should be included in the ISO/IEC TR 15504.

References

[1] ISO, ISO/IEC TR 15504: Information technology – Software process assessment (SPICE).

[2] ISO, ISO/IEC 21827 Information technology – Systems Security Engineering Capability Maturity Model (SSE-CMM).

[3] ISO, ISO/IEC 15408: Information technology - Security techniques - Evaluation criteria for IT security, 1999.

[4] Eun-ser Lee, Kyung-whan Lee, Tai-hoon Kim and Il-hong Jung: Introduction and Evaluation of Development System Security Process of ISO/IEC TR 15504, ICCSA 2004, LNCS 3043, Part 1, 2004.

[5] Tai-hoon Kim and Haeng-kon Kim: A Relationship between Security Engineering and Security Evaluation, ICCSA 2004, LNCS 3046, Part 4, 2004.

VSP International
Science Publishers
P.O. Box 346, 3700 AH Zeist
The Netherlands

Lecture Series on Computer
and Computational Sciences
Volume 1, 2004, pp. 955-957

Advances in Financial Forecasting

Dimitrios D. Thomakos[1]

Department of Economics,
School of Management and Economics,
University of Peloponnese,
GR-221 00 Tripolis, Greece

Accepted: August 8, 2004

Abstract: This Symposium covers a variety of papers that present new methods and applications of existing, state of the art, methods in three broad categories of financial forecasting. In the first category we have five papers that contribute to the theory and applications, to economic and financial data, of forecasting methods, filtering and smoothing. In the second category we have three papers that deal with the construction and evaluation of successful trading rules that can be used in real-life transactions. Finally, in the third category we have seven papers that cover different aspects in the theory and applications of dynamic macroeconomics, asset pricing and risk management.

Keywords: filtering, forecasting, trading methods, asset pricing, risk management.

1 Summary

In the last decade or so the field of financial forecasting has become very active, both in producing new methods but also in incorporating state of the art methods from other fields. In this Symposium we have fifteen papers that cover a variety of methods and applications in financial forecasting. We classified them into three broad categories, based on their content and potential applicability. These categories are methods and applications in forecasting, methods for and evaluation of trading rules, and dynamic macroeconomics, asset pricing and risk management.

The first category includes five papers. Brandl [3] considers the combination of machine learning methods, such as genetic algorithms and neural nets, and economic theory in improving exchange rate forecasts. Heidari [5] provides a comparative analysis of several alternative vector autoregressive models for forecasting inflation, including classical and Bayesian models. Rinderu *et al.* [11] consider the problem of the stability of the monetary transmission mechanism in a developing country and how it can be modeled in the context of a dynamical system. Shmilovici *et al.* [13] use the context tree algorithm of Rissanen, for compression and prediction of a time series to analyze 12 pairs of international intra-day currency exchange rates. Finally, Thomakos [14] proposes a new method for filtering, smoothing and forecasting based on the use of a particular causal filter with time-varying, data-dependent and adaptable weights.

The second category includes three papers. Angelides and Degiannakis [1] test the accuracy of parametric, non-parametric and semi-parametric methods in predicting the one-day-ahead Value-at-Risk (VaR) of perfectly diversified portfolios in three types of markets (stock exchanges, commodities and exchange rates), both for long and short trading positions. Bekiros and Georgoutsos

---

[1]Symposium organizer. E-mail: thomakos@uop.gr

[2] investigate the nonlinear predictability of technical trading rules based on a recurrent neural network as well as a neurofuzzy model, with corresponding trading rules and an application to the NASDAQ, NYSE and NIKKEI markets. Papadamou and Stephanides [9] explore the potential power of digital trading and present a new Matlab tool based on genetic algorithms, which specializes in parameter optimization of technical rules.

The third category includes seven papers. Hardouvelis and Malliaropoulos [4] present evidence that the predictive ability of the yield spread for short-run inflation is related to its predictive ability for economic activity. Kayahan and Stengos [7] tests the conditional version of the Sharpe-Lintner CAPM by adopting Local Maximum Likelihood as its nonparametric methodology. Kottaridi and Siourounis [7] consider an econometric framework where macroeconomic monetary volatility is linked to the probability distribution of liquidity shocks hitting an international investor, providing evidence that for what is called "flight to quality". Koumbouros [8] proposes a new method that decomposes the overall market risk into parts reflecting long-run market uncertainty related to the dynamics of the present value of revisions in expectations about future asset-specific and market cash-flows and discount-rates. Papanastasopoulos and Benos [10] design a hybrid model to act as an early warning system to monitor changes in the credit quality of corporate obligors. Rompolis and Tzavalis [12] propose a new nonparametric approach of estimating the risk neutral density (RND) of asset prices or log-returns, that exploits a relationship between the call and put prices and the conditional characteristic function of the asset price and the log-return. Finally, Wang *et al.* [15] examine the predictability of stock index returns (S&P500, S&P400 and Russell 2000) using short-term interest rates as predictors, and suggest that trading strategies based on their findings can be profitable.

Acknowledgment

The organizer wishes to express his appreciation to Professor T. E. Simos for his encouragement, comments and recommendations in putting together this Symposium. Any errors or omissions are solely the responsibility of the organizer.

References

[1] T. Angelidis and S. Degiannakis. "Modeling Risk in Three Markets: VaR Methods for Long and Short Trading Positions".

[2] S. Bekiros and D. Georgoutsos. "Comparative evaluation of technical trading rules: neurofuzzy models vs. recurrent neural networks".

[3] B. Brandl. "Machine Learning in Economic Forecasting and the Usefulness of Economic Theory: the Case of Exchange Rate Forecasts".

[4] G. A. Hardouvelis and D. Malliaropulos. "The Yield Spread as a Symmetric Predictor of Output and Inflation".

[5] H. Heidari. "An Evaluation of Alternative VAR Models for Forecasting Inflation".

[6] B. Kayahan and T. Stengos. "Testing of Capital Asset Pricing Model with Local Maximum Likelihood Method".

[7] C. Kottaridi and G. Siourounis. "A Flight to Quality! International Capital Structure Under Foreign Liquidity Constraints".

[8] M. Koumbouros. "Temporary and Permanent Long-Run Asset-Specific and Market Risks in the Cross-Section of US Stock Returns".

[9] S. Papadamou and G. Stephanides. "Improving Technical Trading Systems by Using a New Matlab based Genetic Algorithm Procedure".

[10] G. Papanastasopoulos and A. Benos. "Extending the Merton Model: A Hybrid Approach to Assessing Credit Quality".

[11] P.L. Rinderu, Gh. Gherghinescu and O.R. Gherghinescu. "Modeling the Stability of the Transmission Mechanism of Monetary Policy: A Case Study for Romania, 1993-2000".

[12] L. Rompolis and E. Tzavalis. "Estimating Risk Neutral Densities of Asset Prices based on Risk Neutral Moments: An Edgeworth expansion approach".

[13] A. Shmilovici, Y. Kahiri and S. Hauser. "Forecasting with a Universal Data Compression Algorithm: The Forex Market Case".

[14] D. Thomakos. "Functional Filtering, Smoothing and Forecasting".

[15] D. Thomakos, T. Wang and J. T. Wu. "Market Timing and Cap Rotation".

VSP International
Science Publishers
P.O. Box 346, 3700 AH Zeist
The Netherlands

*Lecture Series on Computer
and Computational Sciences*
Volume 1, 2004, pp. 958-960

Modeling Risk in Three Markets: VaR Methods
for Long and Short Trading Positions

Timotheos Angelidis [1]
Stavros Degiannakis

Department of Banking and Financial Management,
University of Piraeus,
GR- 185 34, Piraeus , Greece

Athens Laboratory of Business Administration.
GR-166 71, Vouliagmeni, Greece

Department of Statistics,
Athens University of Economics and Business
GR-104 34, Athens, Greece

Accepted: August 8, 2004

Abstract: The accuracy of parametric, non-parametric and semi-parametric methods in predicting the one-day-ahead Value-at-Risk (VaR) of perfectly diversified portfolios in three types of markets (stock exchanges, commodities and exchange rates) is investigated, both for long and short trading positions. The risk management techniques are designed to capture the main characteristics of asset returns, such as leptokurtosis and asymmetric distribution, volatility clustering, asymmetric relationship between stock returns and conditional variance and power transformation of conditional variance. Based on backtesting measures and a loss function evaluation method, we find out that the modeling of the main characteristics of asset returns produces accurate VaR forecasts. Especially for the high confidence levels, a risk manager must employ different volatility techniques in order to forecast the VaR for the two trading positions.

Keywords: Asymmetric Power ARCH model, Skewed-t Distribution, Value-at-Risk, Volatility Forecasting.

Mathematics SubjectClassification: 62P20.

1. Empirical Investigation

Value-at-Risk (VaR) at a given probability level a, is the predicted amount of financial loss of a portfolio over a given time horizon. Given the fact that the asset returns are not normally distributed, since they exhibit skewness and excess kurtosis, it is plausible to employ volatility forecasting techniques that accommodate these characteristics. The one-step-ahead volatility of daily returns is estimated by a set of Autoregressive Conditional Heteroskedasticity (GARCH, EGARCH, TARCH and APARCH) models by assuming three distribution assumptions (Normal, Student-t and Skewed Student-t), historical and filtered-historical simulations and the commonly used variance-covariance method.

We aim to evaluate the predictive accuracy of various models under a risk management framework. We employ a two stage procedure to investigate the forecasting power of each volatility forecasting technique. In the first stage, two backtesting criteria (Kupiec (1995), Christoffersen (1998)) are implemented to test statistical accuracy of the models, which will serve as the final diagnostic check, in order to judge the "quality" of the VaR forecasts. Moreover, the purpose of the used

---

[1] Corresponding Author. Tel.: +30-210-8964-736.

backtesting measures is twofold. First, to test whether the average number of the VaR violations[2], according to an out-of-sample period, is statistically equal to the expected one. It is important to note that the estimated VaR number must neither overestimate nor underestimate the "true" but unobservable value. In the former case, the financial institution does not use its capital efficiently, while in the latter case it can not cover future losses. Second, given the fact that an adequate model must wide the VaR forecasts during volatile periods and narrow them otherwise, it is necessary to examine if the violations are also randomly distributed. However, in most of the cases there are more than one risk model that satisfies both the backtesting measures and therefore a risk manager can not select a unique volatility forecasting technique. Hence, in order to achieve this goal, we compare the best performed models via a loss function, in the attempt to select one model among the various candidates.

In the second stage, we employ standard forecast evaluation methods in order to examine whether the differences between models (which have converged sufficiently), are statistically significant. We focus on out-of-sample evaluation criteria because we believe that a model that may be inadequate according to some in-sample evaluation criterion, can still yield ``better'' forecasts in an out-of-sample framework than a correctly specified model.

We generate out-of-sample VaR forecasts for two equity indices (S&P500, FTSE100), two commodities (Gold Bullion $/Troy Ounce, London Brent Crude Oil Index US$/BBL) and two exchange rates (US $ to Japanese ¥, US $ to UK £), obtained from Datastream for the period of January 3rd 1989 to June 30th 2003. For all models, we use a rolling sample of 2000 observations in order to calculate the 95% and the 99% $VaR_{t+1|t}$ for long and short trading positions.

Under the framework of the loss function approach, we evaluate all the models with p-value greater than 10% for both unconditional and conditional coverage tests. A high cut-off point is preferred in order to ensure that the successful risk management techniques will not a) over or under estimate statistically the "true" VaR and b) generate clustered violations. In the case of a smaller cut-off point, an incorrect model could not be easily rejected, which might turn to be costly for a risk a manager

Table 1 summarizes the two-stage model selection procedure3. In the first stage (columns 2 and 3) the models that have not been rejected by the statistical backtesting procedures are presented, while in the second stage (column 4), the volatility methods that are preferred over the others, based on the loss function approach, are exhibited. For example, in panel A, for the S&P500 index, the GARCH(1,1)-normal model achieves the smallest value of the loss function, while its forecasting accuracy is not statistically different to that of the EWMA, EGARCH(1,1) and APARCH(1,1) models with normally distributed innovations.

Our study sheds a light on the volatility forecasting methods under a risk management framework, since it juxtaposes the performance of the most well known techniques for different markets (stock exchanges, commodities and exchange rates) and trading positions. Although, the normal distribution produces adequate one-day-ahead VaR forecast at the 95% confidence level, models that parameterize the leverage effect for the conditional variance, the leptokurtosis and the asymmetry of the data, forecast accurate the VaR at the 99% confidence level. Moreover, short-trading positions should be modeled using volatility specifications different to that of portfolios with long trading positions.

Specifically, more sophisticated techniques that accommodate the features of the financial time series are needed, in order to calculate the one-day-ahead VaR. Brooks and Persand (2003) pointed out that models, which do not allow for asymmetries either in the unconditional return distribution or in the volatility specification, underestimate the "true" VaR. Giot and Laurent (2003) proposed the skewed Student-t distribution and pointed out that it performed better than the pure symmetric one, as it reproduced the characteristics of the empirical distribution more accurate. These views are confirmed for both confidence levels and trading positions, as most of the selected models parameterize these features.

At the 95% confidence level, specifications with normally distributed errors achieve the lowest loss function values. In most of the cases, the techniques that produce the most accurate VaR predictions are the same for both long and short trading positions. On the other hand, the volatility specifications, that parameterize the leverage effect for the conditional variance and the asymmetry of

---

2 A violation occurs if the predicted VaR is not able to cover realized loss.
[3] Tables with detailed results are available upon request.

the innovations' distribution, forecast the VaR at the 99% confidence level more adequately. However, the models that must be employed for modeling the short and the long trading positions are not the same. This finding is in contrast with that of Giot and Laurent (2003) who argued that the APARCH model based on the skewed Student-t distribution forecasts the VaR adequate for both trading positions.

Finally, for long position on OIL index (95% VaR), and short position on GOLD index (99% VaR), there are no models that produce adequate VaR forecasts. Given the fact that for these cases they have been rejected by the conditional coverage test, there is evidence that clustered violations were generated.

References

[1] Brooks, C., Persand, G., 2003. The effect of asymmetries on stock index return Value-at-Risk estimates. The Journal of Risk Finance, Winter, 29-42.
[2] Christoffersen, P., 1998. Evaluating interval forecasts. International Economic Review, 39, 841-862.
[3] Giot, P., Laurent, S., 2003. Value-at-Risk for Long and Short Trading Positions. Journal of Applied Econometrics, 18, 641-664.
[4] Kupiec, P.H., 1995. Techniques for Verifying the Accuracy of Risk Measurement Models. Journal of Derivatives, 3, 73-84.

VSP International
Science Publishers
P.O. Box 346, 3700 AH Zeist
The Netherlands

*Lecture Series on Computer
and Computational Sciences*
Volume 1, 2004, pp. 961-963

Comparative Evaluation of Technical Trading Rules: Neurofuzzy Models vs. Recurrent Neural Networks

S. Bekiros[1] & D. Georgoutsos

Department of Accounting and Finance
Athens University of Economics and Business
76 Patission str.,
104 34 Athens, GREECE

Accepted: August 8, 2004

Abstract: This paper investigates the nonlinear predictability of technical trading rules based on a recurrent neural network as well as a neurofuzzy model. The efficiency of the trading strategies was considered upon the prediction of the direction of the market in case of NASDAQ, NYSE and NIKKEI returns. The sample extends over the period 2/8/1971 – 4/7/1998 while the sub-period 4/8/1998 – 2/5/ 2002 has been reserved for out-of-sample testing purposes. Our results suggest that, in absence of trading costs, the return of the proposed neurofuzzy model is consistently superior to that of the recurrent neural model as well as of the buy & hold strategy for bear markets. On the other hand, we found that the buy & hold strategy produces in general higher returns than neurofuzzy model or neural networks for bull periods. The proposed neurofuzzy model which outperforms the neural network predictor allows investors to earn significantly higher returns in bear markets.

Keywords: Technical trading rules, Neurofuzzy models, Neural networks

Mathematics Subject Classification: 89.65, 42.79, 47.52, 05.45

1. Extended Abstract

The recurrent network uses the tansig transfer function (G) in its hidden (recurrent) layer, and the purelin function (S) in its output layer. The output y stands for the forecasted stock index return and is given by:

$$y_t = S\left[\beta_0 + \sum_{i=1}^{q} \beta_i \cdot g(t)\right]$$

where

$$g(t) = G\left(\alpha_{i0} + \sum_{j=1}^{n} \alpha_{ij} x_{j,t} + \sum_{h}^{m} \delta_{ih} g_h(t-1)\right)$$

and $x_{1,t}, \ldots, x_{n,t}$ are past values of the stock index return.

In the neurofuzzy model y also stands for the forecasted stock index return and $x_{1,t}, \ldots, x_{n,t}$ are past values of the stock index return. The output is given by:

---

[1] Corresponding author. E-mail: sbekiros@yahoo.com

$$y = \left[\overline{w}_1 x_1 \quad \overline{w}_1 x_2 \quad \overline{w}_1 \quad \overline{w}_2 x_1 \quad \overline{w}_2 x_2 \quad \overline{w}_2\right] \cdot \left[c_1 \quad d_1 \quad h_1 \quad c_2 \quad d_2 \quad h_2\right]^T$$

where w_i are the firing strengths of the membership function grades and c, d, h coefficients from the general first-order Sugeno model of the form:

$$IF\ x\ is\ A\ AND\ y\ is\ B\ THEN\quad z = h + c \cdot x + d \cdot y$$

References

[1] Elman, J. L. 1990, "Finding structure in time," Cognitive Science, 14, 179-211.

[2] Jang, J., 1993, "Adaptive-Network-Based Fuzzy Inference Systems", IEEE Transactions on Systems, Man and Cybernetics, 23 (3), 665-685.

[3] Masters, T., 1993, "Advanced Algorithms for Neural Networks", John Wiley.

[4] Pesaran, M.H., and Timmermann, A., 1992, "A simple nonparametric test of predictive performance", Journal of Business and Economics Statistics, 10, 461-465

[5] Sugeno, M., 1985, "Industrial applications of fuzzy control", Elsevier Science Publications Co.

[6] Sugeno, M., 1988, "Fuzzy Control", Nikkan Kougyou, Shinbunsha.

TABLE 1: Statistical results for the trading models

Indices	NASDAQ				NYSE				NIKKEI225			
Architecture	RNN		NF		RNN		NF		RNN		NF	
Sub-period	Bull	Bear	Bull	Bear	Bull	Bear	Bull	Bear	Bull	Bear	Bull	Bear
Total Return	0.569	-0.277	0.662	0.512	0.563	-0.317	0.467	-0.048	-0.159	0.216	-0.336	0.543
B&H Return	1.027	-0.982	1.027	-0.982	0.166	-0.190	0.166	-0.190	0.241	-0.773	0.241	-0.773
Sign Rate	0.543	0.507	0.543	0.533	0.514	0.468	0.490	0.465	0.462	0.462	0.451	0.511
PT test	1.665	0.382	1.177	1.362	0.723	-1.275	-0.288	-0.908	-1.708	-1.549	-2.169	0.848
MSE	0.021	0.035	0.018	0.029	0.013	0.014	0.011	0.011	0.018	0.021	0.014	0.016
Sharpe Ratio	0.063	-0.020	0.073	0.037	0.082	-0.080	0.068	-0.012	-0.021	0.027	-0.045	0.069
Ideal Profit	0.081	-0.026	0.094	0.048	0.112	-0.107	0.093	-0.016	-0.029	0.037	-0.062	0.094

Notation:
RNN: Recurrent Neural Network
NF: NeuroFuzzy model
B&H return: Total return from Buy & Hold strategy
Sign Rate: Proportion of the cases that the strategy predicts the correct sign of the market
PT test: Pesaran & Timmermann test (1992)
MSE: Mean Square Error
Sharpe Ratio: Return from the strategy per unit of risk
Ideal Profit: Compares the forecasting system return against the perfect forecaster

VSP International
Science Publishers
P.O. Box 346, 3700 AH Zeist
The Netherlands

*Lecture Series on Computer
and Computational Sciences*
Volume 1, 2004, pp. 964-966

Machine Learning in Economic Forecasting and the Usefulness of Economic Theory: the Case of Exchange Rate Forecasts

B. Brandl[1]

Department of Government,
Faculty of Business, Economics, and Computer Science,
University of Vienna,
A-1210 Vienna, Austria

Accepted: August 8, 2004

Abstract: This paper focuses on an integration of economic theory in a machine learning process for the purpose of exchange rate forecasting. Since the early 1980s literature stresses the weakness of economic theory in forecasting exchange rates. Consistent with these results, in this paper it is asked how machine learning can increase the forecasting performance of theoretical models. Therefore structural exchange rate models are implemented in a machine learning process as a framework in which and among which the possibilities of machine learning are exploited not only to identify further sources of influence but also to test alternative variables of aggregates suggested by exchange rate theory. Thus to serve as a tool for model selection. The applied approach uses a Genetic Algorithm for model selection and Neural Networks for the generation of the forecasts. It is shown that this combination of economic theory and machinery learning not only increases the "fitness" (a popular term in Genetic Algorithm literature) of theoretical exchange rate models but is also fruitful for the effectiveness and correctness of machine learning processes. As experience showed, relationships derived from machine learning techniques are often not convincing as regards their correctness and effectiveness. Most machine learning approaches do not contribute much to persuade otherwise. In this paper we tried to overcome parts of this limitation. The approach is illustrated in some detail for five exchange rates on a monthly frequency.

Keywords: Forecasting, Exchange Rate Theory, Genetic Algorithm, Neural Network

Mathematics Subject Classification: 62P05, 91B64, 62M45

1. Combining machine learning with economic theory

The basic idea of this paper is to combine economic theory with machine learning to forecast exchange rates. Thus to increase the "fitness" of theoretically derived exchange rate models. The reason for this combined approach is that relying merely on economic theory on the one hand and machine learning on the other hand has its problems. Nevertheless, both approaches reveal advantages, but also disadvantages. The main advantage of forecasting exchange rates using economic theory is that relatively stable relationships are considered. Particularly with regard to the forecasting performance a main disadvantage of theoretical approaches is that these stable relationships provide a low statistical fit on actual data. One reason for this is that they abstract from many other (temporary and case specific) sources of influence on current exchange rate fluctuations but also from psychological dynamics which can be modeled by technical indicators. In this paper such indicators (among other fundamentals) are added to the structural models by using machine learning. See the appendix for a discussion on the used models. On the other hand, the problem of most pure machine learning approaches is that they are not able to distinguish between basic and fundamental relationships and temporary relationships. Moreover, machine learning may be "blinded" by spurious causality so that it can only model exchange rate behaviour over a specific (relatively short) time span. Whereas, a principle advantage of machine learning is that over these specific time spans a considerably high goodness of fit can be achieved as such methods have the possibility to detect an trace influences on exchange rates which are not describable by economic theory.

---

[1] Corresponding author. Assistant Professor at the University of Vienna. E-mail: bernd.brandl@univie.ac.at

2. Structuring the GA search space

The combination of theoretical relationships with additional relationships has the advantage that the goodness of fit of the theoretical models can be raised without endangering that relationships offered by economic theory get lost. To search among combinations between those two sorts of influences the GA is applied. However, to constrain the GA, the search space is divided into several clusters or factor groups (F) from each of which the GA has to select variables to build forecast models. Usually one series from each factor group, while those are assembled according to theoretical issues. The advantage of using a GA for this optimization task is that many combinations between series from different factor groups can be evaluated automatically without loosing the structure of the theoretical models. The number of factor groups, depends on the theory used, and the number of series in the factor group depends on how abstract the theory is formulated, as well as on issues such as availability of data on different frequencies. See the appendix for the theoretical aspects of the models. Especially on higher frequencies such as on a monthly frequency many theories are only applicable by using proxies. However, as mentioned, in addition to considering the series in accordance with economic theory, other sources of influence are evaluated. Usually other exchange rates, technical indicators and financial market series such as stock market indices have been taken into account, because they not only are said to have the capacity to proxy (or even anticipate) real economic activity but also are mapping capital movement between countries, which in turn affect exchange rates. This idea of constraining the GA can be expressed analytically. The general representation of the forecast equation to be optimized is of the following form that the search space is divided in several clusters:

$$s_{t+1} = \sum_{j=0}^{k_1} \sum_{i=1}^{n_1} \beta_{ij}^1 F_{i,t-j}^1 + \cdots + \sum_{j=0}^{k_m} \sum_{i=1}^{n_m} \beta_{ij}^m F_{i,t-j}^m + \sum_{j=0}^{k_1} \sum_{i=1}^{n_1} \alpha_{ij} X_{i,t-j} \qquad (1)$$

With F^i indicating series from factor groups according to economic theory and X_i indicating all other series used. Accordingly β denotes the coefficients of the theoretical variables and α that of all others. m stands for the number of theoretical factor groups, n for the number of variables in one group of factors and k for the number of lagged series considered. The GA is used for the model selection, whereas is constrained to select exactly one variable from every theoretical factor group, which means

$$\forall m \; \exists i, j : \beta_{ij}^m \neq 0 . \qquad (2)$$

The number of series that can be selected from cluster X is unconstrained, but usually limited by three series. As heard the number factor groups depends on which theory used. This means for the (general) monetary model $F^1 = M$, $F^2 = M^*$, $F^3 = Y$, $F^4 = Y^*$, $F^5 = r$ and $F^6 = r^*$. M denotes time series for the domestic money supply (M^* for the foreign money supply), Y are time series which express domestic real economic activity (Y^* for foreign real domestic activity) and r, respectively r^* denotes all time series which express the rate of interest in the home country, respectively foreign country. For the Dornbusch Model $F^1 = P_{ratio}$, $F^2 = r^*$ and $F^3 = r$. Series for P_{ratio} all express the ratio between domestic and foreign price level. Last, for the Frankel Model $F^1 = M$, $F^2 = M^*$, $F^3 = Y$, $F^4 = Y^*$, $F^5 = r_{long}$, $F^6 = r_{long}^*$, $F^7 = r_{real}$, $F^8 = r_{real}^*$, $F^9 = r_{short}$ and $F^{10} = r_{short}^*$. Whereas in the Frankel approach the interest rate factor is divided into three groups of factors to consider the real rate of interest (subscript *real*), short term interest rates (subscript *short*) as well as long term interest rates (subscript *long*) separately. The following figure illustrates the allocation time series out of the factor groups to form forecast models using the general monetary model as an example.

3. Conclusions and results

The utilization of machine learning to increase the fitness of theoretical exchange rate models resulted in an increased forecasting performance. It is also shown that the implementation of economic theory (fundamental, structural economic models from exchange rate theory) as a framework for machine learning increased the out-of-sample forecasting performance compared to applying machine learning without a consideration of domain knowledge. To introduce domain knowledge the GA (applied as a tool for model selection) was constrained in a way to consider the demands of economic theory. The reason for doing so was that economic theory is able to provide a stable framework in which the potential of machine learning could be exploited. Additionally to such frameworks the GA was allowed to consider additional sources of influence on future exchange rate behavior. Especially technical indicators and financial market variables. The fitness of theoretical attempts of exchange rate determination was increased considerably by machine learning. By including sensitivity analysis in the fitness function it was able to make such an expansion of theoretical frameworks as the correctness and validity of such frameworks could be considered, i.e. guaranteed. By combining the two approaches (machine learning and economic theory) it was possible to construct forecast models on a monthly frequency which behave stable over time and have a relatively high goodness of fit.

Appendix: The used theoretical exchange rate forecast models

The first structural model to be observed is the so-called flex-price monetary model (FPMM). This model focuses on the current account (rather than the capital account) and assumes flexible prices and exogenous determined output. In the case of floating exchange rates, the FPMM explains a relationship between (rapid) monetary growth and a depreciating exchange rate and vice versa. A second version (regarding underlying assumptions) of the general monetary approach is the sticky-price monetary model (SPMM), which tries to consider expectations by invoking the rational expectations hypothesis. The SPMM is based on the assumption of perfectly mobile capital flows and slow (sticky) adjusting prices on the goods market. Furthermore, in the SPMM framework, the slow adjusting goods market is said to be dominated by excess demand, which works via the price expectations augmented Phillips curve. However, the (reduced form) forecast equation for the (general) monetary approach can be written as given in the equation (3). With s indicating the nominal exchange rate and asterisks variables from foreign countries (for all other models notations are analogous):

$$s_{t+1} = f\left(\overset{(+)}{\Delta M_{t-1}}, \overset{(-)}{\Delta Y_{t-2}}, \overset{(+)}{\Delta r_{l,t}}, \overset{(-)}{\Delta M^*_{t-1}}, \overset{(+)}{\Delta Y^*_{t-2}}, \overset{(-)}{\Delta r^*_{l,t}} \right). \qquad (3)$$

The signs above the variables indicate the (theoretical) expected signs of the first partial derivative. As an ANN is used, the signs are observed by using *sensitivity analysis*. With relative changes in *money supply* denoted by ΔM, relative changes in *national income* by ΔY and relative changes in *long-term interest rates* by $r_{l,t}$. Indices are indicating publication lags. The next model considered is the *Dornbusch overshooting model* with the forecast equation:

$$s_{t+1} = f\left(\overset{(+)}{\Delta \frac{P_{t-1}}{P^*_{t-1}}}, \overset{(-)}{\Delta r_{k,t}}, \overset{(+)}{\Delta r^*_{k,t}} \right). \qquad (4)$$

With P indicating *price levels* and r_k *short-term interest rate*. The next model considered is the *Frankel real interest differential model*. It was shown that the Frankel approach considers money supplies, national incomes and the *expected inflation differentials*. The expected inflation differential is usually approximated by the differential of long-term interest rates or by long-term government Bonds. Contrary to the previous models, the Frankel approach considers countries' *real interest rates* (r_r). Accordingly the forecast equation can be written as follows:

$$s_{t+1} = f\left(\begin{array}{c} \overset{(+)}{\Delta M_{t-1}}, \overset{(-)}{\Delta Y_{t-2}}, \overset{(+)}{\Delta r_{l,t}}, \overset{(-)}{\Delta r_{r,t-1}}, \overset{(-)}{\Delta r_{k,t}}, \\ \overset{(-)}{\Delta M^*_{t-1}}, \overset{(+)}{\Delta Y^*_{t-2}}, \overset{(-)}{\Delta r^*_{l,t}}, \overset{(+)}{\Delta r^*_{r,t-1}}, \overset{(+)}{\Delta r^*_{k,t}} \end{array} \right). \qquad (5)$$

VSP International
Science Publishers
P.O. Box 346, 3700 AH Zeist
The Netherlands

Lecture Series on Computer
and Computational Sciences
Volume 1, 2004, pp. 967-969

The Yield Spread as a Symmetric Predictor of Output and Inflation

Gikas A. Hardouvelis[1]

Department of Banking and Financial Management
University of Piraeus, and CEPR

Dimitrios Malliaropoulos

Department of Banking and Financial Management
University of Piraeus, and National Bank of Greece

Accepted: August 8, 2004

Abstract: We present evidence that the predictive ability of the yield spread for short-run inflation is related to its predictive ability for economic activity. In particular, an increase in the slope of the term structure predicts an increase in output growth and a decrease in inflation of equal magnitude. In order to explain this finding, we develop a monetary asset-pricing model with sticky goods prices. Sticky prices imply that economic disturbances generate predictable changes in output and inflation, thus allowing for intertemporal substitution effects and changes in the slope of the yield curve. We derive analytic solutions of the covariance between the nominal yield spread and future output growth and inflation and show that a moderate degree of price stickiness and relatively high degree of intertemporal substitution can account for the observed correlations in the US data over the period 1960:Q1 - 2003:Q2.

Keywords: general equilibrium, term structure of interest rates.

Mathematics Subject Classification: 91B28, 91B59, 91B64.

1   Summary

Over the past 15 years, a large body of empirical literature has documented that the nominal yield spread - defined as the difference between long-term and short-term interest rates of Treasury securities - is a good predictor of future real GDP growth for horizons up to two years ahead. Related work shows that the yield spread has some predictive power for future changes in the rate of inflation at horizons of two years and beyond. In contrast to previous studies, which examine the predictive ability of the yield spread by regressing the difference of future inflation at long horizons - two years ahead - from current inflation on the current yield spread, in this Paper we examine the relationship between the current yield spread and the future level of inflation and discover a new empirical regularity, which previously went unnoticed: The current level of the yield spread is negatively related to the future level of inflation for horizons between one quarter and one and a half years ahead.

---

[1]Corresponding author. E-mail: hardouvelis@ath.forthnet.gr

We provide evidence that the forecasting ability of the yield spread for future inflation is related to its forecasting ability for output. An increase in the yield spread leads to an increase in future output and a simultaneous drop in future prices of approximately the same percentage as the percentage increase in real output. The predictability of inflation and real output seem to be mirror reflections of the same economic phenomenon! We confirm this finding by testing the hypothesis that the yield spread is a symmetric predictor of future output and inflation. Indeed, this hypothesis cannot be rejected by the data. The symmetry in the predictability of output and inflation is further corroborated by the remarkable finding that during periods when the forecasting ability of the yield spread for output deteriorates (especially after the mid-1980s), its forecasting ability for inflation also deteriorates by a similar amount.

The symmetric predictability of output and inflation via the yield spread is a new stylized fact, which requires an economic explanation. So far the literature has concentrated on providing an economic explanation for the predictability of output. Most authors provide plausible but heuristic economic stories for the predictability of output. Our main challenge in this Paper is to build a general equilibrium model that explains not only output predictability - something the previous literature has so far failed to do - but the symmetric price predictability as well.

We build a parsimonious one-factor general equilibrium model of a monetary economy with short-term price stickiness, which is able to explain the stylized facts as a result of intertemporal smoothing of rational consumers. We make the model as simple as possible and explore how far it can go in explaining the predictive ability of the term structure for output and prices. We derive explicit analytic solutions of the model, which relate the predictive power of the yield spread to two main deep structural economic parameters: the degree of price stickiness and the elasticity of intertemporal substitution of the representative consumer.

We subsequently estimate the model parameters and find that moderate price stickiness and relatively strong intertemporal substitution are sufficient conditions for explaining the stylized facts on the symmetric predictability of the yield spread. We also test the model's over-identifying restrictions. These restrictions cannot be rejected by the data.

One key feature of the model is the simplicity of its dynamics. The dynamics are driven entirely by the nature of price stickiness. This distinguishes our model from the class of affine- yield models, in which the dynamics of the factors driving the economic variables are exogenous. Because prices are sticky, current economic shocks lead to predictable changes in future prices and output. These expectations, coupled with consumption smoothing and arbitrage, lead to contemporaneous changes in real and nominal interest rates. A second key feature of the model is that, with a constant velocity of money and relatively strong intertemporal substitution, economic shocks have opposite effects on real and nominal interest rates. Positive productivity shocks increase real but decrease nominal interest rates. Positive money supply shocks decrease real but increase nominal interest rates. As we explain later, this feature is central in explaining the evidence and distinguishes our model from previous models. Indeed, previous unsuccessful attempts at explaining the predictability of output within a general equilibrium model of an endowment economy focused exclusively on real economies and real magnitudes. Yet, the empirical evidence is based on the nominal yield spread, not the real yield spread. Thus the explanation of the evidence requires a monetary model, which can jointly predict output and prices. To build some intuition on the mechanics of the model and its ability to explain the evidence, consider for example the effects of a permanent positive productivity shock. This shock increases consumption and output and reduces prices contemporaneously, creating the base of comparisons with future levels of consumption, output and prices. Due to price stickiness, prices do not adjust fully to their lower steady state level. In the subsequent periods, they are expected to further decline slowly towards their new steady state. This future further decrease in prices is expected to lead to a symmetric - due to the constant velocity of money - further increase in consumption and output in every future period, albeit at smaller and smaller magnitudes as time goes on, which reflect the ever smaller declines in prices. Real interest rates increase because the expected increase in future consumption relative to today's consumption decreases the marginal utility of future consumption relative to today's marginal utility. This leads rational agents to borrow and consume more today in order to smooth their consumption, hence, pushing upward the real rate of interest. If the elasticity of intertemporal substitution of consumption is high (larger than unity), however, the increase in real interest rates, which is required in order to bring about this consumption smoothing, is not very large and is overwhelmed by the drop in expected inflation. Hence, nominal interest rates decline. All nominal rates decline across the full maturity spectrum, but short rates decline a lot more than long rates. This is because, as was explained above, as time passes, the expected change in prices and output washes out gradually. The largest impact occurs early on and influences current short rates a lot more than expected future short rates. Since long rates are weighted averages of current and expected future short rates, the impact on them will be smaller than the impact on short rates. It follows that the nominal yield spread increases. So we end up with a current increase in the yield spread, a decrease in expected future prices and a symmetric increase in expected future output.

References

[1] Ang, A., M. Piazzesi and Min Wei. What does the Yield Curve Tell us about GDP Growth? *Columbia Business School Discussion Paper* 2003.

[2] De Lint, Christel R. and D. Stolin. The Predictive Power of the Yield Curve: A Theoretical Assessment. *Journal of Monetary Economics*, 2003, vol. 50, pp. 1603-1622.

[3] Estrella, A. Why does the Yield Curve Predict Output and Inflation? *Federal Reserve Bank of New York Discussion Paper*, 2003.

[4] Estrella, A. and G. Hardouvelis. The Term Structure as a Predictor of Real Economic Activity, *Journal of Finance*, 1991, vol. 46, pp. 555-576.

[5] Estrella, A., A. P. Rodrigues and S. Schich. How Stable is the Predictive Power of the Yield Curve? Evidence from Germany and the Uniited States, *The Review of Economics and Statistics*, 2003, vol. 85, pp. 629-644.

[6] Hamilton, J. D. and D. H. Kim. A Reexamination of the Predictability of Economic Activity Using the Yield Spread, *Journal of Money, Credit, and Banking*, 2002, vol. 34, pp. 340-360.

VSP International
Science Publishers
P.O. Box 346, 3700 AH Zeist
The Netherlands

*Lecture Series on Computer
and Computational Sciences*
Volume 1, 2004, pp. 970-973

An Evaluation of Alternative VAR Models
for Forecasting Inflation

H. Heidari[1]

School of Economics
Faculty of Commerce and Economics
The University of New South Wales
Sydney NSW 2052, Australia

Accepted: August 8, 2004

Abstract: This paper provides a comparative analysis of several alternative vector autoregression (VAR) models for forecasting inflation. These include classical and Bayesian framework. Although traditional BVAR models can improve UVAR models forecasts by using extra information as priors, they can not be used to resolve a mixed drift case, which is common in most of the macroeconomic forecasting models. Hence I apply the Bewley transformation for the re-parameterization of the VAR to estimate drift parameters directly by using instrumental variables. As an alternative the g-prior is considered. I assess the performance of the different specifications of BVAR models by one-step ahead forecasts using real data and Monte Carlo experiments.

Keywords: VAR models, Bayesian VAR models, Forecasting, Bewley transformation, Inflation.

Mathematics Subject Classification: 91B64, 62M20, 62M10

PACS: 89.65.Gh, 82.20.Wt, 02.50.Ng

Extended Abstract

Accurate forecasts of variables are made to guide decisions in a variety of fields such as economics, marketing, financial speculation, financial risk management, business and government budgeting and crisis management. Although the forecasting of macroeconomic magnitudes is of great interest, the analysis and prediction of prices and output, under inflation targeting strategy has become even more important from the standpoint of the monetary policy makers and private agents who try understand and react to Central Banks behavior. A significant problem of this regime is the central bank's imperfect control over inflation. It is documented that using conditional inflation forecasts as an intermediate target variable can alleviate this problem, see, e.g. [14]. This makes forecasts as the focal point in monetary policy discussions.

In this paper I will provide a comparative analysis of several alternative VAR models for forecasting of inflation. These include classical and Bayesian approaches. A disadvantage of using an Unrestricted VAR (UVAR) model for forecasting based on unrestricted OLS estimates of the coefficients is the large number of parameters that need to be estimated. In an attempt to restrict the parameters of the UVAR models and improve the forecasting performance of these models, Litterman [8,9] and Doan et al. [3] suggested that they could be estimated using Bayesian techniques, which take into account any prior information available to the forecaster. The Litterman's prior is presented as embodying the idea that series should be random walks. As most of the macroeconomic variables have persistent trends, so the best guesses of the Litterman prior will be the random walk with drift, with a vague prior on the drift. In fact, Litterman [8,9] suggested a class of priors for VAR models that induce a random walk mean for the coefficients and have a parsimonious set of hyperparameters, tightness parameters, which govern their variance.

---

[1] Corresponding author. Email: h.heidari@unsw.edu.au

There is a lot of empirical evidence in the literature, which suggests that Bayesian VAR (BVAR) models with Litterman priors produce forecasts that exhibit a high degree of accuracy relative to UVAR models, see, e.g. [3,6,7,8,9,10,11,12,13,15]. In inflation forecasting, however, the performance of BVAR models with a Litterman prior has been somewhat less impressive, see, e.g. [7,8,9,13,14]. Although BVAR models can improve on UVAR models forecasts by using extra information as priors, they cannot be used to resolve a mixed drift case, which is common in most macroeconomic forecasting models. Bewley [2] argues that further gains in forecast performance of VAR models could follow from restricting drift parameters in a classical or Bayesian approach.

In order to overcome poor forecasting, Hendry and Clements [5] suggested intercept correction with vector error correction models (VECM). One of the ways of restricting the intercept correction is suggested by Bewley [2]. Bewley [2] finds that the BVAR performs better than the VAR mainly because it corrects for the unit root, not because it reduces the over-parameterization, and that its long-run performance for driftless variables is poor. Bewley [2] suggests that Litteman's prior, because of the vague prior on the constant, will not perform well in long-run forecasting of I(1) variables either, particularly if they have no drift.

Most of the macroeconomic forecasting models include time series that demonstrate both drift and no drift. By manipulating a VAR model Bewley [2] has shown that the constant terms in driftless equations also exhibits the non-zeros elements. Moreover the constants are nonlinear functions of these non-zero elements and the other parameters in the system. Hence the BVAR with Litterman prior cannot be used to solve this mixed drift case. Bewley [2] argued that in this condition, the long-run forecast errors of time series without drift in a mixed-drift model are dominated by insignificant drift parameter estimates.

In this paper, I will attempt to apply the Bewley transformation [1] for the re-parameterization of the VAR to estimate drift parameters directly by using instrumental variables. Then I will forecast in this framework, in both the classical and Bayesian frameworks and compare it with some other VAR forecasting models such as a first-Differenced VAR specification, a Bayesian VAR with Litterman prior [9],the Normal-Wishart prior, the VAR specification suggested by Waggoner and Zha [16] and Zha [20], g-prior and Natural- Conjugate prior.

Another way of thinking about mean estimation in the VAR models is using g-priors. Zellner [19] reported that the g-prior is a special form of a Natural Conjugate prior distribution. For the g-prior amounts, Zellner [19] proposed specifying a normal prior for a conditional on variance - covariance matrix, where the mean of the prior is the same suggested by Litterman (random walk), with a Jeffrey's prior for the variance - covariance matrix of residuals. The BVAR with g-prior shrinks the coefficients mean toward the random walk, without any effect on the long-run forecast. Because of this, the reparameterization of BVAR with g-prior may estimate the mean in VAR models much more accurately than the traditional BVAR.

I will investigate the reasons for the differences in the performance of aforementioned models and the forecast accuracy of these models in different monetary regimes. The paper is divided into two parts: First I will carry out some simulations in different situations such as when the data have a unit root, when they do not and also when the variables do or do not have a drift in the forecast period (mixed drift). I will set the DGP to approximately mimic some important macroeconomic series such as inflation, capacity utilization, real GDP, stock prices and unemployment. Moreover I will consider measurement errors and breaks in simulation. I will leave two forecasting periods out; find the value of the hyperparameters that maximize the forecasting performance in the first forecasting period and forecast at several horizons, including rather long horizons, in the second forecasting period. Then at some point (first, second, four and eight quarters ahead) I will evaluate the models in terms of their forecasting performance with different measures of forecast accuracy.

In the second part of this paper, I will use real data of some countries with different monetary policy regimes and economic environments such as USA, Australia, and Islamic Republic of Iran to asses the forecasting performance of the considered models at different forecasting horizons.

Acknowledgments

The author wishes to thanks Paolo Girodani for inspiring discussions and providing some Gauss procedures and Minxian Yang for his careful reading of the manuscript and their fruitful comments and suggestions. However, they cannot be held responsible for any remaining errors which are my own.

References

[1] R. A. Bewley, The Direct Estimation of the Equilibrium Response in a Linear Dynamic Model, *Economics Letters* **3** 357-361(1979).

[2] R. A. Bewley, *Real-time Forecasting with Vector Auto regressions: Spurious Drift, Structural Change and Intercept-Correction*, Working Paper, The University of New South Wales, Sydney, Australia, 2000.

[3] T. R. Doan, R. B. Litterman and C. Sims, Forecasting and conditional projection using realistic prior distributions, *Econometric Reviews*, **3(1)** 1-100(1984).

[4] J. Hamilton, *Time Series Analysis*, Princeton University Press, Princeton, 1994.

[5] D. F. Hendry and M. P. Clements, Economics Forecasting: Some Lessons from Recent Research, *Economic Modeling* **20** 301-329(2003).

[6] K. R. Kadiyala and S. Karlsson, Forecasting with Generalized Bayesian Vector Auto regressions, *Journal of Forecasting* **12** 365-378(1993).

[7] K. R. Kadiyala and S. Karlsson, Numerical Methods for Estimation and Inference in Bayesian VAR- Models, *Journal of Applied Economics* **12(2)** 99-132 (1997).

[8] R. B. Litterman, Forecasting and Policy Analysis with Bayesian Vector Auto regression Models, Federal Reserve Bank of Minneapolis, *Quarterly Review* **4** 30-41(1984).

[9] R. B. Litterman, Forecasting With Bayesian Vector Auto regressions – Five Years of Experience, *Journal of Business and Economic Statistics*, **4(1)** 25- 38(1986).

[10] S. K. McNees, Forecasting Accuracy of Alternative Techniques: A Comparison of US Macroeconomic Forecasts, *Journal of Business and Economic Statistic*, **4(1)** 5-15(1986).

[11] J. C. Robertson and E. W.Tallman, Vector Auto regressions: Forecasting and Reality, Federal Reserve Bank of Atlanta, *Economic Review* **84** 4- 18(1999).

[12] C.A. Sims A Nine – Variable Probabilistic Macroeconomic Forecasting Model, Business *Cycle, Indicators and Forecasting* (Editor: J. Stock and M. Watson),University of Chicago Press, Chicago, (1993), 179-212.

[13] C.A. Sims and T. Zha, Bayesian Methods for Dynamic Multivariate Models, *International Economic Review* **39** 949-968(1998).

[14] L.E.O. Svensson, Inflation forecast targeting: Implementing and monitoring inflation targets, *European Economic Review* **41** 1111-1146(1997).

[15] R. M. Todd, Improving Economic Forecasting With Bayesian Vector Auto regression, Federal Reserve Bank of Minneapolis, *Quarterly Review* **8** 18-29 (1984).

[16] D. Waggoner and T. Zha, *Conditional Forecasts in Dynamic Multivariate Models*, Working Paper 98-22, Federal Reserve Bank of Atlanta, USA, 1998.

[17] R. H. Webb, Forecast of inflation from VAR models, *Journal of Forecasting*, **14(3)** 268-285(1995).

[18] V. Zarnowitz and P. Braun, *Twenty-two years of the NBER-ASA quarterly economic outlook surveys: Aspects and comparisons of forecasting performance*, Working Paper, No. 3965, NBER, USA, 1992.

[19] A. Zellner, On assessing prior distributions and Bayesian regression analysis with g-piror distributions, *Bayesian Inference and Decision Techniques* (Editor: P. Goel and A. Zellner), Elsevier Science Publishers B.V.,(1986), 233-243.

[20] T. A. Zha, A Dynamic Multivariate Model for Use in Formulating Policy, Federal Reserve Bank of Atlanta, *Economic Review* **83** 16-29 (1998).

VSP International
Science Publishers
P.O. Box 346, 3700 AH Zeist
The Netherlands

Lecture Series on Computer
and Computational Sciences
Volume 1, 2004, pp. 974-975

Testing of Capital Asset Pricing Model with Local Maximum Likelihood Method

Burç Kayahan
Thanasis Stengos

Department of Economics
University of Guelph
Guelph, Ontario N1G 2W1
Canada

Accepted: August 8, 2004

Abstract: This paper follows the approach of Wang (2003) in order to test the conditional version of Sharpe-Lintner CAPM by adopting Local Maximum Likelihood as its nonparametric methodology. This methodology does not only avoid the misspeficication of betas, risk premiums and the stochastic discount factor but, is also expected to perform better when compared with other more traditional methods such as the constant Nadaraya-Watson kernel estimator due to its superior performance at the sample extremes.

Keywords: CAPM, Local Maximum Likelihood, nonparametric estimation.

Mathematics Subject Classification: 62G, 62P20.

1. Extended Abstract

Asset pricing models have attracted considerable attention in finance literature due to the ambition of researchers to predict the asset returns. Various models such as Arbitrage Pricing Theory (APT) or Consumption-based General Equilibrium models have been developed in the finance literature to forecast asset returns, however, none of these models have been as popular as the captial asset pricing model (CAPM) of Sharpe (1964) and Lintner (1965). The central assumption of CAPM is that market portfolio of invested wealth[1] is mean-variance efficient and under this assumption it predicts that the risk premiums on individual assets will be proportional to risk premium on the market portfolio and beta coefficient of the market security which measures the sensitivity of the market security's return to market fluctuations. This simple version of CAPM is also known as unconditional or static CAPM due to the fact that the relation between individual securities and market portfolio, as implied by the betas, are assumed to be time-invariant and stable in this framework.

The CAPM has been examined and tested by a great number of authors during the past decades and various anomalies of the static CAPM have been documented as a result of these studies. A group of studies led by Fama and French(1992,1995,1996) have found that the static CAPM does not hold empirically. The empirical results in disfavor of unconditional CAPM, were so strong that it led some authors to conclude that CAPM is dead. However, there were also numerous attempts to explore the nature of these anomalies and correct them. In fact, studies by Keim and Stambaugh (1986), and Breen,Glosten and Jagannathan (1989) showed that betas in the CAPM framework, are not time-invariant and they vary during the business cycles as shown by Chen(1991) and Harvey and Ferson(1991). In their paper Jagannathan and Wang (1996) have developed a conditional version of CAPM where the betas of securities are conditioned on the information set available to investors at a particular point in time and are subject to change with the fluctuating economic conditions in time. Dybvig and Ross (1985) and Hansen and Richard (1987) demonstrated that even though static CAPM may commit serious pricing erros, conditional CAPM might still hold.

The development of conditional CAPM, motivated a new literature that focused on the formation and testing of these models. Despite its solution to some of static version's anomalies,

---

[1] See Markowitz (1959) for a detailed description.

conditional CAPM also brought some new obstacles into the literature. One of such problems, was the choice of conditioning variables and the lack of theory in how to form the relationship between the betas and conditioning variables. At first, conditional CAPM that exhibited betas as the linear functions of conditioning variables, were adopted, however, the performance of such specifications were even worse than the static CAPM! It has been argued that the validity of such tests should be suspected due to the fact that their results are strongly influenced by the modelling assumptions inherent in them. In fact, Ghysels (1998) showed that when beta dynamics are misspecified, the pricing errors of conditional CAPM are much likely to be greater than the unconditional version.

This revelation induced researchers to focus on nonparametric estimation techniques in an attempt to avoid specifying certain functional forms in the estimation of betas. Wang (2003) adopted such a nonparametric approach for specifying the beta risk dynamics. In detail, he used the Nadaraya-Watson Kernel estimator in his estimations and used the stochastic discount factor (SDF) framework to apply and test the restrictions implied by conditional CAPM. In his paper, he showed that conditional CAPM is superior to unconditional version despite the fact that pricing errors committed by conditional CAPM, are still statistically significant even under this nonparametric setup.

In this paper, we adopt the testing methodology of Wang, however, we use Local Maximum Likelihood as the nonparametric estimation method. It is known that Nadaraya-Watson Kernel Method performs poorly in tail estimation (also known as boundary effect) and it is a stylized fact that stock returns exhibit fatter tails. Our goal is improving the explanatory power of conditional CAPM model via adopting a more precise estimation technique and compare whether the efficieny gains from such a procedure are significant.

In order to measure the performance of Local Maximum Likelihood in the conditional CAPM context, we conducted a Monte Carlo experiment where Nadaraya-Watson Kernel Method is also included in order to make a relative comparison. Moreover, we also applied these two estimation techniques to the stock returns data to see their performance from an empirical point of view.

The paper proceeds as follows. The nonparametric presentation of conditional CAPM and the testing methodology are presented in Section II. Local Maximum Likelihood technique is briefly explained in Section III. Section IV summarizes the design and results of the Monte Carlo experiment. Section V provides the description of data and empirical results. Section VI concludes the paper.

VSP International
Science Publishers
P.O. Box 346, 3700 AH Zeist
The Netherlands

*Lecture Series on Computer
and Computational Sciences*
Volume 1, 2004, pp. 976-978

A Flight to Quality! International Capital Structure Under Foreign Liquidity Constraints

Constantina Kottaridi[1]

Department of International and European Economic Studies
Athens University of Economics and Business
104-34 Athens, Greece

Gregorios Siourounis

Department of Economics
London Business School
Regent's Park
London NW1 4SA, United Kingdom

Accepted: August 8, 2004

Abstract: In the last decade we have witnessed a significant change in the structure of capital flows to developed as well as to developing countries. We construct a simple econometric framework where country specific random effects and macroeconomic monetary volatility are linked to the probability distribution of liquidity shocks hitting an international investor. A "volatility augmented" gravity equation is then estimated to provide empirical evidence that as the probability of getting a bad liquidity shock increases, investors switch to safer assets but with a pecking order: they seem to damp equities for more bonds and more direct investments. A flight to quality!

Keywords: foreign bond investments (FBI), foreign direct investments (FDI), foreign equity investments (FEI), liquidity shocks, monetary macroeconomic volatility.

Mathematics Subject Classification: 62P20.

1. Extended Abstract

The present paper aims at testing empirically the information-based approach of capital flows, which states that: investors, who know that they are more likely to get a foreign liquidity shock that forces them to sell before maturity of the project, will choose to make portfolio investments, whereas investors who know they are less likely to get a liquidity shock, will choose to make direct investments.[2] We constructed a simple econometric framework where country specific random effects and macroeconomic monetary volatility are linked to the probability distribution of foreign liquidity shocks[3] hitting an international investor, thus, inducing a shift in foreign capital structure. We then estimate a "volatility augmented" gravity equation.

We document that capital structure depends greatly on the volatility of macroeconomic monetary aggregates. We show that monetary macroeconomic volatility increases international investors' probability of receiving a bad foreign liquidity shock. Interestingly, in this event, international investors' gross cross border transactions switch to more direct investments relative to equity investments. Moreover, in the same event, gross cross border transactions in bonds increase relative to equities. Thus foreign liquidity constraints faced by international investors have a direct effect on the external capital structure of nations implying a pecking order of international capital flows.

---

[1] Corresponding author. E-mail: ck@aueb.gr

[2] For a recent theoretical documentation see Goldstein and Razin (2003).

[3] We define "foreign liquidity shock", the shock associated with the investment in the foreign country.

An important issue emanating from worldwide developments within the context of globalization is the large change in the level and composition of capital flows, both among advanced countries and between advanced and developing ones. International capital flows can take various forms, the characteristics of which differ markedly in terms of i.e., cost, risk bearing, vulnerability to capital flows reversal, conditionality, access to intellectual property (Williamson (2000)). Conventional wisdom classifies international capital flows into three major categories, namely, portfolio equity flows (FEI), debt flows (FBI) and foreign direct investment flows (FDI). A sharing feature of these lies in the blending of foreign savings with domestic savings to finance domestic investment (Razin, Sadka and Yuen, 1999). FDI is nonetheless differentiated from the rest in its origin, since, by default, it entails foreign ownership control[4] of a domestic.[5]

Understanding the influencing factors and implications of the structure of capital flows is of particular interest to international economists. While there is a considerable number of theoretical and empirical studies addressing the issue of determinants of debt flows, equity flows and direct investments[6], few attempts have been made to explore the determinants of international capital structure. This work has clear relevance for policymakers. Stocks of foreign assets, liabilities and direct investments represent an important global linkage. Identifying the sources of the composition in world asset trade can contribute to understanding its sustainability and likely future trends as well as its effects on local economies. In this respect, growth that is related to the injection of foreign capital greatly depends on the fluctuations of capital structure. In contrast, growth that is linked to positively trending variables such as output per capita and goods trade can be predicted to continue into the future.

An extensive database of gross cross-border transactions of debt, equity and direct flows among 15 countries representing the global environment over a period of 13 years is used to empirically evaluate the theoretical prediction that international investor's liquidity cantering associated with the foreign investment will have a direct impact on the composition of her international portfolio. Sample coverage both in terms of time series and cross-countries, enables the distinction of alternative capital structures.

Acknowledgments

We thank Wouter Denhaan, Christos Genakos, Itay Goldstein, Stefan Nagel, Sergey Sanshar, Assaf Razin, and the participants of the EEFS 2003 Conference (Bologna-Italy) for useful comments and suggestions.

References

[1] Froot, K., Japanese Foreign Direct Investment, Working Paper No.3737, Published in US-Japan Economic Forum, edited by Martin Feldstein and Kosai, 1991.

[2] Goldstein, and Razin, An Information-Based Trade-off Between Foreign Direct Investment and Foreign Portfolio Investment: Volatility, Transparency and Welfare, CEPR Working Papers, No. DP3747, 2003.

[3] Portes, R., H. Rey and Y. Oh, Information and Capital Flows: the Determinants of transactions in Financial Assets. European Economic Review, Papers and Proceedings, 783-796, 2001.

[4] Portes, R. and H. Rey, The Determinants of Cross-Border Equity Transaction Flows, Working Paper, 2002.

[5] Razin, A, E Sadka and C. Yuen, Capital Flows with Debt and Equity Financed Investment: Equilibrium Structure and Efficiency Implications, IMF Working Paper No. 159, 1998b.

[6] Razin, A., Sadka, E. and Yuen, C-W., An Information-Based Model of Foreign Direct Investment: The Gains from Trade Revisited. NBER Working Paper No. 6884, 1999.

---

[4] Conventionally, the cut-off point which enables control of a domestic firm by a foreigner, is assumed to be the ownership of 10% of its equity

[5] Froot (1991) however, notes that FDI actually requires neither capital flows nor investment in capacity.

[6] See for example Portes and Rey (2002) and Portes, Rey and Oh (2002).

[7]   Williamson, J. Issues Regarding the Composition of Capital Flows. Conference on Capital Account Liberalization: The Developing Country Perspective, 2000.

VSP International
Science Publishers
P.O. Box 346, 3700 AH Zeist
The Netherlands

Lecture Series on Computer
and Computational Sciences
Volume 1, 2004, pp. 979-983

Temporary and Permanent Long-Run Asset-Specific and Market Risks in the Cross-Section of US Stock Returns

M. Koubouros[1], D. Malliaropulos and E. Panopoulou

Accepted: August 8, 2004

Abstract: This paper decomposes the overall market, or CAPM, risk into parts reflecting long-run market uncertainty related to the dynamics of the present value of revisions in expectations about future asset-specific and market cash-flows and discount-rates respectively. We decompose market betas into four sub-betas (associated with assets' and market's cash-flows and discount-rates) and we employ a discrete time version of the Intertemporal-CAPM (I-CAPM) to derive a four-beta asset pricing model. The model performs well in pricing average excess returns on single-sorted US common stock portfolios according to size, book-to-market and dividend-price ratios, and it produces insignificant pricing errors, high estimates for the explained cross-sectional variation in average returns and economically and statistically acceptable estimates for the coefficient of relative risk aversion.

Keywords: CAPM, asset pricing, betas, VAR-GARCH
Mathematics Subject Classification: Here must be added the AMS-MOS or PACS
PACS: Here must be added the AMS-MOS or PACS Numbers

1. Introduction

In this study we extend the CAPM and we decompose the market systematic risk (CAPM beta) into four long-run risk components related to the covariance of unexpected changes in stock-specific cash-flows and discount rates with unexpected changes in market-wide cash-flows and discount rates. We empirically test whether these sources of risk are priced using a discrete time version of the intertemporal asset pricing model of Merton (1973) recently developed by Campbell (1996).

Our study is related to the work of Campbell (1991), Campbell and Mei (1993), Campbell and Vuolteenaho (2004) and Campbell, Polk and Vuolteenaho (2003). In a novel paper, Campbell (1991) shows that unexpected stock returns can be decomposed into the discounted sum of revisions in expectations about future cash flows and future discount rates. Campbell and Mei (1993) extend this analysis by studying the behaviour of asset specific cash-flow and discount-rate components of portfolio betas but do not provide any evidence on whether these parts of systematic risk carry individual risk prices. More recently, Campbell and Vuolteenaho (2004) show that the market beta can be decomposed into a "relatively bad" cash-flow beta, reflecting news about the market's future cash flows, and a "relatively good" discount rate beta, reflecting news about the market's future discount rates. Empirically, they find that small stocks and value stocks have considerably higher cash-flow (bad) betas than growth stocks and large stocks, and this can explain their higher average returns. However, they restrict their analysis by assuming that good and bad betas are independent of whether the innovation in individual returns is due to unexpected changes in future cash-flows or discount rates of the company. In a recent study closest to ours, Campbell, Polk and Vuolteenaho (2003) decompose the overall market beta into four betas which reflect the covariance of unexpected changes in stock-specific cash-flows and discount rates with unexpected changes in market-wide cash-flows and discount rates. However, they do not test the asset pricing implications of this decomposition.

Our equilibrium four-beta model generates low and insignificant pricing errors, high estimates for the explained cross-sectional variation in average excess returns and statistically and economically acceptable estimates for the degree of relative risk aversion. We find that, as predicted by economic theory, risks associated with permanent shocks to total wealth are the main determinants of the overall

---
[1] Corresponding author, Department of Banking and Financial Management, University of Piraeus, and Department of Economics, University of Peloponnese. Karaiskakis' End, 22 100 Tripolis, Greece, phone: (+3) 2710.230129, fax: (+3) 2710.230139, e-mail: mcoum@uop.gr.

risk premium; covariances with both portfolio cash-flow and discount-rate dynamics earn equilibrium risk premia that are indistinguishable from zero, but the premia associated with asset-specific cash-flow news are greater than those linked to asset-specific discount-rate news. More importantly, we provide some evidence that the coefficient of proportionality between the two premia is equal to the constant coefficient of relative risk aversion, as predicted by theory.

2. Decomposing Risk and Return

2.1 Cash-Flow and Discount-Rate Dynamics

Using the approximation of unexpected return originally developed by Campbell and Shiller (1988), we decompose market risk and into four components each reflecting the sensitivity of asset-specific cash-flow and discount-rate news to market return innovations respectively. More specifically, using the definition of the market beta ($\beta_{im,t} = Cov_t(e_{i,t+1}, e_{m,t+1})/Var_t(e_{m,t+1})$) and the decomposition of the unexpected component of the return:

$$e_{i,t+1} = N_{i,t+1}^C - N_{i,t+1}^D = (E_{t+1} - E_t) \sum_{j=0}^{\infty} \rho_i^j \Delta d_{i,t+1+j} - (E_{t+1} - E_t) \sum_{j=1}^{\infty} \rho_i^j r_{i,t+1+j}, \tag{1}$$

We break the quantity of market systematic risk as follows:

$$\beta_{im,t} = \frac{Cov_t(N_{i,t+1}^C - N_{i,t+1}^D, N_{m,t+1}^C - N_{m,t+1}^D)}{Var_t(e_{m,t+1})} = \beta_{i,CC,t} + \beta_{i,CD,t} + \beta_{i,DC,t} + \beta_{i,DD,t}, \tag{2}$$

where the individual components $\beta_{i,CC,t}, \beta_{i,CD,t}, \beta_{i,DC,t}$, and $\beta_{i,DD,t}$, in (2) are defined as:

$$\beta_{i,CC,t} = \frac{Cov_t(N_{i,t+1}^C, N_{m,t+1}^C)}{Var_t(e_{m,t+1})}, \beta_{i,CD,t} = \frac{Cov_t(N_{i,t+1}^C, -N_{m,t+1}^D)}{Var_t(e_{m,t+1})}, \tag{3}$$

and

$$\beta_{i,DC,t} = \frac{Cov_t(-N_{i,t+1}^D, N_{m,t+1}^C)}{Var_t(e_{m,t+1})}, \beta_{i,DD,t} = \frac{Cov_t(-N_{i,t+1}^D, -N_{m,t+1}^D)}{Var_t(e_{m,t+1})}, \tag{4}$$

respectively. In equations (1) to (4), $Var_t(.)$ and $Cov_t(.)$ denote the conditional, at time t, variance and covariance operators respectively, $r_{i,t+1} = \log(P_{i,t+1} + D_{i,t+1}) - \log P_{i,t}$ is the real continuously compounded return with $D_{i,t+1}$ being the real dividend, $(E_{t+1} - E_t)x_{t+1+j} = E_{t+1}[x_{t+1+j}] - E_t[x_{t+1+j}]$, and finally $\bar{\delta}_i = E[\log(d_{i,t} - p_{i,t}]$ and $\rho_i = 1/[1 + \exp(\bar{\delta}_i)]$ are asset specific constants.

2.2 Pricing of Cash-Flow and Discount-Rate Risks

In order to derive testable restrictions on the premia associated with the cash-flow and discount rate risks in (3) we need a risk story. For this purpose, we employ the recursive utility framework provided by Epstein and Zin (1989, 1991) and Weil (1989) and the log-linear approach employed by Campbell (1996). This model derives the following cross-sectional linear restrictions on assets' risk premia:

$$E_t[R_{i,t+1}] - R_{f,t+1} = \gamma Cov_t(e_{i,t+1}, e_{m,t+1}) + \gamma Cov_t(e_{i,t+1}, N_{m,t+1}^D) \tag{5}$$

where γ is the coefficient of relative risk aversion (CRRA). Using equation (1) for any individual asset as well as the market portfolio, m, gives:

$$E_t[R_{i,t+1}] - R_{f,t+1} = \gamma Cov_t(N_{i,t+1}^C, N_{m,t+1}^C) + Cov_t(N_{i,t+1}^C, -N_{m,t+1}^D)$$
$$+ \gamma Cov_t(-N_{i,t+1}^D, N_{m,t+1}^C) + \gamma Cov_t(-N_{i,t+1}^D, -N_{m,t+1}^D) \qquad (6)$$

Multiplying and dividing each conditional covariance term in (6) by the conditional variance of market's unexpected returns, $Var_t(e_{m,t+1})$, we obtain the following representation for the risk premium on any risky asset i :

$$E_t[R_{i,t+1}] - R_{f,t+1} = \lambda_{0,t} + \lambda_{CC,t}\beta_{i,CC,t} + \lambda_{CD,t}\beta_{i,CD,t} + \lambda_{DC,t}\beta_{i,DC,t} + \lambda_{DD,t}\beta_{i,DD,t}, \qquad (7)$$

where $\lambda_{0,t}$ represents the conditional Jensen's alpha, the rest of the λs represent time-varying prices of beta risks, defined as $\lambda_{CC,t} = \lambda_{DC,t} = \gamma Var(e_{m,t+1})$ and $\lambda_{CD,t} = \lambda_{DD,t} = Var(e_{m,t+1})$ respectively, and the betas are defined similarly to (3) and (4). Equation (7) states that the required risk premium on asset i is jointly determined by the covariances of asset's shocks to cash flows and discount rates with the corresponding decomposed components of the total market innovation. Similarly to Campbell and Vuolteenaho (2004), a conservative risk-averse investor with $\gamma > 1$ demands a higher risk price for risks associated with market cash flow uncertainty ($\beta_{CC,t}$ and $\beta_{i,DC,t}$) rather than for risks linked to shocks to market returns ($\beta_{CD,t}$ and $\beta_{DD,t}$), since any positive (negative) shock to wealth discount rates is at a benefit (cost) of worse future investment opportunities, whereas the investor is never compensated later for every positive (negative) shock to dividends. Since our four-beta model in (7) can be written in the restricted from as follows:

$$E_t[R_{i,t+1}] - R_{f,t+1} = \lambda_{0,t} + \gamma\lambda_t\beta_{i,CC,t} + \lambda_t\beta_{i,CD,t} + \gamma\lambda_t\beta_{i,DC,t} + \lambda_t\beta_{i,DD,t}, \qquad (8)$$

with $\lambda_t = Var(e_{m,t+1})$, the beta prices of market cash-flow risk, λ_{CC} and λ_{DC}, are a γ multiple of the beta risk prices of market discount-rate risk, λ_{CD} and λ_{DD}, respectively.

3. Data, Methodology and Empirical Analysis

3.1 Data and Methodology

In order to derive the cash-flow and discount-rate "news" series N_{t+1}^C and N_{t+1}^D we employ a VAR-GARCH(1,1) methodology. A VAR-GARCH(1,1) system is estimated for each asset i and the for the market portfolio m. Our monthly data consists of 3 sets of 10 portfolios sorted according to market value, book-to-market and dividend-price ratio, and 3 US-economy-wide variables that serve as instruments, from December 1928 to December 2001. Following the common practice, the state variables have been selected under the assumption that they can successfully forecast future portfolio returns, and are (a) the log market price-earnings ratio, p-e, (b) the term yield spread, TY, and (c) the small-stock value spread, VS. The VAR-GARCH approach enables us to construct the news series as linear combinations of the standardized innovations of the state variables. Then using the sample covariances of these standardized residuals we calculate the cash-flow and discount-rate betas in (3) and (4), and we estimate the prices of beta risks by running an cross-sectional OLS regression of an unconditional version of (7) and (8).

3.2 Empirical Analysis

Table 1 report the empirical findings. Our extended model improves the ability of the disappointing static CAPM and the two-beta I-CAPM provided by Campbell and Vuolteenaho (2004) to capture the spread in mean asset premia. The proportion of cross-sectional variability explained increases from 46.4% (two-beta model) to the high 83%, while the pricing error is still highly insignificant ($\hat{t} = -0.566$). Most importantly, and even when the insignificant constant is included in the regression, all except λ_{CD} ($\hat{t} = 1.69$) the slope coefficients are significant indicating that the approach of decomposing cash flow and discount rate market risks into two components each reflecting sensitivities of asset's dividend growth news and future return news yields interesting insights for the determination of average risk premia. Once the insignificant λ_0 is removed all four risk prices are highly significant and the high estimated values for λ_{CC} and λ_{DC} (0.019 and 0.023 respectively) provide further support on the results presented by Campbell and Vuolteenaho (2004) and Campbell, Polk and Vuolteenaho

(2004). They argued that value and small stocks have considerably higher market cash-flow betas than large and growth stocks. We extend their results by showing that the sign and magnitude of our estimated beta-risk prices of the decomposed cash-flow market risk are in line with a rational asset pricing model for a long-lived investor. This investor requires a higher premium per unit of market cash-flow risk than for market discount-rate risk and the factor of proportionality restricted to be equal to the coefficient of relative risk aversion is both economically and statistically significant. For this group of portfolios although we reject the equality hypothesis for the market discount rate premia (λ_{CD} and λ_{DD}) we can safely accept if for the market cash flow premia (λ_{CC} and λ_{DC}). Overall, for our four-beta specification in (7) and (8), the spread in the average excess returns of size, value portfolios and dividend-price sorted portfolios seems to be not puzzling.

Table 1: Asset pricing results: Cross-sectional regressions of average excess returns on cash-flow and discount-rate betas of 25 book-to-market, 10 size and 10 dividend-price portfolios.

		Two-beta	Two-beta	Four-beta	Four-beta
	CAPM	I-CAPM	I-CAPM*	I-CAPM	I-CAPM*
λ_0	0.010* (0.006) [1.799]	−0.004 (0.005) [−0.766]		−0.002 (0.003) [−0.566]	
λ_m	−0.003 (0.006) [−0.470]				
λ_C		0.071*** (0.015) [4.668]	0.0061*** (0.009) [6.519]		
λ_D		0.016** (0.006) [2.747]	0.012*** (0.001) [18.042]		
λ_{CC}				0.024* (0.029) [2.627]	0.019*** (0.009) [2.305]
λ_{CD}				0.006 (0.004) [1.690]	0.004*** (0.001) [3.806]
λ_{DC}				0.028* (0.014) [2.030]	0.023*** (0.011) [2.179]
λ_{DD}				0.009** (0.004) [2.549]	0.008*** (0.001) [5.606]
adj. − R^2	-2.7%	45.6%	46.4%	82.7%	83.1%
F - test (all zero) (p-value)		13.178 (0.000)		35.733 (0.000)	
χ^2 - test (p-value)		$\lambda_0 = \lambda_D = 0$ 321.302 (0.000)		$\lambda_0 = \lambda_{CD} = 0$ 14.429 (0.001)	$\lambda_{CC} = \lambda_{DC}$ 0.846 (0.357)
χ^2 - test (p-value)					$\lambda_{CD} = \lambda_{DD}$ 17.017 (0.000)
CRRA (γ) (t-stat)					5.304*** (9.713)

					0.012^{***}
$\tilde{v}$					(0.001)
					[18.042]

Acknowledgments

The author wishes to thank Dimitrios Thomakos (University of Peloponnese, Dept. of Economics) for his careful reading of the manuscript and his helpful comments.

References

[1] J. Campbell, A variance decomposition of stock returns, *Economic Journal* 101 157-159(1991).

[2] J. Campbell, Understanding risk and return, *Journal of Political Economy* 104 298-345(1996).

[3] J. Campbell and J. Mei, Where do betas come from? Asset pricing dynamics and the sources of systematic risk, *Review of Financial Studies* 6 567-592(1993).

[4] J. Campbell, J. Polk and T. Vuolteenaho, Growth or glamour, unpublished paper, Harvard University (2003)

[5] J. Campbell and R. Shiller, Stock prices, earnings and expected dividends, *Journal of Finance*, 43 661-676(1988).

[6] J. Campbell and T. Vuolteenaho, Bad beta good beta, forthcoming *American Economic Review* (2004).

[7] R. Merton, An Intertemporal capital asset pricing model, *Econometrica* 41 867-887(1973).

VSP International
Science Publishers
P.O. Box 346, 3700 AH Zeist
The Netherlands

Lecture Series on Computer
and Computational Sciences
Volume 1, 2004, pp. 984-987

Improving Technical Trading Systems by Using a New Matlab based Genetic Algorithm Procedure

S. Papadamou[1]
G. Stephanides

Department of Economics,
University of Thessaly,
382 21Volos, Greece,

+Department of Applied Informatics,
University of Macedonia Economic and Social Sciences,
54006, Thessaloniki, Greece

Accepted: August 8, 2004

Abstract: Recent studies in financial markets suggest that technical analysis can be a very useful tool in predicting the trend. Trading systems are widely used for market assessment however parameter optimization of these systems has adopted little concern. In this paper, to explore the potential power of digital trading, we present a new Matlab tool based on genetic algorithms, which specializes in parameter optimization of technical rules. It uses the power of genetic algorithms to generate fast and efficient solutions in real trading terms. Our tool was tested extensively on historical data of a UBS fund investing in Emerging stock markets through a specific technical system. Results show that our proposed GATradeTool outperforms commonly used, non-adaptive, software tools with respect the stability of return and time saving over the whole sample period.

Keywords: financial markets; prediction; genetic algorithms; investment, technical rules

Mathematics SubjectClassification: 91B99, 91B28, 46N60, 92D15.

1. Introduction

In our day's traders and investment analysts require fast and efficient tools in a ruthless financial market. Battles in trading are now mainly waged at computer speed. The development of new software technology and the appearance of new software environments like Matlab provide the basis for solving difficult financial problems in real times.

Technical analysis has been a part of financial practice for many decades, but in many studies researchers have ignored the issue of parameter optimization, leaving them open to criticism of data snooping and the possibility of survivorship bias (Lo and MacKinley [1] , Brock et al [4] and Papadamou et al.[6]). Traditionally researchers used ad hoc specification of trading rules.

Papadamou and Stephanides [5], implemented a new Matlab based toolbox for computer aided technical trading that have included a procedure for parameter optimization problem. However, the weak point of their optimization procedure is time: the objective function (e.g. profit) isn't a simple squared error function but a complicated one (each optimization iteration goes through the data, generates trading signals, calculates profits etc.). When the data sets are large and you would like to re-optimize your system often and you need a solution as soon as possible, then try out all the possible solutions and get the best one would be a very tedious task.

The aim of this study is to show how genetic algorithms (GAs) (Holland [3], Goldberg [2]), a class of algorithms in evolutionary computation, can be employed to improve the performance and the efficiency of computerized trading systems. Bauer [8, 9] in his research offered practical guidance

---

[1] Corresponding author. Lecturer in Monetary & Banking Economics, Department of Economics, University of Thessaly, Argonauton & Filelinon, Volos, Greece. E-mail: stpapada@econ.uth.gr, stefpap3@yahoo.com

concerning how GAs might be used to develop attractive trading strategies based on fundamental information. According to Allen and Karjalainen [10], genetic algorithm is an appropriate method to discover technical trading rules. Fernandez-Rodriguez et al [7] by adopting genetic algorithms optimization in a simple trading rule provide evidence for successful use of GAs from the Madrid Stock Exchange.

2. Methodology

Our methodology is conducted in several steps. Firstly, we have to implement our trading system based on technical analysis. In developing a trading system, you need to determine when to enter and when to exit the market. If the trader is in the market the binary variable F_t is equal to one otherwise is zero. As position traders we base the majority of our entry and exit decisions on daily charts by constructing a trend following indicator (Dimbeta). This indicator calculates the deviation of current prices from its moving average of $\theta 1$ length. The indicators used in our trading system can be formalized as below:

$$Dimbeta_t = \frac{Close_t - MovAv_t(Close, \theta_1)}{MovAv_t(Close, \theta_1)} \tag{1}$$

Where $Close_t$ is the closing price of the fund at time t and function MovAv calculates the simple moving average of the variable Close with time length $\theta 1$.

$$MovAv_t(Close, \theta_1) = \frac{1}{\theta_1} \sum_{i=0}^{\theta_1 - 1} Close_{t-1}, \qquad t = \theta_1, \theta_1 + 1, ..., N \tag{2}$$

Our trading system consists of two indicators, the Dimbeta indicator and the Moving Average of Dimbeta given by the following equation.

$$MovAv_t(Dimbeta, \theta_2) = \frac{1}{\theta_2} \sum_{i=0}^{\theta_2 - 1} Dimbeta_{t-1}, \qquad t = \theta_2, \theta_2 + 1, ..., N \tag{3}$$

If $MovAv_t(Dimbeta, \theta_2)$ cross upward the $Dimbeta_t$ then enter long into the market (i.e. buy signal). If $MovAv_t(Dimbeta, \theta_2)$ cross-downward $Dimbeta_t$ then close the long position in the market (i.e. sell signal).

Secondly, we have to optimize our trading strategy. It is well known that maximizing objective functions such as profit or wealth can optimize trading systems. The most natural objective function for a risk-insensitive trader is profit. In our software no short sales are allowed and the leverage factor is set fixed at $v=1$, the wealth at time T is given by the following formula:

$$W_{(T)} = W_o \prod_{t=1}^{T} (1 + F_{t-1} \cdot r_t) \cdot \{1 - \delta|F_t - F_{t-1}|\} \tag{4}$$

Where $r_t - (Close_t / Close_{t-1}) - 1$ is the return realized for the period ending at time t, δ are the transaction costs and Ft is the binary dummy variable indicating a long position or not (i.e. 1 or 0). The profit is given by subtracting from the final wealth the initial wealth, $P_T = W_{(T)} - W_0$.

Optimizing a system involves performing multiple tests while varying one or more parameters ($\theta 1$, $\theta 2$) within the trading rules. In this paper we investigate the possibility of solving the optimization problem by using genetic algorithms.

Our proposed GATradeTool, operates on a population of candidate solutions encoded. Each decision variable in the parameter set is encoded as a binary string and these are concatenated to form a chromosome. It begins with a randomly constructed population of initial guesses. These solution candidates are evaluated in terms of our objective function (equation 4). In order to obtain optimality each chromosome exchanges information by using operators (i.e. crossover and mutation) borrowed from natural genetic to produce the better solution (Whitley [11]).

3. Empirical Results

We apply our methodology in a UBS Mutual Fund investing in emerging stock markets3. The data analyzed consists of 2800 observations on daily closing prices of that fund for the period 1/5/98 – 25/6/04.The optimization period is defined between 1/5/98 to 25/6/03. The optimized system was evaluated through the extended period 1/5/98 to 25/6/04.

Firstly, the effect of different GA parameter configurations will be studied by changing one parameter while keeping the remaining one's fixed at default values. More specifically we are interested to measure the effect of the population size and the crossover parameter in the performance of the genetic algorithm based optimization procedure. Secondly, we compared the solutions of optimization problem conducted by different software tools in order to measure the validity of the GATradeTool proposed.

By looking in table 1 we can say that as long as you increase the population size the best and the average solutions are higher. However, after a population size of 30 the performance decreased. In order to take into consideration the computational costs involved since increase in population size, we calculate the time needed for solving the problem. Low population size leads to low performance and low completion time. According to the efficiency index the best solution is that given by the population size 20. The results were analogous in case of crossover parameter investigation.

Table 1: Population Size Effect

	5	10	20	30	40
Dimbeta/MADimbeta	53/197	71/141	72/135	138/206	203/202
Completion Time	4,68	9,00	17,57	26,66	36,97
Max Return	88,83%	95,35%	121,09%	126,39%	108,87%
Min Return	-40,00%	-0,34%	-53,00%	-87,52%	-15,26%
Avg Return	59,72%	68,25%	69,74%	76,07%	71,77%
St. Dev. Of Returns	29,08%	33,03%	36,97%	42,33%	39,09%
Max Ret./St. Dev	3,05	2,89	3,28	2,99	2,79

By looking in table two you can compare the results of optimization of our trading system by using three different software tools. The first row gives the result for the GATradeTool against the Metastock and the FinTradeTool (Papadamou and Stephanides, [5]). Our proposed software tool (GATradeTool) can solve the optimization problem very fast without any specific restrictions about the number of total tests. The maximum number of test that can be performed in Metastock software is 32000. The FinTradeTool needs much more time in order to find the optimal solution, that is closed to the best solution provided by the GATradeTool. The optimum parameters that have been found in period 1/5/98-25/6/03 were used in the extended period 5/1/98-25/6/04.

Table 2: Comparison of three different software tools

Software Tool	Optimized Parameters	Total Tests	Completion Time (Min)	Optimization Period Return (1/5/98-25/6/03)	Ext. Evaluation Period Return (1/5/98-25/6/04)	%Change In Return	Index Return/ Time
GATrade	(73,135)	-	17,81	120,0%	127,0%	5,9%	6,7%
FinTrade	(75,129)	39601	67,15	126,4%	141,2%	11,7%	1,9%
Metastock	(60,111)	32000	30,3	116,9%	122,1%	4,5%	3,9%

Note: the optimized parameters are θ_1 and θ_2.

Figure1 depicts the evolution of the maximum, minimum and average return across the 300 generations for the Dimbeta trading system. It can be observed that the maximum return has a positive trend. It appears to be relatively stable after 150 generations and moves in the range between 1.2 and 1 (ie. 120%-100% return). For the minimum fitness no pattern seems to exist. For the average population return a clear upward trend can be found in the first 180 generations, this is an indication that the overall fitness of the population improves over time. Concerning the volatility of the solutions, standard deviation of solutions after an increase in the first generations stabilizes in a range between 0.3 and 0.6 providing evidence of a stable and efficient set of solutions.

In conclusion, Genetic algorithms are better suited since they perform random searches in a structured manner and converge very fast to populations of near optimal solutions. The GA will give you a set (population) of "good" solutions. Analysts are interested to get a few good solutions as fast as possible rather than the globally best solution. New trading system can easily implemented in our procedure by making only few changes in the code producing the trading signals.

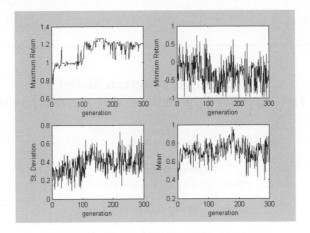

Figure 1: Evolution of Several statistics over 300 generations

References

[1] A.W. Lo, and A.C. MacKinlay When are contrarian profits due to stock market overreaction? *Review of Financial Studies* 3 175-206 (1990).

[2] D.E. Goldberg, *Genetic Algorithms in Search, Optimization and Machine Learning* Addison-Wesley 1989.

[3] J.H. Holland *Adaptation in natural and artificial system* University of Michigan Press 1975.

[4] W. Brock, J. Lakonishok and B. LeBaron Simple technical trading rules and the stochastic properties of stock returns *Journal of Finance* 47 1731-1764 (1992).

[5] S. Papadamou and G. Stephanides A New Matlab-Based Toolbox For Computer Aided Dynamic Technical Trading, *Financial Engineering News* May/June Issue No. 31 (2003).

[6] S. Papadamou and S. Tsopoglou Investigating the profitability of Technical Analysis Systems on foreign exchange markets, *Managerial Finance* 27(8) 63-78 (2001).

[7] F. Fernández-Rodríguez C. González-Martel and S. Sosvilla-Rivero Optimisation of Technical Rules by Genetic Algorithms: Evidence from the Madrid Stock Market Working Papers 2001-14 FEDEA (ftp://ftp.fedea.es/pub/Papers/2001/dt2001-14.pdf)

[8] R.J. Bauer and G.E. Liepins Genetic Algorithms and Computerized trading strategies *Expert Systems in Finance* (Editors D.E. O'Leary and P.R. Watkins) Amsterdam The Netherlands: Elsevier Science Publishers 1992.

[9] R.J. Bauer *Genetic Algorithms and Investment Strategies.* New York John Wiley & Sons Inc. 1994

[10] F. Allen and R. Karjalainen Using genetic algorithms to find technical trading rules *Journal of Financial Economic* 51 245-271 (1999).

[11] D. Whitley The Genitor algorithm and selection pressure: why rank-based allocations of reproductive trials are best *In Proceedings of the third International Conference on Genetic Algorithms* 116-121 (1989).

VSP International
Science Publishers
P.O. Box 346, 3700 AH Zeist
The Netherlands

Lecture Series on Computer
and Computational Sciences
Volume 1, 2004, pp. 988-990

Extending the Merton Model:
A Hybrid Approach to Assessing Credit Quality

George A. Papanastasopoulos[1]
Alexandros V. Benos

Department of Economics,
School of Management and Economics
University of Peloponnese,
GR-221 00 Tripolis Campus, Greece

Department of Banking and Financial Management,
University of Piraeus,
GR-18534 Piraeus, Greece

Accepted: August 8, 2004

Abstract: In this paper we have designed a model to act as an early warning system to monitor changes in the credit quality of corporate obligors. The structure of the model is hybrid in that it combines the two credit risk modeling approaches: (a) a structural model based on Merton's contingent claim view of firms, and (b) a statistical model determined through empirical analysis of historical data. Specifically, we extend the standard Merton approach to estimate a new risk-neutral distance to default metric, allowing liabilities and the corresponding default point to be stochastic. Then, using financial ratios, other accounting based measures and the risk-neutral distance to default metric from our structural model as explanatory variables we calibrate the hybrid model with an ordered – probit regression method. Using the same econometric method, we calibrate a model using our risk-neutral distance to default metric as unique explanatory variable. Then, using cumulative accuracy plots we have test the classification power of those models to predict default events out of sample. Finally, we use the optimized model to predict expected default probabilities for industrial companies listed in the Athens Stock Exchange.

Keywords: credit risk, distance to default, financial ratios, accounting variables, financial ratios

Mathematics Subject Classification: 62P20, 62H30.

1. Extended Abstract

Credit risk refers to the risk due to unexpected changes in the credit quality of a counter party or issuer and its quantification is one of the major frontiers in modern Finance. Credit risk measurement depends on the likelihood of default of a firm to meet its a required or contractual obligation and on what will be lost if default occurs. In this paper we have designed a model to act as an early warning system to monitor changes in the credit quality of corporate obligors. The structure of the model is hybrid in that it combines the two credit risk modeling approaches: (a) a structural model based on Merton's contingent claim view of firms, and (b) a statistical model determined through empirical analysis of historical data.

Central to our hybrid model is a variant of Merton's analytical model of firm value. Fundamental, to Merton's model is the idea that corporate liabilities (equity and debt) could be considered as contingent claims on the value of the firm's assets. To see this, consider the case of a simple firm with market value assets equal to A, representing the expected discounted future cash flows and a capital structure with two classes of liabilities: equity with market value equal to E and zero-coupon debt with face value D^T, maturity at time T. The issue of the debt prohibits the payment of dividends δ until the face value is paid at maturity T. Now consider the position of the

---

[1] Corresponding author. Lecturer of the University of Peloponnese. E-mail: papanast@uop.gr.

equityholders and the debtholders of this simple firm. On debt's maturity T, if the market value of firm's assets A exceeds the face value of debt, the debtholders will receive the promised payment and the equityholders will receive the residual claim $A - D^T$. If the market value of firm's assets does not exceeds the face value of debt the equityholders will find it preferable to exercise their limited liability rights, default on the promised payment and surrender the firm's ownership to its debtholders. These payoffs imply that equity and debt possesses option like features with respect to the solvency of the firm. The payoff of equityholders is equivalent to that of a European Call option with underling asset the firm's asset value A, strike price equal to D^T (firm's default boundary) and maturity T and thus its value at time T equal to $E_T = \max(V_T - D^T, 0)$. The payoff of debtholders is equivalent to that of a portfolio composed of default-free debt with face value D and maturity at T plus a European Put Position on the assets of the firm's A with strike D^T and maturity T $D_T = \min(D^T, A_T) = D^T - \max(D^T - A_T, 0)$. Therefore, the equilibrium theory of option pricing proposed by Black and Scholes (1973), can be used to price corporate liabilities (equity and debt) and to estimate the underlying default probability of a public.

However, some of the underlying assumptions of the original model serve to facilitate its mathematical representation and can be considerable weakened[2]. Specifically, in the original model and most of its modified version the default boundary beneath which promised payments to debtholders are not made and default occurs is assumed to be constant. Hence, the estimated risk – neutral expected default probabilities cannot capture changes in the relationship of asset value to the firm's default point that caused from changes in firm's leverage. However, these changes are critical in the determination of actual default probability. Moreover, the simplifying assumption of a constant default boundary is the major reason that the model results in unrealistic estimated short-term credit spreads that differ from those observed empirically. It is common that firms adjust their liabilities as they are near default. Empirical studies have showed, that the liabilities of commercial and industrial firms increase as they are near default while the liabilities of financial institutions often decrease. This difference reflects the ability of firms to liquidate their assets and adjust their leverage as they encounter difficulties. In order to capture the uncertainty associated with leverage, we have introduced randomness to the default boundary. Therefore, we have developed a new Merton-Type approach to estimate default probabilities and value corporate liabilities.

Our modified version of Merton Model, is based like all structural default risk models on the idea that corporate liabilities (debt & equity) can be valued as contingent claims on the firm's assets. The model is forward looking since it uses current market information regarding the future prospects of the underlying firm. Moreover, it relates different credit risk factors in an analytical way and allows non-linear effects and interaction among them. Its basic output the risk-neutral distance to default equals:

$$RNDD_T = \left(\frac{\ln\dfrac{A - \delta}{DP} - \dfrac{\sigma_A{}^2 \cdot T - \lambda^2}{2}}{\sqrt{\sigma_A{}^2 \cdot T + \lambda^2}} \right)$$

It is straightforward, that risk-neutral distance to default measure and the risk-neutral expected default probability depends on:
- The current market value of firm's assets A
- The asset volatility σ_A, which is a measure of business risk.
- The initial level of the default boundary DP
- The default boundary volatility λ, which captures the uncertainty about changes on firm's leverage.
- The continuously compounded risk free rate r
- The stream of expected cash dividends δ
- The length of time horizon T.

Our risk-neutral distance to default metric does not take into account credit risk factors such as cash flow adequacy, asset quality, earning performance, capital adequacy which are evaluated by

---

[2] Merton 1974, p.450

financial ratios and accounting-based measures. Therefore, we calibrate a hybrid model with an ordered-probit regression to explain credit ratings and rating transitions using our risk-neutral distance to default metric, financial ratios and other accounting based measures as explanatory variables. Using the same econometric method, we calibrate a model using our risk-neutral distance to default metric as unique explanatory variable. Then, using cumulative accuracy plots we have test the classification power of those models to predict default events out of sample and out-of time. We have found that by enriching the risk-neutral distance to default metric with financial ratios and accounting variables into the hybrid model, we can improve both in of-sample fit of credit ratings and out-of-sample predictability of default events. Finally, we have used the hybrid model to predict expected default probabilities for industrial companies listed in the Athens Stock Exchange.

Acknowledgments

The authors gratefully acknowledge professor Gikas A. Hardouvelis, associate professor Dimitrios Thomakos, lecturer George Skiadopoulos and Elias Panagiotidis for their helpful comments and suggestions. All errors are our own.

References

[1] Altman E., "Financial ratios, discriminant analysis and the prediction of corporate bankruptcy", *Journal of Finance*, (1968), vol.23, pp.589-609.

[2] Crouhy M., Galai D., Mark R., "A Comparative Anatomy of Current Credit Risk Models" *Journal of Banking and Finance*, (2000), vol.24, pp.57-117.

[3] Greene H. W., *"Econometric Analysis"*, (Princeton Education 2000).

[4] Harrison J.M., Kreps D., "Martingales and arbitrage in multi-period securities markets", *Journal of Economic Theory*, (1979), vol.20, pp.381-408.

[5] Merton C. R, "Theory of Rational Option Pricing", *Bell Journal of Economics and Management Science*, (1973), vol.4, pp.141-183.

[6] Merton C. R ,." On the Pricing of Corporate Debt: The Risk Structure of Interest Rates ", *Journal of Finance*, (1974), vol. 29, pp.449-470.

[7] Rubinstein M., "Great Moments in financial economics: II. Modigliani-Miller Theorem", *Journal of Investment Management*, (2003), vol.1(2).

[8] Vassalou Maria, Xing Yuhang, "Default Risk in Equity Returns", *University of Columbia Working Paper*, (2003).

VSP International
Science Publishers
P.O. Box 346, 3700 AH Zeist
The Netherlands

Lecture Series on Computer
and Computational Sciences
Volume 1, 2004, pp. 991-993

Modeling the Stability of the Transmission Mechanism of Monetary Policy A Case Study for Romania, 1993-2000

P.L. Rinderu[1]
Gh. Gherghinescu
O.R. Gherghinescu

University of Craiova,
200585 13 Al.I.Cuza Street, Craiova, Romania

Accepted: August 8, 2004

1. Introduction

Most of the mathematical models used for approaching problems in the field of economics should be considered and analyzed under the assumption of dynamic systems. In systems theory the concept of *dynamic system* relates input functions to output functions via the concept of *state*. The most often used examples of dynamic systems are finite-dimensional linear systems and finite-dimensional Gaussian systems. There are though economic problems that cannot be modeled in an adequate way by linear systems. Given the fact that the use of non-linear approaches implies difficulties in obtaining the solution, the most appropriate tool under these circumstances is to create a linear model and to use it as first and acceptable proxy for the real system.

2. Theoretical assumptions

The chosen model should meet some requirements:
- To be realistic for the phenomenon to be modeled. This requires deep understanding of and experience with the phenomenon.
- The models are analytically, algebraically and computationally tractable.

Some other points should be also taken into consideration:
- Characterization of finiteness properties of outcomes.
- Characterization of decidability and complexity of solving problems regarding system identification and control.
- Decomposition of dynamic systems into irreducible components.
- Generating abstractions of systems for distinguishing the levels in hierarchical systems.
- Characterization of classes of control laws for specific control objectives, when appropriate.
Generally speaking, the existence of a control law for a particular control objective may be formulated if a recognition problem is not well understood.

We assume that the control objective is described as a subset of input–output functions of another dynamic system to be called the control objective system. The control objective is then to establish the existence of a controller in a prescribed pattern of dynamic systems such that the set of input–output functions of the associated closed-loop system would be equal to or included in that of the control-objective system.
The formulation of such a problem using a mathematical formalism has the advantage of ensuring a high degree of generality for almost all classes of deterministic systems. The basic formulation, considering a dynamic system, has the form:

$$\dot{x}(t) = f(x(t), u(t)), \ x(t_0) = x_0,$$

---

[1] Corresponding author. E-mail: prinderu@central.ucv.ro

$$y(t) = h(x(t), u(t)),$$
$$z(t) = j(x(t), u(t)),$$

where:

$y(t)$ - the observation function;

$z(t)$ - the function that should be controlled.

If considering a dynamic feedback based on partial observations, the following relation represents, *de facto*, a control law:

$$\dot{x}_c(t) = f_c(x_c(t), y(t)), \; x_c(t_0) = x_{c,0},$$

where the function $v(t)$ represents the external input.

A basic problem for the correctness of the model is represented by the identification of different characteristics, process that requires the availability of an observation-based realization or filter systems with specified finiteness properties. System identification therefore requires restriction to a class of dynamic systems for which the associated class of observation-based realizations or filter systems has a specified finiteness property.

3. Methods

The proposed model aims at simulating the transmission mechanism of monetary policy in Romania in the framework of *monetary targeting* (in which the central bank has been exerting its control over the monetary basis as operational objective in order to attain a certain level for broad money as intermediary objective and thus contribute to price stability, as fundamental objective of its policy). The proposed interval for the analysis is 1993-2000, given the fact that before 1993 a direct control has been exerted upon broad money, whereas after 2003 the role of monetary aggregates in the transmission mechanism has been continuously decreasing in favor of that of the interest rate.

We propose a non-linear model with double feed-back reaction loops for the transmission mechanism of monetary policy during the above-mentioned interval, as presented in fig. 1.

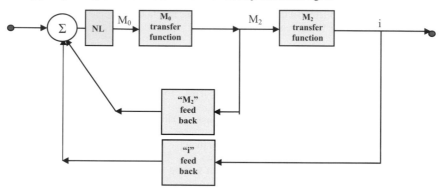

Fig. 1 Block-diagram of the model

where:
NL - non linear block;
M_0 - monetary basis as operational objective for monetary policy;
M_2 - broad money as intermediary objective for monetary policy;
i - inflation rate measured by the consumer price index as a proxy for price stability.

Using linearizations and taking into account the qualitative aspects of the inner processes, after simple algebraic calculus we deduct the partial transfer functions that link M_2 to M_0 via the money multiplier and i to M_2 via money velocity, respectively. The next step is to determine the global transfer function of the system, which allows for considerations over the stability of the simulated process of regulating i as function of M_0 and M_2.

4. Conclusions

The study presents a double feed-back model for controlling price stability via appointing monetary basis as operational objective and broad money as intermediary objective for monetary policy. The model allows for quantifying the stability of the transmission mechanism of monetary policy within the monetary targeting regime in Romania between 1993 and 2000.

Acknowledgments

The authors wish to thank the anonymous referees for their careful reading of the manuscript and their fruitful comments and suggestions.

References

[1] D. Antohi, I. Udrea and H. Braun, *Transmission Mechanism of Monetary Policy in Romania*, Research Papers of the National Bank of Romania, No.13, Bucharest, 2003.
[2] C. Belea, *Non-linear Automation*, Technical Publishing House, Bucharest, 1983.
[3] R.E. Kalman, P.L. Falb and M.A. Arbib, *Topics in Mathematical Systems Theory*, McGraw-Hill Book Co., New York, 1969.
[4] J.H. van Schuppen, *System theory for system identification*, Journal of Econometrics 118, 313-339, 2004.
[5] -, *The Transmission Mechanism of Monetary Policy*, Bank of England, 2000.

VSP International
Science Publishers
P.O. Box 346, 3700 AH Zeist
The Netherlands

*Lecture Series on Computer
and Computational Sciences*
Volume 1, 2004, pp. 994-997

Estimating Risk Neutral Densities of Asset Prices based on Risk Neutral Moments: An Edgeworth Expansion Approach

Leonidas S. Rompolis[1]
Elias Tzavalis

Department of Accounting and Finance,
Athens University of Economics and Business

Department of Economics,
Queen Mary & Westfield College,
University of London,
London E1 4NS

Accepted: August 8, 2004

Abstract: In this paper we suggest a new nonparametric approach of estimating the risk neutral density (RND) of asset prices or log-returns. Our approach exploits a relationship between the call and put prices and the conditional characteristic function of the asset price and the log-return. The latter is used to nonparametrically estimate the risk neutral moments of the underlying asset price or its log-return. These moments' estimates can be employed to estimate the RND using the generalized Edgeworth series expansion of a density. We then evaluate the performance of our approach by estimating the RND for the asset price and/or its associated log-return for three popular option pricing models: the Black-Scholes model, the stochastic volatility and the stochastic volatility jump diffusion model. Using S&P 500 option prices and implied volatilities we estimate the implied RND.

Keywords: risk neutral densities, conditional characteristic function, risk neutral moments, Edgeworth series expansion.

Mathematics Subject Classification: 60E10, 60G44, 91B28.

1. Introduction

In the last several years many studies tried to investigate the expectations and preferences of the investors that are implied by the option prices given by the market. The most important feature, concerning these studies, was the estimation of the risk neutral density (RND). Under the no-arbitrage condition there exists an equivalent probability measure under which all assets (including options) discounted at the risk free rate of interest are martingales. Options are valued under risk-neutrality, therefore the RND is the density implied by the option prices.

The existing methods for estimating the RND can be distinguished in three categories. The former category of methods claims that asset returns are governed by a known parametric model, which implies a specific RND.[2] The second category of methods make use of the Cox and Ross formula (1976), typically using a nonlinear optimization method to find the exact form of the RND that produces the predicted option prices that are "close" to the observed option prices. Among these techniques same place more structure on the RND to be derived,[3] others place less structure, and therefore they can be viewed as nonparametric.[4] The last category is consisted of methods that first try to construct an option pricing formula, from the observed option prices, and then exploit the Breeden-Litzenberger formula (1978) to derive the RND. Shimko (1993), Malz (1997) and Campa *et.al.* (1998)

---

[1] E-mail: rompolis@aueb.gr, rompolis@in.gr
[2] See for example Heston (1993) for a stochastic volatility model and Bates (1996) for a model with stochastic volatility and jumps.
[3] See for example Madan and Milne (1994) and Melick and Thomas (1997).
[4] See for example Rubistein (1994) and Buchen and Kelly (1996).

use tools of numerical analysis, like quadratic polynomials and cubic splines, in order to construct the option pricing function. On the other hand, Ait-Sahalia and Lo (1998) and Ait-Sahalia and Duarte (2003) utilize kernel regression to estimate the functional form that relates the option price to the strike price.

In this paper we propose a new nonparametric method for estimating the RND based on option data. Based on Bakshi and Madan's (2000) fundamental theorem stating that put and call option prices span the conditional characteristic function (CCF) of the underlying asset price (or return), we derive closed-form equations for the conditional risk neutral moments (RNM), based on the out-of-the money (OTM) call and put option prices. We then employ the generalized Edgeworth series expansion in order to approximate the implied RND. This approach gives us the opportunity, without estimating the RND, to obtain the moments of the asset price distribution or of the asset price return. We can therefore estimate the mean (considered as the expected value of the asset return), the variance, the skewness (measuring the asymmetry of the market expectations) and the kurtosis.

2. Extracting the RNM general formula from option prices

Our analysis is motivated by the result of Bakshi and Madan's (2000) fundamental theorem stating that the continuum of characteristic functions and the continuum of options are equivalent classes of spanning securities, under the risk-neutral measure. We can therefore calculate the risk-neutral CCF of the asset price (or return) as a function of option prices. This result can be directly used to extract the RND by the inverse Fourier transform. However, this may lead us to problems in estimating the RND, as the CCF implied by option prices is defined only in the neighborhood of zero. We, thus, follow another approach which relies on the RNM of the RND. We write the CCF as a Taylor series expansion around zero, and we then extract the coefficient of the polynomial, which are exactly the RNM. This approach has its own merits as, apart from estimating the RND, it enables us to directly compare the risk-neutral and physical moments [see Bakshi, Kapadia and Madan (2003)].

The above general formula for the RNM cannot be directly used when we have market data. First, an option pricing function must be constructed from the available option data. To do so, we follow the technique of Campa *et. al.* (1998). Option prices are first converted to implied volatilities, using the Black-Scholes formula. Then the new data are interpolated by cubic splines. Finally the continuous implied volatility function is converted back into a continuous option pricing function. The second path of the RNM numerical approximation is the calculation of the integrals involved in the formula. A Gaussian quadrature procedure is used to evaluate the numerical integration.

3. Estimating the RND from the RNM through the Edgeworth series expansion

Using the RNM that we calculate in the previous section we can approximate the RND exploiting the Edgeworth series expansion up to order four. In order to apply the Edgeworth expansion an approximating density must be chosen. The natural choice would be the log-normal density when we approximate the RND of the asset price, and the normal density when we approximate the RND of the asset price log-return. The coefficients of these "prior" densities are given by the first two RNM. Namely, what we really do is to adjust the approximating density for the risk-neutral skewness and excess kurtosis, appeared in the RND [see Ait-Sahalia and Lo (1998)].

The method that we just described has its own advantages. It is a fully nonparametric approach because it does not assume any additional structure on the strike-call/put pricing relationship to estimate the RND and it is not data intensive. The main disadvantage is that the Egdeworth expansion can gives us negative values. Nevertheless, we can control that before applying the method using the ranges derived by Jondeau and Rockinger (2001).

4. Numerical evaluation and empirical example

We evaluate the performance of the method suggested in the previous section, to accurately approximate the RND. To this end, we conduct a set of numerical exercises evaluating the ability of the method to accurately approximate the theoretical RNM and RND for the log-returns implied by the Black-Scholes, stochastic volatility and stochastic volatility jump diffusion model. The last two models are used extensively in the literature to improve upon the empirical performance of the Black-Scholes model. In all the examples we conducted the theoretical and estimated RNM are very close, and so are the approximated RND and the density implied by the model.

To assess the empirical relevance of our method we present an application to the estimating of the log-return distribution of the S&P 500 index using the S&P 500 European option prices obtained from the CBOE. We calculate the RND at two different days of 2002, and for different maturities.

Acknowledgments

The authors wish to thank the anonymous referees for their careful reading of the manuscript and their fruitful comments and suggestions.

References

[1] Ait-Sahalia, Yacine and Andrew W. Lo, Nonparametric estimation of state-price densities implicit in financial asset prices, *Journal of Finance* 53 499-548(1998).

[2] Ait-Sahalia, Yacine and Jefferson Duarte, Nonparametric option pricing under shape restrictions, *Journal of Econometrics* 116 9-47(2003).

[3] Bakshi, Gurdip and Dilip Madan, Spanning and derivative-security valuation, *Journal of Financial Economics* 55 205-238(2000).

[4] Bakshi, Gurdip, Nikunj Kapadia and Dilip Madan, Stock return characteristics, skew laws, and the differential pricing of individual equity options, *The Review of Financial Studies* 16 101-143(2003).

[5] Bates, David S., Jumps and stochastic volatility: exchange rate processes implicit in Deutsche Mark options, *Journal of Financial Studies* 9 69-107(1996).

[6] Breeden, Douglas and Robert H. Litzenberger, Prices of state-contingent claims implicit in option prices, *Journal of Business* 51 621-651(1978).

[7] Buchen, Peter W. and Michael Kelly, The maximum entropy distribution of an asset inferred from option prices, *Journal of Financial and Quantitative Analysis* 31 143-159(1996).

[8] Campa, Jose M., P.H Kevin Chang and Robert L. Reider, Implied exchange rate distributions: Evidence from OTC option markets, *Journal of International Money and Finance* 17 117-160(1998).

[9] Cox, John and Stephen Ross, The valuation of options for alternative stochastic processes, *Journal of Financial Economics* 3 145-166(1976).

[10] Heston, Steven L., A closed-form solution for options with stochastic volatility with applications to bond and currency options, *Review of Financial Studies* 6 327-343(1993).

[11] Jondeau, Eric and Michael Rockinger, Gram-Charlier densities, *Journal of Economic Dynamics and Control* 25 1457-1483(2001).

[12] Madan, Dilip B. and Frank Milne, Contingent claims valued and hedged by pricing and investing in a basis, *Mathematical Finance* 4 223-245(1994).

[13] Malz, Allan, Estimating the probability distribution of future exchange rate from option prices, *The Journal of Derivatives* 20-36(1997).

[14] Melick, William R. and Charles P. Thomas, Recovering an asset's implied PDF from option prices: An application to crude oil during the Gulf crisis, *Journal of Financial and Quantitative Analysis* 32 91-115(1997).

[15] Rubistein, Mark, Implied binomial trees, *Journal of Finance* 49 771-818(1994).

[16] Shimko, David, Bounds of probability, *Risk* 6 33-37(1993).

VSP International
Science Publishers
P.O. Box 346, 3700 AH Zeist
The Netherlands

*Lecture Series on Computer
and Computational Sciences*
Volume 1, 2004, pp. 998-1001

Forecasting with a Universal Data Compression Algorithm: The Forex Market Case

A. Shmilovici[1],
Y. Kahiri
S. Hauser

Dept. of Information Systems Eng.
Ben-Gurion University
P.O.Box 653, Beer-Sheva, Israel

School of Management
Ben-Gurion University
P.O.Box 653, Beer-Sheva, Israel

Accepted 8 August, 2004

Abstract: We use the context tree algorithm of Rissanen, for compression and prediction of time series. The weak form of the EMH is tested for 12 pairs of international intra-day currency exchange rates for one year series of 1,5,10,15,20,25 and 30 minutes. Statistically significant compression is detected in all the time-series, yet, the Forex market turns out to be efficient most of the time, and the short periods of inefficiency are not sufficient generating excess profit.

Keywords: Efficient Market Hypothesis, Context Tree, Forex Intra-day Trading, Stochastic Complexity

Mathematics Subject Classification: 62P05, 91B84, 62M20

1. Introduction

The Efficient Market Hypothesis (EMH) states that the current market price fully reflects all available information. Assuming that all information is widely spread and accessible to investors, no prediction of future changes in prices can be made. New information is *immediately* discovered and quickly disseminated to produce a change in the market price. The *weak* form of the EMH considers only past price data and rules out predictions based on the price data only. The prices follow a random walk, where successive changes have zero correlation.

The EMH has been investigated in numerous papers. Thorough surveys, such as [1] and [2], present conflicting results. Most arguments in favor of the EMH rely on statistical tests that show no predictive power for the tested models. However, claims for successful predictions with nonlinear models such as neural networks may not contradict the EMH if the trading community does not immediately assimilate new useful methods. In any event, predicting stock prices is generally accepted to be a difficult task, since stock prices do behave very much like a random-walk process most of the time.

Universal coding methods [3],[4] were developed within the context of coding theory to compress a data sequence without any prior assumptions about the statistics of the generating process. The universal coding algorithms - typically used for file compression - constructs a model of the data that will be used for coding it in a less redundant representation. The crucial part in coding is to come up with a conditional probability for the next outcome given the past. This can be done, for example, by calculating sequentially empirical probabilities in each "context" of the data. A prediction algorithm that uses these conditional probabilities to make predictions will work, due to the statistical analysis performed by the universal data compression algorithm.

---

[1] Corresponding author. E-mail: armin@bgumail.bgu.ac.il

Connection between compressibility and predictability exists in the sense that sequences, which are compressible, are easy to predict and conversely, incompressible sequences are hard to predict. Merhav and Feder [4] found an upper and a lower bound to the connection between compressibility and predictability. Since the compressibility of a data set indicates that the data is not random, we can use the model that was used for compressing the data, for prediction of a future outcome.

Here we use the context tree algorithm of Rissanen [5], which is a universal coding method, for measuring the *stochastic complexity* of a time series. The advantage of this specific algorithm, in contrast to other universal algorithms known only to have asymptotic convergence, is that Rissanen's context-tree algorithm has been shown [6] also to have the best asymptotic convergence rate. Thus, it can be used to compress even relatively short data sets - like the ones available from economic time series. We use this algorithm as described [7] to examine the weak form of the market efficiency hypothesis. The idea is that, *in an efficient market, compression of the time series is not possible*, because there are no patterns and the stochastic complexity is high. Periods of reduced stochastic complexity indicate times where a model (or patterns) is found, potentially allowing prediction and obtaining abnormal financial gains. This would indicate periods of potential inefficiency in a financial market.

2. Numerical Experiments

The weak form of the EMH is tested for one year for 12 pairs of international intra-day currency exchange rates. The currencies are described in table 1. The intra-day currency exchange rates were encoded for series of 1,5,10,15,20,25 and 30 minutes.to a tri-nary string indicating a {low, stable, high} trend. A context tree was computed for each sliding window.

First, using the context tree, the compression was computed for each sliding window. Statistically significant compression above random is detected in all of those series for all the different periods. The difference between the compression of different series indicates that it takes some time for the forex market to absorb new information. This indicates potential market short term inefficiency that can be used for forecasting.

Second, using the context tree, forecasts were computed for each sliding window, e.g. a forecast of the 51st day was generated based on the previous 50 days. The forecasts were compared with the actual next trends in the time series. Attempts to use the compression for prediction of the future tri-nary trend {low, stable, high} such that the {high, low} trends are above the trading costs, failed to produce consistent results.

Third, the forecasts were compared only in to values following a compressible window to asses the theory that compressibility indicates the possibility for prediction. Several statistical tests were used to check the quality of the prediction:

The *confusion matrix* measures the match between the forecast and the real data. The context tree supply a forecast with a probability the forecast is correct. Therefore, the prediction quality was measured for the 0%, 50%, 70% and 80% estimated probability for a correct forecast. Then, we used the sign test to verify that a prediction that was made with 80% chance to be correct produces better results then other predictions with lower probability to be accurate. For example, table 1 below presents the sign test between forecast with 70% probabilities and no probabilities to be correct for 1 and 5 minute intervals. We did not find that forecasts given with a higher probability of being correct were significantly better than forecast that were given with lower probabilities.

The Kappa statistic coefficient was computed to measure the degree of compatibility between the forecasted data and the real data. Tables 2 below illustrate the results of this test on forecast that were given with 0% or more chance to an accurate prediction. The bold numbers indicate the significant of the coefficient to be bigger then zero The Kappa coefficient we found can be describe as poor to medium correspondence.

The final part of the research was a simulation of opening and closing positions. The rule for opening a position was given by the context tree, the position was closed in the end of the period the forecast was given for. We found that in some of the currencies there are profit before paying a commission, but in almost all the cases there was loss after paying commission. Table 3 below illustrates an example of the simulation results when the position was taken for 80% chance to be right in the forecast.

Table 1: The sign test 70% and 0% prediction accuracy for 1 and 5 minute

Currency	1	5
EURUSD		+
GBPUSD	+	
USDJPY	+	
AUDUSD		+
CHFJPY		
EURCAD		+
EURCHF		+
EURGBP		
EURJPY	+	+
GBPCHF	+	
GBPJPY	+	
USDCHF	+	+
Sum	**6**	**6**
Statistic	**0.387**	**0.387**

Table 2: The Kappa coefficient for several currency series

Currency	1 Minute				10 Minute		
	Moving Window	Kappa	Standard Erorr	95% CI	Kappa	Standard Erorr	95% CI
EURUSD	100	0.01	0.015	-0.015 to 0.04	0.056	0.005	**0.046 to 0.065**
	75	0.014	0.015	-0.015 to 0.043	0.052	0.005	**0.042 to 0.062**
	50	0.021	0.014	-0.007 to 0.05	0.054	0.005	**0.045 to 0.064**
GBPUSD	100	0.007	0.013	-0.019 to 0.033	0.039	0.005	**0.03 to 0.049**
	75	0.011	0.013	-0.011 to 0.036	0.027	0.005	**0.018 to 0.036**
	50	0.017	0.013	-0.008 to 0.043	0.034	0.005	**0.024 to 0.043**
AUDUSD	100	0.011	0.025	-0.038 to 0.059	0.044	0.007	**0.03 to 0.058**
	75	0.012	0.024	-0.036 to 0.06	0.04	0.007	**0.026 to 0.054**
	50	0.021	0.024	-0.026 to 0.068	0.044	0.007	**0.031 to 0.058**
CHFJPY	100	0.003	0.017	-0.03 to 0.037	0.019	0.005	**0.009 to 0.029**
	75	0.007	0.017	-0.026 to 0.041	0.015	0.005	**0.006 to 0.025**
	50	0.015	0.017	-0.018 to 0.047	0.015	0.006	**0.006 to 0.025**

Table 3: summary of a trading strategy for some currency series

Minutes		1	5	10
EURUSD	Profit	17	4	40
	loss	11	20	40
	commission	34	22	22
Summary		**-28**	**-38**	**-22**
GBPUSD	Profit	8	39	51
	loss	4	12	18
	commission	20	32	30
Summary		**-16**	**-5**	**3**

3. Conclusions

Statistically significant compression is detected in all the time-series. The compression is detected even for 5 minutes and 30 minutes interval series, though to a less extent. It seems that it takes some time for the forex market to absorb new information. This indicates potential market short term inefficiency.

We found that though a good compression capability should indict a good prediction capability, the accuracy of the prediction was poor. There was no significant difference between the quality of prediction between predictions with high degree of success (as predicted by the context tree) and predictions with a low degree of success.

A simulation of opening and closing positions demonstrated no profit beyond the commission for the intra-day trade.

Our conclusion is that though the context tree is a useful tool for forecasting time series, the Forex market is efficient most of the time, and the short periods of inefficiency are not sufficient generating excess profit.

References

[1] E.F. Fama, "Efficient Capital Markets: II", *Journal of Finance*, **46** 1575- 1611(1991).

[2] T. Hellstrom, , Holmstrom, K., "Predicting the Stock Market", Technical Report Ima-TOM-1997-07, Umea University, Sweden - *available on the web*

[3] M. Feder, Merhav N., Gutman M, "Universal Prediction of Individual Sequences", *IEEE Transactions on Information Theory*, **38**(4)1258- 1270(1992).

[4] N. Merhav, Feder M., "Universal Prediction", *IEEE Transactions on Information Theory,* IT-44, 2124-2147(1998).

[5] J. Rissanen, "A Universal Data Compression System", *IEEE Transactions on Information Theory*, **29** (5), 656- 664(1983).

[6] J. Ziv, "A Universal Prediction Lemma and Applications to Universal Data Compression and Prediction", *IEEE Transactions on Information Theory*, **47**(4)1528- 1532(. 2000).

[7] M. Weinberger, Rissanen J., Feder M., "A Universal Finite Memory Source", *IEEE Transactions on Information Theory*, **41**(3)643- 653(1995).

[8] F.M.J. Willems, Shtarkov Y.M. and Tjalkens T.J., "The context-tree weighting method: Basic properties", *IEEE Trans. On Information Theory*, **41**(3) 653-664(1995).

VSP International
Science Publishers
P.O. Box 346, 3700 AH Zeist
The Netherlands

*Lecture Series on Computer
and Computational Sciences*
Volume 1, 2004, pp. 1002-1005

Functional Filtering, Smoothing and Forecasting

Dimitrios D. Thomakos[1]

Department of Economics,
School of Management and Economics,
University of Peloponnese,
GR-221 00 Tripolis, Greece

Accepted: August 8, 2004

Abstract: A new method for filtering and forecasting univariate time series is proposed, which is based on a particular type of a causal filter with functional, time-varying coefficients. The main novelty of the proposed method is the nature of the filtering coefficients, which are data-dependent and local in character. The filtering and forecasting procedures can incorporate two different delay parameters, one applied to the functional coefficients and the other to the data used in the filter; a data-dependent methodology is also presented for selecting these delays and the order of the filter. Various simpler cases of the general method are presented and the potential of the method is illustrated using simulated and real-world time series.

Keywords: filtering, forecasting, functional coefficients, nonparametric.

Mathematics Subject Classification: 60G35, 62G99, 62M10, 62M20.

PACS: to be completed

1   Presentation of the Problem

Suppose that we are given a univariate, real-valued time series x_t and we are interested in filtering and forecasting based on a causal filter of the generic form:

$$f_t \stackrel{\text{def}}{=} \sum_{j=0}^{k-1} \tau_j(\boldsymbol{x}_{t-d_1}) x_{t-d_2-j} \tag{1}$$

where $k \in \mathbb{N}_+$ is the order of the filter, $\boldsymbol{x}_{t-d_1} \stackrel{\text{def}}{=} (x_{t-d_1}, x_{t-d_1-1}, \ldots, x_{t-d_1-\ell+1})^\top$ is an $(\ell \times 1)$ vector of past observations with $\ell \in \mathbb{N}$, and $(d_1, d_2) \in \mathbb{N}$ are two delay parameters. The filtering coefficients $\tau_j(\cdot)$ are time-varying functions of the vector $\boldsymbol{x}_{t-d_1}$ and we call them the filtering functionals. Note that the filtering coefficients differ in two respects (besides being time-varying) from the coefficients found in other filtering equations: first, they incorporate delays and second, they are assumed to be functions of a subset of recent observations. Our aim is to use sample information in computing these coefficients.

In particular, we are interested in adjusting the filtering coefficients based on the spatial distance between values of the time series that are apart h time units, where h is to be determined by the

---

[1]Corresponding author. E-mail: thomakos@uop.gr

model. Therefore, in computing the filtering coefficients we take into account both the time distance as well as the spatial distance of the observations, in anticipation that the coefficients will be more adaptable to changes in the evolution of the time series. To address these requirements we propose that the filtering functionals are constructed based on kernel functions that are widely used in nonparametric time series analysis.

We note that the proposed model would belong in the class of functional coefficient autoregressive models if the data generating process was of the form:

$$x_t = f_t + n_t \tag{2}$$

where n_t is a zero-mean (white) "noise" time series. In this case we would have that f_t would be the minimum mean-squared error forecast for x_t: assuming that $d_2 > 0$, letting $\Omega_t \overset{\text{def}}{=} \sigma(s \le t; x_s)$ be the information set based on observations up to and including time $t - d_2$ we would have $\mathsf{E}\,[x_t | \Omega_t] \overset{\text{def}}{=} f_t$.

2   A Set of Filtering Functionals

Define the sets $A \subset \mathbb{R}^\ell$ and $B \subset (0, +\infty)$; for an $(\ell \times 1)$ vector $\boldsymbol{\psi} \overset{\text{def}}{=} (\psi_1, \psi_2, \dots, \psi_\ell)^\top$ let the function $w(\boldsymbol{\psi}) : A \to B$ denote a kernel function such that it satisfies the conditions $w(\mathbf{0}) \ge w(\boldsymbol{\psi})$, $w(\boldsymbol{\psi}) = w(-\boldsymbol{\psi})$ and $\int_A w(\boldsymbol{\psi}) d\boldsymbol{\psi} = 1$. An example of such a kernel function is the multivariate Gaussian kernel given by:

$$w(\boldsymbol{\psi}) \overset{\text{def}}{=} \prod_{i-1}^{\ell} \frac{1}{\sqrt{2\pi}} \exp(-\frac{1}{2}\psi_i^2) \equiv \left(\frac{1}{\sqrt{2\pi}}\right)^\ell \exp(-\frac{1}{2}\boldsymbol{\psi}^\top \boldsymbol{\psi}) \tag{3}$$

Next, define the $(\ell \times 1)$ vector $\boldsymbol{\psi}_{t-d_2-j} \equiv \boldsymbol{\psi}_{t-d_2-j}(\boldsymbol{x}_{t-d_1}) \overset{\text{def}}{=} (x_{t-d_2-j} - \boldsymbol{x}_{t-d_1})/\sigma$, where $\sigma > 0$ is a scaling factor and $j = 0, 1, \dots, k - 1$. Note that the sth element of the vector $\boldsymbol{\psi}_{t-d_2-j}$, say $\psi_{t-d_2-j,s}$, measures the spatial distance between the observation x_{t-d_2-j} and the observation $x_{t-d_1,s}$ in terms of the scaling factor σ. If $\boldsymbol{\psi}_{t-d_2-j}$ was used as input in the kernel function $w(\cdot)$ above, then the value of $w(\boldsymbol{\psi}_{t-d_2-j})$ would change with the distance between the observations x_{t-d_2-j} and $x_{t-d_1,s}$ and with the scaling factor σ. The scaling factor σ essentially controls the height of the kernel function, a feature which is useful in controlling the degree of smoothing, and therefore allow for explicit training in filtering and forecasting.

Using the above kernel function we can now construct filtering functionals that satisfy the necessary conditions of all filtering coefficients, namely that they are positive and that they sum-up to unity. We thus assume that the filtering fuctionals $\tau_j(\cdot)$ take the following form:

$$\tau_j(\boldsymbol{x}_{t-d_1}) \overset{\text{def}}{=} \frac{w(\boldsymbol{\psi}_{t-d_2-j})}{\sum_{i=0}^{k-1} w(\boldsymbol{\psi}_{t-d_2-i})} \tag{4}$$

and we clearly have that $\tau_j(\cdot) \ge 0$, for all $j = 0, 1, \dots, k - 1$, and $\sum_{j=0}^{k-1} \tau_j(\cdot) = 1$. The filtered value f_t is then given by:

$$f_t = \sum_{j=0}^{k-1} \frac{w(\boldsymbol{\psi}_{t-d_2-j})}{\sum_{i=0}^{k-1} w(\boldsymbol{\psi}_{t-d_2-i})} \cdot x_{t-d_2-j} \tag{5}$$

The weight that is being attached to the value x_{t-d_2-j} depends on the spatial distance between the value itself and the values in the evaluating vector $\boldsymbol{x}_{t-d_1}$, based on observations that are apart $h = |d_1 - d_2 - j|$ time units. It is clear from the above formula that the proposed filtering functionals are similar (but not identical) to a *local* nonparametric regression; note that, in this context, the epithet *local* refers both to the local character of the filtering functionals and to the use of a limited number of observations in forming the filtered value f_t; this is in contrast to standard nonparametric regression.

3 Selecting the Model's Parameters

There are five parameters in the model, that need to be determined by the data: the order of the filter k, the order of the evaluating vector ℓ, the delay parameters (d_1, d_2) and the scaling factor σ. Let $\boldsymbol{\theta} \overset{\text{def}}{=} (k, \ell, d_1, d_2, \sigma)^\top$ denote the (5×1) vector containing these parameters. The choice of the delay parameters and the order of the filter depends on the problem at hand. For smoothing and filtering problems the order of the filter will depend on the degree of smoothing that is required. For forecasting problems the order of the filter as well as the delays will be chosen to optimize the predictive ability of the model. For all of the above, the performance of the smoothing, filtering or forecasting operation will depend on the selected value of the scaling factor σ: for a given sample of observations, as σ increases the filtering coefficients become less and less varying and approach $w(\mathbf{0})$ (i.e. larger and constant, equal weights); conversely, as σ decreases the filtering coefficients become less and less varying and approach zero. Therefore, a balance between larger and smaller values of σ is required. For filtering and smoothing problems a simple grid search over the allowable range of the elements of $\boldsymbol{\theta}$ should be sufficient. However, for forecasting problems (where the forecasting performance could be explicitly evaluated) selection of the parameters can be successfully achieved by training the model using a variant of the generalized cross-validation (GCV) approach.

The generic GCV approach that we propose is the following: suppose that we have available a sample of T observations and let $T_1 < T$ be a subset of training observations. For a suitably chosen objective function that evaluates forecasting performance, say $g(\boldsymbol{\theta})$ use the first T_1 training observations to compute the ν-step ahead forecast, say $\hat{x}_{t+\nu}^t$. Then, advance the time index by one and use the next T_1 training observations to compute the next ν-step ahead forecast, and so on until we have a sequence of $T - T_1$ forecasts. Using these forecasts evaluate $g(\boldsymbol{\theta})$ for the chosen $\boldsymbol{\theta}$. Repeating the above procedure for a sequence of N values of $\boldsymbol{\theta}$, say $\boldsymbol{\Theta} \overset{\text{def}}{=} (\boldsymbol{\theta}_1, \cdots, \boldsymbol{\theta}_N)$ select the "optimal" value of $\boldsymbol{\theta}$ as:

$$\boldsymbol{\theta}^* \overset{\text{def}}{=} \underset{\boldsymbol{\theta} \in \boldsymbol{\Theta}}{\operatorname{argmin}} g(\boldsymbol{\theta}) \qquad (6)$$

Acknowledgment

The author wishes to thank the anonymous referees for their careful reading of the manuscript and their fruitful comments and suggestions.

References

[1] Z. Cai, J. Fan and Q, Yao, 2000. "Functional coefficient models for nonlinear time series", *Journal of the American Statistical Association*, 95, pp. 941-956.

[2] R. Chen and R. S. Tsay, 1993. "Functional coefficient autoregressive models", *Journal of the American Statistical Association*, 88, pp. 298-308.

[3] J. Fan and E. Masry, 1997. "Local polynomial estimation of regression functions for mixing processes", *Scandinavian Journal of Statistics*, 24, pp. 165-179.

[4] X. Li and N. E. Heckman, 1996. "Local linear forecasting", *Technical Report 167*, Department of Statistics, University of British Columbia, Vancouver.

[5] D. Thomakos and J. Guerard, 2004. "Naïve, ARIMA, Nonprametric, Transfer Function and VAR Models: A Comparison of Forecasting Performance", *International Journal of Forecasting*, 20, pp. 53-67.

[6] Q. Yao and H. Tong, 1998. "Cross-validatory bandwidth selections for regression estimation based on dependent data", *Journal of Statistical Planning and Inference*, 68, pp. 387-415.

VSP International
Science Publishers
P.O. Box 346, 3700 AH Zeist
The Netherlands

*Lecture Series on Computer
and Computational Sciences*
Volume 1, 2004, pp. 1006-1008

Market Timing and Cap Rotation

Dimitrios Thomakos
Department of Economics,
University of Peloponnese,
End of Karaiskaki Street, 22100, Tripolis
Greece

Tao Wang[1]
Department of Economics,
Queens College,
65-30 Kissena Blvd,
Flushing, NY 11367
U.S.A.

Jingtao Wu
Department of Economics
260 Heady Hall
Iowa State University
Ames, IA 50011
U.S.A.

Accepted: August 8, 2004

Abstract: We examine the predictability of stock index returns (S&P500, S&P400 and Russell 2000) using short-term interest rate. We find that the short-term interest rate has predictive power over the relative performance of stock index returns. Trading strategies based on the finding prove to be profitable.

Keywords: Market timing, short-term interest rate, trading strategy

Mathematics Subject Classification: 62P20, 62J05.

1. Introduction

Both practitioners and academics have devoted considerable time to study the effect of market timing for returns of different asset classes in asset allocation strategies as well as long-short market neutral strategies. By now, few would disagree that returns are at least predictable, although there is generally no agreement on the reasons for predictability.

One of the most commonly used input variables in market-timing techniques is the short-term interest rate. Among the studies that document predictability of returns by short-term interest rates, Breen, Glosten, and Jagannathan (1989) investigate the performance of a market-timing strategy in shifting funds between Treasury bills and stocks. They find that during 1954 to 1987, a portfolio managed by predictions of a three-year rolling regression of excess stock return on the one-month risk-free rate has a variance of monthly returns about 60% of the variance of monthly returns on the value-weighted stock index, with an average return 2 basis points higher. Such a strategy is found to be worth an annual management fee of 2% of the value of the assets managed. On the other hand, Lee (1997) found that the stock return (S&P500) becomes insensitive to the risk-free rate over time, especially after April 1989. However, the reason for the disappearance of the interest rate effect is not clear.

In this paper, we examine the relationship between indices of large cap, mid cap and small cap stocks and short-term interest rate. In particular, we are interested in the relation between relative performance of different cap indices and short-term interest rate.

---

[1] Corresponding author. E-mail: tao_wang@qc.edu, phone: +212-718-997-5445

One development for the past 20 years is the growth of corporate bonds market. For large corporations, most of their financing need can be fulfilled through the bonds market by using corporate bonds or commercial papers. The cost of external finance through the bonds market in some ways could be cheaper than that from the bank-lending channel. For small firms, they are still affected by changes in real short-term interest rate since it is still more difficult for them to finance their investment through the bonds market, especially from the short-rate commercial paper market. Based on this development, large cap stocks should be affected by the bonds yield while small cap stocks are affected by the real short-term interest rate. This observation is consistent with the Lee (1997) study that S&P500 return becomes insensitive to the interest rate over time, as S&P500 is basically a large-cap index. Therefore, a market-timing strategy based purely on short-term interest rate would not be profitable for large-cap stocks but could be profitable for small-cap stocks. It is based on this argument that we want to study the relative performance of these three cap-stock indices. We seek to identify which cap is likely to outperform the others, which cap is likely to under perform the others, and eventually create a managed portfolio to make full use of our results.

2. Data

We use the S&P500, the S&P Mid-Cap 400 and the Russell 2000 to represent the large, mid and small-cap stock indices. We collect the one-month T-bill return as the short-term interest rate from Ibbotson and Associates

3. Methodology

Our base regression includes the following specification using S&P500 and Russell 2000 as an example:

$$r_t^{SP500} - r_t^{Rus2000} = \alpha_0 + \alpha_1 r_{t-1}^{rf} + \varepsilon_t \tag{1}$$

where r_t^{SP500} is the return for S&P500, $r_t^{Rus2000}$ is the return for Russell 2000 and r_{t-1}^{rf} is the risk-free rate as represented by Ibbotson and Associates. We also run regressions using returns from S&P500, S&P400 and returns from S&P400 and Russell 2000. The data period is from December 1978 to February 2004. In the regression, we split our sample into two periods: 1978:12-1987:12 and 1988:1 – 2004:2, and present our results in the next section.

4. Results

In the following table, we present our results from the above regression.

Table 1. Regression Coefficients from Equation 1 (t-statistics are in the parentheses)

	$r_t^{SP500} - r_t^{Rus2000}$		$r_t^{SP500} - r_t^{SP400}$		$r_t^{SP400} - r_t^{Rus2000}$	
	1978:12-1987:12	1988:1 – 2004:2	1978:12-1987:12	1988:1 – 2004:2	1978:12-1987:12	1988:1 – 2004:2
α_1	-0.7108 (-0.54)	3.4446 (2.79)	-0.3440 (-0.43)	1.9619 (2.38)	-0.3667 (-0.34)	1.4827 (1.85)
R^2	0.0027	0.0246	0.0018	0.0164	0.0013	0.0136

The above results support the argument that after 1980s, because of the change in the financial sector, it becomes easier for large-sized firms to finance through different channels, therefore they do not rely on the bank lending channel as much as before. As a result, α_1 all become positive and statistically significant. Large cap tends to over-perform mid-cap while mid-cap over-performs small cap stocks when interest rate rises.

We then estimated the three-year rolling regression from the above specification. From rolling regression results, we make prediction of the relative performance of three index portfolios. From the prediction, we construct two different managed portfolios. One is the long-only portfolio, which we only invest in the index portfolio with the highest forecasted performance. The other one is the market-neutral portfolio that is constructed by longing 100% of the highest forecasted performance index portfolio and short 100% of the combination of the rest two assets. For example, if the model predicts S&P500 will outperform the rest two portfolio in the next month, and is neutral on the prospect of S&P

400 and Russell 2000, we would long 100% S&P 500 and short 50% S&P 400 and 50% of Russell 2000.

In the whole sample period 1982:1-2004:2, our long-only portfolio has an average monthly return of 1.4% with standard deviation 4.85%. It outperforms all the index portfolios in terms of return and return-standard deviation ratio. The market-neutral portfolio has an average monthly return of 0.52% with standard deviation 2.54%. Its *correlation* with the S&P500 return is 0.032. In the sub-sample period 1987:1-2004:2, our long-only portfolio also outperforms the other index portfolio. It has an average monthly return of 1.36% with standard deviation 4.85%. The market-neutral portfolio has an average monthly return of 0.6% with standard deviation 2.66%.

5. Conclusion

In this paper, we examine whether the short-term interest rate can be used to predict the relative performance of three capitalization indices: S&P500, S&P400 and Russell 2000. We find some evidence that short-term interest rate has some predictive power of the relative performance of these indices. Trading strategies based on this finding prove to be profitable.

Table 2: Summary Statistics for Monthly Returns
1982:1-2004:2

	Mean	Standard Deviation	Mean/S.D.
Market Neutral Portfolio	0.52%	2.54%	0.20
Long-Only Portfolio	1.40%	4.85%	0.29
Russell 2000	0.93%	5.72%	0.16
SP 400	1.03%	4.70%	0.22
SP 500	1.20%	4.56%	0.26

1987:1-2004:2

	Mean	Standard Deviation	Mean/S.D.
Market Neutral Portfolio	0.60%	2.66%	0.22
Long-Only Portfolio	1.36%	4.85%	0.28
Russell 2000	0.85%	5.92%	0.14
SP 400	0.95%	4.87%	0.19
SP 500	1.08%	4.68%	0.23

Acknowledgments

The authors wish to thank the anonymous referees for their careful reading of the manuscript and their fruitful comments and suggestions.

References

[1] Breen, W., L. R. Glosten, and R. Jagannathan, ``Economic significance of predictable variations in stock index returns,'' *Journal of Finance*, 44, 1177-1189 (1989).

[2] Lee W., ``Market timing and short-term interest rate,'' *Journal of Portfolio Management*, spring, 35-46 (1997).

VSP International
Science Publishers
P.O. Box 346, 3700 AH Zeist
The Netherlands

Lecture Series on Computer
and Computational Sciences
Volume 1, 2004, pp. 1009-1010

Computational Molecular Science: From Atoms and Molecules to Clusters and Materials

The aim of this Symposium is to bring forth the essential contribution of Computational Science to modern Molecular Science. The contributed papers represent computational efforts in a wide spectrum of systems and phenomena. The relevance of the reported results extends form atoms and small molecules to biomolecules, clusters and materials. A strikingly large variety of subjects is present: theoretical calculations of electric properties, intermolecular interactions and the structure of weakly bonded molecules, collision-induced phenomena, solvent effects on molecular properties, molecular dynamics and more.

A total of thirty-three papers are presented: eight invited lectures, seventeen oral presentations and eight posters.

Farantos proposes a new approach to the analysis of molecular reaction pathways and elementary bifurcation tracks of periodic orbits. This point of view opens the way to the rigorous study of highly excited vibrational states. The analysis of Saddle-node (SN) bifurcations of periodic orbits is expected to offer new insights into fundamental spectroscopic observations of chemical processes.

Fournier presents a new method, Tabu Search in Descriptor Space (TSDS), for the investigation of potential surfaces of homoatomic clusters. The determination of stable molecular geometries for clusters of some size has attracted particular attention in late years. This is mainly due to the emergence of important applications in Materials Science. The study of chemical bonding and basic physicochemical properties of clusters represents a major challenge to fundamental Molecular Science. Results of the application of the TSDS method to the study of Ar, Li and Si clusters are presented and discussed.

Kusalik presents and discusses a new method for the determination of the properties of water in the liquid state. A mean-field approach is used to describe the electrostatic environment acting on the water molecule. This allows the calculation of the electric multipole moments and (hyper)polarizability. A new technique, Centroid Molecular Dynamics (CMD), is applied to the study of quantum dynamics in liquid water leading to interesting new results. This includes a novel characterization of the phenomenon of "effective tunneling" in liquid water.

Pal presents a multireference version of Coupled-cluster theory, MRCC. This version is expected to perform better than standard Single-reference Coupled-cluster theory in the description of near-degenerate states and molecules away from their equilibrium geometry. Some explicit results are presented for the dipole moment of test systems.

Papadopoulos focuses on the calculation of vibrational polarizabilities and hyperpolarizabilities of pyrrole and HArF. The accurate determination of these vibrational contributions is of capital importance to our understanding the nonlinear optics of molecules. The successful design of NLO materials depends includes fundamental estimations of the relative magnitude of the electronic and vibrational contributions to the electric (hyper)polarizability.

Rode's contribution demonstrates fully the power and significance of modern QM/MM molecular dynamics simulation techniques applied to problems pertaining to the determination of the microscopic structure and ultrafast dynamics of ions in liquids. The presented examples are of importance to both Chemistry and Biology. They include the solvation structure and dynamics of several hydrated ions. Particular attention is paid to the visualization of these effects via the newly designed MOLVISION software tool.

Szalewicz presents new density functional theory (DFT) based methods for the calculation of intermolecular forces. Conventional **ab initio** methods are of high predictive capability when applied to similar problems but their applicability is limited to relatively small systems. DFT offers an attractive alternative. The methods have been successfully applied to the interaction of two molecules containing twelve atoms each.

Thakkar presents an analysis of the computational aspects of the determination of molecular structures and energetics in small hydrogen-bonded clusters. The systems of interest are the formic acid, nitric acid, glycolic acid and the water molecule. The quantum chemical methods used in the calculation extend from semiempirical approaches and density functional theory to Møller-Plesset perturbation theory and Coupled-cluster methods. The study of these systems reveals important challenges for computational Quantum Chemistry.

George Maroulis
Department of Chemistry
University of Patras
Greece

VSP International
Science Publishers
P.O. Box 346, 3700 AH Zeist
The Netherlands

*Lecture Series on Computer
and Computational Sciences*
Volume 1, 2004, pp. 1011-1014

Molecular Mechanics Model for Second Hyperpolarizabilities and Macroscopic Susceptibilities: Applications to Nanostructures and Molecular Materials

Per-Olof Åstrand[1]

Department of Chemistry,
Norwegian University of Science and Technology,
7491 Trondheim, Norway

Received 7 March, 2004; accepted in revised form 10 March, 2004

Abstract: The point-dipole interaction model for the dipole-dipole polarizability and its second hyperpolarizability is discussed. Extensions regarding frequency-dependence, damping of interatomic interactions and macroscopic polarization are included. Results for carbon fullerenes and nanotubes are discussed.

Keywords: Polarizability, hyperpolarizability, point-dipole interaction, fullerene, nanotube

PACS: 31.70-f, 33.15-e, 36.40-c, 41.90-e, 78.20-e, 82.90-j

1 Introduction

The electronic structure of molecular systems may be represented in different ways. In molecular orbital and density functional theory, the wave function or the electronic density are normally expanded in terms of atomic basis function and the resulting orbital coefficients are obtained by solving the Schrödinger or Kohn-Sham equations. On the other hand, in molecular mechanics the electronic structure is represented by atom-type parameters such as atomic charges and van der Waals parameters, and the interaction energy is calculated by adopting the atom-type parameters in analytic function of interatomic distances [1, 2]. Atom-type parameters may be determined either empirically by fitting to experimental data or from quantum chemical calculations. In principle, to calculate the interaction energy by molecular mechanics force fields only takes a small fraction of the time it takes to carry out a quantum chemical calculation. Therefore, quantum chemical calculations are normally carried out on single molecules or relatively small molecular systems whereas force fields are adopted in simulations of liquids and solutions. It should be noted, however, that Car-Parrinello simulations propagate the electronic density as described by density functional theory and such simulations have been adopted for example for liquid water [3]. In quantum chemistry, molecular response properties such as (hyper)polarizabilities are calculated conveniently by adopting response theory.

It should also be expected that the atom-type parameters in a force field mimic the corresponding molecular properties [2]. For example, if the electrostatics is represented by atomic charges, the correct molecular charge, dipole moment and possibly also the molecular quadrupole moment should be described by the atomic charges. In most force fields, however, the atomic charges are

---

[1]Corresponding author. E-mail: per-olof.aastrand@chem.ntnu.no

obtained by parametrizing the electrostatic potential around the molecule. The representation of the electrostatics may be extended to atomic dipole moments, which has been investigated in detail recently [4]. Equivalently, atomic polarizabilities should represent the molecular polarizability and possibly also higher-order quadrupole polarizabilities. One way to model the molecular polarizability according to classical electrostatics is the point-dipole interaction (PDI) model [5, 6]. This model will be described here with extensions to frequency-dependence [8], damping of interatomic interactions [9, 10, 11], to molecular second hyperpolarizabilites [12, 13], and to macroscopic polarization [14].

2   Theoretical Background

The PDI model was first introduced by Silberstein [5] and to a large extent exploited by Applequist and coworkers [6]. For a system of N atomic polarizabilities, $\alpha_{I,\alpha\beta}$, in an external electric field, $E^{\text{ext}}_{I,\beta}$, the atomic induced dipole moment, $\mu^{\text{ind}}_{I,\alpha}$, for atom I is given as

$$\mu^{\text{ind}}_{I,\alpha} = \alpha_{I,\alpha\beta} \left(E^{\text{ext}}_{I,\beta} + \sum_{J \neq I}^{N} T^{(2)}_{IJ,\beta\gamma} \mu^{\text{ind}}_{J,\gamma} \right) , \tag{1}$$

where the second term on the right-hand side is the electric field from all other atomic induced dipole moments. Here and in the following, the Einstein summation convention is adopted for repeated Greek subscripts. This results in $3N$ coupled equations, which may be solved by standard matrix techniques as [6, 9]

$$\boldsymbol{\mu}^{\text{ind}} = \left(\boldsymbol{\alpha}^{-1} - \mathbf{T} \right)^{-1} \mathbf{E}^{\text{ext}} \tag{2}$$

where $\boldsymbol{\mu}^{\text{ind}}$ and $\mathbf{E}^{\text{ext}}$ are $3N$ vectors and $\boldsymbol{\alpha}$ and $\mathbf{T}$ are $3N \times 3N$ matrices. A two-atom relay tensor, $\mathcal{B}^{(2)}$ may be defined as

$$\mathcal{B}^{(2)} = \left(\boldsymbol{\alpha}^{-1} - \mathbf{T} \right)^{-1} , \tag{3}$$

resulting in that the molecular polarizability, $\alpha^{\text{mol}}_{\alpha\beta}$, is given as

$$\alpha^{\text{mol}}_{\alpha\beta} = \sum_{I,J}^{N} \mathcal{B}^{(2)}_{IJ,\alpha\beta} . \tag{4}$$

In our work, a spherically symmetric polarizability was included for each element, and these atom-type polarizabilities were parametrized against quantum chemical calculations of molecular polarizabilities [8]. The results are improved dramatically by including a damping of the interatomic interactions by modifying the T-tensor in Eq. (1) [9]. We regarded the overlap between two classical charge distributions, which resulted in that the distance between two atoms in the system is replaced by a scaled distance [11]. Furthermore, it was assumed that the frequency-dependence of the molecular polarizability may be modelled with an Unsöld approximation as [8]

$$\alpha_P \left(-\omega; \omega \right) = \alpha_P \left(0; 0 \right) \times \left(\frac{\bar{\omega}_P^2}{\bar{\omega}_P^2 - \omega^2} \right) , \tag{5}$$

where ω is the frequency and $\bar{\omega}_P$ is an atom-type parameter to be determined. It is obviously a crude approximation only valid far from absorption since it is the atomic parameter $\bar{\omega}_P$ with the lowest value that determines where the first excitation is located irrespective of the surroundings in the molecule. Hyperpolarizabilities may be obtained by extending Eq. (1) with higher order terms,

$$\mu_{I,\alpha}^{\text{ind}} = \alpha_{I,\alpha\beta}E_{I,\beta}^{\text{tot}} + \frac{1}{6}\gamma_{I,\alpha\beta\gamma\delta}E_{I,\delta}^{\text{tot}}E_{I,\gamma}^{\text{tot}}E_{I,\beta}^{\text{tot}} , \tag{6}$$

where $E_{I,\beta}^{\text{tot}}$ is the total electric fied given as the sum of the external field and the field from the other atomic induced dipole moments in the system. $\gamma_{I,\alpha\beta\gamma\delta}$ is an atomic second hyperpolarizability and an atomic first hyperpolarizability, $\beta_{I,\alpha\beta\gamma}$, has not been included because it is zero for spherically symmetric particles. By adopting the approach by Sundberg [12], the molecular second hyperpolarizability, $\gamma_{\alpha\beta\gamma\delta}^{\text{mol}}$, is given in terms of a four-atom relay tensor, $\mathcal{B}_{IJKL,\alpha\beta\gamma\delta}^{(4)}$, as

$$\gamma_{\alpha\beta\gamma\delta}^{\text{mol}} = \sum_{I,J,K,L}^{N} \mathcal{B}_{IJKL,\alpha\beta\gamma\delta}^{(4)} , \tag{7}$$

where the four-atom relay tensor may be expressed as [13]

$$\mathcal{B}_{IJKL,\alpha\beta\gamma\delta}^{(4)} = \sum_{M}^{N} \gamma_{M,\lambda\mu\nu\xi}\tilde{\mathcal{B}}_{ML,\xi\delta}^{(2)}\tilde{\mathcal{B}}_{MK,\nu\gamma}^{(2)}\tilde{\mathcal{B}}_{MJ,\mu\beta}^{(2)}\tilde{\mathcal{B}}_{MI,\lambda\alpha}^{(2)} , \tag{8}$$

where

$$\tilde{\mathcal{B}}_{IJ,\alpha\beta}^{(2)} = \delta_{IJ}\delta_{\alpha\beta} + \sum_{K\neq I}^{N} T_{IK,\alpha\gamma}^{(2)}\mathcal{B}_{KJ,\gamma\beta}^{(2)} . \tag{9}$$

The Lorentz local-field model for macroscopic polarization has been extended by including also the electric fields of the surrounding atomic induced dipole moments according to Eq. (1). The Lorentz-Lorenz equation is not modified, but the polarizability included in the equation becomes the effective polarizability of a particle in a cluster instead of the polarizability of of the isolated particle [14].

3   Results

The model as developed by us has been used for the frequency-dependent polarizability tensors for carbon nanotubes [15] and boron nitride nanotubes [16] as well as the polarizability of molecular clusters [11]. The extension of the model to second hyperpolarizabilities has been used to study the properties of carbon fullerenes and nanotubes [17, 18]. In particular, the saturation length of γ for carbon nanotubes has been studied for tubes up to a length of 75 nm [17]. It is found that carbon nanotubes are comparable to conjugated polymers with respect to the magnitude of γ, which indicates that they are promising candidates for future optical materials. Finally, the macroscopic polarization of C_{60} fullerenes have been calculated [14].

Acknowledgment

P.-O.Å. has received support from the Norwegian Research Council (NFR) through a Strategic University Program (Grant no 154011/420), a NANOMAT program (Grant no 158538/431) and a grant of computer time from the Norwegian High Performance Computing Consortium (NOTUR)

References

[1] A. K. Rappé, C. J. Casewit, Molecular mechanics across chemistry, University Science Books, Sausalito, 1997.

[2] O. Engkvist, P.-O. Åstrand, G. Karlström, Accurate intermolecular potentials obtained from molecular wave functions: Bridging the gap between quantum chemistry and molecular simulations, Chem. Rev. 100 (2000) 4087–4108.

[3] P. L. Silvestrelli, M. Parrinello, Structural, electronic, and bonding properties of liquid water from first principles, J. Chem. Phys. 111 (1999) 3572–3580.

[4] H. Solheim, K. Ruud, P.-O. Åstrand, Atomic dipole moments calculated using analytical molecular second-moment gradients, J. Chem. Phys. 120 (2004) 10368–10378.

[5] L. Silberstein, Molecular refractivity and atomic interaction, Phil. Mag. 33 (1917) 92–128.

[6] J. Applequist, J. R. Carl, K.-F. Fung, An atom dipole interaction model for molecular polarizability. Application to polyatomic molecules and determination of atom polarizabilities, J. Am. Chem. Soc. 94 (1972) 2952–2960.

[7] J. Applequist, An atom dipole interaction model for molecular optical properties, Acc. Chem. Res. 10 (1977) 79–85.

[8] L. Jensen, P.-O. Åstrand, K. O. Sylvester-Hvid, K. V. Mikkelsen, Frequency-dependent molecular polarizability calculated within an interaction model, J. Phys. Chem. A 104 (2000) 1563–1569.

[9] B. T. Thole, Molecular polarizabilities calculated with a modified dipole interaction, Chem. Phys. 59 (1981) 341–350.

[10] R. R. Birge, Calculation of molecular polarizabilities using an anisotropic atom point dipole interaction model which includes the effect of electron repulsion, J. Chem. Phys. 72 (1980) 5312–5319.

[11] L. Jensen, P.-O. Åstrand, A. Osted, J. Kongsted, K. V. Mikkelsen, Polarizability of molecular clusters as calculated by a dipole interaction model, J. Chem. Phys. 116 (2002) 4001–4010.

[12] K. R. Sundberg, A group-dipole interaction model of the molecular polarizability and the molecular first and second hyperpolarizabilities, J. Chem. Phys. 66 (1977) 114–118.

[13] L. Jensen, K. O. Sylvester-Hvid, K. V. Mikkelsen, P.-O. Åstrand, A dipole interaction model for the molecular second hyperpolarizability, J. Phys. Chem. A 107 (2003) 2270–2276.

[14] L. Jensen, P.-O. Åstrand, K. V. Mikkelsen, Microscopic and macroscopic polarization in C_{60} fullerene clusters as calculated by an electrostatic interaction model, J. Phys. Chem. B 108 (2004) 8226–8233.

[15] L. Jensen, O. H. Schmidt, K. V. Mikkelsen, P.-O. Åstrand, Static and frequency-dependent polarizability tensors for carbon nanotubes, J. Phys. Chem. B 104 (2000) 10462–10466.

[16] J. Kongsted, A. Osted, L. Jensen, P.-O. Åstrand, K. V. Mikkelsen, Frequency-dependent polarizability of boron nitride nanotubes: A theoretical study, J. Phys. Chem. B 105 (2001) 10243–10248.

[17] L. Jensen, P.-O. Åstrand, K. V. Mikkelsen, Saturation of the third-order polarizability of carbon nanotubes characterized by a dipole interaction model, Nano Lett. 3 (2003) 661–665.

[18] L. Jensen, P.-O. Åstrand, K. V. Mikkelsen, The static polarizability and second hyperpolarizability of fullerenes and carbon nanotubes, J. Phys. Chem. A xxx (2004) yy–zz, ASAP article.

VSP International
Science Publishers
P.O. Box 346, 3700 AH Zeist
The Netherlands

*Lecture Series on Computer
and Computational Sciences*
Volume 1, 2004, pp. 1015-1021

Lowest Energy Path of Oxygen near CH:
A Combined Configuration Interaction and Tight-Binding Approach

N.C. Bacalis and A. Metropoulos

Theoretical and Physical Chemistry Institute, National Hellenic Research Foundation,
Vasileos Constantinou 48, GR - 116 35 ATHENS, Greece

D.A. Papaconstantopoulos

Center for Computational Materials Science, Naval Research Laboratory, Washington DC
20375-5345 USA

Abstract: It is demonstrated that the lowest energy path for the formation of a polyatomic molecule (applied to the HC-O formation) is easily calculated via a geometry-independent tight binding Hamiltonian fitted to accurate *ab-initio* configuration interaction (CI) total energies. This Hamiltonian not only reproduces the CI calculations accurately and efficiently, but also effectively identifies any CI energies happening to erroneously converge to excited states.

PACS: 31.10.+z, 31.15.-p, 31.50.-x, 82.20.Kh

The question

At present, an accurate *ab-initio* determination of the reaction path in a chemical reaction, needing the detailed knowledge of the pertinent potential energy surface (PES) (diabatic or adiabatic), may be inhibitively time consuming. For the ground state, the time problem is already traditionally overcome via the density functional theory (DFT) [1], which self-consistently approximates the many electron by a one-electron problem. However, DFT calculations sometimes fail to explain experimentally observed features of the PES [2]. Thus, accurate CI calculations are more or less indispensable, even if performed in a rather limited, but representative, set of molecular geometries. Therefore, a reliable interpolation scheme for the pertinent PES, based on CI calculations, and overcoming the problem of wrong CI convergence, is desirable.

The purpose

It is shown that such an interpolation scheme is possible, based on a spin-polarized [3] geometry-independent [4] Slater - Koster (SK) parametrization [5] of *ab-initio* CI total energies [6]. As a demonstration, the method is applied to the construction of the potential energy surface (PES) of the HCO(X^2A') state [7] (a state without a barrier). The lowest energy path of the formation of HCO (in $^2A'$ symmetry), while O approaches HC, is also computed using the interpolated PES.

The procedure

First several (724 - compared to 71969 of H_3 [8]) accurate CI total energies, based on (less accurate) multi-configuration self-consistent field (MCSCF) orbitals, are calculated at selected geometries of the H,C,O atoms in the A′ symmetry of the Cs group. Most of them (508) are fitted to the interpolation scheme, the remaining serving to check the quality of the fit. For the fit a non-orthogonal spin-polarized tight binding (TB) Hamiltonian is formed, whose matrix elements, along with those of the overlap matrix, are expressed as *functions of the bond direction*, according to the SK scheme [5], *and of the bond length*, according to the Naval Research Laboratory (NRL) technique [4], i.e.: The functions are generally polynomials of the interatomic distance, within exponential envelopes, the coefficients and the exponents being varied as parameters. For two adiabatic states near some (avoided) crossing the TB Hamiltonian naturally produces two diabatic PESs in nearby extrapolation, and predicts to which diabatic PES, ground-state or excited, nearby CI energies belong. Among these, the appropriate ones can be used to extend the fit beyond the (avoided) crossings, around which two sets of parameters are needed for the two PES's. If it happens, as with HCO, that the ground and excited state energies beyond the crossing lie close to each other, the adiabatic PES can be fitted as well, with comparable accuracy.

Using at each point of the DOF space the lowest lying TB-fitted PES, the adiabatic path can be found: For each value of a desired degree of freedom (in our case for each C-O distance) the energy minimum is searched [9] in the space of the remaining degrees of freedom (C-H distance and H-C-O angle). Having the parametrized tight binding Hamiltonian, any property can be trivially computed.

Methodology

For the CI energies the correlation consistent aug-cc-pVTZ basis set was used [10, 11] in conjunction with the complete active space self-consistent field (CASSCF) + 1 + 2 multi-reference CI method (MRCI) employed in the MOLPRO package [6] (the four electrons in the 1s orbitals of C and O were unexcited). The CASSCF calculations were state-averaged, and the active space was limited to the 9 valence orbitals among which the remaining 11 electrons were distributed. In the subsequent MRCI calculations the uncontracted configurations were around 50 million internally contracted to about one million. Calculations between C-O distances of 1.7 and 6 bohr were done for several H-C-O angles between $50°$ and $180°$ and several C-H distances between 1.7 and 4.5 bohr, most around the C-H equilibrium distance of 2.12 bohr. The three lowest roots of the secular equation were computed to increase the accuracy of the calculation. By an analytic gradient optimization at the MCSCF level, an approximate (MCSCF) equilibrium geometry was found at the DOF space point $(\tilde{r}_{HC}, \tilde{r}_{CO}, \tilde{\theta}_{H-C-O}) = (2.12, 2.2, 126°)$ (in a.u.). Because it is not evident whether the aforementioned points are beyond any avoided crossing, where the role of the ground and the excited states would be interchanged, first several DOF points near equilibrium were obtained by employing a generalization of the 3-dimensional sphere to the generally multi-dimensional (in this case also 3-dimensional) DOF space: $x_i = r_i/\tilde{r}_i - 1$, i = {HC, CO}, $x_3 = \theta/\tilde{\theta} - 1$, where generally for n degrees of freedom, points belonging to a n-dimensional hypersphere of radius r and center $(\tilde{x}_i, i = 1,...,n)$ are obtained by

$$
\begin{aligned}
x_n - \tilde{x}_n &= r \, cos\theta_n \\
x_{n-1} - \tilde{x}_{n-1} &= r \, sin\theta_n \, cos\theta_{n-1} \\
&\cdots \\
x_1 - \tilde{x}_1 &= r \, sin\theta_n \, sin\theta_{n-1}...cos\theta_1
\end{aligned}
\tag{1}
$$

where the 1st $\theta_1 = 0$ or 180°, the two points of a "1-dimensional sphere", and the other $0 < \theta_i < 180^\circ$ are the "azimuthal" hypersphere angles (incidentally, a *variable dimensional* do-loop code was invented, needed to treat any larger molecule). Thus, first points with small r were fitted, and gradually the fit was extended to more remote DOF points.

The formalism of the NRL geometry - independent TB parametrization is described in detail in Ref. [4]; here an essential summary is only presented. The total energy is written as

$$
\begin{aligned}
E[n(\vec{r})] &= \sum_{i\ ;\ s=1,2} f(\frac{\mu - \epsilon_{i\ s}}{T})\ \epsilon_{i\ s} + F[n(\vec{r})] \\
&\equiv \sum_{i\ ;\ s=1,2} f(\frac{\mu' - \epsilon'_{i\ s}}{T})\ \epsilon'_{i\ s}
\end{aligned}
\tag{2}
$$

where [12] $f(x) = 1/(1 + e^x)$, T=0.005 mRy, and

$$
\epsilon'_{i\ s} = \epsilon_{i\ s} + V_0 \quad ; \quad \mu' = \mu + V_0 \quad ; \quad V_0 = F[n(\vec{r})]/N_e
\tag{3}
$$

with $N_e = \sum_{i\ ;\ s=1,2} f((\mu - \epsilon_{i\ s})/T)$ being the number of electrons, i counts the states, $s = 1, 2$ counts the spin. Since the total energy is independent of the choice of zero of the potential, the shift V_0 is sufficient to be determined by the requirement that $\epsilon'_{i\ s}$ are the eigenvalues of the generalized eigenvalue problem $(\mathbf{H} - \mathbf{S}\ \epsilon'_{i\ s})\ \psi_{i\ s} = 0$, where $\mathbf{H}$ is the TB Hamiltonian and $\mathbf{S}$ is overlap matrix in an atomic s- and p-orbital basis representation $\{\phi_a\}$. Thus, a non-orthogonal TB calculation uses on-site, hopping and overlap parameters. Demanding that only the on-site SK parameters are affected by the shift V_0, for atom I in a spin-polarized structure the matrix elements are expressed as

$$
h^I_{l\ s} = \sum_{n=0}^{3} b^I_{l\ n\ s}\ \varrho_{I\ s}^{2n/3} \quad ; \quad l = s, p
\tag{4}
$$

where

$$
\varrho_{I\ s} = \sum_{J \neq I} e^{-\lambda^2_{\tilde{I}\ \tilde{J}\ s} R_{I\ J}}\ f(\frac{R_{I\ J} - R_0}{r_c})
\tag{5}
$$

is a generalized pair potential ("density"), with $R_0 = 15$ bohr, $r_c = 0.5$ bohr, $R_{I\ J}$ is the internuclear distance between atoms I and J, $\tilde{I}(\tilde{J})$ denote the type of atom on the site $I(J)$ while $\lambda_{\tilde{I}\ \tilde{J}\ s}$, depending on the atom type, and $b^I_{l\ n\ s}$ are the on-site NRL geometry-independent parameters (GIP). It is found sufficient to keep hopping and overlap parameters spin independent, of the form

$$
P_\gamma(R) = (\sum_{n=0}^{2} c_{\gamma\ n}\ R^n)\ e^{-g^2_\gamma R}\ f(\frac{R - R_0}{r_c})
\tag{6}
$$

where γ indicates the type of interaction (i.e. $ss\sigma$, $sp\sigma$, $pp\sigma$, $pp\pi$ and $ps\sigma$). The NRL GIPs are $c_{\gamma\ n}$ and g_γ, R is the interatomic distance, and R_0 and r_c are as in eq. 5.

Within the context of the NRL code [4], written primarily for solids, the molecule was treated as a base to a large cubic lattice unit cell (lattice constant $= 100$ a.u.) ensuring vanishing interaction between atoms in neighboring cells. Thus, the PES was described in terms of the following NRL GIPs for each spin polarization. On-site: s: H, C, O, (H depending on C), (C on H), (H on O), (O on H), (C on O), and (O on C); p: C, O, (C on H), (O on H), (C on O), and (O on C). Hopping and overlap parameters: $ss\sigma$: H-C, H-O, C-O; $sp\sigma$: H-C, H-O, C-O and O-C (denoted as $ps\sigma$); $pp\sigma$ and $pp\pi$: C-O. For HCO, since similar atoms are well separated, the H-H, C-C and O-O parameters vanish. We fitted 508 CI points and checked the resulting PES against 216 more CI energies not included in the fit. The error was less than 10^{-3} a.u., which is within the *ab-initio* PES

Table 1: Geometric characteristics of HCO around equilibrium, along the reaction path, in a.u. (H-C-O angle in degrees). The last three columns indicate the minimum energy molecular geometry.

C-O distance	Total Energy	C-H distance	H-C-O angle
2.6	-113.6328	2.069	117.53
2.5	-113.6485	2.071	118.69
2.4	-113.6610	2.077	119.77
2.3	-113.6685	2.088	120.84
2.2	-113.6687	2.107	121.91
2.1	-113.6583	2.130	122.98
2.0	-113.6326	2.153	124.09

accuracy (starting from different initial guesses the MCSCF calculation may converge to slightly different results by 10^{-3} a.u.). To ensure obtaining physically meaningful TB parameters, for a very limited number of molecular geometries the Hamiltonian eigenvalues were also fitted, while the total energy was fitted for all 508 structures.

Finally, for the reaction path we used a non-linear energy minimization technique employing Powell's conjugate directions method [13] modified to be restricted to closed intervals of the DOF space [9].

For comparison, each of the 724 *ab-initio* CI calculations needs 3 hours of CPU time, each n-dimensional hypersphere radius r-increase, to fit more remote points (with 10 such hypersphere radial extensions all points can be covered) needs 2-3 hours, and each 2-dimensional energy minimization, using the final TB parameters (i.e. the reaction path determination), needs a few seconds.

Results

The fitted TB Hamiltonian could predict correctly total energy curves for points not included in the fit as shown for example in Fig. 1. Since it produces naturally the diabatically extended branch of the energy, it could distinguish to which adiabatic state near an avoided crossing the CI values belong. Classifying such CI points may sometimes be misleading or unrecognizable by mere observation of the MCSCF orbitals. An example is shown in Fig. 2. However, the most impressive aspect was that we realized, through the fit, that at some points (about 10 in 700) the CI calculation had converged to *excited* energies (which ought to be disregarded, otherwise they would destroy the fit). An example is given in Fig. 3. Finally, Fig. 4 shows the lowest energy path for the formation of HCO, as HC approaches O. For a triatomic molecule the figure contains the whole information: For each C-O distance the minimum energy and the corresponding C-H distance and H-C-O angle are displayed via two arrows starting at C (on the curve) and ending one at O (horizontal) and the other at H (oblique). As seen from the figure, at large C-O distances, O is more attracted toward H, but, in approaching equilibrium, O binds mainly with C, the H-C-O angle gradually becoming $\simeq 122^o$ (representing the CI value). Around equilibrium (c.f. Table 1), the angle changes slightly monotonically by 1-2°, but because, in increasing the C-O distance, the C-H distance decreases, predominantly an antisymmetric stretching vibration occurs. To our knowledge there is no experimental confirmation of the reaction path of this intermediate molecule.

Acknowledgment

We wish to thank Dr. M.J. Mehl for many useful discussions.

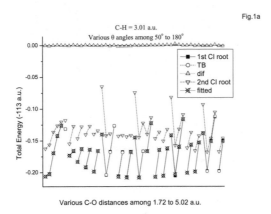

Fig.1a

Fig.1b

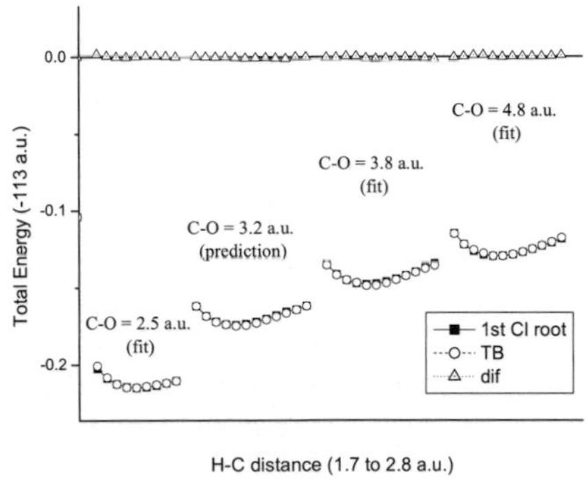

Figure 1: Predicted total energy E in a.u. (Above:) vs C-O distance for C-H distance = 3.01 bohr, and various H-C-O angles. (Below:) vs C-H distance for various C-O distances, and H-C-O angle = 100°.

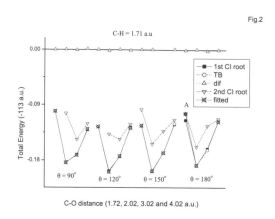

Figure 2: The CI point A (excited) in E vs C-O distance for C-H distance = 1.71 bohr and H-C-O angle =180° is predicted by the fit to belong to the diabatic branch of the curve beyond the avoided crossing. (Inclusion of the lower value to the fit, destroys it.)

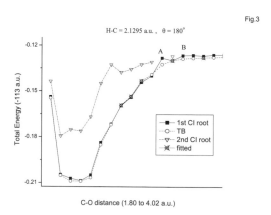

Figure 3: The CI points A and B clearly belong to the excited state as shown by the TB prediction. The CI calculation could not converge to the correct values. The discontinuity can be verified by observing the corresponding MCSCF orbitals.

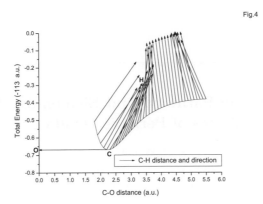

Figure 4: The reaction path for the formation of HCO. Details are described in the text.

References

[1] P. Hohenberg and W. Kohn, Phys. Rev. **136**, B 864 (1964); W. Kohn and L.J. Sham, Phys. Rev. **140**, A1133 (1965); **145**, A561 (1966); U. von Barth and L. Hedin, J. Phys. C **5**, 1629 (1972).

[2] Y. Yourdshahyan, B. Razaznejad, and B.I. Lundqvist, Phys. Rev. B **65**, 075416 (2002).

[3] N.C. Bacalis, D.A. Papaconstantopoulos, M.J. Mehl and M. Lach-hab, Physica B **296**, 125 (2001).

[4] D.A. Papaconstantopoulos and M.J. Mehl, J. Phys.: Condens. Matter **15**, R413 (2003).

[5] J.C. Slater G.F. and Koster, Phys Rev. **94**, 1498 (1954).

[6] Molpro is a package of *ab-initio* programs written by H.-J. Werner and P.J. Knowles, with contributions from J. Amlöf et al.

[7] A. Metropoulos and A. Mavridis, J. Chem. Phys. **115**, 6946, (2001)

[8] Y.-S.M. Wu, A. Kuppermann, and J.B. Anderson, Phys. Chem. Chem. Phys. **1**, 929 (1999).

[9] N. C. Bacalis, J. Phys. B, **29**, 1587 (1996).

[10] T.H. Dunning Jr., J. Chem. Phys. **90** 1007 (1989).

[11] A.K. Wilson, T.V. Maurjk, and T.H. Dunning Jr., J. Mol. Struct.: THEOCHEM **388**, 339 (1997).

[12] M.J. Gillan, J. Phys. Condens. Matter **1**, 689 (1989)

[13] W.H. Press, S.A. Teukolsky, W.T. Vetterling and B.P. Flannery, Numerical Recipes in FOR-TRAN, 2nd ed. (Cambridge University Press, 1992).

VSP International
Science Publishers
P.O. Box 346, 3700 AH Zeist
The Netherlands

Lecture Series on Computer
and Computational Sciences
Volume 1, 2004, pp. 1022-1024

Molecular Reaction Pathways and Elementary Bifurcation Tracks of Periodic Orbits

S. C. Farantos[1]

Institute of Electronic Structure and Lasers, Foundation for Research and Technology-Hellas, and
Department of Chemistry, University of Crete, Iraklion 711 10, Crete, Greece

Spectroscopic techniques such as the Dispersed Fluorescence (DF), and Stimulated Emission Pumping (SEP) [1] have revealed a new dynamical picture of small polyatomic molecules excited at very high vibrational states. Spectra of high complexity at high resolution may show regularities at low resolution, intense regular progressions of spectral lines may coexist with congested bands, and even irregular spectra may be replaced by regular ones as energy increases. Vibrational spectra of excited molecules are the fingerprints of the nonlinear mechanical behaviour of the molecule. The assignment of the spectral lines particularly of the regular ones, and thus the extraction of dynamics, is a challenging task, since the picture which emerges defies the previous simple ideas that a molecule shows regular behaviour at low energies and chaotic one above a threshold energy. Furthermore, when elementary chemical processes such as isomerization and dissociation are involved the understanding of how a bond is breaking and a new one is formed brings back fundamental questions of chemical dynamics.

The established theoretical methods based on a normal mode description of the molecular vibrations, applied at energies close to the equilibrium point, are not any longer valid for vibrationally excited molecules. The departure from the harmonic approximation of the potential energy surface imposes the need for the construction of accurate potential functions far from the equilibrium point and the application of nonlinear mechanics to understand the dynamics of the molecule. Polyatomic molecules stimulate new computational challenges in solving accurately the Schrödinger equation and obtaining hundreds of vibrational states. Nowadays, triatomic molecules can be treated with fully *ab initio* methods both, in their electronic and nuclear part. Tetratomic molecules are more difficult to deal with, in spite of the progress which has recently been achieved. For example, six-dimensional calculations up to energies of the isomerization of acetylene to vinylidine have been published [2].

Apart from the computational challenges small polyatomic molecules expose conceptual and physical interpretation problems. A result of the nonlinear mechanical behaviour of a dynamical system is the simultaneous appearance of ordered motions and chaos, as well as the genesis of new type of motions via bifurcation phenomena. What are the quantum mechanical counterparts of these classical behaviours? What are the spectroscopic signatures of the nonlinear dynamics of the molecules? In cases where the vibrational spectra depict isomerization and dissociation processes how can we identify them in the spectra? As a matter of fact the progress of nonlinear mechanics forces us to reexamine the mechanisms of the breaking and/or forming a single chemical bond as it happens in the elementary chemical processes. To answer the above questions new assignment schemes which allow the classification of quantum states in a meaningful way are required and such novel methods have been developed and applied by our group. We use periodic orbits and

---

[1]E-mail: farantos@iesl.forth.gr

bifurcation theory to explore the complex structure of the molecular phase space and from it to deduce the quantum dynamics.

The theoretical background of our approach stems in several advances of semiclassical theory in the past years. Gutzwiller [3] using Feynman's path integral formulation has derived a semiclassical expression for the trace of the resolvent (Green operator) of the quantum Hamiltonian operator as a sum over all isolated periodic orbits, taking into account their linear stability indices. This formula provides approximate values for the quantum eigenenergies of classically chaotic systems. Berry and Tabor [4] derived in a different way the trace formula based on the Einstein-Brillouin-Keller quantization rule. Their formula gives the quantum density of states as a coherent summation over resonant tori, and therefore it is applicable to integrable systems. Finally, a uniform result bridging the Berry-Tabor and Gutzwiller trace formulas for the case of a resonant island chain was derived by Ozorio de Almeida [5]. Last but not least, the importance of periodic orbits for a qualitative understanding of the localization of quantum mechanical eigenfunctions came from the scarring theory of Heller [6]. It turns out, that for small polyatomic molecules the probability density of eigenfunctions is accumulated along short period stable or the least unstable periodic orbits.

Periodic orbits (POs) evolve with the energy of the system or any other parameter in the Hamiltonian, bifurcate and produce new periodic orbits which portrait the resonances among the vibrational degrees of freedom. Generally, POs reveal the structure of phase space at different energies. For about fifteen years we study families of POs in molecular systems, using model and realistic potentials. We compare the classical and quantum mechanical behaviours of the molecule by constructing bifurcation diagrams of periodic orbits. Locating POs in multidimensional systems is a two point boundary value problem. We apply advanced shooting methods to locate periodic orbits and to continue them in their parameter space [7].

We have systematical study the POs networks for different type of molecules; triatomic [8] and tetratomic [9] molecules, van der Waals [10] molecules and at energies below and above dissociation [11]. The bifurcation theory of Hamiltonian dynamical systems has mainly been developed in the last half of twentieth century. One important outcome of the theory is the identification of the elementary bifurcations which are described by simple Hamiltonians and they are valid for a broad class of Hamiltonian systems. The elementary bifurcations are the saddle-node, transcritical, pitchfork and Hopf bifurcations.

Bifurcation phenomena, i.e. the change of the structure of the orbits by varying one or more parameters, are well known in vibrational spectroscopy. For example, the transition from normal to local mode oscillations is related to the elementary pitchfork bifurcation. A number of studies, classical and quantum mainly in small molecules, revealed that isomerization and dissociation reactions are driven by another type of elementary bifurcation, the saddle-node (SN). Periodic orbits which emerge from SN bifurcations appear abruptly at some critical value of the energy, usually in pairs, and penetrate in regions of nuclear phase space where the normal mode motions can not reach. Saddle-node bifurcations are of generic type, i.e. they are robust and remain for small (perturbative) changes of the potential function [9].

We initially determined the importance of SN bifurcations of periodic orbits in studies of the isomerization dynamics in double well potential functions [12]. These POs connect the two minima and scar the isomerizing wavefunctions, i.e. eigenfunctions with significant probability density in both wells. Their birth is due to the unstable periodic orbit which emanates from the saddle point of the potential energy surface. However, even below the potential barrier a series of SN bifurcations of periodic orbits pave the way to the isomerization process.

The spectroscopic signature of SN bifurcations has been found in a number of triatomic molecules [13, 14, 15]. HCP was the first molecule where experimental evidence for SN bifurcations was given. In the other extreme, infinite dimensional systems, such as periodic or random lattices, show spatially localized and periodic in time motions, called discrete breathers, and it has been shown that they can also be associated with saddle-node bifurcations [16, 17]. Spectroscopic evidence for the existence of discrete breathers can be found among biomolecules [18]. The emanation of SN bifurcations as a generic phenomenon in elementary chemical processes is the main theme of the talk.

References

References

[1] H.-L. Dai and R. W. Field, *Molecular Dynamics and Spectroscopy by Stimulated Emission Pumping*, in *Advanced Series in Physical Chemistry*, **4**, (World Scientific Publ. Co., Singapore, 1995).

[2] S. Zou and J. M. Bowman, J. Chem. Phys. **117**, 5507 (2002).

[3] M. C. Gutzwiller, *Chaos in Classical and Quantum Mechanics,* of *Interdisciplinary Applied Mathematics,* **1,** (Springer-Verlag, New York, 1990).

[4] M. V. Berry and M. Tabor, *J. Phys. A* **10**, 371 (1977).

[5] A. M. Ozorio de Almeida, in T. Seigman, *Quantum Chaos and Statistical Nuclear Physics*, in *Lecture Notes in Physics*, **263**, 197, (Springer-Verlag, New York, 1986).

[6] E. J. Heller, *Phys. Rev. Lett.* **53,** 1515 (1984).

[7] S. C. Farantos, *Comp. Phys. Comm.* **108**, 240 (1998).

[8] S. C. Farantos, *Int. Rev. Phys. Chem.* **15**, 345 (1996).

[9] R. Prosmiti and S. C. Farantos, *J. Chem. Phys.* **118**, 8275 (2003).

[10] R. Guaṅtes, A. Nezis and S. C. Farantos. *J. Chem. Phys.* **111**, 10835 (1999).

[11] M. Founargiotakis, S. C. Farantos, H. Skokos and G. Contopoulos, *Chem. Phys. Letters* **277,** 456 (1997).

[12] S. C. Farantos, *Laser Chemistry* **13,** 87 (1993).

[13] M. Joyeux, S. C. Farantos and R. Schinke, Highly Excited Motion in Molecules: Saddle-Node Bifurcations and their Fingerprints in Vibrational Spectra, *J. Phys. Chem. A* (feature article) **106**, 5407 (2002).

[14] H. Ishikawa, R. W. Field, S. C. Farantos, M. Joyeux, J. Koput, C. Beck and R. Schinke, *HCP - CPH Isomerization: Caught in the Act, Annu. Rev. Phys. Chem.* **50**, 443 (1999).

[15] M. Joyeux, S. Yu. Grebenshchikov, J. Bredenbeck, R. Schinke and S. C. Farantos, *Intramolecular Dynamics Along Isomerization and Dissociation Pathways, Adv. Chem. Phys.*, in press, 2004.

[16] S. Flach and C. R. Willis, *Phys. Rep.* **295,** 181 (1998).

[17] G. Kopidakis and S. Aubry, *Physica D* **130**, 155 (1999).

[18] A. Xie, L. van der Meer, W. Hoff and R. H. Austin, *Phy. Rev. Lett.* **84**, 5435 (2000).

VSP International
Science Publishers
P.O. Box 346, 3700 AH Zeist
The Netherlands

Lecture Series on Computer
and Computational Sciences
Volume 1, 2004, pp. 1025-1028

Global Optimization Methods for Studying Clusters

R. Fournier[1]

Department of Chemistry, York University
4700 Keele Street, M3J 1P3 Toronto, Canada

Received 30 June, 2004; accepted in revised form , 2004

Abstract:
We devised a method, Tabu Search in Descriptor Space (TSDS), for global search of energy minima on potential surfaces of atomic clusters. In each cycle TSDS generates many structures at random, calculates their structural descriptors, and screens them using energy predictions based on descriptors. Only a small fraction (typically less than 10%) of the clusters are retained for energy evaluation. This cycle is repeated many times. In the final step, clusters are sorted and only the best few undergo local optimization. The TSDS method requires between ten and a hundred times fewer energy evaluations than a good genetic algorithm for locating the global minimum of n-atom clusters ($n < 35$) described by a Lennard-Jones potential. It is a promising method for global optimization of functions that are computationally expensive, for example, energy surfaces calculated by first-principles. We will discuss results obtained by combining TSDS with empirical potentials and Kohn-Sham theory that model clusters of Ar, Li, Si, and a few other elements.

Keywords: global optimization, tabu search, clusters, lithium, geometric structure

PACS: 02.60.Pn 36.40.-c 73.22.-f

1 The Cluster Structure Optimization Problem

Small clusters have properties that differ markedly from those of bulk materials and depend on cluster size[1]. It is necessary to elucidate the structure of clusters in order to understand their properties. If real clusters adopt structures that are favored by kinetics, global optimization may not be relevant. Instead, one would have to model in detail the experimental conditions. But that is not our goal. We aim to devise general methods for finding the lowest energy minima on complicated energy surfaces. For clusters, this is a difficult problem for three reasons. First, the number of distinct minima is on the order of thousands or millions even for clusters with only 10 to 20 atoms[2]. Second, most clusters can not be modelled reliably with empirical potentials, and the computer time for calculating an energy by a first-principles method is very high, on the order of seconds or minutes. Third, there are only few structural principles to guide the search in most cases, C_n fullerenes being one exception.

Many global optimization methods have been used for clusters: simulated annealing[3] (SA), genetic algorithms[4] (GA), and several others[5]. Even the best of those methods require on the order of 10^5 energy evaluations for finding the global minimum of a 30-atom Lennard-Jones (LJ) cluster. As a result, global optimizations have largely been limited to empirical or semi-empirical energy surfaces. Our goal is to develop more efficient global optimization techniques and study

---

[1] Corresponding author. E-mail: renef@yorku.ca

energy surfaces calculated by Kohn-Sham Density Functional Theory (KS-DFT). This would make it possible to investigate clusters of almost any composition: bimetallics, oxides, III-V alloys, *etc.* It is probably impossible to devise a global optimization method that is quite general *and* much better than current ones (GA, SA, ...). But we hope that a hybrid algorithm that uses problem-specific knowledge will outperform general methods.

We denote the energy by U, atomic coordinates by $\vec{X}$, and descriptors by $\vec{D}$. What we mean by a descriptor "D_j" is simply some function of atomic positions $D_j(\vec{X})$ that is invariant under rotations and translations and is computationally inexpensive. Ideally, descriptors should correlate with $U(\vec{X})$ and be independent of each other. TSDS has interesting features not normally present in other algorithms. (1) It uses the entire set of known energy points $\{U(\vec{X}_i)\}, i = 1, n$, at every stage in the search, unlike SA and GA. (2) It avoids calculating energy of nearly identical geometries with a tabu condition[6], by forbidding to calculate $U(\vec{X}_n)$ if $\vec{D}(\vec{X}_n) \approx \vec{D}(\vec{X}_j)$, where $\vec{X}_j$ ($j < n$) is an earlier structure. (3) It introduces expert knowledge, mainly through the choice of descriptors. For instance, a descriptor we use for metal clusters is the mean atomic coordination c. Higher values of c normally imply smaller surface area and lower energy. Good descriptors effectively reduce the dimensionality of the problem from ($3n$-6) to 5 or 10 typically, and allow quicker discovery of low energy regions in descriptor space. A very poor choice of descriptors would reduce TSDS to something only marginally better than a random search.

2 Tabu Search in Descriptor Space

Here is a brief summary of the sequence of steps in a TSDS, details can be found elsewhere[7].
1) Input the cluster composition and parameters that control the optimization.
2) Generate many initial random cluster structures.
3) Adjust atomic positions in each initial cluster so that every interatomic distance becomes either equal to d_0 or greater than d_{max}, where d_0 and d_{max} are defined by the user. This makes an initial set $\mathcal{M}$ of clusters.
4) Update a cycle counter variable and the control parameters in such a way that the search progressively becomes more confined to low energy regions of descriptor space.
5) Calculate or get the energy U_i ($i = 1, M$) of all clusters and set U_0=Min$\{U_i\}$.
6) Assign a Boltzmann weight $w_i = exp((U_0 - U_i)/\delta)$ to each cluster
7) Generate many candidate clusters by this procedure:

      7.1) Select one of the M clusters with probability equal to $w_i / \sum_i^M w_i$.
      7.2) Make a candidate cluster by a random geometric modification of cluster i.
      7.3) Calculate the descriptors of the candidate cluster.
      7.4) Calculate a model energy $\tilde{U}$ by interpolation in descriptor space.
8) Select *one* of the candidate clusters, the best, on the basis of how low $\tilde{U}$ is and how dissimilar its descriptors are from those of previous clusters. Calculate the energy for the best cluster "o" and add $(\vec{X}_o, \vec{D}_o, U_o)$ to the main set of clusters.
9) If the cycle counter I is less than I_{max} go to 4, otherwise, go to 10.
10) Put all clusters in increasing order using criteria similar to those in step 8, and do local optimization for each cluster "i" that satisfies two conditions: (a) $U_i < U_0 + U_t$, where U_t is a tolerance, and (b) it is sufficiently different (by descriptors) from all previous structures that underwent local optimization. This final screening typically limits the number of times we rediscover local minima to 3 or 4, and yields 10 to 20 distinct local minima.

3   Lennard-Jones Clusters

We did 40 optimization runs for each n-atom LJ clusters (n =10–40) using TSDS, a random search (RS), and a genetic algorithm (GA), and compared to the best known structures[8]. The parameters that control the RS, GA, and TSDS optimizations were optimized to give the fastest convergence to the global minimum. For clusters with $n \leq 24$ all three methods find the global minimum in at least 80% of the runs. The average number of energy evaluations needed to find the global minimum by TSDS varies from 369 (at $n = 10$) to 735 (at $n = 24$): the number required by GA is, on average, 60 times larger than that, and for RS it is 300 times larger. In the range $25 \leq n \leq 40$, our TSDS runs were too short and often failed to locate the global minimum, whereas GA runs are longer but always succeed in finding the global minimum. In order to allow meaningful comparisons, we looked at both the fraction s of runs where the global minimum is found, and the average number of energy evaluations N required to find the lowest minimum in a run. We define

$$N(0.9) = N \left(ln(1 - 0.9)/ln(1 - s) \right) \tag{1}$$

which is an approximation to the number of energy evaluations required to discover the global minimum with a 0.9 probability. The $N(0.9)$ value is between 5 and 200 times larger in GA than in TSDS over most of the range $26 \leq n \leq 40$, and it is larger still for RS. But there are exceptions: TSDS failed to find the global minimum even once in 40 runs for $n = 35$ and $n = 38$. So it appears that our current implementation of TSDS is very efficient but it is not guaranteed to work in tough cases, whereas GA[4] is slower but more robust.

4   Lithium Clusters Studied by KS-DFT

While we were still developing TSDS, we used a primitive version of it to optimize KS-DFT Li_n (n=7–20) structures. Because of the high computational cost (5 to 100 minutes per energy calculation) we limited the number of energy evaluations for each size to 250 in the global search stage, and roughly 500 in the local optimization stage. For the same reason, we departed from the TSDS algorithm in a few instances by adding specially selected structures. The global minima were discovered by TSDS at most sizes.

Many of the minima we found resemble those reported by Sung *et al.*[9]. The building block for many of these structures is a 9-atom centered square antiprism (CSAP) with D_{4d} symmetry. For example, Li_{11} is found to be a D_{4d} symmetry doubly capped CSAP; Li_{14} is a D_{4d} double CSAP; and Li_{15} is like Li_{14} but with an extra capping atom on the rotation axis. For n=16–20 the structures are somewhat distorted but can be analyzed in terms of fused or capped distorted CSAP building blocks. There are a few exceptions: Li_{13} is an icosahedron, and Li_{10} and Li_{12} contain pentagonal bipyramid motifs like in small LJ clusters. The CSAP structures are reminiscent of the bcc bulk structure of Li and contains two peaks in the pair distribution function at short r, one near 2.60Å and the other near 3.05Å. The ratio 3.05/2.60=1.17 is quite close to the ratio of second and first neighbour distances in bcc, 1.155.

The global minima generally have singlet or doublet ground states, but with some remarkable exceptions: Li_{13} (sextet), Li_{16} (triplet), Li_{17} (quartet), and Li_{18} (triplet). The high-spin state of Li_{13} had been suggested by Gardet and co-workers[10], but the unusual spin states of Li_{16}, Li_{17}, Li_{18} are new. In Li clusters we discovered stable high spin states somewhat by accident, but in magnetic clusters, like Ni_n, it would make sense to use spin like an extra descriptor.

5   Outlook

TSDS is still under development. We obtained encouraging results in structural optimization of LJ[7], Li[11], and Si clusters[12]. But it is too early to give a definite assessment of its applicability.

We are currently using TSDS in structure optimization of KS-DFT Al_n and Be_n, and with an empirical model of covalently bonded clusters. In the future, special attention will have to be given to the choice of descriptors and how they affect performance. Very different descriptors would be needed for optimization of peptides, for instance, the number and position of hydrogen bonds defined in a geometric way. There could be interesting applications of TSDS to alloy clusters, such as Si_xO_y or Ga_xAs_y, for which empirical potentials are often unreliable, and where the standard GA operation[4] for creating new structures may not work because of the requirement of constant chemical composition. By contrast, the geometric operations in TSDS, for example small random displacements of every atom followed by interatomic distance adjustments, are easy to generalize to clusters of any composition. A very different kind of application would be to optimize functions other than energy with respect to both structure and composition. For instance, one could maximize the HOMO-LUMO gap or similar function[13] in a series of related compounds, allowing only limited variations in structure. It is easy to vary chemical composition in TSDS. But a problem where both composition and structure vary freely is obviously much harder than optimizing structure only. However, it is in this kind of computationally difficult problem that hybrid algorithms containing expert knowledge, like TSDS, can be most useful.

References

[1] M. B. Knickelbein Annu Rev Phys Chem **50**, 79 (1999); J. Jortner *Physics and Chemistry of Finite Systems: From Clusters to Crystals* (Kluwer, Netherlands) Vol. 1: 1–17 (1992).

[2] S. F. Chekmarev Phys. Rev. E **64** 036703 (2001); F. H. Stillinger Phys. Rev. E **59** 48 (1999)

[3] R. A. Donnelly Chem Phys Lett **136** 274 (1987).

[4] C. Roberts, R. L. Johnston, N. T. Wilson, Theor. Chem. Acc. **104** 123 (2000); D. M. Deaven and K. M. Ho, Phys. Rev. Lett. **75** 288 (1995). Deaven DM, Tit N, Morris JR, Ho KM (1996) Chem Phys Lett 256: 195

[5] J.A. Northby, J. Chem. Phys. **87** 6166 (1987); P. Amara, J.E. Straub Phys. Rev. B **53** 13857 (1996); J. Pillardy and L. Piela, J. Comp. Chem. **18** 2040 (1997); D.J. Wales and J.P.K. Doye, J. Phys. Chem. A **101** 5111 (1997); H. Leary and J.P.K. Doye, Phys. Rev. E **60** 6320 (1999); D.J. Wales and H.A. Scheraga, Science **285** 1368 (1999); P. Liu and B.J. Berne J. Chem. Phys. **118** 2999 (2003).

[6] D. Cvijovic and J. Klinowski, Science **267** 664 (1995); S.D. Hong and M.S. Jhon MS, Chem. Phys. Lett. **267** 422 (1997).

[7] J.B.Y. Cheng and R. Fournier, Theor. Chem. Acc. **112** 7 (2004).

[8] D.J. Wales, J.P.K. Doye, A. Dullweber, and F.Y. Naumkin, *The Cambridge Cluster Database* http://brian.ch.cam.ac.uk/CCD.html (1997).

[9] M.W. Sung, R. Kawai, and J.H. Weare, Phys. Rev. Lett. **73**, 3552 (1994).

[10] G. Gardet, F. Rogemond, and H. Chermette, J. Chem. Phys. **105**, 9933 (1996).

[11] R. Fournier, J.B.Y. Cheng, and A. Wong, J. Chem. Phys. **119**, 9444 (2003).

[12] J.B.Y. Cheng, MSc Thesis, York University (2002).

[13] P. Jaque and A. Toro-Labbé, J. Phys. Chem. A **104** 995 (2000); J. Aihara, Theor. Chem. Acc. **102** 134 (1998).

VSP International
Science Publishers
P.O. Box 346, 3700 AH Zeist
The Netherlands

Lecture Series on Computer
and Computational Sciences
Volume 1, 2004, pp. 1029-1032

A Quantum Chemical Study of Doped CaCO$_3$ (calcite)

M. Menadakis and G. Maroulis[1]

Department of Chemistry,University of Patras,
GR-26500 Patras, Greece

P. G. Koutsoukos[2]

Department of Chemical Engineering, University of Patras and FORTH-ICEHT,
PO Box 1414 GR 26504 Patras, Greece

Received 7 March, 2004; accepted in revised form 10 March, 2004

Abstract: We have investigated the presence of foreign ions into calcite bulk structure. Four cations isovalent to Ca^{2+} were studied: Mg^{2+}, Sr^{2+}, Ba^{2+} and Zn^{2+}. Our calculations show that the incorporation of these ions into calcite is size dependent. Mg^{2+} and Zn^{2+} can enter into the host lattice while Sr^{2+} and Ba^{2+} cannot at the studied degrees of contamination.

Keywords: Calcite, crystal impurities.

PACS: 61.50.Ah, 61.72.-y

1   Introduction

Calcium carbonate is the most common of all carbonate minerals. It is known to exist abundantly in natural systems and has three polymorphic forms: calcite, aragonite and vaterite. The first two are abundant in geological and biological systems while the third one is the least stable form in ambient conditions. The precise control of crystallization of polymorphs is important to obtain the highly functional crystals for industrial applications. This control may be achieved by structured organic surfaces templates such as self assembled monolayers, biomacromolecules and functionalized polymers[1]. Morphology variations and polymorph selectivity can be also obtained by growth in solution in the presence of growth modifiers such as ions, molecules and polymers.

All these processes involve interactions between the crystal structure (bulk or surface) with foreign ions, molecules and polymers. In order to understand the factors that control crystal growth, dissolution and reactivity there must exist an accurate description of the crystal structure in an atomic scale. There is a number of theoretical and computer simulation studies concerning relevant physical-chemical properties of calcium carbonate. These studies include surface-water interactions, elastic constants, electrical and optical properties.

2   Theory

In the LCAO approxdimation, Crystalline Orbitals are expressed as linear combination of Blöch functions. Blöch functions can be based on localized functions (atomic orbitals) or can be plane

---

[1]Corresponding author. E-mail:maroulis@upatras.gr
[2]E-mail:pgk@chemeng.upatras.gr

waves. In the CRYSTAL98[5] code, which was used for all applications in this work, Blöch functions are based on Gaussian Type Orbitals

$$\psi_i = \sum_\mu \alpha_{\mu,i}(\mathbf{k})\phi_\mu(\mathbf{r};\mathbf{k}) \tag{1}$$

$$\phi_\mu(\mathbf{r};\mathbf{k}) = \sum_T \varphi_\mu(\mathbf{r} - \mathbf{A}_\mu - \mathbf{T})e^{i\mathbf{k}\cdot\mathbf{T}} \tag{2}$$

where A_μ defines the nucleus coordinates in the reference cell on which φ_μ is centered. If translational symmetry is taken into account and Blöch functions are used to form representative sets, which are the bases for the irreducible representation of the translation group, the problem is factorized into n problems of dimension m where n is the number of the irreducible representations of the translational group and m is the number of basis functions in the unit cell[2].

The experimental structural data for calcite were taken from Markgraf[3]. The basis adopted for calcite is that used by Catti[4]. The outer sp and d orbitals were optimized in the crystalline environment. All calculations were performed at the HF level of theory. The irreducible part of the Brillouin zone was sampled at 32 points using a shrinking factor of 6. Other computational parameters such as the overlap and the penetration tolerance for the coulomb and exchange integrals were kept to their default values[5]. The basis sets used for the substitutional cations were among those implemented in the CRYSTAL98 code and have been tested in various crystalline compounds.

3   Results

We have used the supercell approach to study the incorporation of foreign ions into the calcite bulk structure. As a consequence all symmetry operations with translational components were removed thus reducing the symmetry and increasing the computational cost. Four cations isovalent to Ca^{2+} were use as impurities, namely Mg^{2+}, Sr^{2+}, Ba^{2+} and Zn^{2+}. The exact position of the impurity ions was the Ca site. The degree of contamination ranged down to 25%. All structures were allowed to relax by changing the position of the oxygen atom. Table 1 summarizes the effect of the presence of a foreign ion into calcite.

Table 1: Energetic and structural data for the substitution of Ca^{2+} of calcite with a foreign isovalent cation. Unrelaxed and Relaxed refer to the defect formation energy of the corresponding structures given in eV. X-O is the distance in Å between the substitutional ion and the O atom in the relaxed structure. ΔR is the difference in X-O distance between the relaxed and unrelaxed structure.

	Mg		Sr		Ba		Zn	
	50%	25%	50%	25%	50%	25%	50%	25%
Unrelaxed	-2.8744	-2.8740	2.3005	2.3011	6.4686	6.4707	-3.2430	-3.2428
Relaxed	-3.3507	-3.4278	1.9887	1.9775	4.3665	4.3068	-3.6994	-3.7182
X-O	2.2467	2.2457	2.4717	2.4689	2.7161	2.7173	2.2597	2.2559
ΔR	-0.1252	-0.1262	0.0998	0.0970	0.3442	0.3454	-0.1122	-0.1160

It is obvious from table 1 that the substitution of a Ca atom by Mg^{2+} and Zn^{2+} is energetically favourable. For Mg^{2+} this could be expected since the existance of dolomites may suggest that type of substitution. On the other hand the addition of Mg^{2+} ions into a growth solution lead in the formation of aragonite. At this point it must be noticed that the degrees of contamination are relative high. The reported in the literature [6] structures of biogenic Mg-calcites contain 6-13% Mg^{2+}. In addition as the percent of Mg^{2+} in the host lattice change there is also a change in the size of the cell, which in our calculation was kept fixed. All these could explain the overestimated Mg-O distance (approximately 2.10Å in MgCO$_3$ and CaMg(CO$_3$)$_2$).

The incorporation of Zn^{2+} cation into bulk calcite has been confirmed by Reeder[7] using XAFS spectroscopy. The observed Zn-O distance in their experiments is 2.14-2.15Å to be compared with our calculated value of 2.25Å corresponding to 25% of contamination. For Sr^{2+} and Ba^{2+} our calculations show that they cannot incorporate into bulk calcite. Contrary to our results Reeder[7] showed that Ba^{2+} can enter into calcite lattice occupying Ca site. The Ba-O distance observed is 2.68Å which compares well to our calculated value. So the positive value of the defect formation energy could be attributed to relatively small Ba-Ba distance compared to the large size of Ba^{2+} cation. In fact table 1 suggest that the incorporation of foreign ions into calcite is size dependent. This is evident in figs 1 and 2 where the effect of the cation size to the defect formation energy and the local environment is clear.

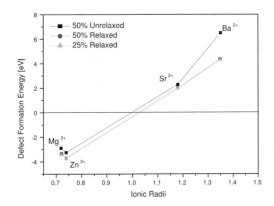

Figure 1: Defect formation energy versus ionic radii for the doped calcite.

Ions smaller than Ca^{2+} like Mg^{2+} and Zn^{2+} (ionic radii 0.72Å and 0.74Å) can be incorporated into the host lattice while larger ions like Sr^{2+} and Ba^{2+} (ionic radii 1.18Å and 1.35Å) cannot. Also the key factor that controls the distortion of the lattice and the local environment around the impurity is the cation size. However in order to have the complete picture of the modes of incorporation lower degrees of contamination and variation of the size of the cell have to be considered.

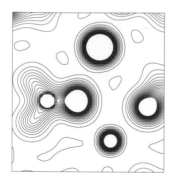 

Figure 2: Left: Electron density map on the Ca-Mg-O plane of a relaxed Mg-doped calcite. Right: Electron density map on the Ca-Ba-O plane of a relaxed Ba-doped calcite. Contour lines [0.005,0.5] ea_0^{-3} are separated by 0.025 ea_0^{-3}.

Acknowledgment

M. Menadakis gratefully acknowledges a scholarship form the Institute of Chemical Engineering and High Temperature Chemical Processes of the Foundation of Research and Technology-Hellas (FORTH/ICE-HT).

References

[1] Y. J. Han and J. Aizemberg, Effect of Magnesium Ions on Oriented Growth of Calcite on Carboxylic Acid Functionalized Self Assembled Monolayers, *JACS* **125** 4032-4033(2003).

[2] C. Picani, *Quantum-Mechanical Ab-initio calculation of the properties of Crystalline Materials*, Springer-Verlag, Berlin Heidelberg, 1996.

[3] S. A. Markgraf R. J. Reeder, High-temperature structure refinements of calcite and magnesite, *Am. Mineral.* **70** 590-600(1985).

[4] M. Catti A. Pavese E. Apra and C. Roetti, Quantum Mechanical Hartree-Fock Study of Calcite ($CaCO_3$) at Variable Pressure and Comparison With Magnesite ($MgCO_3$), *Physics and Chemistry of Minerals* **20** 104-110(1993).

[5] V. R. Saunders R. Dovesi C. Roetti M. Causa N. M. Harrison R. Orlando C. M. Zicovich-Wilson, *CRYSTAL98 User's Manual*, Torino: Universita di Torino 1999.

[6] J. Paquette and R. J. Reeder, Single-crystal X-ray structure refinements of two biogenic magnesian calcite crystals, *Amer. Mineral.* **75** 1151-1158(1990).

[7] R. J. Reeder G. M. Lamble and P. A. Northrup, XAFS study of the coordination and local relaxation around $Co^{2+}, Zn^{2+}, Pb^{2+}$ and Ba^{2+} trace elements in calcite, *Amer. Mineral.* **84** 1049-1060(1999).

VSP International
Science Publishers
P.O. Box 346, 3700 AH Zeist
The Netherlands

Lecture Series on Computer
and Computational Sciences
Volume 1, 2004, pp. 1033-1036

Density-functional-based methods for calculations of intermolecular forces

Krzysztof Szalewicz[1] and Rafal Podeszwa

Department of Physics and Astronomy, University of Delaware, Newark, Delaware 19716

Alston Misquitta

Cambridge University, Chemical Laboratories, Cambridge, CB2 1EW England

Bogumil Jeziorski

Department of Chemistry, University of Warsaw, Pasteura 1, 02-093 Warsaw, Poland

Abstract: Ab initio wave-function-based methods can predict intermolecular force fields very accurately for monomers containing a few atoms. These methods are, however, too time consuming for larger molecules, in particular for any molecules of biological interest or molecules forming energetic solid state materials. The density-functional theory (DFT) methods would be fast enough, but are currently not able to predict the very important dispersion component of the force field. A new perturbational method that is capable of computing the force fields accurately will be presented. This method is based on a DFT description of isolated molecules but computes intermolecular forces using expressions beyond DFT. Calculations for model compounds have shown that the new method reproduces all components of the intermolecular force, including dispersion, extremely well, in fact challenging the accuracy of wave-function-based methods. At the same time, the computer resources required by this method are similar to those of the standard DFT. The method has already been applied to interactions of two monomers containing 12 atoms each, and it should be able to handle interactions of molecules containing 20 and more atoms.

Keywords: Intermolecular forces; Density-functional theory; Dispersion energy; Perturbation theory

PACS: 34.20.Gj

Symmetry-adapted perturbation theory (SAPT) provides not only a conceptual basis for understanding intermolecular interactions but also an efficient computational framework for accurate predictions of interaction energies [1, 2]. The pair interactions and three-body nonadditive interactions of arbitrary closed-shell molecules can now be computed using the program SAPT2002 [3]. This approach has provided some of the most accurate intermolecular potentials for various dimers and trimers, as confirmed by comparisons of the computed spectra with experiment. In particular, the water dimer and trimer spectra agreed with experiment very well [4] and the predictions of the SAPT potential are competitive to those provided by empirical potentials fitted to the spectra [5]. Simulations of liquid water with SAPT potentials [6, 7] provided the first quantitative determination of the role that the three-body effects play in this system.

---

[1] e-mail: szalewic@udel.edu

Applications of wave-function-based *ab initio* methods to interactions of molecules containing ten or more atoms have not been possible since calculations employing SAPT or any other electronic structure method that includes correlation effects at a level adequate for describing intermolecular interactions require relatively significant computer resources. On the other hand, although the existing density-functional theory (DFT) methods are fast enough for such calculations, these methods are known to fail to describe an important part of the van der Waals forces, the dispersion interaction. In fact, two of us have shown [8] that supermolecular DFT calculations lead to large errors also in other interaction energy components (the electrostatic, induction, and exchange interactions) due to an incorrect behavior of electron densities at distances from nuclei that are relevant for intermolecular interactions.

Table 1: Interaction energies (in kcal/mol) for the DMNA dimer at a near minimum geometry.

Hartree-Fock	2.25
frozen-core	
MP2	-7.90
MP4 (SDTQ)	-7.85
CCSD	-5.31
CCSD(T)	-6.85
full-core	
CCSD(T)	-6.86
SAPT	-7.36
SAPT(DFT)/PBE0	-6.22
SAPT(DFT)/B97-2	-6.56

It has recently been shown that a solution to this difficulty is a SAPT approach utilizing the DFT description of monomers [8, 9]. Such a method, now called SAPT(DFT), has been first proposed by Williams and Chabalowski [10] and later developed by some of the present authors [8, 9] and independently by Hesselmann and Jansen [11]. The method does not relay on asymptotic expansions and therefore is applicable for all separations between the interacting molecules. The SAPT(DFT) approach avoids the problems of supermolecular DFT by using this method only to describe each monomer, but calculating the interaction energies from expressions beyond DFT. In addition, the wrong long-range behavior of monomer densities is fixed by applying an asymptotic correction to the exchange-correlation potential of DFT. SAPT(DFT) calculations require only a small fraction of computer resources used by the regular SAPT and converge much faster in the size of the basis sets. Moreover, although initially SAPT(DFT) was expected to be a method providing medium quality results for very large molecules, it turned out that at least in some cases the accuracy of SAPT(DFT) surpasses that which can be reached with the currently programmed regular SAPT and basis sets of a reasonable size. Our most recent results for several dimers show that in all cases when there were significant discrepancies between the results from the two approaches, these were resolved in favor of SAPT(DFT), i.e., were resulting from theory level truncations and basis set incompleteness in the regular SAPT calculations. All the individual physical components of the intermolecular force, including the dispersion energy, have been reproduced by SAPT(DFT) very well.

The initial SAPT(DFT) calculations were performed for fairly small systems, such as He, Ne, H_2O, and CO_2 homogenous dimers. Recently, we were able to compute interaction energies for the dimethylnitramine (DMNA) dimer containing 24 atoms. DMNA is an important model compound for energetic materials and was investigated by SAPT in the past [12]. In Table 1, we show the

Table 2: Individual components of the interaction energy for SAPT and SAPT(DFT) with PBE0 and B97-2 functionals. Energies in kcal/mol. The value in parentheses following the SAPT dispersion energy is this quantity computed with neglect of intramonomer correlation effects.

Component	SAPT		PBE0	B97-2
electrostatic	-10.51		-10.25	-10.05
1-st order exchange	18.28		17.43	16.85
induction	-6.07		-6.54	-6.35
exchange-induction	4.49		5.02	4.82
dispersion	-13.50	(-12.83)	-11.83	-11.75
exchange-dispersion	1.34		1.34	1.30
δ^{HF}	-1.38		-1.38	-1.38
total	-7.36		-6.22	-6.56

total interaction energies at a near minimum geometry computed using SAPT, SAPT(DFT), and several supermolecular methods. We have used the M1 geometry defined in Table 3 of Ref. [12] and the basis set was also taken from that reference. This basis set is of double-zeta quality and includes bond functions. The monomer-centered "plus" basis sets (MC$^+$BS) were used in the SAPT and SAPT(DFT) calculations, whereas the dimer-centered "plus" basis sets (DC$^+$BS) were used in the supermolecular calculations. The latter calculations were performed in the counterpoise corrected way. The "plus" denotes the use of bond function in both approaches and in the MC$^+$BS case also the use of the isotropic part of the basis set of the interacting partner. The supermolecular calculations employ the many-body perturbation theory with the Møller-Plesset decomposition of the Hamiltonian (results denoted by MP) or the coupled-cluster (CC) methods with various levels of electron excitations: single (S), double (D), triple (T), quadruple (Q). The regular SAPT results given in Table 1 employ the complete standard set of corrections, in contrast to the calculations of Ref. [12] which used the sum of the Hartree-Fock interaction energy and of the dispersion energy with neglect of intramonomer correlation effects. Table 1 shows first that the higher-order terms neglected in Ref. [12] are quite important and decrease the magnitude of the interaction energy by more than 3 kcal/mol. SAPT(DFT) gives interaction energies within about 1 kcal/mol of the regular SAPT and about 0.5 kcal/mol of the CCSD(T) method, the most advanced of practically applicable electronic structure approaches. This is an excellent agreement taking into account that both the regular SAPT and CCSD(T) are much more computer intensive than SAPT(DFT). The SAPT(DFT) calculations were performed with two very different functionals: PBE0 [15, 16] and B97-2 [13, 14], which give results within 0.3 kcal/mol of each other, showing that SAPT(DFT) is only weakly dependent on the choice of the functional.

The framework of SAPT provides insights into the physical structuture of the interaction energy. In Table 2, the individual components of the interaction energy: the electrostatic, induction, dispersion, and exchange contributions are shown. The interaction energy at the same level as used in Ref. [12] can be obtained by adding the Hartree-Fock interaction energy from Table 1 and the dispersion energy with neglect of intramonomer correlation effects listed in Table 2. The sum of these two quantities, equal to -10.58 kcal/mol, differs from the minimum energy of -11.06 kcal/mol given in Table 3 of Ref. [12] since the latter result was computed in the DC$^+$BS approach whereas the present results used the MC$^+$BS scheme. It can be seen that, as already pointed out in Ref. [12], our results do not support the conventional description of interactions of large molecules which includes only the electrostatic component. Clearly, the first-order exchange and the dispersion energies are actually larger in magnitude than the electrostatic interactions. An attempt to described the DMNA dimer at the Hartree-Fock level, as it is often done for large molecules, would lead to completely wrong conclusions as the interaction energy at this level is positive. One can see in Table 2 a generally good agreement between the individual SAPT and SAPT(DFT) components. If the findings of Ref. [17] extend to the DMNA dimer, the SAPT(DFT) components can actually be more accurate than the SAPT ones.

This research was supported by grants from the Army Research Office and the National Science Foundation.

References

[1] B. Jeziorski, R. Moszynski, and K. Szalewicz, Chem. Rev. **94**, 1887 (1994).

[2] B. Jeziorski and K. Szalewicz, in *Handbook of Molecular Physics and Quantum Chemistry*, edited by S. Wilson, Wiley, 2002, Vol. 3, Part 2, Chap. 9, p. 232.

[3] *SAPT2002: An Ab Initio Program for Many-Body Symmetry-Adapted Perturbation Theory Calculations of Intermolecular Interaction Energies* by R. Bukowski *et al.*, University of Delaware and University of Warsaw: http://www.physics.udel.edu/~szalewic/SAPT/SAPT.html.

[4] G. C. Groenenboom, E. M. Mas, R. Bukowski, K. Szalewicz, P. E. S. Wormer, and A. van der Avoird, Phys. Rev. Lett. **84**, 4072 (2000).

[5] F.N. Keutsch, N. Goldman, H.A. Harker, C. Leforestier, and R.J. Saykally, Mol. Phys. **101**, 3477 (2003).

[6] E. M. Mas, R. Bukowski, and K. Szalewicz, J. Chem. Phys. **118**, 4386 (2003).

[7] E. M. Mas, R. Bukowski, and K. Szalewicz, J. Chem. Phys. **118**, 4404 (2003).

[8] A.J. Misquitta and K. Szalewicz, Chem. Phys. Lett. **357**, 301 (2002).

[9] A.J. Misquitta, B. Jeziorski, and K. Szalewicz, Phys. Rev. Lett. **91**, 033201 (2003).

[10] H.L. Williams and C.F. Chabalowski, J. Phys. Chem. A **105**, 646 (2001).

[11] A. Hesselmann and G. Jansen, Chem. Phys. Lett. **357**, 464 (2002); *ibid.* **362**, 319 (2002); *ibid.* **367**, 778 (2003).

[12] R. Bukowski, K. Szalewicz, and C. Chabalowski J. Phys. Chem. A, **103**, 7322 (1999).

[13] A.D. Becke, J. Chem. Phys. **107**, 8554 (1997).

[14] P.J. Wilson, T.J. Bradley, and D.J. Tozer, J. Chem. Phys. **115**, 9233 (2001).

[15] J.P. Perdew, K. Burke, and M. Ernzerhof, Phys. Rev. Lett. **77**, 3865 (1996).

[16] C. Adamo, M. Cossi, and V. Barone, J. Mol. Struct. (Theochem) **493**, 145 (1999).

[17] A.J. Misquitta *et al.*, to be published.

VSP International
Science Publishers
P.O. Box 346, 3700 AH Zeist
The Netherlands

Lecture Series on Computer
and Computational Sciences
Volume 1, 2004, pp. 1037-1041

Computational challenges in the determination of structures and energetics of small hydrogen-bonded clusters

Ajit J. Thakkar[1]

Department of Chemistry,
University of New Brunswick,
Fredericton, New Brunswick E3B 6E2,
Canada

Received 25 July, 2004

Abstract: Our laboratory has been involved in the determination of the structures and energetics of small hydrogen-bonded clusters by various quantum chemical methods including semiempirical, density functional, and ab initio Møller-Plesset perturbation theory and coupled cluster methods. Clusters involving formic acid, nitric acid, glycolic acid and water molecules have been studied. The computational challenges encountered, the strategies used to face them, and some of the results obtained are surveyed.

Keywords: Hydrogen-bonded clusters, energy minimization, Gaussian-type functions, Hartree-Fock method, density functional theory, Møller-Plesset perturbation theory, coupled cluster approach.

PACS: 36.40.-c, 31.25.Qm, 31.15.Ew

1   Cluster chemistry

The properties of a piece of bulk crystal do not change dramatically as we repeatedly subdivide it until the piece reaches molecular dimensions or, in other words, the nanometer scale. Particles of a material consisting of a few to a few thousand atoms are called clusters. The properties of clusters often show dramatic size and shape dependence. Clusters of metals, semiconductors, ionic solids, rare gases, and small molecules have been studied using both theoretical and experimental methods. Intense interest in clusters arises because they can be used to investigate surface properties including mechanisms of heterogeneous catalysis [1], and because clusters can serve as building blocks for new materials and electronic devices. An outstanding example of the fruits of cluster chemistry is fullerene chemistry [2, 3] which grew out of the study of carbon clusters, and has become important in nanotechnology [4]. Recent books on clusters include a monograph on metal clusters [5], and edited collections on molecular clusters [6] and metal nanoparticles [7] .

Molecular clusters are held together by relatively weak intermolecular forces [8, 9] or by hydrogen bonds [10]. Hydrogen-bonded clusters are an important class of molecular clusters. Small clusters of water molecules have received a lot of attention; see, for example, the experimental work of Saykally and coworkers [11], and the computational investigations of Xantheas and coworkers [12]. We have studied several types of hydrogen-bonded clusters as outlined next.

---

[1] E-mail: ajit@unb.ca, Web page: http://www.unb.ca/chem/ajit/

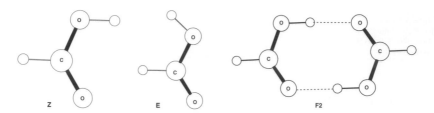

Figure 1: The Z and E rotamers of formic acid, and the gas-phase dimer structure F2.

2 Hydrogen-bonded clusters

Formic acid is present in clouds and fog, and plays an important role in human metabolism. Formic acid exhibits rotational isomerism between the Z and E forms shown in Fig. 1. Both rotamers have been well characterized through spectroscopic techniques. Both experimental and theoretical studies indicate that the Z rotamer is more stable than the E rotamer by about 4.0 kcal/mol [13, 14]. The structure of formic acid differs markedly in different phases. In the gas phase, formic acid forms the cyclic C_{2h} dimer, depicted as D in Fig. 1, with two strong, nearly linear, equivalent O-H$\cdots$O=C H-bonds as shown by spectroscopic [15, 16, 17] and quantum chemical methods [18, 19]. Like acetic and glycolic acid but unlike many other carboxylic acids whose crystal structures consist of associated dimers, formic acid crystallizes in long catameric chains [20, 21] in which H-bonds link each molecule to two neighbors. Chains of the Z form are found in the low temperature (4.5°K) crystal structure [21] whereas chains of the E form are found at higher temperatures [20]. The structure of liquid formic acid remains a subject of debate but probably consists of short chains similar to those observed in the solid [22]. Hence it is of some interest to determine how the structures of small clusters of formic acid molecules evolve with cluster size. Roy, McCarthy and I have performed density functional theory (DFT) computations to study the structures of trimers [23], tetramers [24], pentamers [25] and hexamers [26] of formic acid.

Glycolic acid, $CH_2OHCOOH$, plays an important role in dermatology and the cosmetics industry [27, 28], and it is involved in several life processes [29]. The glycolic acid molecule has rich functionality that allows it to simultaneously form intra- and intermolecular O-H$\cdots$O hydrogen bonds, and also allows for weaker C-H$\cdots$O interactions. The growth of clusters of glycolic acid is of interest for the same reasons as apply to clusters of formic acid. Kassimi and I have investigated the competition between various types of hydrogen bonding in dimers of glycolic acid [30]. Raman and infrared spectroscopy studies [31] suggest that glycolic acid exists in a monomeric form in dilute aqueous solution. A first step towards understanding such a solution is to consider small gas-phase clusters consisting of a glycolic acid molecule and a few water molecules. My group has performed semiempirical, DFT, Møller-Plesset perturbation theory and coupled cluster computations on clusters of glycolic acid with 1–6 [32, 33], and 16 and 28 water molecules [34].

Nitric acid, HNO_3, is widely used to manufacture explosives such as nitroglycerin and fertilizers such as ammonium nitrate. Nitric acid plays an important role in atmospheric chemistry because it acts as a stratospheric reservoir for NO_x. Hart and I have examined the structural isomers of the dimers of nitric acid [35] with DFT and Møller-Plesset perturbation theory.

The computational challenges involved in these studies include the choice of computationally efficient but sufficiently accurate quantum chemical methods and basis sets. Since the number of structural isomers grows exponentially with cluster size, a major challenge is to find all low-lying structures including the global minimum. The purpose of this talk is to survey the strategies we have adopted to face these challenges, and some of the results we have obtained. As an example of

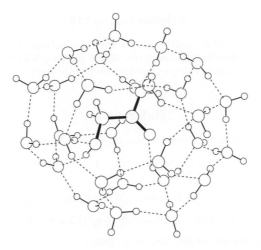

Figure 2: A quasi-spherical, structural isomer of a cluster of glycolic acid and 28 water molecules.

our results, a quasi-spherical isomer of a cluster of glycolic acid and 28 water molecules is shown in Figure 2. It fits standard conceptions of a solvated glycolic acid molecule. A stacked isomer of a cluster of glycolic acid and 28 water molecules is shown in Figure 3. It can be described as a glycolic acid molecule attached to the side of a small crystal of ice made up of a stack of cubic arrangements of water molecules!

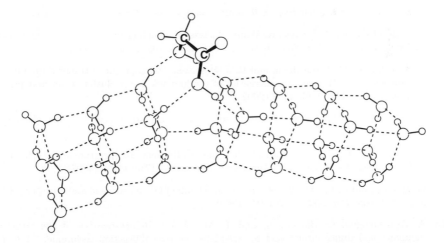

Figure 3: A stacked, structural isomer of a cluster of glycolic acid and 28 water molecules.

Acknowledgments

This lecture would not have been possible without the contributions of my coworkers: Amlan K. Roy, Noureddin El Bakali Kassimi, Shaowen Hu, James R. Hart, Edet F. Archibong and Shane P. McCarthy. I thank the Natural Sciences and Engineering Research Council of Canada for their continuing support.

References

[1] G. Ertl and H. J. Freund. Catalysis and surface science. *Physics Today* **52**, 32–38 (1999).

[2] R. Taylor. *Lecture Notes on Fullerene Chemistry: A Handbook for Chemists* (Imperial College, London, 1999).

[3] A. Hirsch. *The Chemistry of the Fullerenes* (Wiley-VCH, New York, 2002).

[4] E. Osawa. *Perspectives of Fullerene Nanotechnology* (Kluwer Academic, New York, 2002).

[5] W. Ekardt. *Metal Clusters* (Wiley, New York, 1999).

[6] M. Driess and H. Nöth (Eds.). *Molecular Clusters of the Main Group Elements* (Wiley, New York, 2004).

[7] D. L. Feldheim and C. A. Foss, Jr. (Eds.). *Metal Nanoparticles* (Marcel Dekker, New York, 2001).

[8] A. J. Stone. *The Theory of Intermolecular Forces* (Oxford, New York, 1996).

[9] A. J. Thakkar. Intermolecular interactions. In *Encyclopedia of Chemical Physics and Physical Chemistry*, (eds.) J. Moore and N. Spencer (Institute of Physics Publishing, Bristol, 2001), vol. I. Fundamentals, chap. A1.5, pp. 161–186.

[10] S. Scheiner. *Hydrogen bonding: A theoretical perspective* (Oxford, New York, 1997).

[11] F. N. Keutsch and R. J. Saykally. Water clusters: Untangling the mysteries of the liquid, one molecule at a time. *Proc. Nat. Acad. Sci. U.S.A.* **98**, 10533–10540 (2001).

[12] S. S. Xantheas, C. J. Burnham, and R. J. Harrison. Development of transferable interaction models for water. II. Accurate energetics of the first few water clusters from first principles. *J. Chem. Phys.* **116**, 1493–1499 (2002).

[13] W. H. Hocking. The other rotamer of formic acid, cis-formic acid. *Z. Naturforsch.* **31A**, 1113–1121 (1976).

[14] M. Pettersson, J. Lundell, L. Khriachtchev, and M. Räsänen. IR spectrum of the other rotamer of formic acid, *cis*-HCOOH. *J. Am. Chem. Soc.* **119**, 11715–11716 (1997).

[15] G. H. Kwei and R. F. Curl, Jr. Microwave spectrum of O^{18} formic acid and structure of formic acid. *J. Chem. Phys.* **32**, 1592–1594 (1960).

[16] A. Almenningen, O. Bastiansen, and T. Motzfeldt. Reinvestigation of the structure of monomer and dimer formic acid by gas-phase electron diffraction technique. *Acta Chem. Scand.* **23**, 2848–2864 (1969).

[17] A. Almenningen, O. Bastiansen, and T. Motzfeldt. Influence of deuterium substitution on the hydrogen bond of dimer formic acid. *Acta Chem. Scand.* **24**, 747–748 (1970).

[18] L. Turi. Ab initio molecular orbital analysis of dimers of cis-formic acid. Implications for condensed phases. *J. Phys. Chem.* **100**, 11285–11291 (1996).

[19] W. Qian and S. Krimm. Spectroscopically determined molecular mechanics model for the intermolecular interactions in hydrogen-bonded formic acid dimer structures. *J. Phys. Chem. A* **105**, 5046–5053 (2001).

[20] I. Nahringbauer. Hydrogen bond studies: CXXVII. A reinvestigation of the structure of formic acid (98K). *Acta Crystallogr. B* **34**, 315–318 (1978).

[21] A. Albinati, K. D. Rouse, and M. W. Thomas. Neutron powder diffraction analysis of hydrogen-bonded solids. II. Structural study of formic acid at 4.5K. *Acta Crystallogr. B* **34**, 2188–2190 (1978).

[22] P. Jedlovszky, I. Bakó, G. Pálinkas, and J. C. Dore. Structural investigation of liquid formic acid. X-ray and neutron diffraction, and reverse Monte Carlo study. *Mol. Phys.* **86**, 87–105 (1995).

[23] A. K. Roy and A. J. Thakkar. Structures of the formic acid trimer. *Chem. Phys. Lett.* **386**, 162–168 (2004).

[24] A. K. Roy and A. J. Thakkar. Formic acid tetramers: A structural study. *Chem. Phys. Lett.* **393**, 347 354 (2004).

[25] A. K. Roy and A. J. Thakkar. Pentamers of formic acid. *To be published* (2005).

[26] A. K. Roy, S. P. McCarthy, and A. J. Thakkar. Hexamers of formic acid. *To be published* (2005).

[27] L. S. Moy, K. Howe, and R. L. Moy. Glycolic acid modulation of collagen production in human skin fibroblast cultures in vitro. *Dermatal. Surg.* **22**, 439–441 (1996).

[28] R. G. Males and F. G. Herring. A ^{1}H-NMR study of the permeation of glycolic acid through phospholipid membranes. *Biochim. Biophys. Acta-Biomembr.* **1416**, 333–338 (1999).

[29] I. Zelitch. Plant respiration. In *McGraw-Hill Encyclopedia of Science and Technology* (McGraw-Hill, New York, 1992), vol. 13, pp. 705–710. 7th ed.

[30] N. E.-B. Kassimi, E. F. Archibong, and A. J. Thakkar. Hydrogen bonding in the glycolic acid dimer. *J. Mol. Struct. (Theochem)* **591**, 189–197 (2002).

[31] G. Cassanas, M. Morssli, E. Fabreque, and L. Bardet. Étude spectrale de l'acide glycolique, des glycolates et du processus de polymérisation. (Spectral study of glycolic acid, glycolates and of the polymerization process). *J. Raman Spectrosc.* **22**, 11–17 (1991).

[32] A. J. Thakkar, N. E.-B. Kassimi, and S. Hu. Hydrogen-bonded complexes of glycolic acid with one and two water molecules. *Chem. Phys. Lett.* **387**, 142–148 (2004).

[33] A. K. Roy, S. Hu, and A. J. Thakkar. Clusters of glycolic acid with three to six water molecules. *To be published* (2004).

[34] A. K. Roy, J. R. Hart, and A. J. Thakkar. Clusters of glycolic acid with 16 and 28 water molecules. *To be published* (2005).

[35] J. R. Hart and A. J. Thakkar. Nitric acid dimer structures. *To be published* (2004).

VSP International
Science Publishers
P.O. Box 346, 3700 AH Zeist
The Netherlands

Lecture Series on Computer
and Computational Sciences
Volume 1, 2004, pp. 1042-1045

Numerical solution of the Schrödinger equation for two-dimensional double-well oscillators

Amlan K. Roy, Ajit J. Thakkar[1]

Department of Chemistry,
University of New Brunswick,
Fredericton, New Brunswick E3B 6E2,
Canada

B. M. Deb

Department of Chemistry,
Panjab University,
Chandigarh 160 014, India

Received 27 July, 2004

Abstract: Wave functions, energies and selected expectation values of the low-lying stationary states of two-dimensional double well potentials are obtained from the long-time solutions of the corresponding time-dependent Schrödinger equation. The latter is transformed to a diffusion-like equation which is then solved by an alternating-direction, implicit, finite-difference method.

Keywords: Double-well oscillators, two-dimensional Schrödinger equation, numerical solution, finite differences.

PACS: 36.40.-c, 31.25.Qm, 31.15.Ew

1 Introduction

A double-well oscillator is described by a potential function that has two minima separated by a barrier. Problems which are modeled with the help of double-well potentials include the inversion of ammonia, tunneling of protons in hydrogen bonded systems, structural phase transitions, and quantum coherence in Josephson junction superconductors. Thus, it is not surprising that one-dimensional quantum systems with double-well potentials, particularly the anharmonic potential function $V(x) = -Z^2 x^2 + \lambda x^4$, have been studied extensively.

Relatively little work has been done on double-well potentials in two and three dimensions. Progress has been made by Witwit and coworkers [1, 2, 3, 4, 5] on the computation of energy levels for such potentials but wave functions and properties other than the energy remain unexamined to our knowledge. In this work we examine energies, wave functions and properties of the three lowest states of the two-dimensional double-well potential given by

$$V(x,y) = -Z_x^2 x^2/2 - Z_y^2 y^2/2 + \lambda(a_{xx}x^4 + 2a_{xy}x^2 y^2 + a_{yy}y^4)/2. \qquad (1)$$

Atomic units ($\hbar = m_e = e = 1$) are used throughout.

---

[1]Presenting author. E-mail: `ajit@unb.ca`, Web page: `http://www.unb.ca/chem/ajit/`

2 Method

The quantities of interest are solutions of the time-independent Schrödinger equation

$$\hat{H}\phi_{m,n}(x,y) = E_{m,n}\phi_{m,n}(x,y) \tag{2}$$

where the time-independent Hamiltonian is given by

$$\hat{H} = -\frac{1}{2}\left(D_x^2 + D_y^2\right) + V(x,y) \tag{3}$$

in which $D_x^2 = \partial^2/\partial x^2$, $D_y^2 = \partial^2/\partial y^2$, the potential V is given by Eq. (1), and two quantum numbers $\{m,n\}$ are used to label the solutions of the two-dimensional Schrödinger equation (2). In this work, we find the stationary state wave functions $\phi_n(x,y)$ as long-time limits of solutions of the time-dependent Schrödinger equation:

$$\hat{H}\psi(x,y;t) = i\,\frac{\partial\psi(x,y;t)}{\partial t} \tag{4}$$

where $\hat{H}$ is the time-independent Hamiltonian of Eq. (3). As in the work of Anderson [6], we write Eq. (4) in imaginary time τ, substitute $\tau = -it$, and let $D_t = \partial/\partial t$, to obtain a diffusion-like equation in two space dimensions

$$\hat{H}\psi(x,y;t) = -D_t\psi(x,y;t) \tag{5}$$

or simply $\hat{H} = -D_t$ in operator form. Eq. (5) resembles a diffusion quantum Monte Carlo equation. One may express $\psi(x,y;t)$ as [7]

$$\psi(x,y;t) = C_{0,0}\phi_{0,0}(x,y) + \sum_{m,n>0}^{\infty} C_{m,n}\phi_{m,n}(x,y)\,e^{-(E_{m,n}-E_{0,0})t} \tag{6}$$

from which it is apparent that

$$\lim_{t\to\infty}\psi(x,y;t) = C_{0,0}\phi_{0,0}(x,y) \tag{7}$$

so that numerically propagating $\psi(x,y;t)$ to a sufficiently long time t will give us the ground state time-independent wave function apart from a normalization constant. Expectation values of properties, including the energy, can be obtained as mean values of the pertinent operator $\hat{A}$:

$$\langle A\rangle = \lim_{t\to\infty}\langle\psi(x,y;t)|\,\hat{A}\psi(x,y;t)\rangle \tag{8}$$

where the angular brackets indicate integration over the entire domain of the spatial variables $\{x,y\}$. Excited states can be treated in the same way provided that one ensures they stay orthogonal to all lower states at each time step.

A finite-difference method is developed for the numerical solution of Eq. (5) using Peaceman-Rachford splitting [8]. This is an unconditionally stable and convergent method that falls in the category of alternating-direction, implicit, finite-difference methods [9].

In the method, if the spatial grid consists of $N_x \times N_y$ points, then each time step requires the solution of $N_y - 2$ sets of tridiagonal linear equations of dimension $N_x - 2$ followed by the solution of $N_x - 2$ sets of tridiagonal linear equations of dimension $N_y - 2$. A standard LU decomposition technique [10] is used for the solution of the tridiagonal systems. $N_x = N_y = 1951$ was used in our final computations. The energy calculated at each time step is used to monitor convergence of the time-dependent wave function to the desired stationary state wave function. The number of time steps required for convergence varied from a few hundred to a few thousand. Details of the method will be presented in the lecture and will also be published in full elsewhere.

3   Sample Results

Our results are illustrated in Figure 1 for two choices of the potential parameters. Both potentials have four wells separated by ridges near the perimeter, and a large maximum centered at the origin. The wells on the left are relatively shallow and the ground state wave function is peaked at

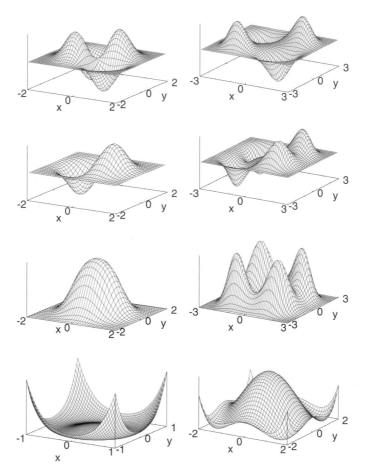

Figure 1: The bottom row shows potentials and the next three rows show the corresponding wave functions for the $(0,0)$, $(1,0)$ and $(1,1)$ states respectively as one moves up the page. The left panels correspond to $Z_x^2 = Z_y^2 = \lambda = 5$, and $a_{xx} = a_{yy} = a_{xy} = 1$. The right panels correspond to $Z_x^2 = Z_y^2 = 10$, $\lambda = 3/2$, $a_{xx} = a_{yy} = 1$, and $a_{xy} = 1/2$.

the origin, whereas the wells on the right are relatively deep and so show localization at the wells manifested as four maxima in the ground state wave function. This trend continues in the $(1,0)$ state with only two extrema seen on the left but four on the right. In the $(1,1)$ state, we see that the probability density for crossing between the peaks is higher for the shallower wells.

Acknowledgments

The Natural Sciences and Engineering Research Council of Canada provided support for this work.

References

[1] M. R. M. Witwit and J. P. Killingbeck. A Hill-determinant approach to symmetric double-well potentials in two dimensions. *Can. J. Phys.* **73**, 632–637 (1995).

[2] M. R. M. Witwit. Energy levels for nonsymmetric double-well potentials in several dimensions: Hill determinant approach. *J. Comp. Phys.* **123**, 369–378 (1996).

[3] M. R. M. Witwit. Application of the Hill determinant approach to several forms of potentials in 2-dimensional and 3-dimensional quantum systems. *J. Math. Chem.* **20**, 273–283 (1996).

[4] M. R. M. Witwit and N. A. Gordon. Calculating energy levels of a double-well potential in a two-dimensional system by expanding the potential function around its minimum. *Can. J. Phys.* **75**, 705–714 (1997).

[5] M. R. M. Witwit. Inner-product perturbation theory. Energy levels of double-well potentials for 2-dimensional quantum systems by expanding the potential functions around their minima. *J. Math. Chem.* **22**, 11–23 (1997).

[6] J. B. Anderson. A random walk simulation of the Schrödinger equation: H_3^+. *J. Chem. Phys.* **63**, 1499–1503 (1975).

[7] B. L. Hammond, W. A. Lester Jr., and P. J. Reynolds. *Monte Carlo Methods in Ab Initio Quantum Chemistry* (World Scientific, Singapore, 1994).

[8] D. W. Peaceman and H. H. Rachford. The numerical solution of parabolic and elliptic differential equations. *J. Soc. Ind. Appl. Math.* **3**, 28–41 (1955).

[9] J. W. Thomas. *Numerical Partial Differential Equations: Finite Difference Methods* (Springer, New York, 1995).

[10] G. Dahlquist and Å. Björck. *Numerical Methods* (Prentice-Hall, London, 1974).

VSP International
Science Publishers
P.O. Box 346, 3700 AH Zeist
The Netherlands

Lecture Series on Computer
and Computational Sciences
Volume 1, 2004, pp. 1046-1050

Applications of the orbital-free embedding formalism to study the environment-induced changes in the electronic structure of molecules in condensed phase

T.A. Wesolowski[1]

Department of Physical Chemistry
University of Geneva
30, quai Ernest-Ansermet, CH-1211 Geneva 4, Switzerland

Received 31 July, 2004

Abstract: We outline the key elements of the universal orbital-free first-principles based embedding formalism applicable in theoretical studies of the electronic structure of atoms, molecules, intermolecular complexes, etc. in the presence of the environment [Wesolowski and Warshel, *J. Phys. Chem.*, **97** (1993) 8050]. So far, most of the applications of this formalism concerned studies of potential energy surface (geometries, IR spectra, etc.) of embedded molecules. Here, we review the past and current applications of the orbital-free embedding formalism to study electronic structure of embedded molecules.

Keywords: Density functional theory, embedding, electronic structure, kinetic energy functional

PACS: 31.15.Ew, 31.15.Bs, 71.10.Ca, 71.70.-d,

1 Introduction

Embedded systems (called also confined systems) are becoming object of great interest in physics, chemistry, and materials science. The theoretical description of an ion, atom, molecule, etc. embedded in a microscopic environment in condensed matter represents a serious challenge for theory (for review of various quantum mechanical approaches to study confined systems, see Ref. [1] for instance). One of the strategies is based on the idea of an embedding (or confining) potential. It traces its origin to the ideas of Sommerfeld and Welker [2]. Although the basic physical laws are known, a first-principles based and universally applicable, embedding potential applicable in practical computer modelling studies has not been formulated yet. Instead, another category of methods using system-tailored empirical embedding potentials is widely used in studies of liquid, solids, and biomolecules [3]. In this work, we outline the density functional theory route to derive the first-principles based embedding potential. It was initially used by Wesolowski and Warshel to study solvated molecules[4]. Its applications to study potential energy surface related properties (physisorption, proton transfer reactions in liquids and enzymes, IR spectra of probe molecules in zeolites, etc.) were reviewed elsewhere [5]. This work, concerns the methodological and computer implementation issues relevant it its applications in studies of electronic structure of embedded systems. They are illustrated by the past and current applications.

---

[1]Corresponding author. E-mail: Tomasz.Wesolowski@chiphy.unige.ch

2 Hohenberg-Kohn theorems in the subsystem based formulation of density functional theory

In this section, we outline the basic ideas of the subsystem-based formulation of DFT originally proposed by Cortona [6]. It will be given for the case of two interacting subsystems relevant for the subsequent considerations concerning orbital-free embedding [4]. For the sake of simplicity, the formalism is given for the spin-compensated case in all formulas throughout this work.

Following Hohenberg-Kohn theorems [7], the ground-state electron density of a given system comprising N_{AB} electrons can be derived from the minimization of the following total-energy functional:

$$E_o = \min_{\rho \longrightarrow 2N_{AB}} E_v^{KS}[\rho] \tag{1}$$

where

$$E_v^{KS}[\rho] = T_s[\rho] + \frac{1}{2} \int \int \frac{\rho(\mathbf{r}')\rho(\mathbf{r})}{|\mathbf{r}' - \mathbf{r}|} d\mathbf{r}' d\mathbf{r} + \int v(\mathbf{r})\rho(\mathbf{r}) \, d\mathbf{r} + E_{xc}[\rho] \tag{2}$$

where v is the external potential, $E_{xc}[\rho]$ is the exchange-correlatiuon functional defined by Kohn and Sham [8], and $T_s[\rho]$ is the kinetic energy functional in a reference system of non-interacting electrons ($T_s[\rho]$) defined in the Levy "constrained search" [9]:

$$T_s[\rho] = \min_{\Psi_s \longrightarrow \rho} \{< \Psi_s|\hat{T}|\Psi_s >\}] \tag{3}$$

where Ψ_s denotes the trial functions of the single determinant form.

Alternatively, E_o can be obtained from a step-wise search:

$$E_o = \min_{\rho_A \longrightarrow 2N_A} \min_{\rho_B \longrightarrow 2N_B} E_v[\rho_A, \rho_B] \tag{4}$$

where

$$E_v[\rho_A, \rho_B] = T_s[\rho_A] + T_s[\rho_B] + T_s^{nadd}[\rho_A, \rho_B]$$

$$+ \frac{1}{2} \int \int \frac{(\rho_A(\mathbf{r}') + \rho_B(\mathbf{r}'))(\rho_A(\mathbf{r}) + \rho_B(\mathbf{r}))}{|\mathbf{r}' - \mathbf{r}|} d\mathbf{r}' d\mathbf{r}$$

$$- \int v(\mathbf{r})\rho(\mathbf{r})(\rho_A(\mathbf{r}) + \rho_B(\mathbf{r})) \, d\mathbf{r} + E_{xc}[\rho_A + \rho_B] \tag{5}$$

In the above formula, the kinetic energy $T_s[\rho_A + \rho_B]$ is is expressed in a hybrid way:

$$T_s[\rho_A + \rho_B] = T_s[\rho_A] + T_s[\rho_B] + T_s^{nad}[\rho_A, \rho_B] \tag{6}$$

3 Orbital-free embedding

Following the same ideas as that leading to Kohn-Sham equations, leads to the one-electron equations allowing one to derive ρ_A for a given ρ_B (electron density of the environment) [4].

$$\left[-\frac{1}{2}\nabla^2 + V_{eff}^{KSCED} \left[\mathbf{r}, \rho_A, \rho_B\right] \right] \phi_{(A)i} = \epsilon_{(A)i}\phi_{(A)i} \tag{7}$$

where $(\rho_A = \sum_i |\phi_{(A)i}|^2)$. The superscript KSCED (Kohn-Sham Equations with Constrained Electron Density) is used to indicate the difference between the effective potential in Eq. 7 and that in the Kohn-Sham formalism [8]. Eq. 8 leads not only to the electron density which minimizes the bi-functional $E_v[\rho_A, \rho_B]$ for a given ρ_B but also to a set of one-electron functions (*embedded orbitals* $\{\phi_{(A)i}\}$) and the corresponding eigenvalues ($\{\epsilon_{(A)i}\}$). If the two sets of orbitals ($\{\phi_{(A)i}\}$ and $\{\phi_{(B)i}\}$), one for each subsystem, are obtained from a coupled calculations in which ρ_A and ρ_B switch their role [18], they are not orthogonal except for some particular cases. Therefore, they cannot be generally considered to be equivalent to the Kohn-Sham orbitals for the whole system.

The total effective potential $V_{eff}^{KSCED}[\rho_A, \rho_B, \mathbf{r}]$ can be conveniently split into one component, the Kohn-Sham effective potential for the isolated subsystem ($V^{KS}[\mathbf{r}, \rho_A]$), and the remaining part representing the environment ($V_{emb}^{eff}[\mathbf{r}, \rho_A, \rho_B]$) which reads:

$$V_{emb}^{eff}[\rho_A, \rho_B, \mathbf{r}] = \sum_{A_B} -\frac{Z_{A_B}}{|\mathbf{r} - \mathbf{R}_{A_B}|} + \int \frac{\rho_B(\mathbf{r}')}{|\mathbf{r}' - \mathbf{r}|} d\mathbf{r}'$$

$$+ \frac{\delta E_{xc}[\rho_A(\mathbf{r}) + \rho_B(\mathbf{r})]}{\delta \rho_A} - \frac{\delta E_{xc}[\rho_A(\mathbf{r})]}{\delta \rho_A} + \frac{\delta T_s^{nad}[\rho_A, \rho_B]}{\delta \rho_A} \tag{8}$$

where the first term shows the explicit form of the contributions of the atomic nuclei to the effective potential and $E_{xc}[\rho]$. Eq. 8 provides a general form of the Hohenberg-Kohn theorem based embedding potential. Several authors adapted the orbital-free embedding potential of the form given in Eq. 8 and used in in various contexts [10].

It is worthwhile to note that using explicit orbitals to evaluate $T_s[\rho_A]$ and $T_s[\rho_B]$ and an explicit but approximate bi-functional of the two electron densities ρ_A and ρ_B to evaluate $T_s^{nad}[\rho_A, \rho_B]$ in this approach, situates it somewhere between the Kohn-Sham formalism where orbitals are used for the whole system and the orbital-free strategy originally introduced to quantum mechanics by Thomas and Fermi [11].

4 Approximating the total energy bi-functional $E[\rho_A, \rho_B]$

Applications of Eqs. 7-8 in computer modelling rely on the approximations to the relevant functionals: $T_s^{nad}[\rho_A, \rho_B]$ and $E_{xc}[\rho]$. In this work, we use the approximate functionals chosen based on dedicated studies [12, 13]. The chosen approximations lead to an approximate bi-functional $E[\rho_A, \rho_B]$ which was shown to reproduce very accurately the energetics of weakly overlapping electron densities ρ_A and ρ_B such as those occurring for weak intermolecular complexes close to the equilibrium geometry [14].

5 Applications

The orbital-free embedding potential of Eq. 8 was originally used to derive the difference of the solvatation free-energy difference between water and methane [15]. The overall accuracy of the results in such calculations depends on both the accuracy of the used approximated functional-derivative $\frac{\delta T_s^{nad}[\rho_A, \rho_B]}{\delta \rho_A}$ and that of the used approximate bi-functional $T_s^{nad}[\rho_A, \rho_B]$. In these applications, they were approximated using gradient expansion approximation of the kinetic energy functional [16] truncated to the second order. Although this approximation proved to be quite satisfactory for evaluating the energies, the potential was found to have a serious flaw. It manifests itself in a significant underestimation of the effect of the solvent on the dipole moment of the solute.

The magnitude of the instantaneous value of the dipole moment of the embedded water molecule evaluated along the molecular dynamics trajectory, was shown to oscillate with the amplitude of about 0.05 Debye around the mean value which was about 0.2 Debye larger than that for the isolated water molecule. The orbital-free embedding reproduced thus qualitatively the effect of the solvent but the solvatation induced increase of the dipole moment was underestimated by a factor of 3-4.

This result indicated that for the studies of the electronic structure related properties the approximation to the embedding potential based on gradient expansion approximation is inadequate. Indeed, dedicated studies on the accuracy of various approximations to $T_s^{nad}[\rho_A, \rho_B]$ resulted in a construction of an approximate bi-functional $T_s^{nad}[\rho_A, \rho_B]$ of the generalized gradient approximation form [13] which was shown to be significantly superiour than the one derived from gradient expansion approximation[12]. In particular, the interaction induced shifts of the dipole moment in such systems as hydrogen-bonded intermolecular complexes were improved considerably. This methodological development, together, with a new implementation of Eqs. 7-8 [18] based on the computer code deMon [19] made it possible to turn back to our interest in electronic structure of embedded systems. However, the new implementation has been used mainly in various studies related to potential energy surface. Such studies have been reviewed elsewhere [5]. The first, practical application of this new implementation to study electronic-structure dependent properties was the evaluation of the effect of the noble gas (either Ne or Ar) matrix on the isotropic component of the hyperfine tensor (A_{iso}) of the embedded magnesium cation [17]. In these studies, the electron density of the noble gas atoms of the first coordination shell was considered as ρ_B in Eq. 8. The numerical results showed that the used approximations in Eq. 8 were indeed encouraging For instance, the calculated values: 222.9 gauss (Ar) and 210.4 gauss (Ne) where in excellent agreement with the exerimental measurements (222.4 gauss for Ar and 211.6 gauss for Ne). It is worthwhile to note that these results were obtained with a rather large basis set. For smaller basis sets, the numerical values agree less good with experiment but the relative shifts of A_{iso} (free cation vs embedded cation) were found to depend only negligibly on the size of the used basis sets. Since, the contribution of the electrostatic components of the embedding potential in Eq. 8 can be expected to be negligible in the cas of the embedding density being that of one or more noble gas atoms, the excellent agreement of the calculated shifts of A_{iso} with experiment indicates that the used approximation for the other terms in Eq. 8 are sufficiently accurate to describe the sum of the remaining non-electrostatic terms.

Recently, we started several new types of applications of the orbital-free embedding formalism to study such a electronic-structure dependent properties as: *i*) hyperfine structure of the Mn impurities in perovskites, *ii*) orbital-level splitting of lanthanide ions in crystalline environment, *iii*) complexation induced shifts of the localized excitations in hydrogen-bonded dimers, *iv*) spin-state of the transition metal centres in metalloenzymes.

The last type of applications became possible owing to the merge of two formalisms [20] the orbital-free embedding and the linear response time-dependent density-functional theory route to investigate electronic excitations.

Acknowledgment

This work is supported by the Swiss National Science Foundation (Project 21-63645.00).

References

[1] W. Jaskólski, *Phys. Rep.*, **271** (1966) 1.

[2] A. Sommerfeld and H. Welker, *Ann. Phys.*, **32** (1938) 56.

[3] There is a vast literature concerning the 'embedded molecule approach'. For pioneering papers in chemistry and solid state physics, see: (a) A. Warshel and M. Karplus, *J. Am. Chem. Soc.*, **94** (1972) 5612; (b) J.L. Whitten and T.A. Pekkanen, *Phys. Rev.*, **B21** (1980) 4357; For review, see for instance: (c) J. Gao, in "Reviews in Computational Chemistry" vol. 7, Eds. K.B. Lipkowitz and B. Boyd, VCH, New York, (1995) p. 119

[4] T.A Wesolowski and A. Warshel, *J. Phys. Chem.*, **97** (1993) 8050.

[5] T.A. Wesolowski, *CHIMIA*, **58** (2004) 311; **56** (2002) 707.

[6] P. Cortona, *Phys. Rev. B.*, **44** (1991) 8454.

[7] P. Hohenberg and W. Kohn, *Phys. Rev.*, **B136** (1964) 864.

[8] W. Kohn and L.J. Sham, *Phys. Rev.* **140** (1965) A1133

[9] M. Levy, *Proc. Natl. Acad. Sci. USA*, **76** (1979) 6062

[10] (a) H.T. Stokes, L.L. Boyer, M.J. Mehl, *Phys. Rev. B*, **54** (1996) 7729; (b) E.V. Stefanovitch, T.N. Truong, *J. Chem. Phys.*, **104** (1996) 2946; (c) N. Govind, Y.A. Wang, E.A. Carter, *J. Chem. Phys.*, **110** (1999) 7677; (d) O. Warschkow, J.M. Dyke, D.E. Ellis, *J. Comput. Phys.*, **143** (1998) 70; (e) J.R. Trail and D.M. Bird, *Phys. Rev. B*, **62**, (2000) 16402

[11] (a) L.H. Thomas, *Proc. Camb. Phil. Soc.*, **23** (1927) 542; (b) E. Fermi, *Z. Physik*, **48** (1928) 73

[12] T.A. Wesolowski, *J. Chem. Phys.*, **106** (1997) 8516.

[13] T.A. Wesolowski, H. Chermette, and J. Weber, *J. Chem. Phys.*, **105** (1996) 9182.

[14] T.A. Wesolowski and F. Tran *J. Chem. Phys.*, **118** (2003) 2072.

[15] T.A. Wesołowski and A. Warshel, *J. Phys. Chem.*, **98** (1994) 5183.

[16] D. A. Kirzhnits, *Sov. Phys. JETP*, **5**, (1957) 64.

[17] T.A. Wesołowski, *Chem. Phys. Lett.*, **311** (1999) 87.

[18] T.A. Wesołowski, J. Weber, *Chem.Phys.Lett.*, **248** (1996), 71.

[19] (a) A. St-Amant and D. R. Salahub, Chem. Phys. Lett. 169 (1990) 387 ; (b) Alain St-Amant, Ph.D. Thesis, University of Montreal (1992); (c) deMon-KS version 3.5, M.E. Casida, C. Daul, A. Goursot, A. Koester, L.G.M. Pettersson, E. Proynov, A. St-Amant, and D.R. Salahub principal authors, S. Chretien, H. Duarte, N. Godbout, J. Guan, C. Jamorski, M. Leboeuf, V. Malkin, O. Malkina, M. Nyberg, L. Pedocchi, F. Sim, and A. Vela contributing authors, deMon Software, 1998.

[20] M. Casida and T.A. Wesolowski, *Intl. J. Quant. Chem.*, (2004) **96** 577.

VSP International
Science Publishers
P.O. Box 346, 3700 AH Zeist
The Netherlands

*Lecture Series on Computer
and Computational Sciences*
Volume 1, 2004, pp. 1051-1053

Application of Purpose-Oriented Moderate-Size GTO Basis Sets to Nonempirical Calculations of Polarizabilities and First Hyperpolarizabilities of Conjugated Organic Molecules in Dielectric Medium

M.Yu. Balakina.[1a], S.E. Nefediev[b]

[a] A.E. Arbuzov Institute of Organic and Physical Chemistry of Kazan Scientific Center,
Russian Academy of Sciences, Arbuzov str., 8, 420088, Kazan, Russia;
[b] Kazan State Technological University, K.Marx str., 68, 420015, Kazan, Russia

Received 7 March, 2004; accepted in revised form 10 March, 2004

Abstract: In the present work the moderate-size basis sets constructed earlier are used for the SCF level calculations of static molecular polarizabilities and hyperpolarizabilities of conjugated organic molecules in dielectric medium. It is shown that studying the solvent effect on the nonlinear optical properties of the conjugated organic molecules with dominant π-electron polarization at the [4s3p2d/3s] level of theory one can obtain rather reliable results.

Keywords: ab initio calculations, basis sets, molecular polarizabilities and hyperpolarizabilities, solvent effect

PACS: 31.15.Ne, 31.15.Ar, 31.70.Dk, 42.70.Jk

1. Introduction

Molecular design of promising nonlinear optical (NLO) chromophores should incorporate the medium effects in the theoretical modeling, as the dielectric environment affects significantly the electric properties of the chromophores [1-3]. The additional motivation for studying the medium effects is conditioned by the fact that the experimental measurements are mostly performed in the condensed phase. Quantum chemical calculations present a natural alternative to the experimental investigation of the chromophores, allowing the estimation of their NLO characteristics both in the gas phase and in the dielectric environment, and providing the insight into the origin of their NLO activity. However, the theoretical estimation of chromophore electric properties, particularly molecular hyperpolarizability, even in the gas phase, is a nontrivial task demanding a special methodology, primarily the adequate basis set [4-6]. The choice of the adequate basis set for the estimation of the NLO properties in solution calls additional examination. It is obvious that for the computational reasons moderate size basis sets are preferable. Recently the purpose-oriented basis sets were suggested in [7,8] for the calculations of molecular polarizability and hyperpolarizability at the SCF level. The aim of the present work is to study the adequacy of the basis sets elaborated in [8] for the calculation of static molecular polarizabilities and hyperpolarizabilities of conjugated organic molecules in solution. For this purpose we apply Time-Dependent Hartree–Fock (TDHF) method [9] for the calculation of electric properties (α_{ij} and β_{ijk}), the solvent effect being taken into account with the Polarizable Continuum Model (PCM) [10,11]. The basis sets are tested here by the example of *p*-nitroaniline (PNA) molecule - the NLO chromophore, widely studied both theoretically and experimentally (see, for example [12, 13]).

2. Results and Discussion

---

[1] Corresponding author. E-mail: marina@iopc.knc.ru

The geometrical parameters of PNA were optimized in 6-31G** basis set both in the gas phase and in two solvents with different dielectric constants (chloroform with $\varepsilon=4.9$ and acetone with $\varepsilon=20.7$). The effect of the solvent on the geometrical parameters was shown to be insignificant: (CN bond lengths are shortened by 0.01 Å, ONC valence angles are increased by 0.3-0.4°, the change in other parameters being negligible). Thus, for the basis set tests we have used the gas-phase geometry of PNA. The electric properties are calculated for the PNA molecule embedded into chloroform. According to the PCM methodology, chromophore occupies in the dielectric medium a cavity, which is defined in terms of interlocking spheres with radii equal to 1.2 times the corresponding van der Waals atomic radii [14], the resulting values used here are: $R_N=1.52$Å, $R_C=2.04$Å, $R_O=1.8$Å, $R_H=1.44$Å.

The energy-optimized (9s5p/4s) Huzinaga atomic basis set [15] was used in [8] as the sp-substrate, for the first row atoms it was augmented with diffuse functions of s- and p- types and two sets of primitive d-functions to account for the polarization of electron density in covalent bonds and in the outer region of the atomic space, the exponent of the corresponding function being optimized in the presence of the applied electric field. The similar assumptions were used for the construction of Hydrogen basis functions. Thus, with various contractions by Dunning [16] the following basis set families were generated: [4s3pNd/3sNp], [5s3pNd/3sNp] and [6s4pNd/4sNp], for N=2 the exponents of the 2d/2p polarization functions being the following: $\alpha_1=0.72$, $\alpha_2=0.115$ for C; $\alpha_1=0.98$, $\alpha_2=0.144$ for N; $\alpha_1=1.28$, $\alpha_2=0.225$ for O; $\alpha_1=1.0$, $\alpha_2=0.125$ for H [7,8]. To provide the better description of the atomic outer region mostly influenced by the polarization effects of interest, the 3d/3p polarization set was also used here, the exponent values being as follows: $\alpha_1=0.72$, $\alpha_2=0.182$, $\alpha_3=0.073$ for C; $\alpha_1=0.98$, $\alpha_2=0.233$, $\alpha_3=0.089$ for N; $\alpha_1=1.28$, $\alpha_2=0.347$, $\alpha_3=0.146$ for O; $\alpha_1=1.0$, $\alpha_2=0.210$, $\alpha_3=0.074$ for H [7,8]. The dependence of the calculated (hyper)polarizabilities on the contraction flexibility of the basis and the dimension of the polarization set is studied.

The analysis of the results of the calculations demonstrate that, as one would expect, the β_{ijk} values are much more sensible to the extension of the basis set than the α_{ij} ones. For the given basis family the convergence of the values of the components of (hyper)polarizability tensors is shown to be achieved, the uncertainties in the results (estimated as half a maximal difference between the data falling in the convergence domain [8]) not exceeding $0.1 \cdot 10^{-24}$ esu for α_{ij} and $0.1 \cdot 10^{-30}$ esu for β_{ijk}, that is notably larger than in the case of the gas phase calculations. The use of contraction of different flexibility is shown to affect the values of both α_{ij} and β_{ijk} values only insignificantly, slightly increasing the calculated data with the increase of the contraction flexibility of the sp-substrate, this tendency is particularly noticeable for the absolute value of the dominant longitudinal component β_{zzz}. The effect of additional polarization function on the first-row atoms is slightly pronounced, causing a small decrease of β_{ijk} values ($\sim 0.3 \cdot 10^{-30}$ esu for β_{zzz}). The effect of the polarization functions on Hydrogen on both polarizability and hyperpolarizability values is negligible. So here, much as in the case of the gas phase, even the [4s3p2d/3s] basis set provides the data fairly well reproducing the results obtained with the [6s4p3d/4s3p] basis, which is the largest basis set examined here, the error being 0.5% for α_{ij} and 1% for dominant longitudinal hyperpolarizability β_{zzz}. Thus we conclude that studying the solvent effect on the NLO properties of the conjugated organic molecules with dominant π-electron polarization at the [4s3p2d/3s] level of theory one can obtain rather reliable SCF results.

Acknowledgments

The support of the Russian Foundation for Basic Research (Project № 02-03-32897) is greatfully acknowledged.

References

[1] A.Willetts, J.E.Rice, D.M.Burland, D.P.Shelton, Problems in the comparison of theoretical and experimental hyperpolarizabilities, *Journal of Chemical Physics* **97** 7590-7599(1992).

[2] K.V.Mikkelsen, Y.Luo, H. Agren, P.Jorgensen, Solvent induced polarizabilities and hyperpolarizabilities of para-nitroaniline, *Journal of Chemical Physics* **100** 8240 (1994).

[3] Y.Luo, P.Norman, H.Agren, A semiclassical approximation model for properties of molecules in solution, *Journal of Chemical Physics* **109** 3589-3595(1998).

[4] P.R.Taylor, T.J.Lee, J.E.Rice, J. Almlöf, The polarizabilities of Ne, *Chemical Physics Letters* **163**, 359-365 (1989).

[5] G.Maroulis. Accurate electric multipole moment, static polarizability and hyperpolarizability derivatives for N_2, *Journal of Chemical Physics* **118** 2673-2687(2003).

[6] G. Maroulis, C. Pouchan, Molecules in static electric fields: linear and nonlinear polarizability of HC≡N and HC≡P, Phys. Rev. A 57 2440-2447(1998).

[7] M.B. Zuev and S.E. Nefediev, Rational Design of Atomic Gaussian Basis Sets for Ab Initio Calculations of the Dipole Polarizabilities and Hyperpolarizabilities. I. Optimized Polarization Sets for the First-Row Atoms from B to F *Journal of Computational Methods in Sciences and Engineering*; Special issue (2003).

[8] M.B. Zuev, M Yu. Balakina, S.E. Nefediev, Rational Design of Atomic Gaussian Basis Sets for Ab Initio Calculations of the Dipole Polarizabilities and Hyperpolarizabilities. II. Moderate size Optimized Sets for the First-Row Atoms from B to F, *Journal of Computational Methods in Sciences and Engineering* Special issue (2003).

[9] S.P. Karna and M. Dupuis, Frequency Dependent Nolnlinear Optical Properties of Molecules: Formulation and Implementation in the HONDO Program, *Journal of Computational Chemistry* **12** 487-504(1991).

[10] R. Cammi, M. Cossi, J. Tomasi, Analytical derivatives for molecular solutes. III. Hartree-Fock static polarizability and hyperpolarizabilities in the polarizable continuum model, *Journal of Chemical Physics* **104** 4611-4620(1996).

[11] R. Cammi, M. Cossi,B. Mennucci, J. Tomasi, Analytical Hartree-Fock calculation of the dynamical polarizabilities α,β, and γ of molecules in solution, *Journal of Chemical Physics* **105** 10556-10564(1996).

[12] M. Stahelin, D.M. Burland, J.E. Rice Solvent dependence of the second order hyperpolarizability in *p*-nitroaniline, *Chemical Physics Letters* **191**, N3/4, 245-250 (1992).

[13] P. Salek, O. Vahtras, T. Helgaker, H. Agren, Density-functional theory of linear and nonlinear time-dependent molecular properties, *Journal of Chemical Physics* **117** 9630-9645(2002).

[14] A. Bondi, van der Waals Volumes and Radii, *Journal of .Physical Chemistry* **68** 441-451(1964).

[15] S. Huzinaga, Gaussian-type functions for polyatomic systems. I, *Journal of Chemical Physics* **42** 1293-1302(1965).

[16] T.H. Dunning, Gaussian basis functions for use in molecular calculations. I. Contraction of (9s5p) atomic basis sets for the first-row atoms, *Journal of Chemical Physics* **53** 2823-2833(1970).

VSP International
Science Publishers
P.O. Box 346, 3700 AH Zeist
The Netherlands

*Lecture Series on Computer
and Computational Sciences*
Volume 1, 2004, pp. 1054-1056

Electronic Polarizabilities of Atoms, Molecules and Crystals, Calculated by Crystal Chemical Method

S.S. Batsanov[1]

Center for High Dynamic Pressure,
Mendeleevo, Moscow Region 141570, Russia

Received 1 August, 2004; accepted 27 August, 2004

Abstract: New empirical methods are proposed for calculating the refractions of free atoms, atoms involved in covalent bonds, atoms in elemental solids and in inorganic compounds with polymeric structures.

Keywords: electronic polarizability, refraction, atom, electronegativity

PACS: 32.10.Dk, 33.15.Kr

The full system of electronic polarizabilities (refractions) of free atoms has been derived from direct measurements and quantum-mechanical calculations (Table 1, upper lines). Refractions of atoms in covalent bonds have been experimentally determined only for nonmetals and metals of the $1a$ and $4b$-$7b$ subgroup of the Periodic Table. In addition, polarizabilities of some metals were have been calculated from the experimental molecular refractions of their organometallic compounds by additive method. However, these data are not available for all elements. To fill all the gaps in Table 1, it is necessary to develop new methods for determining the covalent refraction of atoms.

Hohm [1] and Maroulis et al. [2] have shown that at the dissociation of a molecule the change of the enthalpy D and the refraction ΔR are related by the equation (1), which can be used to calculate atomic refractions.

$$D = A + B\Delta R \tag{1}$$

R is proportional to the atomic volume (and hence to r^3) and D is proportional to Z^*/r, where Z^* is the effective nuclear charge and r is the covalent radius, hence $D/\Delta R$ is proportional to Z^*/r^4. If D is half the dissociation energy of the M–M bond and $\Delta R = R(M) - \frac{1}{2} R(M_2)$, then

$$Z^*/r^4 = k\, D/\Delta R \tag{2}$$

Using the experimentally measured values of the refractions of free atoms and molecules of alkali metals and copper, the present author has calculated the factor k of the equation (2) and found it to be fairly stable for these elements. Then, assuming k to be constant for *all* elements, one can obtain the values of ΔR and, using the known magnitudes of R for the isolated atoms, derive the covalent refractions of metals [3]. The results of these calculations are listed in the middle lines of Table 1.

Taking into account that the refractive indices n of metals for $\lambda \rightarrow \infty$ are very high and hence the the Lorentz–Lorenz function in the formula (3) is close to unity, we can assume the refractions of metals R_M to be simply equal to their atomic volumes [4]. Refractions of atoms in solid metals are presented in the lower lines of Table 1.

$$R = V \frac{n^2 - 1}{n^2 + 2} \tag{3}$$

---

[1] Corresponding author. E-mail: Batsanov@gol.ru

Table 1: Refractions (cm^3) of free atoms, atoms involved in covalent bonds, and atoms in elemental solids

Li	Be	B	C	N	O	F	Ne		
58.4	14.0	7.6	4.4	3.05	1.97	1.40	1.00		
41.4	10.8	4.3	2.07	2.20	1.99	1.45			
13.0	4.9	3.5	2.07						
Na	Mg	Al	Si	P	S	Cl	Ar		
60.8	27.5	17.1	13.6	9.2	7.3	5.5	4.14		
49.9	19.3	11.5	9.05	8.57	7.7	5.69			
23.6	14.0	10.0	9.05	8.76	7.7				
K	Ca	Sc	Ti	V	Cr	Mn	Fe	Co	Ni
110	61.6	44.9	36.8	31.3	29.3	23.7	21.2	18.9	17.1
93.1	46.2	32.5	27.3	21.1	20.9	19.5	14.0	13.3	12.5
45.6	26.3	15.0	10.6	8.4	7.2	7.3	7.1	6.7	6.6
Cu	Zn	Ga	Ge	As	Se	Br	Kr		
15.4	14.5	20.5	15.3	10.9	9.5	7.7	6.27		
12.0	12.9	17.2	11.3	10.9	10.8	8.17			
7.1	9.2	11.7	11.3	10.3	11.6	8.75			
Rb	Sr	Y	Zr	Nb	Mo	Tc	Ru	Rh	Pd
121	72.2	57.2	45.1	39.6	32.3	28.8	24.2	21.7	20.1
99.6	55.0	38.0	27.6	23.8	21.4	19.0	13.3	14.0	14.7
55.9	33.9	20.0	14.0	10.9	9.4	8.2	8.2	8.2	8.8
Ag	Cd	In	Sn	Sb	Te	I	Xe		
18.2	18.55	25.7	19.4	16.6	13.9	12.5	10.20		
13.5	14.5	21.6	16.3	17.7	14.4	13.03			
10.3	12.9	15.8	16.3	17.7	15.4				
Cs	Ba	La	Hf	Ta	W	Re	Os	Ir	Pt
155	95.6	78.4	40.9	33.0	28.0	24.5	21.4	19.2	16.4
131	71.8	55.3	26.7	16.7	15.6	15.5	11.2	11.0	10.0
69.7	37.9	22.5	13.5	10.8	9.6	8.8	8.4	8.5	9.1
Au	Hg	Tl	Pb	Bi	Th	U	Rh		
14.6	12.7	19.2	18.5	18.7	81.0	51.2	13.4		
11.5	12.8	16.7	18.4	21.3	57.3	38.2			
10.2	13.9	17.1	18.3	21.3	19.8	12.5			

However, the coordination numbers N_c of metal atoms in elemental structures and in compounds are different. Therefore one cannot use the R_M values to calculate refractions of crystalline compounds in a straightforward manner. However, we can adjust the refractions of metals to such N_c as are characteristic of the structures of their compounds [5]. The last proposition amounts to calculating the refractions by the equation (4),

$$R_M = V_o / \rho \tag{4}$$

where V_o is the intrinsic atomic volume and ρ is the packing density. The factors by which r^3 should be multiplied for the structures with $N_c = 4$, 6, and 8 are 7.419, 4.472, and 3.710, respectively. The average value for the most common tetrahedral and octahedral structures is 5.9, which practically coincides with the factor of 5.8 proposed by Atoji [6] for conversion of the r^3 of atoms into their refractions.
Using crystalline covalent radii, we obtain the crystalline refractions of metal atoms listed (for the most common coordination numbers) in Table 2. To calculate the molar refractions of crystalline compounds with high accuracy, it is necessary to take into account the polar and metallic character of the chemical bonds, as well as the mutual polarizing influence of ions. (The latter, of course, is the oldest problem in the ionic theory of the chemical bond.) We propose [5] the following formulae for calculating the change of the atomic refractions due to the mutual effects of atoms participating in an A–B bond, viz.

$$\Delta R_a = q \, (Z i)^2 / d \tag{5}$$

where Z is the valence of an atom, i is the bond ionicity, d is the bond distance, and

$$q = [(R_A - R_B)/R_A]^2 \tag{6}$$

is the measure of the mutual polarizing interaction of atoms, which has proved its utility in diverse areas of structural chemistry.

Finally, it is necessary to estimate the bond metallicity m, i.e., the fraction of covalent electrons that form A–A bonds in an AB compound. The simplest method to estimate m consists in calculating the ratio:

$$m = cX_A / (X_A + X_B)$$ (7)

where X_A and X_B are the electronegativities (in Pauling's scale) of the atoms A and B. The abovementioned adjustments allow one to calculate the refractions with the accuracy of ca. 5%.

Table 2: Crystalline covalent atomic refractions (cm^3) for N_c=4 (asterisked) and N_c=6.

Li	Be	B	C	N	O	F			
15	7	4	2.4	2.1	2.0	1.5			
Na	Mg	Al	Si	P	S	Cl			
27	17	11.5	7	8.3	7.6	5.4			
		18*	9*						
K	Ca	Sc	Ti	V	Cr	Mn	Fe	Co	Ni
46	27	17	13	13	13.5	13	13	12.5	12
Cu	Zn	Ga	Ge	As	Se	Br			
16	15	10.5	8.5	10.3	8.5	7.8			
		16*	11*						
Rb	Sr	Y	Zr	Nb	Mo	Tc	Ru	Rh	Pd
55	35	23	19	14	14	12.5	23	23	25
					16.5*				
Ag	Cd	In	Sn	Sb	Te	I			
17	16	16	16	16	14	13			
22*	21*	24*	20*	21*					
Cs	Ba	La	Hf	Ta	W	Re	Os	Ir	Pt
68	44	28	16	15	15	11.5	24	24	24
					17				
Au	Hg	Tl	Pb	Bi	Th	U			
21	21	25	19	18	25	25			
			23*	23*					

References

[1] U. Hohm, *J. Chem. Phys.* **101** 6362 (1994).

[2] G. Maroulis, C. Makris, U. Hohm and D. Goebel, *J. Phys. Chem.* **101** 953 (1997).

[3] S.S. Batsanov, *Rus. J. Phys. Chem.* **77** 1374 (2003).

[4] S.S. Batsanov, *Refractometry and Chemical Structure,* D.Van Nostrand Co., Princeton, 1966.

[5] S.S. Batsanov, *Russ. J. Inorg. Chem.* **49** 560 (2004).

[6] M. Atoji, *J. Chem. Phys.* **25** 174 (1956).

VSP International
Science Publishers
P.O. Box 346, 3700 AH Zeist
The Netherlands

*Lecture Series on Computer
and Computational Sciences*
Volume 1, 2004, pp. 1057-1060

A Quantitative Study of the Rydberg States
of NaAr$_n$(n=1-12) Clusters

H. Berriche[1,a] F. Spiegelman[b] and M. B. El Hadj Rhouma[c]

[a] Laboratoire de Physique et Chimie des Interfaces, Département de Physique,
Faculté des Sciences de Monastir, Avenue de l'Environnement, 5019 Monastir, Tunisia.
[b] Laboratoire de Physique quantique, IRSAMC, Université Paul Sabatier,
118 Route de Narbonne, Toulouse, France.
[c] IPEIM, Monastir, Tunisia.

Received 7 March, 2004; accepted in revised form 10 March, 2004

Abstract: The electronic energy of the excited states of the metal valence electrons in the environment of the rare gas atoms is determined via a procedure close to the ab-initio methodology involving e-Na$^+$ and e-Ar semi-local pseudopotentials complemented by core polarization operators. The geometries of the ground states of the ionic clusters Na$^+$Ar$_n$ were used to determine the energy of the excited states of the neutral NaAr$_n$ clusters. The variation of the cluster energy as a function of the number of argon atoms has shown a stabilization of the Rydberg states in the cluster ionic geometries.

Keywords: NaAr$_n$ clusters, Rhydberg states

PACS: Here must be added the AMS-MOS or PACS Numbers

1. Introduction

The spectroscopy of elementary systems such as atoms or molecules trapped in (or interacting with) clusters has been the object of continuous investigations [1-12]. Their interest is explained by the need to understand the properties of the system from small size clusters to macroscopic states. Most of them were focused on the ground state. The full study of the excited states of the neutral clusters is very complex. It requires an optimisation of the geometries of the excited states. The aim of this study is to investigate the excited states of the NaAr$_n$(n=1-12) clusters using a simple approach based on an analogy between the Rydberg states of the NaAr dimer and NaAr$_n$ clusters. The excited states of the neutral NaAr molecule, starting from the first excited state, are more bound than the ground state and behave like the ground state of the ionic Na$^+$Ar system. Their equilibrium distances are shorter than that of the ground state. They are very close to the equilibrium distance of the ionic molecule in its ground state. This particular property can be used to make a qualitative study of the Rydberg states of the neutral clusters NaAr$_n$. We think, by analogy with the dimers, that the excited states of the neutral clusters are more bound than the ground state. Their geometries approach those of the ionic clusters in their ground states. In this context, the calculation of the energy of the neutral clusters in the geometry of ionic can inform us on the stability of the excited states. The optimised geometries of the ionic clusters Na$^+$Ar$_n$ (n=1-12) where used to calculate the energy and the molecular orbital of the Rydberg states of NaAr$_n$ clusters. In the next section we present the general methodology. The results on NaAr$_n$ will be presented in section 3. Finally we conclude.

2. Summary of the method

The methodology relies on the explicit description of the active electrons and on the representation of the rare gas atoms and the Na$^+$ core via pseudopotentials technique. Using this method we reduce the number of electrons of NaAr$_n$ system to only one valence electron that of Na atom.

---

[1] Corresponding author. E-mail: hamid.berriche@fsm.rnu.tn, hamidberriche@yahoo.fr

The total Hamiltonian for NaAr$_n$ is approximated as:

$$H = -\frac{\Delta}{2} - \frac{1}{r} + W_{Na} + V_{Na}^{pol} + \sum_{Ar}(W_{Ar} + V_{Ar}^{pol} + V_{NaAr^+}) + \sum_{Ar} V_{Ar-Ar} \quad (1)$$

where the sum runs over argon atoms and r stands for electronic coordinates.

The Hartree-Fock extracted pseudopotential, W_{Na} for sodium according to the Barthelat and Durand formulation [13, 14], is completed by core-polarisation pseudo-potential operator V_{Na}^{pol}. For the argon atom we have used a pseudo-potential and a core-polarisation pseudo-potential obtained by fitting differential phase shifts of electron-Ar elastic scattering [15] data for angular moments up to l=2.

The first sum in V^{pol} is over all polarizable cores (sodium and rare gas atoms). α_λ is the dipole polarisability of the core λ and $f_{\alpha\lambda}$ is the electrostatic field at the centre λ produced by the valence electrons and cores of all other centres damped by an cut-off function:

$$f_{\lambda 1} = \sum_i \frac{r_{i\lambda}}{r_{i\lambda}^3} F(r_{i\lambda}) - \sum_{\lambda,\lambda'=Na+} \frac{R_{\lambda\lambda'}}{R_{\lambda\lambda'}^3} \quad (2)$$

We use polarizability values for sodium core and argon atoms of, respectively, 0.995 a$_0^3$ and 11.08 a$_0^3$ respectively. The total Hamiltonian also includes the interaction V_{NaAr^+} between the core electrons of sodium and the electrons of argon, taken from the works of Ahmadi et al [16] on NaAr$^+$ and the pair-interaction between Ar-atoms V_{Ar-Ar} taken from the multi-property fit of Aziz et al [17]. In order to provide accurate transitions of the Na atomic spectra we have used a large uncontracted gaussian basis set 12s/12p/10d.

3. Results

The model has been tested for the diatomic molecules LiAr, NaAr and KAr [18] van der Waals dimers. We produced for them the ground and many exited Rydberg electronic states. Their spectroscopic constants, for the ground and the two first excited states, were in good agreement with both theoretical and experimental previous works. The same accuracy can be expected for complex systems with more argon atoms. The optimised ground state structures of NaAr$_n$ and Na$^+$Ar$_n$ clusters were realized in a previous work [19] using the so-called basin-hopping method. The optimised equilibrium structures of the Na$^+$Ar$_n$ (n=1-12) are used to calculate the energy of the excited states of the neutral clusters NaAr$_n$ (n=1-12). The calculated equilibrium structures are in good agreement with existing previous results [2, 10, 11, 20, 21]. Figure 1 presents the optimised geometries of some ionic and neutral NaAr$_n$ clusters. The Na$^+$ in the ionic geometry is trapped by the argon cluster while the alkaline atom interacts on the surface. Figure 2 presents the energy of the ground state and excited states of the neutral clusters in the geometry of ionic compired with the energy obtained in the neutral geometry. The energy of the clusters in the ionic geometry is lower than that in neutral geometry. This reduction is more marked while going towards increasingly excited states and also by increasing the number of argon atoms. This shows the tendency to stabilize the neutral excited clusters in the ionic geometry. In fact, the excited electron tends to be outside the clusters leading to a reduction in energy. Thus leads to a change in the geometry of the excited neutral cluster to the geometry of the ionic one. The instability of the excited states of the neutral clusters in the neutral geometry can be due to the fact that the orbital of the valence electron inside the cluster is bothered by the argon atoms around. On the other hand, in the ionic geometry, the orbital of the excited metal valence electron tends to include the geometry of the cluster leading to the minimization of the energy. This leads to a stabilization of the excited state of the neutral cluster in the corresponding ionic geometry.

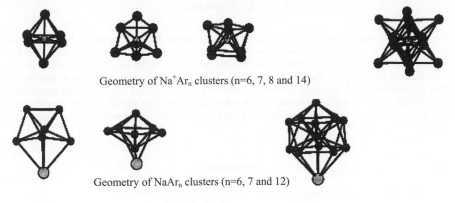

Geometry of Na$^+$Ar$_n$ clusters (n=6, 7, 8 and 14)

Geometry of NaAr$_n$ clusters (n=6, 7 and 12)

Figure 1: Geometries of some neutral and ionic clusters

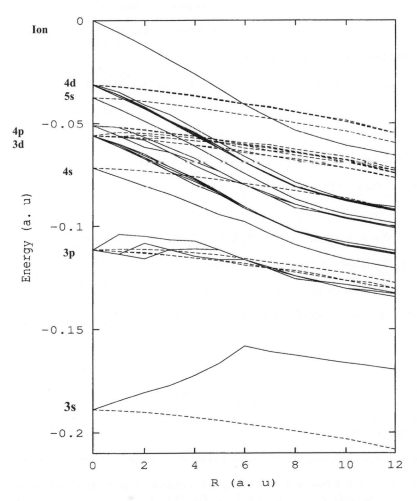

Figure 2: Energy of the electronic states of NaAr$_n$ (n=1-12) clusters in the neutrals (-----) and the ions (——)geometries.

4. Conclusion

Using a simple approach based on the analogy between the excited states of the NaAr dimer and $NaAr_n$ clusters, we have studied the Rydberg states of neutral $NaAr_n$ (n=1-12) small clusters. We have computed the energy of the $NaAr_n$ excited states in the geometries of the ionic corresponding clusters. The variation of the energy of the Rhydberg states as a function of argon atoms shows a minimisation leading to a stabilisation in the ionic geometries where the sodium atom is trapped by the argon cluster and the orbital of the excited valence electron includes the cluster. The stability of the excited neutral cluster in the ionic geometry is more marked for higher excited states and large number of argon atoms.

References

[1] D. J. Wales and J.P.K. Doyle, J. Phys. Chem. A 101 (1997) 5111

[2] C. Tsoo, D. A. Estrin, and S.J. Singer, J. Chem. Phys. 93 (1990) 7187

[3] J. Maclyn, McCarty and G.W. Robinson, Mol. Phys. 2 (1959) 415

[4] B. Meyer, J. Chem. Phys, 43 (1965) 2986

[5] L. C. Balling, M. D. Havey, and J. F. Dawson, J. Chem. Phys. 69 (1978) 1670

[6] L. C. Balling, J. F. Dawson, M. D. Havey, and J. J. Wright, Phys. Rev. Lett. 43 (1979) 435

[7] J. J. Wright and L. C. Balling, J. Chem. Phys. 73 (1978) 994

[8] L. C. Balling and J. J. Wright J. Chem. Phys. 79 (1983) 2941

[9] L. C. Balling and J. J. Wright J. Chem. Phys. 81 (1984) 675

[10] J.A. Boatz and M.E. Fajardo. J. Chem.Phys, 101 (1994)

[11] G. Martyna, C. Cheng, and M. L. Klein, J. Chem. Phys. 95 (1991) 1318

[12] K. T. Tang and J. P. Toennies, J. Chem. Phys. 80 51984) 3726

[13] J. C. Barthelat, Ph. Durand, Theoret. Chim. Acta 38 (1975) 283.

[14] J. C. Barthelat and Ph. Durand, Chim. Ital.108 (1978) 225.

[15] F. Spiegelman, G. Durand, and P. Duplaa, Z. Phys. D 40 (1997) 177.

[16] G. Reza Ahmadi and I. Røeggen, J. Phys. B. At. Mol. Opt. Phys. 27 (1994) 5603

[17] R. A. Aziz and H. H. Chen, J. Chem. Phys, 67 (1988) 5719.

[18] M. B. El Hadj Rhouma, H. Berriche, Z. Belakhdar and F. Spiegelman, J. Chem. Phys. 116 (2002) 1839.

[19] M. B. El Hadj Rhouma, PhD Thesis, the Faculty of Sciences of Tunis, 2002 (unpublished).

[20] J. P. Visticot, P.Depujo, J. M. Mestdagh, A. Lallement, J. Berlande, O. Sublemontier, P. Meynadier, and J. Cuvelier, J. Chem, Phys 100 (1994) 158.

[21] C. Tsoo, D. A.Estrin, and S. J. Singer, J. Chem. Phys. 96 (1992) 7977.

VSP International
Science Publishers
P.O. Box 346, 3700 AH Zeist
The Netherlands

Lecture Series on Computer
and Computational Sciences
Volume 1, 2004, pp. 1061-1064

Evaluation of the Atomic Polarisabities of Li And K using the Long Range LiK$^+$ Electronic States Behaviour

C. Ghanmi, H. Berriche[1] and H. Ben Ouada

Laboratoire de Physique et Chimie des Interfaces, Département de Physique
Faculté des Sciences de Monastir, Avenue de l'Environnement, 5019 Monastir, Tunisia.

Received 7 March, 2004; accepted in revised form 10 March, 2004

Abstract: A vibrational level spacing analysis of the lowest electronic states of the LiK$^+$ molecule dissociating into Li(2s, 2p, 3s, 3p, 3d, 4s and 4p) + K$^+$ and Li$^+$ + K(4s, 4p, 5s, 3d, 5p, 4d, 6s) has been performed using the usual WKB semiclassical approximation. Accurate long range potentials have been determined and allowing for the determination of all vibrational levels near the dissociation limit, the Li and K atomic polarisabilities, the number of trapped vibrational levels, and the vibrational turning points.

Keywords: Polarisability, Li, K, WKB approximation, vibrational analysis

PACS: 33.15.Kr, 32.10.Dk, 33.20.Tp

1. Introduction

In our previous work [1] we have determined the potential energy curves of 25 electronic states of the LiK$^+$ molecule dissociating into Li(2s, 2p, 3s, 3p, 3d, 4s and 4p) + K$^+$ and Li$^+$ + K(4s, 4p, 5s, 5p, 4d and 6s). The use of the pseudopotentials technique [2] for Li and K cores has reduced the number of active electrons to only one valence electron, where the SCF calculation has produced the exact energy in the basis set. In this context, the main source of errors corresponds to the basis-set limitations. Furthermore, we have corrected the energy by taking into account the core-core and core-electron correlation following the formalism of Foucrault et al [3]. The nonempirical pseudopotentials permit the use of very large basis sets for the valence and Rydberg states and allow accurate descriptions for the highest excited states.

The LiK$^+$ ionic system presents an interesting long-range behaviour dominated by charge-dipole interactions. In this study, we use the accuracy of the *ab initio* calculation of the potential energy at long-range distances, to perform a vibrational spacing analysis for the ground and numerous excited states of $^2\Sigma$, $^2\Pi$ *and* $^2\Delta$ symmetries. The paper is structured as follows. In section 2, the numerical method is presented. Section 3 is devoted to the results. Finally we conclude.

2. Summary of the method

An excellent review of the semiclassical approximation applied to the vibrational diatomic molecules can be found in Ref. [4]. In this section we present briefly the used formalism to perform a long-range potential analysis.

The WKB semiclassical approximation applied to a potential of the following expression:

$$V(R) = D - \frac{C_n}{R^n} \tag{1}$$

where D is the asymptotic limit,
leads to an analytical expression for the vibrational energy levels:

---

[1] Corresponding author. E-mail: hamid.berriche@fsm.rnu.tn, hamidberriche@yahoo.fr

$$E(v) = D - \left((v_D - v)\frac{n-2}{2n}K_n\right)^{2n/(n-2)} \tag{2}$$

v_D is the real (non-integer) number corresponding to the last vibrational level near the dissociation energy limit and K_n is given by:

$$K_n = \hbar\sqrt{\frac{2\Pi}{\mu}} \frac{\Gamma(1+1/n)}{\Gamma(1/2+1/n)} \frac{n}{(C_n)^{1/n}} \tag{3}$$

In our case, at long-range distance, the potential energy behaves as a charge-dipole interaction and n=4. Using the last expression for vibrational energy levels, the $(E_v-D)^{1/4}$ can be written as following.

$$(E_v - D)^{1/4} = -(v_D - v)\frac{K_n}{4} \tag{4}$$

The plot of $(E_v-D)^{1/4}$ should be purely linear. A least square fit of this linear law allows us to determine rather accurately the v_D and C_4 constants from which all vibrational levels near the dissociation limit and also the related classical turning points ($R_c(v)$) can be easily determined according to

$$E(v) = D - 11.817045(v_D - v)^4 C_4^{-1}\mu^{-2} \tag{5}$$

and

$$R_c(v) = C_4^{1/4}(D - E(v))^{-1/4} \tag{6}$$

3. Results

Interestingly, the use of the semiclassical approximation finally give us additional information on the long range potential through an analysis of the preliminary vibrational level progression and based on the remarkable adequacy of the linear fit. According to the relation (4), a plot of $(E_v - D)^{1/4}$ versus v should be purely linear. Some plots are reported in Fig. 1 for 4-6, 8, 10-12 $^2\Sigma^+$ states. These plots are remarkably linear. The C_4 and v_D constants and also the number of trapped vibrational levels for $N^2\Sigma^+$ (N=1-8, 10-14) and $M^2\Pi^+$ (M=4, 7) are reported in table 1. Some of the C_4 constants are compared with a result of a previous work [5, 6]. There is a good agreement between the two works. Here, we report the polarisabilities ($C_4=2\alpha$) for more excited levels of the lithium atom and we present for the first time the polarisabilities of the ground and many excited levels of the potassium atom. Furthermore, using this fit we find the total number of vibrational level trapped by each state and specially the last excited vibrational levels near the dissociation limit. For such levels the use of the cubic spline interpolation, in Numerov propagation method used to find the wave function, leads to serious numerical difficulties in that region where the potential energy difference between two *ab inito* calculated points tends to zero. By an inspection of potential energy data the difference between the energy of two closed points at large distances and for higher excited states is about 10^{-6} a u which is the precision of the *ab initio* calculation.

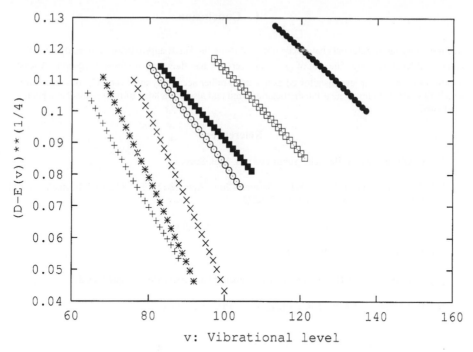

Figure 1: Plot of $(E_v - D)^{1/4}$ versus v for some $^2\Sigma^+$ electronic states of the LiK+ molecule.

Table 1: v_D and C_4 coefficients involved in the semicallsical analysis of the vibrational levels near the related dissociation limit.

State	v_D	C_4	C_4 [5]	Dissociation
$1\,^2\Sigma^+$	73.393	80.303	80.855	$Li(2s) + K^+$
$2\,^2\Sigma^+$	51.82	134.111		$Li^+ + K(4s)$
$3\,^2\Sigma^+$	77.732	2066.033	2029.254	$Li(2p) + K^+$
$4\,^2\Sigma^+$	113.609	5542.288		$Li^+ + K(4p)$
$5\,^2\Sigma^+$	116.524	1906.046	1912.495	$Li(3s) + K^+$
$6\,^2\Sigma^+$	109.649	2050.347		$Li^+ + K(5s)$
$7\,^2\Sigma^+$	82.477	8551.347		$Li^+ + K(3d)$
$8\,^2\Sigma^+$	187.684	37305.430	43494.160	$Li(3p) + K^+$
$10\,^2\Sigma^+$	166.813	30183.313		$Li^+ + K(5p)$
$11\,^2\Sigma^+$	153.185	17105.636		$Li(4s) + K^+$
$12\,^2\Sigma^+$	233.880	82932.323		$Li^+ + K(4d)$
$14\,^2\Sigma^+$	101.525	123564.235		$Li(4p) + K^+$
$4\,^2\Pi^+$	135.013	6355.971	6516.144	$Li(3p) + K^+$
$7\,^2\Pi^+$	192.169	33814.817		$Li^+ + K(4d)$

4. Conclusion

Accurate long-range potentials have been obtained using the WKB approximation by analyzing the vibrational levels spacing. The plot of $(E_v - D)^{1/4}$ versus v has shown a pure linear behaviour. A least squares fit of this linear law has allowed us to extract rather accurately the v_D and C_4 constants from which all vibrational levels near the dissociation limit and also the related turning points have been calculated.

References

[1] C. Ghanmi and H. Berriche, submitted to J. Mol. Spec.

[2]   Ph. Durand and J. C. Barthelat, Theoret. Chim. Acta 38 (1975) 283 ; J. C. Barthelat and Ph. Durand, Gazz Chim. Ital. 108 (1978) 225.

[3] M. Foucrault, Ph. Millié and J. P. Daudey, J. Chem. Phys. 96 (1992) 1257.

[4] J. Vigué, Ann. Phys. (Paris) 3 (1982) 155.

[5] H. Berriche, PhD Thesis from the Paul Sabatier University 1995 (unpublished)

[6] H. Berriche and F. X. Gadea, Chem. Phys. 191 (1995) 119.

VSP International
Science Publishers
P.O. Box 346, 3700 AH Zeist
The Netherlands

Lecture Series on Computer
and Computational Sciences
Volume 1, 2004, pp. 1065-1068

Electric Properties of Boron and Aluminum Trihalides

A. Chrissanthopoulos and G. Maroulis[1]

Department of Chemistry,
University of Patras,
GR-26504 Patras, Greece

Abstract: We have calculated electric multipole moments and dipole polarizabilities for the boron and aluminum trihalides MX_3 (M= B, Al; X= F, Cl, Br, I). The electric properties were obtained from self-consistent field and second-order Møller-Plesset perturbation theory calculations with especially designed Gaussian-type basis sets. Our investigation brings forth interesting patterns for the evolution of electric properties in these molecules. Our results for BI_3 support the recent experimental estimates for the mean and the anisotropy of the dipole polarizability proposed by Keir and Ritchie [Chem.Phys.Lett. **290**, 409 (1998)].

Keywords: Boron trihalides, aluminum trihalides, electric multipole moments, polarizability.

PACS: 32.10.Dk.

1. Introduction

Boron and aluminum halides are of interest both from the experimental and theoretical point of view. Many of the neutral members are powerful Lewis acids and are frequently used as catalysts in synthetic chemistry, principally for Friedel-Crafts reactions. The study of the physicochemical properties of these molecules present a particular challenge as they are highly corrosive, hygroscopic and tend to form polymeric species. In a recent experimental study Keir and Ritchie [1] deduced experimental values for the dipole polarizability anisotropy of BX_3, X = F,Cl and Br, from Rayleigh depolarization ratios. In addition, they estimated the mean and the anisotropy of the dipole polarizability of BI_3. It would be interesting to compare the findings of these authors to theoretical findings.

In this study, we present a theoretical investigation of the electric properties of the boron and aluminum trihalides MX_3, M = B, Al and X = F, Cl, Br and I. We have designed new basis sets for all these molecules. Their construction follows a computational philosophy outlined in previous papers [2,3].

2. Computational strategy and results

The basis sets used in this study are given in Table 1. They are built on substrates of TZV quality [4] for all atoms while a small basis set was used for iodine [5]. All optimizations were performed at the experimental molecular geometry for all molecules. We have also obtained **ab initio** values for the molecular geometries at the MP2 level of theory. We follow Buckingham's definitions and terminology for the electric properties [6]. We have obtained SCF and MP2 values for the electric properties. All calculations were performed with GAUSSIAN 94 [7].

---

[1] Corresponding author, e-mail: maroulis@upatras.gr

Table 1: Basis set composition.

	Uncontracted	Contracted	Final
B	(11s6p)	[5s3p]	[6s4p2d]
Al	(14s9p)	[5s4p]	[6s5p2d]
F	(11s6p)	[5s3p]	[6s4p2d]
Cl	(14s9p)	[5s4p]	[6s5p2d]
Br	(17s13p6d)	[6s5p2d]	[7s6p3d]
I	(13s10p6d)	[5s4p2d]	[6s5p3d]

The molecular orientation adopted in this work is shown in figure 1. There is only one independent component for the quadrupole ($\Theta_{\alpha\beta}$), octopole ($\Omega_{\alpha\beta\gamma}$) and hexadecapole ($\Phi_{\alpha\beta\gamma\delta}$) moment tensors for molecules of D_{3h} symmetry. There are two independent components for the dipole polarizability $\alpha_{\alpha\beta}$ [6]. Thus, by symmetry [6]

$$\alpha_{yy} = \alpha_{\|} \text{ and } \alpha_{xx} = \alpha_{zz} = \alpha_{\perp}$$
$$\Theta_{xx} = \Theta_{zz} \text{ and } \Theta_{yy} = -2\,\Theta_{zz}$$
$$\Omega_{xxz} = -\Omega_{zzz}$$
$$\Phi_{xxxx} = \Phi_{zzzz} = 3/8\,\Phi_{yyyy}$$

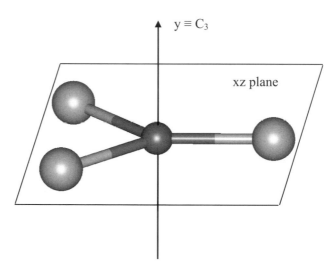

Figure 1. Adopted orientation of the MX_3 (D_{3h}) molecules.

Table 2: M-X bond lengths (in Å), charges from an NBO analysis and electric properties (in atomic units) for boron and aluminum trihalides calculated at the MP2 level of theory

Property	BF_3	BCl_3	BBr_3	BI_3	AlF_3	$AlCl_3$	$AlBr_3$	AlI_3
R_{M-X}/Å	1.321	1.749	1.901	2.119	1.652	2.093	2.252	2.481
$q(M)$	1.438	0.270	-0.044	-0.626	2.240	1.529	1.230	0.675
$q(X)$	-0.479	-0.090	0.015	0.209	-0.747	-0.510	-0.410	-0.225
Θ_{xx}	-1.505	-0.390	0.078	0.622	-3.714	-2.556	-1.845	-0.841
Θ_{zz}	-1.505	-0.390	0.078	0.622	-3.714	-2.556	-1.845	-0.841
Θ_{yy}	3.011	0.779	-1.155	-1.244	7.428	5.112	3.689	1.682
Ω_{xxz}	5.545	-13.599	-25.826	-47.124	21.606	2.417	-12.387	-38.382
Ω_{zzz}	-5.545	13.599	25.826	47.124	-21.606	-2.417	12.387	38.382
Φ_{xxxx}	-0.982	18.444	33.101	58.741	-9.707	16.945	40.763	83.637
Φ_{yyyy}	-2.618	49.183	88.270	156.641	-25.883	45.188	108.702	223.033
$\alpha_\perp$	17.34	62.63	89.63	143.11	21.77	69.61	99.27	155.33
$\alpha_\parallel$	13.63	41.19	56.34	85.09	19.30	50.71	67.95	99.66
$\bar{\alpha}$	16.11	55.48	78.53	123.77	20.95	63.61	88.83	136.77
$\Delta\alpha$	-3.71	-21.44	-33.29	-58.02	-2.47	-18.90	-31.32	-55.66

We give in Table 2 MP2 bond lengths, charges obtained from a NBO analysis and electric properties for all molecules. Our results show very interesting patterns. The total charge of the central atom decreases with the size of the halogen. Monotonic changes are observed for the electric moments as well. We have traced in figure 2 the dependence of the charges and the invariants of the dipole polarizability on the atomic number Z of the halogen. The changes are very similar for the mean and the anisotropy of the polarizability of BX_3 and AlX_3.

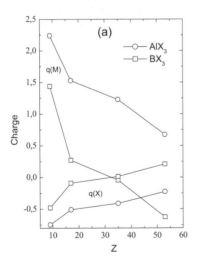
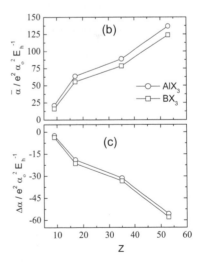

Figure 2: Plot of (a) metal and halogen charges, (b) the electric dipole polarizability and (c) the polarizability anisotropy of the boron and aluminum trihalides, at the optimized geometry and at the MP2 level of theory.

We have obtained for BI_3 MP2 values of $\bar{\alpha}$ = 123.77 and $\Delta\alpha$ = -58.02 $e^2a_0{}^2E_h{}^{-1}$. These values are in reasonable agreement with the estimates of Keir and Ritchie [1], who proposed $\bar{\alpha}$ = 120.7 and $\Delta\alpha$ = -54.6 $e^2a_0{}^2E_h{}^{-1}$. Our values for BF_3, BCl_3 and BBr_3 agree well with the dipole polarizability anisotropies extracted from Rayleigh depolarization ratios by the same authors.

Appendix

Atomic units are used throughout this paper. Conversion factors to SI units are, Length, $1\ a_0$ = $0.529177249 \times 10^{-10}$ m, Θ, $1\ ea_0{}^2$ = $4.486554 \times 10^{-40}\ Cm^2$, Ω, $1\ ea_0{}^3$ = $2.374182 \times 10^{-50}\ Cm^3$, Φ, $1\ ea_0{}^4$ = $1.256363 \times 10^{-60}\ Cm^4$ and α, $1\ e^2a_0{}^2E_h{}^{-1}$ = $1.648778 \times 10^{-41}\ C^2m^2J^{-1}$.

References

[1] R.I.Keir, G.L.D.Ritchie, Chem. Phys. Lett. **290**, 409(1998).

[2] G.Maroulis, J.Chem.Phys. **108**, 5432(1998).

[3] G.Maroulis, J.Chem.Phys. **118**, 2673 (2003).

[4] A. Schäfer, C.Huber and R.Ahlrichs, J.Chem.Phys. **100** 5829(1994).

[5] R. Poirier, R. Kari, I.G. Csizmadia, *Handbook of Gaussian Basis Sets*, Elsevier, Amsterdam, 1985.

[6] A.D.Buckingham, Adv.Chem.Phys. **12**, 107 (1967).

[7] M. J. Frisch, G. W. Trucks, H. B. Schlegel, P. M. W. Gill, B. G. Johnson, M. A. Robb, J. R. Cheeseman, T.Keith, G. A. Petersson, J. A. Montgomery, K. Raghavachari, M. A. Al-Laham, V. G. Zakrzewski, J. V. Ortiz, J. B. Foresman, J. Cioslowski, B. B. Stefanov, A. Nanayakkara, M. Challacombe, C. Y. Peng, P. Y. Ayala, W. Chen, M. W. Wong, J. L. Andres, E. S. Replogle, R. Gomperts, R. L. Martin, D. J. Fox, J. S. Binkley, D. J. Defrees, J. Baker, J. P. Stewart, M. Head-Gordon, C. Gonzalez, and J. A. Pople, GAUSSIAN 94, Revision E.1 (Gaussian, Inc., Pittsburgh PA, 1995.

VSP International
Science Publishers
P.O. Box 346, 3700 AH Zeist
The Netherlands

*Lecture Series on Computer
and Computational Sciences*
Volume 1, 2004, pp. 1069-1072

Collision-Induced Light Scattering Spectra of Mercury Vapor at Different Temperatures

M.S.A.El-Kader and S.M.El-Sheikh[1]

Department of Engineering Mathematics and Physics,
Faculty of Engineering, Giza , 12211, Egypt

Received 7 March, 2004; accepted in revised form 10 March, 2004

Abstract: The two-body collision-induced light scattering spectra of mercury vapor are analyzed in terms of recent interatomic potential and interaction-induced pair polarizability anisotropy model, using line-shape computations. The comparison of the computed spectra with a measurement at different temperatures permits improvements on the existing anisotropy model of the pair polarizability.

Keywords: polarizability, interatomic potential

PACS: 32.10.Dk

1. INTRODUCTION

In this paper we present a new analysis of the depolarized pair Raman spectrum of mercury at different temperatures, based on fitting the spectral profile rather than just the spectral moments of the measurements. Spectral profiles have a number of characteristic features, such as the (discernible) bound dimer contributions, logarithmic slopes and curvatures of the line core and the wings, etc., which may all be utilized in line-shape analyses. A moment analysis, on the other hand, reduces the information available from such spectra to just two numbers: the total intensity and a mean width, ignoring all the rest of the information that may be exploited with line-shape analyses [1]. Spectral profiles are calculated numerically with the help of a quantal computer program, and compared with the measured spectrum. The comparison of calculated and measured spectra provides valuable clues concerning the quality of existing models of both the interaction-induced pair polarizability anisotropy and the interatomic potential; new parameters of the pair polarizability anisotropy may thus be obtained from high-quality measurements.

2. LINE-SHAPE CALCULATIONS

In order to calculate the line profile and the associated zeroth and second moments, the pair polarizability anisotropy and interatomic potential are needed. At moderate densities, the DILS spectrum is determined by binary interactions. It consists of purely translational scattering due to the transitions between the free states and the other is due to the transitions between the bound dimers. The pair polarizability giving rise to the translational scattering in the case of pure mercury vapor is given by the following formula [2]

$$\beta(r) = \frac{6\alpha_0^2}{r^3} + (6\alpha_0^3 + \frac{\gamma C_6}{3\alpha_0})\frac{1}{r^6} + B\exp(-r/r_0) \tag{1}$$

Where α_0 is the polarizability of a single atom, γ is the second hyperpolarizability, and C_6 is the dispersion coefficient. The first term is the first order dipole-induced-dipole contribution. The dispersion formula [3] is

---

[1] Corresponding author. E-mail: lsheikh@aucegypt.edu

$$\beta(r) = (\frac{6\alpha_0^2}{r^3} + (6\alpha_0^3 + \frac{\gamma C_6}{3\alpha_0})\frac{1}{r^6})F(r) - B\exp(-r/r_0) \qquad (2)$$

with the damping function F given by

$$F(r) = \begin{cases} \exp[-(4.4/r - 1)^2] & \text{if } r < 4.4 \\ 1 & \text{otherwise.} \end{cases} \qquad (3)$$

The computation of the bound-state and of the free-state contributions to the DILS spectral profile within the framework of quantum mechanics has been presented only recently [4,5]. The wave function are built step by step, according to the Fox-Goodwin propagative method, through outward propagation of the wave-function ratio at every pair of adjacent points defined on a spatial grid.

Once a model for the pair potential and pair polarizability anisotropy are given, theoretical calculations of the zeroth and second moments can easily be performed within the framework of classical mechanics, which is appropriate for this case at high temperature, by means of the standard expressions [6]

$$M_0^{pair}(\nu) = \int_{-\infty}^{\infty} I_{\parallel}(\nu)\,d\nu = \frac{2}{15}k_0^4 \int_V g(r)\beta^2(r)\,4\pi\,r^2 dr \qquad (4)$$

$$M_2^{pair}(\nu) = \int_{-\infty}^{\infty} \nu^2 I_{\parallel}(\nu)\,d\nu = \frac{2k_B T}{15\,\mu}k_0^4 \int_V g(r)\left((\frac{d\beta(r)}{dr})^2 + 6\frac{\beta^2(r)}{r^2}\right)4\pi\,r^2 dr \qquad (5)$$

Here g(r) is the pair distribution function given by g(r) = exp (-V(r)/ k_B T)   where V(r) is the interatomic potential.

3. ANALYSIS OF SPECTRAL PROFILES TO DETERMINE $\beta(r)$

As a first step in our analysis we calculate the pair spectrum, using the same potentials [7-10] and polarizability models used previously in the moment analysis [11]. While the experimental spectrum is roughly reproduced with that calculation, differences between the computed and measured spectral profiles remain using the full expression Eq.(1), which are clearly greater than the combined uncertainties of the numerical calculation and the experiment as can be seen in Figs.1 and 2. In other words, the combination of the previously used potentials and polarizability model does not reproduce the measurement within the given uncertainties, despite the fact that consistency of measured and computed zeroth and second spectral moments was observed in Table 1.

We therefore decided to calculate the spectrum using the dispersion formula Eq.(2) at different temperatures and to study the variation of the spectral profiles in response to the variation of r_0 and B values.

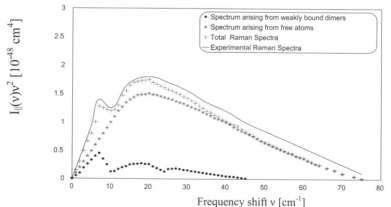

Figure 1. Comparison of computed and measured collision-induced Raman spectra of mercury vapor at T=793 K. For the computed spectrum, the potential from Ref. [9] and the polarizability model of Eq.(1) are used.

Table 1: Moments for the mercury spectra.

Study	Method	M_0 $(10^{-49})cm^5$					M_2/M_0 (cm^{-2})				
		T(K)					T(K)				
		633	793	873	973	1073	633	793	873	973	1073
[12]	Experimental	6.1±.2	5.31±.1	5.81±.1	6.21±.2	5.81±.2	182±2	223±20	233±20	252±4	290±3
Present work	Calculation										
	1st-order DID	3.23	2.898	2.79	2.7	2.62	143.8	179.2	197	218	242
	Full expression (1)	7.46	6.662	6.41	6.18	6.01	170.7	213.98	235.9	263.6	291.5
LJ(12-6)[7]	Dispersion formula (2)	6.95	6.35	5.98	5.75	5.63	159.3	194.6	218.4	245.9	267.9
Ab initio [8]	1st-order DID	3.345	3.1	2.9	2.8	2.73	144.25	180.77	200.0	224.0	247.0
	Full expression	7.6	6.85	6.57	6.35	6.15		218.75	241.0	268.0	297.5
	Dispersion formula	7.28	6.57	6.23	5.92	5.87	162.7	199.4	225.7	247.8	273.8
MSKAK [9]	1st-order DID	2.97	2.73	2.64	2.58	2.53	142.68	178.88	199.0	222.0	245.0
	Full expression	6.85	6.28	6.01	5.94	5.83	172.26	216.65	242.0	267.0	297.0
	Dispersion formula	6.54	6.12	5.74	5.62	5.45	161.15	196.8	220.7	250.8	271.9
Analytical potential [10]	1st-order DID	3.82	3.48	3.37	3.28	3.22	157.6	199.45	220.9	248.3	276.2
	Full expression	6.67	6.48	6.30	6.11	5.85	190.35	226.78	247.43	265.6	295.6
	Dispersion formula	6.29	6.18	6.12	6.05	5.95	179.12	217.05	232.36	252.1	288.5

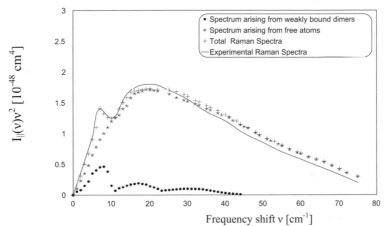

Figure 2. Comparison of computed and measured collision-induced Raman spectra of mercury vapor at T=793 K. For the computed spectrum, the empirical potential from Ref. [10] and the polarizability model of Eq.(1) are used.

The suitable B values are determined by variation of the parameter r_o, forcing the theoretical second spectral moment to reproduce the experimental one. The line- shape calculations are continued for different values of r_o, until a good fit of the line-shape at T=793 K is obtained, with B=45.7 $\overset{\circ}{A}^3$ and

$r_o = 1.2$ $\overset{\circ}{A}$. As can be seen in Fig.3, we now observe agreement of the computed and measured profiles over the complete frequency range The agreement is much closer than anything that was seen before with mercury spectra, particularly when we compare the far wings of the spectra. Agreement at that level apparently requires damping of the long-range part at small separations, presumably in the polarizability model.

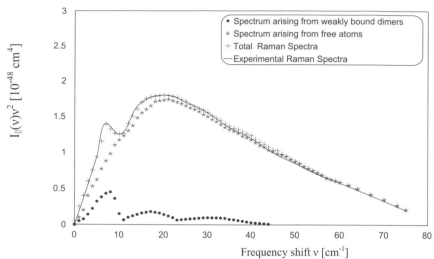

Figure 3. Comparison of computed and measured collision-induced Raman spectra of mercury vapor at T=793 K. For the computed spectrum, the empirical potential from Ref. [10] and the polarizability model of Eq.(2) are used.

Comparison between calculations and experiments for the zeroth and second moments of bound and free states of the depolarized light scattering profile at different temperatures using dispersion model of the polarizability anisotropy is also shown in Table 1. The agreement is excellent in the frequency range of interest and the intensity of the profile calculated by this model of the polarizability anisotropy is the same as the measured one as seen in Fig .3.

References

[1] L.Frommhold; Adv.Chem.Phys.46, 1, 1981.

[2] F.Barocchi and M.Zoppi, in: Intermolecular spectroscopy and dynamical properties of dense systems, J.van Kranendonk (Ed.), Proceeding of the International School of Physics, E.Fermi, Course LXXV, Varenna, North-Holland Pub. Co., 1978.

[3] A.Bonechi, F.Barocchi, M.Moraldi, C.Biermann, R.Winter and L.Frommhold; J.Chem.Phys. 109, 5880, 1998.

[4] M.Chrysos, O.Gaye, and Y.Le Duff, J.Phys.B29, 583, 1996.

[5] O.Gaye, M.Chrysos, V.Teboul, and Y.Le Duff, Phys. Rev. A55, 3484, 1997.

[6] F.Barocchi and M.Zoppi, in Phenomena Induced by intermolecular Interactions, ed.G.Birnbaum, (Plenum, New York), P.311, 1985.

[7] L.E.Epstein and M.D.Power, J.Phys.Chem.336, 57, 1953.

[8] W.E.Baylis, J.Phys.B.10, L583, 1977.

[9] J.Koperski, J.Atkinson and L.Krause; Chem.Phys.Lett.219, 161, 1994.

[10] J.Koperski, J.Atkinson and L.Krause; J.Mol.Spec.184, 300, 1997.

[11] F.Barocchi, F.Hensel and M.Sampoli; Chem.Phys.Lett.232, 445, 1995.

[12] J.Greif, M.Sampoli, F.Barocchi and F.Hensel, Int.Conf. on Spectral Line Shape: Volume11, 15th ICSLS, edited by J.Seidel (American Institute of Physics), 408, 2001.

VSP International
Science Publishers
P.O. Box 346, 3700 AH Zeist
The Netherlands

*Lecture Series on Computer
and Computational Sciences*
Volume 1, 2004, pp. 1073-1076

Empirical Pair- Polarizability Models from Collision-Induced Light Scattering Spectra for Gaseous Argon

S.M.El-Sheikh[1]

Department of Engineering Mathematics and Physics,
Faculty of Engineering, Giza , 12211, Egypt

Received 7 March, 2004; accepted in revised form 10 March, 2004

Abstract: A method based on classical physics, utilizing the first few two moments of the collision-induced light scattering spectra at room temperature to derive an empirical model for the pair-polarizability trace and anisotropy of interacting atoms, with adjustable free parameters, is described and applied to the spectra of argon. Good agreement with *ab initio* results in the literature is obtained and profiles calculated with these models are in excellent agreement with experiments.

Keywords: polarizability, collision-induced light scattering spectra

PACS: 32.10.Dk

1. INTRODUCTION

The polarized and depolarized light which are scattered by a fluid or dense gas due to collisional interactions have a power spectra which are shaped by two functions of the intermolecular separation r, the interaction potential V(r), and, respectively, the trace $\alpha(r)$ and the anisotropy $\beta(r)$ of the induced polarizability [1-4]. In the case of gases consisting of optically isotropic atoms, pure collision-induced depolarized spectra are observed in the vicinity of the Rayleigh line where no monoatomic scattering is allowed [3,4]. Information on the atomic interactions may be obtained from these spectra. For the lower-frequency part of these collision-induced spectra, the dipole-induced-dipole (DID) interaction accounts for most of the observed scattering intensities, whereas, at high frequency range (the well region of the interatomic potential), electron exchange contributions have to be taken into account and can thus be measured for all rare gas diatoms against the dominating classical DID background [3]. For isotropic atoms the excess polarizability induced by interactions in a pair is a tensor that has two invariants: its trace $\alpha(r)$ and its anisotropy $\beta(r)$ related to the isotropic and the anisotropic collision-induced light scattering (CILS), respectively. Recently, the spectral properties of isotropic and anisotropic interaction-induced light scattering could be calculated for both the gaseous and liquid state of monoatomic fluids, on the basis of classical, empirical or *ab initio* models of the induced trace and anisotropy, and of the interaction interatomic potential.

2. ANALYSIS OF MOMENTS TO DETERMINE $\alpha(r)$ and $\beta(r)$

The method of detailed analysis of the first even moments of the polarized and depolarized light scattering spectra has been used by Barocchi and Zoppi [5] for the determination of the extra-dipole-induced dipole (DID) contribution to the pair- polarizability trace and anisotropy of argon. This consists of establishing an appropriate parameterized models and then searching by means of a computer for the sets of parameters that are consistent with the experimental values of the moments. In order to proceed it is convenient to rewrite $\alpha(r)$ and $\beta(r)$ [6] in terms of the reduced variable

---

[1] Corresponding author. E-mail: lsheikh@aucegypt.edu

$x = r/r_m$ where r_m is the separation at the minimum of the interatomic potential $V(x)$ [7]. In this case one has:

$$\alpha(x) = \left(A_6^* x^{-6} + B^* x^{-8} - t_o^* \exp\left(-\frac{x - \sigma^*}{x_t}\right) \right)$$ (1)

$$\beta(x) = \frac{6\alpha_o^2}{r_m^3} \left(x^{-3} + A^* x^{-6} + D^* x^{-8} - g_o^* \exp\left(-\frac{x - \sigma^*}{x_o}\right) \right)$$ (2)

where the different coefficients are defined in Ref.[8]. The substitutions of (1) into the moment expressions [9] for the depolarized spectrum and of (2) into the moment expressions [10] for the depolarized spectrum make it possible to rewrite them in the form of quadratic equations for the unknown t_o^* and g_o^* with coefficients which are parametric functions of x_t and x_o. The coefficients are given by simple expressions involving about 32 integrals over the interatomic separations of the form

$$\int_0^\infty x^{-m} \exp(-n\, x/x_o) \left(\frac{d^2 V}{dx^2}\right)^p \exp[-V(x)/k_B T]\, dx$$ (3)

The great advantages of this method for calculating the parameters of the models for $\alpha(r)$ and $\beta(r)$ are the speed of computation and that the trail-and-error approach is avoided.

For the calculation of the theoretical values of the moments with the empirical models of the trace and anisotropy polarizabilities the most recent pair potential [7] and ten different literature potentials [11-20], are used. Also, for the sake of comparison and discussion, for the present calculations we considered three models of the pair-polarizability trace and anisotropy which are DID [21], PKF [22] and SCF [23] models.

The comparison between the experimental moments and the calculations performed with the two-term dipole-induced dipole (DID) approximation, the empirical PKF and the *ab initio* SCF models of the trace and anisotropy respectively using different potentials mentioned before are shown in Table 1.

Table 1: A comparison between theoretical and experimental moments of the polarized and depolarized light scattering spectra of argon using the different potential models and different trace and anisotropy models of polarizabilities at T=295.0 K.

Potential	M_0 (Å⁹)						M_2 (10^{25} Å⁹/s²)					
	Polarized			Depolarized			Polarized			Depolarized		
	DID	PKF	SCF	DID	PKF	SCF	DID	PKF	SCF	DID	PKF	SCF
LJ (9-6) [11]	0.016	0.135	0.147	40.2	35.9	38.9	0.061	0.164	0.173	45.2	33.0	35.0
LJ(12-6) [12]	0.017	0.141	0.157	45.2	40.6	41.8	0.061	0.167	0.178	50.9	38.0	40.0
LJ(12-6) [13]	0.016	0.137	0.146	44.6	40.1	42.0	0.058	0.154	0.163	49.6	37.4	39.3
LJ(12-6) [14]	0.017	0.137	0.148	45.2	40.6	41.7	0.061	0.162	0.171	50.7	38.0	39.5
LJ(12-6) [15]	0.023	0.143	0.162	51.9	45.9	46.5	0.087	0.219	0.32	62.2	44.7	46.2
LJ(12-6) [16]	0.028	0.151	0.169	56.7	49.5	51.3	0.109	0.247	0.294	71.0	49.2	51.7
MB (exp-6) [17]	0.013	0.121	0.128	30.5	35.6	38.4	0.047	0.137	0.142	34.5	37.0	38.2
Kihara [18]	0.014	0.124	0.131	31.7	36.1	39.2	0.049	0.139	0.144	35.3	37.4	38.7
Aziz [19]	0.064	0.137	0.142	48.8	44.5	43.2	1.275	1.95	2.25	45.2	41.2	40.8
BFW [20]	0.052	0.129	0.138	46.4	41.9	40.6	1.24	1.84	2.19	44.1	41.8	41.2
M3SV [7]	0.071	0.144	0.162	45.7	42.1	39.9	1.31	2.57	2.82	43.2	40.5	39.4
Experiment [24,25]		0.196			42.8±3.8			3.495			42.3±5.2	

All polarizability models give values for these moments those are well outside the experimental uncertainties at room temperature. These results indicate that, in the case of argon, an additional polarizability at long and intermediate range must be added to the DID one in order to verify the experimental zeroth and second moments of the trace and anisotropy. These pair polarizabilities, which are suggested by El-Sheikh *et al.* [6], are the analytical models for the hyperpolarizability contributions of noble gases.

The agreement between the experimental values and the theoretical ones using M3SV [7] is excellent and we have verified that it remains acceptable for $t_o = 0.0301 \pm 0.0004 \ \overset{o}{A}^3$,

$r_t = 0.316 \pm 0.002 \ \overset{o}{A}$ and $g_o = 0.215 \pm 0.02 \ \overset{o}{A}^3$, $r_o = 0.43 \pm 0.015 \ \overset{o}{A}$ in the case of the trace and anisotropy models respectively.

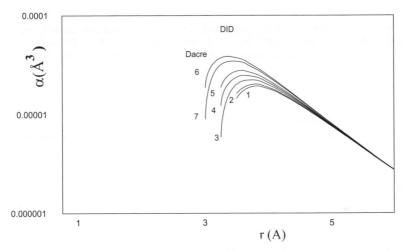

Figure 1: Empirical models (1-7) of the trace $\alpha(r)$. Dot dashed curve represents Dacre's *ab initio* result [23]. The dashed line is the two-term DID function [21].

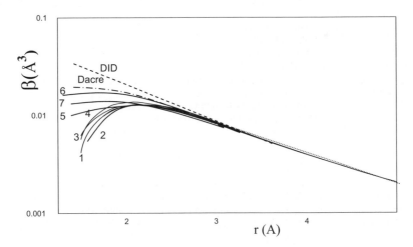

Figure 2: Empirical models (1-7) of the polarizability anisotropy $\beta(r)$. Dot dashed curve represents Dacre's *ab initio* result [23]. The dashed line is the two-term DID function [21].

In figures 1 and 2 the corresponding empirical models for $\alpha(r)$ and $\beta(r)$ are compared with the functions obtained by *ab initio* methods by Dacre [23] for Ar system. In the range of intermolecular separations probed by the molecules, Dacre's *ab initio* models fall between the bounds of the empirical models, determined by the range of acceptable t_o and g_o values.

An effective means for checking the validity of the different models of induced trace and anisotropy $\alpha(r)$ and $\beta(r)$ is to compare the computed second dielectric constant B_ε and second virial kerr coefficient B_K with the experimental ones [26,27]. An excellent agreement with the experimental results for both the second dielectric constant and the second virial kerr coefficient.

3. RESULTS

The result of our analysis is therefore that argon pairs develop an incremental polarizability trace and anisotropy during collisions, besides the DID one, which contributes substantially at intermediate-range distances and can be ascribed to other mechanisms of electron cloud distortion, such as overlap and electron-correlation effects.

References

[1] W.M.Gelbart, Adv.Chem.Phys, 26, 1, 1974.
[2] G.C.Tabisz, Mol.Spectrosc. (Specialist Periodical Reports, The Chemical Societies, London), 6, 136, 1979.
[3] L.Frommhold, Adv.Chem.Phys., 46, 1, 1981.
[4] G.Birnbaum, B.Guillot and S.Bratos, Adv.Chem.Phys., 51, 49, 1982.
[5] U.Bafile, R.Magli, F.Barocchi, M.Zoppi and L.Frommhold, Molec.Phys., 49, 1149, 1983.
[6] S.M.El-Sheikh, G.C.Tabisz and A.D.Buckingham, Chem.Phys., 247, 407, 1999.
[7] M.S.A.El-Kader, J.Phys. B, 35, 4021, 2002.
[8] M.S.A.El-Kader, Z.Phys.Chem., 218, 293, 2004.
[9] H.B.Levine and G.Birnbaum, J.Chem.Phys., 55, 2914, 1971.
[10] F.Barocchi and M.Zoppi in Phenomena Induced by intermolecular Interactions, ed.G.Birnbaum (Plenum ,New York), 311, 1985.
[11] E.Walley and W.G.Schneider, J.Chem.Phys., 23, 1644, 1955.
[12] J.O.Hirshfielder, C.F.Curtiss and R.B.Bird, Molecular theory of gases and liquids (Wiley, New York), 1954.
[13] S.Toxwaerd and E.Praestgaad, Acta Chem.Scand. 22, 1873, 1968.
[14] A.Michels and H.Wijker, Physica (Utr.), 15, 629, 1949.
[15] H.L.Johnston and E.R.Grilly, J.Chem.Phys., 46, 948, 1942.
[16] B.E.F.Fender and G.D.Halsey, J.Chem.Phys., 36, 1881, 1962.
[17] R.DiPippo and J.Kestin, Proceedings Fourth Symposium on Thermophysical Properties (Am.Soc.Mech.Eng., New York), P.304, 1968.
[18] A.E.Sherwood and J.M.Prausnitz, J.Chem.Phys., 41, 429, 1964.
[19] R.A.Aziz and M.J.Slaman, J.Chem.Phys., 92, 1030, 1990.
[20] J.A.Barker, R.A.Fisher and R.O.Watts, Mol.Phys., 21, 657, 1971.
[21] L.Silberstein, Philos.Mag.33, 521, 1917.
[22] M.H.Proffitt, J.W.Keto and L.Frommhold, Can.J.Phys., 59, 1459, 1981.
[23] P.D.Dacre, Mol.Phys., 45, 1, 1982.
[24] O.Gaye, M.Chrysos, V.Teboul and Y.Le Duff, Phys.Rev. A55, 3484, 1997.
[25] F.Chapeau-Blondeau, V.Teboul, J.Berrue and Y.Le Duff, Phys.Lett., 173A, 153, 1993.
[26] H.J.Achtermann, G.Magnus and T.K.Bose, J.Chem.Phys., 94, 5669, 1991.
[27] A.D.Buckingham and D.A.Dunmur, Trans.Faraday Soc. 64, 1776, 1968.

VSP International
Science Publishers
P.O. Box 346, 3700 AH Zeist
The Netherlands

Lecture Series on Computer
and Computational Sciences
Volume 1, 2004, pp. 1077-1080

Interaction Dipole Moment and (Hyper)Polarizability in Rg-Xe

A.Haskopoulos and G. Maroulis[1]

Department of Chemistry,
University of Patras,
GR-26504 Patras, Greece

Received 7 March, 2004; accepted in revised form 10 March, 2004

Abstract: We have obtained interaction dipole moments and (hyper)polarizabilities for the complexes
Rg···Xe, Rg = He, Ne, Ar, Kr and Xe. Relying on finite-field second-order Møller-Plesset perturbation
theory and density functional theory calculations with large, carefully optimized basis sets, the interaction
properties are tabulated for a wide range of interatomic separation distances.

Keywords: (hyper)polarizabilities, van der Waals complexes

1. Introduction

Rare gas dimers have served as model systems to study the van der Waals bonding with both
experimental [1] and theoretical methods [2]. The very weak binding energy of these complexes arise
difficulties which, from a theoretical point of view, necessitate special consideration with respect to the
employed basis set and to the level of theory used in the calculations[3].

The majority of the published work on the interaction of the rare gas atoms has been devoted to the
determination of accurate potential energy curves [4-7]. On the contrary experimental data concerning
fundamental properties, such as the dipole moment and (hyper)polarizability, are notably few. For the
case of Rg···Xe we mention the calculation of the dipole moment of RgRg' systems [8,9] and the
determination of the anisotropy of the dipole polarizability of Rg_2 [10].

The interaction-induced properties of these weakly bound complexes are directly related to the
collision- and interaction-induced spectroscopies. The interaction dipole moment in rare gas
heterodiatoms is linked to collision-induced absorption (CIA) and the mean and anisotropy of the
dipole polarizability are linked to collision-induced light-scattering (CILS) [11].

In previous work on the rare gas diatoms, He···He, Ne···Ne, Ar···Ar and Kr···Kr [12] and on the Ne···Ar
heterodiatom [13] we presented a complete description of the electric properties of the referenced
systems.

In the present study we turn our attention to the interaction properties of the van der Waals complexes
Rg···Xe, Rg = He, Ne, Ar, Kr and Xe. In addition to the conventional ab initio calculations we have also
employed two widely used DFT methods, namely B3LYP and B3PW91. We have calculated the
interaction dipole moment and (hyper)polarizability relying on large, flexible, near-Hartree-Fock
quality basis sets. The above properties are calculated for a broad range of interatomic distances, $3 <
R/\alpha_0 < 20$.

2. Theory

When an uncharged molecule experiences an external, weak, homogeneous electric field, its energy
may be written as [14] :

$$E^p = E^0 - \mu_\alpha F_\alpha - \frac{1}{2}\alpha_{\alpha\beta}F_\alpha F_\beta - \frac{1}{6}\beta_{\alpha\beta\gamma}F_\alpha F_\beta F_\gamma - \frac{1}{24}\gamma_{\alpha\beta\gamma\delta}F_\alpha F_\beta F_\gamma F_\delta + ... \qquad (1)$$

where F_α is the electric field, E^0 is the energy of the free molecule, μ_α is the dipole moment, $\alpha_{\alpha\beta}$ is the
dipole polarizability, $\beta_{\alpha\beta\gamma}$ is the first dipole hyperpolarizability and $\gamma_{\alpha\beta\gamma\delta}$ is the second dipole
hyperpolarizability. The subscripts denote Cartesian components and a repeated subscript implies

---

[1] Corresponding author, e-mail: maroulis@upatras.gr

summation over x, y and z. The number of independent components needed to specify the dipole moment and (hyper)polarizability tensors is regulated by the symmetry of the system. For systems of $C_{\infty v}$ symmetry, as the complexes Rg···Xe, Rg = He, Ne, Ar and Kr, the dipole moment tensor has one, the dipole polarizability two and the first dipole hyperpolarizability two independent components. With z as the molecular axis those are: μ_z, α_{xx} and α_{zz}, β_{zzz} and β_{zxx}. For the complex Xe···Xe, μ and β are zero ($D_{\infty h}$ symmetry). In addition to the Cartesian components we also calculate the following quantities:

$$\bar{\alpha} = (\alpha_{zz} + 2\alpha_{xx})/3 \tag{2}$$

$$\Delta\alpha = \alpha_{zz} - \alpha_{xx} \tag{3}$$

$$\bar{\beta} = \frac{3}{5}(\beta_{zzz} + 2\beta_{zxx}) \tag{4}$$

The interaction properties are obtained via the Boys-Bernardi counterpoise-correction (CP) method [15]. For a defined configuration of system A···B the interaction property $P_{int}(A···B)$ is computed as:

$$P_{int}(A···B) = P(A···B) - P(A···X) - P(X···B) \tag{5}$$

where P(A···X) etc. denotes calculations of the property for the subsystem A in the presence of the ghost orbitals of subsystem B. We have used Eq. (5) to determine the interaction dipole moments, polarizabilities and hyperpolarizabilities.

The ab initio calculations reported in the present study rely on self-consistent-field (SCF), second order Møller-Plesset perturbation theory (MP2) [16]. We also employ the DFT methods, B3LYP and B3PW91.

3. Computational Details

The quality of the employed basis sets is of primary importance in the derivation of interaction properties. We based our study on the basis sets used in previous works [12, 13].

Table 1: Basis sets used in the calculations

System	Basis set	GTF
He···Xe	[6s4p3d/9s8p7d5f]	136
Ne···Xe	[7s5p4d1f/9s8p7d5f]	152
Ar···Xe	[8s6p5d4f/9s8p7d5f]	182
Kr···Xe	[8s7p6d5f/9s8p7d5f]	197
Xe···Xe	[9s8p7d5f/9s8p7d5f]	206

All calculations were performed with GAUSSIAN 94 and GAUSSIAN 98.
Atomic units are used through this work. Conversion factors to SI units are: Energy, 1 E_h = 4.3597482 x 10^{-18} J, Length, 1 a_0 = 0.529177249 x 10^{-10} m, dipole moment ,μ , 1 ea_0 − 8.478358 x 10^{-30} Cm, dipole polarizability, α, 1 $e^2a_0^2E_h^{-1}$ = 1.648778 x 10^{-41} $C^2m^2J^{-1}$ and dipole hyperpolarizability, β, 1 $e^3a_0^3E_h^{-2}$ = 3.206361 x 10^{-53} $C^3m^3J^{-2}$.

4. Results and Discussion

Table 2: Dipole moment and (hyper)polarizability invariants for the complexes Rg⋯Xe, Rg = He, Ne, Ar, Kr and Xe at R = 5 α_0 calculated with the SCF, MP2, B3LYP and B3PW91 method.

System	Method	μ_z	$\overline{\alpha}$	$\Delta\alpha$	$\overline{\beta}$
He⋯Xe	SCF	-0.0890	-0.48	-0.07	-32.3
	MP2	-0.0875	-0.48	-0.04	-36.9
	B3LYP	-0.0813	-0.47	0.18	-40.1
	B3PW91	-0.0801	-0.45	0.22	-38.1
Ne⋯Xe	SCF	-0.1532	-0.52	0.45	-35.4
	MP2	-0.1453	-0.49	0.87	-39.6
	B3LYP	-0.1363	-0.4	1.29	-36.4
	B3PW91	-0.1392	-0.41	1.21	-38.4
Ar⋯Xe	SCF	-0.2388	-0.51	8.37	-63.6
	MP2	-0.2217	-0.21	9.45	-67.2
	B3LYP	-0.2193	-0.03	10.72	-63.8
	B3PW91	-0.2217	0.01	10.54	-67.0
Kr⋯Xe	SCF	-0.1972	-0.05	14.88	-50.0
	MP2	-0.1834	0.53	16.56	-48.1
	B3LYP	-0.1760	0.84	18.73	-36.9
	B3PW91	-0.1792	0.86	18.36	-42.6
Xe⋯Xe	SCF		1.69	27.98	
	MP2		2.85	30.92	
	B3LYP		3.32	34.08	
	B3PW91		3.23	33.32	

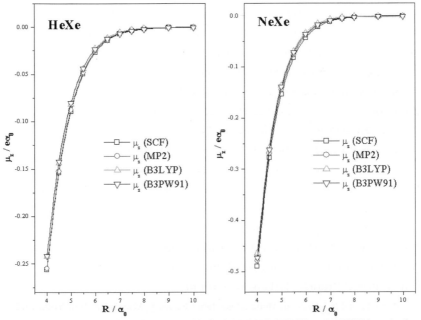

Figure 1: μ_z of the He⋯Xe and Ne⋯Xe with the SCF, MP2, B3LYP and B3PW91 method.

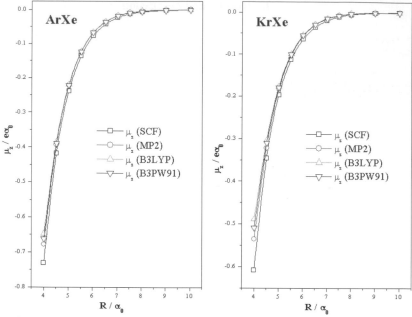

Figure 2: μ_z of the Ar···Xe and Kr···Xe with the SCF, MP2, B3LYP and B3PW91 method.

References

[1] C.E.H. Dessent and K.Müller-Dethlefs, Chem.Rev 100, 3999 (2000).

[2] G.Chałasiński and M.M.Szczęśniak, Chem.Rev. 100, 4227 (2000).

[3] G.Chałasiński and M.M.Szczęśniak, Chem.Rev. 94, 1723 (1994).

[4] T.P.Haley and S.M.Cybulski, J.Chem.Phys. 119, 5487 (2003).

[5] S.Faas, J.H.Van Lenthe and J.G.Snijders, Mol.Phys. 9, 1467 (2000).

[6] K.T.Tang and J.P.Toennies, J.Chem.Phys. 118, 4976 (2003).

[7] T.J.Giese and D.M.York, J.Chem.Phys. 120, 590 (2004).

[8] E.Bar-Ziv and S.Weiss, J.Chem.Phys. 64, 2417 (1976).

[9] W.Jäger, Y.Xu and M.C.Gerry, J.Chem.Phys. 99, 919 (1993).

[10] S.Minemoto, H.Tanji and H.Sakai, J.Chem.Phys. 119, 7737 (2003).

[11] G.C. Tabisz, M.N. Neuman (Eds), Collision- and Interaction-Induced Spectroscopy, Kluwer, Dordrecht, 1995.

[12] G.Maroulis, J.Phys.Chem. A 104, 4772 (2000).

[13] G.Maroulis and A.Haskopoulos, Chem.Phys.Lett. 358, 64 (20002)

[14] A.D.Buckingham, Adv.Chem.Phys. 12, 107 (1967).

[15] S.F. Boys, F. Bernardi, Mol.Phys. 19, 55 (1970).

[16] A. Szabo and N.S.Ostlund, Modern Quantum Chemistry, MacMillan, New York, 1982.

VSP International
Science Publishers
P.O. Box 346, 3700 AH Zeist
The Netherlands

Lecture Series on Computer
and Computational Sciences
Volume 1, 2004, pp. 1081-1084

A Database of Gaussian-Type Basis Sets Applied to the Calculation of Electric Properties of the Azide Anion

A.Hatzis, A.Haskopoulos and G. Maroulis[1]

Department of Chemistry,
University of Patras,
GR-26500 Patras, Greece

Abstract: We have designed a systematic sequence of gaussian-type basis sets for the calculation of electric properties of the azide anion. The basis sets are subsequently used for the determination of the quadrupole and hexadecapole moment and the electric dipole polarizability. Our study brings forth the basis set dependence of these important properties both at the Hartree-Fock and post-Hartree-Fock level of theory. The basis sets would be suitable for intermolecular interaction studies involving the azide anion.

Keywords: Azide, quadrupole moment, hexadecapole moment, polarizability, basis set.

PACS: 32.10.Dk

1. Introduction and theoretical considerations

There is considerable interest in the physicochemical properties of the azide anion N_3^-. A theoretical study of its spectroscopic properties was published by Botschwina [1]. Recent work by Morita et al. [2] and Yang et al [3] has shown that the accurate determination of the quadrupole moment and dipole polarizability of this anion is of primary importance for the interpretation of its physicochemical behaviour in gaseous phase or in solution. Attempts to calculate the quadrupole moment appeared early enough in the literature [4]. Experimental estimates of the mean and the anisotropy of the dipole polarizability have been published by Le Borgne et al. [5].

The aim of this study is to present a systematic sequence of Gaussian-type basis sets especially designed for theoretical calculation of electric properties for N_3^-. We follow a computational philosophy presented in detail in previous work [6]: a) the initial substrate is augmented with diffuse s- and p-GTF b) tight d-GTF are added and their exponent is chosen to minimize the energy of the free molecule c) diffuse d-GTF are added and their exponent is chosen to maximize the mean dipole polarizability. Larger basis sets are created with addition of more polarization functions.

We obtain theoretical values of the quadrupole ($\Theta_{\alpha\beta}$) and hexadecapole ($\Phi_{\alpha\beta\gamma\delta}$) moments and the dipole polarizability ($\alpha_{\alpha\beta}$). We adopt Buckingham's notation and conventions in all cases [7]. A centrosymmetric linear system, as the azide anion, has one independent component for the quadrupole and hexadecapole moment and two for the dipole polarizability. In addition to the Cartesian components of $\alpha_{\alpha\beta}$ we calculate also the mean and the anisotropy defined as

$$\bar{\alpha} = \left(2\alpha_{xx} + \alpha_{zz}\right)/3 \tag{1}$$

$$\Delta\alpha = \alpha_{zz} - a_{xx} \tag{2}$$

The methods used in this study are self-consistent field (SCF), second-order Møller-Plesset perturbation theory (MP2) [8] and the widely used B3LYP density functional theory method (as implemented in the GAUSSIAN 98 program [9]). To analyze the convergence of our values we rely on the following definition of similarity: similarity for two property values Q_i and Q_j is defined as [6]

---

[1] Corresponding author, e-mail: maroulis@upatras.gr

$$S_{ij} \equiv S(i,j) = 1 - \frac{|Q_i - Q_j|}{\max_{ij}|Q_i - Q_j|} \tag{3}$$

2. Results and discussion

The starting point for the construction of basis set is an Ahlrichs TZV basis set TZV (11s6p) contracted to [5s3p] as {62111/411}. The smallest basis set designed in this study is [7s5p2d]. This was augmented systematically to [7s5p5d3f2g]. Thus the total sequence comprises nine basis sets. 5D, 7F and 9G GTF were used on all basis sets.

We give in Tables 1 and 2 SCF and MP2 values for the calculated properties. We observe that convergence is not uniform for all properties. Using as reference our largest basis set, we show in figures 1, 2, 3 and 4 the evolution of $S(k,9)$, $k=1,\ldots,9$. We have included both SCF and MP2 results in all figures.

Table 1: Polarizabilities and multipole moments of N3- calculated with the SCF method.

Basis Set	Θ_{zz}	Φ_{zzzz}	α_{xx}	α_{zz}	$\bar{\alpha}$	$\Delta\alpha$
[7s5p2d]	-8.5265	3.57	24.90	61.66	37.15	36.76
[7s5p3d]	-8.5931	2.21	25.17	62.37	37.57	37.21
[7s5p3d1f]	-8.5740	-1.22	25.24	62.55	37.68	37.31
[7s5p4d1f]	-8.4290	0.27	25.69	62.14	37.84	36.45
[7s5p4d2f]	-8.4286	0.24	25.70	62.14	37.85	36.44
[7s5p5d2f]	-8.4284	0.11	25.70	62.13	37.84	36.43
[7s5p5d3f]	-8.4659	0.70	25.77	62.29	37.94	36.52
[7s5p5d3f1g]	-8.4643	0.49	25.77	62.28	37.94	36.51
[7s5p5d3f2g]	-8.4608	0.54	25.77	62.27	37.94	36.50

Table 2:Polarizabilities and multipole moments of N_3^- calculated with the MP2 method.

Basis Set	Θ_{zz}	Φ_{zzzz}	α_{xx}	α_{zz}	$\bar{\alpha}$	$\Delta\alpha$
[7s5p2d]	-8.6927	-3.72	28.08	67.63	41.26	39.55
[7s5p3d]	-8.7515	-4.32	28.77	68.40	41.98	39.63
[7s5p3d1f]	-8.7168	-6.53	29.07	68.53	42.22	39.46
[7s5p4d1f]	-8.6422	-6.26	28.99	67.81	41.93	38.82
[7s5p4d2f]	-8.6406	-6.37	29.00	67.80	41.93	38.80
[7s5p5d2f]	-8.6400	-6.58	28.99	67.75	41.91	38.77
[7s5p5d3f]	-8.6075	-6.98	28.65	67.02	41.44	38.38
[7s5p5d3f1g]	-8.6027	-6.89	28.69	67.05	41.47	38.36
[7s5p5d3f2g]	-8.5953	-7.50	28.99	67.23	41.74	38.24

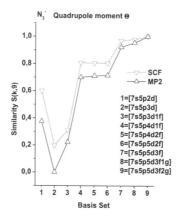

Figure 1: Convergence of the SCF and MP2 values of the quadrupole moment.

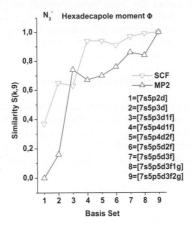

Figure 2: Convergence of the SCF and MP2 values of the hexadecapole moment.

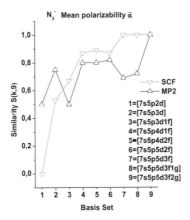

Figure 3: Convergence of the SCF and MP2 values of the mean polarizability.

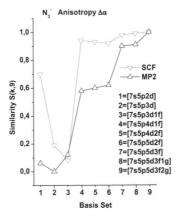

Figure 4: Convergence of the SCF and MP2 values of the dipole polarizability anisotropy.

Comparing SCF and MP2 values, we observe that electron correlation effects are very important for all properties. The second order correction $D2 \equiv MP2-SCF$ is very large for the hexadecapole moment.

We have calculated B3LYP values for basis sets [7s5p2d] to [7s5p5d3f1g]. Comparing our B3LYP/[7s5p5d3f1g] values to the analogous MP2/[7s5p5d3f1g] we see that the DFT method overestimates both Cartesian components of the polarizability. Consequently, the mean is substantially higher. The B3LYP quadrupole moment is in fair agreement with the MP2 one. The two methods diverge considerably over the hexadecapole moment.

Concluding remarks

We have found basis set effects to be important at all levels of theory employed in this study. Higher level methods seem to be necessary for the determination of reliable values. We expect the designed basis sets to be useful in calculations of electric properties for the azide anion. They should also be useful in intermolecular studies involving this anion. Work is currently in progress in our laboratory on the intermolecular interactions of N_3^- with simple atomic and molecular systems.

References

[1] P.Botschwina, *J.Chem.Phys.*, **1986,** 85, 4591.

[2] A.Morita and S.Kato, J.Chem.Phys. 109, 5511 (1998).

[3] X.Yang, B.Kiran, X.B.Wang, L.S.Wang, M.Mucha and P.Jungwirth, J.Phys.Chem A xxx, xxx (2004).

[4] T.Gora and P.J.Kemmey, *J.Phys.Chem.,* **1972,** 57, 3579.

[5] C.LeBorgne, B.Illien, M.Meignon, and M.Chabanel, *Phys. Chem. Chem. Phys.,* **1999,** 1, 4701.

[6] G.Maroulis, J.Chem.Phys. **108**, 5432 (1998).

[7] A.D.Buckingham, Adv.Chem.Phys. **12**, 107 (1967).

[8] A. Szabo and N.S.Ostlund, Modern Quantum Chemistry, MacMillan, New York, 1982.

[9] M.J.Frisch *et al.*, GAUSSIAN 98, Revision A.7 (Gaussian, Inc., Pittsburgh PA, 1998).

VSP International
Science Publishers
P.O. Box 346, 3700 AH Zeist
The Netherlands

*Lecture Series on Computer
and Computational Sciences*
Volume 1, 2004, pp. 1085-1087

Static (Hyper)Polarizability of CH₃X, X=F, Cl, Br and I

P. Karamanis and G. Maroulis[1]

Department of Chemistry,
University of Patras,
GR-26504 Patras, Greece

Received 7 March, 2004; accepted in revised form 10 March, 2004

Abstract: We present a systematic study of static dipole polarizability ($\alpha_{\alpha\beta}$) and static first ($\beta_{\alpha\beta\gamma}$) and second ($\gamma_{\alpha\beta\gamma\delta}$) hyperpolarizability of the monohalogenated methanes, CH₃X, X=F, Cl, Br, I. All properties have been obtained from finite-field Møller-Plesset Perturbation theory and Coupled Cluster calculations with flexible, especially designed gaussian-type basis sets. Good agreement is observed between experimental and theoretical values of the dipole polarizability. In the other hand the comparison in the case of the hyperpolarizabilities is not obvious.

Keywords: Electric polarizability, electric hyperpolarizability, halomethanes

PACS: 32.10.Dk.

1. Introduction

The conjugation effect in the design of new materials that exhibit high non-linear optical character has been proven that has its limitations [1]. Novel molecular architectures have been tested in experimental and theoretical investigations. The combination of highly polarizable atoms with organic molecules may provide new, improved non-linear optical (NLO) materials [2].

In this work we study systematically the evolution of the static electric (hyper)polarizability in the series: CH₃F, CH₃Cl, CH₃Br, CH₃I. The electric properties of halomethanes have been well characterized experimentally and for this reason we have chosen them as fundamental molecular models in our study. Although the dipole polarizability of small molecules can be to a certain degree routinely predicted by **ab initio** methods, the prediction of the hyperpolarizability is no mean task. Basis set effects are important. On the other hand post-Hartree-Fock methods of high predictive potential are required in order to obtain realistic estimates of electron correlation effects. A full presentation of the methods employed in this study can be found in standard textbooks [3,4] We rely on flexible basis sets specially designed for (hyper)polarizability calculations. The computational philosophy underlying our approach has been presented in detail in previous work [5].

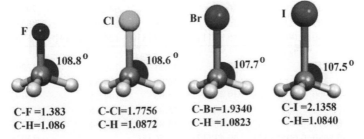

F 108.8°	Cl 108.6°	Br 107.7°	I 107.5°
C-F =1.383	C-Cl=1.7756	C-Br=1.9340	C-I =2.1358
C-H=1.086	C-H =1.0872	C-H =1.0823	C-H=1.0840

Figure 1: Experimental geometries used in all calculations. Bond distances are in Å and angles in degrees

---

[1] Corresponding author, e-mail: maroulis@upatras.gr

2. Theory and Computational Details

The energy of an uncharged molecule interacting with a weak homogeneous electric field can be written as:

$$E(F_a) = E^0 - \mu_a F_a - (1/2)\alpha_{\alpha\beta}F_a F_\beta - (1/6)\beta_{\alpha\beta\gamma}F_a F_\beta F_\gamma - (1/24)\gamma_{\alpha\beta\gamma\delta}F_a F_\beta F_\gamma F_\delta + ... \qquad (1)$$

where F_a is the field E^0 is the energy of the free molecule and $\alpha_{\alpha\beta}$, $\beta_{\alpha\beta\gamma}$, $\gamma_{\alpha\beta\gamma\delta}$ are the dipole (hyper)polarizabilities. The subscripts denote Cartesian components and the repeated subscript implies summation over x, y and z.

For the H, C, F and Cl atoms all basis sets were built upon the D95 [6]. For Br and I we used the substrates $(13s10p4d)[4s3p1d]$ and $(16s13p7d)[5s4p2d]$, respectively [6]. All substrates were systematically augmented in order to obtain near-Hartree-Fock quality basis sets. Experimental molecular geometries were used for all molecules. All optimizations and subsequent calculations were performed with GAUSSIAN 94 and GAUSSIAN 98.

Atomic units are used throughout this study. Conversion factors to SI units are, Length, 1 a_0 = 0.529177249 x 10^{-10} m, dipole (μ), 1 ea_0 = 8.478358 x 10^{-30} Cm, dipole polarizability (α), 1 $e^2 a_0^2 E_h^{-1}$ = 1.648778 x 10^{-41} $C^2 m^2 J^{-1}$, first hyperpolarizability (β), 1 $e^3 a_0^3 E_h^{-2}$ = 3.206361 x 10^{-53} $C^3 m^3 J^{-2}$ and second hyperpolarizability (γ), 1 $e^4 a_0^4 E_h^{-3}$ = 6.235378 x 10^{-65} $C^4 m^4 J^{-3}$.

3. Results

Table 1: Electric properties of monosubstituted halomethanes. All properties in atomic units.

	μ	$\bar{\alpha}$	$\Delta\alpha$	$\bar{\beta}$	$\bar{\gamma}$
CH₃F					
SCF	-0.8107	15.79	1.12	36.24	1239
MP2	-0.7306	16.85	1.53	40.32	1801
MP4	-0.7210	16.94	1.60	41.51	1962
CCSD	-0.7384	16.59	1.48	38.86	1765
CCSD(T)	-0.7265	16.77	1.52	40.64	1816
Exp.	-0.7278 [7]	17.32 [8]	1.57 [9]	59±31 [10]	2875±230 [10]
CH₃Cl					
SCF	-0.8328	28.47	9.74	-11.69	4240
MP2	-0.7483	29.48	9.55	-17.08	5715
MP4	-0.7354	29.43	9.50	-18.03	5702
CCSD	-0.7461	29.12	9.48	-17.87	5230
CCSD(T)	-0.7378	29.38	9.47	-18.15	5602
Exp.	-0.7353±0.0039 [11]	29.80 [12]	9.76 [12]	-13.3±1.4 [13]	6860±360 [13]
CH₃Br					
SCF	-0.8461	35.62	13.28	-53.23	6568
MP2	-0.7337	36.47	12.57	-61.98	8623
MP4	-0.7185	36.51	12.43	-65.46	8658
CCSD	-0.7311	36.16	12.50	-64.21	7991
CCSD(T)	-0.7207	36.46	12.40	-65.59	8531
Exp.	-0.7070±0.0059 [11]	36.81 [11]	13.04 [8]	-124.7±31 [14]	8576 [15]
CH₃I					
SCF	-0.8315	48.60	17.31	-136.26	12201
MP2	-0.7115	49.43	16.00	-137.83	15451
MP4	-0.7000	49.47	15.83	-144.25	15495
CCSD	-0.7073	49.06	15.84	-144.15	14447
CCSD(T)	-0.7010	49.40	15.84	-144.54	15227
Exp.	-0.6480±0.0055 [11]	49.43 [12]	16.56 [8]		22037 [15]

We show in Table 1 electric dipole moments and dipole (hyper)polarizabilities of the monosubstituted halomethanes. A more detailed presentation of experimental data will be given elsewhere. A direct comparison with experiment would necessitate an estimation zero-point vibrational corrections (ZPVC) for all properties. This is far from trivial for polyatomic systems. In addition, the experimental data for the (hyper)polarizability (usually) pertain to opical frequencies. Estimates for the static limit are not always available. It is more instructive to analyze the relative quality of the results obtained through the available methods. Work is now in progress in our laboratory towards this direction.

References

[1] S. R. Marder, J. E. Sohn (Eds), Materials for Nonlinear Optics: Chemical Perspectives, ACS Symposium Series 455 (ACS, Washington DC, 1991)

[2] H.S.Nalwa and S.Miyata (Eds), Nonlinear Optics of Organic Molecules and Polymers (CRC Press, Boca Raton (1997).

[3] A. Szabo and N.S.Ostlund, Modern Quantum Chemistry, MacMillan, New York, 1982.

[4] T.Helgaker, P.Jørgensen, J.Olsen, Molecular Electronic-Structure Theory (Wiley, Chichester, 2000).

[5] G.Maroulis, J.Chem.Phys. 108, 5432 (1998).

[6] Full basis set data are available from the authors.

[7] Marshall, M. D., Muenter and J. S., J. Molec. Spectrosc. **83**, 279 (1980)

[8] M. P. Bogaard, A D Buckingham R. K. Picrens and Λ. H. White, J. Chem. Soc., Faraday Trans 1 **74**, 3008 (1978)

[9] T. N. Olney, N.M. Cann, G.Cooper and C.E. Brion, Chem Phys. Lett. **59-58**, 223 (**1997**)

[10] A. D. Buckingham and B. J. Orr, J. Trans. Faraday Soc. **65**, 673 (1969)

[11] R. G. Shulman, B. P. Dailey and C. H. Townes, Phys. Rev. **78**, 145 (1950)

[12] K. L. Ramaswamy, Proc. Indian Acad. Sci. A **4**, 675 (1936)

[13] J. F. Ward and K. J. Miller, Phys. Rev. A **16**, 1179 (1975)

[14] E. W. Blanch, R. I. Keir, G L. D. Ritchie, J. Phys. Chem. A **106**, 4257 (2002)

[15] T. Lundeen, S. Y. Hou and J. W. Nibler, J. Chem. Phys. **79**, 6301 (1983)

VSP International
Science Publishers
P.O. Box 346, 3700 AH Zeist
The Netherlands

*Lecture Series on Computer
and Computational Sciences*
Volume 1, 2004, pp. 1088-1091

The Interplay Between Molecular Properties and the Local Environment in Liquid Water

P.G. Kusalik[1], A.V. Gubskaya and L. Hernández de la Peña

Department of Chemistry,
Dalhousie University,
Halifax, Nova Scotia, B3H 4J3
Canada

Received 3 August, 2004; accepted in revised form 10 August, 2004

Abstract: The paper explores the interplay between the local environment in liquid water and two key properties of the water molecule, namely its average electronic distribution and the quantum rotational uncertainty of the molecule. A mean-field approach is presented describing the electrostatic environment experienced by water molecule in liquid state, which is used to extract the corresponding hyper- and high-order polarizabilities. It is then shown that the average total dipole moment for the water molecule in the liquid state can be linked to experimental refractive index data by means of a formal framework that relates the temperature dependence of the effective molecular polarizability to the average local electric field experienced by a liquid water molecule over a chosen temperature range. The local environment in liquid water is found to rise to substantially elevated dipole moments, with an almost 10% variation in the dipole being observed over the temperature range 273 to 373 K. The extension of the centroid molecular dynamics (CMD) method to rotational motion of a molecule is also discussed. An algorithm is presented that homogeneously samples the orientational neighborhood associated with the quantum degrees of freedom of a specified orientational centroid; as a critical component of this development a general definition for an orientational centroid (or average rotation) is presented. The application of this methodology in quantum simulations of liquid water is discussed. It is shown that while quantization significantly impacts the bulk properties of water, it is also demonstrated that the local environment in liquid water can influence the quantum dispersion of the molecule in unexpected ways.

Keywords: water, mean-field methods, local electric field, electric response properties, polarizabilities, local liquid structure, quantum rotational uncertainty, rotational centroid.

PACS: 31.15.Ar; 31.15.Fx; 31.15.Qg; 31.25.Qm; 33.15.Kr

1. Introduction

As a substance water is unprecedented in the research interest it has attracted. Still, some very basic properties of the liquid state system and its molecules have remained poorly understood or unknown. For example, the dipole moment of this polar liquid is of fundamental importance and is a basic ingredient for characterizing its local environment, yet an experimental determination of this value and its dependence on temperature had not been achieved previously. The molecular dipole moment of water has received considerable attention as it is known that the gas and liquid state values differ significantly and the apparently elevate values for liquid state molecules are important in our understanding the properties of liquid water. Clearly this elevate dipole moment reflects a perturbation of the electron density of the individual molecules by the local environment. It would be very interesting then to determine the influence of the molecular structure in the liquid upon electrical response properties of the water molecule.

Many of the unique properties of liquid water can be understand in term of the interplay between the local structure and the dynamics within system. Yet the explanations of some behaviour, for example the well-documented difference between the bulk properties of light water and heavy water, have remained incomplete or untested. A possible explanation for isotopic dependence in the properties of liquid water is that even at room temperature the thermal wavelength of a hydrogen atom is non-

---

[1] Corresponding author. E-mail: peter.kusalik@dal.ca

negligible. Hence, we might expect quantum dynamics to play an important role in determining the properties of (bulk) liquid water. While there have been several computer simulation studies, both path integral and centroid molecular dynamics techniques, that have attempted to explore isotopic (quantum) effects in liquid water, none have explored in detail the impact on local water structure, nor the influence of the local environment of the extent of the quantum dispersion.

2. The Molecular Dipole Moment in Liquid Water

The approach used for this work that can be viewed as an extension of mean-field methods; in mean field theories the interactions of a central particle with all its neighbors are collapsed into the influence of a local field. Clearly such a local field will be sensitive to the details of the local structure in the liquid. Then within the *ab initio* calculations that were performed, the influence of the surrounding water molecules will be modeled by a local electric field experienced by the (central) molecule of interest. The nature and strength of this local field are determined from average distributions obtained from classical simulations of liquid water using standard classical water potentials (since they provide a reasonable representation of the local electrostatic environment in liquid water). Three different models of the local environment are examined, where both homogeneous and non-homogeneous local electric fields are imposed using fixed sets of charges. The *ab initio* calculations utilized the MP2 and MP4 levels of theory and large customized basis sets to examine both isolated and liquid-state molecules. A finite field technique (and its charge perturbation variant) is used to calculate molecular polarizabilities, hyper- and high-order polarizabilities (up to fourth-order) for liquid water. For a liquid phase water molecule the first hyperpolarizability (β) and first higher polarizability (A) are observed increase markedly, actually changing sign. The second hyperpolarizability (γ) also increases but much less dramatically, and components of the second high-order polarizability tensor (B) demonstrate a rearrangement of contributions. The excellent agreement of our gas-phase values with experimental results and the most accurate previous theoretical predictions is evident of the quality of our higher order polarizabilities and theoretical models. [1]

In order to extract the average total dipole moment for the water molecule from very reliable experimental data, in this case refractive index data together with densities, three critical components are required: a general formalism that relates an effective molecular polarizability to the average local electric field experienced by the water molecule in the liquid state, a qualitative description of the distributions of the local electric field and field gradient, and accurate results for the molecular response properties of a liquid phase water molecule. Within the general formalism, the perturbing local environment affecting a liquid water molecule is represented with a local electric field. The change in the effective polarizability of the molecule, which can be extracted from refractive index data and characterizes non-linear behavior, is used to quantify the local electric field and its temperature dependence. The required characteristics of the distributions of electric fields and field gradients experienced by a water molecule in the liquid were obtained from MD simulations employing the SPC/E and TIP4P water models. It was found that the contributions of field gradient terms were significant due to near cancellation of the larger contributions from first and second hyperpolarizabilities. The total molecular dipole moment for liquid water as extract from the temperature dependence of experimental refractive index data exhibits an almost 10% decrease in going from 273 to 373K [2]. The consistent behavior observed in our results gives us considerable confidence in our ability to quantify this trend. The present approach also provides estimates for the absolute values for the total dipole moment. At ambient conditions our analysis yields a dipole moment of 2.95 0.2 D, in excellent agreement with other studies.

3. Quantum Effects in Liquid Water

Feynman's path integral representation of statistical mechanics1 has proven to be a very useful tool for studying quantum effects in condensed matter. This approach allows the use of Monte Carlo (MC) and molecular dynamics (MD) methods to study quantum mechanical effects in equilibrium and dynamical properties of complex systems, for example liquid water, excess electrons in liquids, etc. Within this context, centroid molecular dynamics (CMD) has been developed into a very successful technique for examining (approximate) quantum dynamics in condensed phases, becoming exact in the classical limit. CMD has been justified formally and applied in the context of translational degrees of freedom (Cartesian coordinates) in atomic approaches. In this paper we also discuss an approach that extends the CMD method to 3-dimensional rotational dynamics by utilizing an algorithm that in a context of a molecular dynamics simulation samples homogeneously the orientational neighborhood associated with their quantum degrees of freedom. This algorithm ensures that the rotational centroid (or the average orientation of the set of beads that samples the quantum dispersion) remains constant at each step. To help achieve this, we have developed a general definition for the average orientation (rotational centroid) and a straightforward procedure for determining it numerically [3]. Moreover, the requisite definition of an orientational centroid utilized in our approach may have numerous other applications that have explicit orientational dependence.

The application of this CMD methodology to a system that can be modeled as a collection of rigid bodies is demonstrated through the investigation of liquid water. Using a simple (rigid) water model, the TIP4P potential, we have systematically examined isotopic effects in the bulk properties of water. With this approach we have successfully reproduced results for the equilibrium and dynamical properties of light and heavy water as obtained from experiment and from previous path integral simulations. Considerable attention is paid to the parameterization and characterization of the methodology; it is also shown that the removal of the vibrational degrees of freedom, accompanied by the relatively low value of the discretization parameter (P=5) required for convergence, allow the present approach to be significantly (at least 20 times) faster than standard CMD. Additionally, its validity was confirmed by the fact that the rotational CMD technique recovers at a quantitative level essentially the full effect of quantization observed previously in quantum simulations with flexible models. As a result of the considerably enhanced computational efficiency of the rigid-body CMD simulations, it was possible in our simulations to generate significantly longer real-time trajectories with the accompanying reduction in the statistical errors in the properties of interest.

Explicit inclusion of the orientational degrees of freedom additionally allows a quantitative analysis of the rotational uncertainty of the H_2O and D_2O water molecules and its effect on the components of the angular velocity time correlation function; as expected, a notably larger impact of quantization is found in H_2O than in D_2O. Its influence on rotational dynamics, particularly librational motion and dipole relaxation times, is consistent with a "softening" of the intermolecular interactions. The enhancement of the linear self-diffusion coefficient (by roughly 50% at room temperature) is an

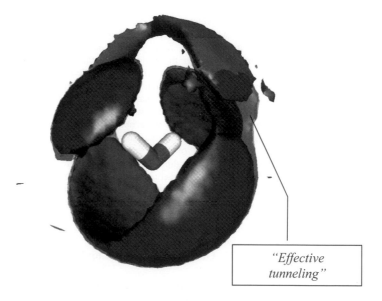

"Effective tunneling"

Figure 1: Oxygen-oxygen spatial distribution functions of classical (red) and quantum (semi-transparent blue) liquid H_2O at 50°C represented by the isosurface $g_{OO}(\mathbf{r}_{ij})=1.4$.

indirect effect of quantization and reflects the importance of rotational-translational coupling in the dynamics of liquid water. Model comparisons further verify the general applicability of the results obtained for our model liquids; after all, the quantum mechanical uncertainty of the protons is an inherent property of nature that manifests itself in liquid water and its effects should be reproducible for any significant water-like molecular model.

Finally, and most importantly for this paper, the molecular uncertainty of water is found to exhibit an unexpected behaviour with temperature. The relationship of this behaviour to the specific local molecular environment found in water is demonstrated. In addition, the phenomenon of "effective tunneling" in liquid water was characterized in an unambiguous manner, and its relation with the water diffusion mechanism is revealed (see Figure 1). The distinction between the impact of quantum effects and the increase in the temperature is also established.

Acknowledgments

We are thankful to the Natural Science and Engineering Research Council of Canada for the financial support.

References

[1] A.V. Gubskaya and P.G. Kusalik, *Mol. Phys.*, **99**, 1107 (2001).

[2] A.V. Gubskaya and P.G. Kusalik, *J. Chem. Phys.*, **117**, 5290 (2002).

[3] L. Hernández de la Peña and P. G. Kusalik, *Mol. Phys.*, to appear (2004).

[4] L. Hernández de la Peña and P.G. Kusalik, *J. Chem. Phys.*, to appear (2004).

VSP International
Science Publishers
P.O. Box 346, 3700 AH Zeist
The Netherlands

*Lecture Series on Computer
and Computational Sciences*
Volume 1, 2004, pp. 1092-1095

Electric Properties of F_2, ClF, BrF and IF

C. Makris and G. Maroulis[1]

Department of Chemistry,
University of Patras,
GR-26 500 Patras, Greece

Abstract: The electric multipole moments and dipole (hyper)polarizability of F_2, ClF, BrF and IF have been obtained from Møller-Plesset perturbation theory and coupled cluster calculations with medium sized, especially designed property-adapted basis sets. Our results represent the first systematic study of the higher electric properties of this sequence.

Keywords: interhalogens, ab-initio, electric multipole moments, polarizability, hyperpolarizability.

PACS: 32.10.Dk

1. Introduction

The interhalogen sequence F_2, ClF, BrF and IF constitutes a privileged ground for fundamental observations on the evolution of electric properties. Figure 1 offers an eloquent picture of the regularities in the electronic structure of these diatomics. Previous systematic work on their electric properties has focused on the dipole moment and polarizability [1]. In this work we extend previous efforts by adding the higher multipole moments and the first and second hyperpolarizability. We have designed flexible Gaussian-basis sets for all molecules. The computational philosophy underlying their construction has been presented elsewhere [2]. We have used Møller-Plesset perturbation theory and coupled cluster techniques for the calculation of electron correlation effects on the molecular properties. Comprehensive presentations of these methods may be found in standard textbooks [3,4]. We follow Buckingham's conventions and terminology throughout [5].

Atomic units are used throughout this study. Conversion factors to SI units are, Length, 1 $a_0 =$ 0.529177249 x 10^{-10} m, dipole (μ), 1 $ea_0 = 8.478358$ x 10^{-30} Cm, quadrupole (Θ), 1 $ea_0^2 = 4.486554$ x 10^{-40} Cm^2, octopole (Ω), 1 $ea_0^3 = 2.374182$ x 10^{-50} Cm^3 and hexadecapole (Φ) moment, 1 $ea_0^4 =$ 1.256363 x 10^{-60} Cm^4, dipole polarizability (α), 1 $e^2a_0^2E_h^{-1} = 1.648778$ x 10^{-41} $C^2m^2J^{-1}$, first hyperpolarizability (β), 1 $e^3a_0^3E_h^{-2} = 3.206361$ x 10^{-53} $C^3m^3J^{-2}$ and second hyperpolarizability (γ), 1 $e^4a_0^4E_h^{-3} = 6.235378$ x 10^{-65} $C^4m^4J^{-3}$.

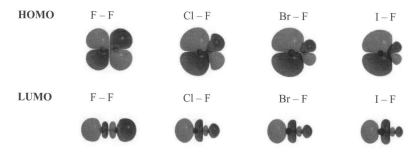

Figure 1: HOMO and LUMO obtained at the MP2 level of theory.

---

[1] Corresponding author, e-mail: maroulis@upatras.gr

2. Results

Electric property values for the sequence of diatomics calculated at the experimental bond length [6] are given in table 1 (See [5] for the definition of the (hyper)polarizability invariants). In all cases the molecule is on the z axis with the F centre on the positive part. The basis sets used in the calculations are augmented, property adapted versions of the widely used split-valence 3-21G basis [7]. We expect the designed basis sets to be of further use in theoretical studies of intermolecular interactions involving the four diatomics.

Table 1: Electric properties of F_2, ClF, BrF and IF.

Molecule	Method	μ_z	Θ_{zz}	Ω_{zzz}	Φ_{zzzz}	$\bar{\alpha}$	$\Delta\alpha$	$\bar{\beta}$	$\bar{\gamma}$
F_2 F:[4s3p2d]	SCF		0.5744		12.60	8.40	8.98		182
	MP2		0.8013		14.62	8.15	5.32		348
	MP4					8.48	6.16		330
	CCSD					8.25	6.02		309
	CCSDT					8.35	6.03		309
ClF Cl:[5s4p2d] F:[4s3p2d]	SCF	-0.4690	0.9341	-8.26	27.74	17.29	7.07	46.5	1013
	MP2	-0.3629	1.1120	-6.82	28.93	18.15	6.28	36.0	1261
	MP4	-0.3661				18.32	6.28	38.4	1342
	CCSD	-0.3800				18.16	6.45	40.5	997
	CCSDT	-0.3658				18.31	6.39	40.8	1026
BrF Br:[6s5p3d] F:[4s3p2d]	SCF	-0.6859	0.3811	-11.82	12.28	22.78	7.25	79.9	2031
	MP2	-0.5485	0.7627	-9.30	18.43	23.96	6.76	64.3	2359
	MP4	-0.5504				24.20	6.67	67.6	2545
	CCSD	-0.5710				23.97	6.92	72.1	2486
	CCSDT	-0.5534				24.18	6.84	71.9	2611
IF I:[7s6p4d] F:[4s3p2d]	SCF	-0.8770	-0.0595	-20.36	-8.00	33.00	4.96	122.7	4415
	MP2	-0.7268	0.4928	-17.41	2.74	35.31	4.85	116.2	5211
	MP4	-0.7173				35.83	4.73	120.2	5540
	CCSD	-0.7463				35.50	4.62	124.9	5339
	CCSDT	-0.7292				35.74	4.66	124.2	5407

Interesting patterns are emerging from an analysis of the contents of Table 1. The dipole moment increases with size in ClF, BrF and IF. Electron correlation reduces the size of the SCF property. As regards the higher electric moments, F_2 is present as an anomaly. Restricting our interest in the subsequence ClF, BrF and IF we see that the value of the property decreases with size. The trend is present in both SCF and MP2 results. F_2 has a relative large $\Delta\alpha$ anisotropy. The effect is less pronounced for the larger diatomics. Important electron correlation effects are observed in the case of the second dipole hyperpolarizability. Their relative size is largest for F_2. Little can be said at this stage on the performance of the MP2 and MP4 methods for $\bar{\gamma}$. It should be noted that for F_2 MP2 > MP4 but this is reversed for the polar diatomics. Note also that for $\bar{\gamma}$ we have MP2 > CCSD(T) for F_2 and ClF but this is reversed for BrF and IF.

We have traced in Figure 2 the dependence of the mean (hyper)polarizability on the size of the diatomic for three key methods: SCF, MP2 and CCSD(T). From a qualitative point of view, all three methods show the same trend in the evolution of these important properties.

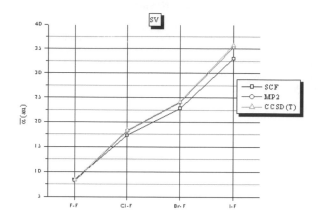

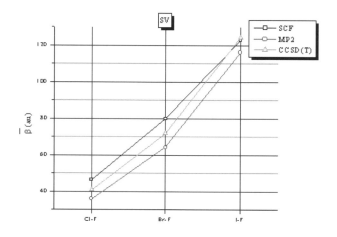

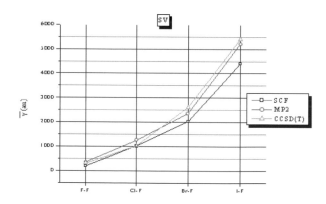

Figure 2: Dependence of the mean (hyper)polarizability of FX, X = F, Cl, Br and I on the size of X.

References

[1] A.J.Sadlej, J.Chem.Phys. 96, 2048 (1992).

[2] G.Maroulis, J.Chem.Phys. 108, 5432 (1998).

[3] A. Szabo and N.S.Ostlund, Modern Quantum Chemistry, MacMillan, New York, 1982.

[4] T.Helgaker, P.Jørgensen, J.Olsen, Molecular Electronic-Structure Theory (Wiley, Chichester, 2000).

[5] A. D. Buckingham, Adv. Chem. Phys. 12, 107 (1967)

[6] K.P. Huber and G. Herzberg, Molecular Spectra and Molecular Structure: IV. Constants of Diatomic Molecules, Van Nostrand, New York, 1979.

[7] M. S. Gordon, J. S. Binkley, J. A. Pople, W. J. Pietro and W. J. Hehre, J. Amer. Chem. Soc. 104, 2797 (1982).

VSP International
Science Publishers
P.O. Box 346, 3700 AH Zeist
The Netherlands

*Lecture Series on Computer
and Computational Sciences*
Volume 1, 2004, pp. 1096-1100

Electric Hyperpolarizability of Small Copper Clusters. The Tetramer Cu₄ as a Test Case

G. Maroulis[1] and A.Haskopoulos

Department of Chemistry,
University of Patras,
GR-26504 Patras, Greece

Abstract: We report an extensive investigation of the static electric polarizability and hyperpolarizability of the copper tetramer. We have designed Gaussian-type basis sets that could be useful in polarizability calculations on small and medium copper clusters. Our results show that it is particularly difficult to obtain reliable values of the differential per atom (hyper)polarizability for Cu₄.

Keywords: Polarizability, hyperpolarizability, copper clusters

PACS: 31.25.Qm, 32.10.Dk.

1. Introduction and Theory

Extensive theoretical work has been reported on the structure and properties of copper clusters Cu_n [1,2]. Of particular importance are studies related to the physicochemical properties of copper clusters. The static dipole polarizability of small Cu_n has attracted some attention [3-5]. Remarkably little is known about their hyperpolarizability. A reference hyperpolarizability value for the copper atom has been calculated by Stiehler and Hinze [6] by a numerical Hartree-Fock method. A detailed study of the polarizability and hyperpolarizability of the dimer [7] showed that the calculation of these properties for higher clusters is a particularly difficult task.

In this work we have recourse to conventional **ab initio** and density functional theory methods in order to determine the static (hyper)polarizability of the copper tetramer. The first category includes Møller-Plesset perturbation theory and coupled cluster techniques [8]. The second category includes the widely used methods B1LYP, B3LYP, P3P86, B3PW91 and MPW1PW91 as implemented in the GAUSSIAN programmes [9,10].

The energy of an uncharged molecule interacting with a weak, homogeneous electric field, can be written as [11] :

$$E^p = E^0 - \mu_\alpha F_\alpha - \frac{1}{2}\alpha_{\alpha\beta}F_\alpha F_\beta - \frac{1}{6}\beta_{\alpha\beta\gamma}F_\alpha F_\beta F_\gamma - \frac{1}{24}\gamma_{\alpha\beta\gamma\delta}F_\alpha F_\beta F_\gamma F_\delta + ... \tag{1}$$

where F_α is the electric field, E^0 is the energy of the free molecule, μ_α is the dipole moment, $\alpha_{\alpha\beta}$ is the dipole polarizability, $\beta_{\alpha\beta\gamma}$ is the first dipole hyperpolarizability and $\gamma_{\alpha\beta\gamma\delta}$ is the second dipole hyperpolarizability. For a molecule of D_{2h} symmetry, as Cu_4, $\mu_\alpha = \beta_{\alpha\beta\gamma} = 0$ [11]. Detailed descriptions to the calculation of the Cartesian components of the (hyper)polarizability of similar systems can be found elsewhere [12]. We also calculate the following invariants:

$$\bar{\alpha} = \left(\alpha_{xx} + \alpha_{yy} + \alpha_{zz}\right)/3$$

$$\Delta\alpha = 2^{-1/2}\left(\left(\alpha_{xx} - \alpha_{yy}\right)^2 + \left(\alpha_{yy} - \alpha_{zz}\right)^2 + \left(\alpha_{zz} - \alpha_{xx}\right)^2\right)^{1/2}$$

$$\bar{\gamma} = \frac{1}{5}\left(\gamma_{xxxx} + \gamma_{yyyy} + \gamma_{zzzz} + 2(\gamma_{xxyy} + \gamma_{yyzz} + \gamma_{zzxx})\right)$$

$$\tag{1}$$

Of interest to us is also the differential per atom (hyper)polarizability defined as:

---

[1] Corresponding author, e-mail: maroulis@upatras.gr

$$\overline{\alpha}_{diff} / 4 \equiv [\overline{\alpha}(Cu_4(D_{2h})) / 4 - \overline{\alpha}(Cu(^2S))]$$

$$\overline{\gamma}_{diff} / 4 \equiv [\overline{\gamma}(Cu_4(D_{2h})) / 4 - \overline{\gamma}(Cu(^2S))] \tag{2}$$

These properties have been found particularly useful in electric property calculations of clusters [12,13].

2. Computational details

An essential part of our effort was directed to the construction of reasonably sized, flexible basis sets. In addition to the basis sets designed for the tetramer we have also obtained basis sets for calculations on the copper atom. All calculations reported in this study pertain to a theoretical geometry obtained with the B3LYP method and a large [9s7p4d1f] basis set. The respective geometrical parameters for the planar $Cu_4(D_{2h})$ are shown in fig 1.

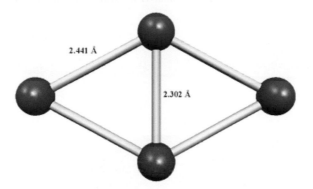

Figure 1: Molecular geometry of Cu_4 (D_{2h}) adopted in this work.

All calculations were performed with GAUSSIAN 94 [9] and GAUSSIAN 98 [10].

Atomic units are used through this work. Conversion factors to SI units are: Length, 1 α_0 = 0.529177249 x 10^{-10} m, dipole polarizability, α, 1 $e^2 a_0^2 E_h^{-1}$ = 1.648778 x 10^{-41} $C^2 m^2 J^{-1}$ and second dipole hyperpolarizability, hyperpolarizability, 1 $e^4 a_0^4 E_h^{-3}$ = 6.235378 x 10^{-65} $C^4 m^4 J^{-3}$.

3. Results and Discussion

We show in figs 2 and 3 the HOMO and LUMO of the planar $Cu_4(D_{2h})$ molecule.

The calculated values for the two Cu-Cu bond lengths agree quite well with previous theoretical results.

We have collected in Table 1 some results on the electric polarizability of the copper atom. It seems particularly difficult to obtain basis sets that yield property values close to the NHF reference ones. Electron correlation reduces considerably both the polarizability and the hyperpolarizability of the atom.

We show in Table 2 **ab initio** and DFT values for the tetramer calculated with a rather small AVDZ1 ≡ [5s3p3d] basis set. Electron correlation effects are not particularly important for the dipole polarizability. Our best AVDZ1/CCSD(T) values for the dipole polarizability $\alpha_{\alpha\beta}/e^2 a_0^2 E_h^{-1}$ are $\overline{\alpha}$ = 192.47 and $\Delta\alpha$ = 169.56. For the mean hyperpolarizability we find $\overline{\gamma}$ = 401 x 10^3 $e^4 a_0^4 E_h^{-3}$

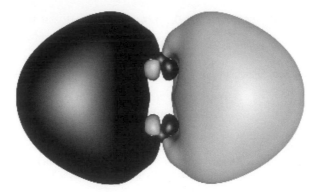

Figure 2: Highest occupied molecular orbital (HOMO) of Cu$_4$ (molecular plane view).

Figure 3: Lowest unoccupied molecular orbital (LUMO) of Cu$_4$ (molecular plane view).

Table 1: Static electric dipole polarizability and hyperpolarizability of the copper atom calculated with near-Hartree-Fock quality basis sets

Basis set	Method	$\bar{\alpha}$	$10^{-3} \times \bar{\gamma}$
[25s22p14d4f]	ROHF	76.39	201
	UHF	77.09	173
	CCSD(T)[a]	59.01	93
	B3LYP	44.65	85
[25s22p14d4f1g]	ROHF	76.39	200
	UHF	77.09	173
	CCSD(T)[a]	59.02	94
	B3LYP	44.65	85
[25s22p15d5f]	ROHF	76.39	201
	B3LYP	44.65	82
[25s22p15d5f3g]	ROHF	76.39	189
	B3LYP	44.73	85
	NHF	77.19	191

[a] The nine innermost orbitals were kept frozen.

Table 2: Static electric dipole polarizability and hyperpolarizability and differential per atom (hyper)polarizability of the copper tetramer calculated with a small AVDZ1 ≡ [5s3p3d] basis set.

Method[a]	$\bar{\alpha}$	$\bar{\alpha}_{diff}$	$\Delta\alpha$	$10^{-3} \times \bar{\gamma}$	$10^{-3} \times \bar{\gamma}_{diff}$
SCF	201.59	-23.98	176.36	616	40
MP2	184.49	-19.92	162.81	425	2
MP4	177.69	-20.21	152.15	217	-40
CCSD	193.36	-20.74	172.20	438	6
CCSD(T)	192.47	-21.06	169.56	401	2
ECC[b]	**-9.12**	**2.92**	**-6.80**	**-216**	**-38**
B1LYP	161.39	-5.49	144.24	307	0
B3LYP	159.75	-5.10	142.15	291	-3
B3P86	160.09	-6.51	141.35	287	-16
B3PW91	161.84	-7.00	140.76	315	-16
MPW1PW91	162.23	-7.73	139.94	322	-19

[a] The thirty-six innermost MO were kept frozen in the correlated calculations.
[b] Defined as ECC = CCSD(T) - SCF

We observe that all DFT predict mean and anisotropy values for the dipole polarizability significantly lower than the conventional **ab initio** methods. The same is clearly true for the mean hyperpolarizability.

Acknowledgments

This work is part of the COST action **D26/0013/02** project of the EEC "Development of Density Functional Theory models for an accurate description of electronic properties of materials possessing potential high non-linear optical properties".

References

[1]   R.J.Van Zee and W.Weltner, J.Chem.Phys. **92**, 6976 (1990).

[2]   P.Calaminici, A.M.Köster, N.Russo and D.R.Salahub, J.Chem.Phys. **105**, 9546 (1996).

[3]   P.Calaminici, A.M.Köster, A.Vela and K.Jug, J.Chem.Phys. **113**, 2199 (2000).

[4]   P.Jaqué and A.Toro-Labbé, J.Chem.Phys. **117**, 3208 (2002).

[5]   Z.Cao, Y.Wang, J.Zhu, W.Wu and Q.Zhang, J.Phys.Chem B **106**, 9649 (2002).

[6]   J.Stiehler and J.Hinze, J.Phys. B **28**, 4055 (1995).

[7]   G.Maroulis, J.Phys.Chem A **107**, 6495 (2003).

[8]   A. Szabo and N.S.Ostlund, **Modern Quantum Chemistry**, MacMillan, New York, 1982.

[9]   M.J.Frisch, *et al.*, GAUSSIAN 94, Revision E.1 (Gaussian, Inc., Pittsburgh PA, 1995)

[10] M.J.Frisch *et al.*, GAUSSIAN 98, Revision A.7 (Gaussian, Inc., Pittsburgh PA, 1998).

[11] A.D.Buckingham, Adv.Chem.Phys. **12**, 107 (1967).

[12] G.Maroulis and D.Xenides, J.Phys.Chem. A **103**, 4590 (1999).

[13] G.Maroulis, D.Begué and C.Pouchan, J.Chem.Phys. **119**, 794 (2003).

VSP International
Science Publishers
P.O. Box 346, 3700 AH Zeist
The Netherlands

*Lecture Series on Computer
and Computational Sciences*
Volume 1, 2004, pp. 1101-1104

The Quantum Mechanical Investigation of Various I-Containing Polyoxides

Agnie M. Kosmas[1] and Evangelos Drougas,

Laboratory of Physical Chemistry, Department of Chemistry
University of Ioannina,
GR-451 10 Ioannina, Greece

Received 5 July, 2004; accepted 25 July, 2004

Abstract: The various isomers of the XOOI, XOOOI, X=H, CH_3 families are investigated using the MP2 methodology combined with two types of basis sets, the 6-311G basis set for all atoms involved and the 6-311G basis set for C, H and O combined with the LANL2DZ ECP procedure for I, augmented with extra polarization functions. The results indicate the improvement accomplished in the quantum treatment of I when additional polarization functions and relativistic corrections are taken into account. In addition, the variations of the I-O bond distances in all types of calculations reveal the differences in the nature of I-O bonding that occur in the various isomeric species. The hypervalent character of I in some isomeric forms is found to have a profound effect in the relative stability order among the members of each family.

Keywords: iodine polyoxides, relativistic effective-core-potential, polarization functions

PACS: 31.15.-p

1. Introduction

Iodine polyoxides, IO_nX, X=H, CH_3, n=1, 2, 3, have long been known to play an important role in the atmospheric chemistry of iodine [1, 2], being the interesting intermediates in the reactions of atomic iodine and iodine oxide with the hydroxy HO, hydroperoxy HO_2 and methoxy CH_3O, methylperoxy CH_3O_2 radicals. Due to their atmospheric significance many experimental measurements of the kinetics of these reactions have been carried out [3-8] and several theoretical investigations of the intermediate complexes have been performed at varying levels of theory [9-12]. Indeed, there has always been a considerable interest in the development of effective computational techniques for the quantum mechanical treatment of I- containing compounds, in which the necessity of including d and f polarization functions for the proper description of I has been stressed repeatedly [13-15].

In addition to their atmospheric significance, the quantum mechanical investigation of iodine polyoxides presents an intrinsic interest because of the hypervalent bonding that iodine assumes in many such species and of its effect in the structural features and the resulting relative stability order among the various isomeric forms. In the present work we carry out a detailed investigation of the iodine polyoxides using two basis sets, i.e., with and without extra polarization and relativistic corrections. From the results a comparative study is performed of the various members of the IO_nX, X=H, CH_3, n=2, 3 families at the same level of theory for all species involved. Both series of results allow us to demonstrate very clearly the interesting overall variation of specific structural features of the various I-O bonds, originating from the hypervalent character and the electropositive behaviour that iodine exhibits in several of the isomeric forms. Furthermore, the second series of calculations indicate most positively the considerable improvement obtained when additional polarization functions and relativistic corrections are taken into account.

---

[1] Corresponding author. E-mail: amylona@cc.uoi.gr

2. Computational details

Two series of calculations have been performed at the MP2 level of theory [16] using two types of basis sets for the optimization procedure. In the first series, denoted as basis A, the standard 6-311G basis set has been used for all atoms involved, C,H,O and I. In the second series the same treatment is used for C, H and O but the iodine atom has been handled in a different way : The effective core potential (ECP) LANL2DZ relativistic procedure has been employed [17], augmented with additional d and f polarization functions (exponents 0.292 and 0.441 respectively) taken from the extensible computational chemistry environment basis set database [18]. Thus, the second series of calculations, denoted hereafter as basis B, allows us to evaluate the improvement in the calculated optimized structures that can be achieved by the use of the additional polarization functions and the relativistic corrections for I that are taken into account implicitly in the LANL2DZ methodology.

3. Results and Discussion

The two families of I- polyoxides examined, XOOI and XOOOI, X=H, CH$_3$, involve two and three oxygen atoms respectively and present three and four isomeric forms i.e., XOOI, XOIO, XIO$_2$ and XOOOI, XOOIO, XOIO$_2$ and XIO$_3$. Thus, we readily see that the structures contain two kinds of I-O bonds depending on the position of the oxygen atom, namely with the   oxygen either being a terminal atom or being bridged to another atom. The chain type XOOI, XOOOI isomers contain only brigded I-O bonds, the XIO$_2$, XIO$_3$ structures contain only terminal I-O bonds and XOIO, XOOIO and XOIO$_2$ contain both bridged and terminal I-O bonds. The calculated O-I equilibrium bond distances are listed in Table 1.

Table 1: The variations in I-O and I-X bond distances (in Å) , depending on the basis set and the position of the oxygen atom

Species	I-O bridged		I-O terminal		I-X	
	Basis A	Basis B	Basis A	Basis B	Basis A	Basis B
HOOI	2.114	2.058				
HOIO	2.048	1.992	1.842	1.815		
HIO$_2$			1.831	1.784	1.748	1.724
CH$_3$OOI	2.103	2.005				
CH$_3$OIO	2.038	1.948	1.829	1.777		
CH$_3$IO$_2$			1.807	1.749	2.120	2.091
HOOOI	2.108	2.012				
HOOIO	2.025	1.982	1.835	1.783		
HOIO$_2$	1.998	1.919	1.798	1.745		
HIO$_3$			1.797	1.794	1.715	1.647
CH$_3$OOOI	2.095	2.003				
CH$_3$OOIO	2.006	1.992	1.815	1.786		
CH$_3$OIO$_2$	2.001	1.994	1.777	1.762		
CH$_3$IO$_3$			1.786	1.738	2.125	2.108

The first observation that can be made from Table 1 reflects the significant improvement in the optimization results accomplished from the use of basis set B which includes extra polarization functions and relativistic corrections for I. Although the same level of theory is used, the calculated geometries are tighter with an average I-O bond distance decrease around 0.04 Å . The same effect is shown in the H-I and C-I equilibrium bond distances which are also found to be systematically shorter when the basis set B is employed.

The second   interesting finding is the   overall variation, regardless of the basis set, that is demonstrated in the I-O bond distances, depending on the position of the oxygen atom within the molecule, i.e., whether it is a bridged O or a terminal O. The bridged O-I distances in XOOI and XOOOI are the largest in each family. In XOIO, XOOIO and XOIO$_2$ iodine is bonded to both bridged and terminal O atoms and exhibits hypervalent character. As a result, the bridged O-I bond distance decreases slightly but the terminal O-I bond length suffers a considerable shrinkage of about 0.2 Å due to the multiple bond characteristics that the terminal I-O bond acquires. The effect is more dramatically presented in the XIO_2, XIO_3 structures when only I-O terminal bonds exist and which assume the shortest I-O distances, 1.738 Å in CH_3IO_3 with basis set B.

The hypervalent nature of terminal I-O bonding increases severely the ionic character of the bond and iodine acquires a considerable positive charge which has interesting consequences in the relative stability order among the various isomers in each family. In the XOOI series, the XOIO isomer is either the most stable form followed very closely by XOOI (case X=H) or it is similar in stability with XOOI (case X=CH₃). The XIO_2 isomers are in both cases very high located in energy because the bonding between the highly electropositively charged I and the electropositive H or CH₃ partners decreases substantially the stability of these structures. In the XOOOI cases analogous effects take place. The structures XOOOI and XOOIO for both H and CH₃ cases are sililar in stability with XIO_3 adducts being very unstable and located very high in energy. The striking feature in these families is the very high stability of $XOIO_2$ structures, X=H, CH₃. Here, the electropositive character of I combined with the highly electronegative HO and CH₃O partners results in considerable stabilization of these structures. Thus, both iodic acid, $HOIO_2$ and methoxy iodate, CH_3OIO_2, are very stable species. In fact, iodic acid, $HOIO_2$, is a white solid, stable at room temperature [19].

3. Summary

The various isomeric forms of XOOI, XOOOI, X=H, CH₃ families have been examined using ab initio theoretical techniques. The hypervalent character of I and the resulting strong ionic nature of I-O bonding is examined and it is found to have a profound effect in the equilibrium I-O bond distances and the relative stability of each adduct. The results also indicate the importance of including extra polarization functions and relativistic corrections for the proper treatment of I.

Acknowledgments

The authors wish to thank the University of Ioannina Computer Center for the computing time generously provided by the Center.

References

[1] S. Solomon, R.R. Garcia and A.R. Ravishankara, J. Geophys. Res. D, **99** 20491 (1994).

[2] S. Solomon, J.B. Burkholder, A.R. Ravishankara and R.R. Garcia, J. Geophys. Res. D, **99** 20929 (1994).

[3] M.E. Jenkin, R.A. Cox and G.D. Hayman, Chem. Phys. Lett. **177** 272 (1991).

[4] F. Maguin, G. Leverdet, G. Le Bras and G. Poulet, J. Phys. Chem. **96** 1775 (1992).

[5] M.K. Gilles, A.A. Turnipseed, R.K. Talukdar, Y. Rudich, P.W. Villalta, L.G. Huey, J.B. Burkholder and A.R. Ravishankara, J. Phys. Chem. **100** 14005 (1996).

[6] J.M. Cronkite, R.E. Stickel, J.M. Nicovich and P.H. Wine, J. Phys. Chem. A **103** 322 (1999).

[7] G.P. Knight and J.N. Crowley, PCCP **3** 393 (2001).

[8] D. Shah, C.E. Canosa-Mas, N.J. Hendy, M.J. Scott, A. Vipond and R.P. Wayne, PCCP 4932 (2001).

[9] A. Misra, R.J. Berry and P. Marshall, J. Phys. Chem. A **101** 7420 (1997).

[10] A. Misra, R.J. Berry and P. Marshall, J. Phys. Chem. A **102** 9056 (1998).

[11] N. Begovic, Z. Markovic, S. Anic and L. Kolar-Anic, J. Phys. Chem. A **108** 651 (2004).

[12] E. Drougas and A.M. Kosmas, Can. J. Chem. accepted.

[13] M.N. Glukhovtsev, A. Pross, M.P. McGrath and L. Radom, J. Chem. Phys. **103** 1878 (1995).

[14] M.N. Glukhovtsev, A. Pross and L. Radom, J. Am. Chem. Soc. **117** 2024 (1995).

[15] A.J. Sadlej, Theor. Chim. Acta **81** 339 (1992).

[16] C. Møller and M.S. Plesset, Phys. Rev. **46** 618 (1934).

[17] W.R. Wadt and P.J. Hay, J. Chem. Phys. **82** 284 (1985).

[18] Extensible Computational Chemistry Environment Basis Set Database, Version 1.0, Molecular Science Computing Facility, Environmental and Molecular Science Laboratory, Pacific Northwest Laboratory, Richland, WA.

[19] A.F. Cotton and G. Wilkinson, Advanced Inorganic Chemistry, Interscience Publishers, 2nd Edition (1967).

VSP International
Science Publishers
P.O. Box 346, 3700 AH Zeist
The Netherlands

*Lecture Series on Computer
and Computational Sciences*
Volume 1, 2004, pp. 1105-1107

Molecular Electric Properties using Coupled-Cluster Method

Sourav Pal[1] , Nayana Vaval, K. R. Shamasundar and D. Ajitha

Physical Chemistry Division,
National Chemical Laboratory,
Pune 411008,
India

Received 7 March, 2004; accepted in revised form 10 March, 2004

Keywords: Multireference Coupled-Cluster Theory, Molecular Properties

PACS: 31.15.Dv, 32.10.Dk

Coupled-cluster (CC) method has been established as a state-of-the-art technique in the calculation of electronic structure, spectra and properties [1]. The single-reference version of the theory in singles and doubles approximation (CCSD) is presently used as a routine black box method for structure determination of primarily non-degenerate situations, e.g. ground closed shell systems [2]. Higher approximations of the theory have also been developed and tested for small molecular cases. CC theory, due to the non-linear structure of the wave-operator, is computationally demanding. However, excellent progress in algorithms coupled with developments in computers, has made CC an affordable method for electronic structure calculation for small to medium size molecules. The development of linear response of the single reference CC wave function has made it possible to use the theory for a large number of chemical applications, namely geometry optimization, electric and magnetic properties, linear as well as nonlinear [3]. The formulation of time-dependent response allows calculation of dynamic properties from which excitation energies can also be computed [4]. Related equation-of-motion based single reference CC theory has been developed in recent years for calculation of direct vertical difference energies [5]. More recently, multi-reference CC (MRCC) theory has been studied as an extremely accurate method for incorporation of electron correlation for near-degenerate situations, like ionized, excited states and away from equilibrium [6]. Linear response of this is a very recent idea [7]. This presentation will be devoted to the computation of electric response of some novel single-reference based CC methods as well as initial results of electric properties using MRCC theory.

The initial development of the linear response formulation of CC theory used a non-variational method of projection, which involved the nth derivative of the cluster amplitudes in the computation of the nth order energy derivatives [8]. Subsequently, with the help of Z-vector technique [9], dependence of cluster derivatives was reduced to make the method more efficient for computation of first and second energy derivatives. However, this algebraic method is quite laborious for extension to progressively higher derivatives. A variation using the constraint of non-variational CC equations using Lagrange multipliers results in the same equations obtained by non-variational Z-vector technique [10]. This variation can be viewed as a stationary condition of the similarity-transformed energy functional pre-multiplied with a linear space consisting of de-excitation amplitudes (left vectors). This de-couples the equations for the right vector amplitudes (normal CC excitation amplitudes) from the ones of the left vectors. The set of amplitudes of the left vector is the set of Lagrange multipliers or the Z-vector amplitudes. Use of a full exponentiation of the left vector results in a structure where the equations for the left and right vector amplitudes are fully coupled. Such a structure was shown to be double-similarity transformed energy functional by Arponen and co-workers [11]. A further transformation of this leads to a double-linked form of he energy functional, in which the excitation amplitudes are directly connected to the Hamiltonian and the left de-excitation amplitudes are either directly

---

[1] Corresponding author. E-mail: pal@ems.ncl.res.in

connected to the Hamiltonian or to two different excitation right vector amplitudes. This is known as the extended CC (ECC) functional. The functional is formally terminating and the variation of the functional leads to a fully extensive results. Linear response of this functional has been studied by us for evaluation of molecular properties [12]. Fully stationary condition ensures (2n+1)-rule and the extensivity coupled with terminating nature of the series lead to much superior values compared to the stationary approach using functionals, e.g. expectation-value functional (XCC) [13]. ECC functional, truncated to a total of cubic terms in the cluster amplitudes in singles and doubles approximation (ECCSD), has been used for our study. We have developed the response using both relaxed and non-relaxed orbitals. In this presentation, results of properties using both the versions of ECC response will be presented for some test molecules, e.g. HF, CO and BH in different basis sets. Although single reference based CC theories are ideally suited to the closed shell ground state description, we have studied at geometry stretched from equilibrium. We show that without relaxation of orbitals, the property results deteriorate progressively at large bond distances. The use of relaxed orbitals improves the results significantly at large bond distances. We also discuss the computational aspects and algorithm involved in the calculations.

While single reference based CC theories can be used away from equilibrium, a more desirable way to include the effects of quasi-degeneracy is to use an appropriate multi-determinantal starting space instead of Hartree-Fock single determinant. This space is called model space. Subsequent cluster expansion on this model space to describe dynamic correlation forms the basis of what are known as "multi-reference coupled-cluster (MRCC) theories" [6]. Effective Hamiltonian version of the theory, which can generate multiple roots, has been studied more extensively and is beginning to be used for potential energy surface and spectroscopic difference energies. Hilbert space MRCC, where each model space determinant is treated as vacuum, is believed to be more suitable for potential energies and Fock space version, which is based on a common vacuum concept and valence-universal wave-operator, is regarded as tailor-made for spectroscopic difference energies. The latter can thus suitably describe excited states or ionized states/ open shell radicals [14]. The class of MRCC theories is much more demanding in terms of computations and is beset with the problem of intruder states due to the multi-root nature causing serious convergence problem at times. A judicious choice of the model space can obviate the problem to some extent. Some serious applications of the MRCC theory, especially Fock space version, have been made in recent times. We have made the first formulation of linear response to the MRCC class of theories [7]. In particular, a satisfactory response idea using the Z-vector type technique of single-reference CC has been formulated by us just recently for both Hilbert and Fock space MRCC theories [15-16]. However, this can be done only for one root at a time. Even without the use of Z-vector technique, some pilot results of first-order electric property have been obtained for multiple excited states or ground and excited states of radicals [17]. In this presentation, we will give a brief overview of the present status of the MRCC linear response with results of dipole moments of test systems.

References

[1]   J. Cizek, *J. Chem. Phys*, **45**, 4256 (1966)

[2]   R.J. Bartlett, *Ann. Rev. Phys. Chem*, **32**, 359 (1981)

[3]   R. J. Bartlett and J. F. Stanton, in *Reviews in Computational Chemistry*, edited by K. B. Lipkowitz and D. B. Boyd (VCH, New York),Vol. 5, pp. 65-169 (1994)

[4]   D. Ajitha and S. Pal, Phys.Rev A 56, 2658 (1997); D. Mukherjee and P.K. Mukherjee, *Chem. Phys*.**37**, 325 (1979)

[5]   S. A. Kucharski, M. Wloch, M. Musial, and R. J. Bartlett, *J. Chem. Phys*. **115**, 8263 (2001); M. Musial, S. A. Kucharski, and R. J. Bartlett, *J. Chem. Phys*. **118**, 1128 (2003); M. Nooijen and R. J. Bartlett, *J. Chem. Phys*., **106**, 6441 (1997).

[6]   D. Mukherjee and S. Pal, *Adv. Quantum Chem*. **20**, 292 (1989). J. Paldus, in *Methods in Computational Molecular Physics*, NATO, edited by S. Wilson and G. H. F. Diercksen (Plenum, New York), Vol. **293**, pp. 99–194 (1992)

[7] S. Pal, *Phys. Rev A* **39**, 39 (1989)

[8] H. J. Monkhorst, *Int. J. Quantum Chem.* **S11**, 421 (1977)

[9] E. A. Salter, G. W. Trucks, and R. J. Bartlett, *J. Chem. Phys.* **90**, 1752 (1989)

[10] P. Jorgensen and T. Helgaker, *J. Chem. Phys.* **89**, 1560 (1988)

[11] J. Arponen, *Annals Phys (NY)*, **151**, 311 (1983); R. Bishop, J. Arponen and P. Pajanne, *Phys. Rev. A*, **36**, 2915 (1987)

[12] N. Vaval, K. B. Ghose and S.Pal, *J. Chem. Phys*, **101**, 4914 (1994); N. Vaval, *Chem. Phys. Lett*, **318**, 168 (2000); A. B. Kumar, N. Vaval, and S. Pal, *Chem. Phys. Lett*, **295**, 189 (1998)

[13] R. J. Bartlett and J. Noga, *Chem. Phys. Lett*, **150**, 29 (1998)

[14] S. Pal, M. Rittby, R. J. Bartlett, D. Sinha, and D. Mukherjee, *Chem. Phys. Lett.* **137**, 273 (1987); S. Pal, M. Rittby, R. J. Bartlett, D. Sinha, and D. Mukherjee, *J. Chem. Phys.* **88**, 4357 (1988); R. Chaudhuri, D. Mukhopadhyay, and D. Mukherjee, *Chem. Phys. Lett.* **162**, 393 (1989); N.Vaval, S. Pal, and D. Mukherjee, *Theor. Chem. Acc.* **99**, 100 (1998); M. Musial and R.J. Bartlett, *J. Chem. Phys.* **121**, 1670 (2004)

[15] K. R. Shamasundar and S. Pal, *J. Chem.. Phys.* **114**, 1981(2001); *ibid* **115**, 1979 (2001) [E] ;K. R. Shamasundar and S. Pal, *Intern. J. Mol. Sci.* **3**, 710 (2002)

[16] K. R. shamasundar and S. Pal, *J. Chem. Phys.* **120**, 6381 (2004)

[17] D. Ajitha, N. Vaval, and S. Pal, *J. Chem. Phys.* **110**, 2316 (1999); D. Ajitha and S. Pal, *ibid.* **114**, 3380 (2001). D. Ajitha and S. Pal, *Chem. Phys. Lett*, **309**, 457 (1999)

VSP International
Science Publishers
P.O. Box 346, 3700 AH Zeist
The Netherlands

*Lecture Series on Computer
and Computational Sciences*
Volume 1, 2004, pp. 1108-1111

On the Vibrational Polarizabilities and Hyperpolarizabilities. Analysis of Some Specific Examples: Pyrrole and Harf

M. G. Papadopoulos[*], A. Avramopoulos and H. Reis

Institute of Organic and Pharmaceutical Chemistry
National Hellenic Research Foundation
48 Vas. Constantinou Ave.
Athens 116 35
Greece

Received 7 March, 2004; accepted in revised form 10 March, 2004

Abstract: In the last few years we have computed both the electronic and vibrational contributions to polarizabilities and hyperpolarizabilities of a large number of molecules, employing a variety of methods. We have found that the static vibrational hyperpolarizabilities have usually large values, while the frequency dependent vibrational contributions are much smaller, but not negligible. Here we demonstrate this point by considering two characteristic recent applications: the properties of pyrrole and HArF.

Keywords: polarizability, hyperpolarizability, electronic contribution, vibrational contribution.

PACS: 33.15.Kr, 42.65.An

1. Introduction

The clumped nucleus approximation allows the resolution of the electric properties to electronic and vibrational contributions. The latter have two components, the zero-point vibrational averaging term, P^{zpva}, and the pure vibrational contribution, P^{pv}. Thus the total electric property may be written as:

$$P^t = P^{el} + P^{pv} + P^{zpva} \tag{1}$$

The pv and zpva contributions to the polarizabilities and hyperpolarizabilities are given by [1]:

$$\alpha^{pv}=[\mu^2]^{(0,0)}+[\mu^2]^{(2,0)}+[\mu^2]^{(1,1)}+[\mu^2]^{(0,2)} \tag{2}$$

$$\beta^{pv}=[\mu\alpha]^{(0,0)}+[\mu\alpha]^{(2,0)}+[\mu\alpha]^{(1,1)}+[\mu\alpha]^{(0,2)}+[\mu^3]^{(1,0)}+[\mu^3]^{(0,1)} \tag{3}$$

$$\gamma^{pv}=[\alpha^2]^{(0,0)}+[\alpha^2]^{(2,0)}+[\alpha^2]^{(1,1)}+[\alpha^2]^{(0,2)}+[\mu\beta]^{(0,0)}+[\mu\beta]^{(2,0)}+[\mu\beta]^{(1,1)}+[\mu\beta]^{(0,2)}+[\mu^2\alpha]^{(1,0)}$$
$$+[\mu^2\alpha]^{(0,1)}+[\mu^4]^{(2,0)}+[\mu^4]^{(1,1)}+[\mu^4]^{(0,2)} \tag{4}$$

$$P^{zpva}=[P^e]^{(1,0)}+[P^e]^{(0,1)} \tag{5}$$

$$[P^e]^{(0,1)}= -\frac{\hbar}{4}\sum_{\alpha}\frac{1}{\omega_\alpha^2}\left(\sum_b \frac{F_{\alpha bb}}{\omega_b}\right)\left(\frac{\partial P^e}{\partial Q_\alpha}\right) \tag{6}$$

and

$$[P^e]^{(1,0)} = \frac{\hbar}{4}\sum_{\alpha}\frac{1}{\omega_\alpha}\left(\frac{\partial^2 P^e}{\partial Q_\alpha^2}\right) \tag{7}$$

where, ω_α is the harmonic frequency, $F_{\alpha bb}$ is the cubic force constant and Q_α is the normal coordinate. Analytical expressions for $[A]^{n,m}$ are given in Ref. 1; n and m are the orders of the electrical and mechanical anharmonicity, respectively. The order of the employed derivatives is described in Refs. 2 and 3, as well as the computational packages we have used.

2. Pyrrole

We have studied the electronic and vibrational contributions to dipole moment, plarizabilites, first and second hyperpoalrizabilites of a series of azoles [2]. Here we shall focus our attention on the

---

[*] Corresponding author: mpapad@eie.gr

vibrational properties of pyrrole, which has been found to have the largest vibrational properties of the considered azoles. Table 1 involves the electronic and vibrational contributions (pv and zpva) to α, β and γ pyrrole. Static and frequency dependent properties are presented. We observe from the static property values (Table 1) that α^{zpva} is small but not negligible; β^{pv} is more than an order of magnitude larger than β^{el}, while γ^{pv} is considerably larger than γ^{el}. The frequency dependent vibrational properties are much smaller than the static ones. However, except of $\alpha(-\omega;\omega)^{pv}$, they are not negligible. For example one may consider the pv contribution to $\beta(-\omega;\omega,\omega)$ or $\gamma(-\omega;\omega,0,0)$.

It has been found that the static pv contribution to α_{xx}, α_{yy} and α_{zz} of pyrrole are 18.55, 0.87 and 2.67 a.u., respectively. Thus in the following analysis we will focus on α_{xx}^{pv} (Table 2).

Table 1. Electronic and vibrational contributions to pola- rizabilities and hyperpolarizabilities of pyrrole. Method : HF/Pola.

Property	P^{el}	P^{pv}	P^{zpva}
α	52.45	7.37	1.51
β	28.75	-305.1	
γ	15322	25588	
$\alpha(-\omega;\omega)$	53.01	-0.06	
$\beta(-\omega;\omega,0)$	29.71	2.7	
$\beta(-2\omega;\omega,\omega)$	31.66	-1.9	
$\gamma(-\omega;\omega,0,0)$	16396	3207	
$\gamma(-2\omega;\omega,\omega,0)$	18961	382	
$\gamma(-3\omega;\omega,\omega,\omega)$	24241	-47	

a Molecular plane : YOZ; the dipole moment (0.774 a.u) is along Zaxis; λ=1064 nm. The properties are presented in atomic units.

Table 2. Analysis of α_{xx}^{pv} (in a.u.) of pyrrole.

	$[\mu^2]^{(0,0)}$	$[\mu^2]^{(2,0)}$	$[\mu^2]^{(1,1)}$	$[\mu^2]^{(0,2)}$
α_{xx}^{pv}	17.13	-0.88	0.49	1.81
18.55				

We observe that by far the largest contribution is made by $[\mu^2]^{(0,0)}$ (Table 1.) Thus the double harmonic approximation is satisfactory for α_{xx}^{pv}. A similar observation has been made for α_{yy}^{pv} [2]. However, the double harmonic approximation is less satisfactory for α_{zz}^{pv} (the double harmonic approximation gives 1.66 a.u., while the other three components, eq. 2 give 1.01. a.u.)[2]. It has been found that β_{xxz}^{pv}, β_{yyz}^{pv}, and β_{zzz}^{pv} of pyrrole are –499.54, 5.55 and –14.52 a.u., respectively [2]. Thus we shall focus our discussion on β_{xxz}^{pv}.

Table 3. Analysis of β^{pv}_{xxz} of pyrrole. The values are in a.u.

	$[\mu\alpha]^{(0,0)}$	$[\mu^3]^{(0,1)}$	$[\mu^3]^{(1,0)}$	$[\mu\alpha]^{(1,1)}$	$[\mu\alpha]^{(2,0)}$	$[\mu\alpha]^{(0,2)}$
β^{pv}_{xxz}	28.50	-213.06	-307.33	-13.32	-0.81	6.48
-499.54						

It is observed that $[\mu\alpha]^{(0,0)}$ contributes 28.5 a.u. which is a small part of –499.54 a.u. Thus the double harmonic approximation is not satisfactory for β^{pv}_{xxz}. The failure of this approximation is due to the very large contribution of $[\mu^3]^{(0,1)}$ and $[\mu^3]^{(1,0)}$ (Table 3).

It has been found that γ^{pv}_{xxxx} = -73501, γ^{pv}_{yyyy} = 5763.9, γ^{pv}_{zzzz} = 3606.8, γ^{pv}_{xxzz} =60368.6, γ^{pv}_{yyzz} =810.6 and γ^{pv}_{xxyy} =1780.5 a.u. [2]. The terms γ^{pv}_{xxxx} and γ^{pv}_{xxzz} are at least an order of magnitude larger than the other ones. Thus we shall present only them (Table 4). The dominant contributions to γ^{pv}_{xxzz} and γ^{pv}_{xxzz} are made by $[\mu^4]^{(2,0)}$ and $[\mu^4]^{(1,1)}$.

3. The HArF paradigm

Räsänen et al. studied HRgY, where Rg is a rare gas atom and Y an electronegative element or group (e.g. OH, CN) [4]. Most of these species have been discussed by using both experimental and theoretical techniques. Perhaps the most well known of these hydrides is HArF (argon flurohydride), the synthesis of which has been reported by employing photolysis of hydrogen fluoride in a solid argon matrix [4]. This is the first experimentally observed covalent neutral condensed phase argon derivative.

We observe that the pv contribution to α_{zz} is approximately equal to the electronic contribution (Table 5). The static and frequency dependent zpva contributions are small, but not negligible. However, the frequency dependent pv contribution to $\alpha_{zz}(-\omega;\omega)$ is negligible. The pv contribution to $\beta_{zzz}(0;0,0)$ is large;. the frequency dependent pv contribution to $\beta_{zzz}(-\omega;\omega,0)$ is smaller than the static one, but it is not negligible. Similarly the zpva contribution to $\beta_{zzz}(-\omega;\omega,0)$ is small but not negligible. It is noted that pv contribution to $\beta_{zzz}(-2\omega;\omega,\omega)$ is negligible. This demonstrates that the frequency dependent pv contribution depends on the process. The first hyperpolarizability of HArF has been interpreted by using the two-state model [3]. The difference between the ground state and the first dipole allowed excited state of HArF has been clarified in terms of resonance structures computed in terms of a complete active space valence bond method [3].

Table 4. Analysis of γ_{xxzz}^{pv} and γ_{xxzz}^{pv} for pyrrole (results in a.u.). Method: HF/Pol.

	γ_{xxzz}^{pv}	γ_{xxzz}^{pv}
$[\mu\beta]^{(0,0)}$	2366.3	1894.3
$[\alpha^2]^{(0,0)}$	686.3	499.8
$[\mu^2\alpha]^{(1,0)}$	2576.5	7011.2
$[\mu^2\alpha]^{(0,1)}$	-5118.9	-5397.1
$[\alpha^2]^{(1,1)}$	1.5	-4.1
$[\mu\beta]^{(1,1)}$	6.3	42.7
$[\mu^4]^{(1,1)}$	15319.4	27382.8
$[\mu\beta]^{(0,2)}$	3200.1	-1559.1
$[\mu\beta]^{(2,0)}$	-61.6	-46.3
$[\alpha^2]^{(0,2)}$	-38.5	18.2
$[\alpha^2]^{(2,0)}$	15.7	29.1
$[\mu^4]^{(0,2)}$	-1268.5	9191.1
$[\mu^4]^{(2,0)}$	-25034.7	21306.0
γ_{iijj}^{pv}	-73501	60368.6

Table 5. Static and frequency dependent properties (in a.u.) of HArF. Method: HF/Pol.

	p^{el}	p^{pv}	p^{zpva}
μ_z	3.473		0.034
$\alpha_{zz}(0;0)$	34.25	34.31	1.43
$\alpha_{zz}(-\omega;\omega)$	35.69	-0.03	1.68
$\beta_{zzz}(0;0,0)$	-561.5	285.9	-68.4
$\beta_{zzz}(-\omega;\omega,0)$	-637.3	-68.5	-88.0
$\beta_{zzz}(-2\omega;\omega,\omega)$	-835.4	1.50	

References

[1] D. Bishop, J. M. Luis and B. Kirtman, *J. Chem. Phys.*, **108**, 10013 (1998); D. M. Bishop and P. Norman, *"Handbook of Advanced Electronic and Photonic Materials"*, (Editor: H. S. Nalwa) Vol.9, p.1. Academic Press, San Diego, (2001).

[2] K. Jug, S. Chiodo, P. Calaminici, A. Avramopoulos and M. G. Papadopoulos, *J. Phys. Chem. A,* **107,** 4172 (2003).

[3] A. Avramopoulos, H. Reis, J. Li and M.G. Papadopoulos, *J. Am. Chem. Soc.,* **126,** 6179 (2004).

[4] L. Khriachtchev, M. Pettersson, N. Runeberg, J. Lundel and M. Räsänen, *Nature*, **406**, 874 (2000), and references therein.

VSP International
Science Publishers
P.O. Box 346, 3700 AH Zeist
The Netherlands

*Lecture Series on Computer
and Computational Sciences*
Volume 1, 2004, pp. 1112-1114

Ab initio QMM/MM Simulations: A Powerful Tool for the Study of Structure and Ultrafast Dynamics of Electrolyte Solutions

Bernd M. Rode[1]

Department of Theoretical Chemistry,
Faculty of Chemistry and Pharmacy,
University of Innsbruck,
A-6020 Innsbruck, Austria

Abstract: The predictive power of modern *ab initio* QM/MM molecular dynamics simulation techniques and the present limits will be demonstrated by a variety of examples dealing with the microscopic structure and ultrafast dynamics of ions in pure and mixed liquids and ligand exchange processes at these ions. The results show that several properties of such systems which are hardly accessible through any experimental technique can be evaluated by the use of simulation techniques, where at least the "chemically important" subsystem is fully treated by *ab initio* quantum mechanics.

Keywords: Molecular Dynamics – Simulations – Femto- and Picosecond Dynamics – Structure of Solvated Ions

PACS: 71.15.Pd

1. Computational Background

For the liquid state - the most important for chemists, but also the most complicated one for theoretical treatment - classical statistical methods employing molecular mechanics have been developed parallel to the generation of a framework of modern quantum chemical programs. While quantum chemical methods have already been used for quite some time to obtain energy surfaces for intermolecular interactions and the subsequent construction of analytical functions describing these interactions, it was not until recently that both approaches have converged to methods efficiently combining the advantages of the statistical approaches with the accuracy of quantum mechanical evaluation of forces in the course of computer simulations of liquids, in particular electrolyte solutions.

The main reason that this approach could not be implemented in a wider scale in the past is the enormous amount of computational power required for the quantum chemical part of the method, increasing the CP time of a classical molecular dynamics simulation by a factor of 50 to 300, depending on the degree of parallelisability and the size of the subsystem to be treated quantum mechanically. The usage of computer clusters with 100 2-3 Ghz processors and a flexible assignement of 4 – 32 processors per simulation has promoted *ab initio* QM/MM simulations towards technical and economical feasibility and thus allowed investigations of larger series of systems.

2. Ion Solvation and its Relevance for Solution Chemistry and Biology

Solvated metal cations and simple anions are not only of particular importance in determining properties and reactivity of aqueous and nonaqueous solutions, they also represent a most interesting research subject due to their structural variety and their ligand exchange dynamics, spanning a range from picoseconds to several hundred years of exchange time. As many of these ions play crucial parts in biological processes or act as toxines for the organism, the understanding of the physico-chemical behaviour of the ions is a prerequisite for the elucidation of numerous biochemical reactions and the complicated interplay of ions, metal complexes and all the organic components in a living organism.

---

[1] E-mail: Bernd.M.Rode@uibk..ac.at

Even simple model systems such as ions with 1 or 2 heteroligands or in mixed solvents can represent useful model systems to understand parts of this interplay.

3. Simulations versus Experimental Investigations of Solutions

Structural investigations of liquids, in particular of solutions with several components, are a complicated and fastidious task, implying not only highly sophisticated experimental equipment, but also much skill in finding the "right" model to interpret the measured data – and every experimental result can be only as good as the theoretical model employed for its deduction! Among the diffraction methods X-ray and Neutron Diffraction do not allow to study solutions at the usual biological concentrations. EXAFS is more advantageous in this aspect, but does not offer other advantages as the possibility of isotope substitution beneficially instrumentalised in Neutron Diffraction. Further, these methods usally give only time-averaged results and, therefore, structural pictures often do not correspondi to the manifold species actually formed in the course of solution dynamics. On the other hand, NMR is not a strong tool to determine structural details of solvated ions, but it can be well utilised to determine exchange rates and thus dynamical data, although only down to the nanosecond range. Femtosecond laser pulse spectroscopy, which could make faster reactions accessible, has not been sufficiently developed yet to investigate ionic solvation phenomena.

In comparison to these experimental methods, molecular dynamics simulations offer many advantages: A complete structrual analysis in terms of all atom-atom pair radial distribution functions, angular distribution functions and coordination number distributions give access to all details of even complicated solutions and reveal all species formed simultaneously or subsequently as intermediates. Dynamical processes can be studied in detail even on the femtosecond scale, the upper limitation set, however, by the feasible simulation time due to computational restrictions. The crucial problem of simulations is the accuracy of intermolecular potentials used to describe the interactions of particles within the solution. It has been generally recognised that simple pair potentials, even three-body corrected, cannot give much more than simple structural data for solvated metal ions, and that many-body and quantum effects play an important role for the "fine-tuning" of structures and conseqently also the dynamics of these ions.

4. *Ab initio* QM/MM Simulations

Classical simulation reach their limits, when n-body effects with n>3 are required, and whenever quantm effects play a role. Therefore, a quantum mechanical treatment of all forces in the elementary box of the simulation would be the simplest solution, but with a number of usually 200 – 500 molecules within this box exceeds any presently affordable computational effort by far. For this reason, a partition of the system into a physico-chemically "more relevant" region (e.g. the ion and its immediate surronnding), treated by means of quantum mechanics, and an "outer" region comprising the rest of the elementary box and treated by classical molecular mechanics based on ab initio generated 2-and 3-body functions, has been introduced. Both regions are separated by a "smoothing region", which guarantees continuous transitions of particles between QM region and MM region. The level of theory to be applied within the QM region is another important decision, and numerous investigations have shown that for solvated ions pure DFT methods are not suitable, and that even hybrid density functional methods as B3LYP do not always produce satisfactory results. Hartree-Fock level *ab initio* calculation of the forces with double zeta + polarisation basis sets appear to be the lower limit for sufficiently accurate results, but they still imply some methodical limitations. While electron correlation has only minor effects in the case of solvated cations, its neglect in the case of solvent-solvent interactions and anion solvation can be quite an error source. However, for both basis set size and level of theory, it is again the present computational facilities setting the limits of performance.

5. Examples to be Presented

Results from *ab initio* QM/MM simlations for solvation structure and dynamics of a large nmber of hydrated ions will be presented, ranging from simple alkali ions including the structure-breaking Cs(I) ion to transition metal ions and ions of heavy elements such as Hg(II), Pb(II) and Au(I). Attention will be focused on ions displaying ultrafast ligand exchange and other experimentally hardly accessible data such as the Jahn-Teller effect. The influence of heteroligands on the reaction rate of ligand exchange will be discussed, and the dynamics of the systems will be illustrated by video clips produced by MOLVISION, a new software tool designed for this purpose (www.molvision.com). The figure below illustrates the powerful abilities of simulations to analyse ionic solvation structures in composite systems (example: Ca(II) ion in aqueous ammonia) and at the same time shows the difficulty to resolve the experimentally observed diffraction pattern for such systems, which will be the sum of all atom pair distributions.

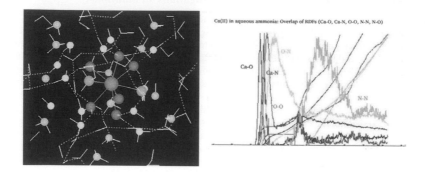

Ca(II) in aqueous ammonia: Overlap of RDFs (Ca-O, Ca-N, O-O, N-N, N-O)

Acknowledgments

The author wishes to thank the Austrian Science Foundation (FWF) for financial support (project no. P16221-NO8)

VSP International
Science Publishers
P.O. Box 346, 3700 AH Zeist
The Netherlands

*Lecture Series on Computer
and Computational Sciences*
Volume 1, 2004, pp. 1115-1118

Two-State Reaction Paradigm in
Transition Metals Mediated Reactions

I. Rivalta, N. Russo[1] and E. Sicilia

Dipartimento di Chimica and Centro di Calcolo ad Alte Prestazioni per Elaborazioni
Parallele e Distribuite-Centro d'Eccellenza MURST, Universita' della Calabria,
I-87030 Arcavacata di Rende (CS), Italy.

Received 7 March, 2004; accepted in revised form 10 March, 2004

Abstract: A brief review of our recent work on the activation of X-H (X=C, N, O) prototypical bonds by
transition metal containing systems is presented.

Keywords: Two-state reactivity paradigm, transition metals, potential energy surfaces

PACS: 31.50.Bc

1. Intoduction, Results and Discussion

Following the interpretation of Arrhenius of reaction rates, the chemical community has largely
accepted a picture of a reaction proceeding along a potential energy path and surmounting a barrier
Since for collisions involving molecules the potential energy depends on the molecular orientations as
well as on the internal geometries of the molecules themselves, the simple potential energy path that
describes the interaction between two atoms becomes a more complicate potential energy surface (PES)
often called potential energy hypersurface. The major part of the chemical reactions could be
considered as the motion of atoms on a single surface that evolves from reactants minimum to
intermediates and products minima passing throughout saddle points that by Evans and Polanyi [1]
were called transition states (TS). These processes are known as adiabatic ones. There are reactions that
can only be interpreted by considering jumps between surfaces. Often, in these non-adiabatic processes
transition metal atoms participate to the reactions and the crossings are due to the presence of excited
spin states close in energy to the ground state. This kind of behaviour, experimentally evidenced by a
series of state selective studies of organometallic reactions [2, 3], has been classified introducing the so
called Two-state Reactivity (TSR) Paradigm [4]. The possibility to follow the elementary step of a
reaction mechanism in going from reactants to products is experimentally enabled, whereas exhaustive
information on these aspects can be theoretically obtained, computing potential energy surfaces of
different spin multiplicities. In Scheme 1 are depicted the possible crossing between two PES's. In the
first case (a) the jump occurs before, while in the last case (c) after the passage of the TS. The case (b)
is the most intriguing since two crossings occur before and after the TS formation.

a)   b)   c)

ΔE

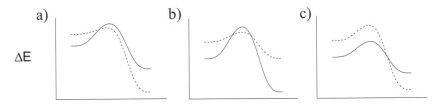

Reaction Coordinate

Scheme 1

[1] Corresponding author, e-mail: nrusso@unical.it

Only some works dedicated to this subject published in the last decades demonstrating that all these two-state reactivity typologies can be met along the PES's for reactions involving transition metal containing species can be cited [5-10]. We report here our recent work [11] on the activation of X-H (X=C, N, O) prototypical bonds mediated by bare (M^+) and ligated (MO^+ and MO_2^+) first row transition metal ions studied in the light of TSR.

The possibility to build up a reliable PES involving transition metal atoms or ions depends on the ability of the used computational tool to correctly assign the electronic ground state and to reproduce the energy gap between ground and first excited states of the considered metal. This is not a trivial question because the density functional bases methods, that we have used, often do not reproduce the experimental data for this fundamental property. In Table 1, we have collected and compared with experimental data [12] the results for the first-row transition metal ions obtained using DFT in its B3LYP [13] formulation and two basis sets of different size (DZVP [14] and TZVP+G(3df,2p) [15]). For the sake of brevity the B3LYP/TZVP+G(3df,2p) computations will be referred as B3LYP/TZVP. As can be realized from the data reported in Table 1, the most dramatic case is that of iron cation for which the order of the ground and first excited states is inverted no matter which basis set, less or more extended, is used. In parenthesis is reported the value of the gap calculated using a DZVP basis set optimized *ad hoc* by us for non local functionals [16]. Even if the gap value is overestimated with respect to the experimental value the correct ordering of the states is reproduced.

For the elimination of two hydrogen atoms as molecular hydrogen or water, the proposed reaction mechanism, in agreement with the experimental findings, involves as main steps the formation of: a stable ion-dipole complex; an insertion intermediate; dehydrogenation products. After the exothermic formation of the first ion-molecule complex, hydrogen migration to give the insertion complex occurs through a first three- or four-centered transition state. In the same way a second hydrogen shift from the X atom generates the last intermediate, in which hydrogen or water are practically formed.

Table 1: Relative energies, in kcal/mol, of the M^+ excited states with respect to ground state.

Cation	Configurations (State)	B3LYP/DZVP[a]	B3LYP/TZVP[a]	Exp[b]
Sc^+	sd (^{3}D) – sd (^{1}D)	4.6	3.8	7.4
Ti^+	sd^2 (^{4}F) – sd^2 (^{2}F)	12.9	12.6	13.7
V^+	d^4 (^{5}D) – sd^3 (^{3}F)	22.5	21.3	25.4
Cr^+	d^5 (^{6}S) – sd^4 (^{4}D)	53.1	52.2	56.7
Mn^+	sd^5 (^{7}S) – sd^5 (^{5}S)	15.1	19.7	27.0
Fe^+	sd^6 (^{6}D) – d^7 (^{4}F)	-10.5 (12.4)	-4.2	5.8
Co^+	d^8 (^{3}F) – sd^7 (^{5}F)	21.6	16.7	9.9
Ni^+	d^9 (^{2}D) – sd^8 (^{4}F)	31.9	28.0	25.0
Cu^+	d^{10} (^{1}S) – sd^9 (^{3}D)	69.1	67.1	64.8

It is obtained surmounting an energy barrier corresponding to a second four-centered transition state.

As an example of the PES's that we have characterized in detail, in Figure 1 are reported the energetical diagrams for the dehydrogenation and dehydration reactions of ammonia by Fe^+ bare cation and FeO^+ oxide. Along the path for the activation of the N-H bond by Fe^+ a surface crossing is localized at the entrance channel of the reaction and a second one before the formation of the last hydrogen-ion complex, after the passage of the TS corresponding to the migration of the second hydrogen atom. When the activation of the N-H bond occurs thanks to the corresponding oxide the driving force of the reaction is the formation of a strong O-H bond. The presence of the oxygen ligand favourably influences the reactivity of the iron cation since all the minima and transition states along the whole reaction path lie below the energy of the reactants dissociation limit. Multiple spin crossing are observed between the surfaces for the sextet ground state of the monoxide cation and that for the quartet excite state. More details on the PES's will be given.

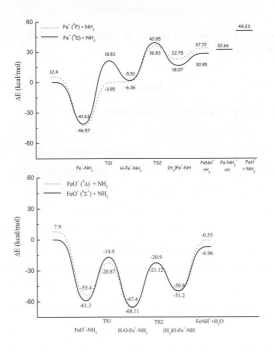

Figure 1

The picture appears to be much more complicated when dioxide cations are considered since an increasing complexity of the reactivity pattern is encountered besides that at the electronic structure level. In Scheme 2 is shown a reaction pattern which accounts for all the experimentally detected [17] products for the activation of methane mediated by CrO_2^+ cation.

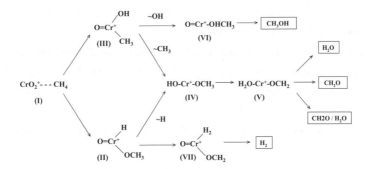

Scheme 2

It seems that TSR plays a crucial role in determining products distribution. Work is in progress to elucidate the mechanistic details of the interaction of chromium dioxide with methane lacking any previous theoretical investigation of this subject.

Acknowledgments

The author wishes to thank the anonymous referees for their careful reading of the manuscript and their fruitful comments and suggestions.

References

[1] Evans M. G.; Polanyi, Trans. Faraday Soc. **1935**, 31, 857

[2] Armentrout, P. B. Science **1991**, *41*, 175.

[3] Armentrout, P. B; Beauchamp, J. L. *Acc. Chem. Res.* **1989**, *22*, 315.

[4] Schröder, D.; Shaik, S.; Schwarz, H. *Acc. Chem. Res.* **2000**, *33*, 139.

[5] Elkind, J. L.; Armentrout P. B. *J. Am. Chem. Soc.* **1986**, *108*, 2765.

[6] Fisher, E. R.; Armentrout P. B. *J. Am. Chem. Soc.* **1992**, *114*, 2039.

[7] Fiedler, A.; Schröder, D.; Shaik, S.; Schwarz, H. *J. Am. Chem. Soc.* **1994**, *116*, 10734.

[8] Danovich, D.; Shaik, S. *J. Am. Chem. Soc.* **1997**, *119*, 1773.

[9] Irigoras, A.; Elizalde, O; Silanes, I.; Fowler, J. E.; Ugalde, J. M. *J. Am. Chem. Soc.* **2000**, *122*, 114.

[10] Schwarz, H. .; Schröder, D. *Angew. Pure Appl. Chem.* **2000**, *72*, 2319

[11] a) Russo, N.; Sicilia, E. *J. Am. Chem. Soc.* **2001**, *123*, 2588; b) Russo, N.; Sicilia, D. *J. Am. Chem. Soc.* **2002**, *124*, 1471; c) Michelini, M. C.; Sicilia, E.; Russo, N. *J. Phys. Chem. A* **2002**, *106*, 8937; d) Michelini, M. C.; Sicilia, E.; Russo, N.; *Inorg. Chem.*, **2003**, *42*, 8773. e) Kondakova, O.; Michelini, M. C.; Sicilia, E.; Russo, N.; *Inorg. Chem.*, **2003**, *42*, 8773. f) Chiodo, S.; Kondakova, O.; Irigoras, A.; Michelini, M. C.; Russo, N.; Sicilia, E.; Ugalde, J. M. *J. Phys. Chem. A,* **2004**, *108*, 1069. g) M. C.; Sicilia, E.; Russo, N.; *Inorg. Chem.*, **2004**, *0*, 00.

[12] Moore, C. E. Atomic Energy Levels; NSRD-NBS, USA; U.S. Government Printing Office: Washington, DC, 1991; Vol 1.

[13] a) Becke, A. D. *J. Chem. Phys.* **1993**, *98*, 5648. b) Stephens P. J.; Devlin, F. J.; Chabalowski, C. F.; Frisch, M. J. *J. Phys. Chem.* **1994**, *98*, 11623.

[14] Goudbout, N.; Salahub, D. R. ; Andzelm, J.; Wimmer, E. *Can. J. Chem.* **1992**, *70*, 560.

[15] Irigoras, A.; Elizalde, O; Silanes, I.; Fowler, J. E.; Ugalde, J. M. *J. Am. Chem. Soc.* **1989**, *111*, 8280 and references therein.

[16] Chiodo, S; Russo, N.; Sicilia, E *J. Comp. Chem.,* in press.

[17] Fiedler, A.; Kretzschmar, I.; Schröder, D.; Schwarz, H. *J. Am.Chem. Soc.* **1996**, *118*, 9941.

VSP International
Science Publishers
P.O. Box 346, 3700 AH Zeist
The Netherlands

*Lecture Series on Computer
and Computational Sciences*
Volume 1, 2004, pp. 1119-1120

Computations of Endohedral Fullerenes as Agents of Nanoscience: Gibbs Energy Treatment for Ca@C72, Ca@C82, La@C82, N2@C60, and La@C60

Z. Slanina, K. Kobayashi and S. Nagase [1]

Department of Theoretical Studies,
Institute for Molecular Science,
Myodaiji, Okazaki 444-8585, Japan

Received 7 March, 2004; accepted in revised form 10 March, 2004

Abstract: The paper surveys ongoing computations on endohedral fullerene systems, combining the treatments of quantum chemistry and statistical mechanics. Relative concentrations of isomers of Ca@C72, Ca@C82, and La@C82 are evaluated using the Gibbs energy based on density-functional theory computations. The results illustrate the enthalpy-entropy interplay in the systems produced under high temperatures. Similar computations are also reported for encapsulation of N2 and La into the C60 cage.

Keywords: Fullerenes and metallofullerenes; Electronic properties; Gibbs energy of encapsulation.

PACS: 31.15.Ew; 36.40.Cg; 74.70.Wz.

1. Computational Outline

Various endohedral cage compounds have been suggested as possible candidate species for molecular memories. One approach is built on endohedral species with two possible location sites of the encapsulated atom while another concept of quantum computing aims at a usage of spin states of N@C60. In this work, several systems related to the first approach are simulated computationally, combining the treatments of quantum chemistry and statistical mechanics. For example, relative concentrations of five isomers of Ca@C72, nine isomers of Ca@C82, and four isomers of La@C82 are computed using the Gibbs energy.

Evaluations of fullerenes require entropy consideration owing to very high synthetic temperatures. An illustration is supplied on two recently observed endohedral systems - Ca@C72 and Ca@C82. Five isomers computed previously [1] are considered for Ca@C72: IPR-related cage (a), two non-IPR C72 cages (b) and (c), a structure with one heptagon (d), and a species with two heptagons (e). For Ca@C82, the isomers [2] derived from the nine IPR structures are treated: C3v(a), C3v(b), C2v, C2(a), C2(b), C2(c), Cs(a), Cs(b), Cs(c). The analytical vibrational analysis is carried out at the B3LYP level in a combined basis set: 3-21G for C atoms and a dz basis set with the effective core potential on Ca. Electronic excitation energies are evaluated by TD DFT at the same level and also by the ZINDO method. The separation energetics is computed at the B3LYP/6-31G* level.

Relative concentrations (mole fractions) in a set of isomers can be expressed through their partition functions and the enthalpies at the absolute zero temperature or ground-state energies (i.e., the relative potential energies corrected for the vibrational zero-point energies) by a compact master formula [3]. This master formula is an exact formula that can be directly derived from the standard Gibbs energies of the isomers, supposing the conditions of the inter-isomeric thermodynamic equilibrium. Rotational-vibrational partition functions are constructed from the calculated structural and vibrational data using the rigid rotator and harmonic oscillator approximation. No frequency scaling is applied as it is not significant for the mole fractions at high temperatures. The electronic partition function was constructed by directed summation. The symmetry and chirality contributions are included accordingly.

---

[1] Corresponding author. E-mail: zdenek@ims.ac.jp

In the Ca@C72 system [4], the (b) and (c) non-IPR species represent major isomers while the IPR-related structure (a) comes as a minor species. Five structures show significant populations at higher temperatures for Ca@C82: C2v > Cs > C2 > C3v > Cs.

The problem of the relative isomeric stabilities is somewhat simpler [5] than the problem of the absolute cluster stabilities as pressure/concentration is avoided and there is only one controlling parameter - temperature. The Gibbs energy for encapsulation has for the first time been evaluated for N2 and La incorporations into C60. It is shown that the term is quite temperature dependent.

Acknowledgments

The reported research has been supported by a Grant-in-aid for NAREGI Nanoscience Project, Scientific Research on Priority Area (A), and Scientific Research (B) from the Ministry of Education, Culture, Sports, Science and Technology of Japan.

References

[1] K. Kobayashi, S. Nagase, M. Yoshida and E. Osawa, *J. Am. Chem. Soc.* 119, 12693 (1997).

[2] K. Kobayashi and S. Nagase, *Chem. Phys. Lett.* 274, 226 (1997).

[3] Z. Slanina, *Contemporary Theory of Chemical Isomerism.* D. Reidel, Dordrecht, 1985.

[4] Z. Slanina, K. Kobayashi and S. Nagase, *Chem. Phys. Lett.* 372, 810 (2003).

[5] Z. Slanina, X. Zhao, F. Uhlik and S.-L.Lee, *Int. J. Quantum Chem.* 99, 640 (2004).

VSP International
Science Publishers
P.O. Box 346, 3700 AH Zeist
The Netherlands

Lecture Series on Computer
and Computational Sciences
Volume 1, 2004, pp. 1121-1125

NMR Solution Structure Determination of Biomolecules and NMR-driven Docking Simulations of Biomolecular complexes

G. Vlachopoulos, G.A. Spyroulias[1], P. Cordopatis

Department of Pharmacy,
School of Health Sciences, University of Patras,
GR-265 04 Patras, Greece

Received 5 July, 2004; accepted in revised form 25 July, 2004

Abstract: Development on technology and methodology in NMR Spectroscopy has resulted in the detailed analysis and exploitation of NMR-derived geometrical information, such as NOEs (*Nuclear Overhausser Enhancement*), paramagnetism, relaxation times, $^3J_{H^N H^\alpha}$ couplings constants. These data are used as structural constrains and implemented in structure calculation through molecular dynamics protocols for the determination of 3D structures of biomolecules in solution. These structures are characterized by resolution comparable to that provided by X-ray in solid state. The main volume of information is extracted from proton-proton through space, dipolar, interaction when they are found in a distance up to 5.0-5.5 Å. The strategy of solution structure determination of biomolecules is presented below and the various stages of structural calculation using NMR derived constraints are analyzed for the case of the 36-residue synthetic peptide which represents the amino acid sequence of the Angiotensin-I Converting Enzyme catalytic site (ACE). Docking simulations are also being performed in order to set the structural basis for the ACE – substrate interaction using the Xray structure of the C-terminal ACE catalytic site and the NMR structure of the decapeptide hormone substrate, Angiotensin I.

Keywords: Structure Calculation, NMR Spectroscopy, Torsion Angle Dynamics, Docking Simulations

PACS: 71.15.Pd

1. Introduction

The detailed analysis of the way that a polypeptide chain of a biomolecule folds in order to adopt a well-defined tertiary structure is an essential step for the profound understanding of its function. The two major techniques that could afford the high resolution 3D structure of a bio-macromolecule are the X-ray crystallography in the solid state and nuclear magnetic resonance (NMR) spectroscopy in solution. NMR overcomes the drawback of obtaining suitable crystals for diffraction (time-scales, preparation of heavy-atoms derivatives, etc.). At the same time the size of macromolecular structures that can be solved by NMR has been dramatically increased over the past few years. At Brookhaven Protein Data Bank (PDB) the coordinates of more than 1100 NMR structures of proteins and nucleic acids with molecular weight which reach up to 35 kDa have been deposited [1].

2. NMR Spectroscopy : Application to Structure Determination

NMR Spectroscopy is based on the property of every proton, which is known as magnetization. When a proton is found in a static magnetic field B_0, the magnetization lies parallel to B_0 (defined as z direction). To record a conventional NMR spectrum, a radio-frequency pulse B_1 is applied, which rotates the magnetization away from z-axis towards the x-y plane. Free-induction decay is recorded immediately after the pulse and yields the conventional spectrum after Fourier transformation. In 2D NMR experiments limitations due to signal overlapping in 1D spectra, are resolved by extending the measurements into a second dimension (or even into a third/fourth). In most homonuclear 2D experiments the symmetrically placed cross peaks on either side of the diagonal indicate the existence of an interaction between two spins while the type of this interaction depends on the type of the experiment. Thus, in a correlation experiment (COSY, TOCSY), the cross peak arises from through-

---

[1] Corresponding author. E-mail: G.A.Spyroulias@upatras.gr

bond scalar correlations, while in a nuclear Overhauser enhancement (NOE) experiment they arise from through-space correlations [2].

These through-space correlations are the main source of the NMR-derived structural restrains. The dipolar cross-relaxation rate constant is proportional to the inverse sixth power of the distance between the interacting protons. Thus, if one inter-proton distance, r_{ref} is known (from covalent geometry), then another, unknown inter-proton distance r_i, is determined by the relationship:

$$r_i = r_{ref} (S_{ref}/S_i)^{1/6} \tag{1}$$

where S_{ref} and S_i are the integrated cross peak volumes/intensities. The extraction of geometrical data is carried through the use of appropriate algorithms and complex calculations, which could precisely determine the ^1H-^1H distance from the NOE cross-peak intensities. For this reason, the unambiguous assignment of the observed resonances in NMR spectra to the corresponding nuclei is the first and the most important step. This assignment procedure is called *Sequence-Specific Resonance Assignment* and applies to the complete resonance assignment of backbone and side chain nuclei of any polypeptide prior to the assignment of through-space nuclei interaction [2]. The strategy is illustrated in Figure 1.

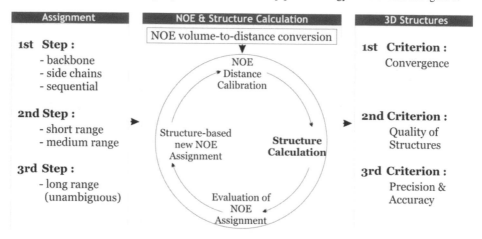

Figure 1: The Strategy of High-Resolution NMR Structure Determination.

NOE cross peaks are usually grouped into three (according to their intensities; strong, medium, and weak) or into five (according to the nature of proton-proton interactions; intra-residue, sequential, medium range, backbone long-range and long-range) categories. Each category is associated with an upper-bound separation between the interacting spins. The cross-peak volume limits and the upper-bound distances are calibrated according to the same distances found in Xray structure or to the NOE intensities observed in protons of known covalent geometry (geminal methylene protons) [2,3].

3. Calculation of NMR Models and Criteria for high-resolution structures

The most applied protocol in calculating structure from NMR data involves the use of restrained molecular dynamics (MD) using the simulating annealing (SA) protocol where the MD force fields are supplemented by square-well pseudo-energy terms based on NMR-derived restraints [4-6]. The generated structure is restrained according to the experimental geometric restraints toward a conformation that satisfies the experimental data while reduces the appeared violations (and consequently the overall energy of the system/molecule) imposed from the same experimental data during the control heat-up – cool-down procedure of the SA protocol. The software used in this study for the calculation of the 36-residue peptide of the Angiotensin-I Converting Enzyme active site (Figure 2), is called DYANA [6] (DYnamic Algorithm for Nmr Applications). DYANA protocol aims to determine coordinates for the protein atoms that will satisfy the input distance and angular restraints in an unbiased fashion while avoiding local energy minimum, and exploring all regions of conformational space, compatible with the observed NMR parameters.

The entire procedure includes structure calculation, evaluation of the calculated models in terms of energy terms and violated restraints, assignment of new NOE cross-peaks from NOESY spectra, conversion of peak intensities to upper distance limits. This procedure is repeated to determine an ensemble of (low-energy) structures consistent with the input data (see Figure 2). A high quality ensemble of structures should exhibit the minimum of violations (where no consisted violation -

violation observed in all the calculated structures - observed) of the input restraints expressed as a penalty term (the *target function*). The *target function V* is $V \geq 0$ with $V = 0$ if all experimental distance constraints and torsion angle constraints are fulfilled and all non-bonded atom pairs satisfy a check for the absence of steric overlap. The target function is related to the potential energy through the equation $E_{pot} = w_0 V$ where w_0 is an overall weighting factor equal to $10 \text{kJmol}^{-1} \text{ Å}^{-2}$.
Additionally, the RMSD (root-mean-square-deviation) among the members of the family should be below 1.0 Å for the backbone and below 2.0 Å for all heavy atoms.

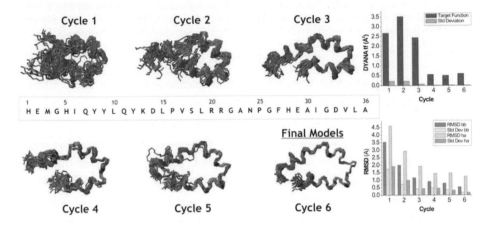

Figure 2: ACE 36-residue peptide model ensemble [7] selected from various cycles of the DYANA calculation procedure using NMR constraints. Statistics related with DYANA target function and RMSD values are also presented.

Convergence to a certain conformation, large number (>15 per residue) and variety of geometric constraints (NOEs, torsion angles, H-bonds, stereospecific assigned protons/methyls), and tight upper distance limits are among the criteria for the precision and accuracy of the calculated structure.
According to the procedure illustrated at Figure 2, the initial structure calculation (Cycle 1) yield low resolution model ensemble with high values of RMSD and target function (due to the constraint violations). Evaluation of the NOE-based distance limits and tight calibration of volume-to-distance ratio results to increase of target function (as new constraint violation appeared) and low RMSD values since the molecular conformational freedom is restrained (Cycle 2). By repeating this procedure, NOE constraint violations are eliminated while convergence to a favourable conformation is achieved. As a result target function and RMSD values drop to acceptable values (Cycles 3 to 6).

4. Mapping the Interaction Interface through NMR – Docking Simulation

The original view that protein-protein interactions originate from a large intramolecular contact surface (600-2000 Å^2) with 10-30 contact side-chains on each side of the protein interfaces [8] ensued from the idea that protein-protein interactions are energetically driven by a limited number or group of residues or hot spots, localized within the contact interfaces [9]. Thus it could be realistic to mimic the binding interaction of large proteins by small peptides. In the case of ACE well-established mechanistic studies on dipeptidyl-carboxypeptidadses in concert with experimental data have postulated that substrate forms a complex coordinated to the metal ion [10-11], providing evidence for an unambiguous interaction site. NMR is applied in order to monitor the effect of binding either on the target (ACE 36-residue constructs) or on the ligand (peptide substrate) through the : (i) *identification of binding residues by differential perturbation of NMR signals* (such as chemical shift perturbation mapping for ^{1}H, ^{15}N and ^{13}C, line-broadening effects caused by T_2-relaxation rate contribution originating from the bound state, etc.), (ii) *determination of the peptide or substrate conformational changes by transferred nuclear Overhauser effects,* (iii) *analysis of molecular diffusion, etc.* [12-14]. Such data extracted from 2D NOESY and HSQC experiments are currently implemented to the appropriate docking simulation programs (HADDOCK [15]) in order to build the peptide-peptide complexes.
However, there are also other software packages (AUTODock & AUTODock Tools [16]) which are using shape complementarity, electrostatic potentials, possible ionic interaction and H-bonds through Simulating Annealing or Genetic algorithms in constructing biomolecular complexes. Such an

approach is currently applied in our laboratory in order to determine the ACE catalytic site – Angiotensin-I complex. For this purpose the NMR structure of ACE presented above and the solution model of Angiotensin-I [17] are implemented in AUTODock program and results suggest an ACE-substrate interaction model with the C-terminal oriented towards the bottom of ACE substrate channel.

Acknowledgments

The authors wish to acknowledge *EU Access to Large Scale Infrastructure Program* (HPRI-CT-1999-00009) and Center of Magnetic Resonance in Florence for access to NMR Instruments and powerful computational systems. GAS and PC thank U of Patras for a *K. Karatheodoris* Research Grant.

References

[1] G. Wider, K. Wüthrich, NMR spectroscopy of large molecules and multimolecular assemblies in solution, *Curr. Opin. Struct. Biol.* 9, 594-601(1999).

[2] K. Wüthrich, *NMR of Proteins and Nucleic Acids.* Wiley, New York, 1986.

[3] P. Güntert, W. Braun, K. Wüthrich, Efficient computation of three-dimensional protein structures in solution from nuclear magnetic resonance data using the program DIANA and the supporting programs CALIBA, HABAS and GLOMSA, *J. Mol. Biol.* 217, 517-530(1991).

[4] A.T. Brünger, G.M. Clore, A.M. Gronenborn, M. Karplus, Three-dimensional structure of proteins determined by molecular dynamics with interproton distance restraints: application to crambin, *Proc. Natl. Acad. Sci. U.S.A.* 83 3801-3805(1986).

[5] G.M. Clore, A.T. Brünger, M. Karplus, A.M. Gronenborn, Application of molecular dynamics with interproton distance restraints to three-dimensional protein structure determination. A model study of crambin, *J. Mol. Biol.* 191, 523-551(1986).

[6] Güntert, P.; Mumenthaler, C.; Wüthrich, K. Torsion angle dynamics for NMR structure calculation with the new program DYANA. *J. Mol. Biol.* 273, 283-298(1997).

[7] G.A. Spyroulias, A.S. Galanis, G. Pairas, E. Manessi-Zoupa, P. Cordopatis *Cur. Top. Med. Chem.* 4, 403-429(2004).

[8] J. Janin and C. Chothia, The structure of protein–protein recognition sites. *J. Biol. Chem.* 265, 16027-16030(1990).

[9] Clackson, T.; Wells, J. A. A hot spot of binding energy in a hormone-receptor interface. *Science (Washington, D.C.)* 267, 383–386(1995).

[10] Lipscomb W. N.; Sträter N. Recent Advances in Zinc Enzymology. *Chem. Rev.* 96, 2375-2433(1996).

[11] Matthews, B. W. Structural Basis of the Action of Thermolysin and Related Zinc Peptidases. *Acc. Chem. Res.* 21, 333-340(1988).

[12] Zuiderweg, E. R. Mapping protein-protein interactions in solution by NMR spectroscopy. *Biochemistry* 41, 1-7(2002).

[13] Ni, F. Recent developments in transferred NOE methods. *Prog. Nucl. Magn. Reson. Spectrosc.* 26, 517–606(1994).

[14] Ni, F.; Scheraga, H. A. Use of the transferred nuclear Overhauser effect to determine the conformations of ligands bound to protein. *Acc. Chem. Res.* 27, 257–264(1994).

[15] C. Dominguez, R. Boelens and A.M.J.J. Bonvin. HADDOCK: a protein-protein docking approach based on biochemical and/or biophysical information. *J. Am. Chem. Soc.* 125, 1731-1737(2003).

[16] G.M. Morris, D.S. Goodsell, R.S. Halliday, R. Huey, W.E. Hart, R.K. Belew and A.J. Olson Automated Docking Using a Lamarckian Genetic Algorithm and and Empirical Binding Free Energy Function *J. Computational Chemistry* 19, 1639-1662(1998).

[17] G.A. Spyroulias, P. Nikolakopoulou, A. Tzakos, I.P. Gerothanassis, V. Magafa, E. Manessi-Zoupa, P. Cordopatis. Comparison of the solution structures of angiotensin I & II. Implication for structure-function relationship. *Eur. J. Biochem.* 270, 2163-2173(2003).

VSP International
Science Publishers
P.O. Box 346, 3700 AH Zeist
The Netherlands

*Lecture Series on Computer
and Computational Sciences*
Volume 1, 2004, pp. 1126-1128

Solvent Effects on the B-Cyclodextrin Inclusion Complexes with *M*-Cresol and Dynamic Hydrophobicity: Molecular Dynamics

Kailiang Yin[1,2,*], Duanjun Xu[1], Chenglung Chen[3]

1. Department of chemistry, Zhejiang University, Hangzhou 310027, China
2. Department of chemical engineering, Jiangsu Polytechnic University, Changzhou 213016, China
3. Department of Chemistry, National Sun Yat-sen University, Kaohsiung, 80424, Taiwan, China

Received 7 March, 2004; accepted in revised form 10 March, 2004

Abstract: Three systems included 1:1 beta-cyclodextrin inclusion complex with m-cresol in vacuum, nano water drop and nano acetone drop have been studied using molecular dynamics with pcff force field and without periodic boundary condition. Solvent effects have been discussed. The pre-equilibrated modelling inclusion complex in nano water drop and acetone drop contained two water and two acetone molecules respectively, and during the simulation one water molecule was observed to dissociate from the host in water drop while no dissociation was observed in acetone case. This result well validates the dynamic hydrophobicity of the beta-cyclodextrin. The researches on the movement of guest molecules, the dynamic structures of host β-CD and the time average configurations of the inclusion complexes indicate that solvent has an important effect on the stability and dynamic structure of the beta-cyclodextrin inclusion complex.

Keywords: β-Cyclodextrin; *m*-cresol; inclusion complex; molecular dynamics; hydrophobicity; solvent effect

PACS: 82.70.Uv

1. Experimental Method Section

1. Simulation method

Molecular dynamics (MD) simulations were carried out using the MATERIALS STUDIO 2.2 software package (Accelrys Co.) [13]. PCFF (Polymer Consistent Force Field) which is conjunction with the SPC water model [14] and well suitable for organic substances was applied in the simulation. The equations of motion were solved with velocity verlet algorithm. The time step of integration was held 0.5 fs in water case and 1 fs in other cases. The summation methods for van der Waals and Coulomb are all atom based (no cutoff). The simulation ensemble was set to *NVE* microcanonical in which number of atoms, volume and energy of system are all kept constant during the simulation. No boundary condition was used for MD run. The system temperature was set to 298 K. After the MD trajectory of one system being collected, analysis was made by the Analysis module in the package and/or our analysis programs compiled by Fortran 90.

2. Modelling and simulation

Three systems were modeled for MD simulation. In all system, the host is β-cyclodextrin and the guest is *m*-cresol which is a common chemical agent, and the initial configuration of the inclusion complex is uniformly as shown in fig. 1a.

In system I, there is no solvent molecule, so the β-CD inclusion can be seen as in the vacuum. After energy minimizing, the system was pre-equilibrated for 500 ps at 298 K by MD simulation which ensemble was *NVE* (the MD within *NVE* ensemble usually be called classical MD) and step time was 1 fs. Then finally classical MD was carried out at 298 K. Iteration step time was chosen as 1 fs, trajectory was saved every 50 fs (0.05 ps), that can also be said sampling time was 50 fs, and 200 ps trajectory in total was collected for the analysis.

In system II, the solvent is water. First, we modeled an energy-minimized nano water drop contained 427 H_2O (the aim we modeled nano drop not periodic system is to save the CPU time) and pre-equilibrated it by the classical MD at 298 K for 200 ps with 0.5 fs step time. Choosing smaller step

---

* Corresponding author. Tel. +86-519-3290253. *E-mail address:* mat_studio@jpu.edu.cn

time – 0.5 fs for water was based on the higher vibrating frequencies of O-H bonds in water. Second, we soaked the β-CD inclusion into the nano water drop and minimized the system's energy to reduce the high energy which arose from the overlap of atoms. Third, we pre-equilibrated the system by the classical MD at 298 K for 500 ps with 1 fs as time step. This well relaxed system is shown in fig. 2. Finally, a classical MD run was made at 298 K with 0.5 fs as time step. Total 100 ps trajectory with every 50 fs as sampling time was collected for the analysis.

In system III, we select a common organic solvent — acetone. Similar modelling and pre-equilibration process to system II were made but the number of acetone was 238 and the step time was chosen as 1 fs. The finally classical MD run was performed on this nano acetone drop at 298 K with 1 fs as time step and total 100 ps trajectory with every 100 fs sampling was collected for the analysis.

Table 1: Some time average structural properties of β-CD complex

system	$r_b(\text{Å})$	$d_{35}(\text{Å})$	$d_{po}(\text{Å})$	$d_{pz}(\text{Å})$	$\cos\theta$	$O_{w1}(\text{Å})$	$mc_{a1}(\text{Å})$	$mc_{a2}(\text{Å})$
I	5.106	2.538	-0.431	0.443	0.164			
II^a	5.057	2.477	-1.340	0.547	0.509	2.685		
II^b	5.089	2.570	-1.252	0.639	0.722	2.619		
III	5.107	2.545	1.381	0.430	0.598		6.572	4.886

a: average before 68 ps, b: average after 68 ps;
r_b: average over seven $r_{b,i}$; d_{35}: distance between the mass centers of H3 and H5 planes; d_{po}: position of the mass center of phenyl ring of *m*-cresol relative to O*b* plane, the same sign definition as D_{po} in fig.5; d_{pz}: distance between the mass center of the phenyl ring and the CD axis which was defined as the line passed through the mass centers of H3 and H5 planes; θ: angle between the vector of the phenyl hydroxyl oxygen point to the mass center of the phenyl ring and the vector of the mass center of the phenyl ring point to that of the O*b* plane; O_{w1}, mc_{a1}, mc_{a2}: distance between the oxygen in w1 or the mass center of a1 or a2 and the mass center of O*b* plane.

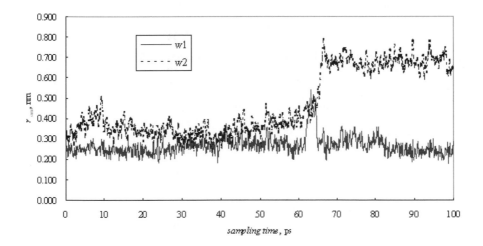

Figure1: Trajectories of the distances r_{om} between the mass center of O*b* plane and oxygen atoms of w1 and w2

Acknowledgments

The author wishes to thank the anonymous referees for their careful reading of the manuscript and their fruitful comments and suggestions. This work was supported JSNFS BK2003402.

References

[1] L. H. Tong, *Chemistry of Cyclodextrin: Fundamental and Application*. Science Publishing, Beijing, 2001.

[2] W. Saenger, *Angrew. Chem. Int. Ed. Engl.* 19 (1980) 344.

[3] L. Catoire, V. Michon, L. Monville, A. Hocquet, L. Jullien, J. Canceill, J-M Lehn, M. Piotto and C. H. du Penhoat, *Carbohydr. Res.* 303 (1997) 379.

[4] K. B. Lipkowitz, *Chem. Rev.* 98 (1998) 1829.

[5] J. R. Grigera, E. R. Caffarena and S. de Rosa, *Carbohydr. Res.* 310 (1998) 253.

[6] E. Cervelló, F. Mazzucchi and C. Jaime, *J. Mol. Struct. (Theochem).* 530 (2000) 155.

[7] J. Varady, X. W. Wu and S. M. Wang, *J. Phys. Chem. B.* 106 (2002) 4863.\

[8] K. V. Damodaran, S. Banba and C. L. Brooks, *J. Phys. Chem. B.* 105 (2001) 9316.

[9] K. L. Yin, D. J. Xu and C. L. Chen, *Chinese J. Inorg. Chem.* 19 (2003) 480.

[10]   K. L. Yin, Q. Xia, D. J. Xu and C. L. Chen, *Chinese J. Chem. Phys.* 2004, in press.

[11]   K. Braesicke, T. Steiner, W. Saenger and E. W. Knapp, *J. Mol. Graphics Mod.* 18 (2000) 143.

[12]   E. B. Starikov, K. Bräcke, E. W. Knapp and W. Saenger, *Chem. Phys. Lett.* 336 (2001) 504.

[13]   Materials Studio 2.0; Discover/ Accelrys: San Diego, CA, 2001.

[14]   H. J. C. Berendsen, J. R. Grigera and T. P. Straatsma, *J. Phys. Chem.* 91 (1987) 6169.

[15]   K. H. Jogun, J. M .Maclennan and J. J. Stczowski, *Eur. Cryst. Meeting* 5 (1979) 34.

[16]   M. Onda, Y. Yamamoto, Y. Inoue and R. Chûjô, *Bull. Chem. Soc. Jpn.* 61 (1988) 4015.

VSP International
Science Publishers
P.O. Box 346, 3700 AH Zeist
The Netherlands

*Lecture Series on Computer
and Computational Sciences*
Volume 1, 2004, pp. 1129-1131

Theoretical Study on the Coumarins for Designing Effective Second-Order Nonlinear Optical Materials

Bo Zhao[1], Zhihua Zhou

Department of Computer Science and Technology,
Faculty of Sciences and Technology,
University of Peloponnese,
GR-221 00 Tripolis, Greece

Received 7 March, 2004; accepted in revised form 10 March, 2004

Abstract: A systematically TDHF calculation study on the molecular nonlinear properties of coumarins were carried out. The results reveal that these compounds have relatively strong molecular nonlinearity. Concerning substituent positions the 7-substitution was much effective in enhancing the β values while the position 4 was inert. The 4- and 7-substituted positions in the molecules show obviously active for the macroscopic nonlinearities especially for the different groups. How to design the excellent coumarin second-order NLO materials is also discussed.

Keywords: molecular second-order nonlinear optical property; Coumarin; molecular design.

PACS: 4265

1. Introduction

There are still have been considerable interests in organic materials with high second-order nonlinear optical (NLO) properties in recent years because of their potential applications in optoelectronics and photonics fields.[1-3] Besides the strong nonlinearity a useful NLO material must satisfy many other qualities, such as short cut-off wavelength and high thermal stability, ect. The greatest challenge in designing excellent NLO materials is to optimize these parameters in a single substance. Usually a material possesses some of these critical requirements but fall short in the others, and rendering it useless for practical applications. Comparing to the currently studied inorganic materials the organic NLO materials have large nonlinear response, extremely fast switching time and high laser damage threshold but usually their transparency and melting points needs improving. So it is important to improve these aspects for the organic materials on the basis of remaining their large nonlinearity.

Often organic compounds have low melting points and fairly long absorption cut-off wavelength, which is the fatal shortcoming of this kind of materials. However, some substituted coumarins have fairly high thermal stability. On the other hand, the conjugation system of the coumarins is relatively small, which favors the good transparency of the solid samples. That is to say this kind of compounds is potentially excellent NLO candidate materials for good transparency and high thermal stability. But unfortunately the macroscopic nonlinearities of the reported coumarin NLO materials were usually weak although the transparency and melting points were much better than the common organic NLO materials. Moreover, most of the coumarin derivatives crystallize in centrosymmetric style and therefore a zero $\chi^{(2)}$ occur, which is no any use for the NLO materials.

As well known that second-order nonlinearity can be described by the molecular hyperpolarizability β at the molecular level and the second-order susceptibility $\chi^{(2)}$ at the macroscopic level. The optical nonlinear response of the bulk materials is related to both the magnitude of the molecular hyperpolarizability and the alignment of the molecules in the medium.[4] So it is important to firstly study the molecular structure dependence of the hyperpolarizability β and the crystal structure for the effective NLO materials design. In order to provide insight into the designing effective NLO materials

---

[1] Corresponding author. E-mail: zhaobo@njnu.edu.cn

using the advantages of this kind of compounds a systematically TDHF calculation study on the molecular nonlinear properties of coumarins were carried out.

2. Results and Discussion

Because the strong donor and acceptor groups usually cause strong red-shift we chose the coumarins substituted with relatively weak substituents for this calculation (at 1064nm) and the results are list in table 1. The calculation gives us an unexpected result that this kind of compound has big molecular hyperpolarizabilities but the β values of some substituted molecules decrease comparing to that of coumarin. Usually the CH_3 group can improve the molecular nonlinearity as a dornor in the organic molecules but it is amphoteric here. When the methyl group is attached on the position 7 of the benzene ring the β value will be enhanced obviously but hardly increase on the position 4 of the pyrone moiety. The other substituents have the same effects on the molecular hyperpolarizabilities of these compounds. So we can say that position 7 in coumarin molecule is the active point for the molecular nonlinearity. The monosubstituted on position 7 can greatly enhance the β values, especially for the 6-methoxy coumarin has the biggest β value of the monosubstituted molecules. On the contrary, the 4- and 6-substituted molecules have less hyperpolarizabilities than coumarin, including the chloro, methyl, methoxy and hydroxy substituted derivatives, these two attached position in the coumarin molecules are certainly inactive points for the molecular nonlinearity.

The molecular hyperpolarizabilities of the disubstituted molecules are not bigger than that of the pertinent monosubstituted ones as expected, in fact they have less β values. Of the groups studied in this work chloro unlike the other groups is an acceptor but it shows the same characteristics as the other donors. The introduction of the hydroxyl group in coumarin molecules is very important for improving the thermal stability of the materials, usually the hydroxyl substituted derivatives have high melting points, perhaps it is derived from the intermolecular hydrogen bond between the hydroxyl groups in the solid materials.

To know the origin of the molecular polarizability a calculation on the charge distributions of 7-hydroxyl-4-methyl coumarin (HMC) was carried out for this study. It is well known that the molecular nonlinearity has close relationship with the intromolecular charge transfer and the charge transfer usually occur between the acceptors and the donors. The calculated results in table 2 obviously reveal that the methyl group attached on 4-position is a weak acceptor, which is different from that it is usually act as a donor. The carbonyl is a relatively weak acceptor group in coumarin molecule, and thus as another weaker acceptor in the pyrone ring the CH_3 can weaken the intromolecular charge transfer and therefore the molecular hyperpolarizability. The charge transfer route in HMC molecule is $O_{13} \rightarrow C_7 \rightarrow C_6 \rightarrow C_5 \rightarrow C_{10} \rightarrow C_4 \rightarrow C_3 \rightarrow C_2 \rightarrow O_{11}$ and $O_{13} \rightarrow C_7 \rightarrow C_8 \rightarrow C_9 \rightarrow C_{10} \rightarrow C_4 \rightarrow C_3 \rightarrow C_2 \rightarrow O_{11}$. Obviously it is interrupt at O_1 atom. Thus the 4-substituted groups counteract some of the intromolecular electron transfer from the benzene ring and so weaken the hyperpolarizabilities.

As known the noncentrosymmetric crystal structure is necessary for the excellent second-order nonlinear optical crystal materials. And so predicting and controlling crystal patterns is very important for the NLO material design although it is very difficult and elusive. In absence of the knowledge that complete understanding the factors control molecular packing in crystals it is maybe an effective and practical approach to study a series of closely related compounds so that a semi-empirical crystal structural principle can be deduced. Primary study reveals that the coumarin derivatives seem to easily crystallize in noncentrosymmetric styles when the substituents at 4 and 7 position are different. Such as 4-methyl-7-chlorocoumarin crystallize in $P2_1$ space group,[5] 4-methyl-7-hydroxycoumarin in $P2_12_12_1$ space group[5] and 4-methyl-7-diethylaminocoumarin in $P2_1$ space group,[5] etc. Here the position 4 and 7 seems to be the active substitution points for noncentrosymmetric crystal structure and therefore the macroscopic nonlinearity of the materials, it will be discussed in another full paper.

Table 1: The substituted location numbering and the calculated results of the coumarins

No.	molecule	$\beta(10^{-30}esu)$	No.	Molecule	$\beta(10^{-30}esu)$
1	coumarin	5.12	9	4,7-2CH$_3$	6.99
2	7-Cl	11.10	10	4-OH	4.24
3	4-Cl	4.09	11	6,7-2OH	13.86
4	6-Cl	4.73	12	4-CH$_3$,7-OH	11.70
5	4-CH$_3$,7-Cl	9.22	13	4-CH$_3$,7-NH$_2$	25.36
6	4-CH$_3$,6-Cl	3.37	14	4-OCH$_3$	4.34
7	7-CH$_3$	8.39	15	7-OCH$_3$	16.16
8	4-CH$_3$	5.37	16	6-OCH$_3$	3.57

Table 2: The charge distribution of HMC at the ground state and the first excited state

No.	atom	The ground state	The first excited state
1	O	-0.161	-0.160
2	C	0.393	0.356
3	C	-0.229	-0.101
4	C	0.045	-0.120
5	C	-0.041	-0.085
6	C	-0.139	-0.103
7	C	-0.022	-0.062
8	C	-0.137	-0.107
9	C	0.116	0.048
10	C	-0.150	-0.034
11	O	-0.336	-0.297
12	C	-0.084	-0.054
13	O	-0.074	-0.067

References

[1] J L Brédas. Molecule Geometry and Nonlinear Optics. Science, 1994,263 :487-488
[2] L Jensen, K O Sylvester-Hvid, K V Mikkelsen. A Dipole Interaction Model for the Molecular Second Hyperpolarizability. J Phys Chem, A 2003,107:2270-2276
[3] Zhao B., Wu Y., Zhou Z. H., et al. The important roles of the bromo group in improving the properties of organic nonlinear optical materials. J Mater Chem, 2000,10:1513-1517.
[4] Oudar J L, Zyss J. Structural dependence of nonlinear-optical properties of methyl-(2,4-dinitrophenyl)- aminopropanoate crystals. Phys Rev, A, 1982, 26:2016-2027.
[5] Gnanaguru K, Ramasubbu N, Venkatesan K, et al. A study on the photochemical dimerization of coumarins in the solid. J Org Chem, 1985, 50:2337-2346.

VSP International
Science Publishers
P.O. Box 346, 3700 AH Zeist
The Netherlands

*Lecture Series on Computer
and Computational Sciences*
Volume 1, 2004, pp. 1132-1133

Preface for the Symposium:
Intelligent Information Systems

Vassilis Kodogiannis
University of Westminster, UK

Ilias Petrounias
UMIST, UK

This special session encompasses papers devoted to the recent developments in the applications of soft computing (SC) techniques to information systems (IS).

The term SC refers to a family of computing techniques that originally comprised four different partners: fuzzy logic, evolutionary computation, neural networks and probabilistic reasoning. The term SC distinguishes these techniques from hard computing that is considered less flexible and computationally demanding. The key point of the transition from hard to SC is the observation that the computational effort required by conventional computing techniques sometimes not only makes a problem intractable, but is also unnecessary as in many applications precision can be sacrificed in order to accomplish more economical, less complex and more feasible solutions. Imprecision results from our limited capability to resolve detail and encompasses the notions of partial, vague, noisy and incomplete information about the real world. In other words, it becomes not only difficult or even impossible, but also inappropriate to apply hard computing techniques when dealing with situations in which uncertainty and imprecision are involved. All the methodologies that constitute the realm of SC (the four above mentioned and some others that have been incorporated in the last few years such as rough sets or chaotic computing) are considered complementary as desirable features lacking in one approach are present in another. Hence, the SC framework is put into effect by hybrid systems combining two or more of the constituent technologies with complementary characteristics.

However, as we strive to design and build *Intelligent Information Systems*, we must take into account developing trends in how these systems will be used. We live in a world that is becoming increasingly mobile and service-oriented, and our information systems must support new forms of interaction and collaboration. There is increased emphasis on the search, acquisition, interchange and sharing of data, information and knowledge. Rather than building large, monolithic systems, we are moving towards agent-based and mediated architectures in which software components are given responsibility to search for goods and services, negotiate for them, acquire them, and provision for their delivery by either electronic or physical means. New requirements from application domains, such as healthcare and web usage, have demonstrated that traditional information systems offer inadequate solutions to a number of problems. More and more scientific and engineering applications make use of intelligent (hybrid) SC information systems.

Advances in knowledge discovery and data mining, neural networks, genetic algorithms as well as in intelligent agents and intelligent temporal databases have made this possible. Such approaches can be applied to the storage, management and retrieval of large collections of data in scientific, engineering and organisational domains. The idea of developing such approaches is both appealing and intuitive, but technically it is significantly challenging and difficult, and especially inappropriate for the conventional information systems.

In particular, the investigation and development of techniques in these areas and their systematic evaluation in various applications domains is becoming more important as collections of diverse data continue to grow exponentially in size. At the same time there has been a lot of research work on systems, methodologies or techniques that integrate more than one intelligent technology. Issues such as intelligent decision support and systems' self learning behaviour become fundamental, as are the ways that this support and learning are manifested within an organisational context.

Six contributions were selected in order to cover as much as possible the range of the different branches of SC in their application to IS. The papers in this session are divided into two main groups. The first block, composed of two papers, corresponds to relevant researchers on the topic and outlines a general framework of the application of different SC branches to Healthcare. Then, the second group, composed of four contributions, provides an additional insight of other aspects of Intelligent IS.

In the first paper, entitled "Artificial Neural Networks and Cardiology: a Fertile Symbiosis?", by M. De Beule et al., the technique of Artificial Neural Networks is applied to model the risk stratification. The performance of the network has proved its ability to find non-linear relationships in (medical) data and some important factors in accomplishing an accurate and reliable network have been derived. In the second paper, entitled "Gluing Web Services through Semantics: The COCOON Project", by E. Della Valle et al. a set of Web Services to support medical decision for risk management is being developed, along with a method for gluing them together through semantics with existing healthcare services from external providers.

The second block of contributions, start with a paper entitled "The Forward Web: A New Web Architecture for Organizing Expanding Knowledge" by S. Courtenage et al. Here, the authors consider a radically different model of web publishing in which the author of a web page is not the user who makes links from the web page to other web pages. Instead, the page author expresses an interest in the kind of content the page should link to and as new content comes online that matches that interest, links are inserted automatically into the original page to point to the new content. The fourth paper entitled "Frameworks for intelligent shopping support" by W. Lin, et al., where the main aim is to reduce the gap between e-marketing strategies and agent technology by presenting a research framework to design an intelligent shopping agent's knowledge base. In the fifth paper "A Survey of Multi-Agent System Architectures", M. Chen et al., a review of multi-agent system architectures is presented. The importance of these architectures is discussed first and a set of functional and non functional criteria are identified. Various architectures from the literature are presented and evaluated against this set of criteria. The last paper in this session entitled "On Probability, Null Values, Intuitionistic Fuzzy, & Value Imperfection", by P. Chountas et al., describes an Intuitionistic fuzzy relational repository, which covers the need for storing and querying imprecise information. The model uses the terms and operations defined in the Intuitionistic fuzzy sets theory in order to describe the relational algebra model with Intuitionistic fuzzy terms.

In conclusion, we believe that the authors have responded magnificently to the challenge of this session, by showing how the ideas of SC have kept pace with new developments, and are providing ideas, approaches, and tools to address problems in Information Systems.

VSP International
Science Publishers
P.O. Box 346, 3700 AH Zeist
The Netherlands

Lecture Series on Computer
and Computational Sciences
Volume 1, 2004, pp. 1134-1138

Gluing Web Services through Semantics:
The COCOON Project

Emanuele Della Valle[1], Nahum Korda[2], Stefano Ceri[1], and Dov Dori[2]

[1]CEFRIEL – Politecnico di Milano
[2]Technion – Israel Institute of Technology
dellava@cefriel.it, korda@tx.technion.ac.il, ceri@cefriel.it, dori@ie.technion.ac.il

Abstract: COCOON is a 6th Framework EU project aimed at setting up a semantics-based healthcare information infrastructure with the goal of reducing medical errors. To this end, COCOON offers a set of Web Services to support medical decision for risk management, along with a method for gluing them together through semantics with existing healthcare services from external providers. This paper presents COCOON's architectural design and describes the two main types of COCOON services: Semantic Information Retrieval, which employs concept-based indexing, and Decision Support, which is based on selected clinical guidelines. Semantic Web Service technologies serve to glue these services in existing healthcare information systems.

Keywords: risk management, Web Services, Semantic Web Services, semantic information retrieval, medical decision support, clinical guidelines

1. Introduction

Presenting the right knowledge to the right medical personnel in the right place and at the right time is of paramount importance in making critical medical decision. Driven by this paradigm, COCOON, a European Community[1] funded research project, concerns the development of a knowledge-based approach to diagnostic and therapeutic risk management. A semantics-based healthcare information infrastructure is designed to offer interactive support to healthcare professionals, assist them to efficiently handle complex medical cases, such that diagnosis and treatment errors are minimized, potentially saving lives and the rising costs of healthcare and malpractice suits. Providing infrastructure for proficient knowledge sharing and transfer, this approach can significantly improve knowledge-driven collaborative practices among healthcare communities in Europe and provide a test bed for similar efforts elsewhere.

Web Services provide standard technologies and protocols to exchange data and services, so they seem to be a proper solution to Internet-wide scale integration problems, However, while Web Services are a major contribution to Service Oriented Architecture, the problem "is not in the plumbing, it's in the semantics" [Brodie, 2003]. Semantic Web Services (SWS) are better suited than non-semantic ones to solve the integration problem, as they enable dynamic, scalable and reusable cooperation between different systems and organizations that employ different syntax, formats, and schemata.

State-of-the-art SWS cater mainly to e-commerce and e-work. The development directions of the three major approaches to SWS, namely OWL-S [DAML, 2004], IRS-II [KMI 2004], and WSMO [WSMO 2004], described briefly below, indicate progression towards accepted standards for Web Services and the Semantic Web.

OWL-S supplies Web Service providers with a core set of markup language constructs for describing the properties and capabilities of their Web Services in unambiguous, computer-intepretable form.

The Internet Reasoning Service (IRS) is a Semantic Web Services framework, which allows applications to semantically describe and execute Web Services. The IRS supports the provision of semantic reasoning services within the context of the Semantic Web.

Web Service Modeling Ontology (WSMO) working group includes the WSML working group, which aims at developing a language called Web Service Modeling Language (WSML) that formalizes WSMO.

---

[1] IST-2002-507126. http://www.cocoon-health.com/site/index.php

These three approaches are based on different logic and ontology frameworks, and they provide different reasoning support. No approach provides a complete solution, but they show complementary strengths that, at present, are subject of industry research efforts.

The Semantic Web Services Initiative (SWSI) is an ad hoc initiative of academic and industrial researchers to create infrastructure that combines Semantic Web and Web Service technologies to "enable maximal automation and dynamism in all aspects of Web service provision and use, including (but not limited to) discovery, selection, composition, negotiation, invocation, monitoring and recovery" [SWSI 2004]. While OWL-S, IRS-II and WSMO may converge to an accepted standard at some time point in the future, the current state of affairs, as levels of industry acceptance and adoption to date indicate, is that SWS are far from maturity.

The evolution of SWS technologies shows that WSMO is a promising initiative to provide mature results in the area of SWS languages, architectures and platforms. WSMO inherits the two fundamentals design principles of WSMF: loose-coupling, which enables scalable solution, and mediation, which bypasses heterogeneity problems. IRS-II, the next IRS version, will be compatible with WSMO, indicating a direction of integration of European research efforts and cooperation with the US-based DAML initiative towards international standardization in SWSI [SWASI 2004]. We base our development on WSMO since it appears to be the most mature and adequate option for the needs of COCOON.

2. COCOON Architecture

COCOON's architecture, depicted by the Object-Process Diagram [Dori, 2002] in Figure 1 is designed to provide two main functions: (1) *Semantic Information Retrieval*, i.e., search that employs concept-based indexing, and (2) *Decision Support*, which makes use of selected clinical guidelines. A set of Web services exposes these two fundamental functions to the rest of the system. This design follows two principles:

1. Web services are the building blocks for the various applications, and
2. The design of applications is carried out by gluing these Web services in a variety of combinations.

COCOON's Semantic Web Services Glue module allows application developers to explore offered Web Services, select the most suitable ones, and define the conditions and restrictions by which they are to be glued together. Using the tools provided by COCOON's Knowledge Management Platform, users will be able to develop a variety of applications that respond to specific information needs and requirements of different national health systems, regional practices, or even individual preferences. Within COCOON, each such application will utilize the most suitable features from the same basic set of fundamental features provided as Web services.

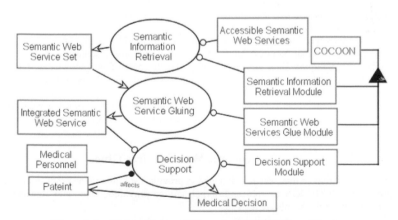

Figure 1 **An Object-Process Diagram of COCOON architecture**

3. Decision Support

COCOON's Decision Support module utilizes clinical guidelines that have been issued by medical authorities as a basis for designing workflows. A clinical guideline is transformed into series of steps that may include diagnostic and treatment activities. Based on the patient's digital health record and results of previous steps, the decision support module offers a set of recommendations regarding the next step(s) to be taken.

Clinical guidelines can differ significantly for various patients' profiles, depending on such factors as demographics, medical history, age, and gender. Selecting the most appropriate medical guidelines for each individual patient is thus of utmost importance. The decision support module accomplishes this through a categorization engine that systematically compares personal background information with medical profiles, and employs complex heuristics in order to evaluate which medical profile is most suitable for the patient being treated. The Decision Support module also consults pharmaceutical databases to match medications to patients, and detect individual counter-indications.

4. Semantic Information Retrieval

The Semantic Information Retrieval module is designed to provide meta-search functionality, i.e., retrieving information from multiple content sources in parallel. These include both internal content sources, such as personal file systems, intranet-based document repositories, corporate document management systems, and external ones, such as Web search engines, digital libraries, and document databases.

The main problem in such meta search lies in analyzing and integrating the search results originating from these various content sources so they support differing ranking strategies, querying syntaxes, and meta-data schemas. A typical solution to this problem is post-retrieval processing of the retrieved documents. During this stage, the results are normalized to a canonical form and integrated into a single presentational structure. They can then be uniformly re-ranked, categorized into a pre-existing classification schema, or serve as a basis for generating an automatically extracted set of cluster labels. COCOON employs GRACE[2] for its meta-search.

Unlike meta search systems that perform post-retrieval integration of the search results, GRACE pre-indexes the documents from the multiple content sources by utilizing LinKBase®[L&C 2004], a state-of-the-art medical ontology. Unlike regular document indices based merely on terminology, LinKBase® is concept-based, and allows querying and browsing based on the semantic relationships of the concepts.

GRACE fuses information resources, knowledge representation system, and the community of users into a single conceptual frame called Knowledge Domain.

Knowledge Domain is the basis for GRACE's pre-indexing. It is based on ontology—a conceptualization of the reality specific to a particular group of end-users. Knowledge Domain operates in three distinct steps:

1. *COCOON index updating*: Concepts from LinKBase® are used to form queries that are periodically submitted to the multiple content sources in order to retrieve the relevant documents and ensure frequent update of the COCOON index.
2. *Concept-based indexing*: The retrieved documents are analyzed, and, based on their lexical content, are assigned to concepts in the LinKBase® index. This is a computationally intensive processing effort that requires significant computational and storage resources, so GRACE utilizes the emerging Data Grid technology. To the best of our knowledge, GRACE and COCOON, which uses GRACE, are the first worldwide Grid-based Information Retrieval systems.
3. *Browsing and querying*: The updated LinKBase® index is used for browsing and querying, as it points to the relevant documents, regardless of the content source from which they originated.

5. Semantic Web Services Glue

The Semantic Web Services Glue, which is designed to provide a solution to the problem of integrating services provided by COCOON as well as external service providers, takes into consideration the specific requirements of the healthcare domain.

---

[2] IST-2001-38100. http://www.grace-ist.org

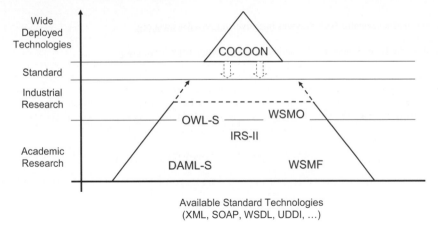

Figure 2 **The relationships between various SWS efforts and the COCOON Glue**

Figure 2 shows schematically the relationships between various SWS efforts and the COCOON Glue. This Glue aims to fill the gap between WSMO and technologies employed in COCOON that cater to the requirements of the healthcare domain.

6. Future Work

Filling the gap is an effort that implies carrying out the following tasks:

1. Selecting and adjusting the ontologies required for describing the healthcare services offered in COCOON through the development of (possibly ad hoc) mediation services to overcome heterogeneity of the various health-care related ontologies,
2. Developing guidelines for selecting and adjusting additional medical ontologies in order for them to be useful in describing healthcare services offered by external providers,
3. Extending and, if necessary, adjusting WSMO to the specific needs and requirements of healthcare systems,
4. Integrating editing facilities in an ontology engineering environment to semantically describe healthcare services using WSMO and medical ontologies,
5. Evaluating different publishing mechanisms to gather and maintain up-to-date semantic descriptions of Web services, and finally
6. Evaluating different storing facilities and inferring their level of support in order to find, orchestrate[3] and choreograph[4] Web services in an efficient and scalable way.

These tasks are currently being the subject of our research in the COCOON project.

References

Brodie, M.L. (2003). The long and winding road to industrial strength (semantic) Web services. 2nd International Semantic Web Conference (ISWC2003). http://iswc2003.semanticweb.org/brodie.pdf

DAML (2004). DAML Services. http://www.daml.org/services/owl-s/

Dori, D. (2002). Object-Process Methodology: A Holistic Systems Paradigm. Springer Verlag, Berlin, Heidelberg, New York. http://www.ObjectProcess.org

KMI (2004). Internet Reasoning Service. http://kmi.open.ac.uk/projects/irs/

L&C (2004). Language and Computers Products. http://www.landcglobal.com/pages/linkbase.php

----

[3] Refers to the interaction among multiple Web Services, including business logic and execution order.
[4] Refers to the sequence of messages exchanges that occur between multiple Web Services.

SWSI (2004). Semantic Web Services Initiative. http://www.swsi.org/

WSMO (2004). Web Services Modeling Ontology. http://www.wsmo.org/

VSP International
Science Publishers
P.O. Box 346, 3700 AH Zeist
The Netherlands

Lecture Series on Computer
and Computational Sciences
Volume 1, 2004, pp. 1139-1143

On Probability, Null Values, Intuitionistic Fuzzy, & Value Imperfection

Panagiotis Chountas[1], Vassilis Kodogiannis[1], Ilias Petrounias[2], Boyan Kolev[3], Krassimir T. Atanassov[3]

[1] – Health Care Computing Group, School of Computer Science, University of Westminster,

Northwick Park, London, HA1 3TP, UK

[2] – Department of Computation, UMIST PO Box 88, Manchester M60 1QD, UK

[3] – CLBME – Bulgarian Academy of Sciences, Bl. 105, Sofia-1113, BULGARIA

Received 7 March, 2004; accepted in revised form 10 March, 2004

Abstract: This paper describes an Intuitionistic fuzzy relational repository, which covers the need for storing and querying imprecise information. The model uses the terms and operations defined in the Intuitionistic fuzzy sets theory in order to describe the relational algebra model with Intuitionistic fuzzy terms.

1. Introduction

Many real world applications, e.g. genome, geophysical and biological systems, must deal with imprecise or vague data. Imperfect information is the partial knowledge of the true value of the real world. It is an epistemic property caused by lack of information. Work in this area is considering imperfect information arising from different elements of the enterprise ontology but always in isolation form each other [1], [2, [3], [4]. This paper is suggesting that the elements involved in information imperfection are in reality related and are affecting each other. The premises on which this paper is based are precise enterprises, where information is unknown. Unknown information can be described as part of a data repository either with the aid of null values with respect to unweighted approaches, or with the support of probabilistic values. In this an analogy between a probability distribution, and null markers with constraints (as enumerated data types), based on the idea of possible worlds is shown. We further delivered the case of intuitionistic fuzzy logic as a more representative way for entailing and retrieving unknown information in data repositories.

2. Associating Null Values & Value Imperfection

The study of incomplete information has led to a body of work on null values. [5] provides a unified framework for treating null values as constraints and a complete algebra. Incomplete information is expressed quite often as either a probability or possibility distribution over a set of values belongs to the domain of an attribute. A probabilistic database represents a set of possible states or worlds. Assuming the probabilistic table, Supplies.

Table 1: Supplies

Supplies	
Supplier	*Item-No*
IBM	*IBM-IDE / 1.0*
INTEL	*INTEL-M3CPU / 0.6*
	INTEL-M2CPU / 0.3
	INTEL-M1CPU / 0.1

The semantics of the "Supplies" probabilistic relation can be expressed in terms of possible worlds as follows

$$W_1= \{(IBM, IBM\text{-}IDE), (INTEL, INTEL\text{-}M3CPU)\}$$
$$P (W_1)= 0.6$$
$$W_2= \{(IBM, IBM\text{-}IDE), (INTEL, INTEL\text{-}M2CPU)\}$$
$$P (W_2)= 0.3$$
$$W_3= \{(IBM, IBM\text{-}IDE), (INTEL, INTEL\text{-}M1CPU)\}$$
$$P (W_3)= 0.1$$

Note that the probabilities add up to one $P (W_1) + P (W_2) + P (W_3)= 1$. Instead of emphasising the additive property of the probability measure, it will be far more interesting to stress the partial order of the possible worlds, implied by the a probability distribution. This is $W_3 \subseteq W_2 \subseteq W_1$, since $P (W_3) \subseteq P (W_2) \subseteq P (W_1)$.

In relational databases with null values it is recognised that there are many different types of null values, each of which reflects different intuitions about why a particular piece of information is unknown. In [11] five different types of nulls are suggested. The labels and semantics of them are defined as follows. Let V be a function, which takes a label and returns a set of possible values that the label may have.

Table 2: Null Markers

Label (X)	V(X)
Ex-mar	D
Ma-mar	$D \cup \{\perp\}$
Pl-mar	$\{\perp\}$
Par-mar (V_s)	V_s
Pm-mar (V_s)	$V_s \cup \{\perp\}$

Intuitively, V (Ex-mar) = D says that the actual value of an existential marker can be any member of the domain D. Likewise, V (Ma-mar) = $D \cup \{\perp\}$ says that the actual value of a maybe marker can be either any member of D, or the symbol $\perp$, denoting a non-existent value. Similarly, V (Par-mar (V s)) = V_s says that the actual value of a partial null marker of the form pa mar (V_s) lies in the set V_s, a subset of the domain D. A controversial issue is the use of the $\perp$, which denotes that an attribute is inapplicable. Certainly this is the interpretation of an algebraic manipulation of the unknown information, instead of a conceptual manipulation and interpretation. Assuming the sample fact spouse, the individual, Tony, is a bachelor, and hence, the wife field is inapplicable to him, $\perp$. Conceptually the issue can be resolved with the use of the subtypes (e.g. married, unmarried) as part of the entity class Person. A subtype is introduced only when there is at least one role recorded for that subtype. In the general case the algebraic issue under the use of subtypes is whether the population of the subtypes in relationship to the supertype is

- Total and Disjoint: Populations are mutually exclusive and collectively exhaustive.
- Non-Total and Disjoint: Populations are mutually exclusive but not exhaustive.
- Total and Overlapping: Common members between subtypes and collectively exhaustive, in relationship to supertype.
- Non-Total and Overlapping: Common members between subtypes and not collectively exhaustive, in relationship to supertype.

The conceptual treatment of null will permit us to reduce the table in "Table 2" using only two types of null markers.

Table 3: Eliminated Null Markers

Label (X)	V(X)
V-mar (V)	$\{V\}$
P-mar (V_s)	$\{V_s\}$
Π-mar $(D\text{-}V_s)$	$\{D - V_s\}$

In [6] a connection between the measure of randomness (p) or observation and compatibility (Π) can be achieved. In this way a fact is presenting information that is observed and testified by one or more information sources therefore a set of alternatives is defined with p constrained by $0 \le p \le 1$ and $\Pi = 1$. In this case the P-mar, marker is used to denote that one of the members of the restricted set V_s is the actual value, of a label value. In this latter case the Π-mar, marker to denote that a label value can be any value derived from the label domain D, excluding the set of probable values V_s. In the case where information is explicit a value marker is any single element (V) from the domain D of the label value. The probability measure stress a partial order of the possible worlds. This is $W_3 \subseteq W_2 \subseteq W_1$, since P ($W_3$) $\subseteq$ P (W_2) $\subseteq$ P (W_1) "Table 1". The challenge is to keep this partial ordering as part of P-mar, marker which a set of alternatives but with no explicit ordering. A set type is defined as following: *set type = Set of Base Type*. The identifier set type is defined over the values specified in base type. The base type can be for example Boolean, character, an enumeration type or a sub-range. Choosing the base type for the P-Mar constraint to be an enumeration type (a_1, a_2, a_3) a partial order is established since $a_1 \subseteq a_2 \subseteq a_3$, therefore the ordering defined by the probability distribution is assured and value imperfection can be expressed and accommodated in the database in a simple way.

Depending on the modelled enterprise modellers may decide to use either the weighted approach (probabilistic values) or the unweighted approach. Both approaches are not negating each other on the contrary a parallelism can be drawn through the partial order property. Further more probabilistic repositories are implying the semantics of the CWA, since every fact instance of fact F not explicitly represented in the relation F, has probability 0. In an effort to null values, maybe fact instances or tuples are utilised. In this way a relation F is composed of a sure component and a maybe component, which may hold, or not. The CWA is utilised with the formulation of the possible world, where in each situation every tuple is certain.

However the use of either probabilistic or null values does not let the modeller to express indefiniteness, to put it differently certainty or uncertainty with respect to likelihood of occurrence of a possible event. For example an information source may not be reliable, i.e. the expert, who enters the information, has a particular reliability coefficient, or the data is not quite actual, etc. That is why we develop a model of an intuitionistic fuzzy repository, which can store data with certain degree of reliability or non-reliability.

3. An Intuitionistic Fuzzy Relational Repository

An Intuitionistic fuzzy relational repository for treating unknown information with respect to the semantics of intuitionistic validity $< \mu^F >$ and falsity $< v^F >$ is defined. Using the ideas for intuitionistic fuzzy expert systems [7] we can estimate any fact F and it can obtain intuitionistic fuzzy truth-values $V(F) = < \mu^F, v^F >$, such that $\mu^F, v^F \in [0,1]$ and $\mu^F + v^F \le 1$. Therefore, the above fact can be represented in the form $< F, \mu_F, v_F >$. Let R be an Intuitionistic Snapshot fuzzy relation:

$R = \{<x, \mu_R(x), v_R(x)> / x \in X\}$, where $x = <col_1, ..., col_n>$ is an ordered tuple belonging to a given universe X.

A *projection* operation over R defines a relation, which is a vertical subset of R, containing the values of the specified attributes:

$\Pi_{col1,...,coln}(R) = \{<<x.col_1, ..., x.col_n>, \mu_R(x), v_R(x)> / x \in X \}$, where $x.col_k$ is the k-th column (attribute) of the ordered couple x.The projection retains the degrees of membership and non-membership of R.

A *selection* operation defines a relation, which contains only those tuples from R for which a certain predicate is satisfied. We can say that the selection modifies the degrees of membership and non-membership of R depending on the corresponding value of the predicate:

$\sigma_P(R) = \{<x, \min(\mu_R(x), \mu(P(x))), \max(v_R(x), v(P(x)))> / x \in X \}$,

where P is the predicate, i.e. the elements of the result relation have degree of membership, which is logically *AND*-ed with the corresponding value of the predicate P.

A *Cartesian product* of two relations $R \times S$ is identical to the Cartesian product operation defined in the Intuitionistic fuzzy sets theory [1], which uses the logical *AND* between the degrees of membership:

Let S be another intuitionistic fuzzy relation: $S = \{<y, \mu_S(y), v_S(y)> / y \in Y\}$, then:

$R \times S = \{ <<x, y>, \min(\mu_R(x), \mu_S(y)), \max(v_R(x), v_S(y),)> / <x, y> \in X \times Y\}$.

If we further wish to represent temporal information about a particular enterprise then a temporal relation can be defined as follows

The elements of the temporal information (we can call them temporal data, or temporal facts, too) can be represented in the form $< F, t_L^F, t_R^F >$, where $[t_L^F, t_R^F]$ is a time interval. Using the ideas for intuitionistic fuzzy expert systems [7] we can estimate any fact F and it can obtain intuitionistic fuzzy truth-values $V(F) =< \mu^F, v^F >$, such that $\mu^F, v^F \in [0,1]$ and $\mu^F + v^F \leq 1$. Therefore, the above fact can be represented in the form $< F, t_L^F, t_R^F, \mu_F, v_F >$.

This form of the fact corresponds to the case in which the fact is valid in interval $t^F = [t_L^F, t_R^F]$ and at every moment of that interval the fact has the truth-value $< \mu^F, v^F >$.

Each of the functions μ, v, and will be regarded as having two arguments i, t, where i is the index of the corresponding fact instance, and t is an element in the interval $T = [t_1, t_2]$. Now, the truth and falsity degrees of the fact instances can be estimated as above, but the pair $<\mu'_i, v'_i>$ converts to

$$< \mu'_i(t), v'_i(t) >=< \frac{\int_{t_1}^{t_2} \mu_i(t)\,dt}{t_2 - t_1}, \frac{\int_{t_1}^{t_2} v_i(t)\,dt}{t_2 - t_1} >$$

Now, if the fact is valid in a set of intervals $t_1^F, t_2^F, ..., t_n^F$ with the forms $t_i^F = [t_{i,L}^F, t_{i,R}^F]$, and if its truth-values for these intervals are, respectively, $< \mu_1^F, v_1^F >, < \mu_2^F, v_2^F >, ..., < \mu_n^F, v_n^F >$, $(1 \leq i \leq n)$, then our fact will have the representation

$$< F, t_1^F, \mu_1^F, v_1^F, t_2^F, \mu_2^F, v_2^F, ... t_n^F, \mu_n^F, v_n^F >$$

Now, for the set of so constructed facts we can define quantors $\forall$ and $\exists$ having truth-values estimations as follows

$$V(\forall t\ F) =< \min_{i=1,2,...,n} \mu_i^F, \max_{i=1,2,...,n} v_i^F >,$$

$$V(\exists t\ F) =< \max_{i=1,2,...,n} \mu_i^F, \min_{i=1,2,...,n} v_i^F >.$$

The first estimation determines the truth-value of the expression "Up to now F has always been valid" and the second one – of the expression "In some interval of time F has been valid".

The advantage of this alternative is that temporal ignorance can be expressed using the generic semantics " it is true up to now" or " is true at some point" recording both the belief and falsity of these propositions. It is assumed that negative information will be stored by default to the equivalent database representation. In this sense it is expected that the database will obey the open world assumption (OWA). This implies that integrity constraints have to be redefined, a problem that still remains unsolved for the database community.

4. Conclusions

In this paper three distinct architectural approaches for building a data repository to accommodate unknown information have been presented. The first architecture is based on a unweighted approach and it is utilised with the aid of null values. The second approach is abased on a weighted approach and it is utilized with the aid of probabilistic values. Both approaches do obey the CWA assumption.

We finally proposed a new architectural approach for building a data repository to accommodate unknown information based on Intuitionistic Fuzzy logic that follows the OWA. The Intuitionistic model has two basic goals: first, to be able to store data with a particular degree of reliability (degree of membership-non membership of the elements) and second, to be able to process queries, which contain indefiniteness.

References

[1]   C.E. Dyreson, R.T. Snodgrass, Support Valid-Time Indeterminacy, ACM Transactions on Database Systems, Vol. 23, No. 1, pp. 1-57, 1998

[2]   F. Kabanza, J-M. Stevenne, P. Wolper, Handling Infinite Temporal Data, Proceedings of ACM Symposium on Principles of Database Systems (PODS), 1990

[3]   D. Dey, S. Sarkar, A Probabilistic Relational Model and Algebra, ACM Transactions on Database Systems, Vol. 21, No. 3, 1996

[4]   L. Lakshmanan, N. Leone, R. Ross, V. Subrahmanian, ProbView: A Flexible Probabilistic Database System, ACM Transactions on Database Systems, Volume 22, No. 3 (Sep. 1997).

[5]   K. Candan, J. Grant, V. Subrahmanian A Unified Treatment of Null Using Constraints, Computer Science Technical Report Series, University of Maryland, USA CS-TR-3456, UMIACS-TR-95-47, April 1995

[6]   M. Delgado, S. Moral, On the concept of Possibility-Probability Consistency in Fuzzy Sets For Intelligent Systems, D.Dubois, H.Prade and R. Yager (eds), Morgan Kaufman Publishers, pp 247-250, 1993

[7]   Atanassov K. Intuitionistic Fuzzy Sets, Springer-Verlag, Heidelberg, 1999.

VSP International
Science Publishers
P.O. Box 346, 3700 AH Zeist
The Netherlands

*Lecture Series on Computer
and Computational Sciences*
Volume 1, 2004, pp. 1144-1147

Artificial Neural Networks and Cardiology:
a Fertile Symbiosis?

M. De Beule[o][1], E. Maes[o], R. Van Impe[o], W. Vanlaere[o] and O. De Winter*

[o]Department of Structural Engineering,
Faculty of Engineering,
Ghent University,
9052 Zwijnaarde, Belgium

*Department of Radiotherapy and Nuclear Medicine,
Faculty of Medicine and Health Sciences,
Ghent University,
9000 Ghent, Belgium

Abstract: The technique of Artificial Neural Networks (ANN's) is applied to model the risk stratification according to d'Agostino et al. [1]. The performance of the network proves its ability to find non-linear relationships in (medical) data and some important factors in accomplishing an accurate and reliable network are derived. At the end an ANN is designed to investigate the predictive quality of certain well chosen risk factors for secondary prevention. The performance of the resulting network is put in the right perspective and some aspects that need further study are mentioned.

Keywords: Neural networks, diagnosis, cardiology, Framingham and d'Agostino.

1. Introduction

Epidemiologists seek to understand the predictive quality of certain risk factors in risk prevention, using - in general - classical determination methods. Since the 1960's a lot of mathematical predictive models are created for primary prevention (prediction of the risk of a new cardiovascular disease (CHD)) based on the Framingham study. In secondary prevention, the risk for a new CHD-event is calculated for persons with a history of coronary or cardiovascular disease. By evaluating afterwards which risk factor affected the global risk for each patient the most, it is possible to support the secondary prevention in an effective way. The risk factors considered in the analyses of d'Agostino are age, log ratio of total to HDL cholesterol and diabetes for men. Log transformed Systolic Blood Pressure (SBP) and smoking are also included in the model for women in addition to the risk factors for men. Some important factors in accomplishing an accurate and reliable network are derived from a basic ANN to model the risk stratification according to d'Agostino. Some of those recommendations are used in the design of a second network. Using classical epidemiological insights, the authors determined 10 (secondary and possible) risk factors: age, diabetes, smoking, personal history, heart rate at rest, systolic blood pressure at rest, total cholesterol value, left ventricular ejection fraction (LVEF), end systolic volume (together with LVEF determined by QGS myocardial gated SPECT software) and defect extent in stress minus defect extent at rest. The idea grew to see whether these factors could be used in a neural network, to predict the outcome of the patients. The reader is presumed to have some basic knowledge of the technique of ANN's. If there would be a lack of this knowledge the authors strongly recommend to consult the work of e.g. Rojas [2].

---

[1] Corresponding author. E-mail: Matthieu.DeBeule@Ugent.be

2. Artificial Neural Networks and d'Agostino

One way to prove the ability of ANN's to deal with complex non-linear datasets, is to see whether an ANN is able to predict the calculated risk scores (in %) reasonably accurate (e.g. Absolute Error < 3 %). For this reason a dataset of 273 patients of the Ghent University Hospital is studied. The data are collected in a time period of about four years and all patients had a history of ischemic heart disease and an ejection fraction of less than 40 %. In a first phase the secondary risk is calculated for each patient and then a network is trained, validated and tested (with the software packet JavaNNS [3]) to see whether the network's risk score (output) corresponds with the calculated risk values. The total database is divided randomly in a training set of 200 persons and a validation set of 73 patients. The application range for the network for the non-discrete parameters is as follows: age [39,7;89,1] (years), TC/HDL [2,017;10,05], SBP [80;190] (mmHg) and secondary risk [4,02;33,49] (%). All data are scaled as symmetrically as possible within the range [-1;1]. In the training phase there are two possible options: searching a network with an acceptable accuracy or seeking the optimal network with the highest accuracy possible. For both options it is necessary to vary all parameters (e.g. number of hidden nodes, number of hidden layers, training algorithm, learning rate, ...) when searching for the lowest error on the output. To find the optimal network, it is necessary to take a lot more combinations of those parameters in consideration than with respect to the search for an acceptable network. For this study we have chosen to search for an acceptable network as it gave already reasonably accurate results. In order to have an independent performance check, it is necessary to test the network with 'virgin' data (not used for training and validation). As all patient data from the dataset have already been used, the authors chose to generate the test set as follows. For every parameter two values (laying within the application range) are considered, as can be seen in Table 1. By combining all possible combinations of these parameters and calculating the secondary risk, a test set with 64 cases is generated.

Table 1: Values for the parameters used in the test set.

Gender	Male	Female
Age (year)	50	70
Diabetes	Yes	No
Smoking	Yes	No
TC / HDL	4	6
SBP (mmHg)	120	140

In Figure 1 the target values are compared to the output values of the network. In this figure the bisector represents the perfect match between the target- and output values. The figure clearly proves that the network performs very accurately in predicting the secondary risk at risks lower than 17 %. For higher risks the maximum absolute error is about 3 % and the resulting network can thus be stated as acceptable for the prediction of the secondary risk according to d'Agostino's method.

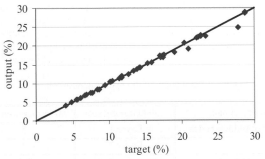

Figure 1: Performance of the resulting network.

The authors would like to draw the attention to the fact that the absolute errors of the test set are much higher than those for the validation set. A plausible cause for this discrepancy is that the data are not very well spread over the whole range (e.g. 14,5 % women in the training set, 2,8 % women in the validation set and 50 % women in the test set). To make an accurate prediction with ANN's, it is necessary to train the network with examples which are distributed evenly over the range of application for each input parameter. If this is not the case, the network can generate results which cannot be trusted in areas where few examples were available. A possibility to cover that lack in distribution is a profound study of the spreading of the data and a good (re)defining the range of applicability of the network (e.g. exclude all women from the data set).

3. Artificial Neural Networks and Risk Stratification

A second ANN is developed to predict the occurrence of a total event (i.e. cardiac death, non fatal myocardial infarct, Coronary Artery Bypass Grafting, Percutaneous Transluminal Coronary Angioplasty or hospitalisation after heart failure). Using this ANN, each patient can be categorised as a high risk (possible occurrence of an event) or a low risk person (no predicted event). Considering the determinant factors for the collected database of the Ghent University Hospital, the designed network is a tool for secondary prevention. To make an accurate prediction with ANN's, it is necessary to train the network with examples which are evenly distributed over the range of application for each input parameter as was mentioned earlier. For this reason 200 male patients with an age between 45 and 85 years were selected from the 273 persons database. Their heart rhythm (in rest) varies from 40 to 100 beats/minute, the systolic blood pressure has a minimum value of 90 and a maximum value of 180 mmHg and the total cholesterol value lies between 120 and 280 mg/dl. All men have an ejection fraction between 15 and 40 % and an end systolic volume of 50 to 300 ml. These ranges determine the field of applicability of the network. A useful tool to define those ranges are scatterplots. In Figure 2 such a plot is given to determine the range of applicability for the ejection fraction (E F). The different slopes of the curve are an indication of the spreading of the data and are an explanation for the exclusion of the patients with an EF less than 15%.

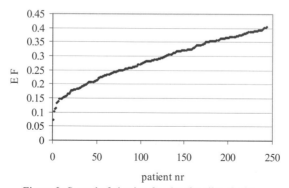

patient nr

Figure 2: Spread of ejection fraction for all male data.

The feedforward network is trained with the backpropagation algorithm with a momentumterm using the software packet JavaNNS. In the quest for a network with an acceptable accuracy, numerous training procedures are carried out while varying several parameters (e.g. number of hidden layers, nodes per hidden layer, momentum term, …). The characteristics of the most accurate network (that produces the highest sensitivity and specificity for the validation set) are shown in Table 2.

Table 2: Parameters of the resulting network.

Number of data	150 (Training set)	50 (Validation set)
Number of hidden nodes	Layer 1	5
	Layer 2	2
Parameters	Learning Rate	0,3
	Momentum Term	0,1
Number of Training cycles		1500

The ratio of the number of training examples to the degrees of freedom for the designed network is 2,5. This approaches the recommended value of 3 in Freeman et al. [4]. Since all data are used for the training and validation set, the network can not be validated with an independent test set. The performance (in %) of the network is summarised in Table 3. The accuracy (Acc) is the percentage of correct classifications, the sensitivity is the fraction of high risk persons who are tested positive (i.e. classified as high risk or occurrence of an event), the specificity is the fraction of low risk persons who are tested negative (i.e. classified as low risk or no event). The Positive Predictive Value (PPV) is the ratio of the number of correct high risk classifications to all high risk classifications, the relation of the number of correct low risk classifications to all low risk classifications is given by the negative predictive value (NPV).

Table 3: Performance of the trained and validated network.

Dataset	Acc (%)	Se (%)	Sp (%)	PPV (%)	NPV (%)
Training set	89,33	66,67	98,15	93,33	88,33
Validation set	82,00	64,29	88,89	69,23	86,49

The designed ANN recognises with success the low risk patients with an acceptable accuracy (i.e. 88,89 %). The support for the remaining group could then be intensified in secondary prevention. In this way the "health care resources" could be spent in a more efficient and economical way. In order to use this model in the clinical practice, a reliable evaluation of the network should be carried out. This involves an external and a temporal evaluation. The external check can be done by testing the performance using another database. This can be an indication of whether the network remains valuable for slightly different populations. A temporal evaluation means that the network has to be tested after a certain period of time to see whether the model remains valid. In addition a comparison of the network with existing classical methods for secondary prevention has to be made. If all tests are positive, the effect of the network on the decision making strategy of the physicians has to be investigated. When this last evaluation is also positive, one could start thinking of implementing such a network in the clinical practice.

The obtained sensitivity of our network is rather low. Maybe the technique of (k-fold) Cross Validation can offer a solution to increase the performance as the number of training examples is rather low (compared to the number of input parameters). As mentioned in the introduction, the input parameters (or risk factors) of our network are chosen using classical epidemiological insights. An alternative method would be to do a full investigation (e.g. Principal Component Analysis) on the complete dataset to see whether the same risk factors are obtained. If this study would reveal new risk factors, their predictive value can be tested with ANN's. Another possibility would be to decrease the number of input parameters and to study the effect on the network. Further research can lead to a network with two output nodes: one parameter that predicts the occurrence of total events and another one that determines the time to this event.

References

[1] R.B. D'Agostino, M.W. Russell and D.M. Huse, Primary and subsequent coronary risk appraisal: new results from the Framingham study, *American Heart Journal* **139** 272–281 (2000) .

[2] R. Rojas: *Neural networks: a systematic introduction*. Springer-Verslag, Berlin Heidelberg, 1996.

[3] Java Neural Network Simulator, User Manual, Version 1.1, Fischer, Hennecke, Bannes and Zell, University of Tübingen.

[4] R.V. Freeman, K.A. Eagle and E.R. Bates, Comparison or artificial neural networks with logistic regression in prediction of in-hospital death after percutaneous transluminal coronary angioplasty, *American Heart Journal* **140** 511-520(2000).

VSP International
Science Publishers
P.O. Box 346, 3700 AH Zeist
The Netherlands

*Lecture Series on Computer
and Computational Sciences*
Volume 1, 2004, pp. 1148-1152

A Short Survey of Multi-Agent System Architectures

Mingwei Chen, Ilias Petrounias[1]

Department of Computation, UMIST,
PO Box 88, Manchester M60 1QD, UK

Received 3 September, 2004; accepted in revised form 4 September, 2004

Abstract: This paper presents a review of multi-agent system architectures. The importance of these architectures is discussed first and a set of functional and non functional criteria are identified. Different architectures from the literature are presented and evaluated against this set of criteria.

Keywords: multi-agent systems, intelligent systems, survey, architectures

1. Introduction

First, we clarify the meaning of terminologies used in MAS research [2]: **Agent Architectures** analyze agents as independent reactive/proactive entities. Agent architectures conceptualize agents as being made of perception, action, and reasoning components [3]. Agent architectures concern how individual agents are constructed internally. **MAS Infrastructure:** An infrastructure is a technical and social substrate that stabilizes and enables instrumental (domain-centric, intentional) activity in a given domain [1]. We can generalize MAS infrastructure as composed of a series of regulations, specifications, standardizations, implemented system components and agent services that serve to solve the "typical, costly, commonly-accepted" [1], non-domain-specific issues in most MASs. **MAS Development Frameworks** are building toolkits for constructing MASs. They should provide support for basic agent communication and for higher-level agent interaction needs, agent management, security and coordination services. **Multi-Agent System Architectures** analyze agents as interacting service provider/consumer entities. Architectures facilitate agent operations and interactions under environmental constraints, and allow them to take advantage of available services and facilities [3].

1.2 Criteria for MAS architectures

On the functional side, multi-agent system architectures are responsible for providing those typical, commonly-used, non-domain-specific functionalities, which are also called basic agent services. The lack of basic agent services in development frameworks are the major obstacle to the long expected prosperity of multi-agent applications. Examples of these, as shown in Fig 1, include facilitation, coordination, message transport, agent management, security etc. The criteria for good MAS architectures are classified into two categories: functional and non-functional. *Functional criteria:* Facilitation capabilities, Coordination capabilities, Agent management capabilities, Security services. *Non Functional criteria:* Modelling capabilities, Naming & addressing dependency, Performance influence, Socialization capabilities, Extensibility, Scalability. The functional criteria consist of the basic agent services that an MAS architecture is expected to offer or is capable of incorporating. Here the following questions should always be asked: what agent services specified by the particular criterion does the system architecture provide? How good are they in terms of their effectiveness and efficiency? What are the pros and cons in terms of the way they are incorporated or implemented?

1.3 Classification of MAS architectures

We classify MAS architectures into four categories according to the coordination (especially facilitation) approaches they use: Federation, Matchmaker, Shared Data Space Coordination Models, Organisational Structuring Architectures. The reasons for choosing coordination approaches as a classifying criterion are twofold. Coordination services, (especially facilitation), are the most basic,

---

[1] E-mail: ilias@co.umist.ac.uk

indispensable functionalities in an MAS architecture. Any MAS architecture would have to provide at least facilitation services in one way or another. The coordination model of system architecture effectively determines the fundamental organisational structure of any MAS based on that architecture. It must be noted that this classification is not disjoint.

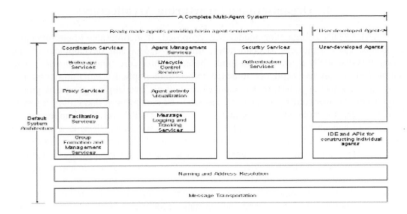

Figure 1: The architectural anatomy of a typical framework

2. Federation Architecture

Federation architecture was first proposed by Genesereth in 1994. It has shaped the developing trend of MAS architectures. The central point [4], [5] is a special system agent, called a facilitator. Every agent in the system must connect to a facilitator through which it interacts with other agents. A facilitator acts like a single gateway for all the agents by handling all their requests. Facilitators from the same or different platforms can be connected together in a peer to peer relation. These architectures offer the most sophisticated and powerful facilitation services. As a coordination model, although claiming to organise agents as peers, it is actually similar to the classic client/server model. A federation architecture based MAS would depend on facilitators to do agent management jobs. When it comes to security concerns, it is the facilitators' responsibility to protect its affiliated agents. On the non-functional side, the advantage of federation architecture is obvious: it enables anonymous interaction between agents. In some early MASs, either agents have to know each other to engage in any interaction, or they have to broadcast their requests or capabilities to other agents, which costs in network bandwidth. The architecture also has some fundamental weaknesses. It is very possible a facilitator could become a "communication bottleneck" or a "point of failure" Its fixed structure means that it is difficult to extend and lacks of the ability to model various organisational structures.

3. Shared Data Spaces Coordination Models for Mobile Agents

Coordination models based on shared data spaces such as Blackboards and Linda have been exploited in distributed applications. Client-server or message-passing models used in most MAS usually require both communication parties to explicitly name each other and to synchronize their interactions in some way (spatial and temporal coupling [6]). Researchers have argued that this spatial and temporal coupling is something that should be avoided in mobile computing environments due to the dynamic nature of mobile agents in terms of their schedules and locations. Instead, they believe blackboard-based or Linda-like coordination models are more suitable in this scenario. A blackboard [7] is a shared data space where distributed or mobile objects can exchange information and get access to resources. Blackboard models enforce temporal uncoupling in terms of that mobile agents can use a blackboard as a common repository to leaving messages for other agents without worrying when they will be read or to retrieve messages that have been left by other agents long ago. Linda [8] went further by organising shared information in tuples and allowing retrieval of tuples via a pattern matching mechanism. Linda-like shared data space coordination models add a new dimension to coordination mechanisms in MAS. However, due to their way of communication, they are only suitable for the coordination of mobile agents. For static agents, message-passing would still be a better way of communication. Therefore, the best use of Linda-like coordination models is to integrate them with other coordination models to form an architecture suitable for both mobile and static agents.

4. Organisational Structuring Architectures

In [9], the AALAADIN coordination model was proposed. It incorporates organisational concepts, such as groups, roles, structures, dependencies, into the design and development of multi-agent applications. It is essentially a generic meta-model based on three core concepts: *agent*, *group* and *role*. Organisational structures are explicitly imposed rather than implicitly emerging from interactions between agents. A whole organisation is composed of multiple overlapping groups. A group contains one or more roles. An agent must play a role in a group to be a member of that group. A role in a group not only defines the functions or services that an agent must fulfil, but also implies the interaction pattern that could happen between it and other roles. For every group, there is a special role name group manager role. An agent can play more than one role at the same time and thus belongs to more than one group. However, agents are not supposed to access these services directly except for the local message transferring. AALAADIN is indeed one of very few efforts that has endeavoured to attack the issue of organisational architecture in MAS. AALAADIN presents quite a few advantages. It makes possible the capability of modelling complex, hierarchical organisational architectures. Groups in AALAADIN are not allowed to contain sub-groups recursively. Nevertheless, incorporation of concepts such as group, role and organisation is a step towards the right direction. By relating interaction patterns to roles, it provides a way for organisation rules or conventions to be imposed on MAS with extremely dynamic nature. Groups effectively provide an additional security layer for agents. Facilitation mechanisms are not specified in AALAADIN. However, it would not be a problem for AALAADIN based MASs to incorporate mediating agents offering brokerage services. Facilitation services can be largely enhanced by grouping of individual agents. Besides, some facilitators with primitive searching algorithms would probably always give the same agent to certain service finding request. This would result in the situation that one agent gets too many requests while others get very few. A group can be set up to recruit agents capable of providing certain services. In a facilitator (or a matchmaker), only one entry would be needed for one group of such agents. With the group manager doing load-balance, the situation that one agent gets too many service requests while others get few can be avoided. The above solution is made on the assumption that agent groups can be treated equally as individual agents and agents in a group can be accessed from the outside by their role names. Unfortunately, AALAADIN does not satisfy those two assumptions at the present.

5. Architectures Using Matchmakers

Agents that work to provide facilitation services are called middle agents. Currently, there are two main types of middle agents: facilitators and matchmakers [13]. Facilitators in the federated architecture, have two fundamental disadvantages: they are potential performance bottlenecks and make architectures inflexible and difficult to be extended. Many MAS architectures have chosen to use matchmakers for middle agents. Compared to facilitators, matchmakers focus on providing yellow page services. They do not require that every communication goes through them. Hence, they are much less likely to become performance bottlenecks and impose less restriction on architectures. The Foundation for Intelligent Physical Agents (FIPA) has made an effort to achieve the interoperation of heterogeneous agent systems by making a series of standards. It has produced a series of specifications that provide regulations for the development of MAS. The FIPA Agent Management Reference Model [10] is actually a minimal system architecture that every MAS must implement to be FIPA-compliant. According to it an Agent Platform contains multiple agents, agent management components (Directory Facilitator, Agent Management System and Message Transport System) as well as environment components. Agent Platforms connect to one another through the Message Transport System. The Agent Management System acts as "the managing authority" of an agent platform having supervisory control over the lifecycle of all the agents on a platform. It also controls access to resources on the local platform. Every agent must register with the agent management system upon creation to get its identification. One Agent Platform can only have one Agent Management System inside it. FIPA suggests that the federation of directory facilitators can be achieved by directory facilitators registered with each other using a special word "fipa-df" as the service description. Java Agent Development Framework (JADE) [11], [12] is one of a few frameworks claimed to be FIPA-Compliant. The directory facilitator in FIPA model is actually a kind of matchmaker. A directory facilitator in a FIPA-compliant multi-agent system would only supply yellow page functions as matchmakers do. The reference model specifies that agent life cycle services must be provided by the agent management system, which would most likely be implemented as a special system agent. Coordination (apart from facilitation) and security mechanisms are not specified and left open to individual implementations, since the reference model is only a minimal architecture to ensure interoperability among multi-agent systems. The FIPA agent management reference model is one of the architectures that supply only facilitation services as coordination mechanisms. Agents are coordinated based on peer to peer relationships. No support is provided for socialization capabilities. As it uses matchmaker type of middle agent instead of facilitators, there is no apparent potential performance bottleneck. It does have good extensibility as it is designed to be extended by concrete implementations. Its naming convention tries to achieve naming & addressing independency by separating agent names from their transport addresses explicitly. However, an independent name & address resolution layer is not specified in the architecture. Furthermore, both the name of an agent and its transport address are required to be included in one object called aid (agent id). This design effectively cancels the advantages of naming and addressing independency as agent names are still forced to associate with their transport addresses on the user level.

6.   Conclusions

We have reviewed eight multi-agent system architectures. We believe that they indicate the state of the art of MAS architectures in that they effectively represent all the important design ideas, techniques and mechanisms that have been used in this area. MAS architectures were first used to fulfil the need for agents to find each other by capabilities. Early architectures offer nothing more than some kind of message transport service and simple facilitating services. With the increasing complexity of MAS, researchers realises that architectures actually possess the solutions to many problems faced by multi-agent applications and should carry more responsibilities. The following are some of the fundamental issues that we believe system architectures should have solutions. Most of them are reflected in our non-functional criteria. Modelling hierarchical organizational structures is a scenario that future multi-agent applications can hardly avoid. However, the idea that "agents in the system should form a community of peers that engage in peer to peer relations" [14] has been prevailing in the industry. As a consequence, most MAS that have been implemented so far fall short in the capabilities to model the complex, hierarchical structures in real world applications. So, how do system architectures support the capability of modelling hierarchical organisational structures? Future multi-agent applications are expected to be running in an extremely dynamic, open environment where self-interested agents compete with each other for resources. In such a situation, global rules or organizational constraints need to be imposed to prevent fraudulent or malicious behaviours. It would be easier to do so if the organizational structure of a system is imposed. Unfortunately, many MAS architectures were designed based on the belief that organisational structures should only emerge from the interactions among agents instead of being specified explicitly. So should the organisational structures of multi-agent systems be explicitly specified and how can global rules and constraints be imposed? Due to their dynamic and unpredictable nature, mobile agents are difficult to incorporate. One of the issues that MAS must face is to deliver messages to mobile agents as their locations are hard to be followed or predicted. Independency between agent naming and addressing would help to solve this problem. But how do system architectures support naming & address independency? Services provided by individual agent are normally unstable and can be in bad quality. Ordinary facilitators (or matchmakers) only show what individual agents claim to be able to offer. There is no credible authentication service for this. One just cannot know how well an agent actually does what it claims to be able to do. Furthermore, individual agents could stop their services at any time due to some unexpected problems and fail to inform facilitators. So how do system architectures avoid such dangers? Sometimes, for agents to be running in resource-limited environments, they need to be really light-weighted (small in size). In such a situation, to implement all kinds of complex functions such as security measures in every individual agent would very likely prove too much for the environment. So what can system architecture do to solve this problem? With increasing number of software agents on the internet, facilitators' performance might be affected by the vast number of services and agents, since normal facilitators have to create an entry for every service offered by every agent. Furthermore, a facilitator might give the same agent to any request for a certain service (i.e. always the first one encountered in its registry). This situation might cause one agent to get too many requests while others get very few. What should system architectures be designed to avoid such problem? None of the existing architectures has been set up to address all the issues we listed above. However, some solutions can actually be found in them. Therefore, it is logical to expect that the next generation of MAS architectures will be the product of integrating the advantages from existing architectures. We believe that group formation capability, separate address resolution layer and the concepts of roles, groups and recursively containing agency will be included in future MAS architectures.

References

[1] L. Gasser, MAS Infrastructure Definitions, Needs, Prospects, *Infrastructure for Agents, MAS and Scalable MAS,* LNAI, v.1887, Berlin: Springer-Verlag, 2000, pp. 1-11.

[2] R. A. Flores-Mendez, Towards the Standardization of Multi Agent Systems Architectures: An Overview, *ACM Crossroads - Special Issue on Intelligence Agents*, Vol. 5(4), ACM, 1999.

[3] M.N. Huhns, M.P. Singh, Agents and Multi-agent Systems: Themes, Approaches, and Challenges, *Readings in Agents, Huhns,* M.N. and Singh, M.P. (Eds.), San Francisco, Morgan Kaufmann, 1998, pp. 1-23.

[4] M. Genesereth, S. Ketchpel, Software Agents, *Communications of the ACM*, July 1994.

[5] M. R. Genesereth, N. Singh and M. Syed, *A* Distributed and Anonymous Knowledge Sharing Approach to Software Interoperation, *International Journal of Cooperative Information Systems*, 4(4), 1995, pp. 339-367.

[6] D. Gelernter, N. Carriero, Coordination Languages and Their Significance, *Communications of the ACM*, Vol. 35, No. 2, February 1992, pp. 96-107.

[7] B. Hayes-Roth, A Blackboard Architecture for Control, *Artificial Intelligence*, 26, 1985, pp.251-321.

[8] S. Ahuja, N. Carriero, D. Gelernter, Linda and Friends, *IEEE Computer,* Vol. 19, No. 8, 1986, pp. 26-34.

[9] J. Ferber, O. Gutknecht, A Meta-Model for the Analysis and Design of Organizations in Multi-Agent Systems, 3rd *International Conference on Multi-Agent Systems,* IEEE, 1998, pp. 128-135.

[10] FIPA, SC00023 FIPA Agent Management Specification,http://www.fipa.org/specs/fipa00023/SC00023J.html.

[11] F. Bellifemine, A. Poggi, G. Rimassa, JADE -- A FIPA-compliant agent framework, *CSELT internal technical report,* 1999, pp.97-108.

[12] G. Vitaglione, F. Quarta, E. Cortese, Scalability and Performance of JADE Message Transport System, A*AMAS Workshop on AgentCities*, Bologna, 2002.

[13] S. Jha, P. Chalasani, O. Shehory, and K. Sycara, A formal treatment of distributed matchmaking, *Agents*, ACM, 1998.

[14] K. Sycara, M. Paolucci, M. van Velsen, and J. Giampapa, RETSINA MAS Infrastructure, CMU-RI-TR-01-05, *Robotics Institute, Carnegie Mellon*, 2001.

VSP International
Science Publishers
P.O. Box 346, 3700 AH Zeist
The Netherlands

*Lecture Series on Computer
and Computational Sciences*
Volume 1, 2004, pp. 1153-1155

The Forward Web: A New Web Architecture for Organizing Expanding Knowledge

Simon Courtenage[1] and Steven Williams
Cavendish School of Computer Science,
University of Westminster,
London, United Kingdom

Received 7 March, 2004; accepted in revised form 10 March, 2004

Abstract: The World-Wide Web allows users to quickly and easily publish information in the form of web pages, and to link their pages to other pages already on the web. This model of web publishing has many advantages, among them are its simplicity and e±ciency. It also has the disadvantage, particularly in areas where knowledge is incomplete and expanding, that its method of organization does not ¯t well with the way knowledge expands.

In this paper, we look at a radically di®erent model of web publishing in which the author of a web page is not the user who makes links from the web page to other web pages. Instead, the page author expresses an interest in the kind of content the page should link to and as new content comes online that matches that interest, links are inserted automatically into the original page to point to the new content. This points to the possibility, in this reverse web, that a hyperlink from a particular location in a web page can lead to multiple destinations, something we call a multi-valued hyperlink.

Keywords: Web architecture, hyperlinks, publish/subscribe, peer-to-peer

1 Introduction

The web has been a source of enormous benefit to a great many people. The great advantages of the web are, of course, the ease with which information can be published and made available to a wide audience, and the ability to organize and connect different documents in a graph-based structure using hyperlinks. However, the way in which the web is constructed, through the addition of new documents, is at odds with the way in which knowledge expands. New pages can only link to pages that already exist in the web graph, but are not themselves linked to by other pages as yet, because they are new.

The web graph, therefore, grows backwards, rather than forwards. We cannot write a web page with links in the page to other pages that do not as yet exist. For example, we do not write a web page containing information about a particular concept, such as number theory, with links from key sub-concepts to pages that have yet to be written. But this is a model of how human knowledge grows, both in terms of a particular individuals understanding of a subject and in terms of research in general. There is a disparity, therefore, between how the web is constructed and how human knowledge is constructed. The web is a great tool for organizing and connecting existing knowledge, but not for knowledge that is incomplete and expanding.

In this paper, we describe a new architecture for the web that models how human knowledge expands when that knowledge is incomplete. The key feature of this web architecture is that authors of web pages do not insert links to other pages as they create or edit the page. Instead they indicate which parts of the content of the page from which links should be established as soon as pages with matching content become available.

---

[1] Corresponding author. Email: courtes@wmin.ac.uk

The basis for this feature is a distributed content-based publish-subscribe system [1] [2] [3]. We show how this publish-subscribe system can implement our forward web over a peer-to-peer network of web servers.

2 Growing the Web Forwards

At present, the web allows pages to be written containing hyperlinks to pages so long as the targets of those hyperlinks (as specified by their URLs) already exist. If a URL is specified for a page that does not exist, then this is considered by HTTP [ref] as an error (code 404 - File Does Not Exist).

If we consider, therefore, that following hyperlinks takes us in a forwards direction, then the expansion of the web, through the addition of new pages or the editing of existing pages to include new hyperlinks, is in a backwards direction, from the pages that will be targets of hyperlinks to the pages that will refer to them. Where the information to be published on the web already exists in some form, the web provides an ideal way to organize and structure it using documents and hyperlinks. However, when the information to be published is incomplete and likely to grow in the future as new information comes to light, the way in which the web expands does not fit well with the way in which the information expands.

We propose, therefore, a new web infrastructure that allows the web to grow forwards, in the same direction as knowledge. However, the current web grows backwards as a consequence of the fact that pages must exist in order to be targeted by an anchor in another tag. To grow forwards, therefore, we must allow anchors to pages that do not yet exist but which may exist at some point in the future.

Clearly, we cannot target specific pages that do not yet exist, since their URL is unknown. Instead, we indicate what kind of pages we want to link to. Web authors, therefore, need to be able to state where links should be inserted into their documents, as and when new and relevant content is made available. We add to HTML a new type of anchor tag that simply specifies the type of content new pages need to contain in order to be linked-to from the page (the existing anchor tag used to create web hyperlinks is retained to maintain interoperability with the existing web). We also include a new SUMMARY tag, to be included in the header of a page and which acts as an aid to matching page content to link requirements. The SUMMARY tag is similar to the use of the META tag in existing HTML, which has been used in the past to advertise a page's content to search engines. The syntax of the SUMMARY tag is simple:

<SUMMARY> *keyword-list* </SUMMARY>

and is placed between the <HEAD> and </HEAD> tags in an HTML document.

A web document author can indicate where new links should be inserted into a document as and when new and relevant content is found by using a <LINKTO KEY="..."> ... </LINKTO tag, rather than an HTML anchor tag. The <LINKTO> tag is used around text which the author would like to act as a hyperlink to other documents (that may or may not exist in the web graph when the document is created). The KEY attribute of the tag contains those keywords to be used in finding matching content on other pages. At present, the matching between the content of the KEY attribute in a <LINKTO> tag and the SUMMARY content in other web documents is done using simple keyword matching.

Using our extended HTML (called F-HTML), the author of a web document does not have complete control over which documents are linked to. In fact, using the <LINKTO> tag in a web document creates the distinct possibility that the anchor text in a <LINKTO> tag can refer to more than one web document. (This is in contrast with the current web which allows only single-valued hyperlinks, i.e., hyperlinks that can only refer to one document by specifying its URL.)

3 A Publish/Subscribe Infrastructure for the Forward Web

The Forward Web demands more from the components of its infrastructure than the current web.

Links between pages must be found and created automatically when a match is found between the content of a <LINKTO> tag in one document and the content of a <SUMMARY> tag in another document. The infrastructure must be aware of the contents of both tags in all documents, therefore, in order to infer which documents should refer to which other documents. Our solution to this problem is to use the publish/subscribe communications paradigm for connecting the components of the web infrastructure, speci¯cally, content-based publish/subscribe, over a peer-to-peer (P2P) network. The use of a publish/subscribe network overcomes the problems in matching <LINKTO> keywords with web page SUMMARY content posed by the asynchronous nature of web publishing, while the use of a P2P network enables document repositories to provide services to each other to allow new hyperlinks to be formed. Publish/Subscribe systems [4][2] [1] [3] form an important communications paradigm in distributed systems, one in which servers (or producers of messages) are decoupled from clients (or consumers) by the network. Instead of clients contacting servers directly to request services or information, clients register a subscription with the network to receive messages satisfying certain criteria. Servers publish information onto the network, without knowing who will receive it, and the network undertakes to route messages to the appropriate clients based on the set of subscriptions currently in effect.

In the Forward Web's publish/subscribe based-architecture, the content of a <LINKTO> tag becomes a subscription. When a new page is added to a web server's document repository, the web server extracts the content of each <LINKTO> tag in the page and advertises it as a subscription to other web server peers. Similarly, it publishes the content of the page's <SUMMARY> tag to other peers. When a <LINKTO> subscription and a <SUMMARY> publication are matched, then the subscribing and publishing web servers are noti¯ed so that the appropriate hyperlink can be inserted into the subscribing web page with the published web page's details. Once the link has been inserted into the page, then following the link in a browser uses the standard client-server model of communication between the browser and the referred-to web page's web server.

4 Conclusions and Further Work

This paper has presented a proposal for a new architecture for the web to allow it to grow progressively as new knowledge is added in the form of new web pages. The architecture envisages web servers acting as peers in a peer-to-peer publish/subscribe-based network, and has been designed to interoperate seamlessly with the existing web.

We are currently developing a testbed platform for the Forward Web, building on our existing research on publish/subscribe architectures, including the development of new peer-to-peer web server networks capable of supporting the Forward Web. We are also investigating more intelligent means of matching forward links with page summaries, including automatic page summarization, other than simple keyword matching.

References

[1] Carzaniga, A., Rosenblum, D., Wolf, A.: Content-based addressing and routing: A general model and its application (2000)

[2] Carzaniga, A., Rosenblum, D.S., Wolf, A.L.: Design and evaluation of a wide-area event noti¯cation service. ACM Transactions on Computer Systems 19 (2001) 332- 383

[3] Pietzuch, P.R., Bacon, J.M.: Hermes: A Distributed Event-Based Middleware Architecture. In: Proc. of the 1st Int. Workshop on Distributed Event-Based Systems (DEBS'02), Vienna, Austria (2002) 611-618

[4] Eugster, P., Felber, P., Guerraoui, R., Kermarrec, A.M.: The many faces of publish/subscribe (2001)

VSP International
Science Publishers
P.O. Box 346, 3700 AH Zeist
The Netherlands

Lecture Series on Computer
and Computational Sciences
Volume 1, 2004, pp. 1156-1156

Preface of the Symposium :
Multimedia Synchronization Model and The Security in the Next Generation Mobile Information Systems

T.E. Simos
Chairman of ICCMSE 2004

Department of Computer Science and Technology,
Faculty of Sciences and Technology,
University of Peloponnese,
GR-221 00 Tripolis, Greece

This symposium has been created after a proposal of Dong Chun Lee of the Dept. of Computer Science, Howon University (Korea). After a successful review, we have accepted her proposal.

The organizer of the symposium, Dong Chun Lee, has selected, after international peer review nine papers:

- *Design and Implementation of Session Initiation Protocol (SIP) Based Multi-Party Secure Conference System* by Hyun-Cheol Kim, Jongkyung Kim, Senog-Jin Ahn, Jin-wook Chung

- *Multimedia Synchronization Algorithm in All-IP Networks* by Gi Sung Lee

- *Design and Implementation of Mobile-Learning System Using Learner-oriented Course Scheduling Agent* by Keunwang Lee, Jonghee Lee, HyeonSeob Cho

- *An Extension of BGP Algorithm for End-to-End Traffic Engineering* by Hyonmin Kong, Youngmi Kwon

- *Real Time Face Detection in Mobile Environments* by Yong-Hwan Lee, Yu-Kyoung Lee, Young-Seop Kim, Sang-Burm Rhee

- *New Mobility Management for Reducing Location Traffic in the Next Generation Mobile Networks* by Dong Chun Lee, Jung-Doo Koo, Hongjin Kim

- *Automatic Identification of Road Signs for Mobile Vision Systems* by Hong-Gyoo Sohn, Yeong-Sun Song, Gi-Hong Kim, Hanna Lee, and Bongsoo Son

- *Mobile-Based Healthcare B2B Workflow System for Efficient Communication* by Sang-Young Lee, Cheol-Jung Yoo, Dong Chun Lee

- *Intelligent All-Way Stop Control System at Unsignalized Intersections* by Sang K.Lee, Young j. Moon

I want to thank the symposium organizer for the activities and excellent editorial work.

VSP International
Science Publishers
P.O. Box 346, 3700 AH Zeist
The Netherlands

Lecture Series on Computer
and Computational Sciences
Volume 1, 2004, pp. 1157-1160

Design and Implementation of Session Initiation Protocol (SIP) Based Multi-Party Secure Conference System

Hyuncheol Kim[1],Jongkyung Kim,Seongjin Ahn,Jinwook Chung

Dept. of Electrical and Computer Engineering,Sungkyunkwan University,
300 Chunchun-Dong Jangan-Gu,Suwon, Korea, 440-746
Dept. of Computer Education, Sungkyunkwan University,
53 Myungryun-Dong Jongro-Gu, Seoul, Korea, 110-745

Received 7 March, 2004; accepted in revised form 10 March, 2004

Abstract: Session Initiation Protocol (SIP) is currently receiving much attention and seems to be the most promising candidate as signaling protocol for the future IP based multimedia services. While the problem of QoS mostly refers to network layer, the problem of security is strictly related to the signaling mechanisms and the service provisioning model. This paper presents, a new SIP based multi-party secure closed conference system. In the traditional participants, except of captain, conference participants doesn't have a privilege that accepts or declines new participants. On the other hand, in closed conference system, adding new participants needs unanimity of opinion of existing participants. We also propose a new authentication procedure for closed conference system. By means of a real testbed implementation, we provide an experimental performance analysis of the closed conference system.

Keywords: Session Initiation Protocol, Secure multimedia conference system, QoS

Mathematics Subject Classification:

PACS:

1   introduction

Along with rapid development of wire and wireless communication technologies, internet service providers face on exorbitant demands of various high quality multimedia services from their subscribers. These requirements cause a migration to the packet-based integrated networks, which are followed by all IP-based converged networks. Session Initiation Protocol (SIP) is the Internet Engineering Task Force (IETF) standard for IP telephony. It seems to be the most promising candidate for call setup signaling for future IP-based telephony services, and it has been chosen by Third-Generation Partnership Project (3GPP) as the protocol for multimedia applications in 3G mobile networks [1][2][3].

Along with such a tendency to be standardized, the next generation communications network is becoming at certitude providing vocal and image services based on IP and SIP. So, development, construction and effective operation of SIP environment are essential. Recently, there have been many detailed investigations within the IETF toward establishing a generic signaling framework for

---

[1]Corresponding author. E-mail: hckim@songgang.skku.ac.kr

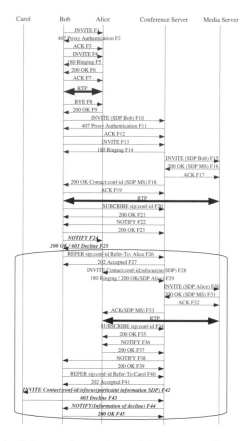

Figure 1: Call control procedures of the secure conference system

NGcNs. However, most of attention has been focused on the possibility of providing new dynamic and powerful services. Less attention has been paid to security feature.

In this paper, we propose a new SIP based secure closed conference system. A discussion of security aspects in SIP is given; we focus on the authentication procedure. We describe a possible SIP based multi-party conference system architecture and related SIP security procedure. The testbed and methodology for evaluating the processing cost of SIP security are reported with experimental results. In case of the existing conference system, only the captain of conference session has the right of invitation of additional participants. However, closed conference system needs all participants' anonymous agreement is needed to invite additional participants. In this paper, the closed conference model and the extended header will be proposed.

2   Proposed authentication procedure for closed conference system

Conference System fundamentally needs that some participants connect each other at the same time. However, the simple expansion of 1:1 connection isn't able to guarantee satisfactory confidence. For instance, when a new participant join conference session, existing participants and new one have to know the information of who has joined conference. That is why any one among them doesn't want to join conference together. Therefore, we need to know all participants of conference. We use the method that is the usage of INVITE message's SDP (Session Description Protocol) from the conference server [4][5].

Fig.1 shows the procedure of the invitation of an additional participant in the closed conference system. To begin with Alice and Bob setup call session through authentication procedure. When the captain, Bob, invite Carol, Bob's UA sends NOTIFY message to Alice in order to agree an additional invitation. When Alice agrees to invite Carol, his UA sends 200 OK message (F25). However, when he doesn't want to do that, it sends 603 declined message (F25). In this case, the invitation is canceled [9]. If Alice agrees the invitation of Carol, next time, Conference Server asks Bob to join conference. Bob checks the received information of participants. If he doesn't want to join that conference, he sends 603 messages (F43). The conference server lets Bob know the information of decline. This procedure presents with gray box in Fig.1. If Bob agrees to join conference session, his UA send 180 Ringing and 200 OK message. It's the same procedure as for conference server to invite Alice. Additional messages are presented in an italic font in the Fig.1.

3   Methodology for the Evaluation of Processing Cost and Experimental Results

In order to experiment with advanced feature in SIP, we have developed an open source VOCAL 1.4.0 and realized a testbed. The testbed consists of PCs equipped with Solaris 9 and Windows XP operating systems, acting as SIP clients, SIP servers. Fig. 10 shows the layout of the testbed [10]. The goal of this testbed was twofold: firstly verification of the function behavior of the various elements and their interoperability; secondly the possibility of making some performance analysis. In particular regarding performance analysis, an interesting point is the evaluation of the cost to be paid in terms of performance for in the introduction of security mechanisms in SIP, like the authentication procedure.

We followed pure experimental approach, trying to evaluate the processing costs of different procedures in the elements of our testbed. Of course, we are conscious that this approach has some limitation, since the results can be severely depend on the specific characteristics of the implemented modules and the testbed. It is not possible to derive result of general validity; however, we found that it is useful to have a first estimate of the relative costs of different procedure in the implementation. In particular, we focused on the SIP proxy servers, which are potential bottlenecks in possible SIP-based multimedia service. With this goal in mind, we have defined a methodology to evaluate the processing cost of SIP procedures in the proxy server [6][7].

The results of our evaluation are reported in Table 1. The Third column reports the experiential average cost of twenty times in terms of second. This includes all the processing that the system performs from the setup of conference session to its terminating. Note that this throughput corresponds to 100 percent utilization of elements processing resources. The two rightmost columns are the most important ones and report the throughput value converted in a relative processing cost. In the first one the reference value of 100 has been assigned to the first procedure/scenario. The result show that the introduction of SIP security accounts for nearly 30 percent of processing cost of no authentication procedure. This increase can be explained with the increase in number of exchanged SIP messages with the actual processing cost of security including the process of the cryptographic algorithm. Furthermore, extra SIP messages include the network delay. Another interesting finding is that the incensement of processing time cost increase accidentally, when third attendant joins conference. The conference mode of testbed is Ad-hoc model. Therefore, the third attendant spends time of disconnecting the session of two attendants.

Table 1: Experimental results

	Procedure/Scenario	Processing Cost(Seconds)	Relative Cost
1	No authentication, 2 attendants	$2.873 * 10^{-2}$	100
2	No authentication, 3 attendants	$4.022 * 10^{-2}$	140
3	No authentication, 4 attendants	$4.252 * 10^{-2}$	148
4	No authentication, 5 attendants	$4.453 * 10^{-2}$	155
5	No authentication, 6 attendants	$4.683 * 10^{-2}$	163
6	Authentication, 2 attendants	$3.735 * 10^{-2}$	130
7	Authentication, 3 attendants	$5.200 * 10^{-2}$	181
8	Authentication, 4 attendants	$5.545 * 10^{-2}$	193
9	Authentication, 5 attendants	$5.746 * 10^{-2}$	200
10	Authentication, 6 attendants	$6.119 * 10^{-2}$	213

4   Conclusion

In this study, the security aspects related to SIP based multi-party closed conference system are discussed. The authentication procedure, based on HTTP Digest authentication, is described. The mapping of authentication mechanism into a possible several service scenarios is provided. Finally, the performance aspects of SIP authentication for closed conference system are considered with pure experimental approach. The processing costs of different security procedure/scenario are compared under a reference implementation. Although the performance results are obviously conditioned by the specific implementation aspects, they can be rough idea of relative processing cost of SIP security procedures.

References

[1] J. Rosenberg, H. Schulzrinne, G. Camarillo , A. Johnston, J. Peterson , R. Sparks, M. Handley , E. Schooler Peterson, *SIP Session Initiation Protocol*, RFC 3261, Jun. 2002.

[2] M. Faccin, Poornima Lalwaney, Basavaraj Patil, *IP Multimedia Services: Analysis of Mobile IP and SIP Interaction in 3G Networks*, IEEE Communication Magazine, Jan. 2004.

[3] Janet R. Dianda, Vijay K. Gurbani, Mark H. Jones, *Session Initiation Protocol*, Bell Labs Technical Journal 7(1), 3-23 2002.

[4] Jonathan Rosenberg, Henning Schulzrinne, *Models for Multi Party Conferencing in SIP*, Jan. 2003.

[5] A. Johnson, *Session Initiation Protocol Call Control-Conferencing for User Agents*, Oct. 2003.

[6] Rohan Mahy, *A Call Control and Multi-party usage Framework for the Session Initiation Protocol (SIP)*, Sep. 2003.

[7] A. B. Roach, *Session Initiation Protocol (SIP)-Specific Event Notification*, RFC 3265 , Jun. 2002.

VSP International
Science Publishers
P.O. Box 346, 3700 AH Zeist
The Netherlands

*Lecture Series on Computer
and Computational Sciences*
Volume 1, 2004, pp. 1161-1164

Multimedia Synchronization Algorithm in All-IP Networks

Gi Sung Lee

Dept. of Computer Science, Howon Univ.,
727 Wolhari, Impi, Gunsan, Chonbuk, Korea
ygslee@sunny.howon.ac.kr

Received 7 March, 2004; accepted in revised form 10 March, 2004

Abstract: This paper proposes multimedia synchronization algorithm that ensures effective presentations of MPEG data in AII-IP networks. While media are being transferred for presentations from any multimedia server in Mobile Switching Center (MSC), this system checks network status and buffering state at Base Station (BS) and gives multimedia server feedback message. Then, the multimedia server offers a scheme that supports synchronization through adjustment of the amount of data being transferred.

Keywords: MPEG data, Synchronization, Mobile switching center, Base station, All-IP networks

1. Introduction

In a mobile environment, play-out of multimedia data on a Mobile Host (MH) is difficult due to inherent characteristics in wireless networks such as the high data loss rate, long delays and low network bandwidth. The distributed multimedia systems which are connected to the mobile networks in large numbers use buffers to overcome network delays and unpredictable losses.

Current research has reached a level in which synchronization schemes based on wireless networks are incorporated into conventional ones. M. Woo, N. U. Qazi, and A. Ghafoor [5] defined a BS as an interface between wired and wireless networks. For wired networks, the interface defines buffering in the BS to reduce jitter delay between packets. One possible shortcoming of this approach is an attempt to apply buffering to synchronization by assigning existing wireless communication channels. Azzedine Boukerche [1] proposed an efficient distributed synchronization solution in mobile multimedia systems to ensure and facilitate MH access to multimedia objects. He also proposed synchronization and handoff management schemes that allow MHs to receive time-dependant multimedia streams without delivery interruption while moving from one cell to another [2].

2. Multimedia Synchronization System

2.1. System Configuration

This system consists of k multimedia servers, m BS, and n MHs in Fig. 1. The BS communicates with i^{th} MH in the any cell. These MHs access the server via the BS. This system monitors variations in the start time for transmission as well as buffers at the BS, using variables such as arrival time of subframes transmitted from multimedia servers and delay jitter. Some advantages of this system include its capability to overcome the limitations of mobile communications like small memory size and low bandwidths

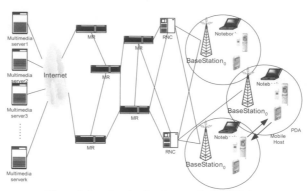

Figure 1 Synchronization System for MPEG data

2.2. Network State for Multimedia Synchronization

To monitor network status and reflect it in current network status, this method uses a DSM (i.e., Difference arrive time from a server to the MH) that is defined as the time difference between sending a Group of Picture (GOP) from a multimedia server and its arrival at a BS. In order to avoid dense networks and reduce data loss rates, a DSM will be sought every time a GOP is transferred, and an optimal DSM that reflects the current network status will be determined. When a GOP located on a media's server arrives at a BS, a current DSM is sought and a data loss rate is then calculated, as a means of finding an optimal DSM, by checking whether the GOP has arrived early or belatedly, as well as whether the buffering state is full. If the current DSM value is either greater than the maximum allowable network delay time, or smaller than the minimum allowable network delay time, the corresponding packet is thrown away as undeliverable and calculated as a data loss rate in order to check if the GOP has arrived belatedly. A DSM with a value greater than the maximum allowable network delay time means that the corresponding packet has arrived too late; whereas a DSM with a value smaller than the minimum allowable network delay time means that the corresponding packet has arrived too early.

2.3 Feedback of Buffering State for Multimedia Synchronization

Due to a dense network, if the difference between the Current DSM (CDSM) and Best DSM (BDSM) exceeds the threshold value, some feedback is required. If the difference between the CDSM and BDSM goes below the threshold value, the rate of data consumption will be higher than the rate of data creation; however, if the buffer level falls below Bn after the buffer status is checked, some feedback is transferred to the server in order to increase data transmission rates. The buffer status is also checked on a regular basis, and a check is made if the buffer level goes beyond low buffering state and high buffering state. Feedback message is given to the server in MSC according to the ordinal occurrence in those two check requirements. In general, DSM-based signals immediately reflect network status compared to buffer-based signals. As a result, DSM-based signals trigger feedback before buffer-based signals. As described above, this scheme can effectively improve our grasp of network status by sending feedback message to multimedia server that MPEG data is transferred properly to the BS. The proposed algorithm describes as the following thing.

Algorithm Multimedia synchronization

```
Procedure
Compute_Best_DSM ( )
Δt = CDSM-BDSM;   /* BDSM: Best DSM,   CDSM: Current DSM*/
check the buffering state;
if (Δt<δ and L_buffer<Bn) or (L_buffer≤Bl and ΔR<0) then
   /*L_buffer: level of buffer */
   send a feedback for speed up to server;
   set the time of feedback to L_f_time;
else if (Δt<δand L_buffer<Bn) or (L_buffer≤Bl and ΔR < 0) then
   /*Δt: difference between CDSM and BDSM.
```

ΔR: difference between production rate and consumption rate */
    send a feedback for slow down to server;
    set the time of feedback to L_f_time;
elseif
while (true) do
Compute_Best_DSM ( )
Δt = CDSM-BDSM;
check the buffering state;
if (Δt<δand L_buffer<Bn) or (L_buffer≤Bl and ΔR < 0) then
    set the Current time to C_time;
    A_time=C_time – L_f_time;
    If A_time > φ then
        send a feedback for speed up to server;
        set C_time to L_f_time;
    endif
else if (Δt > δ and L_buffer < Bn) or (L_buffer ≥ Bh and ΔR > 0) then
    set the Current time to C_time;
    A_time=C_time – L_f_time;
    If A_time > φ then
        send a feedback for slow down to server;
        set C_time to L_f_time;
    end if
  end if
end if

3. Performance Evaluation

Presented are the simulation experiments we carried out to evaluate the performance of our synchronization scheme. We developed a distributed discrete-event model to simulate a cellular wireless multimedia system. Table 1 displays the simulation parameters we have used in our experiments.

Table 1. Simulation Parameters

Number of cells	60	Forward time to BS	100ms
Number of servers	4	Rate of handoff	5% of MMUs
Buffer size of a MH	1 Jitter	RTT to request/deliver a GOP	100ms
Buffer size of a BS	2 Jitter	Average Jitter	200ms
Play-out times/GOP	250ms	Maximum Jitter	600ms

Figure 2 shows the results of 31 tests in which the arrival time was changed for each experiment. In Test 2 and 8, minimum delay and maximum delay were adjusted to 50ms and 600ms each; compared with the no control case, the play-out rate was improved by 6% for the Chen scheme, by 8% for the Azzedin-Boukerche scheme and by about 10% for the scheme proposed in this paper. In Test 4, 9 and 26, minimum delay and maximum delay were adjusted to 20ms and 800ms, respectively, to induce overflow and starvation. In these cases, while network traffic conditions worsened, the play-out rate was improved by 5% for the first variance and by 8-9% for the second variance. In average, the play-out rate was about 79% for the conventional method, 85% for the first variance and 91% for the second variance.

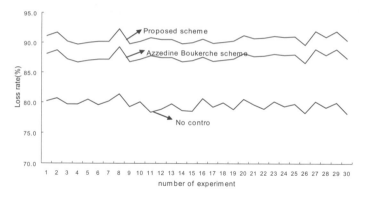

Figure 2. Comparison of Play-out Rate

4. Conclusion

We have proposed multimedia synchronization algorithm that control play-out time reflect two factor which it is DSM, defined as the time difference between sending a GOP from a server and its arrival at a BS, and buffering status. This proposed algorithm overcomes that play-out of multimedia data on a MH is high data loss rate, long delays and low network bandwidth. In the simulation, it showed that the proposed scheme offers continuous MPEG data play-out, higher packet play-out rates, and lower packet loss rates, relative to the previous schemes.

Acknowledgments

This work was supported by Howon University Fund, 2003.

References

[1] Azzedine Boukerche, Sungbum Hong and Tom Jacob, "MoSync: A Synchronization Scheme for Cellular Wireless and Mobile Multimedia System", Proc. of the Ninth International Symposium on Modeling, Analysis and Simulation of Computer and Telecommunication Systems IEEE 2001.

[2] Ernst Biersack, Werner Geyer, "Synchronization Delivery and Play-out of Distributed Stored Multimedia Streams", Multimedia Systems, Vol.7 No.1, pp 70-90, 1999.

[3] T. D. C. Little, and Arif Ghafoor, "Multimedia Synchronization Protocols for Broadband Integrated Services," IEEE J. on Selected Areas in Comm., Vol. 9, No.9, Dec. 1991.

[4] Sook-Young Choi, "A Feedback Scheme for Synchronization in a Distributed Multimedia", KIPS, Vol 9-B, No 1, pp 47-56, Feb., 2002.

[5] Gi-Sung Lee, "Synchronization Scheme of Multimedia Streams," LNCS Vol. 3036, pp 389-396, 2004.

VSP International
Science Publishers
P.O. Box 346, 3700 AH Zeist
The Netherlands

*Lecture Series on Computer
and Computational Sciences*
Volume 1, 2004, pp. 1165-1168

Design and Implementation of Mobile-Learning system Using Learner-oriented Course Scheduling Agent

Keunwang Lee, Jonghee Lee, HyeonSeob Cho

Department of Multimedia Science, Chungwoon University, Chungnam, Korea

Received 7 March, 2004; accepted in revised form 10 March, 2004

Abstract: The demand for the customized courseware which is required from the learners is increased, the needs of the efficient and automated education agents in the mobile learning system are recognized. But many Mobile(m)-learning systems that had been studied recently did not service fluently the courses which learners had been wanting and could not provide the way for the learners to study the learning weakness which is observed in the continuous feedback of the course. In this paper we propose design of m-learning system for learner-oriented course scheduling using weakness analysis algorithm. The proposed system monitors learner's behavior constantly, and evaluates them and calculates his accomplishment. From this accomplishment the multi-agent schedules the suitable course for the learner. The learner achieves an active and complete learning from the repeated and suitable course.

Keywords: m-Learning System, Multi-agent, Course Scheduling

1. Introduction

The fast development of Internet has recently enabled the on-line lecture through the e-learning system, which is now became popular topics in the area of computer education system industry. As this e-learing system is spread widely to the public, the users demand more diverse education service, and that results facilitating study on applied education service being very active [1-3].

Since the agent and broker for the domestic and foreign education software are organized to meet the demands of the average public more rather than customized service for individual learner, it is very difficult to accommodate the various needs for knowledge and evaluated level for each and every individual[4,5]. Although tools to help interaction between learners had been supported in many ways, in instructor's perspective, it is very hard to provide the right course schedule and combinations by analyzing each learner status after facing all registered learners. Hence, agent who can deliver feedback such as effective way of learning, course formation and course schedule to learners is needed in this e-learning system [6-7].

Multi-agent will be proposed in this thesis, who can provides the appropriate active course scheduling and feed back to learner after evaluating the learners education level and method. By developing agent who provides the fast and suitable feedback to learners learning status, we are going to recompose course that is suitable for each learner to increase the effectiveness in learning.

2. Course Scheduling Multi-Agent

2.1. m-Learning System Structure

In this m-learning system, learner and Course Scheduling Multi-Agent (CSMA) are connected via mobile interface, and through Mobile Interface (MI) the request and transfer for course scheduling occurs between learners and CSMA. Learners study the course provided by CSMA in this system.

All the information created by CSMA will be stored into the database and if required, it will be loaded by CSMA and used for reorganizing course. Learner's profile and information obtained by their learning activity as well will be stored into database via MI, then by CSMA it will be regenerated and stored again as necessary information to the learner such as learning achievement level, course, scheduling, evaluation data, feedback and etc. Figure 1 shows the structure for proposed m-learning system.

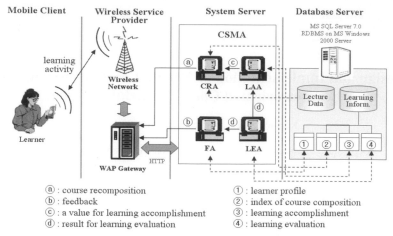

Figure 1. m-Learning system structure

The key component of CSMA consists of four agents, Course Re-composition Agent (CRA), Learning Accomplishment Agent (LAA), Learning Evaluation Agent (LEA) and Feedback Agent (FA). The CRA is delivered the information on the degree of accomplishment of learning from the LAA and creates and provides a new and most customized learner-oriented course. The LAA estimates the degree of learning accomplishment based on the test results from the LEA and tracks the effectiveness of learning. The LEA is carrying out learning evaluation at every stage. The FA provides relevant feedback to learners in accordance with the learners profile and calculated degree of accomplishment of learning.

2.2. Course Scheduling Scheme

The degree of learning accomplishment can be calculated by the comparison between the current test result and the previous test result, and the analysis of the learning effectiveness growth. Let the maximum degree of learning accomplishment be 1 and we can give certain amount as a degree of weakness. Therefore, 1 minus the amount is the degree of learning accomplishment. This can be defined by the following equation.

$$A\,(I,\,i) = 1 - W\,(I,\,i) \tag{1}$$

where $A\,(I,\,i)$ is achievement degree for each subsection and $W\,(I,\,i)$ is weakness for each subsection. The reason the degree of learner's weakness is under 1 is to represent it with percentage. It has observation from 0 to 1. Evaluation agent decides the degree of learning accomplishment by test. Evaluation agent also estimates the weighted value of the weak problems by calculating the marking time of individual question. Figure 2 is a diagram in which section test is inserted into the subsection test.

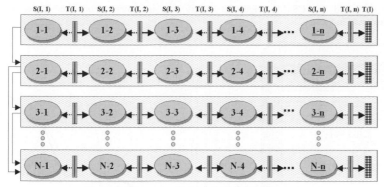

Figure 2. Diagram of subsection and section test

The degree of weakness of subsections W_{tr} (*I,i*) can be represented based on the marking time and ratio of correct answers as following equation.

$$W_t (I, i) = 0 \qquad\qquad : \text{when } t_d (I, i) < t_a (I, i)$$
$$1 \qquad\qquad : \text{when } t_d(I, i) \geq (4 * t_a(I, i))$$
$$(t_d (I, i) - t_a (I, i)) / (3 * t_a (I, i)) \qquad : \text{when } t_d (I, i) < (4 * t_a (I, i)) \qquad (2)$$

$$W_{tR}(I, i) = W_t(I, i) * \alpha_t + (1 - R(I, i)) * (1 - \alpha_t) \qquad (3)$$

where $t_d(I, i)$ is the needed time for solving subsection question in section test, $t_a(I, i)$ is average required time for solving one subsection question in section test, $R(I, i)$ is average required time for solving one subsection question in section test, $W_t(I, i)$ is weakness of solving time for each subsection, $W_{tR}(I, i)$ is weakness of solving time and correct answer for each subsection and α_t is Weight value of time weakness applied. The equations to calculate the weakness of subsection analyzing repeated learning is defined by:

$$W_r (I, i) = (L_c (I, i) - 1) * 0.3 \qquad (4)$$

where $Lc (I, i)$ is count of repeat for subsection learning. Accordingly the degree of learning weakness according to the course test can be calculated as follows equation.

$$W(I, i) = W_{tR}(I, i) * (1 - \alpha_r) + W_r(I, i) * \alpha_r \qquad (5)$$

where $W(I, i)$ is weakness for each subsection and α_r is weight value of repeat learning weakness applied. The degree of learning weakness from the analysis of learning repetition represents the weakness of total subsection along with the marking time. Therefore the degree of learning weakness at each subsection is calculated by the weighted value at between the subsection weakness analyzed by the marking time and the rate of correct answer and the learning weakness analyzing the repetition of subsection study. With this learning weakness, we can estimate the degree of learning accomplishment. And by the degree of learning accomplishment, we track the subsection showing weakness and recompose the course.

3. Experiments

For experimenting the course scheduling multi-agents, 48 people who studied ordinary learning method were extracted randomly from unspecified persons and the same courseware was offered. And another 50 people were extracted with the m-learning system using CSMA. The subsection-specific weakness is demonstrated with the graph and figures and the final test degree so that learners can compare it with their target score. The learners under the target degree can begin the repetition program by the course schedule provided by CSMA. Figure 3 offers the information on the degree of learning accomplishment in PDA.

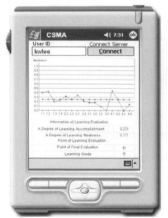

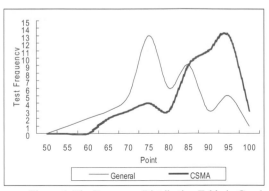

Figure 3. User Interface in PDA   Figure 4. The Frequency Distribution Table in Graph

Figure 4 represents the frequency distribution table in graph. The group that used CSMA system shows higher frequency with the higher marks and group that used general system shows relatively lower marks. The average score of B group is much higher than that of the A group.

4. Conclusion

In this paper we proposed m-learning system using multi-agent for the learner's courseware scheduling creating the courseware customized for the individual learner by evaluating the learning. These agents continuously learn the individual learning course and the feedback on the learning, offering customized scheduling course and giving the maximum learning effectiveness. Accordingly the course ordered by learner can be fittest to the learner with the help of course scheduling agent. The learners can continuously interact with agents until completing the course. If the agents judge the course schedule for learners not to be effective, they recompose the course schedule and offer the new course to learners.

References

[1] Moore, M.G and Kearsley, G., "Distance Education," Wadsworth Publishing Company, 1998.

[2] Hamalainen, M, Whinston, A, and Vishik, S., "Electronic Markets for Learning: Education Brokerages on the Internet," *Communications of the ACM*, vol. 39, No. 6, pp. 51-58, 1996.

[3] Eyong B. Kim, Marc J. Schniederjans, "The role of personality in Web-based distance education courses," *Communications of the ACM*, Vol. 47, No. 3, pp. 95-98, 2004.

[4] Jason A. Brotherton, Gregory D. Abowd, "Lessons learned from eClass: Assessing automated capture and access in the classroom," *ACM Trans on Computer-Human Interaction*, Vol. 11, No. 2, pp. 121-155, 2004.

[5] Hal Berghel, David L. Sallach, "A paradigm shift in computing and IT education," *Communications of the ACM*, Vol. 47, No. 6, pp. 83-88, 2004.

[6] Whinston, A., "Re-engineering MIS Education.", *Journal of Information Science Education*, Fall 1994, 126-133, 1994.

[7] Badrul H. Khan, Web-Based Instruction (WBI): What Is It and Why Is It?, Education Technology Publications, Inc., 1997.

VSP International
Science Publishers
P.O. Box 346, 3700 AH Zeist
The Netherlands

*Lecture Series on Computer
and Computational Sciences*
Volume 1, 2004, pp. 1169-1172

An Extension of BGP Algorithm for
End-to-End Traffic Engineering

Hyonmin Kong[1], Youngmi Kwon[1], Geuk Lee[2]

[1]Dept. of Information and Communications Eng., Chungnam National Univ., Korea
220 Gung-dong, Yuseong-gu, Daejeon, 305-764, Korea

[2] Dept. of Computer Eng., Hannam Univ., Korea

Received 7 March, 2004; accepted in revised form 10 March, 2004

Abstract: To have an effective end-to-end QoS capability in the Internet, it's necessary for the Border Gateway Protocol (BGP) to provide QoS enforcement among the inter-BGPs as well as IGPs. Current BGP4 operates based on the policy, and policies are represented by attributes such as LOCAL_PREF and MED to influence incoming/outgoing traffic flows on purpose. These attributes are optional nontransitive or well known discretionary. So these attributes can influence the preferences between directly-connected neighbor BGPs, but are not sufficient to provide general end-to-end Traffic Engineering. This paper proposes an extended BGP routing algorithm which can provide an end-to-end Traffic Engineering capability. This new method doesn't require additional BGP message type, nor change a structure of existing routing tables. It only uses a slightly changed operation of AS_PATH attribute, which is well-known mandatory. Simulation shows that the new BGP algorithm increases link utilization among ASs and can be used as compulsive load balancing mechanism.

Keywords: BGP, Traffic Engineering (TE), Policy

1. Introduction

BGP is a path vector protocol used to carry routing information between Autonomous Systems (AS). The path vector means that BGP routing information carries a sequence of AS numbers that identifies the path of ASs that identifies the path of ASs that a network prefix has traversed. BGP follows policy-based routing principles [1, 2]. Routes are advertised between a pair of BGP routers in UPDATE messages. The UPDATE message contains the path attributes, which include such information as the degree of preference for a particular route and the list of ASs that the route has traversed. The UPDATE message is the form of {AS_PATH, NEXT_ HOP, Network Layer Reachability Information (NLRI)}. The BGP router receiving the message appends its AS number to the UPDATE message. AS_PATH information prevents the infinite loop of UPDATE message traversal. The other attributes which affects the operation of BGP routing are LOCAL_PREF, ORIGIN, MED, AGGREGATOR, etc. Among these, the LOCAL_PREF and MED attributes are important in TE in BGP. But, those attributes and operations have some limitations to apply the end-to-end traffic engineering in BGP.

In this paper, we explain the limitation of traffic engineering in existing BGP. We, then propose the extended BGP algorithm (BGP$_{new}$) that affords an end-to-end traffic engineering capability to BGP. It doesn't change the existing BGP messages. By just slightly changing its operation to attributes, the end-to-end traffic engineering can be enhanced. After describing a simulation models and results, we have some concluding remarks.

2. The Lack of End-to-End Traffic Engineering Capabilities in BGP

Traffic Engineering is concerned with the performance optimization of operational networks. Its main objective is to reduce congestion hot spots and improve resource utilization across the network through carefully managing the traffic distribution inside a network [3]. BGP routing is generally based on destination address and uses simple metrics such as hop count (AS_PATH length) or delay / capacity for routing decisions. In this routing principle, the poor utilization of network resources can be occurred which is called "fish problem." This problem is primarily caused by two properties of current BGP. First, BGP routing is destination based: all packets whose destination addresses share the same prefix has the same next hop. Second, the decision making in current routing is based on local optimization – any node simply selects a path that is best from its own perspective.

Figure 1 shows the fish problem. The links between routers R_D-R_C and R_C-R_A has higher capacity than the links between routers R_D-R_B and R_B-R_A. Because the length of AS_PATH of two paths from R_D to R_A is same, the policy of R_D may be set to select the R_D-R_C-R_A path for the traffics whose destination is R_A. Even if connections between N_X – N_A wants to be routed via R_D-R_B-R_A path and connections between N_Y-N_A wants to be routed via R_D-R_C-R_A path, current BGP can not support such an end-to-end traffic engineering and load balancing. Router R_D will apply one policy for all the traffics with same destination, R_A. Even though BGP uses LOCAL_PREF or MED attribute for inbound traffic engineering, its effect is bound to local and so cannot set the global policy. Splitting the traffics with the same destination to the distinct paths is basically impossible.

Router R_A can split the outgoing traffics to the router R_X and to the router R_Y using different LOCAL_PREF values. But, R_D can not split the outgoing traffics destined to R_A. One solution for end-to-end traffic splitting is to have the router R_D record the source/destination information in its BGP routing table. Then it will be able to apply individual policies to each traffic. But this method requires for the existing BGP routers to be adjusted to the new algorithm, so it is very expensive solution.

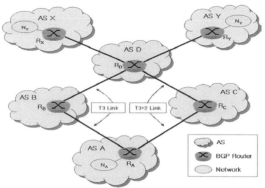

Figure 1. Fish Problem Topology

The new BGP routing algorithm proposed in this paper permits BGP router to transmit the traffic individually according to the link policy of the destination even if the traffic is sent to the same destination.

3. The New BGP Algorithm: BGP$_{NEW}$

3.1. The Extension of Routing Table

To embody the idea which we propose, extension of routing table is necessary, but we have embodied the end-to-end traffic engineering without any structural change of routing table and without the addition of new messages or the change of existing BGP messages. Our method is to send global UPDATE message from source router to destination router by slightly changing the meaning of AS_PATH attribute.

BGP standard specifies that AS numbers are assigned between the numbers from 1 to 65535. The "zero" value is not used as AS number. In our BGP$_{new}$ algorithm, if the source router wants to set the end-to-end traffic engineering policy, it inserts the AS number "zero" in AS_PATH field when it issues an UPDATE message. The source router makes an UPDATE message with "zero" and the destination network prefix and sends it to the only one path which it wants to receive the traffic from the destination. As the UPDATE message traverses the BGP routers, the AS numbers are appended in AS_PATH as normal way regardless of the existence of "zero." During this process, the BGP$_{NEW}$ routers receive the UPDATE message with "zero" field, it records its PATH information into routing table, and applies the maximum precedence weight regardless of the real metric value. Figure 2 describes the end-to-end traffic engineering process issued by the router R_A. R_A sends an UPDATE message {(AS A, (0, N_Y)), R_A, N_A} to the path 2 and {(AS A, (0, N_X)), R_A, N_A} to the path 1. Table 1 shows the eventual routing table for router R_D. The zero valued AS_PATH is appeared in AS_PATH field.

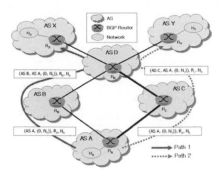

Figure 2. Global UPDATE message

Table 1. Routing Table for Router R_D

Destination Network	NEXT HOP	LOCAL_PREF	AS PATH	Etc.
N_A	R_B	100	AS B, AS A, (0, N_x)	....
N_A	R_C	100	AS C, AS A, {0, N_y}	....
N_A	R_B	100	AS B, AS A	
N_A	R_C	200	AS C, AS A	
N_x	R_x	100	AS X	
N_y	R_y	100	AS Y	
....				

3.2. New Routing Algorithm

When the traffic arrives to the BGP router, the following Routing algorithm should be executed.

A. Search the unanimous record(s) by comparing Destination Network field in its routing table and the destination of the arriving traffic.

B. Check if AS_PATH of searched record(s) includes "zero" value. If there is no "zero" then that is same as the BGP_{old} routing algorithm.

C. If there is a record which has AS number of "zero," then check if the prefix next to "zero" and packet source are unanimous. If they don't match each other, the remaining process is same with general BGP routing algorithm.

D. If they match, ignore the LOCAL_PREF value of the searched record, and give priority. If there are more than two unanimous, follow defaults setting (minimum AS_PATH).

The BGP_{NEW} algorithm can be operated with BGP_{OLD} algorithm without any problem in the network. If the BGP_{OLD} router doesn't recognize the meaning of the "zero," it will simply use the LOCAL_PREF as route selection criteria. If the router recognizes the meaning of the "zero," it will operate the BGP_{NEW} algorithm and so the good load balancing and end-to-end traffic engineering will be obtained. In the worst case, the operation will be same as a normal BGP network.

4. Simulation and Results

Simulation topology is as shown in Figure 3. It is similar to the fish problem topology. To estimate the performance of BGP_{NEW} algorithm, we compared the amount of packet loss and the average packet delay time when the router R_D is old-type and new-type. We used an SSFNet [4] based on Java as a simulation tool. Our assumptions are: (1) The link capacity of the path between router D, C, B and A is T3 * 2. (2) The other links have a capacity of T3. (3) Generated traffic is self-similar traffic [5] and the traffic source has packets to transmit at any time. (4) Buffer size of intermediate routers is 10 Mbytes each other. (5) The ASs V ~ Z generates the unlimited traffic continuously and the traffics are all destined to router A. (6) For simplicity, we assumed that the intermediate routers between router D and A doesn't generate their own traffic. They only play a role of BGP routing.

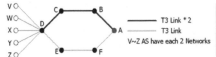

		Traffic generation	Routing	Traffic destination
——— T3 Link * 2	●	No	Yes	No
——— T3 Link	○	Yes	Yes	No
V~Z AS have each 2 Networks	●	No	Yes	Yes

Figure 3. Network Topology for Simulation

Figure 4 (a) and (b) are the results of the packet loss and packet delay comparison. In this figure, we can see that the suggested algorithm enhances a utilization of network resources and reduces the packet loss and delay when compared with a pure BGP.

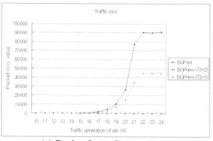
(a) Packet Loss Comparison

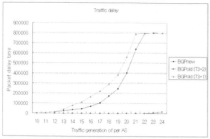
(b) Transmission Delay Comparison

Figure 4. Simulation Result

5. Conclusion

We proposed a new method ensuring the end-to-end traffic engineering capability in the BGP. This algorithm forces the ASs located at the outside of the Neighbor AS to cooperate with the intension of any BGP routers. The proposed BGP algorithm can be operated without any structural change in the existing routing table of the conventional BGP. It uses conventional UPDATE message without addition of the new type of message for the communication between the routers, and it does not have any problem in the interlocking with the other routers.

The proposed method has a less traffic losses than that of BGP_{old} algorithms, especially when the time–critical application traffics are exchanged in the BGP. In addition to that, the resulting traffic delay is far less than that of the conventional BGP.

References

[1] P. Traina, RFC 1965, "Autonomous System Confederations for BGP," Request for Comments 1655, June 1996.

[2] Y. Rekhter and T.Li, RFC 1771, "A Border Gateway Protocol 4 (BGP-4)," March 1995.

[3] Z. Wang, *Internet QoS: Architectures and Mechanisms for Quality of Service*, Morgan Kaufmann Publishers, San Francisco, 2001.

[4] http://www.ssfnet.org.

[5] A. Adas, "Traffic models in broadband networks," IEEE Communications Magazine, 35(7), pp.82-89, 1997.

VSP International
Science Publishers
P.O. Box 346, 3700 AH Zeist
The Netherlands

*Lecture Series on Computer
and Computational Sciences*
Volume 1, 2004, pp. 1173-1176

Real Time Face Detection in Mobile Environments

Yong-Hwan Lee[1], Yu-Kyong Lee[1], Young-Seop Kim[2], Sang-Burm Rhee[1]

[1]Dept. of Electronics & Computer Engineering, Dankook University,
147, Hannam-dong, Yongsan-gu, Seoul, 140-714, Korea
[2]Dept. of Computer Science & Electronics, Dankook University, Korea

Received 7 March, 2004; accepted in revised form 10 March, 2004

Abstract: We propose a new algorithm to implement fast and accurate search for the facial feature in a still color image without constraint background under mobile environments. To improve the performance of detection, Triangle-Square (TS) transformation, as a simple geometric mathematics, is applied. This is of benefit to much computation time are saved, with the evidence of our experimental results. We have similar ratio of face detection to other methods and fast speed, enough to use on real-time system for identification in mobile device. As the results, our proposed algorithm show that it will be more suitable for a real-time application in undefined environments, where varying lighting conditions with complex backgrounds, according to be fast and accurate than any other previous method.

Keywords: Face Detection, Mobile Environment, Skin Color Model, TS , Image Processing

1. Introduction

Digital contents usage by mobile device such as mobile camera phone or PDA equipped with camera has recently increased, and many researches for face verification have become more attractive than before, because of the increasing requirements for security and convenience of users. Since the face detection scheme is always the first step in the processes of face verification or authentication, its performance would be very important to make the performance of whole system better [1]. In case of recognizing a face for mobile device, the detected face should be acknowledged in consideration of variety of illumination variations and camera phone locations [2]. In addition to the accuracy, more important concern is the detection speed, to accept to the real-time system.

Many methods [1, 3, 4] have been published to solve the face detection in a single image, approached with the pattern recognition. These have been best summarized by [3], which classified face detection into four main methods: knowledge-based methods, feature-invariant approaches, template-matching and appearance-based methods. However, all of them have some weakness, and many face detectors [1, 3] deal with only frontal faces in well-constrained environments.

In this paper we propose a novel algorithm to implement fast and accurate search for detection of human faces in a digital still color images, obtained from the mobile device under unconstraint environments, using advanced skin color model and simple geometry mathematics, considered to performance of detection speed improved for real time processing. We transmit an image from the mobile device, and then detect the face by our detector. This process is more efficient to minimize the expensive operations in mobile device with constraint computational power (i.e., battery consumption), and to be independent of the mobile phone regardless of time and place.

2. Face Detection

We design a face detection application, composed of two major steps, depicted in Figure 1: step 1 is a pre-processing for lighting compensation on the input images, transmitted from mobile device, and step 2 is face detection in an image using *EyeMap* and *MouthMap* [4] and TS transformation.

---

[1] Corresponding author. E-mail: hwany1458@empal.com

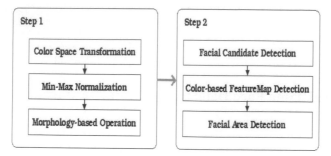

Fig. 1. Block Diagram of Face Detection

2.1. Detection of Skin Color Using *YCbCr*

A first step our scheme is color space transformation and lighting compensation. In this case, we have adopted to use a skin color-based approach using the *YCbCr* Model [3, 5].

The *YCbCr* color space is used widely in digital video, commonly used by European television [5]. In this color model, luminance information is represented by a single component, *Y*, and color information is stored as two color difference components, *Cb* and *Cr*. The transformation used to convert from *RGB* to *YCbCr* is as follow;

$$\begin{bmatrix} Y \\ C_b \\ C_r \end{bmatrix} = \begin{bmatrix} 16 \\ 128 \\ 128 \end{bmatrix} + \begin{bmatrix} 65.481 & 128.553 & 24.966 \\ -37.797 & -74.203 & 112.000 \\ 112.000 & -93.786 & -18.214 \end{bmatrix} \begin{bmatrix} R \\ G \\ B \end{bmatrix} \tag{1}$$

After transformed the color model, the illumination calibration is pre-requisite for the accurate face detection during the pre-processing, especially under mobile environment, where lighting condition and other image features such as distance and pose are un-predefined. So, we solve these problems by using a simple image processing approaches, considered by performance, which are called min-max normalization and histogram equalization [5]. These attempts to equalize the intensity values in a whole image and enhance the performance of the images, which brightness is secund into the one direction.

2.2. Eye/Mouth Detection and Face Boundary Map

After reducing the illumination impact to the brightness component, we attempt to extract the region of eye and mouth on a whole image using *EyeMap* and *MouthMap*, and detect the facial area using TS (Triangle-Square) transformation. We modify the approach of R.L.Hsu [4], which uses Hough transformation to extract the fitting ellipse for detection of the facial feature that wasted almost of processing time to detection. To improve the performance of detection, especially speed which is a strong requirement for real-time, a simple geometric mathematics, called TS transformation, is applied.

The eye region would be easily found due to its intrinsic feature, namely symmetry. So, the method we propose has a limitation whether both eyes are present in the skin region or not. As observed in [4], the eyes are characterized in the *CbCr* planes by a low red component and a high blue one; *EyeMap* transformation [4] is constructed by

$$EyeMap = \frac{1}{3}\left((\alpha \cdot C_b)^2 + \beta \cdot (\hat{C}_r)^2 + (\frac{C_b}{C_r}) \right) \tag{2}$$

where $(C_b)^2$, $(\hat{C}_r)^2$ and C_b / C_r all are normalized to the range [0,255], $\hat{C}_r$ is the negative of C_r (i.e., 255 - C_r) and α is greater than 1, β is less than 1 of positive constant which are emphasized to increase or decrease the color component because of Asian skin color, which generally have $R > G > B$ pattern.

The mouth is characterized by a high red component and a low blue one; *MouthMap* transformation [4] is constructed by

$$MouthMap = \left\{255 - \left(\alpha \cdot C_r - \beta \cdot C_b\right)\right\} \cdot \left(\alpha \cdot C_r\right)^2 \tag{3}$$

After extracting an eye-mouth triangle with the points of two eyes and one mouth, the final extracted face region is easily determined with TS transformation, described in Figure 2. Given an eye-mouth triangle $\Delta T_1 T_2 T_3$, we get the two line segments, D_{ee} and D_{em}, where D_{ee} is the distance between the two eyes, and D_{em} is the distance between the center point of the D_{ee} and the mouth.

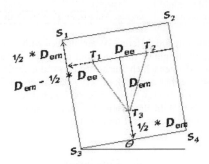

Fig. 2. Determination of Face Region

Now, we can compute the points of square $\square S_1 S_2 S_3 S_4$, using a simple vector calculation and scalar multiplication [6]. Notice that the square can be rotated according to tilted angle of the line segment connecting two eyes. We rotate the bounding box to angle θ, and it must be a shape with four straight sides of equal length forming four right angles.

3. Experimental Results

In experiment environment, we use machines with Pentium IV 1.8 GHz CPU, 512MB memory, running Windows XP Professional OS, SK Teletech's IM-7200 and KTFT's X3600 mobile camera phone as input device. And we implement with MATLAB 6.5 as programming language using QCIF (320x240) image size.

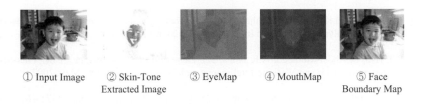

| ① Input Image | ② Skin-Tone Extracted Image | ③ EyeMap | ④ MouthMap | ⑤ Face Boundary Map |

Fig. 3. Some Experimental Results

Figure 3 demonstrates that our proposed algorithm could successfully work with some facial variations applied on the test data set. We evaluate our algorithm on several face images, including multiple faces of different lighting condition and shadows, different positions, different poses(i.e., be happy, be smiling and do nothing) and images with non-face.

To estimate the performance, shown in Table 1, our detector calculated in terms of detection rate (DR) and average run time, used to test on [4], to detect for faces in a still image.

Table 1. Face Detection Results

		Proposed Method
Total number of test images		200
Face Region Detection	DR(%)	94%
	Average Time (sec)	1.02
Facial Feature Detection	DR(%)	93%
	Average Time (sec)	2.38

We tested the implementation in 200 different Asians images with simple and complex background, indoor and outdoor. The reported time of the algorithm presented in Table 1 on a Pentium IV. The algorithm detects 188 of the 200 faces which means a successful detection rate of 94%. Among the 12 false dismissals cases, they were either very dark or very small to detect the face in an image.

On a personal computer, the average run time is 1.02 secs with detecting the face region and 2.38 sec with detecting the facial feature. With the results, we have similar ratio of face detection to other method and fast speed, enough to use on real-time system for identification or verification in mobile device.

4. Conclusion

In this paper we have presented a new face detection algorithm for color images, based on skin color detection to extract the face region from still images of the mobile device with camera phone. For the numerous problems of detection in mobile environments, where varying lighting conditions with complex backgrounds, the good results derived from lighting compensation and simple geometry mathematics in respect of speed with similar detection rate. This is to say that our proposed algorithm is more suitable for real time application such as identification for on-line financial , surveillance system and mobile multimedia processing, according to be fast and accurate than any other previous researches.

References

[1] C.C.Chiang, W.K.Tai, M.T.Yang, Y.T.Huang, C.J.Huang: *A Novel Method for Detecting Lips, Eyes and Faces in Real Time*, Real-Time Imaging 9, pp.277-287, 2003.

[2] D.Masip, J.Vitria, *Real Time Face Detection and Verification for Uncontrolled Environments*, 2nd COST 275 Workshop – Biometrics on the Internet Vigo, 25-26 Mar., 2004.

[3] M.H.Yang, D.J.Kriegman, N.Ahuja: *Detecting Faces in Images: A Survey*, IEEE Transactions on Pattern Analysis and Machine Intelligence, vol.24, no.1, Jan., 2002.

[4] R.L.Hsu, M.Abdel-Mottaleb, A.K.Jain: *Face Detection in Color Images*, IEEE Transactions on Pattern Analysis and Machine Intelligence, vol.24, no.5, pp.696-706, May, 2002.

[5] R.C.Gonzalez, R.E.Woods, S.L.Eddins: *Digital Image Processing using MATLAB*, Prentice Hall, 2004.

[6] Gilbert Strang, *Introduction to Linear Algebra 3rd Edition*, Wellesley-Cambridge Press, 2003.

VSP International
Science Publishers
P.O. Box 346, 3700 AH Zeist
The Netherlands

*Lecture Series on Computer
and Computational Sciences*
Volume 1, 2004, pp. 1177-1180

New Mobility Management for Reducing Location Traffic in the Next Generation Mobile Networks

Dong Chun Lee[1] , Jung-Doo Koo[2], Hongjin Kim[3]

[1]Dept. of Computer Science, Howon Univ., Korea
[2]Dept. of Computer Science, Hanyang Univ., Korea
[3]Dept. of Computer Information, KyungWon College, Korea

Abstract: In this paper, we propose a new location management scheme where the trace of terminals is left in the VLRs, so that a call can be connected by querying only to the VLRs rather than to the HLR when the terminal-terminated- call occurs. To estimate overall mobility management cost, the simulation model is based on the Jackson's network, and makes it possible to estimate mobility management cost of IMT-2000 networks. Considering both location registration cost and call tracking cost, the proposed scheme shows the improvement.

Keywords: location management, location registration, call tracking, IMT-2000 networks

1. Introduction

Two standards exist for carrying out two-level hierarchical strategies using Home Location Register (HLR) and Visitor Location Register (VLR) to support users with mobility in mobile communications. The standard commonly used in North America is the EIA / TIA Interim Standard 95 (IS-95), and in Europe the Global System for Mobile Communications (GSM)[2]. IS-95 and GSM have a structural drawback: as the number of users increase, HLR becomes the bottleneck.

A number of works have been reported to reduce the bottleneck of the HLR. A caching strategy is introduced in [3] which reduce the cost for call tracking by reusing the cached information about a called user's location from a previous call. In [4,5], a location forwarding strategy is proposed to reduce the signaling costs for location registration. A local anchoring scheme is introduced in [6]. Under these schemes, signaling traffic due to location registration is reduced by eliminating the need to report location changes to the HLR. Location update and paging subject to delay constraints is considered in [6]. Hierarchical database system architecture is introduced in [7]. A queuing model of three-level hierarchical database system is illustrated in [8]. These schemes can reduce both signaling traffics due to location registration and call delivery using the properties of call locality and local mobility. A user profile replication scheme is proposed in [6-8]. Based on this scheme, user profiles are replicated at selected local databases. If a replication of the called terminal's user profile is available locally, no HLR query is necessary. When the terminal moves to another location, the network updates all replications of the terminal's user profile.

In this paper we propose a new mobility management method, which effectively reduces mobility management cost in the 3G mobile networks.

2. The Proposed Scheme

To implement the proposed scheme, it is necessary for each terminal to store the IDs of the VLRs in the visited order. We assume each terminal has a queue of length where the VLR IDs are stored.

2.1 The Location Registration
1.   The terminal stores the ID of the new VLR into the tail of its queue. If the number of IDs stored in the queue exceeds , the terminal deletes the VLR ID stored in the head of the

---

[1] Corresponding author. E-mail: ldch@sunny.howon.ac.kr

queue in order to keep the trace length of the terminal   and generates a pointer cancellation (POCANC) message which will be sent to the VLR whose ID is just deleted from the queue.

2. The REGREQ message and the POCANC message (if exists) are sent to the new VLR.
3. The new VLR checks whether the terminal is already registered. If not, it sends the REGNOT message to the HLR and the POCANC message to the new VLR to the old VLR.
4. The HLR sends a REGCANC message including the ID of the new VLR to the old VLR.
5. The old VLR generates a pointer composed of the terminal ID and the ID of the new VLR.

2.2 The Call Tracking

If the VLR of the caller does not contain the pointer to the callee, the call tracking process is performed as follows:

1. The VLR of the caller is queried and a pointer to the callee is found. The VLR sends a ROUTREQ message including its ID to the pointed VLR.
2. The VLR which received the ROUTREQ message sends this message to the next VLR until the pointer ends.
3. If the callee is found in the VLR where it currently resides, a TLDN is assigned to the callee and sent to the VLR of the caller by using the VLR ID included in the ROUTREQ message.
4. The call is established between caller and callee using the TLDN.

3. Analytical Model

To evaluate this end-to-end delay, we treat the database system as Jackson network. The service time of each database operation is assumed to be a major delay. We assume that there are n VLRs and one HLR in the system. The HLR is assumed to have an infinite buffer and single exponential server with the average service time $\frac{1}{uh}$. Likewise, the VLR is assumed to have an infinite buffer and single exponential server with the average service time $\frac{1}{uv}$. We assume that within a RA, the location registration occurs in a Poisson distribution with rate λ_u and the call origination occurs in a Poisson distribution with rate λ_c. λ_{lr} and λ_{tt} represent the average arrival rate of REGCANC message and the average arrival rate of ROUTREQ message, respectively. λ_h represents the average arrival rate of messages to the HLR from other VLRs. P_{vo} is the probability the departure message from the VLR leaves the system. P_{vh} is the probability the departure message from the VR enters the HLR. P_{hv} is the probability the departure message from the HLR enters one of n VLRs. From the definition of λ_{lr} and λ_{tt}, we know that these messages get out of the system after going through the VLR. λ_u enters the VLR in the form of REGREQ message, and after receiving services from the VLR it is delivered to the HLR in the form of REGNOT message. After receiving services from the VLR, $\frac{1}{n}\lambda_c$ get out of the system because the probability the callee is in the same RA with the caller is $\frac{1}{n}$. In the proposed scheme, we denote these messages by λ_{ovt} and λ_{ovrc}, respectively.

Here λ_{ovt} denotes the average arrival rate of the ROUTREQ message from other VLRs. λ_{ovrc} denotes the average arrival rate of the POCANC message from other VLRs.

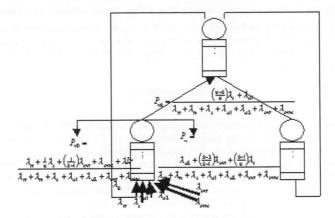

Fig. 1 Jackson network modeling the proposed scheme.

Let W'_v and W'_h denote the average system time in the VLR and in the HLR, respectively, From the Little's result [16], W'_v and W'_h are:

$$W'_v = \frac{1}{\mu_v - \left(\lambda_{tt} + \lambda_{lr} + \lambda_c + \lambda_{u1} + \lambda_{u2} + \lambda_{ovt} + \lambda_{ovrc}\right)} \tag{1}$$

$$W'_h = \frac{1}{\mu_h - n\left(\dfrac{n-k}{n}\lambda_c + \lambda_u\right)} \tag{2}$$

We define the costs for mobility management as follows.

● **Mobility Management Cost**
  Location registration rate × location registration cost + call tracking rate × Call tracking cost.

● **Mobility Management Cost in IS-95 Scheme**
  W_{IS-95M} (Mobility management cost) $= \frac{\lambda_u}{\lambda_c + \lambda_u} \times W_{IS-95L} + \frac{\lambda_c}{\lambda_c + \lambda_u} \times W_{IS-95T}$

● **Mobility Management Cost in the Proposed Scheme**
  W_{PSM} (Mobility management cost) $= \frac{\lambda_u}{\lambda_c + \lambda_u} \times W_{PSL} + \frac{\lambda_c}{\lambda_c + \lambda_u} \times W_{PST}$

To get numerical results, we use the same values of system parameters as those in [4] for example. From these parameters, the average occurrence rate of location registration in an RA, λ_u, is calculated as the performance parameters.

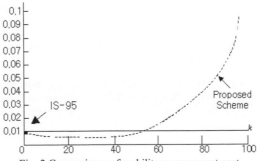

Fig. 2 Comparisons of mobility management cost

We assume that the average service rates of HLR and VLR μ_h are =2000/s, μ_v =1000/s. Fig.2 shows the value of W_{PSM} against k, and compares it with $W_{IS-95\,M}$. W_{PSM} becomes smallest when k=20. The costs for using the proposed scheme show that the proposed scheme reduces the mobility management cost by 34% percent of that of IS-95. It also shows the proposed scheme reduces the average number of arrival messages to the HLR.

4. Conclusions

In this paper we proposed a new mobility management method with the objective of improving IS-95 standard. The proposed scheme performs better than IS-95 because it efficiently utilizes the characteristics of the third generation mobile networks as follows:
Since it is expected that everyone have one terminal in the third generation mobile communications, an enhanced mobility management scheme to reduce the bottleneck of the HLR, should be used in the third generation mobile communications. The proposed scheme can connect a call by querying to only the VLRs when terminal-terminated-call originates in the RA whose VLR has the trace of the callee terminal. And so, the proposed scheme results in distributing messages from the HLR to the VLRs.
Since many calls are likely to be local [6], we could increase the call origination probabilities in the RAs where the trace of the callee terminal is left. Then, the proposed scheme would produce a better result than what we have obtained. As the proposed scheme is based on IS-95, it can be easily implemented with the small modification to the protocol of IS-95.

Acknowledgments

This work was supported by The Korea Sanhak Foundation, and ITRC, 2004.

References

[1] F. Akyildiz, J, McNair, J, Ho, H. Uzunalioglu, W. Wang, "Mobility Management in Current and Future Communication Networks", IEEE Network, Jul./Aug., 1998, pp.39-49

[2] EIA/TIA IS-95.3, "Cellular Radio Telecommunications Intersystem Operations", Technical Report (Revision B), July 1995.

[3] R. Jain, Y. B. Lin, C. N. Lo. and S. Mohan, "A caching Strategy to Reduce Network Impacts of PCS", IEEE Jour. on Selected Areas in Comm., Vol. 12, No. 8, 1994, pp. 1434-1445

[4] J. S. M. Ho and I. F. Akyildiz, "Local Anchor Scheme for Reducing Location Tracking Costs in CNs", Proceeding of ACM MOBICOM'95, Nov. 1995, pp. 181-194

[5] J. Z. Wang, "A Fully Distributed Location Registration Strategy for Universal Personal Communication Systems", IEEE Personal Comm., First Quarter 1994, pp. 42-50

[6] C. Eynard, M. Lenti, A. LOmbardo, O. Marengo, S. Palazzo, "Performance of Data Querying Operations in Universal Mobile Telecommunication System (UMTS)", Proceeding of IEEE INFOCOM '95, Apr. 1995, pp. 473-480

[7] Y. Bing Lin and S. Y. Hwang, "Comparing the PCS Location Tracking Strategies", IEEE Trans. Vehicle. Tech., Vol.45, No. 1, Feb. 1996

[8] R. Jain and Y. -B. Lin, "An Auxiliary User Location Strategy Employing Forwarding Pointers to Reduce Network Impacts of PCS", ACM-Baltzer J. of Wireless Network, July 1997

VSP International
Science Publishers
P.O. Box 346, 3700 AH Zeist
The Netherlands

*Lecture Series on Computer
and Computational Sciences*
Volume 1, 2004, pp. 1181-1184

Automatic Identification of Road Signs for Mobile Vision Systems

Hong-Gyoo Sohn[1], Yeong-Sun Song, Gi-Hong Kim, Hanna Lee, and Bongsoo Son[2]

[1]School of Civil and Environmental Engineering, Yonsei University, Seoul, Korea
[2]School of Urban Planning and Engineering, Yonsei University, Seoul, Korea

Abstract: With the growing need for vision-based vehicle guidance system, this paper is concentrated on gathering information about road signs. A robust algorithm for automatic detecting and identifying road signs from color images of outdoor road scenes is proposed. Usually, urban road and roadsides are so congested and colorful that we need to try more complicated method which considers line and color information in images. The proposed detection algorithm includes edges and color region extraction from images and the Hopfield neural networks. The identification algorithm uses seven invariant moments and parameters that present geometric characteristics. With this combined method we could successfully detect and identify the road signs.

Keywords: Mobile Vision System, Road sign, Hopfield neural networks

1.   Introduction

Many intelligent vision systems have been developed for driver support system and traffic automation. They have many applications such as road center line detection, obstacle detection, traffic control and analysis, license plate finding and reading, and toll correction. Road signs provide drivers with very valuable information about the road for traffic safety. This study is concentrated on automatic detection of road signs and identification of their contents from color road images taken with Charge-Coupled Device (CCD) cameras mounted on a moving vehicle at regular interval.

Road signs are designed and erected at specific points to be easily recognized by drivers. Different techniques have been proposed for detecting road signs from terrestrial images. Habib [2] used color information to extract road signs and tried to identify them with detecting corner points. Corner points are, however, very sensitive to noise and may be difficult to apply to curved or complex shapes. Hsu and Huang [3] proposed a road sign detection and recognition system. They used template-matching methods to detect the road signs, and used matching pursuit filters to recognize. Shape detection methods based on template-matching is very sensitive to image scale and is much slower than methods based on image edges. Urban roads are generally congested and roadsides are unbroken successions of a lot of colorful signboards. With simple algorithms, it was difficult to extract road signs from images. That is why we tried more complicated method which uses line information together with color information. A wide experimentation on a lot of images taken in urban environments indicates that proposed approach is effective and useful for the realization of driver support system.

2.   Computational Procedure

Figure 1 schematically depicts the procedure of detecting and identifying road signs. As the diagram shows the algorithms start with the color road imagery and end up with the identification of road sign. The first step is to detect the road sign. In this step, two parallel processes proceed concurrently to give two input sources, line information and color information, to the neural networks. Then identification of its contents is conducted using seven invariant moments and geometric characteristics of the road signs. Road signs have several types of shape. In this study we concentrated on the identification of triangular-shaped road signs, since it give early "warning" information to the drivers.

---

[1] Corresponding author. E-mail: sohn1@yonsei.ac.kr

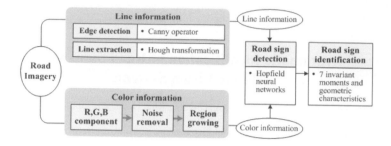

Figure 1 Schematic diagram for detection and identification

3. Road Sign Detection

After color imagery was transformed to gray scale imagery, Canny operator was used to detect edges from road images. Modified Hough transform was applied to extract line components from the detected edges. Modified Hough transform differs from the original Hough transform in that it keeps the coordinates of the start and end point of the detected lines to distinguish lines on the same route. Lines shorter than predefined threshold are eliminated.

Figure 2 shows an example of the line extraction procedure. Figure 2 (a) shows the original color image obtained in the vehicle. Figure 2 (b) represents the extracted edges after applying Canny operator. Figure 2 (c) shows the extracted lines after applying modified Hough transform to Figure 2 (b). In this example, sixty-seven lines remained after the whole line extraction procedure and those line segments are input data for Hopfield neural networks to detect outline of road signs.

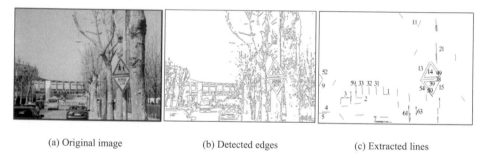

| (a) Original image | (b) Detected edges | (c) Extracted lines |

Figure 2 Results from edge detection and line extraction

A color image can be decomposed into the three R, G, and B components. In other words a specific color can be obtained by mixing R, G, and B components. By assigning intensity ranges to each band, we can extract a specific region from the color images. Unfortunately the condition of a road images is not consistent. The brightness of the color can be influenced by the weather condition, time of the day, shadows, and so on. Sometimes the color of a certain road sign can be faded away. All these conditions make it difficult to extract interesting regions satisfactorily. So, we used a ratio between each color component to adapt to these changing conditions. The color threshold has the following expression:

$$b(i,j) = \begin{bmatrix} 1 & for & \begin{cases} \theta_{1(RG)} \le R(i,j)/G(i,j) \le \theta_{2(RG)} \\ \theta_{1(GB)} \le G(i,j)/B(i,j) \le \theta_{2(GB)} \\ \theta_{1(BR)} \le B(i,j)/R(i,j) \le \theta_{2(BR)} \end{cases} \\ 0 & for & \qquad\qquad otherwise \end{bmatrix} \qquad (1)$$

where, i, j are the row and column number of the image, $b(i, j)$ is the binary value of the pixel at the given coordinates, and $R(i, j)$, $G(i, j)$, and $B(i, j)$ are the functions that give R, G, B levels of each pixel of the image respectively. θs are thresholds of the color ratio. We extracted regions by applying Equation (1) to color image. However, the result image still had lots of noises or excluded wanted region. With this problem, we suggest a step-by-step solution which includes noise removal and region growing steps.

At first, we extracted color region by applying strict tolerance to an image. Then noise removal is accomplished by using a 7x7 modal filter. The filer calculates the number of white pixels versus the number of black pixels. When the total number of white pixel is smaller than the threshold, the pixel is regarded as noise and is removed. After noise is removed, a region growing algorithm is used to enlarge homogeneous regions using higher tolerance than that used in the previous step. Even if the road sign has partially lost color, the proposed method can extract the desired region successfully.

Road signs in images can be modeled and detected by the network with neurons formed by line edge features from the model and road images [4]. The triangular road sign model is represented as two intersecting edges. We used five measuring features defined in Figure 3. Measuring features include angles of lines α_1, α_2, gradient G_1, G_2 (calculated from extracted color regions), and distance D between the upper-right point of the left edge and the upper-left point of the right edge. Equation for interconnection strength between neuron (i, k) and neuron (j, l) is as follows;

$$C_{ikjl} = W_1 \cdot F(\alpha_{Mi}, \alpha_{Ik}) + W_2 \cdot F(\alpha_{Mj}, \alpha_{Il})$$
$$W_3 \cdot F(G_{Mi}, G_{Ik}) + W_4 \cdot F(G_{Mj}, G_{Il}) + W_5 \cdot F(D_{MiMj}, D_{IkIl}) . \tag{2}$$

where, M is the number of model edges and I is the number of edge features in the image. Equal weight factors are applied to our model ($W_1=W_2=W_3=W_4=W_5=0.2$). Note that $F(x, y) = 0$ if $i=j$ or $k=l$.

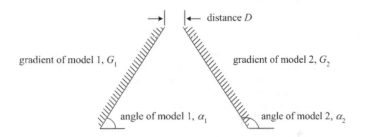

Figure 3 Model of triangular-shaped road sign

The identification of road sign edges subdivides the process into five stages: (1) creating a matrix with a dimension of M by I; (2) Setting the initial neuron states; (3) Building the C coefficients; (4) Updating the values of neuron states iteratively; (5) Calculating the final states. The states of neurons (extracted lines) which match the conditions of the model converge to 1 and otherwise to 0. We concluded that line number 14 and 49 (in Figure 2c) are road sign edges. We confirmed that we could efficiently detect triangular-shaped road signs using Hopfield neural networks. After finding the location of a road sign, we proceeded to identify the contents of road sign.

4.   Road Sign Identification

Our goal is to identify the contents of road signs. For this, we applied seven invariant moments [1] at first. To verify this method, one hundred seven images of triangular-shaped road signs were acquired at various sites. The road signs are detected and extracted from images through above procedure. The verification of seven invariant moment's method showed lower accuracy than expected. One of the plausible reasons is that each member of the same class has different shape. Even though each road sign has its own standard dimension, the design seems to vary from manufacturer to manufacturer. To improve the accuracy of identification of road signs, we devised geometric characteristics of road signs.

They are difference between the coordinates of the center and those of the gravity center, the degree of vertical direction, density, and the density of the holes in the content area. The hole is defined as black pixels surrounded by white pixels. After seven invariant moments and geometric characteristics are calculated from a road sign image, the road sign is classified into the class which has the minimum distance d_i. The distance between reference values of road sign classes and those of input road sign is calculated in Equation (3).

$$d_i = \sum_{j=1}^{7} \frac{\left| m_{ij(\varphi)} - \varphi_j \right|}{\sigma_{ij(\varphi)}} \quad + \quad \sum_{j=1}^{7} \frac{\left| m_{ij(g)} - g_j \right|}{\sigma_{ij(g)}} \tag{3}$$

where i represents the class number of road signs, is the number of invariant moments or geometric characteristics. φ_j and g_j are the jth invariant moment and geometric characteristic of the input road sign, m is the mean of the invariant moments or geometric characteristics of reference road signs, and σ is standard deviation. To analyze the accuracy of suggested identification method, an error matrix was built. When seven invariant moments were only used, the overall accuracy was 0.804. Additional consideration of the geometric characteristics brings the improvement of overall accuracy to 0.991.

5.   Conclusion

We have proposed a robust algorithm to detect and identify road signs from congested and colorful imagery of urban roads. In this sturdy, edge extraction, color threshold, Hopfield neural networks, seven invariant moments, and geometric characteristics of the road signs were used. Color region extraction is one of the most important steps in whole processes. This step can improve not only the accuracy of the road sign detection but also that of the identification of road sign contents. We proposed the use of color ratio, which was more efficient than using the range of each R, G, and B color component. Noise removal and region growing method were efficiently applied to extract the road sign in a variety of conditions. Outline of road signs could be exactly extracted from images in one pixel resolution using Hopfield neural networks. If the GPS, INS, and stereo vision system are mounted on a moving vehicle, the proposed algorithm can be used to point out the accurate location of road signs. The identification of the road sign contents is greatly influenced by small variations in their shapes. The seven invariant moments for identification of contents of road signs are not effective in some cases, but with consideration of the additional geometric constraints the road signs selected for the test were successfully identified.

References

[1] Dai, X. and S. Khorram, A feature-based image registration algorithm using improved chain-cod representation combined with invariant moments, _IEEE Transactions On Geoscience and Remote Sensing_, 37(5), 2351-2362(1999).

[2] Habib, A.F., R. Uebbing, and K. Novak, Automatic Extraction of Road Signs from Terrestrial Color Imagery, _Photogrammetric Engineering & Remote Sensing_, 65(5), 597-601(1999).

[3] Hsu, S.H., and C.L. Huang, Road sign detection and recognition using matching pursuit method, _Image and Vision Computing_, 19, 119-129(2001).

[4] Li, R., W. Wang, and H.Z. Tseng, Detection and Location of Object from Mobile Mapping Image Sequences by Hopfield Neural Network, _Photogrammetric Engineering & Remote Sensing_, 65(10), 1199-1205(1999).

VSP International
Science Publishers
P.O. Box 346, 3700 AH Zeist
The Netherlands

*Lecture Series on Computer
and Computational Sciences*
Volume 1, 2004, pp. 1185-1188

Mobile-Based Healthcare B2B Workflow System for Efficient Communication

Sang-Young Lee[1], Cheol-Jung Yoo[1], Dong Chun Lee[2]

[1]Department of Computer Science, Chonbuk National Univ., Korea
[2]Department of Computer Science, Howon Univ., Korea

Received 7 March, 2004; accepted in revised form 10 March, 2004

Abstract: The recent push for healthcare reform has caused healthcare organizations to focus on ways to streamlined processes in order to secure high quality care as well as reducing costs. Healthcare enterprises involve complex processes that span diverse groups and organizations. These processes involve clinical and administrative tasks, large quantities of data, and large number of patients and personnel. We propose the mobile-based workflow system of passable communication as an important factor in the B2B healthcare. Based on the above proposal the workflow system of business process was designed and implemented on the basis of Java, UML and XPDL.

Keywords: Mobile, workflow system, healthcare B2B, communication

1. Introduction

Business process means tasks, limited conditions and a serial processes including resources to achieve the goal. The business process is important because it enhances the effective activities of organization through the structure provided 1, 2, 3]. The most important information technology in the business process is the workflow consisted of process, information and organization as basic elements. And Workflow Management System (WfMS) instruct, mediate and control the tasks to be done successfully [4, 5].

The healthcare business process applied as shown in this paper consisted of properties related to complicated and diverse organizations. Especially, business processes change dynamically according to the behaviors of each healthcare sectors [6]. Moreover, for the case of hospital material purchasing task the connection forms between supplier and hospital are established through many wholesale merchants participated. Currently, by introduction of e-commerce system brought the hub site applied healthcare B2B [7]. Efforts focused on the workflow introduction for the entire processes optimized to enhance productivity. In this paper, for the process of hospital materials purchasing, healthcare B2B workflow system was implemented. The modeling tool proposed enables important communication in the process of hospital materials purchasing. And the modeling tool was applied for the mobile-based workflow engine.

2. Healthcare B2B Workflow Modeling Tool

The healthcare B2B workflow demands an effective working environment to define and accomplish the diverse business processes generated among sectors. A workflow management system supports the definition and execution of workflow processes that model business applications through a coordinated set of process activities. A process activity may be a manual activity or an automated activity. A

---

[1] E-mail: {sylee230, cjyoo}@chonbuk.ac.kr

manual activity can be represented as a work item in a work list pending completion by a workflow participant. In this paper, the specification [8, 9] of WfMC based workflow participants were divided into manual and automated process. The workflow modeling method on the basis of this conception has a structure divided into two layers of task and actor (see Figure 1).

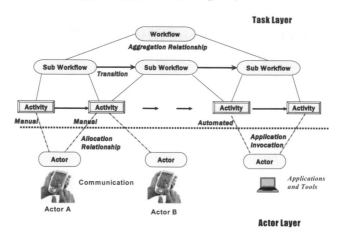

Figure 1: Structure of Workflow Modeling

The workflow modeling method adopted in this study is hybrid concept of the activity-based modeling method and the communication-based modeling method. An advantage exists in practical possibility of healthcare B2B by systematic combination of the mobile-based task and information flow. Referring some studies on workflow pattern [10] the present paper was proposed focusing on the communication pattern [see Table 1].

Table 1: Communication Patterns

Pattern		Notation	Description
Communication Pattern	Static Configuration	A → B → C	Information passed unilateral direction between activity, no feedback is necessitated
	Dynamic Configuration	A → B → C	Process progressing between activities bilaterally, feedback is necessitated.
	Multi-Dynamic Configuration	D / A → B → C	Process progressing between activity, bring about information communications and the information from another section of activity is required.

In this paper, UML was expanded and diagrammatized to apply for business progress and in respect of workflow process. The workflow progress definition language XPDL was used and defined for the exchange of process. The healthcare B2B workflow modeling method implemented with above mentioned are as shown in the Figure 2.

3. Mobile-Based Workflow Engine for Healthcare B2B

The mobile-based workflow engine is used to manage the workflows or business processes created by the business process modeling tool and simulate the executions of them.

The main functions of the system include:

- The user management including new user registry and user login.
- The project management includes project creation and display of projects information.
- The simulation of the workflow or business process execution.
- Automatic Data storage function

1) User login

The "User login" Window will turn up every time when a user initiates the system. The user must input the proper user name and password to login before using the system. The following figure 3 shows the UI of "User login".

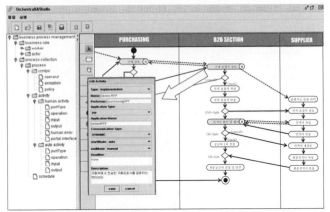

Figure 2: Workflow Modeling Tool for Healthcare B2B

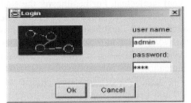

Figure 3: User Login Window

2) Project management function

- Project creation

After a user enters the system, the user can create a project related to a workflow or a business process. The following figure 4 shows the main UI of the system after a user enters the system.

Figure 4: Main UI of the System

-Display of projects information

After a user login to the system, the user can check up the all projects information by selecting "Projects Information menu item in the "Manger" menu. Then an "All Projects Information" dialog will turn up (see figure 5). The dialog shows the information of all projects created in the system. There are four columns in the table of the dialog, which are "Projects", "State", "Activities" and "Users". The Activity List will list all activity names and the current states associated to the project which the user selects in the table.

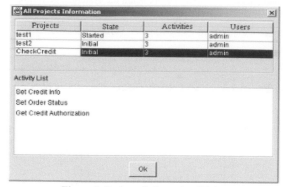

Figure 5: Projects Information Dialog

4. Conclusion
5.

Currently, the principal infrastructure has seldom provided for the effective operation of B2B to drive healthcare business, moreover, the purchasing process between hospital and supplier in electronic procurement system has been found ineffectual. For healthcare, the successful human communication creates superb work efficiency, and provides clients satisfaction consequently.

This paper exploited the mobile based healthcare B2B workflow system for the effective communication. For the first, the workflow modeling method originated from WfMC, the application method of property and the pattern of workflow was proposed. In addition, in the field of healthcare which requires a sound communication, the communication pattern was emphasized to create an efficient workflow modeling. The mobile-based workflow engine was proposed for the workflow process modeled to be exchanged and executed. For the optimum process in the material purchasing business, an efficient communication could be provided by the workflow system in the field of health as shown in this paper.

References

[1] Jutla D., Making Business Sense of Electronic Commerce, IEEE Computer, 32(5), 67-75(1999).
[2] Bussler C., B2B Protocol Standards and their Role in Semantic B2B Integration Engines, IEEE Computer Society, 24(3), 35-43(2002).
[3] Ruth Sara Aguilar-Savlen, Business Process Modeling: Review and Framework, International Journal of Production Economics, 50(5), 235-256(2003).
[4] Layna Fischer, 2003 workflow handbook. Workflow Management Coalition, 2003.
[5] Akhil Kumar, Leon Zhao, Workflow Support for Electronic Commerce Applications. Decision Support Systems, 32(4), 265-278(2002).
[6] Shrivastava S. K, Wheater S. M., Architectural Support for Dynamic Reconfiguration of Distributed Workflow applications, Proc. of IEEE Software Engineering, 155-162(1998).
[7] E. Maij, V.E. van Reijswoud, P.J. Toussaint, E.H. Harms, J.H.M. Zwetsloot-Schonk, A Process View of Medical Practice by Modeling Communicative Acts, Methods of Information in Medicine, 39(1), 56-62(2000).
[8] WFMC-TC-1011, The Workflow Reference Model, Workflow Management Coalition Brussel. Belgium, 1999.
[9] WfMC-TC-1012, Workflow Standard - Interoperability Abstract Specification, Workflow Management Coalition, Winchster. United Kingdom, 1999.
[10] Stephen A. White, Process Modeling Notations and Workflow patterns, The Workflow Management Coalition Terminology & Glossary, 1999.

VSP International
Science Publishers
P.O. Box 346, 3700 AH Zeist
The Netherlands

*Lecture Series on Computer
and Computational Sciences*
Volume 1, 2004, pp. 1189-1192

Intelligent All-Way Stop Control
System at Unsignalized Intersections

Sang K. Lee[1] , Ph.D.
Infrastructure and Construction Economics Research Division,
Korea Research Institute for Human Settlements, Korea

and
Young J. Moon[2], Ph.D.
Advanced Transportation Technology Research Center,
The Korea Transport Institute, Korea

Abstract: This study introduces intelligent all-way stop control system utilizing intelligent transport systems (ITS) technologies to enhance efficiency and safety at unsignalized intersections. An operational and functional concept of intelligent all-way stop controlled intersection is demonstrated based on the intelligent system utilizing loop detectors, LED traffic signs, and a controller specifically designed for right of ways assignments. A prototype and warrants related to traffic volumes for the implementation of intelligent all-way stop control system are presented in this study with a detail procedure and guidelines for the revisions of the roadway and/or traffic law in Korea. Findings from this study suggest that the proposed intelligent all-way stop control system would be implemented in improving both efficiency and safety at unsignalized intersections with the goal of higher driver compliance towards stopping at the intersection and having the right of ways.

Keywords: intelligent transport systems, unsignalized intersections, all way stop control, two way stop control, and right of ways

1. Introduction

Unsignalized intersections, the most common intersection type give no positive indication or control to the driver. Although their capacities may be lower than other important intersection types, they do play an important role in the control of traffic in urban street networks. The driver approaching the unsignalized intersection must decide when it is safe to enter the intersection by looking for a safe opportunity or "gap" in the traffic to enter the intersection. At unsignalized intersections a driver must also respect the priority of other drivers or the right of ways. There may be other vehicles that will have priority over the driver trying to enter the traffic stream and the driver must yield to these drivers. According to the traffic flows and geometry conditions at the unsignalized intersection, three types of control, i.e. two-way stop control (TWSC), all-way stop control (AWSC), and roundabout are typically applied to provide drivers with the right of ways (or priority) to enter the intersection. Under certain traffic volume and geometric conditions, AWSC provides safe and efficient traffic control compared to TWSC and signalized intersections [1]. However, unsignalized intersections in Korea have been operated uncontrolled for a long time, which means no existence of control devices and/or signs for drivers to stop at the intersections and have or yield the right of ways [2].

This study introduces intelligent all-way stop control utilizing intelligent transport systems (ITS) technologies to provide the right of ways for enhancing efficiency and safety at unsignalized intersections.

---

[1] Research Fellow. Korea Research Institute for Human Settlements. E-mail: sklee@krihs.re.kr
[2] Corresponding Author. Research Fellow. The Korea Transport Institute. E-mail: yjmoon@koti.re.kr

2. Analysis of Unsignalized Intersections

Sites Selection and Data Collection

For the analysis of the capacity, service time, delay, etc. at unsignalized intersections, five sites were selected based on the geometry, where the proposed system would be able to be implemented. Selected sites are four-leg intersections with one approach lane allowing through, left turn, and right turn traffics. Stop signs and/or stop bars are installed as a guidance system for drivers to stop first at the stop bar before entering the intersections. Data was collected at the sites during 2 hours each for rush and non-rush traffic conditions for 7 days in a week. Table 1 shows the volume and flow rate collected at the five single lane AWSC intersections.

Table 1: Volume and Flow Rate Collected at the Intersections

Characteristics	Site A	Site B	Site C	Site D	Site E
Volume (vph)	482	614	970	674	482
Flow Rate (vph)	574	654	1032	732	548
Volume Split	63:37	75:25	80:20	84:16	88:12

As shown in Table 1, all sites can be categorized as unsignalized intersections with major and minor roads based on the volume splits on each direction. It is noted that there is no intersection with the conditions where traffic volume is equal to capacity.

Average Service Times

Average service time means the time required for a vehicle to complete the service due to traffic conflict at the intersections. Average service time would be determined as follows:

$$t_s = h_d - t_m \tag{1}$$

where

t_s = average service time (sec);
h_d = departure headway (sec); and
t_m = move-up time (sec).

Average service time calculated according to the collected data of average delay ranges 2.0 – 7.4 seconds with over 90% of confidence level.

It is indicated that service time is generally proportional with traffic volume at the intersections. Site C shows the highest service time compared that of other sites with respect to the traffic volume as shown in Table 1. In order to analyze the service time more detail, it needs to be divided by each direction at the intersections, i.e. major and minor roads. At the Site C, the average service time of east-west bound on major road would 2.35 seconds. However for the south-north bound on minor road, average service time was 8.76 seconds. This means that service time in approach lane with high volume was lower than that with low volume. This would be a typical driving pattern at uncontrolled intersections in Korea that drivers would not be expected to show compliance with the traffic signs at the intersections with control devices and related rules of right of ways.

Traffic conflicts at the intersections are investigated for analyzing average service time and capacity. Four cases of traffic conflicts are considered by the degree of traffic conflict which means the existence of conflict vehicles at other approaches when a vehicle stops and waits for right of ways. They are:

Case 1: No conflict vehicle at any other approaches;
Case 2: One conflict vehicle at the opposing approach;
Case 3: One or two conflict vehicles at left and/or right approaches; and
Case 4: Three conflict vehicles at all other approaches.

In terms of the degree of traffic conflict at the intersections, average service time with high degree of conflict at all approach lanes is 8.9 seconds in Case 4. However, it is decreased to be 1.3 seconds with no traffic conflict in Case 1 at any leg in the intersection.

Service Times at Unsignalized Intersections

As discussed previously, the average service times analyzed from the data at each intersection ranged 3.8 sec to 7.4 sec excepting the Site E. These values are based on the typical driving behavior at unsignalized intersections in Korea. For the purpose of comparing these values with the service times from highway capacity manual [4, 5], an assumption can be made that the sites discussed in this paper are operated by AWSC. The service time can be estimated in terms of control delay of the intersection based on control type, i.e. TWSC and AWSC and driving behavior at the intersection.

Control delay includes initial deceleration delay, queue move-up time, stopped delay, and final acceleration delay. For TWSC, the control delay is as following:

$$d = \frac{3600}{c_{m,n}} + 900T \left[\frac{v_n}{c_{m,n}} - 1 + \sqrt{\left(\frac{v_n}{c_{m,n}} - 1 \right)^2 + \frac{\left(\frac{3600}{c_{m,n}} \right)\left(\frac{v_n}{c_{m,n}} \right)}{450T}} \right] + 5 \tag{2}$$

where
  d = control delay (sec/veh);
  v_n = flow rate for movement n (vph);
  $c_{m,n}$ = capacity of movement n (vph); and
  T = analysis time period (h) (T = 0.25 for a 15-min period).

For AWSC, the control delay is:

$$d = t_s + 900T \left[(x-1) + \sqrt{(x-1)^2 + \frac{h_d x}{450T}} \right] + 5 \tag{3}$$

where
  d = average control delay (sec/veh);
  x = degree of utilization (vh_d/3600);
  t_s = service time (sec);
  h_d = departure headway (sec); and
  T = length of analysis period (h).

Highway capacity software was utilized to calculated service times with the control delay defined in the equation (3) based on the assumption of AWSC, i.e. Site A 2.70, Site B 3.10, Site C, 3.40, and Site D 2.70 sec. These service times expected according to the assumption of AWSC are much lower than those values obtained from the data analysis of the sites. It is shown that AWSC intersections should be more effective on service times than uncontrolled intersections. This indicates the implementation of AWSC system would be recommended at uncontrolled intersections to improve the efficiency in terms of service times.

3. Intelligent All-Way Stop Controlled System

Concept of Intelligent AWSC System
A proposed intelligent all-way stop controlled system is an integrated system of hardware and software. Hardware includes four loop detectors for sensing the stopped vehicle and the order of stopping in each leg, a stop/go sign controller to allocate the priority of right-of-way by the rule of FIFO, and four LED sign board to display "STOP" and "GO" signs by the control signal from the controller. Software represents an algorithm and related programs for the controller to be able to collect the detector data of vehicles' stopping, to make a judgment of what order the vehicles stopped, and to display "GO" sign on each sign board by the priority from the controller.

Warrants for Intelligent AWSC System
Since unsignalized intersections in Korea have been operated uncontrolled due to the driver behavior as indicated previously, no efforts has been fulfilled on the study for establishing warrants of the installation of all-way stop control signs.

Warrants for unsignalized intersections can be derived from that for signalized intersections. According to the MUTCD (6), the minimum vehicular volume warrants for signalized intersections with one approach lane each is satisfied when the traffic volume for each of any 8 hours of an average day is 500 vph on major road and 150 vph on minor road. It is indicated that the values of vehicular volume warrants would be maximum in terms of warrants for unsignalized intersections with one approach lane each. Korean Highway Capacity Manual (2) analyzed capacity for four-legged unsignalized intersections in terms of the number of conflicts and level of service. A guideline was given by the directional split of traffic volume as shown in Table 2.

Table 2: Guideline for Level of Service at AWSC Unsignalized Intersections

Level of Service	Number of Conflicts (vph)	Traffic Volume (vph) By Directional Split (major : minor)		
		50 : 50	60 : 40	70 : 30
A	<= 60	400	440	440
B	<= 120	800	880	880
C	<= 180	1,200	1,320	1,440
D	<= 240	1,600	1,760	1,920
E	<= 300	2,000	2,200	2,400
F	> 301	2,000	2,200	2,400

Based on the consideration of those guidelines discussed previously, the warrants related to traffic volumes are provided for the implementation of intelligent AWSC system as shown in Figure 1.

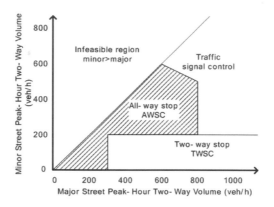

Figure 1: Warrants for Intelligent AWSC System at Unsignalized Intersections.

As shown in the figure, 800 vph of volume would be the maximum volume on the major road with the minimum volume of 300 vph. Whereas in the minor road, 600 vph of the maximum in the case of same volume in the major road would be decreasing with maintaining 1,200 vph as the maximum volume of the intersection until the minimum volume of 200 vph.

4. Conclusions

Feasibility on introduction of AWSC utilizing ITS technologies was demonstrated to provide the right of ways for enhancing efficiency and safety at unsignalized intersections. Findings from this study suggest that the proposed intelligent AWSC system would be implemented for improving both efficiency and safety at unsignalized intersections with the goal of higher driver compliance towards stopping at the intersection and having the right of ways.

References

[1] Marek, J., et al. Determining Intersection Traffic Control Type Using the 1994 Highway Capacity Manual. *ITE Journal*, August 1997.

[2] Transportation Research Board. *Highway Capacity Manual*, National Research Council, Washington, D.C., 2000.

[3] Kyte, M. and Marek, J. Estimating Capacity and Delay at a Single-Lane Approach, All-Way Stop-Controlled Intersection. *Transportation Research Record 1225*, pp. 73-82 (1994).

[4] Kyte, M., et al. Analysis of Traffic Operations at All-Way Stop-Controlled Intersections by Simulation. *Transportation Research Record 1555*, pp. 65-73, (1997).

[5] Kaysi, I. and Alam, G. Driver Behavior and Traffic Stream Interactions at Unsignalized Intersections. *Journal of Transportation Engineering*, ASCE, Vol. 126, No. 6, 2000.